MW00844033

FUNDAMENTALS OF HEAT EXCHANGER DESIGN

FUNDAMENTALS OF HEAT EXCHANGER DESIGN

Ramesh K. Shah

Rochester Institute of Technology, Rochester, New York
Formerly at Delphi Harrison Thermal Systems, Lockport, New York

Dušan P. Sekulić

University of Kentucky, Lexington, Kentucky

WILEY

JOHN WILEY & SONS, INC.

Copyright © 2003 by John Wiley & Sons, Inc. All rights reserved

Published by John Wiley & Sons, Inc., Hoboken, New Jersey
Published simultaneously in Canada

For general information on our other products and services or for technical support, please contact our Customer Care Department within the United States at (800) 762-2974, outside the United States at (317) 572-3993 or fax (317) 572-4002.

Wiley also publishes its books in a variety of electronic formats. Some content that appears in print may not be available in electronic books. For more information about Wiley products, visit our web site at www.wiley.com

Library of Congress Cataloging-in-Publication Data:
Shah, R. K.
    Fundamentals of heat exchanger design / Ramesh K. Shah, Dušan P. Sekulić.
      p. cm.
Includes index.
    ISBN 0-471-32171-0
   1. Heat exchangers–Design and construction.   I. Sekulić, Dušan P. II. Title.
    TJ263 .S42 2003
    621.402′5–dc21
    ISBN 13: 978-0-471-32171-2   2002010161

Contents

Preface

Over the past quarter century, the importance of heat exchangers has increased immensely from the viewpoint of energy conservation, conversion, recovery, and successful implementation of new energy sources. Its importance is also increasing from the standpoint of environmental concerns such as thermal pollution, air pollution, water pollution, and waste disposal. Heat exchangers are used in the process, power, transportation, air-conditioning and refrigeration, cryogenic, heat recovery, alternate fuels, and manufacturing industries, as well as being key components of many industrial products available in the marketplace. From an educational point of view, heat exchangers illustrate in one way or another most of the fundamental principles of the thermal sciences, thus serving as an excellent vehicle for review and application, meeting the guidelines for university studies in the United States and oversees. Significant advances have taken place in the development of heat exchanger manufacturing technology as well as design theory. Many books have been published on the subject, as summarized in the General References at the end of the book. However, our assessment is that none of the books available seems to provide an in-depth coverage of the intricacies of heat exchanger design and theory so as to fully support both a student and a practicing engineer in the quest for creative mastering of both theory and design. Our book was motivated by this consideration. Coverage includes the theory and design of exchangers for many industries (not restricted to, say, the process industry) for a broader, in-depth foundation.

The objective of this book is to provide in-depth thermal and hydraulic design theory of two-fluid single-phase heat exchangers for steady-state operation. Three important goals were borne in mind during the preparation of this book:

1. To introduce and apply concepts learned in first courses in heat transfer, fluid mechanics, thermodynamics, and calculus, to develop heat exchanger design theory. Thus, the book will serve as a link between fundamental subjects mentioned and thermal engineering design practice in industry.

2. To introduce and apply basic heat exchanger design concepts to the solution of industrial heat exchanger problems. Primary emphasis is placed on fundamental concepts and applications. Also, more emphasis is placed on analysis and less on empiricism.

3. The book is also intended for practicing engineers in addition to students. Hence, at a number of places in the text, some redundancy is added to make the concepts clearer, early theory is developed using constant and mean overall heat transfer coefficients, and more data are added in the text and tables for industrial use.

To provide comprehensive information for heat exchanger design and analysis in a book of reasonable length, we have opted not to include detailed theoretical derivations of many results, as they can be found in advanced convection heat transfer textbooks. Instead, we have presented some basic derivations and then presented comprehensive information through text and concise tables.

An industrial heat exchanger design problem consists of coupling component and system design considerations to ensure proper functioning. Accordingly, a good design engineer must be familiar with both system and component design aspects. Based on industrial experience of over three decades in designing compact heat exchangers for automobiles and other industrial applications and more than twenty years of teaching, we have endeavored to demonstrate interrelationships between the component and system design aspects, as well as between the needs of industrial and learning environments. Some of the details of component design presented are also based on our own system design experience.

Considering the fact that heat exchangers constitute a multibillion-dollar industry in the United States alone, and there are over 300 companies engaged in the manufacture of a wide array of heat exchangers, it is difficult to select appropriate material for an introductory course. We have included more material than is necessary for a one-semester course, placing equal emphasis on four basic heat exchanger types: shell-and-tube, plate, extended surface, and regenerator. The choice of the teaching material to cover in one semester is up to the instructor, depending on his or her desire to focus on specific exchanger types and specific topics in each chapter. The prerequisites for this course are first undergraduate courses in fluid mechanics, thermodynamics, and heat transfer. It is expected that the student is familiar with the basics of forced convection and the basic concepts of the heat transfer coefficient, heat exchanger effectiveness, and mean temperature difference.

Starting with a detailed classification of a variety of heat exchangers in Chapter 1, an overview of heat exchanger design methodology is provided in Chapter 2. The basic thermal design theory for recuperators is presented in Chapter 3, advanced design theory for recuperators in Chapter 4, and thermal design theory for regenerators in Chapter 5. Pressure drop analysis is presented in Chapter 6. The methods and sources for obtaining heat transfer and flow friction characteristics of exchanger surfaces are presented in Chapter 7. Surface geometrical properties needed for heat exchanger design are covered in Chapter 8. The thermal and hydraulic designs of extended-surface (compact and noncompact plate-fin and tube-fin), plate, and shell-and-tube exchangers are outlined in Chapter 9. Guidelines for selecting the exchanger core construction and surface geometry are presented in Chapter 10. Chapter 11 is devoted to thermodynamic analysis for heat exchanger design and includes basic studies of temperature distributions in heat exchangers, a heuristic approach to an assessment of heat exchanger effectiveness, and advanced topics important for modeling, analysis, and optimization of heat exchangers as components. All topics covered up to this point are related to thermal–hydraulic design of heat exchangers in steady-state or periodic-flow operation. Operational problems for compact and other heat exchangers are covered in Chapters 12 and 13. They include the problems caused by flow maldistribution and by fouling and corrosion. Solved examples from industrial experience and classroom practice are presented throughout the book to illustrate important concepts and applications. Numerous review questions and problems are also provided at the end of each chapter. If students can answer the review questions and solve the problems correctly, they can be sure of their grasp of the basic concepts and material presented in the text. It is hoped that readers will develop good understanding of the intricacies of heat exchanger design after going through this material and prior to embarking on specialized work in their areas of greatest interest.

For the thermal design of a heat exchanger for an application, considerable intellectual effort is needed in selecting heat exchanger type and determining the appropriate value of the heat transfer coefficients and friction factors; a relatively small effort is needed for executing sizing and optimizing the exchanger because of the computer-based calculations. Thus, Chapters 7, 9, and 10 are very important, in addition to Chapter 3, for basic understanding of theory, design, analysis, and selection of heat exchangers.

Material presented in Chapters 11 through 13 is significantly more interdisciplinary than the rest of the book and is presented here in a modified methodological approach. In Chapter 11 in particular, analytical modeling is used extensively. Readers will participate actively through a set of examples and problems that *extend* the breadth and depth of the material given in the main body of the text. A number of examples and problems in Chapter 11 require analytical derivations and more elaborate analysis, instead of illustrating the topics with examples that favor only utilization of the formulas and computing numerical values for a problem. The complexity of topics requires a more diverse approach to terminology, less routine treatment of established conventions, and a more creative approach to some unresolved dilemmas.

Because of the breadth of the subject, the coverage includes various design aspects and problems for indirect-contact two-fluid heat exchangers with primarily single-phase fluids on each side. Heat exchangers with condensing and evaporating fluids on one side can also be analyzed using the design methods presented as long as the thermal resistance on the condensing or evaporating side is small or the heat transfer coefficient on that side can be treated as a constant. Design theory for the following exchangers is not covered in this book, due to their complexity and space limitations: two-phase and multiphase heat exchangers (such as condensers and vaporizers), direct-contact heat exchangers (such as humidifiers, dehumidifiers, cooling towers), and multifluid and multistream heat exchangers. Coverage of mechanical design, exchanger fabrication methods, and manufacturing techniques is also deemed beyond the scope of the book.

Books by M. Jakob, D. Q. Kern, and W. M. Kays and A. L. London were considered to be the best and most comprehensive texts on heat exchanger design and analysis following World War II. In the last thirty or so years, a significant number of books have been published on heat exchangers. These are summarized in the General References at the end of the book.

This text is an outgrowth of lecture notes prepared by the authors in teaching courses on heat exchanger design, heat transfer, and design and optimization of thermal systems to senior and graduate students. These courses were taught at the State University of New York at Buffalo and the University of Novi Sad, Yugoslavia. Over the past fifteen years or more, the notes of the first author have been used for teaching purposes at a number of institutions, including the University of Miami by Professor S. Kakaç, Rensselaer Polytechnic Institute by Professors A. E. Bergles and R. N. Smith, Rochester Institute of Technology by Professor S. G. Kandlikar, Rice University by Professor Y. Bayazitoğlu, University of Tennessee Space Center by Dr. R. Schultz, University of Texas at Arlington by Professor A. Haji-Sheikh, University of Cincinnati by Professor R. M. Manglik, Northeastern University by Professor Yaman Yener, North Carolina A&T State University by Professor Lonnie Sharpe, Auburn

University by Dr. Peter Jones, Southern Methodist University by Dr. Donald Price, University of Tennessee by Professor Edward Keshock, and Gonzaga University by Professor A. Aziz. In addition, these course notes have been used occasionally at a number of other U.S. and foreign institutions. The notes of the second author have also been used for a number of undergraduate and graduate courses at Marquette University and the University of Kentucky.

The first author would like to express his sincere appreciation to the management of Harrison Thermal Systems, Delphi Corporation (formerly General Motors Corporation), for their varied support activities over an extended period of time. The second author acknowledges with appreciation many years of support by his colleagues and friends on the faculty of the School of Engineering, University of Novi Sad, and more recently at Marquette University and the University of Kentucky. We are also thankful for the support provided by the College of Engineering, University of Kentucky, for preparation of the first five and final three chapters of the book. A special word of appreciation is in order for the diligence and care exercised by Messrs. Dale Hall and Mack Mosley in preparing the manuscript and drawings through Chapter 5.

The first author is grateful to Professor A. L. London of Stanford University for teaching him the ABCs of heat exchangers and for providing constant inspiration and encouragement throughout his professional career and particularly during the course of preparation of this book. The first author would also like to thank Professors Sadik Kakaç of the University of Miami and Ralph Webb of the Pennsylvania State University for their support, encouragement, and involvement in many professional activities related to heat exchangers. The second author is grateful to his colleague and friend Professor B. S. Bačlić, University of Novi Sad, for many years of joint work and teaching in the fields of heat exchanger design theory. Numerous discussions the second author have had with Dr. R. Gregory of the University of Kentucky regarding not only what one has to say about a technical topic, but in particular how to formulate it for a reader, were of a great help in resolving some dilemmas. Also, the continuous support and encouragement of Dr. Frederick Edeskuty of Los Alamos National Laboratory, and Professor Richard Gaggioli of Marquette University were immensely important to the second author in an effort to exercise his academic experience on both sides of the Atlantic Ocean. We appreciate Professor P. V. Kadaba of the Georgia Institute of Technology and James Seebald of ABB Alstom Air Preheater for reviewing the complete manuscript and providing constructive suggestions, and Dr. M. S. Bhatti of Delphi Harrison Thermal Systems for reviewing Chapters 1 through 6 and Dr. T. Skiepko of Bialystok Technical University for reviewing Chapter 5 and providing constructive suggestions. The constructive feedback over a period of time provided by many students (too numerous to mention by name) merits a special word of appreciation.

Finally, we must acknowledge the roles played by our wives, Rekha and Gorana, and our children, Nilay and Nirav Shah and Višnja and Aleksandar Sekulić, during the course of preparation of this book. Their loving care, emotional support, assistance, and understanding provided continuing motivation to compete the book.

We welcome suggestions and comments from readers.

Ramesh K. Shah

Dušan P. Sekulić

NOMENCLATURE

The dimensions for each symbol are represented in both the SI and English systems of units, where applicable. Note that both the hour and second are commonly used as units for time in the English system of units; hence a conversion factor of 3600 should be employed at appropriate places in dimensionless groups.

A total heat transfer surface area (both primary and secondary, if any) on one side of a direct transfer type exchanger (recuperator), total heat transfer surface area of all matrices of a regenerator,[†] m^2, ft^2

A_c total heat transfer area (both primary and secondary, if any) on the cold side of an exchanger, m^2, ft^2

A_{eff} effective surface area on one side of an extended surface exchanger [defined by Eq. (4.167)], m^2, ft^2

A_f fin or extended surface area on one side of the exchanger, m^2, ft^2

A_{fr} frontal or face area on one side of an exchanger, m^2, ft^2

$A_{fr,t}$ window area occupied by tubes, m^2, ft^2

$A_{fr,w}$ gross (total) window area, m^2, ft^2

A_h total heat transfer surface area (both primary and secondary, if any) on the hot fluid side of an exchanger, m^2, ft^2

A_k fin cross-sectional area for heat conduction in Section 4.3 ($A_{k,o}$ is A_k at the fin base), m^2, ft^2

A_k total wall cross-sectional area for longitudinal conduction [additional subscripts c, h, and t, if present, denote cold side, hot side, and total (hot + cold) for a regenerator] in Section 5.4, m^2, ft^2

A_k^* ratio of A_k on the C_{min} side to that on the C_{max} side [see Eq. (5.117)], dimensionless

A_o minimum free-flow (or open) area on one fluid side of an exchanger, heat transfer surface area on tube outside in a tubular exchanger in Chapter 13 only, m^2, ft^2

$A_{o,bp}$ flow bypass area of one baffle, m^2, ft^2

$A_{o,cr}$ flow area at or near the shell centerline for one crossflow section in a shell-and-tube exchanger, m^2, ft^2

$A_{o,sb}$ shell-to-baffle leakage flow area, m^2, ft^2

$A_{o,tb}$ tube-to-baffle leakage flow area, m^2, ft^2

$A_{o,w}$ flow area through window zone, m^2, ft^2

A_p primary surface area on one side of an exchanger, m^2, ft^2

A_w total wall area for heat conduction from the hot fluid to the cold fluid, or total wall area for transverse heat conduction (in the matrix wall thickness direction), m^2, ft^2

a short side (unless specified) of a rectangular cross section, m, ft

$\mathbf{a}$ amplitude of chevron plate corrugation (see Fig. 7.28), m, ft

---

[†] Unless clearly specified, a regenerator in the nomenclature means either a rotary or a fixed-matrix regenerator.

B	parameter for a thin fin with end leakage allowed, h_e/mk_f, dimensionless
Bi	Biot number, $\text{Bi} = h(\delta/2)/k_f$ for the fin analysis; $\text{Bi} = h(\delta/2)/k_w$ for the regenerator analysis, dimensionless
b	distance between two plates in a plate-fin heat exchanger [see Fig. 8.7 for b_1 or b_2 (b on fluid 1 or 2 side)], m, ft
b	long side (unless specified) of a rectangular cross section, m, ft
$\mathscr{C}$	some arbitrary monetary unit (instead of \$, £, etc.), money
C	flow stream heat capacity rate with a subscript c or h, $\dot{m}c_p$, W/K, Btu/hr-°F
C	correction factor when used with a subscript different from c, h, min, or max, dimensionless
$\mathbf{C}$	unit cost, $\mathscr{C}/\text{J}(\mathscr{C}/\text{Btu})$, $\mathscr{C}/\text{kg}$ ($\mathscr{C}/\text{lbm}$), $\mathscr{C}/\text{kW}$ [$\mathscr{C}/(\text{Btu/hr})$], $\mathscr{C}/\text{kW}\cdot\text{yr}(\mathscr{C}/\text{Btu}$ on yearly basis), $\mathscr{C}/\text{m}^2(\mathscr{C}/\text{ft}^2)$
C	annual cost, $\mathscr{C}/\text{yr}$
C^*	heat capacity rate ratio, $C_{\min}/C_{\max}$, dimensionless
$\bar{C}$	flow stream heat capacitance, Mc_p, $C\tau_d$, W $\cdot$ s/K, Btu/°F
C_D	drag coefficient, $\Delta p/(\rho u_\infty^2/2g_c)$, dimensionless
$C_{\max}$	maximum of C_c and C_h, W/K, Btu/hr-°F
$C_{\min}$	minimum of C_c and C_h, W/K, Btu/hr-°F
C_{ms}	heat capacity rate of the maldistributed stream, W/K, Btu/hr-°F
C_r	heat capacity rate of a regenerator, $M_w c_w \text{N}$ or $M_w c_w/P_t$ [see Eq. (5.7) for the hot- and cold-side matrix heat capacity rates $C_{r,h}$ and $C_{r,c}$], W/K, Btu/hr-°F
C_r^*	total matrix heat capacity rate ratio, $C_r/C_{\min}$, $C_{r,h}^* = C_{r,h}/C_h$, $C_{r,c}^* = C_{r,c}/C_c$, dimensionless
$\bar{C}_r$	total matrix wall heat capacitance, $M_w c_w$ or $C_r P_t$ [see Eq. (5.6) for hot- and cold-side matrix heat capacitances $\bar{C}_{r,h}$ and $\bar{C}_{r,c}$], W $\cdot$ s/K, Btu/°F
$\bar{C}_r^*$	ratio of $\bar{C}_r$ to $\bar{C}_{\min}$, dimensionless
C_{UA}	cost per unit thermal size (see Fig. 10.13 and Appendix D), $\mathscr{C}/\text{W/K}$
C_{us}	heat capacity rate of the uniform stream, W/K, Btu/hr-°F
C_w	matrix heat capacity rate; same as C_r, W/K, Btu/hr-°F
$\bar{C}_w$	total wall heat capacitance for a recuperator, $M_w c_w$, W $\cdot$ s/K, Btu/°F
$\bar{C}_w^*$	ratio of $\bar{C}_w$ to $\bar{C}_{\min}$, dimensionless
CF	cleanliness factor, U_f/U_c, dimensionless
c	specific heat of solid, J/kg $\cdot$ K,[†] Btu/lbm-°F
$\mathbf{c}$	annual cost of operation percentile, dimensionless
c_p	specific heat of fluid at constant pressure, J/kg $\cdot$ K, Btu/lbm-°F
c_w	specific heat of wall material, J/kg $\cdot$ K, Btu/lbm-°F
$\mathscr{D}$	exergy destruction rate, W, Btu/hr
D_{baffle}	baffle diameter, m, ft
D_{ctl}	diameter of the circle through the centers of the outermost tubes, $D_{\text{otl}} - d_o$, m, ft
D_h	hydraulic diameter of flow passages, $4r_h$, $4A_o/\mathbf{P}$, $4A_o L/A$, or $4\sigma/\alpha$, m, ft

[†] J = joule = newton × meter = watt × second; newton = N = kg $\cdot$ m/s^2.

$D_{h,w}$	hydraulic diameter of the window section, m, ft
D_{otl}	diameter of the outer tube limit (see Fig. 8.9), m, ft
D_p	port or manifold diameter in a plate heat exchanger, m, ft
D_s	shell inside diameter, m, ft
d	differential operator
d_c	collar diameter in a round tube and fin exchanger, $d_o + 2\delta$, m, ft
d_e	fin tip diameter of a disk (radial) fin, m, ft
d_i	tube inside diameter, m, ft
d_o	tube (or pin) outside diameter, tube outside diameter at the fin root for a finned tube after tube expansion, if any, m, ft
d_w	wire diameter, m, ft
d_1	tube hole diameter in a baffle, m, ft
$\dot{\mathscr{E}}$	exergy rate, W, Btu/hr
E	energy, J, Btu
E	activation energy in Chapter 13 [see Eq. (13.12)], J/kg · mol, Btu/lbm-mole
E	fluid pumping power per unit surface area, $\dot{m}\,\Delta p/\rho A$, W/m^2, hp/ft^2
Eu	row average Euler number per tube row, $\Delta p/(\rho u_m^2 N_r/2g_c)$ or $\Delta p/(G^2 N_r/2g_c\rho)$, dimensionless
e	surface roughness size, m, ft
e^+	roughness Reynolds number, eu^*/ν, dimensionless
F	log-mean temperature difference correction factor [defined by Eq. (3.183)], dimensionless
f	Fanning friction factor, $\tau_w/(\rho u_m^2/2g_c)$, $\Delta p\rho g_c D_h/(2LG^2)$, dimensionless
f_D	Darcy friction factor, $4f$, dimensionless
f_{tb}	row average Fanning friction factor per tube for crossflow to tubes, used in Chapter 7, $\Delta p/(4G^2 N_r/2g_c\rho)$, Eu/4, dimensionless
G	fluid mass velocity based on the *minimum* free area, $\dot{m}/A_o$ (replace A_o by $A_{o,c}$ for the crossflow section of a tube bundle in a shell-and-tube heat exchanger), kg/m^2 · s, lbm/hr-ft^2
Gr	Grashof number [defined by Eq. (7.159)], dimensionless
Gz	Graetz number, $\dot{m}c_p/kL$ [see Eqs. (7.39) and (12.53)], dimensionless
Gz$_x$	local Graetz number, $\dot{m}c_p/kx$, dimensionless
g	gravitational acceleration, m/s^2, ft/sec^2
g_c	proportionality constant in Newton's second law of motion, $g_c = 1$ and dimensionless in SI units, $g_c = 32.174$ lbm-ft/lbf-sec^2
H	head or velocity head, m, ft
H	fluid enthalpy, J, Btu
$\dot{\mathbf{H}}$	enthalpy rate, used in Chapter 11, W, Btu/hr
Hg	Hagen number, defined by Eq. (7.23), dimensionless
Ⓗ	thermal boundary condition referring to constant axial as well as peripheral wall heat flux; also constant peripheral wall temperature; boundary condition valid only for the circular tube, parallel plates, and concentric annular ducts when symmetrically heated

H1	thermal boundary condition referring to constant axial wall heat flux with constant peripheral wall temperature
H2	thermal boundary condition referring to constant axial wall heat flux with constant peripheral wall heat flux
h	heat transfer coefficient [defined by Eqs. (7.11) and (7.12)], $\text{W/m}^2 \cdot \text{K}$, Btu/hr-ft^2-°F
h	specific enthalpy, J/kg, Btu/lbm
h_e	heat transfer coefficient at the fin tip, $\text{W/m}^2 \cdot \text{K}$, Btu/hr-ft^2-°F
h$_{\ell g}$	specific enthalpy of phase change, J/kg, Btu/lbm
$\dot{I}_{\text{irr}}$	irreversibility rate (defined in Table 11.3), W, Btu/hr
$I_n(\cdot)$	modified Bessel function of the first kind and nth order
i_j	flow direction indicator, $i_j = +1$ or -1, fluid $j = 1$ or 2, dimensionless
J	mechanical to thermal energy conversion factor, $J = 1$ and dimensionless in SI units, $J = 778.163$ lbf-ft/Btu
J_i	correction factors for the shell-side heat transfer coefficient for the Bell–Delaware method [see Eq. (9.50)]; $i = c$ for baffle cut and spacing; $i = \ell$ for baffle leakage effects, including both shell-to-baffle and tube-to-baffle leakage; $i = b$ for the bundle bypass flow (C and F streams); $i = s$ for variable baffle spacing in the inlet and outlet sections; $i = r$ for adverse temperature gradient buildup in laminar flow, dimensionless
j	Colburn factor, St Pr$^{2/3}$, (h/Gc_p)Pr$^{2/3}$, dimensionless
K	pressure loss coefficient, $\Delta p/(\rho u_m^2/2g_c)$; subscripts: b for a circular bend, s for a miter bend, and v for a screwed valve in Chapter 6, and br for branches in Chapter 12, dimensionless
$K(\infty)$	incremental pressure drop number for fully developed flow (see Table 7.2 for the definition), dimensionless
K_c	contraction loss coefficient for flow at heat exchanger entrance, dimensionless
K_e	expansion loss coefficient for flow at heat exchanger exit, dimensionless
$K_n(\cdot)$	modified Bessel function of the second kind and nth order
k	fluid thermal conductivity for fluid if no subscript, $\text{W/m} \cdot \text{K}$, Btu/hr-ft-°F
k_f	thermal conductivity of the fin material in Chapter 4 and of the foulant material in Chapter 13, $\text{W/m} \cdot \text{K}$, Btu/hr-ft-°F
k_w	thermal conductivity of the matrix (wall) material, $\text{W/m} \cdot \text{K}$, Btu/hr-ft-°F
L	fluid flow (core) length on one side of an exchanger, m, ft
L_f	fin flow length on one side of a heat exchanger, $L_f \leq L$, m, ft
L_h	plate length in a PHE for heat transfer (defined in Fig. 7.28), m, ft
L_p	plate length in a PHE for pressure drop (defined in Fig. 7.28), m, ft
L_1	flow (core) length for fluid 1 of a two-fluid heat exchanger, m, ft
L_2	flow (core) length for fluid 2 of a two-fluid heat exchanger, m, ft
L_3	noflow height (stack height) of a two-fluid heat exchanger, m, ft
L_q	Lévêque number, defined by Eq. (7.41), dimensionless

ℓ	fin height or fin length for heat conduction from primary surface to either fin tip or midpoint between plates for symmetric heating, $\ell = (d_e - d_o)/2$ for individually finned tubes, ℓ with this meaning used only in the fin analysis and in the definition of η_f, m, ft
ℓ_c	baffle cut, distance from the baffle tip to the shell inside diameter (see Fig. 8.9), m, ft
ℓ_{ef}	effective flow length between major boundary layer disturbances, distance between interruptions, m, ft
ℓ_s	strip length of an offset strip fin, m, ft
ℓ^*	flow length between interruptions, $\ell_{ef}/(D_h \cdot \mathrm{Re} \cdot \mathrm{Pr})$, dimensionless
ℓ_c^*	baffle cut, ℓ_c/D_s, dimensionless
$\mathscr{M}$	molecular weight (molar mass) of a gas, kg/kmol, lbm/lb mole
M_A	foulant material mass per unit heat transfer surface area in Chapter 13, m/A, kg/m^2, lbm/ft^2
M_w	mass of a heat exchanger core or the total mass of all matrices of a regenerator, kg, lbm
m	fin parameter [defined by Eqs. (4.62) and (4.65); see also Table 4.5 for other definitions], 1/m, 1/ft
m	mass of a body or fluid in a control volume, kg, lbm
$\dot{m}$	fluid mass flow rate, $\rho u_m A_o$, kg/s, 1bm/hr
$\dot{m}_n$	fluid mass flow rate for nominal flow passages in Chapter 12, kg/s, 1bm/hr
N	number of subexchangers in gross flow maldistributed exchanger or a number of differently sized/shaped passages in passage-to-passage nonuniformity, used in Chapter 12
N	rotational speed for a rotary regenerator, rev/s, rpm
N_b	number of baffles in a plate-baffled shell-and-tube exchanger
N_c	number of fluid channels in a plate heat exchanger
N_f	number of fins per unit length in the fin pitch direction, 1/m, 1/ft
N_p	number of fluid 1 passages in a two-fluid heat exchanger
N_p	number of pass divider lanes through the tube field that are parallel to the crossflow stream in a shell-and-tube exchanger
N_p'	number of separating plates in a plate-fin exchanger, number of pass divider lanes in a shell-and-tube exchanger
N_r	number of tube rows in the flow direction
$N_{r,c}$	number of effective tube rows crossed during flow through one baffle section, $N_{r,cc} + N_{r,cw}$
$N_{r,cc}$	number of effective tube rows crossed during flow through one crossflow section (between baffle tips)
$N_{r,cw}$	number of effective tube rows crossed during flow through one window zone in a segmental baffled shell-and-tube heat exchanger
N_t	total number of tubes in an exchanger, total number of holes in a tubesheet, or total number of plates in a plate heat exchanger
$N_{t,b}$	total number of tubes associated with one segmental baffle

$N_{t,c}$	number of tubes at the tube bundle centerline cross section
$N_{t,p}$	number of tubes per pass
$N_{t,w}$	number of tubes in the window zone
N_t'	number of tubes in a specified row
NTU	number of exchanger heat transfer units, UA/C_{min} [defined by Eqs. (3.59) through (3.64)], it represents the total number of transfer units in a multipass unit, dimensionless
NTU_1	number of exchanger heat transfer units based on fluid 1 heat capacity rate, UA/C_1; similarly, $NTU_2 = UA/C_2$, dimensionless
NTU_c	number of exchanger heat transfer units based on C_c, UA/C_c, dimensionless
NTU_h	number of exchanger heat transfer units based on C_h, UA/C_h, dimensionless
NTU_o	modified number of heat transfer units for a regenerator [defined by Eq. (5.48)], dimensionless
NTU*	number of heat transfer units at maximum entropy generation, dimensionless
Nu	Nusselt number [defined by Eqs. (7.26) and (7.27)], dimensionless
n, n_p	number of passes in an exchanger
n_c	number of cells of a regenerator matrix per unit of frontal area, $1/m^2$, $1/ft^2$
n_f	total number of fins on one fluid side of an extended-surface exchanger
n_t	number of tubes in each pass
ntu_c	number of heat transfer units based on the cold fluid side, $(\eta_o hA)_c/C_c$, dimensionless
ntu^*_{cost}	reduction in ntu [defined by Eq. (12.44)], dimensionless
ntu_h	number of heat transfer units based on the hot fluid side, $(\eta_o hA)_h/C_h$, dimensionless
$\mathcal{P}$	fluid pumping power, $\dot{m}\,\Delta p/\rho$, W, hp
P	temperature effectiveness for one fluid stream [defined by Eqs. (3.96) and (3.97)], dimensionless
P	wetted perimeter of exchanger passages on one fluid side, $\mathbf{P} = A/L = A_{fr}\beta$, m, ft
$\wp$	deposition probability function, dimensionless
P_c	cold-gas flow period, duration of the cold-gas stream in the matrix or duration of matrix in the cold-gas stream, used in Chapter 5, s, sec
P_h	hot-gas flow period, duration of the hot-gas stream in the matrix or duration of matrix in the hot-gas stream, used in Chapter 5, s, sec
P_r	reversal period for switching from hot- to cold-gas stream, or vice versa, in a fixed-matrix regenerator, used in Chapter 5, s, sec
P_t	total period between the start of two successive heating (or cooling) periods in a regenerator, used in Chapter 5, $P_t = P_h + P_c + P_r \approx P_h + P_c$, s, sec
Pe	Péclet number, $Re \cdot Pr$, dimensionless
Pr	Prandtl number, $\mu c_p/k$, $u_m D_h/\alpha$, dimensionless
p	fluid static pressure, Pa, lbf/ft^2 (psf) or lbf/in^2 (psi)[†]

[†] $Pa = Pascal = N/m^2 = kg/m \cdot s^2$; $N = newton = kg \cdot m/s^2$; $psf = lbf/ft^3$; $psi = lbf/in^3$.

p	porosity of a matrix, a ratio of void volume to total volume of a matrix, $r_h\beta$, dimensionless
p^*	ratio of cold-fluid inlet pressure to hot-fluid inlet pressure, $p_{c,i}/p_{h,i}$, dimensionless
p_d	fin pattern depth, peak-to-valley distance, excluding fin thickness (see Fig. 7.30), m, ft
p_f	fin pitch, $1/N_f$, m, ft
p_t	tube pitch, center-to-center distance between tubes, m, ft
Δp	fluid static pressure drop on one fluid side of a heat exchanger core [see Eq. (6.28)], Pa, psf (psi)
Δp^*	$= \Delta p/(\rho u_m^2/2g_c)$, dimensionless
Δp_b	fluid static pressure drop associated with a pipe bend, Pa, psf (psi)
$\Delta p_{b,i}$	fluid static pressure drop associated with an ideal crossflow section between two baffles, Pa, psf (psi)
Δp_c	fluid static pressure drop associated with the tube bundle central section (crossflow zone) between baffle tips, Pa, psf (psi)
Δp_{gain}	pressure drop reduction due to passage-to-passage nonuniformity [defined by Eq. (12.36)], Pa, psf (psi)
Δp_s	shell-side pressure drop, Pa, psf (psi)
$\Delta p_{w,i}$	fluid static pressure drop associated with an ideal window section, Pa, psf (psi)
Q	heat transfer in a specified period or time, J, Btu
q	total or local (whatever appropriate) heat transfer rate in an exchanger, or heat "duty," W, Btu/hr
q^*	normalized heat transfer rate, $q/[(\dot{m}c_p)(T_{2,i} - T_{1,i})]$, dimensionless
q'	heat transfer rate per unit length, q/L, W/m, Btu/hr-ft
q''	heat flux, heat transfer rate per unit surface area, q/A, W/m^2, Btu/hr-ft^2
q_e	heat transfer rate through the fin tip, W, Btu/hr
q_0	heat transfer rate at the fin base, W, Btu/hr
q_{max}	thermodynamically maximum possible heat transfer rate in a counterflow heat exchanger as expressed by Eq. (3.42), and also that through the fin base as expressed by Eq. (4.130), W, Btu/hr
$\Re$	universal gas constant, 8.3143 kJ/kmol · K, 1545.33 1bf-ft/1b mole-°R
R	heat capacity rate ratio [defined by Eqs. (3.105) and (3.106)], dimensionless
R	thermal resistance based on the surface area A; $\mathbf{R} = 1/UA$ = overall thermal resistance in a two-fluid exchanger, $\mathbf{R}_h = 1/(hA)_h$ = hot-side film resistance (between the fluid and the wall), $\mathbf{R}_c$ = cold-side film resistance, $\mathbf{R}_f$ = fouling resistance, and $\mathbf{R}_w$ = wall thermal resistance [definitions found after Eq. (3.24)], K/W, hr-°F/Btu
$\hat{\mathbf{R}}$	unit thermal resistance, $\hat{\mathbf{R}} = \mathbf{R}A = 1/U$, $\hat{\mathbf{R}}_h = 1/(\eta_o h)_h$, $\hat{\mathbf{R}}_w = 1/(\eta_o h)_c$, $\hat{\mathbf{R}}_w = \delta_w/A_w$, m^2 · K/W, hr-ft^2 °F/Btu
$\mathbf{R}^*$	ratio of thermal resistances on the C_{min} to C_{max} side, $1/(\eta_o hA)^*$; it is also the same as the ratio of hot to cold reduced periods, Π_h/Π_c, Chapter 5, dimensionless

R*	total thermal resistance (wall, fouling, and convective) on the enhanced (or plain with subscript p) "outside" surface side normalized with respect to the thermal resistance $[1/(hA_{i,p})]$ of "inside" plain tube/surface (see Table 10.5 for explicit formulas), dimensionless
$\tilde{R}$	gas constant for a particular gas, $\Re/\mathcal{M}$, J/kg · K, 1bf-ft/1bm-°R
$\hat{\mathbf{R}}_f$	fouling factor or unit thermal resistance ("fouling resistance"), $1/h_f$, m² · K/W, hr-ft²-°F/Btu
R_i	pressure drop correction factor for the Bell–Delaware method, where $i = b$ for bundle bypass flow effects (C stream), $i = \ell$ for baffle leakage effects (A and E streams), $i = s$ for unequal inlet/outlet baffle spacing effects, dimensionless
Ra	Rayleigh number [defined by Eq. (7.160)], dimensionless
Re	Reynolds number based on the hydraulic diameter, GD_h/μ, dimensionless
Re_d	Reynolds number based on the tube outside diameter and mean velocity, $\rho u_m d_o/\mu$, dimensionless
Re_{dc}	Reynolds number based on the collar diameter and mean velocity, $\rho u_m d_c/\mu$, dimensionless
Re_o	Reynolds number based on the tube outside diameter and free stream (approach or core upstream) velocity, $\rho u_\infty d_o/\mu$, dimensionless
r	radial coordinate in the cylindrical coordinate system, m, ft
r_c	radius of curvature of a tube bend (see Fig. 6.5), m, ft
r_f	fouling factor or fouling resistance $r_f = \hat{\mathbf{R}}_f = 1/h_f = \delta_f/k_f$, m² · K/W, hr-ft²-°F/Btu
r_h	hydraulic radius, $A_o L/A$ or $D_h/4$, m, ft
r_i	tube inside radius, m, ft
S	entropy, J/K, Btu/°R
S^*	normalized entropy generation rate, $\dot{S}_{\text{irr}}/C_2$ or $\dot{S}_{\text{irr}}/C_{\max}$, dimensionless
$\dot{S}_{\text{irr}}$	entropy generation rate, W/K, Btu/hr-°R
St	Stanton number, h/Gc_p, $\text{St}_o = U/Gc_p$, dimensionless
s	specific entropy in Chapter 11, J/kg · K, Btu/lbm-°R
s	complex Laplace independent variable with Laplace transforms only in Chapter 11, dimensionless
s	spacing between adjacent fins, $p_f - \delta$, m, ft
T	fluid static temperature to a specified arbitrary datum, except for Eqs. (7.157) and (7.158) and in Chapter 11 where it is defined on an absolute temperature scale, °C, °F
Ⓣ	thermal boundary condition referring to constant wall temperature, both axially and peripherally
$T_{c,o}$	flow area average cold-fluid outlet temperature unless otherwise specified, °C, °F
$T_{h,o}$	flow area average hot-fluid outlet temperature unless otherwise specified, °C, °F
T_ℓ	temperature of the fin tip, °C, °F
T_m	fluid bulk mean temperature, °C, °F

T_s	steam temperature, °C, °F
T_w	wall temperature, °C, °F
T_∞	ambient fluid temperature, free stream temperature beyond the extent of the boundary layer on the wall, °C, °F
T^*	ratio of hot-fluid inlet temperature to cold-fluid inlet temperature, $T_{h,i}/T_{c,i}$, dimensionless
T_c^*	$= (T_c - T_{c,i})/(T_{h,i} - T_{c,i})$, dimensionless
T_h^*	$= (T_h - T_{c,i})/(T_{h,i} - T_{c,i})$, dimensionless
T_w^*	$= (T_w - T_{c,i})/(T_{h,i} - T_{c,i})$, dimensionless
T_0	temperature of the fin base, °C, °F
ΔT	local temperature difference between two fluids, $T_h - T_c$, °C, °F
ΔT_c	temperature rise of the cold fluid in the exchanger, $T_{c,o} - T_{c,i}$, °C, °F
ΔT_h	temperature drop of the hot fluid in the exchanger, $T_{h,i} - T_{h,o}$, °C, °F
ΔT_{lm}	log-mean temperature difference [defined by Eq. (3.172)], °C, °F
ΔT_m	true (effective) mean temperature difference [defined by Eqs. (3.9) and (3.13)], °C, °F
ΔT_{max}	inlet temperature difference (ITD) of the two fluids, $(T_{h,i} - T_{c,i})$, $(T_{w,i} - T_{a,i})$ in Section 7.3.1, °C, °F
U, U_m	overall heat transfer coefficient [defined by Eq. (3.20) or (3.24)], subscript m represents mean value when local U is variable (see Table 4.2 for the definitions of other U's), W/m² · K, Btu/hr-ft²-°F
u, u_m	fluid mean axial velocity, u_m occurs at the minimum free flow area in the exchanger unless specified, m/s, ft/sec
u_c	fluid mean velocity for flow normal to a tube bank based on the flow area of the gap $(X_t - d_o)$; evaluated at or near the shell centerline for a plate-baffled shell-and-tube exchanger, m/s, ft/sec
u_{cr}	critical gap velocity for fluidelastic excitation or critical axial velocity for turbulent buffeting, m/s, ft/sec
u_z, u_w	effective and ideal mean velocities in the window zone of a plate-baffled shell-and-tube exchanger [see Eq. (6.41)], m/s, ft/sec
u_∞	free stream (approach) velocity, m/s, ft/sec
u^*	friction velocity, $(\tau_w g_c/\rho)^{1/2}$, m/s, ft/sec
V	heat exchanger total volume, V_h = volume occupied by the hot-fluid-side heat transfer surface area, V_c defined similarly for the cold fluid side, m³, ft³
V^*	ratio of the header volume to the matrix total volume, dimensionless
$\dot{V}$	volumetric flow rate, $\dot{V} = \dot{m}/\rho = u_m A_o$, m³/s, ft³/sec
V_m	matrix or core volume, m³, ft³
V_p	heat exchanger volume between plates on one fluid side, m³, ft³
V_v	void volume of a regenerator, m³, ft³
v	specific volume, $1/\rho$, m³/kg, ft³/lbm
W	plate width between gaskets (see Fig. 7.28), m, ft
w_p	width of the bypass lane (see Fig. 8.9), m, ft
X^*	axial distance or coordinate, x/L, dimensionless

X_d	diagonal pitch, $(X_t^2 + X_\ell^2)^{1/2}$, m, ft
X_d^*	ratio of the diagonal pitch to the tube outside diameter in a circular tube bank, X_d/d_o, dimensionless
X_ℓ	longitudinal (parallel to the flow) tube pitch (see Table 8.1), m, ft
X_ℓ^*	ratio of the longitudinal pitch to the tube outside diameter in a circular tube bank, X_ℓ/d_o, dimensionless
X_t	transverse (perpendicular to the flow) tube pitch, m, ft
X_t^*	ratio of the transverse pitch to the tube diameter in a circular tube bank, X_t/d_o, dimensionless
x	Cartesian coordinate along the flow direction, m, ft
x^+	axial distance, $x/D_h \cdot \mathrm{Re}$, dimensionless
x^*	axial distance, $x/D_h \cdot \mathrm{Re} \cdot \mathrm{Pr}$, dimensionless
x_f	projected wavy length for one-half wavelength (see Fig. 7.30), m, ft
y	transverse Cartesian coordinate, along the matrix wall thickness direction in a regenerator, or along fluid 2 flow direction in other exchangers, m, ft
Z	capital investment or operating expenses in Chapter 11, $\mathscr{C}$/yr
z	Cartesian coordinate, along the noflow or stack height direction for a plate-fin exchanger, m, ft
α	fluid thermal diffusivity, $k/\rho c_p$, m^2/s, ft^2/sec
α	ratio of total heat transfer area on one fluid side of an exchanger to the total volume of an exchanger, A/V, m^2/m^3, ft^2/ft^3
α_w	thermal diffusivity of the matrix material, $k_w/\rho_w c_w$, m^2/s, ft^2/sec
α^*	aspect ratio of rectangular ducts, ratio of the small to large side length, dimensionless
α_f^*	fin aspect ratio, $2\ell/\delta$, dimensionless
β	chevron angle for a PHE chevron plate measured from the axis parallel to the plate length ($\beta \le 90°$) (see Fig. 1.18c or 7.28), rad, deg
β	heat transfer surface area density: ratio of total transfer area on one fluid side of a plate-fin heat exchanger to the volume between the plates on that fluid side, $A/A_{\mathrm{fr}}L$, packing density for a regenerator, m^2/m^3, ft^2/ft^3
β^*	coefficient of thermal expansion, $1/T$ for a perfect gas, 1/K, 1/°R
γ	unbalance factor, $(\Pi_c/\Lambda_c)/(\Pi_h/\Lambda_h)$ or C_c/C_h in Chapter 5 [see Eq. (5.92)], dimensionless
γ	specific heat ratio, c_p/c_v, dimensionless
Δ	denotes finite difference
∂, δ	partial and finite differential operators
δ	fin thickness, at the root if the fin is not of constant cross section, m, ft
δ_b	segmental baffle thickness, m, ft
δ_{bb}	shell-to-tube bundle diametral clearance, $D_s - D_{\mathrm{otl}}$, m, ft
δ_c	channel deviation parameter [defined in Eqs. (12.44), (12.46), and (12.47)], dimensionless
δ_f	fouling film thickness, m, ft
δ_h	header thickness, m, ft

δ_ℓ	laminar (viscous) sublayer thickness in a turbulent boundary layer, m, ft
δ_{otl}	shell-to-tube outer limit diameter clearance, $D_o - D_{\mathrm{otl}}$, m, ft
δ_s	leakage and bypass stream correction factor to the true mean temperature difference for the stream analysis method [defined by Eq. (4.170)], dimensionless
δ_{sb}	shell-to-baffle diametral clearance, $D_s - D_{\mathrm{baffle}}$, m, ft
δ_{tb}	tube-to-baffle hole diametral clearance, $d_1 - d_o$, m, ft
δ_t	thermal boundary layer thickness, m, ft
δ_v	velocity boundary layer thickness, m, ft
δ_w	wall or primary surface (plate) thickness, m, ft
ε	heat exchanger effectiveness [defined by Eq. (3.37) or (3.44) and Table 11.1]; represents an overall exchanger effectiveness for a multipass unit, dimensionless
ε_c	temperature effectiveness of the cold fluid [defined by Eq. (3.52)], also as the exchanger effectiveness of the cold fluid in Appendix B, dimensionless
ε_{cf}	counterflow heat exchanger effectiveness [see Eq. (3.83)], dimensionless
ε_h	temperature effectiveness of the hot fluid [defined by Eq. (3.51)], also as the exchange effectiveness of the hot fluid in Appendix B, dimensionless
$\varepsilon_{h,o}$	temperature effectiveness of the hot fluid when flow is uniform on *both* fluid sides of a two-fluid heat exchanger (defined the same as ε_h), dimensionless
ε_p	heat exchanger effectiveness per pass, dimensionless
ε_r	regenerator effectiveness of a single matrix [defined by Eq. (5.81)], dimensionless
$\Delta\varepsilon^*$	effectiveness deterioration factor, dimensionless
ζ	Cartesian coordinate, $(y/L_2)C^* \cdot \mathrm{NTU}$ [see Eq. (11.21)], dimensionless
ζ_i	correction factors for shellside pressure drop terms for the Bell–Delaware method [see Eq. (9.51)]; $i = \ell$ for tube-to-baffle and baffle-to-shell leakage; $i = b$ for bypass flow; $i = s$ for inlet and outlet sections, dimensionless
η	reduced time variable for a regenerator [defined by Eq. (5.69)] with subscripts $j = c$ and h for cold- and hot-gas flow periods, dimensionless
η	exergy efficiency [defined by Eq. (11.60)], dimensionless
η_f	fin efficiency [defined by Eq. (4.129)], dimensionless
η_o	extended surface efficiency on one fluid side of the extended surface heat exchanger [see Eqs. (4.158) and (4.160) for the definition], dimensionless
$(\eta_o hA)^*$	convection conductance ratio [defined by Eq. (4.8)], dimensionless
η_p	pump/fan efficiency, dimensionless
η_ε	fin effectiveness [defined by Eq. (4.156)], dimensionless
Θ	$= 1 - \theta = (T - T_{1,i})/(T_{2,i} - T_{1,i})$ in Chapter 11 only, dimensionless
θ	angular coordinate in the cylindrical coordinate system, rad, deg
θ	excess temperature for the fin analysis in Chapter 4 [defined by Eq. (4.63)]; $\theta_0 = T_0 - T_\infty$ at the fin base, °C, °F
θ	$= (T - T_{2,i})/(T_{1,i} - T_{2,i})$ in Chapter 11 only, dimensionless

θ_b angle between two radii intersected at the inside shell wall with the baffle cut (see Fig. 8.9), rad unless explicitly mentioned in degrees

θ_b bend deflection angle (see Fig. 6.5), deg

θ_c disk sector angle for the cold-fluid stream in a rotary regenerator, rad, deg

θ_h disk sector angle for the hot-fluid stream in a rotary regenerator, rad, deg

θ_r disk sector angle covered by the radial seals in a rotary regenerator, rad, deg

θ_t $= \theta_h + \theta_c + \theta_r = 2\pi = 360°$, rad, deg

ϑ fluid temperature for internal flow in Chapter 7, $(T - T_{w,m})/(T_m - T_{w,m})$, dimensionless

ϑ ratio of fluid inlet temperatures, $T_{1,i}/T_{2,i}$ in Chapter 11 where temperatures are on absolute temperature scale, dimensionless

ϑ^* fluid temperature for external flow, $(T - T_w)/(T_\infty - T_w)$ or $(T - T_w)/(T_e - T_w)$, dimensionless

κ length effect correction factor for the overall heat transfer coefficient [see Eqs. (4.32) and (4.33)] dimensionless

κ_T isothermal compressibility, 1/Pa, ft^2/lbf

Λ reduced length for a regenerator [defined by Eqs. (5.84), (5.102), and (5.103)], dimensionless

Λ_m mean reduced length [defined by Eq. (5.91)], dimensionless

Λ^* $= \Lambda_h/\Lambda_c$, dimensionless

$\boldsymbol{\Lambda}$ wavelength of chevron plate corrugation (see Fig. 7.28), m, ft

λ longitudinal wall conduction parameter based on the total conduction area, $\lambda = k_w A_{k,t}/C_{min}L$, $\lambda_c = k_w A_{k,c}/C_c L_c$, $\lambda_h = k_w A_{w,h}/C_h L_h$, dimensionless

μ fluid dynamic viscosity, Pa $\cdot$ s, 1bm/hr-ft

ν fluid kinematic viscosity μ/ρ, m^2/s, ft^2/sec

ξ reduced length variable for regenerator [defined by Eq. (5.69], dimensionless

ξ axial coordinate in Chapter 11, x/L, dimensionless

Π reduced period for a regenerator [defined by Eqs. (5.84), (5.104), and (5.105)], dimensionless

Π_m harmonic mean reduced period [defined by Eq. 5.90)], dimensionless

ρ fluid density, kg/m^3, 1bm/ft^3

σ ratio of free flow area to frontal area, A_o/A_{fr}, dimensionless

τ time, s, sec

τ_d delay period or induction period associated with initiation of fouling in Chapter 13; dwell time, residence time, or transit time of a fluid particle in a heat exchanger in Chapter 5, s, sec

$\tau_{d,min}$ dwell time of the C_{min} fluid, s, sec

τ_s fluid shear stress, Pa, psf

τ_w equivalent fluid shear stress at wall, Pa, psf (psi)

τ^* time variable, $\tau/\tau_{d,min}$, dimensionless

τ_c^*, τ_h^* time variable for the cold and hot fluids [defined by Eq. (5.26)], dimensionless

$\phi(\cdot)$ denotes a functional relationship

χ axial coordinate, (x/L_1)NTU, in Chapter 11 only, dimensionless

χ_i	fractional distribution of the ith shaped passage, dimensionless
ψ	$\Delta T_m/(T_{h,i} - T_{c,i})$, dimensionless
ψ	removal resistance [scale strength factor; see Eq. (13.12)], dimensionless
Ω	water quality factor, dimensionless

Subscripts

A	unit (row, section) A
a	air side
B	unit (row, section) B
b	bend, tube bundle, or lateral branch
c	cold-fluid side, clean surface in Chapter 13
cf	counterflow
cp	constant properties
cr	crossflow section in a segmental baffled exchanger
cv	control volume
cu	cold utility
d	deposit
df	displaced fluid
eff	effective
f	fouling, fluid in Section 7.3.3.2
g	gas side
H	constant axial wall heat flux boundary condition
h	hot-fluid side
hu	hot utility
hex	heat exchanger
$H1$	thermal boundary condition referring to constant axial wall heat flux with constant peripheral wall temperature
i	inlet to the exchanger
i	inside surface in Chapter 13
id	ideal
iso	isothermal
L	coupled liquid
leak	caused by a leak
lm	logarithmic mean
m	mean or bulk mean, manifold (in Chapter 12)
max	maximum
min	minimum
mixing	caused by mixing
ms	maldistributed fluid
n	nominal or reference passage in Chapter 12
o	overall
o	outside surface in Chapter 13

o	outlet to the exchanger when used as a subscript with the temperature
opt	optimal
otl	outer tube limit in a shell-and-tube heat exchanger
p	pass, except for plain surface in Section 10.3
pf	parallelflow
r	reentrainment
ref	referent thermodynamic conditions
s	shellside; steam; structural
std	arbitrarily selected standard temperature and pressure conditions
T	constant wall temperature boundary condition
t	tubeside, tube
tot	total
v	viscous
w	wall or properties at the wall temperature, window zone for a shell-and-tube exchanger
w	water
x	local value at section x along the flow length
1	fluid 1; one section (inlet or outlet) of the exchanger
2	fluid 2; other section (outlet or inlet) of the exchanger
∞	free stream

1 Classification of Heat Exchangers

A variety of heat exchangers are used in industry and in their products. The objective of this chapter is to describe most of these heat exchangers in some detail using classification schemes. Starting with a definition, heat exchangers are classified according to transfer processes, number of fluids, degree of surface compactness, construction features, flow arrangements, and heat transfer mechanisms. With a detailed classification in each category, the terminology associated with a variety of these exchangers is introduced and practical applications are outlined. A brief mention is also made of the differences in design procedure for the various types of heat exchangers.

1.1 INTRODUCTION

A *heat exchanger* is a device that is used to transfer thermal energy (enthalpy) between two or more fluids, between a solid surface and a fluid, or between solid particulates and a fluid, at different temperatures and in thermal contact. In heat exchangers, there are usually no external heat and work interactions. Typical applications involve heating or cooling of a fluid stream of concern and evaporation or condensation of single- or multicomponent fluid streams. In other applications, the objective may be to recover or reject heat, or sterilize, pasteurize, fractionate, distill, concentrate, crystallize, or control a process fluid. In a few heat exchangers, the fluids exchanging heat are in direct contact. In most heat exchangers, heat transfer between fluids takes place through a separating wall or into and out of a wall in a transient manner. In many heat exchangers, the fluids are separated by a heat transfer surface, and ideally they do not mix or leak. Such exchangers are referred to as *direct transfer type*, or simply *recuperators*. In contrast, exchangers in which there is intermittent heat exchange between the hot and cold fluids—via thermal energy storage and release through the exchanger surface or matrix— are referred to as *indirect transfer type*, or simply *regenerators*. Such exchangers usually have fluid leakage from one fluid stream to the other, due to pressure differences and matrix rotation/valve switching. Common examples of heat exchangers are shell-and-tube exchangers, automobile radiators, condensers, evaporators, air preheaters, and cooling towers. If no phase change occurs in any of the fluids in the exchanger, it is sometimes referred to as a *sensible heat exchanger*. There could be internal thermal energy sources in the exchangers, such as in electric heaters and nuclear fuel elements. Combustion and chemical reaction may take place within the exchanger, such as in boilers, fired heaters, and fluidized-bed exchangers. Mechanical devices may be used in some exchangers such as in scraped surface exchangers, agitated vessels, and stirred tank reactors. Heat transfer in the separating wall of a recuperator generally takes place by

1

conduction. However, in a heat pipe heat exchanger, the heat pipe not only acts as a separating wall, but also facilitates the transfer of heat by condensation, evaporation, and conduction of the working fluid inside the heat pipe. In general, if the fluids are immiscible, the separating wall may be eliminated, and the interface between the fluids replaces a heat transfer surface, as in a direct-contact heat exchanger.

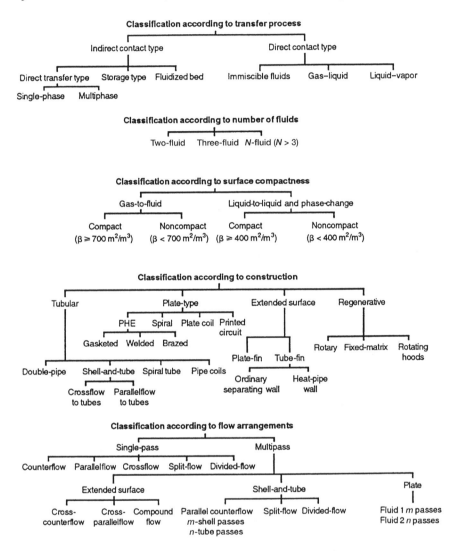

FIGURE 1.1   Classification of heat exchangers (Shah, 1981).

A heat exchanger consists of *heat transfer elements* such as a core or matrix containing the heat transfer surface, and *fluid distribution elements* such as headers, manifolds, tanks, inlet and outlet nozzles or pipes, or seals. Usually, there are no moving parts in a heat exchanger; however, there are exceptions, such as a rotary regenerative exchanger (in which the matrix is mechanically driven to rotate at some design speed) or a scraped surface heat exchanger.

The heat transfer surface is a surface of the exchanger core that is in direct contact with fluids and through which heat is transferred by conduction. That portion of the surface that is in direct contact with both the hot and cold fluids and transfers heat between them is referred to as the *primary* or *direct surface*. To increase the heat transfer area, appendages may be intimately connected to the primary surface to provide an *extended, secondary,* or *indirect surface*. These extended surface elements are referred to as *fins*. Thus, heat is conducted through the fin and convected (and/or radiated) from the fin (through the surface area) to the surrounding fluid, or vice versa, depending on whether the fin is being cooled or heated. As a result, the addition of fins to the primary surface reduces the thermal resistance on that side and thereby increases the total heat transfer from the surface for the same temperature difference. Fins may form flow passages for the individual fluids but do not separate the two (or more) fluids of the exchanger. These secondary surfaces or fins may also be introduced primarily for structural strength purposes or to provide thorough mixing of a highly viscous liquid.

Not only are heat exchangers often used in the process, power, petroleum, transportation, air-conditioning, refrigeration, cryogenic, heat recovery, alternative fuel, and manufacturing industries, they also serve as key components of many industrial products available in the marketplace. These exchangers can be classified in many different ways. We will classify them according to transfer processes, number of fluids, and heat transfer mechanisms. Conventional heat exchangers are further classified according to construction type and flow arrangements. Another arbitrary classification can be made, based on the heat transfer surface area/volume ratio, into compact and noncompact heat exchangers. This classification is made because the type of equipment, fields of applications, and design techniques generally differ. All these classifications are summarized in Fig. 1.1 and discussed further in this chapter. Heat exchangers can also be classified according to the process function, as outlined in Fig. 1.2. However, they are not discussed here and the reader may refer to Shah and Mueller (1988). Additional ways to classify heat exchangers are by fluid type (gas–gas, gas–liquid, liquid–liquid, gas two-phase, liquid two-phase, etc.), industry, and so on, but we do not cover such classifications in this chapter.

1.2 CLASSIFICATION ACCORDING TO TRANSFER PROCESSES

Heat exchangers are classified according to transfer processes into indirect- and direct-contact types.

1.2.1 Indirect-Contact Heat Exchangers

In an indirect-contact heat exchanger, the fluid streams remain separate and the heat transfers continuously through an impervious dividing wall or into and out of a wall in a transient manner. Thus, ideally, there is no direct contact between thermally interacting fluids. This type of heat exchanger, also referred to as a *surface heat exchanger*, can be further classified into direct-transfer type, storage type, and fluidized-bed exchangers.

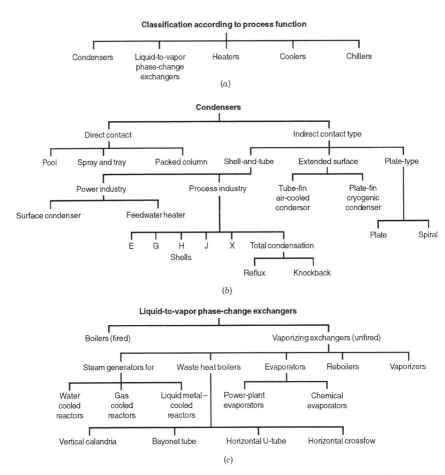

FIGURE 1.2   (a) Classification according to process function; (b) classification of condensers; (c) classification of liquid-to-vapor phase-change exchangers.

1.2.1.1   Direct-Transfer Type Exchangers. In this type, heat transfers continuously from the hot fluid to the cold fluid through a dividing wall. Although a simultaneous flow of two (or more) fluids is required in the exchanger, there is no direct mixing of the two (or more) fluids because each fluid flows in separate fluid passages. In general, there are no moving parts in most such heat exchangers. This type of exchanger is designated as a recuperative heat exchanger or simply as a *recuperator*.[†] Some examples of direct-transfer type heat exchangers are tubular, plate-type, and extended surface exchangers. Note that the term *recuperator* is not commonly used in the process industry for shell-

---

[†] In vehicular gas turbines, a stationary heat exchanger is usually referred to as a *recuperator*, and a rotating heat exchanger as a *regenerator*. However, in industrial gas turbines, by long tradition and in a thermodynamic sense, a stationary heat exchanger is generally referred to as a regenerator. Hence, a gas turbine *regenerator* could be either a recuperator or a regenerator in a strict sense, depending on the construction. In power plants, a heat exchanger is not called a recuperator, but is, rather, designated by its function or application.

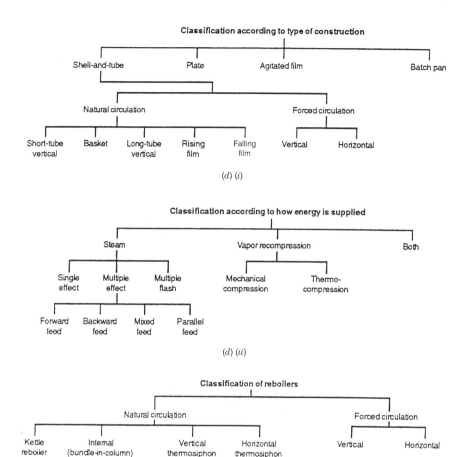

FIGURE 1.2   (*d*) classification of chemical evaporators according to (i) the type of construction, and (ii) how energy is supplied (Shah and Mueller, 1988); (*e*) classification of reboilers.

and-tube and plate heat exchangers, although they are also considered as recuperators. Recuperators are further subclassified as prime surface exchangers and extended-surface exchangers. *Prime surface exchangers* do not employ fins or extended surfaces on any fluid side. Plain tubular exchangers, shell-and-tube exchangers with plain tubes, and plate exchangers are good examples of prime surface exchangers. Recuperators constitute a vast majority of all heat exchangers.

1.2.1.2   Storage Type Exchangers.   In a storage type exchanger, both fluids flow alternatively through the same flow passages, and hence heat transfer is intermittent. The heat transfer surface (or flow passages) is generally cellular in structure and is referred to as a *matrix* (see Fig. 1.43), or it is a permeable (porous) solid material, referred to as a *packed bed*. When hot gas flows over the heat transfer surface (through flow passages), the thermal energy from the hot gas is stored in the matrix wall, and thus the hot gas is being cooled during the matrix heating period. As cold gas flows through the same passages later (i.e., during the matrix cooling period), the matrix wall gives up thermal energy, which is absorbed by the cold fluid. Thus, heat is not transferred continuously through the wall as in a direct-transfer type exchanger (recuperator), but the corresponding thermal energy is alternately stored and released by the matrix wall. This storage type heat exchanger is also referred to as a *regenerative heat exchanger*, or simply as a *regenerator.*[†] To operate continuously and within a desired temperature range, the gases, headers, or matrices are switched periodically (i.e., rotated), so that the same passage is occupied periodically by hot and cold gases, as described further in Section 1.5.4. The actual time that hot gas takes to flow through a cold regenerator matrix is called the *hot period* or *hot blow*, and the time that cold gas flows through the hot regenerator matrix is called the *cold period* or *cold blow*. For successful operation, it is not necessary to have hot- and cold-gas flow periods of equal duration. There is some unavoidable carryover of a small fraction of the fluid trapped in the passage to the other fluid stream just after switching of the fluids; this is referred to as *carryover leakage*. In addition, if the hot and cold fluids are at different pressures, there will be leakage from the high-pressure fluid to the low-pressure fluid past the radial, peripheral, and axial seals, or across the valves. This leakage is referred to as *pressure leakage*. Since these leaks are unavoidable, regenerators are used exclusively in gas-to-gas heat (and mass) transfer applications with sensible heat transfer; in some applications, regenerators may transfer moisture from humid air to dry air up to about 5%.

For heat transfer analysis of regenerators, the ε-NTU method of recuperators needs to be modified to take into account the thermal energy storage capacity of the matrix. We discuss the design theory of regenerators in detail in Chapter 5.

1.2.1.3 Fluidized-Bed Heat Exchangers.

In a fluidized-bed heat exchanger, one side of a two-fluid exchanger is immersed in a bed of finely divided solid material, such as a tube bundle immersed in a bed of sand or coal particles, as shown in Fig. 1.3. If the upward fluid velocity on the bed side is low, the solid particles will remain fixed in position in the bed and the fluid will flow through the interstices of the bed. If the upward fluid velocity is high, the solid particles will be carried away with the fluid. At a "proper" value of the fluid velocity, the upward drag force is slightly higher than the weight of the bed particles. As a result, the solid particles will float with an increase in bed volume, and the bed behaves as a liquid. This characteristic of the bed is referred to as a *fluidized condition*. Under this condition, the fluid pressure drop through the bed remains almost constant, independent of the flow rate, and a strong mixing of the solid particles occurs. This results in a uniform temperature for the total bed (gas and particles) with an apparent thermal conductivity of the solid particles as infinity. Very high heat transfer coefficients are achieved on the fluidized side compared to particle-free or dilute-phase particle gas flows. Chemical reaction is common on the fluidized side in many process applications, and combustion takes place in coal combustion fluidized beds. The common applications of the fluidized-bed heat exchanger are drying, mixing, adsorption, reactor engineering, coal combustion, and waste heat recovery. Since the

---

[†] Regenerators are also used for storing thermal energy for later use, as in the storage of thermal energy. Here the objective is how to store the maximum fraction of the input energy and minimize heat leakage. However, we do not concentrate on this application in this book.

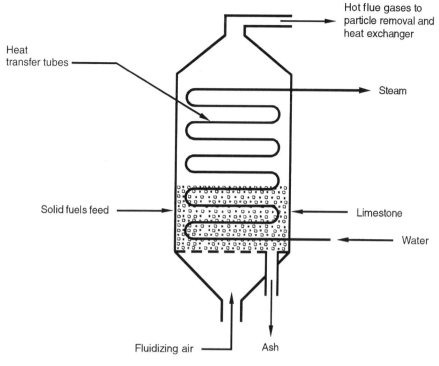

FIGURE 1.3   Fluidized-bed heat exchanger.

initial temperature difference $(T_{h,i} - T_{f,i})^{\dagger}$ is reduced due to fluidization, the exchanger effectiveness is lower, and hence ε-NTU theory for a fluidized-bed exchanger needs to be modified (Suo, 1976). Chemical reaction and combustion further complicate the design of these exchangers but are beyond the scope of this book.

1.2.2   Direct-Contact Heat Exchangers

In a direct-contact exchanger, two fluid streams come into direct contact, exchange heat, and are then separated. Common applications of a direct-contact exchanger involve mass transfer in addition to heat transfer, such as in evaporative cooling and rectification; applications involving only sensible heat transfer are rare. The enthalpy of phase change in such an exchanger generally represents a significant portion of the total energy transfer. The phase change generally enhances the heat transfer rate. Compared to indirect-contact recuperators and regenerators, in direct-contact heat exchangers, (1) very high heat transfer rates are achievable, (2) the exchanger construction is relatively inexpensive, and (3) the fouling problem is generally nonexistent, due to the absence of a heat transfer surface (wall) between the two fluids. However, the applications are limited to those cases where a direct contact of two fluid streams is permissible. The design theory for these

---

† $T_{h,i}$, inlet temperature of the hot fluid to the fluidized bed; $T_{f,i}$, temperature of the fluidized bed itself at the inlet.

exchangers is beyond the scope of this book and is not covered. These exchangers may be further classified as follows.

1.2.2.1 Immiscible Fluid Exchangers. In this type, two immiscible fluid streams are brought into direct contact. These fluids may be single-phase fluids, or they may involve condensation or vaporization. Condensation of organic vapors and oil vapors with water or air are typical examples.

1.2.2.2 Gas–Liquid Exchangers. In this type, one fluid is a gas (more commonly, air) and the other a low-pressure liquid (more commonly, water) and are readily separable after the energy exchange. In either cooling of liquid (water) or humidification of gas (air) applications, liquid partially evaporates and the vapor is carried away with the gas. In these exchangers, more than 90% of the energy transfer is by virtue of mass transfer (due to the evaporation of the liquid), and convective heat transfer is a minor mechanism. A "wet" (water) cooling tower with forced- or natural-draft airflow is the most common application. Other applications are the air-conditioning spray chamber, spray drier, spray tower, and spray pond.

1.2.2.3 Liquid–Vapor Exchangers. In this type, typically steam is partially or fully condensed using cooling water, or water is heated with waste steam through direct contact in the exchanger. Noncondensables and residual steam and hot water are the outlet streams. Common examples are desuperheaters and open feedwater heaters (also known as *deaeraters*) in power plants.

1.3   CLASSIFICATION ACCORDING TO NUMBER OF FLUIDS

Most processes of heating, cooling, heat recovery, and heat rejection involve transfer of heat between two fluids. Hence, two-fluid heat exchangers are the most common. Three-fluid heat exchangers are widely used in cryogenics and some chemical processes (e.g., air separation systems, a helium–air separation unit, purification and liquefaction of hydrogen, ammonia gas synthesis). Heat exchangers with as many as 12 fluid streams have been used in some chemical process applications. The design theory of three- and multifluid heat exchangers is algebraically very complex and is not covered in this book. Exclusively, only the design theory for two-fluid exchangers and some associated problems are presented in this book.

1.4   CLASSIFICATION ACCORDING TO SURFACE COMPACTNESS

Compared to shell-and-tube exchangers, compact heat exchangers are characterized by a large heat transfer surface area per unit volume of the exchanger, resulting in reduced space, weight, support structure and footprint, energy requirements and cost, as well as improved process design and plant layout and processing conditions, together with low fluid inventory.

A gas-to-fluid exchanger is referred to as a *compact heat exchanger* if it incorporates a heat transfer surface having a surface area density greater than about $700 \, \text{m}^2/\text{m}^3$

$(213 \, \text{ft}^2/\text{ft}^3)^\dagger$ or a hydraulic diameter $D_h \le 6 \, \text{mm}$ ($\frac{1}{4}$ in.) for operating in a gas stream and $400 \, \text{m}^2/\text{m}^3$ ($122 \, \text{ft}^2/\text{ft}^3$) or higher for operating in a liquid or phase-change stream. A *laminar flow heat exchanger* (also referred to as a *meso heat exchanger*) has a surface area density greater than about $3000 \, \text{m}^2/\text{m}^3$ ($914 \, \text{ft}^2/\text{ft}^3$) or $100 \, \mu\text{m} \le D_h \le 1 \, \text{mm}$. The term *micro heat exchanger* is used if the surface area density is greater than about $15{,}000 \, \text{m}^2/\text{m}^3$ ($4570 \, \text{ft}^2/\text{ft}^3$) or $1 \, \mu\text{m} \le D_h \le 100 \, \mu\text{m}$. A liquid/two-phase fluid heat exchanger is referred to as a *compact heat exchanger* if the surface area density on any one fluid side is greater than about $400 \, \text{m}^2/\text{m}^3$. In contrast, a typical process industry shell-and-tube exchanger has a surface area density of less than $100 \, \text{m}^2/\text{m}^3$ on one fluid side with plain tubes, and two to three times greater than that with high-fin-density low-finned tubing. A typical plate heat exchanger has about twice the average heat transfer coefficient h on one fluid side or the average overall heat transfer coefficient U than that for a shell-and-tube exchanger for water/water applications. A compact heat exchanger is *not* necessarily of small bulk and mass. However, if it did not incorporate a surface of high-surface-area density, it would be much more bulky and massive. Plate-fin, tube-fin, and rotary regenerators are examples of compact heat exchangers for gas flow on one or both fluid sides, and gasketed, welded, brazed plate heat exchangers and printed-circuit heat exchangers are examples of compact heat exchangers for liquid flows. Basic flow arrangements of two-fluid compact heat exchangers are single-pass crossflow, counterflow, and multipass cross-counterflow (see Section 1.6 for details); for noncompact heat exchangers, many other flow arrangements are also used. The aforementioned last two flow arrangements for compact or noncompact heat exchangers can yield a very high exchanger effectiveness value or a very small temperature approach (see Section 3.2.3 for the definition) between fluid streams.

A spectrum of surface area density of heat exchanger surfaces is shown in Fig. 1.4. On the bottom of the figure, two scales are shown: the heat transfer surface area density β (m^2/m^3) and the hydraulic diameter D_h,‡ (mm), which is the tube inside or outside diameter D (mm) for a thin-walled circular tube. Different heat exchanger surfaces are shown in the rectangles. When projected on the β (or D_h) scale, the short vertical sides of a rectangle indicate the range of surface area density (or hydraulic diameter) for the particular surface in question. What is referred to as β in this figure is either β_1 or β_2, defined as follows. For plate heat exchangers, plate-fin exchangers, and regenerators,

$$\beta_1 = \frac{A_h}{V_h} \quad \text{or} \quad \frac{A_c}{V_c} \tag{1.1}$$

For tube-fin exchangers and shell-and-tube exchangers,

$$\beta_2 = \frac{A_h}{V_{\text{total}}} \quad \text{or} \quad \frac{A_c}{V_{\text{total}}} \tag{1.2}$$

---

† The unit conversion throughout the book may not be exact; it depends on whether the number is exact or is an engineering value.

‡ The hydraulic diameter is defined as $4A_o/\mathbf{P}$, where A_o is the minimum free-flow area on one fluid side of a heat exchanger and $\mathbf{P}$ is the wetted perimeter of flow passages of that side. Note that the wetted perimeter can be different for heat transfer and pressure drop calculations. For example, the hydraulic diameter for an annulus of a double-pipe heat exchanger for q and Δp calculations is as follows.

$$D_{h,q} = \frac{4(\pi/4)(D_o^2 - D_i^2)}{\pi D_i} = \frac{D_o^2 - D_i^2}{D_i} \qquad D_{h,\Delta p} = \frac{4(\pi/4)(D_o^2 - D_i^2)}{\pi(D_o + D_i)} = D_o - D_i$$

where D_o is the inside diameter of the outer pipe and D_i is the outside diameter of the inside pipe of a double-pipe exchanger. See also Eq. (3.65) for a more precise definition of the hydraulic diameter.

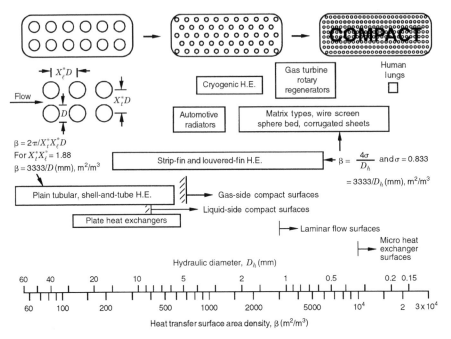

FIGURE 1.4   Heat transfer surface area density spectrum of exchanger surfaces (Shah, 1981).

Here A is the heat transfer surface area, V the exchanger volume, and the subscripts h and c denote hot and cold fluid sides, respectively. V_h and V_c are the volumes individually occupied by the hot- and cold-fluid-side heat transfer surfaces. From this point on in the book, β_1 is simply designated as β and β_2 is designated as α:

$$\beta = \beta_1 \qquad \alpha = \beta_2 \tag{1.3}$$

Note that both β and α (with the definitions noted above) are used in defining the surface area densities of a plate-fin surface; however, only α is used in defining the surface area density of a tube-fin surface since β has no meaning in this case. The following specific values are used in preparing Fig. 1.4:

- For a shell-and-tube exchanger, an inline arrangement[†] is considered with $X_t^* X_l^* = 1.88$.
- For plate and plate-fin exchangers, the porosity between plates is taken as 0.8333; and for a regenerator, the porosity of matrix surface is taken as 0.8333. With these values, β (m^2/m^3) and D_h (mm) are related as $\beta = 3333/D_h$.

[†] The tube array is idealized as infinitely large with thin-walled circular tubes. X_t^* and X_l^* are the transverse and longitudinal tube pitches normalized with respect to the tube outside diameter. Refer to Table 8.1 for the definitions of tube pitches.

Note that some industries quote the total surface area (of hot- and cold-fluid sides) in their exchanger specifications. However, in calculations of heat exchanger design, we need individual fluid-side heat transfer surface areas; and hence we use here the definitions of β and α as given above.

Based on the foregoing definition of a compact surface, a tube bundle having 5 mm (0.2 in.) diameter tubes in a shell-and-tube exchanger comes close to qualifying as a compact exchanger. As β or α varies inversely with the tube diameter, the 25.4 mm (1 in.) diameter tubes used in a power plant condenser result in a noncompact exchanger. In contrast, a 1990s automobile radiator [790 fins/m (20 fins/in.)] has a surface area density β on the order of 1870 m^2/m^3 (570 ft^2/ft^3) on the air side, which is equivalent to 1.8 mm (0.07 in.) diameter tubes. The regenerators in some vehicular gas turbine engines under development have matrices with an area density on the order of 6600 m^2/m^3 (2000 ft^2/ft^3), which is equivalent to 0.5 mm (0.02 in.) diameter tubes in a bundle. Human lungs are one of the most compact heat-and-mass exchangers, having a surface area density of about 17,500 m^2/m^3 (5330 ft^2/ft^3), which is equivalent to 0.19 mm (0.0075 in.) diameter tubes. Some micro heat exchangers under development are as compact as the human lung (Shah, 1991a) and also even more compact.

The motivation for using compact surfaces is to gain specified heat exchanger performance, $q/\Delta T_m$, within acceptably low mass and box volume constraints. The heat exchanger performance may be expressed as

$$\frac{q}{\Delta T_m} = UA = U\beta V \tag{1.4}$$

where q is the heat transfer rate, ΔT_m the true mean temperature difference, and U the overall heat transfer coefficient. Clearly, a high β value minimizes exchanger volume V for specified $q/\Delta T_m$. As explained in Section 7.4.1.1, compact surfaces (having small D_h) generally result in a higher heat transfer coefficient and a higher overall heat transfer coefficient U, resulting in a smaller volume. As compact surfaces can achieve structural stability and strength with thinner-gauge material, the gain in a lower exchanger mass is even more pronounced than the gain in a smaller volume.

1.4.1 Gas-to-Fluid Exchangers

The heat transfer coefficient h for gases is generally one or two orders of magnitude lower than that for water, oil, and other liquids. Now, to minimize the size and weight of a gas-to-liquid heat exchanger, the thermal conductances (hA products) on both sides of the exchanger should be approximately the same. Hence, the heat transfer surface on the gas side needs to have a much larger area and be more compact than can be realized practically with the circular tubes commonly used in shell-and-tube exchangers. Thus, for an approximately balanced design (about the same hA values), a compact surface is employed on the gas side of gas-to-gas, gas-to-liquid, and gas-to-phase change heat exchangers.

The unique characteristics of compact extended-surface (plate-fin and tube-fin) exchangers, compared to conventional shell-and-tube exchangers (see Fig. 1.6), are as follows:

- Availability of numerous surfaces having different orders of magnitude of surface area density

- Flexibility in distributing surface area on the hot and cold sides as warranted by design considerations
- Generally, substantial cost, weight, or volume savings.

The important design and operating considerations for compact extended-surface exchangers are as follows:

- Usually, at least one of the fluids is a gas having a low h value.
- Fluids must be clean and relatively noncorrosive because of low-D_h flow passages and no easy techniques for cleaning.
- The fluid pumping power (and hence the pressure drop) is often as important as the heat transfer rate.
- Operating pressures and temperatures are somewhat limited compared to shell-and-tube exchangers, due to joining of the fins to plates or tubes by brazing, mechanical expansion, and so on.
- With the use of highly compact surfaces, the resulting shape of the exchanger is one having a large frontal area and a short flow length; the header design of a compact heat exchanger is thus important for achieving uniform flow distribution among very large numbers of small flow passages.
- The market potential must be large enough to warrant the sizable initial manufacturing tooling and equipment costs.

Fouling is a major potential problem in compact heat exchangers (except for plate-and-frame heat exchangers), particularly those having a variety of fin geometries or very fine circular or noncircular flow passages that cannot be cleaned mechanically. Chemical cleaning may be possible; thermal baking and subsequent rinsing are possible for small units.[†] Hence, extended-surface compact heat exchangers may not be used in heavy fouling applications. Nonfouling fluids are used where permissible, such as clean air or gases, light hydrocarbons, and refrigerants.

1.4.2 Liquid-to-Liquid and Phase-Change Exchangers

Liquid-to-liquid and phase-change exchangers are gasketed plate-and-frame and welded plate, spiral plate, and printed-circuit exchangers. Some of them are described in detail in Section 1.5.2.

1.5 CLASSIFICATION ACCORDING TO CONSTRUCTION FEATURES

Heat exchangers are frequently characterized by construction features. Four major construction types are tubular, plate-type, extended surface, and regenerative exchangers. Heat exchangers with other constructions are also available, such as scraped surface exchanger, tank heater, cooler cartridge exchanger, and others (Walker, 1990). Some of these may be classified as tubular exchangers, but they have some unique features compared to conventional tubular exchangers. Since the applications of these exchangers

---

[†] Some additional techniques for cleaning and mitigation of fouling are summarized in Section 13.4.

are specialized, we concentrate here only on the four major construction types noted above.

Although the ε-NTU and MTD methods (see end of Section 3.2.2) are identical for tubular, plate-type, and extended-surface exchangers, the influence of the following factors must be taken into account in exchanger design: corrections due to leakage and bypass streams in a shell-and-tube exchanger, effects due to a few plates in a plate exchanger, and fin efficiency in an extended-surface exchanger. Similarly, the ε-NTU method must be modified to take into account the heat capacity of the matrix in a regenerator. Thus, the detailed design theory differs for each construction type and is discussed in detail in Chapters 3 through 5. Let us first discuss the construction features of the four major types.

1.5.1 Tubular Heat Exchangers

These exchangers are generally built of circular tubes, although elliptical, rectangular, or round/flat twisted tubes have also been used in some applications. There is considerable flexibility in the design because the core geometry can be varied easily by changing the tube diameter, length, and arrangement. Tubular exchangers can be designed for high pressures relative to the environment and high-pressure differences between the fluids. Tubular exchangers are used primarily for liquid-to-liquid and liquid-to-phase change (condensing or evaporating) heat transfer applications. They are used for gas-to-liquid and gas-to-gas heat transfer applications primarily when the operating temperature and/ or pressure is very high or fouling is a severe problem on at least one fluid side and no other types of exchangers would work. These exchangers may be classified as shell-and-tube, double-pipe, and spiral tube exchangers. They are all prime surface exchangers except for exchangers having fins outside/inside tubes.

1.5.1.1 Shell-and-Tube Exchangers. This exchanger, shown in Fig. 1.5, is generally built of a bundle of round tubes mounted in a cylindrical shell with the tube axis parallel to that of the shell. One fluid flows inside the tubes, the other flows across and along the tubes. The major components of this exchanger are tubes (or tube bundle), shell, front-end head, rear-end head, baffles, and tubesheets, and are described briefly later in this subsection. For further details, refer to Section 10.2.1.

A variety of different internal constructions are used in shell-and-tube exchangers, depending on the desired heat transfer and pressure drop performance and the methods employed to reduce thermal stresses, to prevent leakages, to provide for ease of cleaning, to contain operating pressures and temperatures, to control corrosion, to accommodate highly asymmetric flows, and so on. Shell-and-tube exchangers are classified and constructed in accordance with the widely used TEMA (Tubular Exchanger Manufacturers Association) standards (TEMA, 1999), DIN and other standards in Europe and elsewhere, and ASME (American Society of Mechanical Engineers) boiler and pressure vessel codes. TEMA has developed a notation system to designate major types of shell-and-tube exchangers. In this system, each exchanger is designated by a three-letter combination, the first letter indicating the front-end head type, the second the shell type, and the third the rear-end head type. These are identified in Fig. 1.6. Some common shell-and-tube exchangers are AES, BEM, AEP, CFU, AKT, and AJW. It should be emphasized that there are other special types of shell-and-tube exchangers commercially available that have front- and rear-end heads different from those in Fig. 1.6. Those exchangers may not be identifiable by the TEMA letter designation.

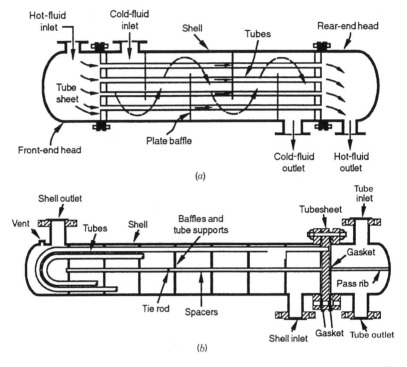

FIGURE 1.5   (*a*) Shell-and-tube exchanger (BEM) with one shell pass and one tube pass; (*b*) shell-and-tube exchanger (BEU) with one shell pass and two tube passes.

The three most common types of shell-and-tube exchangers are (1) fixed tubesheet design, (2) U-tube design, and (3) floating-head type. In all three types, the front-end head is stationary while the rear-end head can be either stationary or floating (see Fig. 1.6), depending on the thermal stresses in the shell, tube, or tubesheet, due to temperature differences as a result of heat transfer.

The exchangers are built in accordance with three mechanical standards that specify design, fabrication, and materials of unfired shell-and-tube heat exchangers. Class R is for the generally severe requirements of petroleum and related processing applications. Class C is for generally moderate requirements for commercial and general process applications. Class B is for chemical process service. The exchangers are built to comply with the applicable *ASME Boiler and Pressure Vessel Code*, Section VIII (1998), and other pertinent codes and/or standards. The TEMA standards supplement and define the ASME code for heat exchanger applications. In addition, state and local codes applicable to the plant location must also be met.

The TEMA standards specify the manufacturing tolerances for various mechanical classes, the range of tube sizes and pitches, baffling and support plates, pressure classification, tubesheet thickness formulas, and so on, and must be consulted for all these details. In this book, we consider only the TEMA standards where appropriate, but there are other standards, such as DIN 28 008.

Tubular exchangers are widely used in industry for the following reasons. They are custom designed for virtually any capacity and operating conditions, such as from high

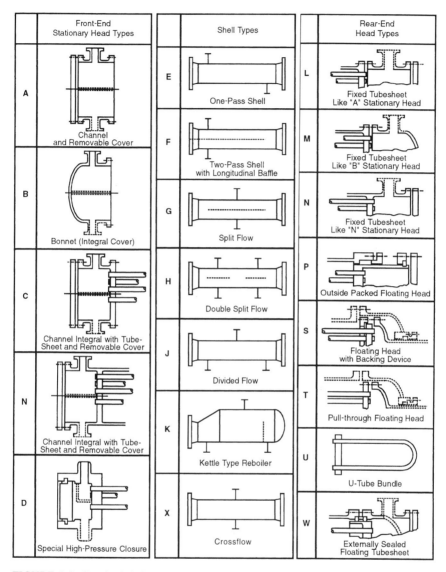

Front-End Stationary Head Types		Shell Types		Rear-End Head Types	
A	Channel and Removable Cover	E	One-Pass Shell	L	Fixed Tubesheet Like "A" Stationary Head
B	Bonnet (Integral Cover)	F	Two-Pass Shell with Longitudinal Baffle	M	Fixed Tubesheet Like "B" Stationary Head
C	Channel Integral with Tube-Sheet and Removable Cover	G	Split Flow	N	Fixed Tubesheet Like "N" Stationary Head
		H	Double Split Flow	P	Outside Packed Floating Head
N	Channel Integral with Tube-Sheet and Removable Cover	J	Divided Flow	S	Floating Head with Backing Device
				T	Pull-through Floating Head
		K	Kettle Type Reboiler	U	U-Tube Bundle
D	Special High-Pressure Closure	X	Crossflow	W	Externally Sealed Floating Tubesheet

FIGURE 1.6   Standard shell types and front- and rear-end head types (From TEMA, 1999).

vacuum to ultrahigh pressure [over 100 MPa (15,000 psig)], from cryogenics to high temperatures [about 1100°C (2000°F)] and any temperature and pressure differences between the fluids, limited only by the materials of construction. They can be designed for special operating conditions: vibration, heavy fouling, highly viscous fluids, erosion, corrosion, toxicity, radioactivity, multicomponent mixtures, and so on. They are the most versatile exchangers, made from a variety of metal and nonmetal materials (such as graphite, glass, and Teflon) and range in size from small [0.1 m² (1 ft²)] to supergiant [over 10^5 m² (10^6 ft²)] surface area. They are used extensively as process heat exchangers in the petroleum-refining and chemical industries; as steam generators, condensers, boiler feedwater heaters, and oil coolers in power plants; as condensers and evaporators in some air-conditioning and refrigeration applications; in waste heat recovery applications with heat recovery from liquids and condensing fluids; and in environmental control.

Next, major components of shell-and-tube exchangers are briefly described.

Tubes. Round tubes in various shapes are used in shell-and-tube exchangers. Most common are the tube bundles[†] with straight and U-tubes (Fig. 1.5) used in process and power industry exchangers. However, sine-wave bend, J-shape, L-shape or hockey sticks, and inverted hockey sticks are used in advanced nuclear exchangers to accommodate large thermal expansion of the tubes. Some of the enhanced tube geometries used in shell-and-tube exchangers are shown in Fig. 1.7. Serpentine, helical, and bayonet are other tube shapes (shown in Fig. 1.8) that are used in shell-and-tube exchangers. In most applications, tubes have single walls, but when working with radioactive,

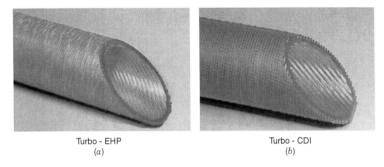

Turbo - EHP   Turbo - CDI
(a)   (b)

FIGURE 1.7   Some enhanced tube geometries used in shell-and-tube exchangers: (*a*) internally and externally enhanced evaporator tube; (*b*) internally and externally enhanced condenser tube. (Courtesy of Wolverine Tube, Inc., Decatur, AL.)

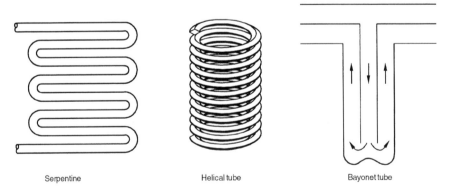

Serpentine   Helical tube   Bayonet tube

FIGURE 1.8   Additional tube configurations used in shell-and-tube exchangers.

[†] A *tube bundle* is an assembly of tubes, baffles, tubesheets and tie rods, and support plates and longitudinal baffles, if any.

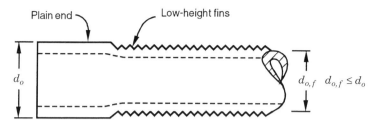

FIGURE 1.9   Low-finned tubing. The plain end goes into the tubesheet.

reactive, or toxic fluids and potable water, double-wall tubing is used. In most applications, tubes are bare, but when gas or low-heat-transfer coefficient liquid is used on the shell side, low-height fins (*low fins*) are used on the shell side. Also, special high-flux-boiling surfaces employ modified low-fin tubing. These are usually integral fins made from a thick-walled tube, shown in Fig. 1.9. Tubes are drawn, extruded, or welded, and they are made from metals, plastics, and ceramics, depending on the applications.

Shells.   The shell is a container for the shell fluid.[†] Usually, it is cylindrical in shape with a circular cross section, although shells of different shapes are used in specific applications and in nuclear heat exchangers to conform to the tube bundle shape. The shell is made from a circular pipe if the shell diameter is less than about 0.6 m (2 ft) and is made from a metal plate rolled and welded longitudinally for shell diameters greater than 0.6 m (2 ft). Seven types of shell configurations, standardized by TEMA (1999), are E, F, G, H, J, K, and X, shown in Fig. 1.6. The E shell is the most common, due to its low cost and simplicity, and has the highest log-mean temperature-difference correction factor F (see Section 3.7.2 for the definition). Although the tubes may have single or multiple passes, there is one pass on the shell side. To increase the mean temperature difference and hence exchanger effectiveness, a pure counterflow arrangement is desirable for a two-tube-pass exchanger. This is achieved by use of an F shell having a longitudinal baffle and resulting in two shell passes. Split- and divided-flow shells, such as G, H, and J (see Fig. 1.6), are used for specific applications, such as thermosiphon boiler, condenser, and shell-side low pressure drops. The K shell is a kettle reboiler used for pool boiling applications. The X shell is a crossflow exchanger and is used for low pressure drop on the shell side and/or to eliminate the possibility of flow-induced vibrations. A further description of the various types of shell configurations is provided in Section 10.2.1.4.

Nozzles.   The entrance and exit ports for the shell and tube fluids, referred to as *nozzles*, are pipes of constant cross section welded to the shell and channels. They are used to distribute or collect the fluid uniformly on the shell and tube sides. Note that they differ from the nozzle used as a fluid metering device or in jet engines, which has a variable flow area along the flow length.

[†] The fluid flowing in the tubes is referred to as the *tube fluid*; the fluid flowing outside the tubes is referred to as the *shell fluid*.

Front- and Rear-End Heads. These are used for entrance and exit of the tube fluid; in many rear-end heads, a provision has been made to take care of tube thermal expansion. The front-end head is stationary, while the rear-end head could be either stationary (allowing for no tube thermal expansion) or floating, depending on the thermal stresses between the tubes and shell. The major criteria for selection of the front-end head are cost, maintenance and inspection, hazard due to mixing of shell and tube fluids, and leakage to ambient and operating pressures. The major criteria for selection of the rear-end head are the allowance for thermal stresses, a provision to remove the tube bundle for cleaning the shell side, prevention of mixing of tube and shell fluids, and sealing any leakage path for the shell fluid to ambient. The design selection criteria for the front- and rear-end heads of Fig. 1.6 are discussed in Sections 10.2.1.5 and 10.2.1.6.

Baffles. Baffles may be classified as transverse and longitudinal types. The purpose of longitudinal baffles is to control the overall flow direction of the shell fluid such that a desired overall flow arrangement of the two fluid streams is achieved. For example, F, G, and H shells have longitudinal baffles (see Fig. 1.6). Transverse baffles may be classified as plate baffles and grid (rod, strip, and other axial-flow) baffles. Plate baffles are used to support the tubes during assembly and operation and to direct the fluid in the tube bundle approximately at right angles to the tubes to achieve higher heat transfer coefficients. Plate baffles increase the turbulence of the shell fluid and minimize tube-to-tube temperature differences and thermal stresses due to the crossflow. Shown in Fig. 1.10 are single- and multisegmental baffles and disk and doughnut baffles. Single- and double-segmental baffles are used most frequently due to their ability to assist maximum heat transfer (due to a high-shell-side heat transfer coefficient) for a given pressure drop in a minimum amount of space. Triple and no-tubes-in-window segmental baffles are used for low-pressure-drop applications. The choice of baffle type, spacing, and cut is determined largely by flow rate, desired heat transfer rate, allowable pressure drop, tube support, and flow-induced vibrations. Disk and doughnut baffles/ support plates are used primarily in nuclear heat exchangers. These baffles for nuclear exchangers have small perforations between tube holes to allow a combination of crossflow and longitudinal flow for lower shell-side pressure drop. The combined flow results in a slightly higher heat transfer coefficient than that for pure longitudinal flow and minimizes tube-to-tube temperature differences. Rod (or bar) baffles, the most common type of grid baffle, used to support the tubes and increase the turbulence of the shell fluid, are shown in Fig. 1.11. The flow in a rod baffle heat exchanger is parallel to the tubes, and flow-induced vibrations are virtually eliminated by the baffle support of the tubes. One alternative to a rod baffle heat exchanger is the use of twisted tubes (after flattening the circular tubes, they are twisted), shown in Fig. 1.12. Twisted tubes provide rigidity and eliminate flow-induced tube vibrations, can be cleaned easily on the shell side with hydrojets, and can be cleaned easily inside the tubes, but cannot be retubed. Low-finned tubes are also available in a twisted-tube configuration. A helical baffle shell-and-tube exchanger with baffles as shown in Fig. 1.13 also has the following advantages: a lower shell-side pressure drop while maintaining the high heat transfer coefficient of a segmental exchanger, reduced leakage streams (see Section 4.4.1), and elimination of dead spots and recirculation zones (thus reducing fouling). Every shell-and-tube exchanger has transverse baffles except for X and K shells, which have support plates because the sole purpose of these transverse baffles is to support the tubes. Baffle types and their design guidelines are described further in Section 10.2.1.3.

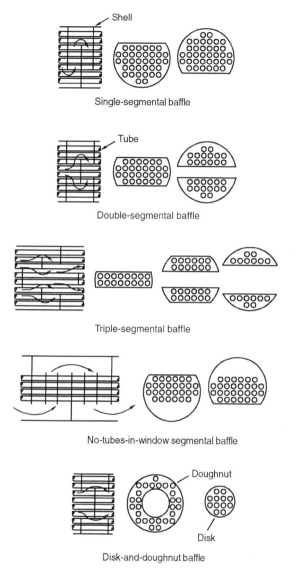

FIGURE 1.10   Plate baffle types, modified from Mueller (1973).

Butterworth (1996) provides further descriptions of these designs, and they are compared in Table 1.1.

Tubesheets. These are used to hold tubes at the ends. A tubesheet is generally a round metal plate with holes drilled through for the desired tube pattern, holes for the tie rods (which are used to space and hold plate baffles), grooves for the gaskets, and bolt holes for flanging to the shell and channel. To prevent leakage of the shell fluid at the

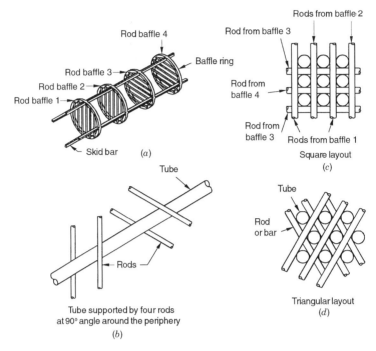

FIGURE 1.11   (*a*) Four rod baffles held by skid bars (no tubes shown); (*b*) tube in a rod baffle exchanger supported by four rods; (*c*) square layout of tubes with rods; (*d*) triangular layout of tubes with rods (Shah, 1981).

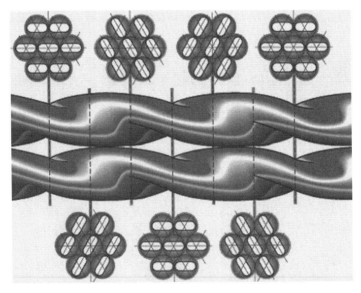

FIGURE 1.12   Twisted tube bundle for a shell-and-tube exchanger. (Courtesy of Brown Fintube Company, Houston, TX.)

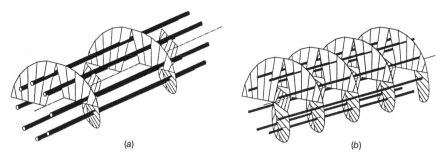

(a)   (b)

FIGURE 1.13   Helical baffle shell-and-tube exchanger: (a) single helix; (b) double helix. (Courtesy of ABB Lumus Heat Transfer, Bloomfield, NJ.)

tubesheet through a clearance between the tube hole and tube, the tube-to-tubesheet joints are made by many methods, such as expanding the tubes, rolling the tubes, hydraulic expansion of tubes, explosive welding of tubes, stuffing of the joints, or welding or brazing of tubes to the tubesheet. The leak-free tube-to-tubesheet joint made by the conventional rolling process is shown in Fig. 1.14.

1.5.1.2 Double-Pipe Heat Exchangers. This exchanger usually consists of two concentric pipes with the inner pipe plain or finned, as shown in Fig. 1.15. One fluid flows in the inner pipe and the other fluid flows in the annulus between pipes in a counterflow direction for the ideal highest performance for the given surface area. However, if the application requires an almost constant wall temperature, the fluids may flow in a parallelflow direction. This is perhaps the simplest heat exchanger. Flow distribution is no problem, and cleaning is done very easily by disassembly. This configuration is also suitable where one or both of the fluids is at very high pressure,

TABLE 1.1   Comparison of Various Types of Shell-and-Tube Heat Exchangers

Characteristic	Segmental Baffle	Rod Baffle	Twisted Tube	Helical Baffle
Good heat transfer per unit pressure drop	No	Yes	Yes	Yes
High shell-side heat transfer coefficient	Yes	No	No	Yes
Tube-side enhancement	With inserts	With inserts	Included	With inserts
Suitable for very high exchanger effectiveness	No	Yes	Yes	No
Tends to have low fouling	No	Yes	Yes	Yes
Can be cleaned mechanically	Yes, with square pitch	Yes	Yes	Yes, with square pitch
Low flow-induced tube vibration	With special designs	Yes	Yes	With double helix
Can have low-finned tubes	Yes	Yes	Yes	Yes

Source: Data from Butterworth with private communication (2002).

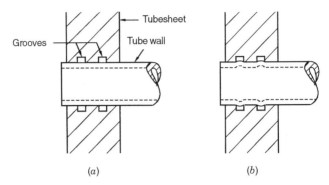

FIGURE 1.14   Details of a leak-free joint between the tube and tube hole of a tubesheet: (*a*) before tube expansion; (*b*) after tube expansion.

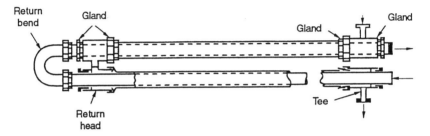

FIGURE 1.15   Double-pipe heat exchanger.

because containment in the small-diameter pipe or tubing is less costly than containment in a large-diameter shell. Double-pipe exchangers are generally used for small-capacity applications where the total heat transfer surface area required is $50\,m^2$ ($500\,ft^2$) or less because it is expensive on a cost per unit surface area basis. Stacks of double-pipe or multitube heat exchangers are also used in some process applications with radial or longitudinal fins. The exchanger with a bundle of U tubes in a pipe (shell) of 150 mm (6 in.) diameter and above uses segmental baffles and is referred to variously as a *hairpin* or *jacketed U-tube exchanger*.

1.5.1.3 Spiral Tube Heat Exchangers. These consist of one or more spirally wound coils fitted in a shell. Heat transfer rate associated with a spiral tube is higher than that for a straight tube. In addition, a considerable amount of surface can be accommodated in a given space by spiraling. Thermal expansion is no problem, but cleaning is almost impossible.

1.5.2   Plate-Type Heat Exchangers

Plate-type heat exchangers are usually built of thin plates (all prime surface). The plates are either smooth or have some form of corrugation, and they are either flat or wound in an exchanger. Generally, these exchangers cannot accommodate very high pressures,

temperatures, or pressure and temperature differences. Plate heat exchangers (PHEs)[†]
can be classified as gasketed, welded (one or both fluid passages), or brazed, depending
on the leak tightness required. Other plate-type exchangers are spiral plate, lamella, and
platecoil exchangers. These are described next.

1.5.2.1 Gasketed Plate Heat Exchangers

Basic Construction. The plate-and-frame or gasketed plate heat exchanger (PHE) consists of a number of thin rectangular metal plates sealed around the edges by gaskets
and held together in a frame as shown in Fig. 1.16. The frame usually has a fixed end
cover (headpiece) fitted with connecting ports and a movable end cover (pressure plate,
follower, or tailpiece). In the frame, the plates are suspended from an upper carrying
bar and guided by a bottom carrying bar to ensure proper alignment. For this purpose,
each plate is notched at the center of its top and bottom edges. The plate pack with
fixed and movable end covers is clamped together by long bolts, thus compressing the
gaskets and forming a seal. For later discussion, we designate the resulting length of the
plate pack as L_{pack}. The carrying bars are longer than the compressed stack, so that
when the movable end cover is removed, plates may be slid along the support bars for
inspection and cleaning.

Each plate is made by stamping or embossing a corrugated (or wavy) surface pattern
on sheet metal. On one side of each plate, special grooves are provided along the periphery of the plate and around the ports for a gasket, as indicated by the dark lines in
Fig. 1.17. Typical plate geometries (corrugated patterns) are shown in Fig. 1.18, and over
60 different patterns have been developed worldwide. Alternate plates are assembled such

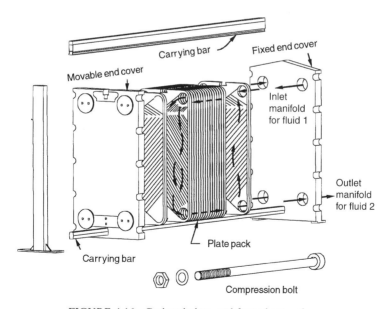

FIGURE 1.16   Gasketed plate- and-frame heat exchanger.

[†] Unless explicitly mentioned, PHE means *gasketed plate heat exchanger.*

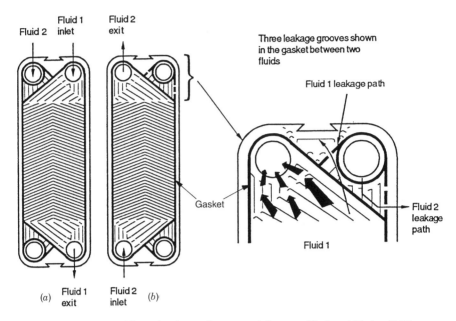

FIGURE 1.17   Plates showing gaskets around the ports (Shah and Focke, 1988).

that the corrugations on successive plates contact or cross each other to provide mechanical support to the plate pack through a large number of contact points. The resulting flow passages are narrow, highly interrupted, and tortuous, and enhance the heat transfer rate and decrease fouling resistance by increasing the shear stress, producing secondary flow, and increasing the level of turbulence. The corrugations also improve the rigidity of the plates and form the desired plate spacing. Plates are designated as *hard* or *soft*, depending on whether they generate a high or low intensity of turbulence.

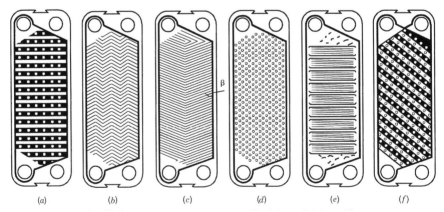

FIGURE 1.18   Plate patterns: (*a*) washboard; (*b*) zigzag; (*c*) chevron or herringbone; (*d*) protrusions and depressions; (*e*) washboard with secondary corrugations; (*f*) oblique washboard (Shah and Focke, 1988).

Sealing between the two fluids is accomplished by elastomeric molded gaskets [typically, 5 mm (0.2 in.) thick] that are fitted in peripheral grooves mentioned earlier (dark lines in Fig. 1.17). Gaskets are designed such that they compress about 25% of thickness in a bolted plate exchanger to provide a leaktight joint without distorting the thin plates. In the past, the gaskets were cemented in the grooves, but now, snap-on gaskets, which do not require cementing, are common. Some manufacturers offer special interlocking types to prevent gasket blowout at high pressure differences. Use of a double seal around the port sections, shown in Fig. 1.17, prevents fluid intermixing in the rare event of gasket failure. The interspace between the seals is also vented to the atmosphere to facilitate visual indication of leakage (Fig. 1.17). Typical gasket materials and their range of applications are listed in Table 1.2, with butyl and nitrile rubber being most common. PTFE (polytetrafluoroethylene) is not used because of its viscoelastic properties.

Each plate has four corner ports. In pairs, they provide access to the flow passages on either side of the plate. When the plates are assembled, the corner ports line up to form distribution headers for the two fluids. Inlet and outlet nozzles for the fluids, provided in the end covers, line up with the ports in the plates (distribution headers) and are connected to external piping carrying the two fluids. A fluid enters at a corner of one end of the compressed stack of plates through the inlet nozzle. It passes through alternate channels[†] in either series or parallel passages. In one set of channels, the gasket does not surround the inlet port between two plates (see, e.g., Fig. 1.17a for the fluid 1 inlet port); fluid enters through that port, flows between plates, and exits through a port at the other end. On the same side of the plates, the other two ports are blocked by a gasket with a double seal, as shown in Fig. 1.17a, so that the other fluid (fluid 2 in Fig. 1.17a) cannot enter the plate on that side.[‡] In a 1 pass–1 pass[§] two-fluid counterflow PHE, the next channel has gaskets covering the ports just opposite the preceding plate (see, e.g., Fig. 1.17b, in which now, fluid 2 can flow and fluid 1 cannot flow). Incidentally, each plate has gaskets on only one side, and they sit in grooves on the back side of the neighboring plate. In Fig. 1.16, each fluid makes a single pass through the exchanger because of alternate gasketed and ungasketed ports in each corner opening. The most conventional flow arrangement is 1 pass–1 pass counterflow, with all inlet and outlet connections on the fixed end cover. By blocking flow through some ports with proper gasketing, either one or both fluids could have more than one pass. Also, more than one exchanger can be accommodated in a single frame. In cases with more than two simple 1-pass-1-pass heat exchangers, it is necessary to insert one or more intermediate headers or connector plates in the plate pack at appropriate places (see, e.g., Fig. 1.19). In milk pasteurization applications, there are as many as five exchangers or sections to heat, cool, and regenerate heat between raw milk and pasteurized milk.

Typical plate heat exchanger dimensions and performance parameters are given in Table 1.3. Any metal that can be cold-worked is suitable for PHE applications. The most

---

[†] A channel is a flow passage bounded by two plates and is occupied by one of the fluids. In contrast, a plate separates the two fluids and transfers heat from the hot fluid to the cold fluid.

[‡] Thus with the proper arrangement, gaskets also distribute the fluids between the channels in addition to providing sealing to prevent leakage.

[§] In a plate heat exchanger, a *pass* refers to a group of channels in which the flow is in the same direction for one full length of the exchanger (from top to bottom of the pack; see Fig. 1.65). In an m pass – n pass two-fluid plate heat exchanger, fluid 1 flows through m passes and fluid 2 through n passes.

TABLE 1.2   Gasket Materials Used in Plate Heat Exchangers

Gasket Material	Generic Name	Maximum Operating Temperature (°C)	Applications	Comments
Natural rubber	*cis*-1,4-polyisoprene	70	Oxygenated solvents, acids, alcohols	
SBR (styrene butadiene)		80	General-purpose aqueous, alkalies, acids, and oxygenated solvents	Has poor fat resistance
Neoprene	*trans*-1,4-polychloroprene	70	Alcohols, alkalies, acids, aliphatic hydrocarbon solvents	
Nitrile		100–140	Dairy, fruit juices, beverage, pharmaceutical and biochemical applications, oil, gasoline, animal and vegetable oils, alkalies, aliphatic organic solvents	Is resistant to fatty materials; particularly suitable for cream
Butyl (resin cured)		120–150	Alkalies, acids, animal and vegetable oils, aldehydes, ketones, phenols, and some esters	Has poor fat resistance; suitable for UHT milk duties; resists inorganic chemical solutions up to 150°C
Ethylene propylene (EDPM) rubber		140	Alkalies, oxygenated solvents	Unsuitable for fatty liquids
Silicone rubber	Polydimethyl-siloxane	140	General low-temperature use, alcohols, sodium hypochlorite	
Fluorinated rubber		175	High-temperature aqueous solutions, mineral oils and gasoline, organic solvents, animal and vegetable oils	
Compressed asbestos fiber		200–260	Organic solvents, high-operating-temperature applications	

common plate materials are stainless steel (AISI 304 or 316) and titanium. Plates made from Incoloy 825, Inconel 625, and Hastelloy C-276 are also available. Nickel, cupronickel, and monel are rarely used. Carbon steel is not used, due to low corrosion resistance for thin plates. Graphite and polymer plates are used with corrosive fluids. The heat transfer surface area per unit volume for plate exchangers ranges from 120 to 660 m^2/m^3 (37 to 200 ft^2/ft^3).

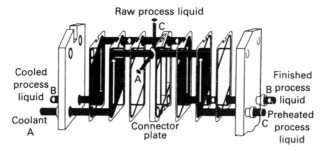

FIGURE 1.19   A three-fluid plate heat exchanger. (Courtesy of Alfa Laval Thermal, Inc., Lund, Sweden.)

Flow Arrangements. A large number of flow arrangements are possible in a plate heat exchanger (shown later in Fig. 1.65), depending on the required heat transfer duty, available pressure drops, minimum and maximum velocities allowed, and the flow rate ratio of the two fluid streams. In each pass there can be an equal or unequal number of thermal plates.[†] Whether the plate exchanger is a single- or multipass unit, whenever possible, the thermodynamically superior counterflow or overall counterflow arrangement (see Sections 1.6.1 and 1.6.2) is used exclusively.

One of the most common flow arrangements in a PHE is a 1-pass–1-pass U configuration (see Fig. 1.65*a*). This is because this design allows all fluid ports to be located on

TABLE 1.3   Some Geometrical and Operating Condition Characteristics of Plate-and-Frame Heat Exchangers

Unit		Operation	
Maximum surface area	2500 m^2	Pressure	0.1 to 3.0 MPa
Number of plates	3 to 700	Temperature	−40 to 260°C
Port size	Up to 400 mm	Maximum port velocity	6 m/s (for liquids)
	(for liquids)	Channel flow rates	0.05 to 12.5 m^3/h
		Maximum unit flow rate	2500 m^3/h
Plates		Performance	
Thickness	0.5 to 1.2 mm	Temperature approach	As low as 1°C
Size	0.03 to 3.6 m^2	Heat exchanger efficiency	Up to 93%
Spacing	1.5 to 7 mm	Heat transfer coefficients	3000 to 8000 W/m^2 · K
Width	70 to 1200 mm	for water–water duties	
Length	0.4 to 5 m		
Hydraulic diameter	2 to 10 mm		
Surface area per plate	0.02 to 5 m^2		

Source: Data from Shah (1994).

[†]In the plate exchanger, the two outer plates serve as end plates and ideally do not participate in heat transfer between the fluids because of the large thermal resistance associated with thick end plates and air gaps between the end plates and the header/follower. The remaining plates, known as *thermal plates*, transfer heat between the fluids.

the fixed end cover, permitting easy disassembly and cleaning/repair of a PHE without disconnecting any piping. In a multipass arrangement, the ports and fluid connections are located on both fixed and movable end covers. A multipass arrangement is generally used when the flow rates are considerably different or when one would like to use up the available pressure drop by multipassing and hence getting a higher heat transfer coefficient.

Advantages and Limitations. Some advantages of plate heat exchangers are as follows. They can easily be taken apart into their individual components for cleaning, inspection, and maintenance. The heat transfer surface area can readily be changed or rearranged for a different task or for anticipated changing loads, through the flexibility of plate size, corrugation patterns, and pass arrangements. High shear rates and shear stresses, secondary flow, high turbulence, and mixing due to plate corrugation patterns reduce fouling to about 10 to 25% of that of a shell-and-tube exchanger, and enhance heat transfer. Very high heat transfer coefficients are achieved due to the breakup and reattachment of boundary layers, swirl or vortex flow generation, and small hydraulic diameter flow passages. Because of high heat transfer coefficients, reduced fouling, the absence of bypass and leakage streams, and pure counterflow arrangements, the surface area required for a plate exchanger is one-half to one-third that of a shell-and-tube exchanger for a given heat duty, thus reducing the cost, overall volume, and space requirement for the exchanger. Also, the gross weight of a plate exchanger is about one-sixth that of an equivalent shell-and-tube exchanger. Leakage from one fluid to the other cannot take place unless a plate develops a hole. Since the gasket is between the plates, any leakage from the gaskets is to the outside of the exchanger. The residence time (time to travel from the inlet to the outlet of the exchanger) for different fluid particles or flow paths on a given side is approximately the same. This parity is desirable for uniformity of heat treatment in applications such as sterilizing, pasteurizing, and cooking. There are no significant hot or cold spots in the exchanger that could lead to the deterioration of heat-sensitive fluids. The volume of fluid held up in the exchanger is small; this feature is important with expensive fluids, for faster transient response, and for better process control. Finally, high thermal performance can be achieved in plate exchangers. The high degree of counterflow in PHEs makes temperature approaches of up to 1°C (2°F) possible. The high thermal effectiveness (up to about 93%) facilitates economical low-grade heat recovery. The flow-induced vibrations, noise, thermal stresses, and entry impingement problems of shell-and-tube exchangers do not exist for plate heat exchangers.

Some inherent limitations of the plate heat exchangers are caused by plates and gaskets as follows. The plate exchanger is capable of handling up to a maximum pressure of about 3 MPa gauge (435 psig) but is usually operated below 1.0 MPa gauge (150 psig). The gasket materials (except for the PTFE-coated type) restrict the use of PHEs in highly corrosive applications; they also limit the maximum operating temperature to 260°C (500°F) but are usually operated below 150°C (300°F) to avoid the use of expensive gasket materials. Gasket life is sometimes limited. Frequent gasket replacement may be needed in some applications. Pinhole leaks are hard to detect. For equivalent flow velocities, pressure drop in a plate exchanger is very high compared to that of a shell-and-tube exchanger. However, the flow velocities are usually low and plate lengths are "short," so the resulting pressure drops are generally acceptable. The normal symmetry of PHEs may make phase-change applications[†] more difficult, due to large differences in volumetric flows. For some cases, heat exchanger duties with widely different fluid flow rates and depending on the allowed pressure drops of the two fluids, an arrangement of a different number of passes for the two fluids may make a PHE advantageous. However, care must be exercised to take full advantage of available pressure drop while multi-passing one or both fluids.

Because of the long gasket periphery, PHEs are not suited for high-vacuum applications. PHEs are not suitable for erosive duties or for fluids containing fibrous materials. In certain cases, suspensions can be handled; but to avoid clogging, the largest suspended particle should be at most one-third the size of the average channel gap. Viscous fluids can be handled, but extremely viscous fluids lead to flow maldistribution problems, especially on cooling. Plate exchangers should not be used for toxic fluids, due to potential gasket leakage. Some of the largest units have a total surface area of about 2500 m^2 (27,000 ft^2) per frame. Some of the limitations of gasketed PHEs have been addressed by the new designs of PHEs described in the next subsection.

Major Applications. Plate heat exchangers were introduced in 1923 for milk pasteur-ization applications and now find major applications in liquid–liquid (viscosities up to 10 Pa · s) heat transfer duties. They are most common in the dairy, juice, beverage, alcoholic drink, general food processing, and pharmaceutical industries, where their ease of cleaning and the thermal control required for sterilization/pasteurization make them ideal. They are also used in the synthetic rubber industry, paper mills, and in the process heaters, coolers, and closed-circuit cooling systems of large petrochemical and power plants. Here heat rejection to seawater or brackish water is common in many applications, and titanium plates are then used.

Plate heat exchangers are not well suited for lower-density gas-to-gas applications. They are used for condensation or evaporation of non-low-vapor densities. Lower vapor densities limit evaporation to lower outlet vapor fractions. Specially designed plates are now available for condensing as well as evaporation of high-density vapors such as ammonia, propylene, and other common refrigerants, as well as for combined evaporation/condensation duties, also at fairly low vapor densities.

1.5.2.2 Welded and Other Plate Heat Exchangers. One of the limitations of the gasketed plate heat exchanger is the presence of gaskets, which restricts their use to compatible fluids (noncorrosive fluids) and which limits operating temperatures and pressures. To overcome this limitation, a number of welded plate heat exchanger designs have surfaced with welded pairs of plates on one or both fluid sides. To reduce the effective welding cost, the plate size for this exchanger is usually larger than that of the gasketed PHE. The disadvantage of such a design is the loss of disassembling flexibility on the fluid sides where the welding is done. Essentially, laser welding is done around the complete circumference, where the gasket is normally placed. Welding on both sides then results in higher limits on operating temperatures and pressures [350°C (660°F) and 4.0 MPa (580 psig)] and allows the use of corrosive fluids compatible with the plate material. Welded PHEs can accommodate multiple passes and more than two fluid streams. A *Platular heat exchanger* can accommodate four fluid streams. Figure 1.20 shows a pack of plates for a conventional plate-and-frame exchanger, but welded on one

---

† Special plate designs have been developed for phase-change applications.

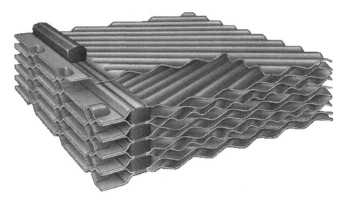

FIGURE 1.20   Section of a welded plate heat exchanger. (Courtesy of Alfa Laval Thermal, Inc., Richmond, VA.)

fluid side. Materials used for welded PHEs are stainless steel, Hastelloy, nickel-based alloys, and copper and titanium.

A *Bavex welded-plate heat exchanger* with welded headers is shown in Fig. 1.21. A *Stacked Plate Heat Exchanger* is another welded plate heat exchanger design (from Packinox), in which rectangular plates are stacked and welded at the edges. The physical size limitations of PHEs [1.2 m wide × 4 m long maximum (4 × 13 ft)] are considerably extended to 1.5 m wide × 20 m long (5 × 66 ft) in Packinox exchangers. A maximum surface area of over 10,000 m^2 (100,000 ft^2) can be accommodated in one unit. The potential maximum operating temperature is 815°C (1500°F) with an operating pressure of up to 20 MPa (3000 psig) when the stacked plate assembly is placed in a cylindrical pressure vessel. For inlet pressures below 2 MPa (300 psig) and inlet temperatures below 200°C (400°F), the plate bundle is not placed in a pressure vessel but is bolted between two heavy plates. Some applications of this exchanger are for catalytic reforming, hydrosulfurization, and crude distillation, and in a synthesis converter feed effluent exchanger for methanol and for a propane condenser.

A vacuum *brazed plate heat exchanger* is a compact PHE for high-temperature and high-pressure duties, and it does not have gaskets, tightening bolts, frame, or carrying and guide bars. It consists simply of stainless steel plates and two end plates, all generally copper brazed, but nickel brazed for ammonia units. The plate size is generally limited to 0.3 m^2. Such a unit can be mounted directly on piping without brackets and foundations. Since this exchanger cannot be opened, applications are limited to negligible fouling cases. The applications include water-cooled evaporators and condensers in the refrigeration industry, and process water heating and heat recovery.

A number of other plate heat exchanger constructions have been developed to address some of the limitations of the conventional PHEs. A double-wall PHE is used to avoid mixing of the two fluids. A wide-gap PHE is used for fluids having a high fiber content or coarse particles/slurries. A graphite PHE is used for highly corrosive fluids. A flow-flex exchanger has plain fins on one side between plates and the other side has conventional plate channels, and is used to handle asymmetric duties (a flow rate ratio of 2 : 1 and higher). A PHE evaporator has an asymmetric plate design to handle mixed process flows (liquid and vapors) and different flow ratios.

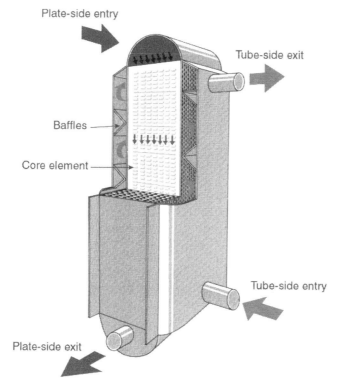

Plate-side entry

Tube-side exit

Baffles

Core element

Tube-side entry

Plate-side exit

FIGURE 1.21   Bavex welded-plate heat exchanger (From Reay, 1999).

1.5.2.3 Spiral Plate Heat Exchangers. A spiral plate heat exchanger consists of two relatively long strips of sheet metal, normally provided with welded studs for plate spacing, wrapped helically around a split mandrel to form a pair of spiral channels for two fluids, as shown in Fig. 1.22. Alternate passage edges are closed. Thus, each fluid has a long single passage arranged in a compact package. To complete the exchanger, covers are fitted at each end. Any metal that can be cold-formed and welded can be used for this exchanger. Common metals used are carbon steel and stainless steel. Other metals include titanium, Hastelloy, Incoloy, and high-nickel alloys. The basic spiral element is sealed either by welding at each side of the channel or by providing a gasket (non–asbestos based) at each end cover to obtain the following alternative arrangements of the two fluids: (1) both fluids in spiral counterflow; (2) one fluid in spiral flow, the other in crossflow across the spiral; or (3) one fluid in spiral flow, the other in a combination of crossflow and spiral flow. The entire assembly is housed in a cylindrical shell enclosed by two (or only one or no) circular end covers (depending on the flow arrangements above), either flat or conical. Carbon steel and stainless steel are common materials. Other materials used include titanium, Hastelloy, and Incoloy.

A spiral plate exchanger has a relatively large diameter because of the spiral turns. The largest exchanger has a maximum surface area of about 500 m² (5400 ft²) for a maximum shell diameter of 1.8 m (72 in.). The typical passage height is 5 to 25 mm (0.20 to 1.00 in.) and the sheet metal thickness range is 1.8 to 4 mm (0.07 to 0.16 in.).

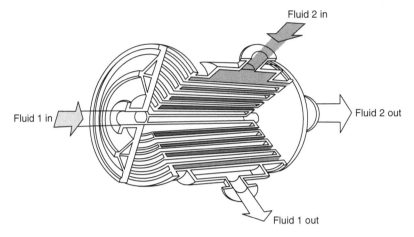

FIGURE 1.22   Spiral plate heat exchanger with both fluids in spiral counterflow.

The heat transfer coefficients are not as high as in a plate exchanger if the plates are not corrugated. However, the heat transfer coefficient is higher than that for a shell-and-tube exchanger because of the curved rectangular passages. Hence, the surface area requirement is about 20% lower than that for a shell-and-tube unit for the same heat duty.

The counterflow spiral unit is used for liquid–liquid, condensing, or gas cooling applications. When there is a pressure drop constraint on one side, such as with gas flows or with high liquid flows, crossflow (straight flow) is used on that side. For condensation or vaporization applications, the unit is mounted vertically. Horizontal units are used when high concentrations of solids exist in the fluid.

The advantages of this exchanger are as follows: It can handle viscous, fouling liquids and slurries more readily because of a *single passage*. If the passage starts fouling, the localized velocity in the passage increases. The fouling rate then decreases with increased fluid velocity. The fouling rate is very low compared to that of a shell-and-tube unit. It is more amenable to chemical, flush, and reversing fluid cleaning techniques because of the single passage. Mechanical cleaning is also possible with removal of the end covers. Thus, maintenance is less than with a shell-and-tube unit. No insulation is used outside the exchanger because of the cold fluid flowing in the outermost passage, resulting in negligible heat loss, if any, due to its inlet temperature closer to surrounding temperature. The internal void volume is lower (less than 60%) than in a shell-and-tube exchanger, and thus it is a relatively compact unit. By adjusting different channel heights, considerable differences in volumetric flow rates of two streams can be accommodated.

The disadvantages of this exchanger are as follows: As noted above, the maximum size is limited. The maximum operating pressure ranges from 0.6 to 2.5 MPa gauge (90 to 370 psig) for large units. The maximum operating temperature is limited to 500°C (930°F) with compressed asbestos gaskets, but most are designed to operate at 200°C (392°F). Field repair is difficult due to construction features.

This exchanger is well suited as a condenser or reboiler. It is used in the cellulose industry for cleaning relief vapors in sulfate and sulfite mills, and is also used as a thermosiphon or kettle reboiler. It is preferred especially for applications having very viscous liquids, dense slurries, digested sewage sludge, and contaminated industrial effluents. A spiral version free of welded studs for plate spacing on one or both fluid sides but with reduced width is used for sludge and other heavily fouling fluids. It is also used in the treatment of bauxite suspensions and mash liquors in the alcohol industry.

1.5.2.4 Lamella Heat Exchangers. A lamella heat exchanger consists of an outer tubular shell surrounding an inside bundle of heat transfer elements. These elements, referred to as *lamellas*, are flat tubes (pairs of thin dimpled plates, edge welded, resulting in high-aspect-ratio rectangular channels), shown in Fig. 1.23. The inside opening of the lamella ranges from 3 to 10 mm (0.1 to 0.4 in.) and the wall thickness from 1.5 to 2 mm (0.06 to 0.08 in.). Lamellas are stacked close to each other to form narrow channels on the shell side. Lamellas are inserted in the end fittings with gaskets to prevent the leakage from shell to tube side, or vice versa. In a small exchanger, lamellas are of increasing width from either end to the center of the shell to fully utilize the available space, as shown in Fig. 1.23a. However, in a larger exchanger, lamellas consist of two (see Fig. 1.23b) or more flat tubes to contain operating pressures. There are no baffles. One end of the tube bundle is fixed and the other is floating, to allow for thermal

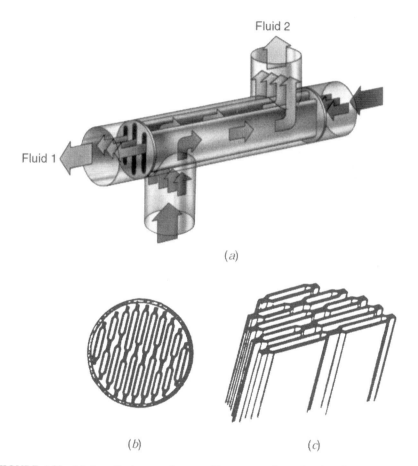

FIGURE 1.23   (*a*) Lamella heat exchanger; (*b*) cross section of a lamella heat exchanger; (*c*) lamellas. (Courtesy of Alfa Laval Thermal, Inc., Lund, Sweden.)

expansion. Thus, this exchanger is a modified floating-head shell-and-tube exchanger. One fluid (tube fluid) flows inside the lamellas and the other fluid (shell fluid) flows longitudinally in the spaces between them, with no baffles on the shell side. The exchanger thus has a single pass, and the flow arrangement is generally counterflow. The flat tube walls have dimples where neighboring tubes are spot-welded. High-heat-transfer coefficients are usually obtained because of small hydraulic diameters and no leakage or bypass streams as encountered in a conventional shell-and-tube exchanger. Also, possible point dimples increase the heat transfer coefficient and pressure drop in the same way as do corrugated plate channels. It can handle fibrous fluids and slurries with proper plate spacing. The large units have surface areas up to $1000\,m^2$ $(10,800\,ft^2)$. A lamella exchanger weighs less than a shell-and-tube exchanger having the same duty. A lamella exchanger is capable of pressures up to 3.45 MPa gauge (500 psig) and temperature limits of 200°C (430°F) for PTFE gaskets and 500°C (930°F) for nonasbestos gaskets. This exchanger is used for heat recovery in the pulp and paper industry, chemical process industry, and for other industrial applications, in competition with the shell-and-tube exchanger.

1.5.2.5 Printed-Circuit Heat Exchangers. This exchanger shown in Fig. 1.24 has only primary heat transfer surface, as do PHEs. Fine grooves are made in the plate by using

FIGURE 1.24 Printed-circuit crossflow exchanger. (Courtesy of Heatric Division of Meggitt (UK) Ltd., Dorset, UK.)

the same techniques as those employed for making printed circuit boards. A block (stack) of chemically etched plates is then diffusion bonded, and fluid inlet/outlet headers are welded to make the exchanger. For the two fluid streams, there are different etching patterns, as desired to make a crossflow, counterflow, or multipass cross-counterflow exchanger. Multiple passes and multiple fluid streams can be made in a single block. Several blocks are welded together for large heat duty applications. The channel depth is 0.1 to 2 mm (0.004 to 0.08 in.). High surface area densities, 650 to 1300 m^2/m^3 (200 to 400 ft^2/ft^3), are achievable for operating pressures 50 to 10 MPa (7250 to 290 psi),[†] and temperatures 150 to 800°C (300 to 1500°F). A variety of materials, including stainless steel, titanium, copper, nickel, and nickel alloys, can be used. It has been used successfully with relatively clean gases, liquids, and phase-change fluids in the chemical processing, fuel processing, waste heat recovery, power and energy, refrigeration, and air separation industries. They are used extensively in offshore oil platforms as compressor aftercoolers, gas coolers, cryogenic processes to remove inert gases, and so on. Having a small channel size, the fluid pressure drop can be a constraint for low-to-moderate pressure applications. However, the main advantage of this exchanger is high pressure/strength, flexibility in design, and high effectivenesses.

1.5.2.6   Panelcoil Heat Exchangers.

The basic elements of this exchanger are called panelcoils, platecoils, or embossed-panel coils, as shown in Fig. 1.25. The panelcoil serves as a heat sink or a heat source, depending on whether the fluid within the coil is being cooled or heated. As a result, the shape and size of the panelcoil is made to fit the system, or flat panelcoils are immersed in a tank or placed in the atmosphere for heat transfer. Basically, three different methods have been employed to manufacture panelcoils: a die-stamping process, spot-weld process, and roll-bond process. In the die-stamping process, flow channels are die-stamped on either one or two metal sheets. When one sheet is embossed and joined to a flat (unembossed sheet), it forms a one-sided embossed panelcoil. When both sheets are stamped, it forms a double-sided embossed panelcoil. The two plates are joined by electric resistance welding of the metal sheets. Examples are shown in Fig. 1.25a and b.

In the spot-weld process, two flat sheets are spot-welded in a desired pattern (no die stamping), and then are inflated by a fluid under high pressure to form flow passages interconnected by weld points. An example is shown in Fig. 1.25d.

In a roll-bond process, two sheets of metal (copper or aluminum) are bonded with a true metallurgical bond, except in certain desired "specified channels," where a special stopweld material is applied. On one of the metal sheets, the stopweld material is applied in the desired flow pattern. This sheet is then stacked with another plain sheet without stopweld material on it. The sheets are then heated and immediately hot-rolled under high pressure to provide a metallurgical bond. Subsequent cold rolling follows to provide an appropriate increase in length. After annealing the panelcoil, a needle is inserted at the edge, exposing stopweld material, and high-pressure air inflates the desired flow passages when the panelcoil is placed between two plates in a heavy hydraulic press. The roll-bond process limits the panelcoils to a flat in shape.

The most commonly used materials for panelcoils are carbon steel, stainless steel, titanium, nickel and its alloys, and monel. The panelcoil sheet metal gauges range between 1.5 and 3.0 mm (0.06 to 0.12 in.) depending on the materials used and whether

---

[†] Note that operating pressures for 650- and 1300-m^2/m^3 surface area densities are 50 and 10 MPa, respectively.

FIGURE 1.25   Die-stamped plate coils: (*a*) serpentine, (*b*) multizone, (*c*) a vessel; (*d*) spot-welded Econocoil bank. (Courtesy of Tranter PHE, Inc., Wichita, TX.)

or not the panels are single or double embossed. The maximum operating pressure ranges from 1.8 MPa (260 psig) for double-embossed and 1.2 MPa (175 psig) for single-embossed carbon steel, stainless steel, and monel panelcoils to 0.7 MPa (100 psig) for double-embossed titanium panelcoils.

Panelcoil heat exchangers are relatively inexpensive and can be made into desired shapes and thicknesses for heat sinks and heat sources under varied operating conditions. Hence, they have been used in many industrial applications, such as cryogenics, chemicals, fibers, food, paints, pharmaceuticals, and solar absorbers.

1.5.3 Extended Surface Heat Exchangers

The tubular and plate-type exchangers described previously are all prime surface heat exchangers, except for a shell-and-tube exchanger with low finned tubing. Their heat exchanger effectiveness (see Section 3.3.1 for the definition) is usually 60% or below, and the heat transfer surface area density is usually less than 700 m^2/m^3 (213 ft^2/ft^3). In some applications, much higher (up to about 98%) exchanger effectiveness is essential, and the box volume and mass are limited so that a much more compact surface is mandated. Also, in a heat exchanger with gases or some liquids, the heat transfer coefficient is quite low on one or both fluid sides. This results in a large heat transfer surface area requirement. One of the most common methods to increase the surface area and exchanger

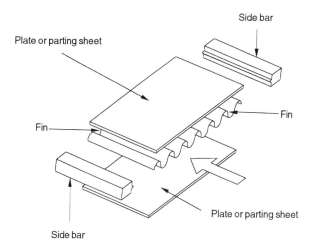

FIGURE 1.26   Basic components of a plate-fin heat exchanger (Shah and Webb, 1983).

compactness is to add the extended surface (fins) and use fins with the *fin density* (*fin frequency*, fins/m or fins/in.) as high as possible on one or both fluid sides, depending on the design requirement. Addition of fins can increase the surface area by 5 to 12 times the primary surface area in general, depending on the design. The resulting exchanger is referred to as an *extended surface exchanger*. Flow area is increased by the use of thin-gauge material and sizing the core properly. The heat transfer coefficient on extended surfaces may be *higher* or *lower* than that on unfinned surfaces. For example, interrupted (strip, louver, etc.) fins provide both an increased area and increased heat transfer coefficient, while internal fins in a tube increase the tube-side surface area but may result in a slight reduction in the heat transfer coefficient, depending on the fin spacing. Generally, increasing the fin density reduces the heat transfer coefficient associated with fins. Flow interruptions (as in offset strip fins, louvered fins, etc.) may increase the heat transfer coefficient two to four times that for the corresponding plain (uncut) fin surface. Plate-fin and tube-fin geometries are the two most common types of extended surface heat exchangers.[†]

1.5.3.1 Plate-Fin Heat Exchangers. This type of exchanger has corrugated fins (most commonly having triangular and rectangular cross sections) or spacers sandwiched between parallel plates (referred to as plates or parting sheets), as shown in Fig. 1.26. Sometimes fins are incorporated in a flat tube with rounded corners (referred to as a *formed tube*), thus eliminating the need for side bars. If liquid or phase-change fluid flows on the other side, the parting sheet is usually replaced by a flat tube with or without inserts or webs (Fig. 1.27). Other plate-fin constructions include drawn-cup

---

[†] If the heat transfer surface in a prime surface heat exchanger is rough (due either to the manufacturing process or made artificially) or small-scale fins (the fin height is approximately 5% or less of the tube radius) are made integral to the prime surface, the exchanger is sometimes referred to as a *microfin heat exchanger*.

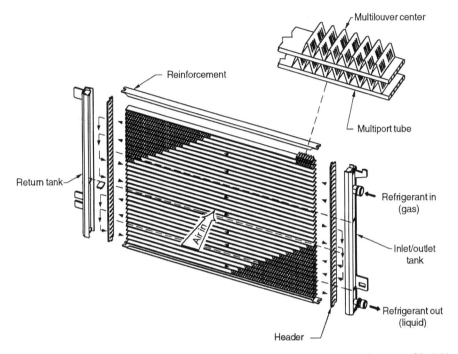

FIGURE 1.27   Flat webbed tube and multilouver fin automotive condenser. (Courtesy of Delphi Harrison Thermal Systems, Lockport, NY.)

(Fig. 1.28) and tube-and-center[†] configurations. The plates or flat tubes separate the two fluid streams, and the fins form the individual flow passages. Alternate fluid passages are connected in parallel by suitable headers to form the two or more fluid sides of the exchanger. Fins are die or roll formed and are attached to the plates by brazing,[‡] soldering, adhesive bonding, welding, mechanical fit, or extrusion. Fins may be used on both sides in gas-to-gas heat exchangers. In gas-to-liquid applications, fins are generally used only on the gas side; if employed on the liquid side, they are used primarily for structural strength and flow-mixing purposes. Fins are also sometimes used for pressure containment and rigidity. In Europe, a plate-fin exchanger is also referred to as a *matrix heat exchanger*.

Plate fins are categorized as (1) plain (i.e., uncut) and straight fins, such as plain triangular and rectangular fins, (2) plain but wavy fins (wavy in the main fluid flow direction), and (3) interrupted fins, such as offset strip, louver, perforated, and pin fins. Examples of commonly used fins are shown in Fig. 1.29. Louver form of the multilouver fin is shown in Fig. 7.29, along with a sketch of its louver form at section AA in

---

[†] In the automotive industry, corrugated fins in the plate-fin unit are referred to as *centers*, to distinguish them from the flat fins outside the tubes in a tube-fin exchanger. The latter are referred to simply as *fins* in the automotive industry.

[‡] In the automotive industry, the most common brazing technique is controlled atmosphere brazing (CAB; brazing at atmospheric pressure in a nitrogen environment and with noncorrosive flux; also known as a Nocolok process), and sometimes vacuum brazing is used. In the cryogenics industry, only vacuum brazing is used.

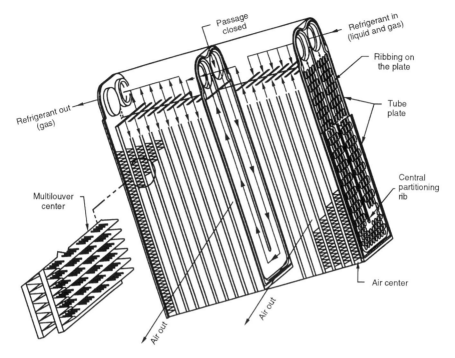

FIGURE 1.28   U-channel ribbed plates and multilouver fin automotive evaporator. (Courtesy of Delphi Harrison Thermal Systems, Lockport, NY.)

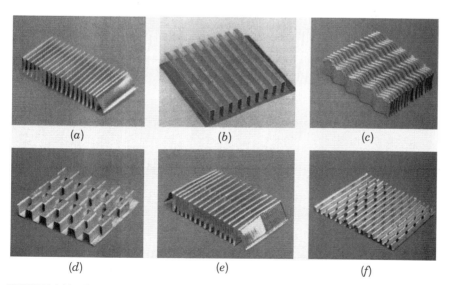

FIGURE 1.29   Corrugated fin geometries for plate-fin heat exchangers: (*a*) plain triangular fin; (*b*) plain rectangular fin; (*c*) wavy fin; (*d*) offset strip fin; (*e*) multilouver fin; (*f*) perforated fin. (Courtesy of Delphi Harrison Thermal Systems, Lockport, NY.)

Fig. 7.29c. Strip fins are also referred to as *offset fins, lance-offset fins, serrated fins,* and *segmented fins.* Many variations of interrupted fins are used in industry since they employ the materials of construction more efficiently than do plain fins and are therefore used when allowed by the design constraints.

Plate-fin exchangers are generally designed for moderate operating pressures [less than about 700 kPa gauge (100 psig)], although plate-fin exchangers are available commercially for operating pressures up to about 8300 kPa gauge (1200 psig). Recently, a condenser for an automotive air-conditioning system (see Fig. 1.27) using carbon dioxide as the working fluid has been developed for operating pressures of 14 MPa (2100 psia). A recently developed titanium plate-fin exchanger (manufactured by superelastic deformation and diffusion bonding, shown in Fig. 1.30) can take 35 MPa (5000 psig) and higher pressures. The temperature limitation for plate-fin exchangers depends on the method of bonding and the materials employed. Such exchangers have been made from metals for temperatures up to about 840°C (1550°F) and made from ceramic materials for temperatures up to about 1150°C (2100°F) with a peak temperature of 1370°C (2500°F). For ventilation applications (i.e., preheating or precooling of incoming air to a building/room), the plate-fin exchanger is made using Japanese treated (hygroscopic) paper and has the operating temperature limit of 50°C (122°F). Thus,

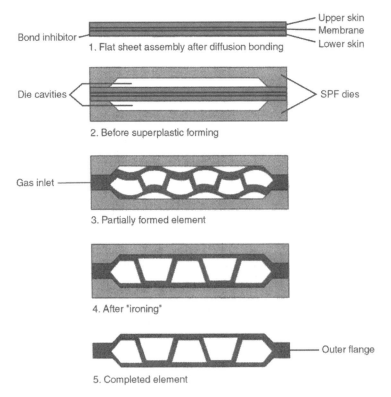

FIGURE 1.30 Process of manufacturing of a super elastically deformed diffusion bonded plate-fin exchanger (From Reay, 1999).

plates and fins are made from a variety of materials, metals, ceramics, and papers. Plate-fin exchangers have been built with a surface area density of up to $5900 \, m^2/m^3$ ($1800 \, ft^2/ft^3$). There is total freedom in selecting the fin surface area on each fluid side, as required by the design, by varying the fin height and fin density. Although typical fin densities are 120 to 700 fins/m (3 to 18 fins/in.), applications exist for as many as 2100 fins/m (53 fins/in.). Common fin thickness ranges between 0.05 and 0.25 mm (0.002 to 0.01 in.). Fin heights may range from 2 to 25 mm (0.08 to 1.0 in.). A plate-fin exchanger with 600 fins/m (15.2 fins/in.) provides about $1300 \, m^2$ ($400 \, ft^2/ft^3$) of heat transfer surface area per cubic meter of volume occupied by the fins. Plate-fin exchangers are manufactured in virtually all shapes and sizes and are made from a variety of materials. A cryogenic plate-fin exchanger has about 10% of the volume of an equivalent shell-and-tube exchanger (Reay, 1999).

Plate-fin exchangers have been produced since the 1910s in the auto industry (copper fin–brass tubes), since the 1940s in the aerospace industry (using aluminum), and in gas liquefaction applications since the 1950s using aluminum because of the better mechanical characteristics of aluminum at low temperatures. They are now used widely in electric power plants (gas turbine, steam, nuclear, fuel cell, etc.), propulsive power plants (automobile, truck, airplane, etc.), systems with thermodynamic cycles (heat pump, refrigeration, etc.), and in electronic, cryogenic, gas-liquefaction, air-conditioning, and waste heat recovery systems.

1.5.3.2 Tube-Fin Heat Exchangers. These exchangers may be classified as conventional and specialized tube-fin exchangers. In a conventional tube-fin exchanger, heat transfer between the two fluids takes place by conduction through the tube wall. However, in a heat pipe exchanger (a specialized type of tube-fin exchanger), tubes with both ends closed act as a separating wall, and heat transfer between the two fluids takes place through this "separating wall" (heat pipe) by conduction, and evaporation and condensation of the heat pipe fluid. Let us first describe conventional tube-fin exchangers and then heat pipe exchangers.

Conventional Tube-Fin Exchangers. In a gas-to-liquid exchanger, the heat transfer coefficient on the liquid side is generally one order of magnitude higher than that on the gas side. Hence, to have balanced thermal conductances (approximately the same hA) on both sides for a minimum-size heat exchanger, fins are used on the gas side to increase surface area A. This is similar to the case of a condensing or evaporating fluid stream on one side and gas on the other. In addition, if the pressure is high for one fluid, it is generally economical to employ tubes.

In a tube-fin exchanger, round and rectangular tubes are most common, although elliptical tubes are also used. Fins are generally used on the outside, but they may be used on the inside of the tubes in some applications. They are attached to the tubes by a tight mechanical fit, tension winding, adhesive bonding, soldering, brazing, welding, or extrusion.

Depending on the fin type, tube-fin exchangers are categorized as follows: (1) an individually finned tube exchanger or simply a *finned tube exchanger*, as shown in Figs. 1.31*a* and 1.32, having normal fins on individual tubes; (2) a tube-fin exchanger having flat (continuous) fins, as shown in Figs. 1.31*b* and 1.33; the fins can be plain, wavy, or interrupted, and the array of tubes can have tubes of circular, oval, rectangular, or other shapes; and (3) longitudinal fins on individual tubes, as shown in Fig. 1.34. A *tube-fin exchanger with flat fins* has been referred to variously as a *plate-fin and tube, plate*

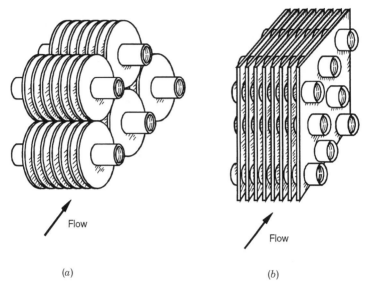

Flow

Flow

(a)   (b)

FIGURE 1.31   (a) Individually finned tubes; (b) flat (continuous) fins on an array of tubes. The flat fins are shown as plain fins, but they can be wavy, louvered, or interrupted.

finned tube, and *tube in-plate fin exchanger* in the literature. To avoid confusion with a plate-fin exchanger defined in Section 1.5.3.1, we refer to it as a tube-fin exchanger having flat (plain, wavy, or interrupted) fins. A tube-fin exchanger of the aforementioned categories 1 and 2 is referred to as a *coil* in the air-conditioning and refrigeration industries and has air outside and a refrigerant inside the tube. Individually finned tubes are probably more rugged and practical in large tube-fin exchangers. The exchanger with flat fins is usually less expensive on a unit heat transfer surface area basis because of its simple and mass-production construction features. Longitudinal fins are generally used in condensing applications and for viscous fluids in double-pipe heat exchangers.

Shell-and-tube exchangers sometime employ low finned tubes to increase the surface area on the shell side when the shell-side heat transfer coefficient is low compared to the tube-side coefficient, such as with highly viscous liquids, gases, or condensing refrigerant vapors. The low finned tubes are generally helical or annular fins on individual tubes; the fin outside diameter (see Fig. 1.9) is slightly smaller than the baffle hole. Longitudinal fins on individual tubes are also used in shell-and-tube exchangers. Fins on the inside of the tubes are of two types: integral fins as in internally finned tubes, and attached fins. Internally finned tubes are shown in Fig. 1.35.

Tube-fin exchangers can withstand ultrahigh pressures on the tube side. The highest temperature is again limited by the type of bonding, materials employed, and material thickness. Tube-fin exchangers usually are less compact than plate-fin units. Tube-fin exchangers with an area density of about $3300 \, \text{m}^2/\text{m}^3$ ($1000 \, \text{ft}^2/\text{ft}^3$) are available commercially. On the fin side, the surface area desired can be achieved through the proper fin density and fin geometry. Typical fin densities for flat fins vary from 250 to 800 fins/m (6 to 20 fins/in.), fin thicknesses vary from 0.08 to 0.25 mm (0.003 to 0.010 in.), and fin flow lengths vary from 25 to 250 mm (1 to 10 in.). A tube-fin exchanger having flat fins with 400 fins/m (10 fins/in.) has a surface area density of about $720 \, \text{m}^2/\text{m}^3$ ($220 \, \text{ft}^2/\text{ft}^3$).

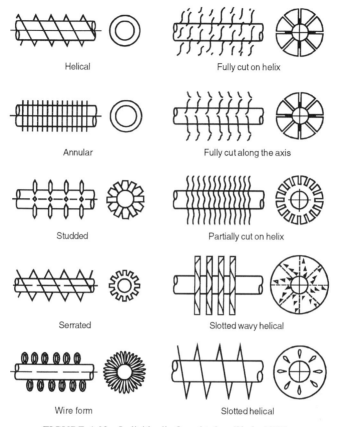

Helical

Fully cut on helix

Annular

Fully cut along the axis

Studded

Partially cut on helix

Serrated

Slotted wavy helical

Wire form

Slotted helical

FIGURE 1.32   Individually finned tubes (Shah, 1981).

Tube-fin exchangers are employed when one fluid stream is at a higher pressure and/ or has a significantly higher heat transfer coefficient than that of the other fluid stream. As a result, these exchangers are used extensively as condensers and evaporators in air-conditioning and refrigeration applications, as condensers in electric power plants, as oil coolers in propulsive power plants, and as air-cooled exchangers (also referred to as *fin-fan exchangers*) in process and power industries.

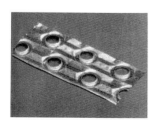
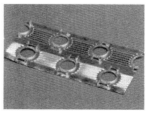

FIGURE 1.33   Flat fins on an array of round, flat, or oval tubes: (*a*) wavy fin; (*b*) multilouver fin; both fins with staggered round tubes; (*c*) multilouver fin with inline elliptical tubes. (Courtesy of Delphi Harrison Thermal Systems, Lockport, NY.)

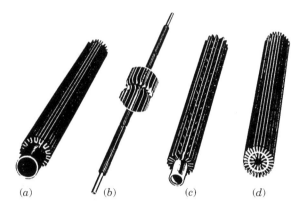

FIGURE 1.34   Longitudinal fins on individual tubes: (*a*) continuous plain; (*b*) cut and twisted; (*c*) perforated; (*d*) internal and external longitudinal fins. (Courtesy of Brown Fintube Company, Houston, TX.)

An *air-cooled exchanger* is a tube-fin exchanger in which hot process fluids (usually liquids or condensing fluids) flow inside the tubes, and atmospheric air is circulated outside by forced or induced draft over the extended surface. If used in a cooling tower with the process fluid as water, it is referred to as a *dry cooling tower*. Characteristics of this type of exchanger are shallow tube bundles (short airflow length) and large face area, due to the design constraint on the fan power.

Heat Pipe Heat Exchangers. This type of exchanger is similar to a tube-fin exchanger with individually finned tubes or flat (continuous) fins and tubes. However, the tube is a

FIGURE 1.35   Internally finned tubes. (Courtesy of Forged-Fin Division, Noranda Metal Industries, Inc., Newtown, CT.)

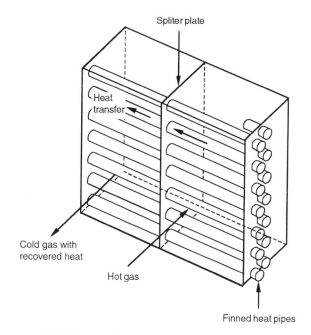

FIGURE 1.36   Heat pipe heat exchanger (Reay, 1979).

heat pipe, and hot and cold gases flow continuously in separate parts of the exchanger, as shown in Fig. 1.36. Heat is transferred from the hot gas to the evaporation section of the heat pipe by convection; the thermal energy is then carried away by the vapor to the condensation section of the heat pipe, where it transfers heat to the cold gas by convection.

As shown in Fig. 1.37, a heat pipe is a closed tube or vessel that has been evacuated, partially filled with a heat transfer fluid (a working fluid sufficient to wet the entire wick),

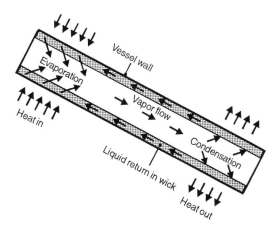

FIGURE 1.37   Heat pipe and its operation.

and sealed permanently at both ends. The inner surfaces of a heat pipe are usually lined with a capillary wick (a porous lining, screen, or internally grooved wall). The wick is what makes the heat pipe unique; it forces condensate to return to the evaporator by the action of capillary force. In a properly designed heat pipe, the wick is saturated with the liquid phase of the working fluid, while the remainder of the tube contains the vapor phase. When heat is applied at the evaporator by an external source, the working fluid in the wick in that section vaporizes, the pressure increases, and vapor flows to the condenser section through the central portion of the tube. The vapor condenses in the condenser section of the pipe, releasing the energy of phase change to a heat sink (to a cold fluid, flowing outside the heat pipe; see Fig. 1.37). The heat applied at the evaporator section tries to dry the wick surface through evaporation, but as the fluid evaporates, the liquid–vapor interface recedes into the wick surface, causing a capillary pressure to be developed. This pressure is responsible for transporting the condensed liquid back to the evaporator section, thereby completing a cycle. Thus, a properly designed heat pipe can transport the energy of phase change continuously from the evaporator to the condenser without drying out the wick. The condensed liquid may also be pumped back to the evaporator section by the capillary force or by the force of gravity if the heat pipe is inclined and the condensation section is above the evaporator section. If the gravity force is sufficient, no wick may be necessary. As long as there is a temperature difference between the hot and cold gases in a heat pipe heat exchanger, the closed-loop evaporation–condensation cycle will be continuous, and the heat pipe will continue functioning. Generally, there is a small temperature difference between the evaporator and condenser section [about 5°C (9°F) or so], and hence the overall thermal resistance of a heat pipe in a heat pipe exchanger is small. Although water is a common heat pipe fluid, other fluids are also used, depending on the operating temperature range.

A heat pipe heat exchanger (HPHE), shown in Fig. 1.36 for a gas-to-gas application, consists of a number of finned heat pipes (similar to an air-cooled condenser coil) mounted in a frame and used in a duct assembly. Fins on the heat pipe increase the surface area to compensate for low heat transfer coefficients with gas flows. The fins can be spirally wrapped around each pipe, or a number of pipes can be expanded into flat plain or augmented fins. The fin density can be varied from side to side, or the pipe may contain no fins at all (liquid applications). The tube bundle may be horizontal or vertical with the evaporator sections below the condenser sections. The tube rows are normally staggered with the number of tube rows typically between 4 and 10. In a gas-to-gas HPHE, the evaporator section of the heat pipe spans the duct carrying the hot exhaust gas, and the condenser section is located in the duct through which the air to be preheated flows. The HPHE has a splitter plate that is used primarily to prevent mixing between the two gas streams, effectively sealing them from one another. Since the splitter plate is thin, a heat pipe in a HPHE does not have the usual adiabatic section that most heat pipes have.

Unit size varies with airflow. Small units have a face size of 0.6 m (length) by 0.3 m (height), and the largest units may have a face size up to 5 m × 3 m. In the case of gas-to-liquid heat exchangers, the gas section remains the same, but because of the higher external heat transfer coefficient on the liquid side, it need not be finned externally or can even be shorter in length.

The heat pipe performance is influenced by the angle of orientation, since gravity plays an important role in aiding or resisting the capillary flow of the condensate. Because of this sensitivity, tilting the exchanger may control the pumping power and ultimately the heat transfer. This feature can be used to regulate the performance of a heat pipe heat exchanger. For further details on the design of a HPHE, refer to Shah and Giovannelli (1988).

Heat pipe heat exchangers are generally used in gas-to-gas heat transfer applications. They are used primarily in many industrial and consumer product–oriented waste heat recovery applications.

1.5.4 Regenerators

The regenerator is a storage-type heat exchanger, as described earlier. The heat transfer surface or elements are usually referred to as a *matrix* in the regenerator. To have continuous operation, either the matrix must be moved periodically into and out of the fixed streams of gases, as in a *rotary regenerator* (Figs. 1.38 through 1.40), or the gas flows must be diverted through valves to and from the fixed matrices as in a *fixed-matrix regenerator* (Fig. 1.41). The latter is also sometimes referred to as a *periodic-flow regenerator*,[†] a *swing regenerator*, or a *reversible heat accumulator*. Thus, in a rotary

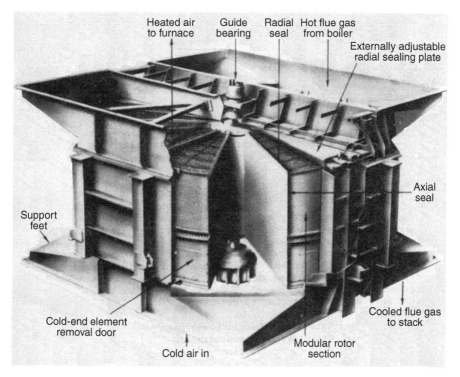

FIGURE 1.38   Ljungstrom air preheater. (Courtesy of ABB Alstom Power Air Preheater, Inc., Wellsville, NY.)

[†] Both rotary matrix and fixed-matrix regenerators have been designated as periodic-flow heat exchangers by Kays and London (1998), because from the viewpoint of an observer riding on the matrix, periodic conditions are experienced in both types of regenerators.

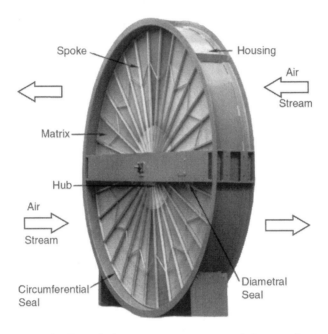

FIGURE 1.39   Heat wheel or a rotary regenerator made from a polyester film.

regenerator, the matrix (disk or rotor) rotates continuously with a constant fraction of the core (having disk sector angle θ_h) in the hot-fluid stream and the remaining fraction (having the disk sector angle θ_c) in the cold-fluid stream; the outlet fluid temperatures vary across the flow area and are independent of time. The two fluids generally flow in the opposite directions and are separated by some form of ductwork and rubbing seals on

FIGURE 1.40   Rotary regenerator made from a treated Japanese paper.

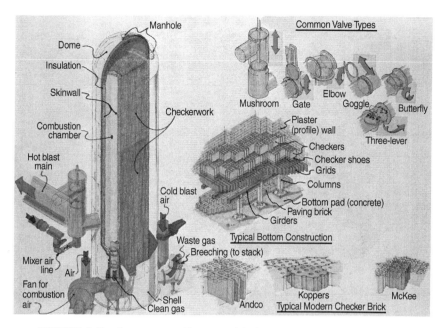

FIGURE 1.41   Cowper stove. (Courtesy of Andco Industries, Inc., Buffalo, NY.)

the matrix. In a fixed-matrix regenerator, the hot and cold fluids are ducted through the use of valves to the different matrices (with a minimum of two identical matrices for continuous operation) of the regenerator in alternate operating periods P_h and P_c; the outlet fluid temperatures vary with time. Here again, the two fluids alternately flow in opposite directions in a given matrix.

A third type of regenerator has a fixed matrix (in disk form) and fixed streams of gases, but the gases are ducted through rotating hoods (headers) to the matrix as shown in Fig. 1.42. This Rothemuhle regenerator is used as an air preheater in some power-generating plants. Since the basic thermal design theory of all types of regenerators is the same, no specific attention will be given to the Rothemuhle regenerator for the thermal design.

The desired material properties for the regenerator are high volumetric heat capacity (high ρc_p) and low effective thermal conductivity in the longitudinal (gas flow) direction. It should be noted that at very low temperatures, 20 K (36°R) and below, the specific heat of most metals decreases substantially, thus affecting the regenerator performance significantly.

The thermodynamically superior counterflow arrangement is usually employed for storage type heat exchangers by introducing gases successively at the opposite ends. When the rotational speed or frequency of switching hot and cold fluids through such a regenerator is increased, its thermal performance ideally approaches that of a pure counterflow heat exchanger; but in reality, the carryover leakage may become significant with increased speed, thus reducing the regenerator performance. For some applications, a parallelflow arrangement (gases introduced successively at the same end) may be used, but there is no counterpart of the single- or multipass crossflow arrangements common in recuperators. For a rotary regenerator, the design of seals to prevent leakages of hot

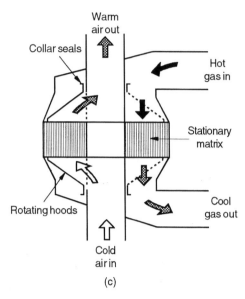

FIGURE 1.42   Rothemuhle regenerator. (Courtesy of Babcock and Wilcox, New Orleans, LA.)

to cold fluids, and vice versa, becomes a difficult task, especially if the two fluids are at significantly different pressures. Rotating drives also pose a challenging mechanical design problem. For a fixed-matrix regenerator operating at high temperatures, due to thermal distortion of housing and valves, various large and small cracks occur in the matrix housing and the valves do not seal the flow of gases perfectly, resulting in pressure leakages.

Major advantages of the regenerators are the following. A much more compact surface may be employed than in a recuperator, thus providing a reduced exchanger volume for given exchanger effectiveness and pressure drop and thereby making a regenerator economical compared to an equivalent recuperator. The major reason for having a much more compact surface for a regenerator is that the hot and cold gas streams are separated by radial seals or valves, unlike in a recuperator, where the primary surface is used to separate the fluid streams. The cost of manufacturing such a compact regenerator surface per unit of heat transfer area is usually substantially lower than that for the equivalent recuperator. Similarly, material cost could be lower in a regenerator than in a recuperator. Hence, a compact regenerator usually has a smaller volume and is lower in weight than an equivalent recuperator. Effectively, many fin configurations of plate-fin exchangers and any finely divided matrix material (high specific heat preferred) that provides high surface area density may be used. However, the leakproof core required in a recuperator is not essential in a regenerator, due to the mode of operation. Regenerators have been made from metals, ceramics, nylon, plastics, and paper, depending on the application. Another important advantage of a counterflow regenerator over a counterflow recuperator is that the design of inlet and outlet headers used to distribute hot and cold gases in the matrix is simple. This is because both fluids flow in different sections (separated by radial seals) of a rotary regenerator, or one fluid enters and leaves one matrix at a time in a fixed-matrix regenerator. In contrast, the header design to separate two fluids at the inlet and outlet in a counterflow recuperator is complex and costly (see Fig. 1.49 for possible header arrangements). Also, in a rotary regenerator, the flow sectors for the hot and cold gases can be designed to optimize the pressure drop on the hot and cold gases; and the critical pressure drop (usually on the hot side) in a rotary regenerator is lower than that in a comparable recuperator. The matrix surface has self-cleaning characteristics, resulting in low gas-side fouling and associated corrosion, if any, because the hot and cold gases flow alternately in opposite directions in the same fluid passage. Hence, regenerators are used with particulate-laden gases that promote surface fouling in a recuperator. Compact surface area density and the counterflow arrangement make the regenerator ideally suited for gas-to-gas heat exchanger applications requiring high exchanger effectiveness, generally exceeding 85%.

A major disadvantage of a rotary regenerator is that an unavoidable carryover of a small fraction of one fluid stream trapped in the flow passages under the radial seal is pushed out by the other fluid stream just after the periodic flow switching. Similar unavoidable carryover of the fluid stream trapped in the void volume of a given matrix of a fixed-matrix regenerator occurs when it is pushed out by the other fluid stream just after valve switching. Where fluid contamination (small mixing) is prohibited as with liquid flows, a regenerator cannot be used. Hence, regenerators are used exclusively for gas-to-gas heat and/or energy transfer applications, primarily for waste heat recovery applications, and are not used with liquid or phase-changing fluids. Other disadvantages are listed separately in the following subsections for rotary and fixed-matrix regenerators.

1.5.4.1 Rotary Regenerators. Rotary regenerators are shown in Figs. 1.38 through 1.40. Depending on the applications, rotary regenerators are variously referred to as a *heat wheel, thermal wheel, Munter wheel,* or *Ljungstrom wheel.* When the gas flows are laminar, the rotary regenerator is also referred to as a *laminar flow wheel.*

In this exchanger, any of the plain plate-fin surface geometries could be used in the matrix made up of thin metal sheets. Interrupted passage surfaces (such as strip fins, louver fins) are not used because a transverse (to the main flow direction) flow leakage is present if the two fluids are at different pressures. This leak mixes the two fluids (contaminates the lower pressure fluid) and reduces the exchanger effectiveness. Hence, the matrix generally has continuous (uninterrupted) flow passages. Flat or wavy spacers are used to stack the "fins"[†] (see Fig. 1.43). The fluid is unmixed at any cross section for these surfaces. Two examples of rotary regenerator surfaces are shown in Fig. 1.43. The herringbone or skewed passage matrix does not require spacers for stacking the "fins". The design Reynolds number range with these types of surfaces is 100 to 1000.

The matrix in the regenerator is rotated by a hub shaft or a peripheral ring gear drive. Every matrix element is passed periodically from the hot to the cold stream and back again. The time required for a complete rotation of the matrix is equivalent to the total period of a fixed-matrix regenerator. In a rotary regenerator, the stationary radial seal locations control the desired frontal areas for each fluid and also serve

---

[†] It should be emphasized that in a regenerator matrix, however, the entire surface acts as a direct heat-absorbing and heat-releasing surface (a primary surface); there is no secondary surface or fins, although the surface between spacers is usually referred to as fins.

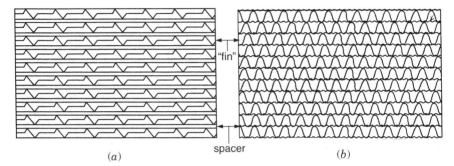

FIGURE 1.43 Continuous-passage matrices for a rotary regenerator: (a) notched plate; (b) triangular passage.

to minimize the primary leakage from the high-pressure fluid to the low-pressure fluid.

A number of seal configurations are used in rotary regenerators. Two common shapes are shown in Fig. 1.44. For the annular sector–shaped seals shown in Fig. 1.44a, flow passages at every radial location experience the same flow exposure and seal-coverage histories. For the uniform-width seals in Fig. 1.44b, flow passages at different radial locations experience different flow exposure and seal coverage. For regenerators with seals of equal area but arbitrary shape, the regenerator effectiveness is highest for annular sector–shaped seals (Beck and Wilson, 1996).

Rotary regenerators have been designed for surface area densities of up to about $6600 \, \text{m}^2/\text{m}^3$ ($2000 \, \text{ft}^2/\text{ft}^3$). They can employ thinner stock material, resulting in the lowest amount of material for a given effectiveness and pressure drop of any heat exchanger known today. Metal rotary regenerators have been designed for continuous operating inlet temperatures up to about 790°C (1450°F). For higher-temperature applications, ceramic matrices are used. Plastics, paper, and wool are used for regenerators operating below 65°C (150°F). Metal and ceramic regenerators cannot withstand large pressure differences [greater than about 400 kPa (60 psi)] between hot and cold gases, because the design of seals (wear and tear, thermal distortion, and subsequent leakage) is the single most difficult problem to resolve. Plastic, paper, and wool regenerators operate approxi-

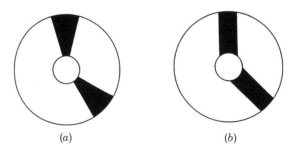

FIGURE 1.44 Seals used in rotary regenerators: (a) annular sector shaped; (b) uniform width shape (Beck and Wilson, 1996).

mately at atmospheric pressure. Seal leakage can reduce the regenerator effectiveness significantly. Rotary regenerators also require a power input to rotate the core from one fluid to the other at the rotational speed desired.

Typical power plant regenerators have a rotor diameter up to 10 m (33 ft) and rotational speeds in the range 0.5 to 3 rpm (rev per min). Air-ventilating regenerators have rotors with diameters of 0.25 to 3 m (0.8 to 9.8 ft) and rotational speeds up to 10 rpm. Vehicular regenerators have diameters up to 0.6 m (24 in.) and rotational speeds up to 18 rpm.

Ljungstrom air preheaters for thermal power plants, commercial and residential oil- and coal-fired furnaces, and regenerators for the vehicular gas turbine power plants are typical examples of metal rotary regenerators for preheating inlet air. Rotary regenerators are also used in chemical plants and in preheating combustion air in electricity generation plants for waste heat utilization. Ceramic regenerators are used for high-temperature incinerators and the vehicular gas turbine power plant. In air-conditioning and industrial process heat recovery applications, heat wheels are made from knitted aluminum or stainless steel wire matrix, wound polyester film, plastic films, and honeycombs. Even paper, wettable nylon, and polypropylene are used in the enthalpy or hygroscopic wheels used in heating and ventilating applications in which moisture is transferred in addition to sensible heat.

1.5.4.2 Fixed-Matrix Regenerator.

1.5.4.2 Fixed-Matrix Regenerator. This type is also referred to as a *periodic-flow, fixed-bed, valved,* or *stationary regenerator*. For continuous operation, this exchanger has at least two identical matrices operated in parallel, but usually three or four, shown in Figs. 1.45 and 1.46, to reduce the temperature variations in outlet-heated cold gas in high-temperature applications. In contrast, in a rotary or rotating hood regenerator, a single matrix is sufficient for continuous operation.

Fixed-matrix regenerators have two types of heat transfer elements: checkerwork and pebble beds. Checkerwork or thin-plate cellular structure are of two major categories: (1) noncompact regenerators used for high-temperature applications [925 to 1600°C (1700 to 2900°F)] with corrosive gases, such as a Cowper stove (Fig. 1.41) for a blast furnace used in steel industries, and air preheaters for coke manufacture and glass melting tanks made of refractory material; and (2) highly compact regenerators used for low- to high-temperature applications, such as in cryogenic process for air separation, in refrigeration, and in Stirling, Ericsson, Gifford, and Vuileumier cycle engines. The regenerator, a key thermodynamic element in the Stirling engine cycle, has only one matrix, and hence it does not have continuous fluid flows as in other regenerators. For this reason, we do not cover the design theory of a Stirling regenerator.

Cowper stoves are very large with an approximate height of 35 m (115 ft) and diameter of 7.5 m (25 ft). They can handle large flow rates at inlet temperatures of up to 1200°C (2200°F). A typical cycle time is between 1 and 3 h. In a Cowper stove, it is highly desirable to have the temperature of the outlet heated (blast) air approximately constant with time. The difference between the outlet temperatures of the blast air at the beginning and end of a period is referred to as a *temperature swing*. To minimize the temperature swing, three or four stove regenerators, shown in Figs. 1.45 and 1.46, are employed. In the *series parallel arrangement* of Fig. 1.45, part of the cold air (blast) flow is bypassed around the stove and mixed with the heated air (hot blast) leaving the stove. Since the stove cools as the blast is blown through it, it is necessary constantly to decrease the amount of the blast bypassed while increasing the blast through the stove by a corre-

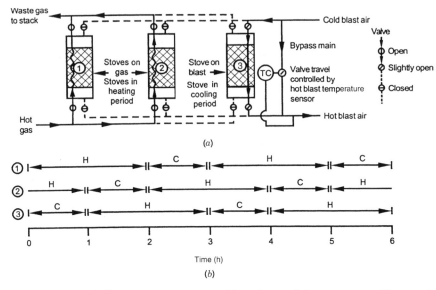

FIGURE 1.45 (a) Three-stove regenerator with series–parallel arrangement; (b) operating schedule. H, hot-gas period; C, blast period (Shah, 1981).

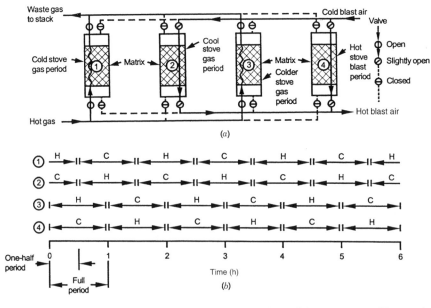

FIGURE 1.46 (a) Four-stove regenerator with staggered parallel arrangement; (b) operating schedule. H, hot-gas period; C, blast period (Shah, 1981).

sponding amount to maintain the hot blast temperature approximately constant. In the *staggered parallel arrangement* of Fig. 1.46, two stoves on air are maintained out of phase by one-half period. In this arrangement, cold blast is routed through a "hot" stove and a "cool" stove (i.e., through which a cold blast has blown for one-half period) rather than being bypassed. The amount of blast through the hot stove is constantly increased while that through the cool stove is decreased by the same amount to maintain the hot blast air temperature approximately constant. At the end of one-half period, the hot stove's inlet valve is fully open and the cool stove's inlet valve is fully closed. At this point, the cool stove is put "on gas," the hot stove becomes the cool stove, and a new hot stove is switched in.

The heat transfer surface used in the aforementioned high-temperature fixed-matrix regenerator is made of refractory bricks, referred to simply as *checkerwork*. The commonly used checker shapes' surface area density range is 25 to $42 \, \text{m}^2/\text{m}^3$ (8 to $13 \, \text{ft}^2/\text{ft}^3$), as shown in Fig. 1.47. The checker flow passage (referred to as a *flue*) size is relatively large, primarily to accommodate the fouling problem associated with highly corrosive hot exhaust gases coming to the regenerator. A typical heat transfer coefficient in such a passage is about $5 \, \text{W/m}^2 \cdot \text{K}$ (1 Btu/hr-ft^2-°F).

The surface geometries used for a compact fixed-matrix regenerator are similar to those used for rotary regenerators. The surface geometries used for packed beds are quartz pebbles, steel, copper, or lead shots, copper wool, packed fibers, powders, randomly packed woven screens, and crossed rods. Heat transfer surface area densities of $82{,}000 \, \text{m}^2/\text{m}^3$ ($25{,}000 \, \text{ft}^2/\text{ft}^3$) are achievable; the heat transfer coefficient range is 50 to $200 \, \text{W/m}^2 \cdot \text{K}$ (9 to 35 Btu/hr-ft^2-°F).

The design flexibility of selecting different frontal areas is not possible for a fixed-matrix regenerator having multiple matrices, but instead, different hot and cold flow periods are selected. The pressure leakage in a fixed-matrix regenerator is through the "imperfect" valves after wear and tear and through the cracks of matrix walls. Fixed-matrix regenerators can be used for large flow rates and can have very large surface areas and high-heat-capacity material, depending on the design requirements.

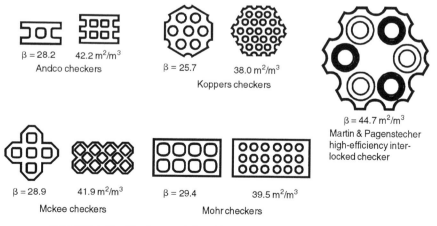

FIGURE 1.47   Checkers used for a blast furnace regenerator (Shah, 1981).

1.6   CLASSIFICATION ACCORDING TO FLOW ARRANGEMENTS

Common flow arrangements of the fluids in a heat exchanger are classified in Fig. 1.1. The choice of a particular flow arrangement is dependent on the required exchanger effectiveness, available pressure drops, minimum and maximum velocities allowed, fluid flow paths, packaging envelope, allowable thermal stresses, temperature levels, piping and plumbing considerations, and other design criteria. Let us first discuss the concept of multipassing, followed by some of the basic ideal flow arrangements for a two-fluid heat exchanger for single- and multipass heat exchangers.

Multipassing. The concept of multipassing applies separately to the fluid and heat exchanger. A *fluid* is considered to have made one pass if it flows through a section of the heat exchanger through its full length. After flowing through one full length, if the flow direction is reversed and fluid flows through an equal- or different-sized section, it is considered to have made a second pass of equal or different size. A *heat*

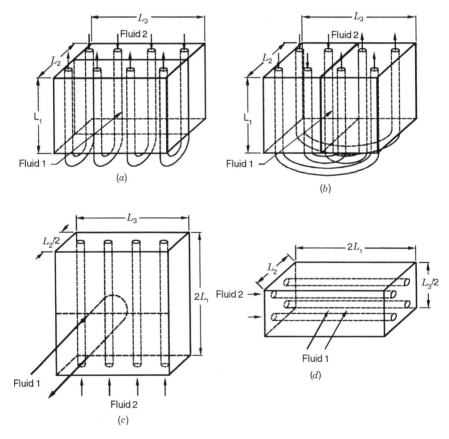

FIGURE 1.48   (*a*) Two-pass cross-counterflow exchanger; (*b*) single-pass crossflow exchanger; (*c*, *d*) unfolded exchangers of (*a*) and (*b*), respectively.

exchanger is considered as a single-pass unit if both fluids make one pass in the exchanger or if it represents any of the single-pass flow arrangements when the multi-pass fluid side is unfolded (note that the folding is used to control the envelope size). To illustrate the concept, consider one exchanger with two different designs of inlet headers for fluid 2 as shown in Fig. 1.48a and b; fluid 1 makes a single pass, and fluid 2 makes two passes in both exchangers. If the exchanger of Fig. 1.48b with fluid 2 unmixed in the headers is unfolded to the horizontal direction (the exchanger length for fluid 2 will be $2L_1$), as in Fig. 1.48d,[†] the resulting geometry is a single-pass exchanger having the same inlet temperatures as fluids 1 and 2 of Fig. 1.48b. Hence, the exchanger of Fig. 1.48b is considered a *single-pass exchanger* from the exchanger analysis point of view. In contrast, the temperature of fluid 1 at the inlet to the first and second pass of fluid 2 is different in Fig. 1.48a. Hence, when it is unfolded vertically as in Fig. 1.48c, the inlet temperature of fluid 1 to each half of the exchanger will be different, due to the presence of two passes, each with one-half of the original flow length L_2. This does not correspond to a single-pass exchanger of the unfolded exchanger height. Therefore, the exchanger of Fig. 1.48a is considered as a *two-pass exchanger*. An additional degree of freedom is introduced by unfolding. This degree of freedom describes how to lead a fluid between the passes (see, e.g., the case considered in Fig. 1.48c, fluid 1). Depending on how the fluid is routed from the exit of one pass to the inlet of the following pass, several distinct flow arrangements can be identified (see Section 1.6.2.1 for further details).

1.6.1 Single-Pass Exchangers

1.6.1.1 Counterflow Exchanger.
In a counterflow or countercurrent exchanger, as shown in Fig. 1.49a, the two fluids flow parallel to each other but in opposite directions within the core.[‡] The temperature variation of the two fluids in such an exchanger may be idealized as one-dimensional, as shown in Fig. 1.50. As shown later, the counterflow arrangement is thermodynamically superior to any other flow arrangement. It is the most efficient flow arrangement, producing the highest temperature change in each fluid compared to any other two-fluid flow arrangements for a given overall thermal conductance (UA), fluid flow rates (actually, fluid heat capacity rates), and fluid inlet temperatures. Moreover, the maximum temperature difference across the exchanger wall thickness (between the wall surfaces exposed on the hot and cold fluid sides) either at the hot- or cold-fluid end is the lowest, and produce minimum thermal stresses in the wall for an equivalent performance compared to any other flow arrangements. However, with plate-fin heat exchanger surfaces, there are manufacturing difficulties associated with the true counterflow arrangement. This is because it is necessary to separate the fluids at each end, and the problem of inlet and outlet header design is complex. Some header arrangements are shown in Fig. 1.49b–f. Also, the overriding importance of other design factors causes most commercial heat exchangers to be designed for flow arrangements different from single-pass counterflow if extremely high exchanger effectiveness is not required.

[†] In unfolded exchangers of Fig. 1.48c and d, the U-bend lengths of the tubes are neglected for the present discussion since they do not take an active part in heat transfer between two fluids.
[‡] This flow arrangement can be rigorously identified as a *countercurrent parallel stream*. However, based on Kays and London's (1998) terminology, used widely in the literature, we use the term *counterflow* for this flow arrangement throughout the book.

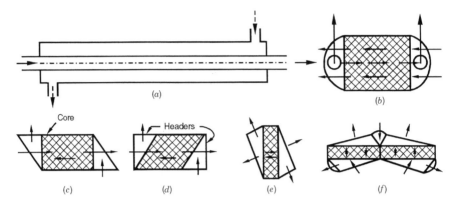

FIGURE 1.49   (a) Double-pipe heat exchanger with pure counterflow; (b–f) plate-fin exchangers with counterflow core and crossflow headers (Shah, 1981).

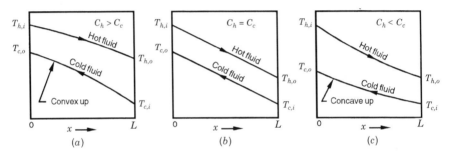

FIGURE 1.50   Temperature distributions in a counterflow heat exchanger of single-phase fluids (no boiling or condensation). Here $C_h = (\dot{m}c_p)_h$ is the heat capacity rate of the hot fluid, C_c is the heat capacity rate of the cold fluid, and specific heats c_p are treated as constant. The symbol T is used for temperature; the subscripts h and c denote hot and cold fluids, and subscripts i and o represent the inlet and outlet of the exchanger (Shah, 1981).

Typical temperature distributions for a counterflow regenerator are shown in Fig. 1.51. Note that the wall temperature fluctuates periodically between the solid line limits shown. Also compare the similarity between the fluid temperature distributions of Fig. 1.50 for $C_h = C_c$ and those of Fig. 1.51*b*.

1.6.1.2 Parallelflow Exchanger. In a parallelflow (also referred to as *cocurrent* or *cocurrent parallel stream*) exchanger, the fluid streams enter together at one end, flow parallel to each other in the same direction, and leave together at the other end. Figure 1.49*a* with the dashed arrows reversed would then depict parallelflow. Fluid temperature variations, idealized as one-dimensional, are shown in Fig. 1.52. This arrangement has the lowest exchanger effectiveness among single-pass exchangers for given overall thermal conductance (*UA*) and fluid flow rates (actually, fluid heat capacity rates) and fluid inlet temperatures; however, some multipass exchangers may have an even lower effectiveness, as discussed later. However, for low-effectiveness exchangers, the difference in parallelflow and counterflow exchanger effectiveness is small. In a parallelflow exchanger, a large temperature difference between inlet temperatures of hot and cold fluids

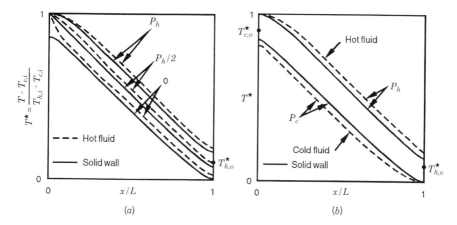

FIGURE 1.51   (*a*) Hot-side solid and fluid temperature excursion; (*b*) balanced ($C_h = C_c$) regenerator temperature distributions at the switching instant (Shah, 1991b).

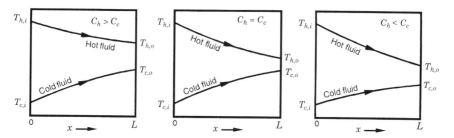

FIGURE 1.52   Temperature distributions in a parallelflow heat exchanger (Shah, 1981).

exists at the inlet side, which may induce high thermal stresses in the exchanger wall at the inlet. Although this flow arrangement is not used for applications requiring high-temperature effectiveness, it may be used in the following applications:

1. It often produces a more uniform longitudinal tube wall temperature distribution and not as high or low tube wall temperature as in a counterflow arrangement at the same surface area (NTU),[†] fluid flow rates (fluid heat capacity rates or C^*), and fluid inlet temperatures (see Example 3.2). For this reason, a parallelflow exchanger is sometimes used with temperature-sensitive materials, highly viscous liquids, and metal recuperators having inlet temperatures in excess of $1100°C$ ($2000°F$).

2. The lowest wall temperature for the parallelflow exchanger is *higher* than that for the counterflow or other flow arrangements for the same NTU, C^*, and fluid inlet temperatures, although the exchanger effectiveness will also be lower. Thus, if acid vapors are present in the exhaust gas, the parallelflow arrangement minimizes or

---

[†] See Sections 3.3.2 and 3.3.3 for definitions of C^* and NTU.

avoids condensation of acid vapors and hence corrosion of the metal surface. The parallelflow exchanger may be preferred when there is a possibility that the temperature of the warmer fluid may reach its freezing point.

3. The highest wall temperature for the parallelflow exchanger is *lower* than that for the counterflow or other flow arrangements for the same NTU, C^*, and inlet temperatures. This may eliminate or minimize the problems of fouling, wall material selections, and fluid decomposition.

4. It provides early initiation of nucleate boiling for boiling applications.

5. A large change in NTU causes a relatively small change in ε for NTU > 2, as shown in Fig. 3.8. Thus a parallelflow exchanger is preferred if the desired exchanger effectiveness is low and is to be maintained approximately constant over a large flow rate range (e.g., for NTU $\sim$ 1 to 5 or higher at $C^* = 1$, see Fig. 3.8).

6. The application allows piping suited only to parallelflow.

1.6.1.3 Crossflow Exchanger. In this type of exchanger, as shown in Fig. 1.53, the two fluids flow in directions normal to each other. Typical fluid temperature variations are idealized as two-dimensional and are shown in Fig. 1.54 for the inlet and outlet sections only. Thermodynamically, the effectiveness for the crossflow exchanger falls in between that for the counterflow and parallelflow arrangements. The largest structural temperature difference exists at the "corner" of the entering hot and cold fluids, such as point *a* in Fig. 1.54. This is one of the most common flow arrangements used for extended-surface heat exchangers, because it greatly simplifies the header design at the entrance and exit of each fluid. If the desired heat exchanger effectiveness is high (such as greater than 80%), the size penalty for the crossflow exchanger may become excessive. In such a case, a counterflow unit is preferred. This flow arrangement is used in a TEMA X shell (see Fig. 1.6) having a single tube pass. The length L_3 (or the "height" in the x direction)

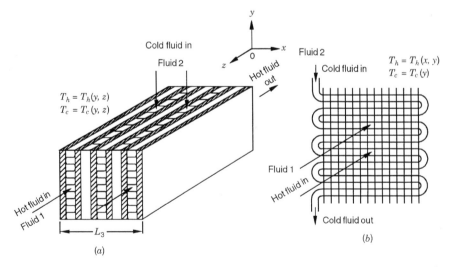

FIGURE 1.53   (*a*) Plate-fin unmixed–unmixed crossflow heat exchanger; (*b*) serpentine (one tube row) tube-fin unmixed–mixed crossflow heat exchanger (Shah, 1981).

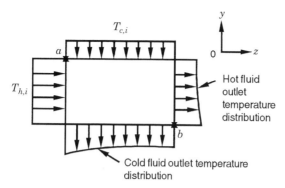

FIGURE 1.54　Temperature distributions at inlets and outlets of an unmixed–unmixed crossflow heat exchanger (Shah, 1981).

in Fig. 1.53*a* does not represent the flow length for either fluid 1 or fluid 2. Hence, it is referred to as *noflow height* or *stack height* since the fins are stacked in the L_3 direction.

In a crossflow arrangement, *mixing* of either fluid stream may or may not occur, depending on the design. A fluid stream is considered *unmixed* when it passes through individual flow channels or tubes with no fluid mixing between adjacent flow channels. In this case within the exchanger, temperature gradients in the fluid exist in at least one direction (in the transverse plane) normal to the main fluid flow direction. A fluid stream is considered completely *mixed* when no temperature gradient exists in the transverse plane, either within one tube or within the transverse tube row within the exchanger. Ideally, the fluid thermal conductivity transverse to the flow is treated as zero for the unmixed-fluid case and infinity for the mixed-fluid case. Fluids 1 and 2 in Fig. 1.53*a* are unmixed. Fluid 1 in Fig. 1.53*b* is unmixed, while fluid 2 is considered mixed because there is only one flow channel. The temperature of an unmixed fluid, such as fluid 1 in Fig. 1.53, is a function of two coordinates z and y within the exchanger, and it cannot be treated as constant across a cross section (in the y direction) perpendicular to the main flow direction x. Typical temperature distributions of the unmixed fluids at exchanger outlet sections are shown in Fig. 1.54. The outlet temperature from the exchanger on the unmixed side is defined as a mixed mean temperature that would be obtained after complete mixing of the fluid stream at the exit. For the cases of Fig. 1.53, it is idealized that there is no variation of the temperature of either fluid in the x direction. The temperature of a mixed fluid (fluid 2 in Fig. 1.53*b*) is mainly dependent on the coordinate y. The temperature change per pass (in the x direction) of fluid 2 in Fig. 1.53*b* is small compared to the total.

In a multiple-tube-row tubular crossflow exchanger, the tube fluid in any one tube is considered mixed at any cross section. However, when split and distributed in different tube rows, the incoming tube fluid is considered unmixed between the tube rows. Theoretically, it would require an infinite number of tube rows to have a truly unmixed fluid on the tube side. In reality, if the number of tube rows is greater than about four, it will practically be an unmixed side. For an exchanger with fewer than about four or five tube rows, the tube side is considered partially unmixed or partially mixed. Note that when the number of tube rows is reduced to one, the tube fluid is considered mixed.

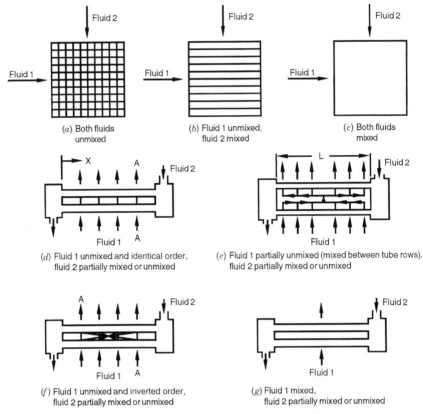

(a) Both fluids unmixed

(b) Fluid 1 unmixed, fluid 2 mixed

(c) Both fluids mixed

(d) Fluid 1 unmixed and identical order, fluid 2 partially mixed or unmixed

(e) Fluid 1 partially unmixed (mixed between tube rows), fluid 2 partially mixed or unmixed

(f) Fluid 1 unmixed and inverted order, fluid 2 partially mixed or unmixed

(g) Fluid 1 mixed, fluid 2 partially mixed or unmixed

FIGURE 1.55   Symbolic presentation of various degrees of mixing in a single-phase crossflow exchanger.

Mixing thus implies that a thermal averaging process takes place at each cross section across the full width of the flow passage. Even though the truly unmixed and truly mixed cases are the extreme idealized conditions of a real situation in which some mixing exists, the unmixed condition is nearly satisfied in many plate-fin and tube-fin (with flat fins) exchanger applications. As will be shown in Section 11.3 and in the discussion of Example 3.5, for the same surface area and fluid flow rates, (1) the exchanger effectiveness generally decreases with increasing mixing on any fluid side, although counter examples can be found in the multipass case; and (2) if the $C_{\max}$ fluid is placed on the unmixed fluid side, the exchanger effectiveness and performance will be higher than that for placing $C_{\max}$ on the mixed fluid side.

Seven idealized combinations of flow arrangements for a single-pass crossflow exchanger are shown symbolically in Fig. 1.55. The flow arrangements are:

(a) *Both fluids unmixed.* A crossflow plate-fin exchanger with plain fins on both sides represents the "both fluids unmixed" case.

(b) *One fluid unmixed, the other mixed.* A crossflow plate-fin exchanger with fins on one side and a plain gap on the other side would be treated as the unmixed–mixed case.

(c) *Both fluids mixed.* This case is practically less important, and represents a limiting case of some multipass shell-and-tube exchangers (e.g., 1–∞ TEMA E and J), as presented later.

(d) *One fluid unmixed and coupled in identical order, the other partially mixed.* Here *identical order* refers to the fact that a fluid coupled in such order leaves the first row at the point where the other fluid enters (leaves) the first row, and enters the other row where the second fluid enters (leaves) that row (see the stream AA in Fig. 1.55d). A tube-fin exchanger with flat fins represents the case of tube fluid partially mixed, the fin fluid unmixed. When the number of tube rows is reduced to one, this exchanger reduces to the case of out-of-tube (fin) fluid unmixed the tube fluid mixed (case b). When the number of tube rows approaches infinity (in reality greater than four), the exchanger reduces to the case of both fluids unmixed (case a).

(e) *One fluid partially unmixed, the other partially mixed.* The case of one fluid (fluid 1) partially unmixed (i.e., mixed only between tube rows) and the other (fluid 2) partially mixed (see Fig. 1.55e) is of less practical importance for single-pass crossflow exchangers. However, as mentioned later (see the middle sketch of Fig. 1.58b with the notation of fluids 1 and 2 interchanged),[†] it represents the side-by-side multipass crossflow arrangement. When the number of tube rows is reduced to one, this exchanger is reduced to the case of out-of-tube fluid unmixed, the tube fluid mixed. When the number of tube rows approaches infinity, the exchanger reduces to the case of out-of-tube fluid mixed, the tube fluid unmixed.

(f) *One fluid unmixed and coupled in inverted order, the other partially mixed.* Here, the term *inverted order* refers to the fact that a fluid coupled in such order leaves the first row at the point where the other fluid enters (leaves) the first row and enters the other row where the second fluid leaves (enters) that row (see the stream AA in Fig. 1.55f). This case is also of academic interest for single-pass crossflow exchangers.

(g) *One fluid mixed, the other partially mixed.* This is the case realized in plain tubular crossflow exchangers with a few tube rows.

1.6.1.4   Split-Flow Exchanger, TEMA G Shell. In this exchanger, shown in Fig. 1.56a, the shell fluid stream enters at the center of the exchanger and divides into two streams. These streams flow in longitudinal directions along the exchanger length over a longitudinal baffle, make a 180° turn at each end, flow longitudinally to the center of the exchanger under the longitudinal baffle, unite at the center, and leave from the central nozzle. The other fluid stream flows straight in the tubes. Typical temperature distributions for the two fluids in this exchanger are shown in Fig. 1.56. This single-pass flow arrangement is found in the TEMA G shell (see Fig. 1.6). Another variant is a double-split flow arrangement, as found in the TEMA H shell (see Fig. 1.6), again having a single tube pass.

[†] In Fig. 1.58b, this means fluid 1 partially mixed (considering mixed in each individual passes) and fluid 2 partially unmixed (i.e., unmixed within a pass and mixed between passes).

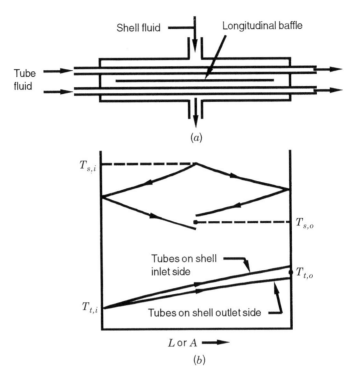

FIGURE 1.56 (*a*) Single-pass split-flow (TEMA G) exchanger; (*b*) idealized shell fluid and tube fluid temperature distributions.

1.6.1.5 Divided-Flow Exchanger, TEMA J Shell. In this exchanger as shown in Fig. 1.57*a*, the shell fluid stream enters at the center of the exchanger and divides into two streams. These streams flow ideally in longitudinal directions along the exchanger length and exit from two nozzles, one at each end of the exchanger. The other fluid stream flows straight in the tubes. Typical temperature distributions for the two fluids are shown in Fig. 1.57*b*. This flow arrangement is found in the TEMA J shell having a single tube pass.

1.6.2 Multipass Exchangers

When the design of a heat exchanger results in either an extreme length, significantly low fluid velocities, or a low effectiveness (sometimes maybe other design criteria), a multi-pass heat exchanger or several single-pass exchangers in series, or a combination of both, is employed. Heat exchangers in any of the five basic flow arrangements of Section 1.6.1 can be put into series to make a multipass unit. In addition, there exist other multipass flow arrangements that have no single-pass counterpart. One of the major advantages of proper multipassing is to increase the exchanger overall effectiveness over the individual pass effectivenesses, but with increased pressure drop on the multipass side. If the overall direction of the two fluids is chosen as counterflow (see Figs. 1.58*a* left and 1.62), the exchanger overall effectiveness approaches that of a pure counterflow exchanger as

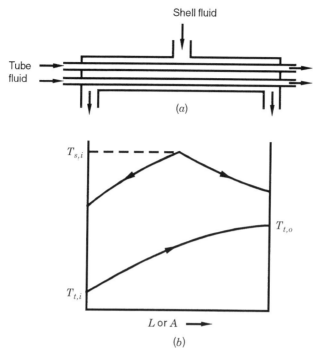

FIGURE 1.57 (*a*) Single-pass divided-flow (TEMA J) exchanger with shell fluid mixed; (*b*) idealized shell and tube fluid temperature distributions.

the number of passes increases. The multipass arrangements are classified according to the type of construction: for example, extended surface, shell-and-tube, or plate exchangers (see Fig. 1.1).

1.6.2.1 Multipass Crossflow Exchangers.

This arrangement is the most common for extended surface exchangers; two or more passes are put in series, with each pass usually having crossflow, although any one of the single-pass basic flow arrangements could be employed. The flow arrangements could be categorized as (a) a *series coupling* of *n* passes or *over-and-under passes*, (b) a *parallel coupling* of *n* passes or *side-by-side passes*, and (c) a combination of both or a compound arrangement. These are shown in Fig. 1.58. Each module in Fig. 1.58 can be either an individual pass or an individual heat exchanger. In the series coupling of *n* passes, each of the fluid streams is in series; whereas in the parallel coupling of *n* passes, one fluid stream is in series, the other in parallel. The parallel coupling (side-by-side) two-pass arrangement is also referred to as the *face-U flow arrangement*. For the same surface area, fluid flow rates and inlet temperatures, a series-coupled *overall counterflow* multipass exchanger yields higher effectiveness and heat transfer rate than that for a parallel-coupled multipass exchanger, as will be shown in Example 3.5. In a series-coupled multipass exchanger, usually the flow direction is chosen such that an overall counterflow is obtained, as shown in Fig. 1.58a, to obtain higher exchanger effectiveness. This arrangement is then referred to as *n*-pass *cross-counterflow*. If the direction of fluid 2 in Fig. 1.58a is reversed, overall

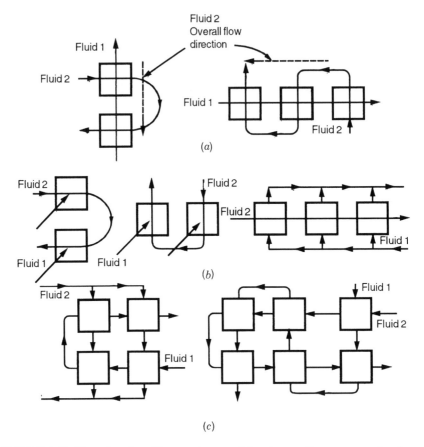

FIGURE 1.58   Examples of multipass exchangers: (*a*) series coupling or over-and-under pass arrangement; (*b*) parallel coupling or side-by-side pass arrangement; (*c*) compound coupling.

parallelflow would be achieved, and it is referred to as an *n-pass cross-parallelflow* arrangement. The latter arrangement is used to prevent freezing of the hot fluid (such as water) in the core near the inlet of the cold fluid (such as air). There are a large number of combinations of the foregoing basic multipass arrangements yielding compound multipass arrangements. No specific broadly accepted classification scheme has emerged for compound multipass arrangements.

Now let us introduce additional basic terminology for series-coupled multipass exchangers. The exchanger effectiveness of any of the foregoing multipass crossflow exchangers depends on whether the fluids are mixed, partially unmixed, or fully unmixed within each pass and mixed or unmixed between passes on each fluid side, in addition to independent variables for single-pass exchangers. When the fluids are unmixed between passes (in the pass return or header), the exchanger effectiveness is dependent on how the fluid is distributed in the header, whether in identical or inverted order, and which fluid has larger heat capacity rate. As shown in Fig. 1.59*a*, consider two rows per pass for the tube fluid and the air flows across the tubes in sequence of rows 1 through 4 in the direction of arrows. The tube fluid in row 1 (the first tube in pass 1) is first in contact

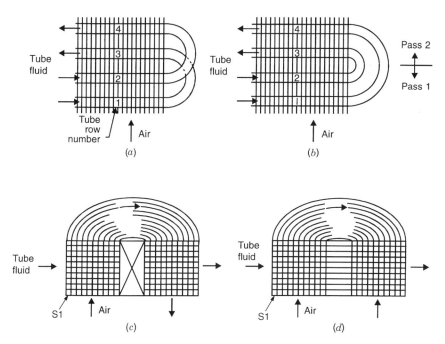

FIGURE 1.59 Two-pass cross-parallelflow exchangers with both fluids unmixed throughout. For the *tube fluid* in the pass return: (*a*) identical order; (*b*) inverted order. Cases (*c*) and (*d*) are symbolic representations of cases (*a*) and (*b*), respectively. In both cases (*a*) and (*b*), air (out-of-tube fluid) is in inverted order.

with the air in pass 1, and the *same* fluid in row 3 (the first tube in pass 2) is first in contact with air in pass 2. The tube fluid between passes for this arrangement is then defined to be in identical order. In Fig. 1.59*b*, the tube fluid stream in row 1 (the first tube in pass 1) is again first in contact with air in pass 1. However, the *different* fluid stream (or row connected to the second tube in pass 1) makes the first contact with air in pass 2. In this arrangement, the tube fluid between passes is said to be in inverted order. In either Fig. 1.59*a* or *b*, the air-side fluid is in inverted order between passes. This is because the airstream that crossed the tube inlet end in the first pass then crosses the tube exit end in the second pass (see the first vertical airstream, S1 in Fig. 1.59*c* and *d*). Figures 1.59*a* and *b* are represented symbolically as Fig. 1.59*c* and *d*, respectively, since one does not have always tubes in the exchanger, for example, in a plate-fin multipass exchanger.

Multipassing of crossflow exchangers retains the header and ducting advantages of a simple crossflow exchanger, while it is possible to approach the thermal performance of a true counterflow heat exchanger using an overall cross-counterflow arrangement. The maximum temperature differences in the wall across the wall thickness direction (sometimes referred to as *structural temperature differences*) are considerably reduced in a multipass cross-counterflow exchanger relative to a single-pass crossflow design for the same terminal temperatures. For high-temperature applications (approximately above 450°C or 850°F), the heat exchanger can be divided into two or more passes having distinct operating temperature ranges. Special metals (such as stainless steel and super-alloys) may be used in passes having high operating temperatures, and ordinary metals (such as aluminum or copper) could be used in moderate to low operating temperatures, thus achieving substantial cost savings and improved durability of the exchanger.

1.6.2.2 Multipass Shell-and-Tube Exchangers. When the number of tube passes is greater than one, the shell-and-tube exchangers with any of the TEMA shell types (except for the F shell) represent a multipass exchanger. Since the shell-side fluid flow arrangement is unique with each shell type, the exchanger effectiveness is different for each shell even though the number of tube passes may be the same. For illustrative purposes, in the following subsections, two tube passes (as in a U-tube bundle) are considered for the multipass shell-and-tube exchangers having E, G, H, and J shells. However, more than two tube passes are also common, as will be mentioned later. The ideal flow arrangement in the F shell with two tube passes is a pure counterflow arrangement as considered with single-pass exchangers and as can be found by unfolding the tubes. Since the liquid is evaporating on the shell side in the K shell (as a kettle reboiler application), it represents the $C^* = 0$ case and can be analyzed using the single-pass exchanger results, as will be explained with Eq. (3.84) and in item 4 after Eq. (3.88). The flow arrangement with the X shell having two or more tube passes represents an overall cross-counterflow or cross-parallelflow arrangement, as described for extended surface exchangers, depending on the overall directions of the two fluids. Hence, only the unique multipass arrangements used in shell-and-tube exchangers are summarized next.

Parallel Counterflow Exchanger, TEMA E Shell. This is one of the most common flow arrangements used in single-phase shell-and-tube heat exchangers, and is associated with the TEMA E shell. One of the simplest flow arrangements is one shell pass and two tube passes, as shown in Fig. 1.60 using a U-tube bundle. A heat exchanger with this arrangement is also simply referred to as a conventional 1–2 *heat exchanger* by industry and in this book, although the more precise terminology would be 1–2 TEMA

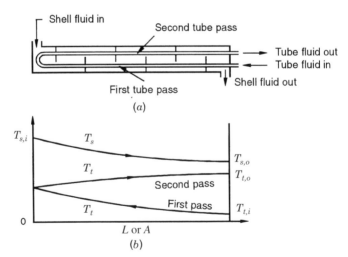

FIGURE 1.60 (*a*) A 1–2 TEMA E heat exchanger (one shell pass and two tube passes); (*b*) corresponding temperature distributions.

E exchanger. Similarly, we simply designate the $1-2n$ TEMA E exchanger as the $1-2n$ exchanger.

As the tubes are rigidly mounted only at one end, thermal expansion is readily accommodated. If the shell fluid is idealized as *well mixed*, its temperature is constant at any cross section but changes from a cross section to another cross section along the shell length direction. In this case, reversing the tube fluid flow direction will not change the idealized temperature distribution of Fig. 1.60b and the exchanger effectiveness.

Increasing the even number of tube passes of a $1-2n$ exchanger from two to four, six, and so on, decreases the exchanger effectiveness slightly, and in the limit when the number of tube passes approaches infinity with one shell pass, the exchanger effectiveness approaches that for a single-pass crossflow exchanger with both fluids mixed. Common tube-side multipass arrangements are shown in Fig. 1.61.[†]

The odd number of tube passes per shell has slightly better effectiveness when the shell fluid flows countercurrent to the tube fluid for more than one half the tube passes. However, this is an uncommon design and may result in structural and thermal problems in manufacturing and design.

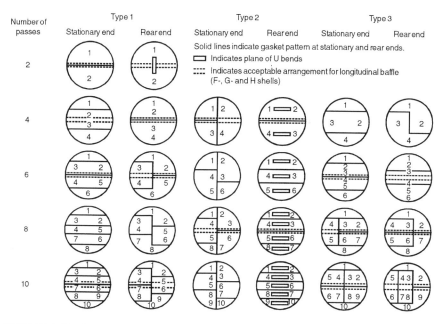

FIGURE 1.61 Common tube-side multipass arrangements in shell-and-tube exchangers (to simplify the sketches, tubes are not shown in the cross section of the exchanger). The solid lines indicate pass ribs in the front header; the dashed lines represent pass ribs in the rear header (From Saunders, 1988).

[†]Each sketch in Fig. 1.61 represents a cross section of the shell and tube fluid nozzles at the inlet and pass partitions. The dashed lines are the pass partitions on the other end of the tube bundle. No tubes or baffles are shown for clarity. Also, no horizontal orientation of the nozzles is shown, although horizontal nozzles are common for some applications.

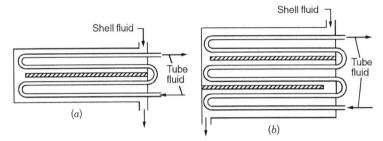

FIGURE 1.62 (a) Two shell pass–four tube pass exchanger; (b) three shell pass–six tube pass exchanger.

Since the 1–2n exchanger has a lower effectiveness than that of a counterflow exchanger, multipassing of the basic 1–2 arrangement may be employed with multiple shells (each shell as a 1–2 exchanger) to approach the counterflow effectiveness. The heat exchanger with the most general flow arrangement would have m shell passes and n tube passes. Figure 1.62 represents two such exchangers.

Split-Flow Exchanger, TEMA G Shell. In this exchanger, there is one central inlet and one central outlet nozzle with a longitudinal baffle, as shown in Fig. 1.63a. Typical temperature distribution is shown in Fig. 1.63b. This arrangement is used in the TEMA

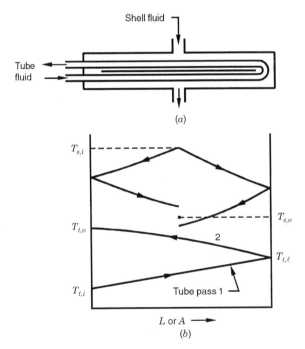

FIGURE 1.63 (a) A 1–2 split flow (TEMA G) exchanger; (b) idealized shell fluid and tube fluid temperature distributions.

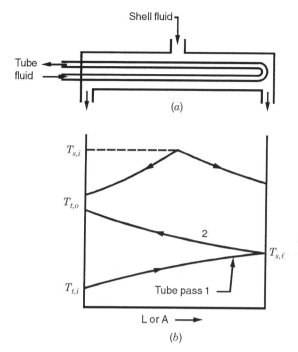

FIGURE 1.64   (a) A 1–2 divided flow (TEMA J) exchanger with shell fluid mixed; (b) idealized shell fluid and tube fluid temperature distributions.

G shell. It is a variant of the conventional 1–2 exchanger. As the "mixing" is less severe than that for the 1–2 exchanger of Fig. 1.60, the exchanger effectiveness is slightly higher, particularly at high NTU values. A double split-flow arrangement is used in the TEMA H shell.

Divided-Flow Exchanger, TEMA J Shell.   In this exchanger, the shell fluid enters at the center, divides into two equal streams, and leaves at both ends, as shown in Fig. 1.64 with typical temperature distributions. The TEMA J shell has this flow arrangement.

1.6.2.3   Multipass Plate Exchanger.   In a plate exchanger, although a single-pass counterflow arrangement is common, there exist a large number of feasible multipass flow arrangements, depending on the gasketing around the ports in the plates. Some of them are shown in Fig. 1.65. Essentially, these are combinations of parallelflow and counterflow arrangements with heat transfer taking place in adjacent channels.

One of the common ways of classifying two-fluid plate exchangers is on the basis of the number of passes on each fluid side. Possible arrangements are 1 pass – 1 pass, 2 pass – 1 pass, and so on, multipass arrangements. Usually, the 1 pass – 1 pass plate exchanger has *looped patterns*, the *m* pass – *n* pass plate exchanger has the *complex flow* arrangement, and the *n* pass – *n* pass plate exchanger has the *series flow* arrangement.

Looped patterns are most commonly used. The flow arrangement represents pure counterflow (although pure parallelflow is also possible) in a single pass. It is used for large flow rates but relatively small temperature drops or rises (ΔT) on each fluid side. Of the two looped patterns, the U-arrangement (Fig. 1.65a) is usually preferred over the Z-arrangement (Fig. 1.65b) since it allows all connections to be made on the same side of the frame. This eliminates the need for disconnecting pipework for maintenance and cleaning purposes.

A complex flow arrangement results by combining Z-arrangements in series with a generally identical number of thermal plates in each pass. Although only three such flow arrangements are shown in Fig. 1.65c–e, many other combinations are possible (see, e.g., Table 3.6). Primarily, these arrangements are used when there is a significant difference in the flow rates of the two fluid streams and the corresponding available pressure drops. Generally, the fluid, having very low permissible pressure drop, goes through the single pass; the other fluid goes through multiple passes in order to utilize the available pressure drop and pumping power. Also, if the flow rates are significantly different, the fluid having the lower flow rate goes through n (> 1) passes such that in each pass the heat capacity rates of both fluid streams are about equal. This would produce approximately equal heat transfer coefficients on each fluid side, resulting in a balanced exchanger (hA values approximately the same). Multipass arrangements always have ports located on fixed and movable end plates.

In the series flow arrangement (Fig. 1.65f), each flow passage represents a pass. The series arrangement is used for small fluid flow rates that must undergo a large temperature difference. It is used for very close temperature approaches. Because of many flow reversals, a significant portion of the available pressure drop is wasted in reversals (i.e., the pressure drop in the series flow arrangement is extremely high). The manifold-

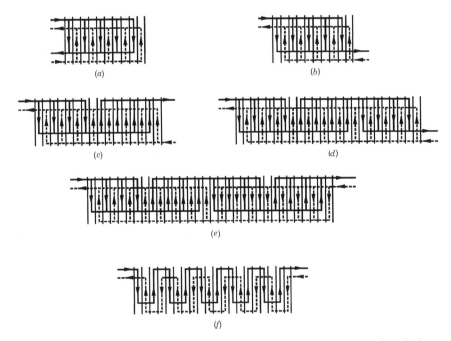

(a)

(b)

(c)

(d)

(e)

(f)

FIGURE 1.65   Single- and multipass plate heat exchanger arrangements. Looped or single-pass arrangements: (a) U arrangement; (b) Z arrangement. Multipass arrangements: (c) 2 pass – 1 pass, (d) 3 pass – 1 pass, (e) 4 pass – 2 pass, and (f) series flow.

induced flow maldistribution (see Section 12.1.3) found in the looped pattern is nonexistent in the series flow arrangement. The series flow is not as effective as pure counterflow because each stream flows parallel to the other fluid stream on one side and counter on the other side. In most pasteurizers, a large section is in series flow.

1.7 CLASSIFICATION ACCORDING TO HEAT TRANSFER MECHANISMS

The basic heat transfer mechanisms employed for transfer of thermal energy from the fluid on one side of the exchanger to the wall (separating the fluid on the other side) are single-phase convection (forced or free), two-phase convection (condensation or evaporation, by forced or free convection), and combined convection and radiation heat transfer. Any of these mechanisms individually or in combination could be active on each fluid side of the exchanger. Such a classification is provided in Fig. 1.1.

Some examples of each classification type are as follows. Single-phase convection occurs on both sides of the following two-fluid exchangers: automotive radiators and passenger space heaters, regenerators, intercoolers, economizers, and so on. Single-phase convection on one side and two-phase convection on the other side (with or without desuperheating or superheating, and subcooling, and with or without noncondensables) occur in the following two-fluid exchangers: steam power plant condensers, automotive and process/power plant air-cooled condensers, gas or liquid heated evaporators, steam generators, humidifiers, dehumidifiers, and so on. Two-phase convection could occur on each side of a two-fluid heat exchanger, such as condensation on one side and evaporation on the other side, as in an air-conditioning evaporator. Multicomponent two-phase convection occurs in condensation of mixed vapors in distillation of hydrocarbons. Radiant heat transfer combined with convective heat transfer plays a role in liquid metal heat exchangers and high-temperature waste heat recovery exchangers. Radiation heat transfer is a primary mode in fossil-fuel power plant boilers, steam generators, coal gasification plant exchangers, incinerators, and other fired heat exchangers.

SUMMARY

Heat exchangers have been classified according to transfer processes, number of fluids, degrees of surface compactness, construction features, flow arrangements, and heat transfer mechanisms. A summary is provided in Fig. 1.1. The major emphasis in this chapter is placed on introducing the terminology and concepts associated with a broad spectrum of commonly used industrial heat exchangers (many specialized heat exchangers are not covered in this chapter). To acquaint the reader with specific examples, major applications of most types of heat exchangers are mentioned. With a thorough understanding of this broad overview of different types of exchangers, readers will be able to apply the theory and analyses presented in the succeeding chapters to their specific needs.

REFERENCES

ASME, 1998, *ASME Boiler and Pressure Vessel Code*, Sec. VIII, Div. 1, *Rules for Construction of Pressure Vessels*, American Society of Mechanical Engineers, New York.

Beck, D. S., and D. G. Wilson, 1996, *Gas Turbine Regenerators*, Chapman & Hall, New York.

Butterworth, D., 1996, Developments in shell-and-tube heat exchangers, in *New Developments in Heat Exchangers*, N. Afgan, M. Carvalho, A. Bar-Cohen, D. Butterworth, and W. Roetzel, eds., Gordon & Breach, New York, pp. 437–447.

Kays, W. M., and A. L. London, 1998, *Compact Heat Exchangers*, reprint 3rd ed., Krieger Publishing, Malabar, FL.

Mueller, A. C., 1973, Heat exchangers, in *Handbook of Heat Transfer*, W. M. Rohsenow and J. P. Hartnett, eds., McGraw-Hill, New York, Chapter 18.

Reay, D. A., 1979, *Heat Recovery Systems—A Directory of Equipment and Techniques*, E. and F.N. Spon, London.

Reay, D. A., 1999, *Learning from Experiences with Compact Heat Exchangers*, CADDET Analyses Series No. 25, Centre for the Analysis and Dissemination of Demonstrated Energy Technologies, Sittard, The Netherlands.

Saunders, E. A. D., 1988, *Heat Exchangers: Selection, Design and Construction*, Wiley, New York.

Shah, R. K., 1981, Classification of heat exchangers, in *Heat Exchangers: Thermal-Hydraulic Fundamentals and Design*, S. Kakaç, A. E. Bergles, and F. Mayinger, eds., Hemisphere Publishing, Washington, DC, pp. 9–46.

Shah, R. K., 1991a, Compact heat exchanger technology and applications, in *Heat Exchanger Engineering*, Vol. 2, *Compact Heat Exchangers: Techniques for Size Reduction*, E. A. Foumeny and P. J. Heggs, eds., Ellis Horwood, London, pp. 1–29.

Shah, R. K., 1991b, Industrial heat exchangers—functions and types, in *Industrial Heat Exchangers*, J-M. Buchlin, ed., Lecture Series No. 1991-04, von Kármán Institute for Fluid Dynamics, Belgium.

Shah, R. K., 1994, Heat exchangers, in *Encyclopedia of Energy Technology and the Environment*, A. Bisio and S. G. Boots, eds., Wiley, New York, pp. 1651–1670.

Shah, R. K., and W. W. Focke, 1988, Plate heat exchangers and their design theory, in *Heat Transfer Equipment Design*, R. K. Shah, E. C. Subbarao, and R. A. Mashelkar, eds., Hemisphere Publishing, Washington, DC, pp. 227–254.

Shah, R. K., and A. D. Giovannelli, 1988, Heat pipe heat exchanger design theory, in *Heat Transfer Equipment Design*, R. K. Shah, E. C. Subbarao, and R. A. Mashelkar, eds., Hemisphere Publishing, Washington, DC, pp. 267–296.

Shah, R. K., and A. C. Mueller, 1988, Heat exchange, in *Ullmann's Encyclopedia of Industrial Chemistry*, Unit Operations II, Vol. B3, Chapt. 2, VCH Publishers, Weinheim, Germany.

Shah, R. K., and R. L. Webb, 1983, Compact and enhanced heat exchangers, in *Heat Exchangers: Theory and Practice*, J. Taborek, G. F. Hewitt, and N. Afgan, eds., Hemisphere/McGraw-Hill, Washington, DC, pp. 425–468.

Suo, M., 1976, Calculation methods for performance of heat exchangers enhanced with fluidized beds, *Lett. Heat Mass Transfer*, Vol. 3, pp. 555–564.

TEMA, 1999, *Standards of TEMA*, 8th ed., Tubular Exchanger Manufacturers Association, New York.

Walker, G., 1990, *Industrial Heat Exchangers: A Basic Guide*, 2nd ed., Hemisphere Publishing, Washington, DC.

REVIEW QUESTIONS

Where multiple choices are given, circle one or more correct answers. Explain your answers briefly.

1.1   Which of the following are compact heat exchangers?

(a) double-pipe exchanger   (b) automobile radiator

(c) plate exchanger   (d) Stirling engine regenerator

1.2 Which of the following are all prime surface heat exchangers?

(a) steam boiler with plain tubes

(b) spiral plate exchanger

(c) automobile radiator

(d) plate exchanger for beer processing

(e) strip-fin gas turbine regenerator

(f) shell-and-tube exchanger with plain tubes

1.3 Fins are used *primarily* to:

(a) increase heat transfer coefficient h

(b) increase surface area A

(c) increase both h and A

(d) increase neither h nor A

1.4 Louver fins (as compared to similar plain uncut fins) are used *primarily* to:

(a) increase heat transfer coefficient h

(b) increase surface area A

(c) increase both h and A

(d) increase neither h nor A

1.5 A finned double-pipe exchanger has fins on the outside of the inner tube(s) for the following reasons:

(a) The tube outside heat transfer coefficient is high.

(b) The tube inside heat transfer coefficient is more than double for tube outside with longitudinal flow.

(c) Fouling is expected on the tube side.

1.6 Which one of the following is *not* a function fulfilled by transverse plate baffles in a shell-and-tube exchanger?

(a) to provide counterflow operation

(b) to support the tubes

(c) to direct the fluid approximately at right angles to the tubes

(d) to increase the turbulence and mixing in the shell fluid

(e) to minimize tube-to-tube temperature differences and thermal stresses

1.7 Which of the following properties of plate heat exchangers, due to their specific construction features, make them particularly suited for the food processing industry?

(a) close temperature control

(b) easy disassembly for cleaning

(c) low probability of one fluid to other fluid contamination

(d) high corrosion resistance

1.8 In which of the following exchangers, is a single-pass crossflow arrangement used?

(a) plate-fin exchanger

(b) Ljungstrom air preheater

(c) gasketed plate heat exchanger

(d) 1–2 TEMA E shell-and-tube exchanger

(e) double-pipe exchanger

1.9 *Commonly* used flow arrangements for a shell-and-tube exchanger are:

(a) parallelflow

(b) cross-counterflow

(c) 1–2 parallel counterflow

1.10 Commonly used flow arrangements in a plate-fin heat exchanger are:

(a) parallel flow

(b) crossflow

(c) counterflow

(d) parallel counterflow

(e) cross-counterflow

(f) split flow

1.11   A single-coolant-tube-row car radiator is a crossflow heat exchanger with following fluid streams:

    (a) mixed–mixed   **(b)** mixed–unmixed   **(c)** unmixed–unmixed

1.12   A truck radiator with six coolant-tube rows and multilouver air centers is a crossflow heat exchanger with following fluid streams:

    (a) mixed–mixed   **(b)** mixed–unmixed   **(c)** unmixed–unmixed

1.13   A multipass exchanger can be identified by:

    (a) inspecting the number of hot-fluid passes

    (b) inspecting the number of cold-fluid passes

    (c) trying to unfold the fluid that travels in series from one pass to the second; this unfolding results in the other fluid traveling in a series (two passes)

    (d) making sure that the number of loops (passes) is greater than one for both fluids

1.14   Identify which of the following are multipass heat exchangers:

    (a) over-and-under multipass arrangement with fluids unmixed between passes

    (b) side-by-side two-pass arrangement with the fluid unmixed between passes

    (c) side-by-side multipass arrangement with fluids mixed between passes

    (d) two-pass cross-parallelflow exchanger with both fluids unmixed and fluids between passes are planar (in inverted order; see Fig. 1.59*b*)

    (e) a 2–2 shell-and-tube exchanger with an F shell

    (f) a 1–2 split-flow exchanger

1.15   Which of the following are possible reasons for using a cross-parallelflow instead of a cross-counterflow multipass exchanger?

    (a) higher effectiveness   **(b)** less prone to core freeze-up near the cold fluid inlet

    (c) reduced thermal stresses   **(d)** reduced size

    (e) reduced higher axial temperature gradient in the wall

1.16   Fill in the blanks.

    (a) A heat exchanger is made up of heat transfer elements called _____ and fluid distribution elements called _____.

    (b) In an extended surface exchanger, the total heat transfer surface consists of _____ and _____.

    (c) A direct-transfer type exchanger is referred to simply as a _____, and a storage type exchanger is referred to simply as a _____.

    (d) Two categories of transverse baffles used for shell-and-tube exchangers are _____ baffle and _____ baffle.

    (e) Thermodynamically, the most efficient single-pass exchanger flow arrangement is _____, and the least efficient flow arrangement is _____.

1.17   Name the specific exchanger construction types used in the following applications:

    (a) milk pasteurizing: _____   **(b)** sulfuric acid cooling: _____

    (c) automotive radiator: _____   **(d)** blast furnace air preheating: _____

    (e) air-cooled condenser: _____

1.18 Circle the following statements as true or false.

(a) T   F   In a well-designed heat exchanger, a significant portion of the total heat transfer takes place in inlet and outlet manifolds/tanks.

(b) T   F   Fins are generally used on the water side of an air-to-water heat exchanger.

(c) T   F   A highly compact rotary regenerator is more compact than human lungs.

(d) T   F   Free convection is more dominant than forced convection in most single-phase two-fluid heat exchangers.

(e) T   F   For highly viscous liquids, a parallelflow exchanger is desirable because the relatively high wall temperature at the inlet reduces liquid viscosity, yielding reduced flow resistance and increased heat transfer coefficient.

(f) T   F   A shell-and-tube exchanger is the most versatile exchanger.

(g) T   F   Tube-fin exchangers are generally more compact than plate-fin exchangers.

(h) T   F   A blast furnace regenerator is generally more compact than a shell-and-tube or plate heat exchanger.

(i) T   F   Figure 1.53b represents a single-pass heat exchanger.

(j) T   F   The heat transfer coefficient for airflow in a compact heat exchanger is higher than that for *high* water flow in a 20 mm diameter tube.

1.19 For the identical average inlet and outlet fluid temperatures, arrange the following exchangers in terms of decreasing largest structural temperature differences across the wall thickness direction:

(a) parallelflow   **(b)** counterflow   **(c)** four-pass overall cross-counterflow

(d) two-pass overall cross-counterflow   **(e)** single-pass crossflow

Now can you tell which exchanger will have the highest thermal stresses in the dividing walls between two fluids and which will have the least thermal stresses? Why? *Hint:* Review the temperature distributions of the hot and cold fluids and of the wall.

1.20 Consider the flow between parallel plates (1 m width × 1 m length) spaced 6 mm apart. Calculate the compactness (m^2/m^3) of the surface exposed to the flow between parallel plates. Now suppose that straight plain fins of 0.05 mm thickness are installed between parallel plates and spaced on 1-mm centers. Calculate the compactness for this plate-fin surface.

1.21 Name five heat exchangers that you are familiar with and classify them in proper subcategories of six major schemes of Fig. 1.1.

2 Overview of Heat Exchanger Design Methodology

An overview of the methodology of heat exchanger design is presented in this chapter. Various quantitative and qualitative design aspects and their interaction and interdependence are discussed, to arrive at an optimum heat exchanger design. Most of these considerations are dependent on each other and should be considered simultaneously to arrive iteratively at the optimum exchanger design based on an optimum system design approach. These are discussed briefly to demonstrate the multidisciplinary approach of heat exchanger design as a component and as part of a system for an overall optimum solution.

2.1 HEAT EXCHANGER DESIGN METHODOLOGY

Design is an activity aimed at providing complete descriptions of an engineering system, part of a system, or just of a single system component. These descriptions represent an unambiguous specification of the system/component structure, size, and performance, as well as other characteristics important for subsequent manufacturing and utilization. This can be accomplished using a well-defined *design methodology*.

From the formulation of the scope of this activity, it must be clear that the design methodology has a very complex structure. Moreover, a design methodology for a heat exchanger as a component must be consistent with the *life-cycle design* of a system. Life-cycle design assumes considerations organized in the following stages.

- Problem formulation (including interaction with a consumer)
- Concept development (selection of workable designs, preliminary design)
- Detailed exchanger design (design calculations and other pertinent considerations)
- Manufacturing
- Utilization considerations (operation, phase-out, disposal)

At the initial stage, an engineer must specify requirements and define the main goal of the system design. This must be based on a good understanding of customer needs. If the problem is clearly formulated, an engineer evaluates alternative concepts of the system design and selects one or more workable design solutions. Based on this analysis, detailed sizing, costing, and optimization have to be completed. This activity leads to a proposed design solution. Simultaneously, project engineering (construction/manufacturing) considerations should be taken into account. The issues related to startups, transients, steady and erratic operations, and ultimately, the retirement, should be considered as well. Through consideration of these steps, a design team reconsiders the conclusions and, in the light of the constraints imposed, iterates one or more steps until all the requirements are met within the tolerable limits. Within the framework of these activities, a particular design methodology has to be developed.

A methodology for designing a new (single) heat exchanger is illustrated in Fig. 2.1; it is based on experience and presented by Kays and London (1998), Taborek (1988), and Shah (1982) for compact and shell-and-tube exchangers. This design procedure may be characterized as a *case study* (one case at a time) method. Major design considerations include:

- Process and design specifications
- Thermal and hydraulic design
- Mechanical design
- Manufacturing considerations and cost
- Trade-off factors and system-based optimization

These design considerations are usually not sequential; there could be strong interactions and feedback among the aforementioned considerations, as indicated by double-sided arrows in Fig. 2.1, and may require a number of iterations before the design is finalized. The overall design methodology is quite complex because of the many qualitative judgments, in addition to quantitative calculations, that must be introduced. It should be emphasized that depending on the specific application, some (but not necessarily all) of the foregoing considerations of heat exchanger designs are applied in various levels of detail during the design process. In the following, these broad considerations are discussed in some detail, accompanied by several examples showing the main features of the design procedures and interactions among them. Refer to appropriate blocks and boxes in Fig. 2.1 for identification of the following specific sections and subsections.

2.1.1 Process and Design Specifications

The process and problem specification (the top dashed block in Fig. 2.1) is one of the most important steps in heat exchanger design. A heat exchanger design engineer can add the most value by working together with a system design engineer to develop "smart" specifications for the heat exchanger that define an *optimum* system. The smart specifications need to be completed based on discussions with the customer, on industry and customer standards, and on design engineer's own experiences.

Process or design specifications include all necessary information to design and optimize an exchanger for a specific application. It includes problem specifications for operating conditions, exchanger type, flow arrangement, materials, and design/manufacturing/operation considerations. In addition, the heat exchanger design engineer provides necessary and missing information on the minimum input specifications required.

2.1.1.1 Problem Specifications. The first and most important consideration is to select the design basis (i.e., design conditions). Next comes an analysis of the performance at the design point and off-design (turndown) conditions. The design basis would require the specification of operating conditions and the environment in which the heat exchan-

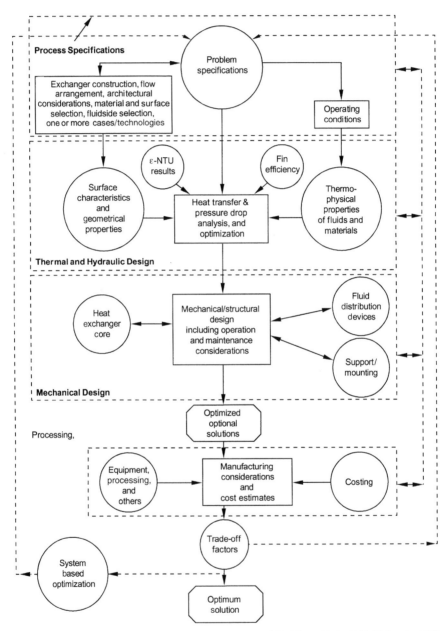

FIGURE 2.1   Heat exchanger design methodology. (Modified from Shah, 1982; Taborek, 1988; and Kays and London, 1998.)

ger is going to be operated. These include fluid mass flow rates (including fluid types and their thermophysical properties), inlet temperatures and pressures of both fluid streams, required heat duty and maximum allowed pressure drops on both fluid sides, fluctuations in inlet temperatures and pressures due to variations in the process or environment parameters, corrosiveness and fouling characteristics of the fluids, and the operating environment (from safety, corrosion/erosion, temperature level, and environmental impact points of view). In addition, information may be provided on overall size, weight, and other design constraints, including cost, materials to be used, and alternative heat exchanger types and flow arrangements. If too many constraints are specified, there may not be a feasible design, and some compromises may be needed for a solution. The heat exchanger designer and system design engineer should work together at this stage to prepare the complete smart specifications for the problem. However, in some industries the heat exchanger designer is constrained by "dumb" specifications that he or she inherits and has little or no opportunity to provide input.

2.1.1.2 Exchanger Specifications. Based on the problem specifications and the design engineer's experience, the exchanger construction type and flow arrangement (see Fig. 1.1) are first selected. Selection of the construction type depends on the fluids (gas, liquid, or condensing/evaporating) used on each side of a two-fluid exchanger, operating pressures, temperatures, fouling and cleanability, fluids and material compatibility, corrosiveness of the fluids, how much leakage is permissible from one fluid to the other fluid, available heat exchanger manufacturing technology, and cost. The choice of a particular flow arrangement is dependent on the required exchanger effectiveness, exchanger construction type, upstream and downstream ducting, packaging envelope/footprint, allowable thermal stresses, and other criteria and design constraints. The orientation of the heat exchanger, the locations of the inlet and outlet pipes, and so on, may be dictated by the system and/or available packaging/footprint space and ducting. Some guidelines on selection of the exchanger type and flow arrangements are provided in Section 10.2.

Next, the core or surface geometry and material are selected. The core geometry (such as shell type, number of passes, baffle geometry, etc.) is selected for a shell-and-tube exchanger, while the surface geometry is chosen for a plate, extended surface, or regenerative heat exchanger. There are several quantitative and qualitative criteria for surface selection. Some of the qualitative and quantitative criteria for compact heat exchanger surfaces are discussed in Sections 10.2 and 10.3. The qualitative criteria for surface selection are the operating temperature and pressure, the designer's experience and judgment, fouling, corrosion, erosion, fluid contamination, cost, availability of surfaces, manufacturability, maintenance requirements, reliability, and safety. For shell-and-tube exchangers, the criteria for selecting core geometry or configuration are the desired heat transfer performance within specified pressure drops, operating pressures and temperatures, thermal/pressure stresses, the effect of potential leaks on the process, corrosion characteristics of the fluids, fouling, cleanability, maintenance, minimal operational problems (vibrations, freeze-up, instability, etc.), and total installed cost. Some of these are discussed in Section 10.2.

For shell-and-tube exchangers, the tube fluid is usually selected as the one having more fouling, high corrosiveness, high pressure, high temperature, increased hazard probability, high cost per unit mass, and/or low viscosity. Maximum allowable pressure drop will also dictate which fluid will be selected for the tube side (high-pressure fluid) and which for the shell side.

For compact heat exchangers, such as for plate-fin exchangers, one may have a choice of considering offset strip fin, louver fin, or other fin geometry. For each fin geometry selected, the thermal/hydraulic design and mechanical design of Fig. 2.1 are carried out. This is what we mean by "one or more cases" in the box. One or more cases also include different heat exchanger technologies.

Example 2.1   A hydrocarbon gas has to be cooled in a chemical plant. A stream of a liquid hydrocarbon is available to be used as a coolant. The gas stream has to change its temperature from 255°C to 30°C. The liquid stream has the inlet temperature of 25°C. The required enthalpy change of the hot gas is smaller than 300 kW (with a small mass flow rate of an order of magnitude 0.01 kg/s). Both fluids are at relatively high pressures (i.e., the respective pressures are of an order of magnitude 10 MPa). Is it possible, using this incomplete set of process data, to offer an unambiguous selection of a feasible heat exchanger type that will be capable of performing the task? Consider the following heat exchanger types: shell-and-tube, double-pipe, welded plate, gasketed plate, lamella, and spiral plate heat exchanger.

SOLUTION

Problem Data and Schematic: An incomplete database is provided to support a process specification. Only information regarding some selected operating conditions is available (a situation often encountered in practice). All the information regarding various heat exchanger types under consideration (shell-and-tube, double-pipe, welded plate, gasketed plate, lamella, and spiral plate heat exchanger) are available (see Sections 1.5.1.1, 1.5.1.2, 1.5.2.2, 1.5.2.1, 1.5.2.4, and 1.5.2.3, respectively).

Determine: Based on the available process specification data, select a feasible heat exchanger type for further design considerations.

Assumptions: Design specifications for the heat exchanger types listed are valid, as discussed in Chapter 1.

Analysis and Discussion: One possible approach to a selection of the feasible heat exchanger type is first to eliminate the types characterized with specifications that conflict with the process conditions. The first important fact to be considered is related to the operating temperature ranges and pressures. A study of various designs (see Chapter 1) leads to the conclusion that lamella, welded, and gasketed plate heat exchangers cannot be used because the allowable operating pressures and temperatures for them are both substantially lower than the process condition data imposed by the problem formulation. More precisely, lamella heat exchangers usually operate at pressures lower than 3.45 MPa and temperatures lower than 200°C (see Section 1.5.2.4). For welded and gasketed plate heat exchangers, these parameters are 4.0 MPa and 350°C, and 3 MPa and 150°C, respectively (see Sections 1.5.2.2 and 1.5.2.1). A spiral heat exchanger can operate at much higher temperatures (up to 500°C; see Section 1.5.2.3), but the pressure limitation is 2 MPa. So, only two remaining types should be considered. This means that only shell-and-tube and double-pipe heat exchangers are feasible candidates.

Both shell-and-tube and double-pipe heat exchangers (see Sections 1.5.1.1 and 1.5.1.2) can easily sustain high pressures and temperatures. Consequently, other criteria should be considered for the selection. These criteria include the required heat exchanger effectiveness, the heat load, fluid characteristics (such as fouling and corrosion ability), cost, and others. For a relatively small heat load (i.e., smaller than 500 kW), a double-pipe heat exchanger would be a cost-effective solution. Also, for higher performance, a multitube double-pipe heat exchanger with or without fins should be considered if cost considerations support this decision. See Example 2.4 for the inclusion of cost considerations in a heat exchanger selection.

Finally, a decision should be made whether to use finned or plain tubes in the double-pipe multitube heat exchanger selected. Due to the fact that the heat exchanger should accommodate both a gas and a liquid, the heat transfer conductance (hA) on the gas side (with low gas mass flow rate) will be low. Hence, employing fins on the gas side will yield a more compact unit with approximately balanced hA values on the gas and liquid sides. Note also that the tube fluid (liquid hydrocarbon) is more prone to fouling. So a double-pipe multitube heat exchanger with finned tubes on the hydrocarbon gas side and liquid hydrocarbon on the tube side has to be suggested as a feasible workable design.

2.1.2 Thermal and Hydraulic Design

Heat exchanger thermal/hydraulic design procedures (the second block from the top in Fig. 2.1) involve exchanger rating (quantitative heat transfer and pressure drop evaluation) and/or exchanger sizing. This block is the heart of this book, covered in Chapters 3 through 9. Only two important relationships constitute the entire thermal design procedure. These are:

1. Enthalpy rate equations

$$q = q_j = \dot{m}_j \, \Delta \mathbf{h}_j \tag{2.1}$$

one for each of the two fluids (i.e., $j = 1, 2$)

2. Heat transfer rate equation or simply the rate equation [see also the equality on the left-hand side in Eq. (1.4)]

$$q = UA \, \Delta T_m \tag{2.2}$$

Equation (2.1) is a well-known relationship from thermodynamics. It relates the heat transfer rate q with the enthalpy rate change for an open nonadiabatic system with a single bulk flow stream (either $j = 1$ or 2) entering and leaving the system under isobaric conditions. For single-phase fluids in a heat exchanger, the enthalpy rate change is equal to $\dot{m}_j \, \Delta \mathbf{h}_j = (\dot{m}c_p)_j \, \Delta T_j = (\dot{m}c_p)_j \left| T_{j,i} - T_{j,o} \right|$. Equation (2.2) reflects a convection–conduction heat transfer phenomenon in a two-fluid heat exchanger. The heat transfer rate is proportional to the heat transfer area A and mean temperature difference ΔT_m between the fluids. This mean temperature difference is a log-mean temperature difference (for counterflow and parallelflow exchangers) or related to it in a way that involves terminal temperature differences between the fluids such as $(T_{h,i} - T_{c,o})$ and $(T_{h,o} - T_{c,i})$. It is discussed in Section 3.7.2. The coefficient of proportionality in Eq. (2.2) is the overall heat transfer coefficient U (for the details, see Section 3.2.4). Solving a design problem means either determining A (or UA) of a heat exchanger to satisfy the required terminal values of some variables (the sizing problem), or determining the terminal values of the variables knowing the heat exchanger physical size A or overall conductance UA (the rating problem). Note that the seven variables on the right-hand side of Eqs. (2.1) and (2.2) are $(\dot{m}c_p)_j$, $T_{j,i}$, $T_{j,o}$, with $j = 1$ or 2 and UA.

Let us define first the heat exchanger thermal design problems. Subsequently, the concept of a design method to solve the problem is introduced. Finally, the main inputs to the design procedures are discussed.

2.1.2.1 Heat Exchanger Thermal Design Problems. From the quantitative analysis point of view, there are a number of heat exchanger thermal design problems. Two of the simplest (and most important) problems are referred to as the *rating* and *sizing problems*.

Rating Problem. Determination of heat transfer and pressure drop performance of either an existing exchanger or an already sized exchanger (to check vendor's design) is referred to as a *rating problem.* Inputs to the rating problem are the heat exchanger construction, flow arrangement and overall dimensions, complete details on the materials and surface geometries on both sides, including their nondimensional heat transfer and pressure drop characteristics (j or Nu and f vs. Re),[†] fluid flow rates, inlet temperatures, and fouling factors. The fluid outlet temperatures, total heat transfer rate, and pressure drops on each side of the exchanger are then determined in the rating problem. The rating problem is also sometimes referred to as the *performance* or *simulation problem.*

Sizing Problem. In a broad sense, the design of a new heat exchanger means the determination/selection of an exchanger construction type, flow arrangement, tube/plate and fin material, and the physical size of an exchanger to meet the specified heat transfer and pressure drops within all specified constraints. However, in a sizing problem for an extended surface exchanger, we will determine the physical size (length, width, height, and surface areas on each side) of an exchanger; we will assume that selection of the exchanger construction type, flow arrangement, and materials was completed beforehand based on some of the selection guidelines presented in Sections 10.1 and 10.2. For a shell-and-tube exchanger, a sizing problem in general refers to the determination of shell type, diameter and length, tube diameter and number, tube layout, pass arrangement, and so on. For a plate exchanger, a sizing problem means the selection of plate type and size, number of plates, pass arrangements, gasket type, and so on. Inputs to the sizing problem are surface geometries (including their dimensionless heat transfer and pressure drop characteristics), fluid flow rates, inlet and outlet fluid temperatures, fouling factors, and pressure drops on each fluid side. The sizing problem is also referred to as the *design problem.* However, in the literature the design problem is variously referred to as a *rating* or *sizing problem.* To avoid confusion with the term *design problem,* we will distinctly refer to heat exchanger thermal design problems as rating and sizing problems. The sizing problem is a subset of the comprehensive design process outlined in Fig. 2.1.

2.1.2.2 Basic Thermal and Hydraulic Design Methods. Based on the number of variables associated with the analysis of a heat exchanger, dependent and independent dimensionless groups are formulated. The relationships between dimensionless groups

---

[†] j, Colburn factor; Nu, Nusselt number; f, Fanning friction factor; Re, Reynolds number. They are defined in Section 7.2.

are subsequently determined for different flow arrangements. Depending on the choice of dimensionless groups, several design methods are being used by industry. These methods include ε-NTU, P-NTU, MTD correction factor, and other methods (see Chapter 3 for details). The basic methods for recuperators are presented in Chapter 3 and for regenerators in Chapter 5. Advanced auxiliary methods for recuperators are presented in Chapter 4. Hydraulic design or pressure drop analyses are presented in Chapter 6. As shown in Fig. 2.1, inputs to the thermal and hydraulic procedures are the surface heat transfer and flow friction characteristics (also referred to as *surface basic characteristics*), geometrical properties, and thermophysical properties of fluids, in addition to the process/design specifications.

2.1.2.3 Surface Basic Characteristics. Surface basic characteristics on each fluid side are presented as j or Nu and f vs. Re curves in dimensionless form and as the heat transfer coefficient h and pressure drop Δp vs. the fluid mass flow rate $\dot{m}$ or fluid mass velocity G in dimensional form. Accurate and reliable surface basic characteristics are a key input for exchanger thermal and hydraulic design. Theoretical solutions and experimental results for a variety of exchanger heat transfer surfaces are presented in Chapter 7 together with the description of experimental techniques for their determinations.

2.1.2.4 Surface Geometrical Properties. For heat transfer and pressure drop analyses, at least the following heat transfer surface geometrical properties are needed on each side of a two-fluid exchanger: minimum free-flow area A_o; core frontal area A_{fr}; heat transfer surface area A which includes both primary and fin area, if any; hydraulic diameter D_h; and flow length L. These quantities are computed from the basic dimensions of the core and heat transfer surface. On the shell side of a shell-and-tube heat exchanger, various leakage and bypass flow areas are also needed. Procedures to compute these quantities for some surface geometries are presented in Chapter 8.

2.1.2.5 Thermophysical Properties. For thermal and hydraulic design, the following thermophysical properties are needed for the fluids: dynamic viscosity μ, density ρ, specific heat c_p, and thermal conductivity k. For the wall, material thermal conductivity and specific heat may be needed. Some information on the thermophysical properties is provided in Appendix A.

2.1.2.6 Thermal and Hydraulic Design Problem Solution. Solution procedures for rating and sizing problems are of an analytical or numerical nature, with empirical data for heat transfer and flow friction characteristics and other pertinent characteristics. These procedures are presented in Chapter 9. Due to the complexity of the calculations, these procedures are often executed using commercial or proprietary computer codes. Since there are many geometrical and operating condition–related variables and parameters associated with the sizing problem, the question is how to formulate the *best possible* design solution (selection of the values of these variables and parameters) among all feasible solutions that meet the performance and design criteria. This is achieved by employing mathematical optimization techniques after initial sizing to optimize the heat exchanger design objective function within the framework of imposed implicit and explicit constraints. A heat exchanger optimization procedure is outlined in Section 9.6. From the viewpoint of a computer code, the thermal and hydraulic design procedures, including optimization (the second major dashed-line block in Fig. 2.1), are summarized in Chapter 9.

Example 2.2 Consider a heat exchanger as a black box with two streams entering and subsequently leaving the exchanger without being mixed. Assume the validity of both Eqs. (2.1) and (2.2). Also take into account that $\dot{m}_j\,\Delta h_j = (\dot{m}c_p)_j \Delta T_j$, where $\Delta T_j = |T_{j,i} - T_{j,o}|$. Note that regardless of the actual definition used, ΔT_m must be a function of terminal temperatures $(T_{h,i}, T_{h,o}, T_{c,i}, T_{c,o})$. With these quite general assumptions, answer the following two simple questions:

(a) How many variables of the seven used on the right-hand sides of Eqs. (2.1) and (2.2) should minimally be known, and how many can stay initially unknown, to be able to determine all design variables involved?

(b) Using the conclusion from question (a), determine how many different problems of sizing a heat exchanger (UA must be unknown) can be defined if the set of variables includes only the variables on the right-hand sides of Eqs. (2.1) and (2.2) [i.e., $(\dot{m}c_p)_j$, $T_{j,i}$, $T_{j,o}$ with $j = 1$ or 2, and UA].

SOLUTION
Problem Data: A heat exchanger is considered as a black box that changes the set of inlet temperatures $T_{j,i}$ of the two fluids [with $(\dot{m}c_p)_j$, $j = 1, 2$] into the set of their respective outlet temperatures $T_{j,o}$ $(j = 1, 2)$ through heat transfer characterized by Eq. (2.2) (i.e., by the magnitude of UA). So in this problem, we have to deal with the following variables: $(\dot{m}c_p)_1$, $(\dot{m}c_p)_2$, $T_{1,i}$, $T_{1,o}$, $T_{2,i}$, $T_{2,o}$, and UA.

Determine: How many of the seven variables listed must be known to be able to determine all the variables involved in Eqs. (2.1) and (2.2)? Note that Eq. (2.1) represents two relationships, one for each of the two fluids. How many different sizing problems (with at least UA unknown) can be defined?

Assumptions: The heat exchanger is adiabatic; the enthalpy changes in enthalpy rate equations, Eq. (2.1), can be determined by $\dot{m}_j\,\Delta h_j = (\dot{m}c_p)_j\,\Delta T_j$; and the heat transfer rate can be determined by Eq. (2.2).

Analysis: (a) The answer to the first question is trivial and can be devised by straightforward inspection of the three equations given by Eqs. (2.1) and (2.2). Note that the left-hand sides of these equalities are equal to the same heat transfer rate. This means that these three equations can be reduced to two equalities by eliminating heat transfer rate q. For example,

$$(\dot{m}c_p)_1 |T_{1,i} - T_{1,o}| = UA\,\Delta T_m$$

$$(\dot{m}c_p)_2 |T_{2,i} - T_{2,o}| = UA\Delta T_m$$

So we have two relationships between seven variables [note that $\Delta T_m = f(T_{1,i}, T_{1,o}, T_{2,i}, T_{2,o})$]. Using the two equations, only two unknowns can be determined. Consequently, the remaining five must be known.

(b) The answer to the second question can now be obtained by taking into account the fact that (1) UA must be treated as an unknown in all the cases considered (a sizing

TABLE E2.2   Heat Exchanger Sizing Problem Types

Problem	UA	$(\dot{m}c_p)_1$	$(\dot{m}c_p)_2$	$T_{1,i}$	$T_{1,o}$	$T_{2,i}$	$T_{2,o}$
1	●	○	○	●	○	○	○
2	●	○	○	○	●	○	○
3	●	○	○	○	○	●	○
4	●	○	○	○	○	○	●
5	●	●	○	○	○	○	○
6	●	○	●	○	○	○	○

●, Unknown variable; ○, known variable.

problem), and (2) only two variables can be considered as unknown since we have only two equations [Eqs. (2.1) and (2.2)] at our disposal. Thus, the number of combinations among the seven variables is six (i.e., in each case the two variables will be unknown and the remaining five must be known). This constitutes the list of six types of different sizing problems given in Table E2.2.

Discussion and Comments: Among the six types of sizing problems, four have both heat capacity rates [i.e., the products $(\dot{m}c_p)_j$ known]; and in addition to UA, the unknown is one of the four terminal temperatures. The remaining two problem types presented in Table E2.2 have one of the two heat capacity rates unknown and the other heat capacity rate and all four temperatures known.

Exactly the same reasoning can be applied to devise a total of 15 rating problems (in each of these problems, the product UA must stay known). A complete set of design problems (including both the sizing and rating problems), devised in this manner, is given in Table 3.11.

2.1.3   Mechanical Design

Mechanical design is essential to ensure the mechanical integrity of the exchanger under steady-state, transient, startup, shutdown, upset, and part-load operating conditions during its design life. As mentioned in the beginning of Chapter 1, the exchanger consists of heat exchanging elements (core or matrix where heat transfer takes place) and fluid distribution elements (such as headers, manifolds, tanks, inlet/outlet nozzles, pipes, and seals, where ideally, no heat transfer takes place). Mechanical/structural design should be performed individually for these exchanger elements. Also, one needs to consider the structural design of heat exchanger mounting. Refer to the third dashed-line block from the top in Fig. 2.1 for a discussion of this section.

The heat exchanger core is designed for the desired structural strength based on the operating pressures, temperatures, and corrosiveness or chemical reaction of fluids with materials. Pressure/thermal stress calculations are performed to determine the thicknesses of critical parts in the exchangers, such as the fin, plate, tube, shell, and tubesheet. A proper selection of the material and the method of bonding (such as brazing, soldering, welding, or tension winding) fins to plates or tubes is made depending on the operating temperatures, pressures, types of fluids used, fouling and corrosion potential, design life, and so on. Similarly, proper bonding techniques should be developed and used for tube-to-header joints,[†] tube-to-tubesheet joints, expansion joints, flanges, and so on. These bonding methods are usually decided upon before conducting the thermal-hydraulic analysis. At this stage, attention should also be paid to operational problems. Thermal stress and fatigue calculations are performed to ensure the durability and desired life of the exchanger for expected startup and shutdown periods and for part-load operating conditions. In addition, some of the less obvious operating problems should also be addressed up front. A check should be made to eliminate or minimize flow-induced vibrations, which may result in erosion, fatigue, and so on, of tubes leading to failure. Flow velocities are checked to eliminate or minimize erosion, corrosion, and fouling. Adequate provisions are also made for thermal expansion. At this stage, considerations are also given to other operating problems, if they exist, such as dynamic instability and freezing. Field experience, if any, becomes quite valuable at this step of design. Fouling and corrosion are covered in Chapter 13.

In addition to the heat exchanger core, proper design of flow distribution devices (headers, tanks, manifolds, nozzles, or inlet–outlet pipes) is made to ensure that there is uniform flow distribution through the exchanger flow passages, and that neither erosion nor fatigue will be a problem during the design life of the exchanger. Header design and flow maldistribution are covered in Chapter 12.

Either the exchanger is mounted on the floor/ground in a room or open environment or in a system along with other components. The structural support for the heat exchanger needs to be designed properly with proper tabs, brackets, and other mounting features on the exchanger to ensure no failure of the supporting structure due to vibration, impact loads, fatigue, and so on. In the mechanical design, consideration is also given to maintenance requirements (such as cleaning, repair, serviceability, and general inspection) and shipping limitations, such as for overall size.

Every heat exchanger must comply with applicable local, state, national, and/or international codes and standards (such as ASME pressure vessel codes, TEMA standards, etc.), and should be designed accordingly for good mechanical design that will also result in good thermal design. Particularly, the exchanger may need extensive structural design to meet the codes and standards for one or more of the following conditions: severe duty (extreme pressures and temperatures); considerable number of pressure and temperature cycles during the design life; earthquake criteria; special application where destructive testing is not feasible, where reliability is critical, or where repair/replacement is not easily effected; or when a customer specifies it. Structural design would include thermal stresses, fatigue, and creep analyses in such a situation to define the life of the exchanger also.

Although some aspects of mechanical design are considered upfront before the thermal design, a common practice in some exchangers is first to design the exchanger to meet the thermal/hydraulic requirements and then check that design from a structural design viewpoint, conducting necessary iterations until the thermal/hydraulic requirements and structural design are both satisfied. Thus, a mechanical design of the exchanger is equally important and probably more difficult than the thermal design primarily because it is not all analytical and is different for each specific case; one must rely on experimentation, prior experience, and good design practice. Many mechanical design criteria should be considered simultaneously or iteratively with thermal design.

[†] The tubesheet in a tube-fin exchanger is referred to as a *header* in the automotive industry. However, conical, triangular, trapezoidal, or other-shaped flow distribution devices at the exchanger inlet and outlet with gas flows are referred to as headers. See, for example, Section 12.4 for further details.

As shown in Fig. 2.1, several optimized optional solutions may be available when the thermal and mechanical designs are completed. The designer addresses, in both parallel and series, manufacturing considerations and cost estimating followed by trade-off factors to arrive at an optimum solution. For shell-and-tube exchangers, since TEMA standards detail most of the mechanical design, the pricing of the exchanger is done before finalizing the mechanical design; the final drawings (solid models) are made after a firm order is placed.

Example 2.3 An engineer has to perform a stress analysis within the scope of a mechanical design of a shell-and-tube heat exchanger for a given task. The following is the available information: A TEMA E shell-and-tube heat exchanger has to be used for cooling oil with water. All process conditions are known (fluid types, thermophysical properties, mass flow rates, temperature and pressure terminal conditions, fouling factors, etc.) The heat load is fully defined. Design specification data (flow arrangement, tube inside and outside diameters, pitch, angle, tube material, shell inside diameter, TEMA designation, etc.) are also known. It is specified that the heat exchanger has to operate indoors and that the plant site is in a seismically inactive area. All data regarding allowable stress limits and fatigue life requirements for the materials used are known. In addition, the engineer has all the data needed to calculate thermal loads caused by differential thermal expansions between the shell and tubes. Enlist the important missing data to perform the stress analysis.

SOLUTION

Problem Data and Schematic: All the process/design specification data, as well as the information about allowable stresses, are known. Specific application of the exchanger is known. The operating environment of the exchanger is specified. The schematic of the TEMA E shell-and-tube heat exchanger (BEM) is given in Fig. 1.5a.

Determine: The missing set of data necessary for performing the stress analysis of the heat exchanger described in the example formulation.

Assumptions: All usual assumptions for the heat exchanger type mentioned are satisfied.

Analysis: Inspection of the data indicates that most of the information needed for stress analysis is available:

- The application of the heat exchanger is known.
- The process and design information and specifications are known.
- The heat load is specified.
- Information about environmental and seismic conditions is available.
- Allowable stress limits and fatigue life data are determined.
- Empty, and static loading can be calculated.
- Vibration can be assessed.

However, very important pieces of information are missing. These are mechanical, superimposed, and operating transient loads. The loads include:

- Mechanical loads caused by pressure and gravity forces
- Superimposed loads caused by piping connections to nozzles (these loads may cause axial, bending, and torsion forces), as well as loads caused by support reactions (consequently, vertical or horizontal orientation must have been specified as well)
- Operating loads under transient conditions (such as startup and shutdown operation), including impact loads caused by eventual erratic fluctuations of pressure and/or temperature

Discussion: This example emphasizes a need for a thorough study of the input data. This is the case not only when a mechanical design is considered, but even more often, that is the case when thermal design is performed. Frequently, an incomplete set of data is available. An engineer must identify the minimum required data set to start the analysis. Subsequently, through numerous additional assumptions, the complete set of data will be defined. This illustrates why two engineers will never provide two exactly equal designs, even if the initial set of data is the same.

2.1.4 Manufacturing Considerations and Cost Estimates

Manufacturing considerations and cost estimates (see the first dashed-line major block from the bottom in Fig. 2.1) are made for those optimized solutions obtained from thermal and mechanical design considerations.

2.1.4.1 Manufacturing Considerations. Manufacturing considerations may be subdivided into manufacturing equipment considerations, processing considerations, and other qualitative criteria. The equipment considerations that may influence which design should be selected include existing tooling versus new tooling; availability and limitations of dies, tools, machines, furnaces, and manufacturing space; production versus offline setup; and funding for capital investment. Processing considerations are related to how individual parts and components of a heat exchanger are manufactured and eventually assembled. This includes manufacturing of individual parts within specified tolerances; flow of parts; stacking of a heat exchanger core and eventual brazing, soldering, welding, or mechanical expansion of tubes or heat transfer surfaces; leak-free mounting (joining) of headers, tanks, manifolds, or return hairpins on the heat exchanger core; mounting of pipes; washing/cleaning of the exchanger; leak testing, mounting of the exchanger in the system; and structural supports. Not only the manufacturing equipment but also the complete processing considerations are evaluated upfront nowadays when a new design of a heat exchanger is being considered, particularly for an extended surface heat exchanger. Other evaluation criteria include the shop workload, delivery dates, company policy, and estimate of the strength of the competition.

2.1.4.2 Costing. The overall total cost, also called *lifetime costs*, associated with a heat exchanger may be categorized as the capital, installation, operating, and sometimes also disposal costs. The capital (total installed) cost includes the costs associated with design, materials, manufacturing (machinery, labor, and overhead), testing, shipping, installation, and depreciation. Installation of the exchanger on the site can be as high as the capital cost for some shell-and-tube and plate heat exchangers. The operating cost consists of the costs associated with fluid pumping power, warranty, insurance, maintenance, repair, cleaning, lost production/downtime due to failure, energy cost associated with the utility (steam, fuel, water) in conjunction with the exchanger in the network, and decommissioning costs. Some of the cost estimates are difficult to obtain and best estimates are made at the design stage.

Example 2.4   A heat exchanger designer needs to make a preliminary selection of a heat exchanger type for known heat transfer performance represented by $q/\Delta T_m$ [Eqs. (1.4) and Eq. (2.2)]. The exchanger should operate with $q/\Delta T_m = 6.3 \times 10^4$ W/K. The criterion for selection at that point in the design procedure is the magnitude of the unit cost per unit of $q/\Delta T_m$. From a preliminary analysis, the engineer has already selected four possible workable design types as follows: (1) a shell-and-tube heat exchanger, (2) a double-pipe heat exchanger, (3) a plate-and-frame heat exchanger, and (4) a welded plate heat exchanger. From the empirical data available, the unit costs (in dollars per unit of $q/\Delta T_m$) for the two values of $q/\Delta T_m$ are given in Table E2.4A. Idealize the dependence of the unit cost vs. $q/\Delta T_m$ as logarithmic. What is going to be the engineer's decision? Discuss how this decision changes with a change in the heat exchanger performance level.

SOLUTION

Problem Data and Schematic: From the available empirical data, the heat exchanger unit cost per unit of heat exchanger performance level is known (see Table E2.4A). Schematics of heat exchanger types selected are given in Figs. 1.5, 1.15, 1.16, and 1.20.

Determine: The heat exchanger type for a given performance. Formulate the decision using the unit cost per unit of $q/\Delta T_m$ as a criterion.

Assumptions: The cost of heat exchangers vs. $q/\Delta T_m$ (W/K) is a logarithmic relationship. All heat exchanger types selected represent workable designs (meet process/design specifications). The heat exchanger performance is defined by $q/\Delta T_m$ as given by Eq. (1.4).

Analysis: The analysis should be based on data provided in Table E2.4A. Because there are no available data for the performance level required, an interpolation should be performed. This interpolation must be logarithmic. In Table E2.4B, the interpolated data are provided along with the data from Table E2.4A.

TABLE E2.4A   Unit Cost for $q/\Delta T_m$

Heat Exchanger Type	$/(W/K) for $q/\Delta T_m$	
	5×10^3 W/K	1×10^5 W/K
Shell-and-tube	0.91	0.134
Double pipe	0.72	0.140
Plate-and-frame	0.14	0.045
Welded plate	1.0	0.108

Source: Modified from Hewitt and Pugh (1998).

TABLE E2.4B   Logarithmically Interpolated Unit Costs for $q/\Delta T_m$

	$/(W/K)$ for $q/\Delta T_m$		
Heat Exchanger Type	5×10^3 W/K	6.3×10^4 W/K	1×10^5 W/K
Shell-and-tube	0.91	0.180	0.134
Double pipe	0.72	0.180	0.140
Plate-and-frame	0.14	0.054	0.045
Welded plate	1.0	0.152	0.108

A comparison of data from Table E2.4B is presented in Table E2.4C in a reduced form. The numbers in Table E2.4C represent dimensionless ratios of the unit cost estimates for the shell-and-tube type versus respective types as listed in Table E2.4B. For example, the number 3.33 in the third column for a plate-and-frame heat exchanger means that a shell-and-tube heat exchanger has a 3.33 times larger unit cost than that of a plate-and-frame heat exchanger for the same performance level. From Table E2.4C, comparing the data in the column for $q/\Delta T_m = 6.3 \times 10^4$ W/K, it becomes clear that the most economical heat exchanger type is the plate-and-frame heat exchanger.

TABLE E2.4C   Reduced Unit Costs Values $q/\Delta T_m$

	$/(W/K)$ for $q/\Delta T_m$		
Heat Exchanger Type	5×10^3 W/K	6.3×10^4 W/K	1×10^5 W/K
Double pipe	1.26	1	0.96
Plate-and-frame	6.5	3.33	2.98
Welded plate	0.91	1.18	1.08

Discussion and Comments: The following conclusions can be formulated. The double-pipe heat exchanger is more economical than the shell-and-tube type only for small performance values. For higher performance levels, the unit cost may be exactly the same for both shell-and-tube and double-pipe heat exchangers (for $q/\Delta T_m = 6.3 \times 10^4$ W/K, see the corresponding column in Table E2.4B), for larger $q/\Delta T_m$, a shell-and-tube exchanger may be even cheaper (compare the numbers in the last column of Table E2.4B). The most economical is the plate-and-frame heat exchanger, regardless of the heat transfer performance level (see Table E2.4B, the last column). A welded plate heat exchanger is the least economical for the small performance level (see the corresponding values in the first column of Table E2.4B). For large duties, the least desirable solution is a double-pipe heat exchanger (it has a reduced unit cost value 0.96 compared to 2.98 and 1.08 for plate-and-frame and welded plate heat exchangers; Table E2.4C). Data presented are based on an approximate costing method developed by Hewitt et al. (1982).

2.1.5   Trade-off Factors

Fairly accurate cost estimates are made as above to the various case studies, after detailed evaluation of thermal, mechanical, and manufacturing design considerations. Now we are in a position to evaluate cost-related trade-off factors. These may be developed to weigh quantitatively the relative costs of pressure drop, heat transfer performance, weight, envelope size, leakage, initial cost versus life[†] of the exchanger for fouling, corrosion, and fatigue failures, and the cost of a one-of-a-kind design versus a design with a large production run. The trade-off factors relate to the physical input of the problem specifications and constraints they impose, including operating conditions. Trade-off analysis may involve both economic considerations and the second law of thermodynamics associated with heat exchanger design.

If the heat exchanger is one component of a system or a thermodynamic cycle, an optimum system design is necessary rather than just an optimum heat exchanger design to arrive at Δp, q, to minimize utilities (pinch analysis), cost, and so on, for an optimum system. In such a case, the problem statement for the heat exchanger design is reformulated after obtaining an optimum design and then applying the trade-off factors. The dashed line from the bottom to top near the left-hand margin of Fig. 2.1 suggests this iterative solution procedure.

2.1.6   Optimum Design

The final output of the quantitative and qualitative analyses is an *optimum design*, or possibly, several such designs (depending on the number of surface or core geometries selected) to submit to the customer.

2.1.7   Other Considerations

If the heat exchanger incorporates new design features, is a critical part of the system, or is going to be mass produced, model and prototype heat exchangers are built and tested in the laboratory to confirm any of the following: its heat transfer and pressure drop performance both as a component and part of a system, fatigue characteristics due to vibration, the quality and life of the fin-to-tube/plate joint, pressure and temperature cycling, corrosion and erosion characteristics, and burst pressure limit.

2.2   INTERACTIONS AMONG DESIGN CONSIDERATIONS

The design methodology presented in Fig. 2.1 could be considered a series solution if we ignore the connecting line branches with double arrows. In this case, one would complete the process specifications and thermal and hydraulic design blocks, followed by the mechanical design, and so on. But in reality, these design considerations are dependent on each other and in many cases cannot be addressed individually without considering the effects on each other. Let us review two examples to illustrate this point.

Consider a shell-and-tube exchanger design with heavy fouling. In the process specifications module, geometry and material are selected properly to minimize the fouling, minimize corrosion, and facilitate cleaning. The heat exchanger is architecturally placed and oriented such that it could provide on-location cleaning or could be removed easily

---

[†]The design life of an exchanger may vary from a few days/weeks operating in a very hostile environment to 30 years or more for nuclear power plant exchangers. For a conventional exchanger, the design life is based on the field experience of similar exchangers. For exchangers with new designs, new heat transfer surfaces or new materials, the design life may be based on accelerated tests in the laboratory.

for external cleaning. During thermal–hydraulic design, the fouling resistance and hence the amount of increased surface area required should be taken into account properly, depending on the cleaning cycle schedule. Also the selection of the core geometry or heat transfer surface should be such that it either minimizes fouling or provides easy cleaning. During mechanical design, the proper material gauge should be selected for the desired life of the exchanger from the fouling and corrosion point of view. The tube gauge could in turn affect both the heat transfer and pressure drop. The decision should be made up front in terms of what type of cleaning technique and maintenance schedule should be employed: online, offline, chemical cleaning, mechanical cleaning, or a throwaway exchanger—since this may affect the exchanger construction, material selection, and thermal/hydraulic design. The material and gauge selection as well as cleaning cycle will affect the fixed and operating cost of the exchanger. For example, two aluminum heat exchangers may be less expensive than one titanium exchanger for the same total service time when using ocean water for cooling. This material choice will affect the geometry, thermal–hydraulic design, and mechanical design.

Consider a plate-fin exchanger for high-temperature (815°C or 1500°F) waste heat recovery. The application will dictate the choice of material as stainless steel or more exotic materials. High-temperature operation means that a special brazing technique will be required, and it needs to be developed for each different fin geometry selected. The resultant contact resistance between the fin and the plate/tube could be finite and not negligible, depending on the brazing material and the technique employed, which in turn will affect the thermal performance. The cost and thermal performance considerations will dictate the selection of material with respect to the desired life of the exchanger. The cost and allowed pressure drop will dictate the choice of whether or not to use fins and which types. The operating thermal stresses and required waste heat recovery will dictate the choice of construction and flow arrangement (crossflow, counterflow, or multipass). So there are many interdependent factors that must be considered while designing and optimizing this exchanger.

SUMMARY

The problem of heat exchanger design is multidisciplinary. Only a part of the total design process consists of quantitative analytical evaluation. Because of a large number of qualitative judgments, trade-offs, and compromises, heat exchanger design is more of an art than a science at this stage. In general, no two engineers will come up with the same heat exchanger design for a given application. Most probably, an experienced engineer will arrive at a "better" design since he or she has a better "feel" for the qualitative considerations.

Heat exchanger design is a complex endeavor and involves not only a determination of one or more feasible solution(s) but also the best possible or nearly optimal design solution. In the chapters that follow, we address systematically the most important quantitative aspects of design methodology outlined here.

REFERENCES

Hewitt, G. F., and S. J. Pugh, 1998, Approximate design and costing methods for heat exchangers, *Proc. Int. Conf. Heat Exchangers for Sustainable Development*, June 14–18, Lisbon, Portugal, pp. 801–820.

Hewitt, G. F., A. R. Guy, and R. H. Marsland, 1982, Heat transfer equipment, in *User Guide on Process Integration for the Efficient Use of Energy*, B. Linhoff et al., eds., Institution of Chemical Engineers, Rugby, UK, pp. 129–186.

Kays, W. M., and A. L. London, 1998, *Compact Heat Exchangers*, reprint 3rd ed., Krieger Publishing, Malabar, FL.

Shah, R. K., 1982, Advances in compact heat exchanger technology and design theory, *Heat Transfer 1982, Proc. 7th Int. Heat Transfer Conf.*, Vol. 1, pp. 123–142.

Taborek, J., 1988, Strategy of heat exchanger design, in *Two-Phase Flow Heat Exchangers: Thermal-Hydraulic Fundamentals and Design*, S. Kakaç, A. E. Bergles, and E. O. Fernandes, eds., Kluwer Academic Publishers, Dordrecht, The Netherlands, pp. 473–493.

REVIEW QUESTIONS

2.1 Establish one-to-one correspondence between design considerations identified within the scope of the heat exchanger design methodology and the process stages within the life-cycle design process.

2.2 An engineer has concluded during a heat exchanger design procedure that three important constraints limit his decision-making process. The first is that the pressure drop of one of the two fluids is larger than the value permitted. The second is that the cost of one of the two fluids used in the exchanger limits the mass flow rate of that fluid through the exchanger. Finally, the third is that the temperature difference between the two fluids at the exit of the exchanger must be reduced due to the need to increase the plant's overall efficiency. From a study of the design methodology chart presented in Fig. 2.1, uncover in which of the suggested segments of the design procedure these constraints are probably identified.

2.3 Two competing feasible heat exchanger types have been identified. List the criteria (as many as you can) to be used to select the heat exchanger type that will suit the imposed requirements (impose your own requirements).

2.4 For a given (known) heat load and inlet temperatures, the outlet temperatures of both fluids can be calculated with the thermophysical properties known. A design engineer does not know either the magnitudes of the required mass flow rates of the fluids or the required size of the heat exchanger. Can he or she determine these three variables using the data available?

2.5 How many distinct rating problems can one identify to be characterized by:
   (a) one known and other unknown mass flow rate?
   (b) one known and other unknown inlet temperature?

2.6 List all the information you need to perform the stress analysis of a given heat exchanger.

PROBLEMS

2.1 Develop a flowchart showing detailed steps for thermal and mechanical design, laboratory, and other tests involved in developing an automotive radiator from an initial concept to the final mass production.

2.2   Develop a flow diagram of various departments involved in carrying out the tasks involved in Problem 2.1.

2.3   Repeat Problems 2.1 and 2.2 for a one-of-a-kind shell-and-tube exchanger for a petroleum refinery.

2.4   Repeat Problems 2.1 and 2.2 for an exchanger either of your interest or for any industry of your interest.

3 Basic Thermal Design Theory for Recuperators

As defined in Chapter 1, in a recuperator, two fluids are separated by a heat transfer surface (wall), these fluids ideally do not mix, and there are no moving parts. In this chapter the thermal design theory of recuperators is presented. In a heat exchanger, when hot and cold fluids are maintained at constant temperatures of T_h and T_c as shown in Fig. 3.1a, the driving force for overall heat transfer in the exchanger, referred to as *mean temperature difference* (MTD), is simply $T_h - T_c$. Such idealized constant temperatures on both sides may occur in idealized single-component condensation on one fluid side and idealized single-component evaporation on the other fluid side of the exchanger. However, a number of heat transfer applications have condensation or evaporation of single-component fluid on one side and single-phase fluid on the other side. In such cases, the idealized temperature distribution is shown in Fig. 3.1b and c. The mean temperature difference for these cases is not simply the difference between the constant temperature and the arithmetic mean of the variable temperature. It is more complicated, as will be discussed. In reality, when the temperatures of both fluids are changing during their passage through the exchanger (see, e.g., Figs. 1.50, 1.52, 1.54, 1.56b, 1.57b, 1.60b, 1.63b and 1.64b), the determination of the MTD is complex. Our objective in this chapter is to conduct the appropriate heat transfer analysis in the exchanger for the evaluation of MTD and/or performance. Subsequently, design methods are outlined and design problems will be formulated.

The following are the contents of this chapter: An analogy between thermal, fluid, and electrical parameters is presented in Section 3.1. Heat exchanger variables and the thermal circuit are presented in Section 3.2. The ε-NTU method is introduced in Section 3.3. Specific ε-NTU relationships for various flow arrangements are summarized in Section 3.4. The P-NTU method is introduced in Section 3.5, and P-NTU relationships for various flow arrangements are summarized in Section 3.6. The mean temperature difference (MTD) method is introduced in Section 3.7. The MTD correction factors F for various flow arrangements are presented in Section 3.8. It is shown in Section 3.9 that the results of applications of ε-NTU and MTD methods are identical, although each method has some limitations. The ψ-P and P_1-P_2 graphical presentation methods, which eliminate some of the limitations of the aforementioned methods, are presented in Section 3.10. A brief description of various methods used to obtain ε-NTU or P-NTU formulas for various exchanger flow arrangements is presented in Section 3.11. Considering seven variables of the heat exchanger design problem, there are a total of 21 design problems possible, as discussed in Section 3.12.

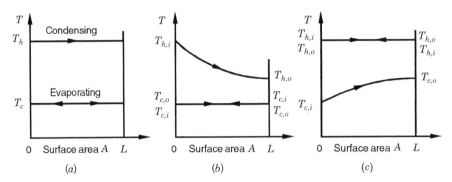

FIGURE 3.1   Idealized temperature distributions of one or both single-component phase-change fluids: (*a*) one fluid condensing, the other evaporating; (*b*) one single-phase fluid cooling, the other fluid evaporating; (*c*) one fluid condensing, the other single-phase fluid heating.

3.1   FORMAL ANALOGY BETWEEN THERMAL AND ELECTRICAL ENTITIES

In heat exchanger analysis, a formal analogy between heat transfer and conduction of electricity is very useful; to understand this analogy clearly, let us start with the definitions. Heat flow q is a consequence of thermal energy in transit by virtue of a temperature difference ΔT. By Ohm's law, the electric current i is a consequence of electrical energy in transit by virtue of an electromotive force (potential) difference ΔE. In both cases, the rate of flow of related entity is inhibited by one or more recognizable resistances acting in series and/or in parallel combinations.

$$\text{Heat Flow (Heat Transfer Rate)} \qquad \text{Electric Current Flow}$$

$$q = \frac{\Delta T}{(UA)^{-1}} = \frac{\Delta T}{\mathbf{R}} \qquad \Delta T = \mathbf{R}q \qquad i = \frac{\Delta E}{\mathbf{R}} \qquad \Delta E = \mathbf{R}i \qquad (3.1)$$

With this notion, the *formal* analogy between various parameters is presented in Table 3.1. It is important to note that the relationships between current, potential, resistance, conductance, capacitance, and time constant terms are analogous for these different physical processes. While the expressions for power and energy are analogous for heat and current flow from the physics point of view, they are not analogous from the resistance circuit point of view as their formulas differ as shown in Table 3.1. Moreover, there is no thermal analogy to electrical inductance or "inertia" in this analogy (not shown in Table 3.1). Note that heat capacity or thermal capacitance energy storage terminology used in heat transfer is used incorrectly as "thermal inertia" in the literature.

Since we know electrical circuit symbolism, we will find it convenient to borrow the symbols for the thermal circuits used to describe the exchanger heat transfer process. These are summarized in Fig. 3.2.

We will also need an analogy between fluid flow in a pipe (exchanger) and electric current for the pressure drop analysis of the exchanger. The basic parameters of pressure drop (head), fluid flow rate, and flow losses are analogous to the voltage potential,

TABLE 3.1   Analogies and Nonalogies between Thermal and Electrical Parameters

Parameter		Electrical		Thermal
		Analogies		
Current	i	ampere, A	q	W, Btu/hr
Potential	E	volts, V	ΔT	°C(K), °F(°R)
Resistance	$\mathbf{R}$	ohms, Ω, V/A	$\mathbf{R} = 1/UA$	°C/W, °F-hr/Btu
Conductance	G	siemens, S, A/V	UA	W/°C, Btu/hr-°F
Capacitance	C	farads, F, A s/V	$\bar{C}$	W · s/°C, Btu/°F
Time constant	$\mathbf{R}C$	s	$\mathbf{R}\bar{C}$	s, hr
		Nonanalogies		
Power	iE	W	q	W, Btu/hr
Energy	$\int_0^\tau iE\,d\tau$	J, W · s	$\int_0^\tau q\,d\tau$	J, Btu

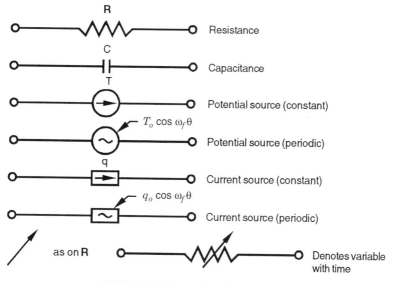

FIGURE 3.2   Thermal circuit symbolism.

current, and resistance.[†] Since the flow losses are measured in terms of the pressure loss or head or fluid column, which have the same units as the potential, this analogy is not as well defined as the analogy between heat transfer and electric current. Again, the relationship between analogous parameters for fluid flow is not linear for transition and turbulent flows or developing laminar flows in pipes.

[†] Pipe and duct design based on one-dimensional lumped parameter analysis typically defines the flow resistance or the flow loss coefficient K as equal to the number of velocity heads lost due to frictional effects [see Eq. (6. 53)].

3.2 HEAT EXCHANGER VARIABLES AND THERMAL CIRCUIT

In this section, starting with the assumptions built into heat exchanger design theory, the basic problem for the exchanger heat transfer analysis is formulated. This includes the differential equations used for the analysis as well as a list of independent and dependent variables associated with heat exchanger design and analysis problems. Next, the basic definitions of important dimensional variables and important terminologies are introduced. Finally, the thermal circuit, and expressions for UA and wall temperatures are presented.

3.2.1 Assumptions for Heat Transfer Analysis

To analyze the exchanger heat transfer problem, a set of assumptions are introduced so that the resulting theoretical models are simple enough for the analysis. The following assumptions and/or idealizations are made for the exchanger heat transfer problem formulations: the energy balances, rate equations, boundary conditions, and subsequent analysis [see, e.g., Eqs. (3.2) and (3.4) through (3.6) in differential or integral form].[†]

1. The heat exchanger operates under steady-state conditions [i.e., constant flow rates and fluid temperatures (at the inlet and within the exchanger) independent of time].

2. Heat losses to or from the surroundings are negligible (i.e. the heat exchanger outside walls are adiabatic).

3. There are no thermal energy sources or sinks in the exchanger walls or fluids, such as electric heating, chemical reaction, or nuclear processes.

4. The temperature of each fluid is uniform over every cross section in counterflow and parallelflow exchangers (i.e., perfect transverse mixing and no temperature gradient normal to the flow direction). Each fluid is considered mixed or unmixed from the temperature distribution viewpoint at every cross section in single-pass crossflow exchangers, depending on the specifications. For a multipass exchanger, the foregoing statements apply to each pass depending on the basic flow arrangement of the passes; the fluid is considered mixed or unmixed between passes as specified.

5. Wall thermal resistance is distributed uniformly in the entire exchanger.

6. Either there are no phase changes (condensation or vaporization) in the fluid streams flowing through the exchanger or the phase change occurs under the following condition. The phase change occurs at a constant temperature as for a single-component fluid at constant pressure; the effective specific heat $c_{p,\text{eff}}$ for the phase-changing fluid is infinity in this case, and hence $C_{\max} = \dot{m} c_{p,\text{eff}} \to \infty$, where $\dot{m}$ is the fluid mass flow rate.

7. Longitudinal heat conduction in the fluids and in the wall is negligible.

8. The individual and overall heat transfer coefficients are *constant* (independent of temperature, time, and position) throughout the exchanger, including the case of phase-changing fluids in assumption 6.

[†] The complete set of differential equations and boundary conditions describing the mathematical models of heat exchangers, based on these assumptions, is presented in Section 11.2.

9. The specific heat of each fluid is constant throughout the exchanger, so that the heat capacity rate on each side is treated as constant. Note that the other fluid properties are not involved directly in the energy balance and rate equations, but are involved implicitly in NTU and are treated as constant.

10. For an extended surface exchanger, the overall extended surface efficiency η_o is considered uniform and constant.

11. The heat transfer surface area A is distributed uniformly on each fluid side in a single-pass or multipass exchanger. In a multipass unit, the heat transfer surface area is distributed uniformly in each pass, although different passes can have different surface areas.

12. For a plate-baffled 1–n shell-and-tube exchanger, the temperature rise (or drop) per baffle pass (or compartment) is small compared to the total temperature rise (or drop) of the shell fluid in the exchanger, so that the shell fluid can be treated as mixed at any cross section. This implies that the number of baffles is large in the exchanger.

13. The velocity and temperature at the entrance of the heat exchanger on each fluid side are *uniform* over the flow cross section. There is no gross flow maldistribution at the inlet.

14. The fluid flow rate is uniformly distributed through the exchanger on each fluid side in each pass i.e., no passage-to-passage or viscosity-induced maldistribution occurs in the exchanger core. Also, no flow stratification, flow bypassing, or flow leakages occur in any stream. The flow condition is characterized by the bulk (or mean) velocity at any cross section.

Assumptions 1 through 5 are necessary in a theoretical analysis of steady-state heat exchangers. Heat losses to the surroundings, if small, may be taken into account approximately by using the effective heat capacity rate C_{eff} for the hot fluid instead of the actual C ($= \dot{m}c_p$) in the analysis. C_{eff} is determined based on the actual heat transfer rate from the hot to cold fluid. Assumption 6 essentially restricts the analysis to single-phase flow on both sides or one side with a dominating thermal resistance. For two-phase flows on both sides, many of the foregoing assumptions are not valid since mass transfer in phase change results in variable properties and variable flow rates of each phase, and the heat transfer coefficients may also vary significantly. As a result, the ε-NTU and other methods presented in Sections 3.3 through 3.11 are not applicable to two-phase heat exchangers. Assumptions 7 through 12 are relaxed in Chapter 4. Assumptions 13 and 14 are addressed in Chapter 12.

If any of the foregoing assumptions are not valid for a particular exchanger application and the sections that cover the relaxation of these assumptions do not provide a satisfactory solution, the approach recommended is to work directly with Eqs. (3.3) and (3.4), or a set of equations corresponding to the model. In this case, modify these differential equations by including a particular effect, and integrate them numerically across sufficiently small segments of the exchanger in which all the assumptions are valid. Refer to Section 4.2.3.2 for an example.

In Sections 3.3 through 3.11, we present ε-NTU, *P*-NTU, MTD, ψ-P, and P_1-P_2 methods of exchanger heat transfer analysis for which the 14 assumptions are invoked. The corresponding model building, based on these assumptions, is discussed in detail in Section 11.2.

3.2.2 Problem Formulation

To perform the heat transfer analysis of an exchanger, our objective is to relate the heat transfer rate q, heat transfer surface area A, heat capacity rate C of each fluid, overall heat transfer coefficient U, and fluid terminal temperatures. Two basic relationships are used for this purpose: (1) energy balance based on the first law of thermodynamics, and (2) rate equations for heat transfer, as outlined by Eqs. (2.1) and (2.2).

Consider a two-fluid exchanger (a counterflow exchanger as an example) shown in Fig 3.3 to arrive at the variables relating to the thermal performance of a two-fluid exchanger. Schematic of Fig. 3.3 and the balance equations for different exchanger flow arrangements may be different, but the basic concept of modeling remains the same. The analysis that will follow is intended to introduce variables important for heat exchanger analysis. Detailed approaches to a general thermodynamic problem formulation are presented in Chapter 11.

Two differential energy conservation (or balance) equations (based on the energy balance implied by the first law of thermodynamics) can be combined as follows for control volumes associated with the differential element of dA area for steady-state flow, an overall adiabatic system, and negligible potential and kinetic energy changes:

$$dq = q'' \, dA = -C_h \, dT_h = -C_c \, dT_c \qquad (3.2)$$

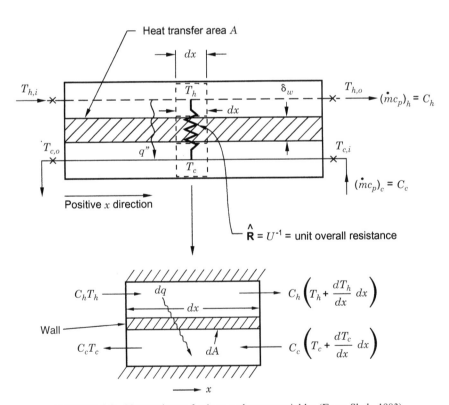

FIGURE 3.3   Nomenclature for heat exchanger variables (From Shah, 1983).

The negative signs in this equation are a result of T_h and T_c decreasing with increasing A (these temperatures decrease with increasing flow length x as shown in Fig. 1.50);[†] also, dq is the heat transfer rate from the hot to cold fluid, $C = \dot{m}c_p$ is the heat capacity rate of the fluid, $\dot{m}$ is the fluid flow rate, c_p is the fluid specific heat at constant pressure, T is the fluid temperature, and the subscripts h and c denote hot and cold fluids, respectively. The heat capacity rate C J/s · °C (Btu/hr · °F) is the amount of heat in joules (Btu) that must be added to or extracted from the fluid stream per second (hour) to change its temperature by 1 °C (°F). The product $\dot{m}c_p = C$ appears in the energy balance [Eq. (3.2)] for constant c_p, and hence C is commonly used in the heat exchanger analysis instead of $\dot{m}$ and c_p as two parameters separately.

In general, for any isobaric change of state, Eq. (3.2) should be replaced by

$$dq = -\dot{m}_h \, d\mathbf{h}_h = -\dot{m}_c \, d\mathbf{h}_c \tag{3.3}$$

where $\mathbf{h}$ is the fluid specific enthalpy, J/kg (Btu/lbm). If the change of state is a phase change, enthalpy differences should be replaced by enthalpies of phase change (either enthalpy of evaporation or enthalpy of condensation). However, c_p can be assumed as infinity for condensing or evaporating single-component fluid streams. Hence, the phase-changing stream can be treated as a "single-phase" fluid having $\Delta T = q/C$ or $dT = dq/C$, with C being infinity for a finite q or dq since the ΔT or $dT = 0$ for isothermal condensing or evaporating fluid streams (see Fig. 3.1). Note that here $\Delta T = T_{h,i} - T_{h,o}$ or $T_{c,o} - T_{c,i}$ as appropriate.

The overall heat transfer rate equation on a differential base for the surface area dA of Fig. 3.3 is

$$dq = U(T_h - T_c)_{\text{local}} \, dA = U \, \Delta T \, dA \tag{3.4}$$

where U is the local overall heat transfer coefficient to be defined in Eq. (3.20).[‡] Thus for this differential element dA, the *driving potential* for heat transfer is the local temperature difference $(T_h - T_c) = \Delta T$ and the *thermal conductance* is $U \, dA$.

Integration of Eqs. (3.2) and (3.4) together over the entire heat exchanger surface for specified inlet temperatures will result in an expression that will relate all important operating variables and geometry parameters of the exchanger. The derivation of such an expression for a counterflow exchanger will be presented in Section 3.3 together with final results for many industrially important exchanger flow arrangements. The common assumptions invoked for the derivation and integration of Eqs. (3.2) and (3.4) are summarized in Section 3.2.1.

Two basic equations, energy conservation (balance) and rate equations, could also be written on an overall basis for the entire exchanger as follows (under the conditions implied by the above-mentioned idealizations):

[†] The sign convention adopted in Eq. (3.2) leads to positive value of heat transfer rate change along each dx element, and should be considered only as formal (i.e., not necessarily in agreement with thermodynamic convention for heat).

[‡] Note that although the overall heat transfer coefficient U is idealized as constant (see assumption 8 in Section 3.2.1), it can vary significantly in a heat exchanger. In that case, the mean overall heat transfer coefficient U_m is obtained from local U data (see Section 4.2.3). Even though $U = U_m = $ constant throughout this chapter, we distinguish between U and U_m in Sections 3.2.3 and 3.2.4 to emphasize how the theory is developed using U and U_m.

$$q = \int C \, dT = C_h(T_{h,i} - T_{h,o}) = C_c(T_{c,o} - T_{c,i}) \tag{3.5}$$

$$q = \int U \, \Delta T \, dA = U_m A \, \Delta T_m \tag{3.6}$$

Here the subscripts i and o denote inlet and outlet, respectively; $T_{h,o}$ and $T_{c,o}$ represent the outlet temperatures; they are bulk temperatures defined by Eq. (7.10) for a nonuniform temperature distribution across a cross section; and U_m and ΔT_m are the mean overall heat transfer coefficient and the exchanger mean temperature difference, respectively, and will be defined shortly.

From Eqs. (3.5) and (3.6) and Fig. 3.3, the steady-state overall-adiabatic heat exchanger behavior can thus be presented in terms of dependent fluid outlet temperatures or heat transfer rate as functions of four operating condition variables and three designer controlled parameters:

$$\underbrace{T_{h,o}, T_{c,o} \text{ or } q}_{\substack{\text{dependent} \\ \text{variables}}} = \phi \, (\underbrace{T_{h,i}, T_{c,i}, C_h, C_c}_{\substack{\text{operating condition} \\ \text{variables}}} \underbrace{U, A, \text{ flow arrangement})}_{\substack{\text{parameters under} \\ \text{designer's control}}} \tag{3.7}$$

$$\underbrace{\phantom{T_{h,i}, T_{c,i}, C_h, C_c \quad U, A, \text{ flow arrangement}}}_{\text{independent variables and parameters}}$$

This equation represents a total of six independent and one or more dependent variables for a given heat exchanger flow arrangement. Of course, any one of the independent variables/parameters in Eq. (3.7) can be made dependent (if unknown); in that case, one of the three dependent variables in Eq. (3.7) becomes an independent variable/parameter. Thus the most general heat exchanger design problem is to determine any two unknown variables from this set when the rest of them are known.

For heat exchanger analysis, it is difficult to understand and work with such a large number of variables and parameters as outlined in Eq. (3.7). From *dimensional* analysis, three dimensionless groups are formulated from six independent and one or more dependent variables of Eq. (3.7). The reduced number of nondimensional variables and parameters simplifies much of the analysis, provides a clear understanding of performance behavior, and the results can be presented in more compact graphical and tabular forms. The specific form of these groups is to some extent optional.

Five such options have been used, depending on which method of heat transfer analysis has been used: the effectiveness–number of heat transfer units (ε-NTU or P-NTU) method, the mean temperature difference (MTD) method, the nondimensional mean temperature difference–temperature effectiveness (ψ-P) method, and the P_1-P_2 method. These methods are discussed in Sections 3.3 through 3.10.

3.2.3 Basic Definitions

The definitions of the mean overall heat transfer coefficient and mean temperature difference are introduced first.

The rate equation (3.4), after rearrangement, is presented in integral form as

$$\int_q \frac{dq}{\Delta T} = \int_A U \, dA \tag{3.8}$$

Now we define the mean temperature difference and mean overall heat transfer coefficient as follows using the terms in Eq. (3.8):

$$\frac{1}{\Delta T_m} = \frac{1}{q} \int_q \frac{dq}{\Delta T} \tag{3.9}$$

$$U_m = \frac{1}{A} \int_A U \, dA \tag{3.10}$$

Substitution of Eqs. (3.9) and (3.10) into Eq. (3.8) results into the following equation after rearrangement:

$$q = U_m A \, \Delta T_m \tag{3.11}$$

Here U_m is the mean overall heat transfer coefficient, and ΔT_m is the true (or effective) mean temperature difference (MTD), also referred to as mean temperature driving potential or force for heat transfer.

Generally, the overall heat transfer coefficient is treated as *constant* in the heat exchanger analysis. It is simply designated as U, without subscripts or overbars, throughout the book except for Section 4.2, where various definitions of mean overall heat transfer coefficients are introduced. Thus the overall rate equation (3.6) is simply

$$q = UA \, \Delta T_m \tag{3.12}$$

Note that if U is treated as a constant, integration of Eq. (3.4) will yield

$$\Delta T_m = \frac{1}{A} \int_A \Delta T \, dA \tag{3.13}$$

Other commonly used important entities for heat exchangers are the inlet temperature difference, temperature range, temperature approach, temperature pinch, temperature gap, temperature meet, and temperature cross. They are discussed below and summarized in Table 3.2.

The *inlet temperature difference* (ITD) is the difference between inlet temperatures of the hot and cold fluids and is designated as $\Delta T_{\max}$ in this book. $\Delta T_{\max} = T_{h,i} - T_{c,i}$ is also sometimes referred to as the *temperature span* or *temperature head* in a heat exchanger.

The *temperature range* for a fluid is referred to as its actual temperature rise or drop ΔT within the exchanger. The temperature ranges for hot and cold fluids in the exchangers are then $\Delta T_h = T_{h,i} - T_{h,o}$ and $\Delta T_c = T_{c,o} - T_{c,i}$, respectively.

The *temperature approach* for exchangers with single-phase fluids is defined as the difference between outlet fluid temperatures $(T_{h,o} - T_{c,o})$ for all single-pass and multipass flow arrangements except for the counterflow exchanger. For the latter, it is defined as the smaller of $(T_{h,i} - T_{c,o})$ and $(T_{h,o} - T_{c,i})$. The temperature approach for multiphase multicomponent streams is defined as the *minimum* local temperature difference between hot and cold fluid streams. This could occur anywhere in the exchanger, depending on the flow arrangement, heat duty, and so on. It can be shown that the temperature approach for single-phase exchangers is related to the exchanger effectiveness ε defined by Eq. (3.44) later as follows:

TABLE 3.2   Expressions for Temperature Span, Range, Approach, Pinch, Gap, Meet, and Cross

Item	Expression
Inlet temperature difference, ITD, maximim temperature span or temperature head	$T_{h,i} - T_{c,i}$
Temperature range for hot fluid	$T_{h,i} - T_{h,o}$
Temperature range for cold fluid	$T_{c,o} - T_{c,i}$
Temperature approach, counterflow exchanger	$\min |(T_{h,i} - T_{c,o}), (T_{h,o} - T_{c,i})|$
Temoerature approach, all other exchangers	$T_{h,o} - T_{c,o}$
Temperature pinch at a local point in an exchanger	$T_h - T_c$ with $(T_h - T_c) \ll (T_{h,i} - T_{c,o})$ or $(T_{h,o} - T_{c,i})$
Temperature gap	$T_{h,o} - T_{c,o}$ with $T_{h,o} > T_{c,o}$
Temperature meet, counterflow sigle-phase exchanger	$T_{h,i} = T_{c,o}$ or $T_{h,o} = T_{c,i}$
Temperature meet, all other single-phase exchangers	$T_{h,o} = T_{c,o}$
Temperature cross, single-pass exchangers	$T_{c,o} - T_{h,o}$ with $T_{c,o} > T_{h,o}$
Temperature cross, multipass exchangers	$T_c - T_h$ with $T_c > T_h$ in one of the passes

$$\frac{\text{temperature}}{\text{approach}} = \begin{cases} (1 - \varepsilon)\, \Delta T_{\max} & \text{for a counterflow exchanger} \\ [1 - (1 + C^*)\varepsilon]\, \Delta T_{\max} = (T_{h,o} - T_{c,o}) & \text{for other exchangers} \end{cases}$$
$$(3.14)$$

where $\Delta T_{\max} = T_{h,i} - T_{c,i}$ and $C^* = C_{\min}/C_{\max}$. For some shell-and-tube, multipass, and two-phase exchangers, it may not be either easy or possible to determine quantitatively the magnitude of the temperature approach. In that case, while the foregoing definition is valid, it loses its usefulness.

A *temperature pinch* refers to a local temperature difference within an exchanger (or an array of exchangers) that is substantially less than either of two terminal temperature differences and is minimum in the exchanger. In the limit, it can approach zero, which is referred to as temperature meet defined below. The temperature pinch usually occurs in an exchanger with two-phase multicomponent fluids. In reality, a temperature pinch close to zero will require a very large (approaching infinity) surface area for a single-pass exchanger. Hence, for a finite-size heat exchanger, the exchanger would cease to function efficiently beyond the temperature pinch point (i.e., resulting in a more significant reduction in heat transfer than justified). However, for a multipass exchanger, the temperature pinch could occur in one pass, and in that case, the performance of *that* pass beyond the temperature pinch is reduced significantly.

A *temperature gap* refers to the temperature difference between hot and cold fluid outlet temperatures provided that $T_{h,o} > T_{c,o}$.

A *temperature meet* refers to the case when the temperature pinch is zero or the hot and cold fluid temperatures are the same locally somewhere in the exchanger or at outlets. This is an idealized condition and does not occur in a single-pass heat exchanger, but may occur in one of the passes of a multipass exchanger.

A *temperature cross* refers to the case when the cold fluid temperature becomes equal or greater than the hot fluid temperature within the exchanger. *External temperature cross* refers to the case when the cold fluid outlet temperature $T_{c,o}$ is greater than the hot fluid outlet temperature $T_{h,o}$. There is no physical (actual) crossing of hot and cold fluid temperature distributions within an exchanger. This is quite common in a counterflow exchanger (see Fig. 1.50a) and other single-pass and multipass exchangers having high

NTUs (see Section 3.3.3 for the definition of NTU). The magnitude of the external temperature cross is $T_{c,o} - T_{h,o}$. *Internal temperature cross* refers to the case when locally somewhere within the exchanger T_c becomes equal to T_h (within a pass or in a segment of the exchanger), and beyond that point along the flow length, $T_c > T_h$ in that pass or segment; subsequently, reverse heat transfer takes place (original cold fluid transferring heat to the original hot fluid). The external temperature cross can occur with or without the internal temperature cross; similarly an internal temperature cross can occur with or without an external temperature cross (see Section 11.4.1).

3.2.4   Thermal Circuit and UA

To understand the exchanger overall heat transfer rate equation [Eq. (3.12)], consider the thermal circuit model of Fig. 3.4. Scale or fouling deposit layers are also shown on each side of the wall.

In the steady state, heat is transferred from the hot fluid to the cold fluid by the following processes: convection to the hot fluid wall, conduction through the wall, and subsequent convection from the wall to the cold fluid. In many heat exchangers, a fouling film is formed as a result of accumulation of scale or rust formation, deposits from the fluid, chemical reaction products between the fluid and the wall material, and/or biological growth. This undesired fouling film usually has a low thermal conductivity and can increase the thermal resistance to heat flow from the hot fluid to the cold fluid. This added thermal resistance on individual fluid sides for heat conduction through the fouling film is taken into account by a *fouling factor*[†] $r_f = 1/h_f$, where the subscript f denotes fouling (or scale); it is shown in Fig. 3.4. Thus, the heat transfer rate per unit area at any section dx (having surface areas dA_h, dA_c, etc.) can be presented by the appropriate convection and conduction rate equations as follows:

$$dq = \frac{T_h - T_{h,f}}{d\mathbf{R}_h} = \frac{T_{h,f} - T_{w,h}}{d\mathbf{R}_{h,f}} = \frac{T_{w,h} - T_{w,c}}{d\mathbf{R}_w} = \frac{T_{w,c} - T_{c,f}}{d\mathbf{R}_{c,f}} = \frac{T_{c,f} - T_c}{d\mathbf{R}_c} \tag{3.15}$$

Alternatively

$$dq = \frac{T_h - T_c}{d\mathbf{R}_o} = U\,dA(T_h - T_c) \tag{3.16}$$

where the overall differential thermal resistance $d\mathbf{R}_o$ consists of component resistances in series (similar to those shown in Fig. 3.4b for a heat exchanger):

$$\frac{1}{U\,dA} = d\mathbf{R}_o = d\mathbf{R}_h + d\mathbf{R}_{h,f} + d\mathbf{R}_w + d\mathbf{R}_{c,f} + d\mathbf{R}_c \tag{3.17}$$

or

$$\frac{1}{U\,dA} = \frac{1}{(\eta_o h\,dA)_h} + \frac{1}{(\eta_o h_f\,dA)_h} + d\mathbf{R}_w + \frac{1}{(\eta_o h_f\,dA)_c} + \frac{1}{(\eta_o h\,dA)_c} \tag{3.18}$$

[†] We also refer to the fouling factor as the *unit fouling thermal resistance* or concisely as *fouling resistance* $r_f = \hat{\mathbf{R}}_f = 1/h_f = \delta_f/k_f$ where δ_f is the thickness of fouling layer and k_f is the thermal conductivity of the fouling material. Refer to Section 13.1 for more details on the fouling resistance concept.

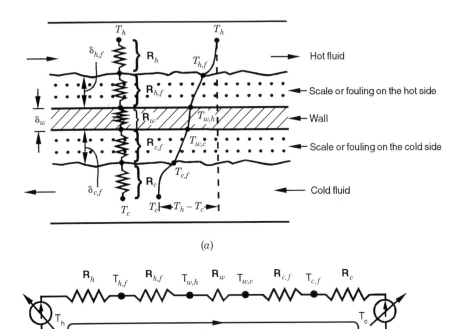

FIGURE 3.4   (a) Thermal resistances; (b) thermal circuit for a heat exchanger (From Shah, 1983).

Various symbols in this equation are defined after Eq. (3.24). If we idealize that the heat transfer surface area is distributed uniformly on each fluid side (see assumption 11 in Section 3.2.1), the ratio of differential area on each fluid side to the total area on the respective side remains the same; that is,

$$\frac{dA}{A} = \frac{dA_h}{A_h} = \frac{dA_c}{A_c} = \frac{dA_w}{A_w} \tag{3.19}$$

Replacing differential areas of Eq. (3.18) by using corresponding terms of Eq. (3.19), we get

$$\frac{1}{UA} = \frac{1}{(\eta_o h A)_h} + \frac{1}{(\eta_o h_f A)_h} + \mathbf{R}_w + \frac{1}{(\eta_o h_f A)_c} + \frac{1}{(\eta_o h A)_c} \tag{3.20}$$

It should be emphasized that U and all h's in this equation are assumed to be *local*. Using the overall rate equation [Eq. (3.6)], the total heat transfer rate will be

$$q = U_m A \, \Delta T_m = U_m A (T_{h,e} - T_{c,e}) = \frac{1}{\mathbf{R}_o}(T_{h,e} - T_{c,e}) \tag{3.21}$$

and a counterpart of Eq. (3.15) for the entire exchanger is

$$q = \frac{T_{h,e} - T_{h,f}}{\mathbf{R}_h} = \frac{T_{h,f} - T_{w,h}}{\mathbf{R}_{h,f}} = \frac{T_{w,h} - T_{w,c}}{\mathbf{R}_w} = \frac{T_{w,c} - T_{c,f}}{\mathbf{R}_{c,f}} = \frac{T_{c,f} - T_{c,e}}{\mathbf{R}_c} \qquad (3.22)$$

where the subscript e denotes the effective value for the exchanger, or $(T_{h,e} - T_{c,e}) = \Delta T_m$. To be more precise, all individual temperatures in Eq. (3.22) should also be mean or effective values for respective fluid sides. However, this additional subscript is not included for simplicity. In Eq. (3.21), the overall thermal resistance $\mathbf{R}_o$ consists of component resistances in series as shown in Fig. 3.4b.

$$\frac{1}{U_m A} = \mathbf{R}_o = \mathbf{R}_h + \mathbf{R}_{h,f} + \mathbf{R}_w + \mathbf{R}_{c,f} + \mathbf{R}_c \qquad (3.23)$$

$$\frac{1}{U_m A} = \frac{1}{(\eta_o h_m A)_h} + \frac{1}{(\eta_o h_{m,f} A)_h} + \mathbf{R}_w + \frac{1}{(\eta_o h_{m,f} A)_c} + \frac{1}{(\eta_o h_m A)_c} \qquad (3.24)$$

For constant and uniform U and h's throughout the exchanger, Eqs. (3.24) and (3.20) are identical. In that case, $U_m = U$ and we will use U throughout the book except for Section 4.2. In Eqs. (3.20) and (3.24), depending on the local or mean value, we define

$$\mathbf{R}_h = \text{hot-fluid-side convection resistance} = \frac{1}{(\eta_o h A)_h} \text{ or } \frac{1}{(\eta_o h_m A)_h}$$

$$\mathbf{R}_{h,f} = \text{hot-fluid-side fouling resistance} = \frac{1}{(\eta_o h_f A)_h} \text{ or } \frac{1}{(\eta_o h_{m,f} A)_h}$$

$$\mathbf{R}_w = \text{wall thermal resistance expressed by Eq. (3.25) or (3.26)}$$

$$\mathbf{R}_{c,f} = \text{cold-fluid-side fouling resistance} = \frac{1}{(\eta_o h_f A)_c} \text{ or } \frac{1}{(\eta_o h_{m,f} A)_c}$$

$$\mathbf{R}_c = \text{cold-fluid-side convection resistance} = \frac{1}{(\eta_o h A)_c} \text{ or } \frac{1}{(\eta_o h_m A)_c}$$

In the foregoing definitions, h is the heat transfer coefficient, discussed in detail in Section 7.1.4.3; h_f is referred to as the fouling coefficient (inverse of fouling factor), discussed in Chapter 13; A represents the total of primary and secondary (finned) surface area; and η_o is the extended surface efficiency of an extended (fin) surface defined in Section 4.3.4. In the literature, $1/(\eta_o h)_f = \mathbf{R}_f A = \hat{\mathbf{R}}_f$ is referred to as *unit fouling resistance*. Note that no fins are shown in the upper sketch of Fig. 3.4; however, η_o is included in the aforementioned various resistance terms in order to make them most general. For all prime surface exchangers (i.e., having no fins or extended surface), $\eta_{o,h}$ and $\eta_{o,c}$ are unity.

The wall thermal resistance $\mathbf{R}_w$ for flat walls is given by

$$\mathbf{R}_w = \begin{cases} \dfrac{\delta_w}{k_w A_w} & \text{for flat walls with a single layer wall} \\[2em] \displaystyle\sum_j \left(\dfrac{\delta_w}{k_w A_w} \right)_j & \text{for flat walls with a multiple-layer wall} \end{cases} \qquad (3.25)$$

For a cylindrical (tubular) wall, it is given as

$$
\mathbf{R}_w =
\begin{cases}
\dfrac{\ln(d_o/d_i)}{2\pi k_w L N_t} & \text{for } N_t \text{ circular tubes with a single-layer wall} \\[4mm]
\dfrac{1}{2\pi L N_t} \sum_j \dfrac{\ln(d_{j+1}/d_j)}{k_{w,j}} & \text{for } N_t \text{ circular tubes with a multiple-layer wall}
\end{cases}
$$

$$(3.26)$$

where δ_w is the wall plate thickness, A_w the total wall area of all flat walls for heat conduction, k_w the thermal conductivity of the wall material, d_o and d_i the tube outside and inside diameters, L the tube or exchanger length, and N_t the number of tubes. A flat (or plain) wall is generally associated with a plate-fin or an all-prime-surface plate exchanger. In this case,

$$
A_w = L_1 L_2 N_p \tag{3.27}
$$

Here L_1, L_2, and N_p are the length, width, and total number of separating plates, respectively. The wall thickness δ_w is then the length for heat conduction.

If there is any contact or bond resistance present between the fin and tube or plate on the hot or cold fluid side, it is included as an added thermal resistance on the right-hand side of Eq. (3.23) or (3.24). For a heat pipe heat exchanger, additional thermal resistances associated with the heat pipe need to be included on the right-hand side of Eq. (3.23) or (3.24); these resistances are evaporation resistance at the evaporator section of the heat pipe, viscous vapor flow resistance inside the heat pipe (very small), internal wick resistance at the condenser section of the heat pipe, and condensation resistance at the condenser section.

If one of the resistances on the right-hand side of Eq. (3.23) or (3.24) is significantly higher than other resistances, it is referred to as the *controlling resistance*. It is considered significantly dominant when it represents more than 80% of the total resistance. For example, if the cold side is gas (air) and the hot side is condensing steam, the thermal resistance on the gas side will be very high (since h for air is very low compared to that for the condensing steam) and will be referred to as the controlling resistance for that exchanger. However, for a water-to-water heat exchanger, none of the thermal resistances may be dominant if the water flow rates are about the same.

The lowest overall thermal resistance in a heat exchanger can be obtained by making the hot- and cold-side thermal resistances about equal (considering wall and fouling resistances is negligible or low). Hence, a low h is often compensated by a high A to make $(\eta_o hA)_h \approx (\eta_o hA)_c$. This is the reason the surface area on the gas side is about 5 to 10 times higher than that on the liquid-side when the liquid side heat transfer coefficient h is 5 to 10 times higher than the h on the gas side. This would explain why fins are used on the gas sides in a gas–to–liquid or gas–to–phase change exchanger.

In Eq. (3.24) or (3.12), the overall heat transfer coefficient U may be defined optionally in terms of the surface area of either the hot surface, the cold surface, or the wall conduction area. Thus,

$$
UA = U_h A_h = U_c A_c = U_w A_w \tag{3.28}
$$

Thus in specifying UA as a product, we don't need to specify A explicitly. However, the option of A_h, A_c, or A_w *must* be specified in evaluating U from the product UA since $U_h \neq U_c$ if $A_h \neq A_c$. It should be mentioned that the value of $\mathbf{R}_o = 1/UA$ will always be *larger* than the largest thermal resistance component of Eq. (3.23). This means that UA will always be *smaller* than the minimum thermal conductance component [a reciprocal of any one term of the right-hand side of Eq. (3.24)]. UA is referred to as the *overall thermal conductance*.

If the overall rate equation is based on a unit surface area

$$\frac{q}{A} = q'' = U \, \Delta T_m \tag{3.29}$$

the unit overall thermal resistance is $\hat{\mathbf{R}}_o = 1/U$. In this case, individual components of resistances are also on a unit area basis, all based on either A_h or A_c explicitly as follows:

$$\frac{1}{U_h} = \frac{1}{(\eta_o h)_h} + \frac{1}{(\eta_o h_f)_h} + \mathbf{R}_w A_h + \frac{A_h/A_c}{(\eta_o h_f)_c} + \frac{A_h/A_c}{(\eta_o h)_c}$$

$$= \hat{\mathbf{R}}_h + \frac{1}{\eta_{o,h}}\hat{\mathbf{R}}_{h,f} + \hat{\mathbf{R}}_w \frac{A_h}{A_w} + \frac{1}{\eta_{o,c}}\hat{\mathbf{R}}_{c,f}\frac{A_h}{A_c} + \hat{\mathbf{R}}_c \frac{A_h}{A_c} \tag{3.30a}$$

$$\frac{1}{U_c} = \frac{A_c/A_h}{(\eta_o h)_h} + \frac{A_c/A_h}{(\eta_o h_f)_h} + \mathbf{R}_w A_c + \frac{1}{(\eta_o h_f)_c} + \frac{1}{(\eta_o h)_c}$$

$$= \hat{\mathbf{R}}_h \frac{A_c}{A_h} + \frac{1}{\eta_{o,h}}\hat{\mathbf{R}}_{h,f}\frac{A_c}{A_h} + \hat{\mathbf{R}}_w \frac{A_c}{A_w} + \frac{1}{\eta_{o,c}}\hat{\mathbf{R}}_{c,f} + \hat{\mathbf{R}}_c \tag{3.30b}$$

where $1/U_h$ is the unit overall thermal resistance based on the hot-fluid-side surface area. Similarly, $1/U_c$ is defined. Also $\hat{\mathbf{R}}_j = 1/(\eta_o h)_j$, $j = h$ or c are unit thermal resistances for hot or cold fluids, $\hat{\mathbf{R}}_f = 1/h_f$ = unit thermal fouling resistance, and $\hat{\mathbf{R}}_w = \delta_w/k_w$ = unit wall thermal resistance. For a plain tubular exchanger, $\eta_o = 1$; then from Eq. (3.30), U_o based on the tube outside surface area is given as follows after inserting the value of $\mathbf{R}_w$ from the first equation of Eq. (3.26):

$$\frac{1}{U_o} = \frac{1}{h_o} + \frac{1}{h_{o,f}} + \frac{d_o \ln(d_o/d_i)}{2k_w} + \frac{d_o}{h_{i,f}d_i} + \frac{d_o}{h_i d_i} \tag{3.31a}$$

$$\frac{1}{U_i} = \frac{d_i}{h_o d_o} + \frac{d_i}{h_{o,f}d_o} + \frac{d_i \ln(d_o/d_i)}{2k_w} + \frac{1}{h_{i,f}} + \frac{1}{h_i} \tag{3.31b}$$

Here the subscripts o and i denote the tube outside and inside, respectively; $1/U_o$ and $1/U_i$ are the unit overall thermal resistances based on the tube outside and inside surface area, respectively.

Knowledge of wall temperature in a heat exchanger is essential to determine the localized hot spots, freeze points, thermal stresses, local fouling characteristics, or boiling/condensing coefficients. In this case, $T_{w,h}$ and $T_{w,c}$ can be computed from Eq. (3.22) on a *local* basis as follows:

$$\frac{T_h - T_{w,h}}{\mathbf{R}_h + \mathbf{R}_{h,f}} = \frac{T_{w,c} - T_c}{\mathbf{R}_c + \mathbf{R}_{c,f}} \tag{3.32}$$

Based on the thermal circuit of Fig. 3.4, when $\mathbf{R}_w$ is negligible, $T_{w,h} = T_{w,c} = T_w$, and Eq. (3.32) reduces to

$$T_w = \frac{T_h + [(\mathbf{R}_h + \mathbf{R}_{h,f})/(\mathbf{R}_c + \mathbf{R}_{c,f})]/T_c}{1 + [(\mathbf{R}_h + \mathbf{R}_{h,f})/(\mathbf{R}_c + \mathbf{R}_{c,f})]} \tag{3.33}$$

When there is no fouling on either fluid side ($\mathbf{R}_{h,f} = \mathbf{R}_{c,f} = 0$), this reduces further to

$$T_w = \frac{(T_h/\mathbf{R}_h) + (T_c/\mathbf{R}_c)}{(1/\mathbf{R}_h) + (1/\mathbf{R}_c)} = \frac{(\eta_o hA)_h T_h + (\eta_o hA)_c T_c}{(\eta_o hA)_h + (\eta_o hA)_c} \tag{3.34}$$

Equations (3.32) through (3.34) are also valid for the entire exchanger if all temperatures are used as mean or effective values on respective fluid sides.

Example 3.1   In a shell-and-tube feedwater heater, cold water at 15°C flowing at the rate of 180 kg/h is preheated to 90°C by flue gases from 150°C flowing at the rate of 900 kg/h. The water flows inside the copper tubes ($d_i = 25\,\text{mm}$, $d_o = 32\,\text{mm}$) having thermal conductivity $k_w = 381\,\text{W/m} \cdot \text{K}$. The heat transfer coefficients on gas and water sides are 120 and 1200 $\text{W/m}^2 \cdot \text{K}$, respectively. The fouling factor on the water side is $0.002\,\text{m}^2 \cdot \text{K/W}$. Determine the flue gas outlet temperature, the overall heat transfer coefficient based on the outside tube diameter, and the true mean temperature difference for heat transfer. Consider specific heats c_p for flue gases and water as 1.05 and 4.19 J/g · K respectively, and the total tube outside surface area as 5 m². There are no fins inside or outside the tubes, and there is no fouling on the gas side.

SOLUTION

Problem Data and Schematic: Fluid flow rates, inlet temperatures, and cold fluid outlet temperature are provided for a shell-and-tube exchanger of prescribed tube inner and outer diameters (Fig. E3.1). Also, the thermal conductivity of the tube and the thermal resistance on the cold fluid side are given. There are no fins on either side of the tubes.

Determine: Hot fluid outlet temperature $T_{h,o}$, overall heat transfer coefficient U, and true mean temperature difference ΔT_m.

Assumptions: The assumptions invoked in Section 3.2.1 are valid. Hot-fluid-side fouling is negligible.

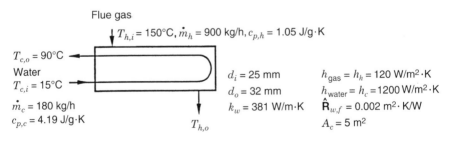

FIGURE E3.1

Analysis: The required heat transfer rate may be obtained from the overall energy balance for the cold fluid [Eq. (3.5)].

$$q = C_c(T_{c,o} - T_{c,i}) = (\dot{m}c_p)_c(T_{c,o} - T_{c,i})$$

$$= \left(\frac{180 \text{ kg/h}}{3600 \text{ s/h}}\right)(4.19 \text{ J/g} \cdot \text{K})(1000 \text{ g/kg})(90 - 15)°\text{C} = 15{,}713 \text{ W}$$

Apply the same Eq. (3.5) on the hot fluid side to find the outlet temperature for flue gas:

$$T_{h,o} = T_{h,i} - \frac{q}{(\dot{m}c_p)_h}$$

Since

$$\dot{m}_h = \frac{900 \text{ kg/h}}{3600 \text{ s/h}} = 0.25 \text{ kg/s}$$

$$c_{p,h} = (1.05 \text{ J/g} \cdot \text{K}) \cdot (1000 \text{ g/kg}) = 1050 \text{ J/kg} \cdot \text{K}$$

we get

$$T_{h,o} = 150°\text{C} - \frac{15{,}713 \text{ W}}{0.25 \text{ kg/s} \times 1050 \text{ J/kg} \cdot °\text{C}} = 90.1°\text{C} \qquad \textit{Ans.}$$

Since U is based on $A = A_h = \pi d_o L N_t$, Eq. (3.31) reduces to the following form after substituting the hot-fluid-side fouling factor $(1/h_f)$ as zero, and replacing the subscripts o and i of U and h with h and c, respectively.

$$\hat{\mathbf{R}} = \frac{1}{U_h} = \frac{1}{h_h} + \frac{1}{h_{h,f}} + \frac{d_o \ln(d_o/d_i)}{2k_w} + \frac{d_o}{h_{c,f}d_i} + \frac{d_o}{h_c d_i}$$

$$= \frac{1}{120 \text{ W/m}^2 \cdot \text{K}} + \frac{0.032 \text{ m}[\ln(32 \text{ mm}/25 \text{ mm})]}{2 \times 381 \text{ W/m} \cdot \text{K}} + \frac{0.002 \text{ m}^2 \cdot \text{K/W} \times 0.032 \text{ m}}{0.025 \text{ m}}$$

$$+ \frac{0.032 \text{ m}}{120 \text{ W/m}^2 \cdot \text{K} \times 0.025 \text{ m}}$$

$$= (0.00833 + 0.00001 + 0.00256 + 0.00107) \text{ m}^2 \cdot \text{K/W} = 0.01197 \text{ m}^2 \cdot \text{K/W}$$
$$\;\; (69.6\%) \quad\;\; (0.1\%) \quad\;\; (21.4\%) \quad\;\; (8.9\%)$$

Hence,

$$U_h = 83.54 \text{ W/m}^2 \cdot \text{K} \qquad \textit{Ans.}$$

Note that the controlling thermal resistance for this feedwater heater is 69.6% on the flue gas side. Now the mean temperature difference can be determined from Eq. (3.12) as

$$\Delta T_m = \frac{q}{U_h A_h} = \frac{15{,}713 \text{ W}}{83.54 \text{ W/m}^2 \cdot \text{K} \times 5 \text{ m}^2} = 37.6°\text{C} \qquad \textit{Ans.}$$

Discussion and Comments: Since the heat transfer coefficient on the cold side is greater than that on the hot side and the hot- and cold-side surface areas are about the same, the hot side becomes the controlling resistance side. This can be seen from the unit thermal resistance distribution as 69.6% of the total unit thermal resistance on the hot-gas side. The tube wall, made of copper, turned out to be a very good conductor with very small thermal resistance. Notice that the fouling resistance on the water side contributes about one-fifth (21.4%) of the total unit thermal resistance and hence about 21% surface area penalty. If there had been no fouling on the water side, we would have reduced the heat transfer surface area requirement by about one-fifth. Hence, if it is desired to make a single important improvement to reduce the surface area requirement in this exchanger, the best way would be to employ fins on the gas side (i.e., employing low finned tubing in the exchanger).

3.3   THE ε-NTU METHOD

In the ε-NTU method, the heat transfer rate from the hot fluid to the cold fluid in the exchanger is expressed as

$$q = \varepsilon C_{min}(T_{h,i} - T_{c,i}) = \varepsilon C_{min}\, \Delta T_{max} \qquad (3.35)$$

where ε is the heat exchanger effectiveness,[†] sometimes referred to in the literature as the *thermal efficiency*, C_{min} is the minimum of C_h and C_c; $\Delta T_{max} = (T_{h,i} - T_{c,i})$ is the fluid *inlet temperature difference* (ITD). The heat exchanger effectiveness ε is nondimensional, and it can be shown that in general it is dependent on the number of transfer units NTU, the heat capacity rate ratio C^*, and the flow arrangement for a direct-transfer type heat exchanger:

$$\varepsilon = \phi(\text{NTU}, C^*, \text{ flow arrangement}) \qquad (3.36)$$

Here the functional relationship ϕ is dependent on the flow arrangement. The three nondimensional groups, ε, NTU, and C^* are first defined below. The relationship among them is illustrated next.

3.3.1   Heat Exchanger Effectiveness ε

Effectiveness ε is a measure of thermal performance of a heat exchanger. It is defined for a given heat exchanger of any flow arrangement as a ratio of the actual heat transfer rate from the hot fluid to the cold fluid to the maximum possible heat transfer rate q_{max} thermodynamically permitted:

$$\varepsilon = \frac{q}{q_{max}} \qquad (3.37)$$

[†] It should be emphasized that the term *effectiveness* may not be confused with *efficiency*. The use of the term *efficiency* is generally restricted to (1) the efficiency of conversion of *energy form A* to *energy form B*, or (2) a comparison of actual system performance to the ideal system performance, under comparable operating conditions, from an energy point of view. Since we deal here with a component (heat exchanger) and there is no conversion of different forms of energy in a heat exchanger (although the conversion between heat flow and enthalpy change is present), the term *effectiveness* is used to designate the efficiency of a heat exchanger. The consequence of the first law of thermodynamics is the energy balance, and hence the definition of the exchanger effectiveness explicitly uses the first law of thermodynamics (see Chapter 11 for further discussion).

Here it is idealized that there are no flow leakages from one fluid to the other fluid, and vice versa. If there are flow leakages in the exchanger, q represents the total enthalpy gain (or loss) of the C_{min} fluid corresponding to its actual flow rate in the outlet (and *not* inlet) stream. How do we determine q_{max}? It would be obtained in a "perfect" *counterflow* heat exchanger (recuperator) of *infinite surface area*, zero longitudinal wall heat conduction, and zero flow leakages from one fluid to the other fluid, operating with fluid flow rates and fluid inlet temperatures the same as those of the actual heat exchanger; also, assumptions 8 to 11, 13, and 14 of Section 3.2.1 are invoked for this perfect counterflow exchanger. This perfect exchanger is the "meterbar" (or "yardstick") used in measuring the degree of perfection of actual exchanger performance. The value of ε ranges from 0 to 1. Thus ε is like an efficiency factor and has thermodynamic significance. As shown below, in such a perfect heat exchanger, the exit temperature of the fluid with the smaller heat capacity will reach the entering temperature of the larger heat capacity fluid.[†]

Consider a counterflow heat exchanger having infinite surface area. An overall energy balance for the two fluid streams is

$$q = C_h(T_{h,i} - T_{h,o}) = C_c(T_{c,o} - T_{c,i}) \tag{3.38}$$

Based on this equation, for $C_h < C_c$, $(T_{h,i} - T_{h,o}) > (T_{c,o} - T_{c,i})$. The temperature drop on the hot fluid side will thus be higher, and over the infinite flow length the hot fluid temperature will approach the inlet temperature of the cold fluid as shown by the two bottom curves in Fig. 3.5, resulting in $T_{h,o} = T_{c,i}$. Thus for an infinite area counterflow exchanger with $C_h < C_c$, we get q_{max} as

$$q_{max} = C_h(T_{h,i} - T_{c,i}) = C_h \, \Delta T_{max} \tag{3.39}$$

Similarly, for $C_h = C_c = C$,

$$q_{max} = C_h(T_{h,i} - T_{c,i}) = C_c(T_{h,i} - T_{c,i}) = C \, \Delta T_{max} \tag{3.40}$$

Based on Eq. (3.38), for $C_h > C_c$, $(T_{c,o} - T_{c,i}) > (T_{h,i} - T_{h,o})$. Hence, $T_{c,o}$ will approach $T_{h,i}$ over the infinite length, and therefore

$$q_{max} = C_c(T_{h,i} - T_{c,i}) = C_c \, \Delta T_{max} \tag{3.41}$$

Or, more generally, based on Eqs. (3.39) throught (3.41),

$$q_{max} = C_{min}(T_{h,i} - T_{c,i}) = C_{min} \, \Delta T_{max} \tag{3.42}$$

where

$$C_{min} = \begin{cases} C_c & \text{for } C_c < C_h \\ C_h & \text{for } C_h < C_c \end{cases} \tag{3.43}$$

[†] It should be mentioned here that the second law of thermodynamics is involved implicitly in the definition of the exchanger effectiveness since the "maximum possible heat transfer rate" is limited by the second law. Further discussion of this and related issues is presented in Section 11.2.2.

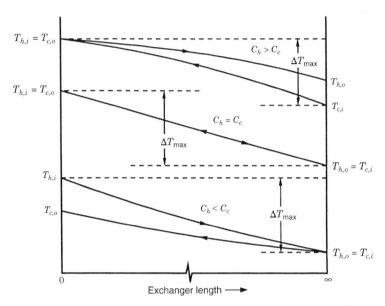

FIGURE 3.5   Temperature distributions in a counterflow exchanger of infinite surface area (From Shah 1983).

Thus q_{max} is determined by Eq. (3.42) for defining the measure of the actual performance of a heat exchanger having *any* flow arrangement. Notice that $\Delta T_{max} = T_{h,i} - T_{c,i}$ in every case and C_{min} appears in the determination of q_{max} regardless of $C_h > C_c$ or $C_h \leq C_c$.

Using the value of actual heat transfer rate q from the energy conservation equation (3.5) and q_{max} from Eq. (3.42), the exchanger effectiveness ε of Eq. (3.37) valid for *all* flow arrangements of the two fluids is given by

$$\varepsilon = \frac{C_h(T_{h,i} - T_{h,o})}{C_{min}(T_{h,i} - T_{c,i})} = \frac{C_c(T_{c,o} - T_{c,i})}{C_{min}(T_{h,i} - T_{c,i})} \tag{3.44}$$

Thus ε can be determined directly from the operating temperatures and heat capacity rates. It should be emphasized here that $T_{h,o}$ and $T_{c,o}$ are the bulk outlet temperatures defined by Eq. (7.10). If there is flow and/or temperature maldistribution at the exchanger inlet, not only the fluid outlet temperatures but also the fluid inlet temperatures should be computed as bulk values and used in Eq. (3.44).

An alternative expression of ε using q from the rate equation (3.12) and q_{max} from Eq. (3.42) is

$$\varepsilon = \frac{UA}{C_{min}} \frac{\Delta T_m}{\Delta T_{max}} \tag{3.45}$$

Now let us nondimensionalize Eq. (3.7). The mean fluid outlet temperatures $T_{h,o}$ and $T_{c,o}$, the dependent variables of Eq. (3.7), are represented in a nondimensional form by the exchanger effectiveness ε of Eq. (3.44). There are a number of different ways to arrive at nondimensional groups on which this exchanger effectiveness will depend. Here we consider an approach in which we list all possible nondimensional groups from visual inspection of Eqs. (3.44) and (3.45) as follows and then eliminate those that are not independent; the exchanger effectiveness ε is dependent on the following nondimensional groups:

$$\varepsilon = \phi\left(\underbrace{\frac{UA}{C_{\min}}, \frac{C_{\min}}{C_{\max}}}_{\text{independent}}, \underbrace{\frac{T_{h,i} - T_{h,o}}{\Delta T_{\max}}, \frac{T_{c,o} - T_{c,i}}{\Delta T_{\max}}, \frac{\Delta T_m}{\Delta T_{\max}}}_{\text{dependent}}, \text{ flow arrangement}\right) \quad (3.46)$$

Note that $\Delta T_{\max} = T_{h,i} - T_{c,i}$ in the last three groups of Eq. (3.46) is an independent parameter. In Eq. (3.46), $C_{\max} = C_c$ for $C_c > C_h$ and $C_{\max} = C_h$ for $C_h > C_c$, so that

$$\frac{C_{\min}}{C_{\max}} = \begin{cases} \dfrac{C_c}{C_h} & \text{for } C_c < C_h \\[2mm] \dfrac{C_h}{C_c} & \text{for } C_h < C_c \end{cases} \quad (3.47)$$

In order to show that the third through fifth groups on the right-hand side of Eq. (3.46) are dependent, using Eqs. (3.5) and (3.44), we can show that the first two of the three groups are related as

$$\frac{T_{h,i} - T_{h,o}}{\Delta T_{\max}} = \frac{C_c}{C_h}\frac{T_{c,o} - T_{c,i}}{\Delta T_{\max}} = \frac{\varepsilon}{C_h/C_{\min}} = \begin{cases} \varepsilon & \text{for } C_h = C_{\min} \\ \varepsilon(C_{\min}/C_{\max}) & \text{for } C_h = C_{\max} \end{cases} \quad (3.48)$$

and using Eq. (3.45), we can show that the fifth group of Eq. (3.46) is

$$\frac{\Delta T_m}{\Delta T_{\max}} = \frac{\varepsilon}{UA/C_{\min}} \quad (3.49)$$

Since the right-hand side of the last equality of Eqs. (3.48) and (3.49) have ε, $C_{\min}/C_{\max}$, and $UA/C_{\min}$ as the only nondimensional groups and they are already included in Eq. (3.46), the dimensionless groups of the left-hand side of Eqs. (3.48) and (3.49) are dependent. Thus Eq. (3.46) can be written in terms of nondimensional groups, without a loss of generality, as follows:

$$\varepsilon = \phi\left(\frac{UA}{C_{\min}}, \frac{C_{\min}}{C_{\max}}, \text{ flow arrangement}\right) = \phi(\text{NTU}, C^*, \text{ flow arrangement}) \quad (3.50)$$

where $UA/C_{\min}$ (number of transfer units = NTU) is a *dimensionless parameter under designer's control*, $C_{\min}/C_{\max}$ (heat capacity rate ratio = C^*) is a *dimensionless operating parameter*, and the heat exchanger flow arrangement built into ϕ is also a designer's parameter. Note that we could have obtained the three nondimensional groups, from the variables and parameters of Eq. (3.7), directly by using Buckingham's π theorem (McAdams, 1954) for a given flow arrangement. In Section 11.2, a rigorous modeling approach is presented to determine the same dimensionless groups.

A comparison of Eqs. (3.50) and (3.7) will support the advantages claimed for the nondimensional analysis approach. For *each flow arrangement*, we have reduced a *seven-parameter problem*[†] [Eq. (3.7)], to a *three-parameter* problem, [Eq. (3.50)] for a given heat exchanger flow arrangement of two fluids.

Before discussing the physical significance of the two independent parameters C^* and NTU, let us introduce the definitions of the *temperature effectiveness* for the hot and cold fluids. The temperature effectiveness ε_h of the hot fluid is defined as a ratio of the temperature drop of the hot fluid to the fluid inlet temperature difference:

$$\varepsilon_h = \frac{T_{h,i} - T_{h,o}}{T_{h,i} - T_{c,i}} = \frac{\Delta T_h}{\Delta T_{max}} \tag{3.51}$$

Similarly, the temperature effectiveness of the cold fluid is defined as a ratio of the temperature rise of the cold fluid to the fluid inlet temperature difference:

$$\varepsilon_c = \frac{T_{c,o} - T_{c,i}}{T_{h,i} - T_{c,i}} = \frac{\Delta T_c}{\Delta T_{max}} \tag{3.52}$$

From an energy balance [Eq. (3.5)], and definitions of ε_h, and ε_c, it can be shown that

$$C_h \varepsilon_h = C_c \varepsilon_c \tag{3.53}$$

A comparison of temperature effectivenesses with Eq. (3.44) for the exchanger (heat transfer) effectiveness reveals that they are related by

$$\varepsilon = \frac{C_h}{C_{min}} \varepsilon_h = \begin{cases} \varepsilon_h & \text{for } C_h = C_{min} \\ \varepsilon_h/C^* & \text{for } C_h = C_{max} \end{cases} \tag{3.54}$$

$$\varepsilon = \frac{C_c}{C_{max}} \varepsilon_c = \begin{cases} \varepsilon_c & \text{for } C_c = C_{min} \\ \varepsilon_c/C^* & \text{for } C_c = C_{max} \end{cases} \tag{3.55}$$

Now let us define and discuss C^* and NTU.

3.3.2 Heat Capacity Rate Ratio C^*

C^* is simply a ratio of the smaller to larger heat capacity rate for the two fluid streams so that $C^* \leq 1$. A heat exchanger is considered *balanced* when $C^* = 1$:

$$C^* = \frac{C_{min}}{C_{max}} = \frac{(\dot{m}c_p)_{min}}{(\dot{m}c_p)_{max}} = \begin{cases} (T_{c,o} - T_{c,i})/(T_{h,i} - T_{h,o}) & \text{for } C_h = C_{min} \\ (T_{h,i} - T_{h,o})/(T_{c,o} - T_{c,i}) & \text{for } C_c = C_{min} \end{cases} \tag{3.56}$$

C^* is a heat exchanger operating parameter since it is dependent on mass flow rates and/or temperatures of the fluids in the exchanger. The C_{max} fluid experiences a smaller temperature change than the temperature change for the C_{min} fluid in the absence of extraneous heat losses, as can be found from the energy balance:

$$q = C_h \Delta T_h = C_c \Delta T_c \tag{3.57}$$

[†] In counting, we have considered only one dependent variable of Eq. (3.7).

where the temperature ranges ΔT_h and ΔT_c are

$$\Delta T_h = T_{h,i} - T_{h,o} \qquad \Delta T_c = T_{c,o} - T_{c,i} \qquad (3.58)$$

Let us reemphasize that for a condensing or evaporating fluid at ideally constant temperature, the ΔT range (rise or drop) is zero, and hence the heat capacity rate C approaches infinity for a finite $q = C\,\Delta T$. Since $C = \dot{m}c_p$, the effective specific heat of condensing or evaporating fluid is hence infinity. The $C^* = 0$ case then represents $C_{\min}$ as finite and $C_{\max}$ approaching infinity in the sense just discussed. The $C^* = 0$ case with $C_{\min} = (\dot{m}c_p)_{\min} = 0$ is not of practical importance, since $\dot{m} = 0$ in that case and hence there is no flow on the $C_{\min}$ side of the exchanger.

3.3.3 Number of Transfer Units NTU

The *number of transfer units* NTU is defined as a ratio of the overall thermal conductance to the smaller heat capacity rate:

$$\mathrm{NTU} = \frac{UA}{C_{\min}} = \frac{1}{C_{\min}} \int_A U\,dA \qquad (3.59)$$

If U is not a constant, the definition of the second equality applies. NTU may also be interpreted as the relative magnitude of the heat transfer rate compared to the rate of enthalpy change of the smaller heat capacity rate fluid. A substitution of the UA magnitude from Eq. (3.24) into Eq. (3.59) for U as constant results in

$$\mathrm{NTU} = \frac{1}{C_{\min}} \frac{1}{1/(\eta_o h_m A)_h + \mathbf{R}_{h,f} + \mathbf{R}_w + \mathbf{R}_{c,f} + 1/(\eta_o h_m A)_c} \qquad (3.60)$$

NTU designates the nondimensional *heat transfer size* or *thermal size* of the exchanger, and therefore it is a design parameter. NTU provides a compound measure of the heat exchanger size through the product of heat transfer surface area A and the overall heat transfer coefficient U. Hence, in general, NTU does not necessarily indicate the physical size of the exchanger. In contrast, the heat transfer surface area designates the physical size of a heat exchanger. A large value of NTU does not necessarily mean that a heat exchanger is large in physical size. As a matter of fact, the automotive gas turbine regenerator at the idle operating point may have $\mathrm{NTU} \approx 10$ and core volume $V \approx 0.01\,\mathrm{m}^3$, whereas a chemical plant shell-and-tube exchanger may have $\mathrm{NTU} \approx 1$ and $V \approx 100\,\mathrm{m}^3$. However, when comparing heat exchangers for a *specific* application, $U/C_{\min}$ approximately remains constant; and in this case, a higher NTU value means a heat exchanger larger in physical size. Hence, NTU is sometimes also referred to as a *heat exchanger size factor*. In general, higher NTU is obtained by increasing either U or A or both or by decreasing $C_{\min}$. Whereas a change in $C_{\min}$ affects NTU directly, a change in $C_{\max}$ (i.e., its flow rate) affects h on the $C_{\max}$ side. This in turn influences U and NTU. Thus, a change in the value of C^* may have direct or indirect effect on NTU.

NTU is also variously referred to as a *performance factor* or *thermal length* θ in the plate heat exchanger[†] literature, and as *reduced thermal flux* in the shell-and-tube

---

[†] In a PHE with chevron plates, a plate with high chevron angle β has a high heat transfer coefficient and high pressure drop in general and is referred to as a *hard plate*; in contrast, a *soft plate* has a low chevron angle β and low heat transfer coefficient and low pressure drop in general.

exchanger literature. Other names are given in the following text with appropriate inter-pretations.

At low values of NTU, the exchanger effectiveness is low. With increasing values of NTU, the exchanger effectiveness generally increases, and in the limit, it approaches a thermodynamic asymptotic value. Note that the perfect exchanger requires that NTU $\to \infty$ (because $A \to \infty$) for $q_{max} = C_{min} \Delta T_{max}$. The following approximate values of NTU will illustrate this point further.

Automobile radiator:	NTU $\approx 0.5 \to \varepsilon \approx 40\%$
Steam plant condenser:	NTU $\approx 1 \to \varepsilon \approx 63\%$
Regenerator for industrial gas turbine engine:	NTU $\approx 10 \to \varepsilon \approx 90\%$
Regenerator for Stirling engine:	NTU $\approx 50 \to \varepsilon \approx 98\%$
Regenerator for an LNG plant:	NTU $\approx 200 \to \varepsilon \approx 99\%$

Another interpretation of NTU as nondimensional residence time is as follows by substituting $C_{min} = \bar{C}_{min}/\tau_d$ in the definition of NTU:

$$\text{NTU} = \frac{1}{(1/UA)C_{min}} = \frac{\tau_d}{(1/UA)\bar{C}_{min}} = \frac{\tau_d}{\mathbf{R}_o \bar{C}_{min}} = \tau_d^* \tag{3.61}$$

Here $\mathbf{R}_o = 1/UA$ is the overall thermal resistance; $\bar{C}_{min} = (Mc_p)_{min} = C_{min}\tau_d$ is the mini-mum-side fluid heat capacitance ($M =$ fluid mass in the exchanger at an instant of time) in the exchanger at any instant of time; and τ_d is the dwell time, residence time, or transit time of a fluid particle passing through the exchanger. Thus, NTU can be interpreted as a *nondimensional residence time* or a ratio of residence time to the time constant of the C_{min} fluid in the exchanger at an instant. NTU might well be expected to play an important part in the transient problem! And from the viewpoint of an observer riding with a particle in the C_{min} stream, he or she would indeed have a transient temperature–time experience.

Yet another interpretation of NTU is as follows. NTU is related to ΔT_m from Eq. (3.45) with the use of Eq. (3.44) as

$$\text{NTU} = \frac{\Delta T_{max}\varepsilon}{\Delta T_m} = \frac{(T_{h,i} - T_{c,i})\varepsilon}{\Delta T_m} = \frac{C_h \Delta T_h}{C_{min} \Delta T_m} = \frac{C_c \Delta T_c}{C_{min} \Delta T_m} \tag{3.62}$$

Therefore, NTU is referred to as a *temperature ratio* (TR), where

$$\text{NTU} = \begin{cases} \dfrac{\Delta T_h}{\Delta T_m} = \dfrac{T_{h,i} - T_{h,o}}{\Delta T_m} & \text{for } C_h = C_{min} \\[4mm] \dfrac{\Delta T_c}{\Delta T_m} = \dfrac{T_{c,o} - T_{c,i}}{\Delta T_m} & \text{for } C_c = C_{min} \end{cases} \tag{3.63}$$

Thus, NTU $= \Delta T_{max,i}/\Delta T_m$, where $\Delta T_{max,i}$ is a maximum of ΔT_h and ΔT_c. When ΔT_m is equal to ΔT_h or ΔT_c whichever is larger, NTU $= 1$. Notice that Eq. (3.63) is convenient for small values of NTU, in which case $\Delta T_m \approx T_{m,h} - T_{m,c}$ is a good approximation; there is no need to calculate ΔT_{lm} or F (see Section 3.7). Here $T_{m,h}$ and $T_{m,c}$ are the arithmetic mean values of the corresponding terminal temperatures.

NTU is also directly related to the overall (total) Stanton number St_o formulated with U in place of h of Eq. (7.28) as

$$NTU = St_o \frac{4L}{D_h} \tag{3.64}$$

Thus, NTU can also be interpreted as a modified Stanton number. Note that here the hydraulic diameter D_h is defined as follows depending on the type of heat exchanger surface geometry involved.

$$D_h = \begin{cases} \dfrac{4 \times \text{flow area}}{\text{wetted perimeter}} = \dfrac{4A_o}{\mathbf{P}} = \dfrac{4A_o L}{A} \\[2ex] \dfrac{4 \times \text{core flow volume}}{\text{fluid contact surface area}} = \dfrac{4\mathbf{p}V}{A} = \dfrac{4\mathbf{p}}{\beta} = \dfrac{4\mathbf{p}}{\alpha} \end{cases} \tag{3.65}$$

where $\mathbf{p}$ is the porosity, a ratio of void volume to total volume of concern. Here the first definition of D_h is for constant cross-sectional flow passages in a heat exchanger. However, when flow area is expanding/contracting across flow cross sections along the flow length as in three-dimensional flow passages (such as in a corrugated perforated fin geometry of Fig. 1.29f), the more general second definition is applicable. In the second definition, $D_h = 4\mathbf{p}/\beta$ for plate-fin type and regenerative surfaces; for tube bundles and tube-fin surfaces, $D_h = 4\mathbf{p}/\alpha$. Note that heat transfer and pressure drop D_h magnitudes will be different if the heated and flow friction perimeters are different, as illustrated in the footnote of p. 9.

Equations (3.63) and (3.59) may also be interpreted as the number of transfer units required by the heat duty ($NTU = \Delta T_{max,i}/\Delta T_m$) and the number of transfer units achieved by the heat exchanger ($NTU = UA/C_{min}$), respectively.

The foregoing definitions and interpretations are for the overall NTU for the exchanger. The number of heat transfer units individually on the hot and cold sides of the exchanger may be defined as:

$$ntu_h = \frac{(\eta_o hA)_h}{C_h} \qquad ntu_c = \frac{(\eta_o hA)_c}{C_c} \tag{3.66}$$

We use ntu_h and ntu_c in Chapter 9 when outlining the solution procedure for the sizing problem for extended surface heat exchangers. The overall thermal resistance equation (3.24), in the absence of fouling resistances, can then be presented in terms of overall and individual number of transfer units as

$$\frac{1}{NTU} = \frac{1}{ntu_h(C_h/C_{min})} + \mathbf{R}_w C_{min} + \frac{1}{ntu_c(C_c/C_{min})} \tag{3.67}$$

3.4 EFFECTIVENESS–NUMBER OF TRANSFER UNIT RELATIONSHIPS

In the preceding section we demonstrated that ε is a function of NTU, C^*, and flow arrangement. We now derive this functional relationship for a single-pass counterflow

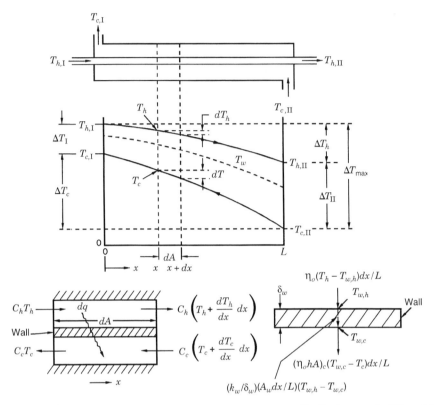

FIGURE 3.6 Counterflow heat exchanger with temperature distributions and differential elements for energy balance and rate equation development.

exchanger and then summarize similar functional relationships for single-pass and multi-pass flow arrangements.

3.4.1 Single-Pass Exchangers

3.4.1.1 Counterflow Exchanger.[†] Consider a counterflow heat exchanger with the temperature distributions for hot and cold fluids as shown in Fig. 3.6. The fluid temperatures on the left-hand end of the exchanger are denoted with the subscript I, and those on the other end with the subscript II.

In the analysis, we consider the overall counterflow exchanger as shown in Fig. 3.3 with only two passages. This is because the idealizations made in Section 3.2.1 (such as uniform velocities and temperatures at inlet, uniform surface area distribution, uniform U, etc.) are also invoked here. Thus, in fact, the hot-fluid passage shown represents all hot-fluid flow passages, and the cold-flow passage shown consists of all cold-fluid flow

---

[†] We derive the heat exchanger effectiveness in this section, and show how to obtain temperature distributions in Chapter 11.

passages. This is the reason that C_h and C_c are associated with the respective single-flow passages shown in Fig. 3.3 and not dC_h and dC_c.

Based on an energy balance consideration on the differential element dx,

$$dq = -C_h \, dT_h = -C_c \, dT_c \qquad (3.68)$$

Here T_h represents the bulk temperature of the hot fluid in the differential element dx of Fig. 3.6. T_c is defined in a similar manner for the cold fluid.

The rate equations applied individually to the dx length elements of the hot fluid, wall, and the cold fluid yield

$$dq = \begin{cases} (\eta_o h A)_h (T_h - T_{w,h}) \dfrac{dx}{L} & \text{for the hot fluid} \qquad (3.69) \\[2ex] \dfrac{k_w}{\delta_w} \dfrac{A_w \, dx}{L} (T_{w,h} - T_{w,c}) & \text{for the wall} \qquad (3.70) \\[2ex] (\eta_o h A)_c (T_{w,c} - T_c) \dfrac{dx}{L} & \text{for the cold fluid} \qquad (3.71) \end{cases}$$

Note that considering a general case, we have included the extended surface efficiency η_o in Eqs. (3.69) and (3.71), although fins are not shown in Fig. 3.6. After getting the expressions for individual temperature differences from Eqs. (3.69)–(3.71), adding them up, and rearranging, we get

$$dq = UA(T_h - T_c) \dfrac{dx}{L} \qquad (3.72)$$

Here U represents the local overall heat transfer coefficient for the element dA or dx. However, we treat this local U the same as U_m for the entire exchanger, and hence U will be treated as a constant throughout the exchanger. Hence UA in Eq. (3.72) is given by Eq. (3.20) or (3.24) without the fouling resistances on the hot and cold sides. If the fouling or other resistances are present, UA in Eq. (3.72) will have those resistances included.

Elimination of dq from Eqs. (3.68) and (3.72) will result in two ordinary differential equations. We need two boundary conditions to solve them and they are

$$T_h(x = 0) = T_{h,\mathrm{I}} \qquad T_c(x = L) = T_{c,\mathrm{II}} \qquad (3.73)$$

Now let us solve the set of Eqs. (3.68), (3.72), and (3.73) as follows to obtain a ratio of terminal temperature differences so that we can determine the exchanger effectiveness directly. We derive the temperature distributions in Section 11.2.1. Here, effectively, we will solve one differential equation that is based on the energy balance and rate equations. Substituting the values of dT_h and dT_c in terms of dq, Eq. (3.68) can be rearranged to

$$d(T_h - T_c) = \left(\dfrac{1}{C_c} - \dfrac{1}{C_h} \right) dq \qquad (3.74)$$

Eliminating dq from Eqs. (3.72) and (3.74) and rearranging yields

$$\dfrac{d(T_h - T_c)}{T_h - T_c} = -\left(1 - \dfrac{C_h}{C_c} \right) \dfrac{UA}{C_h} \dfrac{dx}{L} \qquad (3.75)$$

Integrating this equation from the hot-fluid inlet (section I) to the outlet section (section II) yields

$$\frac{T_{h,\text{II}} - T_{c,\text{II}}}{T_{h,\text{I}} - T_{c,\text{I}}} = \exp\left[-\frac{UA}{C_h}\left(1 - \frac{C_h}{C_c}\right)\right] \qquad (3.76)$$

It can be shown algebraically that the left-hand side of Eq. (3.76) is

$$\frac{T_{h,\text{II}} - T_{c,\text{II}}}{T_{h,\text{I}} - T_{c,\text{I}}} = \frac{1 - (T_{h,\text{I}} - T_{h,\text{II}})/(T_{h,\text{I}} - T_{c,\text{II}})}{1 - (T_{c,\text{I}} - T_{c,\text{II}})/(T_{h,\text{I}} - T_{c,\text{II}})} \qquad (3.77)$$

Now employ the definitions of the temperature effectivenesses from Eqs. (3.51) and (3.52), using the nomenclature of Fig. 3.6:

$$\varepsilon_h = \frac{T_{h,\text{I}} - T_{h,\text{II}}}{T_{h,\text{I}} - T_{c,\text{II}}} \qquad \varepsilon_c = \frac{T_{c,\text{I}} - T_{c,\text{II}}}{T_{h,\text{I}} - T_{c,\text{II}}} \qquad (3.78)$$

Substituting the definitions of ε_h and ε_c in Eq. (3.77), it reduces to

$$\frac{T_{h,\text{II}} - T_{c,\text{II}}}{T_{h,\text{I}} - T_{c,\text{I}}} = \frac{1 - \varepsilon_h}{1 - \varepsilon_c} = \frac{1 - \varepsilon_h}{1 - (C_h/C_c)\varepsilon_h} \qquad (3.79)$$

where Eq. (3.53) is used to obtain the last equality. Substituting Eq. (3.79) into (3.76) and rearranging, we get

$$\varepsilon_h = \frac{1 - \exp[-(UA/C_h)(1 - C_h/C_c)]}{1 - (C_h/C_c)\exp[-(UA/C_h)(1 - C_h/C_c)]} \qquad (3.80)$$

Now the temperature effectiveness of the cold fluid, ε_c, can be determined either directly by employing its definition [Eq. (3.52)], or by substituting Eq. (3.80) into (3.53). Using the second approach, gives us

$$\varepsilon_c = \frac{C_h}{C_c}\varepsilon_h = \frac{1 - \exp[(UA/C_c)(1 - C_c/C_h)]}{(C_c/C_h) - \exp[(UA/C_c)(1 - C_c/C_h)]} \qquad (3.81)$$

Notice that the argument of the exponential terms has been rearranged. By multiplying both numerator and denominator by $\exp\{-(UA/C_c)[1 - (C_c/C_h)]\}$ and rearranging, we get

$$\varepsilon_c = \frac{1 - \exp[-(UA/C_c)(1 - C_c/C_h)]}{1 - (C_c/C_h)\exp[-(UA/C_c)(1 - C_c/C_h)]} \qquad (3.82)$$

A comparison of Eqs. (3.80) and (3.82) reveals that Eq. (3.82) can be obtained directly from Eq. (3.80) by replacing the subscripts h with c and c with h.

To generalize the solution, let $C_{\min} = C_c$, $C^* = C_c/C_h$ and $\text{NTU} = UA/C_{\min} = UA/C_c$. In this case, $\varepsilon = \varepsilon_c$ from Eq. (3.52), and Eq. (3.82) reduces to

$$\varepsilon = \varepsilon_{cf} = \frac{1 - \exp[-\text{NTU}(1 - C^*)]}{1 - C^*\exp[-\text{NTU}(1 - C^*)]} \qquad (3.83)$$

However, if $C_{min} = C_h$, then $C^* = C_h/C_c$, NTU $= UA/C_{min} = UA/C_h$. In this case, $\varepsilon = \varepsilon_h$ from Eq. (3.51), and Eq. (3.80) reduces to the exchanger effectiveness expression of Eq. (3.83). *Thus, regardless of which fluid has the minimum heat capacity rate, the ε-NTU expression for the counterflow exchanger is given by Eq. (3.83).*

Two limiting cases of interest of Eq. (3.83) are $C^* = 0$ and 1. For the special case of $C^* = 0$ (an evaporator or a condenser), the exchanger effectiveness ε, Eq. (3.83), reduces to

$$\varepsilon = 1 - \exp(-\text{NTU}) \tag{3.84}$$

Note that when $C^* = 0$, the temperature of the C_{max} fluid remains constant throughout the exchanger, as shown in Fig. 3.1b and c. In this case, the C_{min} fluid can have any arbitrary flow arrangement. Hence Eq. (3.84) is valid for *all* flow arrangements when $C^* = 0$.

For the special case of $C^* = 1$, Eq. (3.83) reduces to the 0/0 form. Hence, using l'Hospital's rule (taking the derivatives of the numerator and the denominator separately with respect to C^* and substituting $C^* = 1$ in the resultant equation), we get

$$\varepsilon = \frac{\text{NTU}}{1 + \text{NTU}} \tag{3.85}$$

For all $0 < C^* < 1$, the value of ε falls in between those of Eqs. (3.84) and (3.85), as shown in Fig. 3.7.

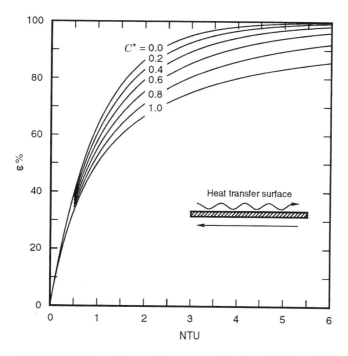

FIGURE 3.7   Counterflow exchanger ε as a function of NTU and C^*.

By inverting Eq. (3.83), NTU for a counterflow exchanger can be expressed explicitly as a function of ε and C^*:

$$\text{NTU} = \frac{1}{1 - C^*} \ln \frac{1 - C^*\varepsilon}{1 - \varepsilon} \tag{3.86}$$

which for $C^* = 0$ and $C^* = 1$ (using l'Hospital's rule) are given by

For $C^* = 0$: $\quad \text{NTU} = \ln \dfrac{1}{1 - \varepsilon} \quad$ or $\quad \text{NTU} = \ln \dfrac{T_{h,i} - T_{c,i}}{T_{h,i} - T_{c,o}} \quad$ for $T_{h,i}$ = constant

$$\tag{3.87}$$

For $C^* = 1$: $\quad \text{NTU} = \dfrac{\varepsilon}{1 - \varepsilon} = \dfrac{T_{h,i} - T_{h,o}}{T_{h,i} - T_{c,o}} = \dfrac{T_{c,o} - T_{c,i}}{T_{h,o} - T_{c,i}} \tag{3.88}$

The ε-NTU results for the counterflow exchanger are presented in Fig. 3.7. The following important observations may be made by reviewing Fig. 3.7:

1. The heat exchanger effectiveness ε increases monotonically with increasing values of NTU for a specified C^*. For all C^*, $\varepsilon \to 1$ as NTU $\to \infty$. Note that this is true for the *counterflow* exchanger, and ε may *not* necessarily approach unity for many other exchanger flow arrangements, as can be found from Tables 3.3 and 3.6.
2. The exchanger effectiveness ε increases with decreasing values of C^* for a specified NTU.
3. For $\varepsilon \lesssim 40\%$, the heat capacity rate ratio C^* does not have a significant influence on the exchanger effectiveness ε. As a matter of fact, it can be shown that when NTU $\to 0$, the effectiveness for *all* flow arrangements reduces to

$$\varepsilon \approx \text{NTU}[1 - \tfrac{1}{2}\text{NTU}(1 + C^*)] \tag{3.89}$$

   This formula will yield highly accurate results for decreasing value of NTU for NTU $\lesssim 0.4$.
4. Although not obvious from Fig. 3.7, but based on Eq. (3.84), the ε vs. NTU curve is identical for *all* exchanger flow arrangements, including those of Tables 3.3 and 3.6 for $C^* = 0$.
5. Heat exchanger effectiveness increases with increasing NTU as noted above in item 1, but at a diminishing rate. For example, increasing NTU from 0.5 to 1 at $C^* = 1$ increases ε from 0.333 to 0.50; a 50% increase in ε for a 100% increase in NTU (or approximately 100% increase in surface area). Increasing NTU from 1 to 2 at $C^* = 1$ increases ε from 0.50 to 0.667, a 33% increase in ε for a 100% increase in NTU (or size). Increasing NTU from 2 to 4 at $C^* = 1$ increases ε from 0.667 to 0.8, a 20% increase in ε for a 100% increase in NTU (or size). This clearly shows a diminishing rate of increase in ε with NTU.
6. Because of the asymptotic nature of the ε-NTU curves, a significant increase in NTU and hence in the exchanger size is required for a small increase in ε at high values of ε. For example, for $C^* = 1$, $\varepsilon = 90\%$ at NTU $= 9$, and $\varepsilon = 92\%$ at NTU $= 11.5$. Thus an increase of 2 percentage points in ε requires an increase

of 28% in NTU and hence in the exchanger size for the same surface area and flow rates. Alternatively, a larger increase in NTU (or the size of the exchanger) is required to compensate for the same (or small) amount of heat loss to the surroundings at high values of ε in comparison to that for a lower ε exchanger.

The counterflow exchanger has the highest exchanger effectiveness ε for specified NTU and C^* of that for all other exchanger flow arrangements. Thus, for a given NTU and C^*, maximum heat transfer performance is achieved for counterflow; alternatively, the heat transfer surface area is utilized most efficiently for counterflow compared to all other flow arrangements.

Should we design heat exchangers at high effectiveness for maximum heat transfer? Let us interpret the results of Fig. 3.7 from the industrial point of view. It should be emphasized that many industrial heat exchangers are *not* counterflow, and the following discussion in general is valid for other flow arrangements.

When the heat exchanger cost is an important consideration, most heat exchangers are designed in the approximate linear range of ε-NTU curves (NTU ≤ 2 or $\varepsilon \leq 60\%$, such as in Fig. 3.7) that will meet the required heat duty with appropriate values of $C_{\min}$ and $\Delta T_{\max}$. The reason for this is that an increase in the exchanger size (NTU) will increase with ε approximately linearly and hence will result in a "good" return on the investment of added surface area. However, when the exchanger is a component in the system, and an increase in the exchanger effectiveness has a significant impact on reducing the system operating cost compared to an increase in the exchanger cost, the exchangers are designed for high effectivenesses. For example, a 1% increase in the regenerator effectiveness approximately increases about 0.5% the thermal efficiency of an electricity generating gas turbine power plant. This increase in the power plant efficiency will translate into millions of dollars' worth of additional electricity generated annually. Hence, the cost of the regenerator represents only a small fraction of the annual savings generated. Therefore, most gas turbine regenerators are designed at about $\varepsilon = 90\%$. A 1% increase in the effectiveness of a cryogenic heat exchanger used in an air separation plant will decrease the compressor power consumption by about 5%. This will translate into an annual savings of more than $500,000 (in 1995 dollars) in operating costs for a 1000-ton oxygen plant. That is why most cryogenic heat exchangers are designed for $\varepsilon \approx 95\%$.

3.4.1.2 Exchangers with Other Flow Arrangements. The derivation of ε-NTU formulas for other heat exchanger flow arrangements becomes more difficult as the flow arrangement becomes more complicated (Sekulić et al., 1999). The solutions have been obtained in the recent past for many complicated flow arrangements, as in Table 3.6 and Shah and Pignotti (1989). Temperature distributions for counterflow, parallelflow, and unmixed–unmixed crossflow exchangers are derived in Sections 11.2.1 and 11.2.4.

ε-NTU Results. Table 3.3 and Figs. 3.8 through 3.11 show the ε-NTU formulas and results for some heat exchanger flow arrangements. NTU can be expressed explicitly as a function of ε and C^* only for some flow arrangements and those formulas are presented in Table 3.4. For all other flow arrangements, NTU is an implicit function of ε and C^* and can be determined iteratively or by using appropriate analytical methods to solve the equation $f(\text{NTU}) = 0$. C^* cannot be expressed explicitly as a function of ε and NTU for any exchanger flow arrangements.

TABLE 3.3 ε–NTU Formulas and Limiting Values of ε for $C^* = 1$ and NTU $\to \infty$ for Various Exchanger Flow Arrangements

Flow Arrangement	ε-NTU Formulas	ε-NTU Formulas for $C^* = 1$	Asymptotic Value of ε When NTU $\to \infty$
Counterflow	$\varepsilon = \dfrac{1 - \exp[-\mathrm{NTU}(1 - C^*)]}{1 - C^* \exp[-\mathrm{NTU}(1 - C^*)]}$	$\varepsilon = \dfrac{\mathrm{NTU}}{1 + \mathrm{NTU}}$	$\varepsilon = 1$ fop all C^*
Parallelflow	$\varepsilon = \dfrac{1 - \exp[-\mathrm{NTU}(1 + C^*)]}{1 + C^*}$	$\varepsilon = \frac{1}{2}[1 - \exp(-\mathrm{NTU})]$	$\varepsilon = \dfrac{1}{1 + C^*}$
Crossflow, both fluids unmixed	$\varepsilon = 1 - \exp(-\mathrm{NTU})$ $\quad - \exp[-(1 + C^*)\mathrm{NTU}]\displaystyle\sum_{n=1}^{\infty} C^{*n} P_n(\mathrm{NTU})$ $P_n(y) = \dfrac{1}{(n+1)!}\displaystyle\sum_{j=1}^{n}\dfrac{(n+1-j)}{j!}\, y^{n+j}$	Same as general formula with $C^* = 1$	$\varepsilon = 1$ for all C^*
Crossflow, one fluid mixed, other unmixed	For $C_{\min}$ mixed, $C_{\max}$ unmixed, $\varepsilon = 1 - \exp\{-[1 - \exp(-\mathrm{NTU}\cdot C^*)]/C^*\}$ For $C_{\max}$ mixed, $C_{\min}$ unmixed, $\varepsilon = \dfrac{1}{C^*}(1 - \exp\{-C^*[1 - \exp(-\mathrm{NTU})]\})$	$\varepsilon = 1 - \exp\{-[1 - \exp(-\mathrm{NTU})]\}$	For $C_{\min}$ mixed, $\quad \varepsilon = 1 - \exp(-1/C^*)$ For $C_{\max}$ mixed, $\quad \varepsilon = [1 - \exp(-C^*)]/C^*$
Crossflow, both fluids mixed	$\varepsilon = \dfrac{1}{\dfrac{1}{1 - \exp(-\mathrm{NTU})} + \dfrac{C^*}{1 - \exp(-\mathrm{NTU}\cdot C^*)} - \dfrac{1}{\mathrm{NTU}}}$	$\varepsilon = \dfrac{1}{\dfrac{2}{1 - \exp(-\mathrm{NTU})} - 1/\mathrm{NTU}}$	$\varepsilon = \dfrac{1}{1 + C^*}$
1-2 shell-and-tube exchanger; shell fluid mixed; TEMA E shell	$\varepsilon = \dfrac{2}{(1 + C^*) + (1 + C^{*2})^{1/2}\coth(\Gamma/2)}$ where $\Gamma = \mathrm{NTU}(1 + C^{*2})^{1/2}$ $\coth(\Gamma/2) = (1 + e^{-\Gamma})/(1 - e^{-\Gamma})$	$\varepsilon = \dfrac{2}{2 + \sqrt{2}\coth(\Gamma/2)}$ where $\Gamma = \sqrt{2}\,\mathrm{NTU}$	$\varepsilon = \dfrac{2}{(1 + C^*) + (1 + C^{*2})^{1/2}}$

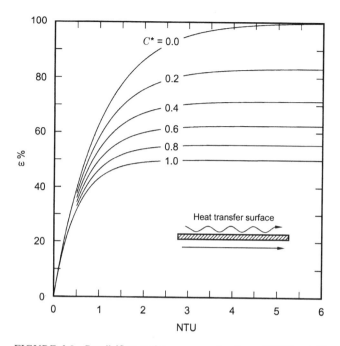

FIGURE 3.8   Parallelflow exchanger ε as a function of NTU and C^*.

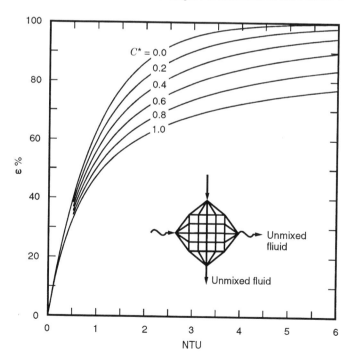

FIGURE 3.9   Unmixed–unmixed crossflow exchanger ε as a function of NTU and C^*.

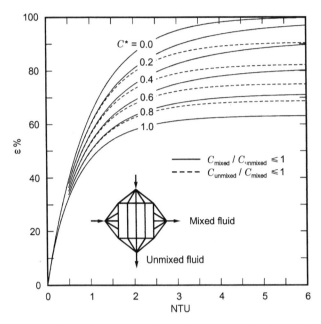

FIGURE 3.10   Unmixed–mixed crossflow exchanger ε as a function of NTU and C^*.

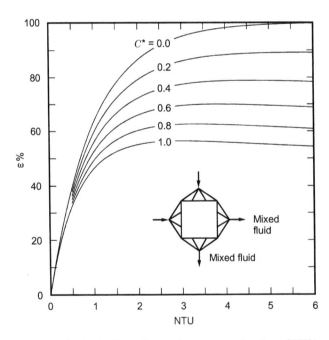

FIGURE 3.11   Mixed–mixed crossflow exchanger ε as a function of NTU and C^*.

TABLE 3.4   NTU as an Explicit Function of ε and C^* for Known Heat Exchanger Flow Arrangements

Flow Arrangement	Formula
Counterflow	$NTU = \dfrac{1}{1 - C^*} \ln \dfrac{1 - C^*\varepsilon}{1 - \varepsilon} \quad (C^* < 1)$ $NTU = \dfrac{\varepsilon}{1 - \varepsilon} \quad (C^* = 1)$
Parallelflow	$NTU = -\dfrac{\ln[1 - \varepsilon(1 + C^*)]}{1 + C^*}$
Crossflow (single pass)	
C_{max} (mixed), C_{min} (unmixed)	$NTU = -\ln\left[1 + \dfrac{1}{C^*}\ln(1 - C^*\varepsilon)\right]$
C_{min} (mixed), C_{max} (unmixed)	$NTU = -\dfrac{1}{C^*}\ln[1 + C^*\ln(1 - \varepsilon)]$
1–2 TEMA E Shell-and-Tube	$NTU = \dfrac{1}{D}\ln \dfrac{2 - \varepsilon(1 + C^* - D)}{2 - \varepsilon(1 + C^* + D)}$ where $D = (1 + C^{*2})^{1/2}$
All exchangers with $C^* = 0$	$NTU = -\ln(1 - \varepsilon)$

From a review of Fig. 3.11, it is found for $C^* > 0$ that the effectiveness of the mixed–mixed crossflow exchanger increases with increasing NTU up to some value of NTU beyond which increasing NTU results in a decrease in ε. The reason for this unexpected behavior is that due to the idealized complete mixing of both fluids, the hot-fluid temperature is lower than the cold-fluid temperature in a part of the exchanger, resulting in a temperature cross (discussed in Section 3.6.1.2) and existence of entropy generation extrema (discussed in Section 11.4). Hence, heat transfer takes place in the reverse direction, which in turn reduces the exchanger effectiveness.

3.4.1.3   Interpretation of ε-NTU Results. As noted in Eq. (3.7), the heat exchanger design problem in general has six independent and one or more dependent variables. For analysis and concise presentation of the results, we arrived at a total of three dimensionless groups (ε, NTU, and C^*) for this method. However, from the industrial viewpoint, the final objective in general is to determine dimensional variables and parameters of interest in any heat exchanger problem. The nondimensionalization is the shortest "road" to obtaining the results of interest. In this regard, let us review Eq. (3.35) again:

$$q = \varepsilon C_{min}(T_{h,i} - T_{c,i}) = \varepsilon(\dot{m}c_p)_{min}(T_{h,i} - T_{c,i}) \tag{3.90}$$

If we increase the flow rate on the $C_c = C_{min}$ side (such as on the air side of a water-to-air heat exchanger), it will increase C^* ($= C_{min}/C_{max}$), reduce NTU ($= UA/C_{min}$), and hence will reduce ε nonlinearly at a lower rate (see Figs. 3.7 to 3.11). Thus from Eq. (3.90), q will increase linearly with increasing $C_{\min}$ and q will decrease nonlinearly at a lower rate with decreasing ε. The net effect will be that q will increase. From Eq. (3.5),

$$q = C_c(T_{c,o} - T_{c,i}) = C_h(T_{h,i} - T_{h,o}) \tag{3.91}$$

Thus, overall, a lower increase in q and a linear increase in C_c will yield reduced $(T_{c,o} - T_{c,i})$; it means that $T_{c,o}$ will be lower for a given $T_{c,i}$. Again from Eq. (3.91), $T_{h,o}$ will also be lower with increased q for given C_h and $T_{h,i}$. These results are shown qualitatively by an arrow up or down in Table 3.5 depending on whether the particular value increases or decreases. It is interesting to note that both $T_{h,o}$ and $T_{c,o}$ in this table can either increase, decrease, or one increases and the other decreases! A thorough understanding of the qualitative trends of the variables of this table is necessary to interpret the experimental/analytical results and anomalies or unexpected test results.

Reviewing Eq. (3.90), the desired heat transfer rate q in a heat exchanger can be obtained by varying either ε of $C_{\min}$ since the ITD ($= T_{h,i} - T_{c,i}$) is generally a fixed design value in many applications. The effectiveness ε can be varied by changing the surface area A (the size of the exchanger) and hence NTU, impacting the capital cost; ε increases with A or NTU nonlinearly and asymptotically. The minimum heat capacity rate $C_{\min}$ ($= \dot{m}c_p$) can be varied by changing the mass flow rate through a fluid pumping device (fan, blower, or pump); the change in $C_{\min}$ has a direct (linear) impact on q. Hence, in general, specified percentage change in $C_{\min}$ will have more impact on q than the change in ε due to the same percentage change in A or the size of the exchanger. Current practice in many industries is to maintain the heat exchanger effectiveness 60% or lower to optimize the heat exchanger cost. The desired heat transfer rate is then obtained by using a proper capacity fluid pumping device, which in turn has a major impact on the operating cost. Thus the heat exchanger and the fluid pumping device are selected individually. However, to get the desired heat transfer rate q in an exchanger, the better cost effective approach would be to select optimally the heat exchanger size (NTU) and the fluid pumping power to get the appropriate flow rate (and hence $C_{\min}$).

Also, a review of Eq. (3.90) indicates that if there is an insufficient fluid flow rate $\dot{m}$, the exchanger will not provide the desired performance even if $\varepsilon \rightarrow 100\%$ for a given

TABLE 3.5 Effect of Increasing One Independent Variable at a Time on the Remaining Heat Exchanger Variables Based on the ε-NTU Relationships[a]

Specific Variable with Increasing Value	Variables Affected					
	ε	NTU	C^*	q	$T_{h,o}$	$T_{c,o}$
$\dot{m}_h$ or C_h	↑	↑	↓	↑	↑	↑
$\dot{m}_c$ or C_c	↓	↓	↑	↑	↓	↓
$T_{h,i}$	—	—	—	↑	↑	↑
$T_{c,i}$	—	—	—	↓	↑	↑
h_c or A_c	↑	↑	—	↑	↓	↑
h_h or A_h	↑	↑	—	↑	↓	↑

[a] $C_{\min} = C_c$ for this table. ↑, increase; —, no change; ↓, decrease.

ΔT_{max}. This is important when the fluid flow is blocked partially or totally in a heat exchanger, due to inadequate manufacturing or specific application/installation in a system.

Equation (3.90) also implies that a perfectly designed exchanger may not operate properly in a system if the heat duty of the exchanger is dictated by the system. For example, if the engine heat rejection is reduced due to the use of a smaller engine in a given automobile, correspondingly, $(T_{h,i} - T_{c,i})$ will reduce for the same fan, and the passenger compartment will not warm up to a comfortable level with the same automotive heater (i.e., given ε). To be more specific, consider the coolant temperature $T_{h,i}$ from the engine lower than the desired value (such as 75°C vs. desired value of 90°C) with a smaller engine. Hence, the ambient air will not be heated to the desired value (such as 50°C vs. the desired value of 62°C) even though the heater is designed properly (i.e., desired ε) due to the reduced ITD.

3.4.1.4 Stream Symmetry. A two-fluid heat exchanger is considered to be *stream symmetric* if the exchanger effectiveness ε (or P) remains invariant by switching the two fluids (such as the tube fluid to the shell side, and the shell fluid to the tube side). Alternatively, the log-mean temperature difference correction factor F (defined in Section 3.7.2) remains the same for the original and stream-switched configurations. Thus, for a stream symmetric exchanger, as shown in Eq. (3.36),

$$\varepsilon = \phi \, (\mathrm{NTU}, C^*, \text{flow arrangement}) \qquad (3.92)$$

regardless of whether the C_{min} fluid is on the tube or shell side in a shell-and-tube exchanger, for example.

All geometrically symmetrical exchangers are stream symmetric such as single-pass counterflow, parallelflow, and mixed–mixed and unmixed–unmixed crossflow exchangers. Some geometrically asymmetric exchangers (such as 1–2 and 1–∞ TEMA E, 1–1 TEMA G, and 1–∞ TEMA J exchangers, all with shell fluid mixed) are also stream symmetric exchangers.

If an exchanger is stream asymmetric, the formula for the exchanger effectiveness ε is different for the two fluid sides of the exchanger (such as the shell side and tube side). This means that

$$\varepsilon = \begin{cases} \phi_1 \, (\mathrm{NTU}, \; C^*, \; \text{flow arrangement}) & \text{for fluid 1 as } C_{min} \text{ fluid} & (3.93a) \\ \phi_2 \, (\mathrm{NTU}, \; C^*, \; \text{flow arrangement}) & \text{for fluid 2 as } C_{min} \text{ fluid} & (3.93b) \end{cases}$$

where ϕ_1 and ϕ_2 are different functional relationships. Some examples of stream asymmetric exchangers are unmixed–mixed or mixed–unmixed crossflow, 1–2 TEMA E with shell fluid unmixed, 1–n TEMA E (with $3 \le n < \infty$) with shell fluid mixed or unmixed, 1–2 TEMA G, 1–1 and 1–2 TEMA H, and 1–1, 1–2 and 1–4 TEMA J exchangers with shell fluid mixed or unmixed. All heat exchanger arrays with parallel coupling and those with compound coupling that include parallel coupling are stream asymmetric.

Example 3.2 In an oil-to-water heat exchanger, the oil enters the exchanger at 100°C with a heat capacity rate of 3700 W/K. Water is available at 15°C and 0.6 kg/s. Determine the exit temperatures in (a) counterflow, and (b) parallelflow arrangements for

$U = 500\,\text{W/m}^2 \cdot \text{K}$ and surface area of $10\,\text{m}^2$. Consider $c_p = 1.88$ and $4.19\,\text{J/g} \cdot \text{K}$ for oil and water, respectively. If the ratio of convection thermal resistances of oil to water is 1.2, and the wall and fouling resistances are negligible, calculate the wall temperature at each end of the counterflow and parallelflow exchangers.

SOLUTION

Problem Data and Schematic: Fluid flow rates, inlet temperatures, and overall heat transfer coefficient are provided for counterflow and parallelflow arrangements (Fig. E3.2A). Also, total heat transfer area and the ratio of thermal resistances of the two fluid sides are given.

Determine: Wall and fluid exit temperatures at each end $(x/L = 0$ and $1)$ of the counterflow and parallelflow arrangements.

Assumptions: The assumptions invoked in Section 3.2.1 are valid, wall and fouling resistances and negligible, the ratio of thermal resistances of oil to water is uniform throughout the exchanger.

Analysis: Knowing the inlet temperatures, we could use the definition of heat exchanger effectiveness ε to find the outlet temperatures of various exchangers. To determine ε, we need to determine C^* and NTU. For the determination of C^*, the heat capacity rates are

$$C_h = 3700\,\text{W/K}$$

$$C_c = (\dot{m}c_p)_c = 0.6\,\text{kg/s} \times (4.19\,\text{J/g} \cdot \text{K} \times 1000\,\text{g/kg}) = 2514\,\text{W/K}$$

where the subscripts h and c denote the hot (oil) and cold (water) sides, respectively. Hence, $C_{\min} = 2514\,\text{W/K}$. Using the definitions of C^* and NTU from Eqs. (3.56) and (3.59), we have

$$C^* = \frac{C_c}{C_h} = \frac{2514\,\text{W/K}}{3700\,\text{W/K}} = 0.679$$

$$\text{NTU} = \frac{UA}{C_{\min}} = \frac{(500\,\text{W/m}^2 \cdot \text{K})(10\,\text{m}^2)}{2514\,\text{W/K}} = 1.989$$

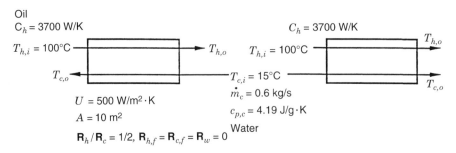

FIGURE E3.2A

(a) *Counterflow Exchanger:* The exchanger effectiveness for the counterflow exchanger, from Eq. (3.83), is

$$\varepsilon = \frac{1 - \exp[-NTU(1 - C^*)]}{1 - C^* \exp[-NTU(1 - C^*)]} = \frac{1 - \exp[-1.989(1 - 0.679)]}{1 - 0.679 \exp[-1.989(1 - 0.679)]} = 0.736$$

Since water is the C_{min} fluid, according to the definition of the effectiveness, Eq. (3.44),

$$\varepsilon = \frac{T_{c,o} - T_{c,i}}{T_{h,i} - T_{c,i}} = \frac{(T_{c,o} - 15)^\circ C}{(100 - 15)^\circ C} = 0.736$$

Hence,

$$T_{c,o} = 15^\circ C + 0.736(100 - 15)^\circ C = 77.6^\circ C \qquad Ans.$$

Employing the energy balance equation (3.5), we could find the heat transfer rate on the water side as

$$q = C_c(T_{c,o} - T_{c,i}) = (2514\,W/K)(77.6 - 15)^\circ C = 157.4 \times 10^3\,W$$

The oil outlet temperature from the energy balance equation (3.5) is then

$$T_{h,o} = T_{h,i} - \frac{q}{C_h} = 100^\circ C - \frac{157.4 \times 10^3\,W}{3700\,W/K} = 57.5^\circ C \qquad Ans.$$

To determine the wall temperature at each end of the exchanger (designated by a second subscript I and II), we use Eq. (3.34) with $\mathbf{R}_{h,f}$ and $\mathbf{R}_{c,f} = 0$, and $\mathbf{R}_h/\mathbf{R}_c = 1.2$.

$$T_{w,I} = \frac{T_{h,i} + (\mathbf{R}_h/\mathbf{R}_c)T_{c,o}}{1 + (\mathbf{R}_h/\mathbf{R}_c)} = \frac{100^\circ C + 1.2 \times 77.6^\circ C}{1 + 1.2} = 87.8^\circ C \qquad Ans.$$

$$T_{w,II} = \frac{T_{h,o} + (\mathbf{R}_h/\mathbf{R}_c)T_{c,i}}{1 + (\mathbf{R}_h/\mathbf{R}_c)} = \frac{57.5^\circ C + 1.2 \times 15^\circ C}{1 + 1.2} = 34.3^\circ C \qquad Ans.$$

(b) *Parallelflow Exchanger:* The heat exchanger effectiveness for a parallelflow exchanger, from Table 3.3, is

$$\varepsilon = \frac{1 - \exp[-NTU(1 + C^*)]}{1 + C^*} = \frac{1 - \exp[-1.989(1 + 0.679)]}{1 + 0.679} = 0.574$$

The water outlet is then calculated from the definition of the effectiveness as

$$\varepsilon = \frac{T_{c,o} - T_{c,i}}{T_{h,i} - T_{c,i}} = \frac{(T_{c,o} - 15)^\circ C}{(100 - 15)^\circ C} = 0.574$$

so that

$$T_{c,o} = 63.8^\circ C \qquad Ans.$$

The heat transfer rate on the water side is then

$$q = C_c(T_{c,o} - T_{c,i}) = (2514\,\text{W/K})(63.8 - 15)°\text{C} = 122.7 \times 10^3\,\text{W}$$

Subsequently, the oil outlet temperature is

$$T_{h,o} = T_{h,i} - \frac{q}{C_h} = 100°\text{C} - \frac{122.7 \times 10^3\,\text{W}}{3700\,\text{W/K}} = 66.8°\text{C} \qquad Ans.$$

The wall temperature at each end of the exchanger is

$$T_{w,\text{I}} = \frac{T_{h,i} + (\mathbf{R}_h/\mathbf{R}_c)T_{c,i}}{1 + (\mathbf{R}_h/\mathbf{R}_c)} = \frac{100°\text{C} + 1.2 \times 15°\text{C}}{1 + 1.2} = 53.6°\text{C} \qquad Ans.$$

$$T_{w,\text{II}} = \frac{T_{h,o} + (\mathbf{R}_h/\mathbf{R}_c)T_{c,o}}{1 + (\mathbf{R}_h/\mathbf{R}_c)} = \frac{66.8°\text{C} + 1.2 \times 63.8°\text{C}}{1 + 1.2} = 65.2°\text{C} \qquad Ans.$$

Discussion and Comments: In this problem, we have compared two thermodynamically extreme performance exchangers (counterflow and parallelflow) at the same NTU, C^*, and fluid inlet temperatures. The following observations may be made based on the results presented graphically in Fig. E3.2B.

- The wall temperature across the exchanger length varies from 87.8 to 34.3°C in the counterflow exchanger and from 53.6 to 65.2°C in the parallelflow exchanger. Thus longitudinal wall temperature distribution is more uniform in the parallelflow exchanger compared to the counterflow exchanger.
- The lowest wall temperatures are 34.3 and 53.6°C in the counterflow and parallelflow exchangers, respectively. Thus, the lowest wall temperature in the parallelflow exchanger is higher than that for the counterflow exchanger.
- Similarly, the highest wall temperature in the parallelflow exchanger is lower than that for the counterflow exchanger (65.2°C vs. 87.8°C).

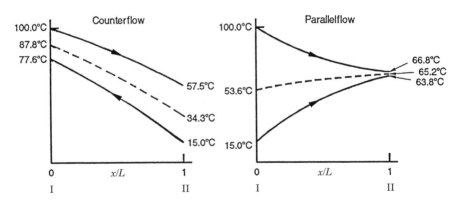

FIGURE E3.2B

It can be shown that these observations are true for comparing a parallelflow exchanger to an exchanger with any other flow arrangement for specified NTU, C^*, and inlet temperatures.

As expected, the parallelflow exchanger provides lower heat transfer rate ($78\% = 122.7\,\mathrm{kW} \times 100/157.4\,\mathrm{kW}$) than that of a counterflow exchanger. However, if the exchanger is designed for the effectiveness lower than 40%, there is not a significant difference in the exchanger effectiveness and heat transfer rate between parallelflow and counterflow exchangers at equal NTU and inlet temperatures. This is an industrially important conclusion for low effectiveness waste heat recovery from exhaust gases having SO_2 as one of the constituents. The sulfuric acid condensation in a heat exchanger can be prevented at atmospheric pressure if the minimum wall temperature is maintained above about 150°C. For this case, the parallelflow exchanger becomes an attractive solution since its lowest wall temperature is higher than that of any other exchanger flow arrangement.

Example 3.3   One important design point for a radiator design is to cool the engine at 50 km/h on a 7% grade road. Your responsibility as a design engineer is to make sure that the coolant (50% water–50% glycol) at the radiator inlet (top tank) does not exceed 120°C temperature at 100 kPa gauge radiator cap pressure. Determine the radiator top tank temperature for the following conditions: engine heat rejection rate $q = 35\,\mathrm{kW}$, airflow rate 0.75 kg/s, air inlet temperature 53°C, and water–glycol flow rate 1.4 kg/s. For this radiator, $UA = 1180\,\mathrm{W/K}$. The specific heats for the air and the water–glycol mixture are 1009 and 3664 J/kg · K respectively. What will be the outlet temperature of the water–glycol mixture? Consider the radiator with both fluids unmixed.

SOLUTION

Problem Data and Schematic: Fluid flow rates, inlet temperature of the cold fluid, heat transfer rate, and the total thermal conductance are given (see Fig. E3.3).

Determine: The inlet temperature of the hot fluid (water–glycol mixture).

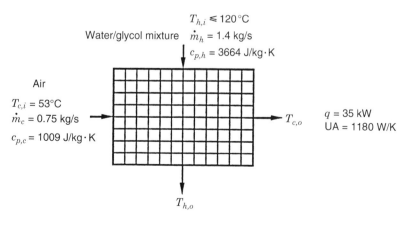

FIGURE E3.3

Assumptions: The fluid properties and UA are constant, and the maximum inlet temperature for the hot fluid is 120°C at 100 kPa, beyond which it will boil in the engine.

Analysis: We could find the NTU from the information given. But first, we have to find C^* and C_{min}:

$$C_{air} = C_c = (\dot{m}c_p)_{air} = 0.75\,\text{kg/s} \times 1009\,\text{J/kg} \cdot \text{K} = 756.75\,\text{W/K} = C_{min}$$

$$C_{liquid} = C_h = (\dot{m}c_p)_{liquid} = 1.4\,\text{kg/s} \times 3664\,\text{J/kg} \cdot \text{K} = 5129.6\,\text{W/K}$$

$$C^* = \frac{C_{air}}{C_{liquid}} = \frac{756.75\,\text{W/K}}{5129.6\,\text{W/K}} = 0.148$$

$$\text{NTU} = \frac{UA}{C_{min}} = \frac{1180\,\text{W/K}}{756.75\,\text{W/K}} = 1.559$$

From Fig. 3.9 or the Table 3.3 formula for an unmixed–unmixed crossflow exchanger, we get

$$\varepsilon = 0.769$$

Hence, $T_{h,i}$ from Eq. (3.35) is given by

$$T_{h,i} = T_{c,i} + \frac{q}{\varepsilon C_{min}} = 53°\text{C} + \frac{35\,\text{kW} \times 1000\,\text{W/kW}}{0.769 \times 756.75\,\text{W/K}} = 113.1°\text{C} \qquad \textit{Ans.}$$

Since this is less than 120°C, the design is safe. If we would have determined $T_{h,i} > 120°\text{C}$, we would have changed the radiator design (such as increasing A and hence UA and NTU) so that $T_{h,i} \leq 120°\text{C}$.

Using the energy balance equation (3.5), we could find the water–glycol mixture outlet temperature as follows:

$$T_{h,o} = T_{h,i} - \frac{q}{C_h} = 113.1°\text{C} - \frac{35\,\text{kW} \times 1000\,\text{W/kW}}{5129.6\,\text{W/K}} = 106.3°\text{C} \qquad \textit{Ans.}$$

Discussion and Comments: As we discussed in Section 2.1.2.1, the two most important heat exchanger design problems are the rating and sizing problems. However, based on Eq. (3.7), the six independent variables of the problem for the specified flow arrangement yields a total of 21 different problems, summarized later in Table 3.11. The problem above is one example beyond the simple sizing problems (numbers 2 and 4 in Table 3.11).

In reality, one needs to design/size a radiator (i.e., UA or NTU to be determined) such that the top tank temperature does not exceed the boiling temperature of the water–glycol mixture at the worst design point (7% grade with air-conditioning on at the highest blower speed for airflow in a desert summer condition); and at the same time, it requires "low" fan power to reduce the total cost of the radiator and fan as well as to reduce a negative impact on the fuel economy. However, to simplify the present problem, the UA and airflow rate were given as part of the input data.

3.5 THE *P*-NTU METHOD

In the *P*-NTU method, the heat transfer rate from the hot fluid to the cold fluid in the exchanger is expressed as

$$q = P_1 C_1 \,\Delta T_{\max} = P_2 C_2 \,\Delta T_{\max} \qquad (3.94)$$

where *P* is the temperature effectiveness for fluid 1 or 2, depending on the subscript 1 or 2, $C = \dot{m} c_p$ is the heat capacity rate for fluid 1 or 2 with the corresponding subscripts, and the inlet temperature difference is $\Delta T_{\max} = T_{h,i} - T_{c,i} = |T_{2,i} - T_{1,i}|$. The temperature effectiveness *P*, similar to ε, is nondimensional and is dependent on the number of transfer units, heat capacity rate ratio, and the flow arrangement as will be shown in Eq. (3.95). We define these dimensionless groups after providing the rationale for this method.

Historically, the *P*-NTU method has been used for designing shell-and-tube exchangers even before the ε-NTU method became widely known in the 1940s. The ε-NTU method represents a variant of the *P*-NTU method. Let us first discuss the reason for use of the *P*-NTU method before presenting the details of the method.

As shown in Table 3.3, the ε-NTU relationship for a mixed–unmixed crossflow exchanger is different depending on whether the mixed fluid is the $C_{\min}$ or $C_{\max}$ fluid, because this exchanger is stream asymmetric, as discussed in Section 3.4.1.4. The other stream asymmetric exchangers are also mentioned in Section 3.4.1.4. Hence, they will have two ε-NTU formulas, depending on which fluid side is the $C_{\min}$ side. Consider that the heat capacity rate ratio (tube-to-shell) $C_t/C_s = 0.9$ at an operating point in a 1–2 TEMA G exchanger. Now, due to a change in the process condition, the tube flow rate is increased in the same exchanger, so that the new $C_t/C_s = 10/9 \approx 1.1$. In this case, $C_t = C_{\min}$ for the first operating point, and $C_s = C_{\min}$ for the second operating point in the *same* exchanger. One needs to use two different ε-NTU formulas for the analysis of these two operating points resulting in two different ε's. Thus one has to keep track of which is the $C_{\min}$ fluid in the calculations.

To avoid possible errors and confusion, an alternative is to present the temperature effectiveness *P* of one fluid stream as a function of NTU based on that side's heat capacity rate and a ratio of heat capacity rates of that side to the other side, *R*. In that case, *R* can vary from 0 to ∞, and only one *P*-NTU formula will suffice for the complete operating range of the exchanger. Of course, we will still have two different formulas (functional relationships as shown below for a stream asymmetric exchanger), and any one will cover the complete operating range $(0 \le R \le \infty)$ in the exchanger.

$$P_1 = \phi_1 \,(\mathrm{NTU}_1, R_1, \text{flow arrangement}) \qquad P_2 = \phi_2 \,(\mathrm{NTU}_2, R_2, \text{flow arrangement})$$
$$(3.95)$$

Note that we use the general subscripts 1 and 2 in the *P*-NTU method for fluids 1 and 2 of the exchanger; and individual fluids 1 and 2 can be hot or cold, or $C_{\min}$ or $C_{\max}$ fluids. Somewhat arbitrarily, we choose the fluid 1 side as the shell fluid side regardless of whether it is the hot- or cold-fluid side in our representation here. For exchangers other than shell-and-tube types, one of the two fluid sides is specifically designated as fluid 1 side, as shown in the corresponding sketches of Table 3.6.

It may not be a problem at all in a computer program to keep track of which fluid is the $C_{\min}$ before using the proper ε-NTU formula for a stream asymmetric exchanger.

However, it may not be possible to determine whether a new but complicated flow arrangement of an industrial heat exchanger is stream symmetric or asymmetric. One does not need to know this fact if one uses the P-NTU method. The P-NTU formulas are flow arrangement dependent (as is the case for the ε-NTU formulas), and *not* on the dimensional value of C_s or C_t (or C_{min}) as an additional variable in the ε-NTU method. Thus, this is where the P-NTU method has an apparent advantage over the ε-NTU method. Note that regardless of how complicated the exchanger flow arrangement is, it may be possible now to obtain a *closed-form* P-NTU formula by the chain rule methodology (Pignotti and Shah, 1992; Sekulić et al., 1999). Now let us define P, NTU, and R of this method.

3.5.1 Temperature Effectiveness P

The temperature effectiveness P is referred to as *thermal effectiveness* in the shell-and-tube heat exchanger literature. It is different for each fluid of a two-fluid exchanger. For fluid 1, it is defined as the ratio of the temperature range (rise or drop) of fluid 1 (regardless of whether it is a cold fluid or a hot fluid) to the inlet temperature difference (ΔT_{max} or ITD) of the two fluids:

$$P_1 = \frac{T_{1,o} - T_{1,i}}{T_{2,i} - T_{1,i}} \tag{3.96}$$

Similarly, the temperature effectiveness of fluid 2 is defined as

$$P_2 = \frac{T_{2,i} - T_{2,o}}{T_{2,i} - T_{1,i}} \tag{3.97}$$

It can be shown that

$$P_1 = P_2 R_2 \qquad P_2 = P_1 R_1 \tag{3.98}$$

where R_1 and R_2 are defined later in Eqs. (3.105) and (3.106). Note that the temperature effectivenesses defined by Eqs. (3.96) and (3.97) are identical to those defined by Eqs. (3.51) and (3.52).

Comparing Eqs. (3.44) and (3.96), it is found that the temperature effectiveness P_1 and the exchanger effectiveness ε are related as

$$P_1 = \frac{C_{min}}{C_1}\varepsilon = \begin{cases} \varepsilon & \text{for } C_1 = C_{min} \\ \varepsilon C^* & \text{for } C_1 = C_{max} \end{cases} \tag{3.99}$$

Similarly,

$$P_2 = \frac{C_{min}}{C_2}\varepsilon = \begin{cases} \varepsilon & \text{for } C_2 = C_{min} \\ \varepsilon C^* & \text{for } C_2 = C_{max} \end{cases} \tag{3.100}$$

Thus, the values of P_1 and P_2 will always be less than or equal to ε.

3.5.2 Number of Transfer Units, NTU

The number of transfer units NTU_1 and NTU_2 are defined as

$$\mathrm{NTU}_1 = \frac{UA}{C_1} \qquad \mathrm{NTU}_2 = \frac{UA}{C_2} \tag{3.101}$$

It can be shown that

$$NTU_1 = NTU_2 R_2 \qquad NTU_2 = NTU_1 R_1 \tag{3.102}$$

These NTU_i's are related to NTU based on C_{min} as

$$NTU_1 = NTU \frac{C_{min}}{C_1} = \begin{cases} NTU & \text{for } C_1 = C_{min} \\ NTU\ C^* & \text{for } C_1 = C_{max} \end{cases} \tag{3.103}$$

$$NTU_2 = NTU \frac{C_{min}}{C_2} = \begin{cases} NTU & \text{for } C_2 = C_{min} \\ NTU\ C^* & \text{for } C_2 = C_{max} \end{cases} \tag{3.104}$$

Thus, NTU_1 or NTU_2 is always less than or equal to NTU.

3.5.3   Heat Capacity Rate Ratio *R*

The heat capacity ratio *R* is defined as

$$R_1 = \frac{C_1}{C_2} = \frac{T_{2,i} - T_{2,o}}{T_{1,o} - T_{1,i}} \tag{3.105}$$

$$R_2 = \frac{C_2}{C_1} = \frac{T_{1,o} - T_{1,i}}{T_{2,i} - T_{2,o}} \tag{3.106}$$

and hence

$$R_1 = \frac{1}{R_2} \tag{3.107}$$

Comparing Eqs. (3.105) and (3.106) with Eq. (3.56), we get

$$R_1 = \frac{C_1}{C_2} = \begin{cases} C^* & \text{for } C_1 = C_{min} \\ 1/C^* & \text{for } C_1 = C_{max} \end{cases} \tag{3.108}$$

$$R_2 = \frac{C_2}{C_1} = \begin{cases} C^* & \text{for } C_2 = C_{min} \\ 1/C^* & \text{for } C_2 = C_{max} \end{cases} \tag{3.109}$$

Thus R_1 and R_2 are always greater than or equal to C^*. Individually, values of R_1 and R_2 range from 0 to ∞, zero being for pure vapor condensation and infinity being for pure liquid vaporization.

3.5.4   General *P*-NTU Functional Relationship

Similar to the exchanger effectiveness ε, the temperature effectiveness P_1 is a function of NTU_1, R_1 and flow arrangement represented by the function ϕ_1:

$$P_1 = \phi_1 (NTU_1, R_1, \text{ flow arrangement}) \tag{3.110}$$

Similarly, P_2 can be expressed as a function ϕ_2 of NTU_2, R_2, and flow arrangement. Hence the functions ϕ_1 and ϕ_2 will be dependent on the exchanger flow arrangement and different for a stream asymmetric exchanger (i.e., the explicit formulas will be different).

The P-NTU results are generally presented on a semilog paper for the following reason. The origin of the P-NTU method is related to shell-and-tube exchangers, and the most useful NTU design range is from about 0.2 to 3.0 for shell-and-tube exchangers. Hence, the NTU scale is chosen as a logarithmic scale to stretch it in the range 0.2 to 3.0 to determine *graphically* a more accurate value of NTU, as was the case before the computer era.

3.6   *P*-NTU RELATIONSHIPS

As demonstrated in Section 3.5, P is a function of NTU, R and flow arrangement; here P, NTU, and R should be defined consistently based on either fluid 1 or 2 side. Since the basic energy balance and rate equations are identical to those mentioned while deriving a ε-NTU relationship, we will not derive any specific P-NTU relationship here. Instead, we illustrate the results for the 1–2 TEMA E exchanger with shell fluid mixed and discuss the concepts of temperature cross and multipassing to emphasize the peculiarity of some results and how to obtain results for more complex flow arrangement. Explicit P-NTU formulas for a large number of exchanger flow arrangements are presented in Table 3.6. It must be emphasized that all formulas are presented in Table 3.6 in terms of P_1, NTU_1 and R_1 where fluid 1 is defined in the accompanying sketch in the first column on the left. If one needs an explicit formula for fluid 2, replace P_1, NTU_1, and R_1 as $P_1 = P_2 R_2$, $NTU_1 = NTU_2 R_2$ and $R_1 = 1/R_2$ to obtain an explicit expression for P_2 in terms of NTU_2 and R_2.

3.6.1   Parallel Counterflow Exchanger, Shell Fluid Mixed, 1–2 TEMA E Shell

3.6.1.1 P-NTU Relationships. This exchanger has one shell pass and an even (rarely, odd) number of tube passes. The simplest arrangement is one shell pass and two tube passes, simply referred to as a 1–2 *exchanger*. Such an exchanger, having two different nozzle arrangements with respect to the tube return end, is shown in Fig. 3.12 with idealized temperature distributions. Note that the tube fluid in one pass is in the counterflow direction and the tube fluid in the second pass is in the parallelflow direction to the shell fluid flow direction.

Although this exchanger is not *geometrically* symmetric, it is highly stream symmetric if the shell fluid is considered perfectly mixed at each cross section (see Section 3.4.1.4). Hence, the temperature effectiveness P is identical for the exchangers of Fig. 3.12a and b for the same NTU and R. In other words, it does not make any difference in the theoretical P as to which end the exchanger inlet nozzle is located on. Also, the flow direction of either the tube or shell fluid or both fluids can be reversed, and P will still be the same for specified values of NTU and R. Note that in all these cases, all four terminal temperatures and the tube fluid temperature at the end of the first tube pass, $T_{t,\ell}$ will be identical, as shown in Fig. 3.12a. The explicit P-NTU formula for this flow arrangement is given in Table 3.6, and the results are shown in Fig. 3.13. From this figure, it is clear that when $NTU_1 \rightarrow \infty$ and $R_1 > 0$, P_1 is less than 1. Also note that P_1 monotonically increases with NTU_1 to asymptotic values for all values of R_1.

The P_1-NTU$_1$ relationship for the 1–n exchanger (one shell pass, even-n tube passes with shell fluid mixed) is also presented in Table 3.6. Note that this 1–n exchanger ($n \geq 4$) is stream asymmetric. The temperature effectiveness P_1 of the 1–n exchanger ($n > 2$ and even) is lower than that of the 1–2 exchanger for specified values of NTU$_1$ and R_1. The maximum reduction in P_1 of the 1–n exchanger over the 1–2 exchanger occurs at $R_1 = 1$. The thermal effectivenesses of the 1–n exchangers at $R_1 = 1$ are shown in Table 3.7. From a review of this table, if the P_1-NTU$_1$ expression of Table 3.6 for the 1–2 exchanger is used for the 1–n exchanger ($n \geq 4$ and even), the error introduced is negligible for low NTU$_1$ (NTU$_1 < 3$), particularly for $R_1 \neq 1$. However, if this error is not acceptable, it is recommended to employ the P_1-NTU$_1$ expression [see Eq. (III.4) in Table 3.6] of the 1–4 exchanger for the 1–n exchanger for $n \geq 4$ and even. When $n \to \infty$, the 1–n exchanger effectiveness approaches that of a crossflow exchanger with both fluids mixed, and the formula is given in Table 3.6 as Eq. (III.5).

The P-NTU results are presented in Figs. 3.14, 3.15, and 3.16 for 1–2 TEMA G, H, and J exchangers for comparisons with those in Fig. 3.13 for the 1–2 TEMA E exchanger.

3.6.1.2 Temperature Cross.[†] As defined earlier, the *temperature cross* refers to the situation where the cold-fluid outlet temperature $T_{c,o}$ is greater than the hot-fluid outlet temperature $T_{h,o}$; and in this case, the magnitude of the temperature cross is $(T_{c,o} - T_{h,o})$. It derives its name from fictitious (e.g., counterflow exchanger; see Fig. 1.50) or actual crossing (e.g., 1–2 TEMA E exchanger as shown in Fig. 3.17b) of the temperature distributions of the hot and cold fluids in heat exchangers. Note that the tube fluid temperature in the second pass beyond point X in Fig. 3.17b is higher than the shell fluid temperature at the corresponding axial position in the exchanger.

To further understand the temperature cross, consider a 1–2 exchanger with two possible shell fluid directions with respect to the tube fluid direction as shown in Fig. 3.17. Since the 1–2 exchanger is symmetric with respect to P or F, when the shell fluid is considered *mixed*, the same overall performance is obtained regardless of the direction of the shell fluid or whether the hot or cold fluid is the shell fluid. In Fig. 3.17 the solid line temperature distributions correspond to a high-NTU$_s$ case; and P_s and F are identical in both cases for the same NTU$_s$ and R_s. Similarly, the dashed line temperature distributions in Fig. 3.17 correspond to a low-NTU$_s$ case, and P_s and F will again be identical, although different from those for the solid lines.

The temperature distributions of Fig. 3.17b reveal that there is a temperature cross of the shell fluid and the tube fluid in the second pass for the high-NTU$_s$ case. In region J of this case, the second tube pass transfers heat to the shell fluid. This is contrary to the original design, in which ideally, heat transfer should have taken place only in one direction (from the shell fluid to the tube fluid in Fig. 3.17) throughout the two passes. This temperature cross can be explained qualitatively as follows: Although an addition of surface area (a high value of NTU$_s$ or a low value of F) is effective in raising the temperature $T_{t,\ell}$ of the tube fluid at the end of the first tube pass, the temperature of the tube fluid rises in the second tube pass up to a point X. Beyond this point, the temperature of the shell fluid is lower than that of the tube fluid, since we have considered the shell fluid mixed at a cross section and it is cooled rapidly by the first tube pass. Thus the

---

[†] In all the discussion of this subsection, we consider only the external temperature cross; refer to Section 11.4.1 for external and internal temperature crosses. Also, throughout this subsection, the shell fluid is considered hot, the tube fluid cold, and heat transfer takes place from the shell fluid to the tube fluid. However, all the discussion and results presented in this subsection are valid if the shell fluid is cold and the tube fluid is hot.

TABLE 3.6 P_1–NTU_1 Formulas and Limiting Values of P_1 for $R_1 = 1$ and $NTU_1 \rightarrow \infty$ for Various Exchanger Flow Arrangements

Flow Arrangement[a]	Eq. No.	General Formula[b]	Value for $R_1 = 1$[c]	Value for $NTU_1 \rightarrow \infty$
Counterflow exchanger, stream symmetric	I.1.1	$P_1 = \dfrac{1 - \exp[-NTU_1(1-R_1)]}{1 - R_1 \exp[-NTU_1(1-R_1)]}$	$P_1 = \dfrac{NTU_1}{1 + NTU_1}$	$P_1 \rightarrow 1 \;\; \text{for } R_1 \leq 1$ $P_1 \rightarrow 1/R_1 \;\; \text{for } R_1 \geq 1$
	I.1.2	$NTU_1 = \dfrac{1}{1 - R_1}\ln\dfrac{1 - R_1 P_1}{1 - P_1}$	$NTU_1 = \dfrac{P_1}{1 - P_1}$	$NTU_1 \rightarrow \infty$
	I.1.3	$F = 1$	$F = 1$	$F = 1$
Parallelflow exchanger, stream symmetric	I.2.1	$P_1 = \dfrac{1 - \exp[-NTU_1(1+R_1)]}{1 + R_1}$	$P_1 = \tfrac{1}{2}[1 - \exp(-2NTU_1)]$	$P_1 \rightarrow \dfrac{1}{1+R_1}$
	I.2.2	$NTU_1 = \dfrac{1}{1 + R_1}\ln\dfrac{1}{1 - P_1(1 + R_1)}$	$NTU_1 = \dfrac{1}{2}\ln\dfrac{1}{1 - 2P_1}$	$NTU_1 \rightarrow \infty$
	I.2.3	$F = \dfrac{(R_1 + 1)\ln[(1 - R_1 P_1)/(1 - P_1)]}{(R_1 - 1)\ln[1 - P_1(1 + R_1)]}$	$F = \dfrac{2P_1}{(P_1 - 1)\ln(1 - 2P_1)}$	$F \rightarrow 0$
Single-pass crossflow exchanger, both fluids unmixed, stream symmetric	II.1	$P_1 = 1 - \exp(-NTU_1)$ $\quad - \exp[-(1 + R_1)NTU_1]$ $\times \displaystyle\sum_{n=1}^{\infty}\dfrac{1}{(n+1)!}R_1^n$ $\times \displaystyle\sum_{j=1}^{n}\dfrac{(n + 1 - j)}{j!}(NTU_1)^{n+j}$	Same as Eq. (II.1) with $R_1 = 1$	$P_1 \rightarrow 1 \;\; \text{for } R_1 \leq 1$ $P_1 \rightarrow \dfrac{1}{R_1} \;\; \text{for } R_1 \geq 1$

$$P_1 \approx 1 - \exp\left[\dfrac{NTU_1^{0.22}}{R_1}\left(e^{-R_1 NTU_1^{0.78}} - 1\right)\right]$$

This approximate equation is accurate within $\pm 1\%$ for $1 < NTU_1 < 7$.

II.2.1

$P_1 = \dfrac{1 - \exp(-KR_1)}{R_1}$

$K = 1 - \exp(-NTU_1)$

$P_1 = 1 - \exp(-K)$

$K = 1 - \exp(-NTU_1)$

$P_1 \to \dfrac{1 - \exp(-R_1)}{R_1}$

$NTU_1 \to \infty$

II.2.2

$NTU_1 = \ln\dfrac{1}{1 + (1/R_1)\ln(1 - R_1 P_1)}$

$NTU_1 = \ln\dfrac{1}{1 + \ln(1 - P_1)}$

II.2.3

$F = \dfrac{\ln[(1 - R_1 P_1)/(1 - P_1)]}{(R_1 - 1)\ln[1 + (1/R_1)\ln(1 - R_1 P_1)]}$

$F = \dfrac{P_1}{(P_1 - 1)\ln[1 + \ln(1 - P_1)]}$

$F \to 0$

II.3.1

$P = 1 - \exp\left(-\dfrac{K}{R_1}\right)$

$K = 1 - \exp(-R_1 \cdot NTU_1)$

$P = 1 - \exp(-K)$

$K = 1 - \exp(-NTU_1)$

$P_1 \to 1 - \exp\left(-\dfrac{1}{R_1}\right)$

II.3.2

$NTU_1 = \dfrac{1}{R_1}\ln\left[\dfrac{1}{1 + R_1\ln(1 - P_1)}\right]$

$NTU_1 = \ln\dfrac{1}{1 + \ln(1 - P_1)}$

$NTU_1 \to \infty$

II.3.3

$F = \dfrac{\ln[(1 - R_1 P_1)/(1 - P_1)]}{(1 - 1/R_1)\ln[1 + R_1\ln(1 - P_1)]}$

$F = \dfrac{P_1}{(P_1 - 1)\ln[1 + \ln(1 - P_1)]}$

$F \to 0$

II.4

$P_1 = \left(\dfrac{1}{K_1} + \dfrac{R_1}{K_2} - \dfrac{1}{NTU_1}\right)^{-1}$

$K_1 = 1 - \exp(-NTU_1)$

$K_2 = 1 - \exp(-R_1 \cdot NTU_1)$

$P_1 = \left(\dfrac{2}{K_1} - \dfrac{1}{NTU_1}\right)^{-1}$

$P_1 \to \dfrac{1}{1 + R_1}$

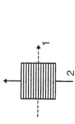

Single-pass crossflow exchanger, fluid 1 unmixed, fluid 2 mixed

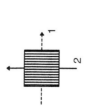

Single-pass crossflow exchanger, fluid 1 mixed, fluid 2 unmixed

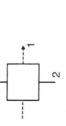

Single-pass crossflow exchanger, both fluids mixed, stream symmetric

(continued over)

TABLE 3.6 (Continued)

Flow Arrangement[a]	Eq. No.	General Formula[b]	Value for $R_1 = 1$[c]	Value for $NTU_1 \to \infty$
1–2 TEMA E shell-and-tube exchanger, shell fluid mixed, stream symmetric	III.1.1	$P_1 = \dfrac{2}{1 + R_1 + E\coth(E\,NTU_1/2)}$ $E = (1+R_1^2)^{1/2}$	$P_1 = \dfrac{1}{1 + \coth(NTU_1/\sqrt{2})/\sqrt{2}}$	$P_1 \to \dfrac{2}{1 + R_1 + E}$
	III.1.2	$NTU_1 = \dfrac{1}{E}\ln\dfrac{2 - P_1(1 + R_1 - E)}{2 - P_1(1 + R_1 + E)}$	$NTU_1 = \ln\dfrac{2 - P_1}{2 - 3P_1}$	$NTU_1 \to \infty$
	III.1.3	$F = \dfrac{E\ln[(1 - R_1 P_1)/(1 - P_1)]}{(1 - R_1)\ln\left[\dfrac{2 - P_1(1 + R_1 - E)}{2 - P_1(1 + R_1 + E)}\right]}$	$F = \dfrac{P_1/(1 - P_1)}{\ln[(2 - P_1)/(2 - 3P_1)]}$	$F \to 0$
1–2 TEMA E shell-and-tube exchanger, shell fluid divided into two streams individually mixed	III.2	$P_1 = \dfrac{1}{R_1}\left[1 - \dfrac{(2 - R_1)(2E + R_1 B)}{(2 + R_1)(2E - R_1/B)}\right]$ $E = \exp(NTU_1)$ $B = \exp\left(-\dfrac{NTU_1 \cdot R_1}{2}\right)$ Same as 1–1 J shell, Eq. (III.10)	$P_1 = \dfrac{1}{2}\left[1 - \dfrac{1 + E^{-2}}{2(1 + NTU_1)}\right]$ for $R_1 = 2$	$P_1 \to \dfrac{2}{2 + R_1}$ for $R_1 \leq 2$ $P_1 \to \dfrac{1}{R_1}$ for $R_1 \geq 2$

III.3

$$P_1 = \frac{1}{R_1}\left(1 - \frac{C}{AC + B^2}\right)$$

$$A = \frac{X_1(R_1+\lambda_1)(R_1-\lambda_2)}{2\lambda_1} - \frac{X_2(R_1+\lambda_2)(R_1-\lambda_1)}{2\lambda_2} - X_3\delta + \frac{1}{1-R_1}$$

$$B = X_1(R_1-\lambda_2) - X_2(R_1-\lambda_1) + X_3\delta$$

$$C = X_2(3R_1+\lambda_1) - X_1(3R_1+\lambda_2) + X_3\delta$$

$$X_i = \frac{\exp(\lambda_i \cdot NTU_1/3)}{2\delta}, \quad i = 1,2,3$$

$$\delta = \lambda_1 - \lambda_2$$

$$\lambda_1 = -\frac{3}{2} + \left[\frac{9}{4} + R_1(R_1-1)\right]^{1/2}$$

$$\lambda_2 = -\frac{3}{2} - \left[\frac{9}{4} + R_1(R_1-1)\right]^{1/2}$$

$$\lambda_3 = R_1$$

Same as Eq. (III.3)
with $R_1 = 1$

$$A = -\frac{\exp(-NTU_1)}{18} - \frac{\exp(NTU_1/3)}{2} + \frac{(NTU_1+5)}{9}$$

$P_1 \to 1$ for $R_1 \le 1$

$P_1 \to \dfrac{1}{R_1}$ for $R_1 \ge 1$

1–3 TEMA E shell-and-tube exchanger, shell and tube fluids mixed, one parallelflow and two counterflow passes

III.4

$$P_1 = 4[2(1+R_1) + DA + R_1B]^{-1}$$

$$A = \coth\frac{D\,NTU_1}{4}$$

$$B = \tanh\frac{R_1\,NTU_1}{4}$$

$$D = (4+R_1^2)^{1/2}$$

$$P_1 = 4(4 + \sqrt{5}A + B)^{-1}$$

$$A = \coth\frac{\sqrt{5}\,NTU_1}{4}$$

$$B = \tanh\frac{NTU_1}{4}$$

$$P_1 \to \frac{4}{2(1+2R_1) + D - R_1}$$

1–4 TEMA E shell-and-tube exchanger, shell and tube fluids mixed

(continued over)

TABLE 3.6 (Continued)

Flow Arrangement[a]	Eq. No.	General Formula[b]	Value for $R_1 = 1$[c]	Value for $NTU_1 \to \infty$
Same as Eq. (III.4) with $n \to \infty$, exchanger stream symmetric	III.5	Eq. (II.4) applies in this limit with $n \to \infty$	Same as for Eq. (II.4)	Same as for Eq. (II.4)
1–1 TEMA G shell-and-tube exchanger, tube fluid split into two streams individually mixed, shell fluid mixed, stream symmetric	III.6	$P_1 = A + B - AB(1+R_1) + R_1AB^2$ $$A = \frac{1}{1+R_1}\left\{1 - \exp\left[-\frac{NTU_1(1+R_1)}{2}\right]\right\}$$ $$B = \frac{1-D}{1-R_1D}$$ $$D = \exp\left[-\frac{NTU_1(1-R_1)}{2}\right]$$	Same as Eq. (III.6) with $$B = \frac{NTU_1}{2+NTU_1}$$ for $R_1 = 1$	$P_1 \to 1$ for $R_1 \le 1$ $P_1 \to \dfrac{1}{R_1}$ for $R_1 \ge 1$
Overall counterflow 1–2 TEMA G shell-and-tube exchanger, shell and tube fluids mixed in each pass at a cross section	III.7	$$P_1 = \frac{B-\alpha^2}{A+2+R_1B}$$ $$A = \frac{-2R_1(1-\alpha)^2}{2+R_1}$$ $$B = \frac{4-\beta(2+R_1)}{2-R_1}$$ $$\alpha = \exp\left[-\frac{NTU_1(2+R_1)}{4}\right]$$ $$\beta = \exp\left[-\frac{NTU_1(2-R_1)}{2}\right]$$	$$P_1 = \frac{1+2NTU_1-\alpha^2}{4+4NTU_1-(1-\alpha)^2}$$ for $R_1 = 2$ $\alpha = \exp(-NTU_1)$	$P_1 \to \dfrac{2+R_1}{R_1^2+R_1+2}$ for $R_1 \le 2$ $P_1 \to \dfrac{1}{R_1}$ for $R_1 \ge 2$

$$B = \frac{NTU_1}{2 + NTU}$$

for $R_1 = 2$

III.8 Same as Eq. (III.8) with

$$P_1 = E[1 + (1 - BR_1/2)(1 - AR_1/2 + ABR_1)]$$
$$-AB(1 - BR_1/2)$$

$$A = \frac{1}{1 + R_1/2}\{1 - \exp[-NTU_1(1 + R_1/2)/2]\}$$

$$B = (1 - D)/(1 - R_1 D/2)$$

$$D = \exp[-NTU_1(1 - R_1/2)/2]$$

$$E = (A + B - ABR_1/2)/2$$

$$P_1 \to \frac{4(1 + R_1) - R_1^2}{(2 + R_1)^2}$$

for $R_1 \le 2$

$$P_1 \to \frac{1}{R_1} \quad R_1 \ge 2$$

1–1 TEMA H shell-and-tube exchanger, tube fluid split into two streams individually mixed, shell fluid mixed

III.9 Same as Eq. (III.11) with

$$P_1 = \frac{1}{R_1}\left[1 - \frac{(1 - D)^4}{B - 4G/R_1}\right]$$

$$B = (1 + H)(1 + E)^2$$

$$G = (1 - D)^2(D^2 + E^2) + D^2(1 + E)^2$$

$$H = \frac{1 - \exp(-2\beta)}{4/R_1 - 1}$$

$$E = \frac{1 - \exp(-\beta)}{4/R_1 - 1}$$

$$D = \frac{1 - \exp(-\alpha)}{4/R_1 + 1}$$

$$\alpha = \frac{NTU_1(4 + R_1)}{8}$$

$$\beta = \frac{NTU_1(4 - R_1)}{8}$$

$$H = NTU_1$$

$$E = \frac{NTU_1}{2}$$

for $R_1 = 4$

$$P_1 \to \left[R_1 + \frac{(4 - R_1)^3}{(4 + R_1)(R_1^3 + 16)}\right]^{-1}$$

for $R_1 \le 4$

$$P_1 \to \frac{1}{R_1} \quad \text{for } R_1 > 4$$

Overall counterflow 1–2 TEMA H shell-and-tube exchanger, shell and tube fluids mixed in each pass at a cross section

(continued over)

TABLE 3.6 (*Continued*)

Flow Arrangement[a]	Eq. No.	General Formula[b]	Value for $R_1 = 1$[c]	Value for NTU$_1 \to \infty$
 1-1 TEMA J shell-and-tube exchanger, shell and tube fluids mixed	III.10	$P_1 = \dfrac{1}{R_1}\left[1 - \dfrac{(2-R_1)(2A+R_1B)}{(2+R_1)(2A-R_1/B)}\right]$ $A = \exp(NTU_1)$ $B = \exp\left(-\dfrac{NTU_1 \cdot R_1}{2}\right)$ Same as Eq. (III.2)	$P_1 = \dfrac{1}{2}\left[1 - \dfrac{1+A^{-2}}{2(1+NTU_1)}\right]$ for $R_1 = 2$	$P_1 \to \dfrac{2}{2+R_1}$ for $R_1 \leq 2$ $P_1 \to \dfrac{1}{R_1}$ for $R_1 \geq 2$
 1-2 TEMA J shell-and-tube exchanger, shell and tube fluids mixed, results remain the same if fluid 2 is reversed	III.11	$P_1 = \left(1 + \dfrac{R_1}{2} + \lambda B - 2\lambda CD\right)^{-1}$ $B = \dfrac{A^\lambda + 1}{A_1^\lambda - 1}$ $C = \dfrac{A^{(1+\lambda)/2}}{\lambda - 1 + (1+\lambda)A^\lambda}$ $D = 1 + \dfrac{\lambda A^{(\lambda-1)/2}}{A^\lambda - 1}$ $A = \exp(NTU_1)$ $\lambda = \left(1 + \dfrac{R_1^2}{4}\right)^{1/2}$	Same as Eq. (III.11) with $R_1 = 1$	$P_1 \to \left(1 + \dfrac{R_1}{2} + \lambda\right)^{-1}$

III.12 — 1–4 TEMA J shell-and-tube exchanger, shell and tube fluids mixed

$$P_1 = \left(1 + \frac{R_1}{4}\frac{1+3E}{1+E} + \lambda B - 2\lambda CD\right)^{-1}$$

$$B = \frac{A^\lambda + 1}{A^\lambda - 1}$$

$$C = \frac{A^{(1+\lambda)/2}}{\lambda - 1 + (1+\lambda)A^\lambda}$$

$$D = 1 + \frac{\lambda A^{(\lambda-1)/2}}{A^\lambda - 1}$$

$$A = \exp(NTU_1)$$

$$E = \exp\left(\frac{R_1 \cdot NTU_1}{2}\right)$$

$$\lambda = \left(1 + \frac{R_1^2}{16}\right)^{1/2}$$

Same as Eq. (III.12) with $R_1 = 1$

$$P_1 \to \left(1 + \frac{3R_1}{4} + \lambda\right)^{-1}$$

III.13 — Limit of 1 – n TEMA J shell-and-tube exchangers for $n \to \infty$, shell and tube fluids mixed, stream symmetric

Eq. (II.4) applies in this limit

Same as for Eq. (II.4)

Same as for Eq. (II.4)

IV.1.1 — Parallel coupling of n exchangers, fluid 2 split arbitrarily into n streams

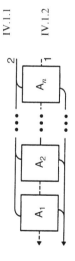

$$P_1 = 1 - \prod_{i=1}^{n}(1 - P_{1,A_i})$$

Same as Eq. (IV.1.1)

Same s Eq. (IV.1.1)

IV.1.2

$$\frac{1}{R_1} = \sum_{i=1}^{n}\frac{1}{R_{1,A_i}}$$

$$1 = \sum_{i=1}^{n}\frac{1}{R_{1,A_i}}$$

Same as Eq. (IV.1.2)

IV.1.3

$$NTU_1 = \sum_{i=1}^{n} NTU_{1,A_i}$$

Same as Eq. (IV.1.3)

$NTU_1 \to \infty$

(*continued over*)

TABLE 3.6 (*Continued*)

Flow Arrangement[a]	Eq. No.	General Formula[b]	Value for $R_1 = 1$	Value for $NTU_1 \rightarrow \infty$
Series coupling of n exchangers, overall counterflow arrangement, stream symmetric if all A_1 are stream symmetric	IV.2.1	$P_1 = \dfrac{\prod_{i=1}^n (1 - R_1 P_{1,A_i}) - \prod_{i=1}^n (1 - P_{1,A_i})}{\prod_{i=1}^n (1 - R_1 P_{1,A_i}) - R_1 \prod_{i=1}^n (1 - P_{1,A_i})}$	$P_1 = \dfrac{\sum_{i=1}^n [P_{1,A_i}/(1 - P_{1,A_i})]}{1 + \sum_{i=1}^n [P_{1,A_i}/(1 - P_{1,A_i})]}$	Same as Eq. (IV.2.1) counterflow
	IV.2.2	$R_1 = R_{1,A_i}, \quad i = 1,\ldots,n$	$1 = R_{1,A_i}, \quad i = 1,\ldots,n$	Same as Eq. (IV.2.2)
	IV.2.3	$NTU_1 = \sum_{i=1}^n NTU_{1,A_i}$	Same as for Eq. (IV.2.3)	Same as Eq. (IV.2.3)
	IV.2.4	$F = \dfrac{1}{NTU_1} \sum_{i=1}^n NTU_{1,A_i} \cdot F_{A_i}$	Same as Eq. (IV.2.4)	Same as Eq. (IV.2.4)
Series coupling of n exchangers, overall parallelflow arrangement, stream symmetric if all A_1 are stream symmetric	IV.3.1.	$P_1 = \dfrac{1}{1 + R_1}\left\{1 - \prod_{i=1}^n [1 - (1 + R_1)P_{1,A_i}]\right\}$	$P_1 = \dfrac{1}{2}\left[1 - \prod_{i=1}^n (1 - 2P_{1,A_i})\right]$	Same as Eq. (IV.3.1)
	IV.3.2	$R_1 = R_{1,A_i}, \quad i = 1,\ldots,n$	$1 = R_{1,A_i}, \quad i = 1,\ldots,n$	Same as Eq. (IV.3.2)
	IV.3.3	$NTU_1 = \sum_{i=1}^n NTU_{1,A_i}$	Same as Eq. (IV.3.3)	$NTU_1 \rightarrow \infty$

In all formulas of plate heat exchangers with the number of thermal plates $N \rightarrow \infty$ (equation numbers starting with V.), the single-pass parallel flow and counterflow temperature effectivenesses are presented in implicit forms. Their explicit forms are as follows with x and y representing the appropriate values of the number of transfer units and heat capacity rate ratio, respectively.

Single-Pass Parallelflow

$$P_p(x,y) = \frac{1 - \exp[-x(1+y)]}{1+y}$$

$$P_p(x,1) = \frac{1}{2}[1 - \exp(-2x)]$$

$$P_p(\infty,y) = \frac{1}{1+y}$$

Single-pass Counterflow

$$P_c(x,y) = \frac{1 - \exp[-x(1-y)]}{1 - y\exp[-x(1-y)]}$$

$$P_c(x,1) = \frac{x}{1+x}$$

$$P_c(\infty,y) = \begin{cases} 1 & \text{for } y < 1 \\ 1/y & \text{for } y > 1 \end{cases}$$

Flow Arrangement[a]	Eq. No.	General Formula[b]	Value for $R_1 = 1$ Unless Specified Differently	Value for $\mathrm{NTU}_1 \rightarrow \infty$
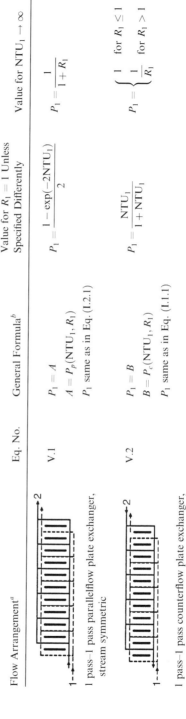 1 pass–1 pass parallelflow plate exchanger, stream symmetric	V.1	$P_1 = A$ $A = P_p(\mathrm{NTU}_1, R_1)$ P_1 same as in Eq. (I.2.1)	$P_1 = \dfrac{1 - \exp(-2\mathrm{NTU}_1)}{2}$	$P_1 = \dfrac{1}{1+R_1}$
1 pass–1 pass counterflow plate exchanger, stream symmetric	V.2	$P_1 = B$ $B = P_c(\mathrm{NTU}_1, R_1)$ P_1 same as in Eq. (I.1.1)	$P_1 = \dfrac{\mathrm{NTU}_1}{1+\mathrm{NTU}_1}$	$P_1 = \begin{cases} 1 & \text{for } R_1 \le 1 \\ \dfrac{1}{R_1} & \text{for } R_1 > 1 \end{cases}$

(continued over)

TABLE 3.6 (*Continued*)

Flow Arrangement[a]	Eq. No.	General Formula[b]	Value for $R_1 = 1$ Unless Specified Differently	Value for $NTU_1 \to \infty$
 1 pass–2 pass plate exchanger	V.3	$P_1 = \frac{1}{2}\left(A + B - \frac{1}{2}ABR_1\right)$ $A = P_p\left(NTU_1, \frac{R_1}{2}\right)$ $B = P_c\left(NTU_1, \frac{R_1}{2}\right)$	Same as Eq. (V.3) with $B = \dfrac{NTU_1}{1 + NTU_1}$ for $R_1 = 2$	$P_1 = \begin{cases} \dfrac{2}{2 + R_1} & \text{for } R_1 \leq 2 \\[2mm] \dfrac{1}{R_1} & \text{for } R_1 > 2 \end{cases}$
 1 pass–3 pass plate exchanger with two end passes in parallelflow	V.4	$P_1 = \frac{1}{3}\left[B + A\left(1 - \frac{R_1 B}{3}\right)\left(2 - \frac{R_1 A}{3}\right)\right]$ $A = P_p\left(NTU_1, \frac{R_1}{3}\right)$ $B = P_c\left(NTU_1, \frac{R_1}{3}\right)$	Same as Eq. (V.4) with $B = \dfrac{NTU_1}{1 + NTU_1}$ for $R_1 = 3$	$P_1 = \begin{cases} \dfrac{9 + R_1}{(3 + R_1)^2} & R_1 \leq 3 \\[2mm] \dfrac{1}{R_1} & R_1 > 3 \end{cases}$
 1 pass–3 pass plate exchanger with two end passes in counterflow	V.5	$P_1 = \frac{1}{3}\left[A + B\left(1 - \frac{R_1 A}{3}\right)\left(2 - \frac{R_1 B}{3}\right)\right]$ $A = P_p\left(NTU_1, \frac{R_1}{3}\right)$ $B = P_c\left(NTU_1, \frac{R_1}{3}\right)$	Same as Eq. (V.5) with $B = \dfrac{NTU_1}{1 + NTU_1}$ for $R_1 = 3$	$P_1 = \begin{cases} \dfrac{9 - R_1}{9 + 3R_1} & R_1 \leq 3 \\[2mm] \dfrac{1}{R_1} & R_1 > 3 \end{cases}$

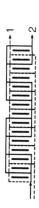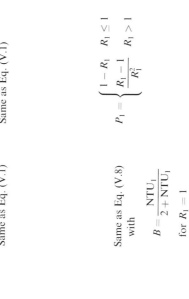1 pass–4 pass plate exchanger	V.6	$$P_1 = \frac{1-Q}{R_1}$$ $$Q = \left(1 - \frac{AR_1}{4}\right)^2 \left(1 - \frac{BR_1}{4}\right)^2$$ $$A = P_p\left(NTU_1, \frac{R_1}{4}\right)$$ $$B = P_c\left(NTU_1, \frac{R_1}{4}\right)$$	Same as Eq. (V.6) with $$B = \frac{NTU_1}{1 + NTU_1}$$ for $R_1 = 4$	$$P_1 = \begin{cases} \dfrac{16}{(4+R_1)^2} & R_1 \le 4 \\[2ex] \dfrac{1}{R_1} & R_1 > 4 \end{cases}$$
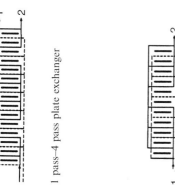2 pass–2 plate exchanger with overall parallelflow and individual passes in parallelflow, stream symmetric	V.7	P_1 same as in Eq. (V.1)	Same as Eq. (V.1)	Same as Eq. (V.1)
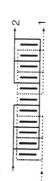2 pass–2 pass plate exchanger with overall parallelflow and individual passes in counterflow, stream symmetric	V.8	$$P_1 = B[2 - B(1 + R_1)]$$ $$B = P_c\left(\frac{NTU_1}{2}, R_1\right)$$	Same as Eq. (V.8) with $$B = \frac{NTU_1}{2 + NTU_1}$$ for $R_1 = 1$	$$P_1 = \begin{cases} 1 - R_1 & R_1 \le 1 \\[2ex] \dfrac{R_1 - 1}{R_1^2} & R_1 > 1 \end{cases}$$

(continued over)

TABLE 3.6 (*Continued*)

Flow Arrangement[a]	Eq. No.	General Formula[b]	Value for $R_1 = 1$ Unless Specified Differently	Value for $NTU_1 \rightarrow \infty$
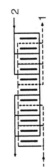 2 pass–2 pass plate exchanger with overall counterflow and individual passes in parallelflow, stream symmetric	V.9	$P_1 = \dfrac{2A - A^2(1+R_1)}{1 - R_1 A^2}$ $A = P_p\left(\dfrac{NTU_1}{2}, R_1\right)$	Same as Eq. (V.9)	$P_1 = \dfrac{1+R_1}{1+R_1+R_1^2}$
 2 pass–2 pass plate exchanger with overall counterflow and individual passes in counterflow, stream symmetric	V.10	P_1 same as Eq. (V.2)	Same as Eq. (V.2)	Same as Eq. (V.2)
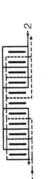  2 pass–3 pass plate exchanger with overall parallelflow	V.11	$P_1 = A + B - \left(\dfrac{2}{9} + \dfrac{D}{3}\right)(A^2+B^2)$ $\quad - \left(\dfrac{5}{9} + \dfrac{4D}{3}\right)AB$ $\quad + \dfrac{D(1+D)AB(A+B)}{3}$ $\quad - \dfrac{D^2 A^2 B^2}{9}$	Same as Eq. (V.11) with $B = \dfrac{NTU_1}{2 + NTU_1}$ for $R_1 = \dfrac{3}{2}$	$P_1 = \begin{cases} \dfrac{9 - 2R_1}{9 + 6R_1} & R_1 \le \dfrac{3}{2} \\[2ex] \dfrac{4R_1^2 + 2R_1 - 3}{2R_1^2(3 + 2R_1)} & R_1 > \dfrac{3}{2} \end{cases}$

V.12 2 pass–3 pass plate exchanger with overall counterflow

$$P_1 = \frac{A + 0.5B + 0.5C + D}{R_1}$$

$$A = \frac{2R_1 EF^2 - 2EF + F - F^2}{2R_1 E^2 F^2 - E^2 - F^2 - 2EF + E + F}$$

$$B = \frac{A(E-1)}{F}; \quad C = \frac{1-A}{E}$$

$$D = R_1 E^2 C - R_1 E + R_1 - \frac{C}{2}$$

$$E = \frac{1}{2R_1 G/3}; \quad F = \frac{1}{2R_1 H/3}$$

$$G = P_c\left(\frac{NTU_1}{2}, \frac{2R_1}{3}\right)$$

$$H = P_p\left(\frac{NTU_1}{2}, \frac{2R_1}{3}\right)$$

Same as Eq. (V.12) with

$$G = \frac{NTU_1}{2 + NTU_1}$$

for $R_1 = \dfrac{3}{2}$

$$P_1 = \begin{cases} \dfrac{27 + 12R_1 - 4R_1^2}{27 + 12R_1 + 4R_1^2} & R_1 \le \dfrac{3}{2} \\[2ex] \dfrac{1}{R_1} & R_1 > \dfrac{3}{2} \end{cases}$$

V.13 2 pass–4 pass plate exchanger with overall parallel flow

$$P_1 = 2D - (1+R_1)D^2$$

$$D = \frac{A + B - ABR_1/2}{2}$$

$$A = P_p\left(\frac{NTU_1}{2}, \frac{R_1}{2}\right)$$

$$B = P_c\left(\frac{NTU_1}{2}, \frac{R_1}{2}\right)$$

Same as Eq. (V.13) with

$$B = \frac{NTU_1}{2 + NTU_1}$$

for $R_1 = 2$

$$P_1 = \begin{cases} \dfrac{4}{(2+R_1)^2} & R_1 \le 2 \\[2ex] \dfrac{R_1 - 1}{R_1^2} & R_1 > 2 \end{cases}$$

(continued over)

TABLE 3.6 (Continued).

Flow Arrangement[a]	Eq. No.	General Formula[b]	Value for $R_1 = 1$ Unless Specified Differently	Value for NTU$_1 \to \infty$
2 pass–4 pass plate exchanger with overall counterflow	V.14	$P_1 = \dfrac{2D - (1 + R_1)D^2}{1 - D^2 R_1}$ $D = \dfrac{A + B - ABR_1/2}{2}$ $A = P_p\left(\dfrac{\mathrm{NTU}_1}{2}, \dfrac{R_1}{2}\right)$ $B = P_c\left(\dfrac{\mathrm{NTU}_1}{2}, \dfrac{R_1}{2}\right)$	Same as Eq. (V.14) with $B = \dfrac{\mathrm{NTU}_1}{2 + \mathrm{NTU}_1}$ for $R_1 = 2$	$P_1 = \begin{cases} \dfrac{4}{(4 + R_1)^2} & R_1 \leq 2 \\[2mm] \dfrac{1}{R_1} & R_1 > 2 \end{cases}$

[a] For those flow arrangements where "stream symmetric" is not mentioned explicitly, they are asymmetric.

[b] All the formulas in this table are based on the fluid 1 side. They can be converted to the fluid side 2 using the following relations: (1) for stream symmetric exchangers, change P_1, NTU$_1$, and R_1 to P_2, NTU$_2$, and R_2. (2) For stream asymmetric exchangers, convert P_1-NTU$_1$-R_1 expressions to P_2-NTU$_2$-R_2 expressions using the following relationships: $P_1 = P_2 R_2$, NTU$_1 =$ NTU$_2 R_2$, and $R_1 = 1/R_2$.

[c] Value for $R_1 = 1$ unless specified differently.

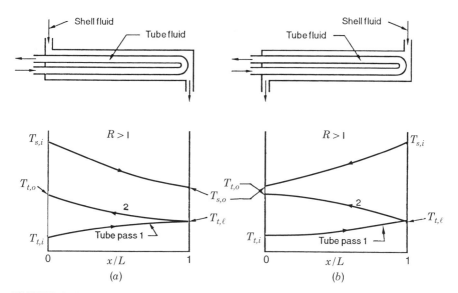

FIGURE 3.12   Idealized temperature distributions in a 1–2 TEMA E exchanger with shell fluid mixed for a low-NTU case.

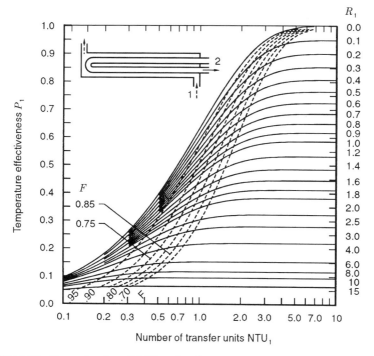

FIGURE 3.13 P_1 as a function of NTU_1 and R_1 for a 1–2 TEMA E exchanger, shell fluid mixed; constant-F factor lines are superimposed (From Shah and Pignotti, 1989).

TABLE 3.7 1–n Exchanger Effectiveness as a Function of NTU_1 for $R_1 = 1$

NTU$_1$	P_1 for the 1–n Exchanger						
	1–2	1–4	1–6	1–8	1–10	1–12	1–∞
1.0	0.463	0.463	0.463	0.463	0.463	0.463	0.462
2.0	0.557	0.553	0.553	0.552	0.552	0.552	0.552
3.0	0.579	0.569	0.567	0.566	0.566	0.566	0.563
4.0	0.584	0.568	0.564	0.562	0.561	0.561	0.559
5.0	0.585	0.564	0.558	0.555	0.554	0.553	0.551
6.0	0.586	0.560	0.552	0.549	0.547	0.546	0.544

addition of the surface area in the second tube pass left of the X is not useful from the thermal design point of view. A review of Fig. 3.13 does reveal that P_s increases monotonically with NTU_s for a specified R_s, although the rate of increase in P_s decreases with increasing values of NTU_s. Thus, increasing NTU_s (or decreasing F) will result in the higher $T_{t,\ell}$ of Fig. 3.17b and higher overall P_s, but with an increased temperature cross (point X will move farther right with increasing NTU_s). Theoretically, the maximum possible temperature cross for a constant U and infinite surface area is

$$T_{t,o,\max} - T_{s,o} = T_{s,o} - T_{t,i} \qquad (3.111)$$

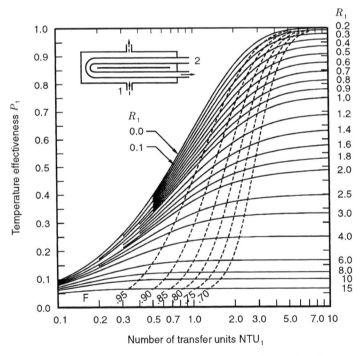

FIGURE 3.14 P_1 as a function of NTU_1 and R_1 for a 1–2 TEMA G (split-flow) exchanger, overall counterflow, shell and tube fluids mixed (From Shah and Pignotti, 1989).

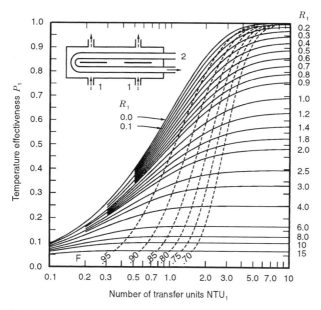

FIGURE 3.15 P_1 as a function of NTU_1 and R_1 for a 1–2 TEMA H (double-split-flow) exchanger, overall counterflow, shell and tube fluids mixed in individual passes (From Shah and Pignotti, 1989).

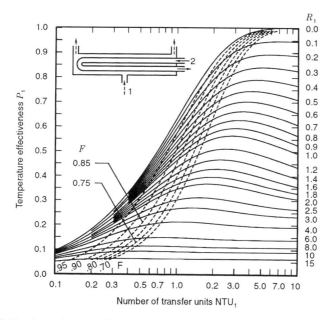

FIGURE 3.16 P_1 as a function of NTU_1 and R_1 for a 1–2 TEMA J (divided-flow) exchanger, shell fluid mixed; overall flow can be either counterflow or parallelflow (From Shah and Pignotti, 1989).

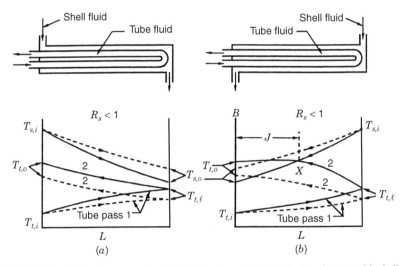

FIGURE 3.17   Idealized temperature distributions in a 1–2 TEMA E exchanger with shell fluid mixed for (*a*) high-NTU$_2$ case with solid lines, and (*b*) low-NTU$_2$ case with dashed lines (From Shah, 1983).

This equation is derived by applying the rate equations at the tube return end (section BB) in Fig. 3.17*b*; the heat transfer rate from the tube fluid of the second pass to the shell fluid is equal to the heat transfer rate from the shell fluid to the tube fluid of the first pass.

Now changing the nozzle orientation of the shell fluid as in Fig. 3.17*a*, we find no apparent crossing of the temperature distributions, although the temperature cross ($T_{t,o} - T_{s,o}$) does exist due to the counterflow direction of the shell fluid and the tube fluid in the second pass. Note that in a counterflow exchanger, the cold-fluid outlet temperature can be higher than the hot-fluid outlet temperature (refer to Fig. 1.50). These cases then have an external temperature cross. It must be emphasized that $T_{t,\ell}$ can never exceed $T_{s,o}$ since the shell fluid and the fluid in tube pass 1 represent a parallel-flow exchanger, as seen in Fig. 3.17*a*.

The temperature cross is undesirable, particularly for shell-and-tube exchangers, because the tube surface area is not utilized cost-effectively. A "good" design avoids the temperature cross in a shell-and-tube exchanger. Theoretically, the optimum design would have the temperature cross (or temperature meet) point just at the end of the second tube pass (Shah, 1983); that is,

$$T_{t,o} = T_{s,o} \quad \text{or} \quad T_{t,o} - T_{s,o} = 0 \tag{3.112}$$

Now

$$\frac{T_{t,o} - T_{s,o}}{T_{s,i} - T_{t,i}} = \frac{T_{t,o} - T_{t,i} + T_{t,i} - T_{s,i} + T_{s,i} - T_{s,o}}{T_{s,i} - T_{t,i}} = P_t - 1 + P_s = P_t - 1 + P_t R_t = 0 \tag{3.113}$$

where P_t and P_s are substituted from their definitions using Eqs. (3.96)–(3.97), and the last equality of 0 comes from Eq. (3.112). Equation (3.113) can be simplified [using Eq. (3.98)] to the following after generalizing the subscripts t and s to 1 and 2:

$$P_1 + P_2 = 1 \qquad P_1 = \frac{1}{1 + R_1} \qquad P_2 = \frac{1}{1 + R_2} \tag{3.114}$$

Thus for a given R, Eq. (3.114) provides the limiting (maximum) value of P; correspondingly NTU computed from the NTU equation for the 1–2 TEMA E exchanger in Table 3.6 provides the limiting (maximum) value of NTU_s beyond which there will be a temperature cross. These limiting values for the 1–2 exchanger are shown in Fig. 3.26 as an F_{min} line. Implications of these limiting values, valid for all exchangers that have temperature crosses, are discussed in detail in Section 11.4.

This concept of the temperature cross (or meet) at the exchanger outlet can readily be utilized to determine whether or not one or more shells in series will be necessary to meet the required heat duty without having a temperature cross in any individual shell. Consider desired inlet and outlet temperatures for the hot fluid to be 310 and 185°C, and those for the cold fluid to be 105 and 275°C. Considering a linear variation in specific heats, the hot- and cold-fluid temperature profiles for an overall counterflow exchanger can be drawn as shown in Fig. 3.18 (the length of the horizontal scale is arbitrary). Now draw a horizontal line from the cold-fluid outlet temperature until it meets the hot-fluid temperature profile,[†] where you draw a vertical line. Continue this process until the horizontal line meets the right-hand-side ordinate. The number of horizontal lines

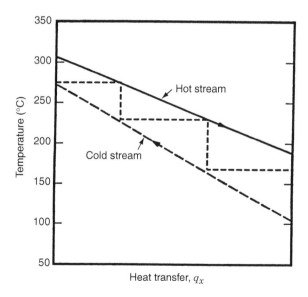

FIGURE 3.18   Estimate of a required number of shells in series to avoid the temperature cross in a 1–2 TEMA E exchanger (From Bell, 1998).

[†] This horizontal line then assures the temperature cross at the outlet section of individual shell-and-tube exchangers.

then indicates the number of shells required (three for Fig. 3.18) to avoid the temperature cross in the exchanger. Thus with this procedure, there is no need to conduct a detailed thermal analysis to find out how many shells are required in series to avoid the temperature cross.

3.6.2 Multipass Exchangers

A multipass exchanger, defined at the beginning of Section 1.6, is a single exchanger having multipassing of one or both fluids in the exchanger. A heat exchanger array consists of a number of individual heat exchangers connected to each other in a specified arrangement. If the two fluids are *the same* in a given exchanger array, the analysis presented below applies. However, in petrochemical and refining applications, there are many heat exchangers interconnected with more than two fluids (although any given exchanger may have only two fluid streams) in such a heat exchanger "train." It is beyond the scope of this book to provide heat transfer analysis of such trains. We will now derive the overall effectiveness and related results for multipass extended surface, shell-and-tube and plate exchangers, and many other construction types, including heat exchanger arrays. An important additional assumption made in the following analyses is: An individual fluid stream is perfectly mixed between passes or between exchangers in an array.

3.6.2.1 Multipass Crossflow Exchangers and Exchanger Arrays. These flow arrangements could be categorized as (1) a series coupling of n passes of a multipass crossflow exchanger or n individual exchangers of an exchanger array, either overall counterflow or parallelflow, (2) a parallel coupling of n passes of a multipass crossflow exchanger or n individual exchangers, and (3) a combination of series and parallel couplings or other compound multipass/array exchangers. The P-NTU expressions are now presented separately for each category.

Series Coupling: Overall Counterflow. Consider exchangers A and B in series coupling as shown in Fig. 3.19a. The results presented below are equally valid if the two exchangers are considered as two passes of a multipass crossflow exchanger. Figure 3.19b shows their combined equivalent exchanger C. In this figure, the temperatures are shown by two subscripts: the first subscript refers to either fluid 1 or fluid 2, and the second subscript denotes the mean terminal temperature at the location outside the passes or exchangers. The temperature effectivenesses of fluid 1 for exchangers A and B from their definitions [see Eq. (3.96)] are given by

$$P_{1,A} = \frac{T_{1,2} - T_{1,1}}{T_{2,2} - T_{1,1}} \qquad P_{1,B} = \frac{T_{1,3} - T_{1,2}}{T_{2,3} - T_{1,2}} \tag{3.115}$$

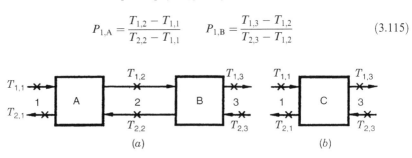

FIGURE 3.19 (*a*) Series coupled exchangers A and B in overall counterflow arrangement; (*b*) combined equivalent exchanger C.

and for the combined equivalent exchanger C,

$$P_{1,C} = \frac{T_{1,3} - T_{1,1}}{T_{2,3} - T_{1,1}} \tag{3.116}$$

Also, we need an expression for the heat capacity rate ratio $R_1 = C_1/C_2$ expressed in the form of temperatures of Fig. 3.19 as given by Eq. (3.105). Since R_1 is the same for exchangers A, B, and C of Fig. 3.19, we get

$$R_1 = \frac{C_1}{C_2} = \frac{T_{2,2} - T_{2,1}}{T_{1,2} - T_{1,1}} = \frac{T_{2,3} - T_{2,2}}{T_{1,3} - T_{1,2}} = \frac{T_{2,3} - T_{2,1}}{T_{1,3} - T_{1,1}} \tag{3.117}$$

$$\begin{array}{ccc} \text{Exchanger} & \text{Exchanger} & \text{Exchanger} \\ \text{A} & \text{B} & \text{C} \end{array}$$

To relate $P_{1,A}$ and $P_{1,B}$ of Eq. (3.115) with the overall $P_1 = P_{1,C}$ of Eq. (3.116), let us first compute the following expression in terms of temperatures for exchanger A by algebraic manipulation using Eqs. (3.115) and (3.117):

$$\frac{1 - P_{1,A}}{1 - R_1 P_{1,A}} = \frac{1 - (T_{1,2} - T_{1,1})/(T_{2,2} - T_{1,1})}{1 - \left(\dfrac{T_{2,2} - T_{2,1}}{T_{1,2} - T_{1,1}}\right)\left(\dfrac{T_{1,2} - T_{1,1}}{T_{2,2} - T_{1,1}}\right)} = \frac{(T_{2,2} - T_{1,1}) - (T_{1,2} - T_{1,1})}{(T_{2,2} - T_{1,1}) - (T_{2,2} - T_{2,1})}$$

$$= \frac{T_{2,2} - T_{1,2}}{T_{2,1} - T_{1,1}} = \frac{(\text{fluid 2 inlet})_A - (\text{fluid 1 outlet})_A}{(\text{fluid 2 outlet})_A - (\text{fluid 1 inlet})_A} \tag{3.118}$$

Similarly, we can arrive at the following expression for exchanger B, using the terminology of the right-hand side of Eq. (3.118):

$$\frac{1 - P_{1,B}}{1 - R_1 P_{1,B}} = \frac{T_{2,3} - T_{1,3}}{T_{2,2} - T_{1,2}} \tag{3.119}$$

and for the combined exchanger C from Fig. 3.19*b* as

$$\frac{1 - P_1}{1 - R_1 P_1} = \frac{T_{2,3} - T_{1,3}}{T_{2,1} - T_{1,1}} \tag{3.120}$$

From Eqs. (3.118)–(3.120), it is obvious that

$$\frac{1 - P_1}{1 - R_1 P_1} = \frac{1 - P_{1,A}}{1 - R_1 P_{1,A}} \frac{1 - P_{1,B}}{1 - R_1 P_{1,B}} = X \tag{3.121}$$

where the right-hand side of the first equality sign is designated as X. Then, from Eq. (3.121),

$$P_1 = \frac{1 - X}{1 - R_1 X} = \frac{(1 - R_1 P_{1,A})(1 - R_1 P_{1,B}) - (1 - P_{1,A})(1 - P_{1,B})}{(1 - R_1 P_{1,A})(1 - R_1 P_{1,B}) - R_1(1 - P_{1,A})(1 - P_{1,B})} \tag{3.122}$$

which upon simplification reduces to

$$P_1 = \frac{P_{1,A} + P_{1,B} - (1 + R_1)P_{1,A}P_{1,B}}{1 - R_1 P_{1,A} P_{1,B}} \qquad (3.123)$$

Equation (3.121) could have been obtained easily without the algebraic manipulations of Eqs. (3.118)–(3.120) by the operating line–equilibrium line approach used in chemical engineering or by the matrix formalism method.

If we refer to exchangers A and B of Fig. 3.19 as A_1 and A_2, Eq. (3.122) can be generalized to n exchangers (or passes) A_i in series in overall counterflow as

$$P_1 = \frac{\prod_{i=1}^{n}(1 - R_1 P_{1,A_i}) - \prod_{i=1}^{n}(1 - P_{1,A_i})}{\prod_{i=1}^{n}(1 - R_1 P_{1,A_i}) - R_1 \prod_{i=1}^{n}(1 - P_{1,A_i})} \qquad (3.124)$$

When $R_1 = 1$, it can be shown that Eq. (3.124) reduces to

$$P_1 = \frac{\sum_{i=1}^{n}[P_{1,A_i}/(1 - P_{1,A_i})]}{1 + \sum_{i=1}^{n}[P_{1,A_i}/(1 - P_{1,A_i})]} \qquad (3.125)$$

These formulas are reported in Table 3.6 as Eq. (IV.2.1). For this case, the overall NTU_1 and R_1 are related to the corresponding individual exchanger/pass quantities by the following relationships:

$$NTU_1 = \sum_{i=1}^{n} NTU_{1,A_i} \qquad (3.126)$$

$$R_1 = R_{1,A_i} \qquad i = 1, 2, \ldots, n \qquad (3.127)$$

Now let us summarize the additional assumptions made to derive the overall P_1 from the individual P_{1,A_i} [i.e., the relationships of Eqs. (3.124) and (3.125)].

1. Both fluids are considered perfectly *mixed* between exchangers or passes. This is the reason we considered the outlet temperatures from individual exchangers, $T_{1,2}$ and $T_{2,2}$ in Fig. 3.19a as mixed mean temperatures.
2. Fluid properties and flow rates are idealized as constant, so that R_1 is the same for each exchanger or each pass.

Note that in the derivation of Eq. (3.124), we did not impose any constraints on the NTUs, flow arrangement, and stream symmetry or asymmetry of individual exchangers. Hence, Eq. (3.124) is valid for:

1. Any values of individual NTU_{1,A_i} may be specified (they need *not* be the same).
2. Individual exchangers can have any flow arrangements, such as counterflow, 1–2 TEMA E, parallelflow, and crossflow, as shown in a hypothetical array of exchangers in Fig. 3.20.
3. Individual exchangers can be stream symmetric or asymmetric.

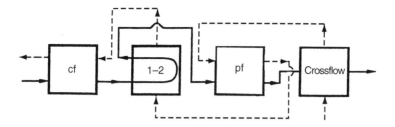

FIGURE 3.20 Hypothetical exchanger array made up of counterflow (cf), 1–2 TEMA E, parallelflow (pf), and crossflow exchangers.

For design purposes, one is also interested in intermediate temperatures between exchangers (or passes). In the example of Fig. 3.19a we want to determine $T_{1,2}$ and $T_{2,2}$. These temperatures expressed as z_1 in dimensionless form are defined as the ratio of the temperature range (rise or drop) of individual fluids 1 and 2 in exchanger A to that in the combined equivalent exchanger C. They are defined by the first two equality signs in the following equation and can be derived as follows:

$$z_1 = \frac{T_{1,2} - T_{1,1}}{T_{1,3} - T_{1,1}} = \frac{T_{2,2} - T_{2,1}}{T_{2,3} - T_{2,1}} = \frac{[(1 - P_{1,A})/(1 - R_1 P_{1,A})] - 1}{[(1 - P_1)/(1 - R_1 P_1)] - 1} \tag{3.128}$$

Here the terms after the first and second equality signs are based on the energy balances on exchangers A and C; the term after the third equality sign can be derived using Eqs. (3.118) and (3.120). If there are n exchangers in series, the temperatures of fluids 1 and 2 after the jth exchanger/pass are $T_{1,j+1}$ and $T_{2,j+1}$, given by

$$z_j = \frac{T_{1,1} - T_{1,j+1}}{T_{1,1} - T_{1,n+1}} = \frac{T_{2,1} - T_{2,j+1}}{T_{2,1} - T_{2,n+1}} = \frac{\prod_{i=1}^{j} [(1 - P_{1,A_i})/(1 - R_1 P_{1,A_i})] - 1}{[(1 - P_1/(1 - R_1 P_1)] - 1} \tag{3.129}$$

For $R_1 = 1$, Eq. (3.129) reduces to

$$z_j = (1 - P_1) \sum_{i=1}^{j} \frac{1}{1 - P_{1,A_i}} \tag{3.130}$$

If all n exchangers (or passes) have identical flow arrangements and identical individual NTU (i.e., NTU_p), Eqs. (3.124) and (3.129) are further simplified as follows:

$$P_1 = \frac{[(1 - R_1 P_{1,p})/(1 - P_{1,p})]^n - 1}{[(1 - R_1 P_{1,p})/(1 - P_{1,p})]^n - R_1} \tag{3.131}$$

$$z_j = \frac{[(1 - P_{1,p})/(1 - R_1 P_{1,p})]^j - 1}{(1 - P_1)/(1 - R_1 P_1) - 1} \tag{3.132}$$

$$z_j = \frac{[(1 - P_1)/(1 - R_1 P_1)]^{j/n} - 1}{(1 - P_1)/(1 - R_1 P_1) - 1} \tag{3.133}$$

Here $P_{1,p}$ is the temperature effectiveness of each pass (or individual identical exchangers). Equation (3.131) reduces to the following forms when $R_1 = 1$ and 0:

$$P_1 = \begin{cases} \dfrac{nP_{1,p}}{1 + (n-1)P_{1,p}} & \text{for } R_1 = 1 \qquad (3.134) \\[4mm] 1 - (1 - P_{1,p})^n & \text{for } R_1 = 0 \qquad (3.135) \end{cases}$$

The temperature effectiveness per pass $P_{1,p}$ can be expressed in terms of P_1, R_1, and n by inverting Eq. (3.131) as follows:

$$P_{1,p} = \frac{[(1 - R_1 P_1)/(1 - P_1)]^{1/n} - 1}{[(1 - R_1 P_1)/(1 - P_1)]^{1/n} - R_1} \qquad (3.136)$$

For the special cases, Eq. (3.136) reduces to

$$P_{1,p} = \begin{cases} \dfrac{P_1}{n - (n-1)P_1} & \text{for } R_1 = 1 \qquad (3.137) \\[4mm] 1 - (1 - P_1)^{1/n} & \text{for } R_1 = 0 \qquad (3.138) \end{cases}$$

Finally, the number of passes n can be presented as a function of P_1, $P_{1,p}$, and R_1 by inverting Eq. (3.131) as

$$n = \frac{\ln[(1 - R_1 P_1)/(1 - P_1)]}{\ln\left[(1 - R_1 P_{1,p})/(1 - P_{1,p})\right]} \xrightarrow{R_1=1} \frac{P_1(1 - P_{1,p})}{P_{1,p}(1 - P_1)} \qquad (3.139)$$

When individual passes of a multipass overall counterflow exchanger are crossflow exchangers (such as exchangers A and B in Fig. 3.19a), the resultant arrangement is referred to as the *multipass cross-counterflow*.

There are many possible combinations of the multipass overall cross-counterflow exchanger, depending on the following: (1) each fluid is mixed or unmixed between passes (in the preceding section we considered the fluids mixed between passes); (2) each fluid is mixed or unmixed within each pass; (3) the fluid that is unmixed between passes has an identical or inverted order (see Fig. 1.55); (4) fluid 1 is the $C_{\min}$ or $C_{\max}$ fluid; and (5) the number of passes. Bačlić (1990) has provided closed-form formulas for 36 such two-pass cross-counterflow exchangers. The simplified relationship of Eq. (3.131) is adequate for the design and analysis of most two- and three-pass industrial exchangers.

Series Coupling. Overall Parallelflow. If the direction of one of the two fluids in Fig. 3.19 is reversed, both fluids will flow in the same overall direction, as shown in Fig. 3.21. The resultant series coupling then has an overall parallelflow arrangement. In this case, the overall temperature effectiveness P_1 of such n exchangers (or passes) in series is given as follows [which is a counterpart of Eq. (3.124) and can be derived similarly].

$$P_1 = \frac{1}{1 + R_1}\left\{1 - \prod_{i=1}^{n}[1 - (1 + R_1)P_{1,A_i}]\right\} \qquad (3.140)$$

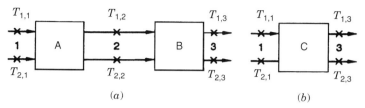

FIGURE 3.21   (*a*) Series-coupled exchangers A and B in overall parallelflow arrangement; (*b*) combined equivalent exchanger C.

The temperatures of fluids 1 and 2 after the *j*th exchanger (pass) are given by

$$z_j = \frac{T_{1,1} - T_{1,j+1}}{T_{1,1} - T_{1,n+1}} = \frac{T_{2,1} - T_{2,j+1}}{T_{2,1} - T_{2,n+1}} = \frac{\prod_{i=1}^{j} P_{1,A_i}}{P_1} \tag{3.141}$$

For this case of series coupling with overall parallelflow, individual NTU_{1,A_i} and R_{1,A_i} are related to the overall NTU_1 and R_1 by Eqs. (3.126) and (3.127).

If all *n* exchangers (or passes) have identical flow arrangements and identical individual NTUs (i.e., NTU_p), Eqs. (3.140) and (3.141) simplify to

$$P_1 = \frac{1}{1 + R_1}\left\{1 - [1 - (1 + R_1)P_{1,p}]^n\right\} \tag{3.142}$$

$$z_j = \frac{P_{1,p}^j}{P_1} = \frac{P_1^{j/n}}{P_1} = P_1^{-(1-j/n)} \tag{3.143}$$

The temperature effectiveness per pass, $P_{1,p}$, can be expressed as follows from Eq. (3.142):

$$P_{1,p} = \frac{1}{1 + R_1}\left\{1 - [1 - (1 + R_1)P_1]^{1/n}\right\} \tag{3.144}$$

It should be emphasized that if *n* exchangers are series coupled in overall parallelflow, their P_1 and R_1 are related as follows for $\text{NTU}_1 \to \infty$ [see Table 3.6, Eq. (I.2.1) for $\text{NTU}_1 \to \infty$]

$$P_1 = \frac{1}{1 + R_1} \tag{3.145}$$

Hence,

$$(1 + R_1)P_1 < 1 \text{ for } \text{NTU}_1 < \infty \tag{3.146}$$

The condition of Eq. (3.146) applies to Eq. (3.144) for its use.

Finally, the number of passes can be expressed from Eq. (3.142) as

$$n = \frac{\ln[1 - (1 + R_1)P_1]}{\ln[1 - (1 + R_1)P_{1,p}]} \tag{3.147}$$

When individual passes of the multipass overall parallelflow exchanger are crossflow exchangers, the resultant arrangement is referred to as the *multipass cross-parallelflow*. There are many possible combinations of this flow arrangement: (1) each fluid is mixed or unmixed between passes (in the preceding section we considered the fluids mixed between passes; (2) each fluid is mixed or unmixed within each pass; (3) the fluid that is unmixed between passes has an identical or inverted order (see Fig. 1.55); (4) fluid 1 is the C_{min} or C_{max} fluid; and (5) the number of passes. Bačlić (1990) has provided closed-form formulas for 36 such two-pass cross-parallelflow exchangers. The simplified relationship of Eq. (3.142) is adequate for the design and analysis of most two- and three-pass industrial exchangers.

Example 3.4   The heat exchanger array shown in Fig. E3.4*a* has hot and cold fluids with equal heat capacity rates and inlet temperatures of 300°C and 100°C, respectively. Heat exchangers A, B, and C are unmixed–unmixed crossflow, counterflow, and parallelflow exchangers, respectively. The corresponding NTUs for each exchanger are 2, 3, and 3. Determine:

(a) the overall exchanger array effectiveness and outlet temperatures at the exit of exchanger C

(b) the outlet temperatures at the exits of exchangers A and B.

SOLUTION

Problem Data and Schematic: Fluid inlet temperatures and each exchanger's NTU are given as shown in Fig. E3.4*a*. Also given are the flow arrangements of each exchanger.

Determine: The overall effectiveness and the outlet temperatures at the exit of each exchanger.

Assumptions: The assumptions invoked in Section 3.2.1 are valid, the hot and cold fluids have equal heat capacity rates, and both fluids are mixed between passes.

Analysis: (a) The overall exchanger array effectiveness will be evaluated first by considering exchangers A and B in the overall counterflow direction. Since the heat capacity rates are equal, $R = 1$. Knowing R, NTU, and the flow arrangement of each exchanger, we could obtain the effectiveness of each exchanger from the formulas in Table 3.6.

$$P_A = 0.6142 \quad \text{for NTU}_A = 2 \text{ from Table 3.6, Eq. (II.1)}$$

$$P_B = 0.7500 \quad \text{for NTU}_B = 3 \text{ from Table 3.6, Eq. (I.1.1)}$$

$$P_C = 0.4988 \quad \text{for NTU}_C = 3 \text{ from Table 3.6, Eq. (I.2.1)}$$

Note that P, NTU, and R are the same for fluids 1 and 2 since $R = 1$. Hence, the subscripts A, B, and C used here for P and NTU designate those values for exchangers A, B, and C for either fluid.

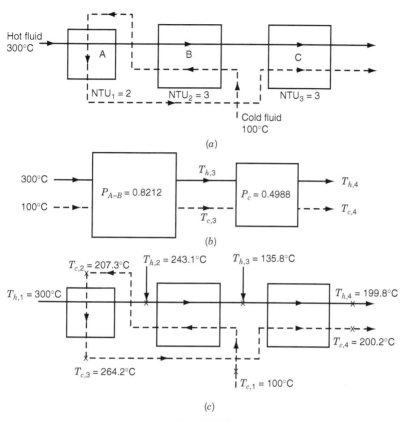

FIGURE E3.4

Considering exchangers A and B in overall counterflow arrangement, we could calculate the overall effectiveness for the two exchangers from Eq. (3.125):

$$P_{\text{A}-\text{B}} = \frac{0.6142/0.3858 + 0.7500/0.2500}{1 + 0.6142/0.3858 + 0.7500/0.2500} = 0.8212$$

So the problem now looks as shown in Fig. E3.4*b*. These exchangers are in overall parallelflow arrangement. Hence, the overall effectiveness from Eq. (3.140) for $R = 1$ is

$$P = \frac{1}{1+R}\{1 - [1 - (1+R)P_{\text{A}-\text{B}}][1 - (1+R)P_{\text{B}}]\}$$

$$= \tfrac{1}{2}\{1 - [1 - (2 \times 0.8212)][1 - (2 \times 0.4988)]\} = 0.5008 \qquad Ans.$$

The definition of the overall P in terms of the temperatures for $R = 1$ are as follows:

$$P = \frac{T_{h,i} - T_{h,o}}{T_{h,i} - T_{c,i}} = \frac{T_{h,1} - T_{h,4}}{T_{h,1} - T_{c,1}} \qquad P = \frac{T_{c,o} - T_{c,i}}{T_{h,i} - T_{c,i}} = \frac{T_{c,4} - T_{c,1}}{T_{h,1} - T_{c,1}}$$

Rearranging the equations above, we could solve for $T_{h,4}$ and $T_{c,4}$:

$$T_{h,4} = T_{h,1} - P(T_{h,1} - T_{c,1}) = 300°C - 0.5008(300 - 100)°C = 199.8°C \qquad Ans.$$

$$T_{c,4} = T_{c,1} + P(T_{h,1} - T_{c,1}) = 100°C + 0.5008(300 - 100)°C = 200.2°C \qquad Ans.$$

(b) Knowing the overall effectiveness of exchangers A and B to be P_{A-B}, we could again use the definition of effectiveness to solve for the cold-fluid outlet temperature at exchanger B and the hot-fluid outlet temperature at exchanger A.

$$T_{h,3} = T_{h,1} - P_{A-B}(T_{h,1} - T_{c,1}) = 300°C - 0.8212(300 - 100)°C = 135.8°C \qquad Ans.$$

$$T_{c,3} = T_{c,1} + P_{A-B}(T_{h,1} - T_{c,1}) = 100°C + 0.8212(300 - 100)°C = 264.2°C \qquad Ans.$$

Thus, we know all temperatures except for the hot-fluid inlet temperature to exchanger B and the cold-fluid inlet temperature to exchanger A. Applying the definition of the temperature effectiveness to exchanger A yields

$$T_{c,3} = T_{c,2} + P_A(300°C - T_{c,2})$$

With known $T_{c,3} = 264.2°C$, we get

$$T_{c,2} = 207.3°C \qquad Ans.$$

Using the definition of the temperature effectiveness again for exchanger A, we get the following equation:

$$T_{h,2} = T_{h,1} - P_A(T_{h,1} - T_{c,2}) = 300°C - 0.6142(300 - 207.3)°C = 243.1°C \qquad Ans.$$

Thus, all temperatures are now known as shown in Fig. E3.4c.

Discussion and Comments: The objective of this example is to demonstrate how to do the analysis of a heat exchanger array when individual exchangers have different flow arrangements and different values of NTU. While the formulas for exchanger arrays derived in the text were used for the analysis, the intermediate temperatures were computed using the definition of the temperature effectiveness for each exchanger rather than using those formulas in the text.

Even though the overall flow arrangement is parallelflow between exchanger C and combined exchangers A and B, the total exchanger effectiveness approaches that of a parallelflow (i.e., 0.5 asymptotically in an oscillatory manner). If we review the inlet temperatures to exchanger C, the original hot fluid is now cold, and the original cold fluid is now hot. The heat transfer occurs in the opposite direction than what one would have thought. Hence, the temperatures at the outlet of exchanger C do make sense (i.e., $T_{h,4} < T_{c,4}$).

Parallel Coupling. Consider exchangers A and B in parallel coupling, as shown in Fig. 3.22a. The results presented below are equally valid if the two exchangers are considered as two passes of a multipass crossflow exchanger with fluids mixed between passes. Figure 3.22b shows their combined equivalent exchanger C. In this figure, the temperatures are shown by two subscripts: The first subscript refers to either fluid 1 or fluid 2, and

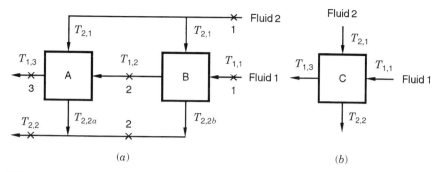

(a)   (b)

FIGURE 3.22   (a) Parallel coupled exchangers A and B; (b) combined equivalent exchanger C.

the second subscript denotes the mean terminal temperature at the location outside the passes or exchangers. The temperature effectiveness of fluid 1 of exchangers A, B, and C from their definitions [see Eq. (3.96)] is given by

$$P_{1,A} = \frac{T_{1,3} - T_{1,2}}{T_{2,1} - T_{1,2}} \qquad P_{1,B} = \frac{T_{1,2} - T_{1,1}}{T_{2,1} - T_{1,1}} \qquad P_{1,C} = \frac{T_{1,3} - T_{1,1}}{T_{2,1} - T_{1,1}} \qquad (3.148)$$

To relate $P_{1,A}$ and $P_{1,B}$ with $P_{1,C}$, let us first compute the following expressions from their definitions of Eq. (3.148):

$$1 - P_{1,A} = 1 - \frac{T_{1,3} - T_{1,2}}{T_{2,1} - T_{1,2}} = \frac{T_{2,1} - T_{1,3}}{T_{2,1} - T_{1,2}} \qquad (3.149)$$

$$1 - P_{1,B} = 1 - \frac{T_{1,2} - T_{1,1}}{T_{2,1} - T_{1,1}} = \frac{T_{2,1} - T_{1,2}}{T_{2,1} - T_{1,1}} \qquad (3.150)$$

$$1 - P_{1,C} = 1 - \frac{T_{1,3} - T_{1,1}}{T_{2,1} - T_{1,1}} = \frac{T_{2,1} - T_{1,3}}{T_{2,1} - T_{1,1}} \qquad (3.151)$$

Hence, from Eqs. (3.149)–(3.151), it can be shown that

$$(1 - P_{1,A})(1 - P_{1,B}) = (1 - P_{1,C}) \qquad (3.152)$$

Rearranging Eq. (3.152), we get

$$P_{1,C} = 1 - (1 - P_{1,A})(1 - P_{1,B}) = P_{1,A} + P_{1,B} - P_{1,A}P_{1,B} \qquad (3.153)$$

For this parallel coupling, the heat capacity rate of fluid 2 is divided into two components:

$$C_2 = C_{2,A} + C_{2,B} \qquad (3.154)$$

Hence R_1 $(= C_1/C_2)$ will be related to $R_{1,A}$ $(= C_1/C_{2,A})$ and $R_{1,B}$ $(= C_1/C_{2,B})$ as

$$\frac{1}{R_1} = \frac{1}{R_{1,A}} + \frac{1}{R_{1,B}} \tag{3.155}$$

If we refer to exchangers A and B of Fig. 3.22 as A_1 and A_2, Eqs. (3.153) and (3.155) can be generalized to n exchangers (or passes) A_i in parallel coupling as

$$P_1 = 1 - \prod_{i=1}^{n} (1 - P_{1,A_i}) \tag{3.156}$$

$$\frac{1}{R_1} = \sum_{i=1}^{n} \frac{1}{R_{1,A_i}} \tag{3.157}$$

and

$$\mathrm{NTU}_1 = \sum_{i=1}^{n} \mathrm{NTU}_{1,A_i} \tag{3.158}$$

In the foregoing derivations, we idealized both fluids perfectly *mixed* between exchangers or passes as well as fluid properties and flow rates as constant. Here again, we did not impose any constraints on the magnitude of individual exchanger NTUs, flow arrangement, and stream symmetry or asymmetry condition.

If all passes (exchangers) have the same NTU_{1,A_i} and R_{1,A_i}, they will have the same temperature effectiveness $P_{1,p}$, and Eq. (3.156) simplifies to

$$P_1 = 1 - (1 - P_{1,p})^n \tag{3.159}$$

or

$$P_{1,p} = 1 - (1 - P_1)^{1/n} \tag{3.160}$$

and

$$n = \frac{\ln(1 - P_1)}{\ln(1 - P_{1,p})} \tag{3.161}$$

Equation (3.159) can be presented in terms of P_2 (of fluid stream 2 of Fig. 3.22) as follows since $P_1 = P_2 R_2$ from Eq. (3.98) and hence $P_{1,A_i} = P_{2,A_i} R_{2,A_i}$:

$$P_2 = \frac{1}{R_2} \left[1 - \left(1 - P_{2,p} R_{2,p} \right)^n \right] = \frac{1}{n R_{2,p}} \left[1 - \left(1 - P_{2,p} R_{2,p} \right)^n \right] \tag{3.162}$$

where $R_2 = n R_{2,p}$ from Eq. (3.157) since $R_1 = 1/R_2$ and $R_{1,A_i} = 1/R_{2,A_i}$.

Exchangers shown in Fig. 1.55e and g are parallel-coupled exchangers, and their temperature effectiveness P_1 is calculated from Eq. (3.159) or Eq. (IV.1.1) of Table 3.6.

Example 3.5   A decision needs to be made on connecting two 1–2 shell-and-tube exchangers so that minimum total surface area is required to heat oil from 15°C to 60°C with a mass flow rate of 1.666 kg/s using hot water at 95°C and 1.000 kg/s flow rate. The overall heat transfer coefficient is 540 W/m² · K based on the shell-side surface area. The specific heats of oil and water are 1675 and 4187 J/kg · K, respectively. The oil flows in the tubes. Determine the *total* surface area required to do the job. Consider fluids mixed between passes and the shell-side fluid mixed at every cross section.

SOLUTION

Problem Data and Schematic: The overall heat transfer coefficient, fluid flow rates, inlet temperatures, and cold-fluid outlet temperature are provided for two 1–2 shell-and-tube exchangers in series (Fig. E3.5). The tube-side fluid is oil and the shell-side fluid is hot water. Consider four possible arrangements for the two exchangers as shown in the figure:

(a) Series coupling and overall counterflow
(b) Series coupling with overall parallelflow
(c) Parallel coupling with the shell fluid in series
(d) Parallel coupling with the tube fluid in series

Determine: The total surface area required to do the job for each case.

Assumptions: The assumptions invoked in Section 3.2.1 are valid, there are no fins on either side of tubes, and the two 1–2 exchangers are identical (i.e., have the same NTU_p).

Analysis: (a) *Two exchangers in series in overall counterflow* (Fig. E3.5a): Since this is an exchanger array problem, we cannot calculate total NTU directly. First, we need to calculate the temperature effectiveness of one of the fluids for one exchanger and then determine NTU_p to come up with total NTU and A. We need to evaluate R_1 and P_1 to proceed. Here we designate the shell fluid (water) as fluid 1 and the tube fluid (oil) as fluid 2.

$$C_{water} = C_s = C_1 = (\dot{m}c_p)_s = 1.000\,\text{kg/s} \times 4187\,\text{J/kg} \cdot \text{K} = 4187.0\,\text{W/K}$$

$$C_{oil} = C_t = C_2 = (\dot{m}c_p)_t = 1.666\,\text{kg/s} \times 1675\,\text{J/kg} \cdot \text{K} = 2790.6\,\text{W/K}$$

$$R_1 = \frac{C_1}{C_2} = \frac{4187.0\,\text{W/K}}{2790.6\,\text{W/K}} = 1.50 = R_{1,p}$$

To calculate the shell fluid temperature effectiveness P_1, we calculate the water outlet temperature based on the energy balance, using Eq. (3.5).

$$(2790.6\,\text{W/K})(60-15)°\text{C} = (4187.0\,\text{W/K})(95 - T_{h,o})°\text{C}$$

Hence,

$$T_{h,o} = T_{s,o} = 65.0°\text{C}$$

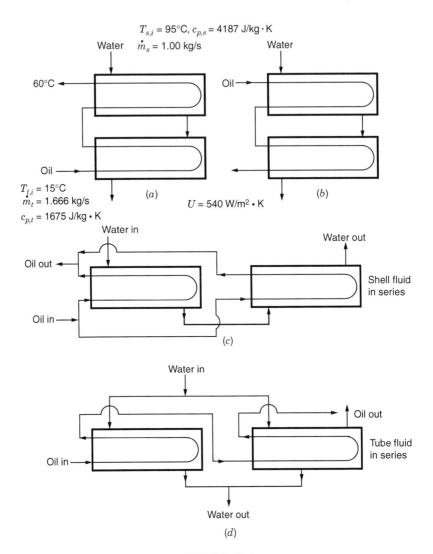

FIGURE E3.5

The shell-side temperature effectiveness, using Eq. (3.96), is

$$P_s = P_1 = \frac{T_{s,i} - T_{s,o}}{T_{s,i} - T_{t,i}} = \frac{95°C - 65°C}{95°C - 15°C} = 0.375$$

Now use Eq. (3.136) to compute the temperature effectiveness of the same fluid per pass:

$$P_{1,p} = \frac{[(1 - R_1 P_1)/(1 - P_1)]^{1/n} - 1}{[(1 - R_1 P_1)/(1 - P_1)]^{1/n} - R_1} = \frac{[(1 - 1.50 \times 0.375)/(1 - 0.375)]^{1/2} - 1}{[(1 - 1.50 \times 0.375)/(1 - 0.375)]^{1/2} - 1.50} = 0.2462$$

Using Eq. (III.1.2) of Table 3.6 or Fig. 3.14, we get $NTU_{1,p} = NTU_{s,p}$:

$$E = [1 + R_1^2]^{1/2} = (1 + 1.50^2)^{1/2} = 1.8028$$

$$NTU_{1,p} = \frac{1}{E} \ln \frac{2 - P_{1,p}(1 + R_{1,p} - E)}{2 - P_{1,p}(1 + R_{1,p} + E)} = \frac{1}{1.8028} \ln \frac{2 - 0.2462(1 + 1.5 - 1.8028)}{2 - 0.2462(1 + 1.5 + 1.8028)}$$
$$= 0.3686$$

Hence, using Eq. (3.126) gives us

$$NTU_1 = \sum_{i=1}^{2} NTU_{1,p} = 0.3686 + 0.3686 = 0.7372$$

and

$$A = \frac{NTU_1 \times C_1}{U} = \frac{0.7372 \times 4187.0 \, \text{W/K}}{540 \, \text{W/m}^2 \cdot \text{K}} = 5.72 \, \text{m}^2 \qquad \textit{Ans.}$$

(b) *Two exchangers in series in overall parallelflow* (Fig. E3.5b). In this case, R_1 and P_1 are specified as in part (a).

$$R_1 = 1.50 \qquad P_1 = 0.375$$

The temperature effectiveness per pass, $P_{1,p}$, is computed from Eq. (3.144) as

$$P_{1,p} = \frac{1}{1 + R_1} \{ 1 - [1 - (1 + R_1)P_1]^{1/n} \}$$
$$= \frac{1}{1 + 1.50} \{ 1 - [1 - (1 + 1.50) \times 0.375]^{1/2} \} = 0.300$$

Using Eq. (III.1.2) of Table 3.6 or Fig. 3.14, we get $NTU_{1,p} = NTU_{s,p}$ as follows with $E = 1.8028$ as before:

$$NTU_{1,p} = \frac{1}{E} \ln \frac{2 - P_{1,p}(1 + R_{1,p} - E)}{2 - P_{1,p}(1 + R_{1,p} + E)} = \frac{1}{1.8028} \ln \frac{2 - 0.300 \times (1 + 1.5 - 1.8028)}{2 - 0.300 \times (1 + 1.5 + 1.8028)}$$
$$= 0.5138$$

Hence, using Eq. (3.126) yields

$$NTU_1 = 2NTU_{1,p} = 2 \times 0.5138 = 1.0277$$

and

$$A = \frac{NTU_1 \times C_1}{U} = \frac{1.0277 \times 4187.0 \, \text{W/K}}{540 \, \text{W/m}^2 \cdot \text{K}} = 7.97 \, \text{m}^2 \qquad \textit{Ans.}$$

(c) *Parallel coupling with the shell fluid in series* (Fig. E3.5c). In this case, for the equivalent combined exchanger, R_1 and P_1 are 1.50 and 0.375. The individual pass temperature effectiveness of fluid 1 is computed from Eq. (3.160) as

$$P_{1,p} = 1 - (1 - P_1)^{1/2} = 1 - (1 - 0.375)^{1/2} = 0.2094$$

The heat capacity rate ratio for the individual pass will change for parallel coupling and can be determined from Eq. (3.157) as

$$R_{1,p} = 2R_1 = 3.00$$

Thus $NTU_{1,p}$ is computed from Eq. (III.1.2) of Table 3.6 with $E = [1 + R_1^2]^{1/2} = (1 + 3.00^2)^{1/2} = 3.1623$ as

$$NTU_{1,p} = \frac{1}{E} \ln \frac{2 - P_{1,p}(1 + R_{1,p} - E)}{2 - P_{1,p}(1 + R_{1,p} + E)} = \frac{1}{3.1623} \ln \frac{2 - 0.2094(1 + 3.00 - 3.1623)}{2 - 0.2094(1 + 3.00 + 3.1623)}$$

$$= 0.4092$$

Hence, using Eq. (3.158) gives us

$$NTU_1 = \sum_{i=1}^{2} NTU_{1,p} = 0.4092 + 0.4092 = 0.8184$$

and

$$A = \frac{NTU_1 \times C_1}{U} = \frac{0.8184 \times 4187.0 \, \text{W/K}}{540 \, \text{W/m}^2 \cdot \text{K}} = 6.35 \, \text{m}^2 \qquad \textit{Ans.}$$

(d) *Parallel coupling with the tube fluid in series* (Fig. E3.5d). For the parallel coupling, consider the temperature effectiveness of the tube fluid in series. Hence, in this case, for the equivalent combined exchanger, *redefine* R_1 and P_1 on the tube side and compute the values as

$$R_1 = R_t = \frac{C_{\text{oil}}}{C_{\text{water}}} = \frac{2790.6 \, \text{W/K}}{4187.0 \, \text{W/K}} = 0.666$$

$$P_1 = P_t = \frac{T_{t,o} - T_{t,i}}{T_{s,i} - T_{t,i}} = \frac{(60 - 15)^\circ\text{C}}{(95 - 15)^\circ\text{C}} = 0.5625$$

The individual pass temperature effectiveness of fluid 1 is computed from Eq. (3.160) as

$$P_{1,p} = 1 - (1 - P_1)^{1/2} = 1 - (1 - 0.5625)^{1/2} = 0.3386$$

The heat capacity rate ratio for the individual pass will change for parallel coupling as in the previous case and can be determined from Eq. (3.157) as

$$R_{1,p} = R_{t,p} = 2R_1 = 1.332$$

Since the 1–2 TEMA E exchanger is a symmetrical exchanger, the formulas presented in Table 3.6, Eq. (III.1), are also valid for the tube side as long as P_1, NTU_1, and R_1 are defined consistently for the tube side. Hence, we compute $NTU_{1,p}$ from Eq. (III.1.2) of Table 3.6 with $E = (1 + R_{1,p}^2)^{1/2} = (1 + 1.332^2)^{1/2} = 1.6656$ as

$$NTU_{1,p} = \frac{1}{E} \ln \frac{2 - P_{1,p}(1 + R_{1,p} - E)}{2 - P_{1,p}(1 + R_{1,p} + E)} = \frac{1}{1.6656} \ln \frac{2 - 0.3386(1 + 1.332 - 1.6656)}{2 - 0.3386(1 + 1.332 + 1.6656)}$$

$$= 0.6062 = NTU_{t,p}$$

Hence, using Eq. (3.158), the total tube side NTU is

$$NTU_t = 2NTU_{1,p} = 2 \times 0.6062 = 1.2124$$

$$UA = NTU_t C_t = 1.2124 \times 2790.6\,\text{W/K} = 3383.3\,\text{W/K}$$

Finally, the shell-side total surface area is

$$A = \frac{UA}{U} = \frac{3383.3\,\text{W/K}}{540\,\text{W/m}^2 \cdot \text{K}} = 6.27\,\text{m}^2 \qquad \textit{Ans.}$$

To compare results with all three previous cases, we redefine all important parameters for this last case based on the shell side.

$$P_s = P_t R_t = 0.5625 \times 0.666 = 0.375$$

$$R_s = \frac{1}{R_t} = \frac{1}{0.666} = 1.50$$

$$R_{s,p} = \frac{1}{R_{t,p}} = \frac{1}{1.332} = 0.75$$

$$P_{s,p} = P_{t,p} R_{t,p} = 0.3386 \times 1.332 = 0.4510$$

$$NTU_{s,p} = NTU_{t,p} R_{t,p} = 0.6062 \times 1.332 = 0.8075$$

$$NTU_s = NTU_t R_t = 1.2124 \times 0.666 = 0.8075$$

We can now summarize the results of four cases analyzed (see Table E3.5).

TABLE E3.5 Results of Analysis

Shell-side Parameter	Series Coupling, Overall Counterflow [Case (a)]	Series Coupling, Overall Parallelflow [Case (b)]	Parallel Coupling, Shell Fluid (C_{max}) in Series [Case (c)]	Parallel Coupling, Tube Fluid (C_{min}) in Series [Case (d)]
P_s	0.375	0.375	0.375	0.375
R_s	1.50	1.50	1.50	1.50
$P_{s,p}$	0.2462	0.300	0.2094	0.4510
$R_{s,p}$	1.50	1.50	3.00	0.75
$NTU_{s,p}$	0.3686	0.5138	0.4092	0.8075
NTU_s	0.7372	1.0277	0.8184	0.8075
$A_s(\text{m}^2)$	5.72	7.97	6.35	6.27

Discussion and Comments: The objective of this example is threefold: compare (1) series coupling arrangements, (2) parallel coupling arrangements, and (3) series vs. parallel coupling arrangements.

- A comparison of cases (a) and (b) of series coupling indicates as expected that the overall counterflow arrangement is superior to the overall parallelflow exchanger requiring $5.72\,m^2$ vs. $7.97\,m^2$ surface area (about 28% less) for the same desired performance. As a matter of fact, the desired overall temperature effectiveness of fluid 1 over 40% cannot be achieved in the overall parallelflow arrangement; NTU_1 or A_1 of infinity should have resulted in $P_1 = 40\%$, as per Eq. (3.145).

- Comparison of cases (c) and (d) indicate that these arrangements feature a minimal difference in overall performance. Far reaching conclusions regarding the performance level of these arrangements vs. (a) and (b) cannot be provided in all details due to specified low exchanger effectiveness. However, what can be stated is that splitting the streams and mixing them again causes a deterioration of the overall performance (for a further discussion see Section 11.5) when compared with a case that does not feature flow splitting and has the same overall flow direction (like in case a). Of course, cases (c) and (d) still have better performance than the overall parallelflow of case (b).

- Finally, a comparison of the best cases of series and parallel coupling, cases (a) and (d), indicate clearly, as expected, that series coupling with overall counterflow arrangement yields smallest area requirement than does parallel coupling with the C_{min} fluid in series. This is due to the fact that overall counterflow leads to smaller temperature differences than in parallelflow of two separated streams (in parallel coupling); this, in turn, results in the smallest surface area requirement for the same temperature effectiveness for the overall counterflow arrangement.

In addition to the considerations above, several other points need to be considered when designing a heat exchanger. From the heat transfer point of view, whenever the fluid velocity is reduced in the exchanger, the corresponding heat transfer coefficient will be lower in turbulent and transition flow regimes than that for the nominal fluid velocity. This, in turn, may reduce the overall heat transfer coefficient and the heat transfer rate in the exchanger. However, the lower velocity will also reduce the pressure drop on that side. Now compare the parallel coupling vs. series coupling. The fluid, which splits into two streams [oil for case (c) and water for case (d)], will have lower heat transfer coefficient and will result in lower overall U and higher required exchanger surface area than those calculated in the example above. However, if the pressure drop specified is severely constrained, parallel coupling may be a choice for the fluid with the limited pressure drop, to split it into two streams in the parallel coupling. Of course, this solution will result in a twofold penalty in surface area requirement for the same heat transfer: (1) an increase in the surface area for parallel coupling vs. that for series coupling with overall counterflow as in the example above, and (2) a reduction in the heat transfer coefficient and U for parallel coupling resulting in increased surface area. Example 7.6 also provides some insight into performance of single-pass vs. two-pass series or parallel coupled exchanges. In addition, there are other considerations, such as fouling, laminar vs. turbulent flow on the shell side, pressure containment, and thermal stresses, that should be considered before the specific choice is made for series vs. parallel coupling for a given application.

Compound Coupling. Many examples are found in industrial applications having compound multipass exchangers and exchanger arrays: (1) a mixed coupling (a series coupling of exchangers with a combination of overall counterflow and parallelflow; Fig. 3.23), (2) a combination of series and parallel couplings (Fig. 3.24), and (3) a compound coupling that cannot be reduced to any simple configuration (Fig. 3.25). Let us illustrate how we can analyze such compound assemblies by three examples.

Series Coupling of a TEMA E Exchanger. Consider a 1–2 TEMA E shell-and-tube exchanger with four transverse (plate) baffles, as shown in Fig. 3.23*a*. The sections between two baffles with pairs of idealized single-pass unmixed–unmixed crossflow exchangers are assumed to have individual fluids mixed between passes, as shown in Fig. 3.23*b*. Using the series coupling of Fig. 3.19, we coalesce exchangers B1 and B2 into a single unit B. Performing an analogous operation with two pairs C1 and C2, D1 and D2, and E1 and E2, we arrive at the configuration shown in Fig. 3.23*c*. The series-coupled exchangers A and B of Fig. 3.23*c* can be combined into one which then can be reduced with *C*, and so on, to end up eventually with one exchanger. Hence, the

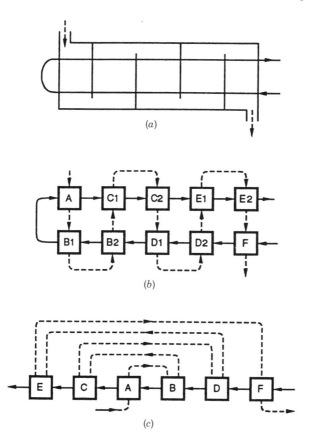

(a)

(b)

(c)

FIGURE 3.23   (*a*) 1–2 TEMA E exchanger with four baffles; (*b*) decomposition into coupled unmixed–unmixed crossflow units mixed between passes; and (*c*) first reduction of the model in (*b*) (From Shah and Pignotti, 1989).

overall combined exchanger temperature effectiveness of fluid 1 of Fig. 3.23a can be computed, making repeated use of the appropriate series coupling formulas given earlier along with the P-NTU formula of the individual exchangers; in this case, they are identical crossflow units.

Series and Parallel Coupled 2 Pass–4 Pass Plate Heat Exchanger. Consider a 2 pass–4 pass plate heat exchanger shown in Fig. 3.24a. We now derive P_1–NTU$_1$ relationship for this exchanger to illustrate how a closed-form formula can be derived for an exchanger that involves series and parallel couplings. A decomposition of this exchanger into counterflow and parallelflow units is shown in Fig. 3.24b. Keeping in mind that exchangers A and D are counterflow and B and C are parallelflow, they can be shown simply as boxes in Fig. 3.24c. We will refer to their temperature effectivenesses as P_{cf} and P_{pf}, respectively, with the subscripts cf and pf here denoting counterflow and parallelflow. Note that for individual exchangers of Fig. 3.24c, fluid 1 heat capacity rate is $C_1/2$ and the surface area is $A/4$. Hence,

$$R_{1,A} = \frac{C_1/2}{C_2} = \frac{1}{2}\frac{C_1}{C_2} = \frac{R_1}{2} = R_{1,B} = R_{1,C} = R_{1,D} \qquad (3.163)$$

$$\mathrm{NTU}_{1,A} = \frac{UA_A}{C_{1,A}} = \frac{UA/4}{C_1/2} = \frac{1}{2}\frac{UA}{C_1} = \frac{\mathrm{NTU}_1}{2} = \mathrm{NTU}_{1,B} = \mathrm{NTU}_{1,C} = \mathrm{NTU}_{1,D} \qquad (3.164)$$

Here A in NTU$_1$ is the total surface area on fluid 1 side of the 2 pass–4 pass exchanger.

Exchangers A and B in Fig. 3.24c are in parallel coupling with fluid 1 divided equally in these exchangers. Note that fluids 1 and 2 in this subassembly correspond to fluids 2 and 1 in Fig. 3.22. Hence the combined temperature effectiveness $P_{2,E}$ of this subassembly (referred to as E in Fig. 3.24d) is given by Eq. (3.153) with the subscript 1 replaced by 2:

$$P_{2E} = 1 - (1 - P_{2,A})(1 - P_{2,B}) \qquad (3.165)$$

Using the relationship of $P_2 = P_1R_1$ of Eq. (3.98), we reduce Eq. (3.165) as

$$P_{1,E}R_{1,E} = 1 - (1 - R_{1,A}P_{1,A})(1 - R_{1,B}P_{1,B}) \qquad (3.166)$$

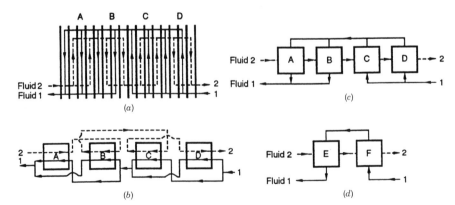

FIGURE 3.24   (a) 2 pass–4 pass plate heat exchanger; (b) decomposition into parallelflow and counterflow units; (c) first reduction of the model in (b); (d) next reduction of the model (c).

Hence, simplify Eq. (3.166) using Eq. (3.163) as

$$P_{1,E} = \frac{P_{cf} + P_{pf} - P_{cf}P_{pf}R_1/2}{2} \tag{3.167}$$

where

$$R_{1,E} = R_1 \qquad P_{1,A} = P_{cf} \qquad P_{1,B} = P_{pf} \tag{3.168}$$

Similarly, the combined $P_{1,F}$ of exchangers C and D of Fig. 3.24c is given by

$$P_{1,F} = \frac{P_{cf} + P_{pf} - P_{cf}P_{pf}R_1/2}{2} \tag{3.169}$$

The resulting exchangers E and F are in series coupling with overall counterflow, similar to Fig. 3.19a. Their combined effectiveness P_1, using Eq. (3.123), is given by

$$P_1 = \frac{P_{1,E} + P_{1,F} - (1 + R_1)P_{1,E}P_{1,F}}{1 - R_1 P_{1,E} P_{1,F}} \tag{3.170}$$

Designating $P_{1,E} = P_{1,F} = P_I$, Eq. (3.170) reduces to

$$P_1 = \frac{2P_I - (1 + R_1)P_I^2}{1 - R_1 P_I^2} \tag{3.171}$$

This is the same expression as Eq. (V.14) in Table 3.6.

Compound Coupling. Consider the 1–4 TEMA E exchanger shown in Fig. 3.25. As was done for the 1–2 TEMA E exchanger in Fig. 3.23b, this exchanger can be modeled, as shown in Fig. 3.25b, consisting of unmixed–unmixed crossflow units with individual fluids mixed between passes. Following the approach of reducing the pairs of exchangers A1–A2, B1–B2, and so on, the model of Fig. 3.25b reduces to that of Fig. 3.25c. No further reduction is possible for the resulting compound assembly (coupling) of Fig. 3.25c. Hence, it can be concluded that *not* all compound assemblies can be reduced using series and parallel coupled exchangers successively. However, it should be emphasized that many compound assemblies, including that of Fig. 3.25c, can be analyzed by the chain rule methodology (Pignotti and Shah, 1992) to come up with a closed-form *P*-NTU expression.

A reasonable estimate of upper and lower bounds for the exchanger temperature effectiveness for a compound multipass exchanger can be made once the results for appropriate series and parallel coupled (over-and-under and side-by-side) passes are known. A procedure for this estimate has been outlined by Pignotti and Cordero (1983).

3.6.2.2 Multipass Shell-and-Tube Exchangers.

A large number of multipass arrangements are possible with shell-and-tube exchangers that employ TEMA E, F, G, H, and J shells. The *P*-NTU relationship for most of these exchangers are provided in Table 3.6, except for the 2–2 TEMA F exchanger. If there is no flow leakage at the baffle-to-shell joints for the longitudinal baffle (see Fig. 1.6), and if there is no heat conduction

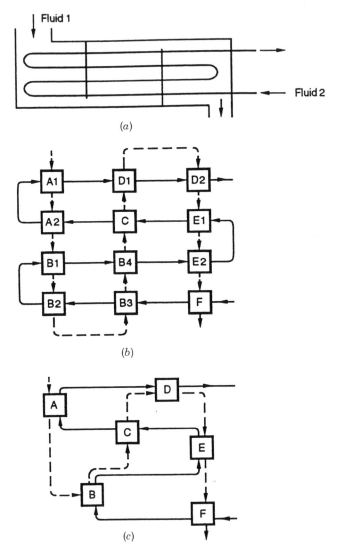

FIGURE 3.25 (*a*) 1–4 TEMA E exchanger with two baffles; (*b*) decomposition into coupled unmixed–unmixed crossflow units mixed between passes; (*c*) irreducible configuration obtained from (*b*) after several reduction steps (From Shah and Pignotti, 1989).

across the longitudinal baffle, this 2–2 F shell exchanger is a true "single-pass" counterflow exchanger. Hence, its P-NTU formula is the same as Eq. (I.1.1) in Table 3.6.

All of the exchangers in the preceding paragraph have only one shell pass (except for the F shell) and one or more tube passes. A careful comparison at the same P_1 and R_1 indicates that NTU_1 and hence the surface area requirement for all these configurations is higher than that for the counterflow exchanger. This fact will be obvious when we compare the log-mean temperature correction factor F, which will be lower than unity for these configurations. An increase in surface area (and hence NTU_1) will increase the temperature effectiveness and hence overall thermal performance of the exchanger for most configurations (except for the J shell; see Fig. 3.16 for P_1-NTU_1 results). However, the gain in performance will be small compared to a large amount of added surface area in the asymptotic region of the *P*-NTU curve.

A remedy for this situation (a lower gain or reduction in P_1 with increasing NTU_1) is to employ multiple shells in series, with one or more tube passes in each shell. If the overall flow direction of the two fluids is chosen as counterflow (that is generally the case), the temperature effectiveness of a given fluid of such an exchanger will approach that of a pure counterflow exchanger as the number of shells is increased. Up to six shells in series have been used in applications; generally, the number of shells in series is limited due to the pressure drop constraint on one of the two fluid streams. With an increased number of shells in series, the temperature change for each stream will represent only a fraction of the total temperature change, and hence the flow arrangement within each shell then has less importance. Note that if the individual exchangers are 1–2 TEMA E exchangers, the number of multiple shells is determined by the procedure outlined in Section 3.6.1.2 to avoid the temperature cross in the exchanger.

Multiple shells in series (with individual identical shell-and-tube exchangers) are used in some applications to increase the overall temperature effectiveness, to obtain high-temperature effectiveness where it is essential, mainly for part-load operation where multiple shells (individual shells of smaller size) are economical, and for shipping/hand-ling considerations. For multiple E, F, and G shells in series, both shell and tube fluids are considered as mixed outside individual shells. In this case, obtain the overall temperature effectiveness as follows:

- For series coupling of exchangers in overall counterflow
  - Use Eq. (3.131) if all individual exchangers are identical and their effectivenesses is known.
  - Use Eq. (3.136) if the overall temperature effectiveness of one fluid is specified and one needs to determine individual exchanger temperature effectiveness of the same fluid for sizing. In this case, all individual exchangers are identical.
  - If the exchangers connected in series are not identical, use Eq. (3.124) to compute the overall temperature effectiveness of one fluid of the exchangers in series.

- For series coupling of exchangers in overall parallelflow
  - Use Eq. (3.142) to compute P_1 if all exchangers are identical and $P_{1,p}$ and R_1 are known.
  - Use Eq. (3.144) to determine $P_{1,p}$ if all exchangers are identical and P_1 and R_1 are known.
  - Use Eq. (3.140) to calculate P_1 if all exchangers are not identical.

- For parallel coupling of exchangers
  - Use Eq. (3.159) to compute P_1 if all exchangers are identical and $P_{1,p}$ is known.
  - Use Eq. (3.160) to determine $P_{1,p}$ if all exchangers are identical and P_1 is known.
  - Use Eq. (3.156) to determine P_1 if all exchangers are not identical.

3.6.2.3 Multipass Plate Exchangers. In a plate exchanger, although a single-pass counterflow arrangement is most common, a large number of multipass arrangements have been used in industry, depending on design criteria. Some of them are shown in

Fig. 1.65. A multipass plate exchanger is designated by the number of passes that each stream makes in the exchanger. For example, in a 2 pass–1 pass plate exchanger, fluid 1 makes two passes and fluid 2 makes one pass in the exchanger. In each pass, there can be any equal or unequal number of thermal plates. Essentially, these are combinations of counterflow and parallelflow arrangements, with heat transfer taking place in adjacent channels. These arrangements can be obtained simply by proper gasketing around the ports in the plates. The single-pass arrangement is used for large flow rates but relatively small ΔT on each fluid side. The n pass–n pass arrangement (with n being a large number) is used for low flow rates and relatively large ΔT on each fluid side. The other flow arrangements are used for intermediate flow rates and ΔT ranges as well as unbalanced flow rates. P-NTU results are presented for 14 flow arrangements for plate exchangers in Table 3.6. For additional 10 flow arrangements (up to 4 pass–4 pass geometry), the P-NTU results are provided by Kandlikar and Shah (1989).

3.7   THE MEAN TEMPERATURE DIFFERENCE METHOD

In this section we introduce the concepts of the log-mean temperature difference LMTD, mean temperature difference MTD, and log-mean temperature difference correction factor F, then implicit and explicit functional relationships among three nondimensional groups of this method.

3.7.1   Log-Mean Temperature Difference, LMTD

The log-mean temperature difference (LMTD or ΔT_{lm}) is defined as

$$\text{LMTD} = \Delta T_{lm} = \frac{\Delta T_I - \Delta T_{II}}{\ln(\Delta T_I / \Delta T_{II})} \tag{3.172}$$

Here ΔT_I and ΔT_{II} are temperature differences between two fluids at each end of a counterflow or parallelflow exchanger. For a counterflow exchanger, from Fig. 1.50,

$$\Delta T_I = T_{h,i} - T_{c,o} \qquad \Delta T_{II} = T_{h,o} - T_{c,i} \tag{3.173}$$

For a parallelflow exchanger, from Fig. 1.52,

$$\Delta T_I = T_{h,i} - T_{c,i} \qquad \Delta T_{II} = T_{h,o} - T_{c,o} \tag{3.174}$$

For *all other flow arrangements*, the heat exchanger is hypothetically considered as a *counterflow* unit operating at the same R (or C^*) value and the same terminal temperatures (or the same effectiveness). Hence, LMTD for all other flow arrangements is evaluated from Eq. (3.172) using ΔT_I and ΔT_{II} of Eq. (3.173). Note that LMTD represents the *maximum* temperature potential for heat transfer that can *only* be obtained in a *counterflow* exchanger.

Some limiting values of ΔT_{lm} defined by Eqs. (3.172) and (3.173) are

$$\Delta T_{lm} = \begin{cases} \dfrac{\Delta T_I + \Delta T_{II}}{2} & \text{for } \Delta T_I \to \Delta T_{II} \\[2mm] \Delta T_I = \Delta T_{II} & \text{for } \Delta T_I = \Delta T_{II} \\[2mm] 0 & \text{for } \Delta T_I \text{ or } \Delta T_{II} = 0 \text{ (NTU} \to \infty) \end{cases} \tag{3.175}$$

It can be shown that when $1 \le \Delta T_I/\Delta T_{II} \le 2.2$, the error introduced by considering the arithmetic mean instead of the log-mean temperature difference is within 5%, i.e., $\Delta T_{am}/\Delta T_{lm} < 1.05$ where

$$\Delta T_{am} = (\Delta T_I + \Delta T_{II})/2 = (T_{h,i} - T_{c,o})/2 + (T_{h,o} - T_{c,i})/2$$

$$= (T_{h,i} + T_{h,o})/2 - (T_{c,i} + T_{c,o})/2 \tag{3.175a}$$

Note that $\Delta T_{lm} \le \Delta T_{am}$.

The log-mean temperature difference ΔT_{lm} normalized with respect to the inlet temperature difference $\Delta T_{max} = T_{h,i} - T_{c,i}$ can be expressed in terms of the temperature (thermal) effectiveness and the exchanger effectiveness:

$$\frac{\Delta T_{lm}}{T_{h,i} - T_{c,i}} = \frac{P_1 - P_2}{\ln[(1 - P_2)/(1 - P_1)]} = \frac{(1 - C^*)\varepsilon}{\ln[(1 - C^*\varepsilon)/(1 - \varepsilon)]} \tag{3.176}$$

This relationship is obtained directly from the definitions of ΔT_{lm}, P_1, P_2, ε, and C^*, and hence is valid for *all* flow arrangements. Following are two limiting forms of Eq. (3.176):

$$\frac{\Delta T_{lm}}{T_{h,i} - T_{c,i}} = \begin{cases} \dfrac{\Delta T_{lm}}{\Delta T_{max}} = 1 - \varepsilon & \text{for } C^* \to 1 \tag{3.177} \\[3mm] \dfrac{\Delta T_{lm}}{\Delta T_{max}} \to 0 & \text{for } \varepsilon \to 1 \tag{3.178} \end{cases}$$

Equation (3.177) is the same as Eq. (3.175) for $\Delta T_I = \Delta T_{II}$. Equations (3.177) and (3.178) clearly show that $\Delta T_{lm} \to 0$ as $\varepsilon \to 1$. Thus a decreasing LMTD means increasing exchanger effectiveness for a given exchanger. An alternative way of interpretation is that ΔT_{lm} decreases with increasing NTU and hence increasing A.

3.7.2 Log-Mean Temperature Difference Correction Factor F

As shown in Eq. (3.12), the heat transfer rate in the exchanger is represented by

$$q = UA \Delta T_m \tag{3.179}$$

Here UA is the exchanger overall thermal conductance defined by Eq. (3.24), and ΔT_m is the true (or effective) mean temperature difference (TMTD), simply referred to as the *mean temperature difference* (MTD). The value of ΔT_m is different for different exchanger flow arrangements at the same inlet and outlet fluid temperatures. In contrast, the LMTD is the same for all exchanger arrangements as given by Eq. (3.172). From Eq.

(3.62) and definitions of ε and ΔT_{max}, the mean temperature difference ΔT_m can be presented in terms of ε, NTU, and temperature drop (or rise) on the hot (or cold) side as

$$\Delta T_m = \frac{\Delta T_{max}\varepsilon}{\text{NTU}} = \frac{T_{h,i} - T_{h,o}}{UA/C_h} = \frac{T_{c,o} - T_{c,i}}{UA/C_c} \qquad (3.180)$$

Alternatively, it can be represented in terms of P_1, and the temperature range (drop or rise) of fluids 1 and 2 or the tube and shell fluids as

$$\Delta T_m = \frac{\Delta T_{max}P_1}{\text{NTU}_1} = \frac{P_1|T_{1,i} - T_{2,i}|}{\text{NTU}_1} = \frac{|T_{1,i} - T_{1,o}|}{UA/C_1} = \frac{|T_{2,o} - T_{2,i}|}{UA/C_2}$$

$$= \frac{|T_{t,o} - T_{t,i}|}{UA/C_t} = \frac{|T_{s,i} - T_{s,o}|}{UA/C_s} \qquad (3.181)$$

As will be shown later for *counterflow, parallelflow, or $C^* = 0$ exchanger* only,

$$\Delta T_m = \Delta T_{\text{lm}} \qquad (3.182)$$

is obtained by eliminating dq from the energy balance equation (3.2) and the rate equation (3.4), and then integrating the resulting expression. For *all other flow arrangements*, integration of these differential energy and rate equations yields a complicated explicit or implicit expression for ΔT_m. Hence, for these flow arrangements it is customary to define a correction factor F as a ratio of the true mean temperature difference to the log-mean temperature difference, or a ratio of the actual heat transfer rate in a given exchanger to that in a counterflow exchanger having the same UA and fluid terminal temperatures, as shown by the following two equalities:

$$F = \frac{\Delta T_m}{\Delta T_{\text{lm}}} = \frac{q}{UA\,\Delta T_{\text{lm}}} \qquad (3.183)$$

Thus,

$$q = UAF\,\Delta T_{\text{lm}} \qquad (3.184)$$

F is referred to as the *log-mean temperature difference correction factor, MTD correction factor*, or *exchanger configuration correction factor*. It is dimensionless. It can be shown that in general it is dependent on the temperature effectiveness P, the heat capacity rate ratio R, and the flow arrangement.

$$F = \begin{cases} \phi_1(P_1, R_1) = \phi_1(P_2, R_2) & \text{for a stream symmetric exchanger} \qquad (3.185) \\ \phi_1(P_1, R_1) = \phi_2(P_2, R_2) & \text{for a stream asymmetric exchanger} \qquad (3.186) \end{cases}$$

As an example, the explicit relationships of Eqs. (3.185) and (3.186) are shown in Table 3.8 for the stream symmetric 1–2 TEMA E exchanger and the stream asymmetric crossflow exchanger with one fluid unmixed and the other mixed.

F is unity for a true counterflow exchanger, and thus the maximum temperature potential (driving force) ΔT_m for any heat exchanger may approach the log-mean temperature difference ΔT_{lm} (computed considering two fluids hypothetically arranged in counterflow). F is generally less than unity for all other flow arrangements provided that

TABLE 3.8 F **as an Explicit Function of** P_1 **and** R_1 **only for the Specific Heat Exchanger Flow Arrangements Listed Here.**

Flow Arrangement	Formula
Counterflow	$F = 1$
Parallelflow	$F = 1$
Crossflow (single-pass)	
fluid 1 unmixed, fluid 2 mixed, stream asymmetric	$F = \dfrac{\ln[(1 - R_1P_1)/(1 - P_1)]}{(R_1 - 1)\ln[1 + (1/R_1)\ln(1 - R_1P_1)]}$
	$= \dfrac{\ln[(1 - R_2P_2)/(1 - P_2)]}{(1 - 1/R_2)\ln[1 + R_2\ln(1 - P_2)]}$
fluid 1 mixed, fluid 2 unmixed, stream asymmetric	$F = \dfrac{\ln[(1 - R_1P_1)/(1 - P_1)]}{(1 - 1/R_1)\ln[1 + R_1\ln(1 - P_1)]}$
	$= \dfrac{\ln[(1 - R_2P_2)/(1 - P_2)]}{(R_2 - 1)\ln[1 + 1/R_2\ln(1 - R_2P_2)]}$
1–2 TEMA E, shell fluid mixed, stream symmetric	$F = \dfrac{D_1\ln[(1 - R_1P_1)/(1 - P_1)]}{(1 - R_1)\ln\dfrac{2 - P_1(1 + R_1 - D_1)}{2 - P_1(1 + R_1 + D_1)}}$
	$= \dfrac{D_2\ln[(1 - R_2P_2)/(1 - P_2)]}{(1 - R_2)\ln\dfrac{2 - P_2(1 + R_2 - D_2)}{2 - P_2(1 + R_2 + D_2)}}$
	where $D_1 = (1 + R_1^2)^{1/2)}$
	and $D_2 = (1 + R_2^2)^{1/2}$
All exchangers with $R_1 = 0$ or ∞	$F = 1$

R_1, and P_1 are both different from zero. F can be explicitly presented as a function of P_1, R_1 and NTU_1 by Eq. (3.202). For these flow arrangements, a limited meaning of F should be clearly understood. It does *not* represent the effectiveness of a heat exchanger. But it represents a degree of departure for the true mean temperature difference from the counterflow log-mean temperature difference. Alternatively, F is a gauge of the actual exchanger performance (in terms of ΔT_m or NTU) in comparison with the counterflow exchanger performance. An F value close to unity does not necessarily mean a highly efficient heat exchanger; it means a close approach to the counterflow performance for the comparable operating conditions of flow rates and fluid inlet temperatures.

It should be emphasized that all idealizations mentioned in Section 3.2.1 for heat exchanger analysis are invoked for the derivation of F and ΔT_m and are *not* built into the concept of the log-mean temperature difference ΔT_{lm} as sometimes referred to in the literature. The definition of ΔT_{lm} is given by Eq. (3.172) for all exchangers, with ΔT_{I} and ΔT_{II} defined by Eq. (3.173) for all exchangers except for the parallelflow exchanger and by Eq. (3.174) for the parallelflow exchanger.

The MTD method is generally used for designing shell-and-tube exchangers because it provides a "feel" to the designer regarding the size if the reduction in effective ΔT (given by ΔT_m) against the best possible ΔT (given by ΔT_{lm}) for the counterflow exchanger. A large reduction in ΔT_m over ΔT_{lm} means a low value of F or a large value of NTU [see Eq. (3.200) or (3.204)] for the same counterflow effectiveness or NTU_{cf}, or a reduction in the temperature effectiveness for the same NTU [see Eq. (3.203) or (3.204)]. In such a case, the exchanger operates in the asymptotic region of the ε-NTU or P-NTU curve; and a large increase in surface area provides only a small increase in heat transfer. This is true in general regardless of whether or not there is a temperature cross for a specified exchanger flow arrangement. Because the capital cost for a shell-and-tube exchanger is an important design consideration, generally it is designed in the steep region of the ε-NTU or P-NTU curve ($\varepsilon < 60\%$); and as a rule of thumb, the F value selected is 0.80 and higher. However, a better guideline for F_{min} is provided by Eq. (3.114) when the temperature meet is at the end of the second tube pass in a 1–2 TEMA E exchanger. Further detailed comparisons between the MTD and ε-NTU methods are provided in Section 3.9.

When the temperatures of the hot and cold fluids remain constant (as in a phase-change condition or $C^* = 0$) in a heat exchanger, Eqs. (3.182) and (3.183) become

$$\Delta T_m = \Delta T_{lm} = T_{h,i} - T_{c,i} = \text{ITD} = \Delta T_{max} \quad \text{and} \quad F = 1 \tag{3.187}$$

This is also a good approximation when condensation takes place on one fluid side and evaporation on the other fluid side (with each fluid side having a single component or an azeotropic fluid), or one of the fluids can be water or other liquids with a high heat capacity rate and having a high flow rate. In this case, Eq. (3.179) becomes

$$q = UA\,\Delta T_m = UA\psi(T_{h,i} - T_{c,i}) = (UA)_{mod}(T_{h,i} - T_{c,i}) \tag{3.188}$$

where ψ is defined by Eq. (3.212). Here $(UA)_{mod} = UA$ when the hot and cold fluid temperatures are truly constant. However, when they are not constant, sometimes in the literature $(UA)_{mod} = UA\psi$ is used to take into account the nonconstancy of the mean temperature difference. Since the MTD method is more commonly used than the ψ-P method (discussed in Section 3.10), it is suggested Eq. (3.184) be used in all cases.

3.8 F FACTORS FOR VARIOUS FLOW ARRANGEMENTS

3.8.1   Counterflow Exchanger

We derive an expression for ΔT_m and hence for F for the counterflow exchanger of Fig. 3.6. Following the same algebraic details starting from Eq. (3.68), and integrating Eq. (3.75) from the hot-fluid inlet (section I) to outlet (section II), we get

$$\ln \frac{\Delta T_{II}}{\Delta T_I} = \left(\frac{1}{C_c} - \frac{1}{C_h} \right) UA \tag{3.189}$$

for constant U. Note that $\Delta T_I = T_{h,i} - T_{c,o}$ represents the temperature difference at *one end* and $\Delta T_{II} = T_{h,o} - T_{c,i}$ at the *other end*. Replacing C_c and C_h of this equation by the values obtained from the energy balances of Eq. (3.5), we get

$$\ln \frac{\Delta T_{II}}{\Delta T_I} = \frac{1}{q}[(T_{c,o} - T_{c,i}) - (T_{h,i} - T_{h,o})]UA = \frac{1}{q}(\Delta T_{II} - \Delta T_I)UA \tag{3.190}$$

A rearrangement of this equation yields

$$q = UA \frac{\Delta T_{\mathrm{I}} - \Delta T_{\mathrm{II}}}{\ln(\Delta T_{\mathrm{I}}/\Delta T_{\mathrm{II}})} \tag{3.191}$$

A comparison of Eq. (3.191) with Eq. (3.179) provides the magnitude for ΔT_m as

$$\Delta T_m = \frac{\Delta T_{\mathrm{I}} - \Delta T_{\mathrm{II}}}{\ln(\Delta T_{\mathrm{I}}/\Delta T_{\mathrm{II}})} = \Delta T_{\mathrm{lm}} \tag{3.192}$$

where ΔT_{lm} after the second equality sign comes from the definition of Eq. (3.172). Thus for a counterflow heat exchanger, from Eq. (3.183),

$$F = 1 \tag{3.193}$$

A comparison of Eq. (3.192) with Eqs. (3.180) and (3.181) yields the following relationship for the *counterflow* exchanger:

$$\Delta T_m = \Delta T_{\mathrm{lm}} = \frac{(T_{h,i} - T_{c,i})\varepsilon}{\mathrm{NTU}} = \frac{\Delta T_{\max}\varepsilon}{\mathrm{NTU}} = \frac{\Delta T_{\max}P_1}{\mathrm{NTU}_1} \tag{3.194}$$

The relationship of Eq. (3.192) or (3.193) is valid for all C^* of a counterflow exchanger and hence also for the special case of $C^* = 0$. As we noted before, when $C^* = 0$, the counterflow ε-NTU relationship of Eq. (3.84) is valid for *all* flow arrangements. Hence, when $C^* = 0$, regardless of the flow arrangement,

$$F = 1 \tag{3.195}$$

This is the case when boiling or condensation takes place on one fluid side in a heat exchanger.

3.8.2 Parallelflow Exchanger

From a derivation similar to the counterflow exchanger, we can show for the parallelflow exchanger that

$$\Delta T_m = \Delta T_{\mathrm{lm}} = \frac{\Delta T_{\mathrm{I}} - \Delta T_{\mathrm{II}}}{\ln(\Delta T_{\mathrm{I}}/\Delta T_{\mathrm{II}})} \tag{3.196}$$

and hence,

$$F = 1 \tag{3.197}$$

Here again ΔT_{I} represents the temperature difference at one end of the exchanger and ΔT_{II} at the other end, and they are defined in Eq. (3.174). Notice that these definitions for the parallelflow exchanger are thus different from those for the counterflow exchanger, Eq. (3.173). If we use the definitions of ΔT_{I} and ΔT_{II} of Eq. (3.173), it can be shown that for the parallelflow exchanger,

$$F = \frac{R_1 + 1}{R_1 - 1} \frac{\ln[(1 - R_1 P_1)/(1 - P_1)]}{\ln[1 - (1 + R_1)P_1]} \tag{3.198}$$

3.8.3 Other Basic Flow Arrangements

As noted in Eq. (3.185) or (3.186), F is a function of P_1 and R_1 for a given flow arrangement. Let us derive this general functional relationship in an explicit form that will have NTU_1 as an additional group. Substituting the equality of Eq. (3.181) for ΔT_m and Eq. (3.194) for ΔT_{lm} for the counterflow exchanger, Eq. (3.183) for F becomes

$$F = \frac{\Delta T_m}{\Delta T_{lm}} = \frac{\Delta T_m}{(\Delta T_{lm})_{cf}} = \frac{\Delta T_{max} P_1}{NTU_1} \left(\frac{NTU_1}{\Delta T_{max} P_1} \right)_{cf} \tag{3.199}$$

To evaluate F, we compare an actual exchanger of any flow arrangement of interest with a reference counterflow exchanger having the same terminal temperatures and heat capacity rates (i.e., the *same* P_1, ΔT_{max}, and R_1). Hence $P_1 = P_{1,cf}$ and $\Delta T_{max} = \Delta T_{max,cf}$, and Eq. (3.199) reduces to

$$F = \frac{NTU_{1,cf}}{NTU_1} \tag{3.200}$$

Here NTU_1 represents the *actual* number of transfer units for a given exchanger. Now $NTU_{1,cf}$ from Eq. (I.1.2) in Table 3.6 can be expressed as

$$NTU_{1,cf} = \begin{cases} \dfrac{\ln[(1 - R_1/P_1)/(1 - P_1)]}{1 - R_1} & \text{for } R_1 \neq 1 \\ \dfrac{P_1}{1 - P_1} & \text{for } R_1 = 1 \end{cases} \tag{3.201}$$

A substitution of Eq. (3.201) into Eq. (3.200) results in the desired explicit relationship among F, P_1, R_1, and NTU_1 valid for *all* flow arrangements (except for parallelflow).

$$F = \begin{cases} \dfrac{\ln[(1 - R_1 P_1)/(1 - P_1)]}{NTU_1(1 - R_1)} & \text{for } R_1 \neq 1 \\ \dfrac{P_1}{NTU_1(1 - P_1)} & \text{for } R_1 = 1 \end{cases} \tag{3.202}$$

Equation (3.202) is *also* valid for parallelflow if the F expression of Eq. (3.198) is considered instead of commonly used $F = 1$.

From Eq. (3.202), we can also express P_1 as a function of F, R_1, and NTU_1 as follows.

$$P_1 = \begin{cases} \dfrac{1 - \exp[F \cdot NTU_1(1 - R_1)]}{R_1 - \exp[F \cdot NTU_1(1 - R_1)]} & \text{for } R \neq 1 \\ \dfrac{F \cdot NTU_1}{1 + F \cdot NTU_1} & \text{for } R = 1 \end{cases} \tag{3.203}$$

The relationship between F and the effectiveness can be obtained by substituting ΔT_m from Eqs. (3.181) and (3.180) with the first equality sign into Eq. (3.183).

$$F = \frac{NTU_{cf}}{NTU} = \frac{\Delta T_{max} P_1}{NTU_1 \Delta T_{lm}} = \frac{\Delta T_{max} \varepsilon}{NTU \Delta T_{lm}} \tag{3.204}$$

Using the relationships of Eqs. (3.99), (3.103), and (3.108), F of Eq. (3.202) can be expressed as a function of ε, NTU, and C^*:

$$
F = \begin{cases} \dfrac{\ln[(1 - C^*\varepsilon)/(1 - \varepsilon)]}{\mathrm{NTU}(1 - C^*)} & \text{for } C^* \neq 1 \\[3mm] \dfrac{\varepsilon}{\mathrm{NTU}(1 - \varepsilon)} & \text{for } C^* = 1 \end{cases} \tag{3.205}
$$

Let us emphasize the interpretation of Eqs. (3.200) and (3.202). Since these equations are based on the premises that $P_1 = P_{1,\mathrm{cf}}$, $R_1 = R_{1,\mathrm{cf}}$, and $\Delta T_{\max} = \Delta T_{\max,\mathrm{cf}}$, it means that to achieve the *same* counterflow effectiveness for a given exchanger, we have $F \cdot \mathrm{NTU}_1 = \mathrm{NTU}_{1,\mathrm{cf}}$; thus the lower the value of the F factor, the higher will be the required NTU_1.

$$
F \propto \frac{1}{\mathrm{NTU}_1} \tag{3.206}
$$

Also, since $F \Delta T_{\mathrm{lm}} = \Delta T_m = (\Delta T_{\mathrm{lm}})_{\mathrm{cf}}$, and $F < 1$ for a noncounterflow exchanger, the log-mean temperature difference for all other exchanger flow arrangements will be higher than that for a counterflow exchanger for specified values of NTU_1 and R_1.

However, if we want to compare a *given* exchanger A to any other exchanger B (e.g., TEMA E vs. TEMA J) at the *same* NTU_1, R_1, and $\Delta T_{\max}$, we obtain the ratio F_A/F_B for the two exchangers using Eq. (3.202) twice (in the ratio form):

$$
\frac{F_A}{F_B} = \begin{cases} \dfrac{\ln[(1 - R_1 P_{1,A})/(1 - P)_{1,A}]}{\ln[(1 - R_1 P_{1,B})/(1 - P)_{1,B}]} & \text{for } R_1 \neq 1 \\[3mm] \dfrac{P_{1,A}(1 - P_{1,B})}{P_{1,B}(1 - P_{1,A})} & \text{for } R_1 = 1 \end{cases} \tag{3.207}
$$

Here the second subscript A or B of P denotes the exchanger A or B, respectively. From this equation it can be shown that for given values of NTU_1, R_1, and $\Delta T_{\max}$, we get

$$
F_A < F_B \qquad \text{if } P_{1,A} < P_{1,B} \tag{3.208}
$$

This means that a reduction in F translates into a reduction in P_1 and vice versa, when comparing two exchangers at the same NTU_1, R_1, and $\Delta T_{\max}$.

Although F in Eq. (3.202) is a function of three nondimensional groups P_1, R_1, and NTU_1, we know that NTU_1 is a function of P_1 and R_1, as can be found from Eq. (3.110). Thus, F is a function of only two independent nondimensional groups, P_1 and R_1 [as noted in Eq. (3.185)], P_1 and NTU_1, or NTU_1 and R_1, for a given flow arrangement.

Based on the results of Table 3.6, P_1 can be presented explicitly as a function of NTU_1 and R_1 for all flow arrangements considered. However, NTU_1 can be presented explicitly as a function of P_1 and R_1 for only a few flow arrangements, as shown in Table 3.4 in terms of the ε-NTU method. Hence, the NTU_1 expression for *these* flow arrangements can be substituted in Eq. (3.204) to obtain F as an explicit function of P_1 and R_1, or ε and C^*. All known explicit formulas for the F factor are presented in Table 3.8. For all other flow arrangements of Table 3.6, since NTU_1 cannot be expressed explicitly as a function

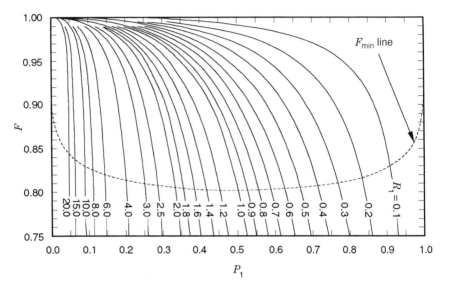

FIGURE 3.26   LMTD correction factor F as a function of P_1 and R_1 for a 1–2 TEMA E shell-and-tube exchanger with the shell fluid mixed (From Shah, 1983).

of P_1 and R_1, it is calculated iteratively for known values of P_1 and R_1. Subsequently, F is computed from Eq. (3.204).

The F factors for the 1–2 TEMA E exchanger (which is stream symmetric) are presented in Fig. 3.26. As noted earlier, Eq. (3.114) represents the temperature cross at the exit of the second pass. The corresponding F factor (computed from the expression in Table 3.8) is designated as F_{min} in Fig. 3.26. There will be a temperature cross in the 1–2 exchanger for F values lower than F_{min}.

The following important observations may be made by reviewing this figure:

- The F factor increases with decreasing R_1 for a specified P_1.
- The F vs. P_1 curves become more steep with increasing R_1.
- The F factor increases with decreasing P_1 for a specified R_1. For all R_1, $F \rightarrow 1$ as $P_1 \rightarrow 0$.
- Although not obvious from Fig. 3.26, it can be shown that when $R_1 > 0$, $F \rightarrow 0$ as $\mathrm{NTU}_1 \rightarrow \infty$ or P_1 approaches the asymptotic value for all $R_1 > 0$. This F vs. P_1 asymptotic trend corresponds to the ε vs. NTU asymptotic behavior at high NTU. A large change in F is required to obtain a slight change in P_1 in this region. A low value of F means that the exchanger will require a large amount of surface area.

It should be pointed out that although the curves at high values of R_1 appear to be too steep compared to the curves for low values of R_1, that steepness, or the asymptotic nature of the curves, is misleading:

1. If we consider that P_1 and R_1 are based on the shell side as 0.2899 and 2.5, the F factor from Fig. 3.26 is 0.7897. If P_1 and R_1 would have been based on the tube

side, they would be 0.7247 ($= 0.2899 \times 2.5$) and 0.4 ($= 1/2.5$), and the corresponding *F* factor would again be 0.7897. A careful review of Fig. 3.26 indeed indicates this fact (although the reading accuracy is only within two digits) for the stream symmetric exchanger considered.[†]

2. A very steep asymptotic curve of *F* vs. P_1 at high values of R_1 implies a large change in *F* for a small change in P_1. However, it does not mean that the exchanger will have large fluctuations in the heat duty since $q \propto F$ (as $q = UAF\,\Delta T_{lm}$). All it means is that UA will have corresponding inverse large fluctuations with P_1 (see the asymptotic behavior of *P*-NTU curves in Fig. 3.13 for the same R_1), and hence the changes in the product UAF will have similar changes in P_1 since $q = P_1 C_1 |T_{1,i} - T_{1,o}|$.

Example 3.6 In a 1–2 TEMA E shell-and-tube exchanger, water enters the shell at 21°C at a rate of 1.4 kg/s. Engine oil flows through the tubes at a rate of 1.0 kg/s. The inlet and outlet temperatures of the oil are 150°C and 90°C, respectively. Determine the surface area of the exchanger by both the MTD and ε-NTU methods if $U = 225\,\text{W/m}^2 \cdot \text{K}$. The specific heats of water and oil are 4.19 and 1.67 J/g · K respectively.

SOLUTION

Problem Data and Schematic: Fluid flow rates, inlet temperatures, and hot fluid outlet temperature are provided for a one shell pass/two tube pass exchanger (Fig. E3.6A). Also given is the overall heat transfer coefficient.

Determine: The surface area of the exchanger by both the MTD and ε-NTU methods.

Assumptions: The assumptions invoked in Section 3.2.1 are valid and these are no fins on either side of the tubes.

Analysis

The MTD Method. We first determine q and ΔT_{lm}. Subsequently, we could determine P, R, and F. Finally, we apply the MTD rate equation, Eq. (3.184) to obtain A. To find the

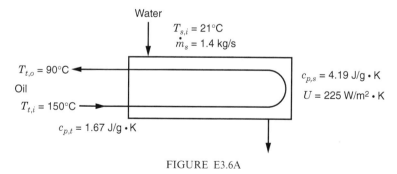

FIGURE E3.6A

[†]Note that the *F* value calculated by the given P_1 and R_1 will not be equal to that for P_2 ($= P_1 R_1$) and R_2 ($= 1/R_1$) for a stream asymmetric exchanger as shown in Eq. (3.186). In that case, compute *F* using Eq. (3.202) by replacing the subscript 1 by 2 at all places.

heat duty, we determine the heat capacity rate for the shell fluid (water) and the tube fluid (oil).

$$C_s = (\dot{m}c_p)_s = 1.4\,\text{kg/s} \times (4.19 \times 10^3\,\text{J/kg}\cdot\text{K}) = 5866\,\text{W/K}$$

$$C_t = (\dot{m}c_p)_t = 1.0\,\text{kg/s} \times (1.67 \times 10^3\,\text{J/kg}\cdot\text{K}) = 1670\,\text{W/K}$$

Therefore, the heat transfer rate from the oil is

$$q = C_t(T_{t,i} - T_{t,o}) = 1670\,\text{W/K}\,(150 - 90)^\circ\text{C} = 100.2 \times 10^3\,\text{W}$$

Using the energy balance equation, we could also find the water outlet temperature:

$$T_{s,o} = T_{s,i} + \frac{q}{C_s} = 21^\circ\text{C} + \frac{100.2 \times 10^3\,\text{W}}{5866\,\text{W/K}} = 38.1^\circ\text{C}$$

Hence all four terminal temperatures are known.

Now let us determine ΔT_{lm}. Using Fig. E3.6B and the definition of ΔT_{lm} of Eq. (3.172), we have

$$\Delta T_{lm} = \frac{\Delta T_{I} - \Delta T_{II}}{\ln(\Delta T_{I}/\Delta T_{II})} = \frac{(111.9 - 69)^\circ\text{C}}{\ln(111.9^\circ\text{C}/69^\circ\text{C})} = 88.74^\circ\text{C}$$

Now the values of tube-side P_1 and R_1 from Eqs. (3.96) and (3.105) are

$$P_1 = \frac{T_{t,i} - T_{t,o}}{T_{t,i} - T_{s,i}} = \frac{(150 - 90)^\circ\text{C}}{(150 - 21)^\circ\text{C}} = 0.4651$$

$$R_1 = \frac{T_{s,o} - T_{s,i}}{T_{t,i} - T_{t,o}} = \frac{C_t}{C_s} = \frac{1670\,\text{W/K}}{5866\,\text{W/K}} = 0.2847$$

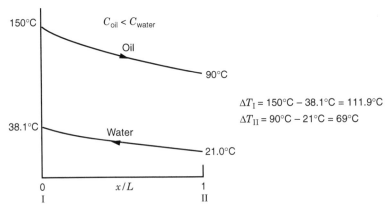

FIGURE E3.6B

Therefore, from Fig. 3.26 or using the formula from Table 3.8, $F = 0.9776$. Thus the heat transfer area from the rate equation is

$$A = \frac{q}{UF \Delta T_{lm}} = \frac{100.2 \times 10^3 \, \text{W}}{225 \, \text{W/m}^2 \cdot \text{K} \times 0.9776 \times 88.74 \, \text{K}} = 5.133 \, \text{m}^2 \qquad Ans.$$

The ε-NTU Method. First, we determine ε and C^*, and subsequently, NTU and A. In this problem, $C_t < C_s$, hence

$$C^* = \frac{C_t}{C_s} = \frac{1670 \, \text{W/K}}{5866 \, \text{W/K}} = 0.2847$$

Using the definition of the effectiveness for the tube side ($C_{\min}$ side), we get

$$\varepsilon = \frac{T_{t,i} - T_{t,o}}{T_{t,i} - T_{s,i}} = \frac{150^\circ \text{C} - 90^\circ \text{C}}{150^\circ \text{C} - 21^\circ \text{C}} = 0.4651$$

Now we could calculate the NTU either from the formula of Table 3.4 for the 1–2 TEMA E exchanger or from Fig. 3.13 with proper interpretation for ε, NTU, and C^*. Therefore, NTU = 0.6916. Hence, the area is

$$A = \frac{C_{\min}}{U} \text{NTU} = \frac{1670 \, \text{W/K}}{225 \, \text{W/m}^2 \cdot \text{K}} \times 0.6916 = 5.133 \, \text{m}^2 \qquad Ans.$$

Discussion and Comments: The solution procedure for the sizing problem is straightforward for both the MTD and ε-NTU methods. Also as expected, the calculated surface area is identical by both methods.

Example 3.7   In an oil-to-water heat exchanger, the oil enters the exchanger at 100°C with a heat capacity rate of 3700 W/K. Water is available at 15°C and 0.6 kg/s. Determine the exit temperatures in a 1–2 TEMA E shell-and-tube exchanger by the MTD method for $U = 500 \, \text{W/m}^2 \cdot \text{K}$ and surface area of 10 m². Consider $c_p = 1.88$ and 4.19 J/g · K for oil and water, respectively.

SOLUTION

Problem Data and Schematic: Fluid flow rates, inlet temperatures, and overall heat transfer coefficient are provided for a 1–2 shell-and-tube exchanger (Fig. E3.7). Also, the total heat transfer area and ratio of thermal resistances of the two fluids are given.

Determine: The outlet temperatures at each end of the exchanger, using the MTD method.

Assumptions: The assumptions invoked in Section 3.2.1 are valid, wall and fouling resistances are negligible, and the ratio of thermal resistances of oil to water is uniform throughout the exchanger.

Analysis: The MTD rate equation is

$$q = UAF \Delta T_{lm}$$

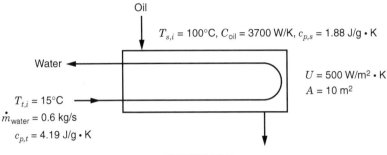

FIGURE E3.7

Since we do not know the outlet temperatures,[†] we cannot calculate directly ΔT_{lm}, q, and F. We have to use an iterative approach to get the solution. Knowing R and NTU, we could estimate P and F to calculate the outlet temperatures from the heat balance.

$$\text{NTU}_s = \frac{UA}{C_s} = \frac{(500\,\text{W/m}^2 \cdot \text{K})(10\,\text{m}^2)}{3700\,\text{W/K}} = 1.351$$

$$R_s = \frac{C_s}{C_t} = \frac{3700\,\text{W/K}}{(0.6\,\text{kg/s})(4190\,\text{J/kg} \cdot \text{K})} = 1.472$$

The first estimate for P from Fig. 3.13 is 0.43. The outlet temperatures are then calculated from the definition of P.

$$P_s = \frac{T_{s,i} - T_{s,o}}{T_{s,i} - T_{t,i}} = \frac{(100 - T_{s,o})°\text{C}}{(100 - 15)°\text{C}} = 0.43$$

Solving for the oil outlet temperature, we get[‡]

$$T_{s,o} = 63.45°\text{C}$$

The water outlet temperature is calculated from the overall energy balance as

$$T_{t,o} = T_{t,i} + \frac{C_s(T_{s,i} - T_{s,o})}{C_t} = 15°\text{C} + 1.472(100 - 63.45)°\text{C} = 68.69°\text{C}$$

The log-mean temperature difference is

$$\Delta T_{lm} = \frac{(100 - 68.79)°\text{C} - (63.45 - 15)°\text{C}}{\ln[(100 - 68.79)°\text{C}/(63.45 - 15)°\text{C}]} = 39.20°\text{C}$$

[†] This rating problem can be solved straightforward, similar to Example 3.2, using the P-NTU method.
[‡] We use here two decimal places for temperatures for iterative calculations although we measure temperatures accuracte to one decimal place.

For $P_s = 0.43$ and $R_s = 1.472$, from Fig. 3.26 or from the F formula in Table 3.8, we get $F = 0.72$. Thus the heat transfer rate is

$$q = UAF\,\Delta T_{\text{lm}} = 500\,\text{W/m}^2 \cdot \text{K} \times 10\,\text{m}^2 \times 0.72 \times 39.20°\text{C} = 141.9 \times 10^3\,\text{W}$$

We could then use the energy balance equations to calculate the outlet temperatures.

$$T_{s,o} = T_{s,i} - \frac{q}{C_s} = 100°\text{C} - \frac{141.9 \times 10^3\,\text{W}}{3700\,\text{W/K}} = 61.66°\text{C}$$

$$T_{t,o} = T_{t,i} + \frac{q}{C_t} = 15°\text{C} + \frac{141.9 \times 10^3\,\text{W}}{2514\,\text{W/K}} = 71.43°\text{C}$$

Since these temperatures are different from those assumed above, let us iterate using the newly calculated temperatures. The new log-mean temperature difference, P_s, and R_s are

$$\Delta T_{\text{lm}} = \frac{(100 - 71.43)°\text{C} - (61.66 - 15)°\text{C}}{\ln[(100 - 71.43)°\text{C}/(61.66 - 15)°\text{C}]} = 36.88°\text{C}$$

$$P_s = \frac{(100 - 61.66)°\text{C}}{(100 - 15)°\text{C}} = 0.451 \qquad R_s = 1.472$$

For $P_s = 0.451$ and $R_s = 1.472$, we get $F = 0.613$ from Fig. 3.26 or from the F formula in Table 3.8. Thus the heat transfer rate is

$$q = 500\,\text{W/m}^2 \cdot \text{K} \times 10\,\text{m}^2 \times 0.613 \times 36.88°\text{C} = 112.9 \times 10^3\,\text{W}$$

Subsequently, new outlet temperatures from the energy balance are

$$T_{s,o} = 100°\text{C} - \frac{112.9 \times 10^3\,\text{W}}{3700\,\text{W/K}} = 69.47°\text{C}$$

$$T_{t,o} = 15°\text{C} + \frac{112.9 \times 10^3\,\text{W}}{2514\,\text{W/K}} = 59.93°\text{C}$$

The outlet temperatures calculated based on the first estimated value of P and then for the first two iterations are summarized as follows:

	$T_{s,o}$ (°C)	$T_{t,o}$ (°C)
Estimation	63.45	68.79
Iteration 1	61.66	71.43
Iteration 2	69.47	59.93
Correct values	63.08	69.35

For this problem, if we had continued iterations, the outlet fluid temperatures would continue to diverge rather than converging to the correct value. Consequently, the application of this approach is not appropriate.

An alternative way for solving the rating problem is to iterate on F rather than on P_s to ensure that the solution will converge. Assume that $F = 0.80$, a generally minimum value for F for a good design. Determine P_1 from Fig. 3.26 for $R_1 = 1.472$ and $F = 0.8$ as $P_1 = 0.4052$ (actually, a computer program was used to determine the exact value of $P_1 = P_s$ here and in the following iterations). Then from the definition of P_s, the oil outlet temperature is given by

$$T_{s,o} = T_{s,i} - P_s(T_{s,i} - T_{t,i}) = 100°C - 0.4052(100 - 15)°C = 65.56°C$$

The heat transfer rate in the exchanger is

$$q = C_s(T_{s,i} - T_{s,o}) = 3700\,W/K \times (100 - 65.56)°C = 127.43\,kW$$

The water outlet temperature is then

$$T_{t,o} = T_{t,i} + \frac{q}{C_t} = 15°C + \frac{127.43\,kW}{(0.6\,kg/s)(4.19\,kJ/kgK)} = 65.69°C$$

The log-mean temperature difference is

$$\Delta T_{lm} = \frac{(100 - 65.69)°C - (65.56 - 15)°C}{ln[(100 - 65.69)°C/(65.56 - 15)°C]} = 41.91°C$$

Subsequently, the new value of the F factor is

$$F = \frac{q}{UA\,\Delta T_{lm}} = \frac{127.43\,kW}{(500\,W/m^2\,K)(10\,m^2)(41.91°C)} = 0.6080$$

With this value of F, determine P_s from Fig. 3.26 for $R_1 = 1.472$. Subsequently, calculate $T_{s,o}$, q, $T_{t,o}$, and ΔT_{lm} as noted above and compute the new value of F. Continue iterations in this manner until the outlet temperatures are converged within the desired accuracy. Following are the first 16 iterations:

F	P_s	$T_{s,o}$	$T_{t,o}$	F	P_s	$T_{s,o}$	$T_{t,o}$
0.8	0.4052	65.56	65.69	0.7243	0.4291	63.53	68.68
0.6080	0.4513	61.64	71.46	0.6868	0.4375	62.81	69.74
0.7701	0.4163	64.61	67.09	0.7173	0.4314	63.33	68.98
0.6435	0.4456	62.13	70.75	0.6949	0.4363	62.92	69.58
0.7477	0.4234	64.01	67.98	0.7125	0.4322	63.26	69.08
0.6672	0.4417	62.46	70.26	0.6979	0.4359	62.95	69.54
0.7328	0.4273	63.68	68.46	0.7113	0.4329	63.20	69.17
0.6806	0.4394	62.65	69.98	0.7005	0.4355	62.98	69.49

The convergence of the iterations is slow and F continues to fluctuate, but it will converge.

This rating problem can be made guaranteed convergent using the Newton–Raphson method [i.e., the solution of $f(x) = 0$ where the function f is nonlinear in the present case] as follows. Since

$$F = \phi_1(P_s, R_s) = \phi_2(T_{s,o}) \tag{1}$$

for this particular problem, and from the overall energy balance between the shell fluid and tube fluid, we get for given P_s and R_s

$$T_{t,o} = \phi_3(T_{s,o}) \tag{2}$$

Hence,

$$\Delta T_{\text{lm}} = \phi_4(T_{s,o}, T_{t,o}) = \phi_5(T_{s,o}) \tag{3}$$

Thus, using the functional relationship of F and ΔT_{lm} with $T_{s,o}$, we obtain the following functional relationship for q from Eq. (3.184):

$$q = UAF \, \Delta T_{\text{lm}} = UA \, \phi_2(T_{s,o}) \phi_5(T_{s,o}) = \phi_6(T_{s,o}) \tag{4}$$

Since

$$q = UAF \, \Delta T_{\text{lm}} = C_s(T_{s,i} - T_{s,o}) \tag{5}$$

Hence,

$$UAF \, \Delta T_{\text{lm}} - C_s(T_{s,i} - T_{s,o}) = 0 = \phi_6(T_{s,o}) - \phi_7(T_{s,o}) = \phi(T_{s,o}) \tag{6}$$

In Eq. (6), the only unknown is $T_{s,o}$, which can be determined accurately by the Newton–Raphson method of solving the $f(x) = 0$ equation. Subsequently, $T_{t,o}$ can be computed from the energy balance.

Discussion and Comments: This example clearly demonstrates that the rating problem may not converge by the simple iterative scheme on P_s. It will converge slowly by iterating on F. One may resort to a more complex numerical convergence scheme for a guaranteed convergence. However, this problem can be solved relatively straightforward by the P-NTU or ε-NTU method, as shown in Example 3.2 for counterflow and parallel-flow arrangements with different operating conditions.

3.8.4   Heat Exchanger Arrays and Multipassing

We consider two cases: (1) determine the F factor for an exchanger array consisting of n identical or nonidentical exchangers, and (2) determine the F factors for an exchanger array for specified performance data (i.e., a combined equivalent exchanger), when the number n of identical individual exchangers is increased so that individual NTU_1/n decreases (where NTU_1 is for the combined equivalent exchanger). As we discussed in Section 3.6.2, our description of an exchanger array also applies to a multipass exchanger having n passes.

For a series-coupled exchanger array in overall counterflow, Fig. 3.19, the F factor for the combined equivalent exchanger is given by (Shah and Pignotti, 1989)

$$F = \frac{1}{\text{NTU}_1} \sum_{i=1}^{n} \text{NTU}_{1,A_i} \cdot F_{A_i} \tag{3.209}$$

where NTU_{1,A_i} and F_{A_i} are the NTU_1 and F factors for individual exchangers in the array. Assume that the array has all identical n exchangers:

$$\text{NTU}_{1,A_i} = \text{NTU}_{1,p} \quad \text{and} \quad F_{A,i} = F_p \tag{3.210}$$

In this case, we find from Eq. (3.209) that

$$F = F_p \tag{3.211}$$

Thus we can conclude that the F factors for individual identical exchangers and the combined equivalent exchangers are the same for series coupling in overall counterflow exchangers.

For exchangers series coupled in overall parallelflow (see Fig. 3.21) and parallel coupled (see Figs. 3.22 and 1.58b), no closed-form relationship exists between the individual and combined equivalent exchanger F factors. For individual exchangers, F factors are determined for known P_{1,A_i} and R_{1,A_i}; and the overall F factor is determined using Eq. (3.202) from the known P_1, NTU_1, and R_1 for the combined equivalent exchanger. The relationships between the individual and combined equivalent exchanger P, NTU, and R are given in Table 3.6: Eq. (IV.1) for parallel coupling and Eq. (IV.3) for series coupling in overall parallelflow.

Compared to series coupling in overall parallelflow or parallel coupling of n exchangers (see Example 3.5), series coupling in overall counterflow direction yields the highest effectiveness and heat transfer rate. Hence, when an exchanger with specified flow arrangement cannot meet the specified heat duty, one alternative is to divide the exchanger into n exchangers, with individual exchangers having $\text{NTU}_{1,p} = \text{NTU}_1/n$, and arrange them such that two fluids have overall counterflow arrangement. In this case, the exchanger effectiveness will approach that of a counterflow exchanger as n increases. Since individual exchanger $\text{NTU}_{1,p}$ will reduce as n increases, F will increase [see Eq. (3.206)] and eventually, $F \to 1$. However, as noted in Eq. (3.211), the F factors for individual and combined equivalent exchangers will be identical for a given n.

Example 3.8   It is desired to preheat the feed stream of a distillation column to $175°C$. It has been proposed to exchange the heat between the column feed and the bottom products in an existing 1–2 TEMA E exchanger. Data for this problem are as follows:

Column feed: shell side, $\dot{m}_s = 725 \, \text{kg}/h, c_s = 3.43 \, \text{kJ}/\text{kg} \cdot \text{K}, \ T_{s,i} = 100°C$

Bottom products: tube side, $\dot{m}_t = 590 \, \text{kg/h}, c_t = 3.38 \, \text{kJ/kg} \cdot \text{K}, T_{t,i} = 235°C$

Surface area $A = 21.58 \, \text{m}^2$, Overall heat transfer coefficient $U = 77 \, \text{W}/\text{m}^2 \cdot \text{K}$

(a) Can the feed be heated to the desired temperature of $175°C$ using this exchanger?

(b) If you cannot heat the feed to the desired temperature, can you explain why?

(c) Now instead of one 1–2 exchanger having NTU calculated in part (a), consider two 1–2 exchangers connected in series in overall counterflow arrangement, each having NTU exactly one-half that of part (a), so that the total NTU remains the same. Now determine whether or not the column feed can be heated to 175°C.

(d) Compare the results of parts (a) and (c) and discuss the implications.

(e) What is the *F* factor for the original 1–2 exchanger? Outline a procedure for how to calculate the *F* factor for the two 1–2 exchangers connected in series in part (c). Will the *F* factor in the latter case be higher or lower than the single 1–2 exchanger of part (a)? Why?

SOLUTION

Problem Data and Schematic: The fluid flow rates, inlet temperatures, total heat transfer area, and total heat transfer coefficient are as given (see Fig. E3.8A).

Determine:

1. Can the feed be heated to the desired temperature? If not, why?
2. Can doubling the number of exchangers heat the feed to the desired temperature?
3. Compare the results and the corresponding *F* factors for the two different types of arrangements.

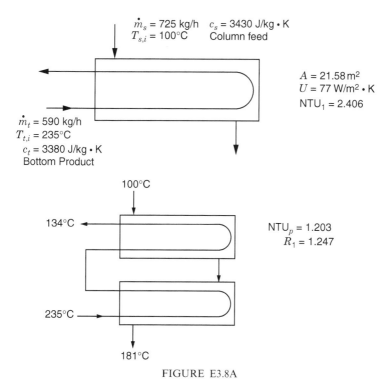

FIGURE E3.8A

Assumptions: The assumptions invoked in Section 3.2.1 are valid, there are no fins on either side of tubes, and there is perfect mixing of fluid streams between exchangers when two exchangers are connected in series.

Analysis: (a)

$$R_1 = \frac{C_s}{C_t} = \frac{725\,\text{kg/h} \times (3.43 \times 10^3)\,\text{J/kg} \cdot \text{K}}{590\,\text{kg/h} \times (3.38 \times 10^3)\,\text{J/kg} \cdot \text{K}} = 1.247$$

$$\text{NTU}_1 = \frac{UA}{(\dot{m}c_p)_s} = \frac{77\,\text{W/m}^2 \cdot \text{K} \times 21.58\,\text{m}^2}{(725/3600)\,\text{kg/s} \times 3430\,\text{kJ/kg} \cdot \text{K}} = 2.406$$

This part could be solved in two alternative ways. As a first alternative, we calculate P_1 and hence outlet temperatures.

From Eq. (III.1.1) of Table 3.6 or Fig. 3.13, for $R_1 = 1.247$ and $\text{NTU}_1 = 2.406$, we get

$$P_1 = 0.5108$$

Hence,

$$T_{s,o} = T_{s,i} + P_1(T_{t,i} - T_{s,i}) = 100°\text{C} + 0.5108(235 - 100)°\text{C} = 169.0°\text{C}$$

$$T_{t,o} = T_{t,i} - P_1 R_1(T_{t,i} - T_{s,i}) = 235°\text{C} - 0.5108 \times 1.247(235 - 100)°\text{C} = 149.0°\text{C}$$

Thus with the 1–2 exchanger, the column feed cannot be heated to the desired $T_{s,o} = 175°\text{C}$ temperature.   *Ans.*

An alternative way to solve this problem is to calculate the required P_1 $(= P_s)$ to get 175°C.

$$P_1 = \frac{T_{s,o} - T_{s,i}}{T_{t,i} - T_{s,i}} = \frac{(175 - 100)°\text{C}}{(235 - 100)°\text{C}} = 0.5556$$

Now, from Eq. (III.1) of Table 3.6, when $\text{NTU}_1 \to \infty$, we get

$$P_{1,\text{max}} = \frac{2}{1 + R_1 + (1 + R_1^2)^{1/2}} = \frac{2}{1 + 1.247 + (1 + 1.247^2)^{1/2}} = 0.5201$$

Thus the maximum shell fluid temperature effectiveness that can ideally be achieved is 0.5201 for $\text{NTU}_1 \to \infty$, while the desired P_1 is 0.5556 for a 175°C outlet temperature. Therefore, we cannot heat the column feed to 175°C even with an infinite surface area.

Note that if we had a true counterflow exchanger with the specified $R_1 = 1.247$ and $\text{NTU}_1 = 2.406$, we get $P_1 = 0.6446$ from Eq. (I.1.1) of Table 3.6. In this case, the shell-side outlet temperature is

$$T_{s,o} = T_{s,i} + P_1(T_{t,i} - T_{s,i}) = 100°\text{C} + 0.6446(235 - 100)°\text{C} = 187.0°\text{C}$$

This means that the feed can be heated to 187.0°C in a pure counterflow exchanger.

(b) The *F* factor for this case is 0.5023 from Eq. (III.1.3) of Table 3.6 or extrapolation in Fig. 3.26 for $P_1 = 0.5108$ and $R_1 = 1.247$. It is too low and there is a *temperature cross* (see F_{min} line in Fig. 3.26). The exchanger flow arrangement is just not efficient to heat the feed to 175°C with the surface area provided or even increasing *A* to ∞.

(c) Now $NTU_{1,p} = 2.406/2 = 1.203$, and $R_1 = 1.247$. Hence, $P_{1,p} = 0.4553$ from Eq. (III.1.1) of Table 3.6 or 0.46 from Fig. 3.13. The overall effectiveness of two 1–2 exchangers in overall counterflow arrangement is determined from Eq. (3.131).

$$P_1 = \frac{[(1 - R_1 P_{1,p})/(1 - P_{1,p})]^2 - 1}{[(1 - R_1 P_{1,p})/(1 - P_{1,p})]^2 - R_1} = \frac{[(1 - 1.247 \times 0.4553)/(1 - 0.4553)]^2 - 1}{[(1 - 1.247 \times 0.4553)/(1 - 0.4553)]^2 - 1.247}$$

$$= 0.5999$$

$$T_{s,o} = T_{s,i} + P_1(T_{t,i} - T_{s,i}) = 100°C + 0.5999(235 - 100)°C = 181.0°C$$

$$T_{t,o} = T_{t,i} - P_1 R_1(T_{t,i} - T_{s,i}) = 235°C - 0.5999 \times 1.247(235 - 100)°C = 134.0°C$$

Using two exchangers, the feed can be heated to 181°C, which is above the required value of 175°C.   *Ans.*

(d) Instead of one 1–2 exchanger having $NTU_1 = 2.406$, if we employ two 1–2 exchangers in series in overall counterflow direction each having $NTU_1 = 1.203$, we can increase the shell fluid temperature effectiveness from 0.5108 to 0.5999. Regardless of the flow arrangement of individual heat exchangers, when two or more exchangers are connected in series in the overall counterflow direction, the overall temperature effectiveness approaches that of a pure counterflow exchanger. Note that in the present problem, part (c), $F = 0.7786$ for $P_{1,p} = 0.4553$ and $R_1 = 1.247$ using Eq. (III.1.3) of Table 3.6 (or see Fig. 3.26). Hence, there is only a small *temperature cross* in individual exchangers, and the surface is used more efficiently.

Note that by making two smaller frontal area 1–2 exchangers, the flow velocities will go up by a factor of 2, due to the reduction in flow area by one-half. This change will also increase individual *h*'s and hence further increase in NTU_1, P_1, and *q* than what is calculated. An important but negative impact of this design will be a significant increase in the pressure drop on both sides (discussed further in Chapter 6).

(e) The *F* factor for the original 1–2 exchanger for $NTU_1 = 2.406$ and $R_1 = 1.247$ is 0.5023, as noted in part (b). When two 1–2 exchangers are connected in overall counterflow direction, the *F* factor for the two exchangers in series is the same as the *F* factor for the individual exchangers [see Eq. (3.211)], since $F = 1$ for counterflow arrangement. Hence, for two 1–2 exchangers in series, $F = 0.7786$ for $NTU_{1,p} = 1.203$ and $R_1 = 1.247$ from Eq. (III.1.3) of Table 3.6 or Fig. 3.26.

An alternative proof can be presented from the general definition of the *F* factor of Eq. (3.204) as follows.

$$F = \frac{\Delta T_{max} P_1}{NTU_1 \Delta T_{lm}}$$

This *F* formula is valid for one exchanger or multiple exchangers in series in overall counterflow direction as long as the appropriate values of P_1, NTU_1, and ΔT_{lm} are used. For two exchangers in series, total $NTU_1 = 2.406$ and $P_1 = 0.5999$, as calculated in part (c), $\Delta T_{max} = 235°C - 100°C = 135°C$, and $\Delta T_{lm} = 43.23°C$ is calculated from the temperatures found in part (c).

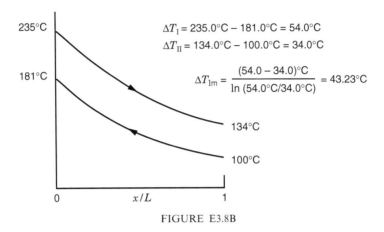

235°C

181°C

$\Delta T_{\mathrm{I}} = 235.0°C - 181.0°C = 54.0°C$

$\Delta T_{\mathrm{II}} = 134.0°C - 100.0°C = 34.0°C$

$$\Delta T_{\mathrm{lm}} = \frac{(54.0 - 34.0)°C}{\ln(54.0°C/34.0°C)} = 43.23°C$$

134°C

100°C

0   x/L 1

FIGURE E3.8B

F in this case is then

$$F = \frac{135°C \times 0.5999}{2.406 \times 43.23°C} = 0.7786$$

Thus, the F factor for two one-half-size 1–2 exchangers connected in series in counterflow will be higher than that for one full-size 1–2 exchanger (i.e., one having the same total NTU).

Discussion and Comments: Several points can be observed from this example:

- When a specified exchanger cannot provide the specified heat duty, one should investigate a higher-performance flow arrangement. One can always approach the effectiveness of a counterflow exchanger by series coupling of exchangers in overall counterflow.
- The F factors for individual identical exchangers and the equivalent combined exchanger are the same for series coupling in overall counterflow. However, the F factor for two 1–2 exchangers in series in overall counterflow is higher than the F factor for one 1–2 exchanger having the same total NTU.
- As one exchanger is divided into n identical exchangers ($\mathrm{NTU}_1 = n\mathrm{NTU}_{1,p}$) and series coupled in overall counterflow, $\mathrm{NTU}_{1,p}$ decreases as n increases. Similarly, with $R_1 = R_{1,p}$, the F of individual exchangers and hence the equivalent combined exchanger increases with increasing n. When $n \to \infty$, $\mathrm{NTU}_{1,p} \to 0$, $\varepsilon_p \to 0$, $F \to 1$, and $P_1 \to P_{1,\mathrm{cf}}$.

The overall temperature effectiveness of the two exchangers exceeded the effectiveness required for this problem. Consequently, the NTUs of the individual exchangers could be reduced. In doing so, the thermal and physical size of the exchanger will be reduced, which in turn may lower the cost; however, the cost of two smaller heat exchangers will probably be higher than the cost of a single larger exchanger. Further investigation into the use of additional exchangers is warranted. The pressure drop constraints and external constraints (e.g., location of pipes, compartment sizes) might limit the number of exchangers allowed.

3.9 COMPARISON OF THE ε-NTU, P-NTU, AND MTD METHODS

The heat transfer analysis of an exchanger can be performed by using any of the ε-NTU, P-NTU, or MTD methods described earlier. Let us first compare the heat transfer rate equation and the relationship of nondimensional groups associated with these methods, presented in Table 3.9. Clearly, there are three nondimensional groups associated with each method, and there is a direct one-to-one correspondence among the three methods; only the algebraic forms of the resulting equations are different.

Now let us discuss the basic steps involved in the solution of the two specific heat exchanger problems, the rating and sizing problems. In a sizing problem, U, C_c, C_h, and the inlet and outlet (terminal) temperatures are specified, and the surface area A is to be determined. The basic steps involved in the ε-NTU and MTD methods are presented below. Since the P-NTU method is closely related to the ε-NTU method, the solution procedure for the P-NTU method will be identical to that for the ε-NTU method by replacing ε, NTU, and C^* by P_1, NTU_1, and R_1, respectively. A computer algorithm can readily be prepared for such solution procedures.

3.9.1 Solutions to the Sizing and Rating Problems

3.9.1.1 Solution to the Sizing Problem

The ε-NTU Method

1. Compute ε from the specified inlet and outlet temperatures. Also calculate $C^* = C_{\min}/C_{\max}$.
2. Determine NTU for known ε and C^* for the given flow arrangement from the ε-NTU plots (similar to Fig. 3.7) or by using analytical expressions and/or numerical routines. If NTU is determined from the ε-NTU formula, it may require iterations to compute NTU for some flow arrangements.
3. Calculate the required surface area A from $A = \mathrm{NTU} \cdot C_{\min}/U$.

TABLE 3.9 Comparison of the ε-NTU, P-NTU, and MTD Methods

ε-NTU	P-NTU	MTD
$q = \varepsilon C_{\min}(T_{h,i} - T_{c,i})$	$q = P_1 C_1 \lvert T_{1,i} - T_{2,i} \rvert$	$q = UAF\,\Delta T_{\mathrm{lm}}$
$\varepsilon = \phi_1\,(\mathrm{NTU}, C^*)$	$P_1 = \phi_2\,(\mathrm{NTU}_1, R_1)$	$F = \phi_3(P_1, R_1)$

$C_c = C_{\min}$	$C_h = C_{\min}$
$P_1 = \varepsilon$	$P_1 = C^*\varepsilon$
$R_1 = C^*$	$R_1 = 1/C^*$

$$F = \frac{\mathrm{NTU}_{cf}}{\mathrm{NTU}} = \frac{(T_{h,i} - T_{c,i})\varepsilon}{\mathrm{NTU}\,\Delta T_{\mathrm{lm}}} = \frac{1}{\mathrm{NTU}(1 - C^*)}\ln\frac{1 - C^*\varepsilon}{1 - \varepsilon} \xrightarrow{C^*=1} \frac{\varepsilon}{\mathrm{NTU}(1 - \varepsilon)}$$

$$F = \frac{(T_{h,i} - T_{c,i})P_1}{NTU_1\,\Delta T_{\mathrm{lm}}} = \frac{(T_{h,i} - T_{c,i})P_1}{\mathrm{NTU}_1\,\Delta T_{\mathrm{lm}}} = \frac{1}{\mathrm{NTU}_1(1 - R_1)}\ln\frac{1 - R_1 P_1}{1 - P_1} \xrightarrow{R_1=1} \frac{P_1}{\mathrm{NTU}_1(1 - P_1)}$$

The MTD Method

1. Compute P_1 and R_1 from the inlet and outlet temperatures specified.
2. Determine F from F–P_1 curves (similar to Fig. 3.26) for known P_1 and R_1 for the flow arrangement given.
3. Calculate the heat transfer rate from $q = \dot{m}c_p|T_i - T_o|$ on either side, and the log-mean temperature difference ΔT_{lm} from the terminal temperatures.
4. Calculate A from $A = q/UF\,\Delta T_{lm}$.

3.9.1.2 Solution to the Rating Problem

The ε-NTU Method

1. Calculate NTU and C^* from the input specifications.
2. Determine ε for the known NTU and C^* for the given flow arrangement from either the ε-NTU graph or the ε-NTU formula.
3. Compute q from $q = \varepsilon C_{\min}(T_{h,i} - T_{c,i})$ and the outlet temperatures from $T_{h,o} = T_{h,i} - q/C_h$ and $T_{c,o} = T_{c,i} + q/C_c$.

The MTD Method

1. Compute R_1 from $R_1 = C_1/C_2$; assume that $F = 0.95$.
2. Determine P_1 from the F–P_1 plot, for the flow arrangement given.
3. Calculate q from $q = P_1 C_1|T_{2,i} - T_{1,i}|$.
4. Calculate $T_{1,o}$ and $T_{2,o}$ from the energy balance for known q. Subsequently, determine ΔT_{lm}.
5. Calculate $F = q/UA\,\Delta T_{lm}$. If F is different from the value in step 4, use the latest value of F and iterate (i.e., go to step 2). In this way, continue iterations until F converges to the degree desired. The outlet temperatures are then known from step 4.

The results for the rating or sizing problem using either method will be identical within the convergence criterion specified. However, when tried, one will appreciate the straightforward solution method for the ε-NTU method compared to the iterative procedure for the MTD method for determining outlet temperatures.

The important points for each method are summarized below.

3.9.2 The ε-NTU Method

- The nondimensional groups involved have thermodynamic significance.
- Energy balance and rate equations are used explicitly in the derivation of ε-NTU formulas, thus carrying the physical significance in heat exchanger design theory.
- Rating and sizing problem solutions are straightforward. If graphical results are not used, NTU is solved from $P_1 = \phi(\mathrm{NTU}_1, R_1)$ by a computer code.
- Determination of the improvement in exchanger performance with increasing surface area is straightforward for this method, to optimize the capital investment. For constant U, it means a direct look-up in the ε-NTU chart and finding an increase in ε for a specified increase in NTU for a given C^*. This is not possible from the F–P charts for the MTD method.

- The effect of change in the inlet temperature on exchanger performance can readily be evaluated from the definition of ε since NTU and C^* are not significantly affected by changes in the inlet temperatures (except through fluid property changes). As a result, one can calculate new outlet temperatures for changes in the inlet temperatures for given ε. This is not readily possible with the MTD method.

- The major drawback of the ε-NTU method is that one needs to keep track of the C_{min} side for a stream asymmetric heat exchanger since it will involve two different formulas for ε, depending on which fluid side is the C_{min} side.

- For the graphical presentation of ε-NTU results, the abscissa ranges from 0 to ∞ and hence is unbounded.

3.9.3 The P-NTU Method

- If fluid 1 side is the C_{min} side, this method is identical to the ε-NTU method for $0 \leq R_1 \leq 1$; otherwise, $P_1 = P_2 R_2$ when $0 \leq R_2 \leq 1$. The first five and last items of the ε-NTU method are also applicable here.

- Since the P-NTU formula is valid for $0 \leq R_1 \leq \infty$, one can use the P-NTU formula for the complete operating range of R_1, regardless of whether the exchanger is symmetric or asymmetric.

3.9.4 The MTD Method

- The F factor represents a penalty paid in the loss of mean temperature difference potential for using a flow arrangement different from the ideal counterflow.

- The F factor is not like an efficiency factor. A value of F close to unity does not represent a highly efficient heat exchanger. All it means is that the exchanger performance is close to that of a counterflow exchanger for the comparable operating conditions of flow rates and inlet fluid temperatures.

- The log-mean rate equation $q = UAF\Delta T_{lm}$ implies that only the rate equation is required for the heat exchanger design theory, whereas, in fact, the energy balance equation is hidden in the F factor.

- The solution to the rating problem is iterative even when the graph of F vs. P_1 with R_1 as a parameter is available.

- The simple rate equation of $q = UAF\Delta T_{lm}$ provides a quick feel for q if ΔT_{lm} is estimated or for ΔT_{lm} if q is known. Here the value of F factor is chosen between 0.8 and 1.0 and UA is known for the application from past experience.

- In the very early design stage of a new exchanger, if the designer has a feel for the values of ΔT_{lm} and the F factor from past practices, one can estimate the surface area required on the "back of an envelope" for known heat duty q and known U. One does not need to know the flow arrangement or exchanger configuration for a rough estimate. These types of estimates are not possible with the ε-NTU or P-NTU method.

Apart from the advantages and disadvantages of each method discussed above, it should again be emphasized that for a given input, all of the methods above will yield identical results within the specified convergence accuracy. Generally, the ε-NTU method is used by automotive, aircraft, air-conditioning, refrigeration, and other industries that design/manufacture compact heat exchangers. The MTD method is used by process, power, and petrochemical industries that design/manufacture shell-and-tube and other noncompact heat exchangers.

The important dimensionless groups of the aforementioned three methods (ε-NTU, P-NTU, and MTD) are P_1 and P_2 (or ε), NTU_1 and NTU_2 (or NTU), R_1 and R_2 (or C^*), and F. Hence, if these groups are presented on one graph, the solutions to the rating and sizing problems described above can be obtained graphically in a straightforward way, without iterations. In this regard, two additional graphical presentation methods are available: the ψ-P method (Mueller charts) and the P_1-P_2 method (Roetzel–Spang charts).[†]

3.10   THE ψ-P AND P_1-P_2 METHODS

3.10.1   The ψ-P Method

For the ψ-P method, a different group ψ is introduced. It is a ratio of the true mean temperature difference (MTD) to the inlet temperature difference (ITD) of the two fluids:

$$\psi = \frac{\Delta T_m}{T_{h,i} - T_{c,i}} = \frac{\Delta T_m}{\Delta T_{max}} \tag{3.212}$$

Using Eqs. (3.180) and (3.181), it can be shown that ψ is related to the parameters of the ε-NTU and P-NTU methods as follows:

$$\psi = \frac{\varepsilon}{NTU} = \frac{P_1}{NTU_1} = \frac{P_2}{NTU_2} \tag{3.213}$$

Also substituting the value of $1/NTU_1$ from Eq. (3.202) into Eq. (3.213), ψ is related to F as

$$\psi = \begin{cases} \dfrac{FP_1(1 - R_1)}{\ln[(1 - R_1 P_1)/(1 - P_1)]} & \text{for } R_1 \neq 1 \\[2ex] F(1 - P_1) & \text{for } R = 1 \end{cases} \tag{3.214}$$

This method was proposed by Mueller (1967) by plotting ψ as a function of P_1 and R_1 with the lines for constant values of NTU_1 and the F factor superimposed, as shown in Fig. 3.27. Note that the constant NTU_1 lines are straight, pass through the (0, 0) point, and have a slope $1/NTU_1$ as found from Eq. (3.213). Hence, any NTU_1 line can be drawn quickly. In this method, the heat transfer rate in the exchanger is given by

$$q = UA\psi(T_{h,i} - T_{c,i}) \tag{3.215}$$

[†] It should be emphasized that all methods (ε-NTU, P-NTU, MTD, ψ-P, and P_1-P_2) will yield identical results for a given set of input variables. Thus they indeed represent different routes to achieving the same final goal. The ψ-P and P_1-P_2 methods are presented here for completeness of literature information. However, the P-NTU or ε-NTU method is more commonly computerized today.

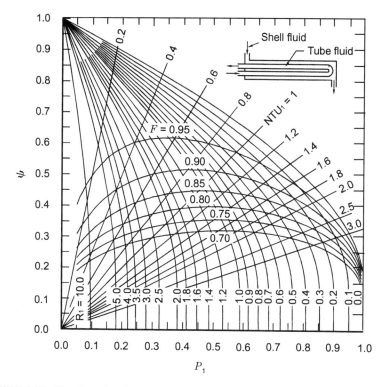

FIGURE 3.27   Nondimensional mean temperature difference ψ as a function of P_1 and R_1 for a 1–2 TEMA E shell-and-tube exchanger with the shell fluid mixed (From Shah, 1983).

3.10.2   The P_1-P_2 Method

In the P_1-P_2 method, P_1 is presented as a function of R_1 with NTU_1 as a parameter on a 45° triangle chart (compared to a 90° rectangle chart such as the P-NTU or F-P chart), and a similar chart of P_2 as a function of R_2 with NTU_2 as a parameter, both put together as shown in Fig. 3.28. Thus straight lines connecting the origin to the R_1 (the top abscissa) or R_2 (the right ordinate) represent constant values of R_1 or R_2. To minimize crowding of many lines, the lines of constant R_1 or R_2 are not shown in Fig. 3.28. The lines of constant F values are also superimposed, thus including all important parameters of heat exchanger theory. The dimensionless mean temperature difference ψ can then be computed from Eq. (3.213). This method of presentation was proposed by Roetzel and Spang (1990). The P_1-P_2 charts for a variety of heat exchanger flow arrangements are provided by Roetzel and Spang (1993).

The advantages of this method are as follows: (1) It includes all major dimensionless heat exchanger parameters. Hence the solution to the rating and sizing problem is noniterative straightforward. (2) One can start with P_1 or P_2 so that one does not need to work with a very small value of P_1 or P_2, thus improving the accuracy of the graphical solution. (3) Since both P_1-NTU_1-R_1 and P_2-NTU_2-R_2 relationships are presented on one graph, symmetry or asymmetry of the exchanger flow arrangement can be determined by visualizing the results, whether or not they are symmetrical about the diagonal

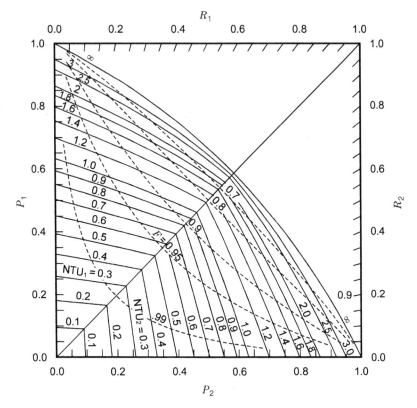

FIGURE 3.28 P_1-P_2 chart for a 1–2 TEMA E shell-and-tube exchanger with shell fluid mixed (From Roetzel and Spang, 1993).

$P_1 = P_2$ axis (the 45° line from the origin). (4) Since the NTU $= \infty$ line is shown, one can easily recognize the region beyond NTU $= \infty$ where a set of mutually consistent heat exchanger parameters does not exist (no design is possible). From the visual inspection point of view, in the P_1-P_2 method, (1) one cannot easily recognize direct asymptotic behavior of P with NTU as in a P-NTU chart, and (2) one cannot recognize P decreasing with increasing NTU after some maximum value as in the temperature cross case.

3.11 SOLUTION METHODS FOR DETERMINING EXCHANGER EFFECTIVENESS

In Section 3.4.1.1, differential energy balance and rate equations were derived for a counterflow exchanger. Subsequently, a procedure for determining heat exchanger effectiveness was illustrated based on the analytical model derived. Although no attempt was made there to derive formulas for temperature distribution in both fluids and the wall, they can be obtained, as presented in Sections 11.2.1 and 11.2.4. All we derived in Section 3.4.1.1 was the outlet temperature difference ratio and subsequently the ε-NTU formula for a counterflow exchanger by integrating compound differential equations across the exchanger surface area (length). In addition to the ε-NTU formulas, expressions for P-NTU and F factors were presented.

A variety of methods have been used to derive ε-NTU or P-NTU formulas for different heat exchanger flow arrangements. The major methods employed are: analytical for obtaining exact solutions, approximate (analytical, seminumerical, and analog), numerical, matrix formalism, and one based on exchanger configuration properties. These methods have been summarized by Sekulic et al. (1999). The ultimate goal of each of these methods is to determine heat exchanger effectiveness. These methods are described briefly next. More details about the temperature distributions and some guidelines for determining thermodynamic properties of heat exchangers are given in Chapter 11.

3.11.1 Exact Analytical Methods

Three major categories of methods are employed:

1. A direct integration of ordinary and/or partial differential equations from application of the energy balances and rate equations. Single-pass parallelflow and counterflow exchangers and 1–$2n$ TEMA E shell-and-tube exchangers have been analyzed by this method.

2. Operational calculus methods. These include the Laplace transform technique and Mikusinski operational calculus. Single-pass and multipass crossflow exchangers and various shell-and-tube heat exchanger flow arrangements have been analyzed by these methods.

3. Transformations of differential equations into integral equations and subsequent solution of integral equations; Volterra integral equations have been solved for analyzing an unmixed–unmixed single-pass crossflow exchanger.

3.11.2 Approximate Methods

Semianalytical/numerical methods used for analyzing single-pass and multipass crossflow exchangers are a collocation method, Galerkin method, and two-dimensional Roesser discrete linear image processing method. In the collocation method applied to a two-pass crossflow exchanger, first the Laplace transforms are applied to independent variables. The resulting equations are coupled at the boundary conditions between passes and lead to integral equations that can be solved exactly or approximately by the collocation and Galerkin methods. When the partial differential equations of a heat exchanger model are discretized and organized in matrix form, they are similar in form to a discrete image-processing model. The important advantage of these models is their straightforward computer implementation for the solution.

3.11.3 Numerical Methods

Finite difference, finite volume, and finite element methods have been used for obtaining ε-NTU or P-NTU relationships for single and multipass crossflow exchangers and some shell-and-tube heat exchangers.

3.11.4 Matrix Formalism

The matrix formalism method uses matrix algebra to obtain the P-NTU formulas for an exchanger array (e.g., see Fig. 3.19 to 3.22) or for a complex exchanger which can be broken into simpler subexchangers whose P-NTU formulas are known. In this method, using the overall energy balance for each fluid, any two fluid temperatures (e.g., outlet temperatures) can be presented in terms of two remaining independent fluid temperatures (e.g., inlet temperatures) (see, e.g., Table 3.10). The coefficients of independent fluid temperatures can then be presented in a 2×2 linear matrix. The overall P value of an exchanger array or a complex exchanger is eventually obtained by proper multiplication of appropriate 2×2 matrices. Thus matrix transformation rules are applied using individual exchanger effectiveness as building blocks to obtain overall effectiveness. Refer to Sekulić et al. (1999) for further details.

3.11.5 Chain Rule Methodology

For a very complex flow arrangement, the matrix formalism method mentioned above becomes too difficult to handle, due to the large number of matrices involved and the associated matrix multiplication and transformation. Our primary interest is in obtaining only P_1 or P_2, rather than performing matrix multiplication (which involves all four terms of each 2×2 matrix) for a complex heat exchanger configuration or an array. Hence, only one term of the 2×2 matrix (P_1 or P_2) for the overall exchanger is evaluated through matrix multiplication of the corresponding terms of N 2×2 matrices. In fact, the chain rule methodology offers a scheme to relate an assembly matrix element M_{ij} (such as P_1) to individual component elements Z_{ij}, Y_{ij}, and so on, *without* the use of direct matrix algebra or matrix multiplication. This is a powerful technique to obtain a closed-form expression for the effectiveness of many highly complex exchanger configurations and assemblies. It can analyze exchangers with multiple (more than two) inlet or

TABLE 3.10 Formulas for Exchanger 2 Terminal Temperatures as Functions of Two Remaining Terminal Temperatures, P_1 and P_2

$T_{1,o} = (1 - P_1)T_{1,i} + P_1 T_{2,i}$	$T_{1,i} = \dfrac{1}{1 - P_1}(T_{1,o} - P_1 T_{2,i})$
$T_{2,o} = P_1 R_1 T_{1,i} + (1 - P_1 R_1)T_{2,i}$	$T_{2,o} = \dfrac{1}{1 - P_1}\{P_1 R_1 T_{1,o} + [1 - P_1(1 + R_1)]T_{2,i}\}$
$T_{1,i} = \dfrac{1}{1 - P_1(1 + R_1)}[(1 - P_1 R_1)T_{1,o} - P_1 T_{2,o}]$	$T_{1,i} = \dfrac{1}{P_1 R_1}[-(1 - P_1 R_1)T_{2,i} + T_{2,o}]$
$T_{2,i} = \dfrac{1}{1 - P_1(1 + R_1)}[-P_1 R_1 T_{1,o} + (1 - P_1)T_{2,o}]$	$T_{1,o} = \dfrac{1}{P_1 R_1}[P_1 T_{2,i} + (1 - P_1)T_{2,o}]$
$T_{1,o} = \dfrac{1}{1 - P_1 R_1}\{[1 - P_1(1 + R_1)]T_{1,i} + P_1 T_{2,o}\}$	$T_{2,i} = \dfrac{1}{P_1}[-(1 - P_1)T_{1,i} + T_{1,o}]$
$T_{2,i} = \dfrac{1}{1 - P_1 R_1}(-P_1 R_1 T_{1,i} + T_{2,o})$	$T_{2,o} = \dfrac{1}{P_1}\{-[1 - P_1(1 + R_1)]T_{1,i} + (1 - P_1 R_1)T_{1,o}\}$

outlet streams (such as 1–2 TEMA J or H exchangers), unmixed streams between exchangers or passes, and the exchanger assembly, which may or may not be coupled only in series and parallel coupling (i.e., connected by some compound coupling). This chain rule methodology with examples is presented by Pignotti and Shah (1992). A number of formulas in Table 3.6 have been determined in the recent past using this methodology.

3.11.6 Flow-Reversal Symmetry

Flow-reversal symmetry refers to identical overall thermal performance (q, ε, P_1, or F) of an exchanger under the inversion of flow directions of both fluids in any one exchanger. Such an exchanger can have one or more inlet and outlet streams for fluid 1 and one or more inlet and outlet streams for fluid 2. This means that q, ε, P_1, or F of a given exchanger and an exchanger with the flow direction of both fluids reversed remains unchanged under the idealizations listed in Section 3.2.1. Thus, the principle of flow-reversal symmetry is valid for *all* two-fluid individual exchangers. Although in some cases, this statement of flow-reversal symmetry is obvious; in other instances it is a useful tool to determine whether or not seemingly different exchanger configurations are truly different or the same. For example, consider the two plate exchangers of Fig. 3.29a and b. At first glance, they appear to be quite different. Reversing the flow directions of both streams of Fig. 3. 29b results in Fig. 3.29c; it will not change the ε-NTU (or P_1-NTU$_1$) relationship, as mentioned above. The mirror image of Fig. 3.29c is then shown in Fig. 3.29d, which will again maintain the ε-NTU (or P_1-NTU$_1$) relationship invariant. Thus the flow arrangements of Fig. 3.29a and d are identical within the exchanger; the only difference is that the directions of fluid 1 in the inlet and outlet manifolds are changed. Since this is outside the active heat transfer region, it does not have any effect on ε or P_1 under the idealizations of Section 3.2.1. Thus, using the principle of flow reversibility, it is shown that the ε-NTU (or P_1-NTU$_1$) relationships of the seemingly different plate exchangers of Fig. 3.29a and b are identical.

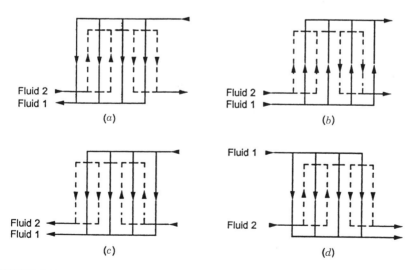

FIGURE 3.29 (a) 1 pass–2 pass plate exchanger; (b) alternative 1 pass– 2 pass plate exchanger; (c) exchanger of (b) with both fluids reversed; (d) mirror image of the exchanger of (c) (From Shah and Pignotti, 1989).

Most of the configurations of Table 3.6 are geometrically symmetric (i.e., after reversal of both fluids, they coincide with the original geometry, except for some trivial transformation, such as rotation, mirror image, etc.). For such cases, flow reversal symmetry does not provide any additional information. The *geometrically* asymmetric cases are III.1, III.2, III.4, III.10, III.11, III.12, V.3, V.6, V.11, V.12, and V.13 in Table 3.6. For these cases, two seemingly different geometries (one shown in the schematics of Table 3.6 and the other with both flows reversed) have the same effectiveness; and hence the effectiveness of the seemingly different configuration is known instantly through the concept of flow reversibility.

3.11.7 Rules for the Determination of Exchanger Effectiveness with One Fluid Mixed

As presented in Section 3.11.6, the principle of flow reversibility indicates that when the directions of both fluids are reversed in any two-fluid individual exchangers (not an array), the exchanger effectiveness ε (and P_1 and F) does not change. In contrast, consider a multipass or complex configuration exchanger having at least *one* fluid side perfectly mixed throughout, the other fluid side being mixed, unmixed, or partially mixed. If the effectiveness of such an exchanger is known for overall parallelflow (or counterflow), the effectiveness of the same exchanger configuration with the direction of only one fluid (which one is immaterial) reversed [i.e., the resulting overall counterflow (parallelflow)], can be found readily as follows. The effectiveness P_1 of the original exchanger and the effectiveness $\hat{P}_1$ of the exchanger with one fluid reversed are related as follows (Sekulić et al., 1999):

$$\hat{P}_1(R_1, \mathrm{NTU}_1) = \frac{P_1(-R_1, \mathrm{NTU}_1)}{1 + R_1 P_1(-R_2, \mathrm{NTU}_1)} \tag{3.216}$$

Here the subscript 1 refers to the fluid side having no restrictions (i.e., it can be mixed, unmixed, or split). The temperature effectiveness of the mixed fluid (the subscript 2) is then given by

$$\hat{P}_2(R_2, \mathrm{NTU}_2) = \frac{-P_2(-R_2, -\mathrm{NTU}_2)}{1 - P_1(-R_2, -\mathrm{NTU}_2)} \tag{3.217}$$

It must be emphasized that Eq. (3.216) or (3.217) is a mathematical relationship between values of P of the original and inverted exchangers. $\hat{P}$ of the inverted exchanger for the physical (positive) values of R is related to P of the original exchanger for the unphysical (negative) values of R as shown in Eq. (3.216) or (3.217).

If one of the fluid flow directions is reversed for the exchangers of Eqs. (III.3), (III.7), and (III.9), Eq. (3.216) can be applied to get the effectiveness of the overall parallelflow exchanger, and hence those configurations are not included in Table 3.6.

3.12 HEAT EXCHANGER DESIGN PROBLEMS

While comparing the ε-NTU and MTD methods in Section 3.9, the solution procedures for two heat exchanger design problems (rating and sizing) were provided. However, considering seven variables of the heat exchanger design problem [Eq. (3.7)], there are a total of 21 problems, as shown in Table 3.11. In this table, the dimensionless parameters

TABLE 3.11 Design Problems in Terms of Dimensional and Dimensionless Parameters

Problem		Dimensional Parameters							Dimensionless Groups					
		UA	$\dot{m}_1$ or C_1	$\dot{m}_2$ or C_2	$T_{1,i}$	$T_{1,o}$	$T_{2,i}$	$T_{2,o}$	P_1	P_2	NTU_1	NTU_2	R_1	R_2
Sizing	1	●	○	○	●	○	○	○	○	○	●	●	○	○
	2	●	○	○	○	●	○	○	○	○	●	●	○	○
	3	●	○	○	○	○	●	○	○	○	●	●	○	○
	4	●	○	○	○	○	○	●	○	○	●	●	○	○
	5	●	●	○	○	○	○	○	○	○	●	●	○	○
	6	●	○	●	○	○	○	○	○	○	●	●	○	○
Rating	7	○	●	●	○	○	○	○	○	●	●	●	○	○
	8	○	○	○	●	●	○	○	●	●	○	○	○	○
	9	○	○	○	●	○	●	○	●	●	○	○	○	○
	10	○	○	○	●	○	●	○	●	●	○	○	○	○
	11	○	○	○	○	●	●	○	●	●	○	○	○	○
	12	○	○	○	●	●	●	○	●	●	○	○	○	○
	13	○	●	○	○	○	○	●	●	●	●	○	●	●
	14	○	●	●	○	○	○	●	●	●	○	○	●	●
	15	○	●	●	○	●	○	○	●	●	○	●	●	●
	16	○	○	●	○	○	●	○	●	●	○	●	●	●
	17	○	○	●	●	○	○	○	○	●	○	●	●	●
	18	○	○	●	○	●	○	●	●	●	●	●	●	●
	19	○	○	○	○	○	○	○	●	●	○	○	●	●
	20	○	●	○	○	○	●	○	●	●	●	○	●	●
	21	○	●	○	○	○	○	●	○	●	●	○	●	●

Source: Data from Bačlić (1990).
[a] ●, Unknown; ○, known parameter.

are also included with known or unknown values based on the dimensional variables. If only one of the temperatures is known, we evaluate the specific heat at that temperature for the determination/estimate of C and hence R; eventually, once we know both temperatures, an iteration may be needed. If specific heats are treated as constant, C_1 and C_2 in the equations above are interchangeable with $\dot{m}_1$ and $\dot{m}_2$. Also, we have presently considered the UA product as one of the design parameters. We discuss separating U and A when we discuss the detailed sizing problem in Chapter 9. In the foregoing count, we have not added q since q can readily be calculated from the overall energy balance [Eq. (3.5)] or the rate equation [Eq. (3.6) or (3.12)]. Alternatively, if q is given, one of the temperatures or the flow rates (heat capacity rate) could be unknown in Table 3.11. We could classify the first six problems as variations of the sizing problem and the next 15 problems as a variation of the rating problem. Using the P-NTU method, these 21 problems can be solved as follows.

1. For problems 1 to 6, P_1 (also P_2) and R_1 are known through the heat balance. Hence, NTU_1 can be calculated from the P-NTU formula for a given flow arrangement, either straightforward or iteratively, depending on whether NTU can be expressed explicitly (see Table 3.4) or implicitly (see Table 3.6).

2. For problem 7, since all four temperatures are known, both P_1 and R_1 are known through their definitions [Eqs. (3.96) and (3.105)]. Hence, NTU_1 can be calculated as mentioned in the preceding paragraph for a given flow arrangement. Then $C_1 = UA/NTU_1$ and $C_2 = C_1/R_1$. Knowing the specific heats of the given fluids, one can determine $\dot{m}_1$ and $\dot{m}_2$.

3. For problems 8 to 13, NTU_1 and R_1 are known. Hence, determine P_1 using the appropriate formula from Table 3.6. Now knowing P_1, R_1, and the definition of P_1, compute the unknown temperatures using the formulas given in Table 3.10.

4. Problems 14 and 16 can only be solved iteratively, with the solution procedure for problem 14 as follows. Assume $\dot{m}_1$ (or C_1). Calculate R_2 $(= C_2/C_1)$. From the problem specifications, $NTU_2 = UA/C_2$ and $q = C_2|T_{2,i} - T_{2,o}|$ are known. Hence, knowing NTU_2 and R_2 for a given flow arrangement, determine P_2 using the appropriate formula from Table 3.6. Subsequently, compute $T_{1,i}$ from the definition of P_2, and C_1 from the energy balance $C_1|(T_{1,i} - T_{1,o})| = q$. With this new value of C_1, repeat all the calculations above. Continue to iterate until the successive values of C_1 converge within the accuracy desired. The solution procedure for problem 16 is identical to that for problem 14 just described, starting with assuming $\dot{m}_2$ (or C_2) and computing R_1 and NTU_1. The solution procedure for problem 15 is relatively straightforward. In this case, P_2 and NTU_2 are given. From the known P-NTU formula from Table 3.6, compute R_2 iteratively (such as using the Newton–Raphson method) since it is an implicit function of P_2 and NTU_2. Similarly, P_1 and NTU_1 are known for problem 17 and determine iteratively R_1 from the known P-NTU formula of Table 3.6.

5. Problems 18 and 20 can only be solved iteratively, with the solution procedure for problem 18 as follows. Assume C_2 and hence determine R_1 $(= C_1/C_2)$. Also, compute NTU_1 $(= UA/C_1)$ from the input. For a given flow arrangement, determine P_1 using the appropriate formula from Table 3.6. Subsequently, compute $T_{1,i}$ from the definition of P_1. Finally, calculate C_2 from the overall energy balance $C_1(T_{1,i} - T_{1,o}) = C_2(T_{2,o} - T_{2,i})$. With this new value of C_2, repeat all the calculations above. Continue to iterate until successive values of C_2 converge within the

desired accuracy. The solution procedures for problem 20 is identical to that for problem 18 just described, starting with assuming C_1 and computing R_2 and NTU_2.

The solution procedure for problem 19 is relatively straightforward. In this case, P_2 and NTU_1 are known. For the given flow arrangement, select the P_1-NTU_1 formula from Table 3.6 and replace P_1 by P_2/R_1 [see Eq. (3.98)]. The resulting equation has only one unknown, R_1, since NTU_1 and P_2 are known; and R_1 is implicit. It can be computed iteratively using, for example, the Newton–Raphson method. Similarly, P_1 and NTU_2 are known in problem 21, and compute R_2 iteratively after replacing P_1 by $P_2 R_2$ in the appropriate P-NTU formula of Table 3.6. See the footnote of Table 3.6 for how to convert P_1-NTU_1-R_1 formulas into P_2-NTU_2-R_2 formulas.

SUMMARY

This chapter is very important to the book. We have presented in considerable detail the basic thermal design theory for recuperators or exchangers with no moving parts or periodic flows as in regenerators. Through the problem formulations, it is shown that there are six independent and one or more dependent variables for the exchanger thermal design problem for any flow arrangement. The solution to this problem can be presented in terms of ε-NTU, P-NTU, MTD, ψ-P and P_1-P_2 methods. The exchanger rating or sizing problem can be solved by *any* of these methods and will yield the *identical* solution within the numerical error of computation. Theoretical details are presented for the ε-NTU, P-NTU, and MTD methods in the text for an understanding of concepts, these methods, and their advantages and disadvantages. Many idealizations are made to simplify the complex heat exchanger design problem to obtain the solution in terms of ε-NTU, P-NTU, and MTD parameters. A thorough understanding of the concepts and results presented in this chapter will provide a strong foundation for rating, sizing, and analysis of industrial heat exchangers.

REFERENCES

Bačlić, B. S., 1990, ε-N_{tu} analysis of complicated flow arrangements, in *Compact Heat Exchangers: A Festschrift for A. L. London,* R. K. Shah, A. D. Kraus, and D. Metzger, eds., Hemisphere Publishing, Washington, DC, pp. 31–90.

Bell, K. J., 1998, Approximate sizing of shell-and-tube heat exchangers, in *Heat Exchanger Design Handbook,* G. F. Hewitt, exec. ed., Begell House, New York, Vol. 3, Sec. 3.1.4.

Kandlikar, S. G., and R. K. Shah, 1989, Asymptotic effectiveness–NTU formulas for multipass plate heat exchangers, *ASME J. Heat Transfer,* Vol. 111, pp. 314–321.

McAdams, W. H., 1954, *Heat Transmission,* 3rd ed., McGraw-Hill, New York, pp. 129–131.

Mueller, A. C., 1967, *New Charts for True Mean Temperature Difference in Heat Exchangers,* AIChE Paper 10, 9th National Heat Transfer Conf., Seattle, WA.

Pignotti, A., and G. O. Cordero, 1983, Mean temperature difference in multipass crossflow, ASME *J. Heat Transfer,* Vol. 105, pp. 584–591.

Pignotti, A., and R. K. Shah, 1992, Effectiveness-number of transfer units relationships for heat exchanger complex flow arrangements, *Int. J. Heat Mass Transfer,* Vol. 35, pp. 1275–1291.

Roetzel, W., and B. Spang, 1990, Verbessertes Diagramm zur Berechnung von Wärmeübertragern, *Wärme-und Stoffübertragung*, Vol. 25, pp. 259–264.

Roetzel, W., and B. Spang, 1993, Design of Heat Exchangers, in *VDI Heat Atlas*, VDI-Verlag GmbH, Düsseldorf, Germany, Sect. Ca, p. 18.

Sekulić, D. P., R. K. Shah, and A. Pignotti, 1999, A review of solution methods for determining effectiveness–NTU relationships for heat exchangers with complex flow arrangements, *Appl. Mech. Rev.*, Vol. 52, No. 3, pp. 97–117.

Shah, R. K., 1983, Heat exchanger basic design methods, in *Low Reynolds Number Flow Heat Exchangers*, S. Kakaç, R. K. Shah, and A. E. Bergles, eds., Hemisphere Publishing, Washington, DC, pp. 21–72.

Shah, R. K., and A. Pignotti, 1989, *Basic Thermal Design of Heat Exchangers*, Report Int-8513531, National Science Foundation, Washington, DC.

REVIEW QUESTIONS

Where multiple choices are given, circle one or more correct answers. Explain your answers briefly.

3.1   Thermal and electrical analogy exists for:

(a) resistances in parallel   (b) resistances in series

(c) power   (d) time constant

3.2   The true mean temperature difference for nonuniform U is defined as:

(a) $\Delta T_m = \dfrac{1}{A} \displaystyle\int_A \Delta T \, dA$ (b) $\Delta T_m = \dfrac{1}{q} \displaystyle\int_q \Delta T \, dq$

(c) $\dfrac{1}{\Delta T_m} = \dfrac{1}{q} \displaystyle\int_q \dfrac{dq}{\Delta T}$

3.3   Fouling generally provides:

(a) an increase in heat transfer coefficient

(b) an increase in thermal resistance to heat flow path

(c) higher outlet temperatures   (d) none of these

3.4   Explicit basic principles used in deriving ε-NTU relationships or F factors are:

(a) second law of thermodynamics   (b) energy balances

(c) rate equation   (d) equation of state

3.5   The explicit rate equation used in the heat exchanger analysis is:

(a) $dq = |C_h dT_h|$ (b) $\varepsilon_h C_h = \varepsilon_c C_c$

(c) $dq = U \, \Delta T \, dA$ (d) $q = C_h(T_{h,i} - T_{h,o})$

(e) $q = \varepsilon C_{\min}(T_{h,i} - T_{c,i})$

3.6   The energy balance equation used in the heat exchanger analysis, from those equations in Question 3.5, is (check as appropriate)

(a) _____   (b) _____   (c) _____   (d) _____   (e) _____

3.7 Consider a heat exchanger for which C^* is near unity. The hot- and cold-fluid temperatures are shown in Fig. RQ3.7 as functions of the position in the exchanger. In this figure,

(a) sketch the variation of *wall* temperature with position (neglecting wall resistance) when $(hA)_h \gg (hA)_c$.

(b) repeat part (a) when $(hA)_c \gg (hA)_h$.

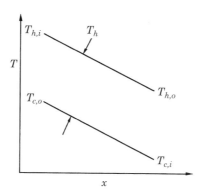

FIGURE RQ3.7

3.8 The ε-NTU and MTD methods employ the following assumptions:

(a) variable heat transfer coefficients   (b) two-dimensional analysis

(c) longitudinal heat conduction negligible

(d) variable velocity and temperature profiles at the entrance

3.9 The perfect (ideal) heat exchanger, providing the maximum possible heat transfer rate, has the following things in common with the actual heat exchanger with constant c_p:

(a) flow arrangement   (b) outlet temperatures   (c) inlet temperatures

(d) surface area   (e) flow rates   (f) none of these

(g) all of these

3.10 A high number of transfer units NTUs (let's say, 100) is generally obtainable in $a(n)$:

(a) physically very large exchanger   (b) exchanger with high flow rates

(c) high effectiveness exchanger   (d) small exchanger

(e) can't tell

3.11 The total NTUs of a two-pass cross-counterflow exchanger is 4. The hot air enters at 315°C and leaves at 150°C on one side. On the other side, 450 kg/h cold air enters at 25°C and leaves at 260°C. The approximate hot-air flow rate should be:

(a) 450 kg/h   (b) 325 kg/h   (c) 640 kg/h   (d) can't tell

3.12 The effectiveness of the exchanger of Question 3.11 is:

(a) 71%   (b) > 85%   (c) 59%   (d) 81%

3.13 A heat exchanger has been designed so that $C_c = 20\,\text{kW/K}$ and $C_h = 40\,\text{kW/K}$. The number of transfer units on the cold side, $\text{ntu}_c = 6$, and that on the hot side, $\text{ntu}_h = 4.5$. Recall $\text{ntu}_{\text{side}} = (\eta_o h A / C)_{\text{side}}$. The total NTUs for the exchanger, neglecting wall resistance and fouling, is:

   (a) 10.5   (b) 1.5   (c) 3.6   (d) 1.33

3.14 What is the ratio of hot- and cold-side convective thermal resistances in Question 3.13?

   (a) 1.33   (b) 2.67   (c) 2   (d) 0.667

3.15 The terminal temperatures of a particular heat exchanger are: hot fluid: 120°C, 50°C; cold fluid: 40°C, 80°C. The effectiveness of this exchanger is approximately:

   (a) 87%   (b) 50%   (c) 75%   (d) 38%

3.16 The heat capacity rate ratio C^* for the heat exchanger of Question 3.15 is:

   (a) 1.75   (b) 0.38   (c) 0.64   (d) 0.57

3.17 The heat exchanger of Question 3.15 may have a parallelflow arrangement.

   (a) true   (b) false   (c) can't tell

3.18 If you consider the heat exchanger of Question 3.15 as counterflow, the temperature distribution in the exchanger will be:

   (a) concave up   (b) convex up   (c) linear

3.19 If you consider the heat exchanger of Question 3.15 as a crossflow unit with both fluids unmixed, the log-mean temperature difference ΔT_{lm} is approximately:

   (a) 70°C   (b) 55°C   (c) 51°C   (d) 40°C   (e) 22°C

3.20 In a hypothetical counterflow gas turbine regenerator, having the same heat capacity rate for both fluids, the design NTU and effectiveness are 0.5 and 33%, respectively. If at part-load operation, NTU doubles, the corresponding effectiveness is:

   (a) 66%   (b) 50%   (c) 83%   (d) 40%

3.21 Other things being equal (including identical heat transfer surfaces), a multipass cross-counterflow heat exchanger designed for a particular set of specifications (i.e., given ε, C^*, flow rates, and flow lengths) will have a shorter noflow (stack) height dimension than a single-pass crossflow heat exchanger.

   (a) It depends on the surface geometries.   (b) true   (c) false
   (d) It depends on the capacity rate ratio.

3.22 The effectiveness of a single-pass crossflow heat exchanger with both fluids unmixed, and with equal heat capacity rates, approaches the following limit as the NTU is increased to infinity:

   (a) 50%   (b) 62%   (c) 67%   (d) 100%

3.23 In a steam condenser, the steam is effectively at a constant temperature of 50°C throughout the heat exchanger, while the temperature of cooling water increases from 20°C to 31°C as it passes through the condenser. The NTU for this heat exchanger is

   (a) 1.00   (b) 0.367   (c) 4.55   (d) 0.457   (e) 2.19

3.24 Suppose that we desire to increase the water outlet temperature to 40°C in the heat exchanger of Question 3.23 at the same flow rate. The surface area of this exchanger required (considering that $U =$ constant) should be increased by a factor of:

    (a) 2.40   **(b)** 1.29   **(c)** 1.0   **(d)** 0.67

3.25 Suppose in Question 3.23 that the flow rate of the cooling water is doubled. Describe qualitatively how the thermal behavior of the exchanger will change. Idealize no subcooling of steam. Consider all the implications you can think of.

3.26 The curves in Fig. RQ3.26a and b represent the temperature profiles in two different counterflow exchangers having the same cold fluids and the same cold fluid flow rate in each exchanger. The heat transfer rate of exchanger A is as follows compared to that of exchanger B:

    (a) higher   **(b)** lower   **(c)** same   **(d)** can't tell

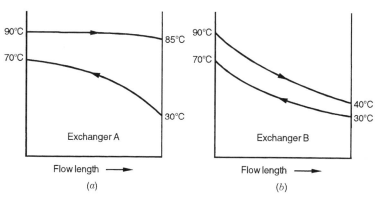

FIGURE RQ3.26

3.27 Circle the following statements as true or false.

    (a) T F   The F factor represents exchanger effectiveness.

    (b) T F   The F factor is a ratio of the true mean temperature difference in a counterflow exchanger to that in the actual exchanger.

    (c) T F   The F factor is a ratio of the counterflow NTU to NTU of the actual exchanger under consideration.

    (d) T F   The higher F factor means generally higher NTU.

    (e) T F   In the asymptotic region of the F-factor curves in Fig. 3.26, a slight change in the exchanger thermal effectiveness means a large change in the F factor, and hence a large change in q and instability in the exchanger performance ($q = UAF \Delta T_{\text{lm}}$).

    (f) T F   The F factor increases with an increasing number of shell passes at the same total NTU for overall counterflow.

    (g) T F   A decreasing LMTD for an exchanger means increasing its heat exchanger effectiveness.

3.28 A value close to unity of the log-mean temperature correction factor F means:
   (a) exchanger effectiveness approaching 100%
   (b) performance approaching that of a crossflow exchanger with both fluids unmixed
   (c) performance approaching that of a counterflow exchanger
   (d) can't tell

3.29 (a) When is it true that ΔT_{lm} for counterflow $= \Delta T_{lm}$ for parallelflow for a given duty?
   (b) When is it true that $\Delta T_{lm} = (\Delta T_1 + \Delta T_2)/2$ for a single-phase counterflow exchanger? (Mention the R value.)
   (c) When is $\Delta T_{lm} = (\Delta T_1 + \Delta T_2)/2$ true for a single-phase parallelflow exchanger?
   (d) When is it true that $\Delta T_a (= T_{a,o} - T_{a,i}) = \Delta T_{lm}$ for a steam condenser where the subscript a denotes the air side?

3.30 TEMA E shell-and-tube exchangers are generally designed with an approximate minimum log-mean temperature difference correction factor F of:
   (a) 0.90   (b) 0.80   (c) 0.40   (d) 0.98   (e) no limit
   Hint: No temperature cross is desired in the exchanger.

3.31 Hot (150 to 100°C) and cold (50 to 75°C) fluids undergo the same temperature changes (as indicated in parentheses) in a counterflow exchanger and a parallel-flow exchanger. For the identical temperature changes and flow rates (same U), the heat transfer surface area required for the counterflow exchanger compared to that for the parallelflow exchanger is:
   (a) 1.00   (b) 1.14   (c) 0.877   (d) 2.00

3.32 Which of the following dimensionless groups can have values ranging from 0 to 1 only?
   (a) ε (b) NTU_1 (c) F (d) ψ
   (e) R_1 (f) NTU (g) P_1 (h) C^*

3.33 Which of the following statements are always true?
   (a) $P_1 \geq \varepsilon$ (b) $R_1 \geq C^*$ (c) $NTU_1 \geq NTU$

3.34 In a heat exchanger, the effectiveness generally increases with:
   (a) increasing NTU   (b) increasing C^*
   (c) increasing the mixing of fluids at a cross section
   (d) increasing F

3.35 For the same surface areas, fluid flow rates, and inlet temperatures, arrange the following flow arrangements from lowest to highest effectiveness:
   (a) unmixed–unmixed   (b) mixed–mixed   (c) mixed–unmixed
   (d) one fluid partially unmixed, other unmixed

3.36 Circle the statements that are true about multipass exchangers.
   (a) Multipass exchangers are always of crossflow type.

(b) A multipass counterflow exchanger can be more efficient than a single-pass counterflow exchanger having the same NTU and C^*.

(c) A multipass overall parallelflow exchanger can be more efficient than a single-pass parallelflow exchanger having the same NTU and C^*.

3.37 The temperature approach for a counterflow exchanger of infinite surface area is:

(a) indeterminate **(b)** zero

(c) a very large value **(d)** can't tell

3.38 In a single-pass *counterflow* heat exchanger with a fixed heat duty, the closer the temperature approach, the more heat transfer surface area is required.

(a) true **(b)** false **(c)** can't tell in general

3.39 The true mean temperature difference in an exchanger can be determined by:

(a) $\Delta T_m = F \Delta T_h$ **(b)** $\Delta T_m = \psi \Delta T_{lm}$

(c) $\Delta T_m = \dfrac{\varepsilon \Delta T_{max}}{\text{NTU}}$ **(d)** $\Delta T_m = \dfrac{\Delta T_h}{(\text{NTU})_c}$

3.40 An $n(>1)$ 1–2 TEMA G exchanger in series in overall counterflow may be analyzed using Eq. (3.131).

(a) true **(b)** false **(c)** can't tell

3.41 The dimensionless temperature difference ψ can be presented as the following functional relationship for a given heat exchanger flow arrangement. Subscripts 1 and 2 refer to fluid 1 and 2 sides, respectively.

(a) $\psi = \phi(P_1, P_2)$ **(b)** $\psi = \phi(\text{NTU}_1, \text{NTU}_2)$ **(c)** $\psi = \phi(R_1, R_2)$
(d) $\psi = \phi(P_1, \text{NTU}_2)$ **(e)** can't tell

3.42 The exchanger effectiveness ε can be presented as the following functional relationship for a given heat exchanger flow arrangement. Subscripts 1 and 2 refer to fluid 1 and 2 sides, respectively.

(a) $\varepsilon = \phi(P_1, P_2)$ **(b)** $\varepsilon = \phi(\text{NTU}_1, \text{NTU}_2)$
(c) $\varepsilon = \phi(R_1, R_2)$ **(d)** $\varepsilon = \phi(P_1, \text{NTU}_2)$
(e) can't tell

3.43 For a given exchanger and specified operating conditions, one can calculate $F_1 = \phi(P_1, R_1)$ or $F_2 = \phi(P_2, R_2)$. Is it correct that $F_1 = F_2$ in general?

(a) yes **(b)** no **(c)** can't tell

3.44 In a heat exchanger, engine oil with a 0.2 kg/s flow rate and 130°C inlet temperature is being cooled by water having a 0.438 kg/s flow rate at 90°C inlet. The engine oil and water specific heats are 2.3 and 4.2 kJ/kg · K, respectively. What is the maximum possible exchanger effectiveness if it is a counterflow exchanger?

(a) 100% **(b)** 25% **(c)** 50% **(d)** 80% **(e)** can't tell

3.45 What is the maximum possible exchanger effectiveness if the exchanger of Question 3.44 is a parallelflow exchanger?

(a) 100% **(b)** 25% **(c)** 50% **(d)** 80% **(e)** can't tell

3.46 In a counterflow double-pipe heat exchanger, cold water enters the annulus at 20°C and leaves at 65°C. In the inner pipe, hot water enters at 110°C and leaves at 65°C. The length of the exchanger is 2 m. We would like to increase the outlet temperature of the cold water to 80°C by increasing the length of the double-pipe exchanger without changing the inlet temperatures and flow rates of both fluids and also keeping constant the tube diameters of the exchanger. Make appropriate idealizations and find the new length of the exchanger as:

(a) 2 m   (b) 3 m   (c) 4 m   (d) 5 m   (e) can't tell

3.47 A 2-ton window air-conditioner needs to remove 8.4 kW of heat from its condenser. In the condenser, the inlet temperature of air is 38°C and the refrigerant condenses at 57°C. Ignore the effect of desuperheating and subcooling in the condenser. UA for this air-conditioner is 700 W/K. Assume the specific heat of air as 1.0 kJ/kg · K. Use any appropriate information from this or any other book, and determine the airflow rate through the condenser as:

(a) 1 kg/s   (b) 0.7 kg/s   (c) 0.5 kg/s   (d) 1.5 kg/s   (e) can't tell

Hint: Assume the exponent to e to be an integer.

3.48 For Question 3.47, the air temperature rise will be:

(a) 10°C   (b) 6°C   (c) 12°C   (d) 20°C   (e) can't tell

3.49 A 1–2 TEMA E shell-and-tube exchanger has the temperature effectiveness of 0.09 and heat capacity rate ratio of 10. At this operating point, the log-mean temperature difference correction factor F will be approximately:

(a) 0.65   (b) 0.75   (c) 0.85   (d) 0.95   (e) can't tell

3.50 A heat exchanger is made of connecting two crossflow subexchangers designed having total $\mathrm{NTU}_1 = 2$ and overall $R_1 = 0.5$. Arrange the following design alternatives from higher to lower effectiveness P_1.

(a) series coupling with overall parallelflow

(b) series coupling with overall counterflow

(c) parallel coupling with fluid 1 in series

(d) can't tell

3.51 An unmixed–unmixed crossflow exchanger has $\varepsilon = 60\%$ and $F = 0.98$ at $\mathrm{NTU} = 1$ for $C^* = 0.2$. To demonstrate how F varies with NTU, we increase $\mathrm{NTU} = 3$ at $C^* = 0.2$. In that case, the exchanger effectiveness ε increases to 90%. The corresponding F value will be:

(a) 0.98   (b) 0.88   (c) 1.00   (d) 0.01

3.52 Consider a clean counterflow heat exchanger with desired $\varepsilon = 85\%$ at $C^* = 1$. If the heat leakage to the ambient is 2%, approximately how much will be an increase in heat transfer surface area on each fluid side to increase the cold-fluid temperature at the same level of the no-heat-leakage case? Consider all other geometric and operating parameters of the exchanger remaining the same and $(\eta_o hA)^* = 1$.

(a) 0%   (b) 2%   (c) 6.4%   (d) 12.8%   (e) can't tell

PROBLEMS

3.1 The typical temperature distributions for hot and cold fluids in a counterflow exchanger are shown in the left-hand sketch of Fig. 1.50 for $C_h > C_c$. Using the energy balance and rate equations, show why the temperature distributions must be convex up. *Hint:* First show that $(T_h - T_c)$ increases as x increases.

3.2 Discuss in detail where each of the assumptions listed in Section 3.2.1 is invoked in the set of Eqs. (3.2), (3.4), (3.5), and (3.6). Alternatively, show how one or more of these equations could be changed if those assumptions were not invoked.

3.3 The required NTU for a given exchanger is 2.0. If $C_c = 10,000\,\text{W/K}$, $C_h = 40,000\,\text{W/K}$, and the thermal resistance $\mathbf{R}_w$ of the wall separating hot and cold fluids is $10^{-5}\,\text{K/W}$, find ntu_h and ntu_c when the convective resistances of hot and cold sides are estimated to be equal. What would the values be if the thermal resistance of the wall is neglected?

3.4 Explain the physical reasoning why $\varepsilon = 1 - e^{-\text{NTU}}$ for all flow arrangements when $C^* = 0$.

3.5 **(a)** For a typical counterflow heat exchanger, explain briefly the change in the effectiveness as the flow rate of the *cold* fluid is decreased slowly from some initial value to near zero. Note that we are not interested in transient effects, just a series of steady-state operating conditions. See Fig. 1.50*b* for the initial temperature profiles.

(b) Explain briefly the effect on the temperature drop of the hot fluid, ΔT_h.

3.6 In a single-pass crossflow exchanger with both fluids unmixed, it has been determined that the hot and cold streams should leave the exchanger at the same mean temperature (i.e. $T_{c,o} = T_{h,o}$). The following data are given for the exchanger: $T_{h,i} = 250°\text{C}$, $T_{c,i} = 30°\text{C}$, $\dot{m}_h = 0.15\,\text{kg/s}$, $\dot{m}_c = 0.60\,\text{kg/s}$, $c_{p,h} = 2000\,\text{J/kg}\cdot\text{K}$, $c_{p,c} = 1000\,\text{J/kg}\cdot\text{K}$, and $U = 1000\,\text{W/m}^2\cdot\text{K}$. Determine **(a)** the required heat exchanger area (m^2); **(b)** the total heat transfer rate between the hot and cold streams (W); **(c)** the outlet temperatures $T_{c,o}$ and $T_{h,o}$ (°C); **(d)** the effectiveness of the heat exchanger; and **(e)** which (if any) of the answers to parts (a) through (d) would be different if the heat exchanger were, instead, a single-pass counterflow. Explain. Be sure to address each of parts (a) through (d).

3.7 Determine the heat transfer surface area required for a heat exchanger constructed from a 25.4 mm OD tube to heat helium from $-7°\text{C}$ to $0°\text{C}$ using hot water at $85°\text{C}$. The water and helium flow rates are 0.6 and 2.4 kg/s. Helium flows within the tubes. The overall heat transfer coefficient is $120\,\text{W/m}^2\cdot\text{K}$. The specific heats for water and helium are 4.18 and 5.20 kJ/kg·K, respectively. Consider **(a)** a counterflow exchanger, and **(b)** a 1–2 TEMA E exchanger. Solve the problem by both the ε-NTU and MTD methods.

3.8 At 80 km/h, the inlet temperature of air to an automobile radiator is $37.8°\text{C}$. Water enters at $98.9°\text{C}$ with a flow rate of 1.89 kg/s and leaves with a temperature of $93.3°\text{C}$. UA for this radiator is $960.6\,\text{W/K}$. Determine the airflow rate and air outlet temperature for this radiator using both the ε-NTU and MTD methods. Consider c_p for air and water as 1.01 and 4.19 kJ/kg·K, respectively. The radiator is unmixed on the air side and mixed on the water side.

3.9 An automobile radiator has a heat transfer rate $q = 98.45\,\text{kW}$. Air and water flow rates are 3.86 and 3.00 kg/s, respectively, and air and water inlet temperatures are 43 and 115°C. The specific heat for air and water are 1.00 and 4.23 kJ/kg · K, respectively. Consider the radiator as a crossflow heat exchanger with both fluids unmixed.

   (a) Determine the exchanger effectiveness ε, the number of transfer units NTU, and UA.

   (b) Determine the true mean temperature difference ΔT_m for heat transfer.

   (c) Determine the F factor. Is the calculated value of F reasonable? If not, why?

   (d) Describe by means of a thermal circuit the makeup of $(UA)^{-1}$, and provide your estimates of the component resistances expressed as a percentage of the overall resistance $(UA)^{-1}$.

3.10 Derive Eq. (3.131) for a four-pass arrangement and demonstrate that it reduces to

$$
P_1 = \begin{cases} \dfrac{4P_p}{1 + 3P_p} & \text{for } C^* = 1 \\[2mm] 1 - e^{-\text{NTU}} & \text{for } C^* = 0 \end{cases}
$$

3.11 Three exchangers identical to the crossflow exchanger of Problem 3.6 are placed in a multipass overall counterflow arrangement with the same inlet temperatures and flow rates. Find the overall effectiveness of this arrangement and the outlet fluid temperatures.

3.12 In the text, the exchanger overall effectiveness ε (P_1) for a multipass overall counterflow exchanger is given by Eq. (3.131) for the case when the NTU per pass is equal. The objective of this problem is to calculate ε for a multipass overall counterflow exchanger when the NTU per pass is not equal. Consider a counterflow exchanger with crossflow headers as a three-pass exchanger (Fig. P3.12).

   (a) Derive an algebraic expression for an overall effectiveness

$$
\varepsilon = \phi(\varepsilon_1, \varepsilon_2, \varepsilon_3, C^*)
$$

Then show that for $C^* = 1$ and $\varepsilon_1 = \varepsilon_3$, it reduces to

$$
\varepsilon = \frac{3\varepsilon_1\varepsilon_2 - 2\varepsilon_1 - \varepsilon_2}{2\varepsilon_1\varepsilon_2 - \varepsilon_1 - 1}
$$

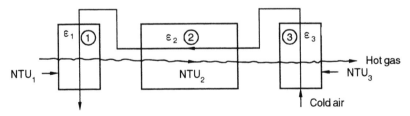

FIGURE P3.12

(b) Harrison Model TR regenerator (for a 12,000-hp GE Frame 3000 marine gas turbine for a Chevron tanker) may be considered as having $NTU = 4$ and $C^* = 1$. Obtain its effectiveness considering it as a true counterflow heat exchanger. Next approximate it as having two crossflow (unmixed–unmixed) headers with a counterflow core, as shown in part (a) with $NTU_1 = NTU_3 = 1$ and $NTU_2 = 2$. Evaluate the overall effectiveness. How good is this approximation?

3.13   Consider a single-pass crossflow exchanger with one fluid mixed and $NTU = 5$ and $C^* = 1$.

(a) Divide this exchanger into two equal passes (so that $NTU_p = 2.5$) and arrange the two passes into overall counterflow arrangement and obtain the exchanger overall effectiveness. Repeat this procedure by dividing the original exchanger into three passes and arranging them in an overall counterflow arrangement, and subsequently obtain the exchanger overall effectiveness. Repeat this procedure by dividing the original exchanger into four and five passes and subsequently, obtain the exchanger effectiveness for overall counterflow arrangement. Make a plot of this crossflow exchanger effectiveness as a function of the number of passes. Compare this curve with the effectiveness of a single-pass counterflow exchanger having $NTU = 5$ and $C^* = 1$. Discuss the results.

(b) Repeat the complete part (a) using an arrangement of n passes ($2 \leq n \leq 5$) in overall parallelflow arrangement. Plot the results on the same graph as that of part (a) and compare them with the effectiveness of a single-pass parallelflow exchanger having $NTU = 5$ and $C^* = 1$. Discuss the results.

3.14   A counterflow heat exchanger is currently used to heat fluid A by fluid C. A need has arisen to heat a second process stream B, and it has been proposed to do this by adding another section of exchanger to the existing unit (see Fig. P3.14). For the data given below and on the figure:

(a) Determine the extra heat exchanger area (i.e., the area of section II) required heating fluid B from 25 to 75°C.

(b) What is the effectiveness of section II?

(c) Suppose that the heat exchanger area of section I was very much larger than $0.75\,m^2$. What would happen to your answer to part (a)? Explain briefly.

Data: The heat exchanger area of section $I = 0.75\,m^2$. For each of sections I and II, $U = 250\,W/m^2 \cdot K$. For fluid A: $\dot{m} = 0.10\,kg/s$, $c_p = 2000\,J/kg \cdot K$. For fluid B: $\dot{m} = 0.30\,kg/s$, $c_p = 1000\,J/kg \cdot K$. For fluid C: $\dot{m} = 0.20\,kg/s$, $c_p = 1250\,J/kg \cdot K$.

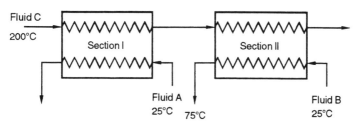

FIGURE P3.14

3.15 Given $T_{h,i} = 60°C$, $T_{h,o} = 43.3°C$, $T_{c,i} = 26.7°C$, and $T_{c,o} = 40.6°C$, calculate the true-mean temperature difference for **(a)** parallelflow, **(b)** counterflow, **(c)** single-pass crossflow with both fluids unmixed, **(d)** single-pass crossflow with cold fluid unmixed and hot fluid mixed, and **(e)** a 1–2 TEMA E exchanger. Also determine the exchanger effectiveness for each case.

3.16 To calculate ΔT_m for any flow arrangement (except for parallelflow), we calculate it from

$$\Delta T_m = F \, \Delta T_{\mathrm{lm}}$$

where F is the correction factor applied to the logarithmic temperature difference ΔT_{lm} for a hypothetical (or actual) counterflow exchanger operating at the same R and the same terminal temperatures. This ΔT_{lm} is obtained from Eq. (3.172) using Eq. (3.173). Now consider that we use the same counterflow ΔT_{lm} definition for a parallelflow exchanger [i.e., Eq. (3.172) in conjunction with Eq. (3.173) and not Eq. (3.174)]. In this case, F will not be unity, but it will be

$$F = \frac{R+1}{R-1} \frac{\ln[(1-RP)/(1-P)]}{\ln[1-(1+R)P]}$$

Derive this equation for F for a parallelflow exchanger.

3.17 Lubricating oil at a temperature of $60°C$ enters a 10 mm diameter tube with a velocity of 2.0 m/s. The tube surface is maintained at $30°C$. Calculate the tube length required to cool the oil to $45°C$. Consider the following properties for oil: $\rho = 865\,\mathrm{kg/m^3}$, $k = 0.14\,\mathrm{W/m \cdot K}$, $c_p = 1780\,\mathrm{J/kg \cdot K}$, and $\mu = 0.0078\,\mathrm{Pa \cdot s}$. For oil flow in the tube, the heat transfer coefficient $h = 51.2\,\mathrm{W/m^2 \cdot K}$.

3.18 A shell-and tube exchanger is required to cool the hot oil from $240°C$ to $90°C$. After some process, the same oil returns to the exchanger at $40°C$ and leaves at $190°C$. Assume a negligible change in the specific heat of oil in the temperature range 110 to $170°C$.

(a) If you design one 1–2 TEMA E shell-and-tube exchanger, would there be a temperature cross?

(b) Is it possible to design one 1–2 TEMA E exchanger to meet the required specifications, and why?

(c) What minimum number of 1–2 TEMA E exchangers in series would you require to eliminate the temperature cross? *Hint:* Use an appropriate equation from Eqs. (3.131)–(3.139). Specify any idealization you may make.

(d) What would be the overall arrangement of two fluids in part (c)? Tell whether or not you could use an alternate (opposite) flow arrangement and why.

(e) What would be the F factor for part (c)?

3.19 Two identical single-pass counterflow exchangers are used for heating water $(c_p = 4.2\,\mathrm{kJ/kg \cdot K})$ at $25°C$ with the hot oil $(c_p = 2.1\,\mathrm{kJ/kg \cdot K})$ at $120°C$. The water and oil flow rates are 1 and 4 kg/s, respectively. The heat exchangers are connected in series on the water side and in parallel on the oil side, as shown in

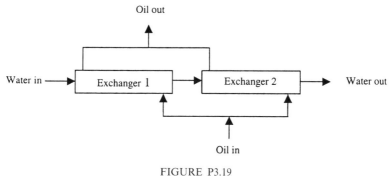

FIGURE P3.19

Fig. P3.19. The oil flow rate splits equally between the two exchangers at the inlet and rejoins at the exit. Use $U = 420 \, \text{W/m}^2 \cdot \text{K}$ and $A_{\text{pass}} = 10 \, \text{m}^2$ for each exchanger. Determine the outlet temperatures of the water and oil.

4 Additional Considerations for Thermal Design of Recuperators

The design theory for heat exchangers developed in Chapter 3 is based on the set of assumptions discussed in Section 3.2.1. That approach allows relatively straightforward solution of the corresponding design problems. In many applications, such design theory suffices and is used extensively. Still, some applications do require inclusion of additional effects i.e., relaxation of a number of assumptions. In all these situations, however, the conventional theory generally fails. So, additional assumptions are necessary to modify the simplified approach or to devise a completely new design methodology.

In industry, heat exchanger design and analysis calculations are performed almost exclusively using commercial and/or proprietary computer software. These tools are equipped with sophisticated routines that can deal with real engineering designs, although they do not possess the transparency necessary for clearly guiding an engineer through the design process. To assess the order of magnitude of various influences, to analyze preliminary designs in a fast and flexible manner, and to involve engineering judgment in a most creative way, analytical and/or *back-of-the-envelope* approaches would be very helpful. These also require, however, insights into the additional influences beyond the basic assumptions mentioned above. Thus, it would be necessary to develop ways of assessing the effects not included in the basic design procedure covered in Chapter 3. In this chapter we consider the following enhancements to the basic design procedure: (1) longitudinal wall heat conduction effects in Section 4.1, (2) nonuniform heat transfer coefficients in Section 4.2, and (3) complex flow distributions in shell-and-tube heat exchangers in Section 4.4.

Additional considerations for completion of either a simplified design approach or one based on the relaxed assumptions are still necessary. Among these, the most important is the need to take into account the fin efficiency of extended heat transfer surfaces commonly used in compact heat exchangers and some shell-and-tube heat exchangers. Hence, considerable theory development and discussion is devoted to fin efficiency in Section 4.3.

4.1 LONGITUDINAL WALL HEAT CONDUCTION EFFECTS

In a heat exchanger, since heat transfer takes place, temperature gradients exist in both fluids and in the separating wall in the fluid flow directions. This results in heat conduction in the wall and in fluids from the hotter to colder temperature regions, which may affect the heat transfer rate from the hot fluid to the cold fluid.

Heat conduction in a fluid in the fluid flow direction is negligible for Pe > 10 and $x^* \geq 0.005$, where the Péclet number Pe = Re $\cdot$ Pr = $u_m D_h / \alpha$ and $x^* = x/(D_h \cdot \text{Pe})$; the significance and meaning of Pe and x^* are presented in Section 7.2. For most heat exchangers, except liquid metal heat exchangers, Pe and x^* are higher than the values indicated above. Hence, longitudinal heat conduction in the fluid is negligible in most applications and is not covered here.

If a temperature gradient is established in the separating walls between fluid flow streams in a heat exchanger, heat transfer by conduction takes place from the hotter to colder region of the wall, flattens the wall temperature distribution, and reduces the performance of the exchanger. For example, let us review typical temperature distributions in the hot fluid, cold fluid, and wall for a counterflow heat exchanger as shown in Fig. 4.1. Dashed lines represent the case of zero longitudinal heat conduction, and solid lines represent the case of finite longitudinal conduction [$\lambda = 0.4$, λ defined in Eq. (4.13)]. From the figure it is clear that longitudinal conduction in the wall flattens the temperature distributions, reduces the mean outlet temperature of the cold fluid, and thus reduces the exchanger effectiveness ε from 90.9% to 73.1%. This in turn produces a penalty in the exchanger overall heat transfer rate. The reduction in the exchanger effectiveness at a specified NTU may be quite significant in a single-pass exchanger having very steep temperature changes in the flow direction (i.e., large $\Delta T_h / L$ or $\Delta T_c / L$). Such a situation arises for a compact exchanger designed for high effectiveness (approximately above 80%) and has a short flow length L. Shell-and-tube exchangers are usually designed for an exchanger effectiveness of 60% or below per pass. The influence of heat conduction in the wall in the flow direction is negligible for such effectiveness.

Since the longitudinal wall heat conduction effect is important only for high-effectiveness single-pass compact heat exchangers, and since such exchangers are usually designed using ε-NTU theory, we present the theory for the longitudinal conduction effect by an extension of ε-NTU theory. No such extension is available in the MTD method.

The magnitude of longitudinal heat conduction in the wall depends on the wall heat conductance and the wall temperature gradient. The latter in turn depends on the thermal conductance on each side of the wall. To arrive at additional nondimensional groups for longitudinal conduction effects, we can work with the differential energy and rate equations of the problem and can derive the same Eq. (4.9) mentioned later. For example, see Section 5.4 for the derivation of appropriate equations for a rotary regenerator. However, to provide a "feel" to the reader, a more heuristic approach is followed here. Let us consider the simple case of a linear temperature gradient in the wall as shown in Fig. 4.1. The longitudinal conduction rate is

$$q_k = k_w A_k \frac{T_{w,1} - T_{w,2}}{L} \tag{4.1}$$

where A_k is total wall cross-sectional area for longitudinal conduction. The convective heat transfer from the hot fluid to the wall results in its enthalpy rate drop as follows, which is the same as enthalpy rate change (convection rate) q_h:

$$(\dot{m}\,\Delta\mathbf{h})_h = C_h(T_{h,1} - T_{h,2}) = q_h \tag{4.2}$$

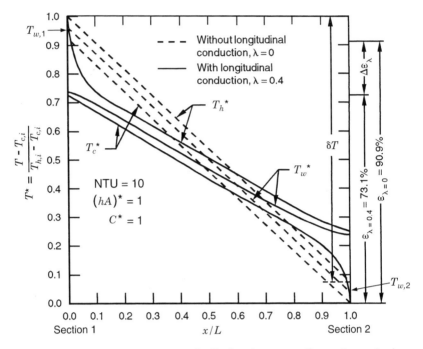

FIGURE 4.1 Fluid and wall temperature distributions in a counterflow exchanger having zero and finite longitudinal wall heat conduction.

Similarly, the enthalpy rate rise of the cold fluid as a result of convection from the wall to the fluid is

$$(\dot{m}\,\Delta h)_c = C_c(T_{c,1} - T_{c,2}) = q_c \tag{4.3}$$

Of course, $q_h = q_c$, with no heat losses to the ambient. Therefore, the ratios of longitudinal heat conduction in the wall to the convection rates in the hot and cold fluids are

$$\frac{q_k}{q_h} = \frac{k_w A_k}{LC_h} \frac{T_{w,1} - T_{w,2}}{T_{h,1} - T_{h,2}} \tag{4.4}$$

$$\frac{q_k}{q_c} = \frac{k_w A_k}{LC_c} \frac{T_{w,1} - T_{w,2}}{T_{c,1} - T_{c,2}} \tag{4.5}$$

The resulting new dimensionless groups in Eqs. (4.4) and (4.5) are defined as

$$\lambda_h = \left(\frac{k_w A_k}{LC}\right)_h \qquad \lambda_c = \left(\frac{k_w A_k}{LC}\right)_c \tag{4.6}$$

where for generality (for an exchanger with an arbitrary flow arrangement), the subscripts h and c are used for all quantities on the right-hand side of the equality sign of λ's.

The individual λ is referred to as the *longitudinal conduction parameter*. It is defined as a ratio of longitudinal wall heat conduction per unit temperature difference and per unit length to the heat capacity rate of the fluid. Keeping in mind that relaxation of a zero longitudinal conduction idealization means the introduction of a new heat transfer process deterioration factor, we may conclude that the higher the value of λ, the higher are heat conduction losses and the lower is the exchanger effectiveness compared to the $\lambda = 0$ case.

Now considering the thermal circuit of Fig. 3.4 (with negligible wall thermal resistance and no fouling), the wall temperature at any given location can be given by Eq. (3.34) as

$$T_w = \frac{T_h + (\eta_o hA)^* T_c}{1 + (\eta_o hA)^*} \tag{4.7}$$

where

$$(\eta_o hA)^* = \frac{(\eta_o hA)_c}{(\eta_o hA)_h} \tag{4.8}$$

Hence, the wall temperature distribution is between T_h and T_c distributions, as shown in Fig. 4.1, and its specific location depends on the magnitude of $(\eta_o hA)^*$. If $(\eta_o hA)^*$ is zero (as in a condenser), $T_w = T_h$. And since T_h is approximately constant for a condenser, T_w will also be a constant, indicating no longitudinal temperature gradients in the wall, even though λ will be finite. Thus, when $(\eta_o hA)^*$ is zero or infinity, the longitudinal heat conduction is zero. Longitudinal heat conduction effects are maximum for $(\eta_o hA)^* = 1$.

Thus, in the presence of longitudinal heat conduction in the wall, we may expect and it can be proven that the exchanger effectiveness is a function of the following groups:

$$\varepsilon = \phi[\text{NTU}, C^*, \lambda_h, \lambda_c, (\eta_o hA)^*, \text{flow arrangement}] \tag{4.9}$$

It should be added that the same holds for a parallelflow exchanger, but as shown in Section 4.1.3, longitudinal wall conduction effects for a parallelflow exchanger are negligible. Note that for a counterflow exchanger,

$$L_h = L_c = L \quad \text{and} \quad A_{k,h} = A_{k,c} = A_k \tag{4.10}$$

Equation (4.6) then becomes

$$\lambda_h = \frac{k_w A_k}{L C_h} \qquad \lambda_c = \frac{k_w A_k}{L C_c} \tag{4.11}$$

and therefore,

$$\frac{\lambda_h}{\lambda_c} = \frac{C_c}{C_h} = \begin{cases} C^* & \text{for } C_c = C_{\min} \\ 1/C^* & \text{for } C_c = C_{\max} \end{cases} \tag{4.12}$$

Thus for a counterflow exchanger, λ_h and λ_c are not both independent parameters since C^* is already included in Eq. (4.9); only one of them is independent. Instead of choosing one of them, alternatively, a unique longitudinal conduction parameter λ is defined as follows for a counterflow exchanger:

$$\lambda = \frac{k_w A_k}{L C_{\min}} \tag{4.13}$$

and Eq. (4.9) takes the following form for a counterflow exchanger:

$$\varepsilon = \phi[\text{NTU}, C^*, \lambda, (\eta_o h A)^*] \tag{4.14}$$

However, for exchangers other than counterflow and parallelflow, Eq. (4.9) is the correct general relationship.

Multiplying by $(T_{h,i} - T_{c,i})$ the numerator and denominator of Eq. (4.13), λ can be interpreted as the ratio of longitudinal heat conduction rate in a counterflow heat exchanger (having NTU $= \infty$ and $C^* = 1$) to the thermodynamically maximum possible heat transfer rate.

Next, we summarize the longitudinal conduction effects for various exchangers.

4.1.1   Exchangers with $C^* = 0$

The operating condition having $C^* = 0$ usually occurs when condensing or boiling takes place on one side of a two-fluid heat exchanger. The thermal resistance $(1/hA)$ in such a case on the two-phase $(C_{\max})$ side is generally very small, and hence the wall temperature is close to the fluid temperature of the $C_{\max}$ side and almost constant (i.e., the wall has a very small temperature gradient, if any, in the longitudinal direction). As a result, longitudinal heat conduction in the wall is negligible and its effect on the exchanger effectiveness is also negligible for the $C^* = 0$ case.

4.1.2   Single-Pass Counterflow Exchanger

The counterflow exchanger problem with finite longitudinal wall conduction has been analyzed by Kroeger (1967), among others, and extensive graphical results are available for the exchanger ineffectiveness $(1 - \varepsilon)$ for a wide range of NTU, C^*, and λ. Kroeger showed that the influence of $(\eta_o h A)^*$ on ε or $(1 - \varepsilon)$ is negligible for $0.1 \leq (\eta_o h A)^* \leq 10$, and hence the longitudinal wall conduction effect can be taken into account by only one additional parameter λ [see Eq. (4.14)] besides NTU and C^*. The penalty to exchanger effectiveness due to longitudinal wall conduction increases with increasing values of NTU, C^*, and λ, and is largest for $C^* = 1$. Kroeger's solution for $C^* = 1$ and $0.1 \leq (\eta_o h A)^* \leq 10$ is

$$\varepsilon = 1 - \frac{1}{1 + \text{NTU}(1 + \lambda\Phi)/(1 + \lambda \cdot \text{NTU})} \tag{4.15}$$

where for NTU ≥ 3,

$$\Phi = \left(\frac{\lambda \cdot \text{NTU}}{1 + \lambda \cdot \text{NTU}}\right)^{1/2} \tag{4.16}$$

For $\lambda = 0$ and ∞, Eq. (4.15) reduces to

$$\varepsilon = \begin{cases} \mathrm{NTU}/(1 + \mathrm{NTU}) & \text{for } \lambda = 0 \\ \frac{1}{2}\left[1 - \exp(-2\,\mathrm{NTU})\right] & \text{for } \lambda = \infty \end{cases} \tag{4.17}$$

Note that for $\lambda \to \infty$, the counterflow exchanger effectiveness ε from Eq. (4.17) is identical to ε for a parallelflow exchanger (see Table 3.4). This is expected since the wall temperature distribution will be perfectly uniform for $\lambda \to \infty$, and this is the case for a parallelflow exchanger with $C^* = 1$.

For $\mathrm{NTU} \to \infty$, Eq. (4.15) reduces to

$$\varepsilon = 1 - \frac{\lambda}{1 + 2\lambda} \tag{4.18}$$

The results from Eq. (4.15) are presented in Fig. 4.2 in terms of ineffectiveness $(1 - \varepsilon)$ as a function of NTU and λ. The concept of ineffectiveness is useful particularly for high values of effectiveness because small changes in large values of effectiveness are magnified. For example, consider an exchanger to be designed for $\varepsilon = 98\%$ and $C^* = 1$. From Fig. 4.2, $\mathrm{NTU} = 49$ when $\lambda = 0$, and $\mathrm{NTU} = 94$ for $\lambda = 0.01$. Thus longitudinal heat conduction increases the NTU required by 92%, a significant penalty in required surface area due to longitudinal conduction. Alternatively, for $\mathrm{NTU} = 49$ and $\lambda = 0.01$, from Fig. 4.2, $(1 - \varepsilon) \times 100 = 3\%$ or $\varepsilon = 97\%$. Thus, the result is a 1% decrease in exchanger effectiveness for $\lambda = 0.01$! A direct expression for the reduction in the exchanger effectiveness due to longitudinal wall heat conduction is

$$\frac{\Delta\varepsilon}{\varepsilon} = \frac{\varepsilon_{\lambda=0} - \varepsilon_{\lambda\neq0}}{\varepsilon_{\lambda=0}} = \frac{(\mathrm{NTU} - \Phi)\lambda}{1 + \mathrm{NTU}(1 + \lambda + \lambda\Phi)} \approx \lambda \tag{4.19}$$

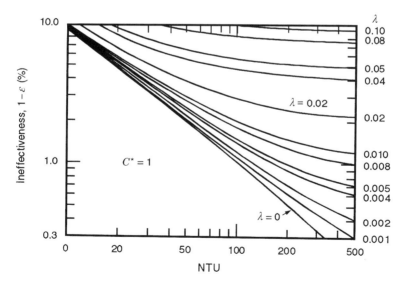

FIGURE 4.2   Counterflow heat exchanger ineffectiveness $(1 - \varepsilon)$ as a function of NTU and λ for $C^* = 1$. (From Kroeger, 1967.)

for NTU large. In Eq. (4.19), the term on the right-hand side of the second equal sign is obtained from using Eq. (4.15) for $\lambda \neq 0$ and Eq. (4.17) for $\lambda = 0$, and the last term on the right-hand side is obtained from Kays and London (1998).

The exchanger ineffectiveness for $C^* < 1$ has been obtained and correlated by Kroeger (1967) as follows:

$$1 - \varepsilon = \frac{1 - C^*}{\Psi \exp(r_1) - C^*} \tag{4.20}$$

where

$$r_1 = \frac{(1 - C^*)\text{NTU}}{1 + \lambda \cdot \text{NTU} \cdot C^*} \tag{4.21}$$

$$\Psi = \frac{1 + \gamma \Psi^*}{1 - \gamma \Psi^*} \qquad \Psi^* = \left(\frac{\alpha}{1 + \alpha}\right)^{1/2} \frac{1 + \gamma}{1/\alpha - \gamma - \gamma^2} \tag{4.22}$$

$$\gamma = \frac{1 - C^*}{1 + C^*} \frac{1}{1 + \alpha} \qquad \alpha = \lambda \cdot \text{NTU} \cdot C^* \tag{4.23}$$

Note that here γ and α are local dimensionless variables as defined in Eq. (4.23). The values of Ψ are shown in Fig. 4.3. The approximate Eq. (4.20), although derived for $(\eta_o hA)^*/C^* = 1$, could be used for values of this parameter different from unity. For $0.5 < (\eta_o hA)^*/C^* \leq 2$, the error introduced in the ineffectiveness is within 0.8% and 4.7% for $C^* = 0.95$ and 0.8, respectively.

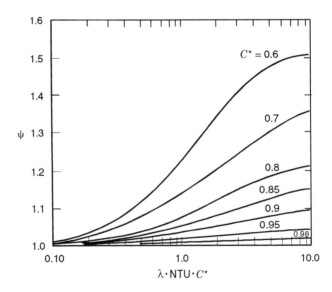

FIGURE 4.3 Function Ψ of Eq. (4.20) for computation of single-pass counterflow exchanger ineffectiveness, including longitudinal wall heat conduction effects. (From Kroeger, 1967.)

In summary, the decrease in exchanger effectiveness due to longitudinal wall heat conduction increases with increasing values of NTU, C^* and λ, and the decrease in ε is largest for $C^* = 1$. Longitudinal wall conduction has a significant influence on the counterflow exchanger size (NTU) for a given ε when NTU > 10 and $\lambda > 0.005$.

4.1.3   Single-Pass Parallelflow Exchanger

In the case of a parallelflow exchanger, the wall temperature distribution is always almost close to constant regardless of the values of C^* and NTU. Since the temperature gradient in the wall is negligibly small in the fluid flow direction, the influence of longitudinal wall conduction on exchanger effectiveness is negligible. Hence, there is no need to analyze or take this effect into consideration for parallelflow exchangers.

4.1.4   Single-Pass Unmixed–Unmixed Crossflow Exchanger

In this case, the temperature gradients in the wall in the x and y directions of two fluid flows are different. This is because $\Delta T_{w,1}$ and $\Delta T_{w,2}$ ($\Delta T_{w,1}$ is the temperature difference in the wall occurring between inlet and outlet locations of fluid 1; similarly $\Delta T_{w,2}$ is defined for fluid 2) are in general different, as well as $L_h \neq L_c$ in general. Hence, λ_h and λ_c are independent parameters for a crossflow exchanger, and ε is a function of five independent dimensionless groups, as shown in Eq. (4.9), when longitudinal wall heat conduction is considered. For the same NTU and C^* values, the crossflow exchanger effectiveness is lower than the counterflow exchanger effectiveness; however, the wall temperature distribution is two-dimensional and results in higher temperature gradients for the crossflow exchanger compared to counterflow. Hence, for identical NTU, C^*, and λ, the effect of longitudinal conduction on the exchanger effectiveness is higher for the crossflow exchanger than that for the counterflow exchanger. Since crossflow exchangers are usually not designed for $\varepsilon \geq 80\%$, the longitudinal conduction effect is generally small and negligible compared to a counterflow exchanger that is designed for ε up to 98 to 99%. Since the problem is more complicated for the crossflow exchanger, only the numerical results obtained by Chiou (published in Shah and Mueller, 1985) are presented in Table 4.1.

4.1.5   Other Single-Pass Exchangers

The influence of longitudinal conduction on exchanger effectiveness is not evaluated for recuperators of other flow arrangements. However, most single-pass exchangers with other flow arrangements are not designed with high effectiveness, hence there seems to be no real need for that information for such industrial heat exchangers.

4.1.6   Multipass Exchangers

The influence of longitudinal conduction in a multipass exchanger is evaluated individually for each pass, depending on the flow arrangement. The results of preceding sections are used for this purpose. Thus although the overall exchanger effectiveness may be quite high for an overall counterflow multipass unit, the individual pass effectiveness may not be high, and hence the influence of longitudinal conduction may not be significant for many multipass exchangers.

TABLE 4.1 Reduction in Crossflow Exchanger Effectiveness ($\Delta\varepsilon/\varepsilon$) Due to Longitudinal Wall Heat Conduction for $C^* = 1$.

$\dfrac{\lambda_x}{\lambda_y}$	$\dfrac{(\eta_0 hA)_x}{(\eta_0 hA)_y}$	NTU	ε for $\lambda_x = 0$	$\Delta\varepsilon/\varepsilon$ for $\lambda_x =$: 0.005	0.010	0.015	0.020	0.025	0.030	0.040	0.060	0.080	0.100	0.200	0.400
0.5	0.5	1.00	0.4764	0.0032	0.0062	0.0089	0.0115	0.0139	0.0162	0.0204	0.0276	0.0336	0.0387	0.0558	0.0720
		2.00	0.6147	0.0053	0.0103	0.0150	0.0194	0.0236	0.0276	0.0350	0.0481	0.0592	0.0689	0.1026	0.1367
		4.00	0.7231	0.0080	0.0156	0.0227	0.0294	0.0357	0.0418	0.0530	0.0726	0.0892	0.1036	0.1535	0.2036
		6.00	0.7729	0.0099	0.0192	0.0279	0.0360	0.0437	0.0510	0.0644	0.0877	0.1072	0.1239	0.1810	0.2372
		8.00	0.8031	0.0114	0.0220	0.0319	0.0411	0.0497	0.0578	0.0728	0.0984	0.1197	0.1377	0.1988	0.2580
		10.00	0.8238	0.0127	0.0243	0.0351	0.0451	0.0545	0.0633	0.0793	0.1066	0.1290	0.1480	0.2115	0.2724
		50.00	0.9229	0.0246	0.0446	0.0616	0.0764	0.0897	0.1017	0.1229	0.1569	0.1838	0.2057	0.2765	0.3427
		100.00	0.9476	0.0311	0.0543	0.0732	0.0893	0.1034	0.1160	0.1379	0.1729	0.2001	0.2223	0.2942	0.3666
	1.0	1.00	0.4764	0.0029	0.0055	0.0079	0.0102	0.0123	0.0143	0.0180	0.0244	0.0296	0.0341	0.0490	0.0634
		2.00	0.6147	0.0055	0.0097	0.0141	0.0182	0.0221	0.0258	0.0327	0.0449	0.0553	0.0643	0.0959	0.1286
		4.00	0.7231	0.0078	0.0151	0.0220	0.0284	0.0346	0.0404	0.0512	0.0702	0.0863	0.1002	0.1491	0.1991
		6.00	0.7729	0.0097	0.0188	0.0273	0.0353	0.0428	0.0499	0.0630	0.0857	0.1049	0.1213	0.1779	0.2344
		8.00	0.8031	0.0113	0.0217	0.0314	0.0404	0.0489	0.0569	0.0716	0.0968	0.1178	0.1356	0.1965	0.2560
		10.00	0.8238	0.0125	0.0240	0.0347	0.0446	0.0538	0.0624	0.0782	0.1052	0.1274	0.1462	0.2096	0.2708
		50.00	0.9229	0.0245	0.0445	0.0614	0.0763	0.0895	0.1015	0.1226	0.1566	0.1834	0.2053	0.2758	0.3405
		100.00	0.9476	0.0310	0.0543	0.0731	0.0892	0.1033	0.1159	0.1378	0.1727	0.1999	0.2221	0.2933	0.3619
	2.0	1.00	0.4764	0.0027	0.0051	0.0074	0.0095	0.0116	0.0135	0.0170	0.0232	0.0285	0.0330	0.0489	0.0652
		2.00	0.6147	0.0048	0.0092	0.0134	0.0173	0.0211	0.0247	0.0313	0.0432	0.0533	0.0621	0.0938	0.1274
		4.00	0.7231	0.0076	0.0147	0.0213	0.0277	0.0336	0.0393	0.0499	0.0685	0.0844	0.0982	0.1468	0.1971
		6.00	0.7729	0.0095	0.0184	0.0268	0.0346	0.0420	0.0490	0.0619	0.0844	0.1033	0.1196	0.1760	0.2328
		8.00	0.8031	0.0111	0.0214	0.0309	0.0398	0.0482	0.0561	0.0706	0.0956	0.1164	0.1342	0.1949	0.2548
		10.00	0.8238	0.0124	0.0238	0.0343	0.0440	0.0532	0.0617	0.0774	0.1041	0.1262	0.1450	0.2083	0.2698
		50.00	0.9229	0.0245	0.0444	0.0613	0.0761	0.0893	0.1013	0.1223	0.1563	0.1831	0.2050	0.2754	0.3401
		100.00	0.9476	0.0310	0.0542	0.0731	0.0891	0.1032	0.1158	0.1377	0.1725	0.1997	0.2219	0.2928	0.3600

1.0	0.5	1.00	0.4764	0.0020	0.0039	0.0057	0.0074	0.0090	0.0106	0.0136	0.0190	0.0237	0.0280	0.0436	0.0609
		2.00	0.6147	0.0034	0.0067	0.0098	0.0128	0.0157	0.0185	0.0238	0.0336	0.0423	0.0501	0.0803	0.1154
		4.00	0.7231	0.0053	0.0103	0.0152	0.0198	0.0243	0.0287	0.0369	0.0520	0.0653	0.0773	0.1229	0.1753
		6.00	0.7729	0.0066	0.0129	0.0189	0.0246	0.0302	0.0355	0.0456	0.0637	0.0797	0.0940	0.1472	0.2070
		8.00	0.8031	0.0076	0.0148	0.0217	0.0283	0.0346	0.0406	0.0520	0.0723	0.0900	0.1057	0.1634	0.2270
		10.00	0.8238	0.0085	0.0165	0.0241	0.0313	0.0382	0.0448	0.0571	0.0790	0.0979	0.1145	0.1751	0.2410
		50.00	0.9229	0.0170	0.0316	0.0445	0.0562	0.0667	0.0765	0.0940	0.1233	0.1474	0.1677	0.2378	0.3092
		100.00	0.9476	0.0218	0.0395	0.0543	0.0673	0.0789	0.0894	0.1080	0.1385	0.1632	0.1840	0.2546	0.3273
	1.0	1.00	0.4764	0.0020	0.0038	0.0055	0.0072	0.0088	0.0103	0.0132	0.0183	0.0228	0.0268	0.0412	0.0567
		2.00	0.6147	0.0034	0.0066	0.0097	0.0127	0.0156	0.0183	0.0236	0.0331	0.0417	0.0493	0.0786	0.1125
		4.00	0.7231	0.0053	0.0103	0.0152	0.0198	0.0243	0.0286	0.0368	0.0517	0.0650	0.0769	0.1220	0.1742
		6.00	0.7729	0.0066	0.0129	0.0189	0.0246	0.0301	0.0354	0.0455	0.0636	0.0795	0.0936	0.1466	0.2064
		0.00	0.8031	0.0076	0.0149	0.0217	0.0283	0.0346	0.0406	0.0519	0.0722	0.0898	0.1054	0.1630	0.2266
		10.00	0.8238	0.0085	0.0165	0.0241	0.0313	0.0382	0.0448	0.0571	0.0789	0.0978	0.1143	0.1749	0.2407
		50.00	0.9229	0.0170	0.0316	0.0445	0.0562	0.0667	0.0765	0.0940	0.1233	0.1473	0.1677	0.2377	0.3090
		100.00	0.9476	0.0218	0.0395	0.0543	0.0673	0.0789	0.0894	0.1080	0.1385	0.1632	0.1840	0.2546	0.3270
	2.0	1.00	0.4764	0.0020	0.0039	0.0057	0.0074	0.0090	0.0106	0.0136	0.0190	0.0237	0.0280	0.0436	0.0609
		2.00	0.6147	0.0034	0.0067	0.0098	0.0128	0.0157	0.0185	0.0238	0.0336	0.0423	0.0501	0.0803	0.1154
		4.00	0.7231	0.0053	0.0103	0.0152	0.0198	0.0243	0.0287	0.0369	0.0520	0.0653	0.0773	0.1229	0.1753
		6.00	0.7729	0.0066	0.0129	0.0189	0.0246	0.0302	0.0355	0.0456	0.0637	0.0797	0.0940	0.1472	0.2070
		8.00	0.8031	0.0076	0.0148	0.0217	0.0283	0.0346	0.0406	0.0520	0.0723	0.0900	0.1057	0.1634	0.2270
		10.00	0.8238	0.0085	0.0165	0.0241	0.0313	0.0382	0.0448	0.0571	0.0790	0.0979	0.1145	0.1751	0.2410
		50.00	0.9229	0.0170	0.0316	0.0445	0.0562	0.0667	0.0765	0.0940	0.1233	0.1474	0.1677	0.2378	0.3092
		100.00	0.9476	0.0218	0.0395	0.0543	0.0673	0.0789	0.0894	0.1080	0.1385	0.1632	0.1840	0.2546	0.3273

(continued)

TABLE 4.1 (*Continued*).

$\frac{\lambda_x}{\lambda_y}$	$\frac{(\eta_o h A)_x}{(\eta_o h A)_y}$	NTU	ε for $\lambda_x = 0$	$\Delta\varepsilon/\varepsilon$ for $\lambda_x =$:											
				0.005	0.010	0.015	0.020	0.025	0.030	0.040	0.060	0.080	0.100	0.200	0.400
2.0	0.5	1.00	0.4764	0.0014	0.0027	0.0039	0.0051	0.0063	0.0074	0.0095	0.0135	0.0170	0.0203	0.0330	0.0489
		2.00	0.6147	0.0024	0.0048	0.0070	0.0092	0.0113	0.0134	0.0173	0.0247	0.0313	0.0375	0.0621	0.0938
		4.00	0.7231	0.0039	0.0076	0.0112	0.0147	0.0181	0.0213	0.0277	0.0393	0.0499	0.0596	0.0982	0.1468
		6.00	0.7729	0.0049	0.0095	0.0141	0.0184	0.0227	0.0268	0.0346	0.0490	0.0619	0.0736	0.1196	0.1760
		8.00	0.8031	0.0057	0.0111	0.0163	0.0214	0.0262	0.0309	0.0398	0.0561	0.0706	0.0837	0.1342	0.1949
		10.00	0.8238	0.0063	0.0124	0.0182	0.0238	0.0291	0.0343	0.0440	0.0617	0.0774	0.0914	0.1448	0.2083
		50.00	0.9229	0.0130	0.0245	0.0349	0.0444	0.0531	0.0613	0.0761	0.1013	0.1223	0.1404	0.2050	0.2754
		100.00	0.9476	0.0168	0.0310	0.0433	0.0542	0.0641	0.0731	0.0891	0.1158	0.1377	0.1563	0.2219	0.2928
	1.0	1.00	0.4764	0.0015	0.0029	0.0042	0.0055	0.0067	0.0079	0.0102	0.0143	0.0180	0.0213	0.0341	0.0490
		2.00	0.6147	0.0026	0.0050	0.0074	0.0097	0.0119	0.0141	0.0182	0.0258	0.0327	0.0391	0.0643	0.0959
		4.00	0.7231	0.0040	0.0078	0.0115	0.0151	0.0186	0.0220	0.0284	0.0404	0.0512	0.0611	0.1002	0.1491
		6.00	0.7729	0.0050	0.0097	0.0144	0.0188	0.0231	0.0273	0.0353	0.0499	0.0630	0.0749	0.1213	0.1779
		8.00	0.8031	0.0057	0.0113	0.0166	0.0217	0.0266	0.0314	0.0404	0.0569	0.0716	0.0848	0.1356	0.1965
		10.00	0.8238	0.0064	0.0125	0.0184	0.0240	0.0295	0.0347	0.0446	0.0624	0.0782	0.0924	0.1462	0.2096
		50.00	0.9229	0.0130	0.0245	0.0350	0.0445	0.0533	0.0614	0.0763	0.1015	0.1226	0.1407	0.2053	0.2758
		100.00	0.9476	0.0169	0.0310	0.0434	0.0543	0.0641	0.0731	0.0892	0.1159	0.1378	0.1564	0.2221	0.2933
	2.0	1.00	0.4764	0.0016	0.0032	0.0047	0.0062	0.0076	0.0089	0.0115	0.0162	0.0204	0.0241	0.0387	0.0558
		2.00	0.6147	0.0027	0.0053	0.0078	0.0103	0.0127	0.0150	0.0194	0.0276	0.0350	0.0418	0.0689	0.1026
		4.00	0.7231	0.0041	0.0080	0.0119	0.0156	0.0192	0.0227	0.0294	0.0418	0.0530	0.0632	0.1036	0.1535
		6.00	0.7729	0.0051	0.0099	0.0146	0.0192	0.0236	0.0279	0.0360	0.0510	0.0644	0.0766	0.1239	0.1810
		8.00	0.8031	0.0058	0.0114	0.0168	0.0220	0.0270	0.0319	0.0411	0.0578	0.0728	0.0862	0.1377	0.1988
		10.00	0.8238	0.0065	0.0127	0.0186	0.0243	0.0298	0.0351	0.0451	0.0633	0.0793	0.0936	0.1480	0.2115
		50.00	0.9229	0.0130	0.0246	0.0350	0.0446	0.0534	0.0616	0.0764	0.1017	0.1229	0.1410	0.2057	0.2765
		100.00	0.9476	0.0169	0.0311	0.0434	0.0543	0.0642	0.0732	0.0893	0.1160	0.1379	0.1566	0.2223	0.2942

Source: Data from Chiou as published in Shah and Mueller (1985).

Example 4.1   A gas-to-air crossflow waste heat recovery exchanger, having both fluids unmixed, has NTU = 6 and $C^* = 1$. The inlet fluid temperatures on the hot and cold sides are 360°C and 25°C, respectively. Determine the outlet fluid temperatures with and without longitudinal wall heat conduction. Assume that $\lambda_c = \lambda_h = 0.04$, and $(\eta_o hA)_h/(\eta_o hA)_c = 1$.

SOLUTION

Problem Data and Schematic: NTU, ratio of heat capacity rates, inlet temperatures, and hot- and cold-side λ's are given for a crossflow gas-to-air waste heat recovery exchanger (Fig. E4.1).

Determine: The outlet temperatures with and without longitudinal wall heat conduction.

Assumptions: Fluid properties are constant and the longitudinal conduction factor λ is also constant throughout the exchanger.

Analysis: In the absence of longitudinal conduction (i.e., $\lambda_x = 0$), we could find the effectiveness $\varepsilon = 0.7729$ from Table 4.1 or Eq. (II.1) of Table 3.6. Using the definition of ε, the outlet temperatures are

$$T_{h,o} = T_{h,i} - \varepsilon(T_{h,i} - T_{c,i}) = 360°C - 0.7729(360 - 25)°C = 101.1°C \qquad Ans.$$

$$T_{c,o} = T_{c,i} + \varepsilon(T_{h,i} - T_{c,i}) = 25°C + 0.7729(360 - 25)°C = 283.9°C \qquad Ans.$$

To take longitudinal conduction into account, we need to find the new effectiveness of the exchanger. Knowing that NTU = 6, $C^* = 1$, $\lambda_c/\lambda_h = 1$, and $\lambda_c = 0.04$, from Table 4.1 we have $\Delta\varepsilon/\varepsilon = 0.0455$. Therefore, the new effectiveness, using the expressions of the first equality of Eq. (4.19), is

$$\varepsilon_{\lambda \neq 0} = \left(1 - \frac{\Delta\varepsilon}{\varepsilon}\right)\varepsilon_{\lambda=0} = (1 - 0.0445)0.7729 = 0.7377$$

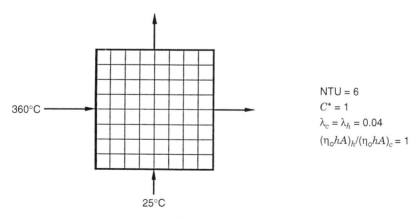

360°C

25°C

NTU = 6
$C^* = 1$
$\lambda_c = \lambda_h = 0.04$
$(\eta_o hA)_h/(\eta_o hA)_c = 1$

FIGURE E4.1

Thus, from the definition of effectiveness, the outlet temperatures are

$$T_{h,o} = T_{h,i} - \varepsilon(T_{h,i} - T_{c,i}) = 360°C - 0.7377(360 - 25)°C = 112.9°C \qquad Ans.$$

$$T_{c,o} = T_{c,i} + \varepsilon(T_{h,i} - T_{c,i}) = 25°C + 0.7377(360 - 25)°C = 272.1°C \qquad Ans.$$

Discussion and Comments: For this particular problem, the reduction in exchanger effectiveness is quite serious, 4.6%, with corresponding differences in the outlet temperatures. This results in a 4.6% reduction in heat transfer or a 4.6% increase in fuel consumption to make up for the effect of longitudinal wall heat conduction. This example implies that for some high-effectiveness crossflow exchangers, longitudinal wall heat conduction may be important and cannot be ignored. So always make a practice of considering the effect of longitudinal conduction in heat exchangers having $\varepsilon \geq 75\%$ for single-pass units or for individual passes of a multiple-pass exchanger.

4.2 NONUNIFORM OVERALL HEAT TRANSFER COEFFICIENTS

In ε-NTU, P-NTU, and MTD methods of exchanger heat transfer analysis, it is idealized that the overall heat transfer coefficient U is constant and uniform throughout the exchanger and invariant with time. As discussed for Eq. (3.24), this U is dependent on the number of thermal resistances in series, and in particular, on heat transfer coefficients on fluid 1 and 2 sides. These individual heat transfer coefficients may vary with flow Reynolds number, heat transfer surface geometry, fluid thermophysical properties, entrance length effect due to developing thermal boundary layers, and other factors. In a viscous liquid exchanger, a tenfold variation in h is possible when the flow pattern encompasses laminar, transition, and turbulent regions on one side. Thus, if the individual h values vary across the exchanger surface area, it is highly likely that U will not remain constant and uniform in the exchanger.

Now we focus on the variation in U and how to take into account its effect on exchanger performance considering local heat transfer coefficients on each fluid side varying slightly or significantly due to two effects: (1) changes in the fluid properties or radiation as a result of a rise in or drop of fluid temperatures, and (2) developing thermal boundary layers (referred to as the *length effect*). In short, we relax assumption 8 postulated in Section 3.2.1 that the individual and overall heat transfer coefficients are constant.

The first effect, due to fluid property variations (or radiation), consists of two components: (1) distortion of velocity and temperature profiles at a given free flow cross section due to fluid property variations (this effect is usually taken into account by the *property ratio method*, discussed in Section 7.6); and (2) variations in the fluid temperature along axial and transverse directions in the exchanger, depending on the exchanger flow arrangement; this effect is referred to as the *temperature effect*. The resulting axial changes in the overall mean heat transfer coefficient can be significant; the variations in U_{local} could be nonlinear, depending on the type of fluid. While both the temperature and thermal entry length effects could be significant in laminar flows, the latter effect is generally not significant in turbulent flow except for low-Prandtl-number fluids.

It should be mentioned that in general the local heat transfer coefficient in a heat exchanger is also dependent on variables other than the temperature and length effects, such as flow maldistribution, fouling, and manufacturing imperfections. Similarly, the overall heat transfer coefficient is dependent on heat transfer surface geometry, individual Nu (as a function of relevant parameters), thermal properties, fouling effects, temperature variations, temperature difference variations, and so on. No information is available on the effect of some of these parameters, and it is beyond the scope of this book to discuss the effect of other parameters. In this section we concentrate on nonuniformities in U due to temperature and length effects.

To outline how to take temperature and length effects into account, let us introduce specific definitions of local and mean overall heat transfer coefficients. The local overall heat transfer coefficient $U(x_1^*, x_2^*, T)$, is defined as follows in an exchanger at a local position $[x^* = x/(D_h \cdot \mathrm{Re} \cdot \mathrm{Pr})$, subscripts 1 and 2 for fluids 1 and 2] having surface area dA and local temperature difference $(T_h - T_c) = \Delta T$:

$$U = \frac{dq}{dA \, \Delta T} \tag{4.24}$$

Traditionally, the mean overall heat transfer coefficient $U_m(T)$ is defined as

$$\frac{1}{U_m A} = \frac{1}{(\eta_o h_m A)_h} + \mathbf{R}_w + \frac{1}{(\eta_o h_m A)_c} \tag{4.25}$$

Here fouling resistances and other resistances are not included for simplifying the discussion but can easily be included if desired in the same way as in Eq. (3.24). In Eq. (4.25), the h_m's are the mean heat transfer coefficients obtained from the experimental/empirical correlations, and hence represent the surface area average values. The experimental/empirical correlations are generally constant fluid property correlations, as explained in Section 7.5. If the temperature variations and subsequent fluid property variations are not significant in the exchanger, the reference temperature T in $U_m'(T)$ for fluid properties is usually the arithmetic mean of inlet and outlet fluid temperatures on each fluid side for determining individual h_m's; and in some cases, this reference temperature T is the log-mean average temperature on one fluid side, as discussed in Section 9.1, or an integral-mean average temperature. If the fluid property variations are significant on one or both fluid sides, the foregoing approach is not adequate.

A more rigorous approach is the area average $\breve{U}$ used in the definition of NTU [see the first equality in Eq. (3.59)], defined as follows:

$$\breve{U} = \frac{1}{A} \int_A U(x, y) \, dA \tag{4.26}$$

This definition takes into account exactly both the temperature and length effects for counterflow and parallelflow exchangers, regardless of the size of the effects. However, there may not be possible to have a closed-form expression for $U(x, y)$ for integration. Also, no rigorous proof is available that Eq. (4.26) is exact for other exchanger flow arrangements.

When both the temperature and length effects are not negligible, Eq. (4.24) needs to be integrated to obtain an overall $\bar{\bar{U}}$ (which takes into account the temperature and length effects) *that can be used in conventional heat exchanger design*. The most accurate approach is to integrate Eq. (4.24) numerically for a given problem. However, if we can come up with some reasonably accurate value of the overall $\bar{\bar{U}}$ after approximately integrating Eq. (4.24), it will allow us to use the conventional heat exchanger design methods with U replaced by $\bar{\bar{U}}$.

Therefore, when either one or both of the temperature and length effects are not negligible, we need to integrate Eq. (4.24) approximately as follows. Idealize local $U(x, y, T) = U_m(T)f(x, y)$ and $U(x_1^*, x_2^*, T) = U_m(T)f(x_1^*, x_2^*)$; here $U_m(T)$ is a pure temperature function and $f(x, y) = f(x_1^*, x_2^*)$ is a pure position function. Hence, Eq. (4.24) reduces to

$$U_m(T)f(x_1^*, x_2^*) = \frac{dq}{dA\,\Delta T} \tag{4.27}$$

and integrate it as follows:

$$\int \frac{dq}{U_m(T)\,\Delta T} = \int f(x_1^*, x_2^*)\,dA \tag{4.28}$$

An overall heat transfer coefficient $\tilde{U}$ that takes the temperature effect into account exactly for a counterflow exchanger is given by the first equality of the following equation, obtained by Roetzel as reported by Shah and Sekulić (1998):

$$\frac{1}{\tilde{U}} = \frac{1}{\ln \Delta T_{\mathrm{II}} - \ln \Delta T_{\mathrm{I}}} \int_{\ln \Delta T_{\mathrm{I}}}^{\ln \Delta T_{\mathrm{II}}} \frac{d(\ln \Delta T)}{U(T)} \approx \frac{1}{\ln \Delta T_{\mathrm{II}} - \ln \Delta T_{\mathrm{I}}} \int_{\ln \Delta T_{\mathrm{I}}}^{\ln \Delta T_{\mathrm{II}}} \frac{d(\ln \Delta T)}{U_m(T)} \tag{4.29}$$

Note that $U(T) = U_m(T)$ in Eq. (4.29) depends only on local temperatures on each fluid side and is evaluated using Eq. (4.25) locally. The approximate equality sign in Eq. (4.29) indicates that the counterflow temperature effect is valid for any other exchanger flow arrangement considering it as hypothetical counterflow, so that ΔT_{I} and ΔT_{II} are evaluated using Eq. (3.173).

The overall heat transfer coefficient $U_m(T)$ on the left-hand side of Eq. (4.28) depends on the temperature only. Let us write the left-hand side of Eq. (4.28) by definition in the following form:

$$\int \frac{dq}{U_m(T)\,\Delta T} = \frac{1}{\tilde{U}} \int \frac{dq}{\Delta T} \tag{4.30}$$

Thus, this equation defines $\tilde{U}$ which takes into account the temperature effect only. The integral on the right-hand side of Eq. (4.30) is replaced by the definition of true mean temperature difference (MTD) as follows:

$$\int \frac{dq}{\Delta T} = \frac{q}{\Delta T_m} \tag{4.31}$$

Integration of the right-hand side of Eq. (4.28) yields the definition of a correction factor κ that takes into account the length effect on the overall heat transfer coefficient.

$$\kappa = \frac{1}{A} \int_A f(x_1^*, x_2^*)\,dA \tag{4.32}$$

where x_1^* and x_2^* are the dimensionless axial lengths for fluids 1 and 2, as noted earlier.

Finally, substituting the results from Eqs. (4.30)–(4.32) into Eq. (4.28) and rearranging yields

$$q = \tilde{U}\kappa A\,\Delta T_m = \bar{\bar{U}}A\,\Delta T_m \qquad (4.33)$$

Thus, an overall heat transfer coefficient $\bar{\bar{U}}$ which takes into account both the temperature $(\tilde{U})$ and length effects (κ) is given by

$$\bar{\bar{U}} = \frac{1}{A}\int U_m(T)f(x,y)\,dA = \tilde{U}\kappa \qquad (4.34)$$

As noted before, $f(x,y) = f(x_1^*, x_2^*)$ is a pure position function. We provide appropriate formulas for $\tilde{U}$ and κ in the following sections. Note that since $\kappa \le 1$ (see Fig. 4.5), $\bar{\bar{U}} \le \tilde{U}$. Also from Eq. (4.30) we find that $\tilde{U} = U_m$ if the temperature effect is not significant (i.e., $U_m(T)$ does not vary significantly with T).

It can be shown that for a counterflow exchanger,

$$\bar{\bar{U}} = \hat{U} = \check{U} \qquad (4.35)$$

with $\check{U}$ defined by Eq. (4.26) and $\hat{U}$ defined by Eq. (4.29) with $U(x,T)$ instead of $U(T)$ or $U_m(T)$. Hence, for evaluating $\bar{\bar{U}}$ for a counterflow exchanger, one can use $\hat{U}$, which is a function of only the area (flow length), such as for laminar gas flows, or $\check{U}$, which is a function of the temperature only such as for turbulent liquid flows. These different definitions of overall heat transfer coefficients are summarized in Table 4.2.

TABLE 4.2   Definitions of Local and Mean Overall Heat Transfer Coefficients

Symbol	Definition	Comments
U	$U = \dfrac{dq}{dA\,\Delta T}$	Basic definition of the local overall heat transfer coefficient.
U_m	$\dfrac{1}{U_m A} = \dfrac{1}{(\eta_o h_m A)_h} + \mathbf{R}_w + \dfrac{1}{(\eta_o h_m A)_c}$	Overall heat transfer coefficient defined using area average heat transfer coefficients on both sides. Individual heat transfer coefficients should be evaluated at respective reference temperatures (usually arithmetic mean of inlet and outlet fluid temperatures on each fluid side).
$\hat{U}$	$\hat{U} = \dfrac{1}{A}\displaystyle\int_A U(A)\,dA$	Mean overall heat transfer coefficient averaged over heat transfer surface area.
$\check{U}$	$\check{U} = (\ln\Delta T_2 - \ln\Delta T_1)\left[\displaystyle\int_{\ln\Delta T_1}^{\ln\Delta T_2}\dfrac{d(\ln\Delta T)}{U(T)}\right]^{-1}$	Mean overall heat transfer coefficient that takes into account the temperature effect only.
$\bar{\bar{U}}$	$\bar{\bar{U}} = \tilde{U}\kappa$	Mean overall heat transfer coefficient that takes into account the temperature and length effects. The correction factor κ takes into account the entry length effect.

Now we discuss methods that take temperature and length effects into account, to arrive at $\bar{\bar{U}}$ for the exchanger analysis.

4.2.1 Temperature Effect

4.2.1.1 Counterflow Exchanger. Consider a single-pass counterflow exchanger in which U varies linearly with the temperature of either fluid stream, $U = a(1 + bT)$, where a and b are constants. In this case, the mean value of $U \Delta T$ (where ΔT is the temperature difference between the hot and cold fluids) is as given by Colburn (1933):

$$\frac{q}{A} = (U \, \Delta T)_m = \frac{U_{\mathrm{I}} \, \Delta T_{\mathrm{II}} - U_{\mathrm{II}} \, \Delta T_{\mathrm{I}}}{\ln(U_{\mathrm{I}} \, \Delta T_{\mathrm{II}} / U_{\mathrm{II}} \, \Delta T_{\mathrm{I}})} \tag{4.36}$$

where U_{I} and U_{II} are overall heat transfer coefficients determined at the exchanger hot and cold terminals, and ΔT_{I} and ΔT_{II} are given by Eq. (3.173). Note that $(U \, \Delta T)_m$ cannot be equal to $\tilde{U} \, \Delta T_m$. In this case, from Eq. (4.36), we get the exchanger heat transfer rate as $q = (U \, \Delta T)_m A$. Equation (4.36) represents a good approximation for very viscous liquids and partial condensation, and further discussion will be provided later with an example.

An alternative approach to take into account the temperature effect on U is to use the approximate method of integration by evaluating local U at specific points in the exchanger or perform a numerical analysis. Since such methods are more general, they are discussed next for all other exchanger flow arrangements.

4.2.1.2 Other Exchangers. We first illustrate the concept of how to include the effect of variable UA for a counterflow exchanger and then extend it to all other flow arrangements. To find out whether or not variations in UA are significant with temperature variations, first evaluate UA at the two ends of a counterflow exchanger or a hypothetical counterflow for all other exchanger flow arrangements. If it is determined that variations in UA are significant, such as shown in Fig. 4.4, the average value $\tilde{U}A$ can be determined by approximate integration of the variations in UA [i.e., Eq. (4.29) with the first equality sign], by the three-point Simpson method as follows (Roetzel and Spang, 1993):

$$\frac{1}{\tilde{U}A} = \frac{1}{6} \frac{1}{U_{\mathrm{I}}A} + \frac{2}{3} \frac{1}{U_{1/2}^* A} + \frac{1}{6} \frac{1}{U_{\mathrm{II}}A} \tag{4.37}$$

where

$$U_{1/2}^* A = U_{1/2} A \frac{\Delta T_{1/2}}{\Delta T_{1/2}^*} \tag{4.38}$$

In Eq. (4.38), $\Delta T_{1/2}$ and $\Delta T_{1/2}^*$ are defined as

$$\Delta T_{1/2} = T_{h,1/2} - T_{c,1/2} \quad \text{and} \quad \Delta T_{1/2}^* = (\Delta T_{\mathrm{I}} \, \Delta T_{\mathrm{II}})^{1/2} \tag{4.39}$$

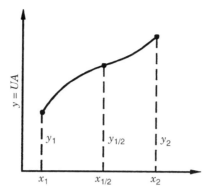

FIGURE 4.4   Variable UA in a counterflow exchanger for the Simpson method.

where the subscripts I and II correspond to terminal points at the end sections, and the subscript $\frac{1}{2}$ corresponds to a point in between defined by the second equation, respectively. Here $T_{h,1/2}$ and $T_{c,1/2}$ are computed through the procedure of Eqs. (4.43)–(4.45).

Usually, uncertainty in the individual heat transfer coefficient is high, so that the three-point approximation may be sufficient in most cases. Note that for simplicity, we have selected the third point as the middle point in the example above. The middle point is defined in terms of ΔT_I and ΔT_{II} [defined by Eq. (3.173)] to take the temperature effect properly into account; it is not a physical middle point along the length of the exchanger. The step-by-step procedure that involves this approach is presented in Section 4.2.3.1.

4.2.2   Length Effect

The heat transfer coefficient can vary significantly in the entrance region of laminar flow. This effect is negligible for turbulent flows. Hence, we associate the length effect to laminar flow. For hydrodynamically developed and thermally developing flow, the local and mean heat transfer coefficients h_x and h_m for a circular tube or parallel plates are related as follows (Shah and London, 1978):

$$h_x = \tfrac{2}{3} h_m (x^*)^{-1/3} \tag{4.40}$$

where $x^* = x/(D_h \cdot \mathrm{Re} \cdot \mathrm{Pr})$. Using this variation in h on one or both fluid sides, counterflow and crossflow exchangers have been analyzed, and the correction factors κ are presented in Fig. 4.5 and Table 4.3 as a function of φ_1 or φ_2, where

$$\varphi_1 = \eta_{o,2} h_{m,2} A_2 \left(\frac{1}{\eta_{o,1} h_{m,1} A_1} + 2\mathbf{R}_w \right) \qquad \varphi_2 = \mathbf{R}_w \left[\frac{1}{(\eta_o h_m A)_1} + \frac{1}{(\eta_o h_m A)_2} \right]^{-1} \tag{4.41}$$

The value of κ is 0.89 for $\varphi_1 = 1$, (i.e., when the exchanger has the hot- and cold-side thermal resistances approximately balanced and $\mathbf{R}_w = 0$). Thus, when a variation in the heat transfer coefficient due to the thermal entry length effect is considered, $\bar{\bar{U}} \le \tilde{U}$ or U_m since $\bar{\bar{U}} = \tilde{U} \kappa$ from Eq. (4.34). This can be explained easily if one considers the thermal resistances connected in series for the problem. For example, consider a very simplified problem with the heat transfer coefficient on each fluid side of a counterflow exchanger

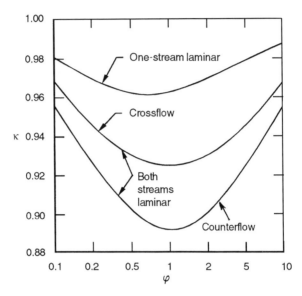

FIGURE 4.5 Length effect correction factor κ for one and both laminar streams based on equations in Table 4.3 (From Roetzel, 1974).

varying from 80 to 40 W/m$^2 \cdot$ K from entrance to exit and $A_1 = A_2$, $\mathbf{R}_w = 0$, $\eta_{o,1} = \eta_{o,2} = 1$, and there is no temperature effect. In this case, the arithmetic average $h_{m,1} = h_{m,2} = 60$ W/m$^2 \cdot$ K and $U_m = 30$ W/m$^2 \cdot$ K. However, at each end of this counterflow exchanger, $U_1 = U_2 = 26.67$ W/m$^2 \cdot$ K (since $1/U = 1/80 + 1/40$). Hence $\bar{\bar{U}} = (U_1 + U_2)/2 = 26.67$ W/m$^2 \cdot$ K. Thus $\bar{\bar{U}}/U_m = 26.67/30 = 0.89$.

TABLE 4.3 Length Effect Correction Factor κ When One or Both Streams Are in Laminar Flow for Various Exchanger Flow Arrangements

One stream laminar counterflow, parallelflow, crossflow, 1–2n TEMA E	$\kappa = (1 + \varphi_1)\left[1 - \frac{4}{3}\varphi_1 + \frac{8}{9}\varphi_1^2 \ln\left(1 + \frac{3}{2\varphi_1}\right)\right]$
Both streams laminar Counterflow	$\kappa = 1 - \dfrac{0.65 + 0.23\mathbf{R}_w(a_1 + a_2)}{4.1 + a_1/a_2 + a_2/a_1 + 3\mathbf{R}_w(a_1 + a_2) + 2\mathbf{R}_w^2 a_1 a_2}$
Crossflow	$\kappa = 1 - \dfrac{0.44 + 0.23\mathbf{R}_w(a_1 + a_2)}{4.1 + a_1/a_2 + a_2/a_1 + 3\mathbf{R}_w(a_1 + a_2) + 2\mathbf{R}_w^2 a_1 a_2}$
Parallelflow	$\kappa = (1 + \varphi_2)\left[1 - \frac{4}{3}\varphi_2 + \frac{8}{9}\varphi_2^2 \ln\left(1 + \frac{3}{2\varphi_2}\right)\right]$

$$\varphi_1 = a_2\left(\frac{1}{a_1} + \mathbf{R}_w\right)_1, \qquad \varphi_2 = \frac{\mathbf{R}_w}{1/a_1 + 1/a_2} \qquad a_1 = (\eta_o h_m A)_1 \qquad a_2 = (\eta_o h_m A)_2$$

4.2.3   Combined Effect

A specific step-by-step procedure is presented below to take into account the combined temperature and length effects on U; the reader may refer to Shah and Sekulić (1998) for further details. First, we need to determine heat transfer coefficients on each fluid side by taking into account fluid property variations due to two effects: (1) distortion of velocity and temperature profiles at a given flow cross section due to fluid property variations, and (2) variations in fluid temperature along the axial and transverse directions in the exchanger. In general, most correlations for the heat transfer coefficient are derived experimentally at almost constant fluid properties (because generally, small temperature differences are maintained during experiments) or are theoretically/numerically obtained for constant fluid properties. When temperature differences between the fluid and wall (heat transfer surface) are large, the fluid properties will vary considerably across a given cross section (at a local x) and will distort both velocity and temperature profiles. In that case, the dilemma is whether to use the fluid bulk temperature, wall temperature, or something in between for fluid properties to determine h's for constant property correlations. Unless a specific heat transfer correlation includes this effect, it is commonly taken into account by a property ratio method using both fluid bulk temperatures and wall temperature, as discussed in Section 7.6. Hence, it must be emphasized that the local heat transfer coefficients at specific points needed in the Simpson method of integration must first be corrected for the local velocity and temperature profile distortions by the property ratio method and then used as local h values for the integration. The net effect on $\bar{U}$ due to these two temperature effects can be significant, and $\bar{U}$ can be considerably higher or lower than U_m at constant properties.

The individual heat transfer coefficients in the thermal entrance region could be generally high. However, in general it will have less impact on the overall heat transfer coefficient. This is because when computing U_{local} by Eq. (4.25), with U_m and h_m's replaced by corresponding local values [see also Eq. (3.20) and the subsequent discussion], its impact will be diminished, due to the presence of the other thermal resistances in the series that are controlling (i.e., having a low hA value). It can also be seen from Fig. 4.5 that the reduction in U_m due to the entry length effect is at the most 11% (i.e., the minimum value of $\kappa = 0.89$). Usually, the thermal entry length effect is significant for laminar gas flow in a heat exchanger.

4.2.3.1   Step-by-Step Procedure to Determine $\bar{\bar{U}}$.

A step-by-step method to determine $\bar{U}A$ for an exchanger is presented below based on the original work of Roetzel and Spang (1993), later slightly modified by Shah and Sekulić (1998). In this method, not only the variations in individual h's due to the temperature effect are taken into account, but the specific heat c_p is considered temperature dependent.

1. Hypothesize the given exchanger as a counterflow exchanger (if it is different from a counterflow exchanger), and determine individual heat transfer coefficients and enthalpies at the inlet and outlet of the exchanger. Subsequently, compute the overall conductances $U_I A$ and $U_{II} A$ at inlet and outlet of the exchanger by using Eq. (3.24).

2. To consider the temperature-dependent specific heats, compute the specific enthalpies **h** of the C_{max} fluid (with a subscript j) at the third point (designated by 1/2 as a subscript, see Section 4.2.1.2) within the exchanger from the following equation

using the known values at each end:

$$\mathbf{h}_{j,1/2} = \mathbf{h}_{j,\text{II}} + \left(\mathbf{h}_{j,\text{I}} - \mathbf{h}_{j,\text{II}}\right) \left(\frac{\Delta T^*_{1/2} - \Delta T_{\text{II}}}{\Delta T_{\text{I}} - \Delta T_{\text{II}}}\right) \tag{4.42}$$

where $\Delta T^*_{1/2}$ is given by

$$\Delta T^*_{1/2} = (\Delta T_{\text{I}} \, \Delta T_{\text{II}})^{1/2} \tag{4.43}$$

Here $\Delta T_{\text{I}} = (T_h - T_c)_{\text{I}}$ and $\Delta T_{\text{II}} = (T_h - T_c)_{\text{II}}$. If $\Delta T_{\text{I}} = \Delta T_{\text{II}}$, (i.e., $C^* = R_1 = 1$), the rightmost bracketed term in Eq. (4.42) becomes $1/2$. If the specific heat is constant, the enthalpies can be replaced by temperatures in Eq. (4.42). If the specific heat does not vary significantly, Eq. (4.42) could also be used for the $C_{\min}$ fluid. However, when it varies significantly as in a cryogenic heat exchanger, the third point calculated for the $C_{\max}$ and $C_{\min}$ fluid by Eq. (4.42) will not be close enough in the exchanger (Shah and Sekulić, 1998). In that case, compute the third point for the $C_{\min}$ fluid by the energy balance as follows:

$$[\dot{m}(\mathbf{h}_i - \mathbf{h}_{1/2})]_{C_{\max}} = [\dot{m}(\mathbf{h}_{1/2} - \mathbf{h}_o)]_{C_{\min}} \tag{4.44}$$

Subsequently, using the equation of state or tabular/graphical results, determine the temperature $T_{h,1/2}$ and $T_{c,1/2}$ corresponding to $\mathbf{h}_{h,1/2}$ and $\mathbf{h}_{c,1/2}$. Then

$$\Delta T_{1/2} = T_{h,1/2} - T_{c,1/2} \tag{4.45}$$

3. For a counterflow exchanger, the heat transfer coefficient $h_{j,1/2}$ on each fluid side at the third point is calculated based on the temperatures $T_{j,1/2}$ determined in the preceding step. For other exchangers, compute $h_{j,1/2}$ at the following corrected reference (Roetzel and Spang, 1993):

$$T_{h,1/2,\text{corr}} = T_{h,1/2} - \frac{3}{2}(T_{h,1/2} - T_{c,1/2}) \frac{1 - F}{1 + R_h^{2/3}} \tag{4.46}$$

$$T_{c,1/2,\text{corr}} = T_{c,1/2} + \frac{3}{2}(T_{h,1/2} - T_{c,1/2}) \frac{1 - F}{1 + R_c^{2/3}} \tag{4.47}$$

In Eqs. (4.46) and (4.47), F is the log-mean temperature difference correction factor and $R_h = C_h/C_c$ or $R_c = C_c/C_h$. The temperatures $T_{h,1/2,\text{corr}}$ and $T_{c,1/2,\text{corr}}$ are used only for the evaluation of fluid properties to compute $h_{h,1/2}$ and $h_{c,1/2}$. The foregoing correction to the reference temperature $T_{j,1/2}$ ($j = h$ or c) results in the cold temperature being increased and the hot temperature being decreased.

Calculate the overall conductance at the third point by

$$\frac{1}{U_{1/2}A} = \frac{1}{\eta_{o,h}h_{h,1/2}A_h} + \mathbf{R}_w + \frac{1}{\eta_{o,c}h_{c,1/2}A_c} \tag{4.48}$$

Note that η_f and η_o can be determined accurately at local temperatures.

4. Calculate the apparent overall heat transfer coefficient at this third point using Eq. (4.38):

$$U_{1/2}^* A = U_{1/2} A \, \frac{\Delta T_{1/2}}{\Delta T_{1/2}^*} \qquad (4.49)$$

5. Find the mean overall conductance for the exchanger (taking into account the temperature dependency of the heat transfer coefficient and heat capacities) from the equation

$$\frac{1}{\tilde{U} A} = \frac{1}{6} \frac{1}{U_{\mathrm{I}} A} + \frac{2}{3} \frac{1}{U_{1/2}^* A} + \frac{1}{6} \frac{1}{U_{\mathrm{II}} A} \qquad (4.50)$$

6. Finally, the true mean heat transfer coefficient $\bar{\bar{U}}$ that also takes into account the laminar flow entry length effect is given by

$$\bar{\bar{U}} A = \hat{U} A \kappa \qquad (4.51)$$

where the entry length effect factor $\kappa \leq 1$ is given in Fig. 4.5 and Table 4.3.

Example 4.2   In a liquid-to-steam two-fluid heat exchanger, the controlling thermal resistance fluid side is the liquid side. Let's assume that the temperature of the steam stays almost constant throughout the exchanger ($T_{\mathrm{steam}} = 108°C$) while the liquid changes its temperature from 26.7°C to 93.3°C. The heat transfer coefficient on the steam side is uniform and constant over the heat transfer surface (12,200 W/m² · K), while on the liquid side, its magnitude changes linearly between 122 W/m² · K (at the cold end) and 415 W/m² · K (at the hot end). Determine the heat transfer surface area if the following additional data are available. The mass flow rate of the liquid is 1.682 kg/s. The specific heat at constant pressure of the liquid is 1,967.8 J/kg · K. The heat exchanger is a double-pipe design with the inner tube inside diameter 52.6 mm and the inner tube outside diameter 60.4 mm. The thermal conductivity of the tube wall is 60.58 W/m · K. Assume that no fouling is taking place.

SOLUTION

Problem Data and Schematics: Data for the double-pipe heat exchanger are provided in Fig. E4.2.

Determine: The heat transfer area of the heat exchanger.

Assumptions: All the assumptions described in Section 3.2.1 hold except for a variable heat transfer coefficient on the liquid side. Also assume that there is no thermal entry length effect on U. To apply a conventional design method (say, the MTD method; Section 3.7) a mean value of the overall heat transfer coefficient should be defined (see Section 4.2.3.1).

Analysis: The heat transfer surface area can be calculated from

$$A = \frac{q}{\bar{\bar{U}} \, \Delta T_{\mathrm{lm}}}$$

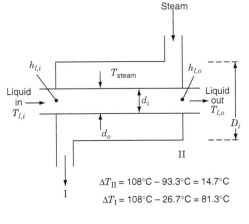

$d_o = 60.4$ mm, $d_i = 52.6$ mm

$\dot{m}_l = 1.682$ kg/s $c_{p,l} = 1967.8$ J/kg $\cdot$ K

$k_w = 60.58$ W/m $\cdot$ K

$h_{l,i} = 122$ W/m^2 $\cdot$ K $h_{l,o} = 415$ W/m^2 $\cdot$ K

$h_{steam} = 12200$ W/m^2 $\cdot$ K

$T_{l,i} = 26.7°$C $T_{l,o} = 93.3°$C

$T_{steam} = 108°$C

$\Delta T_{II} = 108°$C $- 93.3°$C $= 14.7°$C

$\Delta T_I = 108°$C $- 26.7°$C $= 81.3°$C

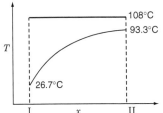

FIGURE E4.2

where $\overline{\overline{U}} = \tilde{U}$ represents an average overall heat transfer coefficient defined by Eq. (4.50). The heat transfer rate can be calculated from the enthalpy change of the cold fluid (liquid):

$$q = \dot{m}c_p \, \Delta T_{liquid} = 1.682 \, \text{kg/s} \times 1967.8 \, \text{J/kg} \cdot \text{K} \times (93.3 - 26.7)°\text{C}$$
$$= 220,435 \, \text{W} = 220.4 \, \text{kW}$$

The log-mean temperature difference by definition [Eq. (3.172)] is equal to

$$\Delta T_{lm} = \frac{\Delta T_I - \Delta T_{II}}{\ln(\Delta T_I / \Delta T_{II})} = \frac{81.3°\text{C} - 14.7°\text{C}}{\ln(81.3°\text{C}/14.7°\text{C})} = 38.9°\text{C}$$

where

$$\Delta T_I = 108°\text{C} - 26.7°\text{C} = 81.3°\text{C} \qquad \Delta T_{II} = 108°\text{C} - 93.3°\text{C} = 14.7°\text{C}$$

and I and II denote the terminal points of the heat exchanger (cold and hot ends of the liquid side, respectively). Now let us calculate U_I and U_{II} (step 1 of Section 4.2.3.1) from the information given.

$$\frac{1}{U_{\text{I}}} = \frac{1}{h_{\text{steam}}(d_o/d_i)} + \frac{d_i \ln(d_o/d_i)}{2k_w} + \frac{1}{h_{\text{liquid},i}}$$

$$= \frac{1}{12200\,\text{W/m}^2 \cdot \text{K}\,(60.4\,\text{mm}/52.6\,\text{mm})} + \frac{(52.6 \times 10^{-3}\,\text{m})\ln(60.4\,\text{mm}/52.6\,\text{mm})}{2 \times 60.58\,\text{W/m} \cdot \text{K}}$$

$$+ \frac{1}{122\,\text{W/m}^2 \cdot \text{K}}$$

$$= (0.7138 + 0.6003 + 81.9672) \times 10^{-4}\,\text{m}^2 \cdot \text{K/W} = 83.2813 \times 10^{-4}\,\text{m}^2 \cdot \text{K/W}$$

Therefore,

$$U_{\text{I}} = 120.1\,\text{W/m}^2 \cdot \text{K}$$

Analogously, $U_{\text{II}} = 393.6\,\text{W/m}^2 \cdot \text{K}$ by changing 122 $\text{W/m}^2 \cdot \text{K}$ to 415 $\text{W/m}^2 \cdot \text{K}$ in the equation above for $1/U_{\text{I}}$. The magnitude of the local overall heat transfer coefficient at the referent temperature $T_{\text{liquid},1/2}$ can be determined using Eq. (4.42), keeping in mind that for a constant specific heat of the fluid, the same form of the equations can be written for both enthalpy and temperature magnitudes:

$$T_{\text{liquid},1/2} = T_{\text{liquid,II}} + (T_{\text{liquid,I}} - T_{\text{liquid,II}})\frac{\Delta T_{1/2}^* - \Delta T_{\text{II}}}{\Delta T_{\text{I}} - \Delta T_{\text{II}}}$$

where

$$\Delta T_{1/2}^* = (\Delta T_{\text{I}}\,\Delta T_{\text{II}})^{1/2} = (81.3°\text{C} \times 14.7°\text{C})^{1/2} = 34.6°\text{C}$$

Therefore,

$$T_{\text{liquid},1/2} = 93.3°\text{C} + (26.7°\text{C} - 93.3°\text{C})\frac{34.6°\text{C} - 14.7°\text{C}}{81.3°\text{C} - 14.7°\text{C}} = 73.4°\text{C}$$

and

$$\Delta T_{1/2} = T_{\text{steam}} - T_{\text{liquid},1/2} = 108°\text{C} - 73.4°\text{C} = 34.6°\text{C}$$

It is specified that the liquid-side heat transfer coefficient varies linearly from 122 to 415 $\text{W/m}^2 \cdot \text{K}$ with the temperature change from 26.7°C to 93.3°C. Hence from a linear interpolation, $h_{\text{liquid},1/2}$ at 73.4°C is

$$h_{\text{liquid},1/2} = 122\,\text{W/m}^2 \cdot \text{K} + \frac{(455 - 122)\,\text{W/m}^2 \cdot \text{K}}{(93.3 - 26.7)°\text{C}}(73.4 - 26.7)°\text{C} = 327.5\,\text{W/m}^2 \cdot \text{K}$$

Now changing 122 $\text{W/m}^2 \cdot \text{K}$ to 327.5 $\text{W/m}^2 \cdot \text{K}$ in the last term of the $1/U_{\text{I}}$ equation above, we get

$$U_{1/2} = 314.0\,\text{W/m}^2 \cdot \text{K}$$

The apparent overall heat transfer coefficient at this third point is given by Eq. (4.49) as

$$U_{1/2}^* = U_{1/2} \frac{\Delta T_{1/2}}{\Delta T_{1/2}^*} = 314.0 \, \text{W/m}^2 \cdot \text{K} \left(\frac{34.6°\text{C}}{34.6°\text{C}} \right) = 314.0 \, \text{W/m}^2 \cdot \text{K}$$

Finally, the mean overall heat transfer coefficient can be calculated, using Eq. (4.50), as

$$\frac{1}{\bar{\bar{U}}} = \frac{1}{6} \left(\frac{1}{120.1 \, \text{W/m}^2 \cdot \text{K}} \right) + \frac{2}{3} \left(\frac{1}{314.0 \, \text{W/m}^2 \cdot \text{K}} \right) + \frac{1}{6} \left(\frac{1}{393.6 \, \text{W/m}^2 \cdot \text{K}} \right)$$

$$= 3.934 \times 10^{-3} \, \text{m}^2 \cdot \text{K/W}$$

Since the thermal entry length is negligible for liquids and zero for condensing steam, $\kappa = 0$. Hence,

$$\tilde{U} = \bar{\bar{U}} = 254.2 \, \text{W/m}^2 \cdot \text{K}$$

The heat transfer surface area of the exchanger is now

$$A = \frac{q}{\bar{\bar{U}} \, \Delta T_{\text{lm}}} = \frac{220,435 \, \text{W}}{254.2 \, \text{W/m}^2 \cdot \text{K} \times 38.9 \, \text{K}} = 22.29 \, \text{m}^2$$

Discussion and Comments: This simple example illustrates how to determine $\bar{\bar{U}}$ and the heat transfer surface area for a counterflow exchanger. For other exchangers, with both fluids as single-phase, the procedure will be more involved, as outlined in Section 4.2.3.1. Calculation of the mean overall heat transfer coefficient by determining the arithmetic mean value of the local values of terminal overall heat transfer coefficients, [i.e., $\tilde{U} = \frac{1}{2}(U_{\text{I}} + U_{\text{II}})$] will result in this case in a heat transfer surface area about 1% smaller than the value determined using the elaborate procedure above. It should be noted that the local heat transfer coefficient on the liquid side changes linearly. Often, the changes of heat transfer coefficients are not linear and the values calculated for the mean overall heat transfer coefficient determined using various approaches may be substantially different (Shah and Sekulić, 1998). In those situations, a numerical approach is the most reliable.

4.2.3.2 Numerical Analysis. In the foregoing section, the methodology was presented to take into consideration variations in U due to the temperature effect, the length effect, or both. As mentioned earlier, other factors also play a role in making U non-uniform. In addition, a number of other factors that could violate the assumptions (see Section 3.2.1) are built into the basic ε-NTU, P-NTU, or MTD methods, such as nonuniform velocity and temperature distributions at the exchanger inlet, and highly variable fluid properties. All these effects can be taken into account by a numerical analysis.

To illustrate the principles, consider an unmixed–unmixed single-pass crossflow exchanger. Divide this exchanger into $m \times n$ segments, as shown in Fig. 4.6, with the hot-fluid passage having m segments, the cold-fluid, n segments. The size of individual segments is chosen sufficiently small so that all fluid properties and other variables/parameters could be considered constant within each segment. Fluid outlet temperatures from each segment are indexed as shown in Fig. 4.6. Energy balance and rate equations for this problem are given in Table 11.2 for an unmixed–unmixed case. For the (j,k)

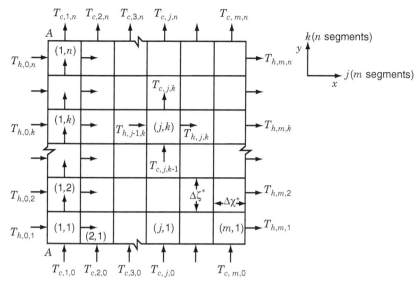

FIGURE 4.6   Numerical modeling of an unmixed–unmixed crossflow exchanger.

element, a set of model equations can be written in finite-difference form as follows:

$$\frac{\theta_{h,j,k} - \theta_{h,j-1,k}}{\Delta\chi^*} = -\left(\frac{\theta_{h,j,k} + \theta_{h,j+1,k}}{2} - \frac{\theta_{c,j,k} + \theta_{c,j,k-1}}{2}\right)\text{NTU}_{h,j,k} \qquad (4.52)$$

$$\frac{\theta_{c,j,k} - \theta_{c,j,k-1}}{\Delta\zeta^*} = +\left(\frac{\theta_{h,j,k} + \theta_{h,j,k-1}}{2} - \frac{\theta_{c,j,k} + \theta_{c,j,k-1}}{2}\right)\text{NTU}_{c,j,k} \qquad (4.53)$$

with the boundary conditions

$$\theta_{h,0,k} = 1 \qquad \theta_{c,j,0} = 0 \qquad (4.54)$$

where $\chi^* = x/L_1$ and $\zeta^* = (y/L_2)C^*$. Equations (4.52) and (4.53) have two unknowns (the two outlet temperatures), and hence the solution is straightforward at each element if the inlet temperatures are known for that element. The NTUs required in Eqs. (4.52) and (4.53) are based on the overall U for individual local segments and are to be evaluated locally taking into account all the effects for the local element, such as fluid property variations and flow maldistribution. Flow maldistribution translates into different local $\Delta\dot{m}$ values for each element in the transverse direction, such as that $\Delta\dot{m}_h$ could be different for each element at the entrance to the first element (left most vertical line AA in Fig. 4.6). Generally, we assume that $\Delta\dot{m}_h$ remains the same all along the row specified. If the inlet temperature is not uniform, the boundary conditions of Eq. (4.54) will need to be changed accordingly. Once the local velocity and temperature distributions are known on each fluid side, individual heat transfer coefficients can be calculated using the appropriate correlations: subsequently, UA and local NTUs for individual elements.

For this particular exchanger, the analysis procedure is straightforward since it represents an explicit marching procedure analysis. Knowing the two inlet temperatures for the element (1,1), two outlet temperatures can be calculated. For the first calculation, all fluid properties can be calculated at the inlet temperature. If warranted, in the next iteration, they can be calculated on each fluid side at the average temperature of the preceding iteration. Once the analysis of the element (1, 1) is completed, analyze element (1, 2) in the same manner since inlet temperatures ($T_{h,0,2}$ and $T_{c,1,1}$) for this element are now known. Continue such analysis for all elements of column 1. At this time, the hot-fluid temperatures at the inlet of the second column are known as well as the cold-fluid outlet temperature from the first column. Continue such analysis to the last column, after which all outlet temperatures are known for both hot and cold fluids.

The example we considered was simple and did not involve any major iteration. If the temperature of one of the fluids is unknown while starting the analysis, the numerical analysis method will become iterative, and perhaps complex, depending on the exchanger configuration, and one needs to resort to more advanced numerical methods. Particularly for shell-and-tube exchangers, not only do the baffles make the geometry much more complicated, but so do the leakage and bypass flows (see Section 4.4.1) in the exchanger. In this case, modeling for evaluating the leakage and bypass flows and their effects on heat transfer analysis needs to be incorporated into the advanced numerical methods.

4.3 ADDITIONAL CONSIDERATIONS FOR EXTENDED SURFACE EXCHANGERS

Extended surfaces or fins are used to increase the surface area[†] and consequently, to increase the total rate of heat transfer. Both conduction through the fin cross section and convection over the fin surface area take place in and around the fin. Hence, the fin surface temperature is generally lower than the base (primary surface) temperature T_0 if the fin is hotter than the fluid (at T_∞) to which it is exposed. This in turn reduces the local and average temperature difference between the fin and the fluid for convection heat transfer, and the fin transfers less heat than it would if it were at the base temperature. Similarly, if the heat is convected to the fin from the ambient fluid, the fin surface temperature will be higher than the fin base temperature, which in turn reduces the temperature differences and heat transfer through the fin. Typical temperature distributions for fin cooling and heating are shown in Fig. 4.13. This reduction in the temperature difference is taken into account by the concept of fin efficiency[‡] η_f and extended surface efficiency η_o for extended surfaces. Once they are evaluated properly, the thermal resistances are evaluated by Eq. (3.24). The heat transfer analysis for direct-transfer type exchangers, presented in Sections 3.2 through 4.2, then applies to the extended surface heat exchangers.

First we obtain the temperature distribution within a fin and the heat transfer through a fin heated (or cooled) at one end and convectively cooled (or heated) along its surface.

[†] As mentioned in Section 1.5.3, the heat transfer coefficient for the fins may be higher or lower than for the primary surface, depending on fin type and density.
[‡] Kays and London (1998) refer to η_f as the fin temperature effectiveness, while Incropera and DeWitt (1996) refer to the fin effectiveness as η_ε, defined by Eq. (4.156). To avoid possible confusion, we refer to η_f as the fin efficiency and η_ε as the fin effectiveness.

Next, we derive an expression for fin efficiency. The analysis presented next is valid for both fin cooling and fin heating situations.

4.3.1 Thin Fin Analysis

4.3.1.1 Thermal Circuit and Differential Equation. Consider a thin fin of variable thickness δ as shown in Fig. 4.7. Its length for heat conduction in the x direction (fin height) is ℓ, its perimeter for surface convection is $\mathbf{P}(x) = 2[L_f + \delta(x)]$ and its cross-sectional area for heat conduction at any cross section x is $A_k(x) = \delta(x)L_f$. Note that throughout this section, $A_k(x)$ will represent the fin cross-sectional area for heat conduction. Note also that both A_f (the fin surface area for heat transfer) and $\mathbf{P}$ can be a function of x (i.e., variable along the fin length ℓ), but they are generally constant with straight fins in heat exchangers so that $A_f = \mathbf{P}\ell$. The fin is considered thin, if $\delta(x) \ll \ell \ll L_f$. Let us invoke the following assumptions for the analysis.

1. There is one-dimensional heat conduction in the fin (i.e., the fin is "thin") so that the temperature T is a function of x only and does not vary significantly in the y and z directions or across A_k. However, A_k can, in general, depend on x.
2. The heat flow through the fin is steady state, so that the temperature T at any cross section does not vary with time.
3. There are no heat sources or sinks in the fin.
4. Radiation heat transfer from and to the fin is neglected.
5. The thermal conductivity of the fin material is uniform and constant.
6. The heat transfer coefficient h for the fin surface is uniform over the surface (except at the fin tip in some cases) and constant with time.
7. The temperature of the ambient fluid T_∞ is uniform.
8. The thermal resistance between the fin and the base is negligible.

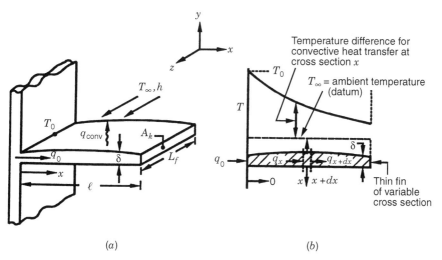

(a)   (b)

FIGURE 4.7   Thin fin of a variable cross section.

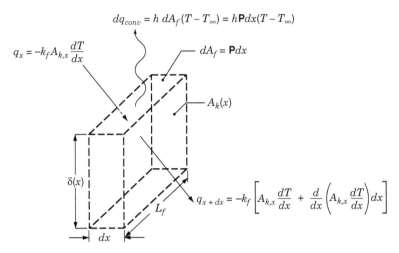

FIGURE 4.8   Energy terms associated with the differential element of the fin.

Although two-dimensional conduction exists at a fin cross section, its effect will be small in most heat exchanger applications. Near the end of Section 4.3.2.2 (p. 285), we discuss what happens if any of assumptions 5 to 8 above are not met.

An energy balance on a typical element between x and $x + dx$ of Fig. 4.7b is shown in Fig. 4.8. Heat enters this element by conduction at x. Part of this heat leaves the cross section at $x + dx$, and the rest leaves by convection through its surface area $dA_f = \mathbf{P}\,dx$. The energy balance on this element of length dx over its full width is

$$q_x - q_{x+dx} - dq_{\text{conv}} = 0 \tag{4.55}$$

The two conduction rate equations associated with conduction through the fin and one convection rate equation for convection to the surroundings for this differential element are given by

$$q_x = -k_f A_{k,x} \frac{dT}{dx} \tag{4.56}$$

$$q_{x+dx} = -k_f \left[A_{k,x} \frac{dT}{dx} + \frac{d}{dx}\left(A_{k,x} \frac{dT}{dx} \right) dx \right] \tag{4.57}$$

$$dq_{\text{conv}} = h\, dA_f\, (T - T_\infty) = h(\mathbf{P}\,dx)(T - T_\infty) \tag{4.58}$$

Substituting these rate equations into Eq. (4.55) and simplifying, we get

$$k_f \frac{d}{dx}\left(A_{k,x} \frac{dT}{dx} \right) dx = h(\mathbf{P}\,dx)(T - T_\infty) \tag{4.59}$$

Carrying out the necessary differentiation and rearranging results in

$$\frac{d^2 T}{dx^2} + \frac{1}{A_{k,x}} \frac{dA_{k,x}}{dx} \frac{dT}{dx} - \frac{h\mathbf{P}}{k_f A_{k,x}}(T - T_\infty) = 0 \tag{4.60}$$

or

$$\frac{d^2 T}{dx^2} + \frac{d(\ln A_{k,x})}{dx}\frac{dT}{dx} - m^2(T - T_\infty) = 0 \tag{4.61}$$

where

$$m^2 = \frac{h\mathbf{P}}{k_f A_{k,x}} \tag{4.62}$$

where both $\mathbf{P}$ and A_k will be functions of x for a variable cross section. Note that m has units of inverse length. To simplify further, define a new dependent variable, excess temperature, as

$$\theta(x) = T(x) - T_\infty \tag{4.63}$$

We assume that the ambient temperature T_∞ is a constant,[†] so that $d\theta/dx = dT/dx$ and Eq. (4.61) reduces to

$$\frac{d^2\theta}{dx^2} + \frac{d(\ln A_{k,x})}{dx}\frac{d\theta}{dx} - m^2\theta = 0 \tag{4.64}$$

This second-order, linear, homogeneous ordinary differential equation with nonconstant coefficients is valid for any thin fins of variable cross section. Once the boundary conditions and the fin geometry are specified, its solution would provide the temperature distribution, and subsequently, the heat transfer rate through the fin, as discussed later.

4.3.1.2 Thin, Straight Fin of Uniform Rectangular Cross Section. Let us derive specific solutions to Eq. (4.64) for a straight fin of uniform thickness δ and constant conduction area A_k of Fig. 4.9, on page 263. This solution will also be valid for pin fins (having a circular cross section) as long as $\mathbf{P}$ and A_k are evaluated properly. For the straight fin of Fig. 4.9,

$$m^2 = \frac{2h(L_f + \delta)}{k_f L_f \delta} \approx \frac{2h}{k_f \delta} \tag{4.65}$$

for $L_f \gg \delta$. Since $A_{k,x}$ is constant, $d(\ln A_{k,x})/dx = 0$, and m^2 is constant. Equation (4.64) reduces to

$$\frac{d^2\theta}{dx^2} - m^2\theta = 0 \tag{4.66}$$

This is a second-order, linear, homogeneous ordinary differential equation. The general solution to this equation is

$$\theta = C_1 e^{-mx} + C_2 e^{mx} \tag{4.67}$$

where C_1 and C_2 (the local nomenclature only), the constants of integration, remain to be established by the boundary conditions discussed in the following subsection.

[†] If T_∞ is not constant, use Eq. (4.61) for the solution.

Boundary Conditions. For a second-order ordinary differential equation [i.e., Eq. (4.66)], we need two boundary conditions to evaluate two constants of integration, C_1 and C_2. The boundary condition at the fin base, $x = 0$, is simply $T = T_0$. Hence,

$$\theta(0) = T_0 - T_\infty = \theta_0 \tag{4.68}$$

At the fin tip ($x = \ell$), there are five possible boundary conditions, as shown in Fig. 4.10.

CASE 1: LONG, THIN FIN. As shown in Fig. 4.10*a* on page 264, the fin is very long compared to its thickness ($\ell/\delta \to \infty$), so that $T \approx T_\infty$ at $x = \ell \to \infty$. Hence,

$$\theta(\infty) = 0 \tag{4.69}$$

CASE 2: THIN FIN WITH AN ADIABATIC TIP. As shown in Fig. 4.10*b* on page 264, the fin tip is considered adiabatic and hence the heat transfer rate through the fin tip is zero. Therefore,

$$q_\ell = q|_{x=\ell} = -k_f A_k \left(\frac{dT}{dx}\right)_{x=\ell} = -k_f A_k \left(\frac{d\theta}{dx}\right)_{x=\ell} = 0 \tag{4.70}$$

or

$$\left(\frac{d\theta}{dx}\right)_{x=\ell} = 0 \tag{4.71}$$

CASE 3: THIN FIN WITH CONVECTIVE BOUNDARY AT THE FIN TIP. As shown in Fig. 4.10*c* on page 264, there is a finite heat transfer through the fin tip by convection and hence

$$q_\ell = -k_f A_k \left(\frac{dT}{dx}\right)_{x=\ell} = h_e A_k (T_\ell - T_\infty) \tag{4.72}$$

Or in terms of θ,

$$\left(\frac{d\theta}{dx}\right)_{x=\ell} = -\frac{h_e}{k_f} \theta_\ell \tag{4.73}$$

Here we have explicitly specified the fin tip convection coefficient h_e as different from h for the fin surface. However, in reality, h_e is not known in most applications and is considered the same as h (i.e., $h_e = h$).

CASE 4: THIN FIN WITH FINITE HEAT TRANSFER AT THE FIN TIP. As shown in Fig. 4.10*d* on page 264, the finite heat transfer through the fin tip is shown as q_ℓ since it could be conduction to the neighboring primary surface (not shown in the figure).

$$-k_f A_k \left(\frac{dT}{dx}\right)_{x=\ell} = q_\ell \tag{4.74}$$

or

$$\left(\frac{d\theta}{dx}\right)_{x=\ell} = -\frac{q_\ell}{k_f A_k} \tag{4.75}$$

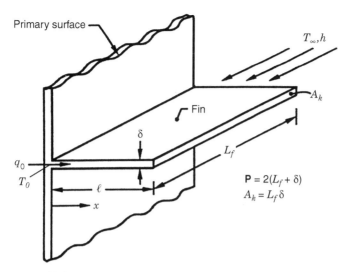

FIGURE 4.9   Straight, thin fin of uniform thickness δ.

CASE 5: THIN FIN WITH FIN TIP TEMPERATURE SPECIFIED. As shown in Fig. 4.10e on page 264, the fin is not very long and the fin tip temperature specified as T_ℓ is constant, so that

$$\theta_\ell = \theta|_{x=\ell} = T_\ell - T_\infty \qquad (4.76)$$

All these boundary conditions are summarized in Table 4.4. In Fig. 4.10, the temperature distribution in the fin near the fin tip region is also shown for the first three boundary conditions. The common trend in all three temperature distributions is that the temperature gradient in the fin decreases continuously with increasing x when the fin convects heat to the ambient ($T > T_\infty$). This is due to less heat available for conduction as x increases because of convection heat transfer from the fin surface. The reverse will be true if the fin receives heat from the ambient. Later we discuss temperature distributions for the last two cases of Fig. 4.10 by Eq. (4.104) and Fig. 4.11, respectively.

Total Fin Heat Transfer. The total convective heat transfer from the fin can be found by computing convective heat transfer from the fin surface once the specific temperature distribution within the fin is found from Eq. (4.67) after applying boundary conditions from the preceding section.

The convective heat transfer rate from the differential element dx, from Fig. 4.8, is dq_{conv}:

$$dq_{\text{conv}} = h\mathbf{P}\, dx\, (T - T_\infty) = h\mathbf{P}\, dx\, \theta \qquad (4.77)$$

Hence, the total convective heat transfer rate q_{conv} from the constant cross section fin, excluding the heat exchanged from the tip (to be included later), is found by integrating

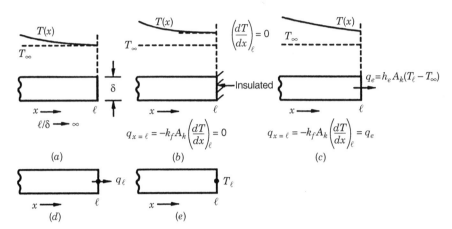

FIGURE 4.10   Five boundary conditions at $x = \ell$ for the thin-fin problem: (a) very long fin; (b) adiabatic fin tip, (c) convection at the fin tip; (d) finite heat leakage at the fin tip, (e) temperature specified at the fin tip.

this equation from $x = 0$ to $x = \ell$.

$$q_{\text{conv}} = h\mathbf{P} \int_0^\ell \theta \, dx \tag{4.78}$$

The conduction heat transfer rate through the base is obtained by

$$q_0 = -k_f A_k \left(\frac{dT}{dx}\right)_{x=0} = -k_f A_k \left(\frac{d\theta}{dx}\right)_{x=0} \tag{4.79}$$

where the temperature gradient $(d\theta/dx)_{x=0}$ is obtained by evaluating the temperature gradient $(d\theta/dx)$ at $x = 0$ from the specific temperature distribution derived. The heat transfer rate between the fin and the environment at the fin tip must be equal to the heat transfer rate conducted through the fin at $x = \ell$; it is given by

$$q_\ell = -k_f A_k \left(\frac{dT}{dx}\right)_{x=\ell} = -k_f A_k \left(\frac{d\theta}{dx}\right)_{x=\ell} \tag{4.80}$$

From the overall energy balance on the fin,

$$q_{\text{conv}} = q_0 - q_\ell \tag{4.81}$$

From here onward, we consider $T_0 > T_\infty$, so that q_0 is coming from the base to the fin (i.e., in the positive x direction) by conduction and q_l is leaving fin (again in the positive x direction). For the boundary conditions of Fig. 4.10a and b, $q_\ell = 0$. Hence

$$q_{\text{conv}} = q_0 \tag{4.82}$$

TABLE 4.4 Boundary Conditions, Temperature Distributions, and Heat Transfer Rates for the Rectangular Profile Thin Fin Problem

Case	Fin Tip Condition	Boundary Condition at $x = \ell$	Temperature Distribution in the Fin	Heat Transfer Rate at $x = 0$ and ℓ, Temperature at $x = \ell$, and Fin Efficiency
1	Long, thin fin	$\theta = 0$	$\dfrac{\theta}{\theta_0} = e^{-mx}$	$q_0 = q_{conv} = \dfrac{h\mathbf{P}}{m}\theta_0 = k_f A_k m\theta_0$ $q_\ell = 0 \quad \theta_\ell = 0 \quad \eta_f = \dfrac{1}{m\ell}$
2	Thin fin with an adiabatic tip	$\dfrac{d\theta}{dx} = 0$	$\dfrac{\theta}{\theta_0} = \dfrac{\cosh m(\ell - x)}{\cosh m\ell}$	$q_0 = \dfrac{h\mathbf{P}}{m}\theta_0 \tanh m\ell = \dfrac{h\mathbf{P}}{m}\theta_0 \dfrac{\cosh(2m\ell) - 1}{\sinh(2m\ell)}$ $q_\ell = 0 \quad \dfrac{\theta_\ell}{\theta_0} = \dfrac{1}{\cosh m\ell} \quad \eta_f = \dfrac{\tanh m\ell}{m\ell}$
3	Thin fin with convection boundary at the fin tip	$\dfrac{d\theta}{dx} = -\dfrac{h_e}{k_f}\theta_\ell$	$\dfrac{\theta}{\theta_0} = \dfrac{\cosh m(\ell - x) + B\sinh m(\ell - x)}{\cosh m\ell + B\sinh m\ell}$ where $B = \dfrac{h_e}{mk_f}$	$q_0 = \dfrac{h\mathbf{P}}{m}\theta_0 \dfrac{\sinh m\ell + B\cosh m\ell}{\cosh m\ell + B\sinh m\ell}$ $q_\ell = h_e A_k \theta_0 \dfrac{1}{\cosh m\ell + B\sinh m\ell}$ $\dfrac{\theta_\ell}{\theta_0} = \dfrac{1}{\cosh m\ell + B\sinh m\ell}$ $\eta_f = \dfrac{\tanh m\ell + B}{(B + m\ell)(1 + B\tanh m\ell)}$
4	Thin fin with finite heat transfer at the fin tip	$\dfrac{d\theta}{dx} = -\dfrac{q_\ell}{k_f A_k}$	$\dfrac{\theta}{\theta_0} = \dfrac{\cosh m(\ell - x) - (q_\ell m/h\mathbf{P}\theta_0)\sinh mx}{\cosh m\ell}$	$q_0 = \dfrac{h\mathbf{P}}{m}\theta_0 \tanh m\ell + q_\ell$ $\dfrac{\theta_\ell}{\theta_0} = \dfrac{1 - (q_\ell m/h\mathbf{P}\theta_0)\sinh m\ell}{\cosh m\ell}$
5	Thin fin with fin tip temperature specified	$\theta = \theta_\ell$	$\dfrac{\theta}{\theta_0} = \dfrac{\sinh m(\ell - x) + (\theta_\ell/\theta_0)\sinh mx}{\sinh m\ell}$	$q_0 = \dfrac{h\mathbf{P}}{m}\theta_0 \dfrac{\cosh m\ell - (\theta_\ell/\theta_0)}{\sinh m\ell}$ $q_\ell = \dfrac{h\mathbf{P}}{m}\theta_0 \dfrac{(\theta_\ell/\theta_0)\cosh m\ell - 1}{\sinh m\ell}$ $= \dfrac{h\mathbf{P}}{m}\theta_\ell \dfrac{\cosh m\ell - (\theta_0/\theta_\ell)}{\sinh m\ell}$

and the fin heat transfer can be obtained either by integrating the temperature distribution along the fin length [Eq. (4.78)] or by differentiating the temperature distribution and evaluating the derivative at the fin base [Eq. (4.79)]!

For the boundary condition of Fig. 4.10c and from Eq. (4.81),

$$q_0 = q_{\text{conv}} + q_e \tag{4.83}$$

If q_e is the convection heat transfer from the fin tip, q_0 represents the total convection heat transfer through the fin surface, including the fin tip.

For the boundary conditions of Fig. 4.10d and e, q_ℓ can be positive, zero, or negative, and one needs to apply Eq. (4.81) to determine q_{conv}.

Now let us derive specific temperature distributions for the five boundary conditions.

CASE 1: LONG THIN FIN $(\ell/\delta \to \infty)$. Substituting the boundary conditions of Eqs. (4.68) and (4.69) into the general solution of Eq. (4.67), we obtain

$$C_1 + C_2 = \theta_0 \tag{4.84}$$

$$C_1 \times 0 + C_2 \times \infty = 0 \tag{4.85}$$

The equality of Eq. (4.85) will hold only if $C_2 = 0$, and hence from Eq. (4.84), $C_1 = \theta_0$. Thus, the specific solution of Eq. (4.67) has

$$C_1 = \theta_0 \qquad C_2 = 0 \tag{4.86}$$

and

$$\frac{\theta}{\theta_0} = e^{-mx} \tag{4.87}$$

As noted before, the total fin heat transfer rate can be obtained by integrating this temperature profile as indicated by Eq. (4.78) or by differentiating it and using Eq. (4.79), and we get

$$q_0 = q_{\text{conv}} = \frac{h\mathbf{P}}{m}\theta_0 = k_f A_k m\theta_0 \tag{4.88}$$

where $m^2 = h\mathbf{P}/k_f A_k$. For this case,

$$\theta_\ell = 0 \qquad q_\ell = 0 \tag{4.89}$$

CASE 2: THIN FIN WITH AN ADIABATIC TIP. Substituting the boundary conditions of Eqs. (4.68) and (4.71) into the general solution, Eq. (4.67), we obtain

$$C_1 + C_2 = \theta_0 \tag{4.90}$$

$$-mC_1 e^{-m\ell} + mC_2 e^{m\ell} = 0 \tag{4.91}$$

Solving for C_1 and C_2 from these equations and substituting them into Eq. (4.67), after simplification, we get

$$\frac{\theta}{\theta_0} = \frac{\cosh m(\ell - x)}{\cosh m\ell} \tag{4.92}$$

and from Eq. (4.79),

$$q_0 = \frac{h\mathbf{P}}{m} \theta_0 \tanh m\ell \tag{4.93}$$

For this case,

$$\frac{\theta_\ell}{\theta_0} = \frac{1}{\cosh m\ell} \qquad q_\ell = 0 \tag{4.94}$$

CASE 3: THIN FIN WITH CONVECTIVE BOUNDARY AT THE FIN TIP. In this case, the boundary conditions are given by Eqs. (4.68) and (4.73). Substituting them into the general solution, Eq. (4.67), we obtain

$$C_1 + C_2 = \theta_0 \tag{4.95}$$

$$-mC_1 e^{-m\ell} + mC_2 e^{m\ell} = -\frac{h_e}{k_f} \theta_\ell \tag{4.96}$$

Solving for C_1 and C_2 from these equations and substituting them into Eq. (4.67), after some algebraic manipulation we get

$$\frac{\theta}{\theta_o} = \frac{\cosh m(\ell - x) + B \sinh m(\ell - x)}{\cosh m\ell + B \sinh m\ell} \tag{4.97}$$

and after finding $d\theta/dx$ from Eq. (4.97), we get q_0 from Eq. (4.79) and q_ℓ from Eq. (4.80):

$$q_0 = \frac{h\mathbf{P}}{m} \theta_0 \frac{\sinh m\ell + B \cosh m\ell}{\cosh m\ell + B \sinh m\ell} \tag{4.98}$$

$$q_\ell = h_e A_k \theta_0 \frac{1}{\cosh m\ell + B \sinh m\ell} \tag{4.99}$$

and from Eq. (4.97) at $x = \ell$,

$$\frac{\theta_\ell}{\theta_0} = \frac{1}{\cosh m\ell + B \sinh m\ell} \tag{4.100}$$

where

$$B = \frac{h_e}{mk_f} = \frac{2}{m\delta} \mathrm{Bi}^* = \frac{\alpha_f^*}{m\ell} \mathrm{Bi}^* \qquad \mathrm{Bi}^* = \frac{h_e \delta}{2k_f} \qquad \alpha_f^* = \frac{2\ell}{\delta} \tag{4.101}$$

Here Bi* is the Biot number at the fin tip; it is the ratio of conduction resistance within the fin $[1/\{k_f/(\delta/2)\}]$ to convection resistance at the fin tip $(1/h_e)$. α_f^* is the fin aspect ratio as defined.

CASE 4: THIN FIN WITH FINITE HEAT TRANSFER AT THE FIN TIP. Substituting the boundary conditions of Eqs. (4.68) and (4.75) into the general solution of Eq. (4.67), we get

$$C_1 + C_2 = \theta_0 \tag{4.102}$$

$$-mC_1 e^{-m\ell} + mC_2 e^{m\ell} = -\frac{q_\ell}{k_f A_k} \tag{4.103}$$

Solving for C_1 and C_2 from these equations and substituting them into Eq. (4.67), after some algebraic manipulation, yields

$$\frac{\theta}{\theta_0} = \frac{\cosh m(\ell - x) - (q_\ell m/h\mathbf{P}\theta_0)\sinh mx}{\cosh m\ell} \tag{4.104}$$

and subsequently, from Eq. (4.79),

$$q_0 = \frac{h\mathbf{P}}{m}\theta_0 \tanh m\ell + q_\ell \tag{4.105}$$

and from Eq. (4.80) at $x = \ell$,

$$\frac{\theta_\ell}{\theta_0} = \frac{1 - (q_\ell m/h\mathbf{P}\theta_0)\sinh m\ell}{\cosh m\ell} \tag{4.106}$$

CASE 5: THIN FIN WITH FIN TIP TEMPERATURE SPECIFIED. Substituting the boundary conditions of Eqs. (4.68) and (4.76) into the general solution, Eq. (4.67), we obtain

$$C_1 + C_2 = \theta_0 \tag{4.107}$$

$$C_1 e^{-m\ell} + C_2 e^{m\ell} = \theta_\ell \tag{4.108}$$

Solving for C_1 and C_2 from these equations and substituting them into Eq. (4.67), after some algebraic manipulation, we get

$$\frac{\theta}{\theta_0} = \frac{\sinh m(\ell - x) + (\theta_\ell/\theta_0)\sinh mx}{\sinh m\ell} \tag{4.109}$$

Subsequently, the heat transfer rates at $x = 0$ and ℓ are obtained for Eqs. (4.79) and (4.80) as

$$q_0 = \frac{h\mathbf{P}}{m}\theta_0 \frac{\cosh m\ell - (\theta_\ell/\theta_0)}{\sinh m\ell} \tag{4.110}$$

$$q_\ell = \frac{h\mathbf{P}}{m}\theta_0 \frac{1 - (\theta_\ell/\theta_0)\cosh m\ell}{\sinh m\ell} = \frac{h\mathbf{P}}{m}\theta_\ell \frac{(\theta_0/\theta_\ell) - \cosh m\ell}{\sinh m\ell} \tag{4.111}$$

Depending on the magnitude of T_ℓ with respect to T_0 and T_∞, four temperature distributions within the fin are possible, as shown in Fig. 4.11 with $T_\ell < T_0$ (if $T_\ell > T_0$, we reverse the notation ℓ and 0 and get $T_\ell < T_0$). The location of zero heat transfer rate in the axial direction of fin then can be determined where the temperature gradient in the fin $(d\theta/dx) = 0$. After obtaining $d\theta/dx$ from Eq. (4.109), equating it to zero and simplifying, we obtain

$$e^{2mX} = \frac{e^{m\ell} - (\theta_\ell/\theta_0)}{(\theta_\ell/\theta_0) - e^{-m\ell}} \tag{4.112}$$

where X denotes the value of x where the temperature gradient within the fin becomes zero, and this location is shown in Fig. 4.11 for four possible temperature distributions (Prasad, 1996). If $\theta_\ell = \theta_0$, then $X = \ell/2$ as expected, and this case is identical to the case of a thin fin with an adiabatic tip discussed previously. From Eq. (4.112), we find the zero heat flux location X as

$$X < \ell \qquad \text{for } \frac{\theta_\ell}{\theta_0} > \frac{1}{\cosh m\ell} \tag{4.113}$$

$$X = \ell \qquad \text{for } \frac{\theta_\ell}{\theta_0} = \frac{1}{\cosh m\ell} \tag{4.114}$$

$$X > \ell \qquad \text{for } \frac{\theta_\ell}{\theta_0} < \frac{1}{\cosh m\ell} \tag{4.115}$$

Equations (4.114) and (4.115) correspond to the temperature distributions of Fig. 4.11d and b, respectively. The location X of zero heat transfer rate for Fig. 4.11a and c can be found from Eq. (4.112) and will depend on the value of θ_ℓ/θ_0 specified. Note that $\theta_\ell/\theta_0 < 1$ for all four cases of Fig. 4.11.

The solutions for the temperature distributions of the foregoing boundary conditions, along with the expressions for q_0, q_ℓ, and θ_ℓ are summarized in Table 4.4.

4.3.1.3   Thin Circular Fin of Uniform Thickness. Next, we derive a solution for the temperature distribution in another important fin geometry, the circular fin, also referred to as the *disk fin, radial fin,* or *circumferential fin,* shown in Fig. 4.12. An energy balance for a differential element dr of the fin of uniform thickness presented in Fig. 4.12 is given by as follows.

$$(q''2\pi r\delta)|_r - (q''2\pi r\delta)|_{r+dr} - 2h(2\pi r)\,dr(T - T_\infty) = 0 \tag{4.116}$$

where $q'' = -k_f(dT/dr)$ represents conduction heat flux at a given location. After simplification, Eq. (4.116) becomes

$$-\frac{d}{dr}\left(r\,\frac{\delta}{2}\,q''\right) - hr(T - T_\infty) = 0 \tag{4.117}$$

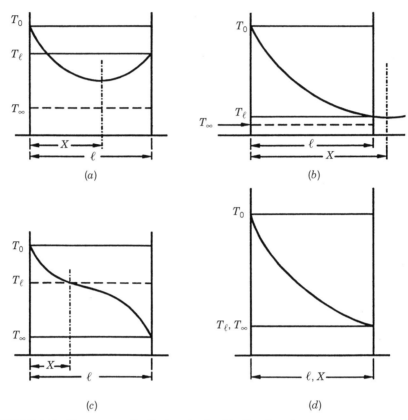

FIGURE 4.11   Temperature distributions in the thin fin when the temperatures are specified at both ends as in Fig. 4.10e: (a) $T_0 > T_\ell > T_\infty$; (b) $T_0 > T_\ell > T_\infty$; (c) $T_0 > T_\infty > T_\ell$; (d) $T_0 > T_\ell = T_\infty$. (From Prasad, 1996).

Replacement of the explicit form of q'' in Eq. (4.117) and rearrangement lead to

$$\frac{d}{dr}\left(r\frac{dT}{dr}\right) - \frac{2hr}{\delta k_f}(T - T_\infty) = 0 \tag{4.118}$$

or

$$\frac{d^2\theta}{dr^2} + \frac{1}{r}\frac{d\theta}{dr} - m^2\theta = 0 \tag{4.119}$$

where $m^2 = h\mathbf{P}/k_f A_k = 2h/k_f\delta = $ constant since $\mathbf{P} = 2(2\pi r)$ and $A_k = 2\pi r\delta$. Equation (4.119) is the modified Bessel equation of order zero. The general solution of this equation is given by

$$\theta = C_3 I_o(mr) + C_4 K_o(mr) \tag{4.120}$$

where C_3 and C_4 are constants of integration, and I_o and K_o are the modified zero-order Bessel functions of the first and second kinds, respectively.

Similar to the uniform thickness straight-fin case, the boundary condition for the circular fin at $r = r_o$ is simply $T = T_0$. There are five boundary conditions at $r = r_e$, similar to those shown in Fig. 4.10. Since the boundary condition of an adiabatic fin tip at $r = r_e$ (similar to that of Fig. 4.10b) represents a good approximation in many practical applications, we will obtain a solution of Eq. (4.120) for this case only.

$$\theta = \theta_0 \qquad \text{at } r = r_o \tag{4.121}$$

$$\frac{d\theta}{dr} = 0 \qquad \text{at } r = r_e \tag{4.122}$$

Substitution of these boundary conditions in Eq. (4.120) results in

$$\theta_0 = C_3 I_o(mr_o) + C_4 K_o(mr_o) \tag{4.123}$$

$$0 = C_3 I_1(mr_e) - C_4 K_1(mr_e) \tag{4.124}$$

where I_1 and K_1 are the modified first-order Bessel functions of the first and second kinds, respectively. Solving for C_3 and C_4 from these two algebraic equations and substituting them into Eq. (4.120) yields

$$\frac{\theta}{\theta_0} = \frac{K_1(mr_e)I_o(mr) + I_1(mr_e)K_o(mr)}{K_1(mr_e)I_o(mr_o) + I_1(mr_e)K_o(mr_o)} \tag{4.125}$$

The heat flow through the base is

$$q_0 = -k_f(2\pi r_o \delta)\left(\frac{d\theta}{dr}\right)_{r=r_o} \tag{4.126}$$

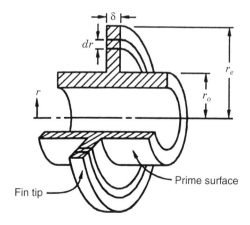

FIGURE 4.12   Thin circular fin of uniform thickness.

After differentiating Eq. (4.125), evaluate $(d\theta/dr)_{r=r_o}$ and substitute it in Eq. (4.126). The result is

$$q_0 = k_f(2\pi r_o \delta)mB_1\theta_0 = \frac{4\pi r_o h}{m}B_1\theta_0 \tag{4.127}$$

where

$$B_1 = \frac{I_1(mr_e)K_1(mr_o) - K_1(mr_e)I_1(mr_o)}{I_o(mr_o)K_1(mr_e) + I_1(mr_e)K_o(mr_o)} \tag{4.128}$$

4.3.2 Fin Efficiency

For extended surface heat exchangers, we are interested in actual heat transfer by the fins. Regardless of whether the fin is heating or cooling the ambient fluid, all heat must pass through the fin base for the first three boundary conditions of Fig. 4.10 or Table 4.4. The heat transfer rate q_0 through the fin base can be calculated in two different ways as shown before, either by integrating or by differentiating the temperature profile for these three boundary conditions.

This heat transfer rate q_0 can be presented in dimensionless form by the concept of fin efficiency η_f, defined as

$$\eta_f = \frac{q_0}{q_{max}} \tag{4.129}$$

Here our meterbar (yardstick) for comparison is a "perfect fin" having (1) the same geometry as the actual fin; (2) the same operating conditions, T_0, T_∞, h, and h_e; and (3) infinite thermal conductivity k_f of the fin material. Under these circumstances, the perfect fin is at the uniform base temperature T_0, as shown in Fig. 4.13 on page 274. The heat transfer from the fin base, q_{max} (the fin is considered to be of uniform cross section), is

$$q_{max} = h\mathbf{P}\ell(T_0 - T_\infty) + h_e A_k(T_0 - T_\infty) = \left(hA_f + h_e A_k\right)(T_0 - T_\infty) = \left(hA_f + h_e A_k\right)\theta_0 \tag{4.130}$$

Thus the heat transfer rate through the fin is then, from Eq. (4.129),

$$q_0 = \eta_f\left(hA_f + h_e A_k\right)(T_0 - T_\infty) \tag{4.131}$$

Thus η_f is a measure of thermal performance of a fin. In the equations above, $A_f = \mathbf{P}\ell$ is the fin convection area and A_k is the fin tip convection area. It will be shown that η_f is independent of q_0 and $\theta_0 = (T_0 - T_\infty)$. The thermal resistance of the fin $\mathbf{R}_f$, based on the temperature difference θ_0, from Eq. (4.131), is

$$\mathbf{R}_f = \begin{cases} \dfrac{1}{\eta_f\left(hA_f + h_e A_k\right)} & \text{if } h_e \neq 0 \tag{4.132a} \\[4mm] \dfrac{1}{\eta_f hA_f} & \text{if } h_e = 0 \tag{4.132b} \end{cases}$$

However, for a finned surface in an exchanger, we need to take the primary surface into consideration, and we will use the thermal resistance of an extended surface given by Eq. (4.164).

We now derive expressions for the fin efficiency for some fin geometries of importance.

4.3.2.1 Thin, Straight Fin of Uniform Rectangular Cross Section. The temperature distributions through this fin for the first three boundary conditions in Table 4.4 are presented by Eqs. (4.87), (4.92), and (4.97) or in Table 4.4. The actual heat transfer rates q_0 through the fin base are given by Eqs. (4.88), (4.93), and (4.98) for these boundary conditions. Using these values of q_0 and q_{max} from Eq. (4.130), the η_f values are obtained as follows using Eq. (4.129). For completeness, the expression for q_0 is also provided below in terms of η_f.

For the *long, thin fin* (case 1),

$$\eta_f = \frac{1}{m\ell} \qquad q_0 = \eta_f\, hA_f(T_0 - T_\infty) \tag{4.133}$$

For the thin fin with an *adiabatic tip* (case 2),[†]

$$\eta_f = \frac{\tanh m\ell}{m\ell} \qquad q_0 = \eta_f\, hA_f(T_0 - T_\infty) \tag{4.134}$$

For the thin fin with a *convection boundary at the fin tip* ($B = h_e/mk_f = 2\mathrm{Bi}^*/m\delta$; case 3),

$$\eta_f = \frac{\tanh m\ell + B}{(B + m\ell)(1 + B\tanh m\ell)} = \frac{m^2\ell^2 \tanh m\ell + m\ell\alpha_f^* \cdot \mathrm{Bi}^*}{(\alpha_f^* \cdot \mathrm{Bi}^* + m^2\ell^2)(m\ell + \alpha_f^* \cdot \mathrm{Bi}^* \tanh m\ell)} \tag{4.135}$$

where $B = \alpha_f^* \cdot \mathrm{Bi}^*/m\ell$ from Eq. (4.101) is substituted to get the expression after the second equality in Eq. (4.135). For this case, q_0 is given by Eq. (4.131). If the convected heat from the fin tip is small, η_f can be approximately computed from Eq. (4.134) with ℓ replaced by $\ell + \delta/2$. This approximation is referred to as the *Harper–Brown approximation*.

The fin efficiency of case 4 of Table 4.4 is the same as that in Eq. (4.135) if we assume that q_ℓ at the fin tip represents convection heat transfer; otherwise, η_f cannot be defined since q_ℓ can be any positive or negative value and q_0 cannot be evaluated explicitly [see Eq. (4.81)]. The fin efficiency for case 5 may not readily be defined; it is discussed later in a subsection on p. 278.

The expression for the fin efficiency becomes increasingly complicated in Eqs. (4.133)–(4.135). It can be shown that

$$\eta_f \text{ of Eq. (4.135)} < \eta_f \text{ of Eq. (4.134)} < \eta_f \text{ of Eq. (4.133)} \tag{4.136}$$

A simple criterion can be set up for the use of Eq. (4.133) for Eq. (4.134) and the use of Eq. (4.134) for Eq. (4.135) as follows.

[†]Note that if q_0 instead of T_0 is specified at $x = 0$ for a case alternative to case 2 (and $q = 0$ at $x = \ell$), it can be shown that the expression for η_f is identical to that in Eq. (4.134).

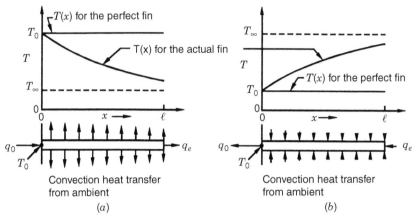

FIGURE 4.13   Temperature distributions for actual and perfect thin fins: (*a*) fin is being cooled; (*b*) fin is being heated (From Shah, 1983).

Since tanh $m\ell \rightarrow 1$ for $m\ell \rightarrow \infty$, η_f of Eq. (4.134) properly reduces to that of Eq. (4.133) in the limit. If we set up a criterion that if η_f of Eq. (4.134) is within 2% of η_f of Eq. (4.133), we designate it as a "long" fin. Now since tanh $m\ell = 0.98$ for $m\ell = 2.30$, we may treat a straight rectangular fin as a "long" fin for $m\ell \geq 2.30$.

Similarly, when B or $\mathrm{Bi}^* = 0$ (i.e., no heat transfer through the fin tip), η_f of Eq. (4.135) reduces to that of Eq. (4.134), as expected. The higher the value of Bi^*, the lower will be η_f of Eq. (4.135) compared to that of Eq. (4.134). If we desire η_f of Eq. (4.135) within 2% of η_f of Eq. (4.134), a simple criterion for the values of B can be set up for each value of $m\ell$. For example, when $m\ell = 1$, tanh $m\ell = 0.761$,

$$\frac{\eta_f \text{ of Eq. } (4.235)}{\eta_f \text{ of Eq. } (4.134)} = \frac{0.761 + B}{(1 + B)(1 + 0.761B)} \times \frac{1}{0.761} = 0.98 \qquad \text{for } B = 0.045 \qquad (4.137)$$

Hence for $m\ell = 1$, $B \leq 0.045$ will provide an error of less than 2% in η_f by the use of Eq. (4.134) instead of Eq. (4.135).

As shown later, the η_f formula of Eq. (4.134) with an adiabatic tip is applicable to most two-fluid plate-fin heat exchangers. This η_f as a function of $m\ell$ is shown in Fig. 4.14 on page 276 for a straight fin designated by $r_e/r_o = 1$. It is found that η_f increases with decreasing $m\ell$, but the variation is nonlinear. Exploitation of this nonlinear behavior by industry is explained in Example 4.3. Since $m^2 = 2h/k_f\delta$, $m\ell$ decreasing means ℓ or h decreasing or k_f or δ increasing. This means that η_f will increase with decreasing fin length ℓ, decreasing heat transfer coefficient h (or decreasing Bi), increasing fin thermal conductivity k_f, or increasing fin thickness δ. From an engineering viewpoint, to attain high fin efficiency: (1) maintain low fin conduction length or plate spacing b (see Fig. 4.15 on page 277); (2) use fins when the heat transfer coefficients are low, such as for forced convection with gases, oils, or a free convection situation; (3) use low height fins when the heat transfer coefficients are high, such as for forced convection with water and other liquids or phase-change fluids; (4) use fin material as aluminum or copper instead of stainless steel and other low-thermal-conductivity materials; and (5) increase the fin thickness to a practical value. However, a decrease in the fin thickness is generally pursued by industry due to the nonlinear behavior of η_f vs. $m\ell$. The decrease in η_f will be much less than the decrease in the fin weight and fin material cost, as illustrated in Example 4.3.

Example 4.3 A plate-fin exchanger has 24-mm-high 0.12-mm-thick rectangular fins (Fig. 4.15a on page 277) with a fin density of 600 fins/m. The heat transfer coefficient for airflow over the fins is 170 $W/m^2 \cdot K$. Determine the fin efficiency if the fins are made from copper with a thermal conductivity of 401 $W/m \cdot K$. If the fin thickness is reduced to 0.06 mm, what happens to the fin efficiency and associated heat transfer rate? How would you change the fin density to bring the same level of heat transfer performance with the thinner fins if you idealize that there is no change in heat transfer coefficient with the change in fin density? Now change the fin material from copper to aluminum with the thermal conductivity as 237 $W/m^2 \cdot K$. Discuss the implication on fin efficiency and fin heat transfer of changing the material from copper to aluminum. The mass densities of these materials are 8933 and 2702 kg/m^3.

SOLUTION

Problem Data and Schematic: The fin is shown in Fig. 4.15a. The following information is provided.

$$h = 170 \text{ W/m}^2 \cdot \text{K} \qquad \ell = 12 \text{ mm } (b = 24 \text{ mm})$$

Copper fins: $\delta = 0.12$ mm $k_f = 401$ W/m·K   fin density = 600 fins/m
$\delta = 0.06$ mm $k_f = 401$ W/m·K
Aluminum fins: $\delta = 0.12$ mm $k_f = 237$ W/m·K   fin density = 600 fins/m

Determine:

(a) The change in the fin efficiency and heat transfer rate if the copper fin thickness is changed from 0.12 mm to 0.06 mm.

(b) The change in fin density for the same heat transfer rate when reducing the fin thickness from 0.12 to 0.06 mm.

(c) How the fin efficiency is changed if the fin material is changed from copper to aluminum, keeping the same fin geometry. What are other design implications?

Assumptions: The heat transfer coefficient does not change with the change in fin density and the assumptions of Section 4.3.1.1 are valid here.

Analysis: Let us compute the fin efficiency using Eq. (4.134) for original copper fins. Using Eq. (4.65) or (4.147) yields

$$m = \left(\frac{2h}{k_f \delta}\right)^{1/2} = \left[\frac{2 \times 170 \text{ W/m}^2 \cdot \text{K}}{401 \text{ W/m} \cdot \text{K} \times (0.12 \times 10^{-3})\text{m}}\right]^{1/2} = 84.0575 \text{ m}^{-1}$$

Therefore,

$$m\ell = 84.0575 \text{ m}^{-1} \times (12 \times 10^{-3})\text{ m} = 1.0087$$

$$\eta_f = \frac{\tanh m\ell}{m\ell} = \frac{\tanh (1.0087)}{1.0087} = \frac{0.7652}{1.0087} = 0.759 \qquad \textit{Ans.}$$

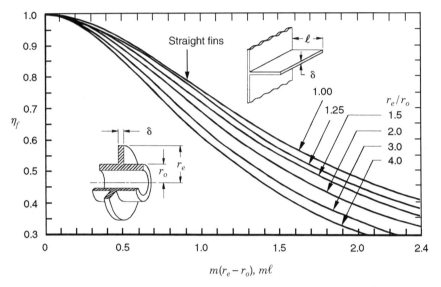

FIGURE 4.14   Fin efficiencies of straight and circular fins of uniform thickness δ (From Kays and London, 1998).

If the fin thickness is reduced to 0.06 mm, we get

$$m = \left(\frac{2h}{k_f\delta}\right)^{1/2} = \left[\frac{2 \times 170\,\text{W/m}^2 \cdot \text{K}}{401\,\text{W/m} \cdot \text{K} \times (0.06 \times 10^{-3})\,\text{m}}\right]^{1/2} = 118.875\,\text{m}^{-1}$$

Therefore,

$$m\ell = 118.875\,\text{m}^{-1} \times (12 \times 10^{-3})\,\text{m} = 1.4265$$

$$\eta_f = \frac{\tanh m\ell}{m\ell} = \frac{\tanh(1.4265)}{1.4265} = 0.625 \qquad\qquad Ans.$$

Thus, the fin efficiency is reduced from 0.759 to 0.625, about 18% reduction. This in turn will reduce fin heat transfer by about 18%. For simplicity here we do not include the effect of the primary surface (which is considered in Section 4.3.4).

Fin heat transfer can be kept approximately constant by increasing the fin surface area by 18% (i.e., increasing the fin density from 600 fins/m to 729 fins/m). Here we have idealized that the heat transfer coefficient h does not change with the change in fin density. In reality, h will decrease with an increase in the fin density in general.

Now changing the fin material from copper to aluminum, m, $m\ell$, and η_f for the 0.12 mm thick fins will be

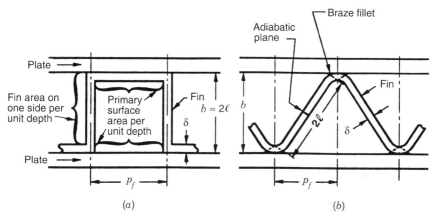

FIGURE 4.15   (a) Idealized plain rectangular fin (b) plain triangular fin (From Shah, 1981).

$$m = \left(\frac{2h}{k_f \delta}\right)^{1/2} = \left[\frac{2 \times 170\,\text{W/m}^2 \cdot \text{K}}{237\,\text{W/m} \cdot \text{K} \times (0.12 \times 10^{-3})\,\text{m}}\right]^{1/2} = 109.339\,\text{m}^{-1}$$

$$m\ell = 109.339\,\text{m}^{-1} \times (12 \times 10^{-3})\,\text{m} = 1.3121$$

$$\eta_f = \frac{\tanh m\ell}{m\ell} = \frac{\tanh(1.3121)}{1.3121} = 0.659 \qquad\qquad Ans.$$

Thus, we can see that by changing the material from copper to aluminum reduces η_f from 0.759 to 0.659. However, if we compare 0.12-mm-thick aluminum fin with 0.06-mm copper fin, the fin efficiency is about the same (0.659 vs. 0.625), and hence from the heat transfer point of view, these two fins are equivalent. Now let us compare the material use for these two fins (0.12 mm aluminum fins vs. 0.06 mm copper fins) at the same fin density. Since the mass densities of aluminum and copper are 2702 and 8933 kg/m³, respectively, the ratio of the fin material is

$$\frac{\text{Al fin material}}{\text{Cu fin material}} = \left(\frac{0.12\,\text{mm}}{0.06\,\text{mm}}\right)\left(\frac{2702\,\text{kg/m}^3}{8933\,\text{kg/m}^3}\right) = 0.60$$

This indicates that aluminum fin material use by weight is about 40% less than that for copper for about the same heat transfer performance despite the double fin thickness for aluminum.

Discussion and Comments: This example clearly indicates that reducing the fin thickness will reduce fin heat transfer as expected, but in a nonlinear way, as shown in Fig. 4.14. In the present case, reducing the fin thickness from 0.12 mm to 0.06 mm (50% reduction) reduces the fin efficiency and heat transfer by 18%. This reduction can be compensated by increasing the fin density by about 18%. Thus there is a net gain of about 32% reduction in weight for the same heat transfer. Although the example above is not for an automobile radiator, this was the direction taken by the automobile industry for the radiator fin. The fin gauge was reduced over the years from 0.20 mm to 0.046 mm as the manufacturing technology improved to make fins with thin gauges, along with increased fin density from about 400 fins/m to 800 fins/m. When the fin thickness was reduced to about 0.046 mm, it was so thin that copper could not handle the corrosion common in the application, and the radiator life was reduced significantly. Changing the material from copper to aluminum and increasing the fin thickness from 0.046 mm to 0.075 mm, the durability was brought back to the desired radiator life with thicker aluminum fins, and at the same time the total material mass was reduced, as was the cost of material. In summary, the material mass for fins can be minimized by reducing the fin thickness as permitted by the design and manufacturing considerations and increasing the fin density to maintain the desired fin heat transfer. This is primarily because fins are used for increasing the surface area for convection heat transfer and thickness comes into picture for conduction heat transfer through the fin, which can be maintained smallest by increasing the fin density somewhat.

To repeat, the important variable for convection heat transfer through the fin surface is the fin surface area, and the important variable for conduction heat transfer from the base along the fin length is the fin thickness. The fin thickness is chosen the thinnest and is of secondary importance for industrial heat exchangers due to a limited range of permissible fin sizes. That is the reason that fins have one-third to one-fourth primary surface thickness to cost effectively utilize the material. Thus the modern fin designs have the largest possible fin density and smallest possible fin thickness permitted by design to make the most use of the fin material for fins in a heat exchanger. This trend will continue with the improvements in material and manufacturing technologies.

Fin Heat Transfer for a Specified Fin Tip Temperature. For this case, the boundary conditions of case 5 shown in Fig. 4.10e, q_0 and q_ℓ are given by Eqs. (4.110) and (4.111). Substituting them in Eq. (4.81) with $\theta_0/\theta_\ell = 1$, we get the total fin transfer rate as (Prasad, 1996)

$$q_{\text{conv}} = q_0 - q_\ell = \frac{h\mathbf{P}}{m}(\theta_0 + \theta_\ell)\frac{\cosh m\ell - 1}{\sinh m\ell} = \frac{h\mathbf{P}}{m}(\theta_0 + \theta_\ell)\tanh\frac{m\ell}{2}$$

$$= h A_{f,1/2}(\theta_0 + \theta_\ell)\frac{\tanh(m\ell/2)}{m\ell/2} = hA_{f,1/2}(\theta_0 + \theta_\ell)\eta_{f,1/2} \qquad (4.138)$$

Here $(\cosh m\ell - 1)/\sinh m\ell = \tanh(m\ell/2)$ is a hyperbolic function identity, and $A_{f,1/2} = \mathbf{P}\ell/2$. Hence, one can see that we can calculate accurately the total fin transfer rate for this case through the half-fin-length idealization (case 2, thin fin with an adiabatic tip in Table 4.4 with ℓ replaced by $\ell/2$). This means that we can consider the fin of Fig. 4.11 made up of two fins of $\ell/2$ length each, one having the heat transfer rate through the fin base as q_0 and the other as q_ℓ and their meeting point at $\ell/2$ as adiabatic. The fin efficiency calculated under this idealization is given by Eq. (4.134) with the fin length as $\ell/2$. For the half-fin idealization, we find $q_{0,1/2}$ from Eq. (4.93) by replacing ℓ with $\ell/2$ as

$$q_{0,1/2} = \frac{h\mathbf{P}}{m}\theta_0 \tanh\frac{m\ell}{2} = \frac{h\mathbf{P}}{m}\theta_0\frac{\cosh m\ell - 1}{\sinh m\ell} \qquad (4.139)$$

As shown clearly shown in Fig. 4.11, the reason is that the adiabatic plane will not be at $x = \ell/2$, but can be found from Eq. (4.112) for the specified temperatures θ_ℓ and θ_0. Thus we find that $q_{0,1/2}$ of Eq. (4.139) is different from the actual q_0 given by Eq. (4.110). Thus, the concept of fin efficiency has real limitations for this fin problem and similarly for the fin problem of case 4, Fig. 4.10d. Further implications are discussed by Prasad (1996).

Dimensionless Groups. A careful review of Eqs. (4.133)–(4.135) reveals that η_f is a function of $m\ell$ and B or Bi*. Employing the definitions of m and B yields

$$\eta_f = \phi(A_k, \mathbf{P}, \ell, k_f, h, h_e) \tag{4.140}$$

Thus η_f for the thin fin is a function of the fin geometry $(A_k, \mathbf{P}, \ell)$, fin material thermal conductivity k_f, heat transfer coefficient h, and the fin tip boundary condition. Based on Eq. (4.140), it should be emphasized that η_f is not a direct function of T_0, T_∞, $(T_0 - T_\infty)$, T_ℓ, q_0, or q_ℓ.

The parameters on the right-hand side of Eq. (4.140) can be cast into dimensionless groups as follows:

$$\eta_f = \phi(m\ell, \text{Bi}^*) = \phi(\alpha_f^* \cdot \text{Bi}^{1/2}, \text{Bi}^*) \tag{4.141}$$

where ϕ is a functional relationship and η_f depends on the fin geometry and fin tip boundary condition. The dimensionless groups $m\ell$ and B have already appeared in Eqs. (4.133)–(4.135). From the definition, $m\ell = \alpha_f^* \cdot \text{Bi}^{1/2}$, where $\text{Bi} = h\delta/2k_f$ is the Biot number at the fin surface and $\alpha_f^* = 2\ell/\delta$ is the aspect ratio of the fin. Also, $\text{Bi}^* = h_e\delta/2k_f$.

In the context of fin heat transfer, the Biot number $\text{Bi} = h\delta/2k_f$ is the ratio of conduction resistance within the fin, $\delta/2k_f = 1/[k_f/(\delta/2)]$, and the convection resistance at the fin surface, $1/h$. A small value of Bi indicates that the conduction resistance is small compared to the convection resistance. Therefore, the temperature gradient within the fin is small compared to that at the fin surface and indicates that the fin may be approximated as a thin fin for the fin efficiency calculation. In contrast, a large value of Bi indicates that the conduction resistance is comparable to the convection resistance; hence, the temperature gradient within the fin may not be negligible for the fin efficiency calculations as in a two-dimensional or "thick" fin problem. For example, an approximate two-dimensional fin efficiency formula for the straight fin of a rectangular profile [a counterpart of Eq. (4.135) for a two-dimensional fin] is given by Huang and Shah (1992):

$$\eta_f = \begin{cases} \dfrac{(\text{Bi}^+)^{1/2}}{\alpha_f^* \cdot \text{Bi} + \text{Bi}^*} \dfrac{\text{Bi}^* \cosh[\alpha_f^*(\text{Bi}^+)^{1/2}] + (\text{Bi}^+)^{1/2} \sinh[\alpha_f^*(\text{Bi}^+)^{1/2}]}{\text{Bi}^* \sinh[\alpha_f^*(\text{Bi}^+)^{1/2}] + (\text{Bi}^+)^{1/2} \cosh[\alpha_f^*(\text{Bi}^+)^{1/2}]} & \text{for } \text{Bi}^* > 0 \\[4pt] & \qquad\qquad (4.142a) \\[6pt] \dfrac{(\text{Bi}^+)^{1/2}}{\alpha_f^* \cdot \text{Bi}} \tanh[\alpha_f^*(\text{Bi}^+)^{1/2}] & \text{for } \text{Bi}^* = 0 \\[4pt] & \qquad\qquad (4.142b) \end{cases}$$

where

$$\text{Bi}^+ = \frac{\text{Bi}}{1 + \text{Bi}/4} \qquad \alpha_f^* = \frac{2\ell}{\delta} \tag{4.143}$$

A comparison of Eqs. (4.134) and (4.142b) reveals that the thin fin approximation of Eq. (4.134) introduces a maximum error of 0.3% for $\text{Bi} \leq 0.01$ and $\alpha_f^* < 100$; otherwise, use Eq. (4.142b) for a thick fin.

Based on Eq. (4.142), we find for a two-dimensional or thick straight fin of rectangular profile,

$$\eta_f = \phi(\alpha_f^*, \text{Bi}, \text{Bi}^*) \tag{4.144}$$

A comparison of Eqs. (4.141) and (4.144) indicates that the fin efficiency is now a distinct function of the fin aspect ratio and Biot number for a thick fin. Based on Eq. (4.142), it can also be shown that η_f increases with decreasing Bi and decreasing α_f^*.

4.3.2.2 Plate-Fin Surfaces.

In most two-fluid plate-fin heat exchangers, heat flow from (or to) both sides of a fin in the interior exchanger flow passages is idealized as symmetrical. For example, see Fig. 4.15a for a typical flow passage in a plate-fin exchanger. Here heat flows from both ends to the center of a fin when the base temperature T_0 is higher than the fluid temperature T_∞. There is no heat transfer through the center of the fin and it is treated as adiabatic. Therefore, the appropriate formula for the fin efficiency is Eq. (4.134), and this formula is one of the most useful formulas for many plate-fin exchangers. For the end passages in a plate-fin exchanger, the heat source (or sink) for the fin is on only one end. For such passages, the appropriate fin efficiency formula is either Eq. (4.135) with finite heat transfer at the fin tip, or the fin tip may again be idealized as adiabatic [use Eq. (4.134)], in which case the fin length is twice the fin length of the central passages.

Consider the two most commonly used fin geometries, rectangular and triangular, shown in Fig. 4.15. The idealized plain rectangular fin geometry has sharp corners instead of actual rounded corners. The fin surface area and primary surface area associated with this fin are shown in Fig. 4.15a. From the review of this figure, the fin length ℓ for heat conduction (up to the adiabatic plane) is

$$\ell = \frac{b - \delta}{2} \approx \frac{b}{2} \quad \text{or} \quad \ell = \frac{b - 2\delta}{2} \approx \frac{b}{2} \tag{4.145}$$

where the last approximate expression in each formula is valid for b (or ℓ) $\gg \delta$. If ℓ is not very large compared to δ, then $\eta_f > 0.95$, and either of the approximations will not significantly affect the value of η_f. The fin efficiency of this plain rectangular fin is then

$$\eta_f = \frac{\tanh m\ell}{m\ell} \tag{4.146}$$

$$m = \left(\frac{hP}{k_f A_k}\right)^{1/2} = \left[\frac{h(2L_f + 2\delta)}{k_f(L_f\,\delta)}\right]^{1/2} = \left[\frac{2h}{k_f\delta}\left(1 + \frac{\delta}{L_f}\right)\right]^{1/2} \approx \left(\frac{2h}{k_f\delta}\right)^{1/2} \tag{4.147}$$

Here review Fig. 4.9 for substituted values of $\mathbf{P}$ and A_k in Eq. (4.147). The last approximate term on the right-hand side is valid when $\delta \ll L_f$.

The plain triangular fin geometry of constant cross section is shown in Fig. 4.15b. Here the corners are shown rounded with a braze or solder fillet. The size of this fillet will depend on the radius of the corner and the manufacturing process. The fin conduction length ℓ for the η_f evaluation is half of the distance 2ℓ shown in this figure. The fin efficiency is computed from Eq. (4.146) with the value of the fin parameter m obtained by Eq. (4.147).

Now let us summarize the η_f evaluation for other plate-fin surfaces of Fig. 1.29. The offset strip fin is similar to the plain rectangular fin except that the fin length L_f is not continuous. The length L_f is the sum of many strip lengths ℓ_s. The fin efficiency is determined by the use of Eq. (4.146) with ℓ and m calculated from Eqs. (4.145) and (4.147), respectively, except that L_f in Eq. (4.147) is replaced by ℓ_s.

In a louver fin, the louver cuts are ideally parallel to the heat conduction path direction from the base to the center of the fin, and the louvers do not extend to the base of the fin in order to maintain the required fin structural strength. Hence, ideal heat transfer through individual louvers is identical to the case of plain triangular fins of Fig. 4.15b, and the applicable fin efficiency formula is Eq. (4.146). The exposed edge area of the louver is part of the fin surface area A_f and should be included in the determination A_f for consistency. However, the common industrial practice is to ignore the louver edge area (see Example 8.2) in the A_f determination and also in the calculation of m from Eq. (4.147). Thus for the η_f evaluation from Eq. (4.146), the louver fin is treated as the plain triangular fin with $m = (2h/k_f\delta)^{1/2}$.

The wavy and corrugated fins are treated as either rectangular or triangular fins, depending on the cross-section shape for the η_f evaluation. Ambiguity exists for the determination of A_f and ℓ for perforated fins for the determination of η_f. The common practice is to ignore the perforations and determine η_f as if the fin were unperforated. As long as the heat transfer coefficient is determined experimentally using this assumption and the fin efficiency is calculated subsequently the same way for its use in the design of a heat exchanger, the error introduced may not be significant. For pin fins, η_f is evaluated from Eq. (4.146) with

$$\ell = \frac{b}{2} - d_0 \approx \frac{b}{2} \qquad m = \left[\frac{h(\pi d_0)}{k_f\left(\pi d_0^2/4\right)}\right]^{1/2} = \left(\frac{4h}{k_f d_0}\right)^{1/2} \tag{4.148}$$

where d_0 is the pin diameter and b is the plate spacing.

In all of the foregoing fin geometries, the fin thickness δ is considered constant, and hence the cross section of the fin in the ℓ (or x) direction is rectangular (see Fig. 4.9). Solutions for η_f have been obtained for this cross-section shape as triangular, concave parabolic, convex parabolic, and so on. We do not consider such fins here, as they are uncommon in two-fluid extended surface exchangers. Refer to Kraus et al. (2001) for further details.

From the foregoing discussion, it is clear that the fin efficiency formula of Eq. (4.134) or (4.146) is the most common expression for plate-fin surfaces, and it is presented in Fig. 4.14 as a straight fin. Fin efficiency formulas for some important fin geometries are summarized in Table 4.5.

Example 4.4   A gas-to-air waste heat recovery heat exchanger has $0.3 \times 0.3 \times 0.6\,\text{m}$ modules with the 0.6 m dimension as the noflow height. Each module is a single-pass crossflow plate-fin exchanger having fluids unmixed on both fluid sides. Each fluid side has plain rectangular aluminum fins of Fig. 4.15a. Consider the plate spacing as 13 mm, fin thickness as 0.15 mm, double fin height 2ℓ as 13 mm, and the fin length as 300 mm. Determine the fin efficiency and heat transfer rate per unit fin surface area for (a) a fin in the center of the core, and (b) a fin at one end of the core for which the heat source is only on one side. Treat the other fin end as having a finite heat transfer with $h = h_e$. Use thermal conductivity of fin material as 190 W/m · K for aluminum and heat transfer coefficient is 120 W/m² · K. Fin base temperature is 200°C and fluid temperature 30°C.

SOLUTION

Problem Data and Schematic: Fin geometry and material properties and heat transfer coefficients are provided for a single-pass crossflow plate-fin exchanger (Fig. E4.4) on page 284. Fluids are unmixed on both sides. The fin base temperature as well as the ambient temperature are also provided.

Determine: The fin efficiency and heat transfer rate per unit area for:

(a) A fin in the center of the core (fin 2 in Fig. E4.4)
(b) A fin at one end of the core (fin 1 in Fig. E4.4) for which heat source is only on one side

Assumptions: The assumptions of Section 4.3.1.1 are invoked here.

Analysis: (a) For a fin at the center of the core, the heat source is on both sides. Hence, the adiabatic plane in the fin will be at the center of the fin, as shown in Fig. 4.15a. For this case, $\ell = 6.5$ mm. Now from Eq. (4.147),

$$m = \left[\frac{2h}{k_f\delta}\left(1 + \frac{\delta}{L_f}\right)\right]^{1/2}$$

$$= \left[\frac{2 \times 120 \text{ W/m}^2 \cdot \text{K}}{(190 \times 10^{-3} \text{ W/m} \cdot \text{K})(0.15 \times 10^{-3} \text{ m})}\left(1 + \frac{0.15 \text{ mm}}{300 \text{ mm}}\right)\right]^{1/2} = 91.79 \text{ m}^{-1}$$

and hence

$$m\ell = 91.79 \text{ m}^{-1} \times 6.5 \times 10^{-3} \text{ m} = 0.5966$$

$$\eta_f = \frac{\tanh m\ell}{m\ell} = \frac{\tanh(0.5966)}{0.5966} = 0.896 \qquad\qquad Ans.$$

The heat transfer rate is just the product of the fin efficiency and the theoretical maximum heat transfer rate:

$$q_0'' = \eta_f q_{max}'' = \eta_f h(T_0 - T_\infty) = 0.896 \times 120 \text{ W/m}^2 \cdot \text{K} \times (200 - 30)°\text{C}$$

$$= 18{,}278 \text{ W/m}^2 \qquad\qquad Ans.$$

(b) When the heat source is only on one side of the fin, the heat conduction length for the fin will be at least 2ℓ (see Fig. 4.15a), and there will be heat transfer from that end to the braze fillet area and the plate. Let us designate the appropriate m and ℓ for this case with the subscript 1:

$$m_1 = m = 91.79 \text{ m}^{-1}$$

$$\ell_1 = 2\ell = 13 \text{ mm} = 0.13 \text{ m}$$

TABLE 4.5   Fin Efficiency for Plate-Fin and Tube-Fin Geometries of Uniform Fin Thickness

Geometry	Fin efficiency formula where $m_i = \dfrac{2h}{k_f \delta_i}\left(1 + \dfrac{\delta_i}{L_f}\right)^{1/2}$ $\quad E_i = \dfrac{\tanh(m_i \ell_i)}{m_i \ell_i}$ $\quad i = 1, 2, 3$
 Plain, wavy, or offset strip fin of rectangular cross section	$\eta_f = E_1$ $\ell_1 = \dfrac{b}{2} - \delta_1 \quad \delta_1 = \delta$
 Triangular fin heated/cooled from one side	$\eta_f = \dfrac{hA_1(T_0 - T_\infty)\dfrac{\sinh(m_1\ell_1)}{m_1\ell_1} + q_e}{\cosh(m_1\ell_1)\left[hA_1(T_0 - T_\infty) + q_e\dfrac{T_0 - T_\infty}{T_1 - T_\infty}\right]}$ $\quad \delta_1 = \delta$
 Plain, wavy, or louver fin of triangular cross section	$\eta_f = E_1$ $\ell_1 = \ell/2 \quad \delta_1 = \delta$
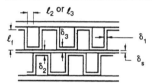 Double sandwich fin	$\eta_f = \dfrac{E_1\ell_1 + E_2\ell_2}{\ell_1 + \ell_2}\dfrac{1}{1 + m_1^2 E_1 E_2 \ell_1 \ell_2}$ $\ell_1 = b - \delta + \delta_s/2 \quad \ell_2 = \ell_3 = p_f/2$ $\delta_1 = \delta \quad \delta_2 = \delta_3 = \delta + \delta_s$
 Pin fin	$\eta_f = \dfrac{\tanh(m\ell)}{m\ell}$ $\ell = \dfrac{b}{2} - d_o \quad m = \left(\dfrac{4h}{k_f d_o}\right)^{1/2} \quad \delta = \dfrac{d_o}{2}$
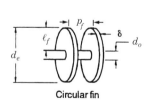 Circular fin	$\eta_f = \begin{cases} a(m\ell_e)^{-b} & \text{for } \Phi > 0.6 + 2.257(\text{r}^*)^{-0.445} \\ \dfrac{\tanh \Phi}{\Phi} & \text{for } \Phi \le 0.6 + 2.257(\text{r}^*)^{-0.445} \end{cases}$ $a = (\text{r}^*)^{-0.246} \quad \Phi = m\ell_e(\text{r}^*)^{\text{n}} \quad \text{n} = \exp(0.13m\ell_e - 1.3863)$ $b = \begin{cases} 0.9107 + 0.0893\text{r}^* & \text{for r}^* \le 2 \\ 0.9706 + 0.17125\,\ell\text{n r}^* & \text{for r}^* > 2 \end{cases}$ $m = \left(\dfrac{2h}{k_f \delta}\right)^{1/2} \quad \ell_e = \ell_f + \dfrac{\delta}{2} \quad \text{r}^* = d_e/o$
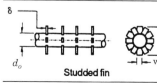 Studded fin	$\eta_f = \dfrac{\tanh(m\ell_e)}{m\ell_e}$ $m = \left[\dfrac{2h}{k_f\delta}\left(1 + \dfrac{\delta}{w}\right)\right]^{1/2} \quad \ell_e = \ell_f + \dfrac{\delta}{2} \quad \ell_f = \dfrac{(d_e - d_o)}{2}$

Source: Data from Shah (1985).

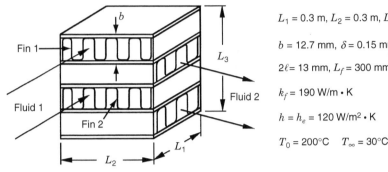

$L_1 = 0.3$ m, $L_2 = 0.3$ m, $L_3 = 0.6$ m

$b = 12.7$ mm, $\delta = 0.15$ mm

$2\ell = 13$ mm, $L_f = 300$ mm

$k_f = 190$ W/m $\cdot$ K

$h = h_e = 120$ W/m$^2 \cdot$ K

$T_0 = 200°$C $T_\infty = 30°$C

FIGURE E4.4

$$m_1\ell_1 = 91.79\,\text{m}^{-1} \times 0.013\,\text{m} = 1.1933$$

$$\tanh m_1\ell_1 = 0.8316$$

$$B = \frac{h_e}{mk_f} = \frac{120\,\text{W/m}^2 \cdot \text{K}}{91.79\,\text{m}^{-1} \times 190\,\text{W/m} \cdot \text{K}} = 0.0069$$

The fin efficiency from Eq. (4.135) is

$$\eta_f = \frac{\tanh m_1\ell_1 + B}{(B + m_1\ell_1)(1 + B\tanh m_1\ell_1)} = \frac{0.8316 + 0.0069}{(0.0069 + 1.1933)\,(1 + 0.0069 \times 0.8316)} = 0.695$$

Ans.

Since $B \ll 0.1$, η_f from Eq. (4.134) is

$$\eta_f = \frac{\tanh m_1\ell_1}{m_1\ell_1} = \frac{0.8316}{1.1933} = 0.697 \qquad\qquad Ans.$$

which is within 0.3% of η_f calculated by Eq. (4.135). The heat transfer rate could be calculated by Eq. (4.131), per unit of fin surface area A_f as

$$q_0'' = \eta_f q_{\max}'' = \eta_f \left(h + h_e \frac{A_k}{A_f} \right)(T_0 - T_\infty)$$

Since A_k/A_f ($\approx \delta/2\ell$) $\ll 1$, we neglect the last term. Therefore, the heat transfer rate is

$$q_0'' = 0.697 \times 120\,\text{W/m}^2 \cdot \text{K} \times (200 - 30)°\text{C} = 14{,}219\,\text{W/m}^2$$

Discussion and Comments: Two points of interest from this example may be observed: (1) decreasing the fin conduction length from 13 mm to 6.5 mm increased the fin efficiency from 0.697 to 0.896 but not by a factor of 2; and (2) the heat transfer rate per unit fin area increased from 14,219 to 18,278 W/m^2. This is because the addition of the fin surface area from the conduction length of 6.5 mm to 13 mm (for only one heat source) is not quite as effective as that from $\ell = 0$ to 6.5 mm (having a heat source on both sides in Fig. 4.15a).

We can consider this problem in a different way: The results obtained in this example are applicable to fins with a heat source on both sides and having fin lengths $2\ell = 13$ mm and 26 mm. In this comparison, the heat transfer rate ratio of the two fins is

$$\frac{q_{\ell=6.5\,\text{mm}}}{q_{\ell=13\,\text{mm}}} = \frac{\left(q_0'' \times 2\ell \times 1\right)_{\ell=6.5\,\text{mm}}}{\left(q_0'' \times 2\ell \times 1\right)_{\ell=13\,\text{mm}}} = \left(\frac{18{,}278 \text{ W/m}^2}{14{,}219 \text{ W/m}^2}\right)\left(\frac{6.5 \text{ mm}}{13 \text{ mm}}\right) = 0.643$$

Thus reducing the fin height by 50% (i.e., reducing the fin material mass by 50%) reduces the fin heat transfer rate by 35.7%. Even increasing the fin density (fins per meter or inch) by 35.7% and assuming no change in the heat transfer coefficient, there will be 14.3% material savings. However, more important is the fact that the shorter fin has much higher column structural strength and that increasing the fin density will increase the column strength. Hence, a shorter fin is more desirable from both the heat transfer and mechanical strength points of view. In reality, reduction in the heat transfer coefficient due to shorter and denser fins should be taken into account as well as a potential fouling problem before making a final selection on the fin height.

Influence of Violation of Basic Assumptions for the Fin Analysis. The basic assumptions made for heat transfer analysis of fins are presented in Section 4.3.1.1. The influence of a violation of assumptions 5 to 8 is found as follows for the thin straight fin of a rectangular profile (Huang and Shah, 1992).

- A 10% linear variation in fin thermal conductivity will increase or decrease η_f by up to 1.7%, depending on whether the fin is being cooled or heated, respectively, when $\eta_f > 80\%$. For a composite fin, the low thermal conductivity layer plays a dominant role in the heat transfer mechanism; a reduction of heat flow by the insulating layer due to frost can be significant, about 8% at a typical $\eta_f = 96\%$.
- The assumption of uniform heat transfer coefficient may lead to gross errors in predicting fin heat transfer. It was found that for a particular linear variation of h, η_f was lower by 6% and 16% for $\eta_f = 90\%$ and 80%. In reality, since h for a given heat transfer surface is determined experimentally based on η_f for constant h, the assumption of constant h for η_f for the design of the exchanger would not introduce a significant error, particularly for high η_f, such as $\eta_f > 80\%$. However, one needs to be aware of the impact of nonuniform h on η_f if the heat exchanger test conditions and design conditions are significantly different.
- Nonuniform ambient temperature T_∞ has less than a 1% effect on the fin efficiency for $\eta_f > 60\%$, and hence this effect can be neglected.
- Longitudinal heat conduction in the fin in the L_f direction in Fig. 4.9 affects η_f less than 1% for $\eta_f > 10\%$, and hence this effect can be neglected.
- The thermal resistance between the fin and the base can have a significant impact on fin heat transfer, and hence care must be exercised to minimize or eliminate it.

It should be emphasized that for most plate-fin heat exchangers, the fin efficiency is maintained at 80% and higher, and more typically at 90% and higher. One way of thinking of a fin having $\eta_f = 80\%$ means that 80% of that fin material is as good as the primary surface from the convection heat transfer point of view; thus, effectively 20% of the material cost is wasted. A typical automotive radiator has a fin efficiency of over 95% at a 80- to 90-km/h automobile speed. If η_f is maintained high, as pointed out above, most of the assumptions made for the fin heat transfer analysis are quite adequate.

4.3.2.3 Circular Fin and Tube-Fin Surfaces

Thin Circular Fin of Uniform Thickness. The temperature distribution for this fin (see Fig. 4.12) is given by Eq. (4.125) and the heat flow q_0 through the base as Eq. (4.127). The fin surface area and q_{max} for this fin are given by

$$A_f = 2\pi\left(r_e^2 - r_o^2\right) \tag{4.149}$$

$$q_{max} = 2\pi\left(r_e^2 - r_o^2\right)h(T_0 - T_\infty) = 2\pi\left(r_e^2 - r_o^2\right)h\theta_o \tag{4.150}$$

where the fin tip surface area is neglected but can be added if desired. Finally, using the definition of Eq. (4.129), the fin efficiency is given by

$$\eta_f = \frac{q_0}{q_{max}} = \frac{2r_o B_1}{m(r_e^2 - r_o^2)} \tag{4.151}$$

where B_1 is given by Eq. (4.128). Since B_1 involves evaluating six Bessel functions, approximations have been proposed using simpler expressions for hand calculations. Such a formula is given in Table 4.5, which is accurate within 1% for most useful ranges of $m(r_o - r_e)$ and r_e/r_o.

If the radial fin tip is not adiabatic and heat transfer at the fin tip is small, it can be taken into account by the Harper–Brown approximation by replacing r_e with $(r_e + \delta/2)$ in Eq. (4.151).

Although it is not obvious from Eq. (4.151), a review of the formula in Table 4.5 shows that for the thin circular fin,

$$\eta_f = \phi\left[m(r_e - r_o), r^*\right] \tag{4.152}$$

for an adiabatic fin tip. Hence, η_f for the thin circular fin is shown in Fig. 4.14 as a function of $m(r_o - r_e)$ and $r^* = r_e/r_o$. Note that for $r_e/r_o = 1$, the circular fin becomes a straight fin. Also note that η_f for the circular fin is less than η_f for a straight fin.

The fin efficiency for most of the fin geometries shown in Fig. 1.32 can be evaluated by Eq. (4.151) except for studded, serrated, and wire form fins. Those fins having slots and cuts are treated as if they were plain circular fins. Studded and serrated fins are treated as straight fins, and the η_f formula is given in Table 4.5 as a last item. A wire form fin is treated as a pin fin with the η_f given by Eq. (4.146) with $\ell = (d_e - d_o)/2$ and m given by Eq. (4.148).

Flat Fins on an Array of Circular Tubes. The fin efficiency for flat plain fins on inline and staggered tube arrangements may be obtained approximately by an *equivalent annulus method* or by a more accurate *sector method*. Based on the arrangement of the tubes, first the idealized adiabatic planes are located. Such planes for inline and staggered tube arrangements are shown in Fig. 4.16a and b by dashed lines, resulting in a rectangle or a hexagon around the tube. In an equivalent annulus method, the rectangular or hexagonal fin around the tube (see Fig. 4.16a and b) is represented hypothetically as a circular fin (see Fig. 4.12) having the same fin surface area. The fin efficiency of this fin is then computed using Eq. (4.151). In the sector method, the smallest representative segment of the fin of Fig. 4.16a is shown in Fig. 4.16c, which is divided into two parts, OAB and OBC. The part OAB (having the subtended angle θ_0) is then divided into m equal-angle ($\Delta\theta = \theta_0/m$) segments. Similarly, the part OBC (having the subtended angle ϕ_0) is then divided into n equal-angle ($\Delta\phi = \phi_0/n$) segments. The outer radius of each circular sector is determined by equating the area of the sector with the area of the equivalent annular sector (Kundu and Das, 2000). Thus for the inline tube arrangement of Fig. 4.16a and c, it is given by

$$r_{e,i} = \frac{X_t}{2}\left[\frac{\tan(i\,\Delta\theta) - \tan[(i-1)\,\Delta\theta]}{\Delta\theta}\right]^{1/2} \qquad r_{e,j} = \frac{X_\ell}{2}\left[\frac{\tan(j\,\Delta\phi) - \tan[(j-1)\,\Delta\phi]}{\Delta\phi}\right]^{1/2}$$

$$(4.153)$$

The smallest representative segment of the staggered tube arrangement of Fig. 4.16b is shown in Fig. 4.16d, which is divided into two parts, OAD and ODF; and ODF is divided into two equal parts, ODE and OEF, as shown in Fig. 4.16e. The part OAD (having the subtended angle θ_0) is then divided into m equal angle ($\Delta\theta = \theta_0/m$) segments. Similarly, each part (ODE and OEF, each having the subtended angle ϕ_0) is then divided into n equal angle ($\Delta\phi = \phi_0/n$) segments. Here again, $r_{e,i}$ and $r_{e,j}$ of ith and jth segments of OAD and ODE are given by

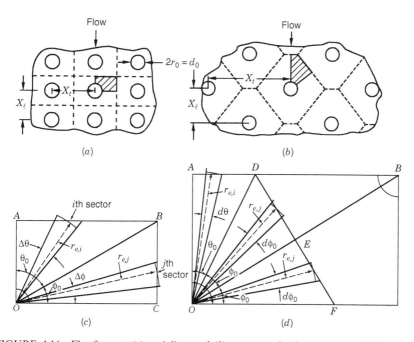

FIGURE 4.16 Flat fin over (a) an inline, and (b) a staggered tube arrangement. The smallest representative shaded segment of the fin for (c) an inline, and (d) a staggered tube arrangement.

$$r_{e,i} = X_\ell \left[\frac{\tan(i\,\Delta\theta) - \tan[(i-1)\,\Delta\theta]}{\Delta\theta} \right]^{1/2}$$

$$r_{e,j} = \frac{\left[X_\ell^2 + (X_t/2)^2 \right]^{1/2}}{2} \left[\frac{\tan(j\,\Delta\phi) - \tan[(j-1)\,\Delta\phi]}{\Delta\phi} \right]^{1/2}$$

$$(4.154)$$

The fin efficiency of each sector is then determined by the circular fin of constant cross section [Eq. (4.151)]. Once η_f for each sector is determined, η_f for the entire fin is the surface-area weighted average of η_f's for each sector.

$$\eta_f = \frac{\sum_{i=1}^m \eta_{f,i}\, A_{f,i} + a \sum_{j=1}^n \eta_{f,j} A_{f,j}}{\sum_{i=1}^m A_{f,i} + a \sum_{j=1}^n A_{f,j}} \tag{4.155}$$

Here $a = 1$ for inline arrangement (Fig. 4.16c) for segment OBC and $n = 2$ for staggered arrangement (Fig. 4.16d) for two equal segments, ODE and OEF. This approximation improves as the number of sectors m, $n \to \infty$. However, in reality, only a few sectors m and n will suffice to provide η_f within the desired accuracy (such as 0.1%). An implicit assumption made in this method is that the heat flow is only in the radial direction and not in the path of least thermal resistance. Hence, η_f calculated by the sector method will be lower than that for the actual flat fin, a conservative value. However, the equivalent annulus method yields an η_f that may be considerably higher than that by the sector method, particularly when the fin geometry around the circular tube departs more and more from a square fin; as a result, the heat transfer computed will be too high.

4.3.3 Fin Effectiveness

Another measure of fin heat transfer performance is the *fin effectiveness*. It is a ratio of actual heat transfer rate through the fin base area to the heat transfer rate through the same fin base area $(A_{k,0})$ if the fin were removed. Thus,

$$\eta_\varepsilon = \frac{q_0}{h A_{k,0} \theta_0} \tag{4.156}$$

This concept is useful for the first three boundary conditions of Fig. 4.10, and particularly for the most common boundary condition of an adiabatic fin tip. In the definition of η_ε, it is idealized that the heat transfer coefficient at the fin base when the fin is removed is identical to the heat transfer coefficient at the fin surface. Using the q_0 expression of Eq. (4.134) for the thin straight fin with an adiabatic fin tip, η_ε of Eq. (4.156) is related to η_f as

$$\eta_\varepsilon = \frac{A_f}{A_{k,0}}\, \eta_f = \frac{2(L_f + \delta)\ell}{L_f \delta}\, \eta_f \approx \frac{2\ell}{\delta}\, \eta_f \tag{4.157}$$

where it is idealized that $L_f \gg \delta$ for the term after the approximate equality sign. This equation is also applicable to the fin of an infinite length. A similar relationship can also be developed between η_ε and η_f using q_0 of Eq. (4.131) for a fin with the convective boundary at the fin tip.

The following observations can be made from Eq. (4.157):

- The explicit formula for η_ε can be obtained when the expression for η_f is available. Of course, Eq. (4.157) needs to be modified by substituting q_0 from Eq. (4.131) in Eq. (4.156) for a thin fin with convective boundary at the fin tip.

- η_ε will generally be much greater than unity. For a typical automotive radiator, $\ell \approx 3\,\text{mm}$ and $\delta \approx 0.075\,\text{mm}$. Hence, $\eta_\varepsilon \approx 75$! In a good design, η_ε must exceed a minimum of 2.

- All parameters that increase η_f [see the discussion in Section 4.3.2.1 after Eq. (4.137)] will also increase η_ε except for ℓ and δ. Note that an increase in ℓ will increase $m\ell$ and reduce η_f (see Fig. 4.14). However, the reduction in η_f is much smaller than the increase in ℓ on a percentage basis. Since η_ε is directly proportional to ℓ, the overall η_ε will increase linearly with an increase in ℓ. Similarly, as noted above, an increase in δ increases η_f, but at a much lower rate, while η_ε is inversely proportional to δ from Eq. (4.157). Hence, overall η_ε will increase with a decrease in δ. Thus, η_ε will increase with an increase in ℓ and k_f and a decrease in δ and h.

4.3.4 Extended Surface Efficiency

An extended surface exchanger has fins attached to the primary surface by brazing, soldering, welding, adhesive bonding, mechanical (press) fit, or extrusion, such as the plate-fin exchanger section shown in Fig. 4.17a. In a plate-fin exchanger, fins are spaced sufficiently far apart to allow desired fluid flow rate (that produces pressure drop within allowed limits) and to have minimum fouling potential. The primary surface A_p is then exposed between fins as shown in Fig. 4.17a. In most tube-fin designs, there is also a primary surface exposed to the fluid along with the fins (secondary surface) as shown in Fig. 4.17b. The heat transfer performance of this extended surface is measured by an *extended* (or *overall*) *surface efficiency* η_o, defined as[†]

$$\eta_o = \frac{q_{\text{total}}}{q_{\max}} = \frac{q_p + q_f}{h(A_p + A_f)(T_0 - T_\infty)} \tag{4.158}$$

In this equation the definition of $q_{\max}$ is similar to that defined by Eq. (4.130) for an adiabatic fin tip with an added heat transfer component for the primary surface. We have also redefined A_f and $q_f\ (= q_0)$ as total values for all finned surfaces; we also redefine η_f using the values of A_f and q_f for all finned surfaces instead of a single fin. Also, A_p in Eq. (4.158) represents the total primary surface area associated with the particular fluid side of concern in the exchanger. We have idealized heat transfer coefficients for the primary and fin surface areas as the same in Eq. (4.158). Now the total heat transfer rate q_{total} is given by

$$q_{\text{total}} = q_p + q_f = hA_p(T_0 - T_\infty) + hA_f\eta_f(T_0 - T_\infty) = h(A_p + \eta_f A_f)(T_0 - T_\infty) \tag{4.159}$$

Substituting Eq. (4.159) into Eq. (4.158), and simplifying yields

$$\eta_o = 1 - \frac{A_f}{A}\left(1 - \eta_f\right) \tag{4.160}$$

[†] For the case of finite heat transfer at the fin tip with h_e as the actual or equivalent heat transfer coefficient, replace A_f with $A_f + (h/h_e)A_k$ in Eqs. (4.158), (4.160), (4.162), and (4.163), and A_f with $(A_f + A_k)$ in Eq. (4.158).

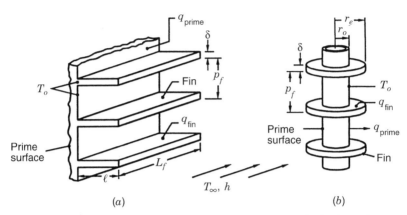

FIGURE 4.17 Extended surface heat exchanger: (a) plate-fin, (b) tube-fin (From Incropera and DeWitt, 2002).

where $A = A_p + A_f$. Note that we considered $\eta_f = 100\%$ for the primary surface in Eq. (4.159). Hence, one can readily see that[†]

$$\eta_o \geq \eta_f \tag{4.161}$$

and

$$q_{\text{total}} = \eta_o h A (T_0 - T_\infty) \tag{4.162}$$

In the derivation of Eq. (4.159), we have neglected the thermal contact resistance that could be present in the exchanger when the fins are attached by the mechanical fit. The thermal circuit for heat transfer in such an extended surface is shown in Fig. 4.18 on page 292. Here h_{cont} is the contact conductance and $A_{k,0}$ is the cross-sectional area of all fins at the fin base. The total thermal resistance for the extended surface from this circuit is given by

$$\frac{1}{\mathbf{R}_t} = hA_p + \frac{1}{1/h_{\text{cont}}A_{k,0} + 1/\eta_f hA_f} = hA_p + \frac{1}{1/h_{\text{cont}}A_{k,0} + \mathbf{R}_f} \tag{4.163}$$

If the thermal contact resistance ($1/h_{\text{cont}}$) is zero,

$$\mathbf{R}_t = \frac{1}{hA_p + \eta_f hA_f} = \frac{1}{\eta_o hA} \tag{4.164}$$

Thus, $\mathbf{R}_t$ represents the combined thermal resistance of the fin and primary surface area. If we want to include the thermal contact resistance, η_o of Eq. (4.164) is not given by

[†] For any finite fin length ℓ, $\eta_o > \eta_f$. When $\ell \to 0$, $\eta_o \to \eta_f$.

Eq. (4.160) but needs to be modified as follows:

$$\eta_{o,\text{cont}} = 1 - \frac{A_f}{A}\left(1 - \frac{\eta_f}{C_1}\right) \tag{4.165}$$

where

$$C_1 = 1 - \frac{\eta_f h A_f}{h_{\text{cont}} A_{k,0}} \tag{4.166}$$

The thermal resistance term of Eq. (4.164) is used for thermal resistances ($\mathbf{R}_h$ and $\mathbf{R}_c$) on the hot- and cold-fluid sides in Eq. (3.23). It must thus be emphasized that the extended surface efficiency η_o appears in the overall resistance equation, such as Eq. (3.24) and *not* η_f. In research articles, η is sometimes used for η_o and is designated as fin efficiency, although it is really the extended surface efficiency.

In some industries, the effective surface area A_{eff} is used instead of the extended surface efficiency, defined as

$$A_{\text{eff}} = A_p + \eta_f A_f = \eta_o A \tag{4.167}$$

so that

$$q = h A_{\text{eff}}(T_o - T_\infty) \tag{4.168}$$

4.4   ADDITIONAL CONSIDERATIONS FOR SHELL-AND-TUBE EXCHANGERS

Any of the basic methods (ε-NTU, P-NTU, MTD, ψ-P and P_1-P_2) can be used for the design of shell-and-tube heat exchangers. However, the construction features of these exchangers may invalidate assumptions 11, 12, and 14 listed in Section 3.2.1. The influence of flow bypassing and leakages, unequal pass area, and a finite number of baffles on the MTD analysis is presented in this section.

4.4.1   Shell Fluid Bypassing and Leakage

Conventional shell-and-tube exchangers have segmental plate baffles. The shell-side flow is very complex in such a heat exchanger, due to a substantial portion of the fluid bypassing the tube bundle through necessary constructional clearances. As a result, the conventional heat transfer correlations and the MTD method applied to a shell-and-tube exchanger generally do not predict the actual performance. To understand the reasons, let us first discuss the shell-side flow phenomenon, and then briefly present two current approaches, the Bell–Delaware method and the stream analysis method, for the determination of exchanger performance.

4.4.1.1   Shell-side Flow Patterns.   Even though one of the major functions of the plate baffle is to induce crossflow (flow normal to the tubes) for higher heat transfer coefficients and hence improved heat transfer performance, this objective is not quite achieved in conventional shell-and-tube heat exchangers. This is because various

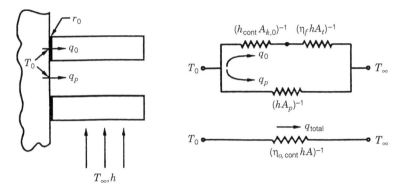

FIGURE 4.18   Thermal circuit for an extended surface with finite thermal contact resistance (From Incropera and DeWitt, 2002).

clearances are required for the construction of the exchanger and the shell fluid leaks or bypasses through these clearances with or without flowing past the tubes (heat transfer surface). Three clearances associated with a plate baffle are tube-to-baffle hole clearance, bundle-to-shell clearance, and baffle-to-shell clearance. In a multipass unit, the tube layout partitions may create open lanes for bypass of the crossflow stream.

The total shell-side flow distributes itself into a number of distinct partial streams due to varying flow resistances, as shown in Fig. 4.19. This flow model was originally proposed by Tinker (1951) and later modified by Palen and Taborek (1969) for a segmental baffle exchanger.

Various streams in order of decreasing influence on thermal effectiveness are as follows:[†]

- *B stream:* crossflow stream flowing over the tubes (and fins, if any) between successive windows. This stream is the "desired" stream and is considered fully effective for both heat transfer and pressure drop.

- *A stream:* tube-to-baffle hole leakage stream through the annular clearance between the tubes and baffle holes of a baffle. This stream is created by the pressure difference on the two sides of the baffle. As heat transfer coefficients are very high in the annular spaces, this stream is considered fully effective.

- *C stream:* bundle-to-shell bypass stream through the annular spaces (clearances) between the tube bundle and shell. This bypass flow area exists because the tube holes cannot be punched close to the tubesheet edge, due to the structural strength requirement. The C stream flows between successive baffle windows. This stream is only partially effective for heat transfer, as it contacts only those tubes near the circumference.

- *E stream:* shell-to-baffle leakage stream through the clearance between the edge of a baffle and the shell. This stream is least effective for heat transfer, particularly in laminar flow, because it may not come in contact with any tube.

[†] Note that there is no D stream since D is used for the shell diameter.

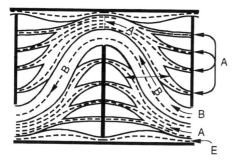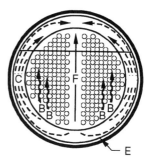

FIGURE 4.19   Shell-side flow distribution and identification of various streams.

- *F stream:* tube-pass partition bypass stream through open passages created by tube layout partitions (when placed in the direction of the main crossflow stream) in a multipass unit. This stream is less effective than the A stream because it comes into contact with less heat transfer area per unit volume; however, it is slightly more effective than the C stream. It is listed last because not all exchangers have two or more passes.

4.4.1.2   Flow Fractions for Each Shell-Side Stream. Each of the streams has a certain flow fraction F_i of the total flow such that the total pressure drop is the same for each stream from the entrance to the exit of the exchanger. Each stream undergoes different acceleration/deceleration and frictional processes and influences heat transfer in different ways.

The design of the plate-baffled shell-and-tube exchanger should be such that most of the flow (ideally, about 80%) represents the crossflow B stream. However, this is rarely achieved in practice. The narrow baffle spacing results in a higher Δp for the B stream and forces more flow into the A, C, and E streams. If the computed values of the B stream are lower than those indicated, the baffle geometry and various clearances should be checked.

Since the A stream is effective from a heat transfer point of view, it is not of great concern if its flow fraction is large, as in the case of a narrow baffle spacing. If the tube-to-baffle hole clearance is plugged due to fouling, the shell-side pressure drop generally increases. The flow fraction of the A stream generally decreases for increasing values of multiple-segmental baffles.

Since C and F streams are only partially effective, the design of the tube bundle should be such that it minimizes the flow fraction for each of these streams to below 10%. Sealing devices are used for this purpose.

The E stream does not contact the heat transfer area and is ineffective from the heat transfer viewpoint. It mixes only poorly with other streams. Since the baffle-to-shell clearances are dictated by TEMA standards, if the flow fraction computed for the E stream is found to be excessive (15% or more), the designer should consider multiple-segmental baffles instead of a single-segmental baffle. This is because the total shell-side pressure drop is lower for the multiple-segmental baffle case, thus forcing more flow to the B, A, and C streams.

Based on extensive test data, Palen and Taborek (1969) arrived at the flow fractions of Table 4.6, on page 296, for various streams. It is surprising to note from this table that the B stream may represent only 10% of the total flow for some exchangers. Even for a good design, it represents only 65% of the total flow in turbulent flow. Hence the performance predicted based on the conventional MTD method will not be accurate in general. As a result, there is no need to compute very accurate values of the MTD correction factor F for various exchanger configurations.

4.4.1.3 The Bell–Delaware Method

In this method the flow fraction for each stream on the shell side is found by knowing the corresponding flow areas and flow resistances. The heat transfer coefficient for ideal crossflow is then modified for the presence of each stream by correction factors. These correction factors are based on experimental data obtained on units of 200 mm diameter TEMA E shell and segmental baffles by Bell (1988) in the 1950s at the University of Delaware. The shell-side heat transfer coefficient h_s is given by

$$h_s = h_{id} J_c J_\ell J_b J_s J_r \qquad (4.169)$$

where h_{id} is the heat transfer coefficient for the pure crossflow stream (B stream) evaluated at a Reynolds number at or near the centerline of the shell. J_c is the correction factor for baffle configuration (baffle cut and spacing) and takes into account the heat transfer in the window. J_ℓ is the correction factor for baffle leakage effects and takes into account both the shell-to-baffle (E stream) and tube-to-baffle hole (A stream) leakages. J_b is the correction factor for bundle and pass partition bypass (C and F) streams and is dependent on the flow bypass area and number of sealing strips. J_s is the correction factor for baffle spacing that is larger at the inlet and outlet sections than in the center. J_r is the correction factor for the adverse temperature gradient in laminar flows (at low Reynolds numbers). These correction factors are discussed further with Eq. (9.50) and their effects are discussed in Section 9.5.1.1. These correction factors in equation form are presented in Table 9.2.

4.4.1.4 The Stream Analysis Method.

The conventional MTD method cannot be applied to a shell-and-tube exchanger to determine the actual heat transfer rate. This is because of the existence of crossflow, leakage, and bypass streams on the shell side; the leakage and bypass streams are idealized as zero in the MTD analysis. Each stream flows through different paths, has different heat transfer rates, and has different temperature profiles, depending on the mixing.

As an illustration, consider the temperature distributions of the tube-side fluid and of various fluid streams on the shell side, shown in Fig. 4.20. The shell-side exit temperature $T_{s,o}$ represents a mixed mean temperature of these streams, depending on the heat transfer effectiveness and heat capacity rate of each stream. Whereas $T_{s,o}$ is used to calculate the log-mean temperature difference, $T_{B,o}$ (the exit temperature of the crossflow B stream) defines the crossflow driving potential for heat transfer.

Since the apparent temperature profile is different from an idealized temperature profile (B stream), this effect must be considered in the determination of the true mean temperature difference. To obtain ΔT_m from ΔT_{lm}, Palen and Taborek (1969) proposed a correction factor δ_s, the delta factor, in addition to the F factor, as follows:

$$\Delta T_m = \delta_s F \, \Delta T_{lm} \qquad (4.170)$$

TABLE 4.6   Flow Fractions (%) for Various Shell-Side Flow Streams

Flow Stream	Turbulent Flow	Laminar Flow
Crossflow stream B	30–65	10–50
Tube-to-baffle leakage stream A	9–23	0–10
Bundle-to-shell bypass stream C	15–33	30–80
Baffle-to-shell leakage stream E	6–21	6–48

Source: Data from Palen and Taborek (1969).

so that

$$q = UA \, \Delta T_m = UA\delta_s F \, \Delta T_{lm} \qquad (4.171)$$

Theoretically, it is possible to derive an expression for δ_s in terms of flow distribution and idealized mixing efficiencies for each stream if these factors are all known. Palen and Taborek derived δ_s empirically as

$$\delta_s = \phi\left(F_E, \frac{T_{t,i} - T_{s,o}}{T_{s,o} - T_{s,i}}, \mathrm{Re}\right) \qquad (4.172)$$

where F_E is the baffle-to-shell leakage stream flow fraction, Re is the Reynolds number for crossflow, and the subscripts t and s for temperatures are for tube and shell fluids, respectively. They found that F_E is the only important flow fraction affecting δ_s in most heat exchangers; all other streams mix with the crossflow stream in the exchanger. The baffle-to-shell leakage stream also becomes well mixed at high Reynolds numbers, and δ_s approaches unity as Re becomes large. The qualitative behavior of δ_s at low Reynolds numbers is shown in Fig. 4.21; δ_s decreases with decreasing $(T_{t,i} - T_{s,o})/(T_{s,o} - T_{s,i})$ and increasing F_E. Palen and Taborek found δ_s varying from 0.4 to 1.0 for their units. Thus, δ_s can be a large correction compared to F, and hence the minimum possible baffle-to-shell clearances are recommended.

In the stream analysis method, first flow fractions for each stream are calculated considering the total pressure drop for each stream the same. Next, the heat transfer

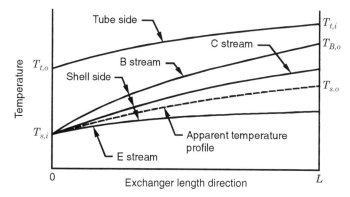

FIGURE 4.20   Temperature profiles of shell-side streams in a single-pass TEMA E exchanger (From Palen and Taborek, 1969).

effectiveness is assigned to each stream and appropriate correction factors are developed. Finally, an equation of the type Eq. (4.171) is used for an overall rate equation for rating or sizing problems.

4.4.2 Unequal Heat Transfer Area in Individual Exchanger Passes

In a multipass exchanger, it may be preferable to have different heat transfer surface areas in different passes to optimize exchanger performance. For example, if one pass has two fluids in counterflow and the second pass has two fluids in parallelflow, the overall exchanger performance for a specified total surface area will be higher if the parallelflow pass has a minimum amount of surface area.

The 1–2, 1–3, and 1–n ($n \geq 4$ and even) TEMA E exchangers for an unequal heat transfer area in counterflow and parallelflow passes have been analyzed with the shell inlet at either the stationary head or floating head by Roetzel and Spang (1989). For a 1–2 TEMA E exchanger, they obtained the following expression for tube-side P_t, NTU_t, and R_t:

$$\frac{1}{P_t} = \nu + R_t + \frac{1}{\text{NTU}_t} \frac{m_1 e^{m_1} - m_2 e^{m_2}}{e^{m_1} - e^{m_2}} \tag{4.173}$$

where

$$m_1, m_2 = \frac{\text{NTU}_t}{2} \{\pm[(R_t + 2\nu - 1)^2 + 4\nu(1 - \nu)]^{1/2} - (R_t + 2\nu - 1)\} \tag{4.174}$$

$$\nu = \frac{\text{NTU}_{pf}}{\text{NTU}_t} \qquad R_t = \frac{C_t}{C_s} \tag{4.175}$$

Here NTU_{pf} represents the NTU on the tube side of the parallelflow pass and NTU_t ($= \text{NTU}_{pf} + \text{NTU}_{cf}$) is the total NTU of the exchanger on the tube side.

Equation (4.173) represents an excellent approximation for a 1–n (n even) exchanger for $\text{NTU}_t \leq 2$, with ν being not close to zero. If ν is close to zero, the appropriate formulas are given by Roetzel and Spang (1989). Refer to Roetzel and Spang (1989) for formulas for unequal passes for 1–3 and 1–n (n even) exchangers. The following are the general observations that may be made from the results above.

- As expected, F factors are higher for $\mathbf{K} > 1.0$ than for the $\mathbf{K} = 1$ (balanced pass) case for given P and R, where $\mathbf{K} = (UA)_{cf}/(UA)_{pf} = (1 - \nu)/\nu$ and the subscripts cf and pf denote counterflow and parallelflow passes, respectively.
- As $\mathbf{K}$ increases, P increases for specified F (or NTU) and R.
- The F factors for the 1–2 exchanger are higher than those for the 1–4 exchanger for specified values of P, R, and $\mathbf{K}$.
- As the number of passes is increased, the F factors (or P) continue to approach that of a crossflow exchanger with both fluids mixed, and the advantage of unbalanced passes over balanced passes becomes negligible.
- Although not evaluated specifically, the unbalanced UA (i.e., $\mathbf{K} > 1$) exchanger will have a higher total tube-side pressure drop and lower tube-side h than those for the balanced UA (i.e., $\mathbf{K} = 1$) exchanger.

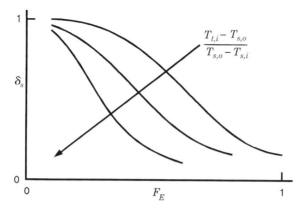

FIGURE 4.21 Temperature distortion correction factor δ_s as a function of the baffle-to-shell stream flow fraction and the temperature ratio indicated (From Palen and Taborek, 1969).

Since the analysis was based on the value of $\mathbf{K} = U_{cf} A_{cf} / U_{pf} A_{pf}$, it means that not only the influence of unequal tube pass area can be taken into account, but also the unequal tube-side overall heat transfer coefficients. Similarly, it should be emphasized that, if properly interpreted, the results for nonuniform UA presented in Section 4.2 can also apply to unequal surface areas in different passes. As noted above, higher exchanger performance can be achieved with higher values of $\mathbf{K} = U_{cf} / U_{pf}$ for equal pass areas. Hence, the shell inlet nozzle should be located at the stationary head when heating the tube fluid and at the floating head when cooling the tube fluid. This is because higher temperatures mean higher heat transfer coefficients. It should be emphasized that U_{cf} and U_{pf} represent mean values of U across the counterflow and parallelflow tube passes and not at the inlet and outlet ends.

4.4.3 Finite Number of Baffles

Assumption 12 in Section 3.2.1 indicates that the number of baffles used is very large and can be assumed to approach infinity. Under this assumption, the temperature change within each baffle compartment is very small in comparison with the total temperature change of the shell fluid through the heat exchanger. Thus the shell fluid can be considered as uniform (perfectly mixed) at every cross section (in a direction normal to the shell axis). It is with this model that the mean temperature difference correction factor for the exchanger effectiveness is normally derived for single-phase exchangers. In reality, a finite number of baffles are used and the condition stated above can be achieved only partially. Shah and Pignotti (1997) have made a comprehensive review and obtained new results as appropriate; they arrived at the following specific number of baffles beyond which the influence of the finite number of baffles on the exchanger effectiveness is not significantly larger than 2%.

- $N_b \geq 10$ for 1–1 TEMA E counterflow exchanger
- $N_b \geq 6$ for 1–2 TEMA E exchanger for $NTU_s \leq 2$, $R_s \leq 5$
- $N_b \geq 9$ for 1–2 TEMA J exchanger for $NTU_s \leq 2$, $R_s \leq 5$
- $N_b \geq 5$ for 1–2 TEMA G exchanger for $NTU_s \leq 3$ for all R_s

- $N_b \geq 11$ for 1–2 TEMA H exchanger for $NTU_s \leq 3$ for all R_s

For 1–n TEMA E exchangers, exchanger effectiveness will depend on the combination of the number of baffles and tube passes, as discussed by Shah and Pignotti (1997).

SUMMARY

Many assumptions are made to simplify the complex heat exchanger design problem to obtain the solution in terms of ε-NTU, P-NTU, and MTD parameters. Sufficient information is provided in this chapter for relaxing these assumptions. These include approaches to design that include longitudinal wall heat conduction, variable local overall heat transfer coefficient, and specific effects (the influence of flow bypassing and leakages, unequal pass areas, and a finite number of baffles) in a shell-and-tube heat exchanger. For an extended heat transfer surface exchanger, careful determination of the fin efficiencies of the extended surfaces must be accomplished. Sufficient details are provided in this chapter for the most important extended surfaces used in heat exchangers by industry. A thorough understanding of concepts and results presented in this chapter and Chapter 3 will provide a strong foundation for rating, sizing, and analysis of industrial heat exchangers.

REFERENCES

Bell, K. J., 1988, Delaware method for shell-side design, in *Heat Transfer Equipment Design*, R. K. Shah, E. C. Subbarao, and R. A. Mashelkar, eds., Hemisphere Publishing, Washington, DC, pp. 145–166.

Colburn, A. P., 1933, Mean temperature difference and heat transfer coefficient in liquid heat exchangers, *Ind. Eng. Chem.*, Vol. 25, pp. 873–877.

Huang, L. J., and R. K. Shah, 1992, Assessment of calculation methods for efficiency of straight fins of rectangular profile, *Int. J. Heat Fluid Flow*, Vol. 13, pp. 282–293.

Incropera, F. P., and D. P. DeWitt, 1996, *Fundamentals of Heat and Mass Transfer*, 4th ed., Wiley, New York.

Kays, W. M., and A. L. London, 1998, *Compact Heat Exchangers*, reprint 3rd ed., Krieger Publishing, Malabar, FL.

Kraus, A. D., A. Aziz, and J. R. Welty, 2001, *Extended Surface Heat Transfer*, Wiley, New York.

Kroeger, P. G., 1967, Performance deterioration in high effectiveness heat exchangers due to axial heat conduction effects, *Adv. Cryogen. Eng.*, Vol. 12, pp. 363–372.

Kundu, B., and P. K. Das, 2000, Performance of symmetric polygonal fins with and without tip loss—a comparison of different methods of prediction, *The Canadian J. Chem. Eng.*, Vol. 78, pp. 395–401.

Palen, J. W., and J. Taborek, 1969, Solution of shell side flow, pressure drop and heat transfer by stream analysis method, *Chem. Eng. Prog. Symp. Ser. 92*, Vol. 65, pp. 53–63.

Prasad, B. S. V., 1996, Fin efficiency and mechanisms of heat exchanger through fins in multi-stream plate-fin heat exchangers: formulation, *Int. J. Heat Mass Transfer*, Vol. 39, pp. 419–428.

Roetzel, W., 1974, Heat exchanger design with variable transfer coefficients for crossflow and mixed arrangements, *Int. J. Heat Mass Transfer*, Vol. 17, 1037–1049.

Roetzel, W., and B. Spang, 1989, Thermal calculation of multipass shell and tube heat exchangers, *Chem. Eng., Res. Des.*, Vol. 67, pp. 115–120.

Roetzel, W., and B. Spang, 1993, Design of heat exchangers, Sect. Cb, Heat transfer, *VDI Heat Atlas*, VDI-Verlag, Dusseldorf, Germany.

Shah, R. K., and A. L. London, 1978, *Laminar Flow Forced Convection in Ducts*, Supplement 1 to *Advances in Heat Transfer*, Academic Press, New York.

Shah, R. K., 1981, Compact heat exchanger design procedures, in *Heat Exchangers: Thermal-Hydraulic Fundamentals and Design*, S. Kakaç. A. E. Bergles and F. Mayinger, eds., Hemisphere Publishing, Washington, DC, pp. 495–536.

Shah, R. K., 1983, Heat exchanger basic design methods, in *Low Reynolds Number Flow Heat Exchangers*, S. Kakaç, R. K. Shah and A. E. Bergles, eds., Hemisphere Publishing, Washington, DC., pp. 21–72.

Shah, R. K., and A. C. Mueller, 1985, Heat exchanger basic thermal design methods, in *Handbook of Heat Transfer Applications*, 2nd ed., W. M. Rohsenow, J. P. Hartnett, and E. N. Ganić, eds., McGraw-Hill, New York, pp. 4–1 to 4–77.

Shah, R. K., and A. Pignotti, 1997, The influence of a finite number of baffles on the shell-and-tube heat exchanger performance, *Heat Transfer Eng.*, Vol. 18, No. 1, pp. 82–94.

Shah, R. K., and D. P. Sekulić, 1998, Nonuniform heat transfer coefficients in conventional heat exchanger design theory, *ASME J. Heat Transfer*, Vol. 119, pp. 520–525.

Tinker, T., 1951, Shell side characteristics of shell and tube heat exchangers, *General Discussion on Heat Transfer*, Proc. Institution of Mechanical Engineers, London, UK, pp. 89–116.

REVIEW QUESTIONS

Where multiple choices are given, circle one or more correct answers. Explain your answers briefly.

4.1 The true mean temperature difference for nonuniform U is defined as:

(a) $\Delta T_m = \dfrac{1}{A}\displaystyle\int_A \Delta T\, dA$ (b) $\Delta T_m = \dfrac{1}{q}\displaystyle\int_q \Delta T\, dq$ (c) $\dfrac{1}{\Delta T_m} = \dfrac{1}{q}\displaystyle\int_q \dfrac{dq}{\Delta T}$

4.2 The mean heat transfer coefficient $\bar{\bar{U}}$ in a counterflow exchanger is exactly defined as:

(a) $\bar{\bar{U}} = \dfrac{h_i + h_o}{2}$ (b) $\bar{\bar{U}} = \dfrac{1}{A}\displaystyle\int_A U\, dA$ (c) $\dfrac{1}{\bar{\bar{U}}} = \dfrac{1}{A}\displaystyle\int_A \dfrac{1}{U}\, dA$

4.3 Longitudinal wall heat conduction effect is more likely important for a:

(a) shell-and-tube exchanger (b) condenser with $C^* \approx 0$

(c) regenerator with $C^* \approx 1$ (d) parallelflow exchanger

4.4 A loss in counterflow exchanger effectiveness (and not the absolute value of ε) due to longitudinal heat conduction increases with:

(a) decreasing value of λ (b) increasing value of C^*

(c) increasing value of NTU

4.5 The analysis for the fins presented in the text is valid for:

(a) a fin with a large square cross section

(b) a fin with a small circular cross section

(c) variable heat transfer coefficients

(d) finite longitudinal heat conduction in the fin

4.6   The following is the differential energy equation applicable to thin fins:

$$\frac{d^2 T}{dx^2} + \frac{d(\ln A_k)}{dx} \frac{dT}{dx} - m^2 (T - T_\infty) = 0$$

It is valid for thin fins of the following cross sections:

(a) triangular straight fins   **(b)** concave parabolic fins

(c) rectangular constant   **(d)** circular constant cross-sectional fin
cross-sectional fins

4.7   The fin efficiency depends on:

(a) fin geometry   **(b)** heat flux level   **(c)** fin base temperature

(d) fin material   **(e)** ambient temperature   **(f)** heat transfer coefficient

(g) fin tip boundary conditions

4.8   The fin efficiency $\eta_f = \tanh(m\ell)/m\ell$ is valid for a straight thin fin of constant cross section and:

(a) heat flux specified at the base and adiabatic fin tip

(b) temperature specified at the base and fin tip heat transfer allowed

(c) temperature specified at the base and adiabatic fin tip

(d) none of these

4.9   For the specified k_f, h, h_e, A_k, $\mathbf{P}$, and ℓ, arrange the following three thin fins in decreasing order of the fin efficiency:

(a) long fin, $\ell \to \infty$ **(b)** fin with an adiabatic end (tip)

(c) fin with finite heat transfer at tip allowed

4.10   For a plate-fin heat exchanger, the total surface efficiency of the finned surface side is given by $\tanh(m\ell)/m\ell$.

(a) true   **(b)** false   **(c)** It depends on the fin geometry.

(d) It depends on the boundary conditions.

4.11   The fin efficiency increases with:

(a) increasing fin height   **(b)** increasing heat transfer coefficient

(c) increasing thermal conductivity of the fin material

(d) increasing fin thickness

4.12   Circle the following statements as true or false.

(a) T   F   Overall extended surface efficiency is always higher than the fin efficiency in a plate-fin exchanger.

(b) T   F   The fin efficiency of low-finned tubes is higher than that of the high-finned tubes if the only difference is the fin height.

(c) T   F   The fin efficiency is determined from the parameter $m\ell$, where $m = (2h/k\delta)^{1/2}$. In this equation, k is the thermal conductivity of the fluid that flows over the finned surface.

(d) T   F   The fin efficiency of a specified fin is higher for water flows compared to air flows at the same velocity.

(e) T   F   For a fin of given geometry and thickness, stainless steel will yield a higher fin efficiency than copper will.

(f) T   F   Doubling the fin thickness doubles the value of the fin efficiency.

(g) T   F   Even though fouling may add a large thermal resistance to primary and secondary surfaces, it has no primary effect on the fin efficiency η_f.

4.13   It is proposed to add pin fins to the outside surfaces of tubes in a tube bank over which air flows. Water flows inside the tubes fast enough to be in turbulent flow, so that $h_{water} \gg h_{air}$.

(a) Sketch the variation of fin efficiency η_f as a function of fin length ℓ for a constant value of m (Fig. RQ4.13a).

(b) For fixed water and air temperatures, sketch the total heat transfer rate as a function of the fin length (Fig. RQ4.13b).

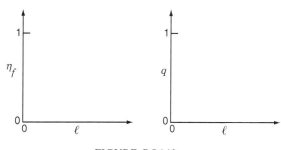

FIGURE RQ4.13

4.14   Plate-fin exchanger A has plain fins whereas exchanger B has offset strip fins. For equal total surface area, frontal area, free-flow area, and flow rates, which has a higher fin efficiency?

(a) A   **(b)** B

4.15   The thermal conductivity of plastics is about 1000 times lower than that for the aluminum. If the fin efficiency is 95% for an aluminum fin of 6 mm height, the height required for a similar (same thickness and cross section) plastic fin for 95% fin efficiency would be:

(a) 6 mm   **(b)** 0.006 mm   **(c)** 0.2 mm   **(d)** 0.5 mm   **(e)** can't tell

Hint: Do not calculate the fin efficiency. Assume the heat transfer coefficient as constant and same.

4.16   In a shell-and-tube exchanger, the following streams are substantially less effective for heat transfer from the hot fluid to the cold fluid:

(a) main crossflow stream   **(b)** baffle-to-shell leakage stream

(c) tube-to-baffle leakage stream   **(d)** all of these   **(e)** none of these

4.17   Which of the following dimensionless parameters have values ranging from 0 to 1 only?

(a) λ **(b)** $(\eta_o hA)^*$ **(c)** κ **(d)** η_f **(e)** η_ε **(f)** η_o

4.18   Which of the following statements are always true?

(a) $\bar{\bar{U}} \geq \tilde{U}$ (b) $\varepsilon_{\lambda=0} \geq \varepsilon_{\lambda \neq 0}$ (c) $\eta_o \geq \eta_f$

(d) $\eta_f \geq \eta_\varepsilon$ (e) $\mathbf{R}_t \geq \mathbf{R}_f$ (f) $h_s \geq h_{id}$

4.19   The Bell–Delaware method for shell-and-tube exchanger design is used to:

(a) determine shell-side heat transfer coefficient and pressure drop

(b) determine tube-side heat transfer coefficient and pressure drop

(c) determine the required number of tube passes

(d) determine the required number of shell passes

4.20   Arrange fluids for case (a) and materials for case (b) below having from the highest to lowest magnitude of the fin efficiency for identical corrugated plain fin geometry in a given heat exchanger.

(a) air, water, viscous oil   (b) stainless steel, copper, aluminum

PROBLEMS

4.1   A counterflow ceramic vehicular gas turbine recuperator operates at the following conditions: Air and gas inlet temperatures are 234 and 730°C, respectively, air and gas flow rates are 0.57 and 0.59 kg/s, respectively, and the product of the overall heat transfer coefficient and heat transfer area is 6650 W/K. The cross-sectional area for longitudinal conduction is 0.03226 m^2 and the core length is 101.6 mm. Calculate the air and gas outlet temperatures with and without longitudinal wall heat conduction. Consider air and gas specific heats as 1.08 and 1.09 kJ/kg · K, respectively and the thermal conductivity of the ceramic material k_w as 3.12 W/m · K. Discuss the results.

4.2   In conventional ε-NTU or F–P theory, heat transfer coefficients on each fluid side are treated as constant and uniform (arithmetic average of inlet and outlet values) throughout the exchanger. If the exchanger is short and has a considerable thermal entrance length effect, the heat transfer coefficient may vary considerably along the flow length. The purpose of this example is to investigate the influence of the thermal entrance length effect on the overall heat transfer coefficient. Consider a double-pipe counterflow heat exchanger with thin walls, negligible wall resistance, and without fins on either fluid side. On the tube side, the heat transfer coefficients on the inlet and outlet sides are 100 and 30 W/m^2 · K, respectively. Similarly, on the annulus side, the heat transfer coefficients on the outlet and inlet locations are 30 and 100 W/m^2 · K, respectively.

(a) Calculate the mean heat transfer coefficient on each fluid side as normally computed as an arithmetic mean. Subsequently, determine the overall heat transfer coefficient (lumped mean value).

(b) Compute the local values of the overall heat transfer coefficient individually at each end (inlet and outlet) of the exchanger. Now consider that the local overall heat transfer coefficient varies linearly from one end of the exchanger to the other end, and determine the integrated average value of the overall heat transfer coefficient.

(c) Based on the results of the foregoing two values of U (lumped mean value and integrated mean value), determine the percentage increase or decrease in the integrated mean value of U compared to the lumped mean value of U.

(d) Now discuss the results. Will the overall heat transfer coefficient increase, decrease, or remain the same when you take into account the influence of thermal entry length effect on U? Explain the physical reason on whatever you found out.

4.3   Heat is transferred from water to air through a $0.305\,\text{m} \times 0.305\,\text{m}$ brass wall (thermal conductivity $77.9\ \text{W/m}\cdot\text{K}$). The addition of straight fins of rectangular cross section is contemplated. These fins are $0.76\,\text{mm}$ thick, $25.4\,\text{mm}$ long and spaced $12.7\ \text{mm}$ apart. Consider the water- and air-side heat transfer coefficients as 170 and $17.0\ \text{W/m}^2\cdot\text{K}$, respectively.

(a) Determine η_f and η_o if the fins are added only on (i) the air side and (ii) the water side.

(b) Calculate the gain in heat transfer rate achieved by adding fins to (i) the air side, (ii) the water side, and (iii) both sides. If you try to make the thermal resistances more balanced, on which side would you add fins? Why? Consider the temperature drop through the wall negligible and the fin tip as adiabatic for η_f calculations.

4.4   A solid copper rod is used to cool a hot spot inside a box of airborne electronic equipment (Fig. P4.4). The shell of the box is double walled and has Freon boiling at $30°C$ serving as a sink for the thermal energy generated inside the box. The ambient temperature inside the box is $90°C$. Because of good thermal contact at the box wall, the rod temperature is $40°C$ at the rod-to-wall junction. The ultimate purpose of the analysis is to establish q_0 and q_ℓ, heat flows in and out of the rod. Do the following to this end:

(a) Picture how you expect $T(x)$ to vary relative to the other temperatures noted in the sketch and text of this problem.

(b) Derive from basic considerations, using a thermal circuit, an appropriate differential equation for $T(x)$ and formulate the boundary conditions explicitly.

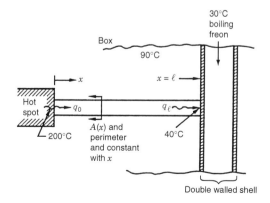

FIGURE P4.4

(c) Write down a general solution for the differential equation and indicate how you would reduce it for this particular problem.

(d) Given the solution of your differential equation and boundary conditions for $T(x)$, how would you evaluate q_0 and q_ℓ?

4.5   In a cryogenics multifluid heat exchanger, offset strip fins are generally used between plates, In neighboring channels, different fluids with different heat transfer coefficients and temperature differences $(T_h - T_c)$ flow. Consider a typical fin of length ℓ as shown in Fig. P4.5. Derive the temperature distribution in this fin as follows.

$$T - T_\infty = \frac{q_0 \cosh[m(\ell - x)] + q_\ell \cosh mx}{k_f A_k \sinh m\ell}$$

Also locate the plane, a value of x, in terms of q_0, q_ℓ, m, and ℓ where the temperature gradient $dT/dx = 0$. Show that $x = \ell/2$ for $q_0 = q_\ell$.

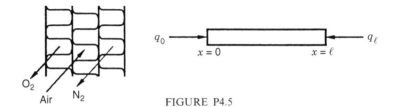

FIGURE P4.5

4.6   As an alternative to the boundary conditions of Problem 4.5, consider the thin fin shown in Fig. P4.6. Derive the temperature distribution for this fin as

$$\frac{T - T_\infty}{T_1 - T_\infty} = e^{-mx} + \left(\frac{T_2 - T_\infty}{T_1 - T_\infty} - e^{-m\ell}\right)\frac{\sinh mx}{\sinh m\ell}$$

Subsequently, derive the heat transfer rate expression at $x = 0$ and ℓ. Also, locate the plane, a value of x, in terms of T_1, T_2, T_∞, m and ℓ where the temperature gradient $dT/dx = 0$. Show that the foregoing temperature distribution reduces to Eq. (4.92) when the value of T_2 is such that $dT/dx = 0$ at $x = \ell$.

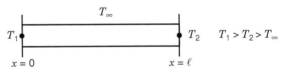

FIGURE P4.6

4.7   Consider the composite thin rectangular fin $(\delta_1 \ll \ell_1, \delta_2 \ll \ell_2, L_f \gg \delta_1, \delta_2)$ shown in Fig. P4.7. Idealize the constant and identical heat transfer coefficients for both fin segments and uniform ambient temperature T_∞. Show the fin efficiency for this fin as

$$\eta_f = \frac{E_1\ell_1 + E_2\ell_2}{\ell_1 + \ell_2}\frac{1}{1 + m_1^2\ell_1\ell_2 E_1 E_2}$$

where

$$E_i = \frac{\tanh m_i \ell_i}{m_i \ell_i} \qquad m_i = \left(\frac{2h}{k_f \delta_i}\right)^{1/2} \qquad i = 1, 2$$

Perform the analysis from the solutions presented in the text for thin fin with **(a)** an adiabatic tip, and **(b)** finite heat transfer at the fin tip without solving any differential equations. Make an appropriate energy balance at $x = \ell_1$. Mention explicitly any additional assumptions that you may make.

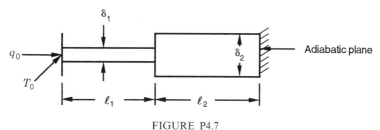

FIGURE P4.7

4.8   A heat exchanger design is generally considered good if $\eta_o hA$ on hot and cold fluid sides are about the same. Because of very low values of heat transfer coefficients with gas flows compared to those for liquid flows, a considerable amount of surface area is needed on the gas side. It can be achieved by increasing either the fin density or the fin height, or both. High fin height may be structurally very weak. An alternative way is to put two layers of fins in between liquid tubes as shown in the Fig. P4.8a. Figure P4.8b represents a general unit fin section. Show that the fin efficiency for this fin is

$$\eta_f = \frac{2E_1 \ell_1 + E_2 \ell_2 + E_3 \ell_3}{2\ell_1 + \ell_2 + \ell_3} \frac{1}{1 + (m_1^2/2)E_1 \ell_1 (E_2 \ell_2 + E_3 \ell_3)}$$

where

$$E_i = \frac{\tanh m_i \ell_i}{m_i \ell_i} \qquad m_i = \left(\frac{2h}{k_f \delta_i}\right)^{1/2} \qquad i = 1, 2, 3$$

Perform the analysis using the solutions presented in the text for thin fins with **(a)** an adiabatic fin tip, and **(b)** finite heat transfer at the fin tip without solving any

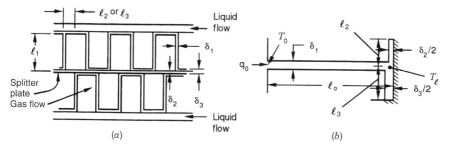

FIGURE P4.8

differential equations. Make an appropriate energy balance at the T_ℓ point. Mention explicitly any additional assumptions that are needed for your analysis.

4.9   Gränges Metallverken Co. of Sweden has developed a compact fin geometry for a crossflow plate-fin exchanger as a car heater. Two layers of air centers (fins), as shown in Fig. P4.8, made from copper (thermal conductivity 380 W/m·K) are sandwiched between the water tubes. The fins and the splitter plate are 0.0254 mm thick. The distance between the water tube and the splitter plate is 3.16 mm ($\ell_1 = 3.175\,\text{mm} - 0.0127\,\text{mm}$). The fin density is 2 fins/mm, so that $\ell_2 = 0.25\,\text{mm}$. Consider the heat transfer coefficient on the air side as 120 W/m²·K. Determine η_f and η_o for this fin geometry using the η_f formula of Problem 4.8. Consider as an approximation only one fin of the full height 6.40 ($3.175 + 3.175 + 2 \times 0.0254$) mm and the same fin density. What is the approximate η_f? How good is this approximation?

4.10   In the conventional fin efficiency analysis, the ambient temperature T_∞ is idealized as constant along the fin length ℓ. However, for the case of laminar flow, the transverse mixing along the fin length ℓ may be negligible after a short distance along the heat exchanger flow length L_f. In such a case, the difference between the fin temperature and the ambient temperature $(T - T_\infty)$ at any x will be constant, independent of x. Consider the straight, thin fin of uniform thickness shown in Fig. P4.10.

(a) Starting with Eq. (4.61), show that the temperature distribution within this fin is given by

$$\frac{T - T_o}{T - T_\infty} = m^2 \left(\frac{x^2}{2} - \ell x \right) \quad \text{with} \quad m^2 = \frac{h\mathbf{P}}{k_f A_k}$$

(b) Derive an expression for the actual heat transfer q_0 through the base.

(c) For the fin efficiency, consider the integrated average temperature $\bar{T}_\infty$ to obtain $q_{\max}$, where

$$\bar{T}_\infty = \frac{1}{\ell} \int_0^\ell T_\infty \, dx$$

Using the results of part (a), show that

$$T_0 - \bar{T}_\infty = (T - T_\infty)\left(1 + \frac{m^2 \ell^2}{3} \right)$$

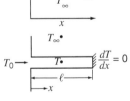

FIGURE P4.10

(d) Define the fin efficiency for this problem and obtain

$$\eta_f = \frac{1}{1 + m^2 \, \ell^2/3}$$

(e) If we would have considered T_∞ as a constant, the fin efficiency would have been

$$\eta_f = \frac{\tanh m\ell}{m\ell}$$

as summarized in Eq. (4.134). For the fin of Example 4.4, $\eta_f = 0.896$. How much error is introduced in this η_f, if we would have considered $(T - T_\infty)$ as constant instead of T_∞ as constant?

4.11 In a round tube and flat fin exchanger, the air on the fin side is heated with water in the tubes. The use of turbulators is contemplated on the tube side to augment the exchanger performance. The following experimental data have been obtained at various air and water flows with and without turbulators.

Test Point	Airflow Rate (m^3/s)	Water Flow Rate (L/s)	$(\eta_o hA)_{air}$ (W/K)	$(\eta_o hA)_{water}$ (W/K) Without Turbulators	$(\eta_o hA)_{water}$ (W/K) With Turbulators
1	0.118	6.3×10^{-5}	395.6	110.8	395.6
2	0.024	6.3×10^{-5}	182.0	110.8	395.6
3	0.118	4.8×10^{-4}	395.6	264.8	817.6
4	0.024	4.8×10^{-4}	182.0	264.8	1345.1

(a) For the same ΔT_m, determine an increase in the heat transfer rate due to the turbulators for each of the test points. Consider the wall thermal resistance as 2.27×10^{-5} K/W.

(b) Discuss at what airflows and water flows the turbulators provide a significant increase in performance, and why.

(c) For what design points would you recommend the use of turbulators?

5 Thermal Design Theory for Regenerators

In this chapter our objective is to present comprehensive thermal design theory for rotary and fixed-matrix regenerators. Definitions, types, operation, and applications of regenerators are described in Sections 1.1 and 1.5.4. Here in Section 5.1, basic heat transfer analysis is presented, including a list of assumptions made in the analysis and development of governing equations. Two methods have been used for the regenerator thermal performance analysis: ε-NTU$_o$ and Λ-Π methods, respectively, for rotary and fixed-matrix regenerators. These are discussed in Sections 5.2 and 5.3. The influence of longitudinal wall heat conduction is important in high-effectiveness regenerators and is discussed in Section 5.4. The influence of transverse conduction may be important in low-thermal-conductivity materials and in thick-walled regenerators and is presented in Section 5.5. The influence of pressure and carryover leakages is important in regenerators, particularly those operating at high effectivenesses. A detailed procedure to take these effects into account is presented in Section 5.6. Finally, the influence of matrix material, size, and arrangement is summarized in Section 5.7.

5.1 HEAT TRANSFER ANALYSIS

For the quasi-steady-state or regular periodic-flow heat transfer analysis of regenerators, the basic assumptions are presented first. Next, the basic variables and parameters are outlined, and then governing equations are derived.

5.1.1 Assumptions for Regenerator Heat Transfer Analysis

The following assumptions are built into the derivation of the governing equations presented in Section 5.1.3 for rotary and fixed-matrix regenerators.

1. The regenerator operates under quasi-steady-state or regular periodic-flow conditions (i.e., having constant mass flow rates and inlet temperatures of both fluids during respective flow periods).
2. Heat loss to or heat gain from the surroundings is negligible (i.e., the regenerator outside walls are adiabatic).
3. There are no thermal energy sources or sinks within the regenerator walls or fluids.

4. No phase change occurs in the regenerator.

5. The velocity and temperature of each fluid at the inlet are uniform over the flow cross section and constant with time.

6. The analysis is based on average and thus constant fluid velocities and the thermophysical properties of both fluids and matrix wall material throughout the regenerator (i.e., independent on time and position).

7. The heat transfer coefficients (h_h and h_c) between the fluids and the matrix wall are *constant* (with position, temperature, and time) throughout the exchanger.

8. Longitudinal heat conduction in the wall and the fluids is negligible.

9. The temperature across the wall thickness is uniform at a cross section and the wall thermal resistance is treated as zero for transverse conduction in the matrix wall (in the wall thickness direction).

10. No flow leakage and flow bypassing of either of the two fluid streams occurs in the regenerator due to their pressure differences. No fluid carryover leakage (of one fluid stream to the other fluid stream) occurs of the fluids trapped in flow passages during the switch from hot to cold fluid period, and vice versa, during matrix rotation or valve switching.

11. The surface area of the matrix as well as the rotor mass is uniformly distributed.

12. The time required to switch the regenerator from the hot to cold gas flow is negligibly small.

13. Heat transfer caused by radiation within the porous matrix is negligible compared with the convective heat transfer.

14. Gas residence (dwell) time in the matrix is negligible relative to the flow period.

The first eight assumptions parallel those generally made in the design theory for recuperators (direct-transfer type exchangers). In the fifth assumption, the fluid velocities and temperatures are considered uniform over the entering cross sections. Generally, a deterioration in heat transfer occurs for nonuniform entering velocity and temperature profiles. In many cases, the temperature is uniform; however, the velocity profile may be nonuniform, due to the header design. The influence of the nonuniform velocity profile at the entrance is considered in Chapter 12.

Saunders and Smoleniec (1951) investigated a part of the sixth assumption. They found the error in the effectiveness less than 1% due to variations in fluid and matrix specific heats. However, if a significant influence of variations in specific heats of the gases and the matrix is anticipated, a numerical solution to the problem is suggested.

Molecular heat conduction in the fluids is generally negligible for the Péclet number $Pe = Re \cdot Pr > 10$, as discussed in Section 7.2.2.5. It is important primarily for the liquid metals having $Pr \leq 0.03$. Since the regenerators are used exclusively for gas-to-gas heat exchanger applications having $Pe > 10$, the assumption of negligible molecular heat conduction in the fluids is very reasonable.

Longitudinal heat conduction in the wall may not be negligible, particularly for metal matrices having continuous flow passages. This effect is considered in Section 5.4. The thermal conductance for transverse heat conduction across intrinsic thick ceramic walls may not be infinity. This effect is considered in Section 5.5.

If there is any pressure difference between the hot and cold fluids in the regenerator, there will be pressure leakage from high- to low-pressure gas across the radial, peripheral, and/or axial seals. This includes flow bypassing from the inlet to the outlet side on each fluid stream. The pressure leakage will depend on the pressure difference between the hot and cold gases. Flow bypassing on each fluid side in the gap between the rotor and the housing will depend on its pressure drop in the matrix. Fluid carryover leakage will occur of the fluids trapped in flow passages during the switch from hot to cold fluid, and vice versa, during matrix rotation or valve switching. This effect will depend on the matrix rotational speed and the void volume of the matrix and headers. Similarly, for a fixed-matrix regenerator, the pressure leakage will occur across the valves and cracks in the housing, and the carryover leakage will occur when switching the fluids. The effects of various pressure and carryover leakages are discussed in Section 5.6.

The influence of matrix material (specific heat c_w and packing density β) and of the size and arrangement of the layers in a multilayer regenerator are discussed in Section 5.7. The governing equations based on the aforementioned assumptions are developed in Section 5.1.3 for the regenerator heat transfer analysis.

The matrix and fluid temperatures depend on x and τ coordinates in a fixed-matrix regenerator. In a rotary regenerator, the fluid temperatures are functions of the axial coordinate x and the angular coordinate θ for a stationary observer looking at the regenerator. Based on the foregoing assumptions, for an observer riding on the matrix, the fluid temperatures in a rotary regenerator are also functions of x and τ. Thus, we consider the fluid temperatures T_h and T_c as functions of x and τ for both types of regenerators.

5.1.2　Definitions and Description of Important Parameters

In this section we define and describe the heat capacity rates, heat capacitances, heat transfer areas, porosity, and volumetric heat capacity for rotary and fixed-matrix regenerators. Let us first define heat capacitance terms ($\bar{C}_h$, $\bar{C}_c$, $\bar{C}_{r,h}$, $\bar{C}_{r,c}$, and $\bar{C}_r$) for the fluids and the matrix, and their relationship to the heat capacity rates (C_h and C_c) and the heat transfer areas A_h and A_c before setting up the energy balance and rate equations. C_h is the hot-fluid *heat capacity rate* and $\bar{C}_h$ is the hot-fluid *heat capacitance* within the regenerator. The same set of entities can be defined for cold fluid as well, i.e., C_c and $\bar{C}_c$. Their definitions and relationships are as follows:

$$C_j = \dot{m}_j c_{p,j} \qquad \bar{C}_j = M_j c_{p,j} = C_j \tau_{d,j} = \left(\frac{CL}{u_m}\right)_j \qquad j = h \text{ or } c \qquad (5.1)$$

Here the subscript $j = h$ for the hot fluid and $j = c$ for the cold fluid; $\dot{m}_j$ is the mass flow rate; M_j is the mass of the j-fluid contained in the regenerator matrix at any instant of time; $c_{p,j}$, $u_{m,j}$, and $\tau_{d,j}$ are the specific heat, mean fluid axial velocity, and fluid dwell time (or residence time), respectively; and L_j is the regenerator matrix length. The regenerator matrix wall heat capacitance $\bar{C}_r$ and the matrix wall heat capacity rate C_r are defined and related as follows:

$$\bar{C}_r = M_w c_w \qquad C_r = \begin{cases} M_w c_w \mathbf{N} = \bar{C}_r \mathbf{N} & \text{rotary regenerator} \\[2mm] \dfrac{M_w c_w}{P_t} = \dfrac{\bar{C}_r}{P_t} & \text{fixed-matrix regenerator} \end{cases} \qquad (5.2)$$

Here M_w is the mass of all matrices (disks), c_w is the specific heat of the matrix material, $\mathbf{N}$ is the rotational speed for a rotary regenerator, and P_t is the total period for a fixed-matrix regenerator. This P_t is the interval of time between the start of two successive heating periods and is the sum of the hot-gas flow period P_h, cold-gas flow period P_c, and reversal period P_r (i.e., the time required to switch from the hot to the cold gas period, and vice versa):

$$P_t = P_h + P_c + P_r \qquad (5.3)$$

Since P_r is generally small compared to P_h or P_c, it is usually neglected. Similarly, for a rotary regenerator, θ_h and θ_c are the disk sector angles through which hot and cold gases flow, and

$$\theta_t = \theta_h + \theta_c + \theta_r = 2\pi \qquad (5.4)$$

with θ_r as the sector angle covered by the radial seals shown Fig. 5.13. The periods and sector angles are related as

$$\frac{P_j}{P_t} = \frac{\theta_j}{\theta_t} \qquad j = h \text{ or } c \qquad (5.5)$$

Now the matrix wall heat capacitances $\bar{C}_{r,j}$, $j = h$ or c, are related to the total matrix heat capacitance as

$$\bar{C}_{r,j} = \bar{C}_r \frac{P_j}{P_t} \quad \text{or} \quad \bar{C}_r \frac{\theta_j}{\theta_t} \qquad j = h \text{ or } c \qquad (5.6)$$

The matrix wall heat capacity rates during the hot and cold periods, using the definition, are

$$C_{r,j} = \frac{\bar{C}_{r,j}}{P_j} = \frac{\bar{C}_r}{P_t} = \bar{C}_r \mathbf{N} \qquad j = h \text{ or } c \qquad (5.7)$$

where the second equality is from Eq. (5.6) and the third equality from Eq. (5.2). Thus,

$$C_{r,j} = C_{r,h} = C_{r,c} = C_r \qquad (5.8)$$

The heat transfer areas A_h and A_c are related to the total heat transfer area A of all matrices of a fixed-matrix regenerator as

$$A_j = \frac{AP_j}{P_t} = \frac{\beta V P_j}{P_t} \qquad j = h \text{ or } c \qquad (5.9)$$

and for a rotary regenerator,

$$A_j = \frac{A\theta_j}{\theta_t} = \frac{\beta V \theta_j}{\theta_t} \qquad j = h \text{ or } c \qquad (5.10)$$

Here β is the heat transfer surface area density or packing density, and V is the total volume of all matrices.

At this time, it may again be pointed out that we have selected the reference coordinate system as (x, τ) for both rotary and fixed-matrix regenerators (see Fig. 5.1). Hence, even for a rotary regenerator, we will use Eq. (5.9) and the pertinent expression in Eqs. (5.2), (5.6), and (5.7) in terms of P_t, P_h, and P_c.

Porosity and Volumetric Heat Capacity of the Matrix. The *core* or *matrix porosity* is a ratio of the void volume to the total core or matrix volume. If the heat transfer surface is made of continuous flow passages (see Fig. 1.43), the porosity is the ratio of the flow area to the frontal area of the core. If the heat transfer surface has interruptions (such as perforations) or is made up of porous materials, the porosity is the ratio of the void volume to the total core volume. Thus, the porosity is defined as follows:

$$\sigma = \begin{cases} \dfrac{A_o}{A_{fr}} & \text{for continuous flow passages} \\[2ex] \dfrac{V_{\text{void}}}{A_{fr}L} & \text{for porous flow passages} \end{cases} \tag{5.11}$$

The porosity σ is related to the flow passage hydraulic diameter as

$$D_h = \frac{4\sigma}{\beta} = \begin{cases} \dfrac{4A_oL}{A} & \text{for continuous flow passages} \\[2ex] \dfrac{4V_{\text{void}}}{A} & \text{for porous flow passages} \end{cases} \tag{5.12}$$

A high-porosity matrix surface is desired because higher porosity means effectively thinner walls, lower wall thermal resistance for transverse conduction to and from the wall and gases, and higher heat transfer performance. High porosity is particularly desired for low-thermal-conductivity materials such as ceramics; for stainless steel and higher-thermal-conductivity materials, the wall thermal resistance is negligibly small. Higher desired porosity means lower solid volume for the matrix, and in order to store the maximum amount of heat within the matrix, the matrix material should have high volumetric heat capacity (i.e., high $\rho_w c_w$).

5.1.3 Governing Equations

On the basis of the foregoing assumptions, let us derive the governing equations and boundary conditions. Consider the counterflow regenerator of Fig. 5.1. A rotary regenerator is shown in Fig. 5.1a. For clarity, only one regenerator elemental flow passage and the associated flow matrix are shown in Fig. 5.1b during the hot-gas flow period and in Fig. 5.1c during the cold-gas flow period. In fact, in the derivation of the governing differential equations, all quantities (surface area, flow area, flow rate, etc.) associated with a complete cross section of the regenerator at x and $x + dx$ are considered. The reference coordinate system considered is (x, τ), so that Figs. 5.1b and c are valid for a rotary regenerator having an observer riding on the matrix. Figure 5.1b and c are also valid for a fixed-matrix regenerator with the observer standing on the stationary matrix. To show clearly that the theoretical analysis is identical for rotary and fixed-matrix regenerators, we consider variables and parameters associated with a complete

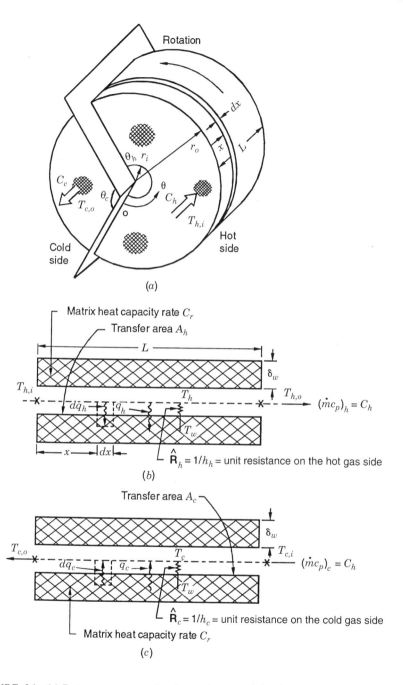

(a)

(b)

(c)

FIGURE 5.1 (a) Rotary regenerator showing sections x and dx; (b) regenerator elemental flow passage and associated matrix during the hot-gas flow period; (c) same as (b) during the cold-gas flow period (From Shah, 1981).

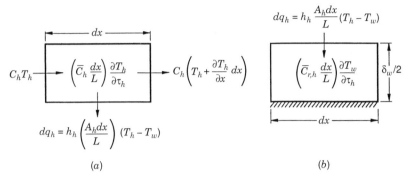

FIGURE 5.2   Energy rate terms associated with the elemental passage dx (a) of fluid, and (b) of matrix at a given instant of time during the hot-gas flow period (From Shah, 1981).

regenerator in this section. That means that we will consider the heat transfer surface area, flow rates, and so on, associated with *all* matrices of a fixed-matrix regenerator.

Hot Period: Fluid. The differential fluid and matrix elements of the hot-gas flow period are shown in Fig. 5.2 with the associated energy transfer terms at a given instant of time. In Fig. 5.2a, during its flow through the elemental passage, the hot gas transfers heat to the wall by convection, resulting in a reduction in its outlet enthalpy and internal thermal energy storage. Applying the energy balance, the first law of thermodynamics, to this elemental passage, we get

$$C_h T_h - C_h \left(T_h + \frac{\partial T_h}{\partial x} dx \right) - h_h \frac{A_h dx}{L} (T_h - T_{w,h}) = \bar{C}_h \frac{dx}{L} \frac{\partial T_h}{\partial \tau_h} \tag{5.13}$$

Substituting the value of $\bar{C}_h$ from Eq. (5.1) into this equation and simplifying, we get

$$\frac{\partial T_h}{\partial \tau_h} + \frac{L}{\tau_{d,h}} \frac{\partial T_h}{\partial x} = \frac{(hA)_h}{C_h \tau_{d,h}} \left(T_{w,h} - T_h \right) \tag{5.14}$$

Hot Period: Matrix. With zero longitudinal and infinite transverse wall heat conduction, the heat transferred from the hot fluid to the matrix wall is stored in the wall in the form of an increase in wall enthalpy. An energy balance on the matrix wall elemental passage is (see Fig. 5.2b)

$$\left(\bar{C}_{r,h} \frac{dx}{L} \right) \frac{\partial T_{w,h}}{\partial \tau_h} = h_h \frac{A_h dx}{L} \left(T_h - T_{w,h} \right) \tag{5.15}^\dagger$$

[†] This equation is accurate for a fixed-matrix wall. For a rotary regenerator, the temperature time derivative ($\partial T_{w,h}/\partial \tau_h$) may be interpreted as a substantial derivative (i.e., a material derivative), $DT_{w,h}/D\tau_h$. This is because a fixed reference frame is used and the wall temperature appears to be a function of both time and angular coordinate, which are not independent variables. Consequently, Eq. (5.15) can be represented accurately as follows for a rotary regenerator matrix:

$$\left(\bar{C}_{r,h} \frac{dx}{L} \right) \frac{DT_{w,h}}{D\tau_h} = h_h \frac{A_h dx}{L} \left(T_h - T_{w,h} \right) \tag{5.16}$$

Because the angular velocity is constant, Eq. (5.16) reverts to Eq. (5.15).

Combining it with Eq. (5.7) and simplifying, we get

$$\frac{\partial T_{w,h}}{\partial \tau_h} = \frac{(hA)_h}{C_{r,h} P_h} (T_h - T_{w,h})$$ (5.17)

Cold Period: Fluid and Matrix. For the cold-gas flow period, a pair of equations similar to Eqs. (5.14) and (5.17) results:

$$-\frac{\partial T_c}{\partial \tau_c} + \frac{L}{\tau_{d,c}} \frac{\partial T_c}{\partial x} = \frac{(hA)_c}{C_c \tau_{d,c}} (T_c - T_{w,c})$$ (5.18)

$$-\frac{\partial T_w}{\partial \tau_c} = \frac{(hA)_c}{C_{r,c} P_c} (T_{w,c} - T_c)$$ (5.19)

The boundary conditions are as follows. The inlet temperature of the hot gas is constant during the hot-gas flow period, and the inlet temperature of the cold gas is constant during the cold-gas flow period:

$$T_h(0, \tau_h) = T_{h,i} = \text{constant} \quad \text{for } 0 \le \tau_h \le P_h$$ (5.20)

$$T_c(L, \tau_c) = T_{c,i} = \text{constant} \quad \text{for } 0 \le \tau_c \le P_c$$ (5.21)

The periodic equilibrium conditions for the wall are

$$T_{w,h}(x, \tau_h = P_h) = T_{w,c}(x, \tau_c = 0) \quad \text{for } 0 \le x \le L$$ (5.22)

$$T_{w,h}(x, \tau_h = 0) = T_{w,c}(x, \tau_c = P_c) \quad \text{for } 0 \le x \le L$$ (5.23)

Since the regenerator is in periodic equilibrium, Eqs. (5.20)–(5.23) are valid for $\tau = \tau + nP_t$, where n is an integer, $n \ge 0$.

The boundary conditions of Eqs. (5.20) and (5.21) are simplest for the analysis. The corresponding analytical models can be solved using analytical and seminumerical methods. In applications, the hot- and cold-fluid inlets to the regenerator may have nonuniform temperature profiles. Then the solution can only be obtained by a numerical analysis.

Based on the foregoing differential equations and boundary conditions [Eqs. (5.14) and (5.17)–(5.23)], the dependent fluid and matrix temperatures are functions of the following variables and parameters:

$$\underbrace{T_h, T_c, T_w}_{\substack{\text{dependent} \\ \text{variables}}} = \phi [\underbrace{x, \tau_h, \tau_c,}_{\substack{\text{independent} \\ \text{variables}}} \underbrace{T_{h,i}, T_{c,i}, C_h, C_c, \tau_{d,h}, \tau_{d,c},}_{\substack{\text{operating condition} \\ \text{variables}}} \underbrace{C_r, (hA)_h, (hA)_c, L, P_h, P_c}_{\text{parameters under designer's control}}]$$ (5.24)

Neither $C_{r,c}$ nor $C_{r,h}$ is included in the foregoing list since $C_{r,c} = C_{r,h} = C_r$ [see Eq. (5.8)].

The regenerator fluid and wall temperatures are thus dependent on 14 independent variables and parameters (since τ is the independent time variable designated with the subscripts h and c for ease of understanding) . Through nondimensionalization we obtain four independent and one dependent dimensionless groups. The specific form of these groups is optional to some extent. Two such sets have been used for regenerator analysis leading to two design methods. The effectiveness–number of transfer units (ε-NTU$_o$) method is generally used for rotary regenerators. The reduced length–reduced period (Λ-Π) method is generally used for fixed-matrix regenerators. It was shown by Shah (1981) that both methods are equivalent. These methods are presented in the following sections, with solutions for counterflow and parallelflow regenerators. Note that for regenerators, there are no counterparts to the other flow arrangements used in recuperators, such as crossflow and multipass cross-counterflow.

5.2 THE ε-NTU$_o$ METHOD

This ε-NTU$_o$ method was developed by Coppage and London (1953). The dimensionless groups in this method are first formulated in Section 5.2.1 such that most of the important groups parallel those of the recuperators discussed in Section 3.3. In Sections 5.2.2 and 5.2.3, the physical significance of the additional dimensionless groups (compared to that for a recuperator) is discussed. In Sections 5.2.4 and 5.2.5, the ε-NTU$_o$ results for counterflow and parallelflow regenerators are presented.

5.2.1 Dimensionless Groups

There are a number of different ways to formulate dimensionless groups. In Section 3.3.1 on recuperators, we made a list of all possible dimensionless groups from the energy balances and rate equations and then eliminated those which were dependent. Now we use a different approach to illustrate an alternative way to formulate dimensionless groups for the regenerator problem. We derive dimensionless groups by making differential equations and boundary conditions nondimensional.

Introduce the following dimensionless independent variables X^* and τ^* as[†]

$$X^* = \frac{x}{L} \tag{5.25}$$

$$\tau_h^* = \frac{\tau_h}{P_h} \qquad \tau_c^* = \frac{\tau_c}{P_c} \tag{5.26}$$

and the following definitions of dependent dimensionless temperatures:

$$T_h^* = \frac{T_h - T_{c,i}}{T_{h,i} - T_{c,i}} \qquad T_c^* = \frac{T_c - T_{c,i}}{T_{h,i} - T_{c,i}} \qquad T_w^* = \frac{T_w - T_{c,i}}{T_{h,i} - T_{c,i}} \tag{5.28}$$

[†] In the case of cryogenics and Stirling engine regenerators, the rotational speed or the valve switching frequency is so high that $\tau_{d,h}$ and $\tau_{d,c}$ may not be so small as to be negligible compared to P_h and P_c. Consideration of this effect is beyond the scope of this chapter. However, in that case, τ_h^* and τ_c^* are defined as

$$\tau_h^* = \frac{1}{P_h}\left(\tau_h - \frac{x}{L}\tau_{d,h}\right) \qquad \tau_c^* = \frac{1}{P_c}\left(\tau_c - \frac{x}{L}\tau_{d,c}\right) \tag{5.27}$$

Define the dimensionless design parameters as

$$\text{ntu}_h = \frac{(hA)_h}{C_h} \qquad \text{ntu}_c = \frac{(hA)_c}{C_c} \tag{5.29}$$

$$C_{r,h}^* = \frac{C_{r,h}}{C_h} \qquad C_{r,c}^* = \frac{C_{r,c}}{C_c} \tag{5.30}$$

With these nondimensional groups, Eqs. (5.14), (5.17), (5.18), and (5.19) reduce as follows:

$$\frac{\partial T_h^*}{\partial X^*} = \text{ntu}_h \left(T_w^* - T_h^* \right) \tag{5.31}$$

$$\frac{\partial T_w^*}{\partial \tau_h^*} = \frac{\text{ntu}_h}{C_{r,h}^*} \left(T_h^* - T_w^* \right) \tag{5.32}$$

$$\frac{\partial T_c^*}{\partial X^*} = \text{ntu}_c \left(T_c^* - T_w^* \right) \tag{5.33}$$

$$\frac{\partial T_w^*}{\partial \tau_c^*} = \frac{\text{ntu}_c}{C_{r,c}^*} \left(T_c^* - T_w^* \right) \tag{5.34}$$

The boundary conditions and periodic equilibrium conditions of Eqs. (5.20)–(5.23) reduce to

$$T_h^* \left(0, \tau_h^* \right) = 1 \qquad \text{for } 0 \le \tau_h^* \le 1 \tag{5.35}$$

$$T_c^* \left(1, \tau_c^* \right) = 0 \qquad \text{for } 0 \le \tau_c^* \le 1 \tag{5.36}$$

$$T_{w,h}^* \left(X^*, \tau_h^* = 1 \right) = T_{w,c}^* \left(X^*, \tau_c^* = 0 \right) \qquad \text{for } 0 \le X^* \le 1 \tag{5.37}$$

$$T_{w,h}^* \left(X^*, \tau_h^* = 0 \right) = T_{w,c}^* \left(X^*, \tau_c^* = 1 \right) \qquad \text{for } 0 \le X^* \le 1 \tag{5.38}$$

It is clear from Eqs. (5.31)–(5.38) that the dependent temperatures are functions of

$$T_h^*, T_c^*, T_w^* = \phi\left(X^*, \tau_h^*, \tau_c^*, \text{ntu}_h, \text{ntu}_c, C_{r,h}^*, C_{r,c}^* \right) \tag{5.39}$$

Thus we are able to reduce independent variables and parameters from 14 to 6 (considering only one period at a time).

For overall regenerator performance, we are interested in determining average fluid outlet temperatures. In a rotary regenerator, the outlet temperatures vary as a function of the angular coordinate θ. If θ_h and θ_c represent the angles for the sectors through which hot and cold gases flow, respectively, the space average outlet temperatures are

$$\bar{T}_{h,o} = \frac{1}{\theta_h} \int_0^{\theta_h} T_{h,o}(\theta)\, d\theta, \qquad \bar{T}_{c,o} = \frac{1}{\theta_c} \int_0^{\theta_c} T_{c,o}(\theta)\, d\theta \tag{5.40}$$

where $T_{h,o}(\theta)$ and $T_{c,o}(\theta)$ represent the angular coordinate-dependent fluid temperatures at the regenerator outlet. However, for an observer riding on the rotary regenerator matrix, the fluid outlet temperatures are functions of the time τ. In this case, the time-averaged outlet temperatures are

$$\bar{T}_{h,o} = \frac{1}{P_h} \int_0^{P_h} T_{h,o}(\tau)\, d\tau, \qquad \bar{T}_{c,o} = \frac{1}{P_c} \int_0^{P_c} T_{c,o}(\tau)\, d\tau \qquad (5.41)$$

Here $T_{h,o}(\tau)$ and $T_{c,o}(\tau)$ are the time-dependent fluid temperatures at the regenerator outlet. $\bar{T}_{h,o}$ and $\bar{T}_{c,o}$ in Eqs. (5.40) and (5.41), respectively, represent the space- and time-averaged fluid temperatures at the regenerator outlet. Equation (5.41) is valid for both the fixed-matrix and rotary regenerators, provided that an observer rides on the matrix for a rotary regenerator.

Thus, the functional relationship of Eq. (5.39) for the dependent regenerator *average* outlet temperatures is

$$T_{h,o}^*, \ T_{c,o}^* = \phi(\mathrm{ntu}_h, \ \mathrm{ntu}_c, \ C_{r,h}^*, \ C_{r,c}^*) \qquad (5.42)$$

A bar on these dimensionless as well as dimensional [of Eq. (5.41)] *average* outlet temperatures is eliminated for convenience (like in a recuperator) in the rest of the chapter, except where such terminology could create ambiguity.

These outlet temperatures are conveniently expressed by the regenerator effectiveness $\varepsilon = q/q_{\max}$. Using the outlet temperatures defined by Eq. (5.41), the actual heat transfer rate in the rotary regenerator is

$$q = C_h(T_{h,i} - \bar{T}_{h,o}) = C_c(\bar{T}_{c,o} - T_{c,i}) \qquad (5.43)$$

In this section we have considered either a rotary regenerator or all matrices of a fixed-matrix regenerator as a system. The hot and cold fluids flow *continuously* in and out of such a system. To determine $q_{\max}$ for such a system, we define a "perfect" heat exchanger as defined for recuperators after Eq. (3.37). This perfect heat exchanger is a counterflow recuperator of infinite surface area, zero longitudinal wall heat conduction, and zero flow leakages from one fluid to another fluid, operating with fluid flow rates and fluid inlet temperatures the same as those of an actual regenerator; fluid properties are considered constant for both exchangers. $q_{\max}$ for this perfect heat exchanger as in Eq. (3.42) is

$$q_{\max} = C_{\min}(T_{h,i} - T_{c,i}) \qquad (5.44)$$

where $C_{\min}$ is the minimum of C_h and C_c. The regenerator effectiveness is thus

$$\varepsilon = \frac{q}{q_{\max}} = \frac{C_h(T_{h,i} - \bar{T}_{h,o})}{C_{\min}(T_{h,i} - T_{c,i})} = \frac{C_c(\bar{T}_{c,o} - T_{c,i})}{C_{\min}(T_{h,i} - T_{c,i})} \qquad (5.45)$$

Then for $C_c = C_{\min}$, comparing ε with $\bar{T}_{h,o}^* = 1 - \varepsilon C^*$ and $\bar{T}_{c,o}^* = \varepsilon$ from the definitions in Eq. (5.28) for the outlet temperatures, we get

$$\varepsilon = \frac{1 - \bar{T}_{h,o}^*}{C^*} = \bar{T}_{c,o}^* \qquad (5.46)$$

and Eq. (5.42) in terms of ε is

$$\varepsilon = \phi(\mathrm{ntu}_h, \ \mathrm{ntu}_c, \ C_{r,h}^*, \ C_{r,c}^*) \qquad (5.47)$$

Since these independent dimensionless groups, defined in Eqs. (5.29) and (5.30), are not parallel to those of a recuperator (a direct-transfer exchanger), let us define a related set as follows:

$$\text{NTU}_o = \frac{1}{C_{\min}} \left[\frac{1}{1/(hA)_h + 1/(hA)_c} \right] \tag{5.48}$$

$$C^* = \frac{C_{\min}}{C_{\max}} \tag{5.49}$$

$$C_r^* = \frac{C_r}{C_{\min}} \tag{5.50}$$

$$(hA)^* = \frac{(hA) \text{ on the } C_{\min} \text{ side}}{(hA) \text{ on the } C_{\max} \text{ side}} \tag{5.51}$$

Since Eqs. (5.48)–(5.51) utilize only four independent dimensionless groups [to be shown related to those of Eq. (5.47)], it is then valid to recast Eq. (5.47) as

$$\varepsilon = \phi[\text{NTU}_o, C^*, \ C_r^*, \ (hA)^*] \tag{5.52}$$

Here NTU$_o$ is the modified number of transfer units. Since there is no direct heat transfer from the hot fluid to the cold fluid in a regenerator (similar to that in a recuperator), UA does not come into the picture directly for the regenerator. However, if the bracketed term of Eq. (5.48) is designated as U_oA, with U_o termed as a *modified* overall heat transfer coefficient, then

$$\frac{1}{U_oA} = \frac{1}{(hA)_h} + \frac{1}{(hA)_c} \quad \text{and} \quad \text{NTU}_o = \frac{U_oA}{C_{\min}} \tag{5.53}$$

A comparison of this expression with Eq. (3.24) or (3.20) reveals that U_oA is the same as UA when the wall thermal resistance and fouling resistances are zero. Note that the entire surface in a regenerator is primary surface (no fins), and hence the overall extended surface efficiency $\eta_o = 1$. Thus the definition of NTU$_o = U_oA/C_{\min}$ parallels that of NTU of Eq. (3.59), and in the limiting case of $C_r^* = \infty$, the numerical solutions demonstrate that the regenerator has the same performance as a recuperator with its NTU$_o$ identical to NTU provided that the pressure and carryover leakage effects are neglected. The newly defined dimensionless groups of Eq. (5.52) are related to those of Eq. (5.47) as follows for $C_c = C_{\min}$:

$$\text{NTU}_o = \frac{1}{(C_{r,h}^*/C_{r,c}^*)/\text{ntu}_h + 1/\text{ntu}_c} \tag{5.54}$$

$$C^* = \frac{C_{r,h}^*}{C_{r,c}^*} \tag{5.55}$$

$$C_r^* = C_{r,c}^* \tag{5.56}$$

$$(hA)^* = \frac{\text{ntu}_c}{\text{ntu}_h} \frac{C_{r,h}^*}{C_{r,c}^*} \tag{5.57}$$

A comparison of Eq. (5.52) with Eq. (3.50) reveals that the effectiveness of a regenerator is dependent on two additional parameters, C_r^* and $(hA)^*$. Since the thermal energy is stored and delivered by the matrix wall periodically, the wall temperature is going to be dependent on (1) the storage heat capacity rate of the matrix wall, and (2) the thermal conductances between the matrix wall and the hot fluid, $(hA)_h$, and between the matrix wall and the cold fluid, $(hA)_c$. And as a result, two additional dimensionless groups, C_r^* and $(hA)^*$, come into the picture; they are discussed further next.

5.2.2   Influence of Core Rotation and Valve Switching Frequency

Heat transfer in the regenerator from the hot gas to the matrix surface is by convection/radiation during the hot gas flow period, depending on the applications, and that from the matrix surface to the cold gas is by convection/radiation during the cold-gas flow period. With all other variables/parameters the same, this heat transfer rate is greatest when the temperature difference between the gas and surface is the greatest. For this reason, the matrix surface is not allowed to be heated to the hot-fluid inlet temperature or cooled to the cold-fluid temperature; otherwise, this would result in zero temperature potential and zero heat transfer near the end of the hot- or cold-gas flow period. The temperature swing at the exit of the regenerator is reduced with large C_r^*, which translates into relatively fast rotational speeds for rotary regenerators or fast valve switching in fixed-matrix regenerators. Hence, from the heat transfer point of view, the dimensionless group C_r^* which takes the rotational speed/valve switching frequency into account, can have a large impact on the design of the regenerator.

When the cold-gas period starts, the cold gas gets heated at the entrance from the heat rejected by the local matrix elements. If the blow period is long (i.e., low values of C_r^*), the local matrix elements at the entrance will be cooled to the cold-fluid inlet temperature, and hence no heat transfer will take place in that region subsequently. A similar phenomenon will take place during the hot-gas period, where the entrance region will be heated to the hot-fluid inlet temperature, and hence no heat transfer will take place in that region subsequently. This effect propagates downstream, depending on the length of the period until the fluid flow switches. This phenomenon is sometimes quantified by an *exhaustion coefficient* which is inversely proportional to C_r^*.

Figure 5.3 clearly demonstrates that the regenerator effectiveness ε increases with C_r^* for given values of NTU_o and C^*. However, higher rotational speeds/valve switching frequency would induce larger carryover leakage and effect higher seal wear and tear (thus increasing the seal leakage), both of which will reduce the regenerator effectiveness. The range of the optimum value of C_r^* is between 2 and 4 for optimum regenerator effectiveness and seal life (Shah, 1988), although many rotary regenerators are designed with C_r^* larger than 4.

5.2.3   Convection Conductance Ratio $(hA)^*$

As mentioned earlier, the wall temperature profile in a regenerator (in the absence of longitudinal wall heat conduction) is going to be dependent on the thermal conductances $(hA)_h$ and $(hA)_c$ between the matrix wall and the hot/cold fluids. For a high-temperature regenerator, the thermal conductance will not only include convection conductance but also radiation conductance. The dimensionless group that takes into account the effect of the convection conductance ratio is $(hA)^*$, as defined by Eq. (5.51). Lambertson (1958) and others have shown through a detailed analysis that $(hA)^*$ has a negligible influence

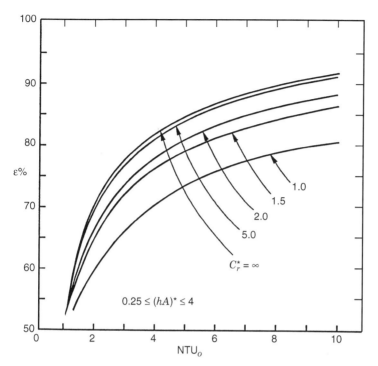

FIGURE 5.3   Counterflow regenerator ε as a function of NTU$_o$ and for $C^* = 1$ (Kays and London, 1998).

on the regenerator effectiveness for the range $0.25 \leq (hA)^* < 4$. Since most regenerators operate in this range of $(hA)^*$, fortunately, the effect of $(hA)^*$ on the regenerator effectiveness can usually be ignored.

5.2.4 ε-NTU$_o$ Results for a Counterflow Regenerator

No closed-form exact solution of the theoretical model [Eqs. (5.31)–(5.38)] is available presently. The solutions mentioned below are obtained to determine the regenerator effectiveness directly, and not the detailed temperature distributions. A numerical solution to Eqs. (5.31)–(5.38) has been obtained by Lamberston (1958) using a finite difference method and a closed-form approximate solution by Bačlić (1985) using a Galerkin method. Lamberston (1958) employed a finite difference method to analyze a rotary regenerator. He presented the solution in terms of ε as a function of the four dimensionless groups of Eq. (5.52). He covered the following range of the parameters: $1 \leq \text{NTU}_o \leq 10$, $0.1 \leq C^* \leq 1.0$, $1 \leq C_r^* \leq \infty$, and $0.25 \leq (hA)^* \leq 1$. His results are presented by Kays and London (1998). A closed-form solution for a *balanced* $(C^* = 1)$ *and symmetric* $[(hA)^* = 1]$ counterflow regenerator has been obtained by Bačlić (1985), valid for all values of $C_r^* < \infty$ as follows:

$$\varepsilon = C_r^* \frac{1 + 7\beta_2 - 24\{B - 2[R_1 - A_1 - 90(N_1 + 2E)]\}}{1 + 9\beta_2 - 24\{B - 6[R - A - 20(N - 3E)]\}} \tag{5.58}$$

where

$$B = 3\beta_3 - 13\beta_4 + 30(\beta_5 - \beta_6)$$

$$R = \beta_2[3\beta_4 - 5(3\beta_5 - 4\beta_6)]$$

$$A = \beta_3[3\beta_3 - 5(3\beta_4 + 4\beta_5 - 12\beta_6)]$$

$$N = \beta_4[2\beta_4 - 3(\beta_5 + \beta_6)] + 3\beta_5^2$$

$$E = \beta_2\beta_4\beta_6 - \beta_2\beta_5^2 - \beta_3^2\beta_6 + 2\beta_3\beta_4\beta_5 - \beta_4^3 \qquad (5.59)$$

$$N_1 = \beta_4[\beta_4 - 2(\beta_5 + \beta_6)] + 2\beta_5^2$$

$$A_1 = \beta_3[\beta_3 - 15(\beta_4 + 4\beta_5 - 12\beta_6)]$$

$$R_1 = \beta_2[\beta_4 - 15(\beta_5 - 2\beta_6)]$$

$$\beta_i = \frac{V_i(2\mathrm{NTU}_o, 2\mathrm{NTU}_o/C_r^*)}{(2\mathrm{NTU}_o)^{i-1}} \qquad i = 2, 3, \dots, 6$$

and

$$V_i(x, y) = \exp[-(x+y)] \sum_{n=i-1}^{\infty} \binom{n}{i-1}\left(\frac{y}{x}\right)^{n/2} I_n\left(2\sqrt{xy}\right) \qquad i = 2, 3, \dots, 6 \qquad (5.60)$$

In these equations, all variables/parameters are local except for NTU_o, C_r^*, and ε. Here I_n represents the modified Bessel function of the first kind and nth (integer) order. The regenerator effectiveness of Eq. (5.58) is presented in Table 5.1 for $0.5 \leq \mathrm{NTU}_o \leq 500$ and $1 \leq C_r^* \leq \infty$, and some typical results are presented in Fig. 5.3. The values for $C_r^* = \infty$ can be calculated from an asymptotic expression $\varepsilon = \mathrm{NTU}_o/(1 + \mathrm{NTU}_o)$.

Now let us discuss further the reasons for choosing the set of dimensionless independent groups of Eq. (5.52) instead of those of Eq. (5.47):

1. For specified NTU_o, C^*, and C_r^*, the effectiveness ε generally decreases with decreasing values of $(hA)^*$, and the reverse occurs for large values of NTU_o and $C^* \approx 1$. However, the influence of $(hA)^*$ on ε is negligibly small for $0.25 \leq (hA)^* \leq 4$, as shown by Lambertson, among others. A maximum error of 0.5 point on % ε occurs for $C^* \geq 0.9$, $C_r^* \geq 1$, and $\mathrm{NTU}_o \leq 9$. A maximum error of 2 points on % ε occurs at $C^* \geq 0.7$, $C_r^* \geq 1$, and $\mathrm{NTU}_o \leq 9$. The maximum error of 5 points on % ε occurs for $C^* = 0.1$, $C_r^* \geq 1$, and $\mathrm{NTU}_o = 3$. Since in most regenerators, $C^* > 0.8$, we can effectively eliminate $(hA)^*$ from Eq. (5.52). Hence,

$$\varepsilon = \phi\,(\mathrm{NTU}_o,\ C^*,\ C_r^*) \qquad (5.61)$$

2. When $C_r^* \to \infty$, the effectiveness ε of a regenerator approaches that of a recuperator. The difference in ε for $C_r^* \geq 5$ and that for $C_r^* = \infty$ is negligibly small and may be ignored for the design purpose.

TABLE 5.1 Counterflow Regenerator ε as a Function of NTU_o and for $C^* = 1.0$ and $(hA)^* = 1$

NTU_o	ε for $C_r^* = :$										
	1.0	1.5	2.0	2.5	3.0	3.5	4.0	4.5	5.0	10.	∞
0.5	0.3321	0.3282	0.3304	0.3315	0.3320	0.3324	0.3326	0.3328	0.3329	0.3332	0.3333
1.0	0.4665	0.4845	0.4912	0.4943	0.4960	0.4971	0.4978	0.4982	0.4986	0.4996	0.5000
1.5	0.5477	0.5757	0.5861	0.5910	0.5937	0.5954	0.5965	0.5972	0.5977	0.5994	0.6000
2.0	0.6006	0.6359	0.6490	0.6553	0.6587	0.6608	0.6622	0.6631	0.6638	0.6659	0.6667
2.5	0.6385	0.6791	0.6940	0.7011	0.7051	0.7075	0.7091	0.7102	0.7109	0.7134	0.7143
3.0	0.6672	0.7117	0.7279	0.7356	0.7399	0.7425	0.7442	0.7454	0.7463	0.7491	0.7500
3.5	0.6900	0.7374	0.7544	0.7625	0.7670	0.7698	0.7716	0.7729	0.7738	0.7768	0.7778
4.0	0.7086	0.7583	0.7757	0.7840	0.7887	0.7916	0.7935	0.7948	0.7958	0.7989	0.8000
4.5	0.7242	0.7756	0.7933	0.8017	0.8065	0.8094	0.8114	0.8128	0.8138	0.8171	0.8182
5.0	0.7375	0.7903	0.8080	0.8165	0.8213	0.8243	0.8264	0.8278	0.8288	0.8322	0.8333
5.5	0.7491	0.8029	0.8206	0.8290	0.8339	0.8369	0.8390	0.8404	0.8415	0.8450	0.8462
6.0	0.7592	0.8139	0.8315	0.8398	0.8447	0.8477	0.8498	0.8513	0.8524	0.8559	0.8571
6.5	0.7682	0.8236	0.8410	0.8492	0.8540	0.8571	0.8592	0.8607	0.8618	0.8654	0.8667
7.0	0.7762	0.8322	0.8494	0.8575	0.8622	0.8653	0.8674	0.8689	0.8701	0.8737	0.8750
7.5	0.7834	0.8399	0.8568	0.8648	0.8695	0.8726	0.8747	0.8762	0.8773	0.8811	0.8824
8.0	0.7900	0.8468	0.8635	0.8713	0.8759	0.8790	0.8811	0.8827	0.8838	0.8876	0.8889
8.5	0.7959	0.8531	0.8695	0.8771	0.8817	0.8848	0.8869	0.8885	0.8896	0.8934	0.8947
9.0	0.8014	0.8588	0.8749	0.8824	0.8869	0.8900	0.8921	0.8937	0.8948	0.8987	0.9000
9.5	0.8065	0.8641	0.8799	0.8872	0.8917	0.8947	0.8968	0.8984	0.8995	0.9034	0.9048
10.0	0.8111	0.8689	0.8844	0.8916	0.8960	0.8990	0.9011	0.9026	0.9038	0.9077	0.9091
10.5	0.8155	0.8733	0.8886	0.8956	0.8999	0.9029	0.9050	0.9065	0.9077	0.9117	0.9130
11.0	0.8195	0.8774	0.8924	0.8992	0.9035	0.9064	0.9086	0.9101	0.9113	0.9153	0.9167
11.5	0.8233	0.8813	0.8960	0.9026	0.9068	0.9097	0.9119	0.9134	0.9146	0.9186	0.9200
12.0	0.8268	0.8848	0.8993	0.9058	0.9099	0.9128	0.9149	0.9165	0.9176	0.9217	0.9231
12.5	0.8302	0.8882	0.9204	0.9087	0.9127	0.9156	0.9177	0.9193	0.9205	0.9245	0.9259

13.0	0.8333	0.8913	0.9053	0.9114	0.9154	0.9182	0.9203	0.9219	0.9231	0.9271	0.9286
13.5	0.8363	0.8942	0.9079	0.9140	0.9178	0.9207	0.9228	0.9243	0.9255	0.9296	0.9310
14.0	0.8391	0.8970	0.9105	0.9163	0.9202	0.9230	0.9250	0.9266	0.9278	0.9319	0.9333
14.5	0.8417	0.8996	0.9128	0.9186	0.9223	0.9251	0.9272	0.9287	0.9299	0.9340	0.9355
15.0	0.8442	0.9020	0.9151	0.9207	0.9243	0.9271	0.9292	0.9307	0.9319	0.9360	0.9375
16.0	0.8489	0.9065	0.9192	0.9245	0.9280	0.9307	0.9328	0.9344	0.9356	0.9397	0.9412
17.0	0.8532	0.9106	0.9229	0.9279	0.9313	0.9340	0.9360	0.9376	0.9388	0.9430	0.9444
18.0	0.8571	0.9143	0.9262	0.9310	0.9343	0.9369	0.9389	0.9405	0.9417	0.9459	0.9474
19.0	0.8607	0.9176	0.9292	0.9338	0.9370	0.9395	0.9415	0.9431	0.9443	0.9485	0.9500
20.0	0.8640	0.9207	0.9320	0.9363	0.9394	0.9419	0.9439	0.9455	0.9467	0.9509	0.9524
25.0	0.8775	0.9328	0.9427	0.9461	0.9487	0.9510	0.9530	0.9545	0.9557	0.9600	0.9615
30.0	0.8874	0.9412	0.9502	0.9529	0.9551	0.9572	0.9591	0.9607	0.9619	0.9662	0.9677
35.0	0.8951	0.9475	0.9558	0.9578	0.9597	0.9617	0.9636	0.9651	0.9663	0.9707	0.9722
40.0	0.9014	0.9524	0.9600	0.9616	0.9632	0.9651	0.9670	0.9685	0.9697	0.9740	0.9756
45.0	0.9065	0.9562	0.9634	0.9646	0.9660	0.9678	0.9696	0.9711	0.9723	0.9767	0.9783
50.0	0.9109	0.9594	0.9662	0.9670	0.9682	0.9700	0.9717	0.9732	0.9744	0.9788	0.9804
60.0	0.9180	0.9642	0.9704	0.9707	0.9716	0.9732	0.9749	0.9764	0.9776	0.9820	0.9836
70.0	0.9235	0.9677	0.9734	0.9733	0.9740	0.9755	0.9772	0.9787	0.9799	0.9843	0.9859
80.0	0.9280	0.9704	0.9758	0.9754	0.9758	0.9773	0.9789	0.9804	0.9816	0.9860	0.9877
90.0	0.9317	0.9725	0.9776	0.9770	0.9773	0.9787	0.9803	0.9817	0.9829	0.9874	0.9890
100.0	0.9348	0.9742	0.9791	0.9782	0.9784	0.9798	0.9814	0.9828	0.9840	0.9885	0.9901
200.0	0.9522	0.9820	0.9860	0.9841	0.9837	0.9848	0.9863	0.9877	0.9889	0.9934	0.9950
300.0	0.9602	0.9847	0.9884	0.9861	0.9854	0.9864	0.9879	0.9894	0.9905	0.9950	0.9967
400.0	0.9650	0.9861	0.9896	0.9871	0.9863	0.9873	0.9888	0.9902	0.9914	0.9959	0.9975
500.0	0.9684	0.9869	0.9903	0.9877	0.9869	0.9878	0.9893	0.9907	0.9919	0.9964	0.9980

[a]Neglecting the effects of longitudinal wall heat conduction, fluid bypass and carryover.

Source: Data from Shah (1988).

324

Thus by the selection of the Eq. (5.52) set, we have demonstrated the similarities and differences between a regenerator and a recuperator in the absence of flow leakages and carryover.

The following observations may be made by reviewing Fig. 5.3, the results of Table 5.1 for $C^* = 1$, and those for $C^* < 1$ by Kays and London (1998):

1. For specified C_r^* and C^*, the heat exchanger effectiveness increases with increasing NTU$_o$. For all C_r^* and C^*, $\varepsilon \to 1$ as NTU$_o \to \infty$.

2. For specified NTU$_o$ and C^*, ε increases with increasing values of C_r^* and approaches asymptotically the value for a counterflow recuperator.

3. For specified NTU$_o$ and C_r^*, ε increases with decreasing values of C^*. The percentage change in ε is largest in the lower NTU$_o$ range, and this percentage change in ε increases with increasing values of C_r^*.

4. For $\varepsilon < 40\%$ and $C_r^* > 0.6$, C^* and C_r^* do not have a significant influence on the exchanger effectiveness.

Now let us present approximate formulas to compute ε for a wide range of C_r^* and C^*. The influence of C_r^* on ε can be presented by an empirical correlation for $\varepsilon \leq 90\%$ by Kays and London (1998) as

$$\varepsilon = \varepsilon_{cf}\left[1 - \frac{1}{9(C_r^*)^{1.93}}\right] \tag{5.62}$$

where ε_{cf} is the counterflow recuperator effectiveness as follows:

$$\varepsilon_{cf} = \frac{1 - \exp[-\text{NTU}_o(1 - C^*)]}{1 - C^*\exp[-\text{NTU}_o(1 - C^*)]} \xrightarrow{C^*=1} \frac{\text{NTU}_o}{1 + \text{NTU}_o} \tag{5.63}$$

Equation (5.62) agrees within 1% with the tabular results of Lambertson (1958) for $C^* = 1$ for the following ranges: $2 < \text{NTU}_o < 14$ for $C_r^* \leq 1.5$, $\text{NTU}_o \leq 20$ for $C_r^* = 2$, and a complete range of NTU$_o$ for $C_r^* \geq 5$. For decreasing values of C^*, the error due to the approximation increases with lower values of C_r^*. For example, to obtain accuracy within 1%, $C_r^* \geq 1.5$ for $C^* = 0.9$, and $C_r^* \geq 2.0$ for $C^* = 0.7$.

The following approximate procedure is proposed by Razelos (1980) to calculate the regenerator effectiveness ε for the case of $C^* < 1$. For the known values of NTU$_o$, C^*, and C_r^*, calculate "equivalent" values of NTU$_o$ and C_r^* for a balanced regenerator ($C^* = 1$), designated with a subscript m, as follows:

$$\text{NTU}_{o,m} = \frac{2\text{NTU}_o \cdot C^*}{1 + C^*} \tag{5.64}$$

$$C_{r,m}^* = \frac{2C_r^*C^*}{1 + C^*} \tag{5.65}$$

With these values of NTU$_{o,m}$ and $C_{r,m}^*$, obtain the value of ε_r using Eq. (5.58) with $\varepsilon = \varepsilon_r$ or from the approximate equation

$$\varepsilon_r = \frac{\text{NTU}_{o,m}}{1 + \text{NTU}_{o,m}}\left[1 - \frac{1}{9(C_{r,m}^*)^{1.93}}\right] \tag{5.66}$$

Subsequently, calculate ε from

$$\varepsilon = \frac{1 - \exp\{\varepsilon_r(C^{*2} - 1)/[2C^*(1 - \varepsilon_r)]\}}{1 - C^* \exp\{\varepsilon_r(C^{*2} - 1)/[2C^*(1 - \varepsilon_r)]\}} \tag{5.67}$$

A comparison of ε from this procedure with that of Eq. (5.62) shows that the Razelos approximation yields more accurate values of ε compared to that from Eq. (5.62) for $C^* < 1$. It can be shown that ε's of Eqs. (5.62) and (5.67) are identical for $C_r^* = \infty$. Note that either by employing the foregoing approximate method or a direct use of Eq. (5.62), we at most need Table 5.1 or Fig. 5.3; thus the tabular data for $C^* < 1$ are not needed.

5.2.5 ε-NTU$_o$ Results for a Parallelflow Regenerator

The differential equations and boundary conditions for the parallelflow regenerator are the same as those of Eqs. (5.31)–(5.38) except for Eq. (5.36). The boundary condition of Eq. (5.36) for this case is

$$T_c^*(0, \tau_c^*) = 0 \qquad \text{for } 0 \le \tau_c^* \le 1 \tag{5.68}$$

The solution may be presented in terms of ε as a function of the same four dimensionless groups as for a counterflow regenerator, Eq. (5.52). Theoclitus and Eckrich (1966) obtained the solution numerically by a finite difference method. They covered the following ranges of the parameters: $1 \le \text{NTU}_o \le 10$, $0.5 \le C^* \le 1.0$, $0.2 \le C_r^* \le \infty$, and $0.25 \le (hA)^* \le 1$. Their results for $C^* = 1$ are presented in Fig. 5.4.

From a review of Fig. 5.4, it is interesting to note that the effectiveness of a parallelflow regenerator oscillates above and below that for a parallelflow recuperator

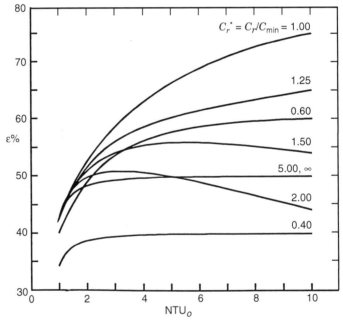

FIGURE 5.4 Parallelflow regenerator ε as a function of NTU$_o$ and for $C_r^* = 1$ and $(hA)^* = 1$.

($\varepsilon_{max} = 50\%$). The oscillations decrease in amplitude as C_r^* increases. The maximum effectiveness for the parallelflow regenerator is reached at $C_r^* \approx 1$, and it exceeds the effectiveness of a parallelflow recuperator. These results contrast with those for the counterflow regenerator where the limiting effectiveness, represented by the counterflow recuperator, is never exceeded, but is approached asymptotically as C_r^* increases.

Similar to a recuperator, parallelflow is sometimes preferred to counterflow in some regenerator applications for the first three reasons mentioned for a recuperator in Section 1.6.1.2.

Example 5.1 A boiler is equipped with a regenerator rotating at 4 rpm and having the flue gas (hot) and air (cold) flow areas with disk sector angles of 195° and 165°, respectively. The rotor with 2 m diameter and 0.4 m height is turned by a shaft of 0.2 m diameter. The matrix material has the following characteristics:

Density = 7800 kg/m^3 Packing density = 1200 m^2/m^3

Specific heat = 0.44 kJ/kg · K   Porosity = 0.76

The flue gas and airstream flowing in counterflow have the following characteristics:

	Flue Gas	Air
Heat transfer coefficient (W/m^2 · K)	60	50
Isobaric specific heat (kJ/kg · K)	1.11	1.005
Mass flow rate (kg/s)	8.3	6.9
Inlet temperature (°C)	320	26

Assuming that 10% of the rotor face is covered by radial seals, calculate the regenerator effectiveness and heat transfer rate from the flue gas stream to the airstream.

SOLUTION

Problem Data and Schematic: The heat transfer coefficients, fluid flow rates, and inlet temperatures are provided in Fig. E5.1 for both the hot and cold fluid streams. In

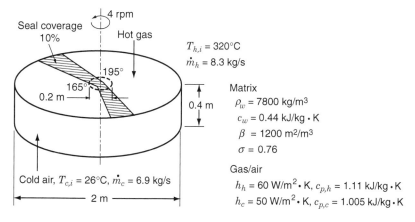

FIGURE E5.1

addition, the regenerator dimensions and rotational speed are specified. Also specified are the physical properties of both hot and cold fluid streams as well as the matrix material.

Determine: The regenerator effectiveness and the heat transfer rate from the flue gas to the air.

Assumptions: The assumptions of Section 5.1.1 are invoked here.

Analysis: The regenerator effectiveness ε is a function of four dimensionless groups as given in Eq. (5.52):

$$\varepsilon = \phi[\text{NTU}_o, C^*, C_r^*, (hA)^*]$$

To determine whether the effect of the convection conductance ratio $(hA)^*$ is negligible, we first establish which side has the lower heat capacity rate. To that end,

$$C_h = \dot{m}_h c_{p,h} = 8.3\,\text{kg}/\text{s} \times 1.11\,\text{kJ}/\text{kg} \cdot \text{K} = 9.21\,\text{kW}/\text{K}$$

$$C_c = \dot{m}_c c_{p,c} = 6.9\,\text{kg}/\text{s} \times 1.005\,\text{kJ}/\text{kg} \cdot \text{K} = 6.93\,\text{kW}/\text{K}$$

Thus,

$$C_{\max} = C_h = 9.21\,\text{kW}/\text{K} \qquad C_{\min} = C_c = 6.93\ \text{kW}/\text{K}$$

$$C^* = \frac{C_{\min}}{C_{\max}} = \frac{6.93\ \text{kW}/\text{K}}{9.21\ \text{kW}/\text{K}} = 0.752$$

Now

$$(hA)^* = \frac{(hA) \text{ on } C_{\min} \text{ side}}{(hA) \text{ on } C_{\max} \text{ side}}$$

Noting that the heat transfer surface areas on the two sides are in proportion to the disk sector angles, we have

$$(hA)^* = \frac{50\,\text{W}/\text{m}^2 \cdot \text{K} \times 165°}{60\,\text{W}/\text{m}^2 \cdot \text{K} \times 195°} = 0.71$$

As pointed out by Lambertson (1958), $(hA)^*$ has a negligible influence on the regenerator effectiveness ε in the range $0.25 \le (hA)^* \le 4$. Hence, in the present case with $(hA)^* = 0.71$, we have

$$\varepsilon = \phi(\text{NTU}_o, C^*, C_r^*)$$

To determine C_r^*, we first determine matrix mass as

M_w = rotor cross-sectional area $\times$ rotor height $\times$ matrix material density

$\times$ matrix solidity

$$= \frac{\pi}{4}(2^2 - 0.2^2)\,\text{m}^2 \times 0.4\,\text{m} \times 7800\,\text{kg}/\text{m}^3 \times (1 - 0.76) = 2328.9\,\text{kg}$$

Knowing the matrix mass, its heat capacity rate is computed as

$$C_r = M_w c_w \mathbf{N} = 2328.9\,\text{kg} \times 0.44\,\text{kJ/kg} \cdot \text{K} \times \frac{4}{60}\,\text{rev/s} = 68.3\,\text{kW/K}$$

Knowing C_r and $C_{\min}$, we get

$$C_r^* = \frac{C_r}{C_{\min}} = \frac{68.3\,\text{kW/K}}{6.93\,\text{kW/K}} = 9.86$$

To find the hot- and cold-side heat transfer surface areas, we note that the total matrix surface area A is given by

A = rotor cross-sectional area × rotor height × matrix packing density β

$\quad$× fraction of rotor face area not covered by radial seals

$$= \frac{\pi}{4}(2^2 - 0.2^2)\,\text{m}^2 \times 0.4\,\text{m} \times 1200\,\text{m}^2/\text{m}^3 \times (1 - 0.1) = 1343.6\,\text{m}^2$$

The hot- and cold-gas-side surface areas are proportional to the respective sector angles, so that

$$A_h = \left(\frac{195°}{360°}\right)A = \frac{195°}{360°} \times 1343.6\,\text{m}^2 = 727.8\,\text{m}^2$$

$$A_c = \left(\frac{165°}{360°}\right)A = \frac{165°}{360°} \times 1343.6\,\text{m}^2 = 615.8\,\text{m}^2$$

Thus,

$$(hA)_h = 60\,\text{W/m}^2 \cdot \text{K} \times 727.8\,\text{m}^2 = 43{,}668\,\text{W/K} = 43.67\,\text{kW/K}$$

$$(hA)_c = 50\,\text{W/m}^2 \cdot \text{K} \times 615.8\,\text{m}^2 = 30{,}790\,\text{W/K} = 30.79\,\text{kW/K}$$

From the foregoing values in conjunction with Eq. (5.48), we get

$$\text{NTU}_o = \frac{1}{C_{\min}}\frac{1}{1/(hA)_h + 1/(hA)_c} = \frac{1}{6.93\,\text{kW/K}}\left[\frac{1}{(1/43.67 + 1/30.79)\,\text{K/kW}}\right] = 2.606$$

Knowing C^* and NTU_o, the counterflow recuperator effectiveness ε_{cf} is determined from Eq. (5.63). Thus

$$\varepsilon_{cf} = \frac{1 - \exp[-\text{NTU}_o(1 - C^*)]}{1 - C^*\exp[-\text{NTU}_o(1 - C^*)]} = \frac{1 - \exp[-2.606(1 - 0.752)]}{1 - 0.752\exp[-2.606(1 - 0.752)]} = 0.7855$$

Knowing C_r^* and ε_{cf}, the regenerator effectiveness ε can be calculated from Eq. (5.62) valid for $\varepsilon \leq 0.9$:

$$\varepsilon = \varepsilon_{cf}\left(1 - \frac{1}{9C_r^{*1.93}}\right) = 0.7855\left(1 - \frac{1}{9 \times 9.86^{1.93}}\right) = 0.7844 \qquad \textit{Ans.}$$

By definition,

$$\varepsilon = \frac{q}{q_{max}} = \frac{q}{C_{min}\left(T_{h,i} - T_{c,i}\right)}$$

Hence,

$$q = \varepsilon C_{min}\left(T_{h,i} - T_{c,i}\right) = 0.7844 \times 6.93 \text{ kW/K} \times (320 - 26) \text{ K} = 1598 \text{ kW} \qquad Ans.$$

We now demonstrate the use of Razelos method (1980) to evaluate ε since C^* is considerably less than unity. First compute the parameters $\text{NTU}_{o,m}$, $C^*_{r,m}$, and ε_r of Eqs. (5.64)–(5.66).

$$\text{NTU}_{o,m} = \frac{2\text{NTU}_o C^*}{1 + C^*} = \frac{2 \times 2.606 \times 0.752}{1 + 0.752} = 2.237$$

$$C^*_{r,m} = \frac{2C^*_r C^*}{1 + C^*} = \frac{2 \times 9.86 \times 0.752}{1 + 0.752} = 8.46$$

$$\varepsilon_r = \frac{\text{NTU}_{o,m}}{1 + \text{NTU}_{o,m}}\left(1 - \frac{1}{9C^{*1.93}_{r,m}}\right) = \frac{2.237}{1 + 2.237}\left(1 - \frac{1}{9 \times 8.46^{1.93}}\right) = 0.6898$$

In terms of these parameters, the regenerator effectiveness ε is given in Eq. (5.67) as

$$\begin{aligned}\varepsilon &= \frac{1 - \exp\{\varepsilon_r(C^{*2} - 1)/[2C^*(1 - \varepsilon_r)]\}}{1 - C^*\exp\{\varepsilon_r(C^{*2} - 1)/[2C^*(1 - \varepsilon_r)]\}}\\[2mm]
&= \frac{1 - \exp\{0.6898(0.752^2 - 1)/[2 \times 0.752(1 - 0.6898)]\}}{1 - 0.752\exp\{0.6898(0.752^2 - 1)/[2 \times 0.752(1 - 0.6898)]\}} = 0.7842 \qquad Ans.\end{aligned}$$

This value of ε is very close to the value (0.7844) calculated by Eq. (5.62). Hence, the value of q calculated above will remain virtually the same.

Discussion and Comments: Determination of the regenerator effectiveness and subsequent heat transfer rate for a regenerator is straightforward, as was the case with the recuperator. The only difference is that the regenerator effectiveness depends on C^*_r and $(hA)^*$ in addition to NTU_o and C^*. Only the latter two dimensionless groups are used for the recuperator effectiveness determination.

A comparison of Kays and London (1998) and Razelos methods for the determination of ε shows that both methods yield virtually the same effectiveness, as expected. The Razelos method would be more accurate for low values of C^*.

As mentioned in the text, the regenerator effectiveness approaches the recuperator effectiveness for higher values of C^*_r. It can be seen here that when $C^*_r = 9.86$, $\varepsilon \approx \varepsilon_{cf}$ (0.784 vs. 0.785). When we consider the seal and carryover leakages, the actual ε will be lower than 0.784, depending on the amount of leakage. These leakage effects are discussed in Section 5.6.

Example 5.2   A rotary regenerator, with a rotational speed of 10 rpm, is used to recover energy from a gas stream at 250°C flowing at 10 kg/s. This heat is transferred to the airstream at 10°C, also flowing at 10 kg/s. The wheel depth is 0.22 m and diameter 1.6 m, so that its face area is approximately 1.8 m². The mass of the matrix is 150 kg with a surface-to-volume ratio of 3000 m²/m³, and the mean specific heat of the matrix material is 0.8 kJ/kg · K. The heat transfer coefficient for both fluid streams is 30 W/m² · K. The mean isobaric specific heat of the gas is 1.15 kJ/kg · K and that of air is 1.005 kJ/kg · K. The flow split gas:air = 50%:50%. For a counterflow arrangement, calculate the following values:

(a) The regenerator effectiveness

(b) The rate of heat recovery and the outlet temperatures of air and gas

(c) The rate of heat recovery and the outlet temperatures of air and gas if the rotational speed of the wheel is increased to 20 rpm

(d) The rate of heat recovery and the outlet temperatures of air and gas if the rotational speed of the wheel is reduced to 5 rpm

SOLUTION

Problem Data and Schematic: The heat transfer coefficients, fluid flow rates, and inlet temperatures are provided in Fig. E5.2 for both fluid streams. In addition, the regenerator dimensions and rotational speed are specified. Also specified are the physical properties of both hot and cold fluid streams as well as the matrix material.

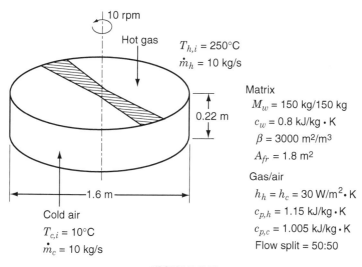

FIGURE E5.2

Determine: (a) The heat exchanger effectiveness, (b) the heat recovery rate and air outlet temperature at the given wheel speed, (c) the heat recovery rate and air outlet temperature at the increased wheel speed, and (d) the heat recovery rate and air outlet temperature at the reduced wheel speed.

Assumptions: The assumptions of Section 5.1.1 are invoked here. Seal face coverage is assumed negligible.

Analysis: (a) The regenerator effectiveness ε can be calculated using Eq. (5.62), valid for $\varepsilon \leq 0.9$, in conjunction with Eq. (5.63). For this purpose, we first need to determine C^*, C_r^*, and NTU_o. Let us first calculate various heat capacity rates.

$$C_h = \dot{m}_h c_{p,h} = 10 \, \text{kg/s} \times 1.15 \, \text{kJ/kg} \cdot \text{K} = 11.5 \, \text{kJ/s} \cdot \text{K} = 11.5 \, \text{kW/K}$$

$$C_c = \dot{m}_c c_{pc} = 10 \, \text{kg/s} \times 1.005 \, \text{kJ/kg} \cdot \text{K} = 10.05 \, \text{kJ/s} \cdot \text{K} = 10.05 \, \text{kW/K}$$

$$C_{\max} = C_h = 11.5 \, \text{kW/K} \qquad C_{\min} = C_c = 10.05 \, \text{kW/K}$$

$$C^* = \frac{C_{\min}}{C_{\max}} = \frac{10.05 \, \text{kW/K}}{11.5 \, \text{kW/K}} = 0.8739$$

$$C_r = M_w c_w \text{N} = 150 \, \text{kg} \times 0.8 \, \text{kJ/kg} \cdot \text{K} \times \frac{10}{60} \, \text{rev/s} = 20 \, \text{kW/K}$$

$$C_r^* = \frac{C_r}{C_{\min}} = \frac{20 \, \text{kW/K}}{10.05 \, \text{kW/K}} = 1.99$$

The modified overall heat transfer coefficient U_o, with negligible wall thermal resistance and fouling resistance, is given as

$$\frac{1}{(U_o A)_h} = \frac{1}{(U_o A)_c} = \frac{1}{h_h A_h} + \frac{1}{h_c A_c}$$

With equal hot- and cold-side heat transfer areas (since flow split $= 50 : 50$; i.e., $A_h = A_c$), U_o is given by

$$\frac{1}{U_o} = \frac{1}{h_h} + \frac{1}{h_c} = \frac{1}{30 \, \text{W/m}^2 \cdot \text{K}} + \frac{1}{30 \, \text{W/m}^2 \cdot \text{K}} = \frac{1}{15} \, \text{m}^2 \cdot \text{K/W}$$

or

$$U_o = 15 \, \text{W/m}^2 \cdot \text{K}$$

The total heat transfer area A is expressible as

$$A = \text{matrix volume} \times \text{surface area density}$$
$$= \text{matrix face area} \times \text{depth} \times \text{surface area density}$$
$$= 1.8 \, \text{m}^2 \times 0.22 \, \text{m} \times 3000 \, \text{m}^2/\text{m}^3 = 1188 \, \text{m}^2$$

Therefore,

$$A_h = A_c = 594 \, \text{m}^2$$

Thus, NTU_o is determined as

$$NTU_o = \frac{U_o A}{C_{\min}} = \frac{15 \, \text{W/m}^2 \cdot \text{K} \times 594 \, \text{m}^2}{10.05 \times 10^3 \, \text{W/K}} = 0.8866$$

Now ε_{cf} and ε are evaluated from Eqs. (5.63) and (5.62) as

$$\varepsilon_{cf} = \frac{1 - \exp[-\mathrm{NTU}_o(1 - C^*)]}{1 - C^* \exp[-\mathrm{NTU}_o(1 - C^*)]} = \frac{1 - \exp[-0.8866(1 - 0.8739)]}{1 - 0.8739 \exp[-0.8866(1 - 0.8739)]} = 0.4840$$

$$\varepsilon = \varepsilon_{cf}\left[1 - \frac{1}{9(C_r^*)^{1.93}}\right] = 0.4840\left(1 - \frac{1}{9 \times 1.99^{1.93}}\right) = 0.4697 \qquad \textit{Ans.}$$

(b) By definition,

$$\varepsilon = \frac{q}{q_{max}} = \frac{q}{C_{min}\left(T_{h,i} - T_{c,i}\right)}$$

Hence,

$$q = \varepsilon C_{min}(T_{h,i} - T_{c,i}) = 0.4697 \times 10.05\,\mathrm{kW/K} \times (250 - 10)\,\mathrm{K} = 1133\,\mathrm{kW} \qquad \textit{Ans.}$$

To find the outlet temperatures of the cold and hot fluid streams, we have from the energy balance

$$q = C_c(T_{c,o} - T_{c,i}) = C_h(T_{h,i} - T_{h,o})$$

Thus,

$$T_{c,o} = T_{c,i} + \frac{q}{C_c} = 10^\circ\mathrm{C} + \frac{1133\,\mathrm{kW}}{10.05\,\mathrm{kW/^\circ C}} = 122.7^\circ\mathrm{C} \qquad \textit{Ans.}$$

$$T_{h,o} = T_{h,i} - \frac{q}{C_h} = 250^\circ\mathrm{C} - \frac{1133\,\mathrm{kW}}{11.5\,\mathrm{kW/^\circ C}} = 151.5^\circ\mathrm{C} \qquad \textit{Ans.}$$

(c) When the rotational speed is increased to 20 rpm,

$$C_r = M_w c_w \mathrm{N} = 150\,\mathrm{kg} \times 0.8\,\mathrm{kJ/kg \cdot K} \times \frac{20}{60}\,\mathrm{rev/s} = 40\,\mathrm{kW/K}$$

$$C_r^* = \frac{C_r}{C_{min}} = \frac{40\,\mathrm{kW/K}}{10.05\,\mathrm{kW/K}} = 3.98$$

In this case, ε_{cf} will still remain the same as in part (a) since it is not affected by the rpm. Then ε is given by

$$\varepsilon = \varepsilon_{cf}\left(1 - \frac{1}{9C_r^{*1.93}}\right) = 0.4840\left(1 - \frac{1}{9 \times 3.98^{1.93}}\right) = 0.4803$$

$$q = \varepsilon C_{min}(T_{h,i} - T_{c,i}) = 0.4803 \times 10.05\,\mathrm{kW/K} \times (250 - 10)\,\mathrm{K} = 1158\,\mathrm{kW} \qquad \textit{Ans.}$$

$$T_{c,o} = T_{c,i} + \frac{q}{C_c} = 10^\circ\mathrm{C} + \frac{1158\,\mathrm{kW}}{10.05\,\mathrm{kW/^\circ C}} = 125.2^\circ\mathrm{C} \qquad \textit{Ans.}$$

$$T_{h,o} = T_{h,i} - \frac{q}{C_h} = 250^\circ\mathrm{C} - \frac{1158\,\mathrm{kW}}{11.5\,\mathrm{kW/^\circ C}} = 149.3^\circ\mathrm{C} \qquad \textit{Ans.}$$

(d) When the rotational speed is reduced to 5 rpm,

$$C_r = M_w c_w \mathrm{N} = 150\,\mathrm{kg} \times 0.8\,\mathrm{kJ/kg \cdot K} \times \frac{5}{60}\,\mathrm{rev/s} = 10\,\mathrm{kW/K}$$

$$C_r^* = \frac{C_r}{C_{min}} = \frac{10\,\mathrm{kW/K}}{10.05\,\mathrm{kW/K}} = 0.995$$

$$\varepsilon = \varepsilon_{cf}\left(1 - \frac{1}{9C_r^{*1.93}}\right) = 0.4840\left(1 - \frac{1}{9 \times 0.995^{1.93}}\right) = 0.4297$$

$$q = \varepsilon C_{min}(T_{h,i} - T_{c,i}) = 0.4297 \times 10.05\,\mathrm{kW/K} \times (250 - 10)\,\mathrm{K} = 1036\,\mathrm{kW} \qquad Ans.$$

$$T_{c,o} = T_{c,i} + \frac{q}{C_c} = 10^\circ\mathrm{C} + \frac{1036\,\mathrm{kW}}{10.05\,\mathrm{kW/^\circ C}} = 113.1^\circ\mathrm{C} \qquad Ans.$$

$$T_{h,o} = T_{h,i} - \frac{q}{C_h} = 250^\circ\mathrm{C} - \frac{1036\,\mathrm{kW}}{11.5\,\mathrm{kW/^\circ C}} = 159.9^\circ\mathrm{C} \qquad Ans.$$

Discussion and Comments: This example illustrates the effect of the rotational speed on the heat recovery rate and the cold- and hot-stream outlet temperatures. As seen from the following list, the higher the rotational speed, the higher the heat recovery rate and the cold-stream outlet temperature and the lower the hot-stream outlet temperature. However, the rate of change decreases with increasing C_r^* when comparing the results for 5, 10, and 20 rpm. The rotational speed (in low C_r^* operating range) is clearly an important variable in controlling all three quantities.

Wheel Rotational Speed (rpm)	Heat Recovery Rate (kW)	Cold Stream (Air) Outlet Temperature (°C)	Hot Stream (Gas) Outlet Temperature (°C)	ε
5	1036	113.1	159.9	0.4297
10	1133	122.7	151.5	0.4697
20	1158	125.2	149.3	0.4803

A review of the regenerator effectivenesses indicates that as the rotational speed (and C_r) increases, ε increases and approaches that of a counterflow recuperator ($\varepsilon_{cf} = 0.4840$).

Example 5.3 Determine the effectiveness and the outlet temperatures of the hot and cold fluids of a two-disk (in parallel) counterflow rotary regenerator for a vehicular gas turbine using the following data:

Operating conditions
Airflow rate = 2.029 kg/s
Gas flow rate = 2.094 kg/s
Disk speed = 15 rpm
Air inlet temperature = 480°C
Gas inlet temperature = 960°C
Flow length = 0.0715 m

Disk geometry (exclusive of rim)
Disk diameter = 0.683 m
Hub diameter = 0.076 m
Seal face coverage = 7%
Matrix effective mass (two disks) = 34.93 kg
Matrix compactness $\beta = 5250\,\mathrm{m^2/m^3}$
Flow split, gas : air = 50:50

Heat transfer coefficients
$h_{\text{air}} = 220.5 \text{ W/m}^2 \cdot \text{K}$
$h_{\text{gas}} = 240.5 \text{ W/m}^2 \cdot \text{K}$

Physical properties
$c_{p,\text{air}} = 1.050 \text{ kJ/kg} \cdot \text{K}$
$c_{p,\text{gas}} = 1.084 \text{ kJ/kg} \cdot \text{K}$
$c_w = 1.130 \text{ kJ/kg} \cdot \text{K}$

SOLUTION

Problem Data and Schematic: The heat transfer coefficients, fluid flow rates, and inlet temperatures are provided in Fig. E5.3 for both hot and cold streams. In addition, the regenerator dimensions and rotational speed are specified. Also specified are the physical properties of both hot and cold fluid streams as well as the matrix material.

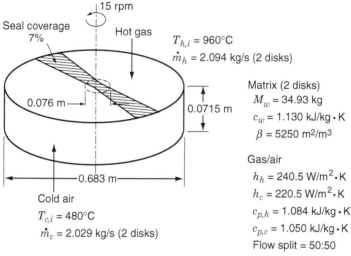

FIGURE E5.3

Determine: The regenerator effectiveness and outlet temperatures of both hot and cold fluids.

Assumptions: The assumptions of Section 5.1.1 are invoked here.

Analysis: Let us first evaluate the dimensionless groups NTU_o, C^*, C_r^*, and $(hA)^*$ for the determination of ε. To that end, first we need to compute the frontal area and volume of both disks:

frontal area = (disk area − hub area) × (1 − seal face coverage) × number of disks

$$= \frac{\pi}{4}\left(0.683^2 - 0.076^2\right) \text{m}^2 \times (1 - 0.07) \times 2 = 0.673 \text{ m}^2$$

matrix volume = frontal area × flow length (disk height)

$$= 0.673 \text{ m}^2 \times 0.0715 \text{ m} = 0.0481 \text{ m}^3$$

The total heat transfer area is given by

$$A = \beta V = 5250 \text{ m}^2/\text{m}^3 \times 0.0481 \text{ m}^3 = 252.53 \text{ m}^2$$

Since the flow split is $50:50$, the hot and cold side areas are equal, so that

$$A_h = A_c = \frac{252.53 \text{ m}^2}{2} = 126.26 \text{ m}^2$$

Knowing the heat transfer areas and heat transfer coefficients for both sides, we have

$$(hA)_h = 240.5 \text{ W/m}^2 \cdot \text{K} \times 126.26 \text{ m}^2 = 30{,}366 \text{ W/K} = 30.37 \text{ kW}$$

$$(hA)_c = 220.5 \text{ W/m}^2 \cdot \text{K} \times 126.26 \text{ m}^2 = 27{,}840 \text{ W/K} = 27.84 \text{ kW/K}$$

$$(hA)^* = \frac{(hA)_c}{(hA)_h} = \frac{27.84 \text{ kW/K}}{30.37 \text{ kW/K}} = 0.917$$

Knowing the flow rates and specific heats of the two fluids, we get

$$C_h = \dot{m}_h c_{p,h} = 2.094 \text{ kg/s} \times 1.084 \text{ kJ/kg} \cdot \text{K} = 2.270 \text{ kW/K}$$

$$C_c = \dot{m}_c c_{p,c} = 2.029 \text{ kg/s} \times 1.050 \text{ kJ/kg} \cdot \text{K} = 2.130 \text{ kW/K}$$

$$C_{\min} = 2.130 \text{ kW/K} \qquad C_{\max} = 2.270 \text{ kW/K}$$

$$C^* = \frac{C_{\min}}{C_{\max}} = \frac{C_c}{C_h} = \frac{2.130 \text{ kW/K}}{2.270 \text{ kW/K}} = 0.9383$$

$$C_r = M_w c_w \mathbf{N} = 34.93 \text{ kg} \times 1.130 \text{ kJ/kg} \cdot \text{K} \times \frac{15}{60} \text{ rev/s} = 9.868 \text{ kW/K}$$

$$C_r^* = \frac{C_r}{C_{\min}} = \frac{9.868 \text{ kW/K}}{2.130 \text{ kW/K}} = 4.63$$

$$\text{NTU}_o = \frac{1}{C_{\min}} \frac{1}{1/(hA)_h + 1/(hA)_c} = \frac{1}{2130 \text{ kW/K}} \left[\frac{1}{1/(30.37 + 1/27.84) \text{ K/kW}} \right] = 6.819$$

With the foregoing values of NTU_o, C^*, C_r^*, and $(hA)^*$, we determine the regenerator effectiveness ε using Eqs. (5.63) and (5.62).

$$\varepsilon_{cf} = \frac{1 - \exp[-\text{NTU}_o(1 - C^*)]}{1 - C^* \exp[-\text{NTU}_o(1 - C^*)]} = \frac{1 - \exp[-6.819(1 - 0.9383)]}{1 - 0.9383 \exp[-6.819(1 - 0.9383)]} = 0.8945$$

Introducing this value in Eq. (5.62), we obtain

$$\varepsilon = \varepsilon_{cf} \left(1 - \frac{1}{9 C_r^{*1.93}} \right) = 0.8945 \left(1 - \frac{1}{9 \times 4.63^{1.93}} \right) = 0.8893 \qquad \textit{Ans.}$$

Knowing the heat exchanger effectiveness ε, the average outlet temperatures of the fluids are readily computed based on the definition of ε:

$$T_{h,o} = T_{h,i} - \varepsilon C^* \left(T_{h,i} - T_{c,i} \right) = 960°\text{C} - 0.8893 \times 0.9383(960 - 480)°\text{C} = 559.5°\text{C} \ \textit{Ans.}$$

$$T_{c,o} = T_{c,i} - \varepsilon \left(T_{h,i} - T_{c,i} \right) = 480°\text{C} - 0.8893(960 - 480)°\text{C} = 906.9°\text{C} \qquad \textit{Ans.}$$

Discussion and Comments: This example is very similar to Example 5.1 or 5.2 except that the regenerator consists of two disks. We have opted for the analysis that includes both disks together as a single entity; we could have performed the analysis considering a single disk and would have obtained the same results. Note that the design rotational speed of the regenerator is sufficiently high, as reflected by $C_r^* = 4.63$ to yield the regenerator effectiveness $\varepsilon = 0.8893$, within 0.58% of the pure counterflow recuperator $\varepsilon_{cf} = 0.8945$.

5.3 THE Λ-Π METHOD

This method for determining regenerator performance is due to Hausen (1929, 1983). He analyzed a fixed-matrix regenerator starting with Eqs. (5.14) and (5.17)–(5.19). He defined spatial and time-independent nondimensional variables such that Eqs. (5.31)–(5.34) became parameter-free. Hausen introduced the following, now referred to as *Schumann dimensionless independent variables*[†]:

$$\xi_j = \left(\frac{hA}{C}\right)_j \frac{x}{L} \qquad \eta_j = \left(\frac{hA}{\bar{C}_r}\right)_j \tau_j \qquad j = h \text{ or } c \tag{5.69}$$

Here A represents the total heat transfer area of the matrix in the jth period, M_w represents its mass, and $\bar{C}_{r,j} = (M_w c_w)_j = M_w c_w$ based on Eqs. (5.2) and (5.8). Substituting these variables into Eqs. (5.14) and (5.17)–(5.19) yield the following. For the hot-gas flow period,

$$\frac{\partial T_h}{\partial \xi_h} = T_w - T_h \qquad \frac{\partial T_w}{\partial \eta_h} = T_h - T_w \tag{5.71}$$

For the cold-gas flow period,

$$\frac{\partial T_c}{\partial \xi_c} = T_c - T_w \qquad \frac{\partial T_w}{\partial \eta_c} = T_c - T_w \tag{5.72}$$

Note that these equations are now parameter-free. The boundary and periodic-flow conditions are still those of Eqs. (5.20)–(5.23).

The independent variables ξ and η of Eq. (5.69) are represented irrespective of the period as

$$\xi = \frac{hA}{CL} x = bx \propto x \qquad \eta = \frac{hA}{\bar{C}_r} \tau = c\tau \propto \tau \tag{5.73}$$

Here b and c are the constants since h, A, C, L, and $\bar{C}_r$ all are constants due to the assumptions listed in Section 5.1.1. For this reason, the variables ξ and η are interpreted as the reduced length and reduced period variables, respectively.

[†] If the dwell periods $\tau_{d,h}$ and $\tau_{d,c}$ are not negligible, η_h and η_c are expressed as

$$\eta_h = \frac{h_h A_h}{\bar{C}_{r,h}} \left(\tau_h - \frac{x}{L} \tau_{d,h}\right) \qquad \eta_c = \frac{h_c A_c}{\bar{C}_{r,c}} \left[\tau_c - \left(1 - \frac{x}{L}\right) \tau_{d,c}\right] \tag{5.70}$$

If the temperatures T_h, T_c, and T_w are made dimensionless as before [Eq. (5.28)], it is evident from Eqs. (5.69)–(5.72) that

$$T_h^*, T_c^*, T_w^* = \phi(\xi_h, \xi_c, \eta_h, \eta_c) \tag{5.74}$$

For overall regenerator performance, we are interested in determining the time-averaged fluid outlet temperatures. These temperatures are obtained from their definitions [Eq. (5.41)] after suitable modifications based on Eq. (5.69), with τ replaced by η_h and η_c for the hot- and cold-gas flow periods, respectively. These temperatures are expressed by the regenerator effectiveness. Since we have considered only one matrix at a time in the foregoing analysis, the hot and cold fluid flows are intermittent. In an ideal steady-state periodic condition, the actual heat transfer (joules or Btu) during one hot- or cold-gas flow period will be

$$Q = C_h P_h \left(T_{h,i} - \bar{T}_{h,o} \right) = C_c P_c \left(\bar{T}_{c,o} - T_{c,i} \right) \tag{5.75}$$

This means that the regenerator is in a cyclic equilibrium because the heat transferred to the matrix during the hot period equals the heat transferred to the cold fluid during the cold period. The maximum possible heat transfer will be in a counterflow regenerator of infinite surface area having the same fluid flow rates and fluid inlet temperatures. Thus this maximum possible heat transfer is

$$Q_{\max} = (CP)_{\min}(T_{h,i} - T_{c,i}) \tag{5.76}$$

Thus, the effectiveness for a fixed-matrix regenerator (with two matrices) is defined as follows:

$$\varepsilon = \frac{Q}{Q_{\max}} = \frac{(CP)_h(T_{h,i} - \bar{T}_{h,o})}{(CP)_{\min}(T_{h,i} - T_{c,i})} = \frac{(CP)_c(\bar{T}_{c,o} - T_{c,i})}{(CP)_{\min}(T_{h,i} - T_{c,i})} \tag{5.77}$$

However, the effectiveness of a single matrix of two or more matrices of a fixed-matrix regenerator can be defined as follows by considering only one matrix going through the hot and cold period (i.e., one complete period). The maximum possible heat transfer during hot- and cold-gas flow periods, respectively, is

$$Q_{\max,h} = C_h P_h \left(T_{h,i} - T_{c,i} \right) \qquad Q_{\max,c} = C_c P_c \left(T_{h,i} - T_{c,i} \right) \tag{5.78}$$

Thus, the regenerator effectiveness during hot- and cold-gas flow periods are

$$\varepsilon_h = \frac{Q_h}{Q_{\max,h}} = \frac{C_h P_h(T_{h,i} - \bar{T}_{h,o})}{C_h P_h(T_{h,i} - T_{c,i})} = \frac{T_{h,i} - \bar{T}_{h,o}}{T_{h,i} - T_{c,i}} = 1 - \bar{T}_{h,o}^* \tag{5.79}$$

$$\varepsilon_c = \frac{Q_c}{Q_{\max,c}} = \frac{C_c P_c(\bar{T}_{c,o} - T_{c,i})}{C_c P_c(T_{h,i} - T_{c,i})} = \frac{\bar{T}_{c,o} - T_{c,i}}{T_{h,i} - T_{c,i}} = \bar{T}_{c,o}^* \tag{5.80}$$

where the last equalities in Eqs. (5.79) and (5.80) are obtained using the definitions of Eq. (5.28). Note that the effectiveness in Eqs. (5.79) and (5.80) is similar to the temperature effectiveness for a recuperator [Eqs. (3.51) and (3.52)]. Razelos (1979) defined the overall effectiveness of a single matrix as

$$\varepsilon_r = \frac{Q_h + Q_c}{Q_{\max,h} + Q_{\max,c}} = \frac{2Q}{Q_{\max,h} + Q_{\max,c}} \tag{5.81}$$

Using Eqs. (5.79)–(5.81), it can be shown that

$$\frac{1}{\varepsilon_r} = \frac{1}{2}\left(\frac{1}{\varepsilon_h} + \frac{1}{\varepsilon_c}\right) \quad \text{therefore,} \quad \varepsilon_r = \frac{2(C_c/C_h)}{1 + (C_c/C_h)}\varepsilon_c \tag{5.82}$$

Comparing Eqs. (5.77) and (5.82), it can be shown for a two-matrix regenerator that

$$\varepsilon = \left(\frac{1 + C^*}{2C^*}\right)\varepsilon_r \tag{5.83}$$

The independent dimensionless variables of Eqs. (5.69), collocated at $x = L$, $\tau_h = P_h$ or $\tau_c = P_c$, become the dimensionless parameters as follows[†]:

$$\Lambda_j = \xi_j(L) = \left(\frac{hA}{C}\right)_j, \quad \Pi_j = \eta_j(P_j) = \left(\frac{hA}{C_r}\right)_j P_j \quad j = h \text{ or } c \tag{5.84}$$

Hence, the regenerator effectiveness is a function of four parameters:

$$\varepsilon = \phi(\Lambda_h, \Lambda_c, \Pi_h, \Pi_c) \tag{5.86}$$

From the first equation of Eq. (5.84), $\Lambda_j = \xi_j(L)$; and from Eq. (5.73), $\xi(L) = bL$, we get

$$\Lambda = bL \tag{5.87}$$

Similarly, from the second equation of Eq. (5.84), $\Pi_j = \eta_j(P_j)$; and from Eq. (5.73), $\eta_j(P_j) = cP_j$. Hence, we get

$$\Pi = cP_h \quad \text{or} \quad cP_c \tag{5.88}$$

Since b and c are constants in Eqs. (5.87) and (5.88), Λ and Π are designated as *reduced length* and *reduced period*, respectively, for the regenerator. The reduced length Λ also designates the dimensionless heat transfer or thermal size of the regenerator [see, e.g., Eq. (5.107)], and this method is referred to as *the Λ-Π method*. It has been used primarily for the design of fixed-matrix regenerators, but of course it can also be used for the design of rotary regenerators. In the Λ-Π method, several different designations are used to classify regenerators, depending on the values of Λ and Π. Such designations and their equivalent dimensionless groups of the ε-NTU$_o$ method are summarized in Table 5.2.

For the effectiveness of the most general unbalanced and unsymmetrical regenerator, Razelos (1979) proposed an alternative set of four dimensionless groups, instead of those of Eq. (5.86):

$$\varepsilon, \varepsilon_r, \varepsilon_h, \varepsilon_c = \phi(\Lambda_m, \Pi_m, \gamma, \mathbf{R}^*) \tag{5.89}$$

[†] If the dwell periods $\tau_{d,h}$ and $\tau_{c,h}$ are not negligible, Π_h and Π_c would be

$$\Pi_h = \eta_h(P_h) = \frac{h_h A_h}{\bar{C}_{r,h}}(P_h - \tau_{d,h}) \quad \Pi_c = \eta_c(P_c) = \frac{h_c A_c}{\bar{C}_{r,c}}(P_c - \tau_{d,c}) \tag{5.85}$$

TABLE 5.2 Designation of Various Types of Regenerators Depending on the Values of Dimensionless Groups

Regenerator Terminology	Λ-Π Method	ε-NTU$_o$ Method
Balanced	$\Lambda_h/\Pi_h = \Lambda_c/\Pi_c$ or $\gamma = 1$	$C^* = 1$
Unbalanced	$\Lambda_h/\Pi_h \neq \Lambda_c/\Pi_c$	$C^* \neq 1$
Symmetric	$\Pi_h = \Pi_c$ or $\mathbf{R}^* = 1$	$(hA)^* = 1$
Unsymmetric	$\Pi_h \neq \Pi_c$	$(hA)^* \neq 1$
Symmetric and balanced	$\Lambda_h = \Lambda_c, \Pi_h = \Pi_c$	$(hA)^* = 1, C^* = 1$
Unsymmetric but balanced	$\Lambda_h/\Pi_h = \Lambda_c/\Pi_c$	$(hA)^* \neq 1, C^* = 1$
Long	$\Lambda/\Pi > 5$	$C_r^* > 5$

Source: Data from Shah (1985).

where Λ_m and Π_m are the *mean reduced length* and *mean reduced period*, respectively. They have been proposed by Hausen (1983) as the harmonic means in the following sense:

$$\frac{1}{\Pi_m} = \frac{1}{2}\left(\frac{1}{\Pi_h} + \frac{1}{\Pi_c}\right) \tag{5.90}$$

$$\frac{1}{\Lambda_m} = \frac{1}{2\Pi_m}\left(\frac{\Pi_h}{\Lambda_h} + \frac{\Pi_c}{\Lambda_c}\right) \tag{5.91}$$

and γ and $\mathbf{R}^*$ are defined as

$$\gamma = \frac{\Pi_c/\Lambda_c}{\Pi_h/\Lambda_h} = \frac{C_c P_c}{M_{w,c}c_w}\frac{M_{w,h}c_w}{C_h P_h} = \frac{C_c}{C_h} \tag{5.92}$$

$$\mathbf{R}^* = \frac{\Pi_h}{\Pi_c} = \frac{(hA)_h}{(hA)_c}\frac{P_h}{M_{w,h}c_w}\frac{M_{w,c}c_w}{P_c} = \frac{(hA)_h}{(hA)_c} \tag{5.93}$$

In Eqs. (5.92) and (5.93), note that $M_{w,h}c_w/P_h = C_{r,h}$, $M_{w,c}c_w/P_c = C_{r,c}$, and $C_{r,h} = C_{r,c} = C_r$, from Eq. (5.8). Razelos (1979) also showed that the influence of $\mathbf{R}^*$ on ε_r is negligible for $1 \leq \mathbf{R}^* \leq 5$.[†] Thus,

$$\varepsilon_r = \phi(\Lambda_m, \Pi_m, \gamma) \tag{5.94}$$

He also pointed out that

$$\varepsilon_r(\Lambda_m, \Pi_m, \gamma) = \varepsilon_r(\Lambda_m, \Pi_m, 1/\gamma) \tag{5.95}$$

and therefore the tabulation of ε_r is needed only for $\gamma \leq 1$.

Schack (1965) and others have characterized the measure of fixed-matrix regenerator performance in terms of the matrix (brick)[‡] utilization coefficient η. It is defined as the

---

[†] Since $\mathbf{R}^* = 1/(hA)^*$ for $C_c = C_{\min}$, it has been already shown by Lambertson (1958) that the influence of $(hA)^*$ is negligible on ε for $0.25 \leq (hA)^* \leq 4$.

[‡] See Fig. 1.47 for brick geometries used in a fixed-matrix regenerator.

ratio of the heat that the matrix (brick) actually absorbs or gives up during one period to the heat it would have absorbed or given up if it had infinite transverse thermal conductivity. The amount of heat that the matrix absorbs or gives up during one period is the same as the amount of heat transferred from the hot gas to the matrix or the amount given up from the matrix to cold gas during one period.

$$\eta = \frac{C_h P_h (T_{h,i} - \bar{T}_{h,o})}{(M_w c_w)_h (T_{h,i} - T_{c,i})} = \frac{C_c P_c (\bar{T}_{c,o} - T_{c,i})}{(M_w c_w)_c (T_{h,i} - T_{c,i})} = \frac{\Pi_c}{\Lambda_c} \varepsilon_c = \frac{\varepsilon}{C_r^*} \qquad (5.96)$$

Thus it is clear that the higher the value of C_r^*, the lower is the utilization coefficient, indicating the lower amount of heat stored in the regenerator.

Before presenting the specific results, let us compare the dimensionless groups of the Λ-Π method with those of the ε-NTU$_o$ method.

5.3.1   Comparison of the ε-NTU$_o$ and Λ-Π Methods

The functional relationships for these methods are given by Eqs. (5.52) and (5.86) or (5.89). For comparison purposes, we consider $C_c = C_{\min}$. The regenerator effectiveness ε is related to ε_r, ε_h, and ε_c as

$$\varepsilon = \begin{cases} \dfrac{(\gamma + 1)\varepsilon_r}{2\gamma} \\ \dfrac{\varepsilon_h}{\gamma} \\ \varepsilon_c \end{cases} \qquad (5.97)$$

The independent variables of Eqs. (5.52) and (5.89) are related as follows for $C_c = C_{\min}$:

$$\text{NTU}_o = \left(\frac{C^*}{\text{ntu}_h} + \frac{1}{\text{ntu}_c} \right)^{-1} = \frac{\Lambda_m (1 + \gamma)}{4\gamma} \qquad (5.98)$$

$$C^* = \gamma \qquad (5.99)$$

$$C_r^* = \frac{\Lambda_m (1 + \gamma)}{2\gamma \Pi_m} \qquad (5.100)$$

$$(hA)^* = \frac{1}{R^*} \qquad (5.101)$$

Similarly, the independent variables of Eqs. (5.86) and (5.47) are related as follows for $C_c = C_{\min}$:

$$\Lambda_h = \left(\frac{hA}{C}\right)_h = \text{ntu}_h = C^*\left[1 + \frac{1}{(hA)^*}\right]\text{NTU}_o \tag{5.102}$$

$$\Lambda_c = \left(\frac{hA}{C}\right)_c = \text{ntu}_c = [1 + (hA)^*]\text{NTU}_o \tag{5.103}$$

$$\Pi_h = \left(\frac{hA}{\bar{C}_r}\right)_h P_h = \frac{\text{ntu}_h}{C^*_{r,h}} = \frac{1}{C^*_r}\left[1 + \frac{1}{(hA)^*}\right]\text{NTU}_o \tag{5.104}^\dagger$$

$$\Pi_c = \left(\frac{hA}{\bar{C}_r}\right)_c P_c = \frac{\text{ntu}_c}{C^*_{r,c}} = \frac{1}{C^*_r}[1 + (hA)^*]\text{NTU}_o \tag{5.105}$$

Noting the relationships of Eqs. (5.98)–(5.105), it is clear that there is a one-to-one correspondence between the dimensionless groups in the ε-NTU$_o$ and Λ-Π methods.

To summarize in concise form, the general functional relationship and basic definitions of dimensionless groups of ε-NTU$_o$, and Λ-Π methods are presented in Table 5.3, the relationship between the dimensionless groups of these methods in Table 5.4, and the definitions of dimensionless groups in Table 5.5.

TABLE 5.3 General Functional Relationships and Basic Definitions of Dimensionless Groups for ε-NTU$_o$ and Λ-Π Methods for Counterflow Regenerators

ε-NTU$_o$ Method	Λ-Π Method[a]
$q = \varepsilon C_{\min}(T_{h,i} - T_{c,i})$	$Q = \varepsilon_h C_h P_h(T_{h,i} - T_{c,i}) = \varepsilon_c C_c P_c(T_{h,i} - T_{c,i})$
$\varepsilon = \phi[\text{NTU}_o, C^*, C^*_r, (hA)^*]$	$\varepsilon_r, \varepsilon_h, \varepsilon_c = \phi(\Lambda_m, \Pi_m, \gamma, \mathbf{R}^*)$
$\varepsilon = \dfrac{C_h(T_{h,i} - T_{h,o})}{C_{\min}(T_{h,i} - T_{c,i})} = \dfrac{C_c(T_{c,o} - T_{c,i})}{C_{\min}(T_{h,i} - T_{c,i})}$	$\varepsilon_h = \dfrac{Q_h}{Q_{\max,h}} = \dfrac{C_h P_h(T_{h,i} - \bar{T}_{h,o})}{C_h P_h(T_{h,i} - T_{c,i})} = \dfrac{T_{h,i} - \bar{T}_{h,o}}{T_{h,i} - T_{c,i}}$
$\text{NTU}_o = \dfrac{1}{C_{\min}}\dfrac{1}{1/(hA)_h + 1/(hA)_c}$	$\varepsilon_c = \dfrac{Q_c}{Q_{\max,c}} = \dfrac{C_c P_c(\bar{T}_{c,o} - T_{c,i})}{C_c P_c(T_{h,i} - T_{c,i})} = \dfrac{\bar{T}_{c,o} - T_{c,i}}{T_{h,i} - T_{c,i}}$
$C^* = \dfrac{C_{\min}}{C_{\max}}$	$\varepsilon_r = \dfrac{Q_h + Q_c}{Q_{\max,h} + Q_{\max,c}} = \dfrac{2Q}{Q_{\max,h} + Q_{\max,c}}$
$C^*_r = \dfrac{C_r}{C_{\min}}$	$\dfrac{1}{\varepsilon_r} = \dfrac{1}{2}\left(\dfrac{1}{\varepsilon_h} + \dfrac{1}{\varepsilon_c}\right) \qquad \varepsilon = \left(\dfrac{1+\gamma}{2\gamma}\right)\varepsilon_r$
$(hA)^* = \dfrac{hA \text{ on the } C_{\min} \text{ side}}{hA \text{ on the } C_{\max} \text{ side}}$	$\dfrac{1}{\Pi_m} = \dfrac{1}{2}\left(\dfrac{1}{\Pi_h} + \dfrac{1}{\Pi_c}\right) \quad \dfrac{1}{\Lambda_m} = \dfrac{1}{2\Pi_m}\left(\dfrac{\Pi_h}{\Lambda_h} + \dfrac{\Pi_c}{\Lambda_c}\right)$
	$\gamma = \dfrac{\Pi_c/\Lambda_c}{\Pi_h/\Lambda_h} \quad \mathbf{R}^* = \dfrac{\Pi_h}{\Pi_c} \qquad \Lambda_h = \dfrac{(hA)_h}{C_h}$
	$\Lambda_c = \dfrac{(hA)_c}{C_c} \quad \Pi_h = \left(\dfrac{hA}{C_r}\right)_h \quad \Pi_c = \left(\dfrac{hA}{C_r}\right)_c$

Source: Data from Shah (1985).
[a] P_h and P_c represent hot- and cold-gas flow periods, respectively, in seconds.

---

† If the hot- and cold-gas dwell times are not neglected in the definition of Π_h and Π_c, the right-hand term of Eqs. (5.104) and (5.105) should be multiplied by $(1 - \tau_{d,h}/P_h)$ and $(1 - \tau_{d,c}/P_c)$, respectively.

TABLE 5.4   Relationship between Dimensionless Groups of ε-NTU$_o$ and Λ-Π Methods for $C_c = C_{min}$ ^{a}

ε-NTU$_o$	Λ-Π
$\text{NTU}_o = \dfrac{\Lambda_m(1+\gamma)}{4\gamma} = \dfrac{\Lambda_c/\Pi_c}{1/\Pi_h + 1/\Pi_c}$	$\Lambda_h = C^*\left[1 + \dfrac{1}{(hA)^*}\right]\text{NTU}_o$
$C^* = \gamma = \dfrac{\Pi_c/\Lambda_c}{\Pi_h/\Lambda_h}$	$\Lambda_c = [1 + (hA)^*]\text{NTU}_o$
$C_r^* = \dfrac{\Lambda_m(1+\gamma)}{2\gamma\Pi_m} = \dfrac{\Lambda_c}{\Pi_c}$	$\Pi_h = \dfrac{1}{C_r^*}\left[1 + \dfrac{1}{(hA)^*}\right]\text{NTU}_o$
$(hA)^* = \dfrac{1}{\mathbf{R}^*} = \dfrac{\Pi_c}{\Pi_h}$	$\Pi_c = \dfrac{1}{C_r^*}[1 + (hA)^*]\text{NTU}_o$

Source: Data from Shah (1985).
^{a} $\varepsilon = \varepsilon_c = \varepsilon_h/\gamma = (\gamma+1)(\varepsilon_r/2\gamma)$ for $C_c = C_{min}$. If $C_h = C_{min}$, the subscripts c and h in this table should be changed to h and c, respectively.

TABLE 5.5   Definitions of Dimensionless Groups for Regenerators in Terms of Dimensional Variables of Rotary and Fixed-Matrix Regenerators for $C_c = C_{min}$ ^{a}

Dimensionless Group	Rotary Regenerator	Fixed-Matrix Regenerator
NTU$_o$	$\dfrac{h_c A_c}{C_c}\dfrac{h_h A_h}{h_h A_h + h_c A_c}$	$\dfrac{h_c A_c}{C_c}\dfrac{h_h P_h}{h_h P_h + h_c P_c}$
C^*	$\dfrac{C_c}{C_h}$	$\dfrac{C_c P_c}{C_h P_h}$
C_r^*	$\dfrac{M_w c_w \mathbf{N}}{C_c}$	$\dfrac{M_w c_w}{C_c P_c}$
$(hA)^*$	$\dfrac{h_c A_c}{h_h A_h}$	$\dfrac{h_c P_c}{h_h P_h}$
$\dfrac{1}{\Lambda_m}$	$\dfrac{C_h + C_c}{4}\left(\dfrac{1}{h_h A_h} + \dfrac{1}{h_c A_c}\right)$	$\dfrac{C_h P_h + C_c P_c}{4A}\left(\dfrac{1}{h_h P_h} + \dfrac{1}{h_c P_c}\right)$
$\dfrac{1}{\Pi_m}$	$\dfrac{M_w c_w \mathbf{N}}{2}\left(\dfrac{1}{h_h A_h} + \dfrac{1}{h_c A_c}\right)$	$\dfrac{M_w c_w}{2A}\left(\dfrac{1}{h_h P_h} + \dfrac{1}{h_c P_c}\right)$
γ	$\dfrac{C_c}{C_h}$	$\dfrac{C_c P_c}{C_h P_h}$
$\mathbf{R}^*$	$\dfrac{h_h A_h}{h_c A_c}$	$\dfrac{h_h P_h}{h_c P_c}$

Source: Data from Shah (1985).
^{a} If $C_h = C_{min}$, the subscripts c and h in this table should be changed to h and c, respectively. The definitions are given for one rotor (disk) of a rotary regenerator or for one matrix of a fixed-matrix regenerator.

5.3.2 Solutions for a Counterflow Regenerator

Hausen (1929) obtained the fluid temperature distributions for a *balanced and symmetric* counterflow regenerator $[C^* = 1$ and $(hA)^* = 1)]$ in the form of infinite series by using the method of separation of variables for solving sets of governing equations (5.71)–(5.72) with the boundary and periodic-flow conditions of Eqs. (5.20)–(5.23). Subsequently, integrating exit temperatures over the entire duration of the period, average exit temperatures and ultimately the effectiveness can be determined. In this case,

$$\varepsilon = \phi(\Lambda, \Pi) \tag{5.106}$$

where $\Lambda = \Lambda_h = \Lambda_c$ and $\Pi = \Pi_h = \Pi_c$. His results are shown in Fig. 5.5. From the relationships of Eqs. (5.102)–(5.105), we have

$$\Lambda = 2\mathrm{NTU}_o \qquad \Pi = \frac{2\mathrm{NTU}_o}{C_r^*} \tag{5.107}$$

Using the results of Table 5.1 and Eq. (5.107), the effectiveness of Fig. 5.5 can be determined for the balanced and symmetric $[C^* = 1, (hA)^* = 1]$ regenerators.

For a general (balanced/unbalanced and symmetric/asymmetric) counterflow regenerator, Dragutinović and Bačlić (1998) have presented exact analytical relations and computational algorithms for evaluation of temperature distributions and regenerator effectiveness in terms of four dimensionless groups; they refer to them as the utilization factor U_1, reduced length Λ_1, unbalance factor β, and asymmetry factor σ, defined as follows.

$$U_1 = \frac{\Pi_1}{\Lambda_1} = \frac{(CP)_1}{M_w c_w} \qquad \Lambda_1 = \left(\frac{hA}{C}\right)_1 \qquad \beta = \frac{U_1}{U_2} = \frac{(CP)_1}{(CP)_2} \qquad \sigma = \frac{\Lambda_1}{\Lambda_2} = \left(\frac{hA}{C}\right)_1 \left(\frac{C}{hA}\right)_2 \tag{5.108}$$

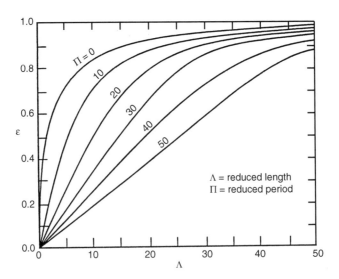

FIGURE 5.5 Effectiveness chart for a balanced and symmetric counterflow regenerator (From Hausen, 1983).

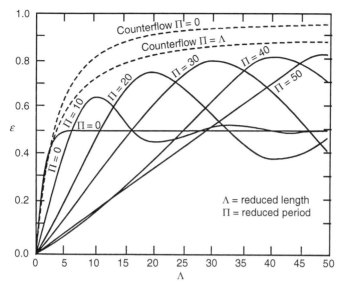

FIGURE 5.6   Effectiveness chart for a balanced and symmetric parallelflow regenerator (From Hausen, 1983).

They provided comprehensive tabular results for the counterflow regenerator effectiveness for the following ranges of the dimensionless groups: $\Lambda_1 \sim 1$ to ∞; $U_1 \sim 0$ to 2, $\beta_1 \sim 0$ to 1, and $\sigma \sim 0$ to 1.

5.3.3   Solution for a Parallelflow Regenerator

Hausen (1983) also obtained the solution for a balanced and symmetric parallelflow regenerator as shown in Fig. 5.6. The oscillations in ε above and below $\varepsilon = 0.5$ are clearly observed in this figure (see Fig. 5.4 for similar trends, although clearly not seen due to the shortened abscissa, $\mathrm{NTU}_o = 10 \rightarrow \Lambda = 20$). Comparing the results for the effectiveness of a counterflow regenerator (Fig. 5.5) and parallelflow (Fig. 5.6), it becomes obvious that parallelflow provides significantly smaller effectiveness for the same regenerator length and period duration. The performance of a regenerator reduces to the performance of a parallelflow recuperator in the limit of an infinitely short period ($\Pi = 0$).

Example 5.4   A rotary regenerator has a circular matrix of 2 m diameter and 1 m depth rotating at 1 rpm. The metal matrix weighs 3500 kg and has a total surface area of 7900 m². Its specific heat is 0.43 kJ/kg · K. Flue gas at 500°C flows through the matrix at the rate of 52 kg/s and air at 20°C flows at the same flow rate in a countercurrent fashion. Determine the air and flue gas outlet temperatures assuming that the isobaric specific heat is 1.05 kJ/kg · K and the heat transfer coefficient is 130 W/m² · K for both the flue gas and air. Idealize the regenerator as symmetric and balanced.

SOLUTION

Problem Data and Schematic: The heat transfer coefficients, fluid flow rates, and inlet temperatures are provided in Fig. E5.4 for both hot and cold streams. In addition, the

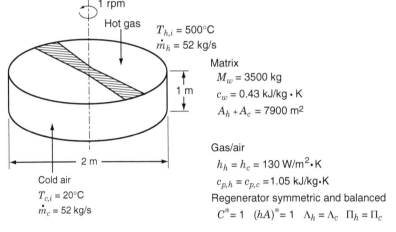

FIGURE E5.4

regenerator dimensions and rotational speed are specified. Also specified are the physical properties of both hot and cold fluid streams as well as the matrix material.

Determine: The cold- and hot-fluid outlet temperatures.

Assumptions: The assumptions of Section 5.1.1 are invoked here.

Analysis: For a symmetric and balanced regenerator, the reduced length Λ and the reduced period Π are equal on the hot and cold sides:

$$\Lambda_h = \Lambda_c = \Lambda = \Lambda_m = \frac{hA}{\dot{m}c_p} = \text{ntu} \tag{1}$$

$$\Pi_h = \Pi_c = \Pi = \Pi_m = \frac{hAP}{M_w c_w} \tag{2}$$

Note that in Eq. (2), P is the total period if M_w is the total matrix mass of the regenerator; otherwise, $P = P_c$ or P_h when $M_w = M_{w,c}$ or $M_{w,h}$.

The heat transfer coefficients, flow rates, and specific heats are equal on the hot and cold fluid sides. Based on Eqs. (1) and (2), for a symmetric and balanced regenerator ($\Lambda_h = \Lambda_c, \Pi_h = \Pi_c$), this implies that the flow split between the hot and cold sides must be 50:50 and the heat transfer surface area must be equal. Now let us first determine the reduced length Λ and the reduced period Π from using Eqs. (1) and (2).

$$\Lambda = \frac{hA}{\dot{m}c_p} = \frac{0.130\,\text{kW/m}^2 \cdot \text{K} \times (7900/2)\,\text{m}^2}{52\,\text{kg/s} \times 1.05\,\text{kJ/kg} \cdot \text{K}} = 9.4$$

$$\Pi = \frac{hAP}{M_w c_w} = \frac{0.130\,\text{kW/m}^2 \cdot \text{K} \times 3950\,\text{m}^2 \times 60\,\text{s}}{3500\,\text{kg} \times 0.43\,\text{kJ/kg} \cdot \text{K}} = 20.47$$

With $\Lambda = 9.4$ and $\Pi = 20.47$, the regenerator effectiveness is obtained as $\varepsilon = 0.42$ from Fig. 5.5. Knowing ε together with the inlet temperatures $T_{h,i}$ and $T_{c,i}$ of the hot and cold fluid streams, the cold-fluid outlet temperature $\bar{T}_{c,o}$ and the hot-fluid temperature outlet $\bar{T}_{h,o}$ are determined from Eqs. (5.79) and (5.80).

$$\bar{T}_{c,o} = T_{c,i} + \varepsilon\left(T_{h,i} - T_{c,i}\right) = 20°C + 0.42(500 - 20)°C = 221.6°C \qquad Ans.$$

$$\bar{T}_{h,o} = T_{h,i} - \varepsilon\left(T_{h,i} - T_{c,i}\right) = 500°C - 0.42(500 - 20)°C = 298.4°C \qquad Ans.$$

Alternatively, the example can be analyzed by the ε-NTU$_o$ method as follows:

$$C_{min} = C_h = C_c = \dot{m}c_p = 52 \text{ kg/s} \times 1.05 \text{ kJ/kg} \cdot \text{s} = 54.6 \text{ kW/K}$$

$$C^* = \frac{C_{min}}{C_{max}} = 1$$

$$C_r = M_w c_w \mathbf{N} = 3500 \text{ kg} \times 0.43 \text{ kJ/kg} \cdot \text{K} \times \tfrac{1}{60} \text{rev/s} = 25.08 \text{ kW/K}$$

$$C_r^* = \frac{C_r}{C_{min}} = \frac{25.08 \text{ kW/K}}{54.6 \text{ kW/K}} = 0.459$$

$$(hA)_h = (hA)_c = 0.130 \text{ kW/m}^2 \cdot \text{K} \times 3950 \text{ m}^2 = 513.5 \text{ kW/K}$$

Now let us calculate NTU$_o$.

$$NTU_o = \frac{1}{C_{min}} \frac{1}{1/(hA)_h + 1/(hA)_c} = \frac{1}{54.6 \text{ kW/K}} \left[\frac{1}{(1/513.5 + 1/513.5) \text{ K/kW}}\right] = 4.702$$

Now the regenerator effectiveness ε could have been determined by using Fig. 5.3 if the graph for $C_r^* = 0.459$ or a close value (0.50) would have been available. Hence, we will resort to Eqs. (5.62) and (5.63) for the determination of ε. Using Eq. (5.63) for $C^* = 1$,

$$\varepsilon_{cf} = \frac{NTU_o}{1 + NTU_o} = \frac{4.702}{1 + 4.702} = 0.8246$$

Subsequently, ε from Eq. (5.62) is

$$\varepsilon = 0.8246\left(1 - \frac{1}{9 \times 0.459^{1.93}}\right) = 0.413$$

The outlet temperatures, using the definition of ε from Eq. (5.45), are

$$\bar{T}_{c,o} = T_{c,i} + \varepsilon\left(T_{h,i} - T_{c,i}\right) = 20°C + 0.413(500 - 20)°C = 218.2°C \qquad Ans.$$

$$\bar{T}_{h,o} = T_{h,i} - \varepsilon\left(T_{h,i} - T_{c,i}\right) = 500°C - 0.413(500 - 20)°C = 301.8°C \qquad Ans.$$

Discussion and Comments: First it is shown how to solve the problem by the Λ-Π method. Since Fig. 5.5 is valid only for the balanced and symmetric regenerator, Π should be evaluated carefully using Eq. (2), where A represents the heat transfer surface area on only one fluid side (hot or cold), *not* the total surface area; however, P/M_w can be evaluated either for the total matrix or for the hot (or cold) gas side. The rest of the procedure for determining ε is straightforward.

The same problem is solved by the ε-NTU$_o$ method for illustration. The regenerator effectiveness is determined graphically for the Λ-Π method and by an empirical formula for the ε-NTU$_o$ method; they are found to be close enough within the reading accuracy of the graphical results. Thus, as expected, the results should be identical regardless of which method is used for the analysis.

5.4 INFLUENCE OF LONGITUDINAL WALL HEAT CONDUCTION

Longitudinal heat conduction in the wall may not be negligible, particularly for a high-effectiveness regenerator having a short flow length L. Longitudinal wall heat conduction reduces the exchanger effectiveness as discussed in Section 4.1 with Fig. 4.1, hence it is important that its influence on ε be determined quantitatively. It should be emphasized that one end of the matrix (regenerator) in a counterflow regenerator is always hotter than the other end during both hot and cold blow periods. Hence, longitudinal heat conduction in the matrix wall occurs in the same direction through both periods.

In Section 4.1, we used a heuristic approach to derive the dimensionless groups associated with the longitudinal wall heat conduction effect; here we use a more rigorous approach. The basic differential energy balance equations, Eqs. (5.14) and (5.18) for the hot and cold fluid sides, do not change, but those for the wall change. For finite axial heat conduction in the wall during the hot-gas flow period, the model is modified by adding the corresponding heat conduction terms as shown in Fig. 5.7 (compare with Fig. 5.2b).

Applying the energy balance to the differential element of the wall in Fig. 5.7, we get

$$-k_w A_{k,h} \frac{\partial T_w}{\partial x} - \left[-k_w A_{k,h} \left(\frac{\partial T_w}{\partial x} + \frac{\partial^2 T_w}{\partial x^2} \, dx \right) \right] + h_h \left(\frac{A_h \, dx}{L} \right)(T_h - T_w) = \left(\bar{C}_{r,h} \frac{dx}{L} \right) \frac{\partial T_w}{\partial \tau_h}$$

(5.109)

Upon simplification,

$$\frac{\partial T_w}{\partial \tau_h} = \frac{(hA)_h}{\bar{C}_{r,h}}(T_h - T_w) + \frac{k_w A_{k,h} L}{\bar{C}_{r,h}} \frac{\partial^2 T_w}{\partial x^2}$$

(5.110)

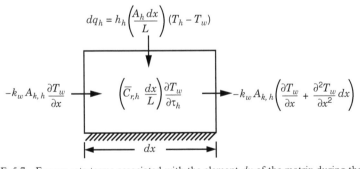

FIGURE 5.7   Energy rate terms associated with the element dx of the matrix during the hot-gas flow period; longitudinal heat conduction has been included (From Shah, 1981).

Using the previous definitions of X^*, τ_h^*, T_h^*, T_w^*, ntu$_h$, and $C_{r,h}^*$ [Eqs. (5.25)–(5.30)], Eq. (5.110) is made dimensionless:

$$\frac{\partial T_w^*}{\partial \tau_h^*} = \frac{\text{ntu}_h}{C_{r,h}^*}(T_h^* - T_w^*) + \frac{\lambda_h}{C_{r,h}^*}\frac{\partial^2 T_w^*}{\partial X^{*2}} \tag{5.111}$$

where

$$\lambda_h = \frac{k_w A_{k,h}}{L C_h} \tag{5.112}$$

Similarly, the governing differential equation for the matrix wall temperature during the cold-gas flow period is

$$\frac{\partial T_w^*}{\partial \tau_c^*} = \frac{\text{ntu}_c}{C_{r,c}^*}(T_w^* - T_c^*) + \frac{\lambda_c}{C_{r,c}^*}\frac{\partial^2 T_w^*}{\partial X^{*2}} \tag{5.113}$$

where

$$\lambda_c = \frac{k_w A_{k,c}}{L C_c} \tag{5.114}$$

Since Eqs. (5.111) and (5.113) are second-order partial differential equations with respect to X^*, we need to define four boundary conditions for the matrix wall temperatures in order to get a particular solution: two during the hot-gas flow period and two during the cold-gas flow period. The realistic boundary conditions are adiabatic boundary conditions for each period at $X^* = 0$ and 1:

$$\left(\frac{\partial T_w^*}{\partial X^*}\right)_{X^*=0} = \begin{cases} \left(\dfrac{\partial T_w^*}{\partial X^*}\right)_{X^*=1} = 0 & \text{for } 0 \le \tau_h^* \le 1 \quad (5.115) \\[3mm] \left(\dfrac{\partial T_w^*}{\partial X^*}\right)_{X^*=1} = 0 & \text{for } 0 \le \tau_c^* \le 1 \quad (5.116) \end{cases}$$

Thus, the inclusion of the effect of longitudinal heat conduction adds two dimensionless groups, λ_h and λ_c, on which the exchanger effectiveness ε would depend. Bahnke and Howard (1964) suggested an alternative set of two dimensionless groups:

$$\lambda = \frac{k_w A_{k,t}}{L C_{\min}} \qquad A_k^* = \frac{A_k \text{ on the } C_{\min} \text{ side}}{A_k \text{ on the } C_{\max} \text{ side}} \tag{5.117}$$

where $A_{k,t}$ is the total area for longitudinal conduction,

$$A_{k,t} = A_{k,h} + A_{k,c} = A_{fr} - A_o = A_{fr}(1 - \sigma) \tag{5.118}$$

Note that λ and A_k^* are related to λ_h and λ_c as follows for $C_c = C_{\min}$:

$$\lambda = \lambda_c + \frac{\lambda_h}{C^*} \qquad A_k^* = \frac{A_{k,c}}{A_{k,h}} = C^*\frac{\lambda_c}{\lambda_h} \tag{5.119}$$

This choice of dimensionless groups offers the advantage that the resulting ε is not affected significantly by A_k^* for $0.25 \le A_k^* \le 1$ (Bahnke and Howard, 1964; Skiepko and Shah, 1994). Thus, the effect of longitudinal heat conduction in the wall is taken into account by λ and is added to the functional relationship for ε of Eq. (5.61):

$$\varepsilon = \phi(\text{NTU}_o, C^*, C_r^*, \lambda) \tag{5.120}$$

In order to obtain an exact solution for this problem, Eqs. (5.31), (5.33), (5.111), and (5.113) need to be solved using the boundary conditions and periodic equilibrium conditions of Eqs. (5.35) through (5.38), (5.115), and (5.116). The first closed-form analytical solution to these equations was obtained by Skiepko (1988). Bahnke and Howard (1964) obtained a numerical solution by a finite difference method. They determined the exchanger effectiveness and longitudinal conduction effect over the following ranges of dimensionless parameters: $1 \le \text{NTU}_o \le 100$, $0.9 \le C^* \le 1$, $1 \le C_r^* \le \infty$, $0.01 \le \lambda \le 0.32$, $0.25 \le (hA)^* \le 1$, $0.25 \le A_k^* \le 1$. The ineffectiveness $(1 - \varepsilon)$ as a function of NTU_o and λ is shown in Fig. 5.8 for $C^* = 0.95$ and $C_r^* > 5$. Similar results for $C^* = 1$ and $C_r^* > 5$ are already shown in Fig. 4.2.

Bahnke and Howard's results are correlated by Shah (1975) as follows[†]:

$$\varepsilon = \varepsilon_{cf} \left[1 - \frac{1}{9(C_r^*)^{1.93}} \right] \left(1 - \frac{C_\lambda}{2 - C^*} \right) \tag{5.121}$$

where

$$C_\lambda = \frac{1}{1 + \text{NTU}_o(1 + \lambda\Phi)/(1 + \lambda\text{NTU}_o)} - \frac{1}{1 + \text{NTU}_o} \tag{5.122}$$

and

$$\Phi = \left(\frac{\lambda\text{NTU}_o}{1 + \lambda\text{NTU}_o} \right)^{1/2} \tanh \left\{ \frac{\text{NTU}_o}{[\lambda\text{NTU}_o/(1 + \lambda\text{NTU}_o)]^{1/2}} \right\} \tag{5.123a}$$

$$\approx \left(\frac{\lambda\text{NTU}_o}{1 + \lambda\text{NTU}_o} \right)^{1/2} \quad \text{for NTU}_o \ge 3 \tag{5.123b}$$

The regenerator effectiveness ε of Eq. (5.121) agrees well within $\pm 0.5\%$ with the results of Bahnke and Howard for the following range of parameters: $3 \le \text{NTU}_o \le 12$, $0.9 \le C^* \le 1$, $2 \le C_r^* \le \infty$, $0.5 \le (hA)^* \le 1$, and $0 \le \lambda \le 0.04$. It agrees within $\pm 1\%$ for the following range of parameters: $1 \le \text{NTU}_o \le 20$, $0.9 \le C^* \le 1$, $2 \le C_r^* \le \infty$, $0.25 \le (hA)^* \le 1$, and $0 \le \lambda \le 0.08$.

The following is a more accurate method for a wider range of $C^* < 1$.

1. Use the Razelos method to compute $\varepsilon_{r,\lambda=0}$ for an equivalent balanced regenerator using the procedure from Eqs. (5.64)–(5.66).
2. Compute C_λ from Eq. (5.122) using computed $\text{NTU}_{o,m}$ and given λ.
3. Calculate $\varepsilon_{r,\lambda\neq0} = C_\lambda \varepsilon_{r,\lambda=0}$.
4. Finally, determine ε from Eq. (5.67) with ε_r replaced by $\varepsilon_{r,\lambda\neq0}$.

[†] The correlation of Shah (1975) has been modified here by replacing Hahnemann's solution with Kroeger's solution (1967) for $\kappa > 0$ because of the ready availability of the later reference. The accuracy of the results is identical.

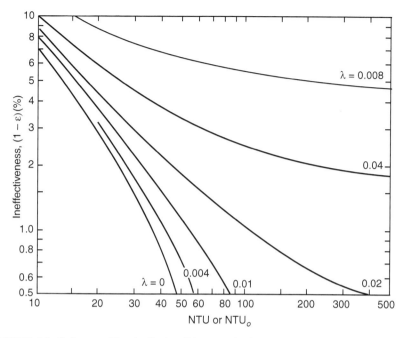

FIGURE 5.8   Influence of longitudinal wall heat conduction on the performance of storage and direct-transfer counterflow exchangers; $C^* = 0.95$. (From Kays and London, 1998.)

This procedure yields ε, which is accurate within 1% for $1 \leq \text{NTU}_o \leq 20$ for $C_r^* \leq 1$ when compared to the results of Bahnke and Howard.

A careful review of Figs 5.8 and 4.2 and Eq. (5.121) reveals that longitudinal heat conduction in the wall reduces the regenerator effectiveness. Thus, similar to a recuperator, due to longitudinal heat conduction in the wall, the regenerator effectiveness ε decreases with increasing values of NTU, C^*, and λ, and the decrease in ε is largest for $C^* = 1$. However, the effect of increasing C_r^* when NTU, C^*, and λ are kept constant is complicated. For given values of NTU_o, C^*, and λ, increasing C_r^* increases $\Delta\varepsilon/\varepsilon$ at small values of NTU_o for $\lambda \leq 0.32$; at intermediate values of NTU_o (≈ 4 to 5), increasing C_r^* from 1 to 5 (and higher) first increases and then decreases $\Delta\varepsilon/\varepsilon$; at large values of NTU_o (≥ 9), increasing C_r^* from 1 to 10 decreases $\Delta\varepsilon/\varepsilon$. Bahnke and Howard's results show that Kays and London's approximation (1998)

$$\frac{\Delta\varepsilon}{\varepsilon} = \lambda \tag{5.124}$$

is a very good engineering approximation for $\text{NTU}_o > 10$ and $\lambda < 0.1$.

Longitudinal heat conduction can have a serious impact on the regenerator effectiveness or NTU for an ultra high-effectiveness regenerator. For example, a Stirling engine regenerator may require 350 ideal NTU to get 200 usable NTU due to longitudinal heat conduction. As a result, such regenerators may require a stack of high thermal conductivity (copper or aluminum) perforated plates (Venkatarathnam, 1996) or wire screens, alternating with low thermal conductivity spacers made up of plastic, stainless steel, and so on. Such a design would significantly reduce longitudinal heat conduction or the *stack conduction.*

No detailed temperature distributions for fluids and wall were obtained by either Lambertson (1958) or Bahnke and Howard (1964). Mondt (1964) obtained these temperature distributions by solving the differential equations numerically for some values of the associated dimensionless groups. Illustrative results are shown in Figs. 5.9–5.11.

In Fig. 5.9, the matrix wall temperatures T_w^* at $x = 0$, $x = L/2$, and $x = L$ are shown as functions of a dimensionless time for the $\lambda = 0$ case. Also shown are the hot- and cold-gas inlet and outlet temperatures. Experimental points shown for $T_{c,o}^*$ are in good agreement with the theoretical predictions. The wall temperatures are linear with time except for the sections of hot- and cold-fluid inlets.

In Fig. 5.10, hot-gas and matrix temperatures are shown as a function of the flow length X^* for $\tau_h = 0$, $P_h/2$, and P_h. Here, again, these temperature distributions are linear except for the regenerator ends.

In Fig. 5.11, the matrix wall temperatures are shown at switching time $\tau = P_h$ and P_c. The reduction in matrix wall temperature gradients at $X^* = 0$ due to longitudinal heat conduction is evident. Note that the time average $T_{c,o}^*$ is reduced, which in turn indicates that the exchanger effectiveness ε is reduced due to longitudinal wall heat conduction, as expected.

Skiepko (1988) presented three-dimensional temperature charts demonstrating how the longitudinal matrix heat conduction affects the matrix as well as the gas temperature distributions, shown as dependent on coordinate x and time τ.

Example 5.5 Determine the reduction in the regenerator effectiveness of Example 5.3 due to longitudinal wall heat conduction given that the thermal conductivity of the

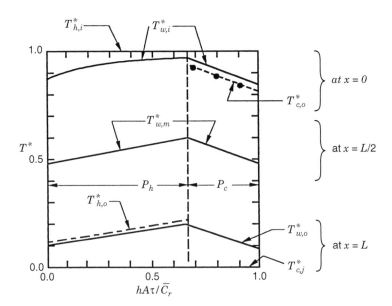

FIGURE 5.9   Cyclic temperature fluctuations in the matrix at the entrance, midway, and exit of a regenerator. (From Mondt, 1964.)

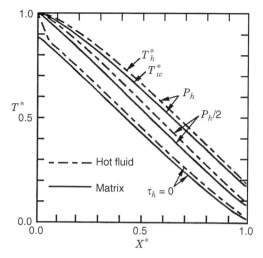

FIGURE 5.10   Fluid and matrix wall temperature excursion during hot-gas flow period. (From Mondt, 1964.)

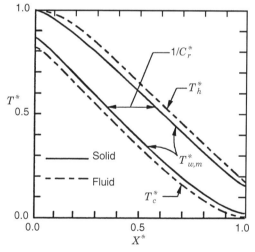

FIGURE 5.11   Balanced regenerator temperature distributions at switching instants. (From Mondt, 1964.)

ceramic matrix is 0.69 W/m · K and its porosity is 0.7. Use all other pertinent information from Example 5.3.

SOLUTION

Problem Data and Schematic: The heat transfer coefficients, fluid flow rates, and inlet temperatures are provided in Fig. E5.5 for both hot and cold streams. In addition, the

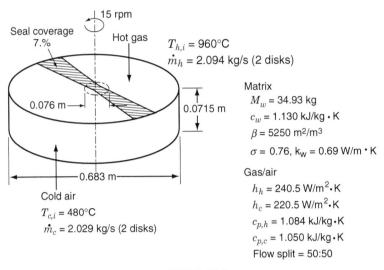

FIGURE E5.5

regenerator dimensions and rotational speed are specified. Also specified are the physical properties of both hot- and cold-fluid streams as well as the matrix material.

Determine: The regenerator effectiveness and outlet temperatures of both hot and cold fluids.

Assumptions: The assumptions of Section 5.1.1 are invoked here.

Analysis: To assess the impact of the longitudinal wall heat conduction on the regenerator effectiveness, we must first compute the total conduction area $A_{k,t}$ as the difference between the effective frontal area and the free-flow area. Thus,

$$A_{k,t} = A_{fr} - A_o = A_{fr}(1 - \sigma) = 0.673\,\text{m}^2 \times (1 - 0.70) = 0.2019\,\text{m}^2$$

Introducing this value of $A_{k,t}$ together with the value of $C_{\min} = 2130$ W/K computed in Example 5.3, we obtain from Eq. (5.117) the value of the longitudinal wall conduction parameter λ as

$$\lambda = \frac{k_w A_{k,t}}{L C_{\min}} = \frac{0.69\,\text{W/m} \cdot \text{K} \times 0.2019\,\text{m}^2}{0.0715\,\text{m} \times 2130\,\text{W/K}} = 9.133 \times 10^{-4}$$

Knowing λ and $\text{NTU}_o = 6.819$ from Example 5.3, the dimensionless parameter Φ, defined in Eq. (5.123b) for $\text{NTU}_o \geq 3$, can be determined as

$$\Phi = \left(\frac{\lambda \text{NTU}_o}{1 + \lambda \text{NTU}_o}\right)^{1/2} = \left(\frac{9.133 \times 10^{-4} \times 6.819}{1 + 9.133 \times 10^{-4} \times 6.819}\right)^{1/2} = 0.0787$$

Knowing λ, NTU_o, and Φ, the parameter C_λ, defined in Eq. (5.122), can be determined as

$$C_\lambda = \frac{1}{1 + \text{NTU}_o(1 + \lambda\Phi)/(1 + \lambda\text{NTU}_o)} - \frac{1}{1 + \text{NTU}_o}$$

$$= \frac{1}{1 + 6.819(1 + 9.133 \times 10^{-4} \times 0.0787)/(1 + 9.133 \times 10^{-4} \times 6.819)} - \frac{1}{1 + 6.819}$$

$$= 6.860 \times 10^{-4}$$

From Example 5.3 we have $\varepsilon_{cf} = 0.8945$, $C_r^* = 4.63$, and $C^* = 0.9383$. Hence, the value of the regenerator effectiveness with longitudinal conduction from Eq. (5.121) is

$$\varepsilon = \varepsilon_{cf}\left(1 - \frac{1}{9C_r^{*1.93}}\right)\left(1 - \frac{C_\lambda}{2 - C^*}\right)$$

$$= 0.8945\left(1 - \frac{1}{9 \times 4.63^{1.93}}\right)\left(1 - \frac{6.860 \times 10^{-4}}{2 - 0.9383}\right)$$

$$= 0.8888$$

On comparing this value with the value of $\varepsilon = 0.8893$ calculated in Example 5.3 without longitudinal heat conduction, we notice that longitudinal heat conduction accounts for only a 0.06% decrease in the regenerator effectiveness.

Discussion and Comments: This example was intended to demonstrate how to evaluate the effect of longitudinal heat conduction on the regenerator effectiveness. The result shows that this effect is negligible for this case primarily because of the very low value of λ. However, this may not be the case for matrices made up of metals, thick walls, or short regenerator flow lengths. The recommended practice is to include the effect of longitudinal wall heat conduction in a computer program and thus always consider it, no matter of how small or large.

5.5 INFLUENCE OF TRANSVERSE WALL HEAT CONDUCTION

One of the idealizations (assumption 9 in Section 5.1.1) made in the foregoing regenerator design theory is that the wall thermal resistance is zero. This assumption is invoked in deriving Eqs. (5.17) and (5.19). The temperature gradient in the wall thickness ($\delta_w/2$) direction in Fig. 5.2b is zero. It was also shown that the wall thermal resistance in U_oA of Eq. (5.53) is zero. The zero wall thermal resistance represents a good approximation for metal matrices having high thermal conductivity. For those matrices having thick walls or low thermal conductivity, such as ceramic matrices for fixed-matrix regenerators, the wall resistance may not be negligible.

5.5.1 Simplified Theory

A simplified method is now outlined to include the influence of wall thermal resistance on regenerator effectiveness. An additional but essential assumption made for the analysis[†] is: The temperatures of hot and cold gases and the wall at any cross section

---

[†] In this section, the analysis is made for finite thermal resistance in the wall thickness direction and infinite thermal resistance in the longitudinal direction (the zero-longitudinal heat conduction case).

in the regenerator are linear with time, and the numerical value of this time derivative of temperature at any point in the wall is the same. For the wall temperature, this means

$$\frac{\partial T_w}{\partial \tau} = \frac{\partial T_{w,o}}{\partial \tau} = \frac{\partial T_{w,m}}{\partial \tau} = \text{constant} \tag{5.125}$$

Here $T_{w,o}$ and $T_{w,m}$ are the surface wall temperature and mean wall temperature at a given instant in time. This linear temperature–time relationship represents a good approximation to the actual wall temperature profile in the greater part of either hot- or cold-gas flow period along most of the regenerator length, as shown in Fig. 5.9.

Now let us consider a differential element of the regenerator matrix wall as shown in Fig. 5.12 during the hot-gas flow period. Here A_w represents the conduction area for the wall for heat conduction in the y direction. For continuous flow passages, $A_w = A$ (the convective heat transfer surface area), and $A_w \approx A$ for noncontinuous flow passages. The energy balance on the element of Fig. 5.12b yields the well-known one-dimensional transient conduction equation valid at each x coordinate.

$$\alpha_w \frac{\partial^2 T_w}{\partial y^2} = \frac{\partial T_w}{\partial \tau} \tag{5.126}$$

where $\alpha_w = k_w / \rho_w c_w$ is the thermal diffusivity of the matrix material. The appropriate boundary conditions are (see Fig. 5.12)

$$T_w = T_{w,o} \quad \text{at } y = 0 \text{ and } \delta_w \tag{5.127}$$

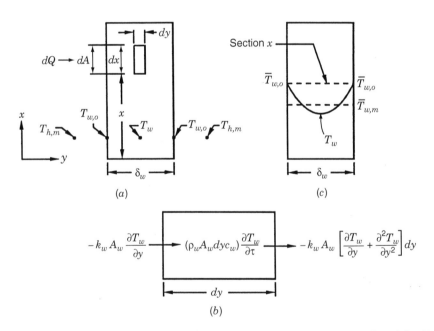

FIGURE 5.12 (a) Matrix wall of thickness δ_w, (b) energy transfer terms across the differential wall element; (c) parabolic temperature distribution in the wall at a given x (From Shah, 1981).

Now let us first obtain the temperature distribution in the wall by a double integration of Eq. (5.126) using the boundary conditions of Eq. (5.127) and the assumption of Eq. (5.125). We get

$$\bar{T}_w - \bar{T}_{w,o} = -\frac{1}{2\alpha_w}\left(\frac{\partial T_w}{\partial \tau}\right)(\delta_w - y)y \tag{5.128}$$

Hence, T_w and $T_{w,o}$ of Eq. (5.128) must be time-averaged temperatures, and they are designated by a bar in Eq. (5.128) to denote them as time-averaged quantities. The temperature profile of Eq. (5.128) is parabolic at each x as shown in Fig. 5.12c. The mean wall temperature $\bar{T}_{w,m}$ is obtained by integrating Eq. (5.128) with respect to y across the wall thickness as

$$\bar{T}_{w,m} - \bar{T}_{w,o} = -\frac{\delta_w^2}{12\alpha_w}\frac{\partial T_w}{\partial \tau} \tag{5.129}$$

Thus, the differences between the mean wall temperature and the surface wall temperature during hot- and cold-gas flow periods are

$$\left(\bar{T}_{w,m} - \bar{T}_{w,o}\right)_h = -\frac{\delta_w^2}{12\alpha_w}\left(\frac{\partial T_w}{\partial \tau}\right)_h \tag{5.130}$$

$$\left(\bar{T}_{w,m} - \bar{T}_{w,o}\right)_c = -\frac{\delta_w^2}{12\alpha_w}\left(\frac{\partial T_w}{\partial \tau}\right)_c \tag{5.131}$$

The mean wall temperatures during the hot- and cold-gas flow periods must by equal for the idealized true periodic flow conditions (i.e., $\bar{T}_{w,m,h} = \bar{T}_{w,m,c}$). Subtracting Eq. (5.131) from Eq. (5.130), we get

$$\left(\bar{T}_{w,o}\right)_h - \left(\bar{T}_{w,o}\right)_c = -\frac{\delta_w^2}{12\alpha_w}\left[\left(\frac{\partial T_w}{\partial \tau}\right)_c - \left(\frac{\partial T_w}{\partial \tau}\right)_h\right] \tag{5.132}$$

To determine the influence of wall thermal resistance on regenerator effectiveness, we want to arrive at an expression for $\mathbf{R}_w$ for a flat (plain) wall. This means that we want to derive an expression for an equivalent UA for the regenerator for the finite wall thermal resistance case. This will now be achieved by writing an energy balance and a rate equation for an element dx during hot- and cold-gas periods.

During the hot-gas flow period, heat will be conducted from both sides of the wall (exposed to convection from the gas) to the center of the wall. In this case, the heat transfer surface area of a matrix element of length dx will be dA, considering both sides of the wall. The heat will be conducted through the wall thickness $\delta_w/2$. The reverse heat transfer by conduction will take place in the wall during the cold-gas flow period. They are given by

$$dQ_h = \rho_w c_w \frac{\delta_w}{2}\,dA\left(\frac{\partial T_w}{\partial \tau}\right)_h P_h \qquad dQ_c = -\rho_w c_w \frac{\delta_w}{2}\,dA\left(\frac{\partial T_w}{\partial \tau}\right)_c P_c \tag{5.133}$$

Substituting values of the temperature gradients from these equations into Eq. (5.132) and noting that $dQ_h = dQ_c = dQ$ for the idealized periodic conditions yields

$$\left(\bar{T}_{w,o}\right)_h - \left(\bar{T}_{w,o}\right)_c = \frac{\delta_w}{6k_w}\left(\frac{1}{P_h} + \frac{1}{P_c}\right)\frac{dQ}{dA} \tag{5.134}$$

Now the rate equations during hot- and cold-gas flow periods are

$$dQ_h = h_h \, dA[\bar{T}_h - (\bar{T}_{w,o})_h]P_h \qquad dQ_c = -h_c \, dA[\bar{T}_c - (\bar{T}_{w,o})_c]P_c \qquad (5.135)$$

where T_h and $\bar{T}_c$ are the hot- and cold-fluid time-averaged temperatures at a section x during the hot- and cold-gas flow periods, respectively. Substituting $(\bar{T}_{w,o})_h$ and $(\bar{T}_{w,o})_c$ from these equations into Eq. (5.134) and noting that $dQ_h = dQ_c = dQ$, rearrangement yields

$$dQ = \left[\frac{1}{h_h P_h} + \frac{1}{h_c P_c} + \frac{\delta_w}{6k_w}\left(\frac{1}{P_h} + \frac{1}{P_c}\right)\right]^{-1} dA(\bar{T}_h - \bar{T}_c) \qquad (5.136)$$

Since $Q = Q_h = Q_c$ represents total heat transfer per cycle in time $(P_h + P_c)$, the average heat transfer rate during one cycle is

$$q = \frac{Q}{P_h + P_c} \qquad (5.137)$$

and hence

$$dq = \frac{dQ}{P_h + P_c} \qquad (5.138)$$

Substituting dQ from Eq. (5.136) into Eq. (5.138), we get

$$dq = U_o \, dA(\bar{T}_h - \bar{T}_c) \qquad (5.139)$$

where

$$\frac{1}{U_o} = \left[\frac{1}{h_h P_h} + \frac{1}{h_c P_c} + \frac{\delta_w}{6k_w}\left(\frac{1}{P_h} + \frac{1}{P_c}\right)\right](P_h + P_c) \qquad (5.140)$$

Integration of Eq. (5.139) along the length of a counterflow regenerator yields

$$q = U_o A \, \Delta T_{\mathrm{lm}} \qquad (5.141)$$

where

$$\Delta T_{\mathrm{lm}} = \frac{(T_{h,i} - \bar{T}_{c,o}) - (\bar{T}_{h,o} - T_{c,i})}{\ln[(T_{h,i} - \bar{T}_{c,o})/(\bar{T}_{h,o} - T_{c,i})]} \qquad (5.142)$$

The bar on T represents the corresponding (hot or cold)-period time-averaged temperatures. A in Eq. (5.141) represents the total surface area $(A_h + A_c)$, *in contrast* to either A_h or A_c for a recuperator. This is because dq in Eq. (5.138) is over one cycle.

Now dividing both sides of Eq. (5.140) by the total surface area A and introducing the definitions of A_h and A_c from Eq. (5.9), we get

$$\frac{1}{U_o A} = \frac{1}{h_h A_h} + \frac{1}{h_c A_c} + \frac{\delta_w}{6k_w}\left(\frac{1}{A_{w,h}} + \frac{1}{A_{w,c}}\right) \qquad (5.143)$$

where the relationships $A_{w,h} = A_h$ and $A_{w,c} = A_c$, as noted just before Eq. (5.126), are used in Eq. (5.143). Now this equation is valid for both the rotary and fixed-matrix regenerators as long as proper values of A_h and A_c from Eqs. (5.9) and (5.10) are used. From a comparison of this equation with that for a recuperator, the equivalent wall thermal resistance for the regenerator is

$$\mathbf{R}_w = \frac{\delta_w}{6k_w}\left(\frac{1}{A_{w,h}} + \frac{1}{A_{w,c}}\right) \tag{5.144}$$

Now consider a rotary or fixed-matrix regenerator with two matrices and the special case of a $50:50$ split of flow areas in a rotary regenerator or $P_h = P_c$ in a fixed-matrix regenerator. For this regenerator, $A_{w,h} = A_{w,c} = A_w$. Thus, for this case, Eq. (5.144) reduces to

$$\mathbf{R}_w = \frac{\delta_w}{3k_wA_w} \tag{5.145}$$

For a comparable recuperator having plain walls of thickness δ_w, conduction area A_w, and thermal conductivity k_w, the wall thermal resistance is

$$\mathbf{R}_w = \frac{\delta_w}{k_wA_w} \tag{5.146}$$

From a comparison of Eqs. (5.145) and (5.146), it is evident that the wall thermal resistance of a regenerator is one-third that of an equivalent recuperator. Alternatively, the regenerator is equivalent to a recuperator of one-third wall thickness. The qualitative reason for the lower wall thermal resistance is that the thermal energy is not transferred through the wall in a regenerator; it is stored and rejected.

One of the basic assumptions made in the foregoing analysis is that the temperatures T_h, T_c, and T_w are all linear with time. Since this is not true at the switching moment and near the regenerator inlet and outlet, Hausen (1983) in 1942 suggested modifying $\mathbf{R}_w$ of Eq. (5.145) by a factor Φ^*, so that

$$\frac{1}{U_oA} = \frac{1}{h_hA_h} + \frac{1}{h_cA_c} + \mathbf{R}_w\Phi^* \tag{5.147}$$

where for a plain wall,

$$\Phi^* = \begin{cases} 1 - \dfrac{\delta_w^2}{60\alpha_w}\left(\dfrac{1}{P_h} + \dfrac{1}{P_c}\right) & \text{for } \dfrac{\delta_w^2}{2\alpha_w}\left(\dfrac{1}{P_h} + \dfrac{1}{P_c}\right) \leq 10 \\[4mm] 2.142\left[0.3 + \dfrac{\delta_w^2}{2\alpha_w}\left(\dfrac{1}{P_h} + \dfrac{1}{P_c}\right)\right]^{-1/2} & \text{for } \dfrac{\delta_w^2}{2\alpha_w}\left(\dfrac{1}{P_h} + \dfrac{1}{P_c}\right) \geq 10 \end{cases} \tag{5.148}$$

Generally, the wall thermal resistance is much smaller than the hot- or cold-gas film resistance. Hence the wall thermal resistance formula of Eq. (5.144) is adequate for rating and sizing problems of most applications. This correction factor Φ^* is not adequate to accurately determine the temperature distribution in the wall immediately after the changeover and near the regenerator ends. In those cases, a numerical method of Heggs and Carpenter (1979) is suggested to take into account the wall thermal resistance effect. Heggs and Carpenter designate this effect the *intraconduction effect*.

5.6 INFLUENCE OF PRESSURE AND CARRYOVER LEAKAGES

In both rotary and fixed-matrix regenerators, flow leakages from cold to hot gas streams, and vice versa, occur due to pressure differences and carryover (as a result of the matrix rotation or valve switching). *Pressure leakage* is defined as any leakage due to the pressure difference between the cold and hot gases in a regenerator. In a rotary regenerator, this leakage occurs at face seals, through the pores in the matrix, or through the circumferential gap between the disk and the housing. In a fixed-matrix regenerator, this leakage occurs at valves as well as to and from the cracks in the housing (which is usually made of bricks) in high-temperature applications. In such an application, the operating pressure on the hot-gas matrix is below ambient pressure (i.e., under vacuum) to avoid the leakage of poisonous flue gas from the matrix to the surrounding room; and in this case, the pressure leakage is from outside to the matrix through cracks in the walls. For further details, refer to Shah and Skiepko (1998).

Carryover leakage in a rotary regenerator is defined as the transport of the trapped gas in the voids of the matrix to the other fluid side just after the switching from fluid 1 stream to fluid 2 stream, and vice versa, due to matrix rotation. The carryover leakage is unavoidable and its mass flow rate is a function of total void volume, rotor speed, and fluid density. In a fixed-matrix regenerator, the carryover leakage in one cycle from fluid 1 to fluid 2 is the gas in the void volume of one matrix of a two-matrix regenerator and associated header volume.

In the following, a model is presented to take into account the pressure and carryover leakages in a rotary regenerator (Shah and Skiepko, 1997). Refer to Shah and Skiepko (1998) for modeling leakages for a fixed-matrix regenerator.

5.6.1   Modeling of Pressure and Carryover Leakages for a Rotary Regenerator

In a rotary regenerator, pressure leakage occurs at face seals separating the hot and cold gas sides, through the pores in the matrix itself (which is neglected in the present analysis), or through the circumferential gap between the disk (rotor) and housing. The circumferential leakage is sometimes referred to as the *side bypass leakage*, and the pressure leakage at the face seals is referred to as the *cross bypass leakage*. The influence of the total pressure leakage on the regenerator effectiveness is modeled and summarized next. However, this net flow leakage also represents a loss in the heated cold-gas flow rate to the process or the thermodynamic system. This loss may also have a substantial influence on the process or cycle efficiency. For example, in a gas turbine power plant, a 6% cold high-pressure air leak to the exhaust gas stream in a regenerator means a 6% reduction in the net power output—a significant penalty! However, unless there are significantly large leakages, the regenerative cycle has substantially higher thermal efficiency than a simple cycle for the gas turbine power plant.

A Ljungstrom rotary regenerator with radial, peripheral, and axial seals is shown in Fig. 5.13. Radial seals prevent leakage of high-pressure gas to low-pressure gas at the inlet and outlet faces of the regenerator. The axial seals prevent the leakage of high-pressure gas to low-pressure gas in the circumferential direction in the gap between the housing and rotor. The peripheral seals prevent the flow bypass from the regenerator inlet to regenerator outlet side on each gas side (in the axial direction) of the regenerator in the gap between the housing and rotor if the axial seals are perfect (zero leakage); the leakage between the axial seals (since there are usually 12 or fewer axial seals along the perimeter of the rotor) also eventually has to go through the peripheral seals. Note that in

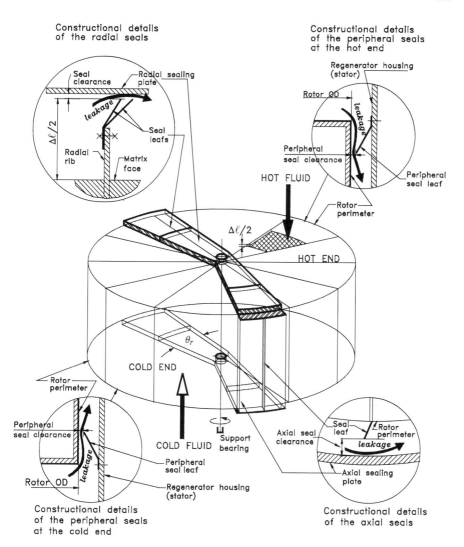

Constructional details
of the radial seals

Constructional details
of the peripheral seals
at the hot end

FIGURE 5.13   A Ljungstrom rotary regenerator with radial, peripheral, and axial seals. (From Shah and Skiepko, 1997.)

some regenerator applications, such as for a gas turbine, the axial seals are not used; only peripheral seals are used to prevent/minimize the flow leakage bypasses in axial and circumferential directions.

A model for regenerator thermal design consists of an *internal (or ideal) regenerator* (represented by the rotating disk with no leakage streams within its boundary as marked by dashed lines in Fig. 5.14), and an *actual regenerator* (that is considered as the internal regenerator with its housing, and radial, peripheral, and axial seals that include *all* leakages and bypass flows). The concept of these two regenerators is used for the thermal design procedure. It is easily seen from Fig. 5.14 that there are no leakages within the

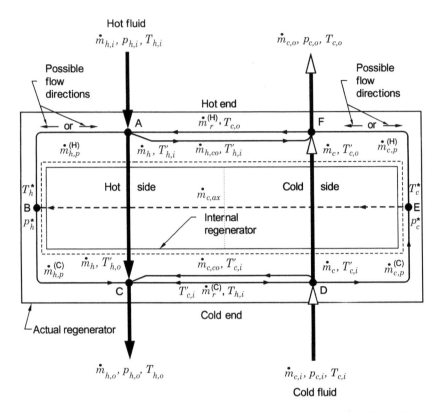

FIGURE 5.14   Regenerator gas flow network with all categories of leakages. (From Shah and Skiepko, 1997.)

region bounded by dashed lines (i.e., the internal regenerator). In Fig. 5.14, all leakage streams are shown by medium-thickness lines with arrows indicating the flow directions; the exception is the axial leakage stream, which is shown as a large-dashed line through the internal regenerator but which in reality flows circumferentially through the gap between the rotor and housing. The flow rates associated with each stream are identified as follows: The first subscript, h or c, designates the hot or cold gas stream. The second subscripts r, p, ax, and co designate the radial, peripheral, axial, and carryover leakage streams. The superscripts (H) and (C) denote the hot end (hot-gas inlet side) and cold end (cold-gas inlet side) of the regenerator disk. The temperatures associated with each stream are identified as follows: The first subscript, h or c, designates the hot or cold stream. The second subscripts, i and o, refer to the inlet and outlet gas sides of the regenerator; those with a prime mark denote inlet and outlet from the internal regenerator, and those without a prime mark denote inlet and outlet from the actual regenerator.

All radial, peripheral, and axial seal leakages are modeled as a flow through an orifice as follows:

$$\dot{m}_{\text{seal}} = C_d A_{o,s} Y \sqrt{2\rho \, \Delta p} \tag{5.149}$$

Here we have assumed that the thermodynamic process involved is isentropic, the pressure leakages through seals can be evaluated by applying Bernoulli and continuity equations, and the thermodynamic relations are used for gas flow through the leakage area. In Eq. (5.149), C_d is the coefficient of discharge, assumed to be 0.80 (Shah and Skiepko, 1997); $A_{o,s}$ is the seal gap flow area; Y is the expansion factor for the compressibility effect, assumed to be 1; ρ is the gas density before the seal; and Δp is the pressure drop in the seal. The leakage flow rate terms, Δp, and inlet density for each leakage stream are summarized in Table 5.6.

The carryover leakage from hot to cold gas in one rotation will correspond to the hot void volume times the average density $\bar{\rho}$ of the gas:

$$\dot{m}_{h,co} = (A_{fr} L\sigma)\bar{\rho}_h \mathbf{N} \tag{5.150}$$

where $\bar{\rho}_h$ is determined at arithmetic averages of inlet and outlet gas temperature and pressure. A similar expression can be written for the carryover leakage $\dot{m}_{c,co}$ from the cold side to the hot side. In some regenerator applications, heat transfer surface of different porosity is used along the flow length, in which case $L\sigma$ in Eq. (5.150) is replaced by $\sum_i L_i \sigma_i$. In the Ljungstrom regenerator and some other regenerators, the matrix flow length is L, and the radial ribs (separating heat transfer baskets) are slightly higher by $\Delta L/2$ on the hot and cold ends on which the radial seals scrub (see Fig. 5.13). Hence, the gas trapped in this header volume is also carried over to the other side. Hence, the carryover leakage term in general for each fluid side is given by

$$\dot{m}_{j,co} = A_{fr} \mathbf{N} \bar{\rho}_j \left[\sum_{i=1}^{n} (L_i \sigma_i) + \Delta L \right] \qquad j = h \text{ or } c \tag{5.151}$$

TABLE 5.6 **Rotary Regenerator Pressure Leakage and Carryover Flow Rates, Pressure Drops, and Inlet Density for the Orifice Analysis**

Leakage Term	Symbols	Pressure Drop	Density
Flows through radial seal clearances			
Flow of the higher-pressure cold gas at the hot end	$\dot{m}_{c,r}^{(H)}$	$p_{c,o} - p_{h,i}$	$\rho_{c,o}$
Flow of the higher-pressure cold gas at the cold end	$\dot{m}_{c,r}^{(C)}$	$p_{c,i} - p_{h,o}$	$\rho_{c,i}$
Flows through peripheral seal clearances			
Flows at the hot end of the disk face			
Flow around the inlet to the lower-pressure hot-gas zone	$\dot{m}_{h,p}^{(H)}$	$p_{h,i} - p_h^*$ $p_h^* - p_{h,i}$	$\rho_{h,i}$ if $p_{h,i} > p_h^*$ ρ_h^* if $p_{h,i} < p_h^*$
Flow around the outlet from the higher-pressure cold-gas zone	$\dot{m}_{c,p}^{(H)}$	$p_{c,o} - p_c^*$ $p_c^* - p_{c,o}$	$\rho_{c,o}$ if $p_{c,o} > p_c^*$ ρ_c^* if $p_{c,o} < p_c^*$
Flows at the cold end of the disk face			
Flow around the outlet from the lower-pressure hot-gas zone	$\dot{m}_{h,p}^{(C)}$	$p_h^* - p_{h,o}$	ρ_h^*
Flow around the inlet to the higher-pressure cold-gas zone	$\dot{m}_{c,p}^{(C)}$	$p_{c,i} - p_c^*$	$\rho_{c,i}$
Flow through axial seal clearances	$\dot{m}_{c,ax}$	$p_c^* - p_h^*$	ρ_c^*
Gas carryover			
Carryover of the lower-pressure hot gas into cold gas	$\dot{m}_{h,co}$		$\bar{\rho}_h$
Carryover of the higher-pressure cold gas into hot gas	$\dot{m}_{c,co}$		$\bar{\rho}_c$

Source: Data from Shah and Skiepko (1997).

Generally, the carryover leakage is very small and its influence on the regenerator effectiveness is also negligibly small for most rotary regenerator applications, except for the cryogenics and Stirling engine regenerators, which have effectively a very high value of switching frequency.

The leakage modeling of an actual regenerator involves the determination of all nine leakage flow rates mentioned in Table 5.6, plus mass flow rates, pressures, and temperatures at the mixing points A, B, ..., F in Fig. 5.14. To be more specific, four additional mass flow rate terms are hot- and cold-gas mass flow rates through the regenerator disk $(\dot{m}_h, \dot{m}_c)$, and hot- and cold-gas mass flow rates at the outlet of the actual regenerator $(\dot{m}_{h,o}, \dot{m}_{c,o})$; four pressures and temperatures at points B and E of Fig. 5.14: p_h^*, p_c^*, T_h^*, T_c^* (note these are dimensional values); and three temperatures from the actual regenerator to the outside or to an internal regenerator: $T_{h,o}, T_{c,o}$, and $T'_{h,i}$. Thus it represents a total of 20 unknowns for the actual regenerator. Determination of various leakage flow rates of Table 5.6 represents nine equations [seven of Eq. (5.149) and two of Eq. (5.150)]. The mass flow rate balances at six junction points A, B, ..., F in Fig. 5.14 represent six equations presented in Table 5.7. The energy balances at five points, A, B, C, E, and F in Fig. 5.14, provide an additional five equations, as summarized in Table 5.8. Thus a total of 20 linear/nonlinear equations are available to determine 20 unknowns mentioned above. If $\dot{m}_{c,r}^{(C)}$ leaks from point C to D in Fig. 5.14, one additional energy balance equation at point D is available in Table 5.8, and the additional unknown is $T'_{c,i}$.

To investigate the influence of leakage distribution on regenerator heat transfer performance, we also need to consider heat transfer and pressure drop modeling of the internal regenerator. Heat transfer analysis includes applying ε-NTU$_o$ theory, including longitudinal and transverse wall heat conduction effects, discussed in Sections 5.2, 5.4, and 5.5 for the internal regenerator. Here the appropriate mass flow rates and inlet temperatures are determined for the internal regenerator from an analysis of the actual regenerator as mentioned above. The internal regenerator heat transfer analysis will then yield temperatures and the internal regenerator effectiveness $\varepsilon_{i,b}$. The pressure drop analysis for the hot- and cold-gas sides of the internal regenerator is also straightforward and is presented in Chapter 6. Appropriate flow rates within the internal regenerator and

TABLE 5.7 Mass Flow Rate Balances at Mixing Locations in the Regenerator

Mixing Location	Pressure Condition	Mass Flow Rate Balance
A	$p_{h,i} > p_h^*$	$-\dot{m}_{h,i} - \dot{m}_{c,r}^{(H)} + \dot{m}_{h,p}^{(H)} + \dot{m}_{h,co} + \dot{m}_h = 0$
	$p_{h,i} < p_h^*$	$-\dot{m}_{h,i} - \dot{m}_{c,r}^{(H)} - \dot{m}_{h,p}^{(H)} + \dot{m}_{h,co} + \dot{m}_h = 0$
B	$p_{h,i} > p_h^*$	$-\dot{m}_{h,p}^{(H)} - \dot{m}_{c,ax} + \dot{m}_{h,p}^{(C)} = 0$
	$p_{h,i} < p_h^*$	$\dot{m}_{h,p}^{(H)} - \dot{m}_{c,ax} + \dot{m}_{h,p}^{(C)} = 0$
C		$-\dot{m}_h - \dot{m}_{c,r}^{(C)} - \dot{m}_{h,p}^{(C)} - \dot{m}_{c,co} + \dot{m}_{h,o} = 0$
D		$-\dot{m}_{c,i} + \dot{m}_{c,co} + \dot{m}_{c,p}^{(C)} + \dot{m}_{c,r}^{(C)} + \dot{m}_c = 0$
E	$p_{c,o} > p_c^*$	$-\dot{m}_{c,p}^{(C)} - \dot{m}_{c,p}^{(H)} + \dot{m}_{c,ax} = 0$
	$p_{c,o} < p_c^*$	$-\dot{m}_{c,p}^{(C)} + \dot{m}_{c,p}^{(H)} + \dot{m}_{c,ax} = 0$
F	$p_{c,o} > p_c^*$	$-\dot{m}_c - \dot{m}_{h,co} + \dot{m}_{c,p}^{(H)} + \dot{m}_{c,r}^{(H)} + \dot{m}_{c,o} = 0$
	$p_{c,o} < p_c^*$	$-\dot{m}_c - \dot{m}_{h,co} - \dot{m}_{c,p}^{(H)} + \dot{m}_{c,r}^{(H)} + \dot{m}_{c,o} = 0$

Source: Data from Shah and Skiepko (1997).

TABLE 5.8 Energy Balances or Temperatures at Mixing Locations in the Regenerator

Mixing Location	Pressure Condition	Energy Balance
A	$p_{h,i} > p_h^*$	$-\dot{m}_{h,i}H_{h,i} - \dot{m}_{c,r}^{(H)}H_{c,o} + (\dot{m}_{h,p}^{(H)} + \dot{m}_{h,co} + \dot{m}_h)H_{h,i}' = 0$
	$p_{h,i} < p_h^*$	$-\dot{m}_{h,i}H_{h,i} - \dot{m}_{c,r}^{(H)}H_{c,o} - \dot{m}_{h,p}^{(H)}H_h^* + (\dot{m}_{h,co} + \dot{m}_h)H_{h,i}' = 0$
B	$p_{h,i} > p_h^*$	$-\dot{m}_{h,p}^{(H)}H_{h,i}' - \dot{m}_{c,ax}H_c^* + \dot{m}_{h,p}^{(C)}H_h^* = 0$
	$p_{h,i} < p_h^*$	$T_h^* = T_c^*$
C		$-(\dot{m}_{c,co} + \dot{m}_{c,r}^{(C)})H_{c,i} - \dot{m}_{c,r}^{(C)}H_h^* - \dot{m}_hH_{h,o}' + \dot{m}_{h,o}H_{h,o} = 0$
D		If $\dot{m}_{c,r}^{(C)}$ from D to C: $T_c' = T_{c,i}$
		If $\dot{m}_{c,r}^{(C)}$ from C to D: $\dot{m}_{c,co}H_{c,i}' - \dot{m}_{c,r}^{(C)}H_{h,o} + \dot{m}_{c,p}^{(C)}H_{c,i}' + \dot{m}_cH_{c,i}' - \dot{m}_{c,i}H_{c,i} = 0$
E	$p_{c,o} > p_c^*$	$-\dot{m}_{c,p}^{(C)}H_{c,i} - \dot{m}_{c,p}^{(H)}H_{c,o} + \dot{m}_{c,ax}H_c^* = 0$
	$p_{c,o} < p_c^*$	$T_c^* = T_{c,i}$
F	$p_{c,o} > p_c^*$	$-\dot{m}_cH_{c,o}' - \dot{m}_{h,co}H_{h,i}' + (\dot{m}_{c,p}^{(H)} + \dot{m}_{c,r}^{(H)} + \dot{m}_{c,o})H_{c,o} = 0$
	$p_{c,o} < p_c^*$	$-\dot{m}_cH_{c,o}' - \dot{m}_{h,co}H_{h,i}' - \dot{m}_{c,p}^{(H)}H_{c,i} + (\dot{m}_{c,r}^{(H)} + \dot{m}_{c,o})H_{c,o} = 0$

Source: Data from Shah and Skiepko (1997).

temperatures (for density calculations) are needed. This analysis yields outlet pressures $p_{h,o}$ and $p_{c,o}$.

Thus heat transfer and pressure drop modeling of the internal regenerator has five unknowns: $\varepsilon_{i,b}$, $T_{h,o}'$, $T_{c,o}'$, $p_{h,o}$, and $p_{c,o}$. Correspondingly, there are five equations: the definition of $\varepsilon_{i,b}$, two equations for outlet temperatures (for known inlet temperatures, $\varepsilon_{i,b}$ and C_{min}), and two pressure drop equations. The modeling of actual and internal regenerators thus requires solving 25 nonlinear equations iteratively for 25 unknowns using, for example, an iterative Newton–Raphson method. Such a large set of equations can only be solved by a computer program. Specific illustrative results for 5 and 10% leakages individually through radial, peripheral, and axial seals for a gas turbine regenerator problem are presented by Shah and Skiepko (1997). The following are specific conclusions from that study.

- Any radial seal leakage location (at the hot or cold end of a regenerator) has a negligible effect on actual heat transfer to the cold stream.[†] Hence, one can assume 50% radial seal leakage at the hot end and 50% at the cold end. For one specific example of a gas turbine regenerator, 5% and 10% radial seal leakages reduced the useful thermal energy transfer (heat recovery) to the cold gas in the outlet duct by 3.2% and 6.9%, respectively. These 5% and 10% leakages reduced the cold-gas side pressure drop by 1.4% and 3.2% and increased the hot-gas side pressure drop by 4.5% and 9.5%, respectively. If a conservative (i.e., high) value of the pressure drop is desired, consider total pressure leakage of the cold gas at the regenerator hot end only.

[†] This is often true for high-effectiveness ε regenerators. If $\varepsilon \leq 40$ to 50%, the hot-end leakage of cold gas can dilute the hot-gas temperature entering the actual regenerator, thus having some impact on $(T_{h,i} - T_{c,i})$ and reduction in heat transfer.

- The effects of the peripheral seal leakage on heat transfer and cold and hot gas side pressure drops are similar to those for radial seal leakage.
- Four cases of leakage across axial seals have been considered for the following flow paths in Fig. 5.14: DEBC, DEBA, FEBC, and FEBA. The effects on heat transfer and pressure drops are very sensitive to the leakage distribution (flow path): from about the same effect as for the radial seals to detrimental effect over a factor of two.
- Since all leakages occur simultaneously in a regenerator, even a moderate or small leak through individual seals can have a significant effect on regenerator performance. It is highly desirable to have all radial, peripheral, and axial seals as tight as possible. However, it is shown that the axial seals should be tightened as much as possible because of the most detrimental effects for the gas turbine regenerator problem considered.

5.7   INFLUENCE OF MATRIX MATERIAL, SIZE, AND ARRANGEMENT

In a fixed-matrix regenerator, the outlet temperature of the cold gas (air) decreases as a function of time during the cold-gas flow period. The difference in the outlet temperature from $\tau = 0$ to that at $\tau = P_c$ is referred to as the *temperature swing*. The swing should be minimized so that the heated air from the regenerator is at a relatively constant temperature for the process downstream. The design of three and four stove systems, as mentioned in Section 1.5.4.2, has been developed to minimize this temperature swing. Since the temperature swing depends on the heat capacity of the matrix material, it can be minimized by employing a high-volumetric-heat-capacity ($\rho_w c_w$) material in the matrix, such as silicon carbide or corundum instead of fireclay. However, the high-heat-capacity material is much more expensive. Use of proper matrix material size and arrangement along with different materials in a two- or three-zoned regenerator results in an optimum regenerator.

Heggs and Carpenter (1978), among others, proposed the use of high-heat-capacity material near the outlet end of the cold-gas flow path and low-heat-capacity material near the inlet end of the cold-gas flow path. Only 10% of the heat transfer surface area is required to have higher-heat-capacity material, the other 90% with lower-heat-capacity material. Heggs and Carpenter also found that thick bricks can be used near the outlet end and thin bricks near the inlet end of the cold-gas flow path in a two-zoned regenerator. Either of the alternatives will reduce the temperature swing. The design of a two-zoned regenerator is carried out by considering two zones in series; the outlet fluid temperature from the first zone is the inlet fluid temperature of the second zone. Alternatively, an equivalent height for one zone is computed in terms of the second-zone heat transfer coefficient, surface area, and packing density, as shown in Example 5.6.

The steam boiler regenerator utilizes two layers (each having different surfaces) in the matrix for the following reasons: (1) To get higher heat transfer, the hot end has a higher performance surface of rather complex geometry, which is unsuitable for the cold end due to the possibility of plugging by particulate (a plain surface is often used so that particulates can flow through easily); and (2) near the cold end, the hot flue gas is cooled enough so that there is a possibility of sulfuric acid corrosion. Hence, the sheet metal thickness of the surface near the cold end is higher than that for the hot matrix layer, where corrosion is not a problem. Also, due to the condensed water vapor near the cold end, the surface is wet and hence fouling is a serious problem. Hence, the matrix surface flow passages are straight (duct flow with lower heat transfer coefficients) to get easier self cleaning and forced blow-off of the matrix. At the cold end, ceramic elements are sometimes applied to the matrix surface, or the steel sheet surface is covered with ceramic material to resist corrosion.

Example 5.6   A rotary regenerator matrix used to preheat air in a thermal power plant is made of two layers (each having different surfaces) in the air/gas flow direction. The upper layer near the hot end of the regenerator has an enhanced surface for higher performance. The lower layer near the cold end has a plain surface with higher sheet metal thickness (compared to that for the upper layer) to minimize and accommodate corrosion and fouling, as mentioned in the text above. The geometries, heat transfer coefficients, and physical properties of the matrix material for the upper and lower layers are as follows:

Regenerator Parameters and Fluid Properties	Upper Layer (Near the Hot End)	Lower Layer (Near the Cold End)
Height of the individual layer (m)	1.300	0.300
Hydraulic diameter (mm)	8.54	7.08
Sheet metal thickness (mm)	0.7	1.0
Packing density (m^2/m^3)	402.3	440.4
Porosity	0.859	0.780
Flue gas–side heat transfer coefficient $(W/m^2 \cdot K)$	82.2	33.0
Air-side heat transfer coefficient $(W/m^2 \cdot K)$	72.7	29.3
Density ρ_w (kg/m^3)	7841	7841
Specific heat c_w $(J/kg \cdot K)$	456	456
Thermal conductivity k_w $(W/m \cdot K)$	50.3	50.3

The frontal area of the rotor, excluding the shaft but including the seal coverage, is 19.1 m^2. The rotor turns at 2.32 rpm with the flue gas (hot) side spanning 195° and the air (cold) side spanning 165° of the rotor face. The radial seals cover 10% of the rotor face area. The mass flow rate of the flue gas stream entering at 339°C is 51.1 kg/s, and it flows in countercurrent to the air stream at 29°C, flowing at 42.4 kg/s. The isobaric specific heat of gas and air are 1.11 and 1.02 kJ/kg · K respectively.

Determine the regenerator effectiveness and the heat transfer rate from the flue gas to the airstream.

SOLUTION

Problem Data and Schematic: The rotor geometry comprising two layers of the same material but with different physical characteristics is specified. In addition, the rotor dimensions and rotational speed are specified. The fluid flow rates and inlet temperatures are provided for both the hot and cold fluid streams. The physical properties of matrix material and fluids are specified (see Fig. E5.6).

Determine: The regenerator effectiveness and the heat transfer rate from the flue gas to the air.

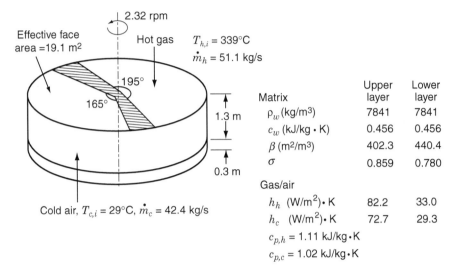

FIGURE E5.6

Assumptions: The assumptions of Section 5.1.1 are invoked here, and the fluid temperature distributions after the first layer and entering the second layer are uniform.

Analysis: To solve this problem, compute the height of the equivalent lower layer to represent it as if it were an equivalent upper layer maintaining the same actual heat transfer rate of the lower layer. Once this equivalent upper layer height is obtained for the lower layer, add it up with the upper layer height to obtain a single height (in terms of the upper layer geometry and other characteristics) of the equivalent regenerator individually for hot gas and cold air. So this equivalent regenerator may have different heights for the gas and air sides for performance calculation purposes. Then the solution procedure of this single-layer regenerator follows the same steps as those presented in Section 5.2.

   Let us calculate the height of the equivalent lower layer. Noting that the heat transfer area A of individual layers is given as

$$A = \text{rotor frontal area } A_{fr} \times \text{layer height } L \times \text{packing density } \beta$$

we can write, based on Eq. (5.141), the following expression for the heat transfer rate q in the lower layer:

$$q = U_{o,\text{lo}} A_{\text{lo}} \, \Delta T_{\text{lo}} = U_{o,\text{lo}} A_{fr} L_{\text{lo}} \, \beta_{\text{lo}} \, \Delta T_{\text{lo}}$$

Similarly, using Eq. (5.141), we can write an expression for the heat transfer rate q in the lower layer in terms of the height of the equivalent upper layer as follows:

$$q = U_{o,\text{up}} A_{fr} L_{eq,\text{up}} \, \beta_{\text{up}} \, \Delta T_{\text{up}}$$

Equating the two equations for the lower layer heat transfer rates and taking cognizance of the fact that the temperature potential for heat transfer in the original and equivalent layers is equal ($\Delta T_{\text{lo}} = \Delta T_{\text{up}}$), we obtain an equivalent upper layer height of the same heat transfer duty as the lower layer:

$$L_{\text{eq,up}} = \frac{U_{o,\text{lo}}}{U_{o,\text{up}}} \frac{\beta_{\text{lo}}}{\beta_{\text{up}}} L_{\text{lo}}$$

We use Eq. (5.143) to calculate the overall heat transfer coefficients $U_{o,\text{lo}}$ and $U_{o,\text{up}}$. Neglecting the heat conduction term in Eq. (5.143), we get

$$\frac{1}{U_o} = \frac{1}{h_h(A_h/A)} + \frac{1}{h_c(A_c/A)}$$

where $A_h = A_{fr,h}L\beta$ and $A_c = A_{fr,c}L\beta$. However, the frontal areas are given as

$$A_{fr,h} = 0.9 A_{fr}\theta_h/360° \quad \text{and} \quad A_{fr,c} = 0.9 A_{fr}\theta_c/360°$$

and, moreover, $A = A_h + A_c$. Thus, we get

$$\frac{A_h}{A} = \frac{\theta_h}{\theta_h + \theta_c} \quad \text{and} \quad \frac{A_c}{A} = \frac{\theta_c}{\theta_h + \theta_c}$$

where θ_h and θ_c are specified as 195° and 165° respectively. Using A_h/A and A_c/A in the formula for U_o, we get the overall heat transfer coefficients in the lower and upper layers as

$$\frac{1}{U_{o,\text{lo}}} = \frac{1}{33.0 \text{ W/m}^2 \cdot \text{K } (195°/360°)} + \frac{1}{29.3 \text{ W/m}^2 \cdot \text{K } (165°/360°)}$$

$$\Rightarrow U_{o,\text{lo}} = 7.67 \text{ W/m}^2 \cdot \text{K}$$

and

$$\frac{1}{U_{o,\text{up}}} = \frac{1}{82.2 \text{ W/m}^2 \cdot \text{K } (195°/360°)} + \frac{1}{72.7 \text{ W/m}^2 \cdot \text{K } (165°/360°)}$$

$$\Rightarrow U_{o,\text{up}} = 19.1 \text{ W/m}^2 \cdot \text{K}$$

Inserting the numerical values in the formula above for $L_{\text{eq,up}}$, we obtain an equivalent upper layer height for the lower layer as

$$L_{\text{eq,up}} = \frac{7.67 \text{ W/m}^2 \cdot \text{K}}{19.1 \text{ W/m}^2 \cdot \text{K}} \times \frac{440.4}{402.3} \times 0.3 \text{ m} = 0.132 \text{ m}$$

Adding the equivalent height to the actual upper layer height of 1.300 m, we obtain the effective height for the purpose of heat transfer as $L_{\text{eff}} = (1.300 + 0.132) \text{ m} = 1.432 \text{ m}$. Note that this effective height is lower than the actual physical height of the two layers, which is $(1.300 + 0.300) \text{ m} = 1.600 \text{ m}$. Now we can treat this two-layer regenerator as a single-layer regenerator with a height of 1.432 m, having all the characteristics of the upper layer only.

Now let us calculate hA values on each fluid side. Frontal areas for the hot and cold sides are calculated knowing the fraction of frontal area not covered by seals, total frontal area, and the sector angle divided by 360°:

$$A_{fr,h} = 0.9 A_{fr}(\theta_h/360°) = 0.9 \times 19.1 \, \text{m}^2 \times (195°/360°) = 9.311 \, \text{m}^2$$

$$A_{fr,c} = 0.9 A_{fr}(\theta_c/360°) = 0.9 \times 19.1 \, \text{m}^2 \times (165°/360°) = 7.879 \, \text{m}^2$$

Knowing individual frontal areas and effective heights, the heat transfer areas for the hot and cold sides are found as

$$A_h = A_{fr,h} L_{\text{eff}} \beta_h = 9.311 \, \text{m}^2 \times 1.432 \, \text{m} \times 402.3 \, \text{m}^2/\text{m}^3 = 5364 \, \text{m}^2$$

$$A_c = A_{fr,c} L_{\text{eff}} \beta_h = 7.879 \, \text{m}^2 \times 1.432 \, \text{m} \times 402.3 \, \text{m}^2/\text{m}^3 = 4539 \, \text{m}^2$$

Knowing the heat transfer areas and heat transfer coefficients, we obtain

$$(hA)_h = 0.0822 \, \text{kW/m}^2 \cdot \text{K} \times 5364 \, \text{m}^2 = 440.9 \, \text{kW/K}$$

$$(hA)_c = 0.0727 \, \text{kW/m}^2 \cdot \text{K} \times 4539 \, \text{m}^2 = 330.0 \, \text{kW/K}$$

Hence, the ratio of the convection conductances is found as

$$(hA)^* = \frac{(hA)_{\min}}{(hA)_{\max}} = \frac{(hA)_c}{(hA)_h} = \frac{330.0 \, \text{kW/K}}{440.9 \, \text{kW/K}} = 0.748$$

Thus $(hA)^*$ is within the range 0.25 to 4.0, and hence its effect is negligible on the regenerator ε.

To determine the regenerator effectiveness, we calculate various heat capacity rates:

$$C_h = \dot{m}_h c_{p,h} = 51.1 \, \text{kg/s} \times 1.11 \, \text{kJ/kg} \cdot \text{K} = 56.72 \, \text{kW/K}$$

$$C_c = \dot{m}_c c_{p,c} = 42.4 \, \text{kg/s} \times 1.02 \, \text{kJ/kg} \cdot \text{K} = 43.25 \, \text{kW/K}$$

$$C_{\max} = C_h = 56.72 \, \text{kW/K} \qquad C_{\min} = C_c = 43.25 \, \text{kW/K}$$

$$C^* = \frac{C_{\min}}{C_{\max}} = \frac{43.25 \, \text{kW/K}}{56.72 \, \text{kW/K}} = 0.7625$$

For the matrix heat capacity rate, we must first determine the matrix mass as follows.

M_w = rotor frontal area × rotor height × matrix solidity × matrix material density

$$= 19.1 \, \text{m}^2 \times [1.3 \, \text{m} \times (1 - 0.859) + 0.3 \, \text{m} \times (1 - 0.780)] \times 7841 \, \text{kg/m}^3 = 37{,}336 \, \text{kg}$$

Knowing the matrix mass, its heat capacity rate is computed as

$$C_r = M_w c_w \text{N} = 37{,}336 \, \text{kg} \times 0.456 \, \text{kJ/kg} \cdot \text{K} \times \frac{2.32}{60} \, \text{rev/s} = 658.3 \, \text{kW/K}$$

$$C_r^* = \frac{C_r}{C_{\min}} = \frac{658.3 \, \text{kW/K}}{43.25 \, \text{kW/K}} = 15.22$$

From the foregoing values, we can compute NTU_o, given in Eq. (5.48), as

$$NTU_o = \frac{1}{C_{min}} \frac{1}{1/(hA)_h + 1/(hA)_c} = \frac{1}{43.25\,kW/K} \left[\frac{1}{(1/440.9 + 1/330.0)\,K/kW} \right] = 4.364$$

Now determine the regenerator ε using Eqs. (5.62) and (5.63):

$$\varepsilon_{cf} = \frac{1 - \exp[-NTU_o(1 - C^*)]}{1 - C^* \exp[-NTU_o(1 - C^*)]} = \frac{1 - \exp[-4.364(1 - 0.7625)]}{1 - 0.7625 \exp[-4.364(1 - 0.7625)]} = 0.8845$$

$$\varepsilon = \varepsilon_{cf} \left(1 - \frac{1}{9C_r^{*1.93}} \right) = 0.8845 \left(1 - \frac{1}{9 \times 15.22^{1.93}} \right) = 0.8840$$

It can be shown that ε calculated by the Razelos method [Eqs. (5.64)–(5.67)] would have been identical to 0.8840. The heat transfer rate is then determined as

$$q = \varepsilon C_{min}(T_{h,i} - T_{c,i}) = 0.8840 \times 43.25\,kW/K\,(339 - 29)\,°C = 11{,}852\,kW \qquad Ans.$$

Knowing the heat transfer rate, the hot- and cold-fluid outlet temperatures are calculated as

$$T_{h,o} = T_{h,i} - \frac{q}{C_h} = 339°C - \frac{11{,}852\,kW}{56.72\,kW/K} = 130.0°C \qquad Ans.$$

$$T_{c,o} = T_{c,i} + \frac{q}{C_c} = 29°C + \frac{11{,}852\,kW}{43.25\,kW/K} = 303.0°C \qquad Ans.$$

Discussion and Comments: This example illustrates the methodology of analyzing a regenerator with multilayer packing. As discussed in the text, the multilayer packing is used to accommodate high temperatures, high fouling and corrosion, and/or high matrix volumetric heat capacity rates in different regions of the regenerator. The analysis above presumes that the fluid temperature distribution is uniform after leaving the first zone and entering the second zone on each fluid side. This is a reasonably good approximation, considering the fact that there are large uncertainties involved in the determination of heat transfer coefficients and various flow leakages associated with regenerator analysis. Once the problem is reformulated having an equivalent single layer, the analysis is straightforward, as outlined in Section 5.2 since we have neglected the longitudinal wall conduction effect.

SUMMARY

Regenerators differ from recuperators in that the heat is transferred intermittently from the hot fluid to the cold fluid via periodic thermal energy storage and release from the heat transfer surface (matrix). As a result, two additional parameters enter in the analysis of regenerators: the storage heat capacity rate of the matrix wall and the ratios of the thermal conductances between the wall and hot and cold fluids [in a dimensionless form C_r^* and $(hA)^*$, respectively], with the latter group of less importance in most industrial regenerators. The two most commonly used methods for the design and analysis of regenerators described are the ε-NTU$_o$ and Λ-Π methods. The details of these methods with the basic concepts and advantages/disadvantages are presented. In addition to longitudinal conduction in the wall (as in recuperators) at high effectivenesses, the transverse conduction in the wall can also be important for ceramic and low-thermal-conductivity materials. The design theory is presented to take these effects into account. In addition, pressure and carryover leakages can reduce the regenerator effectiveness significantly, depending on the operating conditions. A design theory is presented to take these effects into account. Since the last two effects are too complex and interdependent on the regenerator geometry and operating conditions, the design theory involves iterative calculations. The details presented in the text are essential for any computer analysis of regenerator rating and sizing.

REFERENCES

Bačlić, B. S., 1985, The application of the Galerkin method to the solution of the symmetric and balanced counterflow regenerator problem, *ASME J. Heat Transfer*, Vol. 107, pp. 214–221.

Bahnke, G. D., and C. P. Howard, 1964, The effect of longitudinal heat conduction on periodic-flow heat exchanger performance, *ASME J. Eng. Power*, Vol. 86, Ser. A, pp. 105–120.

Coppage, J. E., and A. L. London, 1953, The periodic-flow regenerator: a summary of design theory, *Trans. ASME*, Vol. 75, pp. 779–787.

Dragutinović, G. D., and B. S. Bačlić, 1998, *Operation of Counterflow Regenerators*, Vol. 4, Computational Mechanics Publications, WIP Press, Southampton, UK.

Hausen, H., 1929, Über die Theorie von Wärmeaustauches in Regeneratoren, *Z. Angew. Math. Mech.*, Vol. 9, pp. 173–200.

Hausen, H., 1983, *Heat Transfer in Counterflow, Parallel Flow and Cross Flow*, McGraw-Hill, New York.

Heggs, P. J., and K. J. Carpenter, 1978, The effects of packing material, size and arrangement of the performance of thermal regenerators, *Heat Transfer 1978, Proc. 6th Int. Heat Transfer Conf.*, Vol. 4, Hemisphere Publishing, Washington, DC, pp. 321–326.

Heggs, P. J., and K. J. Carpenter, 1979, A modification of the thermal regenerator infinite conduction model to predict the effects of intraconduction, *Trans. Inst. Chem. Eng.*, Vol. 57, pp. 228–236.

Kays, W. M., and A. L. London, 1998, *Compact Heat Exchangers*, reprint 3rd ed., Krieger Publishing, Malabar, FL.

Kroger, P. G., 1967, Performance deterioration in high effectiveness heat exchangers due to axial heat conduction effects, *Advances in Cryogenic Engineering*, Vol. 12, pp. 363–372.

Lambertson, T. J., 1958, Performance factors of a periodic-flow heat exchanger, *Trans. ASME*, Vol. 80, pp. 586–592.

Mondt, J. R., 1964, Vehicular gas turbine periodic-flow heat exchanger solid and fluid temperature distributions, *ASME J. Eng. Power*, Vol. 86, Ser. A, pp. 121–126.

Razelos, P., 1979, An analytic solution to the electric analog simulation of the regenerative heat exchanger with time-varying fluid inlet temperatures, *Wärme-und Stoffübertragung*, Vol. 12, pp. 59–71.

Razelos, P., 1980, Personal communication, Department of Applied Science, City University of New York, Staten Island, NY.

Saunders, O. A., and S. Smoleniec, 1951, Heat transfer in regenerators, *IMechE–ASME General Discussion on Heat Transfer*, London, pp. 443–445.

Schack, A., 1965, *Industrial Heat Transfer*, Wiley, New York.

Shah, R. K., 1975, A correlation for longitudinal heat conduction effects in periodic-flow heat exchangers, *ASME J. Eng. Power*, Vol. 97, Ser. A, pp. 453–454.

Shah, R. K., 1981, Thermal design theory for regenerators, in *Heat Exchangers: Thermal-Hydraulic Fundamentals and Design*, S. Kakaç, A. E. Bergles, and F. Mayinger, eds., Hemisphere Publishing, Washington, DC, pp. 721–763.

Shah, R. K., 1985, Compact heat exchangers, in *Handbook of Heat Transfer Applications*, 2nd ed., W. M. Rohsenow, J. P. Hartnett and E. N. Ganić, eds., McGraw-Hill, New York, Chapter 4, pp. 174–311.

Shah, R. K., 1988, Counterflow rotary regenerator thermal design procedures, in *Heat Transfer Equipment Design*, R. K. Shah, E. C. Subbarao, and R. A. Mashelkar, eds., Hemisphere Publishing, Washington, DC, pp. 267–296.

Shah, R. K., and T. Skiepko, 1997, Influence of leakage distribution on the thermal performance of a rotary regenerator, in *Experimental Heat Transfer, Fluid Mechanics and Thermodynamics 1997*, Vol. 1, Edizioni ETS, Pisa, Italy, pp. 365–377.

Shah, R. K., and T. Skiepko, 1998, Modeling of leakages in fixed matrix regenerators, *Heat Transfer 1998, Proc. 11th Int. Heat Transfer Conf.*, Vol. 6, pp. 233–238.

Skiepko, T., 1988, The effect of matrix longitudinal heat conduction on the temperature fields in the rotary heat exchanger, *Int. J. Heat Mass Transfer*, Vol. 31, pp. 2227–2238.

Skiepko, T., and R. K. Shah, 1994, Some additional data on longitudinal wall heat conduction in counterflow rotary regenerators, in *Heat Transfer 1994, Proc. 10th Int. Heat Transfer Conf.*, Vol. 4, pp. 429–434.

Theoclitus, G., and T. L. Eckrich, 1966, Parallel flow through the rotary heat exchanger, *Proc. 3rd Int. Heat Transfer Conf.*, Vol. I, pp. 130–138.

Venkatarathnam, G., 1996, Effectiveness–N_{tu} relationship in perforated plate matrix heat exchangers, *Cryogenics*, Vol. 36, pp. 235–241.

REVIEW QUESTIONS

Where multiple choices are given, circle one or more correct answers. Explain your answers briefly.

5.1   What is the difference between heat capacitance and the heat capacity rate of a fluid? Explain with equations.

5.2   How do you define the heat capacity rate of a matrix? How is it related to the heat capacity of the wall for a rotary regenerator and for a fixed-matrix regenerator?

5.3   In a storage-type heat exchanger, when the hot fluid flows through the matrix, the temperature of the entire matrix rises to the hot-fluid temperature. When cold fluid flows through it, its temperature drops to the cold-fluid temperature. Check the appropriate answer for this phenomenon to occur.

(a) It depends on the matrix NTU.   (b) true   (c) false

(d) It depends on the temperature levels.

5.4   Given a number of independent variables associated with differential equations and boundary conditions, independent dimensionless groups can be obtained by:

(a) making differential equations and boundary conditions parameter free

(b) using the Buckingham Π theorem

(c) making differential equations and boundary conditions dimensionless

(d) first making all possible dimensionless groups and then eliminating dependent groups

5.5 Time is an important variable in:

(a) storage-type heat exchanger design theory
(b) shell-and-tube exchanger steady-state theory

(c) recuperator steady-state pressure drop analysis

5.6 In normal steady-state or periodic operation, the cold-fluid outlet temperature varies as a function of time for the following exchanger:

(a) plate exchanger   (b) rotary regenerator   (c) tube-fin exchanger

(d) fixed-matrix regenerator   (e) double-pipe exchanger

5.7 The following factors make analysis of regenerators more complicated than for recuperators:

(a) Thermal resistance of solid material is important.

(b) Thermal capacitance of solid material is important.

(c) Heat transfer coefficients vary with position.

(d) Heat transfer from/to fluid streams occurs independently.

(e) Compactness β (m^2/m^3) is higher.

5.8 Regenerator effectiveness depends on the following dimensionless groups:

(a) NTU   (b) NTU$_o$ (c) F (d) C^* (e) C_r^* (f) P (h) R (h) $(hA)^*$

5.9 Regenerator effectiveness depends on the following dimensionless groups:

(a) τ^* (b) X^* (c) Λ_h (d) Π_c (e) Π_h (f) ξ_c (g) η_h (h) Λ_c (i) T_w^*

5.10 How does the regenerator effectiveness of an ideal counterflow regenerator change with increasing rotational speed **N** (rpm)? Ignore carryover and pressure leakages.

(a) decreases, but does not approach 0 as $\mathbf{N} \to \infty$

(b) decreases, approaches 0 as $\mathbf{N} \to \infty$

(c) increases, but does not approach 1 as $\mathbf{N} \to \infty$ for NTU$_o < \infty$

(d) increases, approaches 1 as $\mathbf{N} \to \infty$ for NTU$_o \to \infty$

5.11 The following things occur when the thickness (flow length) of an ideal counterflow regenerator matrix is doubled with all other operating conditions remaining unchanged:

(a) Wall thermal resistance effects become more important.

(b) Exchanger effectiveness increases (ideal regenerator).

(c) Longitudinal conduction effects become more important.

(d) Capacity rate ratio C^* increases.

(e) Pressure drop increases.

5.12 The counterflow regenerator effectiveness will increase with:

(a) increasing pressure leakage between hot and cold streams

(b) increasing rotational speed and neglecting carryover losses

(c) increasing the specific heat of the matrix

(d) increasing the thermal conductivity of the matrix material

(e) increasing the regenerator flow length

5.13 How would ε change for a counterflow regenerator (having $C_r^* = 5$ and $C^* = 1$) by increasing the disk speed from 3 rpm to 20 rpm?

(a) less than 1%   (b) about 5%   (c) more than 10%

5.14 In a vehicular gas turbine rotary regenerator, the terminal temperatures are: hot fluid: 760°C, 297°C; cold fluid: 204°C, 667°C. The rotational speed is such that $C_r^* = 10$. The effectiveness of this regenerator is approximately:

(a) 83%   (b) 75%   (c) 100%   (d) 64%

5.15 The number of transfer units NTU for the regenerator of Question 5.14 is approximately:

(a) 10   (b) 5   (c) 8   (d) 1   (e) 7.33

5.16 A loss in the counterflow regenerator effectiveness due to longitudinal heat conduction increases with:

(a) decreasing value of NTU_o (b) decreasing value of C^*

(c) decreasing value of λ

5.17 The unbalanced regenerator effectiveness depends on Λ and Π.

(a) false   (b) true   (c) can't say anything

5.18 A greater temperature "swing" is exhibited in a regenerator with $C_r^* = 5$ than in a regenerator with $C_r^* = 1$ for a fixed $C_{\min}$.

(a) true   (b) false   (c) can't say anything

5.19 The transverse conduction in the regenerator matrix wall is important for:

(a) gas turbine rotary regenerators   (b) Ljungstrom air preheaters

(c) Cowper stoves   (d) glass melting furnace air preheaters

(e) thick ceramic walls   (f) thin ceramic walls

5.20 For identical surface geometry, matrix surface thickness and rpm, which matrix material will require the least disk depth for the same ε? Assume fixed frontal area and velocities.

(a) *Aluminum:* $\rho_w = 2702 \text{ kg/m}^3$, $c_w = 903 \text{ J/kg} \cdot \text{K}$, $k_w = 237 \text{ W/m} \cdot \text{K}$

(b) *Stainless steel:* $\rho_w = 8238 \text{ kg/m}^3$, $c_w = 468 \text{ J/kg} \cdot \text{K}$, $k_w = 13.4 \text{ W/m} \cdot \text{K}$

Ignore longitudinal and transverse conduction as well as pressure and carryover leakages.

5.21 Consider two counterflow rotary regenerators, one with $C_r^* = 5$ and the other with $C_r^* = 2$. Otherwise, the exchangers are identical, with the same inlet temperatures and flow rates. Sketch the temperature variation with position of the hot-fluid outlet temperature during one blow period in Fig. RQ5.21a for each exchanger. Repeat in Fig. RQ5.21b for the cold-fluid outlet temperature variation. For convenience, let $C^* = 1$.

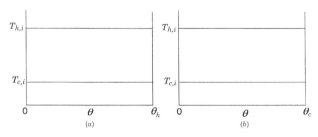

FIGURE RQ5.21

PROBLEMS

5.1 Following are the design data for a ceramic counterflow disk rotary regenerator. Two such disks are used in a truck gas turbine engine.

Regenerator operating conditions
Airflow rate = 1.400 kg/s
Gas flow rate = 1.414 kg/s
Disk speed = 15 rpm
Air inlet temperature = 204°C
Gas inlet temperature = 866°C
Air inlet pressure = 393 kPa
Gas inlet pressure = 105 kPa
Regenerator leakage
  = 5.15% of airflow

Dimensions for one disk (exclusive of rim)
Disk diameter = 597 mm
Hub diameter = 50.8 mm
Flow length = 74.7 mm
Seal face and hub coverage = 10.8%
Flow split, gas : air = 1.2 : 1
Effective total frontal area = 0.2495 m^2
Matrix volume (excluding coverage)
  = 0.01863 m^2
Matrix effective mass = 13.43 kg
Header-to-matrix volume ratio $V^* = 0.5$

Physical properties
$c_{p,\text{air}} = 1093$ J/kg · K
$c_{p,\text{gas}} = 1118$ J/kg · K
c_w at 400°C = 1093 J/kg · K
Matrix wall density = 2214 kg/m^3
Matrix $k = 0.19$ W/m · K

Matrix geometry
Surface area density $\beta = 6463$ m^2/m^3
Hydraulic diameter $D_h = 0.44$ mm
Porosity $\sigma = 0.71$
Cell count $N = 2.17$ mm^2
Wall thickness = 0.076 mm
Triangular passages

Heat transfer coefficients
$h_{\text{air}} = 409$ W/m^2 · K
$h_{\text{gas}} = 425$ W/m^2 · K

Calculate the regenerator effectiveness and outlet temperatures in the absence of longitudinal conduction, wall thermal resistance, and pressure and carryover leakage effects. Then determine separately the influences of longitudinal wall heat conduction and wall thermal resistance, again neglecting the effect of pressure and carryover leakages on regenerator effectiveness. What will be the outlet temperatures in the latter case? Discuss your results.

5.2 The rotary air preheaters at the Moss Landing power plant have the following geometry and operating conditions: 14.63-m disk diameter with $A_h = A_c = 31{,}120$ m^2/unit, $\dot{m}_{\text{gas}} \approx \dot{m}_{\text{air}} = 340$ kg/s per preheater, effectiveness $\varepsilon = 65\%$,

$M_w = 1121\,\text{kg/m}^3$ of total volume, $c_\text{steel} = 460.6\,\text{J/kg}\cdot\text{K}$, $c_\text{air} \approx c_\text{gas} = 1026$ J/kg·K, and the matrix air flow length $= 1.52\,\text{m}$. The preheater rotates at 48 rev/h. What are the disk heat capacity rate ratio C_r^* and the disk NTU_o? What is the average air-side convective conductance? Assume that $h_c \approx h_h$. Explain why you would neglect longitudinal conduction in the wall for this problem. In this case, the C_λ of Eq. (5.122) is assumed as zero.

5.3 Consider a rotary regenerator of compactness $\beta\,(\text{m}^2/\text{m}^3)$, disk diameter D, thickness (flow length) L, hot and cold gas flow ratio $\dot{m}_h$ and $\dot{m}_c$, respectively, and rotating speed **N**. For what flow split (a fraction of a cycle in which an element of matrix is in the hot blow divided by the fraction in the cold blow) will the exchanger effectiveness be maximum? Justify your answer briefly. Assume that the heat transfer coefficients h_h and h_c are equal and do not depend on the flow split. Neglect any seal area. *Hint:* Use Eq. (5.10) and the definition of NTU_o.

5.4 Consider a fixed-matrix regenerator used as a thermal storage device with the following data: $\dot{m}_h = \dot{m}_c = 0.156\,\text{kg/s}$, $c_{p,c} = c_{p,h} = 1011.0\,\text{J/kg}\cdot\text{K}$, $h_h = h_c = 50.23\,\text{W/m}^2\cdot\text{K}$, $A = 5.8\,\text{m}^2$, $M_w = 904.8\,\text{J/kg}\cdot\text{K}$, and $c_w = 920\,\text{J/kg}\cdot\text{K}$. The cycle period P_t is specified as very short. Determine the mean outlet temperatures of the hot and cold gases if $T_{h,i} = 80°\text{C}$ and $T_{c,i} = 10°\text{C}$ and the hot and cold gases are in counterflow.

5.5 An industrial water tube boiler produces 2.52 kg/s of steam at 5.17 MPa. The flue gases leave the boiler at 321°C (55.6°C above the steam saturation temperature). To increase the boiler efficiency, a rotary counterflow regenerator will be installed to recover heat from 321°C gases to preheat $-17.8°\text{C}$ combustion air. The boiler is oil fired, and the combustion products, which contain sulfur, have a dew point of 135°C. To avoid corrosion, the gas temperature leaving the regenerator is limited to 143°C. The following design data are provided: Air and gas flow rates are 2.39 and 2.55 kg/s, respectively, and their specific heats are 1005 and 1089 J/kg·K. As a good design practice, consider $C_r^* = 5$. The regenerator matrix is made up of low-alloy steel with $\rho = 8009\,\text{kg/m}^3$ and $c_w = 461\,\text{J/kg}\cdot\text{K}$. The heat transfer coefficients for air and gas sides are 190 and 207 W/m²·K, respectively, and the split of gas to air flow is 50:50%. Determine the total heat transfer surface area required for the regenerator. Neglect longitudinal conduction, carryover, and pressure leakage effects. Once you determine the total area required, how would you determine the disk diameter and flow length? Answer the last question qualitatively only.

6 Heat Exchanger Pressure Drop Analysis

Fluids need to be pumped through the heat exchanger in most applications. It is essential to determine the fluid pumping power required as part of the system design and operating cost analysis. The fluid pumping power is proportional to the fluid pressure drop, which is associated with fluid friction and other pressure drop contributions along the fluid flow path. The fluid pressure drop has a direct relationship with exchanger heat transfer, operation, size, mechanical characteristics, and other factors, including economic considerations. The objective of this chapter is to outline the methods for pressure drop analysis in heat exchangers and related flow devices. In this chapter, we start with the importance of pressure drop evaluation, pumping devices used for fluid flow, and the major contributions to the total pressure drops associated with a heat exchanger in Section 6.1. Two major contributions to pressure drop are associated with (1) core or matrix and (2) flow distribution devices. Then a detailed derivation for various contributions to the core pressure drop is outlined for a plate-fin exchanger, followed by the pressure drop equations for all other major types of exchangers, in Sections 6.2 through 6.5. For most other exchangers, the flow distribution devices are of varied type (ducting, pipe bends, valves and fitting, etc.), and the pressure drop associated with them is calculated separately; appropriate methods for their computation are summarized in Section 6.6. Next, the presentation of pressure drop data in nondimensional and dimensional forms is outlined for the design and analysis of heat exchangers in Section 6.7. Finally, pressure drop dependence on geometry and fluid properties is outlined in Section 6.8.

6.1 INTRODUCTION

First we outline why pressure drop is important for gases vs. liquids, what devices are used for pumping fluids in the exchanger, and the major components of the pressure drop in a heat exchanger. Assumptions for the pressure drop analysis are also presented in Section 6.1.4 before we start the pressure drop analysis in the following subsections.

6.1.1 Importance of Pressure Drop

The determination of pressure drop Δp in a heat exchanger is essential for many applications for at least two reasons: (1) The fluid needs to be pumped through the exchanger, which means that fluid pumping power is required. This pumping power is proportional to the exchanger pressure drop. (2) The heat transfer rate can be influenced significantly

378

by the saturation temperature change for a condensing/evaporating fluid if there is a large pressure drop associated with the flow. This is because saturation temperature changes with changes in saturation pressure and in turn affects the temperature potential for heat transfer.

Let us first determine the relative importance of the fluid pumping power $\mathcal{P}$ for gas flow vs. liquid flow in a heat exchanger. $\mathcal{P}$ is proportional to Δp in a heat exchanger and is given by

$$\mathcal{P} = \frac{\dot{V}\,\Delta p}{\eta_p} = \frac{\dot{m}\,\Delta p}{\rho\eta_p} \tag{6.1}$$

where $\dot{V}$ is the volumetric flow rate and η_p is the pump/fan efficiency. Now introduce the following relationships:

$$\dot{m} = GA_o \qquad \Delta p \approx f\frac{4L}{D_h}\frac{G^2}{2g_c\rho} \qquad \mathrm{Re} = \frac{GD_h}{\mu} \tag{6.2}$$

where G is referred to as the core *mass velocity* $(G = \rho u_m)$, A_o is the minimum free flow area, f is the Fanning friction factor,[†] and Re is the Reynolds number as defined in Eq. (6.2). The Δp expression in Eq. (6.2) is for the core frictional pressure drop and is derived later as Eq. (6.29). Substituting the expressions of Eq. (6.2) into Eq. (6.1) and simplifying results in

$$\mathcal{P} = \frac{\dot{m}\,\Delta p}{\rho\eta_p} \approx \begin{cases} \dfrac{1}{2g_c\eta_p}\dfrac{\mu}{\rho^2}\dfrac{4L}{D_h}\dfrac{\dot{m}^2}{D_hA_o}(f\cdot\mathrm{Re}) & \text{for fully developed laminar flow} \quad (6.3a) \\[2em] \dfrac{0.046}{2g_c\eta_p}\dfrac{\mu^{0.2}}{\rho^2}\dfrac{4L}{D_h}\dfrac{\dot{m}^{2.8}}{A_o^{1.8}D_h^{0.2}} & \text{for fully developed turbulent flow} \quad (6.3b) \end{cases}$$

Here $f = 0.046\mathrm{Re}^{-0.2}$ [see Eq. (7.72) in Table 7.6] is used in the derivation of the right-hand-side expression of Eq. (6.3b) for fully developed turbulent flow.[‡] Note also that $f\cdot\mathrm{Re}$ in Eq. (6.3a) is constant for fully developed laminar flow, as discussed in Section 7.4.1.1. To determine the order of magnitude for the fluid pumping power requirement for gas vs. liquid flow, let us assume that the flow rate and flow passage geometry are given (i.e., $\dot{m}$, L, D_h, and A_o are specified). It is evident from Eq. (6.3) that $\mathcal{P} \propto 1/\rho^2$ (i.e., strongly dependent on ρ in laminar and turbulent flows); $\mathcal{P} \propto \mu$ (i.e., strongly dependent on μ in laminar flow); and $\mathcal{P} \propto \mu^{0.2}$ (i.e., weakly dependent on μ in turbulent flow). For high-density moderate-viscosity liquids, the pumping power is generally so small that it has only a minor influence on the design. For laminar flow of highly viscous liquids in large L/D_h exchangers, the fluid pumping power is an important constraint. In addition, the pumping power is an important consideration for gases, in both turbulent and laminar flow, because of the great impact of $1/\rho^2$. For example, the density ratio for liquid water vs. air at ambient conditions is approximately 800 : 1, which indicates that the pumping power for airflow will be much higher than that for water if Δp is to be kept

---

[†] The definition of the Fanning friction factor is given in Eq. (6.9).

[‡] It should be emphasized that Eqs. (6.3a) and (6.3b) are presented for demonstrating the importance of Δp, and they do not imply that we avoid the transition flow regime. As a matter of fact, most automotive heat exchangers operate in the transition regime. We discuss the range of Re for laminar, transition, and turbulent flows in heat exchangers in Section 7.1.2.2.

the same. Hence, typical design values of Δp for water and air as working fluids in a heat exchanger are 70 kPa (10 psi) (a typical value in shell-and-tube exchangers) and 0.25 kPa (1 in. H_2O) (for compact exchangers with airflows near ambient pressures), respectively, to maintain the low fluid pumping power requirement for exchanger operation. Refer also to the first footnote in Section 13.1 and the associated discussion.

6.1.2 Fluid Pumping Devices

The most common fluid pumping devices are fans, pumps, and compressors. A *fan* is a low-pressure air- or gas-moving device, which uses rotary motion. There are two major types of fans: axial and radial (centrifugal), depending on the direction of flow through the device. Fans may be categorized as blowers and exhausters. A *blower* is a centrifugal fan when it is used to force air through a system under positive pressure, and it develops a reasonably high static pressure (500 Pa or 2.0 in. H_2O). An *exhauster* is a fan placed at the end of a system where most of the pressure drop is on the suction side of the fan. A *pump* is a device used to move or compress liquids. A *compressor* is a high-volume centrifugal device capable of compressing gases [100 to 1500 kPa (15 to 220 psi) and higher].

Fans and pumps are volumetric devices and are commonly used to pump fluids through heat exchangers. This means that a fan will develop the same dynamic head [pressure rise per unit fluid (gas) weight across the fan; Eq. (6.4)] at a given capacity (volumetric flow rate) regardless of the fluids handled, with all other conditions being equal. This means that the pressure rise across a fan will be proportional to the fluid density at a given volumetric flow rate for all other conditions being equal. Note that the *head, dynamic head or velocity head* is referred to as the kinetic energy per unit weight of the fluid pumped, expressed in units of millimeters or inches (feet). Thus the pressure rise across a fan (which is mainly consumed as the pressure drop across a heat exchanger) can be expressed in terms of the head H as follows:

$$\frac{\Delta p}{\rho g/g_c} = H = \frac{u_m^2}{2g} \tag{6.4}$$

Since fans and pumps are generally head limited, the pressure drop in the heat exchanger can be a major consideration.

6.1.3 Major Contributions to the Heat Exchanger Pressure Drop

The pressure drop associated with a heat exchanger is considered as a sum of two major contributions: pressure drop associated with the core or matrix, and pressure drop associated with fluid distribution devices such as inlet/outlet headers, manifolds, tanks, nozzles, ducting, and so on. The purpose of the heat exchanger is to transfer thermal energy from one fluid to the other; and for this purpose, it requires pressure difference (and fluid pumping power) to force the fluid flow over the heat transfer surface in the exchanger. Hence, ideally most of the pressure drop available should be utilized in the core and a small fraction in the manifolds, headers, or other flow distribution devices. However, this ideal situation may not be the case in plate heat exchangers and other heat exchangers in which the pressure drop associated with manifolds, headers, nozzles, and so on, may not be a small fraction of the total available pressure drop.

If the manifold and header pressure drops are small, the core pressure drop dominates. This results in a relatively uniform flow distribution through the core. All heat transfer and core pressure drop analyses outlined here and in preceding chapters presume that the flow distribution through the core is uniform. A serious deterioration in performance may result for a heat exchanger when the flow through the core is not uniformly distributed. This topic is covered in Chapter 12.

The core pressure drop is determined separately on each fluid side. It consists of one or more of the following contributions, depending on the exchanger construction: (1) frictional losses associated with fluid flow over the heat transfer surface (this usually consists of skin friction plus form drag), (2) momentum effect (pressure drop or rise due to the fluid density changes in the core), (3) pressure drop associated with sudden contraction and expansion at the core inlet and outlet, and (4) gravity effect due to the change in elevation between the inlet and outlet of the exchanger. The gravity effect is generally negligible for gases. For vertical liquid flow through the exchanger, the pressure drop or rise due to the elevation change is given by

$$\Delta p = \pm \frac{\rho_m g L}{g_c} \tag{6.5}$$

where the " + " sign denotes vertical upflow (i.e., pressure drop), the " − " sign denotes vertical downflow (i.e., pressure rise or recovery), g is gravitational acceleration, L is the exchanger length, and ρ_m is the mean fluid mass density calculated at bulk temperature and mean pressure between the two points where the pressure drop is to be determined. The first three contributions to the core pressure drop are presented for extended surface exchangers, regenerators, and tubular and plate heat exchangers in Sections 6.2 through 6.5. Since the manifolds are integral parts of a PHE, the pressure drop associated with manifolds is also included in Δp calculations for a PHE in Section 6.5.

6.1.4 Assumptions for Pressure Drop Analysis

The following are the major assumptions made for the pressure drop analysis presented in this chapter.

1. Flow is steady and isothermal, and fluid properties are independent of time.
2. Fluid density is dependent on the local temperature only or is treated as a constant (inlet and exit densities are separately constant).
3. The pressure at a point in the fluid is independent of direction. If a shear stress is present, the pressure is defined as the average of normal stresses at the point.
4. Body forces are caused only by gravity (i.e., magnetic, electrical, and other fields do not contribute to the body forces).
5. If the flow is not irrotational, the Bernoulli equation is valid only along a streamline.
6. There are no energy sinks or sources along a streamline; flow stream mechanical energy dissipation is idealized as zero.
7. The friction factor is considered as constant with passage flow length.

6.2   EXTENDED SURFACE HEAT EXCHANGER PRESSURE DROP

The pressure drop analysis is presented now for plate-fin and tube-fin heat exchangers.

6.2.1   Plate-Fin Heat Exchangers

One flow passage in a plate-fin heat exchanger is shown in Fig. 6.1 along with fluid flow and static pressure distribution along the flow path. The incoming flow to the passage is assumed to be uniform. As it enters the passage, it contracts due to the free-flow area change. Flow separation takes place at the entrance followed by irreversible free expansion. In the core, the fluid experiences skin friction; it may also experience form drag at the leading and trailing edges of an interrupted fin surface; it may also experience internal contractions and expansions within the core, such as in a perforated fin core. If heating or cooling takes place in the core, as in any heat exchanger, the fluid density and mean velocity change along the flow length. Thus, the fluid within the flow passage accelerates or decelerates depending on whether it is being heated or cooled. At the core exit, flow separation takes place followed by an expansion due to the free-flow area change. Then the total pressure drop on one side of the exchanger, from Fig. 6.1, is

$$\Delta p = \Delta p_{1-2} + \Delta p_{2-3} - \Delta p_{3-4} \tag{6.6}$$

Here the subscripts 1, 2, 3, and 4 represent locations far upstream, passage entrance, passage exit, and far downstream, respectively, as shown in Fig. 6.1. The Δp_{1-2} is the pressure drop at the core entrance due to sudden contraction, Δp_{2-3} the pressure drop within the core (also simply referred to as the core pressure drop), and Δp_{3-4} the pressure rise at the core exit. Usually, Δp_{2-3} is the largest contribution to the total pressure drop, and we evaluate it first before the other two contributions.

6.2.1.1   Core Pressure Drop. The pressure drop within the core consists of two contributions: (1) the pressure loss caused by fluid friction, and (2) the pressure change due to the momentum rate change in the core. The friction losses take into account both skin friction and form drag effects. The influence of internal contractions and

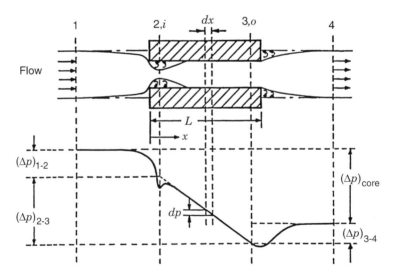

FIGURE 6.1   Pressure drop components associated with one passage of a heat exchanger (From Shah, 1983; modified from Kays and London, 1998).

expansions due to flow area changes, if present, is also lumped into the core friction loss term. Consider a differential element of flow length dx in the core as shown in Fig. 6.1. Various force and momentum rate terms in and out of this element are shown in Fig. 6.2.[†]

Applying Newton's second law of motion, we have

$$\frac{G^2 A_o}{g_c}\left[\frac{1}{\rho} + \frac{d}{dx}\left(\frac{1}{\rho}\right)dx\right] - \frac{G^2 A_o}{g_c \rho} = pA_o - \left(p + \frac{dp}{dx}dx\right)A_o - \tau_w \mathbf{P}\,dx \qquad (6.7)$$

Here τ_w is the *effective* wall shear stress[‡] due to skin friction, form drag, and internal contractions and expansions, if any. $\mathbf{P}$ is the wetted perimeter of the fluid flow passages of heat exchanger surface. Rearranging and simplifying Eq. (6.7), we get

$$-\frac{dp}{dx} = \frac{G^2}{g_c}\frac{d}{dx}\left(\frac{1}{\rho}\right) + \tau_w\frac{\mathbf{P}}{A_o} \qquad (6.8)$$

Note that we use the mass velocity $G = \dot{m}/A_o$ as the flow variable for the exchanger Δp analysis. This is because G is constant for a constant-steady state fluid flow rate $\dot{m}$ and constant A_o[§] even though both ρ and u_m in $G = \rho u_m$ vary along the flow length in the heat exchanger.

Now define the Fanning friction factor f[¶] as the ratio of wall shear stress τ_w to the flow kinetic energy per unit volume.

$$f = \frac{\tau_w}{\rho u_m^2/2g_c} = \frac{\tau_w}{G^2/2g_c\rho} \qquad (6.9)$$

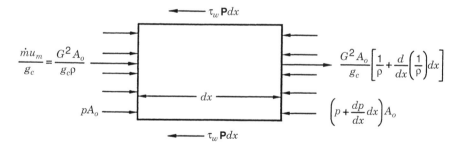

FIGURE 6.2   Force and momentum rate terms for a differential element of a heat exchanger core.

[†] While $\tau_w \mathbf{P}\,dx$ is shown acting on both top and bottom surface in Fig. 6.2, in reality it acts along the entire surface $\mathbf{P}\,dx$.

[‡] τ_w is dependent on the flow passage geometry and size, fluid velocity, fluid density and viscosity, and surface roughness, if any.

[§] The minimum free-flow area A_o is constant in most heat exchangers, including that for flow over the tube banks and flow in the regenerator matrix, made up of solid objects such as bricks, rocks, pebbles, and so on.

[¶] The friction factor is either derived experimentally for a surface or derived theoretically for laminar flow and simple geometries. It is discussed further in Chapter 7.

As discussed in Section 7.2.1.2, τ_w represents the effective wall shear stress, and ρ is the fluid mass density determined at the local bulk temperature and mean pressure. Also define the hydraulic radius r_h as

$$r_h = \frac{A_o}{P} \tag{6.10}$$

Note that the hydraulic diameter $D_h = 4r_h$; it was defined earlier by Eq. (3.65) and in the footnote in Section 1.4.

Substituting Eqs. (6.9) and (6.10) into Eq. (6.8) and using $d(1/\rho) = -(1/\rho^2)\,d\rho$, we get, upon simplification,

$$-\frac{dp}{dx} = \frac{G^2}{2g_c}\left(-\frac{2}{\rho^2}\frac{d\rho}{dx} + f\frac{1}{\rho r_h}\right) \tag{6.11}$$

Integration of this equation from $x = 0$ ($\rho = \rho_i$, $p = p_2$; see Fig. 4.1) to $x = L$ ($\rho = \rho_o$, $p = p_3$) will provide the expression for the core pressure drop $\Delta p_{2-3} = p_2 - p_3$ as

$$\Delta p_{2-3} = \frac{G^2}{2g_c\rho_i}\left[2\left(\frac{\rho_i}{\rho_o} - 1\right) + f\frac{L}{r_h}\rho_i\left(\frac{1}{\rho}\right)_m\right] \tag{6.12}$$

where the mean specific volume with respect to the flow length, $(1/\rho)_m$, is defined as

$$\left(\frac{1}{\rho}\right)_m = \frac{1}{L}\int_0^L \frac{dx}{\rho} \tag{6.13}$$

Here, the fluid mean specific volume $(1/\rho)_m$ can be expressed as follows [see also Eq. (9.18)]. For a liquid with any flow arrangement, or for an ideal gas with $C^* = 1$ and any flow arrangement except for parallelflow,

$$\left(\frac{1}{\rho}\right)_m = v_m = \frac{v_i + v_o}{2} = \frac{1}{2}\left(\frac{1}{\rho_i} + \frac{1}{\rho_o}\right) \tag{6.14}$$

Here v denotes the specific volume in m^3/kg or ft^3/lbm; v_i and v_o are evaluated at inlet and outlet temperatures and pressures, respectively. Note that, in general,

$$\left(\frac{1}{\rho}\right)_m \neq \frac{1}{\rho_m} \qquad \text{where} \quad \rho_m = \frac{\rho_i + \rho_o}{2} \tag{6.15}$$

However, $(1/\rho)_m \approx 1/\rho_m$ is a good approximation for liquids with very minor changes in density with temperatures and small changes in pressure. For a perfect gas with $C^* = 0$ and any exchanger flow arrangement,

$$\left(\frac{1}{\rho}\right)_m = \frac{\tilde{R}}{p_{\text{ave}}}T_{\text{lm}} \tag{6.16}$$

where $\tilde{R}$ is the gas constant in $J/kg\cdot K$ or $lbf\ ft/lbm\text{-}°R$, $p_{\text{ave}} = (p_i + p_o)/2$ and $T_{\text{lm}} = T_{\text{const}} \pm \Delta T_{\text{lm}}$. Here T_{const} is the average temperature of the fluid stream on the other side of the heat exchanger, and ΔT_{lm} is the log-mean temperature difference.

Note the core pressure drop [Eq. (6.12)] has two contributions: The first term represents the momentum rate change or the flow acceleration (deceleration) effects due to the fluid heating (cooling); its positive value represents a pressure drop for flow acceleration and the negative value a pressure rise for flow deceleration. The second term represents the frictional losses and is the dominating term for Δp.

6.2.1.2   Core Entrance Pressure Drop. The core entrance pressure drop consists of two contributions: (1) the pressure drop due to the flow area change, and (2) the pressure losses associated with free expansion that follow sudden contraction. To evaluate the core entrance losses, it will be assumed that the temperature change at the entrance is small and that the fluid velocity is small compared to the velocity of sound. Thus the fluid is treated as incompressible. The pressure drop at the entrance due to the area change alone, for a frictionless incompressible fluid, is given by the Bernoulli equation as

$$p_1 - p_2' = \rho_i \left(\frac{u_2^2}{2g_c} - \frac{u_1^2}{2g_c} \right) = \frac{\rho_i u_2^2}{2g_c} \left[1 - \left(\frac{u_1}{u_2} \right)^2 \right] \tag{6.17}$$

where ρ_i is the fluid density at the core inlet and $\rho_i = \rho_1 = \rho_2$ in Fig. 6.1; and p_2' is the hypothetical static pressure at section 2 in Fig. 6.1 if the pressure drop would have been alone due to the area change. From the continuity equation,

$$\rho_i A_{o,1} u_1 = \rho_i A_{o,2} u_2 \tag{6.18}$$

Introduce σ as the ratio of core minimum free-flow area to frontal area and G as the core mass velocity:

$$\sigma = \frac{A_{o,2}}{A_{o,1}} = \frac{A_{o,3}}{A_{o,4}} \tag{6.19}$$

$$G = \rho_i u_2 = \frac{\dot{m}}{A_{o,2}} \tag{6.20}$$

Substituting Eqs. (6.18)–(6.20) into Eq. (6.17), the pressure drop at the core entrance due to the area change alone is

$$p_1 - p_2' = \frac{G^2}{2g_c \rho_i}(1 - \sigma^2) \tag{6.21}$$

The second contribution to the pressure drop at the entrance is due to the losses associated with irreversible free expansion that follows the sudden contraction. A region of flow separation and secondary flows (as shown in Fig. 6.1 at the vena contracta) produces irreversible pressure losses, and the change in the momentum rate (due to any nonuniform flow) will also produce pressure losses. The resulting pressure change is due to the change in the momentum rate downstream of the vena contracta. Pressure drop due to these losses is taken into account by the contraction loss coefficient K_c multiplied by the dynamic velocity head at the core inlet as follows:

$$\Delta p_{\text{loss}} = K_c \frac{\rho_i u_2^2}{2g_c} = K_c \frac{G^2}{2g_c \rho_i} \tag{6.22}$$

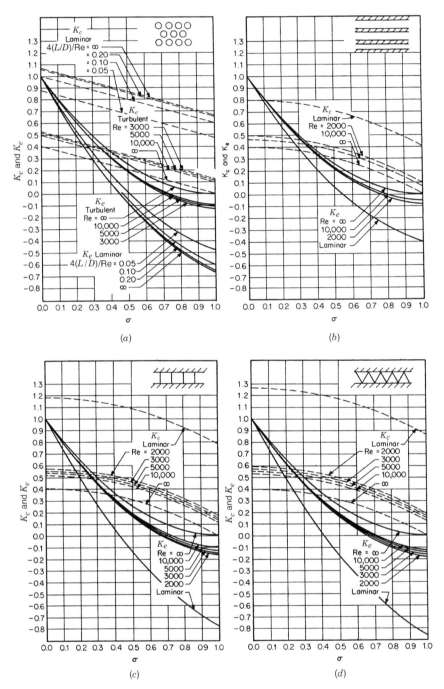

FIGURE 6.3 Entrance and exit pressure loss coefficients for (a) a multiple circular tube core, (b) multiple-tube flat-tube core, (c) multiple square tube core, and (d) multiple triangular tube core with abrupt contraction (entrance) and abrupt expansion (exit). (From Kays and London, 1998.)

K_c is a function of the contraction ratio σ, Reynolds number Re, and flow cross-sectional geometry. Values of K_c for four different entrance flow passage geometries are presented in Fig. 6.3.

It may be mentioned at this stage that if the velocity profile just downstream of the vena contracta is partially or fully developed, it represents a gain in the momentum rate over the flat velocity profile (at the entrance) due to boundary layer displacement effects and the velocity profile shape for the constant flow rate. This gain in the momentum rate results in a decrease in the static pressure or it represents a pressure drop. This pressure drop is also lumped into the Δp loss of Eq. (6.22) that defines K_c. Thus, K_c is made up of two contributions: irreversible expansion after the vena contracta and the momentum rate change due to a partially or fully developed velocity profile just downstream of the vena contracta.

The total pressure drop $\Delta p_{1-2} = p_1 - p_2$ at the core entrance is the sum of those from Eqs. (6.21) and (6.22):

$$\Delta p_{1-2} = \frac{G^2}{2g_c\rho_i}(1 - \sigma^2 + K_c) \tag{6.23}$$

6.2.1.3 Core Exit Pressure Rise. The core exit pressure rise $(p_4 - p_3)$ is divided into two contributions idealizing the fluid as incompressible at the core exit $(\rho_3 = \rho_4 = \rho_o)$ in Fig. 6.1. The first contribution is the pressure rise due to the deceleration associated with an area increase and it is given by an expression similar to Eq. (6.21):

$$\Delta p_{\text{rise}} = \frac{G^2}{2g_c\rho_o}(1 - \sigma^2) \tag{6.24}$$

The second contribution is the *pressure loss* associated with the irreversible free expansion and momentum rate changes following an abrupt expansion, and it is similar to Eq. (6.22).

$$\Delta p_{\text{loss}} = K_e \frac{\rho_3 u_3^2}{2g_c} = K_e \frac{G^2}{2g_c\rho_o} \tag{6.25}$$

Note that K_e is based on the dynamic velocity head at the core outlet. The exit loss coefficient K_e is a function of the expansion ratio $1/\sigma$, the Reynolds number Re, and the flow cross-sectional geometry. Values of K_e for four different flow passage geometries are presented in Fig. 6.3.

It should be emphasized that two effects are lumped into defining K_e: (1) pressure *loss* due to the irreversible free expansion at the core exit, and (2) pressure *rise* due to the momentum rate changes, considering partially or fully developed velocity profile at the core exit and uniform velocity profile far downstream at section 4 in Fig. 6.1. Hence, the magnitude of K_e will be positive or negative, depending on whether the sum of the foregoing two effects represents a pressure loss or a pressure rise.

The net pressure rise at the core exit, $\Delta p_{3-4} = p_4 - p_3$, from Eqs. (6.24) and (6.25), is

$$\Delta p_{3-4} = \frac{G^2}{2g_c\rho_o}\left(1 - \sigma^2 - K_e\right) \tag{6.26}$$

6.2.1.4   Total Core Pressure Drop. The total core pressure drop on one fluid side of a plate-fin exchanger is given by Eq. (6.6) as

$$\Delta p = \Delta p_{1-2} + \Delta p_{2-3} - \Delta p_{3-4} \tag{6.27}$$

Introducing Δp components from Eqs. (6.23), (6.12), and (6.26), we get

$$\frac{\Delta p}{p_i} = \frac{G^2}{2g_c\rho_i p_i}\left[\underbrace{1 - \sigma^2 + K_c}_{\text{entrance effect}} + \underbrace{2\left(\frac{\rho_i}{\rho_o} - 1\right)}_{\text{momentum effect}} + \underbrace{f\frac{L}{r_h}\rho_i\left(\frac{1}{\rho}\right)_m}_{\text{core friction}} - \underbrace{(1 - \sigma^2 - K_e)\frac{\rho_i}{\rho_o}}_{\text{exit effect}}\right] \tag{6.28}$$

Generally, the core frictional pressure drop is a dominating term, about 90% or more of Δp for gas flows in many compact heat exchangers. The entrance effect represents the pressure loss, and the exit effect in many cases represents a pressure rise; thus the net effect of entrance and exit pressure losses is usually compensating.

The entrance and exit losses are important when their values with respect to the core friction term in the brackets of Eq. (6.28) is nonnegligible. Reviewing the terms in the brackets of Eq. (6.28), it is clear that the entrance and exit effects may be nonnegligible when σ and L are small, r_h (or D_h) is large, and f is small. Small values of f for a given surface are usually obtained at high values of Re (such as in turbulent flow). Thus, the entrance and exit losses are important at small values of σ and L (short cores), large values of D_h and Re, and for gases; they are generally negligible for liquids because the total Δp of Eq. (6.28) is small compared to that for gases. Note that the small values of σ are obtained in a plate-fin exchanger (1) if the passages are small and the plates/fins are relatively thick, and/or (2) a large portion of the frontal area (on the fluid side of interest) is blocked by the flow passages of the other fluid.

The values of K_c and K_e presented in Fig. 6.3 apply to long tubes for which flow is fully developed at the exit. For partially developed flows, K_c is lower and K_e is higher than that for fully developed flows, due to the associated momentum rate changes, as discussed before. For interrupted surfaces, flow is hardly ever fully developed but may be periodic. For highly interrupted fin geometries, the entrance and exit losses are generally small compared to a high value of the core pressure drop, and the flow is mixed very well; hence, K_c and K_e for Re $\to \infty$ should represent a good approximation. For many enhanced and compact heat exchangers, flow passages are nonsmooth and uninterrupted (such as wavy, ribbed, stamped, etc.) or interrupted with flows partially developed, periodic or with flow separation, attachment, recirculation, vortices, and so on. For flows through such passages, the estimate of K_c and K_e from Fig. 6.3 may not be accurate. However, if the entrance and exit losses are only a small fraction of the core pressure drop Δp, the error in the calculation of Δp due to large errors in K_c and K_e will still be small.

The core frictional pressure drop, being the major contribution in the total core pressure drop of Eq. (6.28), may be approximated as follows in different forms:

$$\Delta p \approx \frac{4fLG^2}{2g_cD_h}\left(\frac{1}{\rho}\right)_m = \frac{4fL\dot{m}^2}{2g_cA_o^2\rho_mD_h} = f\frac{4L}{D_h}\frac{\rho_mu_m^2}{2g_c} = f\frac{4L}{D_h}\frac{G^2}{2g_c\rho} = \frac{\mu}{2g_c\rho}\frac{4L}{D_h^2}\frac{\dot{m}}{A_o}(f \cdot \text{Re}) \tag{6.29}$$

where it is idealized that $(1/\rho)_m \approx 1/\rho_m \approx 1/\rho$. Corresponding fluid pumping power $\mathcal{P}$ is

$$\mathcal{P} = \frac{\dot{m}\,\Delta p}{\rho} = \frac{GA_o}{\rho}\frac{4fLG^2}{2g_c\rho D_h} = \frac{1}{2g_c\rho^2}fAG^3 \qquad (6.30)$$

where the last term is obtained after substituting $D_h = 4A_oL/A$ in the preceding term and simplifying it. Equations (6.29) and (6.30) will be considered later for comparing, assessing, and evaluating the merits of different heat exchanger surfaces.

We can evaluate the flow area ratio for two different surfaces on one fluid side for a given application (specified mass flow rate and pressure drop on one fluid side) using Eq. (6.29) as follows:

$$\frac{A_{o,1}}{A_{o,2}} = \frac{f_1}{f_2}\frac{(L/D_h)_1}{(L/D_h)_2} \qquad (6.31)$$

Example 6.1   A gas-to-air single-pass crossflow plate-fin heat exchanger has overall dimensions of 0.300 m × 0.600 m × 0.900 m and employs strip fins on the air side. The following information is provided for the air side:

Geometrical properties
Fin density = 0.615 mm^{-1}
Plate spacing = 6.35 mm
Fin offset length = 3.18 mm
Airflow length = 0.6 m
Hydraulic diameter = 0.002383 m
Fin metal thickness = 0.15 mm
Minimum free-flow area = 0.1177 m^2
Free-flow area/frontal area = 0.437

Operating conditions
Volumetric airflow rate = 0.6 m^3/s
Reynolds number = 786
Fanning friction factor = 0.0683
Inlet pressure = 110 kPa
Inlet temperature = 4°C
Outlet temperature = 194.5°C
Gas constant for air
  = 287.04 J/kg · K

Determine the air-side pressure drop.

SOLUTION

Problem Data and Schematic: All necessary geometrical information and operating conditions are given for the air side (Fig. E6.1), as listed above, to compute the pressure drop.

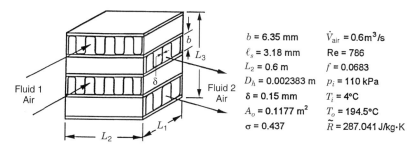

b = 6.35 mm $\dot{V}_{air}$ = 0.6 m^3/s
ℓ_s = 3.18 mm   Re = 786
L_2 = 0.6 m f = 0.0683
D_h = 0.002383 m p_i = 110 kPa
δ = 0.15 mm T_i = 4°C
A_o = 0.1177 m^2 T_o = 194.5°C
σ = 0.437 $\tilde{R}$ = 287.041 J/kg·K

FIGURE E6.1

Determine: The air-side pressure drop for this plate-fin heat exchanger.

Assumptions: The flow distribution through the heat exchanger is uniform, and air is treated as an ideal gas.

Analysis: To compute the pressure drop for a plate-fin heat exchanger using Eq. (6.28), first we need to determine the inlet, outlet, and mean air densities in the core as well as the core mass velocity G. Considering air as an ideal gas, the inlet density is given by

$$\rho_{a,i} = \frac{p_{a,i}}{\tilde{R}T_{a,i}} = \frac{110 \times 10^3 \, \text{Pa}}{287.04 \, \text{J/kg} \cdot \text{K} \times (4.0 + 273.15) \, \text{K}} = 1.3827 \, \text{kg/m}^3$$

Note that we converted the inlet temperature to an absolute temperature scale. Similarly, the air density at the core outlet is given by

$$\rho_{a,o} = \frac{p_{a,o}}{\tilde{R}T_{a,o}} = \frac{110 \times 10^3 \, \text{Pa}}{287.04 \, \text{J/kg} \cdot \text{K} \times (194.5 + 273.15) \, \text{K}} = 0.8195 \, \text{kg/m}^3$$

Note that we have considered here the outlet pressure as 110 kPa since the pressure drop across the core is usually very small, and hence it is neglected in the first trial. The mean density is the harmonic mean value given by Eq. (6.14) as

$$\left(\frac{1}{\rho}\right)_m = \frac{1}{2}\left(\frac{1}{\rho_i} + \frac{1}{\rho_o}\right) = \frac{1}{2}\left(\frac{1}{1.3827 \, \text{kg/m}^3} + \frac{1}{0.8195 \, \text{kg/m}^3}\right) = 0.9717 \, \frac{\text{m}^3}{\text{kg}} = \frac{1}{1.0291 \, \text{kg/m}^3}$$

Since the airflow is given as the volumetric flow rate at the inlet, let us calculate the mass velocity as

$$G = \frac{\dot{V}_i \rho_i}{A_o} = \frac{0.6 \, \text{m}^3/\text{s} \times 1.3827 \, \text{kg/m}^3}{0.1177 \, \text{m}^2} = 7.0486 \, \text{kg/m}^2 \cdot \text{s}$$

Now let us calculate the entrance and exit pressure loss coefficients so that we can compute the pressure drop using Eq. (6.28). Since strip fins are used, the flow is well mixed. Hence, K_c and K_e are evaluated at Re $\to \infty$ using Fig. 6.3. Reviewing Fig. 6.3b and c for parallel plates and square ducts, while it is found that K_c and K_e are dependent on the aspect ratio of the rectangular passages, it is also found that K_c and K_e are identical for Re $\to \infty$ (i.e., independent of the aspect ratio). Hence, even though one can easily calculate the aspect ratio of the rectangular passages formed by the offset strip fin from the given geometry data, there is no need to compute it. Then from Fig. 6.3b or c, for $\sigma_a = 0.437$, we get

$$K_c = 0.33 \qquad K_e = 0.31$$

The core pressure drop for the air side is then given by Eq. (6.28) as

$$\frac{\Delta p}{p_i} = \frac{G^2}{2g_c \rho_i p_i} \left[(1 - \sigma^2 + K_c) + 2\left(\frac{\rho_i}{\rho_o} - 1\right) + f\frac{L}{r_h}\rho_i \left(\frac{1}{\rho}\right)_m - (1 - \sigma^2 - K_e)\frac{\rho_i}{\rho_o} \right]$$

$$= \frac{(7.0486 \text{ kg/m}^2 \cdot \text{s})^2}{2 \times 1 \times 110 \times 10^3 \text{ Pa} \times 1.3827 \text{ kg/m}^3}$$

$$\times \left[(1 - 0.437^2 + 0.33) + 2\left(\frac{1.3827 \text{ kg/m}^3}{0.8194 \text{ kg/m}^3} - 1\right) \right.$$

$$\left. + \frac{0.0683 \times 0.6 \text{ m} \times 1.3827 \text{ kg/m}^3}{(0.002383/4) \text{ m} \times 1.0291 \text{ kg/m}^3} - (1 - 0.437^2 - 0.31)\left(\frac{1.3827 \text{ kg/m}^3}{0.8194 \text{ kg/m}^3}\right) \right]$$

$$= 0.1633 \times 10^{-3}(1.1390 + 1.3745 + 92.4226 - 1.8883)$$
$$\qquad\qquad (1.2\%) \quad (1.5\%) \quad (98.2\%) \quad (0.9\%)$$

$$= 0.01536$$

Hence,

$$(\Delta p)_a = p_i \frac{\Delta p}{p_i} = 110 \text{ kPa} \times 0.01536 = 1.69 \text{ kPa} \qquad\qquad Ans.$$

Note that the pressure drop on the air side (1.69 kPa) is 1.5% of the inlet pressure (110 kPa). Hence, our assumption of $p_o \approx p_i$ to calculate ρ_o is good. Otherwise, once the pressure drop is computed, determine the outlet pressure and then iterate once more with the newly computed outlet density.

Discussion and Comments: As one can see, the determination of the pressure drop is straightforward. In this example, the core pressure drop is dominant, consisting of 98.2% of the total pressure drop. The combined entrance and exit losses are 0.3% ($= 1.2 - 0.9$) of the pressure drop. Since the core frictional pressure drop contribution for this example is so large that reducing the core depth by 50% would not have made any significant difference in the contribution of the combined entrance and exit losses. However, modern automotive compact heat exchangers have an airflow length of about 12 to 50 mm only. In this case, the entrance and exit losses may become a nonnegligible fraction of the total core pressure drop; and the approximation of the entrance and exit losses for Re $\rightarrow \infty$ may not be a good approximation. However, until better information is available, the current approach of using K_c and K_e from Kays and London (1998) is recommended to obtain at least a good approximate correction.

6.2.2 Tube-Fin Heat Exchangers

6.2.2.1 Tube Inside. The pressure drop inside the tubes is determined in the same manner as that for plate-fin surfaces using Eq. (6.28). Appropriate values of the f factor and K_c and K_e are used in this expression for flow inside the tubes with or without fins.

6.2.2.2 Tube Outside with Fins. The three types of fins on tubes (Section 1.5.3.2) are: normal fins on individual tubes, longitudinal fins on individual tubes, and flat fins on an array of tubes. For the first two types of finned tubes, the tube outside flow in each tube row experiences a contraction and an expansion. Thus, the pressure losses associated with a tube row within the core are of the same order of magnitude as those at the entrance with the first tube row and those at the exit with the last tube row. Consequently, the entrance and exit pressure drops are not calculated separately, but they are generally lumped into the friction factor (which is generally derived experimentally) for individually finned tubes and longitudinally finned tubes. Then the total pressure drop associated with the core, from Eq. (6.28), becomes

$$\frac{\Delta p}{p_i} = \frac{G^2}{2g_c\rho_i p_i}\left[f\frac{L}{r_h}\rho_i\left(\frac{1}{\rho}\right)_m + 2\left(\frac{\rho_i}{\rho_o} - 1\right)\right] \tag{6.32}$$

It should be emphasized that the friction factor in Eq. (6.32) is based on the hydraulic diameter. If, instead, the pressure drop correlation is available in terms of an average Euler number Eu per tube row [see Eq. (7.22) for the definition], the pressure drop will be

$$\frac{\Delta p}{p_i} = \frac{G^2}{2g_c\rho_i p_i}\left[EuN_r\rho_i\left(\frac{1}{\rho}\right)_m + 2\left(\frac{\rho_i}{\rho_o} - 1\right)\right] \tag{6.33}^\dagger$$

where N_r represents the number of tube rows. Thus, entrance and exit pressure losses are effectively lumped into the friction factor f by eliminating them from the Δp equation.

For flat fins on an array of tubes (Fig. 1.31b), the components of the total core pressure drop on the fin side are all the same as those for plate-fin surfaces. The only difference is that the flow area at the entrance and exit is between the fins and is independent of the tube arrangement.

To obtain the entrance and exit losses based on the flow area at the leading edge, first apply the continuity equation as follows:

$$\dot{m} = (\rho u_m A_o)_{\text{leading edge}} = (\rho u_m A_o)_{\text{core}} \tag{6.34}$$

Introducing $G' = (\rho u_m)_{\text{leading edge}}$ and $\sigma' = A_{o,\text{leading edge}}/A_{fr}$ in this equation, we get

$$G'\sigma' = G\sigma \tag{6.35}$$

Thus, K_c and K_e are evaluated for σ' from Fig. 6.3. The total pressure drop for this geometry (flat fins on an array of tubes) is then given by

$$\frac{\Delta p}{p_i} = \frac{G^2}{2g_c\rho_i p_i}\left[f\frac{L}{r_h}\rho_i\left(\frac{1}{\rho}\right)_m + 2\left(\frac{\rho_i}{\rho_o} - 1\right)\right] + \frac{G'^2}{2g_c\rho_i p_i}\left[(1 - \sigma'^2 + K_c) - (1 - \sigma'^2 - K_e)\frac{\rho_i}{\rho_o}\right] \tag{6.36}$$

6.3   REGENERATOR PRESSURE DROP

For a rotary regenerator matrix having either continuous cylindrical passages or herringbone (or skewed) passages, the pressure drop consists of the same components

---

†Note that Eq. (6.33) can also be presented in terms of an average Hagen number Hg [see Eq. (7.23) for the definition] by relacing Eu with 2 Hg/Re$_d^2$.

as that for the plate-fin exchanger [Eq. (6.28)]. For a fixed-matrix regenerator matrix made up of any porous material (such as randomly packed screens, cross rods, bricks, tiles, spheres, copper wool, etc.), the entrance and exit pressure drops are included in the experimental friction factors. Thus Eq. (6.32) applies for the pressure drop of fixed-matrix regenerators.

6.4 TUBULAR HEAT EXCHANGER PRESSURE DROP

6.4.1 Tube Banks

The pressure drop on the tube side is determined from Eq. (6.28). The pressure drop associated with flow over the tube banks consists of the same contributions as that for the plate-fin exchanger, except that the entrance and exit pressure drops are included in the friction factors. Hence, the total pressure drop on the outside of a tube bank is given by Eq. (6.32).

6.4.2 Shell-and-Tube Exchangers

6.4.2.1 Tube Side. The pressure drop inside the tube is determined from Eq. (6.28) with proper values of K_c, K_e, and f. However, in shell-and-tube exchangers, the entrance and exit pressure drops for the tube flow are generally neglected since their contribution is small compared to the losses associated with inlet and outlet nozzles and chambers. If U-tubes or hairpins are used in a multipass unit, additional pressure drop due to the 180° bend needs to be included. The pressure drop associated with such a bend is discussed in Section 6.6.3.

6.4.2.2 Shell Side. The pressure drop evaluation on the shell side of a shell-and-tube heat exchanger is complicated due to the presence of bypass and leakage streams (discussed briefly in Section 4.4.1.1) in addition to the crossflow stream. In this case, the pressure drop is evaluated first for an ideal crossflow section and an ideal window section. Correction factors are then applied for the leakage and bypass streams. The total pressure drop is then the sum of the pressure drops for each window section and each crossflow section (Bell, 1988). In this section we just provide the empirical correlations for the pressure drop for the ideal crossflow and window sections. The expression for the total shell-side pressure drop is presented where correction factors for leakage and bypass streams are defined in Section 9.5.1.2.

The pressure drop associated with liquid flow in an ideal crossflow section between two baffles is

$$\Delta p_{b,\mathrm{id}} = \mathrm{Eu}\frac{G_c^2}{2g_c\rho_s}N_{r,\mathrm{cc}}\left(\frac{\mu_w}{\mu_m}\right)^{0.25} = \frac{4f_{\mathrm{id}}G_c^2}{2g_c\rho_s}N_{r,\mathrm{cc}}\left(\frac{\mu_w}{\mu_m}\right)^{0.25} = \frac{\mu^2}{\rho g_c}\frac{N_{r,\mathrm{cc}}}{d_o^2}\,\mathrm{Hg} \qquad (6.37)^{\dagger}$$

[†] The viscosity correction factor $(\mu_w/\mu_m)^{0.25}$ is considered only for liquids. Replace this term from Eq. (6.37) with $(T_w/T_m)^m$ for a gas on the shell side. See Tables 7.12 and 7.13 for the values of m.

where

$$Eu = \frac{\Delta p_{b,id}}{(\rho u_m^2/2g_c)} \frac{1}{N_{r,cc}} \quad Eu = 4f_{id} \quad \text{and} \quad Hg = 32\,Re \tag{6.38}$$

Hence, where Eu is the average Euler number per tube row, $N_{r,cc}$ is the number of effective tube rows crossed during flow through one crossflow section; G_c is the crossflow mass velocity, a ratio of total mass flow rate divided by the minimum free-flow area $A_{o,c}$ over the tubes at or near the *shell centerline* for one crossflow section; μ is the liquid viscosity evaluated at the tube wall (w) or bulk/mean (m) temperatures; and Hg is the Hagen Number per tube row defined by Eq. (7.23). The Euler number and Hagen number are determined from the correlations for flow normal to a tube bank of a specified arrangement. Correlations in terms of the Hagen number are summarized in Section 7.5.1. It should be emphasized that $Eu = 4f_{id}$ in Eq. (6.37), and f_{id} is the ideal Fanning friction factor per tube row as defined by (with data provided by) Bell (1988). This definition of f_{id} is used only here and in Section 9.5.1.2 while discussing the Bell–Delaware method. At all other places in the book, the definition of the Fanning friction factor used is given by Eqs. (7.17) and (7.18).

The pressure drop associated with an ideal *one*-window-section $\Delta p_{w,i}$ depends on the shell-side Reynolds number $Re_d = Gd_o/\mu = \rho u_c d_o/\mu$, where u_c is evaluated at or near the shell centerline in one crossflow section as mentioned above. It is given by

$$\Delta p_{w,id} = \begin{cases} (2 + 0.6 N_{r,cw}) \dfrac{G_w^2}{2g_c \rho_s} & \text{for } Re_d > 100 \tag{6.39a} \\[3mm] \dfrac{26 G_w \mu_s}{g_c \rho_s} \left(\dfrac{N_{r,cw}}{p_t - d_o} + \dfrac{L_b}{D_{h,w}^2} \right) + \dfrac{G_w^2}{g_c \rho_s} & \text{for } Re_d \leq 100 \tag{6.39b} \end{cases}$$

where p_t is the tube pitch, N_w is the number of effective crossflow tube rows in each window, L_b is the baffle spacing, and $D_{h,w}$ (the hydraulic diameter of the window section) and u_z (and G_w) are given by

$$D_{h,w} = \frac{4A_{o,w}}{\pi d_o N_{t,w} + \pi D_s \theta_b/360°} \tag{6.40}$$

$$\frac{G_w^2}{\rho_s^2} = u_z^2 = u_c u_w = \frac{\dot{m}}{A_{o,cr} \rho_s} \frac{\dot{m}}{A_{o,w} \rho_s} \tag{6.41}$$

where $\dot{m}$ is the total shell-side flow rate, $A_{o,cr}$ and $A_{o,w}$ are the flow areas for the crossflow and window sections, respectively, u_c and u_w are the ideal crossflow and window mean velocities, and θ_b in Eq. (6.40) is in degrees. The $\rho_s = \rho_m$ is the mean density of the shell-side fluid.

The combined pressure drop associated with the inlet and outlet sections on the shell side is given by

$$\Delta p_{i-o} = 2 \Delta p_{b,id} \left(1 + \frac{N_{r,cw}}{N_{r,cc}} \right) \zeta_b \zeta_s \tag{6.42}$$

The total pressure drop on the shell side is the sum of the pressure drop associated with each crossflow section between baffles, the pressure drop associated with each window section, and the pressure drop for crossflow sections on each end between the first (and last) baffle and the tubesheet. Since the $\Delta p_{b,\mathrm{id}}$, $\Delta p_{w,\mathrm{id}}$, and Δp_{i-o} of Eqs. (6.37), (6.39), and (6.42) are for ideal conditions, they must be corrected for the presence of the bypass and leakage streams. The total pressure drop on the shell side, excluding the Δp associated with the entrance and exit nozzles and headers, is

$$\Delta p_s = \Delta p_{\mathrm{cr}} + \Delta p_w + \Delta p_{i-o} = \left[(N_b - 1)\, \Delta p_{b,\mathrm{id}} \zeta_b + N_b\, \Delta p_{w,\mathrm{id}} \right] \zeta_\ell$$

$$+ 2\Delta p_{b,\mathrm{id}} \left(1 + \frac{N_{r,\mathrm{cw}}}{N_{r,\mathrm{cc}}} \right) \zeta_b \zeta_s \tag{6.43}$$

where N_b is the number of baffles, ζ_b is the pressure drop correction factor for bypass flows (C and F streams in Fig. 4.19), ζ_l is the pressure drop correction factor for both baffle-to-shell (E stream) and tube-to-baffle (A stream) leakage streams, and ζ_s is the pressure drop correction factor for the unequal baffle spacings for inlet and exit baffle sections. The values of ζ_b, ζ_l, and ζ_s are presented later in Table 9.3.

Example 6.2   A shell-and-tube heat exchanger is designed to cool the shell-side lubricating oil from 65°C to 60°C. Following are the specifications for the shell-and-tube heat exchanger.

Tube outside diameter = 19 mm	Tube wall thickness = 1.2 mm
Tube pitch = 25 mm, square layout	Number of baffles = 14
Crossflow area near the shell centerline	Number of effective tube rows
= 0.04429 m²	crossed in one window zone = 3.868
Flow area through the window zone	Oil flow rate = 36.3 kg/s
= 0.01261 m²	Ideal tubebank friction factor = 0.23
Number of effective tube rows	Shell-side Reynolds number = 242
baffle section = 9	Oil density = 849 kg/m³

Factors for various leakage and bypass flows for the pressure drop correction are (1) 0.59 for baffle-to-shell and tube-to-baffle leakage streams, (2) 0.69 for baffle-to-shell bypass stream, and (3) 0.81 for unequal baffle spacing on inlet and exit baffle sections. Calculate the shell-side pressure drop.

SOLUTION

Problem Data and Schematic: All necessary geometrical information and operating conditions are given for the oil side, as listed below, to compute the shell-side oil pressure drop. A schematic of a shell-and-tube heat exchanger is shown in Fig. 1.5a.

Geometry: $d_o = 19$ mm, $p_t = 25$ mm, $A_{o,c} = 0.04429$ m², $A_{o,w} = 0.01261$ m², $N_{r,\mathrm{cc}} = 9$, $N_{r,\mathrm{cw}} = 3.868$, $\delta_w = 1.2$ mm, $N_b = 14$

Operating conditions and oil density: $\dot{m} = 36.3$ kg/s, $f_{\mathrm{id}} = 0.23$, $\mathrm{Re}_s = 242$, $\rho = 849$ kg/m³

Correction factors: $\zeta_l = 0.59$, $\zeta_b = 0.69$, $\zeta_s = 0.81$

Determine: The oil-side pressure drop for this shell-and-tube heat exchanger.

Assumption: The assumptions adopted for the pressure drop analysis made in Section 6.1.4 are invoked here. Fluid density on hot-water side is treated as constant.

Analysis: To compute the pressure drop for the shell side using Eq. (6.43), let us first compute individual pressure drop components using Eqs. (6.37) and (6.39a). The mass velocity

$$G_c = \frac{\dot{m}}{A_{o,c}} = \frac{36.3 \, \text{kg/s}}{0.04429 \, \text{m}^2} = 819.60 \, \text{kg/m}^2 \cdot \text{s}$$

$$\Delta p_{b,\text{id}} = \frac{4 f_{\text{id}} G_c^2}{2 g_c \rho_s} N_{r,\text{cc}} \left(\frac{\mu_w}{\mu_m}\right)^{0.14} = \frac{4 \times 0.23 \times (819.60 \, \text{kg/m}^2 \cdot \text{s})^2}{2 \times 1 \times 849 \, \text{kg/m}^3} \times 9 \times (1)^{0.14} = 3275.6 \, \text{Pa}$$

Note that we have not included the viscosity correction primarily because no data are given (due to the small temperature drop specified). Since the shell-side Reynolds number is given as 242, the appropriate equation for the window zone pressure drop is Eq. (6.39a). Let us first calculate the velocity u_z using Eq. (6.41):

$$u_z^2 = u_c u_w = \frac{\dot{m}}{A_{o,c} \rho_s} \frac{\dot{m}}{A_{o,w} \rho_s}$$

$$= \frac{36.3 \, \text{kg/s}}{0.04429 \, \text{m}^2 \times 849 \, \text{kg/m}^3} \frac{36.3 \, \text{kg/s}}{0.01261 \, \text{m}^2 \times 849 \, \text{kg/m}^3} = 3.2732 \, \text{m}^2/\text{s}^2$$

The ideal window section pressure drop is then given by

$$\Delta p_{w,\text{id}} = (2 + 0.6 N_{r,\text{cw}}) \frac{\rho_m u_z^2}{2 g_c} = (2 + 0.6 \times 3.868) \frac{849 \, \text{kg/m}^3 \times 3.2732 \, \text{m}^2/\text{s}^2}{2 \times 1} = 6003.6 \, \text{Pa}$$

We are now ready to compute the shellside pressure drop using Eq. (6.43) as

$$\Delta p_s = \left[(N_b - 1) \Delta p_{b,\text{id}} \zeta_b + N_b \Delta p_{w,\text{id}} \right] \zeta_\ell + 2 \Delta p_{b,\text{id}} \left(1 + \frac{N_{r,\text{cw}}}{N_{r,\text{cc}}} \right) \zeta_b \zeta_s$$

$$= \left[(14 - 1) \times 3275.6 \, \text{Pa} \times 0.69 + 14 \times 6003.6 \, \text{Pa} \right] \times 0.59 + 2 \times 3275.6 \, \text{Pa}$$

$$\times \left(1 + \frac{3.868}{9} \right) \times 0.69 \times 0.81$$

$$= (17{,}335 + 49{,}590 + 5235) \, \text{Pa} = 72{,}160 \, \text{Pa} = 72.2 \, \text{kPa} \qquad \qquad Ans.$$
$$\quad (24\%) \qquad (69\%) \qquad (7\%)$$

Discussion and Comments: Since all the data were provided, the computation of the shell-side pressure drop is straightforward. Note that the total pressure drop contribution associated with the crossflow streams is $24 + 7 = 31\%$ and the pressure drop contribution associated with the window zone is 69%. It should be clear that the pressure drop in the window zone can be significant and can also increase substantially with smaller baffle cuts and baffle spacings. In Chapters 8 and 9, we show how to obtain some of the input data of this problem.

6.5 PLATE HEAT EXCHANGER PRESSURE DROP

Pressure drop in a plate heat exchanger consists of three contributions: (1) pressure drop associated with the inlet and outlet manifolds and ports, (2) pressure drop within the core (plate passages), and (3) pressure drop due to the elevation change for a vertical flow exchanger. The pressure drop in the manifolds and ports should be kept as low as possible (generally $< 10\%$, but may be as high as 25 to 30% or higher in some designs). Empirically, it is calculated as approximately 1.5 times the inlet velocity head per pass. Since the entrance and exit losses in the core (plate passages) cannot be determined experimentally, they are included in the friction factor for the given plate geometry. Although the momentum effect [see Eq. (6.28)] is negligibly small for liquids, it is also included in the following Δp expression. The pressure drop (rise) caused by the elevation change for liquids is given by Eq. (6.5). Summing all contributions, the pressure drop on one fluid side in a plate heat exchanger is given by

$$\Delta p = \frac{1.5 G_p^2 n_p}{2 g_c \rho_i} + \frac{4 f L G^2}{2 g_c D_e} \left(\frac{1}{\rho}\right)_m + \left(\frac{1}{\rho_o} - \frac{1}{\rho_i}\right) \frac{G^2}{g_c} \pm \frac{\rho_m g L}{g_c} \qquad (6.44)$$

where $G_p = \dot{m}/(\pi/4) D_p^2$ is the fluid mass velocity in the port, n_p is the number of passes on the given fluid side, D_e is the equivalent diameter of flow passages (usually, D_e equals twice the plate spacing), and ρ_o and ρ_i are fluid mass densities evaluated at local bulk temperatures and mean pressures at outlet and inlet, respectively.

Example 6.3 A 1-pass 1-pass plate heat exchanger with chevron plates is being used to cool hot water with cold water on the other fluid side. The following information is provided for the geometry and operating conditions: number of flow passages 24 on the hot-water side, plate width 0.5 m, plate height 1.1 m, port diameter 0.1 m, channel spacing 0.0035 m, equivalent diameter 0.007 m, hot-water flow rate 18 kg/s, mean dynamic viscosity 0.00081 Pa · s, and mean density 995.4 kg/m^3 for both manifolds and core. The hot water is flowing vertically upward in the exchanger. The friction factor for the plates is given by $f = 0.8 \, \mathrm{Re}^{-0.25}$, where $\mathrm{Re} = G D_e/\mu$ is the Reynolds number. Compute the pressure drop on the hot-water side.

SOLUTION

Problem Data and Schematic: The following information is provided:

$$n_p = 1 \quad N_p = 24 \quad w = 0.5 \, \mathrm{m} \quad L = 1.1 \, \mathrm{m} \quad D_p = 0.1 \, \mathrm{m} \quad b = 0.0035 \, \mathrm{m}$$

$$D_e = 0.007 \, \mathrm{m} \quad \dot{m} = 18 \, \mathrm{kg/s} \quad \mu = 0.00081 \, \mathrm{kg/m \cdot s} \quad \rho = 995.4 \, \mathrm{kg/m^3}$$

$$f = 0.8 \, \mathrm{Re}^{-0.25}$$

A typical plate heat exchanger is shown in Fig. 1.16.

Determine: The hot-water-side pressure drop for this plate heat exchanger.

Assumptions: All assumptions of Section 6.1.4 involved in the pressure drop evaluation are invoked here.

Analysis: We compute the pressure drop using Eq. (6.44). The mass velocity through the port is given by

$$G_p = \frac{\dot{m}}{(\pi/4)D_p^2} = \frac{18\,\text{kg/s}}{(\pi/4)(0.1\,\text{m})^2} = 2291.83\,\text{kg/m}^2 \cdot \text{s}$$

The mass velocity through the core is given by

$$G = \frac{\dot{m}}{A_o} = \frac{18\,\text{kg/s}}{0.042\,\text{m}^2} = 428.57\,\text{kg/m}^2 \cdot \text{s}$$

where

$$A_o = N_p \times w \times b = 24 \times 0.5\,\text{m} \times 0.0035\,\text{m} = 0.042\,\text{m}^2$$

Next, compute the Reynolds number and the friction factor as follows:

$$\text{Re} = \frac{GD_e}{\mu} = \frac{428.57\,\text{kg/m}^2 \cdot \text{s} \times 0.007\,\text{m}}{0.00081\,\text{kg/m} \cdot \text{s}} = 3704$$

$$f = 0.8\,\text{Re}^{-0.25} = 0.8 \times (3704)^{-0.25} = 0.1025$$

Now we are ready to compute the pressure drop on the hot-water side using Eq. (6.44):

$$
\Delta p = \frac{1.5G_p^2 n_p}{2g_c \rho_i} + \frac{4fLG^2}{2g_c D_e}\left(\frac{1}{\rho}\right)_m + \frac{\rho_m g L}{g_c}
$$

$$
= \frac{1.5 \times (2291.83\,\text{kg/m}^2 \cdot \text{s})^2 \times 1}{2 \times 1 \times 995.4\,\text{kg/m}^3} + \frac{4 \times 0.1025 \times 1.1\,\text{m} \times (428.57\,\text{kg/m}^2 \cdot \text{s})^2}{2 \times 1 \times 0.007\,\text{m}}
$$

$$
\times \frac{1}{995.4\,\text{kg/m}^3} + \frac{995.4\,\text{kg/m}^3 \times 9.87\,\text{m/s}^2 \times 1.1\,\text{m}}{1}
$$

$$
= (3957.6 + 5944.2 + 10,807.1)\,\text{kg/m}^2 \cdot \text{s} = 20{,}708.9\,\text{kg/m}^2 \cdot \text{s} = 20.71\,\text{kPa} \qquad Ans.
$$
$$(19.1\%) \quad (28.7\%) \quad (52.2\%)$$

Note that we did not include the momentum effect term of the pressure drop above since the inlet and outlet densities were not given primarily because the inlet and outlet density difference will be negligible for water.

Discussion and Comments: The pressure drop evaluation is straightforward for this exchanger. For this problem, the pressure drop due to the elevation change is quite significant. However, if it had been a two-pass exchanger, the pressure drop and pressure rise due to the elevation change would have been almost canceled out for vertical flows in the exchanger. Note also that the port pressure drop is about the same order of magnitude as the core pressure drop. The port pressure drop is generally not a small fraction of the total pressure drop in plate heat exchangers. Care should be exercised in the design of the exchanger to minimize the port pressure drop so that more allowable pressure drop is available for the core and that will also result in more uniform flow through the plate passages.

6.6 PRESSURE DROP ASSOCIATED WITH FLUID DISTRIBUTION ELEMENTS

Fluid flows into and out of the heat exchanger through pipes, ducts, or nozzles. These are usually connected to manifolds, tanks, or headers for fluid distribution to the heat exchanger flow passages. The pressure drop associated with these components usually consists of wall friction, bend losses, sudden contraction and expansion losses, and branch losses, depending on the geometry. When information on the friction factor is not available, these pressure losses are generally presented in terms of velocity heads, as in the first equality of Eq. (6.53), where the pressure loss coefficient K represents the number of velocity heads. These pressure losses are summarized separately next.

6.6.1 Pipe Losses

The pressure drop associated with a pipe of constant cross section, due to wall friction, is given by Eq. (6.29) as

$$\Delta p = f \, \frac{4L}{D_h} \frac{\rho u_m^2}{2g_c} \tag{6.45}$$

where f is the Fanning friction factor, generally dependent on the Reynolds number and flow cross-section geometry. In turbulent flow, f is also dependent on the surface roughness of the pipe. Fluid mass density is evaluated at local bulk temperature and mean pressure.

The Fanning friction factor as a function of Re and $e/d_i = e/D_h$ for a circular tube is presented in Fig. 6.4. Here e is the surface roughness magnitude (average height) and $d_i (= D_h)$ is the tube inside diameter. The results are valid for fully developed laminar and turbulent flows. Notice that the surface roughness has no influence on the f factors in laminar flow. If the ordinate of Fig. 6.4 is changed to $4f$ = Darcy friction factor [see Eq. (7.20)], the resulting figure is referred to as the *Moody diagram*. The turbulent flow f factors of Fig. 6.4 are also valid for noncircular pipes provided that a proper value of the hydraulic diameter is used in Eq. (6.45). The laminar friction factors f are dependent on the cross-section geometry and are presented in Chapter 7. Further explanation of this figure and the theory are presented in a subsection on Circular Tube with Surface Roughness in Section 7.4.1.3.

6.6.2 Sudden Expansion and Contraction Losses

At the entrance of a heat exchanger, a pipe is generally connected to a manifold. Fluid experiences a sudden expansion during the flow to the manifold. Similarly, it experiences a sudden contraction while flowing from the exit manifold to the exit pipe. Sudden expansion and contraction losses presented in Fig. 6.3 are applicable here. In heat exchangers, these losses are associated with relatively *large* pipes, for which the flow is turbulent. A review of Fig. 6.3 reveals that K_e is highest for Re $= \infty$ for a given σ. By

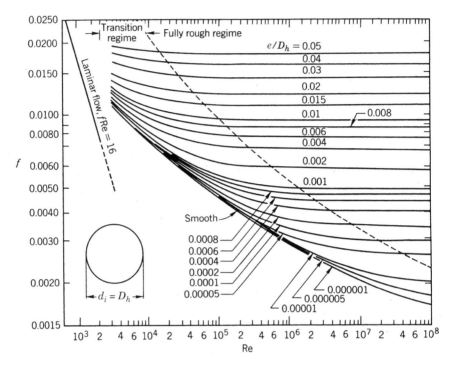

FIGURE 6.4   Fanning friction factors for smooth and rough circular tubes (From Bhatti and Shah, 1987.)

applying the momentum equation and Bernoulli equation across the sudden expansion (sections 3 and 4 in Fig. 6.1), it can be shown rigorously that

$$K_e = (1 - \sigma)^2 \tag{6.46}$$

This equation, referred to as the *Borda–Carnot equation*, is valid for Re = ∞. It is shown in all four geometries of Fig. 6.3 for Re = ∞. This value of K_e is generally used for a sudden expansion in a single pipe.

For a sudden contraction, an experimentally derived ratio of *vena contracta* area to pipe area is required to obtain the value of K_c. The Crane Co. (1976) presents the value of K_c for a single pipe as

$$K_c = 0.5(1 - \sigma) \tag{6.47}$$

This represents a 25% higher value of K_c than the more conservative value of Kays and London (1998) in Fig. 6.3a for Re = ∞. The pressure drops due to sudden contraction and expansion are then determined from Eqs. (6.22) and (6.25), respectively, with the appropriate values of the density and mean velocities.

Example 6.4   Determine the effect of a change in the cross-sectional area on the pressure drop of a square pipe. The original square pipe has a cross section side length of 70.7 mm.

To reduce the pressure drop, an engineer decided to double the pipe cross section with a side length of 141.4 mm over the pipe length of 1 m. Consider air flowing at 0.05 kg/s at 27°C. The air density and dynamic viscosity are 1.1614 kg/m³ and 184.6×10^{-7} Pa · s, respectively. For fully developed turbulent flow for rectangular (and square) ducts, consider

$$f = 0.0791 \text{Re}^{-0.25}(1.0875 - 0.1125\alpha^*)$$

where α^* is the aspect ratio of the rectangular flow passage.

Problem Data and Schematic: The schematics of the pipe cross sections are shown in Fig. E6.4. The following data are provided:

$a_1 = b_1 = 0.0707 \,\text{m}$ $a_2 = b_2 = 0.1414 \,\text{m}$ $L = 1 \,\text{m}$ $\dot{m} = 0.05 \,\text{kg/s}$

$\alpha^* = 1$ $\rho = 1.1614 \,\text{kg/m}^3$ $\mu = 184.6 \times 10^{-7} \,\text{kg/m} \cdot \text{s}$

$f = 0.0791 \text{Re}^{-0.25}(1.0875 - 0.1125\alpha^*)$

where the subscripts 1 and 2 are for the small and large cross-sectional-area (square) pipes.

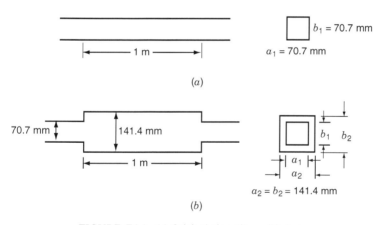

(a)

(b)

FIGURE E6.4   (*a*) Original pipe; (*b*) modified pipe.

Determine: The increase or decrease in pressure drop due to the sudden expansion and sudden contraction associated with the square pipe.

Assumptions: It is assumed that the flow entering the small cross-sectional-area pipe is fully developed turbulent flow and is isothermal throughout the flow length.

Analysis: Let us first compute the necessary geometry and other information to determine the pressure drop desired.

Flow area before the sudden expansion: $A_{o,1} = 0.0707 \,\text{m} \times 0.0707 \,\text{m} = 5 \times 10^{-3} \,\text{m}^2$

Flow area after the sudden expansion: $A_{o,2} = 0.1414 \,\text{m} \times 0.1414 \,\text{m} = 0.02 \,\text{m}^2$

Thus, the area ratio

$$\sigma = \frac{A_{o,1}}{A_{o,2}} = \frac{5 \times 10^{-3}\,\text{m}}{0.02\,\text{m}} = 0.25$$

The hydraulic diameter of the flow passage for a square duct is the length of its side:

$$D_{h,1} = 0.0707\,\text{m} \qquad D_{h,2} = 0.1414\,\text{m}$$

The mass velocities G based on the flow area before and after the sudden expansion with subscripts 1 and 2 are

$$G_1 = \frac{\dot{m}}{A_{o,1}} = \frac{0.05\,\text{kg/s}}{5 \times 10^{-3}\,\text{m}^2} = 10\,\text{kg/m}^2 \cdot \text{s} \quad \text{and} \quad G_2 = 2.5\,\text{kg/m}^2 \cdot \text{s}$$

Flow Reynolds number Re in the small-cross-sectional-area pipe is given by

$$\text{Re}_1 = \left(\frac{GD_h}{\mu}\right)_1 = \frac{10\,\text{kg/m}^2 \cdot \text{s} \times 0.0707\,\text{m}}{184.6 \times 10^{-7}\,\text{kg/m} \cdot \text{s}} = 38{,}304$$

Flow Reynolds number Re in the large-cross-sectional area pipe is given by

$$\text{Re}_2 = \left(\frac{GD_h}{\mu}\right)_2 = \frac{2.5\,\text{kg/m}^2 \cdot \text{s} \times 0.1414\,\text{m}}{184.6 \times 10^{-7}\,\text{kg/m} \cdot \text{s}} = 19{,}152$$

Using the equation given, let us compute the friction factor for the small- and large-cross-sectional-area ducts

$$f_1 = 0.0791\text{Re}^{-0.25}(1.0875 - 0.1125\alpha^*)$$
$$= 0.0791 \times (38{,}304)^{-0.25}\,(1.0875 - 0.1125 \times 1) = 0.005513$$
$$f_2 = 0.0791\text{Re}^{-0.25}(1.0875 - 0.1125\alpha^*)$$
$$= 0.0791 \times (19{,}152)^{-0.25}\,(1.0875 - 0.1125 \times 1) = 0.006556$$

Now we determine the sudden expansion and contraction losses for the large pipe using Fig. 6.3. In both cases, we calculate the mass velocity for the small-cross-sectional-area pipe. The sudden expansion and sudden contraction coefficients for the square pipe for $\text{Re} = 38304$ and $\sigma = 0.25$ from Fig. 6.3 are

$$K_e = 0.55 \qquad K_c = 0.47$$

Hence, the pressure rise due to sudden expansion at the entrance to the large-cross-sectional-area pipe, using Eq. (6.26), is given by

$$\Delta p_{\text{exp}} = \frac{G^2}{2g_c\rho}\left(1 - \sigma^2 - K_e\right) = \frac{\left(10\,\text{kg/m}^2 \cdot \text{s}\right)^2}{2 \times 1 \times 1.1614\,\text{kg/m}^3}\left(1 - 0.25^2 - 0.55\right) = 16.68\,\text{Pa}$$

The pressure drop due to sudden contraction at the exit of the large-cross-sectional-area pipe, from Eq. (6.23), is

$$\Delta p_{\text{con}} = \frac{G^2}{2g_c\rho}\left(1 - \sigma^2 + K_c\right) = \frac{\left(10\,\text{kg/m}^2 \cdot \text{s}\right)^2}{2 \times 1 \times 1.1614\,\text{kg/m}^3}\left(1 - 0.25^2 + 0.47\right) = 60.59\,\text{Pa}$$

The pressure drop due to friction in a large-cross-sectional-area pipe of length 1 m is given by

$$\Delta p_{fr} = f_2 \frac{4L}{D_{h,2}} \frac{G^2}{2g_c\rho} = 0.006556 \times \frac{4 \times 1\,\text{m}}{0.1414\,\text{m}} \times \frac{\left(2.5\,\text{kg/m}^2 \cdot \text{s}\right)^2}{2 \times 1 \times 1.1614\,\text{kg/m}^3} = 0.499\,\text{Pa}$$

Thus the total pressure drop associated with the large cross-sectional area pipe due sudden expansion, friction and sudden contraction is

$$\Delta p_t = -\Delta p_{\text{exp}} + \Delta p_{fr} + \Delta p_{\text{con}} = (-16.68 + 0.499 + 60.59)\,\text{Pa} = 44.41\,\text{Pa}$$

For a straight small cross-sectional area pipe of 1 m length, the associated pressure drop is only for the friction component and is given by

$$\Delta p_{fr} = f_1 \frac{4L}{D_{h,1}} \frac{G_1^2}{2g_c\rho} = 0.005513 \times \frac{4 \times 1\,\text{m}}{0.0707\,\text{m}} \times \frac{\left(10\,\text{kg/m}^2 \cdot \text{s}\right)^2}{2 \times 1 \times 1.1614\,\text{kg/m}^3} = 13.43\,\text{Pa}$$

Thus, the pressure drop for the pipe with large cross-sectional area and 1 m length is 44.41 Pa; while the pressure drop for the straight pipe without any change in cross section is 13.43 Pa. *Ans.*

Discussion and Comments: From this example it is found that the frictional pressure drop in the small-cross-sectional-area pipe is increased by a factor of about 27 (13.43 Pa/0.499 Pa) compared to that for a pipe with four times the flow area and double the hydraulic diameter. Despite this significant increase in the frictional component, there are no other pressure losses, whereas for the large-cross-sectional-area pipe, the sudden expansion and contraction losses are significantly larger than the frictional pressure loss contribution. Hence, the increase in cross-sectional area for supposedly reducing pressure drop, in fact, increases the pressure drop.

However, for this example, if the length of the large cross-sectional pipe were increased over 3.4 m (by comparing the total pressure drop for each pipe), its total pressure drop would be lower than that for the small-cross-sectional-area pipe. This is because the expansion and contraction losses remain constant irrespective of the pipe length.

6.6.3 Bend Losses

In a number of applications, the inlet and outlet pipes that carry the fluids into and out of heat exchangers have various bends. These pipes may have a circular or rectangular cross section with a certain bend deflection angle θ_b (see Fig. 6.5) or a miter bend (see inset in Fig. 6.12) with a circular cross section.

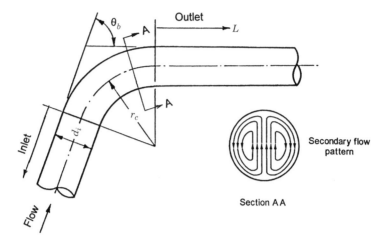

FIGURE 6.5   Circular-cross-section pipe bend with a secondary flow pattern.

When a fluid flows through a bend or a curve, it experiences a variation of centrifugal force across the tube. This results in a pressure gradient between the outer wall (maximum pressure) and the inner wall (minimum pressure) at a given cross section. This results in a secondary flow, as shown at section AA in Fig. 6.5, superimposed on the main flow. The frictional energy loss near the tube walls is thus increased and the pressure drop is greater than that for the corresponding flow in a straight tube.

In the following formulation of expressions for the pipe bend pressure losses, it is assumed that these losses are associated only with the pipe bend and the outlet pipe section of Fig. 6.5. The pressure drop associated with a smooth straight inlet pipe section of the bend is not included (but must be included when calculating the total pressure drop associated with a heat exchanger having pipe bends) because it can be calculated straight-forwardly [e.g., by using Eq. (6.29)] separately. If the bend follows another component (bend, tee, etc.), the pressure loss associated with the inlet pipe section is already included in Eq. (6.49) for specific lengths of the inlet pipe section specified in Sections 6.6.3.1 through 6.6.3.3.

The pressure drop associated with a bend may be represented by

$$\Delta p_b = K_{b,t} \frac{\rho u_m^2}{2g_c} \qquad (6.48)$$

where ρ is the fluid density evaluated at local bulk temperature and mean pressure, u_m the mean axial velocity (both ρ and u_m evaluated at the entrance of the bend), and $K_{b,t}$ is the total pressure drop coefficient due to the bend. The total pressure drop for a bend consists of two contributions: (1) the pressure drop for the bend due to the curvature effect, the flow development effect in the outlet pipe, and the surface roughness effect, and (2) the pressure drop associated with the outlet straight pipe of specified surface roughness. The pressure loss coefficients K_b and K_f for these two contributions to the pressure drop are strongly dependent on the flow Reynolds number. Thus,

$$K_{b,t} = K_b + K_f = K_b + f\frac{4L}{D_h} \qquad (6.49)$$

where the Fanning friction factor f for the outlet pipe is determined from Fig. 6.4 for the appropriate surface roughness size, and the bend pressure loss coefficient K_b is given by Miller (1990) as

$$K_b = K_b^* C_{Re} C_{dev} C_{rough} \qquad (6.50)$$

Here K_b^* is the bend pressure loss coefficient evaluated at $Re = \rho u_m d_i / \mu = 10^6$, C_{Re} is the correction factor for the actual Reynolds number for a given application, C_{dev} is the correction factor for flow development in the outlet pipe, and C_{rough} is the correction factor for the pipe surface roughness. C_{rough} is given by

$$C_{rough} = \frac{f_{rough}}{f_{smooth}} \qquad (6.51)$$

where f_{smooth} is the friction factor for a hydraulically smooth pipe and f_{rough} is the friction factor for the given or assumed roughness for the pipe bend. Both these friction factors can be obtained from Fig. 6.4. Values of K_b^*, C_{Re}, and C_{dev} are provided next for pipe bends of three different geometries.

6.6.3.1 Pipe Bends with a Circular Cross Section. The pipe bend is shown in Fig. 6.5 with important geometrical parameters as the pipe inside diameter d_i, the radius of curvature r_c of the bend, the bend deflection angle θ_b, and the outlet pipe length L. The bend total pressure loss coefficient $K_{b,t}$ is computed using Eq. (6.49) with K_b from Eq. (6.50), where K_b^* at $Re = 10^6$ is given in Fig. 6.6. Since the actual Reynolds number for a given application in general will be different, a correction factor C_{Re} for the Reynolds number is applied. C_{Re} is presented in Fig. 6.7 as a function of Re for various values of r_c/d_i. For bends with $r_c/d_i < 1$, C_{Re} is strongly dependent on Re. It is calculated as follows (Miller, 1990):

1. For $0.7 < r_c/d_i < 1$ or $K_b^* < 0.4$, use C_{Re} from Fig. 6.7 for bends with $r_c/d_i = 1$.
2. For other $r_c/d_i (< 1)$ and $K_b^* \geq 0.4$, compute C_{Re} from the equation

$$C_{Re} = \frac{K_b^*}{K_b^* - 0.2 C_{Re}' + 0.2} \qquad (6.52)$$

where C_{Re}' is C_{Re} from Fig. 6.7 for $r_c/d_i = 1$.

The outlet pipe length correction C_{dev} is presented in Fig. 6.8 as a function of L/d_i and K_b^*. For $r_c/d_i > 3$ and/or $\theta_b > 100°$, $C_{dev} \approx 1$ (Miller, 1990). Refer to Miller (1990) for K_b^* values for short outlet pipes.

The Δp_b of Eq. (6.48) is not underestimated using the pressure loss coefficient $K_{b,t}$ of Eq. (6.49) if the inlet pipe lengths (L_i) comply with the following conditions (Miller, 1990): (1) $L_i/d_i > 2$ when the bend follows another component which has a pressure loss coefficient of less than 0.25 at $Re = 10^6$, and (2) $L_i/d_i > 4$ when the bend follows another component which has a pressure loss coefficient greater than 0.5 at $Re = 10^6$.

Example 6.5 In a 90° circular bend, water at 25°C flows at 2 kg/s. The pipe inside diameter is 25 mm, the radius of curvature of the bend is 150 mm, and the average height of the pipe inside surface roughness is 0.025 mm. The downstream straight pipe length is 0.25 m. Assume that the flow entering the pipe is fully turbulent. Compute the pressure

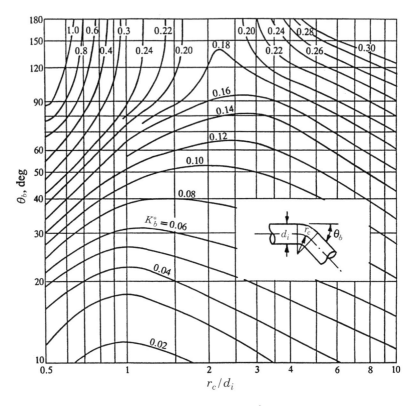

FIGURE 6.6   Bend pressure loss coefficient K_b^* at $Re = 10^6$ for circular cross section bends. (From Miller, 1990.)

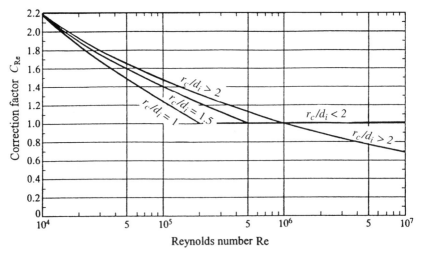

FIGURE 6.7   Reynolds number correction factor C_{Re} as a function of Re and r_c/d_i. (From Miller, 1990.)

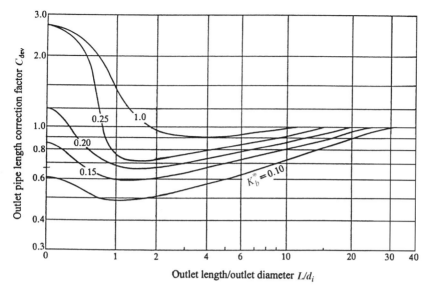

FIGURE 6.8   Outlet pipe length correction factor C_{dev} as a function of the outlet pipe L/d_i and K_b^*. (From Miller, 1990.)

drop associated with the bend and the downstream pipe. Use the following properties for water: density 997 kg/m^3 and a dynamic viscosity of 0.000855 Pa · s.

SOLUTION

Problem Data and Schematic: The schematic of the pipe bend is shown in Fig. E6.5 together with the pipe data.

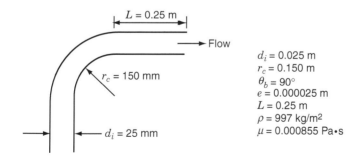

d_i = 0.025 m
r_c = 0.150 m
θ_b = 90°
e = 0.000025 m
L = 0.25 m
ρ = 997 kg/m^2
μ = 0.000855 Pa·s

FIGURE E6.5

Determine: Bend and downstream pipe pressure drops for the specified pipe.

Assumptions: The flow entering the bend is fully developed turbulent flow.

Analysis: Let us first compute the necessary geometry data and other information to determine the desired pressure drops.

Pipe flow area: $A_o = \dfrac{\pi}{4} d_i^2 = \dfrac{\pi}{4}(0.025)^2 = 0.0004909\,\text{m}^2$

Mass velocity: $G = \dfrac{\dot{m}}{A_o} = \dfrac{2\,\text{kg/s}}{0.0004909\,\text{m}^2} = 4074.15\,\text{kg/m}^2 \cdot \text{s}$

Flow velocity: $u_m = \dfrac{\dot{m}}{\rho A_o} = \dfrac{2\,\text{kg/s}}{997\,\text{kg/m}^3 \times 0.0004909\,\text{m}^2} = 4.086\,\text{m/s}$

Reynolds number: $\text{Re} = \dfrac{G d_i}{\mu} = \dfrac{4074.15\,\text{kg/m}^2 \cdot \text{s} \times 0.025\,\text{m}}{855 \times 10^{-6}\,\text{kg/m} \cdot \text{s}} = 0.119 \times 10^6$

Dimensionless pipe surface roughness: $e/d_i = 0.025\,\text{mm}/25\,\text{mm} = 10^{-3}$

From Fig. 6.4 we get the following values of friction factors for smooth and rough pipes at $\text{Re} = 0.119 \times 10^6$ and $e/d_i = 10^{-3}$:

$$f_{\text{smooth}} = 0.0043 \qquad f_{\text{rough}} = 0.0056$$

The bend pressure drop is computed using Eqs. (6.48)–(6.50). The bend loss pressure loss coefficient K_b from Eq. (6.50) is

$$K_b = K_b^* C_{\text{Re}} C_{\text{dev}} C_{\text{rough}}$$

For $r_c/d_i = 0.150\,\text{m}/0.025\,\text{m} = 6$ and $\theta_b = 90°$, we get $K_b^* = 0.20$ from Fig. 6.6. For $\text{Re} = 0.119 \times 10^6$ and $r_c/d_i = 6$, we get $C_{\text{Re}} = 1.48$ from Fig. 6.7. For $K_b^* = 0.20$ and $L/d_i = 0.25\,\text{m}/0.025\,\text{m} = 10$, we get $C_{\text{dev}} = 0.86$ from Fig. 6.8. The correction factor for the pipe surface roughness, from Eq. (6.51), is

$$C_{\text{rough}} = \frac{f_{\text{rough}}}{f_{\text{smooth}}} = \frac{0.0056}{0.0043} = 1.30$$

Thus using Eq. (6.50),

$$K_b = K_b^* C_{\text{Re}} C_{\text{dev}} C_{\text{rough}} = 0.20 \times 1.48 \times 0.86 \times 1.30 = 0.331$$

and

$$K_{b,t} = K_b + K_f = K_b + f\,\frac{4L}{D_h} = 0.331 + 0.0056 \times \frac{4 \times 0.25\,\text{m}}{0.025\,\text{m}} = 0.555$$

Finally, the bend pressure loss is given by Eq. (6.48) as

$$\Delta p_b = K_{b,t}\,\frac{\rho u_m^2}{2g_c} = 0.555 \times \frac{997\,\text{kg/m}^3 \times (4.086\,\text{m/s})^2}{2 \times 1} = 4619\,\text{Pa} = 4.62\,\text{kPa} \qquad \textit{Ans.}$$

For comparison purposely let us compute the pressure drop associated with the straight pipe of the same length and surface roughness. The straight length equals the length of 90° bend with the radius of curvature as 150 mm plus the straight pipe of 0.25 m downstream.

$$L_{eq,st} = \frac{\pi}{2} r_c + L = \frac{\pi}{2} \times 0.150\,\text{m} + 0.25\,\text{m} = 0.486\,\text{m}$$

Using the friction factor for a rough pipe as 0.0056 from the above,

$$\Delta p = \frac{4 L_{eq,st}}{D_h} \times f \times \frac{\rho u_m^2}{2 g_c} = \frac{4 \times 0.486\,\text{m}}{0.025\,\text{m}} \times 0.0056 \times \frac{997\,\text{kg/m}^3 \times (4.086\,\text{m/s})^2}{2 \times 1}$$

$$= 3624\,\text{Pa} = 3.62\,\text{kPa}$$

Thus the increase in the pressure drop due to the bend effect is 27% $[(4619/3624 - 1) \times 100]$.

Discussion and Comments. As shown through this example, the bend effect increases the pressure drop due to the curvature effect and the subsequent effect on the downstream flow development. For this example, the increase is 27%. If heat transfer is taking place, there will also be an increase in heat transfer (though not 27%) for the 90° bend.

6.6.3.2   Pipe Bends with Rectangular Cross Section. The pipe bend with rectangular cross section is shown as an inset in Fig. 6.9, with the geometrical parameters as rectangles of sides a and b as shown (and not necessarily short and long sides), the radius of curvature r_c, the bend deflection angle θ_b, and the outlet pipe length L.

The bend total pressure loss coefficient $K_{b,t}$ is computed using Eq. (6.49) with K_b from Eq. (6.50), where K_b^* is given in Figs. 6.9 through 6.11 for the cross-section aspect ratio $\alpha^* = a/b = 0.5$, 1, and 2, respectively. The magnitude of C_{Re}, the Reynolds number correction factor (for actual $Re \neq 10^6$), is the same as that given for the circular cross-section bend in Section 6.6.3.1. The outlet pipe length correction C_{dev} is determined from Fig. 6.8 for the circular cross section bend with the following modifications:

1. $C_{dev,rect} = 1 - (1 - C_{dev,cir})/2$ for $\alpha^* = a/b < 0.7$ and $L/D_h > 1$.
2. $C_{dev,rect} = C_{dev,cir}$ for $\alpha^* = a/b < 0.7$ and $L/D_h < 1$.
3. $C_{dev,rect} = C_{dev,cir}$ for $\alpha^* = a/b > 1$ and $L/D_h > 1$.
4. C_{dev} for the circular cross section for $\alpha* = a/b > 1.0$ and $L/D_h < 1$ except for r_c/b between 1.5 and 3 when the basic coefficient K_b^* should be multiplied by 2.

The values for K_b^* for short outlet pipes are given by Miller (1990). If the inlet pipe length exceeds $4D_h$, the pressure loss computed for a rectangular bend with the foregoing methodology will probably be conservative.

6.6.3.3   Miter Bends. For the miter bends of circular and rectangular cross sections (see inset in Fig. 6.12), the procedure to determine $K_{b,t}$ is identical to that for a circular-cross-section pipe bend (see Section 6.6.3.1) except that K_b^* is determined from Fig. 6.12; and no correction is applied for the outlet pipe flow development (i.e., $C_{dev} = 1$). The loss coefficients for composite miter bends and other bends are given by Miller (1990) and Idelchik (1994).

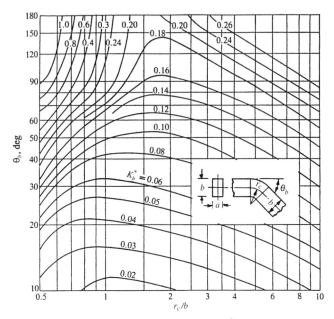

FIGURE 6.9   Bend pressure loss coefficient K_b^* at Re = 10^6 for a rectangular cross section bend with the aspect ratio $\alpha^* = a/b = 0.5$. (From Miller, 1990.)

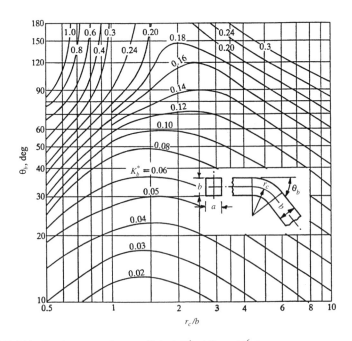

FIGURE 6.10   Bend pressure loss coefficient K_b^* at Re = 10^6 for a square cross section bend. (From Miller, 1990.)

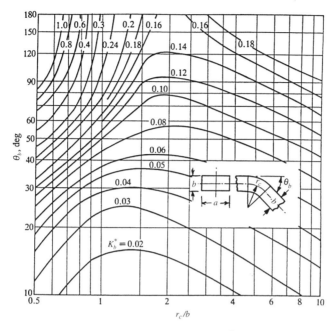

FIGURE 6.11   Bend pressure loss coefficient K_b^* at Re $= 10^6$ for a rectangular cross section bend with the aspect ratio $\alpha^* = a/b = 2$. (From Miller, 1990.)

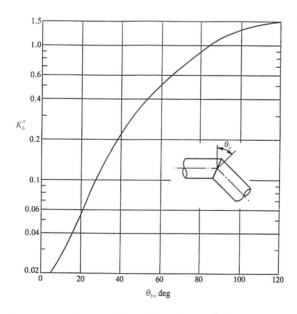

FIGURE 6.12   Bend pressure loss coefficient K_b^* at Re $= 10^6$ for a miter bend. (From Miller, 1990.)

Example 6.6   Determine the pressure drop in a 90° miter bend using all the data given in Example 6.5 for a 90° circular bend.

SOLUTION

Problem Data and Schematic: The schematic of the 90° miter bend is shown in Fig. E6.6 together with the input data.

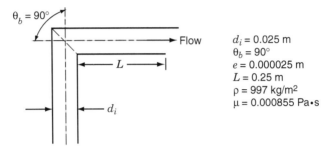

d_i = 0.025 m
θ_b = 90°
e = 0.000025 m
L = 0.25 m
ρ = 997 kg/m²
μ = 0.000855 Pa·s

FIGURE E6.6

Determine: Miter bend and downstream pipe pressure drop for the specified pipe.

Assumptions: The flow entering the bend is fully developed turbulent flow.

Analysis: Most of the information generated for the 90° circular bend is also applicable here. For the miter bend, as mentioned in Section 6.6.3.3, $C_{dev} = 1$ and $K_b^* = 1.2$ from Fig. 6.12. From Example 6.4, $C_{Re} = 1.48$ and $C_{rough} = 1.30$. Hence,

$$K_b = K_b^* C_{Re} C_{dev} C_{rough} = 1.2 \times 1.48 \times 1 \times 1.30 = 2.31$$

and

$$K_{b,t} = K_b + K_f = K_b + f \frac{4L}{D_h} = 2.31 + 0.0056 \times \frac{4 \times 0.25\,\text{m}}{0.025\,\text{m}} = 2.534$$

The pressure drop associated with the 90° miter bend is then

$$\Delta p_b = K_{b,t} \frac{\rho u_m^2}{2g_c} = 2.534 \times \frac{997\,\text{kg/m}^3 (4.086\,\text{m/s})^2}{2 \times 1} = 21{,}090\,\text{Pa} = 21.09\,\text{kPa} \qquad \textit{Ans.}$$

Discussion and Comments: The pressure drop associated with the miter bend is 21.09 kPa; for the same fluid and flow rates, the pressure drop for the circular bend is 4.62 kPa. Thus the miter bend resulted in an increase in the pressure drop by about 4.5 times for this example. Hence, whenever there is an option, a circular bend is generally preferred over a miter bend for a lower pressure drop through the bend.

6.7   PRESSURE DROP PRESENTATION

For most heat exchanger surfaces, the core pressure drop [the core friction term in Eq. (6.28)] for a given heat transfer surface is determined experimentally (using a 'small size' heat exchanger) as a function of the fluid flow rate. Details on the experimental method are presented in Section 7.3.4. These data may then be used for the design and analysis of heat exchangers of different physical size, operating at different temperatures and/or pressures, and operating with different fluids compared to those for the test exchanger, but using the same heat transfer surface. Hence, we need to present the test exchanger core Δp vs. $\dot{m}$ results in a universal form so that they can be used for operating conditions, physical sizes, and fluids beyond those for the test exchanger.

If the detailed geometry (such as the hydraulic diameter D_h, minimum free flow area A_o, etc.) of the heat exchanger surface is known, the best approach is to present the core pressure drop versus mass flow rate results in nondimensional form, such as an f vs. Re curve. If the detailed geometry information is not available, the measured core Δp vs. $\dot{m}$ results are presented in dimensional form. Both approaches are presented next.

6.7.1 Nondimensional Presentation of Pressure Drop Data

The core pressure drop in a dimensionless form is generally presented in two alternative forms,[†] in terms of the pressure loss coefficient K or Euler number Eu and the Fanning friction factor f defined by

$$\Delta p = K\,\frac{\rho u_m^2}{2g_c} = \mathrm{Eu}\,\frac{\rho u_m^2}{2g_c} \tag{6.53}$$

$$\Delta p = f\,\frac{4L}{D_h}\,\frac{\rho u_m^2}{2g_c} \tag{6.54}$$

Thus, the Euler number is the same as the pressure loss coefficient K. If the pressure loss coefficient K is constant along the flow length, such as in tube banks, manifolds, bends, valves, and so on, as a result of turbulent flow, usually K or Eu is used to present the pressure drop. For a tube bank with N_r rows, the Euler number is generally defined as an N_r row average as in Eq. (6.37) for $N_{r,cc} = N_r$.

The Fanning friction factor generally represents primarily the frictional component of the pressure drop and is used when the given heat transfer surface has approximately the same frictional pressure drop per unit length along the flow direction. Thus, use of the friction factor allows pressure drop prediction of different flow lengths of the heat exchanger surface. In fluid dynamics books, generally the Darcy friction factor f_D is used and is related to the Fanning friction factor f as

$$f_D = 4f = \frac{\Delta p}{(\rho u_m^2/2g_c)}\,\frac{D_h}{L} \tag{6.55}$$

A comparison of Eqs. (6.54) and (6.55) indicates that one needs to know the hydraulic diameter and flow length of the exchanger surface if the pressure drop is presented in terms of the friction factor. No such information is needed for Δp to be presented in terms of K or Eu.

[†] See also Eq. (7.115) for presenting pressure drop in terms of the Hagen number.

The fluid flow rate is presented in dimensionless form as the Reynolds number defined as

$$\text{Re} = \frac{\rho u_m D_h}{\mu} = \frac{G D_h}{\mu} = \frac{\dot{m} D_h}{A_o \mu} = \frac{\rho \dot{V} D_h}{A_o \mu} \tag{6.56}$$

The further significance of Re is presented in Section 7.2.1.1.

The greatest advantage of the nondimensional presentation is that for a given Reynolds number, geometrically similar surfaces (regardless of the physical size) have the same friction factor with any fluid flowing through the surface. This means that when converted into terms of f vs. Re (such as Fig. 7.8), the experimental data (Δp vs. $\dot{m}$) can be used for different operating conditions (temperature, pressure, etc.), different physical sizes (different D_h but geometrically similar surfaces), and different fluids from those used in the test conditions. Also, such f vs. Re plots allow one to compare Δp vs. $\dot{m}$ data taken on different surfaces with different fluids or operating conditions so that a heat exchanger design with minimum flow resistance can be selected for a given application.

6.7.2 Dimensional Presentation of Pressure Drop Data

In industry, it is a common practice to present the pressure drop data in a dimensional form for specified heat exchanger surfaces since no geometry information is required for such a presentation. These results are presented in different forms by different industries, such as Δp vs. $\dot{m}$, Δp vs. $\dot{V}$, or Δp vs. G.[†] To correct for an operating/design temperature being different from the test temperature, a density correction is usually applied to the pressure drop by plotting the pressure drop at some standard density. However, let us outline the theory for this pressure drop correction by matching the friction factor and Reynolds number between the actual and standard conditions for Δp vs. $\dot{m}$, $\dot{V}$, or G. The standard pressure and temperature are different for different industries depending on their applications, fluids used, and other factors. For example, the standard conditions for air for some industrial heat exchangers could be 1 atm pressure and 20°C (68°F) temperature.

To obtain a standard flow rate (in terms of $\dot{m}$ or G) from the actual flow rate measured at operating conditions for a given exchanger surface, we need to match

$$\text{Re}_{\text{std}} = \text{Re}_{\text{act}} \Rightarrow \frac{\dot{m}_{\text{std}} D_h}{\mu_{\text{std}} A_o} = \frac{\dot{m}_{\text{act}} D_h}{\mu_{\text{act}} A_o} \tag{6.57}$$

Thus,

$$\dot{m}_{\text{std}} = \dot{m}_{\text{act}} \frac{\mu_{\text{std}}}{\mu_{\text{act}}} \tag{6.58}$$

Since $\dot{m} = \dot{V}\rho$, we get the following relationship between $\dot{V}_{\text{std}}$ and $\dot{V}_{\text{act}}$ using Eq. (6.58):

$$\dot{V}_{\text{std}} = \dot{V}_{\text{act}} \frac{\rho_{\text{act}}}{\rho_{\text{std}}} \frac{\mu_{\text{std}}}{\mu_{\text{act}}} \tag{6.59}$$

[†] Note that the pressure drop is identical for two exchangers having different frontal areas but identical heat transfer surfaces and flow length when the mass velocities G are identical in both exchangers. Thus, different frontal areas are taken into account if Δp is plotted against G rather than $\dot{m}$. However, in this case, the information on the minimum free-flow area is required to determine $G = \dot{m}/A_o$.

Since $G = \dot{m}/A_o$, we get from Eq. (6.58),

$$G_{std} = G_{act} \frac{\mu_{std}}{\mu_{act}} \tag{6.60}$$

To match friction factors, use Eq. (6.54) and $\dot{m} = \rho u_m A_o$ to get

$$f = \Delta p \frac{2g_c}{\rho u_m^2} \frac{D_h}{4L} = \Delta p \frac{2g_c \rho A_o^2}{\dot{m}^2} \frac{D_h}{4L} \tag{6.61}$$

Now let us match the standard and actual friction factors for a given geometry using Eq. (6.61).

$$f_{std} = f_{act} \Rightarrow \frac{\Delta p_{std} \rho_{std}}{\dot{m}_{std}^2} = \frac{\Delta p_{act} \rho_{act}}{\dot{m}_{act}^2} \tag{6.62}$$

Thus

$$\Delta p_{std} = \Delta p_{act} \frac{\rho_{act}}{\rho_{std}} \left(\frac{\dot{m}_{std}}{\dot{m}_{act}} \right)^2 \tag{6.63}$$

Substituting $\dot{m}_{std}$ from Eq. (6.58) into Eq. (6.63), we get

$$\Delta p_{std} = \Delta p_{act} \frac{\rho_{act}}{\rho_{std}} \left(\frac{\mu_{std}}{\mu_{act}} \right)^2 \tag{6.64}$$

This is the most general relationship between Δp_{std} and Δp_{act} that requires both density and viscosity corrections, and should be used in all cases.

To illustrate the aforementioned effects of density and viscosity corrections, the experimental pressure drop Δp with air as a function of the measured mass velocity G is shown in Fig. 6.13a for a heat exchanger. The pressure drop Δp vs. mass velocity G tests were conducted with air at six different inlet temperatures under isothermal conditions. There is about a 17% spread in the experimental pressure drop over the temperature range covered. The same results are replotted in Fig. 6.13b with a traditional density correction to a standard temperature of 15°C (59°F) [i.e., the pressure drop is corrected using Eq. (6.66), and no correction is applied to the measured mass velocity G]. The results collapse to about a 4% spread with the test temperature. Figure 6.13c shows the recommended correction of density and viscosity [i.e., the pressure drop is corrected using Eq. (6.64) and the mass velocity is corrected using Eq. (6.60)]. The results collapse to a single curve, as expected from the foregoing theoretical basis.

However, in industry the viscosity correction in Eq. (6.64) is usually neglected and only the density correction is applied. It can be rationalized as follows. When the friction factor is constant (independent of the Reynolds number for turbulent flow, such as in a rough pipe; see Fig. 6.4) or when f is not strongly variable for turbulent flow in a smooth pipe, we do not need to match Re_{std} and Re_{act} as is done in Eq. (6.57). Hence, in this case,

$$\dot{m}_{std} = \dot{m}_{act} \tag{6.65}$$

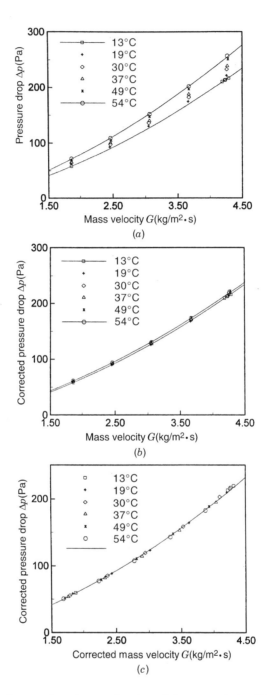

FIGURE 6.13 Pressure drop Δp as a function of the mass velocity G for a heat exchanger: (*a*) measured data; (*b*) density correction applied to Δp; (*c*) density and viscosity correction applied to Δp and the density correction applied to G.

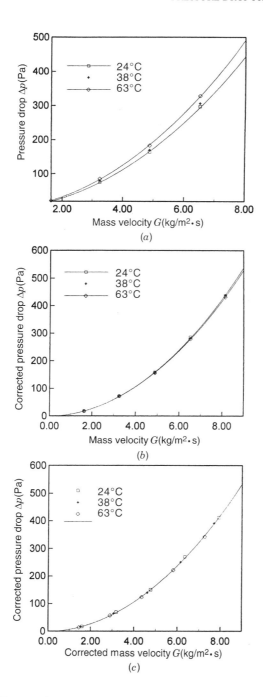

FIGURE 6.14 Pressure drop Δp as a function of the mass velocity G for a perforated plate:
(*a*) measured data; (*b*) density correction applied to Δp; (*c*) density and viscosity correction
applied to Δp and the density correction applied to G.

and Eq. (6.63) simplifies to

$$\Delta p_{\text{std}} = \Delta p_{\text{act}} \frac{\rho_{\text{act}}}{\rho_{\text{std}}} \qquad (6.66)$$

To demonstrate that the viscosity correction is not significant for a turbulent flow (or where the pressure drop is approximately proportional to the velocity square), isothermal tests similar to those of Fig. 6.13a are shown in Fig. 6.14a with air at three temperatures on a perforated plate. The results are shown in Fig. 6.14b with the density correction alone, and in Fig. 6.14c with the density and viscosity corrections together. As it is found, whether only the density correction is applied (Fig. 6.14b) or both the density and viscosity corrections are applied (Fig. 6.14c), the results of these tests collapse to a single curve, as anticipated from the theoretical basis outlined above.

In most situations, whether or not f is constant is not known a priori, and therefore Eq. (6.64) is recommended for correcting Δp_{act} to Δp_{std} and Eqs. (6.58), (6.59), and (6.60) for correcting $\dot{m}$, $\dot{V}$, and G, respectively, for a plot of Δp vs. $\dot{m}$, $\dot{V}$, or G at standard conditions.

6.8   PRESSURE DROP DEPENDENCE ON GEOMETRY AND FLUID PROPERTIES

Pressure drop in a heat exchanger core/matrix is dependent on some fluid properties and geometrical parameters. Since the frictional pressure drop is the dominating contribution to the core pressure drop, we use that term only for the analysis [i.e., Eq. (6.2) or (6.29)]. Substituting $A_o = D_h A/4L$ from the definition of the hydraulic diameter, we get the expression for Δp from Eq. (6.3) as

$$\Delta p = \begin{cases} \dfrac{1}{D_h^3}\left[\dfrac{1}{2g_c}\dfrac{\mu}{\rho}\dfrac{(4L)^2}{A}\dot{m}(f\cdot\text{Re})\right] & \text{for laminar flow} \qquad (6.67a) \\[4mm] \dfrac{1}{D_h^3}\left[\dfrac{0.046}{2g_c}\dfrac{\mu^{0.2}}{\rho}\dfrac{(4L)^{2.8}}{A^{1.8}}\dot{m}^{1.8}\right] & \text{for turbulent flow} \qquad (6.67b) \end{cases}$$

Thus the pressure drop is proportional to D_h^{-3} (for constant $\dot{m}$, L, A, and fluid properties). For a circular tube, since $A = \pi D_h L$, Δp of Eq. (6.67) is proportional to D_h^{-4} and $D_h^{-4.8}$ in laminar and turbulent flows respectively. Hence, based on Eq. (6.67),

$$\Delta p \propto \begin{cases} \dfrac{1}{D_h^3} \text{ to } \dfrac{1}{D_h^4} & \text{for laminar flow} \qquad (6.68a) \\[4mm] \dfrac{1}{D_h^3} \text{ to } \dfrac{1}{D_h^{4.8}} & \text{for turbulent flow} \qquad (6.68b) \end{cases}$$

Reviewing Eq. (6.3) we find that Δp is proportional to L when D_h and A_o are constant, and Δp is proportional to $1/A_o$ and $1/A_o^{1.8}$ for laminar and turbulent flow respectively when D_h and L are constant. In these comparisons, we also keep $\dot{m}$ and fluid properties constant. Note that the surface area A is then not an independent variable since $A = (4A_o L)/D_h$.

$$\Delta p \propto \begin{cases} L, \dfrac{1}{A_o} & \text{for laminar flow} & (6.69a) \\[3mm] L, \dfrac{1}{A_o^{1.8}} & \text{for turbulent flow} & (6.69b) \end{cases}$$

From Eq. (6.67), we find that

$$\Delta p \propto \begin{cases} \dfrac{1}{\rho}, \mu & \text{for laminar flow} & (6.70a) \\[3mm] \dfrac{1}{\rho}, \mu^{0.2} & \text{for turbulent flow} & (6.70b) \end{cases}$$

Thus, Δp is dependent on ρ and μ as shown, but is not directly dependent on c_p and k.

Finally, the dependence of the pressure drop on the surface geometry is given as follows from the second equality of Eq. (6.29):

$$\Delta p \propto \frac{L}{A_o^2 D_h} \qquad (6.71)$$

for a specified fluid and its mass flow rate. Since the friction factors and Colburn factors (or Nu) for enhanced heat transfer surfaces are higher than those for a plain surface, both heat transfer rate and pressure drop for an enhanced surface will be higher than those for a plain surface for a given fluid mass flow rate $\dot{m}$ and the same heat exchanger dimensions. However, it is possible to maintain high-heat-transfer performance with the same pressure drop (particularly in laminar flows) by properly choosing the heat exchanger dimensions and surface geometry [i.e., the proper choice of L, A_o, and D_h in Eq. (6.71) so that Δp remains the same]. Refer to Example 10.3 to gain some insight into this concept. However, if the exchanger frontal area cannot be changed, one will end up with a larger pressure drop with enhanced surface compared to that for the plain surface for a given fluid flow rate. In that case (i.e., for a fixed frontal area), a plain unenhanced surface will have a lower pressure drop but longer exchanger flow length (and hence larger volume and mass of the exchanger) to meet the heat transfer requirement specified.

SUMMARY

In this chapter, important issues related to the pressure drop analysis and data presentation for a heat exchanger as a component are discussed with the following highlights.

- In a heat exchanger, the pressure drop (in addition to heat transfer) becomes an important design consideration for gases in laminar, transition, and turbulent flow and for highly viscous liquids in laminar flow. For other liquids, and particularly in turbulent flow, the pressure drop is not as critical a design consideration as heat transfer.

- Pressure drop associated with a heat exchanger is made up of two contributions: (1) pressure drop associated with a core/matrix/heat transfer surface where heat transfer takes place, and (2) pressure drop associated with flow distribution devices that contribute to the pressure drop without any effective heat transfer. Ideally, most of the design pressure drop allowed should be associated with the first contribution—the heat transfer surface.

- Core pressure drop may consist of the following contributions: (1) core friction, (2) entrance effect, (3) exit effect, (4) momentum effect, and (5) elevation change effect. The first contribution should be ideally above 80 to 90% because that portion of the pressure drop is utilized where heat transfer takes place. The first four contributions are included in Eq. (6.28) and the fifth one is given by Eq. (6.5).

- Pressure drop associated with the following flow distribution devices is presented in Section 6.6: (1) pipes, (2) sudden expansion and contraction at the inlet/outlet, and (3) bends. For other flow distribution devices, refer to Miller (1990) and Idelchik (1994).

- For most heat exchanger surfaces, pressure drop data are obtained experimentally. They are presented in dimensional or dimensionless form as outlined in Section 6.7. When presented in dimensional form, they should be corrected for both density and viscosity changes, as shown in Eq. (6.64).

- Pressure drop is a strong function of the passage hydraulic diameter, and it is also dependent on the exchanger flow length, free-flow area, and heat transfer surface area. The specific functional dependency is shown in Eqs. (6.68) and (6.69). The pressure drop is also dependent on the type of fluid used, particularly its density and viscosity, as shown in Eq. (6.70).

REFERENCES

Bell, K. J., 1988, Delaware Method for shell-side design, in *Heat Transfer Equipment Design*, R. K. Shah, E. C. Subbarao, and R. A. Mashelkar, eds., Hemisphere Publishing, Washington, DC, pp. 145–166.

Crane Co., 1976, *Flow of Fluids through Valves, Fittings, and Pipes*, Technical Paper 410, Crane Co., Chicago.

Idelchik, I. E., 1994, *Handbook of Hydraulic Resistance*, 3rd ed., CRC Press, Boca Raton, FL.

Kays, W. M., and A. L. London, 1998, *Compact Heat Exchangers*, reprint 3rd ed., Krieger Publishing, Malabar, FL.

Miller, D. S., 1990, *Internal Flow Systems*, 2nd ed., BHRA (Information Services), Cranfield, UK.

REVIEW QUESTIONS

Where multiple choices are given, circle one or more correct answers. Explain your answers briefly.

6.1   Pressure drop is more critical and must be allocated carefully for:

   (a) compact heat exchangers   **(b)** double-pipe heat exchangers

   (c) gases   **(d)** moderately viscous liquids

6.2   Heat exchangers are ideally designed to have a significantly higher pressure drop in inlet and outlet headers than in the core.

   (a) depends on the application   **(b)** true

(c) false   (d) shell-and-tube exchangers

(e) plate-fin exchangers

6.3   Entrance and exit losses in a given fluid side of an exchanger are important:

(a) for high core Δp (b) at high Reynolds numbers

(c) for a high ratio of frontal to   (d) for long cores
free-flow area

6.4   Entrance and exit loss terms are usually treated as zero for:

(a) crossflow to tube banks

(b) rotary regenerator with cylindrical flow passages

(c) plate heat exchangers

(d) normal fins on individual tubes

6.5   In total core pressure drop evaluation, the entrance and exit pressure losses are generally:

(a) additive   (b) compensating   (c) can't tell in general

6.6   Two designs meet heat transfer performance requirement for a fluid stream (fluid 1) in a plate heat exchanger: (1) a 1-pass 1-pass design (see Fig. 1.65a) with 10 flow channels for fluid 1 and 11 fluid channels for fluid 2, and (2) a series flow arrangement (see Fig. 1.65f) with seven flow channels for fluid 1 and eight flow channels for fluid 2. If the flow is fully developed turbulent in both designs (i.e., f is about the same), the pressure drop ratio for fluid 1 for design 2 to design 1 is approximately:

(a) 0.7   (b) 0.49   (c) 100

(d) 700   (e) 343   (f) 1.43

6.7   The pressure rise/drop due to elevation change may be important for:

(a) gases   (b) liquids   (c) two-phase fluids

6.8   Pressure loss in a regular 90° bend generally consists of:

(a) frictional loss   (b) elevation change loss

(c) curvature effect loss   (d) sudden contraction loss

6.9   The pressure drop for flow in a rough tube of 25 mm diameter compared to a similar smooth tube at the same Reynolds number in laminar flow is:

(a) lower   (b) higher

(c) the same   (d) can't tell definitely

6.10   The pressure drop in the fluid distribution system is generally presented in terms of:

(a) Euler number   (b) Reynolds number

(c) velocity head   (d) momentum flux correction factor

6.11   The pressure drop constraint is more important for fluids having:

(a) high density   (b) high viscosity

(c) high thermal conductivity   (d) high specific heat

6.12 Arrange the following exchangers in the order of most allowed design pressure drop Δp to least allowed Δp.

(a) noncompact liquid-to-liquid exchanger

(b) highly compact gas-to-gas exchanger

(c) moderately compact gas-to-gas exchanger

(d) compact water-to-water exchanger

6.13 The core pressure drop is generally expressed as follows:

$$\frac{\Delta p}{p_i} = \frac{G^2}{2g_c \rho_i p_i} \left[\underbrace{\left(1 - \sigma^2 + K_c\right)}_{(1)} + \underbrace{2\left(\frac{\rho_i}{\rho_o} - 1\right)}_{(2)} + \underbrace{f \frac{L}{r_h} \rho_i \left(\frac{1}{\rho}\right)_m}_{(3)} - \underbrace{\left(1 - \sigma^2 - K_e\right)\frac{\rho_i}{\rho_o}}_{(4)} \right]$$

Identify which terms of this equation are treated as zero for the following exchangers:

(a) gas-to-gas waste heat recovery exchanger: _____

(b) fin side of a circular finned tube: _____

(c) shell side of a shell-and-tube exchanger for sulfuric acid cooling: _____

(d) plate-fin exchanger during isothermal testing: _____

6.14 Air at a given mass flow rate goes in series through two passes of the same length in a heat exchanger. Now consider an alternative arrangement where those two passes are in parallel coupling and the total airflow rate is the same (i.e., these cases correspond to Fig. 3.21 vs. Fig. 3.22 for fluid 2). The air-side core pressure drop for the second case will be the following factor of the core pressure drop for the first case, assuming fully developed laminar flow through the exchanger. Note that except for different coupling of two passes, both passes of both exchangers are identical.

(a) 1   (b) $\frac{1}{2}$ (c) 2

(d) $\frac{1}{4}$ (e) 4

PROBLEMS

6.1 Determine the pressure drop on the hot and cold sides of the regenerator of Problem 5.1 having $\varepsilon = 95\%$. Compute $\Delta p/p_i$ on each side in percent and determine the total $\Delta p/p$ for this regenerator. Generally, design $(\Delta p/p)_{\text{total}}$ is kept below 4 to 5%. Discuss your results. Treat gas as air. Use $\mu = 0.369 \times 10^{-4}$ and 0.360×10^{-4} Pa · s for the hot- and cold-fluid sides, respectively, and $f = 14/\text{Re}$.

6.2 Determine the pressure drops on the hot and cold sides of the regenerator of Example 5.3. The following are additional data for the surface: $f = 17/\text{Re}$, $D_h = 0.44$ mm, $\mu = 17 \times 10^{-5}$ Pa · s, $p_{a,i} = 442$ kPa and $p_{g,i} = 150$ kPa.

6.3 Explain why K_e can be negative for some values of σ and Re for all geometries of Fig. 6.3. Explain why K_c is not negative for any value of σ and Re.

6.4 The objective is to evaluate the pressure drop for air and water at (1) equal velocity and (2) equal flow rate through a 25-mm-inside-diameter 3-m-long circular tube. Air and water are at 25°C. Assume $f = 16/\mathrm{Re}$ for laminar flow and $f = 0.00128 + 0.1143\mathrm{Re}^{-0.311}$ for turbulent flow. The following fluid properties are specified:

Thermophysical Properties	Air	Water
Specific heat c_p (J/kg · K)	1009	4187
Dynamic viscosity μ (Pa · s)	0.184×10^{-4}	8.853×10^{-4}
Thermal conductivity k (W/m · K)	0.0261	0.602
Prandtl number Pr	0.71	6.16
Density ρ (kg/m^3)	1.183	996.4

(a) Consider air and water velocities at 6 m/s. Determine the following:
  (i)   Air flow and water flow rates.
  (ii) Pressure drop for each fluid, neglecting entrance and exit losses.
  (iii) Fluid pumping power for each fluid.

(b) Now assume that air and water flow rates are the same as the airflow rate in part (a)(i). Determine the following.
  (i) Pressure drop for each fluid, neglecting entrance and exit losses.
  (ii) Fluid pumping power for each fluid.

(c) Discuss the results of parts (a) and (b) from an engineering viewpoint.

6.5 To select an appropriate shell type for a given application (with single-phase fluids on both sides), we want to evaluate all major shell types: E, F, G, H, J, and X. Because of the thermal stress considerations, use U tubes (two tube passes) in a single-shell exchanger for all shell types specified.

(a) For the identical tube fluid flow rate, will the pressure drop on the tube side be the same or different in the types of shells above for the same effective shell length? Why? Assume all fluid properties to be constant.

(b) Estimate the shell-side pressure drop for each shell type as a function of u_m and L, where u_m is the mean shell-side velocity in the E shell and L is the shell length. Note that the shell-side flow rate is the same for all shell types and each has single-segmental baffles/support plates at the same spacing. The shell and tube diameters and number of tubes are the same for all shell types. The shell length is much greater than the shell diameter. Select a shell type for the lowest shell-side pressure drop. *Hints:* Don't forget to add Δp's qualitatively due to 180° bends. Don't use Eqs. (6.37) and (6.39).

6.6 Estimate whether or not the pressure drops through the core are going to be acceptable for the design given below. The inlet temperature, pressure, and mass flow rate of an airstream entering a gas-to-gas two-fluid heat exchanger are: 214°C, 490 kPa, and 21 kg/s, respectively. The outlet temperature of that fluid is 346°C. The mass flow rate of the other airstream is the same as for the first, while the inlet temperature and pressure are 417°C and 103 kPa, respectively. The pressure drops are limited to 4.9 kPa for the first fluid stream and to 2.57 kPa for the second fluid stream. Fanning and Colburn friction factors can be calculated using the following correlations for a plain plate-fin surface with the designation 19.86 of Kays and London (1998):

$$j \text{ or } f = \exp\left[a_0 + r(a_1 + r\{a_2 + r[a_3 + r(a_4 + a_5 r)]\})\right]$$

where a_j, $j = 1, \ldots, 5$ are given below, and $r = \ln(\text{Re})$.

	a_0	a_1	a_2	a_3	a_4	a_5
f	-0.81952×10^3	0.54562×10^3	-0.14437×10^3	0.18897×10^2	-0.12260×10^1	0.31571×10^{-1}
j	-0.26449×10^3	0.16720×10^3	-0.41495×10^2	0.50051×10^1	-0.29456	0.67727×10^{-2}

For both heat transfer surfaces, the hydraulic radius is 0.001875 m and the extended-surface efficiency is 0.9. The number of transfer units for the exchanger is 4.9. The heat transfer surface wall temperature may be assumed to be 317°C. Thermal conductivity of the fin material and fin thickness are 200 W/m · K, and 0.152 mm. The plate spacing is 6.35 mm. The ratio of minimum free-flow area to frontal area for both heat transfer surfaces is 0.3728, while the minimum free-flow areas on both sides are 0.4258 m^2 and 1.5306 m^2. Elaborate all decisions on the assumptions required if any of the input data are missing.

7 Surface Basic Heat Transfer and Flow Friction Characteristics

In exchanger design, the dimensional heat transfer and pressure drop vs. fluid flow test data obtained for one exchanger size cannot be used to size or rate accurately an exchanger of a different size (having the same basic surface geometries) because of the nonlinear relationships among the geometrical and operating parameters of the exchanger. To rate or size an exchanger of different size or different performance requirements, one must have dimensionless heat transfer and fluid friction characteristics[†] of heat transfer surfaces. That is why among the most important inputs to the heat exchanger thermal and hydraulic design and analysis are accurate dimensionless heat transfer and fluid friction characteristics, as shown in Fig. 2.1. These heat transfer characteristics are generally presented in terms of the Nusselt number Nu vs. the Reynolds number Re, the dimensionless axial distance x^*, or the Graetz number Gz. The experimental characteristics are usually presented in terms of the Stanton number St or the Colburn factor j vs. the Reynolds number Re. Flow friction characteristics are generally presented in terms of the Fanning friction factor f vs. the Reynolds number Re or the dimensionless axial distance x^+; alternatively, they are presented for flow over tube banks in terms of the Euler number Eu or Hagen number Hg vs. the Reynolds number Re. Depending on flow and heat transfer conditions, some additional important dimensionless numbers may be involved, as discussed later. The surface characteristics are obtained analytically for simple geometries. The surface characteristics are primarily obtained experimentally for most exchanger surfaces because the flow phenomena are complex due to the geometry features of flow area and/or heat transfer surface. Now more emphasis is placed on numerical (CFD) analysis to obtain the surface characteristics of complex heat exchanger surfaces, but full three-dimensional numerical analyses are yet not practical at a reasonable cost covering a wide range of Re values, due to computing limitations.

In this chapter we start with basic concepts in Section 7.1 for understanding the behavior of the surface characteristics. These will include the concepts of boundary layers, types of flows usually encountered in industrial heat exchangers, convection mechanisms, and the basic definitions of the mean velocity, mean temperature and heat transfer coefficient. In Section 7.2, dimensionless groups used for fluid flow and heat transfer characteristics of heat exchanger surfaces are presented along with an

---

[†] We will not use the terminology *surface performance data* since *performance* in some industries is related to a dimensional plot of the heat transfer rate and one fluid-side pressure drop as a function of its fluid flow rate for an exchanger. Note that we need to distinguish between the performance of a surface geometry (of one side of an exchanger) and the performance of a heat exchanger.

425

illustrative example for airflow through a circular tube. Experimental techniques used for measurement of the heat transfer and flow friction characteristics of heat exchanger surfaces are presented in Section 7.3. These include the steady state, transient, and Wilson plot techniques. In Section 7.4, analytical and semiempirical correlations are presented for heat transfer and flow friction for simple geometries. In Section 7.5, experimental correlations are presented for complex geometries that include tubular surfaces, plate heat exchanger surfaces, plate-fin surfaces, tube-fin surfaces, and regenerator surfaces. The influence of temperature-dependent fluid properties can be significant in a heat exchanger. This is discussed in Section 7.6, particularly the property ratio method for gases and liquids. Although forced convection is the major mode of heat transfer in many heat exchangers, free convection and radiation heat transfer can be important in some applications. These issues are summarized briefly in Section 7.7. Thus, the major motivation of this chapter is to outline most important issues related to the understanding and utilization of available literature information for the determination and minor extrapolation of accurate heat transfer and flow friction characteristics of heat exchanger surfaces.

7.1   BASIC CONCEPTS

Some of the basic concepts needed to understand heat transfer and flow friction characteristics of heat exchanger surfaces are described in this section. These include the concepts of a boundary layer, flow types and convection mechanism, and definitions of the mean velocity, temperature, and heat transfer coefficient. Although these concepts have been introduced in the first courses in fluid mechanics and heat transfer, our emphasis here is to review them from the heat exchanger application point of view for understanding and minor extrapolation of design correlations.

7.1.1   Boundary Layers

The concepts of velocity and thermal boundary layers are first discussed below.

7.1.1.1   Velocity Boundary Layer. The flow field around a body may be divided into two regions for the purpose of analysis. The thin region close to the body surface, where the influence of fluid viscosity becomes increasingly predominant when approaching the surface, is referred to as the *velocity* or *momentum boundary layer*, as shown in Fig. 7.1 for flow over a flat plate. The remainder of the flow field can to a good approximation

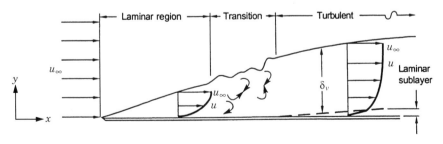

FIGURE 7.1   Velocity boundary layer on a flat plate.

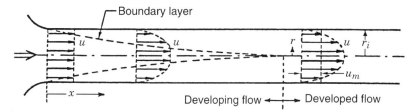

FIGURE 7.2   Hydrodynamically developing and developed isothermal laminar flow in a tube. (From Shah and London, 1978.)

be treated as inviscid and can be analyzed by the potential flow theory. Somewhat arbitrarily, the velocity boundary layer thickness δ_v is defined as the distance at which 99% of the free stream or entrance velocity magnitude is achieved.

For internal flows, such as flow in a circular or noncircular pipe, the velocity boundary layer starts at the pipe entry. It grows along the flow length and eventually fills up the pipe cross section, beyond which the entire pipe flow becomes the fully developed flow, as shown in Fig. 7.2 for laminar flow and in Fig. 7.3 for turbulent flow. Hence, two regions can be identified for the laminar boundary layer in a pipe flow: developing and developed boundary layer regions; and these are discussed later. Similarly, a turbulent boundary layer in a pipe flow has four regions: developing laminar, transition, developing turbulent, and fully developed boundary layer regions, as shown in Fig. 7.3. Notice that if the tube length is ℓ or less (see Fig. 7.3), there will be developing laminar flow regardless of a high value of the operating Reynolds number (such as any value greater than 10^4) based on the tube hydraulic diameter.

The velocity profile for fully developed laminar flow of a power-law fluid $[(\tau = -\mu(-du/dr)^m]$ in a circular tube with the origin at the tube axis and radius r_i is given by

$$\frac{u}{u_m} = \frac{3m+1}{m+1}\left[1 - \left(\frac{r}{r_i}\right)^{(m+1)/m}\right] \tag{7.1}$$

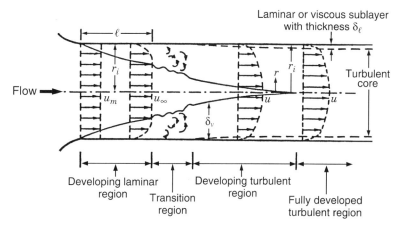

FIGURE 7.3   Flow regimes in developing and developed turbulent flow in a tube at a high Reynolds number (Re $\geq 10^4$). (From Shah, 1983.)

Note that for a Newtonian fluid ($m = 1$), the velocity profile in Eq. (7.1) is governed by $u = u_{max}[1 - (r/r_i)^2]$ and $u_{max} = 2u_m$, where u_m represents mean velocity and u_{max} the maximum, centerline velocity.

The empirical, power-law velocity profile for fully developed turbulent flow in a circular tube is given by

$$\frac{u}{u_{max}} = \left(1 - \frac{r}{r_i}\right)^{1/n} \qquad \frac{u_m}{u_{max}} = \frac{2n^2}{(n+1)(2n+1)} \tag{7.2}$$

where the exponent n varies with Re and the values of n are between 6 and 10 for the range of Re between 10^4 and 10^6 (Hinze, 1975; Bhatti and Shah, 1987). In Eq. (7.2), u, u_m, and u_{max} represent the time-averaged turbulent local, cross-sectional mean, and cross-sectional maximum velocities. Thus as Re increases, n increases, and the velocity profile becomes flatter over most of the tube cross section. An analysis of Eqs. (7.1) and (7.2) reveals that all turbulent velocity profiles are significantly steeper near the wall than the laminar velocity profile, contributing significantly higher pressure drops in turbulent flow, as discussed later.

7.1.1.2   Temperature Boundary Layer. When heat transfer takes place between the fluid and the solid surface, the major temperature change occurs in a region very close to the surface in most cases. This region is referred to as the *temperature* or *thermal boundary layer*, shown in Fig. 7.4 for laminar internal flow. The temperature profile across the cross section can be determined integrating the corresponding energy equation of the boundary layer problem, and it is beyond the scope of this discussion. The solution approach to determining these profiles can be found in any advanced heat convection textbook (see, e.g., Kays and Crawford, 1993). Specific temperature profiles can be found in Shah and Bhatti (1987) for laminar flow and in Bhatti and Shah (1987) for turbulent flow. The temperature boundary layer, also sometimes referred to as fluid film near the surface, may be interpreted in terms of the thermal resistance to heat transfer from the fluid to the solid surface (or vice versa). The heat transfer rate per unit of surface area across the boundary layer in Fig. 7.4 ($T_m > T_w$) is

$$q_w'' = -k\left(\frac{\partial T}{\partial y}\right)_{y=0} = k\,\frac{T_m - T_w}{\delta_t} \tag{7.3}$$

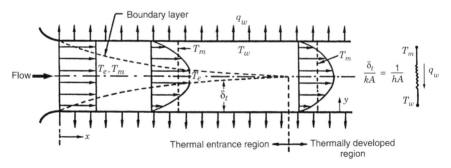

FIGURE 7.4   Laminar thermal boundary layer development in a tube.

Instead of expressing the heat flux in terms of the thermal boundary layer thickness δ_t, generally it is expressed in terms of the *film coefficient*[†] or the *heat transfer coefficient* $h = k/\delta_t$, where k is the fluid thermal conductivity. The thinner the thermal boundary layer (the smaller δ_t), the larger will be the heat transfer coefficient h and the smaller the film resistance. The bulk temperature T_m (for the pertinent definition, see Section 7.1.4.2) for internal flow is replaced by T_∞ for external flow; however, in some published correlations, some other temperatures, such as T_i, replaces T_m, and hence the user should check the definition of the temperature difference used in the Nusselt number evaluation.

The thermal boundary layer is not necessarily of the same thickness as the velocity boundary layer. The boundary layer thickness and the growth rates are dependent on the fluid Prandtl number, discussed in Section 7.1.2.4 (Laminar Flow).

7.1.2 Types of Flows

In heat exchangers, many different types of flows are encountered because a wide variety of heat transfer surfaces and a wide range of mass flow rates are being used. Our objective for describing these flows is (1) to understand the fluid flow and heat transfer behavior of heat transfer surfaces, (2) to develop/apply analytical heat transfer and pressure drop correlations for these surfaces, and (3) to facilitate further enhancement of heat transfer for improved heat exchanger performance. With these objectives in mind, we will classify major types of flows with their specific behavior. We restrict this classification to low Mach number flows as found in most industrial heat exchangers. Most of the information presented in this section is condensed from Jacobi and Shah (1998).

7.1.2.1 Steady and Unsteady Flows. If the velocity and temperature throughout the flow field do not change with time, the flow is said to be *steady*. Such flows are easier to model and analyze. Although in most heat exchanger applications, true steady flow is not encountered, the steady flow approximation is often useful. On the other hand, unsteady effects can be important.

The most obvious case of an *unsteady flow* occurs when the boundary conditions, either velocity (pressure) or temperature, change with time. Such *imposed unsteadiness* can be caused by changes in the fan or pump speed or fluid and/or wall temperature. However, even if the boundary conditions are steady, the flow may exhibit unsteadiness. A simple example is flow normal to a circular cylinder; with a steady approach flow, unsteadiness and vortex shedding occur downstream. Another example is the flow in an offset strip fin geometry where unsteadiness and vortex shedding (such as von Kármán vortices) occur downstream of the strips. In these cases, unsteadiness is self-sustained in the flow. For offset strip fins, the *self-sustained unsteadiness* is periodic in nature along the flow length after a few strips. A rigorous numerical modeling of such a self-sustained flow and its experimental verification for a generic flow geometry of communicating channels indicates a great potential for heat transfer enhancement of such a geometry (Amon et al., 1992). Imposed or self-sustained unsteadiness can be laminar or turbulent and generally causes an early transition to turbulent flow. Laminar unsteady flows are preferred over turbulent flows because the pressure drops in the turbulent flows are higher than those associated with unsteady laminar flows.

[†] It is referred to as the *film* coefficient since the entire thermal resistance is assumed to be in the boundary layer or *film*.

7.1.2.2 Laminar, Transition, and Turbulent Flows. Flow is considered *laminar* when the velocities are free of random fluctuations at every point in the flow field; thus, the fluid motion is highly ordered. Fluid particles flow along nearly parallel streamlines with no rotation. However, in a developing laminar boundary layer (see Figs. 7.1 and 7.2), fluid particles along a streamline have u and v velocity components for a 2D geometry, with the v component being responsible for the momentum or enthalpy transfer across the boundary layer. In laminar flow, viscous forces are dominant over inertia forces. For steady laminar flow, all velocities at a stationary point in the flow field remain constant with respect to time, but the velocities may be different at different points in the flow field. For unsteady laminar flow, all velocities at a stationary point in the flow field will also be time dependent. Laminar flow, also referred to as *viscous* or *streamline flow*, is associated with viscous fluids, low flow velocities, and/or small flow passages.

Flow is considered *turbulent* when the fluid particles do not travel in a well-ordered manner; however, it is difficult to define turbulence in simple terms. Some important characteristics are:

- Turbulent flows have self-sustained, irregular velocity fluctuations (u', v', w') in all directions. In turbulent flows, fluid particles travel in *randomly* moving fluid masses of varying sizes called *eddies*. The irregularity of these fluctuations distinguishes turbulence from laminar self-sustained oscillations, which are usually periodic.
- Turbulent flows have eddies with a broad distribution of sizes. In a turbulent pipe flow, the eddy length scale can range from a few millimeters to the pipe diameter. The range of eddy sizes is the result of the generation of large eddies, their breakup into smaller and smaller eddies, and their eventual dissipation at small scales. This broad eddy-size distribution distinguishes turbulence from laminar self-sustained unsteadiness—laminar unsteadiness usually produces larger eddies (a few length scales or a few pipe diameters) with a small eddy-size distribution.
- A turbulent flow is accompanied by large-scale mixing due to the advective effect of turbulent eddies. Mixing increases heat transfer and wall friction when the near-wall region of the flow is affected (refer to the discussion related to Table 7.1).

If a streak of colored dye is injected in turbulent water flow through a transparent pipe, one will be able to see that the color will immediately disperse throughout the flow field. Thus, the transport of momentum and energy transverse to the main flow direction is greatly enhanced in turbulent flow.

The turbulent boundary layer has a multilayer character. The simplest model includes two layers, as shown in Fig. 7.3; the near-wall region is referred to as the *viscous sublayer* (where molecular diffusion plays a major role), and the outer region is called a *fully turbulent* region or core (where turbulent mixing is dominant). However, molecular diffusion and turbulent mixing phenomena are dependent on the fluid Prandtl number, as summarized in Table 7.1 and discussed as follows. For liquid metals, $Pr < 0.03$, molecular diffusion or heat transfer by conduction within the fluid in the fully turbulent region is important in addition to turbulent mixing. Hence, the thermal resistance is distributed throughout the flow cross section, and the heat transfer coefficient is less dependent on the Reynolds number than for higher Prandtl numbers ($Pr > 0.5$). For all fluids with $Pr > 0.5$, in the fully turbulent region, molecular diffusion plays a minor role and the turbulent mixing is dominant. For fluids with $0 < Pr < 5$, the viscous sub-

TABLE 7.1   Dominance of Molecular Diffusion and Turbulent Mixing in a Two-Layer Turbulent Boundary Layer as a Function of the Fluid Prandtl Number

Sublayer	Pr < 0.03	$0.5 \leq \text{Pr} \leq 5$	Pr > 5
Viscous sublayer	Molecular diffusion	Molecular diffusion	Molecular diffusion and turbulent mixing
Turbulent core	Molecular diffusion and turbulent mixing	Turbulent mixing	Turbulent mixing

layer is dominated by molecular diffusion and turbulent mixing is unimportant in the viscous sublayer. For fluids with Pr > 5, the turbulent mixing in addition to molecular diffusion can be important in the viscous sublayer. Thus for gas and some liquid flows (0.5 < Pr < 5), a two-layer model provides a demarcation of heat transfer by molecular diffusion in the viscous sublayer and by turbulent mixing in the turbulent core. Almost all of the thermal resistance in a turbulent boundary layer for fluids with Pr > 0.5 is due to the thin viscous sublayer; a disruption or further thinning of this sublayer will enhance heat transfer accompanied by large increases in the skin friction.

Flow in the intermediate range of the Reynolds number between fully developed or developing laminar and turbulent flows is designated as *transition flow*. In this region, the laminar boundary layer becomes unstable locally in the presence of small disturbances. Orderly laminar pattern transforms to a chaotic turbulent pattern when the Reynolds number exceeds a certain critical value Re_{cr}. The transition starts in the duct core region rather than at the duct wall and it decays downstream in a relatively short distance compared to a fully developed turbulent flow. This flow instability continues to grow as Re increases until the flow becomes fully turbulent. In a smooth circular pipe, the transition flow exists between $2300 \leq \text{Re} \leq 10,000$. The transition from the laminar to turbulent flow is dependent on the entrance configuration, flow passage geometry, whether or not the surface is interrupted, surface roughness, natural convection effects, flow pulsation, and even the change of viscosity when large heating rates occur. Also, noise and vibration at the exterior of the duct wall could influence Re_{cr}. For a plate heat exchanger, the transition flow initiates between $10 \leq \text{Re} \leq 200$, depending on the corrugation geometry.[†] Under certain conditions (such as an increase in the gas viscosity with temperature in a heat exchanger), a turbulent flow may revert to a laminar flow. This process is known as *reverse transition* or *laminarization*.

Since a basic definition of turbulence is difficult to formulate, it is no surprise that classifying a flow as laminar or turbulent can be difficult. In fact, there may be regions of laminar, laminar unsteady, transitional, and turbulent flows all coexisting in a heat exchanger complex flow passage. The problem arises in deciding where to look in the flow. To clarify the problem, consider a typical finned-tube heat exchanger (see Fig. 1.31a) operating at a Reynolds number of, say, 200, based on the conventional hydraulic diameter for flow normal to the tubes. At this flow rate, it is reasonable to expect that the boundary layer developing on the fins will remain laminar for most flow lengths of interest. Should the flow become fully developed (discussed below), the Reynolds number magnitude suggests that such a duct flow would also be laminar. However, it is easy to imagine that downstream of the first tube row, in the tube-wake region, the wake could

---

[†] See Section 7.1.2.4 for examples of other geometries. In this section and throughout the book, Re without any subscript is consistently defined using the hydraulic diameter D_h and fluid mean velocity u_m.

become turbulent, since wakes, which include a separated shear layer, become turbulent at very low Reynolds numbers (50 or less; see Bejan, 1995). As the flow approaches the next tube row, a favorable pressure gradient will accelerate the flow around the next row of tubes. In this region of favorable pressure gradient, the flow could relaminarize, to take on a turbulent character again in the wake. Thus, in this finned tube heat exchanger, it may be possible to have a confined region of the flow that looks turbulent. A similar situation occurs in offset-strip fin and multilouver fin designs at locations downstream from the fins. How is one to classify such a flow? There is no doubt that a flow with regions of turbulence should be classified as turbulent. Accurate predictions of the local flow and heat transfer cannot be obtained without attention to the relevant turbulent processes. Nevertheless, turbulent mixing is unimportant throughout much of such a flow—locally, the flow appears laminar. This type of a combined-flow situation is common in heat exchangers. Many such flows may be characterized as *low-Reynolds-number turbulent flows* (Re < 3000).

7.1.2.3 Internal, External, and Periodic Flows.

The flows occurring in confining passages of various regular or irregular, singly or doubly connected, constant or variable cross sections such as circular, rectangular, triangular, annular, and the like are referred to as *internal flows*. Associated with these flows is a pressure gradient. The pressure gradient either decreases along the flow length (favorable pressure gradient) or increases in the flow direction (adverse pressure gradient), depending on whether the cross-section size remains constant or increases. In the internal duct flow, the boundary layers are eventually constrained by the surface, and after the development length, the entire cross section represents a boundary layer.

The unconfined flows occurring over surfaces such as flat plates, circular cylinders, turbine blades, and the like are referred to as *external flows*. In the external flow, the boundary layers may continue to grow along the flow length, depending on the surface geometry. Unlike the internal flows, these flows can occur without a sizable pressure gradient in the flow direction as in flow past a flat plate. Frequently, however, these flows occur with pressure gradients that may be positive or negative in the flow direction. For example, in flow over a circular cylinder both negative and positive pressure gradients occur along the main flow direction.

Another distinction between the internal and external flows is that in the latter the effect of viscosity is dominant near the solid wall only, with a potential inviscid flow being away from the wall on the unbounded side. In internal flows, in general, the effect of viscosity is present across the flow cross section. The only exception is the flow near the passage inlet where a potential inviscid core develops near the center of the cross section as shown in Fig. 7.2.

It should be noted that in spite of the enlisted distinctions between internal and external flows, it is very difficult to define the flow either as internal or as external in many practical situations. Heat exchangers are considered to have internal flows—a somewhat artificial view. The classification of flow in complex heat exchanger passages depends on how the boundaries affect the flow. Consider the sketch of flat plates shown in Fig. 7.5. In the first case, (Fig. 7.5a), the flow path appears long compared to the plate spacing, and an internal flow is suggested. In Fig. 7.5b, the flow path appears very small compared to the plate spacing, and an external flow is suggested. The simplified situation reflected in Fig. 7.5c may be difficult to decide since the length scales do not firmly suggest internal or external flow. Spacing and length are important, but only inasmuch as they

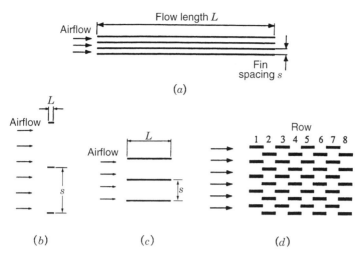

FIGURE 7.5 Length scales important in continuous and interrupted passages: (*a*) flow length much longer than the plate spacing; (*b*) flow length much shorter than the plate spacing; (*c*) flow length and plate spacing comparable; (*d*) periodic array (offset strip fins) that provides an interrupted passage. (From Jacobi and Shah, 1996.)

reflect the effect of the boundary layer on the flow. It would be more precise to consider the boundary layer thickness as a length scale.

When flow develops between neighboring plates (fins), velocity and temperature boundary layers grow. If the neighboring boundary layers do not interact, the flow is considered external; however, if the boundary layers interact, the flow should be considered internal. In cases where the flow is affected by boundary layers on neighboring fins, a blockage effect occurs. That is, as the boundary layers grow (along with the accompanying displacement boundary layers), the effective flow area for the out-of-the-boundary-layer flow is reduced. By continuity, the core flow must accelerate and a favorable pressure gradient (i.e., the static pressure decreasing with increasing flow length x) must be established. The favorable pressure gradient thins the boundary layers, with a commensurate increase in skin friction and heat transfer. As with turbulence, it is useful to admit a paradigm with combined internal and external flows in different regions. Such an approach suggests that flow development is important in classifying an exchanger flow as internal or external; however, before addressing flow development, it is useful to discuss periodic flows.

The last situation depicted in Fig. 7.5*d* shows an array of flat plates in the flow. Such arrays are common in contemporary heat exchanger designs (e.g., in offset strip fin designs). For these extended arrays, it is useful to drop the internal and external classification (view) and adopt the periodic classification, as it reflects the boundary condition. Periodic flows can be developing or fully developed (as discussed in Section 7.1.2.4). In periodic flows, the surface pattern is spatially repeated, and the boundary conditions may be normalized in such a way as to take a periodic structure. For such geometries, it is doubtful that the flow will be steady—instabilities in the wake regions almost ensure unsteadiness. Modeling approaches which assume that the flow to be steady will probably be of little value except at very low Reynolds numbers (Re < 200). It is very common to find that periodic flows exhibit self-sustained, laminar flow oscillations which increase the heat transfer coefficient and the friction factor (Amon et al., 1992). This behavior, along with the inherent boundary layer restarting effect of the interrupted surface, makes such geometries especially attractive for heat exchanger applications.

7.1.2.4 Developed and Developing Flows

Laminar Flow. Four types of laminar duct (internal) flows are: fully developed, hydrodynamically developing, thermally developing (abbreviated for thermally developing and hydrodynamically developed flow), and simultaneously developing (abbreviated for thermally and hydrodynamically developing flow). A further description of these flows is now provided with the aid of Fig. 7.6.

Referring to Fig. 7.6a, suppose that the temperature of the duct wall is held at the entering fluid temperature ($T_w = T_e$) and there is no generation or dissipation of heat within the fluid. In this case, the fluid experiences no gain or loss of heat. In such an isothermal flow, the effect of viscosity gradually spreads across the duct cross section

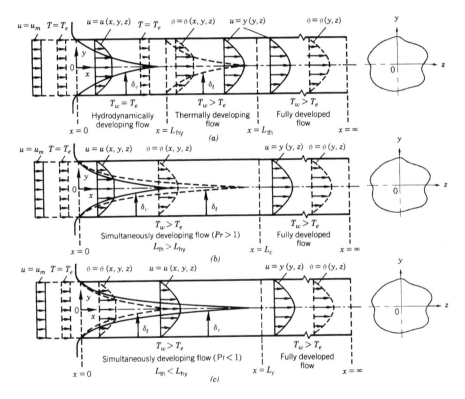

FIGURE 7.6 Types of laminar flows for constant wall temperature boundary condition: (*a*) hydrodynamically developing flow followed by thermally developing and hydrodynamically developed flow; (*b*) simultaneously developing flow, Pr > 1; (*c*) simultaneously developing flow, Pr < 1. Solid lines denote the velocity profiles, dashed lines the temperature profiles. (From Shah and Bhatti, 1987.)

beginning at $x = 0$. The extent to which the viscous effects diffuse normally from the duct wall is represented by the hydrodynamic boundary layer thickness δ_v, which varies with the axial coordinate x. In accordance with Prandtl's boundary layer theory, the hydrodynamic boundary layer thickness divides the flow field into two regions: a viscous region near the duct wall and an essentially inviscid region around the duct axis.

At $x = L_{\mathrm{hy}}$,[†] the viscous effects have spread completely across the duct cross section. The region $0 \leq x \leq L_{\mathrm{hy}}$ is called the *hydrodynamic entrance region*, and the fluid flow in this region is called the *hydrodynamically developing flow*. As shown in Fig. 7.6a, the axial velocity profile in the hydrodynamic entrance region varies with all three space coordinates [i.e., $u = u(x, y, z)$]. For hydrodynamically developed flow, the velocity profile at a given cross section becomes independent of the axial coordinate and varies with the transverse coordinates alone [i.e., $u = u(y, z)$ or $u(r, \theta)$]. The fully developed velocity profile is essentially independent of type of the velocity profile at the duct inlet.

After the flow becomes hydrodynamically developed ($x > L_{\mathrm{hy}}$; Fig. 7.6a), suppose that the duct wall temperature T_w is raised above the fluid entrance temperature T_e (i.e., $T_w > T_e$). In this case, the local temperature changes diffuse gradually from the duct wall, beginning at $x = L_{\mathrm{hy}}$. The extent to which the thermal effects diffuse normally from the duct wall is denoted by the thermal boundary layer thickness δ_t, which also varies with the axial coordinate x. The δ_t is defined as the value of y' (the coordinate measured from the duct wall in Fig. 7.6) for which the ratio $(T_w - T)/(T_w - T_e) = 0.99$. According to Prandtl's boundary layer theory, the thermal boundary layer thickness divides the flow field into two regions: a heat-affected region near the duct wall and an essentially unaffected region around the duct axis. At $x = L_{\mathrm{th}}$,[‡] the thermal effects have spread completely across the duct cross section. The region $L_{\mathrm{hy}} \leq x \leq L_{\mathrm{th}}$ is termed the *thermal entrance region*, and the fluid flow in this region is called the *thermally developing flow*. It may be emphasized that the thermally developing flow is already hydrodynamically developed in Fig. 7.6a. As shown in this figure, in the thermally developing flow region, the local dimensionless fluid temperature $\vartheta = (T_w - T)/(T_w - T_e)$, varies with all three spatial coordinates [i.e., $\vartheta = \vartheta(x, y, z)$].

For $L_{\mathrm{th}} \leq x < \infty$ in Fig. 7.6a, the viscous and thermal effects have completely diffused across the duct cross section. This region is referred to as the *fully developed region*. The fluid flow in this region is termed the *fully developed flow*. In this region, the dimensionless temperature ϑ varies with the transverse coordinates alone, although the local fluid temperature T varies with all three-space coordinates, and fluid bulk temperature T_m varies with the axial coordinate alone. In the fully developed region, the boundary conditions at the wall govern the convective heat transfer process so that the temperature distribution is essentially independent of the duct inlet temperature and velocity distributions.

[†] The hydrodynamic entrance length L_{hy} is defined as the duct length required to achieve a maximum duct cross-section velocity as 99% of that for a fully developed flow when the entering fluid velocity profile is uniform. The maximum velocity occurs at the centroid for the ducts symmetrical about two axes (e.g., circular tube and rectangular ducts). The maximum velocity occurs away from the centroid on the axis of symmetry for isosceles triangular, trapezoidal, and sine ducts (Shah and London, 1978). For nonsymmetrical ducts, no general statement can be made for the location of $u_{\max}$. There are a number of other definitions also used in the literature for L_{hy}.
[‡] The thermal entrance length L_{th} is defined, somewhat arbitrarily, as the duct length required to achieve a value of local Nusselt number equal to 1.05 times the Nusselt number for fully developed flow, when the entering fluid temperature profile is uniform. As discussed in Section 7.4.3.1, theoretically the Nusselt number is infinity at $x = 0$ and reduces asymptotically to a constant value in fully developed flow.

The fourth type of flow, called *simultaneously developing flow*, is illustrated in Fig. 7.6*b* and *c*. In this case, the viscous and thermal effects diffuse simultaneously from the duct wall, beginning at $x = 0$. Depending on the value of the Prandtl number Pr, the two effects diffuse at different rates. In Fig. 7.6*b*, Pr > 1 and $\delta_v > \delta_t$, whereas in Fig. 7.6*c*, Pr < 1 and $\delta_v < \delta_t$. This relationship among Pr, δ_v and δ_t is easy to infer from the definition of the Prandtl number, which for this purpose can be expressed as Pr $= \nu/\alpha$, a ratio of kinematic viscosity to thermal diffusivity. The kinematic viscosity is the diffusion rate for momentum or for velocity in the same sense that the thermal diffusivity is the diffusion rate for heat or for temperature change. For Pr > 1, the velocity profile (hydraulic) development is faster than the temperature profile (thermal) development (so that $L_{hy} < L_{th}$). For Pr < 1, the velocity profile development is slower than the temperature profile development (so that $L_{hy} > L_{th}$). When Pr $= 1$, the viscous and thermal effects diffuse through the fluid at the same rate. This equality of diffusion rates does not guarantee that the hydrodynamic and thermal boundary layers in internal duct flows will be of the same thickness at a given axial location. The reason for this apparent paradox lies in the fact that with Pr $= 1$, the applicable momentum and energy differential equations do *not* become analogous. In external laminar flow over a flat plate, on the other hand, the energy and momentum equations do become analogous when Pr $= 1$. When the boundary conditions for the momentum and thermal problems are also analogous, we get $\delta_v = \delta_t$ for all values of x.

The region $0 \leq x \leq L_c$ in Fig. 7.6*b* and *c* is referred to as the *combined entrance region*. It is apparent that the combined entrance length L_c is dependent on Pr. For Pr > 1, $L_c \approx L_{th}$, and for Pr < 1, $L_c \approx L_{hy}$. It may also be noted that in the combined entrance region, both the axial velocity and the dimensionless temperature vary with all three space coordinates [i.e., $u = u(x, y, z)$ and $\vartheta = \vartheta(x, y, z)$ where $\vartheta = (T_{w,m} - T)/(T_{w,m} - T_m)$]. The region $L_c \leq x < \infty$ is the fully developed region, similar to the one depicted in Fig. 7.6*a* with axially invariant $u(y, z)$ and $\vartheta(y, z)$, satisfying

$$\left(\frac{\partial u}{\partial x}\right)_{x>L_{hy}} = 0 \qquad \left(\frac{\partial \vartheta}{\partial x}\right)_{x>L_{th}} = 0 \qquad (7.4)$$

In a *fully developed laminar flow*, the fluid appears moving by sliding laminae of infinitesimal thickness relative to adjacent layers. If a dye were injected at the centerline of a tube in laminar flow of water, the colored streak will continue to flow without being mixed at the tube centerline. While fully developed laminar flow is obtained for Re ≤ 2300 for a smooth circular tube with smooth tube inlet, the actual value of the Reynolds number to achieve/maintain fully developed or developing laminar flow depends on the tube cross-sectional geometry, straightness, smoothness and constancy of the cross section along the flow length, and the flow inlet geometry. We will discuss this after we define the types of turbulent duct flows.

Turbulent Flow. The turbulent duct flows can also be divided into four categories: fully developed, hydrodynamically developing, thermally developing, and simultaneously developing. This division is identical to the one adopted for laminar duct flows in the preceding subsection. However, there are some important differences:

- The hydrodynamic entrance length and the thermal entrance length for turbulent duct flow are characteristically much shorter than the corresponding lengths in laminar duct flow. Typical values of L_{th}/D are 8 to 15 for air and less than 3 for liquids. Consequently, results for fully developed turbulent fluid flow and heat transfer are frequently used in design calculations without reference to the hydrodynamic and thermal entrance regions. However, caution must be exercised in using the fully developed results for the low Prandtl number liquid metals, since the entrance region effects are quite pronounced for such fluids even in turbulent duct flows ($L_{th}/D_h = 5$ to 30, depending on Re and Pr).

• The turbulent flow development is preceded by the laminar boundary layer development and through the transition region as shown in Fig. 7.3.

• Fully developed turbulent flow generally exists for Re > 10,000 in a circular pipe. For a sharp corner flow passage geometry such as a triangular passage, while the flow is turbulent in the core region for Re ≥ 10,000, it is laminar in the sharp corner region.

• The flow is fully developed turbulent for Re < 10,000 for those flow geometries in which turbulence is frequently generated within the flow passage geometry, such as in a perforated fin exchanger or a corrugated plate heat exchanger. Typically, flow is fully developed turbulent for Re ≥ 200 for plate heat exchanger geometries (see Fig. 1.16) and for Re > 100 on the shell side of a plate baffled shell-and-tube exchanger.

Thus the Reynolds number at which fully developed turbulent flow can be achieved depends on tube cross-section geometry, its variation along the flow length, the surface roughness, and the flow inlet geometry.

Periodic Flow. In the regularly interrupted heat exchanger surfaces, such as a strip fin surface, the conventional fully developed laminar or turbulent flow does not exit [i.e., the velocity gradient of Eq. (7.4) is dependent on x]. Sufficiently downstream, the velocity profile and the dimensionless temperature profile $\vartheta = (T_{w,m} - T)/(T_{w,m} - T_m)$ at a given cross section of each strip fin are identical:

$$u(x_1, y, z) = u(x_2, y, z) \qquad \vartheta(x_1, y, z) = \vartheta(x_2, y, z) \qquad (7.5)$$

where $x_2 = x_1 + \ell$, where ℓ is the length of one spatial period. This condition means that

$$\mathrm{Nu}_x(x_1, y, z) = \mathrm{Nu}_x(x_2, y, z) \qquad (7.6)$$

Usually, two to eight spatial repetitions are required to attain fully developed conditions in a periodic geometry, with the exact number depending on the geometry and the Reynolds number.

7.1.2.5   Flows with Separation, Recirculation, and Reattachment.

The purpose of the surface interruptions, such as in offset strip fins and louver fins, is to break the growth of the boundary layer. This results in thin boundary layers and enhances the heat transfer coefficient (usually accompanied by an increase in the friction factor). However, an undesirable consequence of the surface interruption is often the separation, recirculation, and reattachment features in the flow, which results in more increase in the f factor than in j or Nu. Consider, for example, the flow at the leading edge of a fin of finite thickness. The flow typically encounters such a leading edge at the heat

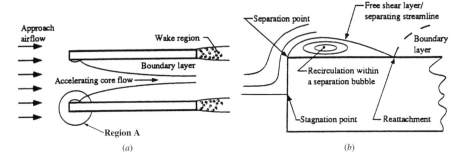

FIGURE 7.7 (*a*) Flow passing between two neighboring fins. Boundary layer separation, reattachment, and growth are shown. Free shear layer and wake regions are also identified. (*b*) Region A. (From Jacobi and Shah, 1996.)

exchanger inlet or at the start of new fins, offset strips, or louvers. For most Reynolds numbers, flow separation will occur at the leading edge because the flow cannot turn the sharp corner of the fin, as shown in Fig. 7.7. Downstream from the leading edge, the flow reattaches to the fin. The streamline emanating from the leading edge and terminating at the reattachment point is called the *separating streamline* (see Fig. 7.7). Fluid between the separating streamline and the fin surface is recirculating; this region is sometimes called a *separation bubble* or *recirculation zone*. Within the recirculation zone, relatively slow-moving fluid flows in a large eddy pattern. The boundary between the separation bubble and the separated flow (along the separation streamline) consists of a free shear layer. Free shear layers are highly unstable, and it can be expected that velocity fluctuations will develop in the free shear layer downstream from the separation point. These perturbations will be advected downstream to the reattachment region, and there they will result in increased heat transfer. The fin surface in contact with the recirculation zone is subject to lower heat transfer because of the lower fluid velocities and the thermal isolation associated with the recirculation eddy. Boundary layer separation occurs only in regions where there is an adverse pressure gradient. As increasing pressure slows the flow, there may be a point at which the velocity gradient at the surface vanishes; thus, along a curve on the fin surface is the *separation line*. Sharp corners and bends, such as those commonly found in interrupted-surface designs, are subject to separation. If the flow does not reattach to the surface from which it separated, a wake results. The separation bubble increases form drag and thus usually represents an increase in fluid pumping power with no heat transfer benefit. Therefore, as a general rule, flow separation should be avoided in surface design, where extremely low fluid pressure drop/pumping power is required.

7.1.3 Free and Forced Convection

In *convection heat transfer*, a combination of mechanisms is present. Pure conduction exists within and near the wall. Fluid particles at a cross section, receiving or rejecting heat at the wall by conduction, are being carried away in the fluid flow direction. This represents internal thermal energy transport from one cross section to another. During the motion, fluid particles will also transfer heat by conduction to neighboring particles in the direction of a negative temperature gradient. This complex phenomenon of heat conduction (molecular diffusion) at the wall and within the fluid and the movement of fluid is referred to as *convection heat transfer.* Energy transport due solely to bulk fluid motion (and having no heat conduction) is referred to as *advection*:

$$\text{convection} = \text{conduction} + \text{advection} \tag{7.7}$$

Thus, knowledge of both heat conduction and fluid mechanics is essential for convection heat transfer. If the motion of fluid arises solely due to external force fields such as gravity, centrifugal, magnetic, electrical, or Coriolis body forces, the process is referred to as *natural* or *free convection.* If the fluid motion is induced by some *external* means such as pump, fan (blower or exhauster), wind, or vehicle motion, the process is referred to as *forced convection.*

7.1.4 Basic Definitions

Now let us introduce some basic, but important definitions that are used in heat exchanger design and analysis. These are mean velocity, mean temperatures, and heat transfer coefficient.

7.1.4.1 Mean Velocity. The fluid mean axial velocity u_m is defined as the integrated average axial velocity with respect to the free-flow area A_o:

$$u_m = \frac{1}{A_o} \int_{A_o} u \, dA \tag{7.8}$$

where u is the local velocity distribution across the flow cross section. In the heat exchanger analysis (ε-NTU, MTD, etc.), not only is the flow assumed to be uniform for all flow channels but also uniform within a channel at velocity u_m. The true velocity distribution in a flow passage is considered only for evaluation of the theoretical/ analytical value of the heat transfer coefficient and friction factor for that flow passage.

7.1.4.2 Mean Temperatures. The peripheral mean wall temperature $T_{w,m}$ and the fluid bulk temperature T_m at an arbitrary duct cross section x are defined as

$$T_{w,m} = \frac{1}{\mathbf{P}} \int_{\mathbf{P}} T_w \, ds \tag{7.9}$$

$$T_m = \frac{1}{A_o u_m} \int_{A_o} u T \, dA \tag{7.10}$$

where $\mathbf{P}$ is the duct perimeter and s is the spatial coordinate at a point on the duct wall (of finite thickness) along the inside perimeter. For a duct with uniform curvature, such as a circular tube, $T_w = T_{w,m}$. However, for a noncircular tube, T_w may not be uniform but will be dependent on the boundary condition.

The fluid bulk temperature T_m in Eq. (7.10) is an enthalpy average over the flow cross section for constant ρc_p. However, if we idealize uniform flow in a cross section with $u = u_m$, the T_m in Eq. (7.10) is the integrated average value over the cross section. The fluid bulk temperature T_m is also referred to as the *mixed mean fluid temperature,* or *mixing cup, mass average,* or *flow average temperature.* Conceptually, T_m is the temperature one would measure if the duct were cut off at a section and escaping fluid were collected and thoroughly mixed in an adiabatic container.

7.1.4.3 Heat Transfer Coefficient.

The complex convection (= conduction + advection) phenomenon described earlier can be simplified by introducing a concept of the heat transfer coefficient as a proportionality factor in *Newton's law of cooling*:

$$q'' = \begin{cases} h(T_w - T_m) & \text{for internal flow or for heat exchangers} \quad (7.11) \\ h(T_w - T_\infty) & \text{for external flow or sometimes for developing} \\ & \text{flow in a duct with } T_\infty \text{ replaced by } T_e \quad (7.12) \end{cases}$$

Hence, the heat transfer coefficient h represents quantitatively the convective heat flux q'' per unit temperature difference $(T_w - T_m)$ or $(T_w - T_\infty)$ between a surface (wall) and a fluid. Thus, the complex flow and heat transfer phenomena for a heat transfer surface are all lumped into the definition of h, making it dependent on many variables or operating conditions. Some of these variables/conditions are: phase condition (single-phase, condensation, boiling/evaporation, multiphase), flow regime (laminar, transition, turbulent), flow passage geometry, fluid physical properties (i.e., the type of the fluid), flow and thermal boundary conditions, convection type (free and forced), heat transfer rate, nonuniformity of wall temperature at individual flow cross sections, viscous dissipation, and other parameters/variables, depending on the flow type. Thus, in general, the concept of h to simplify the convection phenomenon is useful only in a limited number of applications and may not provide an easy solution to a wide range of convective heat transfer problems. In external forced convection flow, q'' is often directly proportional to the temperature difference/potential $\Delta T = (T_w - T_\infty)$, so h is found to be independent of this temperature difference. Moreover, the heat transfer coefficient may be considered as nearly constant in some flow regimes. Only in these situations (linear problem), may Eqs. (7.11) and (7.12) be considered to provide a linear relationship between the driving potential for heat transfer and convective heat flux. For nonlinear problems, h may also be dependent on ΔT, such as in natural convection flows (where convection heat flux is proportional to, say, $\Delta T^{5/4}$ for laminar flow), or it may be dependent on q'' and ΔT, as in boiling. In such situations, the operational convenience of the conductance concept is diminished (since now h is dependent on ΔT and/or q'', such as $\Delta T^{1/4}$ for laminar free convection), although the *definitions* of Eqs. (7.11) and (7.12) are still valid.

Depending on the value of the temperature potential $\Delta T = (T_w - T_m)$ and the heat flux q'' at a local point on the wall, averaged over a cross section or averaged over the flow length, the magnitude of h will be local at a point, peripherally average but axially local, or averaged over the flow length. The h signifies a conductance in the thermal circuit representation of Fig. 7.4 where $q'' = q_w/A$ signifies the current and ΔT is the potential. This h is the characteristic of the heat transfer process, the "property" of a fluid–heat transfer surface thermal interaction, *not* a fluid property. It is also referred to as the *convection conductance, film coefficient, surface coefficient of heat transfer, unit conduction for thermal-convection heat transfer*, and *unit thermal convective conductance*. It is also represented by the symbol α mostly in European literature.

In those applications where the wall temperature is nonuniform at individual cross sections, Moffat (1998) and others have proposed a concept of an adiabatic heat transfer coefficient defined as

$$q'' = h_{\text{adiabatic}}(T_w - T_{\text{adiabatic}}) \tag{7.13}$$

where the adiabatic temperature is the temperature that the heat transfer surface would achieve if there were no heat transfer to or from it by radiation and conduction; thus it represents the effective average fluid temperature near the heat transfer surface. Although this concept may be quite useful for simple geometries used in electronic cooling (Moffat, 1998), where $T_{\text{adiabatic}}$ can be determined readily, it is difficult to determine in complex heat exchanger geometries having nonuniform temperature over the fin cross section.

In the convection heat transfer process using h [Eq. (7.11)], only the wall temperature and fluid bulk temperature (or the free stream temperature) are involved for the determination of convective heat flux. However, this heat flux is actually dependent on the temperature gradient at the wall. Namely, combining Eqs. (7.3) and (7.11), the heat transfer rate per unit surface area can be represented either as conduction or convection heat flux as follows (heat transfer taking place from the wall to the fluid):

$$q''_w = -k\left(\frac{\partial T}{\partial y}\right)_{y=0} = q'' = h(T_w - T_m) \tag{7.14}$$

Consequently, the heat transfer coefficient is given by

$$h = \frac{-k(\partial T/\partial y)_{y=0}}{T_w - T_m} \tag{7.15}$$

i.e., the heat transfer coefficient is, indeed, dependent on the temperature gradient at the wall. If viscous dissipation is significant, a straightforward application of Eq. (7.14) may yield a negative heat transfer coefficient. This is a consequence of an incorrectly assumed driving potential difference for heat transfer (i.e., the imposed temperature difference would not be proportional to the temperature gradient at the wall). In such cases, the heat transfer coefficient must be based on the difference between the given and the adiabatic wall temperatures (Burmeister, 1993).

7.2 DIMENSIONLESS GROUPS

Heat transfer characteristics of an exchanger surface are presented in terms of the Nusselt number, Stanton number, or Colburn factor vs. the Reynolds number, x^*, or Graetz number. Flow friction characteristics are presented in terms of the Fanning friction factor vs. Re or x^+. These and other important dimensionless groups used in heat exchanger design and internal flow forced convection are described next and are also summarized in Table 7.2 with their definitions and physical meanings.

It should be emphasized that in all dimensionless groups, wherever applicable, for *consistency* the hydraulic diameter D_h is used as a characteristic length. However, the hydraulic diameter or any other characteristic length does not represent a universal characteristics dimension. This is because three-dimensional boundary layer and wake

TABLE 7.2   Important Dimensionless Groups for Internal Flow Forced Convection Heat Transfer and Flow Friction, Useful in Heat Exchanger Design

Dimensionless Group	Definitions and Working Relationships	Physical Meaning and Comments
Reynolds number	$\mathrm{Re} = \dfrac{\rho u_m D_h}{\mu} = \dfrac{G D_h}{\mu}$	Flow modulus, proportional to the ratio of flow momentum rate ("inertia force") to viscous force
Fanning friction factor	$f = \dfrac{\tau_w}{\rho u_m^2 / 2 g_c}$ $f = \Delta p^* \dfrac{r_h}{L} = \dfrac{\Delta p}{\rho u_m^2 / 2 g_c} \dfrac{r_h}{L}$	Ratio of wall shear (skin frictional) stress to the flow kinetic energy per unit volume; commonly used in heat transfer literature
Apparent Fanning friction factor	$f_{\mathrm{app}} = \Delta p^* \dfrac{r_h}{L}$	Includes the effects of skin friction and the change in the momentum rates in the entrance region (for developing flows)
Incremental pressure drop number	$K(x) = (f_{\mathrm{app}} - f_{fd}) \dfrac{L}{r_h}$ $K(\infty) = \text{constant for } x \to \infty$	Represents the excess dimensionless pressure drop in the entrance region over that for fully developed flow
Darcy friction factor	$f_D = 4f = \Delta p^* \dfrac{D_h}{L}$	Four times the Fanning friction factor; commonly used in fluid mechanics literature
Euler number	$\mathrm{Eu} = \Delta p^* = \dfrac{\Delta p}{\rho u_m^2 / 2 g_c}$	Pressure drop normalized with respect to the dynamic velocity head; see an alternative definition for a tube bank in Eq. (7.22)
Hagen number	$\mathrm{Hg} = \dfrac{\rho g_c}{\mu^2} D_h^3 \dfrac{\Delta p}{\Delta x}$	Alternative to a friction factor to represent the dimensionless pressure drop. It does not have any velocity explicitly in its definition, so it avoids the ambiguity of velocity definitions
Dimensionless axial distance for fluid flow problem	$x^+ = \dfrac{x}{D_h \mathrm{Re}}$	Ratio of the dimensionless axial distance x/D_h to the Reynolds number; axial coordinate in the hydrodynamic entrance region
Nusselt number	$\mathrm{Nu} = \dfrac{h}{k/D_h} = \dfrac{q'' D_h}{k(T_w - T_m)}$	Ratio of the convective conductance h to pure molecular thermal conductance k/D_h over the hydraulic diameter
Stanton number	$\mathrm{St} = \dfrac{h}{G c_p} = \dfrac{\mathrm{Nu}}{\mathrm{Pe}} = \dfrac{\mathrm{Nu}}{\mathrm{Re} \cdot \mathrm{Pr}}$	Ratio of convection heat transfer (per unit duct surface area) to the enthalpy rate change of the fluid reaching the wall temperature; St does not depend on any geometric characteristic dimension
Colburn factor	$j = \mathrm{St} \cdot \mathrm{Pr}^{2/3} = \dfrac{\mathrm{Nu} \cdot \mathrm{Pr}^{-1/3}}{\mathrm{Re}}$	Modified Stanton number to take into account the moderate variations in the Prandtl number for $0.5 \leq \mathrm{Pr} \leq 10$ in turbulent flow
Prandtl number	$\mathrm{Pr} = \dfrac{\nu}{\alpha} = \dfrac{\mu c_p}{k}$	Fluid property modulus representing the ratio of momentum diffusivity to thermal diffusivity of the fluid

Péclet number	$Pe = \dfrac{\rho c_p u_m D_h}{k} = \dfrac{u_m D_h}{\alpha}$ $= Re \cdot Pr$	Proportional to the ratio of thermal energy transported to the fluid (fluid enthalpy rise) to thermal energy conducted axially within the fluid; the inverse of Pe indicates relative importance of fluid axial heat conduction
Dimensionless axial distance for heat transfer problem	$x^* = \dfrac{x}{D_h \cdot Pe} = \dfrac{x}{D_h \cdot Re \cdot Pr}$	Axial coordinate for describing the thermal entrance region heat transfer results
Graetz number	$Gz = \dfrac{\dot{m} c_p}{kL} = \dfrac{Pe \cdot P}{4L} = \dfrac{P}{4D_h}\dfrac{1}{x^*}$ $Gz = \dfrac{\pi}{4x^*}$ for a circular tube	Conventionally used in the chemical engineering literature; related to x^* as shown when the flow length in Gz is treated as a length variable
Lévêque number	$Lq = \dfrac{1}{2} x_f\, HgPr\, \dfrac{D_h}{L}$ $= x_f\, f\, Re\, \dfrac{1}{x^*}$	Used for the Nusselt number prediction in thermally developing internal flows; related to x^* as shown when the flow length in Lq is treated as a length variable; $x_f \approx 0.5$ for many interrupted surface geometries

Source: Data modified from Shah and Mueller (1988).

effects in noncircular continuous/interrupted flow passages cannot be correlated with a single length dimension D_h or some equivalent diameter. For some of the dimensionless groups of Table 7.2, a number of different definitions are used in the literature; the user should pay particular attention to the specific definitions used in any literature source before using specific results. This is particularly true for the Nusselt number (where many different temperature differences are used in the definition of h), and for f, Re, and other dimensionless groups having characteristic dimensions different from D_h.

7.2.1 Fluid Flow

7.2.1.1 Reynolds Number. The Reynolds number is defined for internal flow as

$$Re = \frac{\rho u_m D_h}{\mu} = \frac{G D_h}{\mu} = \frac{u_m D_h}{\nu} \tag{7.16}$$

It is usually interpreted as a flow characteristic proportional to the ratio of flow momentum rate (ρu_m^2) or inertia force to viscous force $(\mu u_m / D_h)$ for a specified duct geometry, where the mathematical expressions in parentheses are provided as an example for flow in a constant-cross-sectional duct. Note that the inertia force is zero for fully developed internal flow, while the momentum rate is still finite. Hence, the Reynolds number Re is a ratio of flow momentum rate to viscous force, and thus it is a flow modulus. It should be added, though, that the physical interpretation of the Reynolds number as the ratio of inertia force to viscous force usually referred to in textbooks is not necessarily correct.

This is due to the fact that in a boundary layer there is actually always a balance between inertia and friction, which can easily be demonstrated by studying the order of magnitude of respective terms in an energy equation. See Bejan (1995) for further discussion. If Re is the same for two systems that are geometrically and kinematically similar, a dynamic similarity is also realized, irrespective of the fluid.

7.2.1.2 Friction Factor and Related Groups. The ratio of wall shear stress τ_w to the flow kinetic energy per unit volume $\rho u_m^2 / 2g_c$ is defined as the Fanning friction factor:

$$f = \frac{\tau_w}{\rho u_m^2 / 2g_c} \tag{7.17}$$

If f is based strictly on the true wall shear stress, it truly represents the *skin friction*[†] component only. It is the skin friction that relates to the convective heat transfer over a surface in the Reynolds analogy discussed in Section 7.4.5. In a heat exchanger core, depending on the geometry of the heat exchanger surface, there could be form drag and internal contraction/expansion (as in a tube bank or a perforated plate core) included in the experimental value of the f factor; in that case, τ_w in Eq. (7.17) represents the effective wall shear stress.

The friction factor is strongly dependent on the flow passage geometry in laminar flow, and weakly dependent in turbulent flow. The friction factor is inversely proportional to the Reynolds number in fully developed laminar flow and is dependent on $x^+ = x/(D_h \cdot \text{Re})$ in developing laminar flow. The friction factor is proportional to Re^{-n} (where $n \approx 0.20$ to 0.25 in turbulent flow) for smooth tubes. It is dependent on the surface roughness in turbulent flow. In addition to the flow passage geometry and flow regimes, the friction factor could also be dependent on fluid physical properties (ρ and μ), phase condition (single-phase, condensation, and vaporization), and other parameters, depending on the flow type.

In a steady-state isothermal fully developed flow in a constant-cross-sectional geometry, the momentum rate at any cross section is constant. The pressure drop is then a result of the wall friction. In the absence of core entrance and exit losses, it can be shown [see Eq. (6.12)] from the application of Newton's second law of motion that

$$\Delta p^* = \frac{\Delta p}{\rho u_m^2 / 2g_c} = f \frac{L}{r_h} = f \frac{4L}{D_h} \tag{7.18}$$

where L is the duct length in which Δp occurs.

In developing flow, since the velocity profile changes at any cross section of a constant cross-sectional duct, the momentum rate changes. Both the effects of wall friction and changes of the momentum rate in developing flows are incorporated into an *apparent Fanning friction factor*, defined by

$$\Delta p^* = \frac{\Delta p}{\rho u_m^2 / 2g_c} = f_{\text{app}} \frac{L}{r_h} = f_{fd} \frac{L}{r_h} + K(x) \tag{7.19}$$

[†] Skin friction is the flow friction associated with scrubbing the surface (through or over which the fluid flows) by the fluid.

Here f_{fd} is the friction factor for fully developed flow and $K(x)$ is the *incremental pressure drop number* (Shah and London, 1978). In the case of fully developed flow in a duct, there is always an entrance region (see Fig. 7.2) in the beginning. In this case, $K(x)$ of Eq. (7.19) takes its maximum value $K(\infty)$ reported for many duct geometries by Shah and London (1978) and Shah and Bhatti (1987). However, the incremental contribution of $K(\infty)$ to the pressure drop may be negligible and can be ignored in long ducts.

In the literature, several other definitions of the friction factor are also used. Hence one should be careful to distinguish before its use. The other more common definition used in fluid mechanics literature for the friction factor is the "large" or *Darcy friction factor* f_D, also sometimes referred to as *Darcy–Weisbach friction factor*:

$$f_D = 4f \tag{7.20}$$

and Eq. (7.18) becomes

$$\Delta p^* = f_D \frac{L}{D_h} \quad \text{or} \quad \Delta p = f_D \frac{L}{D_h} \frac{\rho u_m^2}{2g_c} \tag{7.21}$$

For flow over a tube bank, the skin friction contribution to the pressure drop may not be major, and no unique flow length can be defined for Δp to be proportional to L. For such geometries, the pressure drop is presented in terms of an average *Euler number* Eu per tube row instead of the friction factor f:

$$\text{Eu} = \frac{\Delta p}{\rho u_m^2/2g_c} \frac{1}{N_r} = \frac{\Delta p^*}{N_r} \tag{7.22}$$

For other external flow geometries, the Euler number is reported in the literature for a complete tube bank by eliminating N_r from Eq. (7.22) or per tube row as in Eq. (7.22), or some other definitions, and hence care should be exercised to note the specific definition of Eu used.

An alternative way of representing the driving force $(\Delta p/\Delta x)$ of internal fluid flow in a dimensionless form is in terms of the *Hagen number* Hg, defined by the first equality in the following equation (Martin, 2002). The second equality relates the Hagen number to the Fanning friction factor and the Euler number.

$$\text{Hg} = \begin{cases} \dfrac{\rho g_c}{\mu^2} D_h^3 \dfrac{\Delta p}{\Delta x} = 2f \cdot \text{Re}^2 & \text{for internal flow} \tag{7.23a} \\[3mm] \dfrac{\rho g_c}{\mu^2} d_o^2 \dfrac{\Delta p}{N_r} = \tfrac{1}{2}\text{Eu} \cdot \text{Re}_d^2 & \text{for flow normal to a tube bundle} \tag{7.23b} \end{cases}$$

where $\text{Re}_d = \rho u_m d_o/\mu$. Thus the Hagen number is an alternative to a friction factor or an Euler number to represent the dimensionless pressure drop, and it is an average value per tube row for flow normal to a tube bundle. Note that the Hagen number does not have any velocity explicitly in its definition, which may be advantageous in flow normal to the tube bank and other external flow geometries having some ambiguity in defining the maximum velocity, as needed in the Fanning friction factor definition. Hence when relating to the Hagen number and friction factor or Euler number of Eq. (7.23), any reference velocity may be used as long it is the same in both the f or Eu and Re number definitions.

For Hagen–Poiseuille flow (fully developed laminar flow in a circular tube), the Fanning friction factor, Darcy–Weisbach friction factor, and Hagen number definitions relate the skin friction and Reynolds number as

$$f \cdot \mathrm{Re} = 16 \qquad f_D \cdot \mathrm{Re} = 64 \qquad \mathrm{Hg} = 32\mathrm{Re} \qquad (7.24)$$

7.2.1.3 Dimensionless Axial Distance. The dimensionless axial distance in the flow direction for the hydrodynamic entrance region is defined as

$$x^+ = \frac{x}{D_h \cdot \mathrm{Re}} \qquad (7.25)$$

The apparent friction factor f_{app} decreases with increasing value of x^+ and asymptotically approaches the fully developed value f as $x^+ \to \infty$.

7.2.2 Heat Transfer

7.2.2.1 Nusselt Number. The Nusselt number is one of the dimensionless representations of the heat transfer coefficient. It is defined for an internal flow as the ratio of the convective conductance h to the pure molecular thermal conductance k/D_h:

$$\mathrm{Nu} = \frac{h}{k/D_h} = \frac{hD_h}{k} = \frac{q'' D_h}{k(T_w - T_m)} \qquad (7.26)$$

The Nusselt number has a physical significance in the sense that the heat transfer coefficient h in Nu represents the convective conductance in a thermal circuit representation (Fig. 7.4) with the heat flux q'' as the current and $(T_w - T_m)$ as the potential. Alternatively, the Nusselt number may be interpreted as the ratio of convection heat transfer to conduction heat transfer.

For external flow or in the thermal entrance region, the Nusselt number is defined by either of the first two equalities in the following equation:

$$\mathrm{Nu} = \frac{hL}{k} = \frac{hD_h}{k} = -\left. \frac{\partial \vartheta^*}{\partial y^*} \right|_{y^*=0} \qquad (7.27)$$

where $\vartheta^* = (T - T_w)/(T_\infty - T_w)$ and $y^* = y/L$ or $\vartheta^* = (T - T_w)/(T_m - T_w)$ and $y^* = y/D_h$. The term after the last equality in Eq. (7.27) is based on Eq. (7.15). Thus, Nu can be interpreted in this case as the dimensionless temperature gradient at the surface.

The Nusselt number is strongly dependent on the thermal boundary condition and flow passage geometry in laminar flow and weakly dependent on these parameters in turbulent flow. The Nusselt number is constant for thermally and hydrodynamically fully developed laminar flow. It is dependent on $x^* = x/(D_h \cdot \mathrm{Pe})$ for developing laminar temperature profiles and on x^* and Pr for developing laminar velocity and temperature profiles. The Nusselt number is dependent on Re and Pr for fully developed turbulent flows. In addition to thermal boundary conditions, flow passage geometry, and flow regimes, the Nusselt number could also be dependent on the phase condition (single-phase, condensation, and vaporization); fluid physical properties, including Pr; and other variables/parameters, depending on the flow type and convection type (free and forced).

7.2.2.2 Stanton Number. The Stanton number is another dimensionless representation of the heat transfer coefficient. It is defined as

$$St = \frac{h}{Gc_p} = \frac{h}{\rho u_m c_p} \tag{7.28}$$

By multiplying both the numerator and denominator by $(T_w - T_m)$, the physical meaning of the Stanton number becomes apparent: namely, St is the ratio of convected heat transfer (per unit duct surface area) to the enthalpy rate change of the fluid reaching the wall temperature (per unit of flow cross-sectional area).

Single-phase heat transfer from the wall to the fluid (or vice versa) is related to its enthalpy rate change:

$$hA(T_w - T_m) = A_o Gc_p(T_o - T_i) = Gc_p A_o \, \Delta T \tag{7.29}$$

When we introduce $T_w - T_m = \Delta T_m$, Eq. (7.29) reduces to

$$\frac{h}{Gc_p} = \frac{A_o \, \Delta T}{A \, \Delta T_m} \tag{7.30}$$

Hence, the Stanton number can also be interpreted as being proportional to the temperature change in the fluid divided by the convective heat transfer driving potential.

When axial fluid heat conduction is negligible (large Pe), St is frequently preferred to Nu as a dimensionless modulus for the convective heat transfer correlation. This is because it relates more directly to the designer's task of establishing the number of exchanger transfer units NTU. Moreover, the behavior of St with Re parallels that of the Fanning friction factor f vs. Re [see Fig. 7.8, in terms of modified St or the $j \, (= St \cdot Pr^{2/3})$ factor and f factor vs. Reynolds number]. For turbulent flow, since $h \propto u_m^{0.2}$ or $Re^{0.2}$, it is not a strong function of Re; it is nearly constant.

The Stanton number is also directly related to the number of heat transfer units on one fluid side of the exchanger, as follows:

$$St = \frac{h}{Gc_p} = \frac{hA}{\dot{m}c_p} \frac{A_o}{A} = \text{ntu} \frac{r_h}{L} = \text{ntu} \frac{D_h}{4L} \tag{7.31}$$

The Nusselt number is related to the Stanton, Prandtl, and Reynolds numbers as follows by *definition*:

$$Nu = St \cdot Re \cdot Pr \tag{7.32}$$

Hence, irrespective of the flow passage geometry, boundary condition, flow types, and so on, Eq. (7.32) is *always* valid.

7.2.2.3 Colburn Factor. The Colburn factor is a modified Stanton number to take into account the moderate variations in the fluid Prandtl number (representing different fluids). It is defined as

$$j = St \cdot Pr^{2/3} = \frac{Nu \cdot Pr^{-1/3}}{Re} \tag{7.33}$$

As the Stanton number is dependent on the fluid Prandtl number, the Colburn factor j is nearly independent of the flowing fluid for $0.5 \leq \text{Pr} \leq 10$ from laminar to turbulent flow conditions. Thus the j vs. Re data obtained for a given heat exchanger surface for air can be used for water under certain flow conditions, as discussed in Section 7.4.6. Note that using Eqs. (7.31) and (7.33), it can be shown that j is related to ntu as follows:

$$j = \text{ntu} \, \frac{r_h}{L} \, \text{Pr}^{2/3} = \text{ntu} \, \frac{D_h}{4L} \, \text{Pr}^{2/3} \tag{7.34}$$

7.2.2.4 Prandtl Number. The Prandtl number is defined as the ratio of momentum diffusivity to the thermal diffusivity of the fluid:

$$\text{Pr} = \frac{\nu}{\alpha} = \frac{\mu c_p}{k} \tag{7.35}$$

The Prandtl number is solely a fluid property modulus. Its range for several fluids is as follows: 0.001 to 0.03 for liquid metals, 0.2 to 1 for gases, 1 to 13 for water, 5 to 50 for light organic liquids, 50 to 10^5 for oils, and 2000 to 10^5 for glycerin.

7.2.2.5 Péclet Number. The Péclet number is defined as

$$\text{Pe} = \frac{\rho c_p u_m D_h}{k} = \frac{u_m D_h}{\alpha} \tag{7.36}$$

On multiplying the numerator and denominator of the first equality of Eq. (7.36) by the axial fluid bulk temperature gradient (dT_m/dx), it can be shown that

$$\text{Pe} = D_h \, \frac{\dot{m}c_p(dT_m/dx)}{kA_o(dT_m/dx)} \tag{7.37}$$

Thus the Péclet number represents the relative magnitude of the thermal energy transported to the fluid (fluid enthalpy change) to the thermal energy axially conducted within the fluid. The inverse of the Péclet number is representative of the relative importance of fluid axial heat conduction. Therefore, Pe is important for liquid metal heat exchangers because of its low Pr values ($\text{Pr} \leq 0.03$). Longitudinal heat conduction within the fluid for all other fluids is negligible for $\text{Pe} > 10$ and $x^* > 0.005$ (Shah and London, 1978). From Eqs. (7.16) and (7.35), Eq. (7.36) becomes

$$\text{Pe} = \text{Re} \cdot \text{Pr} \tag{7.38}$$

7.2.2.6 Dimensionless Axial Distance, Graetz Number, and Lévêque Number. The dimensionless distance x^* in the flow direction for heat transfer in the thermal entrance region and the Graetz number Gz are defined as

$$x^* = \frac{x}{D_h \cdot \text{Pe}} = \frac{x}{D_h \cdot \text{Re} \cdot \text{Pr}} \qquad \text{Gz} = \frac{\dot{m}c_p}{kL} = \frac{\text{Pe} \cdot \mathbf{P}}{4L} = \frac{\text{Re} \cdot \text{Pr} \cdot \mathbf{P}}{4L} \tag{7.39}$$

This definition of the Graetz number is used conventionally in the chemical engineering literature. If the flow length L in Gz is treated as a length variable, x^* is related to Gz as

$$x^* = \frac{P}{4D_h} \frac{1}{Gz} \tag{7.40}$$

which reduces to $x^* = \pi/4Gz$ for the circular tube. Thus, the inverse of the Graetz number is proportional to the dimensionless axial distance for thermal entrance region effects in laminar flow.

Lévêque (1928) obtained a developing thermal boundary layer solution for the fully developed laminar velocity profile in a circular tube of length L for the mean Nusselt number for the ⓉT boundary condition as $Nu = 1.615(x^*)^{-1/3} = 1.615(Re \cdot Pr \cdot D_h/L)^{1/3}$. Since $f \cdot Re = 16$ for a circular tube, this Lévêque solution can be generalized for other flow passage approximately as follows by multiplying $f \cdot Re/16$ by the term in parentheses to get $Nu = 0.641(f \cdot Re^2 \cdot Pr \cdot D_h/L)^{1/3} = 0.404(f_D \cdot Re^2 \cdot Pr \cdot D_h/L)^{1/3} = 0.404(2Hg \cdot Pr \cdot D_h/L)^{1/3}$. Since many of the heat exchanger surfaces have *developing* velocity and temperature profiles, Martin (2002) proposed this generalized Nusselt number expression for chevron plates, tube banks, cross rod matrices, and packed beds in laminar and turbulent flow regimes using only the contribution of the skin friction factor for f or f_D in the equation above. Martin refers to this generalized Nusselt number expression as the *generalized Lévêque equation* and the parenthetic group as the *Lévêque number*, Lq, which is slightly modified as follows:

$$Lq = \frac{1}{2}x_f \cdot Hg \cdot Pr \frac{D_h}{L} = x_f f \cdot Re^2 \cdot Pr \frac{D_h}{L} \tag{7.41}$$

where x_f is the fraction of the total or apparent friction factor that corresponds to the skin friction. As noted near the end of Section 7.4.5, $j/f = 0.25$; hence, $x_f \approx 0.5$ for many interrupted flow geometries.

7.2.3 Dimensionless Surface Characteristics as a Function of the Reynolds Number

Since the majority of basic data for heat exchanger (particularly compact heat exchanger) surfaces are obtained experimentally (because computational fluid dynamics techniques and modeling cannot analyze accurately three-dimensional real surfaces at present), the dimensionless heat transfer and pressure drop characteristics of these surfaces are presented in terms of j and f vs. Re. As an example, basic heat transfer and flow friction characteristics for airflow in a long circular tube are presented in Fig. 7.8. This figure shows three flow regimes: laminar, transition, and turbulent. This is characteristic of fully developed flow in continuous flow passage geometries such as a long circular tube, triangular tube, or other noncircular cross-sectional ducts. Generally, the compact interrupted surfaces do not have a sharp dip in the transition region (Re $\sim$ 1500 to 10,000), as shown for the circular tube. Notice that there is a parallel behavior of j vs. Re and f vs. Re curves, but no such parallelism exists between Nu and f vs. Re curves. The parallel behavior of j and f vs. Re is useful for (1) identifying erroneous test data for some specific surfaces for which a parallel behavior is expected but indicated otherwise by test results (see Fig. 7.11 and related discussion); (2) identifying specific flow phenomena in which the friction behavior is different from the heat transfer behavior (such as rough surface flow for friction and smooth surface flow for heat transfer for highly interrupted fin

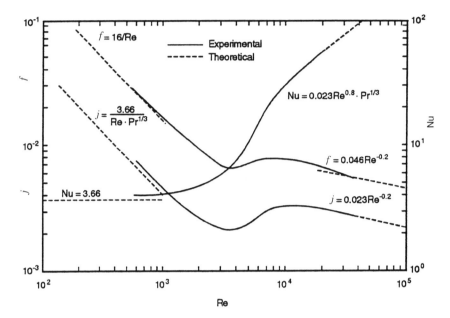

FIGURE 7.8   Basic heat transfer and flow friction characteristics for airflow through a long circular tube. (From Shah, 1981.)

geometries in turbulent flow); and (3) predicting the f factors for an interrupted surface (that has considerable form drag contribution) when the j factors are known by some predictive method. It should be remembered that j vs. Re can readily be converted to Nu vs. Re curves, or vice versa, for given fluids because $j = \text{Nu} \cdot \text{Pr}^{-1/3}/\text{Re}$ by definition. Because the values of j, f, and Re are dimensionless, the test data are applicable to surfaces of any hydraulic diameter, provided that a complete geometric similarity is maintained.

7.3   EXPERIMENTAL TECHNIQUES FOR DETERMINING SURFACE CHARACTERISTICS

The surface basic characteristics of most heat transfer surfaces are determined experimentally. Steady-state techniques are one of the most common test techniques used to establish the j vs. Re characteristics of a recuperator (nonregenerator) surface. Different data acquisition and reduction methods are used, depending on whether the test fluid is primarily a gas (air) or a liquid. Kays and London (1950, 1998) developed the steady-state test technique for gases (also used for oils), where the test fluid has the *controlling* thermal resistance. The Wilson plot technique (Wilson, 1915) and its modifications (Shah, 1990) are usually employed to measure surface heat transfer characteristics for test fluids as liquids (such as water), where the test fluid has *noncontrolling* thermal resistance. The transient test technique is usually employed to establish the Colburn factor versus Reynolds number characteristics of a matrix type or a high-ntu surface. The transient technique may vary the fluid inlet temperature by means of a step change, periodic, or arbitrary rise/drop. The determination of the Fanning friction factor is made under steady fluid flow rates with or without heat transfer, regardless of the core construction and the method of heat transfer testing. A detailed review of thermal performance test methods for industrial heat exchangers is provided by Lestina and Bell (2001).

7.3.1   Steady-State Kays and London Technique

Generally, a crossflow heat exchanger is employed as a test section. On one fluid side (the "known side"), a surface for which the j vs. Re characteristic is known is employed; a fluid with a high heat capacity rate flows on this side. On the other fluid side (the "unknown side") of the exchanger, a surface is employed for which the j vs. Re characteristic is to be determined. The fluid that flows over this unknown-side surface is preferably the one that is used in a particular application of the unknown-side surface. Often, air is used on the unknown side; while steam, hot water, chilled water, or oils (resulting in a high value of hA) are used on the known side. In the following subsections, we describe the test setup, experimental method, data reduction method, and test core design. This method could become quite inaccurate, due to inaccuracies in temperature measurements, and hence it is generally not used for high (> 3) or low (< 0.5) core NTU.

7.3.1.1   Test Setup and Experimental Procedure. In general, the test setup consists of the following basic elements on the unknown side: (1) a test section; (2) a fluid metering device, such as a nozzle, orifice, or rotameter; (3) a fluid pumping device, such as a fan, pump or high-pressure fluid supply; (4) temperature measurement devices, such as thermocouples or resistance thermometers; and (5) pressure measurement devices, such a manometers or pressure transducers. Similar devices are also used on the known side. As an example, Fig. 7.9 shows the air-side schematic of the test rig used by Kays and London (1950) at Stanford University.

In the experiments, the flow rates on both fluid sides of the exchanger are set at constant predetermined values. Once the steady-state conditions are achieved, fluid temperatures upstream and downstream of the test section on both fluid sides are measured, as well as all pertinent measurements for the determination of fluid flow rates. The upstream pressure and pressure drop across the core on the unknown side are also recorded to determine the hot friction factors.[†] The tests are repeated with different

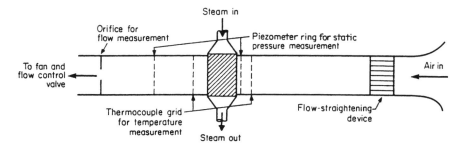

FIGURE 7.9   Schematic of a steam-to-air steady-state heat transfer test rig. (From Shah, 1985.)

[†] The friction factor determined from the Δp measurement taken during heat transfer testing is referred to as the *hot friction factor*.

flow rates on the unknown side to cover the desired range of Reynolds number for the j and f vs. Re characteristics. To assure high-accuracy data, the difference between the enthalpy rate drop of the hot fluid (e.g., steam or water) and the enthalpy rate rise of the cold fluid (e.g., air) should be maintained below $\pm 3\%$; also, the heat capacity rate ratio $C^* = 0$ for condensing steam on one fluid side; C^* should be maintained below 0.2 if a liquid is used on the $C_{\max}$ side to maintain the $C_{\min}$ side as the controlling thermal resistance side.

7.3.1.2   *Theoretical Analysis and Test Data Reduction Methods.*   The determination of the airside film coefficient h, Colburn factor j, and Reynolds number Re is now shown for condensing steam-to-air tests as an illustration. If water or other fluid is used on the known side, the following calculations need to be modified, as discussed below.

The fluid temperature distributions in the core on both fluid sides are shown in Fig. 7.10a. Generally, a slightly superheated steam is entered at the core inlet, to ensure dry steam. However, the limited influence of the degree of superheat on the data reduction is neglected. The steam-side temperature is taken as uniform, corresponding to the average core saturation pressure.

An energy balance equation for a control volume between x and $x + dx$ of an air channel in the heat exchanger core (Fig. 7.10b), can be written as follows:

$$C_a T_{a,x} + U\mathbf{P}\,dx(T_s - T_{a,x}) - C_a\left(T_{a,x} + \frac{dT_{a,x}}{dx}\,dx\right) = 0 \qquad (7.42)$$

Therefore,

$$U\mathbf{P}\,dx(T_s - T_{a,x}) = C_a\,dT_{a,x} \Rightarrow \frac{dT_{a,x}}{T_s - T_{a,x}} = \frac{U\,dA}{C_a} \qquad (7.43)$$

where $C_a = (\dot{m}c_p)_a$ is the heat capacity rate for air (cold fluid), $\mathbf{P}\,dx = dA$, and subscripts a and s denote air and steam, respectively. Idealizing that the local overall heat transfer coefficient U is uniform throughout the channel (i.e., $U = U_m$) and integrating Eq. (7.43), we get

$$\frac{T_s - T_{a,o}}{T_s - T_{a,i}} = e^{-\mathrm{NTU}} = 1 - \varepsilon \qquad (7.44)$$

where $\mathrm{NTU} = UA/C_a$. The second equality on the right-hand side of Eq. (7.44) denotes the exchanger ineffectiveness $(1 - \varepsilon)$, which is obtained directly from the definition of

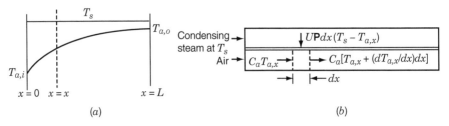

(a) (b)

FIGURE 7.10   (a) Fluid temperature distributions in the test core for steam-to-air tests; (b) control volume for energy balance.

$\varepsilon[= (T_{a,o} - T_{a,i})/(T_s - T_{a,i})]$ or from Eq. (3.84) for $C^* = C_{\min}/C_{\max} = 0$, since condensing steam is used on one fluid side. Subsequently, NTU is computed from Eq. (7.44).

If hot water, chilled water, or some oil is used on the known side instead of steam, the crossflow test core would have a finite value of C^* ($C^* \le 0.2$). In that case, the hot fluid temperature in Fig. 7.10a will not be constant and integration of the energy balance equation(s) would lead to the appropriate ε-NTU relationship, different from Eq. (7.44). These relationships are given in Table 3.6, such as Eq. (II.1) for an unmixed–unmixed crossflow exchanger, or Eq. (II.2) or (II.3) for mixed–unmixed crossflow exchanger. In such a case, NTU must be obtained from the corresponding ε-NTU relationship iteratively (often) or directly (rarely), depending on the pertinent expression. When $C^* \ne 0$, the heat capacity rate is determined from the measured mass flow rates on each fluid side and the specific heats of the fluids at their average temperatures. On the known side, the fluid properties are evaluated at the arithmetic average temperature T_s. On the unknown side, the fluid properties (c_p, μ, Pr, ρ) are evaluated at the log-mean average temperature:

$$T_{a,\mathrm{lm}} = T_s - \Delta T_{\mathrm{lm}} \tag{7.45}$$

where

$$\Delta T_{\mathrm{lm}} = \frac{(T_s - T_{a,i}) - (T_s - T_{a,o})}{\ln[(T_s - T_{a,i})/(T_s - T_{a,o})]} \tag{7.46}$$

The overall heat transfer coefficient U_a based on the total *airside* surface area A_a is then evaluated from NTU as $U_a = \mathrm{NTU} \cdot C_a/A_a$. Now, the reciprocal of U_a, an overall thermal resistance, is considered as having three components in series: (1) air-side thermal resistance, including the extended surface efficiency on the air side; (2) wall thermal resistance; and (3) steam-side thermal resistance, including the extended surface efficiency on the steam side:

$$\frac{1}{U_a} = \frac{1}{\eta_{o,a}h_a} + \frac{A_a}{A_w}\frac{\delta_w}{k_w} + \frac{A_a}{\eta_{o,s}A_s h_s} \tag{7.47}$$

Then

$$h_a = \frac{1}{\eta_{o,a}}\left(\frac{1}{U_a} - \frac{A_a\delta_w}{A_w k_w} - \frac{A_a}{\eta_{o,s}A_s h_s}\right)^{-1} \tag{7.48}$$

The test cores are generally new, and no fouling or scale resistance is on either side, so the corresponding resistance is not included in Eq. (7.47). The wall thermal resistance is constant and is evaluated separately for each test core. The steam side (or liquid side, if liquid is employed) heat transfer coefficient h_s must also be evaluated separately for each core and should be known a priori. However, the steam-side (or liquid-side) resistance is generally a very small percentage of the total resistance, and a reasonable estimate will suffice, even though considerable uncertainty may be involved in its determination.

The term $\eta_{o,a}$ in Eq. (7.48) is the extended surface efficiency (see Section 4.3.4) of the air-side surface and is related to the fin efficiency η_f of the extended surface as follows [see Eq. (4.160)]:

$$\eta_{o,a} = 1 - \frac{A_f}{A}(1 - \eta_f) \tag{7.49}$$

where the fin efficiency η_f for various geometries has been presented in Table 4.5. For many plate-fin surfaces, the relation for the straight fin with constant conduction cross section may be used to a good approximation [see Eq. (4.146)]. In that case,

$$\eta_f = \frac{\tanh m\ell}{m\ell} \tag{7.50}$$

where $m^2 = 2h/k_f\delta$ and ℓ is the fin length from the root to the center of the fin (see Fig. 4.15 as an example). Once the surface area and the geometry are known for the extended surface, h and η_o are computed iteratively from Eqs. (7.48)–(7.50). For example, calculate the value of h_a from Eq. (7.48) for assumed $\eta_{o,a} = 1$, and subsequently determine η_f from Eq. (7.50), $\eta_{o,a}$ from Eq. (7.49), and h_a from Eq. (7.48). With this new value of h_a, compute η_f from Eq. (7.50) and $\eta_{o,a}$ from Eq. (7.49) and subsequently the next value of h_a from Eq. (7.48). Continue such iterations until the desired degree of convergence is achieved. Alternatively, a graph, tabular values, or a curve fit to $\eta_o h$ vs. h is determined for the given extended surface using Eqs. (7.50) and (7.49), and h_a is determined from it once $\eta_{o,a} h_a$ is found from Eq. (7.48) or (7.47).

The Stanton number St and the Colburn factor j are then evaluated from their definitions, knowing the heat transfer coefficient h, the core mass velocity G, the air–water vapor mixture specific heat c_p, and the Prandtl number Pr as follows:

$$\mathrm{St} = \frac{h}{Gc_p} \qquad j = \mathrm{St} \cdot \mathrm{Pr}^{2/3} \tag{7.51}$$

The Reynolds number on the unknown side for the test point is determined from its definition:

$$\mathrm{Re} = \frac{GD_h}{\mu} \tag{7.52}$$

where $G = \dot{m}/A_o$ is the core mass velocity, $D_h = 4A_o/\mathbf{P}$ is the hydraulic diameter, and μ is the dynamic viscosity for the unknown-side fluid evaluated at the temperature given by Eq. (7.45).

Example 7.1   The objective of this example is to determine Colburn factor j and Reynolds number Re for the air-side surface tested by a steady-state technique.[†] The test unit is a crossflow heat exchanger with steam condensing on the known side and air flowing on the unknown side. The air-side surface has the offset strip fin surface shown in Fig. E7.1.

Geometrical properties on the air side of this core for the data reduction are:

$$D_h = 0.00121 \text{ m} \qquad \sigma = 0.3067 \ (K_c = 0.37, \ K_e = 0.48)$$

$$\frac{r_h}{L} = 0.005688 \qquad \delta = 0.102 \text{ mm}$$

$$A_{f,a} = 1.799 \text{ m}^2 \qquad \ell_s = 2.819 \text{ mm}$$

$$A_a = 2.830 \text{ m}^2 \qquad \ell = 0.851 \text{ mm} \ \left(\ell = \frac{b}{2} - \delta\right)$$

[†] The data for this problem are for a test point of core 105 tested by London and Shah (1968).

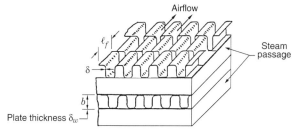

FIGURE E7.1

$A_{o,a} = 0.0161 \text{ m}^2$ $b = 1.905 \text{ mm}$

$\delta_w = 0.406 \text{ mm}$ $\dfrac{A_a \delta_w}{A_w k_w} = 5.68 \times 10^{-5} \text{ m}^2 \cdot \text{K/W}$

The recorded data for a test point are:[†]

Airflow rate $\dot{m}_a = 0.274 \text{ kg/s}$ Air inlet temperature $T_{a,o} = 23.05°\text{C}$

Air inlet pressure $p_i = 101.60 \text{ kPa}$ Air outlet temperature $T_{a,o} = 101.60°\text{C}$

Air-side pressure drop $\Delta p = 1.493 \text{ kPa}$ Condensing steam temperature

$\qquad\qquad\qquad\qquad\qquad\qquad\qquad\quad T_s = 107.87°\text{C}$

Show that the log-mean temperature on the air side is 77.71°C. At this temperature, air properties are $c_p = 1.0166 \text{ kJ/kg} \cdot \text{K}$ for humid air, Pr = 0.697, and $\mu = 2.0822 \times 10^{-5} \text{ Pa} \cdot \text{s}$. Determine Re and j for this test point. Consider the steam-side thermal resistance as zero. Fins are made from aluminum for which $k_f = 192.1 \text{ W/m} \cdot \text{K}$.

SOLUTION

Problem Data and Schematic: The test core geometry information, test point inlet pressure, air-side pressure drop, and temperatures are given for the crossflow test unit with air as the test fluid and steam as the known fluid on the other side. A small section of the test core is shown in Fig. E7.1.

Determine: The Reynolds number and j factor for this test point.

Assumptions: Constant fluid properties apply, the steam-side thermal resistance is zero and there is no fouling on either fluid side.

Analysis: First, we calculate the log-mean average temperature of the air using Eqs. (7.46) and (7.45):

$$\Delta T_{\text{lm}} = \frac{(T_s - T_{a,i}) - (T_s - T_{a,o})}{\ln[(T_s - T_{a,i})/(T_s - T_{a,o})]}$$

$$= \frac{(107.87 - 23.05)°\text{C} - (107.87 - 101.60)°\text{C}}{\ln[(107.87 - 23.05)°\text{C}/(107.87 - 101.60)°\text{C}]} = 30.16°\text{C}$$

[†] The air temperatures at the inlet and outlet of the test core represent an average of 9 and 27 thermocouple readings, respectively; the thermocouple wire spool was calibrated to ±0.05°F accuracy, traceable to the United States National Bureau of Standards.

Hence,

$$T_{a,\text{lm}} = T_s - \Delta T_{\text{lm}} = (107.87 - 30.16)^\circ\text{C} = 77.71^\circ\text{C}$$

Let us calculate the air-side Reynolds number from its definition after calculating the mass velocity G:

$$G = \frac{\dot{m}_a}{A_{o,a}} = \frac{0.274\,\text{kg/s}}{0.0161\,\text{m}^2} = 17.019\,\text{kg/m}^2 \cdot \text{s}$$

$$\text{Re} = \frac{GD_h}{\mu} = \frac{17.019\,\text{kg/m}^2 \cdot \text{s} \times 0.00121\,\text{m}}{2.0822 \times 10^{-5}\,\text{Pa} \cdot \text{s}} = 989$$

To determine the j factor, first compute ε. Since the air side is the C_{min} side,

$$\varepsilon = \frac{T_{a,o} - T_{a,i}}{T_s - T_{a,i}} = \frac{(101.60 - 23.05)^\circ\text{C}}{(107.87 - 23.05)^\circ\text{C}} = 0.9261 \quad \text{therefore,} \quad 1 - \varepsilon = 0.0739$$

From Eq. (7.44), we have

$$\text{NTU} = -\ln(1 - \varepsilon) = -\ln(0.0739) = 2.605$$

Therefore,

$$U_a = \frac{\text{NTU} \cdot C_{\text{min}}}{A_a} = \frac{(\dot{m}c_p)_a}{A_a} = \frac{2.605 \times (0.274\,\text{kg/s} \times 1.0166 \times 10^{-3}\,\text{kJ/kg} \cdot \text{K})}{2.830\,\text{m}^2}$$

$$= 256.38\,\text{W/m}^2 \cdot \text{K}$$

Since the steam-side thermal resistance is zero, h on the air side from Eq. (7.48) is then

$$h_a = \frac{1}{\eta_{o,a}(1/U_a - A_a\delta_w/A_w k_w)} \tag{7.53}$$

Here we need to know $\eta_{o,a}$ to determine h_a. The $\eta_{o,a}$ is calculated from Eq. (7.49) for known η_f, which in turn is dependent on h. Thus, we need to calculate h_a iteratively. Let us assume that $\eta_{o,a} = 1.00$; then, from Eq. (7.53),

$$h_a = \frac{1}{1 \times [1/(256.38\,\text{W/m}^2 \cdot \text{K}) - 5.68 \times 10^{-5}\,\text{m}^2 \cdot \text{K/W}]} = 260.16\,\text{W/m}^2 \cdot \text{K} \tag{7.54}$$

The fin efficiency of the offset strip fin, from Eq. (4.146) with m from Eq. (4.147), is

$$m = \left[\frac{2h_a}{k_f\delta}\left(1 + \frac{\delta}{\ell_s}\right)\right]^{1/2} = \left[\frac{2 \times 260.16\,\text{W/m}^2 \cdot \text{K}}{192.1\,\text{W/mK} \times 0.102 \times 10^{-3}\,\text{m}}\left(1 + \frac{0.102\,\text{mm}}{2.819\,\text{mm}}\right)\right]^{0.5}$$

$$= 165.881\,\text{m}^{-1}$$

$$m\ell = 165.88\,\text{m}^{-1} \times 0.851 \times 10^{-3}\,\text{m} = 0.1412$$

$$\eta_f = \frac{\tanh m\ell}{m\ell} = \frac{\tanh(0.1412)}{0.1412} = 0.993$$

$$\eta_{o,a} = 1 - \frac{(1 - \eta_f)\,A_{f,a}}{A_a} = 1 - \frac{(1 - 0.993) \times 1.799\,\text{m}^2}{2.830\,\text{m}^2} = 0.996$$

Hence, the refined value of h_a using Eq. (7.53) with $\eta_{o,a} = 0.996$ and the rest of the terms from Eq. (7.54) is

$$h_a = \frac{260.16\,\text{W}/\text{m}^2 \cdot \text{K}}{\eta_{o,a}} = \frac{260.16\,\text{W}/\text{m}^2 \cdot \text{K}}{0.996} = 261.21\,\text{W}/\text{m}^2 \cdot \text{K}$$

With this value of h_a, the new values of η_f and $\eta_{o,a}$ are found to be

$$\eta_f = 0.993 \qquad \eta_{o,a} = 0.996 \qquad\qquad\qquad Ans.$$

Thus, the values of $\eta_{o,a}$ and h_a are converged in two iterations for this problem. The Colburn factor is finally calculated from its definition:

$$j = \text{St} \cdot \text{Pr}^{2/3} = \frac{h}{Gc_p}\text{Pr}^{2/3} = \frac{261.21\,\text{W}/\text{m}^2 \cdot \text{K}}{17.019\,\text{kg}/\text{m}^2 \cdot \text{s} \times 1.0166 \times 10^3\,\text{J}/\text{kg} \cdot \text{K}}(0.697)^{2/3}$$

$$= 0.01187 \qquad\qquad\qquad Ans.$$

Discussion and Comments: This example demonstrates how to obtain the j factor for a heat transfer surface when its thermal resistance is dominant in a two-fluid heat exchanger. The methodology is straightforward. If hot or chilled water had been used on the steam side, we would need to make a few changes in the procedure. Instead of using Eq. (7.44) for the ε-NTU relationship for $C^* = 0$, we would have used the appropriate ε-NTU formula from Table 3.6 [such as Eq. (II.1)]. In that case, NTU being implicit in the formula, it would have been computed iteratively for known ε and C^*. The water-side thermal resistance would have been finite and Eq. (7.48) should have been used for the airside h. If the test core were not a new core, fouling on the water side should have been included.

7.3.1.3 Test Core Design.

The test core is designed with two basic considerations in mind to reduce the experimental uncertainty in the j factors: (1) the appropriate magnitudes of thermal resistances on each fluid side as well as of the wall, and (2) the proper range of NTU.

The thermal resistances in a heat exchanger are related by Eq. (7.47) by multiplying $1/A_a$ on both sides. To reduce the uncertainty in the determination of the thermal resistance of the unknown side (with known overall thermal resistance, $1/UA$), the thermal resistances of the exchanger wall and the known side should be kept as small as possible by design. The wall thermal resistance value is usually negligible when one of the fluids in the exchanger is air. This may be minimized further through the use of a thin material with high thermal conductivity. On the known side, the thermal resistance is minimized by the use of a liquid (hot or cold water) at high flow rates, or a condensing steam, to achieve a high h, and also by using an extended surface. Therefore, the thermal boundary condition achieved during steady-state testing is generally a close approach to a uniform wall temperature condition.

The NTU range for testing is generally restricted between 0.5 and 3 or between 40 and 90% in terms of the exchanger effectiveness. To understand this restriction and point out the precise problem areas, consider the test fluid on the unknown side to be cold air being heated in the test section and the fluid on the known side to be hot water (steam replaced by hot water and its flow direction reversed in Fig. 7.9 to avoid air bubbles). The high

NTU occurs at low airflow values for a given test core. Both temperature and mass flow rate measurements become more inaccurate at low airflows, and the resultant heat unbalances $(q_w - q_a)/q_a$ increase sharply at low airflows with decreasing air mass flow rates. In this section, the subscripts w and a denote water and air sides, respectively. Now, the exchanger effectiveness can be computed in two different ways:

$$\varepsilon = \frac{q_a}{C_a(T_{w,i} - T_{a,i})} \quad \text{or} \quad \varepsilon = \frac{q_w}{C_a(T_{w,i} - T_{a,i})} \tag{7.55}$$

In an ideal case, $q_a = q_w$ and the two relationships of Eq. (7.55) must give identical results. However, in reality, $q_a \neq q_w$, as emphasized above. Thus, a large variation in ε will result at low airflows, depending on whether it is based on q_a or q_w. Since ε-NTU curves are very flat at high ε (high NTU), there is a very large error in the resulting NTU, and hence in h and j. The j vs. Re curve drops off consistently with decreasing Re, as shown by a dashed line in Fig. 7.11. This phenomenon is referred to as *rollover* or *drop-off* in j. Some of the problems causing the rollover in j are errors in temperature and air mass flow rate measurements, as follows:

1. A thermocouple measures the junction temperature, not the ambient temperature. Hence, the measured air temperature downstream of the test core, $T_{a,o}$, may be too low, due to heat conduction along the thermocouple wire. This and other heat losses at low airflows result in quite low heat transfer coefficients associated with a thermocouple junction or a resistance thermometer. This heat conduction error is not so pronounced for the upstream temperature measurement since air is at a lower temperature. However, the measured air temperature $T_{a,i}$ upstream of the test core may be too high, due to the radiation effect from the hot core and hot walls of the wind tunnel because of heat conduction in the duct wall from the hot test core. This error is negligible for the core downstream, since the duct walls are at about the same temperature as the outlet air. Both the aforementioned errors in $T_{a,i}$ and $T_{a,o}$ will decrease the q_a calculated.
2. At low airflows, temperature stratification in the vertical direction would be a problem both upstream and downstream of the test core. Thus, it becomes difficult to obtain true bulk temperatures $T_{a,i}$ and $T_{a,o}$.

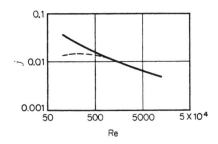

FIGURE 7.11   Rollover phenomenon for j vs. Re characteristic of a heat exchanger surface at low airflows. The dashed curve indicates the rollover phenomenon; the solid curve represents the accurate characteristic. (From Shah, 1985.)

3. On the water side, the temperature drop is generally very small, and hence it will require very accurate instrumentation for ΔT_w measurements; the water flow rate should be adjusted downward from a high value to ensure the magnitude of ΔT_w large enough for an accurate measurement. Also, care must be exercised to ensure good mixing of water at the core outlet before ΔT_w is measured.

4. There are generally some small leaks in the wind tunnel between the test core and the point of air mass flow rate measurement. These leaks, although small, are approximately independent of the air mass flow rate, and they represent an increasing fraction of the measured flow rate $\dot{m}_a$ at low airflows. A primary leak test is essential to ensure negligible air leakage at the lowest encountered test airflow before any testing is conducted.

5. Heat losses to ambient are generally small for a well-insulated test section. However, they could represent a good fraction of the heat transfer rate in the test section at low airflows. A proper calibration is essential to determine these heat losses.

6. For some test core surfaces, longitudinal heat conduction in the test core surface wall may be important and should be taken into account in the data reduction.

The first five factors cause heat imbalances $(q_w - q_a)/q_a$ to increase sharply at decreasing low airflow rates. To minimize or eliminate the rollover in j factors, the data should be reduced based on $q_{ave} = (q_w + q_a)/2$, and whenever possible, by reducing the core flow length by half (i.e., reducing NTU) and then retesting the core.

The uncertainty in the j factors obtained from the steady-state tests ($C^* \approx 0$ case) for a given uncertainty in ΔT_o $(= T_s - T_{a,o}$ or $T_{w,o} - T_{a,o})$ is given as

$$\frac{d(j)}{j} = -\frac{d(\Delta T_o)}{\Delta T_{max}} \frac{ntu_a}{NTU} \frac{e^{NTU}}{NTU} \qquad (7.56)^\dagger$$

Here $\Delta T_{max} = T_{w,i} - T_{a,i}$. In general, $ntu_a/NTU \leq 1.1$ for most testing for j data. Thus, a measurement error in the outlet temperature difference [i.e., $d(\Delta T_o)$] magnifies the error in j by the foregoing relationship both at high NTU (NTU > 3) and low NTU (NTU < 0.5). The error at high NTU is due to the errors in ΔT_o and other factors discussed above as well as due to e^{NTU} in the numerator. The error in the j factor at low NTU can also be significant, as can be found from Eq. (7.56) (NTU^2 in the denominator), and hence a careful design of the test core is essential for obtaining accurate j factors.

In addition to the foregoing measurement errors, inaccurate j data are obtained for a given surface if the test core is not constructed properly. The problem areas are poor thermal bond between the fins and the primary surface, gross blockage (gross flow maldistribution) on the water (steam) or air side, and passage-to-passage nonuniformity (or maldistribution) on the air side. These factors influence the measured j and f factors

---

† This equation is obtained as follows: After multiplying Eq. (7.47) on both sides by C_a/A_a, the resulting equation has NTU in terms of ntu_a and the rest of the terms, which are treated as constant; differentiating this equation, we get $d(NTU)$ in terms of $d(ntu_a)$. Substituting the last relationship in Eq. (7.44) after it is differentiated [i.e., it has the $d(NTU)$ term], we eventually can get a formula for $d(ntu_a)/ntu_a$. Using the definition of j from Eq. (7.34), we finally get Eq. (7.56).

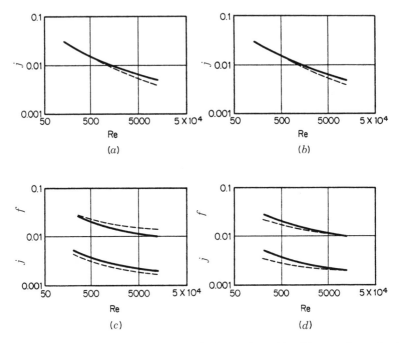

FIGURE 7.12   Influence on measured j data of (a) poor thermal bond between fins and primary surface, (b) water (steam)-side gross blockage, (c) air-side gross blockage, and (d) air-side passage-to-passage nonuniformity. The solid lines are for the perfect core, the dashed lines for the imperfect core specified. (From Shah, 1985.)

differently in different Reynolds number ranges. Qualitative effects of these factors are presented in Fig. 7.12 to show the trends. The solid lines in these figures represent the j data of an ideal core having a perfect thermal bond, no gross blockage, and perfect flow uniformity. The dashed lines represent what happens to j factors when the specified imperfections exist. It is imperative that a detailed air temperature distribution be measured at the core outlet during testing to ensure that none of the foregoing problems are associated with the core.

The experimental uncertainty in the j factor for the foregoing steady-state method is usually within $\pm 5\%$ when the temperatures are measured accurately to within $\pm 0.1°C$ ($0.2°F$) and none of the aforementioned problems exist in the test core. The uncertainty in the Reynolds number is usually within $\pm 2\%$ when the mass flow rate is measured accurately within $\pm 0.7\%$.

7.3.2   Wilson Plot Technique

To obtain highly accurate j factors, one of the requirements for the design of a test core in the preceding method was to have the thermal resistance on the test fluid (gas) side dominant [i.e., the test fluid-side thermal conductance $\eta_o hA$ is significantly lower than that on the other (known) h side]. This is achieved by either steam or hot or cold water at high mass flow rates on the known h side. However, if the fluid on the unknown h side is water or another liquid and it has a high heat transfer coefficient, it may not represent a dominant thermal resistance, even if condensing steam is used on the other side. This is because the test fluid thermal resistance may be of the same order of magnitude as the wall thermal resistance. Hence, for liquids, Wilson (1915) proposed a technique to obtain heat transfer coefficients h for turbulent flow in a circular tube. This technique has been used extensively over the years and its modifications are often used today as well.

In this method, liquid (test fluid, unknown side, fluid 1) flows on one fluid side for which j vs. Re characteristics are being determined; condensing steam, liquid, or air flows on the other fluid side (fluid 2), for which we may or may not know the j vs. Re characteristics. The fluid flow rate on the fluid 2 side and the log-mean average temperature *must* be kept constant (through iterative experimentation) so that its thermal resistance and C_2 of Eq. (7.60) are truly constant. The flow rate on the unknown (fluid 1) side is varied systematically. The fluid flow rates and temperatures upstream and downstream of the test core on each fluid side are measured for each test point. Thus when ε and C^* are known, NTU and UA for the test core are computed. For discussion purposes, consider the test fluid side to be cold and the other fluid side to be hot. UA is given by

$$\frac{1}{UA} = \frac{1}{(\eta_o hA)_c} + \mathbf{R}_{c,f} + \mathbf{R}_w + \mathbf{R}_{h,f} + \frac{1}{(\eta_o hA)_h} \tag{7.57}$$

Note that $\eta_o = 1$ on the fluid side, which does not have fins. For fully developed turbulent flow through constant-cross-sectional ducts, the Nusselt number correlation is of the form

$$\mathrm{Nu} = C_o \cdot \mathrm{Re}^a \cdot \mathrm{Pr}^{0.4} \left(\frac{\mu_w}{\mu_m}\right)^{-0.14} \tag{7.58}$$

where C_o is a constant and $a = 0.8$ for the Dittus–Boelter correlation (see Table 7.6). However, note that a is a function of Pr, Re, and the geometry; it is shown in Fig. 7.19 for a circular tube. Theoretically, a will vary depending on the tube cross-sectional geometry, particularly for augmented tubes or with turbulators, and it is not known a priori. Wilson (1915) used $a = 0.82$. The term $(\mu_w/\mu_m)^{-0.14}$ takes into account the variable property effects for liquids; for gases, it should be replaced by an absolute temperature ratio function [see Eq. (7.157)]. By substituting the definitions of Re, Pr, and Nu in Eq. (7.58) and considering fluid properties as constant we obtain

$$h_c A_c = A_c (C_o k^{0.6} \rho^{0.82} c_p^{0.4} \mu^{-0.42} D_h^{-0.18})_c u_m^{0.82} = C_1' u_m^{0.82} = \frac{C_1 u_m^{0.82}}{\eta_{o,c}} \tag{7.59}$$

The test conditions are maintained such that the fouling resistances $\mathbf{R}_{c,f}$ and $\mathbf{R}_{h,f}$ remain approximately constant, although not necessarily zero; Wilson (1915) had neglected them. Since h is maintained constant on the fluid 2 side, the last four terms on the right-hand side of Eq. (7.57) are constant, let us say equal to C_2. A requirement for this condition (as mentioned above) is to have the log-mean average temperature constant so that the thermal resistance and C_2 remain constant. Now substituting Eq. (7.59) into Eq. (7.57), we get:

$$\frac{1}{UA} = \frac{1}{C_1 u_m^{0.82}} + C_2 \tag{7.60}$$

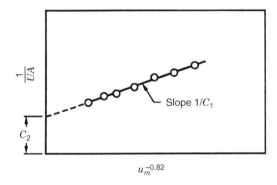

FIGURE 7.13   Original Wilson plot of Eq. (7.60). (From Shah, 1985.)

Equation (7.60) has the form $y = mx + b$ with $y = 1/UA$, $m = 1/C_1$, $x = u_m^{-0.82}$, and $b = C_2$. Wilson plotted $1/UA$ vs. $u_m^{-0.82}$ on a linear scale as shown in Fig. 7.13. The slope $1/C_1$ and the intercept C_2 (at the hypothetical position $x = 0$ on the linear scale axis) are then determined from this plot. Once C_1 is known, h_c from Eq. (7.59) and hence the correlation given by Eq. (7.58) are known. However, replacing $u_m^{0.82}$ with $Re^{0.82}$ and related terms, we get an equation as shown in Example 7.2, which yields C_o directly.

For this method, the Re exponent of Eq. (7.58) should be known, and both terms on the right-hand side of Eq. (7.60) should be of the same order of magnitude. If C_2 is too small, it could end up negative in Fig. 7.13, depending on the slope of the fitted line due to the scatter in the test data; in this case, ignore the Wilson plot technique and use Eq. (7.57) for data reduction using the best estimate of C_2. If C_2 is too large, such that the slope $1/C_1$ approaches close to zero, C_1 will contain a large experimental uncertainty. If $\mathbf{R}_w$ or $\mathbf{R}_{h,f}$ is too high, $\mathbf{R}_h = 1/(\eta_o hA)_h$ must be kept very low (by a very high value of h on the hot side), so that C_2 is not very large. However, if $\mathbf{R}_h$ is too low and the hot fluid is a liquid or gas, its temperature drop may be difficult to measure accurately. Increasing h on that side can reduce C_2.

The limitations of the Wilson plot technique may be summarized as follows:

1. The fluid flow rate and its log-mean average temperature on the fluid 2 side *must* be kept constant, so that C_2 is a constant.
2. The Re exponent in Eq. (7.58) is presumed to be known (such as 0.82 or 0.8). However, in reality it is a function of Re, Pr, and the geometry itself. Since the Re exponent is not known a priori, the classical Wilson plot technique *cannot* be utilized to determine the constant C_o of Eq. (7.58) for most noncircular or enhanced heat transfer surfaces.
3. All the test data must be in one flow region (e.g., turbulent flow) on fluid 1 side, or the Nu correlation of Eq. (7.58) must be replaced by an explicit equation with two (or more) unknown constants, such as Eq. (7.58) or any other variant.
4. Variations in fluid thermophysical properties and the fin thermal resistance are not taken into consideration on the unknown fluid 1 side.
5. The fouling resistance on both fluid sides of the exchanger *must* be kept constant so that C_2 remains constant in Eq. (7.60).

Shah (1990) discusses how to relax all the foregoing limitations of the Wilson plot technique except for the third limitation (one flow region for complete testing), this is discussed later.

The original Wilson plot technique has two unknowns, C_1 and C_2 in Eq. (7.60). In general, the Wilson plot technique determines two unknowns of the overall thermal resistance equation by a linear (or log-linear) plot or by a linear regression analysis. In the problem above, if $\mathbf{R}_{c,f}$, $\mathbf{R}_w$, and $\mathbf{R}_{h,f}$ are known a priori, we can determine the Nu correlation of the type of Eq. (7.58) for unknown constant C_o when its exponent a is known for the fluid 2 side. Thus, the heat transfer correlation on the fluid 2 side can also be evaluated using the Wilson plot technique if the exponents on Re in Eq. (7.58) are known on both fluid sides. Alternatively, we could determine C_o and a of Eq. (7.58) by the Wilson plot technique if C_2 is known. The Wilson plot technique thus represents a problem with two unknowns. Briggs and Young (1963) extended this technique with three unknowns.

For a more general problem (e.g., a shell-and-tube exchanger), consider the Nu correlation on the tube side as Eq. (7.58) with $C_o = C_t'$ and the exponent on Re as a. On the shell side, consider Nu correlation as given by Eq. (7.58) with $C_o = C_s'$ and the Re exponent as d. We can rewrite Eq. (7.57) as follows after neglecting $\mathbf{R}_{t,f}$ and $\mathbf{R}_{s,f}$ for a new/clean exchanger:

$$\frac{1}{UA} = \frac{1}{C_t[\mathrm{Re}^a \cdot \mathrm{Pr}^{0.4} \cdot Ak/D_h]_t(\mu_w/\mu_m)_t^{-0.14}} + \mathbf{R}_w + \frac{1}{C_s[\mathrm{Re}^d \cdot \mathrm{Pr}^{0.4} \cdot Ak/D_h]_s(\mu_w/\mu_m)_s^{-0.14}}$$
$$(7.61)$$

where $C_t = \eta_{o,t}C_t'$ and $C_s = \eta_{o,s}C_s'$. Thus, the more general Wilson plot technique has five unknowns (C_t, C_s, a, d, and $\mathbf{R}_w$); no verified solution procedure is reported in the literature for this problem.

Example 7.2   An air–water round tube and flat louvered fin heat exchanger test core with turbulators in the tubes is tested to determine the Colburn factor vs. Reynolds number correlation on the turbulators (water) side. Since generally the air side has larger (controlling) thermal resistance, the conventional Kays and London method cannot be used for getting the j or Nu vs. Re characteristics for the water side with a turbulator. Hence, the Wilson plot technique has been used to determine the j vs. Re characteristics. Testing is performed for three different air mass flow rates and for a range of mass flow rates for water. The unknown heat transfer coefficient side is the water side. The air-side thermal resistance and its log-mean average temperature are kept constant during testing for a specified nominal airflow rate. The air- and water-side heat transfer surface areas are 2.211 and 0.1522 m^2, respectively. The hydraulic diameter $D_{h,w}$ of the water-side flow channels is 0.005282 m. From the temperature and flow measurements, the partially reduced test data shown in Table E7.2 are determined for three constant values of airflow rates.[†] The calculated values in the sixth column were obtained assuming Nu $= C_o \cdot \mathrm{Re}^{0.85} \cdot \mathrm{Pr}^{0.4}$ correlation for heat transfer on the water side. Determine this correlation based on the data provided using the Wilson plot technique, assuming negligible viscosity variation on the water side. How will this correlation change if the

---

[†] If the test setup and measurements are correct, the water-side heat transfer correlation should be identical for all three airflow rates. From the temperatures measured, the exchanger effectiveness is evaluated and NTU is determined for the known exchanger flow arrangement. Subsequently, $1/U$ is computed for known A and $C_{\min}$.

TABLE E7.2   Partially Reduced Test Data to Determine Water Side j vs. Re Characteristics.

Air Mass Flow Rate (kg/s)	Water Mass Flow Rate (kg/s)	Pr for Water	Re for Water Side	$1/U$ $(m^2 \cdot K/W)$	$(A_a D_{h,w})/$ $(Ak \cdot Re^{0.85} \cdot Pr^{0.4})_w$ $(m^2 \cdot K/W)$
\multicolumn{6}{c}{Test Data for Nominal Air Mass Flow Rate of 0.76 kg/s}
0.757	0.399	2.32	8,960	5.4979×10^{-3}	3.6306×10^{-5}
0.756	0.452	2.30	10,191	5.3426×10^{-3}	3.2578×10^{-5}
0.756	0.624	2.29	14,178	5.1246×10^{-3}	2.4676×10^{-5}
0.755	0.757	2.27	17,327	5.0402×10^{-3}	2.0858×10^{-5}
0.758	1.145	2.25	26,404	4.8165×10^{-3}	1.4617×10^{-5}
0.758	1.517	2.24	35,168	4.7137×10^{-3}	1.1480×10^{-5}
0.757	2.256	2.23	52,610	4.5513×10^{-3}	8.1658×10^{-6}
\multicolumn{6}{c}{Test Data for Nominal Air Mass Flow Rate of 1.13 kg/s}
1.133	2.253	1.88	52,354	3.8866×10^{-3}	8.1922×10^{-6}
1.134	1.537	1.90	35,505	3.9903×10^{-3}	1.1371×10^{-5}
1.135	1.170	1.92	26,935	4.1135×10^{-3}	1.4362×10^{-5}
1.134	0.756	1.94	17,209	4.3221×10^{-3}	2.0940×10^{-5}
1.135	0.651	1.96	14,759	4.3957×10^{-3}	2.3822×10^{-5}
1.135	0.460	1.98	10,319	4.6706×10^{-3}	3.2173×10^{-5}
1.134	0.382	1.97	8,501	4.8089×10^{-3}	3.7829×10^{-5}
\multicolumn{6}{c}{Test Data for Nominal Air Mass Flow Rate of 1.89 kg/s}
1.885	3.864	2.27	8,532	4.0781×10^{-3}	3.7609×10^{-5}
1.882	0.460	2.25	10,243	3.9381×10^{-3}	3.2293×10^{-5}
1.883	0.617	2.22	13,884	3.7570×10^{-3}	2.5026×10^{-5}
1.888	0.777	2.20	17,637	3.6245×10^{-3}	2.0482×10^{-5}
1.887	1.149	2.18	26,328	3.4303×10^{-3}	1.4620×10^{-5}
1.890	1.524	2.17	35,088	3.3350×10^{-3}	1.1471×10^{-5}
1.888	2.209	2.15	51,124	3.2518×10^{-3}	8.3478×10^{-6}

Reynolds number dependence of the Nusselt number is found to be $Re^{0.60}$ instead of $Re^{0.85}$?

SOLUTION

Problem Data and Schematic: The test data for an air–water heat exchanger are determined experimentally and are provided in Table E7.2. $Nu = C_o \cdot Re^{0.85} \cdot Pr^{0.4}$ is given as the correlation form on the water side. The exchanger is made up of round tube and multilouver fins (Fig. 1.33*b*).

Determine: The Colburn factor vs. Re number correlations for three values of the air mass flow rates. Identify the influence of a change of the Re number exponent in the heat transfer correlation on the water side. Evaluate the alternate correlation as $Nu = C_o \cdot Re^{0.6} \cdot Pr^{0.4}$.

Assumptions: Standard assumptions for heat exchanger design as enlisted in Section 3.2.1 are valid. The assumptions adopted in Section 7.3.2 for the Wilson plot technique are also invoked. Fouling is neglected on both fluid sides because of the new test cores.

Analysis: The following Nusselt number correlation on the water side is given:

$$\text{Nu} = \frac{hD_h}{k} = C_o \cdot \text{Re}^a \cdot \text{Pr}^{0.4}$$

where $a = 0.85$ is a known exponent. The Colburn factor by definition is

$$j = \frac{\text{Nu}}{\text{Re} \cdot \text{Pr}^{1/3}} = C_o \cdot \text{Re}^{a-1} \cdot \text{Pr}^{0.07}$$

The overall thermal resistance is related to air, water, and wall thermal resistances as follows [see Eq. (7.57)]:

$$\frac{1}{UA_a} = \frac{1}{\eta_o h_a A_a} + \mathbf{R}_w + \frac{1}{C_o[\text{Re}^{0.85} \cdot \text{Pr}^{0.4}(k/D_h)A]_w}$$

Therefore,

$$\frac{1}{U} = \frac{A_a}{C_o[\text{Re}^{0.85} \cdot \text{Pr}^{0.4}(k/D_h)A]_w} + \left(\frac{1}{\eta_o h_a} + \mathbf{R}_w A_a\right)$$

or

$$y = \frac{1}{U} = mx + b$$

where $m = 1/C_o$, $x = (A_a D_{h,w})/(Ak \cdot \text{Re}^{0.85} \cdot \text{Pr}^{0.4})_w$, and $b = 1/\eta_o h_a + \mathbf{R}_w A_a$. Note that b is a constant for all test points having the same airflow rate. The Wilson plot technique provides the value of m (a slope of the line $y = mx + b$ determined from the regression analysis for which the pairs of y and x values are determined from the experimental data). In the table of the problem statement, the values of y and x are given in the fifth and sixth columns. These data are presented graphically in Fig. E7.2.

The results of the regression analysis are as follows:

Nominal Airflow Rate (kg/s)	m	b (m$^2 \cdot$K/W)	Number of Data Points	Coefficient of Determination	Residual Mean Square Error
1.89	28.58	0.003019	7	0.9976	2.81×10^{-10}
1.13	31.36	0.003646	7	0.9973	3.73×10^{-10}
0.76	31.99	0.004333	7	0.9929	9.91×10^{-10}

Since $m = 1/C_o$, C_o can then be computed for each airflow testing from the results above. The desired correlation for the circular tube with a turbulator is then

$$j = C_o \cdot \text{Re}^{-0.15} \cdot \text{Pr}^{0.07} = 0.0350\text{Re}^{-0.15} \cdot \text{Pr}^{0.07}$$

where in the expression after the second equality, $C_o = 1/28.58 = 0.0350$ is based on the tests for 1.89 kg/s airflow rates. The Colburn factors from the correlation above are presented graphically in Fig. E7.2 for three airflow rates. We find that the Colburn factors for the largest airflow rate are slightly larger than the corresponding values for two smaller airflow rates. The symbols used in the plot correspond to the same air mass flow rates as in the Wilson plot. To determine the cause of this deviation, the

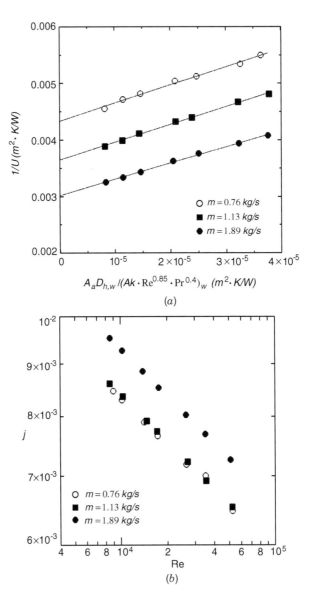

FIGURE E7.2

experimental procedure and equipment operation should be checked and ensure that the log-mean average temperature on the air side is kept constant. A detailed study in this case showed that during the experiments for the latter two set of data, a water valve leak was present but not corrected. The correction of the water mass flow rates would lead to a very good agreement of all three sets of data, as one would expect.

If we had assumed a Reynolds number exponent of 0.60 instead of 0.85, regression analysis on the modified set of data would yield $m = 3.56$ and hence $C_o = 0.2808$ for a nominal airflow rate of 1.89 kg/s. As can be found, the resulting correlation for a circular tube with a turbulator depends strongly on the value of the exponent selected for the Reynolds number.

Discussion and Comments: From the results above, it is clear that the Wilson plot technique yields the heat transfer correlation on the unknown side in a straightforward manner when its thermal resistance is not so dominant. To assure that one gets an accurate correlation for the Nu or j factor vs. Reynolds number, it is highly desirable that the sets of test data be taken at two or more constant thermal resistances on the other fluid side (as was done here for three airflow rates). If there is something wrong with the test setup or measurement errors, they would then show up by having different correlations through the Wilson plot technique.

The calculated numerical value of the correlation coefficient C_o depends strongly on the assumed exponent on the Reynolds number, as one would expect from the theoretical considerations. Although the exponent 0.85 on the Reynolds number would be a good choice for a circular tube, it may not be reasonably accurate for some augmented tubes, tubes with turbulators, those having developing laminar flow, and others. In that case, it is essential either that we know the correct Reynolds number exponent beforehand or consider it as an unknown and may apply the modified Wilson plot technique with three unknowns as developed by Briggs and Young (1963) and summarized by Shah (1990).

7.3.3   Transient Test Techniques

Transient test techniques are used to establish the j vs. Re characteristics for a matrix-type regenerator surface or a high-ntu surface. The most important advantage of most of these techniques is that only one fluid (air) is employed during the testing. Therefore, these techniques are also referred to as *single-blow techniques*. The test section is a single-fluid exchanger (matrix or core) built up from the heat transfer surface for which the j vs. Re characteristics are to be determined. Generally, air is used as a working fluid. Initially, the matrix wall and fluid temperatures are both constant and uniform at the fluid temperature at time equals zero. A known monotonic increase or decrease is imposed on the inlet temperature of the fluid. The resultant temperature–time history of the fluid at the core outlet is recorded. This outlet fluid temperature response is related directly to heat transfer and the average heat transfer coefficient in the matrix. A comparison of theoretical and experimental outlet temperature responses then permits evaluation of the average heat transfer coefficient. The fluid temperature at the core inlet could be a step, exponential, periodic, or ramp function, or any arbitrary variation. The step inlet temperature was utilized most commonly for the single-blow method before the automatic data acquisition by a computer became common in the early 1980s. Heggs and Burns (1988) provided a comparison of four commonly used methods for the single-blow technique (i.e., the direct matching, maximum slope, shape factor, and differential fluid enthalpy methods). They reported the direct matching and differential fluid enthalpy methods to be most accurate. For details of various transient techniques and data reduction methods, refer to Shah and Zhou (1997). The direct curve-matching data reduction method is most accurate for the complete range of ntu and is now the most commonly used and recommended method.

If a single-blow transient technique is used for a recuperator surface of a two-fluid heat exchanger, the experimental surface characteristics will be fine for comparison purposes, but may not be accurate on an absolute basis, for the following reasons:

- The test surface in transient testing acts as a primary surface even if fins or extended surfaces are used. Since the real effect of the fin surface heat transfer is not taken into account (in contrast, it is taken into account in a two-fluid heat exchanger test), one relies on the idealized fin efficiency evaluation as outlined in Section 4.3, and this could introduce an error of unknown magnitude in the measured j factors.

- The blockage of the second fluid side on the frontal area has an effect on the measured j factors in a two-fluid heat exchanger and is included in the measured j factors by the steady-state technique. This effect is absent in the transient technique and can have an effect of unknown magnitude in the measured j factors for a real two-fluid heat exchanger.

- For the steady-state testing of a two-fluid heat exchanger, the boundary condition is generally constant wall temperature. The boundary condition in the transient test technique is between constant wall temperature and constant heat flux (Shah and London, 1970).

As mentioned in the second item above, the single-blow transient technique has a drawback related to the blockage of the second fluid side. To overcome that problem, a new method has been proposed by Gvozdenac (1994). That technique, the *double-blow method*, takes into account the responses of both fluids on an inlet temperature perturbation experienced by one of the two fluid streams. The double-blow method still cannot overcome the problems mentioned in the first and third items; those problems may not be important if one does not need highly accurate j factors.

7.3.3.1   Test Setup and Experimental Procedure.   The test setup consists of the following basic elements: (1) test section, (2) heating device, (3) fluid metering device, (4) fluid pumping device, (5) temperature measurement devices, and (6) pressure measurement devices. A schematic of the transient test rig used at Stanford University is shown in Fig. 7.14.

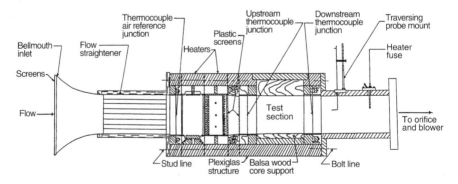

FIGURE 7.14   Schematic of the transient single-blow heat transfer test rig. (From Wheeler, 1968.)

The test section is a single-fluid exchanger in which the heat transfer surface is stacked up with or without plain sheets in between. The test fluid is air. The heating device is generally a heating screen of fine wires (wire mesh) or made up of thin strips. The latter approach is preferred, due to the large surface area of thin strips and hence lower temperature, resulting in less radiation error in the upstream temperature measurements (see Shah and Zhou, 1997). Also, the air temperatures upstream and downstream are measured by point measurements (using a thermocouple grid) or line measurements (such as 10 platinum wires 50 μm in diameter across the test section height arranged at equal distance in the test cross section). The line measurements provide more accurate temperature measurement across the test section with less flow disturbance. There are a number of choices for accurate flow and pressure measurement devices, and the selection is made by the individual investigators.

The following is the experimental procedure used to obtain the air temperature–time history at core outlet during the core heating and cooling. The airflow rate is set at a constant predetermined value. The air is heated with the resistance-heating device to about 11°C (20°F) above the ambient temperature; the heated air in turn heats the matrix. The temperature–time history of the air at the core inlet and outlet is recorded continuously during matrix heating, as shown in Fig. 7.15a. Then heating continues until the core reaches a uniform temperature exhibited by a negligible difference between the air temperature at the inlet and exit of the matrix. Once the stable condition is reached, the power to the heating device is turned off. The temperature–time history of the air leaving the matrix is recorded continuously during the matrix cooling period as shown in Fig. 7.15b. During each heating and cooling period of the matrix, measurements are taken for airflow rates, core upstream and downstream pressures, and core upstream temperature before the heating device. Usually, two heating and two cooling curves are recorded for each flow rate, and the average heat transfer coefficient of four curves is used to determine the j factor. Similar tests are repeated with different airflow rates to cover the desired range of the Reynolds number.

7.3.3.2 Theoretical Model. A brief background on the theoretical analysis of the test point may be given as follows. Pertinent differential equations for the matrix heating case of Fig. 7.15a are identical to those of Eq. (5.71) for the hot-gas flow period in a regenerator presented in Section 5.3. Similarly, the differential equations for the matrix

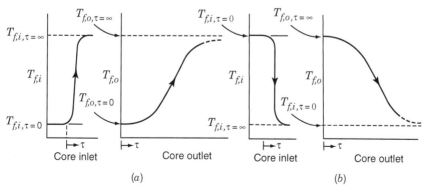

FIGURE 7.15 Exponential temperature change imposed at the inlet and its response at the core outlet: (*a*) matrix heating; (*b*) matrix cooling.

cooling case of Fig. 7.15b are those of Eq. (5.72). Generally, longitudinal heat conduction in the matrix wall is included in the analysis. In this case, the second equations of Eqs. (5.71) and (5.72) are replaced by Eqs. (5.111) and (5.113). Hence, the pertinent differential equations, using the same nomenclature, for the single-blow analysis problem are

$$\frac{\partial T_f}{\partial \xi} = (T_w - T_f) \tag{7.62}$$

$$\frac{\partial T_w}{\partial \eta} = (T_f - T_w) + \lambda \cdot \text{ntu} \frac{\partial^2 T_w}{\partial \xi^2} \tag{7.63}$$

where

$$\xi = \frac{hA}{C} \frac{x}{L} \qquad \eta \approx \frac{hA}{M_c c_w} \tau \qquad \lambda = \frac{k_w A_k}{LC} \qquad \text{ntu} = \frac{hA}{C} \tag{7.64}$$

Here T_f and T_w are the fluid (air) and wall temperatures. The boundary and initial conditions for this problem are

At $\xi = 0$: $T_f(0, \eta) = 1 - e^{-\tau/\tau_H} = 1 - e^{-(M_w c_w/hA\tau_H)\eta} \tag{7.65}$

At $\eta = 0$: $T_w(\xi, 0) = 0 \qquad T_f(\xi, 0) = 0 \tag{7.66}$

At $\xi = 0$ and 1: $\frac{\partial T_w(0, \eta)}{\partial \xi} = 0 \qquad \frac{\partial T_w(1, \eta)}{\partial \xi} = 0 \tag{7.67}$

where the time constant of heater $\tau_H \approx (Mc/hA)_H$ and the subscript H denotes the values for the heater (in contrast to the test core). Note that longitudinal heat conduction in the air is negligible and hence has not been incorporated in Eq. (7.62).

A variety of methods have been used to solve Eqs. (7.62), (7.63), and (7.65)–(7.67) as summarized by Shah and Zhou (1997). One of the most accurate and rapid methods is the numerical analysis of these equations for the measured inlet and outlet air temperatures from the core. In this case, the inlet and outlet temperatures are measured using an online data acquisition system; the temperatures are then digitized as finely as desired and are fed into the data reduction program, which employs a direct curve-matching method. Also input to the data reduction program are the measured airflow rate, core geometrical properties (D_h, A_o, A, σ), and core and air thermophysical properties. For an assumed value of the heat transfer coefficient, Eqs. (7.62)–(7.67) are solved numerically and the outlet air temperature distribution is determined as a function of time. This is then compared with the measured temperature–time distribution. If the measured and numerical outlet temperature distributions do not match within a desired degree of accuracy, iterations are continued on the heat transfer coefficient until the temperature distributions do match. Mullisen and Loehrke (1986) used the *Regula-falsi method* for iterative adjustment on h. Typical curve matching is shown in Fig. 7.16. Once h is known, the j factor is determined from its definition. The Reynolds number is also calculated from the measured flow rate, core geometrical properties, and air thermophysical properties. The j factors determined by this direct curve-matching method are accurate within $\pm 2\%$ for the complete range of NTU. For further details on the data reduction method, refer to Mullisen and Loehrke (1986).

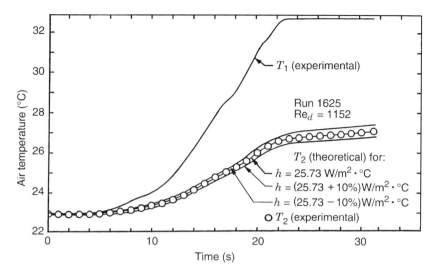

FIGURE 7.16   Typical curve matching of experimental temperature–time history at the core outlet with the numerically predicted curve. (From Mullisen, 1983.)

Bačlić et al. (1986a) proposed an alternate method, a differential enthalpy method, for the data reduction of single-blow test results that can have any arbitrary inlet fluid temperature variation. The percentage error range in prediction of NTU spans between 1.5% and 9.5% for the range of NTU between 0.4 and 10 if the evaluation of the enthalpy change is kept at an experimental error level of 1%. The application of this method for determining j factors of compact heat transfer surfaces is provided by Bačlić et al. (1986b).

7.3.4   Friction Factor Determination

The experimental determination of flow friction characteristics of compact heat exchanger surfaces is relatively straightforward. Regardless of the core construction and the method of heat transfer testing (steady state or transient), determination of the f factor is made under steady fluid flow rates with or without heat transfer. For a given fluid flow rate on the unknown f side, the following measurements are made: core pressure drop, core inlet pressure and temperature, core outlet temperature for hot friction data, fluid mass flow rate, and core geometrical properties. The Fanning friction factor f is then determined:

$$f = \frac{r_h}{L}\frac{1}{(1/\rho)_m}\left[\frac{2g_c\Delta p}{G^2} - \frac{1}{\rho_i}\left(1 - \sigma^2 + K_c\right) - 2\left(\frac{1}{\rho_o} - \frac{1}{\rho_i}\right) + \frac{1}{\rho_o}\left(1 - \sigma^2 - K_e\right)\right] \quad (7.68)$$

This equation is an inverted form of the core pressure drop equation (6.28). For the isothermal pressure drop data, $\rho_i = \rho_o = 1/(1/\rho)_m$. Here K_c and K_e are sudden contraction and expansion pressure loss coefficients presented in Fig. 6.3. The friction factor thus determined includes the effects of skin friction, form drag, and local flow contraction and expansion losses, if any, within the core. Tests are repeated with different flow rates on the unknown side to cover the desired range of the Reynolds number. The experimental uncertainty in the f factor is usually within $\pm 5\%$ when Δp is measured accurately within $\pm 1\%$.

The Reynolds number is determined in the same way as described in Section 7.3.1.2 for heat transfer tests. The uncertainty in Reynolds numbers for both j and f factor testing is $\pm 2\%$ when the fluid flow rate is measured accurately within $\pm 0.7\%$.

Generally, the Fanning friction factor f is determined from isothermal pressure drop data (no heat transfer across the core). The hot friction factor f vs. Re curve should be close to the isothermal f v. Re curve, particularly when variations in the fluid properties are small (i.e., the average fluid temperature for the hot f factors is not significantly different from the wall temperature). Otherwise, the hot f factors must be corrected to take into account the temperature-dependent fluid properties (see Section 7.6.1).

Example 7.3 Calculate the friction factor for the heat exchanger operating point of Example 7.1.

SOLUTION

Problem Data and Schematic: Test core, geometry, operating conditions, and the schematic are the same as those in Example 7.1; the following are specific data for the friction factor evaluation:

$p_i = 101.60\,\text{kPa}$ $\qquad$ $\Delta p = 1.493\,\text{kPa}$ $\qquad$ $T_i = 23.05°\text{C}$ $\qquad$ $T_o = 101.60°\text{C}$

$G = 17.019\,\text{kg/m}^2 \cdot \text{s}$ $\qquad$ $r_h/L = 0.005688$ $\qquad$ $K_c = 0.37$ $\qquad$ $K_e = 0.48$ $\qquad$ $\sigma = 0.3067$

Determine: The friction factor for the given heat exchanger operating point.

Assumptions: Steady-state flow and constant fluid properties apply.

Analysis: We use Eq. (7.68) to determine the friction factor. Let us first calculate the air density ρ_i, ρ_o, and $(1/\rho)_m$. For air as a perfect gas, the gas constant $\tilde{R} = 287.04\,\text{J/kg} \cdot \text{K}$ or $(\text{N} \cdot \text{m/kg} \cdot \text{K})$ for air.

$$\rho_i = \frac{p_i}{\tilde{R}T_i} = \frac{101.60 \times 10^3\,\text{Pa}}{287.04\,\text{J/kg} \cdot \text{K} \times (273.15°\text{C} + 23.05°\text{C})} = 1.1946\,\text{kg/m}^3$$

With $p_o = p_i - \Delta p = (101.60 - 1.493)\,\text{kPa} = 100.11\,\text{kPa}$,

$$\rho_o = \frac{p_o}{\tilde{R}T_o} = \frac{100.11 \times 10^3\,\text{Pa}}{287.04\,\text{J/kg} \cdot \text{K} \times (273.15°\text{C} + 101.60°\text{C})} = 0.9307\,\text{kg/m}^3$$

$$\left(\frac{1}{\rho}\right)_m = \frac{1}{2}\left(\frac{1}{\rho_i} + \frac{1}{\rho_o}\right) = \frac{1}{2}\left(\frac{1}{1.1946\,\text{kg/m}^3} + \frac{1}{0.9307\,\text{kg/m}^3}\right) = 0.9558\,\text{m}^3/\text{kg}$$

Now we compute the friction factor as follows from Eq. (7.68):

$$
f = \frac{r_h}{L} \frac{1}{(1/\rho)_m} \left[\frac{2g_c\,\Delta p}{G^2} - \frac{1}{\rho_i}\left(1 - \sigma^2 + K_c\right) - 2\left(\frac{1}{\rho_o} - \frac{1}{\rho_i}\right) + \frac{1}{\rho_o}\left(1 - \sigma^2 - K_e\right) \right]
$$

$$
= \frac{0.005688}{0.9558\,\text{m}^3/\text{kg}} \left[\frac{2 \times 1 \times 1.493 \times 10^3\,\text{Pa}}{(17.019\,\text{kg/m}^2 \cdot \text{s})^2} - \frac{1 - (0.3067)^2 + 0.37}{1.1946\,\text{kg/m}^3} \right.
$$

$$
\left. - 2\left(\frac{1}{0.9307} - \frac{1}{1.1946}\right)\frac{1}{\text{kg/m}^3} + \frac{1 - (0.3067)^2 - 0.48}{0.9307\,\text{kg/m}^3} \right]
$$

$$
= \; 0.06134 - 0.00636 - 0.00282 + 0.00272 = 0.05488 \qquad\qquad Ans.
$$
$$
\;\;(111.7\%)\;\;\;\;(11.6\%)\;\;\;\;(5.1\%)\;\;\;\;(5.0\%)
$$

The Reynolds number for this point was calculated in Example 7.1 as Re = 989.

Discussion and Comments: This example shows that the friction factor determination is again straightforward if the geometry is known and all required measurements are made. For this particular test point, notice that the core pressure drop contribution is the largest; the entrance and exit loss contributions compensate each other, with the net result as a small effect (6.6%).

7.4 ANALYTICAL AND SEMIEMPIRICAL HEAT TRANSFER AND FRICTION FACTOR CORRELATIONS FOR SIMPLE GEOMETRIES

Analytical correlations[†] for simple geometries are presented in this section with the following objectives:

1. Assess the available analytical correlations for obtaining as much insight as possible into the performance behavior of complex flow passage geometries having insufficient experimental data and empirical correlations. As a result, based on the analytical correlations presented in this section, one would be able to predict *j* and *f* data for some complex flow passage geometries with reasonable accuracy, and obtain some direction toward the goal for heat transfer surface enhancement and for an increase in surface compactness.

2. Provide accurate, concise, and pertinent correlations for important flow passage geometries. These correlations are also important (a) for extrapolating the experimental data when they do not exist in the Re range desired, (b) for getting a first

---

[†] The closed-form or approximate (such as numerical) solutions of governing equations in this book are referred to as *theoretical solutions* (*not* as analytical correlations); for example, the solutions are in terms of the temperature distribution for the energy equation, or the velocity/pressure distribution for the momentum equation. When secondary parameters are computed (such as the heat transfer coefficient or the Nusselt number) from exact or approximate theoretical solutions, we refer to them as analytical (or numerical, if appropriate) correlations. In this case, analytical/numerical results are generally expressed in a closed-form (preferably simplified) equation (with or without curve fitting) *without* any experimental input. When experimental results are expressed in a theory-based equation (with constants modified), we refer to them as *semiempirical correlations*; when the experimental results are expressed in an equation form without much theory base (such as regression), we refer to such an equation as an *empirical correlation*.

approximation to the j and f factors when no experimental data are available, and (c) in the development of semiempirical correlations.

Flow passages in most compact heat exchangers are complex, with frequent boundary layer interruptions; some heat exchangers (particularly the tube side of shell-and-tube exchangers and highly compact regenerators) have continuous flow passages. The velocity and temperature profiles across the flow cross section are generally fully developed in the continuous flow passages, whereas they develop at each boundary layer interruption in an interrupted surface and may reach a periodic fully developed flow. The heat transfer and flow friction characteristics are generally different for fully developed flows and developing flows. Fully developed laminar flow solutions are applicable to highly compact regenerator surfaces or highly compact plate-fin exchangers with plain uninterrupted fins. Developing laminar flow correlations are applicable to interrupted fin geometries and plain uninterrupted fins of short lengths, and turbulent flow solutions for not-so-compact heat exchanger surfaces.

Next, analytical correlations are discussed separately for developed and developing flows for simple flow passage geometries. For complex surface geometries, the basic surface characteristics are primarily obtained experimentally, as discussed in Section 7.3; the pertinent correlations are presented in Section 7.5.

The heat transfer rate in laminar duct flow is very sensitive to the thermal boundary condition. Hence, it is essential to identify thermal boundary conditions carefully in laminar flow. The heat transfer rate in turbulent duct flow is insensitive to the thermal boundary condition for most common fluids ($Pr \geq 0.7$); the exception is liquid metals ($Pr < 0.03$). Hence, there is generally no need to identify thermal boundary conditions in turbulent flow for all *fluids* except for liquid metals. A systematic classification of thermal boundary conditions for internal flow is given by Shah and London (1978). Three important thermal boundary conditions for heat exchangers are Ⓣ, Ⓗ①, and Ⓗ② as shown in Fig. 7.17. The Ⓣ boundary condition refers to constant wall temperature, both axially and peripherally throughout the passage length. This boundary condition is approximated in condensers, evaporators, and liquid-to-gas heat exchangers with high liquid flow rates. The Ⓗ① boundary condition refers to the constant wall heat transfer rate in the axial direction and constant wall temperature at any cross section in the peripheral direction. The Ⓗ② boundary condition refers to constant wall heat transfer rate in the

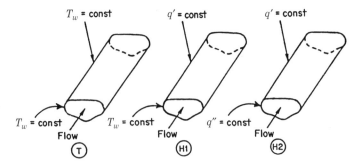

FIGURE 7.17   Thermal boundary conditions for duct flow: Ⓣ constant wall temperature; Ⓗ① constant axial wall heat flux with constant peripheral wall temperature; Ⓗ② constant wall heat flux axially and peripherally. (From Shah, 1983.)

axial direction as well as in the peripheral direction. The Ⓗ① and Ⓗ② boundary conditions may be realized in gas turbine regenerators, counterflow exchangers with $C^* \approx 1$, and nuclear and electric resistance heating. In these applications, the Ⓗ① boundary condition could be realized for highly conductive materials (such as copper, aluminum, etc.) where the temperature gradients in the peripheral direction are at a minimum; the Ⓗ② boundary condition is realized for very poorly conducting materials (such as ceramics, plastics, Teflon, etc.) for which temperature gradients exist in the peripheral direction. For intermediate thermal conductivity values, the boundary condition will be in between that of Ⓗ① and Ⓗ②. It may be noted that the Ⓗ① and Ⓗ② boundary conditions for the symmetrically heated passages with no sharp corners (e.g., circular, flat, and concentric annular ducts) are identical. In general, $Nu_{H1} > Nu_T$, $Nu_{H1} \geq Nu_{H2}$, and $Nu_{H2} \simeq Nu_T$.

7.4.1 Fully Developed Flows

7.4.1.1 Laminar Flow. Nusselt numbers are constant for fully developed laminar flow in ducts of constant cross-sectional area;[†] but they depend on the flow passage geometry and thermal boundary conditions. The product of the Fanning friction factor and the Reynolds number is also constant, but dependent on the flow passage geometry. The fully developed laminar flow problem has been analyzed extensively for many duct geometries. Analytical correlations for some technically important flow passages are presented in Table 7.3. More detailed analytical correlations for some important geometries are also presented in closed-form equations in Table 7.4. The following observations may be made from Table 7.3:

- There is a strong influence of flow passage geometry on Nu and fRe. For micro- or meso-scale flow passages, the surface roughness height profile can be nonnegligible compared to the passage size (D_h) and hence can change the flow passage geometry shape regardless of the original shape, and thus affect Nu and $f \cdot Re$ even in laminar flow. Rectangular passages with a small aspect ratio exhibit the highest Nu and $f \cdot Re$.

- The thermal boundary conditions Ⓗ①, Ⓗ②, Ⓣ have a strong influence on the Nusselt numbers. Depending on the flow geometry, j factors for Ⓗ① boundary condition may be roughly 50% greater than that for the Ⓗ② boundary condition, and about 20% greater than that for the Ⓣ boundary condition.

- As $Nu = hD_h/k$, a constant Nu implies the convective heat transfer coefficient h independent of the flow velocity (Reynolds number) and fluid type (Prandtl number).

- An increase in h can best be achieved either by reducing D_h or by selecting geometry with a low-aspect-ratio rectangular flow passage. Reducing the hydraulic diameter is an obvious way to increase exchanger compactness and heat transfer (a direction adopted for the development of meso and micro heat exchangers, the ultracompact heat exchangers), or D_h can be optimized using well-known heat transfer correlations based on design problem specifications.

[†] Note that for micro- and meso-scale passage geometries, the cross-sectional area of the duct may not be uniform along the flow length due to manufacturing processes. In this case, the analytical correlations presented in Table 7.3 may not be accurate and truly representative for performance of such ducts.

TABLE 7.3 Solutions for Heat Transfer and Friction for Fully Developed Laminar Flow through Specified Ducts

Geometry $(L/D_h > 100)$	Aspect Ratio	Nu_{H1}	Nu_{H2}	Nu_T	$f \cdot Re$	$\dfrac{j_{H1}}{f}$ [a]	$K(\infty)$ [b]	L_{hy}^{+} [c]
	$\dfrac{2b}{2a} = \dfrac{\sqrt{3}}{2}$	3.014	1.474	2.39	12.630	0.269	1.739	0.040
	$\dfrac{2b}{2a} = \dfrac{\sqrt{3}}{2}$	3.111	1.892	2.47	13.333	0.263	1.818	0.040
	$\dfrac{2b}{2a} = 1$	3.608	3.091	2.976	14.227	0.286	1.433	0.090
		4.002	3.862	3.34	15.054	0.299	1.335	0.086
	$\dfrac{2b}{2a} = \dfrac{1}{2}$	4.123	3.017	3.391	15.548	0.299	1.281	0.085
		4.364	4.364	3.657	16.000	0.307	1.250	0.056
	$\dfrac{2b}{2a} = \dfrac{1}{4}$	5.331	2.94	4.439	18.233	0.329	1.001	0.078
	$\dfrac{2b}{2a} = \dfrac{1}{6}$	6.049	2.93	5.137	19.702	0.346	0.885	0.070
	$\dfrac{2b}{2a} = \dfrac{1}{8}$	6.490	2.94	5.597	20.585	0.355	0.825	0.063
	$\dfrac{2b}{2a} = 0$	8.235	8.235	7.541	24.000	0.386	0.674	0.011

Source: Data and recommendations from Shah and London (1978).
[a] $j_{H1}/f = Nu_{H1} \cdot Pr^{-1/3}/(f \cdot Re)$ with $Pr = 0.7$. Similarly, values of j_{H2}/f and j_T/f may be computed.
[b] $K(\infty)$ for sine and equilateral triangular channels may be too high; $K(\infty)$ for some rectangular and hexagonal channels is interpolated.
[c] L_{hy}^{+} for sine and equilateral triangular channels is too low, so use with caution. L_{hy}^{+} for rectangular channels is based on the smoothened curve. L_{hy}^{+} for a hexagonal channel is an interpolated value.

- Since $f \cdot Re = $ constant, $f \propto 1/Re \propto 1/u_m$. In this case, it can be shown that $\Delta p \propto u_m$.

Analytical correlations have been verified by many researchers for flow friction and heat transfer in a single channel. Hence, these results provide a valuable guideline for exchangers that may employ many such channels in parallel. However, passage-to-passage flow nonuniformity could result in significant deviations in Nu and f from the analytical predictions (see Section 12.1.2). Also, the actual thermal boundary conditions for heat transfer in a specific application may not correspond to any of the previously

TABLE 7.4 Laminar Fully Developed Analytical Correlations for Friction Factors and Nusselt Numbers for Some Duct Geometries

Duct Geometry, D_h, α^*, r^*	$f \cdot Re$, Nu_T, Nu_{H1}, Nu_{H2}

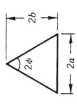

Rectangular:

$$D_h = \frac{4ab}{a+b} = \frac{4b}{1+\alpha^*}$$

$$\alpha^* = 2b/2a$$

For $0 \leq \alpha^* \leq 1$:

$$f \cdot Re = 24(1 - 1.3553\alpha^* + 1.9467\alpha^{*2} - 1.7012\alpha^{*3} + 0.9564\alpha^{*4} - 0.2537\alpha^{*5})$$

$$Nu_T = 7.541(1 - 2.610\alpha^* + 4.970\alpha^{*2} - 5.119\alpha^{*3} + 2.702\alpha^{*4} - 0.548\alpha^{*5})$$

$$Nu_{H1} = 8.235(1 - 2.0421\alpha^* + 3.0853\alpha^{*2} - 2.4765\alpha^{*3} + 1.0578\alpha^{*4} - 0.1861\alpha^{*5})$$

$$Nu_{H2} = 8.235(1 - 10.6044\alpha^* + 61.1755\alpha^{*2} - 155.1803\alpha^{*3} + 176.9203\alpha^{*4} - 72.9236\alpha^{*5})$$

For $0 \leq \alpha^* \leq 1$:

$$f \cdot Re = 12(1 - 0.0115\alpha^* + 1.7099\alpha^{*2} - 4.3394\alpha^{*3} + 4.2732\alpha^{*4} - 1.5817\alpha^{*5} + 0.0599\alpha^{*6})$$

$$Nu_T = 0.943(1 + 4.8340\alpha^* - 2.1738\alpha^{*2} - 4.0797\alpha^{*3} - 2.1200\alpha^{*4} + 11.3589\alpha^{*5} - 6.2052\alpha^{*6})$$

$$Nu_{H1} = 2.059(1 + 0.7139\alpha^* + 2.9540\alpha^{*2} - 7.8785\alpha^{*3} + 5.6450\alpha^{*4} + 0.2144\alpha^{*5} - 1.1387\alpha^{*6})$$

$$Nu_{H2} = \begin{cases} 1.088\alpha^* & \text{for } \alpha^* \leq 0.125 \\ -0.2113(1 - 10.9962\alpha^* - 15.1301\alpha^{*2} + 16.5921\alpha^{*3}) & \text{for } 0.125 < \alpha^* \leq 1 \end{cases}$$

Isosceles triangle:

$$D_h = \frac{4ab}{a + \sqrt{a^2 + 4b^2}}$$

$$\alpha^* = 2b/2a$$

For $1 \leq \alpha^* \leq \infty$:

$$f \cdot Re = 12(\alpha^{*3} + 0.2595\alpha^{*2} - 0.2046\alpha^* + 0.0552)/\alpha^{*3}$$

$$Nu_T = 0.943(\alpha^{*5} + 5.3586\alpha^{*4} - 9.2517\alpha^{*3} + 11.9314\alpha^{*2} - 9.8035\alpha^* + 3.3754)\alpha^{*5}$$

$$Nu_{H1} = 2.059(\alpha^{*5} + 1.2489\alpha^{*4} - 1.0559\alpha^{*3} + 0.2515\alpha^{*2} + 0.1520\alpha^* - 0.0901)/\alpha^{*5}$$

$$Nu_{H2} = \begin{cases} 0.912(\alpha^{*3} - 13.3739\alpha^{*2} + 78.9211\alpha^* - 46.6239)/\alpha^{*3} & \text{for } 1 \leq \alpha^* < 8 \\ 0.312/\alpha^* & \text{for } 8 \leq \alpha^* \leq \infty \end{cases}$$

TABLE 7.4 Continued

Duct Geometry, D_h, a^*, r^*	$f \cdot Re$, Nu_T, Nu_{H1}, Nu_{H2}

Right triangle:

$$D_h = \frac{4ab}{a + b + \sqrt{a^2 + b^2}}$$

$$\alpha^* = 2b/2a$$

For $0 \leq \alpha^* \leq 1$:

$$f \cdot Re = 12(1 + 0.27956\alpha^* - 0.2756\alpha^{*2} + 0.0591\alpha^{*3} + 0.0622\alpha^{*4} - 0.0290\alpha^{*5})$$

$$Nu_T = 1.1731(1 + 3.1312\alpha^* - 3.5919\alpha^{*2} + 1.7893\alpha^{*3} - 0.3189\alpha^{*4})$$

$$Nu_{H1} = 2.0581(1 + 1.2981\alpha^* - 2.1837\alpha^{*2} + 4.3496\alpha^{*3} - 6.2381\alpha^{*4} + 4.3140\alpha^{*5} - 1.0911\alpha^{*6})$$

$$Nu_{H2} = \begin{cases} 0.2299\alpha^* & \text{for } 0 \leq \alpha^* \leq 0.125 \\ 0.4402\alpha^*(1 - 6.8176\alpha^* + 53.2849\alpha^{*2} - 77.9848\alpha^{*3} + 33.5641\alpha^{*4}) & \text{for } 0.125 < \alpha^* \leq 1 \end{cases}$$

Elliptical:

$$D_h = \frac{\pi b}{E(m)}$$

$$\alpha^* = 2b/2a$$

For $0 \leq \alpha^* \leq 1$:

$$f \cdot Re = 2(1 + \alpha^{*2})[\pi/E(m)]^2$$

$$Nu_T = 0.3536(1 + 0.9864\alpha^* - 0.7189\alpha^{*2} + 3.3364\alpha^{*3} - 3.0307\alpha^{*4} + 1.0130\alpha^{*5})[\pi/E(m)]^2$$

$$Nu_{H1} = 9(1 + \alpha^{*2})\left(\frac{\alpha^{*4} + 6\alpha^{*2} + 1}{17\alpha^{*4} + 98\alpha^{*2} + 17}\right)\left(\frac{\pi}{E(m)}\right)^2$$

$$Nu_{H2} = 0.3258\alpha^*(1 + 15.6397\alpha^* - 29.5117\alpha^{*2} + 16.2250\alpha^{*3})[\pi/E(m)]^2$$

$$E(m) = 1 + 0.463015(1 - m) + 0.107781(1 - m)^2 - [0.245273(1 - m) + 0.041250(1 - m)^2]\ln(1 - m)$$

For $0 \leq \alpha^* \leq 2$:

$$f \cdot Re = 9.5687(1 + 0.0772\alpha^* + 0.8619\alpha^{*2} - 0.8314\alpha^{*3} + 0.2907\alpha^{*4} - 0.0338\alpha^{*5})$$

$$Nu_T = 1.1791(1 + 2.7701\alpha^* - 3.1901\alpha^{*2} - 1.9975\alpha^{*3} - 0.4966\alpha^{*4})$$

$$Nu_{H1} = 1.9030(1 + 0.4556\alpha^* + 1.2111\alpha^{*2} - 1.6805\alpha^{*3} + 0.7724\alpha^{*4} - 0.1228\alpha^{*5})$$

$$Nu_{H2} = \begin{cases} 0.76\alpha^* & \text{for } 0 \leq \alpha^* \leq 0.125 \\ -0.0202(1 - 32.0594\alpha^* - 216.1635\alpha^{*2} + 244.3812\alpha^{*3} - 82.4951\alpha^{*4} + 7.6733\alpha^{*5}) & \text{for } 0.125 < \alpha^* \leq 2 \end{cases}$$

For $0 \leq r^* \leq 1$:

$$f \cdot \text{Re} = \frac{16(1 - r^{*})^2}{1 + r^{*2} - 2r_m^{*2}}, \quad r_m = \left(\frac{1 - r^{*2}}{2\ln(1/r^{*})} \right)^{1/2}$$

Sine:

D_h is a function of $\alpha^{*\ddagger}$

$\alpha^* = 2b/2a$

For $0 \leq r^* \leq 0.02$:

$\text{Nu}_T = 3.657 + 98.95r^*$

$\text{Nu}_H = 4.364 + 100.95r^*$

For $0.02 \leq r^* \leq 1$:

$\text{Nu}_T = 5.3302(1 + 3.2904r^* - 12.0075r^{*2} + 18.8298r^{*3} - 9.6980r^{*4})$

$\text{Nu}_H = 6.2066(1 + 2.3108r^* - 7.7553r^{*2} + 13.2851r^{*3} - 10.5987r^{*4} + 2.6178r^{*5} + 0.4680r^{*6})$

Concentric annular:

$D_h = 2(r_o - r_i)$

$r^* = \dfrac{r_i}{r_o}$

Source: Data from Shah and Bhatti (1987).

described boundary conditions. In addition, the developing flow effect may be present if the flow passage is not long enough. Because these and other effects (such as brazing, fouling, fluid property variations, free convection, etc.) could affect the actual Nu and $f \cdot$ Re as outlined in Table 7.5, when very accurate Nu and $f \cdot$ Re are needed for specific applications, they are obtained experimentally even for simple flow passage geometries.

Let us further emphasize the influence of the entrance effect and surface roughness on Nu and f for fully developed laminar duct flow. For most duct geometries, the mean Nu and f will be within 10% of the fully developed value if $L/D_h > 0.2\,\text{Re} \cdot \text{Pr}$. If $L/D_h < 0.2\,\text{Re} \cdot \text{Pr}$, analytical correlations for fully developed flow may not be adequate, since Nu and f are higher in the developing region. A review of analytical correlations indicates that the entrance region effect could be nonnegligible for $L/D_h \approx 100$ for gas flows. Hence, its effect on the pressure drop should not be neglected even for $L/D_h \approx 100$ for gas flows. However, if the passage-to-passage nonuniformity (see Section 12.1.2) exists in a heat exchanger, it reduces Nu substantially, and also reduces f slightly (and the effect on f factors could be neglected for practical purposes). Hence, in compact heat exchanger applications, the increase in Nu due to the entrance-length effect is over-compensated by a reduction in Nu due to the passage-to-passage nonuniformity. Hence, the increase in Nu due to the entrance-length effect for $L/D_h \approx 100$ for gas flows is generally neglected.

The duct surface roughness generally does not affect Nu and f for fully developed laminar flow as long as the height of the surface roughness is negligible compared to the duct hydraulic diameter (i.e., $e/D_h < 0.01$). However, for highly compact flow passages (i.e., passages with small D_h), the surface roughness height may not be negligible in comparison to the passage D_h. In that case, the surface roughness then changes the

TABLE 7.5 **Influence of Increasing Specific Variables on Theoretical Fully Developed *Laminar* Friction Factors and Nusselt Numbers**

Variable	f	Nu
Entrance effect	Increases	Increases
Passage-to-passage nonuniformity	Decreases slightly	Decreases significantly
Gross flow maldistribution	Increases sharply	Decreases
Free convection in a horizontal passage	Increases	Increases
Free convection with vertical aiding flow	Increases	Increases
Free convection with vertical opposing flow	Decreases	Decreases
Property variation due to fluid heating	Decreases for liquids and increases for gases	Increases for liquids and decreases for gases
Property variation due to fluid cooling	Increases for liquids and decreases for gases	Decreases for liquids and increases for gases
Fouling	Increases sharply	Increases slightly
Surface roughness	Affects only if the surface roughness height profile is significant compared to the passage size (D_h)	Affects only if the surface roughness height profile is significant compared to the passage size (D_h)

Source: Data from Shah and Bhatti (1988).

effective flow cross-sectional geometry, which in turn affects Nu and $f \cdot \text{Re}$ (Table 7.3 shows that Nu and $f \cdot \text{Re}$ are geometry dependent). Thus, even in laminar flow, surface roughness can affect Nu and $f \cdot \text{Re}$ for highly compact heat exchangers, due to the change in the resulting flow passage geometry. This is commonly found in some recent research results for micro- and meso-scale heat exchangers ($D_h = 1$ to 1000 μm).

The entrance effects, flow maldistribution, free convection, property variation, fouling, and surface roughness all affect fully developed analytical correlations, as shown in Table 7.5. Hence, to consider these effects in real plate-fin plain fin geometries having fully developed flows, it is best to reduce the magnitude of the analytical Nu by a minimum of 10% and increase the value of the analytical $f \cdot \text{Re}$ by a minimum of 10% for design purposes.

Analytical values of L_{hy}^{+} and $K(\infty)$ are also listed in Table 7.3. The hydrodynamic entrance length L_{hy} [dimensionless form is $L_{hy}^{+} = L_{hy}/(D_h \cdot \text{Re})$] is the duct length required to achieve a maximum channel velocity of 99% of that for fully developed flow when the entering fluid velocity profile is uniform. Since the flow development region precedes the fully developed region, the entrance region effects could be substantial, even for channels having fully developed flow along a major portion of the channel. This increased friction in the entrance region and the change of momentum rate are taken into account by the incremental pressure drop number $K(\infty)$, defined by

$$\Delta p = \left[\frac{4f_{fd}L}{D_h} + K(\infty) \right] \frac{G^2}{2g_c \rho} \tag{7.69}$$

where the subscript fd denotes the fully developed value.

7.4.1.2 Transition Flow.

For the initiation of transition to turbulent flow, the lower limit of the critical Reynolds number (Re_{cr}), depends on the type of entrance (e.g., smooth vs. abrupt configuration at the exchanger flow passage entrance) in smooth ducts. For a sharp square inlet configuration, Re_{cr} is about 10 to 15% lower than that for a rounded inlet configuration. For most exchangers, the entrance configuration would be sharp. Tam and Ghajar (1997) provide some information on Re_{cr} and friction factors for isothermal and nonisothermal transition flow in a horizontal circular tube with different inlet geometries. The lower limits of Re_{cr} for various passages with a sharp square inlet configuration vary from about 2000 to 3100 (Bhatti and Shah, 1987). The upper limit of Re_{cr} may be taken as 10^4 for most practical purposes.

Transition-flow Fanning friction factor and Nusselt number correlations for circular tubes are summarized in Table 7.6. For the transition-flow Nu data, the Gnielinski correlation (1976) has a step function in Nu from transition to laminar flow at Re = 2300. Taborek (1990) proposed a linear proration of Nu_{lam} and Nu_{turb} between Re = 2000 and 8000 as given by Eq. (7.77) in Table 7.6. Ghajar and Tam (1994) provide transition-flow Nusselt numbers for a horizontal circular straight tube with three different inlets for the uniform wall heat flux boundary condition. The transition-flow f and Nu data for noncircular passages are rather sparse; Eqs. (7.70) and (7.77) may be used to obtain fair estimates of f and Nu for noncircular flow passages (having no sharp corners) using the hydraulic diameter as the characteristic dimension.

7.4.1.3 Turbulent Flow.

Analytical and experimental correlations for friction factors and Nusselt numbers are presented for smooth and rough circular tubes here. All these

TABLE 7.6 Some Important Correlations for f and Nu for Transition and Turbulent Flows in Circular and Noncircular Smooth Tubes

Type of Flow and Geometry	Correlations	Eq. No.	Remarks	Reference
Transition and turbulent flow, circular or noncircular duct	$f = A + B \cdot Re^{-1/m}$ 1. $A = 0.0054$; $B = 2.3 \times 10^{-8}$; $m = -2/3$ 2. $A = 0.00128$; $B = 0.1143$; $m = 3.2154$	(7.70)	Bhatti–Shah correlation 1. $2100 \leq Re \leq 4000$ 2. $4000 \leq Re \leq 10^7$ Accuracy ±2%	Bhatti and Shah (1987)
Turbulent flow in a smooth duct	Blasius: $\quad f = 0.0791Re^{-0.25}$ McAdams: $\quad f = 0.046Re^{-0.2}$ Bhatti and Shah: $\quad f = 0.00128 + 0.1143Re^{-0.311}$	(7.71) (7.72) (7.73)	$4000 < Re < 10^5$ $30,000 < Re < 10^6$ $4000 < Re < 10^7$ Accuracy ±2%	Bhatti and Shah (1987)
Turbulent flow in a smooth duct	$Nu = \dfrac{(f/2)Re \cdot Pr}{C + 12.7(f/2)^{1/2}(Pr^{2/3} - 1)}$ $C = 1.07 + \dfrac{900}{Re} - \dfrac{0.63}{1 + 10Pr}$	(7.74) (7.75)	Petukhov–Popov correlation Accuracy ±5% $4000 \leq Re \leq 5 \times 10^6$ $0.5 \leq Pr \leq 10^6$ Obtain f from Eq. (7.70) or (7.73)	Petukhov and Popov (1963)
Transition and turbulent flow, circular or noncircular duct	$Nu = \dfrac{(f/2)(Re - 1000)Pr}{1 + 12.7(f/2)^{1/2}(Pr^{2/3} - 1)}$	(7.76)	Gnielinski correlation Accuracy ±10% $2300 \leq Re \leq 5 \times 10^6$, $0.5 \leq Pr \leq 2{,}000$ Not a good correlation in the transition regime Obtain f from Eq. (7.70) or (7.73)	Gnielinski (1976)
Transition and turbulent flow, circular or noncircular duct	$Nu = \phi Nu_{lam} + (1 - \phi)Nu_{turb}$ $\phi = 1.33 - (Re/6000)$	(7.77) (7.78)	Use Nu_{turb} from Eq. (7.74) or (7.76) and Nu_{lam} from Table 7.3 Applicable for $2000 < Re < 8000$	Taborek (1990)
Turbulent flow in a smooth duct	$Nu = 0.023Re^{0.8} \cdot Pr^{0.4}$	(7.79)	So-called Dittus–Boelter correlation Used for approximate calculations	Winterton (1998)

Turbulent flow in a smooth duct	$\mathrm{Nu} = \begin{cases} 0.024\mathrm{Re}^{0.8} \cdot \mathrm{Pr}^{0.4} & \text{for heating} \\ 0.026\mathrm{Re}^{0.8} \cdot \mathrm{Pr}^{0.3} & \text{for cooling} \end{cases}$	(7.80)	Dittus–Boelter correlation $2500 < \mathrm{Re} < 1.24 \times 10^5$ and $0.7 \leq \mathrm{Pr} \leq 120$ See text for accuracy	Bhatti and Shah (1987)
Turbulent flow in a smooth duct	$\mathrm{Nu} = 0.023\mathrm{Re}^{0.8} \cdot \mathrm{Pr}^{1/3}$	(7.81)	Colburn correlation $10^4 < \mathrm{Re} < 10^5$ and $0.5 \leq \mathrm{Pr} \leq 3$ Accuracy within $+27.6\%$ and -19.8%	Bhatti and Shah (1987)
Liquid metal turbulent flow in a smooth duct	$\mathrm{Nu}_H = 5.6 + 0.0165\mathrm{Re}_f^{0.85} \cdot \mathrm{Pr}_w^{0.86}$	(7.82)	Chen and Chiou correlation	Chen and Chiou (1981)
	$\mathrm{Nu}_T = 4.5 + 0.0156\mathrm{Re}_f^{0.85} \cdot \mathrm{Pr}_w^{0.86}$	(7.83)	$10^4 < \mathrm{Re} < 5 \times 10^6$ and $0 \leq \mathrm{Pr} \leq 0.1$	
Turbulent flow in a smooth duct	$\mathrm{Nu} = 0.023\mathrm{Re}^{n} \cdot \mathrm{Pr}^{0.4}$	(7.84)	Petukhov–Popov correlation with exponent n as in Fig. 7.19	Present authors

correlations are applicable to low-Mach-number flows, as is the case in heat exchanger applications.

Smooth Circular Tube. In fully developed turbulent flow, the *constant-property*[†] Nusselt number is independent of thermal boundary conditions for Pr > 0.7 but is dependent on both Re and Pr. In contrast, the constant-property Nu is independent of Re and Pr in fully developed laminar flow but is dependent on the thermal boundary conditions. In fully developed flow with high-Prandtl-number fluids, the thermal resistance is primarily very close to the wall, and the temperature distribution across the cross section is flat and is insensitive to different thermal boundary conditions at wall. For Pr < 0.7, the turbulent flow Nusselt number is also dependent on the thermal boundary conditions. For very low Prandtl number fluids (liquid metals, Pr < 0.03), the thermal diffusivity of the fluid is very high, resulting in thermal resistance distributed over the entire flow cross section (see also Table 7.1). Different thermal boundary conditions then yield different temperature distributions across a cross section, resulting in different h and Nu. Sleicher and Tribus (1957) computed analytically the ratio of Nu for the Ⓗ boundary condition to that for the Ⓣ boundary condition for turbulent flow. They reported that $Nu_H/Nu_T \geq 1$, as is the case for laminar flow, but the departure from unity is negligible for Pr > 1.0.

A large number of analytical solutions and empirical correlations are available for turbulent flow in a pipe (Bhatti and Shah, 1987). The most commonly referred to and the most accurate correlations for a smooth circular tube are presented in Table 7.6. Equations (7.70)–(7.84) are assigned to these correlations in Table 7.6. Some comments on the correlations of Table 7.6 are now presented.

- The Petukhov and Popov (1963) correlation, Eq. (7.74), is the most accurate. Petukhov and Popov simplified Eq. (7.74) by modifying the expression for C of Eq. (7.74) to simply $C = 1.07$. Note that Gnielinski (1976) further modified C to 1.00 and Re to (Re − 1000), as shown in Eq. (7.76), to extend the validity of Re to 2300. The Nusselt numbers based on Eq. (7.76) for Pr ≥ 0.5 are presented in Fig. 7.18.

- A comparison of the Dittus–Boelter (1930) correlations [Eq. (7.80)] with the Gnielinski correlation, for heating for $10^4 \leq Re \leq 1.24 \times 10^5$, is as follows: (1) 13.5 to 17% higher for air (Pr = 0.7), (2) 15% lower to 7% higher for water (3 ≤ Pr ≤ 10), and (3) 10% lower to 21% higher for oil (Pr = 120). Predictions of the cooling correlation for $10^4 \leq Re \leq 1.24 \times 10^5$ are: (1) 29 to 33% higher for air (Pr = 0.7), (2) 26% lower to 3% higher for water (3 ≤ Pr ≤ 10), and (3) 39 to 18% lower for oil (Pr = 120). The predictions below Re = 10^4 are much worse. Based on the comparisons above, the Dittus–Boelter correlation for air and gases is not very accurate.

- The Nusselt numbers for liquid metals are dependent on thermal boundary conditions, as mentioned earlier. In Eqs. (7.82) and (7.83), Re_f denotes the Reynolds number with the fluid properties evaluated at the *film* temperature $(T_m + T_w)/2$, and Pr_w denotes the Prandtl number with the fluid properties evaluated at the wall temperature. Note that the influence of temperature-dependent fluid properties is thus included in Eqs. (7.82) and (7.83).

[†] Turbulent-flow Nusselt numbers are strongly dependent on the variations in the fluid properties across a cross section when large temperature differences are involved. This subject is covered in Section 7.6.

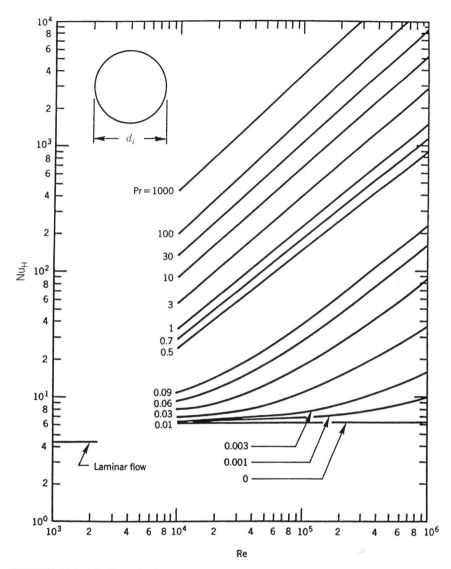

FIGURE 7.18 Nu_H for a circular tube for fully developed turbulent flow. (From Bhatti and Shah, 1987.)

Recently, Churchill and co-researchers (2001, 2002) have arrived at a theory-based correlation that is very accurate for the complete range of Pr $(0 < Pr < \infty)$.

The Petukhov–Popov correlation could be represented in a simplified formula (similar to the Dittus–Boelter correlation) as shown in Eq. (7.84), with the exponent of Re as n. Note that the Prandtl number exponent for fully developed turbulent flow in a circular tube is now accepted as 0.4 (Gnielinski, 1976; Kays and Crawford, 1993). The Reynolds number exponent n is then obtained by comparing Eqs. (7.84) and (7.74) and is dependent on Re as shown in Fig. 7.19.

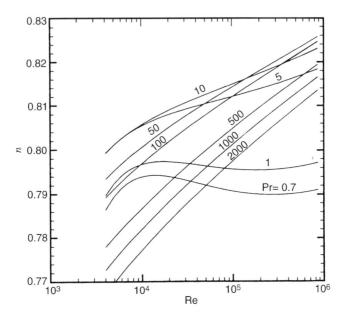

FIGURE 7.19   Dependence of the exponent n of Eq. (7.84) on Re and Pr when comparing Eq. (7.84) with Eq. (7.74).

Most of the thermal or flow resistance is concentrated in the viscous sublayer near the wall for turbulent flow, and the temperature and velocity profiles are relatively flat over most of the channel cross section. Hence, provided that there are no sharp corners, the influence of the channel shape in turbulent flow is not as great as that in laminar flow. A common practice is to employ hydraulic diameter as the characteristic length in the circular tube correlations to predict Nu and f for noncircular channels. Hence, it is generally an accepted fact that the hydraulic diameter correlates Nu and f for fully developed turbulent flow in circular and noncircular ducts. This is true for the results to be accurate to within $\pm15\%$[†] for most noncircular ducts except for those having sharp or acute-angled corners in the flow passage or concentric annuli with the inner wall heated. In these cases, the values of Nu and f factors could be more than 15% lower than the values for the circular tube. Also, the corners and noncircularity of flow passages would affect the flow phenomena, and Nu and f will be somewhat different from those for the circular tube.

A careful observation of accurate experimental friction factors for all noncircular smooth ducts (and the correlations in Table 7.7) reveals that ducts with laminar $f \cdot \mathrm{Re} < 16$ have turbulent f factors lower than those for the circular tube; whereas ducts with laminar $f \cdot \mathrm{Re} > 16$ have turbulent f factors higher than those for the circular tube. Similar trends are observed for the Nusselt numbers. If one is satisfied

---

[†] This order of accuracy is adequate for most engineering calculations for overall heat transfer and pressure drop, although it may not be adequate for detailed flow distribution and local temperature distribution analyses as required, for example, in a nuclear reactor or when mass production of heat exchangers requires accuracy within $\pm5\%$ or lower for cost savings.

TABLE 7.7   Fully Developed Turbulent Flow Friction Factors and Nusselt Numbers (Pr > 0.5) for Some Technically Important Smooth-Walled Ducts

Duct Geometry	Recommended Correlations

Rectangular:

$$D_h = \frac{4ab}{a+b}, \alpha^* = \frac{2a}{2b}$$

$$\frac{D_\ell}{D_h} = \frac{2}{3} + \frac{11}{24}\alpha^*(2 - \alpha^*)$$

f factors: (1) Substitute D_ℓ for D_h in the circular duct correlation, Eq. (7.71), and calculate f from the resulting equation. (2) Alternatively, calculate f from $f = (1.0875 - 0.1125\alpha^*)f_c$, where f_c is the friction factor for the circular duct using D_h. In both cases, predicted f factors are within $\pm 5\%$ of the experimental results.

Nusselt numbers: (1) With uniform heating at four walls, use circular duct Nu correlation for an accuracy of $\pm 9\%$ for $0.5 \leq$ Pr ≤ 100 and $10^4 \leq$ Re $\leq 10^6$. (2) With equal heating at two long walls, use circular duct correlation for an accuracy of $\pm 10\%$ for $0.5 <$ Pr ≤ 10 and $10^4 \leq$ Re $\leq 10^5$. (3) With heating at one long wall only, use circular duct correlation to get approximate Nu values for $0.5 <$ Pr < 10 and $10^4 \leq$ Re $\leq 10^6$. These calculated values may be up to 20% higher than the actual experimental values.

Isosceles triangular:

$$D_h = \frac{4ab}{a + \sqrt{a^2 + 4b^2}}$$

$$\frac{D_g}{D_h} = \frac{1}{2\pi}\left[3\ln\cot\frac{\theta}{2} + 2\ln\tan\frac{\phi}{2} - \ln\tan\frac{\theta}{2}\right]$$

where $\theta = (90° - \phi)/2$

For $0 < 2\phi < 60°$, use circular duct f and Nu correlations with D_h replaced by D_g; for $2\phi = 60°$, replace D_h by D_ℓ ($= \sqrt{3}a$) and for $60° < 2\phi \leq 90°$ use circular duct correlations directly with D_h. Predicted f and Nu are within $+9\%$ and -11% of the experimental values. No recommendations can be made for $2\phi > 90°$ due to the lack of experimental data.

Concentric annular:

$$D_h = 2(r_o - r_i), r^* = \frac{r_i}{r_o}$$

$$\frac{D_\ell}{D_h} = \frac{1 + r^{*2} + (1 - r^{*2})/\ln r^*}{(1 - r^*)^2}$$

f Factors: (1) Substitute D_ℓ for D_h in the circular duct correlation, Eq. (7.71) and calculate f from the resulting equation. (2) Alternatively, calculate f from $f = (1 + 0.0925r^*)f_c$ where f_c is the friction factor for the circular duct using D_h. In both cases, predicted f factors are within $\pm 5\%$ of the experimental results.

Nusselt numbers: In all the following recommendations, use D_h with the *wetted* perimeter in Nu and Re: (1) Nu at the outer wall *can* be determined from the circular duct correlation within an accuracy of about $\pm 10\%$ regardless of the heating/cooling condition at the inner wall. (2) Nu at the inner wall *cannot* be determined accurately regardless of the heating/cooling condition at the outer wall. (3) For the Ⓗ and Ⓣ boundary conditions, see Bhatti and Shah (1987) for correlations.

Source: Data from Bhatti and Shah (1987).

within $\pm 15\%$ accuracy, Eqs. (7.70) and (7.76) in Table 7.6 for f and Nu can be used for noncircular passages with the hydraulic diameter as the characteristic length in f, Nu, and Re; otherwise, refer to Table 7.7 for more accurate results for turbulent flow in those duct geometries.

Example 7.4   A parallelflow exchanger is operating at the following conditions: $\dot{m}_h = 3000\,\text{kg/h}$, $\dot{m}_c = 6000\,\text{kg/h}$, $T_{h,i} = 50°\text{C}$, $T_{h,o} = 30°\text{C}$, $T_{c,i} = 10°\text{C}$. The properties for both streams (water) are $\rho = 1000\,\text{kg/m}^3$, $c_p = 4180\,\text{J/kg}\cdot\text{K}$, $k = 0.59\,\text{W/m}\cdot\text{K}$, and $\mu = 0.001\,\text{Pa}\cdot\text{s}$. The heat transfer coefficients on both sides are $5000\,\text{W/m}^2\cdot\text{K}$. Assume fouling resistances and the wall thermal resistance to be negligible. Determine how much surface area is necessary for this exchanger. Assume water properties constant with temperature and $D_h = 15\,\text{mm}$ on both fluid sides.

The heat transfer coefficients vary with the velocity (or flow rate) according to the Dittus–Boelter correlation:

$$\frac{hD_h}{k} = 0.023 \left(\frac{GD_h}{\mu}\right)^{0.8} \cdot \text{Pr}^{0.4}$$

What are the exit temperatures if (1) the hot fluid rate is doubled; (2) flow rates of both fluids are doubled? Also compute and compare the heat transfer rates for each case.

SOLUTION

Problem Data and Schematic: Flow rates, heat transfer coefficients, and inlet and outlet temperatures as well as fluid properties are given in Fig. E7.4 for a parallelflow exchanger. Also given is the variation of heat transfer coefficients with water velocity in terms of the Dittus–Boelter correlation.

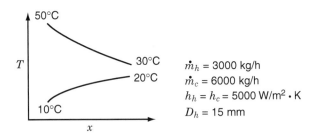

FIGURE E7.4

Determine: The surface areas of on both fluid sides, and the exit temperatures for various specified flow rates.

Assumptions: Fluid properties are constant, wall thermal resistance and fouling resistances are negligible, and both fluids flow through smooth surfaces.

Analysis: In the first portion of the problem, ε and C^* are specified. We determine NTU and subsequently the surface area. Note that the hot fluid is the $C_{\min}$ fluid since its flow rate is considerably lower than that for the cold fluid.

$$\varepsilon = \frac{T_{h,i} - T_{h,o}}{T_{h,i} - T_{c,i}} = \frac{(50 - 30)°C}{(50 - 10)°C} = 0.50$$

$$\dot{m}_h = 3000\,\text{kg/h} = (3000/3600)\,\text{kg/s} = 0.8333\,\text{kg/s}$$

$$\dot{m}_c = (6000/3600)\,\text{kg/s} = 1.6667\,\text{kg/s}$$

$$C_h = \dot{m}_h c_{p,h} = 0.8333\,\text{kg/s} \times 4.180\,\text{kJ/kg}\cdot\text{K} = 3.483\,\text{kJ/s}\cdot\text{K} = 3.483\,\text{kW/K}$$

$$C_c = \dot{m}_c c_{p,c} = 1.6667\,\text{kg/s} \times 4.180\,\text{kJ/kg}\cdot\text{K} = 6.967\,\text{kJ/s}\cdot\text{K} = 6.967\,\text{kW/K}$$

$$C^* = \frac{C_{\min}}{C_{\max}} = \frac{C_h}{C_c} = \frac{3.483\,\text{kW/K}}{6.967\,\text{kW/K}} = 0.5$$

Using Eq. I.2.2 of Table 3.6 for the parallelflow exchanger for the ε-NTU relationship, we have

$$\text{NTU} = \frac{1}{1+C^*}\ln\left[\frac{1}{1-\varepsilon(1+C^*)}\right] = \frac{1}{1+0.5}\ln\left[\frac{1}{1-0.5(1+0.5)}\right] = 0.9242$$

Hence, from the definition of NTU, we get

$$UA = \text{NTU}\cdot C_{\min} = 0.9242 \times 3.483\,\text{kW/K} = 3.219\,\text{kW/K}$$

For thin-walled tubes, $A_h \approx A_c = A$. Also, $\mathbf{R}_w \approx 0$, $h_{h,f} \approx 0$, and $h_{c,f} \approx 0$. Since there are no fins, $\eta_{o,h} = \eta_{o,c} = 1$. Therefore, from Eq. (3.30a),

$$\frac{1}{U_h} = \frac{1}{h_h} + \frac{1}{h_c} = \frac{1}{5000\,\text{W/m}^2\cdot\text{K}} + \frac{1}{5000\,\text{W/m}^2\cdot\text{K}} = \frac{1}{2500\,\text{W/m}^2\cdot\text{K}}$$

$$\Rightarrow U_h = 2500\,\text{W/m}^2\cdot\text{K}$$

Since we have determined UA and U_h separately, the surface area is then computed as

$$A_h = \frac{UA}{U_h} = \frac{3219\,\text{W/K}}{2500\,\text{W/m}^2\cdot\text{K}} = 1.288\,\text{m}^2 \qquad\qquad Ans.$$

Similarly,

$$A_c = 1.288\,\text{m}^2 \qquad\qquad Ans.$$

The heat transfer rate q for this case is

$$q = \varepsilon C_{\min}(T_{h,i} - T_{c,i}) = 0.50 \times 3.483\,\text{kW/K} \times (50 - 10)°C = 69.7\,\text{kW}$$

The cold-fluid outlet temperature can be calculated from an overall energy balance involving enthalpy rate changes of both hot and cold fluids:

$$C_c(T_{c,o} - T_{c,i}) = C_h(T_{h,i} - T_{h,o})$$

or

$$T_{c,o} = T_{c,i} + \frac{C_h}{C_c}(T_{h,i} - T_{h,o}) = 10 + 0.5 \times (50 - 30) = 20°C$$

Case 1. If the hot-fluid flow rate is doubled, the new value of C_h will be

$$C_{h,\text{new}} = 2 \times C_{h,\text{old}} = 2 \times 3.483\,\text{kW/K} = 6.967\,\text{kW/K}$$

and

$$C_h = C_c = C_{\min} = 6.967\,\text{kW/K}$$

Because of the doubling of the hot-fluid flow rate, the heat transfer coefficient on the hot-fluid side will increase but less than doubled. Subsequently, U and NTU will increase, resulting in lower ε. From the Dittus–Boelter correlation, $h \propto G^{0.8} \propto \dot{m}^{0.8}$. Hence, the ratio of the two heat transfer coefficients will be

$$h_{h,\text{new}} = h_{h,\text{old}}\left(\frac{\dot{m}_{\text{new}}}{\dot{m}_{\text{old}}}\right)^{0.8} = 5000\,\text{W/m}^2 \cdot \text{K} \times 2^{0.8} = 8705.5\,\text{W/m}^2 \cdot \text{K}$$

The new overall heat transfer coefficient is

$$\frac{1}{U_{\text{new}}} = \frac{1}{h_{h,\text{new}}} + \frac{1}{h_c} = \frac{1}{8705.5\,\text{W/m}^2 \cdot \text{K}} + \frac{1}{5000\,\text{W/m}^2 \cdot \text{K}} = 0.0003149\,\text{m}^2 \cdot \text{K/W}$$

Therefore, $U_{\text{new}} = 3175.9\,\text{W/m}^2 \cdot \text{K}$. So the new value of NTU will be

$$\text{NTU} = \frac{3175.9\,\text{W/m}^2 \cdot \text{K} \times 1.288\,\text{m}^2}{6.967 \times 10^3\,\text{W/K}} = 0.5871$$

Since $C_h = C_c = 6.967\,\text{kW/K}$, $C^* = 1$. Hence, from Eq. I.2.1 of Table 3.6 or for parallel-flow in Table 3.3,

$$\varepsilon = \frac{1 - e^{-\text{NTU}(1+C^*)}}{1 + C^*} = \frac{1}{2}(1 - e^{-2\text{NTU}}) = 0.5(1 - e^{-2 \times 0.5871}) = 0.3455$$

The new outlet temperatures for this $C^* = 1$ case are

$$T_{h,o} = T_{h,i} - \varepsilon(T_{h,i} - T_{c,i}) = 50°C - 0.3455(50 - 10)°C = 36.2°C \qquad \textit{Ans.}$$

$$T_{c,o} = T_{c,i} + \varepsilon(T_{h,i} - T_{c,i}) = 10°C + 0.3455(50 - 10)°C = 23.8°C \qquad \textit{Ans.}$$

The heat transfer rate is

$$q = \varepsilon C_{\min}(T_{h,i} - T_{c,i}) = 0.3455 \times 6.967\,\text{kW/K} \times (50 - 10)°C = 96.3\,\text{kW}$$

Case 2. If the flow rates of both fluids are doubled, then

$$C_h = 6.967\,\text{kW/K} \qquad C_c = 13.933\,\text{kW/K} \qquad C_{\min} = C_h = 6.967\,\text{kW/K}$$

The new heat capacity rate ratio is

$$C^* = \frac{6000\,\text{kg/h} \times 4180\,\text{J/kg} \cdot \text{K}}{12000\,\text{kg/h} \times 4180\,\text{J/kg} \cdot \text{K}} = 0.5$$

Based on the results of case 1, $h_h = h_c = 8705.5\,\text{W/m}^2 \cdot \text{K}$. Therefore,

$$U = \left(\frac{1}{h_h} + \frac{1}{h_c}\right)^{-1} = \frac{8705.5\,\text{W/m}^2 \cdot \text{K}}{2} = 4352.8\,\text{W/m}^2 \cdot \text{K}$$

and NTU will be

$$\text{NTU} = \frac{UA}{C_{\min}} = \frac{4352.8\,\text{W/m}^2 \cdot \text{K} \times 1.288\,\text{m}^2}{6967\,\text{W/K}} = 0.8047$$

The exchanger effectiveness from Eq. (I.2.1) of Table 3.6 is

$$\varepsilon = \frac{1 - e^{-\text{NTU}(1+C^*)}}{1 + C^*} = \frac{1 - e^{-0.875 \times 1.5}}{1 + 0.5} = 0.4673$$

Finally, the new outlet temperatures are

$$T_{h,o} = T_{h,i} - \varepsilon(T_{h,i} - T_{c,i}) = 50°\text{C} - 0.4673(50 - 10)°\text{C} = 31.1°\text{C} \qquad \textit{Ans.}$$

$$T_{c,o} = T_{c,i} + \varepsilon C^*(T_{h,i} - T_{c,i}) = 10°\text{C} + 0.4673 \times 0.5 \times (50 - 10)°\text{C} = 19.3°\text{C} \qquad \textit{Ans.}$$

and the heat transfer rate is

$$q = \varepsilon C_{\min}(T_{h,i} - T_{c,i}) = 0.4673 \times 6.967\,\text{kW/K} \times (50 - 10)°\text{C} = 130.2\,\text{kW} \qquad \textit{Ans.}$$

The results of three cases are summarized as follows:

Case	$\dot{m}_h$ (kg/h)	$\dot{m}_c$ (kg/h)	h_h (kW/m²·K)	h_c (kW/m²·K)	NTU	C^*	ε	q	$T_{h,o}$ (°C)	$T_{c,o}$ (°C)
Base	3000	6000	5.0	5.0	0.9242	0.5	0.5000	69.7	30.0	20.0
1	6000	6000	8.7	5.0	0.5871	1.0	0.3455	96.3	36.2	23.8
2	6000	12000	8.7	8.7	0.8047	0.5	0.4673	130.2	31.3	19.3

Discussion and Comments: The objective of this example is to demonstrate the effect of increasing flow rates of one or both fluids on exchanger performance. Since the ε-NTU relationship is not linear in general [for this problem, it is given for parallelflow in Table 3.3], one cannot expect that doubling the flow rates on one or both fluid sides will double the heat transfer rates. Also, the heat transfer coefficient will vary nonlinearly with the flow rate; in this example, it will vary according to the Dittus–Boelter correlation mentioned in the problem statement.

Doubling the flow rate on the C_{min} side as in case 1 decreases NTU, increases C^*, and decreases ε as expected. However, the decrease in ε is much less than linear, and C_{min} increases linearly with $\dot{m}_h$. Hence, the net heat transfer rate in the exchanger increases. Notice that despite increasing q, the $T_{h,o}$ increases (and not decreases) due to the increased flow rate. However, since we have not changed the C_{max} ($= C_c$) flow rate, the cold fluid outlet temperature increases due to the increased heat transfer in the exchanger. Thus, this problem clearly demonstrates the peculiar behavior of arrows going up or down for $T_{h,o}$ and $T_{c,o}$ in Table 3.5. However, notice that we cannot use Table 3.5 directly for this problem since $C_{min} = C_h$ for this problem and Table 3.5 is for $C_{min} = C_c$.

Next, when we double both fluid flow rates, the value of NTU is in between the base case and case 1, due to increased overall U. This, along with decreased C^* compared to case 1, produces higher ε and q. However, notice that due to the nonlinear nature of the problem, by doubling the flow rates on both sides, q is still not doubled compared to the base case. As a result, we get $T_{c,o}$ even lower than that for the base case. However, as expected, $T_{h,o}$ will be in between that for the base case and case 1.

Thus, we find that due to the nonlinear relationship between h and $\dot{m}$, and between ε and NTU, we obtain outlet temperatures that are higher or lower than the base case from the exchanger with increasing flow rates. Thus the use of Table 3.5 for $C_{min} = C_c$ or its counterpart for $C_{min} = C_h$ can provide qualitative guidelines on the variations in $T_{h,o}$ and $T_{c,o}$, but actual computation as shown in this example is necessary for quantitative results.

Example 7.5   You as a designer need to decide whether to select a rectangular (aspect ratio $\alpha^* = \frac{1}{8}$) or a square duct ($\alpha^* = 1$) of 5000 mm² cross-sectional area for air flowing at 0.05 kg/s at 27°C for maximum heat transfer and for minimum pressure drop. The duct length is 5 m. The duct wall temperature is 100°C, due to the steam on the other side. For fully developed turbulent flow for rectangular and square ducts, use the following correlations for friction factors and Nusselt numbers:

$$f = 0.0791\,\mathrm{Re}^{-0.25}(1.0875 - 0.1125\alpha^*)$$

$$\mathrm{Nu} = 0.024\,\mathrm{Re}^{0.8} \cdot \mathrm{Pr}^{0.4}$$

where α^* is the aspect ratio of the rectangular flow passage. Use the following properties for air:

$$\rho = 1.0463\,\mathrm{kg/m^3} \qquad \mu = 202.5 \times 10^{-7}\,\mathrm{Pa \cdot s}, \qquad k = 29.1 \times 10^{-3}\,\mathrm{W/m \cdot K}$$

$$c_p = 1008\,\mathrm{J/kg \cdot K} \qquad \mathrm{Pr} = 0.702$$

SOLUTION

Problem Data and Schematic: Two pipes for this problem are shown in Fig. E7.5. The fluid properties are specified in the problem statement.

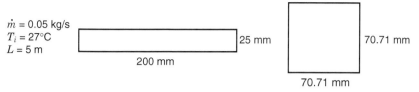

FIGURE E7.5

Determine: The pressure drop and heat transfer for each duct, and decide which geometry will yield a lower pressure drop and higher heat transfer.

Assumptions: Fully developed turbulent flow exists with constant fluid properties.

Analysis: Let us first compute the hydraulic diameters and Reynolds numbers for the rectangular and square ducts.

$\alpha^* = 1/8$	$\alpha^* = 1$
$D_h = \dfrac{4A_o}{P} = \dfrac{4 \times 200\,\text{mm} \times 25\,\text{mm}}{2(200 + 25)\,\text{mm}}$	$D_h = \dfrac{4A_o}{P} = \dfrac{4 \times 70.71\,\text{mm} \times 70.71\,\text{mm}}{2(70.71 + 70.71)\,\text{mm}}$
$= 44.44\,\text{mm}$	$= 70.71\,\text{mm}$
$G = \dfrac{\dot{m}}{A_o} = \dfrac{0.05\,\text{kg/s}}{5000 \times 10^{-6}\,\text{m}^2} = 10\,\text{kg/m}^2 \cdot \text{s}$	$G = \dfrac{\dot{m}}{A_o} = \dfrac{0.05\,\text{kg/s}}{5000 \times 10^{-6}\,\text{m}^2} = 10\,\text{kg/m}^2 \cdot \text{s}$
$\text{Re} = \dfrac{GD_h}{\mu} = \dfrac{10\,\text{kg/m}^2 \cdot \text{s} \times (44.44 \times 10^{-3}\,\text{m})}{202.5 \times 10^{-7}\,\text{Pa} \cdot \text{s}}$	$\text{Re} = \dfrac{GD_h}{\mu} = \dfrac{10\,\text{kg/m}^2 \cdot \text{s} \times (70.71 \times 10^{-3}\,\text{m})}{202.5 \times 10^{-7}\,\text{Pa} \cdot \text{s}}$
$= 21{,}946$	$= 34{,}919$

Heat Transfer. To calculate the air outlet temperature and heat transfer rate, we determine the exchanger effectiveness for computed values of NTU and C^*. For the T boundary condition, the temperature distributions of the wall and air are similar to those of Fig. 3.1c. They correspond to the $C^* = 0$ case. Hence the ε-NTU formula of Eq. (3.84) applies. Since there is no thermal resistance specified for the wall, the other fluid side and fouling, and no fins, we have from Eq. (3.20),

$$UA = hA$$

where h is computed from the Nusselt number correlation given. We now compute q and $T_{a,o}$ separately for $\alpha^* = \frac{1}{8}$ and 1.

$\alpha^* = \frac{1}{8}$ *Case.* Let us first compute the air-side heat transfer coefficient. From the correlation given,

$$\text{Nu} = 0.024\,\text{Re}^{0.8} \cdot \text{Pr}^{0.4} = 0.024(21946)^{0.8}(0.702)^{0.4} = 61.92$$

Hence,

$$h = \text{Nu}\,\frac{k}{D_h} = 61.92 \times \frac{29.1 \times 10^{-3}\,\text{W/m} \cdot \text{K}}{44.44 \times 10^{-3}\,\text{m}} = 40.55\,\text{W/m}^2 \cdot \text{K}$$

To compute $\text{NTU} = UA/C_{\min} = hA/C_{\min}$ for this problem, let us compute A and $C_{\min}$. The surface area A of 5 m long rectangular pipe ($\alpha^* = \frac{1}{8}$) is

$$A = PL = 2(200 + 25) \times 10^{-3}\,\text{m} \times 5\,\text{m} = 2.25\,\text{m}^2$$

$$C_{\min} = \dot{m}c_p = 0.05\,\text{kg/s} \times 1008\,\text{J/kg} \cdot \text{K} = 50.40\,\text{W/K}$$

Therefore,

$$\text{NTU} = \frac{hA}{C_{\min}} = \frac{40.55\,\text{W/m}^2 \cdot \text{K} \times 2.25\,\text{m}^2}{50.40\,\text{W/K}} = 1.810$$

Now

$$\varepsilon = 1 - e^{-\text{NTU}} = 1 - e^{-1.810} = 0.8364$$

$$q = \varepsilon C_{\min}(T_w - T_{a,i}) = 0.8364 \times 50.40\,\text{W/K} \times (100 - 27)^\circ\text{C} = 3077\,\text{W} \qquad Ans.$$

The outlet temperature from the energy balance is

$$T_{a,o} = T_{a,i} + \frac{q}{C_{\min}} = 27^\circ\text{C} + \frac{3077\,W}{50.40\,\text{W/K}} = 88.1^\circ\text{C}$$

$\alpha^* = 1$ *Case.* For this case,

$$\text{Nu} = 0.024\,\text{Re}^{0.8} \cdot \text{Pr}^{0.4} = 0.024(34919)^{0.8}(0.702)^{0.4} = 89.78$$

Hence h, A, and NTU are calculated as follows:

$$h = \text{Nu}\,\frac{k}{D_h} = 89.78 \times \frac{29.1 \times 10^{-3}\,\text{W/m} \cdot \text{K}}{70.71 \times 10^{-3}\,\text{m}} = 36.95\,\text{W/m}^2 \cdot \text{K}$$

$$A = PL = 2(70.71 + 70.71) \times 10^{-3}\,\text{m} \times 5\,\text{m} = 1.414\,\text{m}^2$$

$$C_{\min} = \dot{m}c_p = 0.05\,\text{kg/s} \times 1008\,\text{J/kg} \cdot \text{K} = 50.40\,\text{W/K}$$

$$\text{NTU} = \frac{hA}{C_{\min}} = \frac{36.95\,\text{W/m}^2 \cdot \text{K} \times 1.414\,\text{m}^2}{50.40\,\text{W/K}} = 1.037$$

Now determine ε, q, and $T_{a,o}$ as follows:

$$\varepsilon = 1 - e^{-\text{NTU}} = 1 - e^{-1.037} = 0.6455$$

$$q = \varepsilon C_{\min}(T_w - T_{a,i}) = 0.6455 \times 50.40\,\text{W/K} \times (100 - 27)^\circ\text{C} = 2375\,\text{W} \qquad Ans.$$

The outlet temperature from the energy balance is

$$T_{a,o} = T_{a,i} + \frac{q}{C_{\min}} = 27^\circ\text{C} + \frac{2375\,\text{W}}{50.40\,\text{W/K}} = 74.1^\circ\text{C} \qquad Ans.$$

Pressure Drop for $\alpha^* = \frac{1}{8}$. From the correlation given,

$$f = 0.0791\text{Re}^{-0.25}(1.0875 - 0.1125\alpha^*) = 0.0791(21946)^{-0.25}(1.0875 - 0.1125 \times \tfrac{1}{8})$$

$$= 0.006976$$

For this single duct with $\sigma \approx 1$, entrance and exit losses will be small and the pressure drop due to flow acceleration will also be small. Hence, the core pressure drop for this rectangular duct, from Eq. (6.29), is given by

$$\Delta p = \frac{4fLG^2}{2g_c\rho D_h} = \frac{4 \times 0.006976 \times 5\,\text{m} \times (10\,\text{kg/m}^2 \cdot \text{s})^2}{2 \times 1 \times 1.0463\,\text{kg/m}^3 \times (44.44 \times 10^{-3}\,\text{m})} = 150.0\,\text{Pa} \qquad Ans.$$

Pressure Drop for $\alpha^* = 1$. The friction factor for this case, from the correlation given, is

$$f = 0.0791\text{Re}^{-0.25}(1.0875 - 0.1125\alpha^*)$$
$$= 0.0791(34919)^{-0.25}(1.0875 - 0.1125 \times 1)$$
$$= 0.005642$$

Hence, the frictional pressure drop for this pipe is

$$\Delta p = \frac{4fLG^2}{2g_c\rho D_h} = \frac{4 \times 0.005642 \times 5\,\text{m} \times (10\,\text{kg/m}^2 \cdot \text{s})^2}{2 \times 1 \times 1.0463\,\text{kg/m}^3 \times (70.71 \times 10^{-3}\,\text{m})} = 76.3\,\text{Pa} \qquad Ans.$$

The foregoing results can be summarized as follows:

α^*	q (W)	Δp (Pa)	$q/q_{\alpha^*=1}$	$\Delta p/\Delta p_{\alpha^*=1}$
$\frac{1}{8}$	3077	150.0	1.30	1.97
1	2375	76.3	1.00	1.00

From the results above and intermediate aspect ratios, we find that the lower the aspect ratio of a rectangular duct, the higher is the heat transfer rate and pressure drop in turbulent flows. The increase in Δp is higher than the increase in q. Hence, if the pressure drop is the constraint, choose a square duct in the rectangular duct family. If the higher heat transfer rate is the requirement, choose a rectangular duct with as low an aspect ratio as possible compared to a square duct.

Discussion and Comments: This example demonstrates that the pressure drop increases considerably as the aspect ratio of a rectangular duct decreases in turbulent flow. It can also be shown that this pressure drop increase with decreasing aspect ratio is even more pronounced (>1.97) for laminar flow. Hence, if one is interested in a rectangular duct design for fluid flow only, the square duct will provide the minimum pressure drop. The heat transfer rate increases with decreasing aspect ratio as shown above, but the increase is lower compared to the increase in the pressure drop. Although this problem is meant for the duct design connecting various components of a thermal system, if one is interested in heat exchanger surface design, the exchanger surface selection criteria are presented in Section 10.3.

Circular Tube with Surface Roughness. A roughness element has no effect on laminar flow *unless* the height of the roughness element is not negligible compared to the flow cross-sectional size[†]. However, it exerts a strong influence in transition and turbulent

---

[†] This means that surface roughness does influence j (or Nu) and f factors, even in laminar flow if the surface roughness height is not negligible compared to the flow passage size. This is because the flow cross-section shape is changed due to the surface roughness and hence Nu and f change, as discussed in Section 7.4.1.1.

flow regimes; in these regimes, a roughness element on the surface causes local flow separation and reattachment. This generally results in an increase in the friction factor as well as in the heat transfer coefficient in a certain Reynolds number range. The major thermal resistance to heat transfer in a turbulent boundary layer for liquids is the viscous sublayer. It causes a 60% and 95% temperature drop for liquids with $\text{Pr} = 5$ and 100, respectively (Šlančiauskas, 2001). Hence, if the roughness element height e is of the same order of magnitude as the laminar (viscous) sublayer thickness δ_ℓ in turbulent flow (see Fig. 7.3), the roughness element tends to break up the laminar sublayer thereby, increasing the wall shear stress and heat transfer. In fact, if the surface is sufficiently rough, no viscous sublayer can exist. In that case, the apparent turbulent shear stresses are transmitted directly to the wall as form drag. For air and gases, the surface roughness–induced heat transfer enhancement is due to greater height surface roughness, which creates recirculation (flow mixing) and flow impact (reattachment) downstream, making the viscous sublayer thinner. It is thus apparent that the ratio e/δ_ℓ is a determining factor for the effect of surface roughness. The surface roughness provides heat transfer enhancement in the transition regime only, up to about a surface roughness Reynolds number e^+ [defined by Eq. (7.85)] of about 100 for both liquids and gases.

The roughness elements of practical interest can be divided into two categories: (1) uniformly distributed three-dimensional elements, including sand-grain roughness and roughness imparted by electroetching; and (2) repeated two-dimensional elements such as transverse ribs aligned normal to the flow direction; the helical ribs placed in a spiral fashion over the surface may also be viewed as a two-dimensional type of element. For liquids, the uniformly distributed three-dimensional surface roughness is preferable. For gases, greater height repeated rib roughness induces flow mixing and reattachment downstream, causing an increase in heat transfer. The rib height and angle to flow (most desirable is 45°) are more important, followed by rib spacing and rib shape for heat transfer enhancement with gas flows (Šlančiauskas, 2001).

The fully developed velocity distribution in a rough circular pipe can be described by the power-law formula, Eq. (7.2), with the exponent n ranging between 4 and 5. In the turbulent core region, the velocity profile for smooth and rough pipes are identical, underscoring the fact that the turbulence mechanism in the turbulent core is independent of the conditions at the pipe wall.

For rough pipes, the turbulent flow f factor depends on the roughness type, roughness height e relative to the pipe diameter d_i, and other geometrical dimensions for two- and three-dimensional roughnesses. The results are generally correlated in terms of a *roughness Reynolds number* e^+, defined as

$$e^+ = \frac{eu^*}{\nu} = \frac{e}{\nu}\left(\frac{\tau_w g_c}{\rho}\right)^{1/2} = \frac{e}{d_i}\frac{u_m d_i}{\nu}\left(\frac{\tau_w g_c}{\rho u_m^2}\right)^{1/2} = \frac{e}{d_i}\text{Re}\sqrt{f/2} \qquad (7.85)$$

Here $u^* = (\tau_w g_c/\rho)^{1/2}$ is referred to as the *friction velocity*. The roughness Reynolds number has e as the characteristic dimension and u^* as the velocity. Equation (7.85) also shows the relationship between the roughness Reynolds number and the conventional Reynolds number, based on the hydraulic diameter and u_m through the last equality sign.

Based on the experiments with sand-grain surface roughness in circular tubes, Nikuradse (1933) identified three flow regimes in rough flow: hydraulically smooth regime ($0 \leq e^+ < 5$), transition regime ($5 \leq e^+ \leq 70$), and fully rough regime ($e^+ > 70$). Note that the currently accepted ranges of e^+ for the three regimes defined above are slightly different from those originally proposed by Nikuradse. For the fully rough regime, there will not be a viscous sublayer. The friction factor correlations for these three regimes are presented in Table 7.8.

The roughness obtained by Nikuradse with closely packed sand-grains can be regarded as roughness of the maximum density. The roughness is adequately characterized by the height e or the ratio e/D_h. However, other types of roughness involving uniformly distributed elements of finite sizes such as spheres, spherical segments, cones, and the like, cannot be characterized by e or e/D_h alone. In such cases, it is convenient to determine an *equivalent sand-grain roughness* e_s so that Nikuradse's measurements of *fully rough regime* friction factors and the velocity distribution with sand-grain roughness can be utilized.

When two-dimensional rib roughness with ribs at right angles to the flow direction are fitted to the equivalent sand-grain roughness scale, a very large value of e_s/D_h results compared to the e/D_h value of the ribs. For example, with $e/D_h = 0.001$, a value of $e_s/D_h = 0.025$ is obtained in the fully rough flow regime. This means that a two-dimensional riblike roughness element perpendicular to the flow direction is appreciably more effective in increasing f than is a sand-grain type of element of the same height.

The natural roughness occurring in commercial pipes is three-dimensional (similar to the sand-grain roughness) and has a random distribution and arbitrary shape. The major difference in f vs. Re curves of sand-grain vs. natural roughness is in the transition region of rough flow. For the sand-grain roughness, f reaches a minimum at an intermediate Re and then gradually rises to an asymptotic constant value with increasing Re. For natural roughness, f monotonically decreases with Re and reaches an asymptotic value at high Re. The Fanning friction factors for commercial rough pipes are shown in Fig. 6.4.

Based on Eq. (7.87) in Table 7.8 or Fig. 6.4, the friction factor for the fully rough region is constant, independent of Re. Hence, whenever experimental f factors for a heat exchanger surface are *almost constant*, independent of Re (usually at high Re), we characterize that surface as behaving as a rough surface in that Re range.

Turbulent flow heat transfer studies with sand-grain (two- and three-dimensional) and repeated rib roughnesses have been conducted in detail as summarized by Dipprey and Sabersky (1963) and Webb (1994). Dipprey and Sabersky recommended the follow-

TABLE 7.8   Circular Tube Rough Surface Flow Regimes and Friction Factor Correlations

Flow Regime	Range of e^+	Correlation	Eq. No.
Hydraulically smooth	$0 \leq e^+ < 5$	See Eq. (7.70) in Table 7.6	(7.70)
Transition	$5 \leq e^+ \leq 70$	$\dfrac{1}{\sqrt{f}} = 3.48 - 1.7372 \ln \left(\dfrac{2e}{d_i} - \dfrac{16.2426}{\mathrm{Re}} \ln A_2 \right)$	(7.86)
		where $A_2 = \dfrac{(2e/d_i)^{1.1098}}{6.0983} + \left(\dfrac{7.149}{\mathrm{Re}} \right)^{0.8981}$	
Fully rough	$e^+ > 70$	$\dfrac{1}{\sqrt{f}} = 3.48 - 1.737 \ln \dfrac{2e}{d_i} = 2.28 - 1.737 \ln \dfrac{e}{d_i}$	(7.87)

Source: Data from Bhatti and Shah (1987).

ing correlation based on heat transfer measurements for closed-packed sand-grain type roughness in a circular tube:

$$\text{Nu} = \frac{(f/2)\text{Re} \cdot \text{Pr}}{1 + \sqrt{f/2}[\bar{g}(e^+)\text{Pr}^{-0.44} - B(e^+)]} \qquad (7.88)$$

where and f is given by Eqs. (7.70), (7.86), and (7.87) (in Table 7.8); $B(e^+)$ is given in Table 7.9 for the appropriate ranges of e^+ [obtained based on the approximate straight-line curves given by Dipprey and Sabersky (1963)]; and $\bar{g}(e^+)\text{Pr}^{-0.44}$ is presented in Fig. 7.20 and correlated by[†]

$$g(e^+)\text{Pr}^{-0.44} = \begin{cases} 9.684 + 0.0658e^+ - 0.0003(e^+)^2 & \text{for } 5 < e^+ < 70 \text{ and Pr} = 1.20 \\ 24.5 - 0.24e^+ & \text{for } 5 < e^+ < 50 \text{ and Pr} = 2.79 \\ 2.5 + 0.2e^+ & \text{for } 50 < e^+ < 70 \text{ and Pr} = 2.79 \\ 18.02 - 0.4086e^+ + 0.0101(e^+)^2 & \text{for } 5 < e^+ < 70 \text{ and Pr} = 4.58 \\ \quad -8 \times 10^{-5}(e^+)^3 & \text{and } 5.94 \\ 17.278 - 0.4781e^+ + 0.0131(e^+)^2 & \text{for } 5 < e^+ < 70 \text{ and} \\ \quad -0.000105(e^+)^3 & \text{Pr} = 1.20 - 5.94 \end{cases} \qquad (7.89)$$

$$g(e^+)\text{Pr}^{-0.44} = 16.342 + 0.002e^+ \quad \text{for } e^+ \geq 70 \qquad (7.90)$$

Note that the last expression in Eq. (7.89) covers the complete range of Prandtl numbers from 1.20 to 5.94, but the values for Pr = 1.20 for $e^+ \leq 20$ have an error of up to 25% compared to the data of Dipprey and Sebarsky (1963). The other individual expressions are accurate within ±8% of a one-decimal-point reading of the data from the figure of Dipprey and Sabersky (1963). Equation (7.88) is valid for $1 < \text{Pr} < 6$. A more general equation applicable to the fully rough flow regime ($e^+ > 70$) in the range $0.5 < \text{Pr} < 5000$ is given by Shah and Bhatti (1988) as[‡]

$$\text{Nu} = \frac{(f/2)(\text{Re} - 1000)\text{Pr}}{1 + \sqrt{f/2}[\{17.42 - 13.77\text{Pr}_t^{0.8}\}(e^+)^{0.2}\text{Pr}^{0.5} - 8.48]} \qquad (7.91)$$

TABLE 7.9 Values of $B(e^+)$ for Eq. (7.88)

e^+ Range	$0 \leq e^+ \leq 3$	$3 \leq e^+ \leq 7$	$7 \leq e^+ \leq 14$	$14 \leq e^+ \leq 70$	$e^+ \geq 70$
$B(e^+)$	$5.5 + 2.5 \ln e^+$	$7.30 + 0.318e^+$	9.52	$9.78 - 0.0186e^+$	8.48

[†] These correlations were developed by Ms. Zeng Deng, a Ph.D. student at Oklahoma State University, Stillwater, OK (2002) from the graphical results of Dipperey and Sebarsky (1963).

[‡] In previous publications, there were some errors in this equation; this is the correct version.

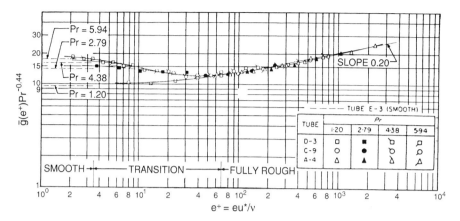

FIGURE 7.20   Heat transfer correlation of sand grain roughness data. (From Dipprey and Sabersky, 1963.)

where the turbulent Prandtl number Pr_t is a function of Pr being given by the following correlation, presented by Malhotra and Kang (1984) for a circular tube in the range $10^4 < Re < 10^6$:

$$Pr_t = \begin{cases} 1.01 - 0.09Pr^{0.36} & \text{for } 1 \leq Pr \leq 145 \\ 1.01 - 0.11\ln Pr & \text{for } 145 \leq Pr \leq 1800 \\ 0.99 - 0.29(\ln Pr)^{1/2} & \text{for } 1800 < Pr \leq 12,500 \end{cases} \qquad (7.92)$$

The prediction of Eq. (7.91) in conjunction with Eq. (7.92) agrees within $\pm 15\%$, with the reported experimental data in the fully rough flow regime for $0.7 \leq Pr \leq 4600$.

In the transition regime of flow ($5 \leq e^+ \leq 70$) over sand-grain surface roughness, the St number can increase by a factor of 2 to 3 with a corresponding increase in the friction factor with St/f approaching 0.25 from lower values. This is the region where the surface roughness can reduce the surface area requirement for a given pressure drop. However, beyond a certain combination of Re and Pr, the surface roughness increases f in fully rough region (see Fig. 6.4) with no significant increase in St. In that region, it is found that the j factor continues to decrease as Re increases for all e/d_i values, while the f factor remains constant. The reason for this behavior is that the surface roughness induces form drag that becomes dominant compared to skin friction at high Re. This results in the flattened (nondecreasing) f vs. Re curves shown in Fig. 6.4. This form drag does not have a heat transfer counterpart, and hence j decreases as Re increases, similar to a smooth surface, thus reducing the relative heat transfer performance of a rough surface.

7.4.2   Hydrodynamically Developing Flows

The hydrodynamically developing flow in an internal passage resembles the flow over an external passage due to the presence of an inviscid potential core around the axis of the passage. The axial distance L_{hy} from the passage inlet to the point where the potential core disappears (see Figs. 7.2 and 7.3) is referred to as the *hydrodynamic entrance length*.

For a circular tube, this distance is given by (Shah and Bhatti, 1987; Bhatti and Shah, 1987)

$$\frac{L_{hy}}{D_h} = \begin{cases} 0.056\mathrm{Re} & \text{for laminar flow } (\mathrm{Re} \leq 2100) \\ 1.359\mathrm{Re}^{1/4} & \text{for turbulent flow } (\mathrm{Re} > 10^4) \end{cases} \tag{7.93}$$

Equation (7.93) shows that $L_{hy}/D_h = 137$ for laminar flow at $\mathrm{Re} = 2100$, whereas $L_{hy}/D_h = 13.6$ for turbulent flow at $\mathrm{Re} = 10^4$. Thus, the hydrodynamic entrance effect persists over a longer distance in laminar flow than in turbulent flow. For practical purposes, the hydrodynamic entrance effect manifests itself as an increase in both the friction factor and heat transfer coefficient.

7.4.2.1 Laminar Flow. The friction factors in the hydrodynamic entrance region are higher than those for the fully developed case. The effects of skin friction and momentum rate change in the entrance region are included in the apparent friction factor [see, Eq. (7.19)]. The $f_{app} \cdot \mathrm{Re}$ factors for four passage geometries are presented as a function of x^+ in Fig. 7.21. For other geometries, refer to Shah and London (1978) and Shah and Bhatti (1987). It can be seen from Fig. 7.21 that for very low values of x^+, the values of $f_{app} \cdot \mathrm{Re}$ are about the same for all four geometries. This is because the effect of the boundary layer development is the same and is not influenced by the boundary layer of the neighboring surface in the very early entrance region.

Based on the solutions for laminar boundary layer development over a flat plate and the fully developed flow in circular and some noncircular ducts, $f_{app} \cdot \mathrm{Re}$ are correlated by the following equation (Shah and London, 1978):

$$f_{app} \cdot \mathrm{Re} = 3.44(x^+)^{-0.5} + \frac{K(\infty)/(4x^+) + f\,\mathrm{Re} - 3.44(x^+)^{-0.5}}{1 + C'(x^+)^{-0.2}} \tag{7.94}$$

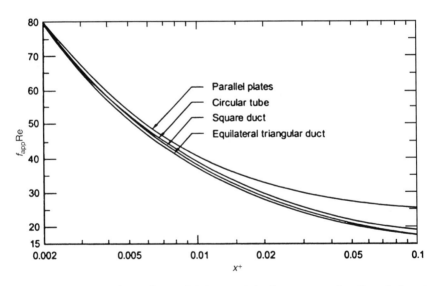

FIGURE 7.21 $f_{app} \cdot \mathrm{Re}$ factors for parallel plates, and circular, square, and equilateral triangular ducts for developing flow. (From Shah and London, 1978.)

TABLE 7.10 $K(\infty)$, $f \cdot$ Re, and C' for Use in Eq. (7.94)

	$K(\infty)$	$f \cdot$ Re	C'
		Rectangular Ducts	
α^*			
1.00	1.43	14.227	0.00029
0.50	1.28	15.548	0.00021
0.20	0.931	19.071	0.000076
0.00	0.674	24.000	0.000029
		Equilateral Triangular Ducts	
2ϕ			
60°	1.69	13.333	0.00053
		Concentric Annular Ducts	
r^*			
0	1.25	16.000	0.000212
0.05	0.830	21.567	0.000050
0.10	0.784	22.343	0.000043
0.50	0.688	23.813	0.000032
0.75	0.678	23.967	0.000030
1.00	0.674	24.000	0.000029

Source: Data from Shah and London (1978).

where $K(\infty)$, $f \cdot$ Re, and C' are given in Table 7.10 for rectangular, equilateral triangular, and concentric annular ducts. Here f_{app} is defined the same way as f [compare Eqs. (7.18) and (7.19)], but it includes the contributions of additional pressure drop due to the momentum rate change and excess wall shear between developing and developed flows. Considering only the first term on the right-hand of Eq. (7.94), it can be shown that $f \propto \mathrm{Re}^{-0.5}$.

It should be emphasized that while Eq. (7.94) will predict the apparent friction factors accurately for a single duct, it generally does not predict the apparent friction factors even fairly approximately for interrupted fin geometries, such as strip fins and louver fins used in compact heat exchangers. This is because it includes only the effect of skin friction. The form drag associated with the smooth and burred edges of surface interruptions and the wake effect may contribute significantly to the pressure drop. Hence, analytical apparent friction factors are generally not used in designing exchangers. But as a rule of thumb, $f \approx 4j$ or a similar relationship may be used to predict f factors for interrupted surfaces for which j factors are already known either from the theory or from experiments.

7.4.2.2 Turbulent Flow. In the entrance region, the turbulent flow friction factors are higher than those in fully developed flow. However, the entrance length is very short, generally less than 10 tube diameters, and other sources of pressure drop at the tube entrance are far more important in most applications. Hence, the influence of the developing turbulent flow region is generally neglected in pressure drop evaluation. Zhi-qing (1982) has developed a closed-form solution for $f_{app} \cdot$ Re for developing turbulent flow in a circular tube and has shown that $f_{app} \cdot$ Re in this case is dependent on both Re and x^+ (Shah and Bhatti, 1988).

7.4.3   Thermally Developing Flows

The temperature profile development in laminar flow with a fully developed velocity profile is shown in Figs. 7.4 and 7.6a for circular and noncircular ducts. No general formula is available to predict the thermal entrance length. For a circular tube having a fully developed velocity profile, $L_{th}^* = L_{th}/(D_h \cdot \text{Re} \cdot \text{Pr})$ for the Ⓗ and Ⓣ boundary conditions are given by Shah and London (1978) as

$$L_{th,H}^* = 0.0431 \qquad L_{th,T}^* = 0.0335 \qquad (7.95)$$

This translates into L_{th}/D_h for air (Pr = 0.7) at Re = 2000 as 60 and 47 and significantly higher (over 100) for water and other liquids with high Prandtl numbers. The reason for the longer thermal entrance length with high-Pr fluids (i.e., fluids with relatively low thermal diffusivity compared to the momentum diffusivity) is that the thermal boundary layer develops gradually (see Fig. 7.6b to get the basic concept).

The turbulent flow thermal entry length for gases and liquids is almost independent of Re and thermal boundary condition. L_{th}/D_h varies from about 8 to 15 for air, and $L_{th}/D_h < 3$ for liquids. For noncircular ducts, L_{th}/D_h may be as high as 30 to 40, due to the coexistence of laminar flow in corner regions.

7.4.3.1   Laminar Flow.   The thermal entrance Nusselt numbers are higher than those for the fully developed case. The local Nusselt numbers for four flow geometries are shown in Figs. 7.22 and 7.23 for the Ⓣ and Ⓗ boundary conditions. Theoretically, the Nusselt numbers are infinity at $x = 0$ and asymptotically approach the fully developed

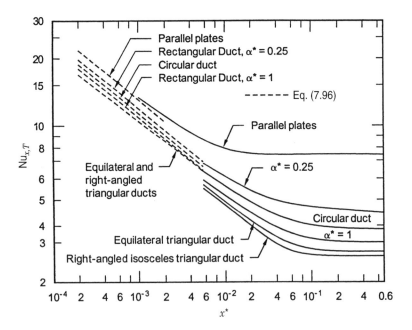

FIGURE 7.22   Comparison of $\text{Nu}_{x,T}$ for parallel plates, and circular, rectangular, and isosceles triangular ducts for developed velocity and developing temperature profiles. (From Shah and London, 1978.)

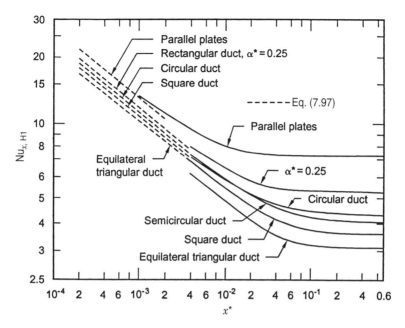

FIGURE 7.23   Comparison of $Nu_{x,H1}$ for parallel plates, and circular, rectangular, equilateral triangular, and semicircular ducts for developed velocity and developing temperature profiles. (From Shah and London, 1978.)

values as x^* increases. The thermal entrance solutions for Nu for many geometries are presented in Shah and London (1978) and Shah and Bhatti (1987).

The thermal entrance local and mean Nusselt numbers for the ⓉⓉ and ⒽⒽ boundary conditions for circular and noncircular ducts having laminar developed velocity profiles and developing temperature profiles are correlated as (Shah and London, 1978)

$$Nu_{x,T} = 0.427(f \cdot Re)^{1/3}(x^*)^{-1/3} \quad Nu_{m,T} = 0.641(f \cdot Re)^{1/3}(x^*)^{-1/3} \quad (7.96)$$

$$Nu_{x,H1} = 0.517(f \cdot Re)^{1/3}(x^*)^{-1/3} \quad Nu_{m,H1} = 0.775(f \cdot Re)^{1/3}(x^*)^{-1/3} \quad (7.97)$$

where f is the Fanning friction factor for fully developed flow, Re is the Reynolds number, and $x^* = x/(D_h \cdot Re \cdot Pr)$. For interrupted surfaces, $x = \ell_s$. The equations above are recommended for $x^* < 0.001$. The following observations may be made from Eqs. (7.96) and (7.97) and other solutions for laminar flow surfaces having developing laminar flows.

- The influence of thermal boundary conditions on thermally developing flow appears to be of the same order as that for thermally fully developed flow; we have considered hydrodynamically fully developed flow in both cases.
- Since $Nu \propto (x^*)^{-1/3} = [x/(D_h \cdot Re \cdot Pr)]^{-1/3}$, $Nu \propto Re^{1/3} \propto u_m^{1/3}$. Therefore, h varies as $u_m^{1/3}$.
- Since the velocity profile is considered fully developed, $\Delta p \propto u_m$, as noted earlier.

- The influence of the duct shape on thermally developing Nu is not as great as that for the fully developed Nu.

Equations (7.96) and (7.97) may not yield accurate values of the Nusselt numbers for interrupted surfaces such as strip fin, and louver fin, due to flow separation, vortex flow, burred edges, and other factors. In experimental data of many interrupted surface geometries, it is found that the exponent on x^* varies from -0.3 to -0.6, depending on the flow type, flow development, and other factors. Using an experimental value of the exponent (which may be different from -0.33), Eqs. (7.96) and (7.97) do provide valuable guidelines for predicting the performance of a surface of a family for which no experimental j and f data are available. For example, consider that we know j_1 and f_1 at some Re (in laminar flow) for an interrupted surface having the interruption length ℓ_1. However, in the design, we would like to employ an interrupted surface of the same type but having the interruption length ℓ_2. As soon as we change the interruption length from ℓ_1 to ℓ_2, the original j and f data for the surface with the interruption length ℓ_1 are no longer valid. In this case, the new j_2 and f_2 at the same Re for the same fluid can be obtained heuristicly as follows.

Since $j = \mathrm{Nu} \cdot \mathrm{Pr}^{-1/3}/\mathrm{Re}$, from Eqs. (7.96) and (7.97), we get

$$j \propto (x^*)^{-1/3} \propto \left(\frac{\ell}{D_h}\right)^{-1/3} \tag{7.98}$$

Therefore,

$$\frac{j_2}{j_1} = \left(\frac{\ell_2/D_{h,2}}{\ell_1/D_{h,1}}\right)^{-1/3} \tag{7.99}$$

Since all other quantities are known, j_2 can readily be calculated from Eq. (7.99). If no other information is available, consider

$$\frac{f_2}{f_1} = \frac{j_2}{j_1} \tag{7.100}$$

to evaluate f_2. This is because the contribution of the form drag is of the same order of magnitude as that of the skin friction. Hence, we cannot use the theory-based developing flow friction solutions for interrupted finned surfaces. It should be emphasized that Eq. (7.100) implies that the relationship between f_2 and j_2 is the same as it is between f_1 and j_1. Instead of changing the interruption length from ℓ_1 to ℓ_2 in the foregoing example, if we had changed the fin density, it would have resulted in the hydraulic diameter changing from $D_{h,1}$ to $D_{h,2}$. In that case, the same foregoing procedure could be used to determine j_2 and f_2.

The theoretical ratio $\mathrm{Nu}_m/\mathrm{Nu}_{fd}$ is shown in Fig. 7.24 as a function of x^* for several passage geometries having the constant wall temperature boundary condition. The following observations may be made from this figure.

- The entrance region Nusselt numbers and hence the heat transfer coefficients could be two to three times higher than the fully developed values, depending on the interruption length $\ell^* = x^*$.

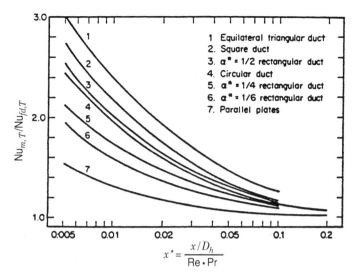

FIGURE 7.24 Ratio of laminar developing to developed flow Nu_T for different ducts; the velocity profile is developed for both Nu's. (From Shah and Webb, 1983.)

- At $x^* \approx 0.1$, although the *local* Nusselt number approaches the fully developed value, the value of the mean Nusselt number can be significantly higher than that for fully developed flow for a channel of length $\ell^* = x^* = 0.1$.

- The ratio $\text{Nu}_m/\text{Nu}_{fd}$ increases from the lowest values for parallel plates to the highest values for the equilateral triangular duct at a given x^*, and thus this order is a function of the channel shape. Notice that it is just the opposite nature of that for Nu_{fd} for fully developed flow in Table 7.3. This is because the duct cross-sectional shape affects $\text{Nu}_{x,T}$ and $\text{Nu}_{m,T}$ marginally until the thermal boundary layers of the neighboring walls start interacting. For a highly interrupted surface, a basic inferior passage geometry for fully developed flow (such as triangular) does not penalize significantly in terms of low values of Nu or h in thermally developing flow.

- A higher value of $\text{Nu}_m/\text{Nu}_{fd}$ at $x^* = 1$ means that the flow channel has a longer entrance region.

7.4.3.2 Turbulent Flow. The Nusselt numbers (Nu_x) in the thermal entrance region are higher than those for the fully developed turbulent flow. This trend is similar to that for the laminar flow case. However, unlike laminar flow, those Nusselt numbers are not dependent on the thermal boundary conditions for $\text{Pr} \geq 0.7$.

The following correlations are for the local and mean Nusselt numbers for the Ⓣ and Ⓗ thermally developing flow in a circular passage (Bhatti and Shah, 1987):

$$\frac{\text{Nu}_x}{\text{Nu}_\infty} = 1 + \frac{C_6}{10(x/D_h)} \qquad \frac{\text{Nu}_m}{\text{Nu}_\infty} = 1 + \frac{C_6}{x/D_h} \qquad (7.101)$$

where Nu_∞ stands for the fully developed Nu_T or Nu_H derived from the formulas recommended in Table 7.7 and

$$C_6 = \frac{(x/D_h)^{0.1}}{\mathrm{Pr}^{1/6}}\left(0.68 + \frac{3000}{\mathrm{Re}^{0.81}}\right) \tag{7.102}$$

These correlations are valid for $x/D_h > 3$, $3500 < \mathrm{Re} < 10^5$, and $0.7 < \mathrm{Pr} < 75$. The value of Nu_m agrees within $\pm 12\%$ with the experimental measurements for $\mathrm{Pr} = 0.7$.

Notice that Nu_x and Nu_m of Eq. (7.101) are functions of x/D_h, Re, and Pr; in contrast, the Nusselt numbers for developing laminar flow depend only on $x^* = x/(D_h \cdot \mathrm{Re} \cdot \mathrm{Pr})$. As an illustration, the turbulent entrance region local Nusselt numbers are presented in Fig. 7.25 for a circular tube having the Ⓗ boundary condition.

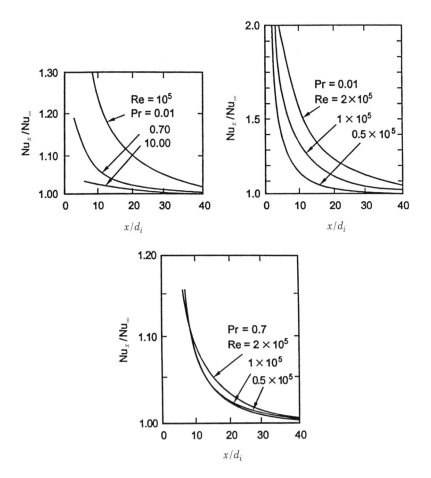

FIGURE 7.25 Nusselt numbers as a function of x/d_i, Re, and Pr for Ⓗ turbulent flow thermal entrance region in a circular tube. (From Kays and Crawford, 1993.)

7.4.4   Simultaneously Developing Flow

7.4.4.1   Laminar Flow. In simultaneously developing flow, the velocity and temperature profiles both develop in the entrance region. If they are uniform at the duct inlet, the fluid velocity, velocity gradients, and temperature gradients near the wall in the entrance region will be higher than those for velocity profiles already developed. The higher velocities near the wall convect more thermal energy in the flow direction, and heat transfer in the thermal entrance region is higher for the case of developing velocity profiles. As an example, Fig. 7.26 compares the circular duct $Nu_{x,H}$ for developed and developing velocity profiles. The curve for $Pr = \infty$ in this figure represents the correlation for *any fluid* ($Pr < \infty$) having its velocity profile fully developed before the temperature profile starts developing. It is clear from Fig. 7.26 that the Nusselt number is higher for the simultaneously developing flow situation ($Pr \leq \infty$).

Additionally, in simultaneously developing flow, the rate of development of the temperature boundary layer relative to the velocity boundary layer does depend on the fluid Prandtl number (refer to Fig. 7.6*b* and *c*). If the velocity and temperature profiles are uniform at the duct entrance, the lower the fluid Prandtl number, the faster the development of the temperature boundary layer will be in comparison to the velocity boundary layer in the entrance region of the duct. This would result in lower temperature gradients at the wall and in turn decrease the Nusselt number and heat transfer at a given $x^{+} = x/(D_{h} \cdot Re)$. Thus, the lower the Prandtl number, the lower the Nusselt number will be at a given x^{+} for specified duct geometry. However, if the axial coordinate is stretched or compressed by considering $x^{*} = x^{+}/Pr$, it is found that the lower the

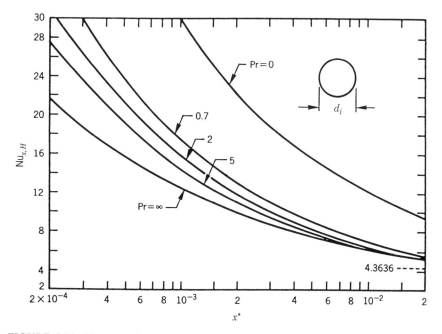

FIGURE 7.26 $Nu_{x,H}$ as a function of x^{*} and Pr for simultaneously developing laminar flow in a circular tube. (From Shah and London, 1978.)

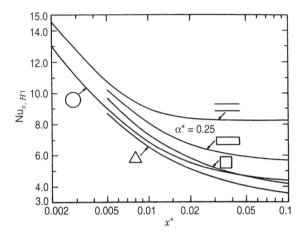

FIGURE 7.27   Simultaneously developing laminar flow $\mathrm{Nu}_{x,\mathrm{H1}}$ for $\mathrm{Pr} = 0.7$ for several constant-cross-section duct geometries. (From Shah and London, 1978.)

Prandtl number, the higher the Nusselt number will be at a given x^*. The effect of the fluid Prandtl number on the entrance region Nusselt numbers is shown in Fig. 7.26 for the circular tube. The $\mathrm{Nu}_{x,\mathrm{H1}}$ for simultaneously developing flow for the circular duct, parallel plates, rectangular ducts, and equilateral triangular ducts are compared in Fig. 7.27 for $\mathrm{Pr} = 0.7$.

The theoretical entrance-region Nusselt numbers for simultaneously developing flow are higher than those for thermally developing and hydrodynamically developed flow. These theoretical solutions do not take into account the wake effect or secondary flow effect present in flow over interrupted heat transfer surfaces. Experimental data indicate that the compact interrupted surfaces *do not* achieve the higher heat transfer coefficients predicted for simultaneously developing flows. The results for thermally developing flows (and developed velocity profiles), such as those in Fig. 7.24 or Eqs. (7.96) and (7.97), are in better agreement with the experimental data for interrupted surfaces, and hence those are recommended for design guidelines; the correct exponent on x^*, instead of $-\frac{1}{3}$, should be used in these equations if available based on experimental data.

7.4.4.2   Turbulent Flow. The Nusselt numbers for simultaneously developing turbulent flow are practically the same as the Nusselt numbers for thermally developing turbulent flow. However, the Nusselt numbers for simultaneously developing flows are quite sensitive to the passage inlet configuration (Bhatti and Shah, 1987).

Table 7.11 summarizes the dependence of Δp and h on u_m for developed and developing laminar and turbulent flows based on the solutions/correlations presented for the specific flow type. Although these results are for the circular tube, the general functional relationship should be valid for noncircular ducts as a first approximation.

7.4.5   Extended Reynolds Analogy

For convective transport of mass, momentum, and heat transfer across a boundary layer, the associated coefficients (mass transfer, skin friction, and heat transfer) in dimension-

TABLE 7.11   Dependence of Pressure Drop and Heat Transfer Coefficient on the Flow Mean Velocity for Internal Flow in a Constant Cross-Sectional Duct

	$\Delta p \propto u_m^p$		$h \propto u_m^q$	
Flow Type	Laminar	Turbulent	Laminar	Turbulent
Fully developed	u_m	$u_m^{1.8}$	u_m^0	$u_m^{0.8}$
Hydrodynamically developing	$u_m^{1.5}$	$u_m^{1.8}$	—	—
Thermally developing	u_m	$u_m^{1.8}$	$u_m^{1/3}$	$u_m^{0.8}$
Simultaneously developing	$u_m^{1.5}$	$u_m^{1.8}$	$u_m^{1/2}$	$u_m^{0.8}$

Source: Data from Shah and Mueller (1988).

less form are related to each other for attached boundary layers over a heat/mass transfer surface. This relationship in its simplest form was discovered by Reynolds in 1874 (see Bejan, 1995). The analysis infers that the differential momentum and energy equations are similar for fully developed turbulent flow over a sharp-edge flat plate; and for the ⓉT boundary condition, the resulting dimensionless velocity and temperature profiles are identical for Pr = 1 and the heat transfer coefficient (i.e., St number) and friction factor are related as

$$St = \frac{f}{2} \qquad (7.103)$$

It has also been shown that the boundary layer equations for convective transport of mass, momentum, and energy have the identical form when the pressure gradient dp/dx along the flow direction is zero and Pr = Sc = 1 for laminar flow (Incropera and DeWitt, 2002). Here Sc, the Schmidt number, denotes a ratio of momentum diffusivity to the mass diffusivity. Thus, the Reynolds analogy is now referred to as

$$\frac{f}{2} Re = Nu = Sc \qquad (7.104)$$

This is valid for laminar and turbulent boundary layer flows with zero axial pressure gradient and the constant wall temperature boundary condition. By converting the Nu or Sc in the corresponding Stanton numbers and extending the Prandtl number range based on the experimental results, the modified Reynolds or Chilton–Colburn analogies are

$$\frac{f}{2} = \begin{cases} St \cdot Pr^{2/3} = j & 0.6 < Pr < 60 \qquad (7.105) \\ St_m \cdot Pr^{2/3} = j_m & 0.6 < Sc < 3000 \qquad (7.106) \end{cases}$$

where St_m and j_m are the Stanton number and Colburn factor for mass transfer. However, for most heat exchanger surfaces, the analogy above is not quite correct. This is because there is always some finite pressure gradient along the flow length (finite pressure drop in the heat exchanger), and the boundary conditions are not necessarily the constant wall temperature. Despite this fact, we know from the empirical evidence that skin friction and convective heat transfer at the wall are related to each other. Whenever there is convective heat transfer at the heat exchanger surface, there will be, correspondingly, associated skin friction. We cannot get the convective heat transfer without the expenditure of skin friction in the boundary layer flows. We refer to it as an *extended Reynolds* analogy. It is formulated as (Shah and Bhatti, 1988)

$$j = \text{St} \cdot \text{Pr}^{2/3} = \frac{\text{Nu}}{\text{Re}} \cdot \text{Pr}^{-1/3} = \frac{f}{2} \phi_w \qquad (7.107)$$

where

$$\phi_w = \phi_w \text{ (duct geometry, flow type, boundary condition, Pr)} \qquad (7.108)$$

Here, ϕ_w is the function that modifies the Reynolds analogy relationship between j and f factors; the flow type in ϕ_w indicates developing or developed laminar or turbulent flow.

Based on a variety of turbulent boundary layer models and experimental data, many correlations have been developed for turbulent flow Nu, and some of them are summarized in Table 7.6. One of the most accurate correlations for fully developed turbulent flow (and also for transition flow) in a circular tube is due to Gnielinski and is given in Eq. (7.76) in Table 7.6. A comparison of the Gnielinski correlation with Eq. (7.107) will then yield the most accurate value of ϕ_w, which is dependent on Pr and Re.

In general, based on the extended Reynolds analogy, skin friction and heat transfer at the wall (heat transfer surface) for boundary layer flow are related to each other by a parameter ϕ_w. It may or may not be possible to have an explicit expression/value of ϕ_w for complex heat transfer surface geometries. However, this analogy clearly indicates that *an increase in heat transfer will be accompanied by an increase in the skin friction factor!* Then for an enhanced heat transfer surface, it is the heat exchanger designer's responsibility to maintain the same pressure drop by adjusting the geometry of the exchanger [i.e., L, A_o, and D_h, see Eq. (6.29)] to take advantage of heat transfer enhancement (and associated increase in the f factor) without increasing the pressure drop and fluid pumping power for the exchanger. It should also be mentioned that the Reynolds analogy in general would not be valid for nonboundary layer flows. For example, form drag associated with the flow normal to a tube bank or form drag associated with the flow in the wake region of an interrupted heat transfer surface does not enhance heat transfer in general. Hence, for such geometries, j/f for gases and other fluids is closer to $\frac{1}{4}$ rather to $\frac{1}{2}$, as indicated by Eq. (7.103), if the contribution of the form drag is of the same order of magnitude as that of the skin friction. Due to the form drag associated with surface roughness in turbulent flow, the friction factor takes on an approximate constant value (see Fig. 6.4). However, there is no counterpart of the form drag for heat transfer, hence the j factor continues to decrease with an increase in Re for a rough surface in the fully rough region as discussed with the results of Fig. 7.20, a behavior similar to that for a smooth surface!

7.4.6   Limitations of *j* vs. Re Plot

The limitations of the j vs. Re plot, commonly used in presenting compact heat exchanger surface basic data, should be understood for various basic flow types as follows.

- In fully developed laminar flow, as discussed in Section 7.4.1.1, the Nusselt number is theoretically constant, independent of Pr (and also Re). Since $j = \text{St} \cdot \text{Pr}^{2/3} = \text{Nu} \cdot \text{Pr}^{-1/3}/\text{Re}$ by definition, the j factor will be dependent on Pr in the fully developed laminar region. Hence, the j factors presented in Chapter 7 of Kays and London (1998) for gas flows in the fully developed laminar region should first be converted to a Nusselt number (using $Pr = 0.70$), which can then be used directly for liquid flows as constant-property results.

- Based on analytical correlations for thermally developing laminar flow in Section 7.4.3.1, $Nu \propto (x^*)^{-1/3}$. This means that $Nu \cdot Pr^{-1/3}$ is independent of Pr and hence j is independent of Pr for thermally developing laminar flows.[†] As a result, the j vs. Re data of Kays and London (1998) could be used for air, water, and other fluids as constant-property data.

- For fully developed turbulent flow, $Nu \propto Pr^{0.4}$, and hence $j \propto Pr^{0.07}$. Thus, j is again dependent on Pr in the fully developed turbulent region.[‡] In this case, if the experimental data/correlations are available for specific geometries in terms of the j vs. Re characteristics, convert those data into Nusselt numbers using the definition $j = Nu \cdot Pr^{-1/3}/Re$ with the properties of the test fluids (i.e., using the same Pr values as reported in the original work). Then use these Nusselt numbers for the fluids of interest for a given application. If the original experimental data/correlations are available in terms of the Nusselt numbers, there is no need to change anything.

All of the foregoing comments apply to either constant-property theoretical solutions or almost-constant-property (low-temperature difference) experimental data. The influence of property variations, discussed in Section 7.6, must be taken into account by correcting the aforementioned constant-property j or Nu when designing a heat exchanger.

7.5 EXPERIMENTAL HEAT TRANSFER AND FRICTION FACTOR CORRELATIONS FOR COMPLEX GEOMETRIES

The analytical correlations presented in Section 7.4 are useful for well-defined constant cross-sectional surfaces with essentially unidirectional flows. The flows encountered in heat exchangers are generally very complex, having flow separation, reattachment, recirculation, and vortices, as discussed earlier. Such flows significantly affect Nu and f for the specific exchanger surfaces. Since no analytical or accurate numerical correlations are available, Nu (or j) and f (or Eu) vs. Re data are obtained experimentally. Kays and London (1998) and Webb (1994) presented experimental results reported in the open literature for many heat exchanger surfaces. In the following, empirical correlations for some important surfaces are summarized.

[†] If a slope of -1 for the log-log j–Re characteristic is used as a criterion for *fully developed laminar flow*, most of the interrupted finned surfaces reported in Chapter 10 of Kays and London (1998) would not qualify as being in a fully developed laminar condition. Data for most of these surfaces indicate thermally developing flow conditions for which j is almost independent of Pr, as indicated (as long as the exponent on Pr is about $-\frac{1}{3}$, although for some surfaces that exponent may be as high as -0.5). Hence, the j–Re characteristic should not be converted to the Nu–Re characteristic for the interrupted finned surface basic data in Chapter 10 of Kays and London (1998).

[‡] Colburn (1933) proposed $j = St \cdot Pr^{2/3}$ as a correlating parameter to include the effect of Prandtl number based on the then available data for turbulent flow. Based on presently available experimental data, however, the j factor is clearly dependent on Pr for fully developed turbulent flow and for fully developed laminar flow but not for ideal thermally developing laminar flows.

To determine the fluid properties for the correlations of this section, refer to Section 9.1 for the appropriate mean temperature on each fluid side.

7.5.1   Tube Bundles

Žukauskas (1987) has presented extensive experimental results *graphically* for flow normal to inline and staggered plain and finned tube bundles. Comprehensive pressure drop correlations for flow normal to inline and staggered *plain* tube bundles have been developed by Gaddis and Gnielinski (1985) and recast in terms of the Hagen number per tube row by Martin (2002) as follows:

$$
\mathrm{Hg} = \begin{cases} \mathrm{Hg_{lam}} + \mathrm{Hg_{turb,i}}\left[1 - \exp\left(1 - \dfrac{\mathrm{Re}_d + 1000}{2000} \right) \right] & \text{inline tube bundles} \\[4mm] \mathrm{Hg_{lam}} + \mathrm{Hg_{turb,s}}\left[1 - \exp\left(1 - \dfrac{\mathrm{Re}_d + 200}{1000} \right) \right] & \text{staggered tube bundles} \end{cases}
$$

$$(7.109)$$

where

$$
\mathrm{Hg_{lam}} = 140\mathrm{Re}_d \frac{(X_\ell^{*0.5} - 0.6)^2 + 0.75}{X_t^{*1.6}(4X_t^* X_\ell^* / \pi - 1)} \tag{7.110}
$$

Equation (7.110) is valid for all inline tube bundles. It is also valid for staggered tube bundles except that the term $X_t^{*1.6}$ needs to be changed to $X_d^{*1.6}$ when the minimum free-flow area occurs in the diagonal planes of the staggered tube bundle [i.e., $X_\ell^* < 0.5(2X_t^* + 1)^{1/2}$].

$$
\mathrm{Hg_{turb,i}} = \left\{ \left[0.11 + \frac{0.6(1 - 0.94/X_\ell^*)^{0.6}}{(X_t^* - 0.85)^{1.3}} \right] \times 10^{0.47(X_\ell^*/X_t^* - 1.5)} + 0.015(X_t^* - 1)(X_\ell^* - 1) \right\}
$$
$$
\times \mathrm{Re}_d^{2 - 0.1(X_\ell^*/X_t^*)} + \phi_{t,n}\mathrm{Re}_d^2 \tag{7.111}
$$

$$
\mathrm{Hg_{turb,s}} = \left\{ \left[1.25 + \frac{0.6}{(X_t^* - 0.85)^{1.08}} \right] + 0.2\left(\frac{X_\ell^*}{X_t^*} - 1 \right)^3 - 0.005\left(\frac{X_t^*}{X_\ell^*} - 1 \right)^3 \right\}
$$
$$
\times \mathrm{Re}_d^{1.75} + \phi_{t,n}\mathrm{Re}_d^2 \tag{7.112}
$$

Equation (7.112) is valid for $\mathrm{Re}_d \leq 250{,}000$. For higher Re_d, correct $\mathrm{Hg_{turb,s}}$ of Eq. (7.112) as follows:

$$
\mathrm{Hg_{turb,s,corr}} = \mathrm{Hg_{turb,s}}\left(1 + \frac{\mathrm{Re}_d - 250{,}000}{325{,}000} \right) \qquad \text{for } \mathrm{Re}_d > 250{,}000 \tag{7.113}
$$

$$
\phi_{t,n} = \begin{cases} \dfrac{1}{2X_t^{*2}}\left(\dfrac{1}{N_r} - \dfrac{1}{10} \right) & \text{for } 5 \leq N_r \leq 10 \text{ and} \\ & \qquad X_\ell^* \geq 0.5(2X_t^* + 1)^{1/2} \\[4mm] 2\left[\dfrac{X_d^* - 1}{X_t^*(X_t^* - 1)} \right]^2 \left(\dfrac{1}{N_r} - \dfrac{1}{10} \right) & \text{for } 5 \leq N_r \leq 10 \text{ and} \\ & \qquad X_\ell^* < 0.5(2X_t^* + 1)^{1/2} \end{cases}
$$

$$(7.114)$$

and $\phi_{t,n} = 0$ for $N_r > 10$. It should be emphasized that $\phi_{t,n}\mathrm{Re}_d^2$ takes into account the influence of tube bundle inlet and outlet pressure drops while the first bracketed term on the right side of Eqs. (7.111) and (7.112) takes into account the frictional pressure drop in the tube bundle. For the total pressure loss of a tube bundle, the complete Eq. (7.111) or (7.112) should be used (i.e., both terms on the right-hand sides of these equations should be included).

The foregoing correlation of Eq. (7.109) is valid for $1 < \mathrm{Re}_d < 300,000$ and $N_r \geq 5$ for both inline and staggered tube bundles: $1.25 \leq X_t^* \leq 3.0$ and $1.2 \leq X_\ell^* \leq 3.0$ for inline tube bundles and $1.25 \leq X_t^* \leq 3.0$, $0.6 \leq X_\ell^* \leq 3.0$, and $X_d^* \geq 1.25$ for staggered tube bundles. The experimental data for this correlation had $7.9 \leq d_o \leq 73$ mm.

The pressure drop for flow normal to the tube bundle is then computed from

$$\Delta p = \frac{\mu^2}{\rho g_c}\frac{N_r}{d_o^2}\,\mathrm{Hg} \tag{7.115}$$

where N_r is the number of tube rows in the flow direction. Note that only when the minimum flow area occurs in the diagonals for a staggered tube bundle should the term N_r be replaced by $N_r - 1$, considering the number of flow resistances in the diagonal flow area.

In all correlations for heat transfer and pressure drop for tube bundles, $\mathrm{Re}_d = \rho u_m d_o/\mu$, where

$$u_m = \begin{cases} u_\infty \dfrac{X_t^*}{X_t^* - 1} & \text{inline tube bundles} \\[2mm] u_\infty \dfrac{X_t^*}{X_t^* - 1} & \text{staggered tube bundles with } X_\ell^* \geq 0.5(2X_t^* + 1)^{1/2} \\[2mm] u_\infty \dfrac{X_t^*}{2(X_d^* - 1)} & \text{staggered tube bundles with } X_\ell^* < 0.5(2X_t^* + 1)^{1/2} \end{cases} \tag{7.116}$$

Martin (2002) developed comprehensive correlations for heat transfer with flow normal to inline and staggered plain tube bundles as follows:

$$\mathrm{Nu} = \begin{cases} 0.404\mathrm{Lq}^{1/3}\left(\dfrac{\mathrm{Re}_d + 1}{\mathrm{Re}_d + 1000}\right)^{0.1} & \text{inline tube bundles} \\[3mm] 0.404\mathrm{Lq}^{1/3} & \text{staggered tube bundles} \end{cases} \tag{7.117}$$

where

$$\mathrm{Lq} = \begin{cases} 1.18\,\mathrm{Hg}\cdot\mathrm{Pr}\left(\dfrac{(4X_t^*/\pi) - 1}{X_\ell^*}\right) & \text{inline tube bundles} \\[3mm] 0.92\,\mathrm{Hg}\cdot\mathrm{Pr}\left(\dfrac{(4X_t^*/\pi) - 1}{X_d^*}\right) & \text{staggered tube bundles with } X_\ell^* \geq 1 \\[3mm] 0.92\,\mathrm{Hg}\cdot\mathrm{Pr}\left(\dfrac{(4X_t^* X_\ell^*/\pi) - 1}{X_\ell^* X_d^*}\right) & \text{staggered tube bundles with } X_\ell^* < 1 \end{cases}$$

$$\tag{7.118}$$

where Hg is obtained from Eq. (7.109). Note that the Lévêque number Lq is defined in Eq. (7.41).

The foregoing heat transfer correlation of Eq. (7.117) is valid for $1 < \mathrm{Re}_d < 2{,}000{,}000$, $0.7 \leq \mathrm{Pr} \leq 700$ (validity expected also for $\mathrm{Pr} > 700$, but not for $\mathrm{Pr} < 0.6$), and $2 \leq N_r \leq 15$ for inline tube bundles (90° tube bundles) and $4 \leq N_r \leq 80$ for staggered tube bundles (30, 45, 60° tube bundles and other staggered tube bundles within the correlation tube pitch ranges); the tube pitch ratio range is $1.02 \leq X_t^* \leq 3.0$ and $0.6 \leq X_\ell^* \leq 3.0$ for both inline and staggered tube bundles. The experimental data for this correlation had $7.9 \leq d_o \leq 73\,\mathrm{mm}$. The Nusselt numbers are predicted accurately within ±20% for inline tube bundles and ±14% for staggered tube bundles using Eq. (7.117), where the Hagen number of Eq. (7.118) is determined from the Gaddis and Gnielinski correlation of Eq. (7.109). The Nusselt number prediction may be better if the experimental friction factors are used. Note that when the Gaddis and Gnielinski correlation is extrapolated outside their ranges of Re_d and N_r for Nu calculations, it has predicted Nu within the accuracy mentioned.

7.5.2 Plate Heat Exchanger Surfaces

One of the most commonly used high-performance plate heat exchanger surfaces has chevron plates with the important geometrical parameters identified in Fig. 7.28. A considerable amount of research has been conducted to determine heat transfer and flow friction characteristics of this geometry. Martin (1996) provides comprehensive correlations for friction factors and Nusselt numbers for this geometry. His correlation for the Fanning friction factors is

$$\frac{1}{\sqrt{f}} = \frac{\cos\beta}{(0.045\tan\beta + 0.09\sin\beta + f_0/\cos\beta)^{1/2}} + \frac{1 - \cos\beta}{\sqrt{3.8 f_1}} \qquad (7.119)$$

where

$$f_0 = \begin{cases} \dfrac{16}{\mathrm{Re}} & \text{for } \mathrm{Re} < 2000 \\[2mm] (1.56\ln\mathrm{Re} - 3.0)^{-2} & \text{for } \mathrm{Re} \geq 2000 \end{cases} \qquad (7.120a)$$

$$f_1 = \begin{cases} \dfrac{149.25}{\mathrm{Re}} + 0.9625 & \text{for } \mathrm{Re} < 2000 \\[2mm] \dfrac{9.75}{\mathrm{Re}^{0.289}} & \text{for } \mathrm{Re} \geq 2000 \end{cases} \qquad (7.120b)$$

$$\mathrm{Re} = \frac{G D_h}{\mu} \qquad (7.121)$$

$$G = \frac{\dot{m}}{2\mathbf{a}W} \qquad D_h = \frac{4\mathbf{a}}{\Phi} \qquad \Phi = \frac{1}{6}\left(1 + \sqrt{1 + X^2} + 4\sqrt{1 + X^2/2}\right) \qquad X = \frac{2\pi\mathbf{a}}{\Lambda} \qquad (7.122)$$

As shown in Fig. 7.28, $\mathbf{a}$ is the amplitude and Λ is the wavelength of chevron corrugations, and W is the plate width between gaskets. An exact formula for Φ is given by Eq. (8.131). The friction factor correlation of Eq. (7.119) is valid for the corrugation angle β within 0 to 80° and is accurate within −50% and +100%. If the model plate data (those having only the central portion of the plates without inlet and outlet ports and

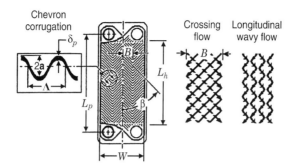

FIGURE 7.28   Geometrical parameters of a chevron plate. (From Martin, 1996.)

distribution regions) are eliminated, the correlation of Eq. (7.119) is based on industrial plates of PHEs and is within ±40% accuracy. Of course, this correlation can be improved further if the actual detailed geometrical information would be available.

Martin (1996) also obtained the Nusselt number correlation as follows, using the momentum and heat transfer analogy from a generalized Lévêque solution in thermal entrance turbulent flow in a circular pipe (Schlünder, 1998):

$$\mathrm{Nu} = \frac{hD_h}{k} = 0.205\mathrm{Pr}^{1/3}\left(\frac{\mu_m}{\mu_w}\right)^{1/6}(f \cdot \mathrm{Re}^2 \sin 2\beta)^{0.374} \qquad (7.123)$$

This correlation is valid for the corrugation angle β within 10 to 80°, and is accurate within ±30%, and within ±13% for industrial plates. Note that if Eq. (7.123) is used for gases, the viscosity correction term $(\mu_m/\mu_w)^{1/6}$ should be omitted. Note that a 100% error in f will translate to 30% error in Nu, due to the exponent 0.374 on f in Eq. (7.123).

7.5.3   Plate-Fin Extended Surfaces

In this section we provide correlations for the offset strip fin and louver fin geometries, and briefly mention other geometries.

A careful examination of all accurate published data has revealed the ratio $j/f \leq 0.25$ for strip fin, louver fin, and other similar interrupted surfaces. This can be justified approximately as follows. The flow develops along each interruption in such a surface. Based on the Reynolds analogy for fully developed turbulent flow over a flat plate, in the absence of form drag, j/f should be 0.5 for $\mathrm{Pr} \approx 1$ (see Section 7.4.5). Since the contribution of form drag is of the same order of magnitude as the skin friction in developing laminar flows for such an interrupted surface, j/f will be about 0.25. Published data for strip and louver fins may be questionable if $j/f > 0.3$, since they would indicate that the contribution of form drag in interrupted fins is small; however, the form drag contribution will not be small in actual fins because of the finite fin thickness and possible burrs at the edges due to the manufacturing process. All pressure and temperature measurements and possible sources of flow leaks and heat losses must be checked thoroughly for all basic surface characteristics having $j/f > 0.3$ for strip and louver fins (refer to some problems identified in Section 7.3.1.3 for the test core design).

7.5.3.1 Offset Strip Fins. This is one of the most widely used enhanced fin geometries (Fig. 1.29d) in aircraft, cryogenics, and many other industries that do not require mass production. This surface has one of the highest heat transfer performances relative to the friction factor. Extensive analytical, numerical and experimental investigations have been conducted over the last 50 years. The most comprehensive correlations for j and f factors for the offset strip fin geometry are provided by Manglik and Bergles (1995) as follows.

$$j = 0.6522 \, \text{Re}^{-0.5403} \left(\frac{s}{h'}\right)^{-0.1541} \left(\frac{\delta}{\ell_s}\right)^{0.1499} \left(\frac{\delta}{s}\right)^{-0.0678}$$

$$\times \left[1 + 5.269 \times 10^{-5} \text{Re}^{1.340} \left(\frac{s}{h'}\right)^{0.504} \left(\frac{\delta}{\ell_s}\right)^{0.456} \left(\frac{\delta}{s}\right)^{-1.055}\right]^{0.1} \quad (7.124)$$

$$f = 9.6243 \, \text{Re}^{-0.7422} \left(\frac{s}{h'}\right)^{-0.1856} \left(\frac{\delta}{\ell_s}\right)^{0.3053} \left(\frac{\delta}{s}\right)^{-0.2659}$$

$$\times \left[1 + 7.669 \times 10^{-8} \text{Re}^{4.429} \left(\frac{s}{h'}\right)^{0.920} \left(\frac{\delta}{\ell_s}\right)^{3.767} \left(\frac{\delta}{s}\right)^{0.236}\right]^{0.1} \quad (7.125)$$

where

$$D_h = \frac{4A_{o,\text{cell}}}{A_{\text{cell}}/\ell_s} = \frac{4sh'\ell_s}{2(s\ell_s + h'\ell_s + h'\delta) + s\delta} \quad (7.126)$$

Geometrical symbols in Eqs. (7.124)–(7.126) are shown in Fig. 8.7; here $s = p_f - \delta$, $h' = b - \delta$, and $b = b_1$. For Eq. (7.126), $A_{o,\text{cell}}$ and A_{cell} are given by Eqs. (8.72) and (8.71), respectively.

These correlations predict the experimental data of 18 test cores within $\pm 20\%$ for $120 \leq \text{Re} \leq 10^4$. Although all experimental data for these correlations are obtained for air, the j factor takes into consideration minor variations in the Prandtl number, and the correlations above should be valid for $0.5 < \text{Pr} < 15$.

The heat transfer coefficients for the offset strip fins are 1.5 to 4 times higher than those of plain fin geometries. The corresponding friction factors are also high. The ratio of j/f for an offset strip fin to j/f for a plain fin is about 0.8. If properly designed [refer to Eq. (6.71) and associated discussion], the offset strip fin would require substantially lower heat transfer surface area than that of plain fins at the same Δp, but about a 10% larger flow area.

7.5.3.2 Louver Fins. Louver or multilouver fins are used extensively in the auto industry due to their mass production manufacturability and hence lower cost. They have generally higher j and f factors than those for the offset strip fin geometry, and also the increase in the friction factors is in general higher than the increase in j factors. However, the exchanger can be designed for higher heat transfer and the same pressure drop compared to that with offset strip fins by a proper selection of exchanger frontal area, core depth, and fin density [refer to the discussion of Eq. (6.71)]. Published early literature and correlations on the louver fins are summarized by Webb (1994) and Cowell et al. (1995), and an understanding of flow and heat transfer phenomena is summarized by Cowell et al. (1995). A correlation for the Colburn factors for the

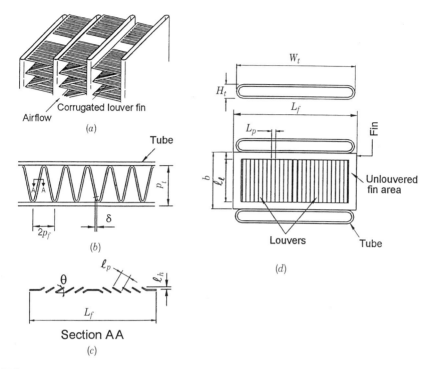

FIGURE 7.29   Definition of geometrical parameters of corrugated louver fins. (From Chang and Wang, 1997.)

corrugated louver fins (see Fig. 7.29a or 1.29e), based on an extensive database for *airflow* over louver fins, is obtained by Chang and Wang (1997) and Wang (2000) as follows:

$$j = \mathrm{Re}_{\ell_p}^{-0.49} \left(\frac{\theta}{90}\right)^{0.27} \left(\frac{p_f}{\ell_p}\right)^{-0.14} \left(\frac{b}{\ell_p}\right)^{-0.29} \left(\frac{W_t}{\ell_p}\right)^{-0.23} \left(\frac{\ell_\ell}{\ell_p}\right)^{0.68} \left(\frac{p_t}{\ell_p}\right)^{-0.28} \left(\frac{\delta}{\ell_p}\right)^{-0.05} \quad (7.127)$$

where $\mathrm{Re}_{\ell_p} = G\ell_p/\mu$ represents the Reynolds number based on the louver pitch ℓ_p. Also, θ is the louver angle (deg), p_f the fin pitch (mm), b the vertical fin height (mm), W_t the tube outside width (= total fin length in the airflow direction if there are no overhangs; mm), ℓ_ℓ the louver cut length (mm), p_t the tube pitch (mm), and δ the fin thickness (mm). These geometrical parameters are shown in Fig. 7.29. Equation (7.127) is valid for the following ranges of the parameters: $0.82 \leq D_h \leq 5.02$ mm, $0.51 \leq p_f \leq 3.33$ mm, $0.5 \leq \ell_p \leq 3$, $2.84 \leq b \leq 20$ mm, $15.6 \leq W_t \leq 57.4$ mm, $2.13 \leq \ell_\ell \leq 18.5$ mm, $7.51 \leq p_t \leq 25$ mm, $0.0254 \leq \delta \leq 0.16$ mm, $1 \leq N_r \leq 2$, and $8.4 \leq \theta \leq 35°$. This correlation predicts 89% of experimental j factors of 91 test cores within $\pm15\%$ for $30 < \mathrm{Re}_{\ell_p} < 5000$ with a mean deviation of 8%.

Chang and Wang (1997) also presented a simplified correlation for Eq. (7.127) as

$$j = 0.425 \, \mathrm{Re}_{\ell_p}^{-0.496} \quad (7.128)$$

They report that this correlation predicts 88% of data points within ±25% with the mean deviation of 13%.

The correlation for the Fanning friction factor based on the same database by Chang et al. (2000) is

$$f = f_1 f_2 f_3 \tag{7.129}$$

where

$$f_1 = \begin{cases} 14.39 \, \mathrm{Re}_{\ell p}^{(-0.805 p_f/b)} \{\ln[1.0 + (p_f/\ell_p)]\}^{3.04} & \mathrm{Re}_{\ell p} < 150 \\ 4.97 \, \mathrm{Re}_{\ell p}^{(0.6049 - 1.064/\theta^{0.2})} \{\ln[(\delta/p_f)^{0.5} + 0.9]\}^{-0.527} & 150 < \mathrm{Re}_{\ell p} < 5000 \end{cases} \tag{7.130}$$

$$f_2 = \begin{cases} \{\ln[(\delta/p_f)^{0.48} + 0.9]\}^{-1.435} (D_h/\ell_p)^{-3.01} [\ln(0.5 \, \mathrm{Re}_{\ell p})]^{-3.01} & \mathrm{Re}_{\ell p} < 150 \\ [(D_h/\ell_p) \ln(0.3 \, \mathrm{Re}_{\ell p})]^{-2.966} (p_f/\ell_\ell)^{-0.7931(p_t/b)} & 150 < \mathrm{Re}_{\ell p} < 5000 \end{cases} \tag{7.131}$$

$$f_3 = \begin{cases} (p_f/\ell_\ell)^{-0.308} (L_f/\ell_\ell)^{-0.308} (e^{-0.1167 p_t/H_t}) \theta^{0.35} & \mathrm{Re}_{\ell p} < 150 \\ (p_t/H_t)^{-0.0446} \{\ln[1.2 + (\ell_p/p_f)^{1.4}]\}^{-3.553} \theta^{-0.477} & 150 < \mathrm{Re}_{\ell p} < 5000 \end{cases} \tag{7.132}$$

Additional parameters for the friction factor correlations are: D_h is the hydraulic diameter of the fin geometry (mm), H_t the tube outside height (mm), and L_f the fin length in the airflow direction (mm). Note that θ in Eqs. (7.130) and (7.132) is in degrees. The conventional definition of the hydraulic diameter [the first definition in Eq. (3.65)] is used in Eq. (7.131) assuming that the louver fins were plain fins (without cuts) for the calculation of A_o and A; the effect of the braze fillets has also been neglected since no such information is available in the open literature. The correlation of Eq. (7.129) predicts 83% of the experimental friction factor data points within ±15%, with a mean deviation of 9% for the parameter ranges the same as those for Eq. (7.127).

Note that the authors included the tube width W_t in the heat transfer correlation (to take into account the correct surface area). However, they used the fin length $L_f^\dagger$ for the friction factor correlation since the friction factor f of Eq. (7.129) is for the fin friction component only (excluding entrance and exit pressure losses from the pressure drops measured).

7.5.3.3 Other Plate-Fin Surfaces.

Perforated and pin fin geometries have been investigated and it is found that they do not have superior performance then that for offset strip and louver fin geometries (Shah, 1985). Perforated fins are now used only in a limited number of applications. They are used as "turbulators" in oil coolers and in cryogenic air separation exchangers as a replacement for the existing perforated fin exchangers; modern cryogenic air separation exchangers use offset strip fin geometries.

[†] $W_t < L_f$ for 15 cores of Achaichia and Cowell (1988); for all other cores in the correlations, $W_t = L_f$.

Considerable research work has been reported on vortex generators using winglets (Jacobi and Shah, 1999), and research continues for heat exchanger applications.

7.5.4 Tube-Fin Extended Surfaces

Two major types of tube-fin extended surfaces are (1) individually finned tubes and (2) flat fins (sometimes referred to as plate fins) with or without enhancements/interruptions on an array of tubes as shown in Fig. 1.31. An extensive coverage of the published literature and correlations for these extended surfaces is provided by Webb (1994), Bemisderfer (1998), Kays and London (1998), and Wang (2000). Empirical correlations for some important geometries are summarized below.

7.5.4.1 Individually Finned Tubes. This fin geometry, helically wrapped (or extruded) circular fins on a circular tube as shown in Fig. 1.31a or 8.5, is commonly used in process and waste heat recovery industries. The following correlation for j factors is recommended by Briggs and Young (1963) for individually finned tubes in a staggered tube bank:

$$j = 0.134 \mathrm{Re}_d^{-0.319} \left(\frac{s}{\ell}\right)^{0.2} \left(\frac{s}{\delta}\right)^{0.11} \tag{7.133}$$

where $\mathrm{Re}_d = \rho u_m d_o / \mu$ (where u_m occurs at the minimum free-flow area defined in Section 8.2.1.2), $\ell [= (d_e - d_o)/2]$ is the radial height of the fin, δ is the fin thickness, $s(= p_f - \delta)$ is the distance between adjacent fins, and p_f is the fin pitch; see the circular fin figure in Table 4.5 for the definitions of $\ell(= \ell_f)$, δ and p_f. Equation (7.133) is valid for the following ranges: $1100 \le \mathrm{Re}_d \le 18{,}000$, $0.13 \le s/\ell \le 0.63$, $1.01 \le s/\delta \le 7.62$, $0.09 \le \ell/d_o \le 0.69$, $0.011 \le \delta/d_o \le 0.15$, $1.54 \le X_t/d_o \le 8.23$, fin root diameter d_o between 11.1 and 40.9 mm, and fin density $N_f (= 1/p_f)$ between 246 and 768 fins/m. All data have been obtained on equilateral triangular pitch tube bundles (30° in Table 8.1). The standard deviation of Eq. (7.133) with experimental results is 5.1%.

For friction factors, Robinson and Briggs (1966) recommended the following correlation:

$$f_{tb} = 9.465 \mathrm{Re}_d^{-0.316} \left(\frac{X_t}{d_o}\right)^{-0.927} \left(\frac{X_t}{X_d}\right)^{0.515} \tag{7.134}$$

Here $X_d = \left(X_t^2 + X_\ell^2\right)^{1/2}$ is the diagonal pitch, and X_t and X_ℓ are the transverse and longitudinal tube pitches, respectively. The correlation is valid for the following ranges: $2{,}000 \le \mathrm{Re}_d \le 50{,}000$, $0.15 \le s/\ell \le 0.19$, $3.75 \le s/\delta \le 6.03$, $0.35 \le \ell/d_o \le 0.56$, $0.011 \le \delta/d_o \le 0.025$, $1.86 \le X_t/d_o \le 4.60$, $18.6 \le d_o \le 40.9$ mm, and $311 \le N_f \le 431$ fins/m. The standard deviation of Eq. (7.134) with correlated data is 7.8%.

For crossflow over low-height finned tubes, Ganguli and Yilmaz (1987) correlated all data of Rabas and Taborek (1987) as well as some additional data published in the literature for air. They correlated the heat transfer result data as follows:

$$j = 0.255 \mathrm{Re}_d^{-0.3} \left(\frac{d_e}{s}\right)^{-0.3} \tag{7.135}$$

where d_e is the fin tip diameter of the radial (low height) fins. This correlation is valid for the following ranges: $\ell \leq 6.35\,\text{mm}$, $800 \leq \text{Re}_d \leq 800{,}000$, $20° \leq \varphi \leq 40°$, $N_r \geq 4$, $0.6 \leq \text{Pr} \leq 0.7$, and $5 \leq d_e/s \leq 60$, where φ is the tube layout angle for a tube bundle (see Table 8.1). All the heat transfer data correlated within $\pm 20\%$, and 95% of the data correlated within $\pm 15\%$. Note that the low-finned tube j factors for a given Re_d are lower than those for a plain tube and the j factors approach the plain tube value asymptotically at high Re_d.

For low-finned tubes, Ganguli and Yilmaz (1987) correlated the plain tube bank friction factor data as follows:

$$f = F_s f_{tb,p} \tag{7.136}$$

where F_s is the surface factor (a ratio of the friction factor for the finned tube to that for the plain tube) and $f_{tb,p}$ is the friction factor for a plain tube bank. They are given by

$$F_s = 2.5 + \frac{3}{\pi}\tan^{-1}\left[0.5\left(\frac{A}{A_p}\right) - 5\right] \tag{7.137}$$

where

$$A = \pi d_o(1 - \delta N_f) + 2N_f \frac{\pi}{4}\left(d_e^2 - d_o^2\right) + \pi d_e \delta N_f \qquad A_p = \pi d_o \tag{7.138}^\dagger$$

$$f_{tb,p} = 0.25 K_p \cdot \text{Re}_d^{-0.25} \tag{7.139}$$

$$K_p = 2.5 + 1.2(X_t^* - 0.85)^{-1.06} + 0.4\left(\frac{X_\ell^*}{X_t^*} - 1\right)^3 - 0.01\left(\frac{X_t^*}{X_\ell^*} - 1\right)^3 \tag{7.140}$$

This friction factor correlation is valid for the same parameter ranges mentioned for the heat transfer correlation. All of the friction data correlated within $\pm 20\%$, and 95% of the data correlated within $\pm 15\%$.

7.5.4.2 Plain Flat Fins on a Tube Array. This geometry, plain flat fins (Fig. 1.31b) on a staggered tube bank, is used in the air-conditioning/refrigeration industry as well as where the pressure drop on the fin side prohibits the use of enhanced/interrupted flat fins. An inline tube bank is generally not used unless very low fin side pressure drop is the essential requirement. A heat transfer correlation for plain flat fins on staggered tube banks (Fig. 1.31b) is provided by Wang and Chi (2000) and summarized by Wang (2000) as follows:

$$j = \begin{cases} 0.108\text{Re}_{dc}^{-0.29}\left(\dfrac{X_t}{X_\ell}\right)^{c_1}\left(\dfrac{p_f}{d_c}\right)^{-1.084}\left(\dfrac{p_f}{D_h}\right)^{-0.786}\left(\dfrac{p_f}{X_t}\right)^{c_2} & \text{for } N_r = 1 \\[3mm] 0.086\text{Re}_{dc}^{c_3}\cdot N_r^{c_4}\left(\dfrac{p_f}{d_c}\right)^{c_5}\left(\dfrac{p_f}{D_h}\right)^{c_6}\left(\dfrac{p_f}{X_t}\right)^{-0.93} & \text{for } N_r \geq 2 \end{cases} \tag{7.141}$$

---

† Here A is the total outside surface area of the finned tube and A_p is the plain tube outside surface area, and both A and A_p are per unit tube length.

where

$$c_1 = 1.9 - 0.23 \ln \text{Re}_{dc} \qquad c_2 = -0.236 + 0.126 \ln \text{Re}_{dc} \qquad (7.142a)$$

$$c_3 = -0.361 - \frac{0.042 N_r}{\ln \text{Re}_{dc}} + 0.158 \ln \left[N_r \left(\frac{p_f}{d_c} \right)^{0.41} \right] \qquad c_4 = -1.224 - \frac{0.076(X_\ell / D_h)^{1.42}}{\ln \text{Re}_{dc}}$$

$$(7.142b)$$

$$c_5 = -0.083 + \frac{0.058 N_r}{\ln \text{Re}_{dc}} \qquad c_6 = -5.735 + 1.21 \ln \frac{\text{Re}_{dc}}{N_r} \qquad (7.142c)$$

where p_f is the fin pitch, d_c is the collar diameter of the fin and $\text{Re}_{dc} = \rho u_m d_c / \mu$. This j factor correlation predicts 89% of the test points of 74 cores within ±15%, with a mean deviation of 8%. Wang and Chi (2000) also provided the following correlation for the friction factors:

$$f = 0.0267 \text{Re}_{dc}^{c_7} \left(\frac{X_t}{X_\ell} \right)^{c_8} \left(\frac{p_f}{d_c} \right)^{c_9} \qquad (7.143)$$

where

$$c_7 = -0.764 + 0.739 \left(\frac{X_t}{X_\ell} \right) + 0.177 \left(\frac{p_f}{d_c} \right) - \frac{0.00758}{N_r} \qquad (7.144a)$$

$$c_8 = -15.689 + \frac{64.021}{\ln \text{Re}_{dc}} \qquad c_9 = 1.696 - \frac{15.695}{\ln \text{Re}_{dc}} \qquad (7.144b)$$

Equations (7.141) and (7.143) are valid for the following ranges of the parameters: $300 \le \text{Re}_{dc} \le 20{,}000$, $6.9 \le d_c \le 13.6$ mm, $1.30 \le D_h \le 9.37$ mm, $20.4 \le X_t \le 31.8$ mm, $12.7 \le X_\ell \le 32$ mm, $1.0 \le p_f \le 8.7$ mm, and $1 \le N_r \le 6$. This friction factor correlation of Eq. (7.143) predicts 85% of experimental friction factors of 74 test cores within ±15%, with a mean deviation of 8%.

7.5.4.3 Corrugated Flat Fins on a Tube Array.

There are a number of variations available for flat fins with a sharp vs. smooth wave. The specific flat fin geometry shown in Fig. 7.30 is designated as a corrugated (herringbone or sharp-wave) fin. The heat transfer and flow friction correlations are developed by Wang (2000) and presented separately for large- and small-diameter tubes as follows. For larger tube diameters ($d_o = 12.7$ and 15.88 mm, before tube expansion), the following are the correlations:

$$j = 1.7910 \text{Re}_{dc}^{c_1} \left(\frac{X_\ell}{\delta} \right)^{-0.456} N_r^{-0.27} \left(\frac{p_f}{d_c} \right)^{-1.343} \left(\frac{p_d}{x_f} \right)^{0.317} \qquad (7.145)$$

$$f = 0.05273 \text{Re}_{dc}^{c_2} \left(\frac{p_d}{x_f} \right)^{c_3} \left(\frac{p_f}{X_t} \right)^{c_4} \left(\ln \frac{A}{A_{p,t}} \right)^{-2.726} \left(\frac{D_h}{d_c} \right)^{0.1325} N_r^{0.02305} \qquad (7.146)$$

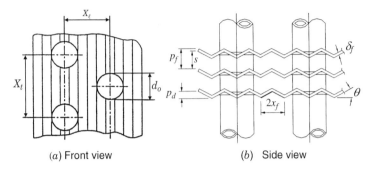

(a) Front view   (b) Side view

FIGURE 7.30   Corrugated fin on a tube array.

where

$$c_1 = -0.1707 - 1.374\left(\frac{X_\ell}{\delta}\right)^{-0.493}\left(\frac{p_f}{d_c}\right)^{-0.886} N_r^{-0.143}\left(\frac{p_d}{x_f}\right)^{-0.0296} \tag{7.147a}$$

$$c_2 = 0.1714 - 0.07372\left(\frac{p_f}{X_\ell}\right)^{0.25}\left(\ln\frac{A}{A_{p,t}}\right)\left(\frac{p_d}{x_f}\right)^{-0.2} \tag{7.147b}$$

$$c_3 = 0.426\left(\frac{p_f}{X_t}\right)^{0.3}\ln\frac{A}{A_{p,t}} \qquad c_4 = -\frac{10.2192}{\ln \mathrm{Re}_{dc}} \tag{7.147c}$$

Here x_f is the projected fin pattern length for one-half wavelength, p_d the fin pattern depth (peak-to-valley distance, excluding fin thickness, as shown in Fig. 7.30), and $A_{p,t}$ the tube outside surface area when there are no fins. Equations (7.145) and (7.146) are valid for the following ranges of the parameters: $500 \leq \mathrm{Re}_{dc} \leq 10{,}000$, $3.63 \leq D_h \leq 7.23$ mm, $13.6 \leq d_c \leq 16.85$ mm, $31.75 \leq X_t \leq 38.1$, $27.5 \leq X_\ell \leq 33$ mm, $2.98 \leq p_f \leq 6.43$ mm, $1 \leq N_r \leq 6$, $12.3 \leq \theta \leq 14.7°$, $6.87 \leq x_f \leq 8.25$ mm, and $p_d = 1.8$ mm. The correlation of Eq. (7.145) predicts 93% of experimental Colburn factors for 18 test cores within $\pm10\%$, with a mean deviation of 4%. Similarly, the correlation of Eq. (7.146) predicts 92% of experimental friction factors for 18 test cores within $\pm10\%$, with a mean deviation of 5%.

For smaller-diameter tubes ($d_o = 7.94$ and 9.53 mm before tube expansion), following are the correlations for the j and f factors:

$$j = 0.324\mathrm{Re}_{dc}^{c_1}\left(\frac{p_f}{X_\ell}\right)^{c_2}(\tan\theta)^{c_3}\left(\frac{X_\ell}{X_t}\right)^{c_4} N_r^{0.428} \tag{7.148}$$

$$f = 0.01915\mathrm{Re}_{dc}^{c_5}(\tan\theta)^{c_6}\left(\frac{p_f}{X_\ell}\right)^{c_7}\left(\ln\frac{A}{A_{p,t}}\right)^{-5.35}\left(\frac{D_h}{d_c}\right)^{1.3796} N_r^{-0.0916} \tag{7.149}$$

where

$$c_1 = -0.229 + 0.115 \left(\frac{p_f}{D_c}\right)^{0.6} \left(\frac{X_\ell}{D_h}\right)^{0.54} N_r^{-0.284} \ln(0.5 \tan \theta) \qquad (7.150a)$$

$$c_2 = -0.251 + \frac{0.232 N_r^{1.37}}{\ln(\mathrm{Re}_{dc}) - 2.303} \qquad c_3 = -0.439 \left(\frac{p_f}{D_h}\right)^{0.09} \left(\frac{X_\ell}{X_t}\right)^{-1.75} N_r^{-0.93} \qquad (7.150b)$$

$$c_4 = 0.502[\ln(\mathrm{Re}_{dc}) - 2.54] \qquad c_5 = 0.4604 - 0.01336 \left(\frac{p_f}{X_\ell}\right)^{0.58} \left(\ln \frac{A}{A_{p,t}}\right) (\tan \theta)^{-1.5}$$
$$\qquad (7.150c)$$

$$c_6 = 3.247 \left(\frac{p_f}{X_t}\right)^{1.4} \ln \frac{A}{A_{p,t}} \qquad c_7 = -\frac{20.113}{\ln \mathrm{Re}_{dc}} \qquad (7.150d)$$

Equations (7.148) and (7.149) are valid for the following ranges of the parameters: $300 \le \mathrm{Re}_{dc} \le 8000$, $1.53 \le D_h \le 4.52\,\mathrm{mm}$, $8.58 \le d_c \le 10.38\,\mathrm{mm}$, $X_t = 25.4\,\mathrm{mm}$, $19.05 \le X_\ell \le 25.04\,\mathrm{mm}$, $1.21 \le p_f \le 3.66\,\mathrm{mm}$, $1 \le N_r \le 6$, $14.5 \le \theta \le 18.5°$, $4.76 \le x_f \le 6.35\,\mathrm{mm}$, and $1.18 \le p_d \le 1.68\,\mathrm{mm}$. The correlation of Eq. (7.148) predicts 95% of experimental Colburn factors of 27 test cores within $\pm 15\%$, with a mean deviation of 6%. Similarly, the correlation of Eq. (7.149) predicts 97% of experimental friction factors for data points of 27 test cores within $\pm 15\%$, with a mean deviation of 5%.

7.5.5 Regenerator Surfaces

The two most common types of regenerator surfaces are (1) continuous cylindrical passages for rotary regenerators and some compact fixed-matrix regenerators, and (2) randomly packed woven screens, crossed rods, and packed beds using a variety of materials.

For compact regenerators, the continuous cylindrical flow passages have simple geometries, such as triangular, rectangular, and hexagonal passages. The Nu and f factors of Tables 7.3 and 7.4 are a valuable baseline for such passages. Due to the entrance length effect, actual Nusselt numbers should be higher than that for fully developed flow. However, the actual flow passages are never ideal and uniform, due to manufacturing processes and tolerances. The passage-to-passage nonuniformity, discussed in Section 12.1.2, reduces heat transfer more than the gain by the thermal entrance effect. The friction factors are generally higher than those for fully developed flow because of the significant effect of the hydrodynamic entrance length. The passage-to-passage nonuniformity reduces the friction factor and Δp only slightly. Thus, generally j or Nu is lower and f is higher than those for fully developed flow. Also, the thermal boundary condition for heat transfer may not exactly correspond to any of the boundary conditions described previously. Hence, accurate j and f versus Re characteristics are generally determined experimentally even for simple geometries in addition to those for complex geometries. As an illustration, London et al. (1970) presented the following correlations for airflow through triangular passages ($40 < \mathrm{Re} < 800$):

$$f = \frac{14.0}{\mathrm{Re}} \qquad j = \frac{3.0}{\mathrm{Re}} \qquad (7.151)$$

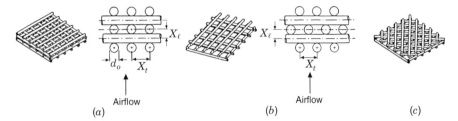

Airflow Airflow
(a) (b) (c)

FIGURE 7.31 Cross rod geometries. (From Kays and London, 1998.)

London and Shah (1973) presented the following correlations for airflow through hexagonal passages ($80 < \mathrm{Re} < 800$):

$$f = \frac{17.0}{\mathrm{Re}} \qquad j = \frac{4.0}{\mathrm{Re}} \qquad (7.152)$$

Crossed rod geometries of Fig. 7.31 have alternate layers of parallel solid rods touching each other and stacked 90° to each other. Martin (2002) has correlated the j factors as follows for these geometries based on the generalized Lévêque solution in thermal entrance turbulent flow in a circular pipe (Schlünder, 1998):

$$\mathrm{Nu} = 0.535[\mathbf{p}(1 - \mathbf{p})^{-0.25}f \cdot \mathrm{Re}^2 \cdot \mathrm{Pr}]^{1/3} \qquad (7.153)$$

where the friction factors of Kays and London (1998) data are correlated by Das (2001) as

$$f = \begin{cases} 0.603\mathrm{Re}^{-0.104}(X_t^*)^{0.136} & \text{for inline arrangement} \\ 0.728\mathrm{Re}^{-0.188}(X_t^*)^{0.913} & \text{for staggered arrangement} \\ 0.475\mathrm{Re}^{-0.108}(X_t^*)^{0.458} & \text{for random arrangement} \end{cases} \qquad (7.154)$$

and d_w is the wire diameter, $X_t^* = X_t/d_w$, and X_t is the transverse pitch of the rods. Note that the hydraulic diameter D_h, the ratio of free-flow area to frontal area σ, the porosity (void fraction) $\mathbf{p}$, and the ratio of heat transfer surface area to volume α of the crossed rod matrices (of square mesh using circular cylinder rods) are related to d_w and X_t as follows:

$$D_h = \frac{4X_t}{\pi} - d_w = \frac{\mathbf{p}d_w}{1 - \mathbf{p}} \qquad \sigma = \frac{(X_t - d_w)^2}{X_t^2} \qquad \mathbf{p} = 1 - \frac{\pi d_w}{4X_t} \qquad \alpha = \frac{\pi}{X_t} \qquad (7.155)$$

The correlations of Eqs. (7.153) and (7.154) are valid for $d_w = 9.53\,\mathrm{mm}$, and $1.571 \le X_t/d_w \le 4.675$ or $0.500 \le \mathbf{p} \le 0.832$. The correlation for Nu (the j factor) is accurate within $\pm 4.8\%$ and that for the f factors is accurate within ± 7.8, 6.7, and 7.1% for inline, staggered, and random arrangements of Kays and London data (1998).

Example 7.6 The purpose of this example is to determine the change in performance of a single-pass unmixed–unmixed crossflow exchanger (Fig. E7.6a) when it is made a two-pass configuration either in series coupling (over and under) Fig. E7.6b), or in parallel

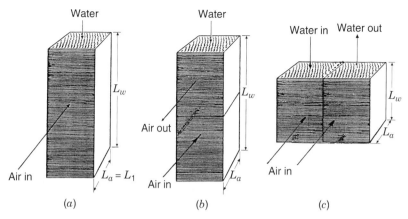

FIGURE E7.6   (a) Single-pass unmixed–unmixed crossflow exchanger; (b) series coupled (over-and-under) two-pass cross-counterflow exchanger; (c) parallel coupled (side-by-side) two-pass exchanger.

coupling (side by side, Fig. E7.6c), with water *mixed* between passes for two cases: 1:1 and 10:1 thermal resistances on air and water sides. Note that the total heat transfer surface area is the same in all three exchangers. The performance specifications for the single-pass exchanger of Fig. E7.6a are: number of heat transfer units, 1.2; ratio of heat capacity rates, 0.8; effectiveness, 0.547; and pressure drop on the air side, 0.25 kPa. All areas, geometrical properties, flow rates, heat transfer coefficients, friction factors, and fluid properties are known (i.e., the exchangers have been designed).

Consider $j_a \propto \mathrm{Re}_a^{-0.4}$ and $f_a \propto \mathrm{Re}_a^{-0.2}$ on the air side and assume fluid properties to be constant. Determine the exchanger effectivenesses and the pressure drops for the exchangers for Fig. E7.6b and c. Evaluate ε and Δp for the following two cases of thermal resistances for the single-pass exchanger of Fig. E7.6a: (a) $\mathbf{R}_a \approx \mathbf{R}_w$ and (b) $\mathbf{R}_a \approx 10\mathbf{R}_w$. Neglect the wall resistances and fouling resistances in all three exchangers.

SOLUTION

Problem Data and Schematic: An exchanger with known thermal capacities is given as in Fig. E7.6a. Also given are two exchangers that vary in overall configuration from the first exchanger as shown in Figs. E7.6b and c. For the exchanger of Fig. E7.6a,

$$\mathrm{NTU} = 1.2 \qquad C^* = \frac{C_{\mathrm{air}}}{C_{\mathrm{water}}} = 0.8 \qquad \varepsilon = 0.547 \qquad \Delta p_a = 0.25 \text{ kPa}$$

Determine: The exchanger effectivenesses and pressure drops for the exchangers of Fig. E7.6b and c.

Assumptions: Constant fluid properties apply; pressure drops due to the entrance, exit, and momentum effects on the air side are neglected, as is the effect of flow turning from pass 1 to pass 2 on heat transfer and pressure drop.

Analysis: Let us use subscripts 1, 2, and 3, respectively, for all pertinent quantities of exchangers of Fig. E7.6a, b, and c. Careful observations of the geometry and flow arrangement will yield the following results:

Variable	Exchanger 1	Exchanger 2	Exchanger 3
Airflow length	L_1	$L_2 = 2L_1$	$L_3 = L_1$
Air-side mass velocity	G_1	$G_2 = 2G_1$	$G_3 = G_1$
Air-side surface area per pass	A_1	$A_2 = 0.5A_1$	$A_3 = 0.5A_1$
Water-side surface area per pass	$A_{w,1}$	$A_{w,2} = 0.5A_{w,1}$	$A_{w,3} = 0.5A_{w,1}$
Air-side heat capacity rate per pass	C_1	$C_2 = C_1$	$C_3 = 0.5C_1$
Heat capacity rate ratio air/water for each pass	C^*	$C_p^* = C^*$	$C_p^* = 0.5C^*$
Air-side heat transfer coefficient	h_1	h_2	$h_3 = h_1$
Water-side heat transfer coefficient	h_w	h_w	h_w
U based on air-side surface area per pass	U_1	U_2 will depend on $\mathbf{R}_a$ and $\mathbf{R}_w$	$U_3 = U_1$
$[(hA)_w]$ per pass	$[(hA)_w]_1$	$0.5[(hA)_w]_1$	$0.5[(hA)_w]_1$
$[(hA)_a]$ per pass	$(hA)_1$	$h_2(0.5A_1)$	$0.5(hA)_1$

We first calculate the change in the heat transfer coefficient on air side for exchanger 2. By definition, $h = j_a G c_p \cdot \mathrm{Pr}^{2/3}$, and $j_a = \mathrm{Re}_a^{-0.4}$ is specified where $\mathrm{Re}_a = G D_h / \mu$. Hence,

$$h = h_a \propto j_a G \propto \mathrm{Re}_a^{-0.4} G \propto G^{-0.4} G = G^{-0.6}$$

Therefore,

$$\frac{h_2}{h_1} = \left(\frac{G_2}{G_1}\right)^{0.6} = (2)^{0.6} = 1.516$$

Now, for the single-pass exchanger (exchanger 1), we designate $(\eta_o hA)_a = (hA)_1$. Then we get the following:

$\mathbf{R}_a = \mathbf{R}_w$ Case	$\mathbf{R}_a = 10\mathbf{R}_w$ Case
$\dfrac{1}{(hA)_1} = \dfrac{1}{[(hA)_w]_1}$ (i.e., $[(hA)_w]_1 = (hA)_1$)	$\dfrac{1}{(hA)_1} = \dfrac{10}{[(hA)_w]_1}$ (i.e., $[(hA)_w]_1 = 10(hA)_1$)
$\dfrac{1}{(UA)_1} \approx \dfrac{1}{(hA)_1} + \dfrac{1}{(hA)_1} = \dfrac{2}{(hA)_1}$	$\dfrac{1}{(UA)_1} \approx \dfrac{1}{(hA)_1} + \dfrac{1}{10(hA)_1} = \dfrac{1.1}{(hA)_1}$
Hence, $(hA)_1 = 2(UA)_1$	Hence, $(hA)_1 = 1.1(UA)_1$

Now let us calculate the exchanger effectiveness for exchanger 2, assuming a two-pass exchanger, for the two cases of thermal resistance distribution. The appropriate relationships derived in Chapter 3 are used here.

$\mathbf{R}_a = \mathbf{R}_w$ Case	$\mathbf{R}_a = 10\mathbf{R}_w$ Case
Since $[(hA)_w]_{2,p} = 0.5[(hA)_w]_1$ and $[(hA)_w]_1 = (hA)_1$, we get $[(hA)_w]_{2,p} = 0.5(hA)_1$. Now, $[(hA)_a]_2 = 1.516h_1(0.5A_1) = 0.758(hA)_1$. Hence, from Eq. (3.24),	Since $[(hA)_w]_{2,p} = 0.5[(hA)_w]_1$ and $[(hA)_w]_1 = 10(hA)_1$, we get $[(hA)_w]_{2,p} = 5(hA)_1$. Now, $[(hA)_a]_2 = 1.516h_1(0.5A_1) = 0.758(hA)_1$. Hence, from Eq. (3.24),
$\dfrac{1}{(UA)_{2,p}} = \dfrac{1}{0.758(hA)_1} + \dfrac{1}{0.5(hA)_1} = \dfrac{3.32}{(hA)_1}$	$\dfrac{1}{(UA)_{2,p}} = \dfrac{1}{0.758(hA)_1} + \dfrac{1}{5(hA)_1} = \dfrac{1.52}{(hA)_1}$

Substituting the value of $(hA)_1 = 2(UA)_1$ for exchanger 1 from above,

$$(UA)_{2,p} = 0.602(UA)_1$$

Since $C_{\min} = C_1$ also for exchanger 2,

$$NTU_{2,p} = 0.602NTU_1 = 0.602 \times 1.2 = 0.722$$

Now, based on the input, $C_p^* = C^* = 0.8$. From Fig. 3.9 (or from the ε-NTU formula in Table 3.3) with $C_p^* = 0.8$ and $NTU_{2,p} = 0.722$, we get $\varepsilon_p = 0.422$. Thus, ε_2 for the two-pass exchanger of Fig. E7.6b, from Eq. (3.131), is

$$\varepsilon_2 = \frac{[(1 - 0.8 \times 0.422)/(1 - 0.422)]^2 - 1}{[(1 - 0.8 \times 0.422)/(1 - 0.422)]^2 - 0.8}$$

$$= 0.610$$

Increase in ε_2 over ε_1 $(= 0.547$ given) is

$$\frac{\varepsilon_2}{\varepsilon_1} = \frac{0.610}{0.547} = 1.12 \quad \text{or} \quad 12\%$$

Substituting the value of $(hA)_1 = 1.1(UA)_1$ for exchanger 1 from above,

$$(UA)_{2,p} = 0.724(UA)_1$$

Since $C_{\min} = C_1$ also for exchanger 2,

$$NTU_{2,p} = 0.724NTU_1 = 0.724 \times 1.2 = 0.869$$

Now, based on the input, $C_p^* = C^* = 0.8$. From Fig. 3.9 (or from the ε-NTU formula in Table 3.3) with $C_p^* = 0.8$ and $NTU_{2,p} = 0.869$, we get $\varepsilon_p = 0.468$. Thus, ε_2 for the two-pass exchanger of Fig. E7.6b, from Eq. (3.131), is

$$\varepsilon_2 = \frac{[(1 - 0.8 \times 0.468)/(1 - 0.468)]^2 - 1}{[(1 - 0.8 \times 0.468)/(1 - 0.468)]^2 - 0.8}$$

$$= 0.657$$

Increase in ε_2 over ε_1 $(= 0.547$ given) is

$$\frac{\varepsilon_2}{\varepsilon_1} = \frac{0.657}{0.547} = 1.20 \quad \text{or} \quad 20\%$$

Thus, the increase in heat transfer rate for a two-pass exchanger is 12% and 20%, respectively, over similar single-pass balanced (the $\mathbf{R}_a = \mathbf{R}_w$ case) and highly unbalanced (the $\mathbf{R}_a = 10\mathbf{R}_w$ case) exchangers.

Now we calculate the effectiveness of exchanger 3, which is a parallel coupled (side-by-side) two-pass exchanger with the fluid (water) mixed between passes. The "unfolding" of this exchanger will not result in exchanger 1 because the water side is not unmixed throughout. As noted earlier, for each pass of exchanger 3, C_{air} is one-half that of the exchanger 1. Hence,

$$C_p^* = \frac{C_{a,3}}{C_w} = \frac{0.5C_{a,1}}{C_w} = 0.5C_1^* = 0.4$$

and the NTU per pass is

$$NTU_{3,p} = \frac{U_1(0.5A_1)}{0.5C_1} = NTU_1 = 1.2$$

Note that the overall U will be the same for exchangers 1 and 3 as noted earlier. Hence, ε_p from Fig 3.9 (or from the ε-NTU formula for an unmixed–unmixed crossflow exchanger in Table 3.3) is 0.6175 for $NTU_p = 1.2$ and $C_p^* = 0.4$. In this case, the parallel airstream is the fluid stream 2 of Fig. 3.22. Hence, the exchanger effectiveness ε_3 for the airside is calculated from modified Eq. (3.162) in terms of $\varepsilon - \varepsilon_p - C_p^*$ formula as follows with $\varepsilon_p = 0.6175$, $C_p^* = 0.4$, and $n = 2$.

$$\varepsilon_3 = \frac{1}{nC_p^*}\left[1 - \left(1 - \varepsilon_p C_p^*\right)^n\right] = \frac{1}{2 \times 0.4}[1 - (1 - 0.6175 \times 0.4)^2] = 0.5412$$

Since the difference between exchanger 1 and exchanger 3 on the water side is only unmixed (exchanger 1) vs. mixed (exchanger 3) between passes (here we considered two-pass on the water side for exchanger 3), we expect the effectiveness of exchanger 3 to be slightly lower,[†] and this is what we found $\varepsilon_3 = 0.541$ vs. $\varepsilon_1 = 0.547$.

Now let us first calculate the air-side pressure drop for exchanger 2. Since $f_a \propto Re_a^{-0.2}$ and $Re_a = GD_h/\mu$,

$$\frac{f_2}{f_1} = \left(\frac{Re_2}{Re_1}\right)^{-0.2} = \left(\frac{G_2}{G_1}\right)^{-0.2} \tag{7.156}$$

because D_h on the air side is the same for both exchangers, and we assume that the fluid properties do not vary with relatively small changes in the air temperature. If we consider the core frictional pressure drop as the main component of the total pressure drop, then, from Eq. (6.29),

$$\Delta p = \frac{4fLG^2}{2g_c D_h}\left(\frac{1}{\rho}\right)_m$$

Therefore, the ratio of the pressure drops is

$$\frac{\Delta p_2}{\Delta p_1} = \frac{f_2}{f_1}\frac{L_2}{L_1}\left(\frac{G_2}{G_1}\right)^2$$

Substituting in this equation, f_2/f_1 of Eq. (7.156), $L_2/L_1 = 2$, and $G_2/G_1 = 2$, and we get

$$\frac{\Delta p_2}{\Delta p_1} = \left(\frac{G_2}{G_1}\right)^{-0.2}\left(\frac{L_2}{L_1}\right)\left(\frac{G_2}{G_1}\right)^2 = 2\left(\frac{G_2}{G_1}\right)^{1.8} = 2(2)^{1.8} = 6.96$$

or

$$\Delta p_2 = 6.96 \times 0.25 = 1.74\,\text{kPa} \qquad\qquad Ans.$$

Thus, the pressure drop on the air side for the two-pass exchanger 2 is about seven times that for a similar single-pass exchanger 1. Thus, the penalty in the air-side pressure drop is substantial compared to the gain in heat transfer (12 to 20%) for a two-pass cross-counterflow exchanger.

For exchanger 3, the pressure drop on the air side will not change and the pressure drop on the water side will be slightly higher, due to the return tank (a 180° bend) after the first pass. The only advantage of exchanger 3 could be different packaging.

Discussion and Comments: The motivation of this problem is to demonstrate the calculation procedure and differences in heat transfer and pressure drop performance of a heat exchanger when it is packaged in different ways, ideally having the same surface areas and the same individual heat transfer surface geometries. Since exchanger 3 is theoretically almost identical to exchanger 1, there is no significant difference in either heat transfer or pressure drop. However, exchanger 2 is two-pass exchanger, and as a result, its effectiveness improves over that for the single-pass exchanger as expected. However,

---

[†] Refer to Chapter 11 for detailed discussion and thermodynamics arguments for the reasons of this performance reduction.

there is a significant penalty in the pressure drop. Hence the design decision on which arrangement to select (exchanger 1 or 2) will depend on the allowable pressure drop on the air side.

Note that if we had made exchanger 2 a two-pass cross-counterflow exchanger with the water side as two-pass and airflow as a straight throughflow, the resulting increase in the water-side pressure drop (about 7- to 8-fold) due to the increased velocity (twice), double the flow length, and a 180° turn may be tolerable due to low fluid pumping power (refer to the end of Section 6.1.1). Such a flow arrangement is used in liquid-to-gas exchangers when the liquid-side heat transfer coefficient is very low (or hA on the liquid side is equal or lower than gas side); a substantial gain in heat transfer performance may be obtained in that case, due to the increased velocity and hence h and U.

7.6 INFLUENCE OF TEMPERATURE-DEPENDENT FLUID PROPERTIES

One of the basic assumptions made in the theoretical correlations for Nu and f of Section 7.4 is that the fluid properties remain constant throughout the flow field. Most of the original experimental j and f data presented in Section 7.5 involve small temperature differences so that the fluid properties generally do not vary significantly. In certain heat exchanger applications, fluid temperatures vary significantly. At least two questions arise: (1) Can we use the j and f data obtained for air at 50 to 100°C (100 to 200°F) for air at 500 to 600°C (900 to 1100°F)? (2) Can we use the j and f data obtained with air (such as all data in Kays and London, 1998) for water, oil, and viscous liquids? The answer is yes, by modifying the constant-property j and f data for the variations in the fluid properties within a heat exchanger, and we discuss this method in this section.

Let us first illustrate how the variation in fluid properties affect the pressure drop and heat transfer in a heat exchanger. Consider a specific case of influence of the variable liquid viscosity on pressure drop and heat transfer for fully developed laminar flow through a circular tube. In absence of heat transfer (isothermal case), the velocity profile will be parabolic (Fig. 7.32). Now consider heat being transferred from the tube wall to the liquid. This will set up a temperature gradient in the radial direction in the liquid at a tube cross section. Liquid will be hotter near the wall and cooler near the tube centerline. Since the liquid viscosity decreases with increasing temperature, the fluid near the wall will have lower viscosity than that of the isothermal case. This will result in increased velocity near the tube wall and a decreased velocity near the tube center to satisfy the continuity equation for the steady-flow case. The resulting velocity profile is "flatter," as shown in Fig. 7.32. The decrease in liquid viscosity near the wall yields lower f and lower Δp. The increased velocity near the wall may mean a more efficient convection heat

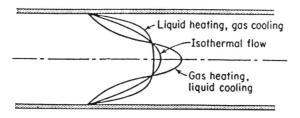

FIGURE 7.32   Influence of heat transfer on laminar velocity distribution.

transfer process, resulting in higher heat transfer coefficient h (and higher Nu or j, depending on the flow type in the exchanger) and higher heat transfer rate.

Now consider cooling of the liquid so that the tube wall is at a lower temperature than that of the liquid. As the liquid near the wall is at a lower temperature it has a higher viscosity and a lower velocity than for the isothermal case (see Fig. 7.32). This results in higher f and higher Δp and a lower heat transfer coefficient h (lower Nu or j) and lower heat transfer rate.

Since the viscosity increases with increasing temperature for gases, the gas heating situation is similar to the liquid cooling situation, resulting in higher f and Δp and lower h, j, and q. Since the viscosity decreases with decreasing temperature for gases, the gas cooling situation is similar to the liquid heating situation. However, it should be emphasized that the variation in viscosity for gases (such as air) is about an order of magnitude lower than that for liquids (such as water) over comparable temperature differences. Hence, the effect of variable fluid properties on *laminar flow* Nusselt numbers is found to be negligible, as will be shown in Table 7.13, as the exponent n of Eq. (7.157) is zero.

7.6.1   Correction Schemes for Temperature-Dependent Fluid Properties

For engineering applications, it is convenient to employ the constant-property analytical solutions, or experimental data obtained with small temperature differences, and then to apply some kind of correction to take into account the effect of fluid property variations. Three options (the first two commonly used) for such a correction are (1) the property ratio method, (2) the reference temperature method, and (3) other methods. In the *property ratio method*, all properties are evaluated at the bulk temperature, and then all the variable-property effects are lumped into a correction factor. This factor is a ratio of some pertinent property evaluated at the surface temperature to that property evaluated at the bulk temperature. Therefore, this correction factor is a function of temperature. In the *reference temperature method*, pertinent groups are evaluated such that the constant-property results could be used directly to evaluate variable property performance. Typically, this may be the film temperature or the surface temperature. The *film temperature* is an arithmetic average of the wall temperature and the fluid bulk temperature. There are a number of other methods reported in the literature to take into account the variable-property effects; one of them used in Russian literature is the Prandtl number ratio instead of the temperature ratio of Eq. (7.157) or the viscosity ratio of Eq. (7.158).

The property ratio method is used extensively in internal flow (such as in heat exchanger design and analysis), while the reference temperature method is the most common for external flow (as in aerodynamic problems). One of the reasons for the selection of the property ratio method for internal flow is that the determination of $G = \rho u_m$ (used in calculations of Re and St or j factor) is straightforward. It is computed as $G = \dot{m}/A_o$, regardless of the variations in the fluid density ρ in the heat exchanger. In the reference temperature method, ρ in ρu_m is evaluated at the reference temperature, while ρ in determining u_m ($u_m = \dot{m}/A_o\rho$) is evaluated at the bulk temperature. Thus two densities are needed for the reference temperature method, which leads to awkwardness and ambiguous interpretations.

For gases, the viscosity, thermal conductivity, and density are functions of the absolute temperature T; they generally increase with T. This absolute temperature dependence is similar for different gases except near the temperature extremes (near the critical

TABLE 7.12   Property Ratio Method Exponents of Eqs. (7.157) and (7.158) for Laminar Flow

Fluid	Heating	Cooling
Gas	$n = 0.0$, $m = 1.00$ for $1 < T_w/T_m < 3$	$n = 0.0$, $m = 0.81$ for $0.5 < T_w/T_m < 1$
Liquid	$n = -0.14$, $m = 0.58$ for $\mu_w/\mu_m < 1$	$n = -0.14$, $m = 0.54$ for $\mu_w/\mu_m > 1$

Source: Data from Shah (1985).

temperature). The temperature-dependent property effects for *gases* are adequately correlated by the following equations for Nusselt numbers and friction factors:

$$\frac{\mathrm{Nu}}{\mathrm{Nu_{cp}}} = \left(\frac{T_w}{T_m}\right)^n \qquad \frac{f}{f_{cp}} = \left(\frac{T_w}{T_m}\right)^m \qquad (7.157)$$

where T_w and T_m are *absolute* temperatures.

For liquids, the viscosity is the only property of importance that varies greatly with the temperature; the variations in the thermal conductivity and specific heat with temperature roughly compensate their effects except near the critical points. Thus the temperature-dependent property effects for *liquids* are adequately correlated by the following equations for Nusselt numbers and friction factors:

$$\frac{\mathrm{Nu}}{\mathrm{Nu_{cp}}} = \left(\frac{\mu_w}{\mu_m}\right)^n \qquad \frac{f}{f_{cp}} = \left(\frac{\mu_w}{\mu_m}\right)^m \qquad (7.158)$$

Here the subscript cp in Eqs. (7.157) and (7.158) refers to constant properties (i.e., Nu and f for constant fluid properties), and all properties in the dimensionless groups of Eqs. (7.157) and (7.158) are evaluated at the *bulk* temperature. The values of the exponents n and m for fully developed laminar and turbulent flows in a circular tube are summarized in Tables 7.12 and 7.13 for heating and cooling situations.

TABLE 7.13   Property Ratio Method Correlations or Exponents of Eqs. (7.157) and (7.158) for Turbulent Flow

Fluid	Heating	Cooling
Gas	$\mathrm{Nu} = 5 + 0.012\mathrm{Re}^{0.83}\,(\mathrm{Pr} + 0.29)(T_w/T_m)^n$ $n = -[\log_{10}(T_w/T_m)]^{1/4} + 0.3$ for $1 < T_w/T_m < 5$, $0.6 < \mathrm{Pr} < 0.9$, $10^4 < \mathrm{Re} < 10^6$, and $L/D_h > 40$ $m = -0.1$ for $1 < T_w/T_m < 2.4$	$n = 0$ $m = -0.1$ (tentative)
Liquid	$n = -0.11^a$ for $0.08 < \mu_w/\mu_m < 1$ $f/f_{cp} = (7 - \mu_w/\mu_m)/6^b$ or $m = 0.25$ for $0.35 < \mu_w/\mu_m < 1$	$n = -0.25^a$ for $1 < \mu_w/\mu_m < 40$ $m = 0.24^b$ for $1 < \mu_w/\mu_m < 2$

Source: Data from Shah (1985).
[a] Valid for $2 \leq \mathrm{Pr} \leq 140$, $10^4 \leq \mathrm{Re} \leq 1.25 \times 10^5$.
[b] Valid for $1.3 \leq \mathrm{Pr} \leq 10$, $10^4 \leq \mathrm{Re} \leq 2.3 \times 10^5$.

These correlations [Eqs. (7.157) and (7.158)] with exponents from Tables 7.12 and 7.13, are derived for the constant-heat flux boundary condition. The variable-property effects are generally not important for fully developed flow having a constant wall temperature boundary condition, since T_m approaches T_w for fully developed flow. Therefore, to take into account the minor influence of property variations for the constant wall temperature boundary condition, the correlations of Eqs. (7.157) and (7.158) are adequate. The values of the exponents n and m may also depend on the duct cross-sectional shape, surface interruptions, developing flows and thermal boundary conditions different from the constant heat flux or constant wall temperature. However, no such information is available in the literature. Hence, Eqs. (7.157) and (7.158) may not be able to fully correct the variable-property effects for many diverse situations in heat exchangers, but they still represent the best primary corrections in open literature.

7.7   INFLUENCE OF SUPERIMPOSED FREE CONVECTION

When the velocities of secondary flows due to free convection effects are not negligible in comparison to the mean velocity of the forced convection flow, the free convection may have a significant influence on heat transfer and pressure drop calculated by a pure forced convection correlation. The influence of superimposed free convection over pure forced convection flow is important when the fluid velocity is low, a high temperature difference $(T_w - T_m)$ is employed, or the passage geometry has a large hydraulic diameter D_h. The effect of the superimposed free convection is generally important in the laminar flow of many shell-and-tube heat exchangers; it is quite negligible for compact heat exchangers.

The effect of free convection is correlated by combinations of the following dimensionless numbers: Grashof number Gr, Rayleigh number Ra, Prandtl number Pr, and L/D along with the Reynolds number Re. The Grashof and Rayleigh numbers are defined as

$$\text{Gr} = \frac{g\rho^2 D_h^3 \beta^*(T_w - T_m)}{\mu^2} \tag{7.159}$$

$$\text{Ra} = \text{Gr} \cdot \text{Pr} = \frac{g\rho^2 D_h^3 \beta^* c_p(T_w - T_m)}{\mu k} \tag{7.160}$$

where β^* is the coefficient of thermal expansion and g is the gravitational acceleration. The Grashof number represents a ratio of the buoyancy force to the viscous force, and the buoyancy force is due to spatial variation in the fluid density as a result of temperature differences. The Rayleigh number indicates that the free convection boundary layer is laminar or turbulent, depending on its value. For vertical plates, the boundary layer is laminar for $\text{Ra}_x < 10^9$ and turbulent for $\text{Ra}_x > 10^9$, where Ra_x is defined by Eq. (7.160) with D_h replaced by x, the axial coordinate along the vertical plate from the point of initiation of free convection. For pure forced convection flow, Gr or Ra approaches zero value.

The problem of free convection over forced convection is referred to as the *combined* or *mixed* convection problem. Here the Reynolds and Grashof numbers are the dimensionless groups characterizing forced convection and free convection *flows*, respectively. From an order-of-magnitude analysis of the boundary layer equations, it can be shown that when Gr/Re^2 is on the order of unity, the problem should be considered as the combined convection problem. If $Gr/Re^2 \ll 1$, free convection is negligible; and if $Gr/Re^2 \gg 1$, forced convection is negligible. Since the combined convection problem is more important for laminar flow, some correlations will be provided next for laminar flow through horizontal and vertical tubes.

7.7.1 Horizontal Circular Tubes

For horizontal tubes, free convection sets up secondary flows at a cross section that aids the convection process. Hence, the heat transfer coefficient and Nusselt number for the combined convection are higher than those for the pure forced convection for flow in a horizontal tube. The maximum heat transfer occurs at the bottom of the tube. When the free convection effect is significant in laminar flow, large temperature gradients exist near the wall, and the temperature variations in the horizontal and vertical directions at a given flow cross section are also markedly different from the parabolic velocity distribution for Poiseuille flow.

Metais and Eckert (1964) recommend a free, mixed, and forced convection regime map, as shown in Fig. 7.33, for a horizontal circular tube with axially constant wall temperature boundary condition. The limits of the forced and free convection regimes are defined in such a manner that free convection effects contribute only about 10% to the heat flux. Figure 7.33 may therefore be used as a guide to determine whether or not free convection is important. Recently, Ghajar and Tam (1995) have presented a flow regime map for a horizontal circular tube with uniform wall heat flux boundary condition and three inlet configurations to the tube. They have developed this chart for a limited range of the Rayleigh number Ra and provided correlations for the Nusselt number.

A number of correlations for developed and developing mixed convection flows are presented by Aung (1987) for a horizontal circular tube with various boundary conditions. As an illustration, Morcos and Bergles (1975) presented the following combined convection correlation in a horizontal circular tube:

$$Nu_{H4} = \{(4.36)^2 + [0.145(Gr^* \cdot Pr^{1.35} \cdot K_p^{0.25})^{0.265}]^2\}^{1/2} \qquad (7.161)$$

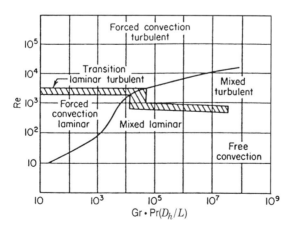

FIGURE 7.33 Horizontal circular tube free, forced, and mixed convection flow regimes for $10^{-2} < Pr(D_h/L) < 1$. (From Metais and Eckert, 1964.)

where

$$\mathrm{Gr^*} = \mathrm{Gr} \cdot \mathrm{Nu} = \frac{g\beta^* D_h^4 q''}{\nu^2 k} \qquad K_p = \frac{k_w \delta_w}{k D_h} \qquad (7.162)$$

Here $\mathrm{Gr^*}$ is the modified Grashof number and K_p is the peripheral heat conduction parameter. The subscript H4 denotes the axially constant wall heat flux boundary condition with finite heat conduction in the peripheral direction (Shah and London, 1978). This correlation is valid for $3 \times 10^4 < \mathrm{Ra} < 10^6$, $4 < \mathrm{Pr} < 175$, and $0.015 < K_p < 0.5$, where Ra is defined in Eq. (7.160) with T_w replaced by the mean wall temperature $T_{w,m}$. All the fluid properties in $\mathrm{Nu_{H4}}$, $\mathrm{Gr^*}$, Pr, and K_p should be evaluated at the *film* temperature [i.e., at $(T_{w,m} + T_m)/2$].

The thermal entrance length is reduced significantly in the presence of the free convection effects. Thus, for most cases when free convection effects are superimposed, the combined convection flow is fully developed. Hence, Eq. (7.161) is sufficient to determine the Nusselt numbers for combined convection.

The friction factors are also *higher* for the combined convection case. Based on the ethylene glycol data, Morcos and Bergles (1975) presented the following correlation:

$$\frac{f}{f_{fc}} = [1 + (0.195\mathrm{Ra}^{0.15})^{15}]^{1/15} \qquad (7.163)$$

where $f_{fc} = 16/\mathrm{Re}$ is the *isothermal* forced convection friction factor. Both f and Ra are based on the fluid properties evaluated at the *film* temperature.

It should be emphasized that the free convection effects are unimportant for compact heat exchanger surfaces due to the small D_h, illustrated in the following example.

Example 7.7 Investigate the influence of superimposed free convection on the air side of the automobile radiator of Problem 3.8. Use the following additional information: $\Delta T_{lm} = 49.9°\mathrm{C}$, $D_h = 0.003\,\mathrm{m}$ for air-side surface, $\mu = 0.20 \times 10^{-4}\,\mathrm{Pa \cdot s}$, $\rho = 1.09\,\mathrm{kg/m^3}$, $\mathrm{Pr} = 0.7$, $\ell_{ef}/D_h = 0.5$, and the operating Reynolds number $\mathrm{Re} = 1500$.

SOLUTION

Problem Data and Schematic: Data are provided for the radiator shown in Fig. E7.7.

Determine: The influence of superimposed free convection.

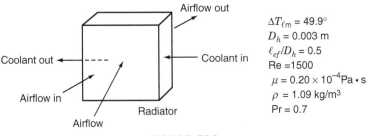

Airflow out

$\Delta T_{\ell m} = 49.9°$
$D_h = 0.003\,\mathrm{m}$
$\ell_{ef}/D_h = 0.5$
$\mathrm{Re} = 1500$
$\mu = 0.20 \times 10^{-4}\,\mathrm{Pa \cdot s}$
$\rho = 1.09\,\mathrm{kg/m^3}$
$\mathrm{Pr} = 0.7$

Coolant out

Coolant in

Airflow in

Radiator

Airflow

FIGURE E7.7

Assumptions: The flow pattern map of Fig. 7.33 is applicable for this problem.

Analysis: Using Fig. 7.33, we determine whether or not the free convection effects are important for this problem. To determine the Grashof number, let us determine $(T_w - T_m)$ and β^*. For this problem,

$$T_w - T_m = \Delta T_{lm} = 49.9°C = 49.9 \text{ K}$$

The mean air temperature is the arithmetic mean water temperature minus the log-mean temperature difference.

$$T_{m,c} = T_{m,h} - \Delta T_{lm} = \frac{(98.9 + 93.3)°C}{2} - 49.9°C = 46.2°C$$

For air as a perfect gas,

$$\beta^* = \frac{1}{T_{m,c}} = \frac{1}{(273.15 + 46.2) \text{ K}} = 0.0031 \frac{1}{\text{K}}$$

The gravitational acceleration at sea level, $g = 9.807 \text{ m/s}^2$. Now

$$\text{Gr} = \frac{g\rho^2 D_h^3 \beta^* (T_w - T_m)}{\mu^2}$$

$$= \frac{9.807 \text{ m/s}^2 \times (1.09 \text{ kg/m}^3)^2 \times (0.003 \text{ m})^3 \times 0.0031 \text{ K}^{-1} \times 49.9 \text{ K}}{(0.20 \times 10^{-4} \text{ Pa} \cdot \text{s})^2} = 121.7$$

Hence

$$\frac{\text{Gr} \cdot \text{Pr} \cdot D_h}{\ell_{ef}} = 121.7 \times \frac{0.7}{0.5} = 170.4$$

Note that we consider the interrupted length (louver length in the flow direction) ℓ_{ef} for ℓ_{ef}/D_h in the correlation of Fig. 7.33. We find from Fig. 7.33 that for $\text{Gr} \cdot \text{Pr} \cdot D_h/\ell_{ef} = 170.4$ and $\text{Re} = 1500$, the flow regime is forced convection laminar flow, and hence the influence of superimposed free convection is negligible.

Discussion and Comments: For the radiator, the mean temperature difference is relatively small (49.9°C for this case), the hydraulic diameter is small ($D_h = 0.003$ m), and the velocity is relatively high compared to the free convection velocities of 1 m/s or less. Hence, we would not expect the free convection effects to be important. And this is what is found from this example.

7.7.2 Vertical Circular Tubes

Unlike horizontal tubes, the effect of superimposed free convection for vertical tubes is dependent on the flow direction and whether or not the fluid is being heated or cooled. For fluid heating with upward flow or fluid cooling with downward flow, free convection aids forced convection since the velocities due to free and forced convection are in the

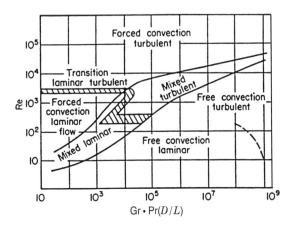

FIGURE 7.34 Vertical circular tube free, forced, and mixed convection flow regimes for $10^{-2} < \mathrm{Pr}(D_h/L) < 1$. (From Metais and Eckert, 1964.)

same direction. The resultant friction factor and heat transfer coefficient are higher than the pure forced convection coefficient. However, for fluid cooling with upward flow or fluid heating with downward flow, free convection counters forced convection, and lower friction factor and heat transfer coefficient result. The flow regime chart of Metais and Eckert (1964) for vertical tubes, as shown in Fig. 7.34, provides guidelines to determine the significance of the superimposed free convection. The results of Fig. 7.34 are applicable for upflow and downflow and both constant heat flux and constant wall temperature boundary conditions.

Chato (1969) summarizes the literature on mixed convection in vertical channels and suggests the following correlations. For upward *liquid flow* (water and oil) with liquid heating in a circular tube with *decreasing density* (and decreasing viscosity) *along the flow direction* ($\partial \rho / \partial x < 0$), the Nusselt numbers and friction factors are given by the following correlations devised by Worsøe-Schmidt as presented by Chato (1969) for fully developed flow:

$$\mathrm{Nu} = \begin{cases} 4.73 & \text{for } \dfrac{\mathrm{Gr}}{\mathrm{Re}} < 12.8 \\[2ex] 2.5\left(\dfrac{\mathrm{Gr}}{\mathrm{Re}}\right)^{0.25} & \text{for } \dfrac{\mathrm{Gr}}{\mathrm{Re}} > 12.8 \end{cases} \tag{7.164}$$

$$f \cdot \mathrm{Re} = \begin{cases} 16 & \text{for } \dfrac{\mathrm{Gr}}{\mathrm{Re}} < 8.4 \\[2ex] 5.8\left(\dfrac{\mathrm{Gr}}{\mathrm{Re}}\right)^{0.475} & \text{for } \dfrac{\mathrm{Gr}}{\mathrm{Re}} > 8.4 \end{cases} \tag{7.165}$$

where Gr is defined by Eq. (7.159). The measured temperature profiles and Nusselt numbers agree quite well with Eq. (7.164) for moderate heat input. Equation (7.165) lies just above the upper range of the friction factor data obtained with water and oil.

For upward *liquid flow* with liquid cooling in a circular tube with *increasing density* (and increasing viscosity) *along the flow direction* ($\partial \rho / \partial x > 0$), the Nusselt numbers and friction factors both decrease rapidly beyond $Gr/Re = 12.8$ and 8.4, respectively, and the friction factors even become negative, due to reverse flows occurring at the wall.

For upward *gas flow* and constant heat flux, the *local* Nusselt numbers and friction factors are given by the following correlations:

$$Nu_x = \begin{cases} Nu_{cp} + 0.025(q^*)^{1/2}\dfrac{(Gz_x - 3)(Gz_x - 20)}{Gz_x^{3/2}} & \begin{array}{l}\text{for } 3 < Gz_x < 1000 \text{ and}\\ 0 < q^* < 5 \text{ for air and helium}\end{array} \\[4mm] Nu_{cp} + 0.07(q^*)^{1/2}\dfrac{Gz_x - 8}{Gz_x^{1/2}} & \begin{array}{l}\text{for } 10 < Gz_x < 1000 \text{ and}\\ 0 < q^* < 5 \text{ for carbon dioxide}\end{array} \end{cases}$$

$$(7.166)$$

where the constant property Nusselt numbers are given by

$$Nu_{cp} = \begin{cases} 1.58\,Gz_x^{0.3} & \text{for } Gz_x > 26 \\ 4.2 & \text{for } Gz_x < 26 \end{cases} \qquad (7.167)$$

and $q^* = qd_i/(2k_eT_eA_w)$, with the subscript e denoting the condition at the channel entrance. In Eqs. (7.166) and (7.167), the Graetz number Gz_x denotes a local Graetz number $Gz_x = (\dot{m}c_p/kx)$, where x is the length along the tube. Similarly, Nu_x represents a local Nu number based on the hydraulic diameter of the channel. For upward-flowing gases with heating, the friction factors are given by

$$f = \begin{cases} \dfrac{16}{Re}\left(\dfrac{T_w}{T_m}\right) & \text{for } \dfrac{T_w}{T_m} < 1.2 \text{ to } 1.5 & \text{for gases} \\[4mm] \dfrac{15.5}{Re}\left(\dfrac{T_w}{T_m}\right)^{1.10} & \text{for } 1.5 < \dfrac{T_w}{T_m} < 3 & \text{for air and helium} \\[4mm] \dfrac{15.5}{Re}\left(\dfrac{T_w}{T_m}\right)^{1.25} & \text{for } 1.2 < \dfrac{T_w}{T_m} < 2 & \text{for carbon dioxide} \end{cases} \qquad (7.168)$$

where Re is the mean Reynolds number with respect to the flow length. For upward-flowing gases with cooling, the friction factors are given by

$$f = \dfrac{16}{Re}\left(\dfrac{T_w}{T_m}\right)^{0.81} \qquad \text{for } 0.5 < \dfrac{T_w}{T_m} < 1 \qquad (7.169)$$

7.8 INFLUENCE OF SUPERIMPOSED RADIATION

The theory and correlations presented in this chapter so far are applicable if the radiation effect is negligible (which is the case in many heat exchanger applications). However, when the heat exchanger wall is at a temperature significantly higher than that of the fluid, or vice versa, heat transfer by radiation may also take place in parallel with heat transfer by convection. Some examples of high-temperature fluids in heat exchangers are (1) the gas turbine exhaust gas flowing over tube bundles in cogeneration applications,

(2) exhaust gases from metal industries used in a fixed-matrix regenerator, (3) liquid metals (e.g., sodium) in a fast breeder reactor heat exchanger, and (4) high-temperature fluids used in some heat exchanger applications. However, liquid metals are opaque and the radiation effect is negligible.

7.8.1 Liquids as Participating Media

Radiative heat transfer between the heat exchanger surface and the liquid is generally negligible for water and many liquids having considerably high density compared to gases such as air. In other words, the effect of radiation would be suppressed due to a too high fluid participation through absorption of any thermal radiation that may take place. However, there may be some wavelength ranges where the transmissivity may not be negligible; this wavelength range is dependent on the particular liquid and the mean beam length for radiation. Most liquids, including water used in heat exchanger applications, are opaque and the radiation heat transfer to and from liquids is negligible, although some of the liquids may have slight absorptivity in some wavelength range.

7.8.2 Gases as Participating Media

We will consider separately whether or not the high-temperature gas is transparent (i.e., does not participate in thermal radiation effects). If the gas is transparent, the net radiation heat transfer will be from the hotter to the colder surfaces only. That is, for nonparticipating (i.e., nonabsorbing/emitting) gases, the radiation heat transfer between the gas and the heat transfer surface is negligible. If the gas is a thermally participating medium and is at a temperature higher or lower than that for the heat transfer surfaces, the non-negligible radiation heat transfer will be from the gas to the surface (wall), or vice versa, depending on the temperatures.

For *nonparticipating gases*, the net radiation heat transfer takes place from a hot surface k at temperature T_k to one or more cold surfaces j at temperature T_j without any thermal radiation absorbed or emitted by the gas. For black, individually isothermal surfaces in an enclosure surrounding the nonparticipating medium completely, the net thermal energy transfer can be computed from (Siegel and Howell, 2002)

$$q_{\text{rad}} = \sigma A_k \sum_{j=1}^{n} F_{kj}(T_k^4 - T_j^4) \tag{7.170}$$

where F_{kj} is the view factor or configuration factor for thermal radiation from the surface k to surfaces j, σ is the Stefan–Boltzmann constant,[†] and T_k and T_j *must* be absolute temperatures (K or °R). Equation (7.170) implies that all n black surfaces in the enclosure are flat. Howell (1982) has provided the view factors for many configurations, and the latest catalog is available in a CD with the book by Siegel and Howell (2002). Some view factors applicable to heat exchanger applications are summarized in Table 7.14.

In the case of *absobing gas* flows, the radiation effect is important only for those components of gas that have significant concentration and also absorb/emit infrared radiation. Water vapor and carbon dioxide often fall in this category. Sulfur and nitrogen oxides, on the other hand, are usually not present in high-enough concentration to warrant their consideration.

[†] $\sigma = 5.6696 \times 10^{-8}$ W/m$^2 \cdot$ K$^4 = 0.1714 \times 10^{-8}$ Btu/hr-ft^2-°R^4.

TABLE 7.14   View Factors and Radiation Heat Transfer between Some Heat Exchanger Surfaces for Transparent Flowing Gas

Large (Infinite) Parallel Planes

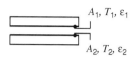

A_1, T_1, ε_1

A_2, T_2, ε_2

$$A_1 = A_2 = A \qquad q_{12} = \frac{A\sigma(T_1^4 - T_2^4)}{1/\varepsilon_1 + 1/\varepsilon_2 - 1}$$

$$F_{12} = 1$$

Long (Infinite) Concentric Cylinders

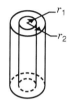

r_1

r_2

$$\frac{A_1}{A_2} = \frac{r_1}{r_2} \qquad q_{12} = \frac{\sigma A_1(T_1^4 - T_2^4)}{(1/\varepsilon_1) + [(1 - \varepsilon_2)/\varepsilon_2](r_1/r_2)}$$

$$F_{12} = 1$$

Concentric Spheres

r_1

r_2

$$\frac{A_1}{A_2} = \frac{r_1^2}{r_2^2} \qquad q_{12} = \frac{\sigma A_1(T_1^4 - T_2^4)}{(1/\varepsilon_1) + [(1 - \varepsilon_2)/\varepsilon_2](r_1/r_2)^2}$$

$$F_{12} = 1$$

Small Convex Object in a Large Cavity

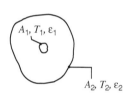

A_1, T_1, ε_1

A_2, T_2, ε_2

$$\frac{A_1}{A_2} \approx 0 \qquad q_{12} = \sigma A_1 \varepsilon_1 (T_1^4 - T_2^4)$$

$$F_{12} = 1$$

Source: Data from Incropera and DeWitt (2002).

Consider a gas component having emissivity ε_g and absorptivity α_g, with a bounding heat transfer surface area A to which the gas will interchange radiation. The wall surface temperature is T_w, and the gas flowing through the heat exchanger is at a mean temperature T_g (which is the same as T_m in the rest of this book). Then the net radiation heat transfer by the gas to a black wall surface is given by

$$q_{\text{rad}} = \sigma A(\varepsilon_g T_g^4 - \alpha_g T_w^4) \qquad (7.171)$$

Here T_g and T_w *must* be absolute temperatures (in K or °R). Since the radiation band structure (e.g., absorption coefficient as a function of wavelength) of the gas is dependent on its temperature, the energy emitted by gas depends on T_g, whereas the energy absorbed by the gas depends as well on the radiation emitted by the wall at T_w in addition to its dependence on T_g. Thus $\varepsilon_g(T_g)$ and $\alpha_g(T_g, T_w)$ depend on the temperatures indicated. Additionally, both ε_g and α_g depend on the partial pressure of the gas component, which participates in radiation heat transfer. Note that if the wall radiates heat to the gas,

q_{rad} in Eq. (7.171) will be negative. We explain how to evaluate ε_g and α_g by the mean beam length treatment of gas properties after we develop the magnitude of the combined heat transfer coefficient in a heat exchanger when radiation cannot be neglected in relation to convection.

In the case of gray walls with an emissivity of ε_w, the evaluation of q_{rad} is more complicated, due to multiple reflections. It can be shown that the net heat transfer from the gas to the wall is given by

$$q_{rad} = \frac{\varepsilon_w \sigma A}{1 - (1 - \varepsilon_w)(1 - \alpha_g)} \left(\varepsilon_g T_g^4 - \alpha_g T_w^4 \right) \tag{7.172}$$

Only if $T_g/T_w \approx 1$, then $\varepsilon_g \approx \alpha_g$, and Eq. (7.172) will simplify to

$$q_{rad} = \frac{\sigma A}{1/\varepsilon_g + 1/\varepsilon_w - 1} \left(T_g^4 - T_w^4 \right) \tag{7.173}$$

Since the convection heat transfer equation is in terms of the temperature difference $(T_w - T_g)$, the *radiation heat transfer coefficient* h_{rad} can be presented similarly using Eq. (7.172):

$$h_{rad} = \frac{q_{rad}}{A(T_g - T_w)} = \frac{\varepsilon_w \sigma}{1 - (1 - \varepsilon_w)(1 - \alpha_g)} \frac{\varepsilon_g T_g^4 - \alpha_g T_w^4}{T_g - T_w} \tag{7.174}$$

In this equation, T_w and T_g *must* be given in K or °R. Note that h_{conv} for forced convection is generally not a strong function of the temperatures T_w and T_g, but h_{rad} is a strong function of T_w and T_g, as shown in Eq. (7.174).

In heat exchangers, the combined convection and radiation effects are then taken care of approximately by considering the convection and radiation heat transfer phenomena in parallel. Hence,

$$h_{combined} = h_{conv} + h_{rad} \tag{7.175}$$

The rest of the heat exchanger analysis for a combined convection and radiation problem is performed the same as before by replacing h (or h_{conv}) with $h_{combined}$, assuming that $h_{combined}$ is defined using $|T_w - T_g| = |T_w - T_m|$.

Now we consider the determination of ε_g and α_g of Eq. (7.174) from the experimental results/correlations of Hottel and Sarofim (1967), and some of them reproduced by Incropera and DeWitt (2002), among others. The gas emissivity ε_g has been correlated in terms of the temperature T_g, the total pressure p of the gas, the partial pressure p_g of the gas species (such as water vapor, carbon dioxide, etc.), and the radius L of an equivalent hemispherical gas mass. More precisely, L_e, the *mean beam length*, is the required radius of a gas hemisphere such that it radiates a heat flux to the center of its base equal to the average flux radiated to the area of interest by the actual volume of gas (Siegel and Howell, 2002). In most exhaust gases, water vapor and carbon dioxide are the most important components from the radiation effect viewpoint, and hence we only consider them here. For other gas components, refer to Hottel and Sarofim (1967).

The water vapor emissivity ε_{H_2O} is presented in Fig. 7.35 as a function of the gas temperature T_g and $p_{H_2O} L_e$. Here p_{H_2O} is the partial pressure of water vapor in the gas mixture at a total pressure of 1 atm; and L_e is the *mean beam length* to take into account

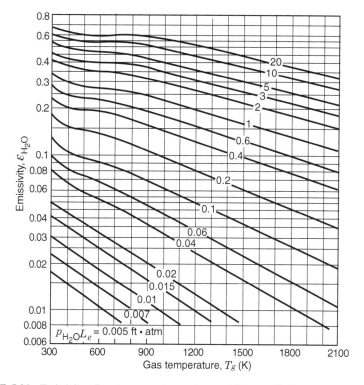

FIGURE 7.35 Emissivity of water vapor in a mixture with nonradiating gases at 1 atm total pressure and of hemispherical shape. (From Hottel, 1954.) (1 ft · atm = 0.305 m · atm).

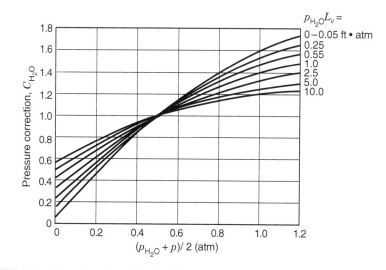

FIGURE 7.36 Correction factor for obtaining water vapor emissivity at pressures other than 1 atm; here, $\varepsilon_{H_2O, p \neq 1\,atm} = C_{H_2O} \varepsilon_{H_2O, p=1\,atm}$. (From Hottel, 1954.) (1 ft · atm = 0.305 m · atm).

TABLE 7.15   Mean Beam Lengths L_e for Various Gas Geometries

Geometry	Characteristic Length	L_e
Sphere (radiation to surface)	Diameter D	$0.65D$
Infinite circular cylinder (radiation to curved surface)	Diameter D	$0.95D$
Semi-infinite circular cylinder (radiation to base)	Diameter D	$0.65D$
Circular cylinder of equal height and diameter (radiation to entire surface)	Diameter D	$0.60D$
Infinite parallel planes (radiation to planes)	Spacing between planes L	$1.8L$
Cube (radiation to any surface)	Side L	$0.66L$
Arbitrary shape of volume V (radiation to surface of area A)	Volume to area ratio V/A	$3.6V/A$

Source: Data from Incropera and DeWitt (2002).

the size and shape of the gas geometry, which is different from a hemispherical geometry. If the total pressure of the gas is different from 1 atm, a correction factor C_{H_2O} (to be multiplied to ε_{H_2O} for 1 atm pressure) is obtained from Fig. 7.36. The mean beam length L_e is provided in Table 7.15 for various gas geometries (Incropera and DeWitt, 2002). Similarly, the carbon dioxide emissivity ε_{CO_2} and corresponding correction factor C_{CO_2} are provided in Figs. 7.37 and 7.38, respectively.

The gas absorptivity α_g can be evaluated using the approximate procedure from the following expressions (Siegel and Howell, 2002), where the emissivity and the correction for total pressure different from 1 atm for a specific component are obtained from Figs. 7.35 through 7.38.

$$\alpha_{H_2O} = C_{H_2O} \left(\frac{T_m}{T_w}\right)^{0.5} \varepsilon_{H_2O}\left(T_w, p_{H_2O} L_e \frac{T_w}{T_m}\right) \tag{7.116a}$$

$$\alpha_{CO_2} = C_{CO_2} \left(\frac{T_m}{T_w}\right)^{0.5} \varepsilon_{CO_2}\left(T_w, p_{CO_2} L_e \frac{T_w}{T_m}\right) \tag{7.116b}$$

The aforementioned emissivity and absorptivity values for water vapor and carbon dioxide apply only when they are alone with other nonradiating components in the gas mixture. However, when both are present in the gas with other nonradiating components, the total gas emissivity and absorptivity are given by

$$\varepsilon_{gas} = C_{H_2O}\varepsilon_{H_2O} + C_{CO_2}\varepsilon_{CO_2} - \Delta\varepsilon \qquad \alpha_{gas} = \alpha_{H_2O} + \alpha_{CO_2} - \Delta\alpha \tag{7.177}$$

Now evaluate $\Delta\alpha = \Delta\varepsilon$ where the correction factor $\Delta\varepsilon$ is given in Fig. 7.39 (see Hottel and Sarofim, 1967).

SUMMARY

The most important inputs for the thermal and hydraulic design of a heat exchanger are heat transfer coefficients and friction factors for the heat transfer surfaces used. The thermal design of the exchanger depends on the accuracy of this information. Except for the major heat exchanger industries having extensive test facilities, such information is not available on modern heat transfer surfaces to most small companies, consultants,

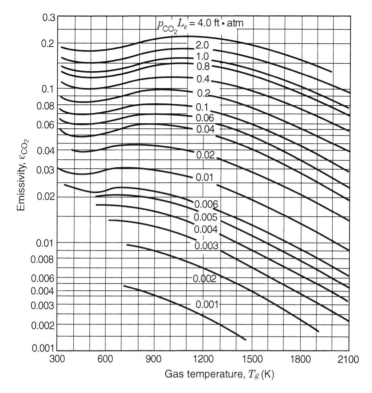

FIGURE 7.37 Emissivity of carbon dioxide in a mixture with nonradiating gases at 1 atm total pressure and of hemispherical shape. (From Hottel, 1954.) (1 ft · atm = 0.305 m · atm).

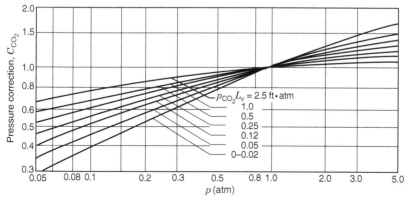

FIGURE 7.38 Correction factor for obtaining carbon dioxide emissivity at pressures other than 1 atm; here, $\varepsilon_{CO_2, p \neq 1\,atm} = C_{CO_2} \varepsilon_{CO_2, p=1\,atm}$. (From Hottel, 1954.) (1 ft · atm = 0.305 m · atm).

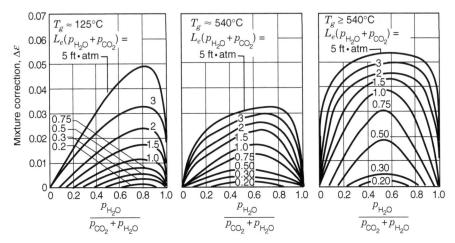

FIGURE 7.39   Correction factors associated with mixtures of water vapor and carbon dioxide. (From Hottel, 1954.) (1 ft · atm = 0.305 m · atm).

and academia. Usually, this information may be obtained from the correlations in the open literature, analytical solutions, and/or experimentation. Hence, in this chapter we have provided sufficient details on all these aspects with background information so that one may have a broad understanding of what to look for, may be able to either obtain accurate j (or Nu) and f data, or be able to use the analytical correlations for approximations and extrapolations with sufficient accuracy. An approximate correlation or methodology is also presented in the chapter to cover situations where fluid properties vary significantly, there is combined free and forced convection, or there is combined convection and radiation in a heat exchanger. With this broad understanding of the topics covered in this chapter, the reader will be able to obtain approximate values of the surface characteristics and be able to design exchangers with reasonably accurate required performance level or the size. Hence, after learning the basic design theory of Chapter 3, the information provided in this chapter is most important for the accurate thermal design of heat exchangers.

REFERENCES

Achaichia, A., and T. A. Cowell, 1988, Heat transfer and pressure drop characteristics of flat tube louvered plate fin surfaces. *Exp. Thermal Fluid Sci.*, Vol. 1, pp. 147–157.

Amon, C. H., D. Majumdar, C. V. Herman, F. Mayinger, B. B. Mikić, and D. P. Sekulić, 1992, Numerical and experimental studies of self-sustained oscillatory flows in communicating channels, *Int. J. Heat Mass Transfer*, Vol. 35, pp. 3115–3129.

Aung, W., 1987, Mixed convection in internal flow, in *Handbook of Single-Phase Convective Heat Transfer*, S. Kakaç, R. K. Shah, and W. Aung, eds., Wiley, New York, Chap. 15.

Bačlić, B. S., P. J. Heggs, and H. Z. Z. A. Ziyan, 1986a, Differential fluid enthalpy method for predicting heat transfer coefficients in packed beds, *Heat Transfer 1986, Proc. 8th Int. Heat Transfer Conf.*, Vol. 5, pp. 2617–2622.

Bačlić, B. S., D. D. Gvozdenac, D. P. Sekulić, and E. J. Becić, 1986b, Laminar heat transfer characteristics of a plate-louver fin surface obtained by the differential fluid enthalpy method, in *Advances in Heat Exchanger Design*, eds. R. K. Shah and J. T. Pearson, HTD-Vol. 66, ASME, New York, pp. 21–28.

Bejan, A., 1995, *Convection Heat Transfer*, 2nd ed., Wiley, New York.

Bemisderfer, C. H., 1998, Contemporary developments in the thermal design of finned-tube heat exchangers, *J. Enhanced Heat Transfer*, Vol. 5, pp. 71–90.

Bhatti, M. S., and R. K. Shah, 1987, Turbulent and transition convective heat transfer in ducts, in *Handbook of Single-Phase Convective Heat Transfer*, S. Kakaç, R. K. Shah, and W. Aung, eds., Wiley, New York, Chap. 4.

Briggs, D. E., and E. H. Young, 1963, Convection heat transfer and pressure drop of air flowing across triangular pitch banks of finned tubes, *Chem. Eng. Prog. Symp. Ser. 41*, Vol. 59, pp. 1–10.

Burmeister, L. C., 1993, *Convective Heat Transfer*, Wiley, New York.

Chang, Y. J., and C. C. Wang, 1997, A generalized heat transfer correlation for louver fin geometry, *Int. J. Heat and Mass Transfer*, Vol. 40, pp. 533–544.

Chang, Y. J., K. C. Hsu, Y. T. Lin and C. C. Wang, 2000, A generalized friction correlation for louver fin geometry, *Int. J. Heat Mass Transfer*, Vol. 43, pp. 2237–2243.

Chato, J. C., 1969, Combined free and forced convection flows in channels, in *Advanced Heat Transfer*, B. T. Chao, ed., University of Illinois Press, Urbana, IL, pp. 439–453.

Chen, C. J., and J. S. Chiou, 1981, Laminar and turbulent heat transfer in the pipe entrance for liquid metals, *Int. J. Heat Mass Transfer*, Vol. 24, pp. 1179–1190.

Churchill, S. W., and S. C. Zajic, 2002, The improved prediction of turbulent convection, *Heat Transfer 2002, Proc. 12th Int. Heat Transfer Conf.*, pp. 279–284.

Colburn, A. P., 1933, A method of correlating forced convection heat transfer data and a comparison with fluid friction, *Trans. Am. Inst. Chem. Eng.*, Vol. 29, pp. 174-210; reprinted in *Int. J. Heat Mass Transfer*, Vol. 7, pp. 1359–1384, 1964.

Cowell, T. A., M. R. Heikal, and A. Achaichia, 1995, Flow and heat transfer in compact louvered fin surfaces, *Exp. Thermal Fluid Sci.*, Vol. 10, pp. 192–199.

Das, S. K., 2001, Private Communication, Department of Mechanical Engineering, IIT Madras, Chennai, India.

Dipprey, D. F., and R. H. Sabersky, 1963, Heat and momentum transfer in smooth and rough tubes at various Prandtl numbers, *Int. J. Heat Mass Transfer*, Vol. 6, pp. 329–353.

Dittus, F. W., and L. M. K. Boelter, 1930, Heat transfer in automobile radiators of tubular type, *University of Calif. Publications in Engineering*, Vol. 2, pp. 443–461.

Gaddis, E. S., and V. Gnielinski, 1985, Pressure drop in cross flow across tube bundles, *Int. Chem. Eng.*, Vol. 25, pp. 1–15.

Ganguli, A., and S. B. Yilmaz, 1987, New heat transfer and pressure drop correlations for crossflow over low-finned tube banks, *AIChE Symp. Ser. 257*, Vol. 83, pp. 9–14.

Ghajar, A. J., and L. M. Tam, 1994, Heat transfer measurements and correlations in the transition region for a circular tube with three different inlet configurations, *Exp. Thermal Fluid Sci.*, Vol. 8, pp. 79–90.

Ghajar, A. J., and L. M. Tam, 1995, Flow regime map for a horizontal pipe with uniform wall heat flux and three inlet configurations, *Exp. Thermal Fluid Sci.*, Vol. 10, pp. 287–297.

Gnielinski, G., 1976, New equation for heat and mass transfer in turbulent pipe and channel flow, *Int. Chem. Eng.*, Vol. 16, pp. 359–368.

Gvozdenac, D. D., 1994, Experimental prediction of heat transfer coefficients by use of double-blow method, *Wärme- und Stoffübertragung*, Vol. 29, pp. 361–365.

Heggs, P. J., and D. Burns, 1988, Single blow experimental prediction of heat transfer coefficients: a comparison of commonly used techniques, *Exp. Thermal Fluid Sci.*, Vol. 1, pp. 243–251.

Hinze, J. O., 1975, *Turbulence*, 2nd ed., McGraw-Hill, New York.

Hottel, H. T., 1954, Radiant-heat transmission, in *Heat Transmission*, 3rd ed., W. H. McAdams, ed., McGraw-Hill, New York.

Hottel, H. C., and A. F. Sarofim, 1967, *Radiative Transfer*, McGraw-Hill, New York.

Howell, J. R., 1982, *A Catalog of Radiation Configuration Factors*, McGraw-Hill, New York.

Incropera, F. P., and D. P. DeWitt, 2002, *Fundamentals of Heat and Mass Transfer*, 5th ed., Wiley, New York.

Jacobi, A. M., and R. K. Shah, 1996, Air-side flow and heat transfer in compact heat exchangers: a discussion of physics, in *Process, Enhanced, and Multiphase Heat Transfer—A Festschrift for A. E. Bergles*, R. M. Manglik and A. D. Kraus, eds., Begell House, New York, pp. 379–390; see also *Heat Transfer Engineering*, Vol. 19, No. 4, pp. 29-41, 1998.

Kays, W. M., and M. E. Crawford, 1993, *Convective Heat and Mass Transfer*, 3rd ed., McGraw-Hill, New York.

Kays, W. M., and A. L. London, 1950, Heat transfer and flow friction characteristics of some compact heat exchanger surfaces, Part I; Test system and procedure, *Trans. ASME*, Vol. 72, pp. 1075–1085.

Kays, W. M., and A. L. London, 1998, *Compact Heat Exchangers*, reprint 3rd ed., Krieger Publishing, Malabar, FL.

Lestina, T., and K. J. Bell, 2001, Thermal performance testing of industrial heat exchangers, *Advances in Heat Transfer*, Academic Press, San Diego, Vol. 35, pp. 1–55.

Lévêque, A., 1928, Les lois de la transmission de chaleur par convection, *Ann. Mines*, Vol. 13, pp. 201–299, 305–362, 381–415.

London, A. L., and R. K. Shah, 1968, Offset rectangular plate-fin surfaces – heat transfer and flow friction characteristics, *ASME J. Eng. Power*, Vol. 90, Ser. A, pp. 218–228.

London, A. L., and R. K. Shah, 1973, Glass-ceramic hexagonal and circular passage surfaces – heat transfer and flow friction design characteristics, *SAE Trans.*, Vol. 82, Sec. 1, pp. 425–434.

London, A. L., M. B. O. Young, and J. H. Stang, 1970, Glass ceramic surfaces, straight triangular passages – heat transfer and flow friction characteristics, *ASME J. Eng. Power*, Vol. 92, Ser. A, pp. 381–389.

Malhotra, A., and S. S. Kang, 1984, Turbulent Prandtl number in circular pipes, *Int. J. Heat Mass Transfer*, Vol. 27, pp. 2158–2161.

Manglik, R. M., and A. E. Bergles, 1995, Heat transfer and pressure drop correlations for the rectangular offset-strip-fin compact heat exchanger, *Exp. Thermal Fluid Sci.*, Vol. 10, pp. 171–180.

Martin, H., 1996, A theoretical approach to predict the performance of chevron-type plate heat exchangers, *Chem. Eng. Processing*, Vol. 35, pp. 301–310.

Martin, H., 2002, The generalized Lévêque equation and its practical use for the prediction of heat and mass transfer rates from pressure drop, *Chem. Eng. Sci.*, Vol. 57, pp. 3217–3223.

Martin, H., 2002, Private communication, Thermische Verfahrenstechnik, Universität Karlsruhe (TH), Karlsruhe, Germany.

Metais, B., and E. R. G. Eckert, 1964, Forced, mixed and free convection regions, *ASME J. Heat Transfer*, Vol. 86, pp. 295–296.

Moffat, R. J., 1998, What's new in convective heat transfer, *Int. J. Heat Fluid Flow*, Vol. 19, pp. 90–101.

Morcos, S. M., and A. E. Bergles, 1975, Experimental investigation of combined forced and free laminar convection in horizontal tubes, *ASME J. Heat Transfer*, Vol. 97, pp. 212–219.

Mullisen, R. S., and R. I. Loehrke, 1986, A transient heat exchanger evaluation test for arbitrary fluid inlet temperature variation and longitudinal core conduction, *ASME J. Heat Transfer*, Vol. 108, pp. 370–376.

Nikuradse, J., 1933, Strömungsgesetze in rauhen Rohren, *Forsch. Arb. Ing.-Wes.*, No. 361; English translation, NACA TM 1292.

Petukhov, B. S., and V. N. Popov, 1963, Theoretical calculation of heat exchange in turbulent flow in tubes of an incompressible fluid with variable physical properties, *High Temp.*, Vol. 1, No. 1, pp. 69–83.

Rabas, T. J., and J. Taborek, 1987, Survey of turbulent forced-convection heat transfer and pressure drop characteristics of low-finned tube banks in cross flow, *Heat Transfer Eng.*, Vol. 8, No. 2, pp. 49–62.

Robinson, K. K., and D. E. Briggs, 1966, Pressure drop of air flowing across triangular pitch banks of finned tubes, *Chem. Eng. Prog. Symp. Series 64*, Vol. 62, pp. 177–184.

Schlünder, E.-U., 1998, Analogy between heat and momentum transfer, *Chem. Eng. Process*, Vol. 37, pp. 103–107.

Shah, R. K., 1981, Compact heat exchangers, in *Heat Exchangers: Thermal-Hydraulic Fundamentals and Design*, S. Kakaç, A. E. Bergles, and F. Mayinger, eds., Hemisphere Publishing Corp., Washington, DC, pp. 111–151.

Shah, R. K., 1983, Fully developed laminar flow through channels, in *Low Reynolds Number Flow Heat Exchangers*, S. Kakaç, R. K. Shah and A. E. Bergles, eds., Hemisphere Publishing Corp., Washington, DC, pp. 75–108.

Shah, R. K., 1985, Compact heat exchangers, in *Handbook of Heat Transfer Applications*, 2nd ed., W. M. Rohsenow, J. P. Hartnett, and E. N. Ganić, eds., McGraw-Hill, New York, pp. 4–174 to 4–311.

Shah, R. K., 1990, Assessment of modified Wilson plot techniques for obtaining heat exchanger design data, *Heat Transfer 1990, Proc. 9th Int. Heat Transfer Conf.*, Vol. 5, pp. 51–56.

Shah, R. K., and M. S. Bhatti, 1987, Laminar convective heat transfer in ducts, in *Handbook of Single-Phase Convective Heat Transfer*, S. Kakaç, R. K. Shah, and W. Aung, eds., Wiley, New York, Chap. 3.

Shah, R. K., and M. S. Bhatti, 1988, Assessment of correlations for single-phase heat exchangers, in *Two-Phase Flow Heat Exchangers: Thermal-Hydraulic Fundamentals and Design*, S. Kakaç, A. E. Bergles, and E. O. Fernandes, eds., Kluwer Academic Publishers, Dordrecht, The Netherlands, pp. 81–122.

Shah, R. K., and A. L. London, 1978, *Laminar Flow Forced Convection in Ducts*, Supplement 1 to *Advances in Heat Transfer*, Academic Press, New York.

Shah, R. K., and A. C. Mueller, 1988, Heat exchange, in *Ullmann's Encyclopedia of Industrial Chemistry*, VCH Publishers, Weinheim, Germany, Unit Operations II, Vol. B3, Chapter 2.

Shah, R. K., and Webb, R. L., 1983, Compact and enhanced heat exchangers, in *Heat Exchangers: Theory and Practice*, J. Taborek, G. F. Hewitt and N. Afgan, eds., Hemisphere/McGraw-Hill, Washington, DC, pp. 425–468.

Shah, R. K., and S. Q. Zhou, 1997, Experimental techniques for obtaining design data for compact heat exchanger surfaces, in *Compact Heat Exchangers for the Process Industries*, R. K. Shah, ed., Begell House, New York, pp. 365–379.

Siegel, R., and J. R. Howell, 2002, *Thermal Radiation Heat Transfer*, Taylor & Francis, New York.

Šlančiauskas, A., 2001, Two friendly rules for the turbulent heat transfer enhancement, *Int. J. Heat Mass Transfer*, Vol. 44, pp. 2155–2161.

Sleicher, C. A., and M. Tribus, 1957, Heat transfer in a pipe with turbulent flow and arbitrary wall-temperature distribution, *Trans. ASME*, Vol. 97, pp. 789–796.

Taborek, J., 1990, Design method for tube-side laminar and transition flow regime with effects of natural convection, paper presented in the Open Forum at the 9th Int. Heat Transfer Conf., Jerusalem, Israel.

Tam, L. M., and A. J. Ghajar, 1997, Effect of inlet geometry and heating on the fully developed friction factor in the transition region of a horizontal tube, *Exp. Thermal Fluid Sci.*, Vol. 15, pp. 52–64.

Wang, C. C., 2000, Recent progress on the air-side performance of fin-and-tube heat exchangers, *Int. J. Heat Exchangers*, Vol. 1, pp. 49–76.

Wang, C. C., and K. U. Chi, 2000, Heat transfer and friction characteristics of plain fin-and-tube heat exchangers; Part 2; Correlation, *Int. J. Heat Mass Transfer*, Vol. 43, pp. 2692–2700.

Webb, R. L., 1994, *Principles of Enhanced Heat Transfer*, Wiley, New York.

Wheeler, A. J., 1968, Single-blow transient testing of matrix-type heat exchanger surfaces at low values of N_{tu}, TR. 68, Department of Mechanical Engineering, Stanford University, Stanford, California.

Wilson, E. E., 1915, A basis for rational design of heat transfer apparatus, *Trans. ASME*, Vol. 37, pp. 47–82.

Winterton, R. H. S., 1998, Where did the Dittus–Boelter equation come from? *Int. J. Heat Mass Transfer*, Vol. 41, pp. 809–810.

Yu, B., H. Ozoe, and S. C. Churchill, 2001, The characteristics of fully developed turbulent convection in a round tube, *Chem. Eng. Sci.*, Vol. 56, pp. 1781–1800.

Zhi-qing, W., 1982, Study on correction coefficients of laminar and turbulent entrance region effect in round pipe, *Appl. Math. Mech.*, Vol. 3, No. 3, pp. 433–446.

Žukauskas, A., 1987, Convective heat transfer in cross flow, *Handbook of Single-Phase Convective Heat Transfer*, S. Kakaç, R. K. Shah, and W. Aung, eds., Wiley, New York, Chap. 6.

REVIEW QUESTIONS

For each question, circle one or more correct answers. Explain your answers briefly.

7.1   Characteristics of steady-state (nonpulsating) laminar flow are:

(a) Flow velocity may have two or three components (u, v, or w).

(b) Velocity at the tube centerline is always constant.

(c) If a color dye is injected in the fully developed flow, it disperses immediately.

(d) Flow velocity components may be time dependent.

(e) None of these.

7.2   In fully developed turbulent flow in a circular tube, the velocity profile is:

(a) parabolic   (b) flat, power-law form

(c) parabolic near the wall and uniform near the centerline   (d) can't tell.

7.3   The boundary layer thickness of fully developed laminar flow in a circular tube at a given axial location is the following fraction of the tube radius:

(a) $\frac{1}{8}$ (b) $\frac{1}{4}$ (c) $\frac{1}{2}$ (d) 1   (e) 2   (f) can't tell

7.4   For constant-property hydrodynamically and thermally developed flow in a tube:

(a) The velocity profile is independent of x.

(b) The temperature profile is independent of x.

(c) The velocity profile is independent of x, but the temperature profile is dependent on x.

7.5 In a *simultaneously developing flow* in a circular tube, consider the velocity boundary layer thickness as 10 mm. The approximate thickness in millimeters of the temperature boundary layer for air, water and sodium flow will be (circle one number from each group):

(a) air: 5, 10, 15 mm   (b) water: 5, 10, 15 mm

(c) liquid sodium: 5, 10, 15 mm

7.6 The heat transfer coefficient h is the proportionality coefficient in the:

(a) conduction rate equation   (b) convection rate equation

(c) energy balance   (d) perfect gas equation

7.7 The heat transfer coefficient for forced convection could be dependent on the:

(a) thermal boundary condition   (b) fluid Prandtl number

(c) flow velocity   (d) wall-to-fluid bulk temperature

(e) flow passage shape   difference

(f) flow passage size

7.8 The heat transfer coefficient in dimensionless form can be presented in terms of the:

(a) Nusselt number   (b) Stanton number   (c) Colburn factor

(d) Prandtl number   (e) Péclet number   (f) Graetz number

(g) Lévêque number

7.9 The groups that designate dimensionless pressure drop characteristics are the:

(a) Fanning friction factor   (b) Reynolds number

(c) axial distance x^+ (d) incremental pressure drop number

(e) Euler number   (f) Hagen number

7.10 The Reynolds number is solely a:

(a) fluid property modulus   (b) flow modulus

(c) heat transfer modulus   (d) pressure drop modulus

(e) all of these   (f) none of these

7.11 In a good design, NTU of the test core for the steady-state technique for determining heat transfer coefficient is usually:

(a) above 3   (b) below 0.4   (c) $0.4 \leq \text{NTU} \leq 3$ (d) can't tell

7.12 For the ⒣ boundary condition, the following are kept uniform and constant for developing flow in a tube:

(a) heat flux along the periphery

(b) wall temperature along the axial direction

(c) heat transfer coefficient

(d) finite wall thermal conductivity ($k_w < \infty$)

(e) none of these

(f) all of these

7.13 Consider a circular tube and a *sharp cornered* quadrilateral duct both having the same flow area. Compared to the circular tube value, the Nusselt number for the quadrilateral duct for fully developed laminar flow will be:

(a) higher   (b) lower   (c) equal   (d) can't tell

7.14 In fully developed laminar flow in constant cross-sectional ducts, it is always true that:

(a) $Nu_T \geq Nu_{H1}$ (b) $Nu_{H1} \geq Nu_{H2}$ (c) $Nu_T \geq Nu_{H2}$ (d) $Nu_T \leq Nu_{H2}$

7.15 In fully developed flow of the following regimes, the Nusselt number is independent of the Prandtl number:

(a) laminar   (b) transition   (c) turbulent   (d) none of these

7.16 In fully developed laminar flow, the pressure drop is approximately proportional to:

(a) u_m (b) u_m^2 (c) $u_m^{1.8}$ (d) $u_m^{0.8}$ (e) none of these

7.17 In fully developed laminar flow, the heat transfer coefficient is approximately proportional to:

(a) u_m (b) u_m^2 (c) $u_m^{1.8}$ (d) $u_m^{0.8}$ (e) none of these

7.18 In fully developed turbulent flow, the pressure drop is approximately proportional to:

(a) u_m (b) u_m^2 (c) $u_m^{1.8}$ (d) $u_m^{0.8}$ (e) none of these

7.19 In fully developed turbulent flow, the heat transfer coefficient is approximately proportional to:

(a) u_m (b) u_m^2 (c) $u_m^{1.8}$ (d) $u_m^{0.8}$ (e) none of these

7.20 If we change from two to four passes on the tube side of a given shell-and-tube exchanger (the same shell diameter and number of tubes) with the same tube flow rate, the tube-side velocity in turbulent flow will:

(a) remain the same   (b) increase by 40%
(c) increase by 74%   (d) increase by a factor of 2
(e) increase by a factor of 4   (f) increase by a factor of 8

7.21 For the case of Question 7.20, the tube-side pressure drop will:

(a) remain the same   (b) increase by 40%
(c) increase by 74%   (d) increase by a factor of 2
(e) increase by a factor of 4   (f) increase by a factor of 8

7.22 For the case of Question 7.20, the tube-side heat transfer coefficient will:

(a) remain the same   (b) increase by 40%
(c) increase by 74%   (d) increase by a factor of 2
(e) increase by a factor of 4   (f) increase by a factor of 8

Hint: Use Dittus–Boelter correlation of Eq. (7.79) for Nu.

7.23 The heat transfer coefficient for fully developed laminar flow increases with:

(a) increase in surface area   (b) decrease in hydraulic diameter
(c) decrease in fluid velocity   (d) increase in fluid thermal conductivity

7.24 For a fully developed laminar flow through a circular tube, identify how the heat transfer coefficient will vary with increasing values of the following variables or fluid properties (circle one for each item):

(a) flow velocity u_m: **(i)** increase   **(ii)** decrease   **(iii)** doesn't change

(b) thermal conductivity k: **(i)** increase   **(ii)** decrease   **(iii)** doesn't change

(c) density ρ: **(i)** increase   **(ii)** decrease   **(iii)** doesn't change

(d) specific heat c_p: **(i)** increase   **(ii)** decrease   **(iii)** doesn't change

(e) dynamic viscosity μ: **(i)** increase   **(ii)** decrease   **(iii)** doesn't change

(f) tube diameter d_i: **(i)** increase   **(ii)** decrease   **(iii)** doesn't change

(g) tube length L: **(i)** increase   **(ii)** decrease   **(iii)** doesn't change

(h) heat flux q'':   **(i)** increase   **(ii)** decrease   **(iii)** doesn't change

7.25 Repeat Question 7.24 by assuming the flow in a circular tube to be fully developed turbulent. *Hint:* Use Eq. (7.79).

7.26 Pressure drop for airflow through a long circular tube at Re = 2000 is measured as 7 kPa and the heat transfer coefficient as $100 \, \text{W/m}^2 \cdot \text{K}$. Neglect the entrance and exit losses and the momentum effect in the pressure drop evaluation. Consider $\Delta p = 16 \mu \dot{m}(f \cdot \text{Re})L/2\pi g_c \rho D_h^4 = (4fL \cdot \text{Re}^2 \mu^2 / 2g_c \rho D_h^3)$. Assume that Nu = 4.36 for laminar flow, Pr = 0.7, and constant fluid properties.

(a) If the tube diameter is doubled, the resulting pressure drop will be the following factor of the original pressure drop (assume that the flow rate remains constant):

(i) $\frac{1}{4}$ **(ii)** $\frac{1}{2}$ **(iii)** 2   **(iv)** $\frac{1}{16}$ **(v)** 1   **(vi)** can't tell

(b) If the tube diameter is doubled, the resulting heat transfer coefficient will be the following factor of the original heat transfer coefficient:

(i) $\frac{1}{4}$ **(ii)** $\frac{1}{2}$ **(iii)** 2   **(iv)** $\frac{1}{16}$ **(v)** 1   **(vi)** can't tell

(c) If the tube length is doubled, the resulting pressure drop will be the following factor of the original pressure drop:

(i) $\frac{1}{4}$ **(ii)** $\frac{1}{2}$ **(iii)** 2   **(iv)** $\frac{1}{16}$ **(v)** 1   **(vi)** can't tell

(d) If the tube length is doubled, the resulting heat transfer coefficient will be the following factor of the original heat transfer coefficient:

(i) $\frac{1}{4}$ **(ii)** $\frac{1}{2}$ **(iii)** 2   **(iv)** $\frac{1}{16}$ **(v)** 1   **(vi)** can't tell

(e) If the flow rate is increased five times, the resulting pressure drop will be the following factor of the original pressure drop:

(i) 5   **(ii)** 25   **(iii)** 24.7   **(iv)** doesn't change   **(v)** can't tell

(f) If the flow rate is increased five times, the resulting heat transfer will be the following factor of the original heat transfer coefficient:

(i) 5   **(ii)** 3.62   **(iii)** 8.64   **(d)** 25   **(e)** doesn't change

Hint: Use Eqs. (7.72) and (7.79), for items (e) and (f).

7.27 Given the physical location x^+ in a circular tube, the Nusselt number in the laminar thermal entrance region will:

(a) increase with increasing Pr   **(b)** decrease with increasing Pr   **(c)** can't tell

7.28 The Nusselt number for fully developed laminar duct flow depends on:

(a) Reynolds number
(b) Prandtl number
(c) cross-sectional geometry
(d) thermal boundary condition
(e) $T_w - T_m$
(f) heat flux
(g) St
(h) none of these
(i) all of these

7.29 The Nusselt number for hydrodynamically developed and thermally developing laminar duct flow depends on:

(a) Reynolds number
(b) Prandtl number
(c) cross-sectional geometry
(d) thermal boundary condition
(e) $T_w - T_m$
(f) heat flux
(g) St
(h) none of these
(i) all of these

7.30 The Nusselt number for simultaneously developing laminar duct flow depends on:

(a) Reynolds number
(b) Prandtl number
(c) cross-sectional geometry
(d) thermal boundary condition
(e) $T_w - T_m$
(f) heat flux
(g) St
(h) none of these
(i) all of these

7.31 The Nusselt number for fully developed turbulent duct flow ($Pr > 0.5$) depends on:

(a) Reynolds number
(b) Prandtl number
(c) cross-sectional geometry
(d) thermal boundary condition
(e) $T_w - T_m$
(f) heat flux
(g) St
(h) none of these
(i) all of these

7.32 In fully developed turbulent flow, the flow passage shape is of:

(a) significant importance   (b) minor importance   (c) can't tell

7.33 For an identical upstream velocity, compared to a two-row staggered tube arrangement, a four-row staggered tube arrangement will have a *mean* heat transfer coefficient that is:

(a) lower   (b) higher   (c) the same

7.34 Decreasing the interrupted flow length will yield:

(a) heat transfer coefficients: higher, lower, same   (circle one)
(b) friction factors: higher, lower, same   (circle one)

7.35 Compared to the constant-viscosity case, a consideration of variable viscosity at a cross section for the liquid heating case results in:

(a) higher heat transfer coefficient   (b) higher pressure drop
(c) higher heat transfer rate   (d) higher friction factor

7.36 Variable fluid properties at a cross section yield higher heat transfer coefficients for:

(a) liquid heating (b) liquid cooling (c) gas heating (d) gas cooling

7.37 In a heat exchanger analysis, heat transfer coefficients are generally considered uniform on each fluid side. Heat transfer coefficients can be nonuniform due to the following reasons:

(a) temperature-dependent fluid properties

(b) thermal entry length effect

(c) distortion of the uniform velocity and temperature profiles at a cross section due to boundary layers

(d) none of these

7.38 Superimposed free convection over forced convection will yield higher heat transfer coefficients for:

(a) gas heating in a horizontal tube

(b) liquid cooling in a horizontal tube

(c) liquid heating in a vertical tube with upflow

(d) gas cooling in a vertical tube with downflow

7.39 The free convection effect in a forced flow may be significant for:

(a) small $(T_w - T_m)$ (b) highly compact exchanger

(c) shell-and-tube exchanger (d) high Reynolds number

PROBLEMS

7.1 A crossflow heat exchanger uses strip fins on the air side with the following geometry: fin pitch $= 549\,\text{m}^{-1}$, plate spacing $= 9.53\,\text{mm}$, fin length $= 3.18\,\text{mm}$, flow passage hydraulic diameter $= 2.68\,\text{mm}$, fin metal thickness $= 0.25\,\text{mm}$, total heat transfer area/volume between plates $= 1250\,\text{m}^2/\text{m}^3$, and fin area/total area $= 0.840$. Fins and parting sheets are made of stainless steel $(k_w = 20.77\,\text{W/m}\cdot\text{K})$. Parting sheet thickness is 0.381 mm. Air flows at $u_\infty = 3.05\,\text{m/s}$ with $\rho = 1.1213\,\text{kg/m}^3$, $\nu = 1.58 \times 10^{-5}\,\text{m}^2/\text{s}$, $\text{Pr} = 0.70$, and $c_p = 1.00\,\text{kJ/kg}\cdot\text{K}$. For the air side, $\sigma = A_o/A_{fr} = 0.402$.

(a) Determine h for Re = 1000.

(b) If the fin thickness is reduced from 0.25 mm to 0.16 mm, how would h and f be affected? Give qualitative reasons only.

(c) If every geometric dimension of the strip fin geometry is scaled up by a factor of 5, estimate j and f factors for operation at Re = 1000. Use $j = 0.0192$ and $f = 0.0927$ at Re = 1000 for this surface.

7.2 A plate-fin condenser is to be designed on the air side with 790 fins/m and a 0.025 mm fin thickness. You are to choose between the triangular and rectangular flow passage surfaces shown in Fig. P7.2. The mass flow rate $\dot{m}$ and the frontal area A_{fr} of the air side are fixed, and the frontal velocity $u_\infty = 3.05\,\text{m/s}$. Use the following fluid properties: $\rho = 1.1213\,\text{kg/m}^3$, $\nu = 1.58 \times 10^{-5}\,\text{m}^2/\text{s}$, $\text{Pr} = 0.70$, and

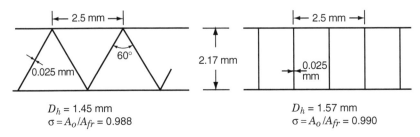

FIGURE P7.2

$c_p = 1.00 \, \text{kJ/kg} \cdot \text{K}$. For the air side, calculate the ratios of **(a)** heat transfer coefficients h_T/h_R, **(b)** surface areas A_T/A_R for the same hA, and **(c)** fluid pumping power $\mathcal{P}_T/\mathcal{P}_R$. Which surface would you select? Why? Note that the subscripts T and R here denote triangular and rectangular passages. *Hint:* Knowing Re for each surface, calculate j (and h), and f from theoretical solutions provided in the text.

7.3   A heat exchanger is constructed so that the hot flue gases at 425°C flow inside a 25.4 mm ID copper tube with 1.6 mm wall thickness. A 51 mm tube is placed around the 25.4 mm diameter tube, and high-pressure water at 150°C flows in the annular space between the tubes. If the mass flow rate of water is 1.51 kg/s and the total heat transfer is 17.6 kW, estimate the length of the heat exchanger for a gas mass flow rate of 0.76 kg/s. Consider the properties of the flue gas to be the same as those of air at atmospheric pressure and 425°C. Consider the exchanger to be counterflow.

7.4   A heat exchanger consists of 300 tubes 1.83 m long and 25.4 mm OD. The tubes are arranged in 15 rows with $X_t = X_\ell = 50.8$ mm. The tube surface temperature is maintained at 93.3°C. Air at 1 atm and 48.9°C flows normal to the tube bank at 6.1 m/s. Calculate the air-side heat transfer coefficient, outlet air temperature, total heat transfer rate, and air-side pressure drop considering **(a)** inline and **(b)** staggered tube arrangements. Compare results and discuss.

7.5   Determine the heat transfer coefficient for air, water, and liquid sodium flow through a tube of circular, rectangular with $\alpha^* = 0.125$, and equilateral triangular cross sections, each with $D_h = 25.4$ mm. First consider Re = 1000 and repeat the calculations at Re = 10,000. Assume fully developed flow for each case and the Ⓗ boundary condition. Which fluid and which cross-section geometry yield the highest heat transfer coefficient? Why? Consider each fluid at T_m of 365 K and at an appropriate pressure having the following fluid properties.

	Air	Water	Sodium
$c_p(\text{kJ/kg} \cdot \text{K})$	1.011	4.209	1.38
$\mu(\text{Pa} \cdot \text{s})$	2.15×10^{-5}	3.06×10^{-4}	6.98×10^{-4}
$k \, (\text{W/m} \cdot \text{K})$	0.0311	0.677	86.2
Pr	0.699	1.90	0.011

7.6 The coolant passages in a reactor core are 0.127 m long and have a rectangular cross section of 19 mm × 25.4 mm. The walls of the passages are to be maintained at a constant temperature of 371°C.

(a) If the coolant is nitrogen gas at 8 atm pressure and a temperature of 260°C, what velocity in the passages will result in a nitrogen discharge temperature of 316°C?

(b) If the passage walls are assumed to have a roughness equivalent to that of commercial steel pipes ($e = 0.046$ mm), what velocity is required for nitrogen discharge temperature of 316°C?

7.7 A compact air-cooled steam condenser employs offset strip fins on the airside as shown in Fig. 8.7, and the important geometrical properties are: 949.6 fins/m, $b = 1.88$ mm, $\ell_s = 2.82$ mm, $D_h = 1.21$ mm, $\delta = 0.10$ mm, $\beta = 2830 \text{ m}^2/\text{m}^3$, $A_f/A = 0.857$, $L = L_f = 53.1$ mm, $A = 2.83 \text{ m}^2$, and $A_o = 0.0161 \text{ m}^2$. The air flows at 8.89 m/s at a log-mean average temperature of 86.1°C. The air properties are: $\rho = 0.982 \text{ kg/m}^3$, $c_p = 1.00 \text{ kJ/kg} \cdot \text{K}$, $\mu = 2.112 \times 10^{-5} \text{ Pa} \cdot \text{s}$, and Pr = 0.696. Calculate the heat transfer coefficient, pressure drop, heat transfer power per unit temperature and unit area, and fluid pumping power due to core friction. Predict the heat transfer coefficient and pressure drop if the strip length is increased from 2.83 mm to 5.66 mm assuming that everything else remains the same. The j and f vs. Re characteristics of the air-side surface are as follows:

Re	j	f	Re	j	f
3000	—	0.0362	600	0.0138	0.0639
2000	0.00965	0.0398	500	0.0152	0.0713
1500	0.0103	0.0432	400	0.0170	0.0828
1200	0.0110	0.0467	300	0.0202	0.102
1000	0.0117	0.0500	200	—	0.137
800	0.0125	0.0551	150	—	0.171

7.8 Consider an automobile air-conditioning/heater duct having a square cross section as shown in Fig. P7.8 with the outside dimension of one side as 100

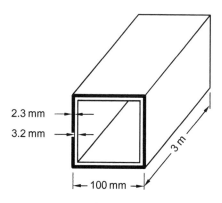

2.3 mm

3.2 mm

3 m

100 mm

FIGURE P7.8

mm, wall thickness 2.3 mm, and duct length as 3 m. The temperature of the air inlet to the duct is 65°C and the ambient temperature is −15°C. The airflow rate is 0.08 kg/s. The purpose of the problem is to minimize heat losses to the ambient and hence to investigate the influence of an insulation layer of 3.2 mm thickness inside this duct. Determine without insulation; (a) the heat transfer coefficient inside the duct, (b) the heat exchange system effectiveness, (c) the outlet air temperature, and (d) the heat loss. Repeat the calculations and determine the same four quantities with insulation. Discuss the results. Use the following air properties: $\rho = 1.058$ kg/m^3, $c_p = 1.008$ kJ/kg · K, $\mu = 2.04 \times 10^{-4}$ Pa · s, $k = 0.0288$ W/m · K, and Pr $= 0.701$. The thermal conductivity of the wall and insulation are 2.5 and 0.045 W/m · K, respectively. Consider the natural convection coefficient on the outside of the duct as 5 W/m^2 · K. Specify clearly any assumptions that you may make.

7.9 Compute the performance of a staggered bank of 19 mm OD plain tubes arranged on an equilateral triangular pitch on 25.4 mm centers. Consider a total of 95 tubes and 10 tube rows. Water flows at $u_\infty = 0.6$ m/s and at 21.1°C normal to the tube bundle. Compute h, hA/L, and Δp. Here L is the tube length. Use the following properties for water: $\rho = 998$ kg/m^3, $\mu = 9.75 \times 10^{-4}$ Pa · s, $k = 0.604$ W/m · K, $c_p = 4182$ J/kg · K, and Pr $= 6.75$.

7.10 A compact air-to-water heat exchanger is to be designed with an air-side flow rate of 0.83 kg/s. The NTU required for the exchanger is 2. We would like to design for an air-side Reynolds number Re$_a$ of 3000 at which $j = 0.006$. The following additional data are available:

Air side: $D_h = 3.475$ mm, $\sigma = 0.48$, $\alpha = 557.7$ m^2/m^3, $\eta_o = 0.8$, Pr $= 0.7$, $c_p = 1005$ J/kg · K, $\mu = 2.07 \times 10^{-5}$ Pa · s

Water side: $h = 1.703$ kW/m^2 · K, $\alpha = 32.8$ m^2/m^3

Note that the air side is the fluid side with the minimum heat capacity rate. Neglect wall thermal resistance and fouling for part (a).

(a) Determine (i) the air-side frontal area, (ii) the air-side heat transfer surface area, and (iii) the air-side flow length.

(b) Now if the water-side fouling factor is 3.52×10^{-4} m^2 · K/W and all other specifications of the problem remain unchanged, what will be your answers to part (a)?

(c) Would you observe the same types of changes, if the same fouling factor were present on the air side rather than the water side? Why?

7.11 The purpose of this problem is to determine the performance of a compact heat exchanger if its width, height, or depth is changed from the original specified size keeping the same fin geometry, fin pitch, and hydraulic diameter. These changes can affect the core NTU, ε or C^*. By evaluating the new values of these parameters properly, a new heat transfer rate can be established. Similarly, for the same velocity, the core pressure drop is proportional to the flow length, and a new Δp can be established with changes in L.

An automotive radiator, shown in Fig. P7.11, has a 0.305 m × 0.305 m frontal area on the air side and 25.4 mm airflow length. At a design point, the airflow rate is 1.05 kg/s, the water flow rate is 1.260 kg/s, and the total heat transfer rate from

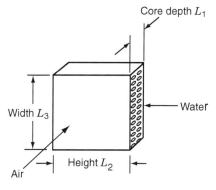

FIGURE P7.11

the water to air is 29.31 kW. Water and air inlet temperatures are 82.2° and 26.7°C, respectively, and the specific heats are 4187 and 1009 J/kg · K, respectively. The air-side pressure drop is 250 Pa, and the water-side pressure drop is 34.5 kPa. The objective of this problem is to determine the new heat transfer rate and pressure drop on both fluid sides (a) when the core width is doubled, (b) when the core height is doubled, (c) when the core depth is doubled, and (d) when the air inlet temperature of the original problem is reduced to 15.6°C. In all cases, assume that the core mass velocities on the air and water sides do not change due to the ram effect on the air side and changing the water pump on the water side.

7.12 Consider a single-pass shell-and-tube air heater with condensing steam outside the tube bundle and turbulent airflow inside the tubes. The following equations are applicable for the design with Nu, f, G, and Δp expressions for the tube/air side:

$$q = (\dot{m}c_p)_a(T_{a,o} - T_{a,i}) = (\dot{m}c_p)_a\,\Delta T_a$$

$$q = U(\pi d_i L N_t)\,\Delta T_{\mathrm{lm}} = \varepsilon(\dot{m}c_p)_a(T_s - T_{a,i})$$

$$\Delta T_{\mathrm{lm}} = \frac{(T_s - T_{a,i}) - (T_s - T_{a,o})}{\ln[(T_s - T_{a,i})/(T_s - T_{a,o})]}$$

$$\varepsilon = \frac{T_{a,o} - T_{a,i}}{T_s - T_{a,i}} = 1 - e^{-\mathrm{NTU}}$$

$$\mathrm{NTU} = \frac{UA}{(\dot{m}c_p)_a} = \frac{\Delta T_a}{\Delta T_{\mathrm{lm}}}$$

$$\mathrm{Nu} = \frac{hD_h}{k} = 0.023\left(\frac{Gd_i}{\mu}\right)^{0.8}\mathrm{Pr}^{0.4}$$

$$U \approx h_i \qquad G = \frac{4\dot{m}_a}{\pi d_i^2 N_t} \qquad \Delta p = \frac{4fLG^2}{2g_c\rho d_i} \qquad f = 0.046\left(\frac{Gd_i}{\mu}\right)^{-0.2}$$

Below is a list of independent variables and some dependent variables. When one independent variable is changed, all other independent variables are assumed to remain constant. Assume constant fluid properties and turbulent flow on the shell side.

Independent Variables	Dependent Variables
1. Inlet temperature of air	A. Heat transfer rate
2. Outlet temperature of air	B. Total heat transfer area
3. Tube diameter	C. Mass velocity of air
4. Tube wall thickness	D. Overall heat transfer coefficient
5. Mass flow rate of air	E. Pressure drop through tubes
6. Steam pressure	F. Ratio of ΔT_a to ΔT_{lm}
7. Total cross-sectional area of tubes	G. Cost of installing the tubes at a fixed charge per tube (labor cost)
8. Heat transfer coefficient on steam side	H. Cost of installing the heater at a fixed charge per meter of length
9. Thermal conductivity of tube material	I. Cost of the tube material at a fixed charge per kilogram
10. Density of tube material	

(a) For constant values of independent variables 1, 3, 4, 5, 6, 7, 8, 9, 10, and if the independent variable 2 is increased, what happens to the dependent variables A through I? Mark your answers in the following table as well as give separately the reasoning using pertinent equations for your answers.

Dependent	Increases	Remains Same	Decreases
A			
B			
C			
D			
E			
F			
G			
H			
I			

Change to the following conditions, and answer as in part (a):

(b) For constant values of independent variables 1, 2, 3, 4, 5, 7, 8, 9, and 10, and if the independent variable 6 is increased.

(c) For constant values of independent variables 1, 2, 3, 4, 5, 6, 8, 9, and 10, and if the independent variable 7 is increased.

(d) For constant values of independent variables 1, 2, 3, 4, 5, 6, 7, 9, and 10, and if the independent variable 8 is increased.

(e) For constant values of independent variables 1, 2, 3, 4, 5, 6, 7, 8, and 9 and if the independent variable 10 is increased. For most materials, the thermal conductivity and density of tube material are related. However, for this problem, we assume them to be independent variables.

7.13 To protect against freezing, a 50%: 50% mixture of water–glycol is used in a radiator instead of pure water. Assume that the velocity of both liquids (water and water–glycol mixture) is the same in the radiator. Determine the following at a given vehicle speed.

(a) Increase or decrease in the heat transfer coefficient for the water–glycol mixture compared to that for pure water. Consider fully developed turbulent flow in the water tubes and use the following Dittus–Boelter correlation for the Nusselt number: $\mathrm{Nu} = 0.023\mathrm{Re}^{0.8} \cdot \mathrm{Pr}^{0.4}$. The physical properties of water and the water–glycol mixture at 93°C are as follows:

Property	Water	Water–Glycol Mixture
Density ρ (kg/m^3)	963.5	1017.6
Specific heat c_p (J/kg · K)	4212	3592
Dynamic viscosity μ (Pa · s)	3.05×10^{-4}	7.50×10^{-4}
Thermal conductivity k (W/m · K)	0.675	0.410
Prandtl number	1.90	6.57

(b) Increase or decrease in UA for the water–glycol mixture compared to that for pure water. Assume that $(hA)_{\mathrm{water}}/(\eta_o hA)_{\mathrm{air}} = 10$, and fouling and wall thermal resistances are negligible. *Hint:* Use Eq. (3.24) and obtain the ratio of two UA's.

(c) Increase or decrease in the heat rejection rate of the radiator due to the use of the water–glycol mixture. Idealize NTU = 0.5 with water, $C^* \approx 0$, $C_{\min} = C_{\mathrm{air}}$, and identical inlet temperature differences $\Delta T_{\max}$'s.

7.14 Design a tubular laminar flow exchanger with small-diameter tubes to replace a tubular turbulent flow exchanger with 20 mm diameter tubes. The tubes are arranged inline with $X^* (= X_\ell^* = X_t^*) = p_t/d_i = 1.25$ and 2.0 for turbulent and laminar flow exchangers, respectively. Here $p_t =$ tube pitch and $d_i =$ inside tube diameter. If the total heat duty, fluid flow rates, and mean temperature difference remain the same for both exchangers, and if we assume equal thermal resistances on both fluid sides of the exchangers, the following ratios are obtained for exchanger volume, pumping power, frontal area, and core lengths for laminar and turbulent flow exchangers.

$$\frac{V_L}{V_T} = \frac{\mathrm{Nu}_T}{\mathrm{Nu}_L} \left(\frac{X_L^*}{X_T^*}\right)^2 \left(\frac{d_{i,L}}{d_{i,T}}\right)^2$$

$$\frac{\mathcal{P}_L}{\mathcal{P}_T} = \frac{\mathrm{Nu}_T}{\mathrm{Nu}_L} \left(\frac{d_{i,T}}{d_{i,L}}\right)^2 \frac{(f\,\mathrm{Re})_L}{(f\,\mathrm{Re})_T} \left(\frac{\mathrm{Re}_L}{\mathrm{Re}_T}\right)^2 \qquad \frac{A_{fr,L}}{A_{fr,T}} = \left(\frac{X_L^*}{X_T^*}\right)^2 \frac{\mathrm{Re}_T}{\mathrm{Re}_L} \frac{d_{i,L}}{d_{i,T}}$$

Compute L_L/L_T from known ratios of V_L/V_T and $A_{fr,L}/A_{fr,T}$. Consider water flowing through the tubes at 310 K with the thermal conductivity $k = 0.628$ W/m · K and Pr = 4.62. Assume that $\mathrm{Re}_T = 5 \times 10^4$ and $\mathrm{Re}_L = 100$. Here the subscripts T and L denote turbulent flow and laminar flow, respectively.

(a) Determine the tube diameter required for a laminar flow exchanger to achieve the same heat transfer coefficient as found in the 20 mm tube. The following correlations are given:

$$\mathrm{Nu}_L = 3.657 \qquad\qquad (f \cdot \mathrm{Re})_L = 16 \text{ for laminar flow}$$

$$\mathrm{Nu}_T = 0.024\mathrm{Re}^{0.8} \cdot \mathrm{Pr}^{0.4} \qquad f_T = 0.046\mathrm{Re}^{-0.2} \text{ for turbulent flow}$$

(b) Determine the ratios of exchanger volume, pumping power, frontal area, and core length for laminar flow to those for turbulent flow.

(c) Discuss the results of part (b), including peculiarities, advantages, and disadvantages of the laminar flow exchanger.

7.15 A water pipe is embedded in an ice–water bath such that the pipe wall temperature is 0°C. The water mass flow rate through the pipe is 1 kg/s. The water inlet temperature is 50°C and the specific heat of water is 4187 J/kg · °C.

(a) If the exchanger effectiveness is 80%, determine the outlet temperature of water.

(b) What is the total heat transfer rate between water and the ice–water bath?

(c) If the pipe diameter is doubled $(D_2 = 2D_1)$, what would be the corresponding new pipe length ratio L_2/L_1 if the exchanger effectiveness remains unchanged? Assume fully developed laminar flow.

7.16 During the performance testing of a compact heat exchanger, if the room air is not conditioned, there is no control on the inlet temperature of air. Similarly, since the fan delivers volumetric flow rate as a function of its speed, the mass flow rate will depend on the air density (inlet temperature). Hence, in general, one cannot attain the exact desired air inlet temperature and airflow rate during testing. The objective of this problem is to correct the performance of a heat exchanger from the measured values to the standard values.

A crossflow heat exchanger has been tested for performance by the steady-state technique using hot water and ambient air as the fluids on the tube and fin sides, respectively. The following data have been measured:

Air inlet temperature = 32.2°C   Air outlet temperature = 60°C

Water inlet temperature = 82.2°C   Water outlet temperature = 76.7°C

Air mass flow rate = 1.57 kg/s   Water mass flow rate = 1.89 kg/s

Use c_p for air and water as 1005 and 4187 J/kg · K. Calculate the heat transfer rate of this exchanger. For the simplicity of calculations, consider $C^* = 0$ for all sections of the problem below. Also note that the mass flow rate is given, so that it will not depend on the inlet temperature.

(a) Now consider the desired inlet air temperature of 26.7°C and no change in air and water mass flow rates as well as water inlet temperature. Assume constant fluid properties. Determine the heat transfer rate of this exchanger. Also compute the percentage change in the heat transfer rate and the percentage change in the inlet temperature difference (ITD).

(b) Instead, now consider the desired air mass flow rate as 1.76 kg/s, and determine the heat transfer rate of the exchanger. Assume no change in air and water inlet temperatures, water mass flow rate, and air-side heat transfer coefficient. For this case, also compute the percentage changes in the heat transfer rate, air mass flow rate, and the exchanger effectiveness.

(c) Next determine the heat transfer rate of this exchanger if the air mass flow rate and air inlet temperature would have been 1.76 kg/s and 26.7°C. Note that the water mass flow rate and water inlet temperature remain the same. Continue all assumptions parts (a) and (b).

(d) Discuss the results of parts (a), (b), and (c). *Hint:* First discuss individual functional relationships for changes.

7.17 You are asked to design an oil cooler to cool the lubricating oil for a 168 kW diesel engine. The oil inlet temperature is 82°C, the oil flow rate is $7.57 \times 10^{-4}\ \mathrm{m^3/s}$ (≈ 0.643 kg/s), and the required heat rejection is 19.78 kW. The allowable pressure drop on the oil side is 3.8×10^4 Pa. Water at 27°C and $1.262 \times 10^{-3}\ \mathrm{m^3/s}$ (≈ 1.285 kg/s) is available as a coolant. You have designed a single-pass counter-flow shell-and-tube heat exchanger with 100 tubes of 3.18 mm inside diameter and 0.3 m length to do the job. Calculate Δp on the oil side to check whether or not it is within specifications. Assume $f \cdot \mathrm{Re} = 18$ for the oil flow through the tubes. Since there was a serious mistake in the design, the oil cooler is unable to deliver the desired flow at design Δp. What is the mistake? For your reanalysis of Δp, consider the mean wall temperature as 35°C since h on the water side is very high. Also use the following fluid properties: $c_p = 4.187$ kW/kg · K for water; for oil $\rho = 855.7\ \mathrm{kg/m^3}$, and $c_p = 2081$ J/kg · K at the exchanger mean temperature. The dynamic viscosity for the engine oil varies with temperature as

$$\mu = 36.87 \times 10^5\ (1.8T + 32)^{-3.59}$$

where T is in °C and μ is in Pa · s. Neglect entrance and exit losses and the flow momentum effect for the Δp calculations.

7.18 The purpose of this problem is to investigate the influence of flow gross maldistribution on an air-cooled air-conditioning condenser with round tubes and flat fins (Fig. P7.18). Because of the specific packaging arrangement, the condenser's face area is blocked by 50%, as shown in Fig. P7.18*b*. Assume that the total airflow rate over the partially blocked condenser of Fig. P7.18*b* is the same as the airflow rate over the unblocked condenser of Fig. P7.18*a*. The Nusselt number and friction factor correlations for the airside surface are as follows:

$$\mathrm{Nu} = 2.0\mathrm{Re}^{1/3} \qquad f = 4.0\mathrm{Re}^{-0.5}$$

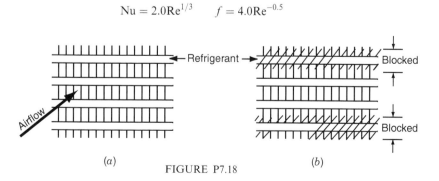

(*a*)   (*b*)

FIGURE P7.18

Determine the following, mentioning clearly any additional assumptions that you may make for the solution during each step.

(a) Percentage increase or decrease in the air-side heat transfer coefficient due to blockage of the face area.

(b) Percentage increase or decrease in the air-side pressure drop.

(c) Percentage increase or decrease in $\eta_o h A$ on the air side. Here assume that η_o is sufficiently high that it changes negligibly with the change in h.

(d) The total thermal resistance for the unblocked condenser in terms of the air-side thermal resistance assuming negligible fouling and wall resistances. For this case, the ratio $\eta_o h A$ on the air side to $h A$ on the refrigerant side is $\frac{1}{3}$. Note that there are no fins on the refrigerant side. *Hint:* Use Eq. (3.24).

(e) The total thermal resistance of the partially blocked condenser in terms of its air-side thermal resistance, again neglecting fouling and wall resistances. Note that the refrigerant-side heat transfer coefficient for unblocked and partially blocked condensers remains the same (since the refrigerant passages are not blocked); only its heat transfer surface area is affected.

(f) The NTU of the partially blocked condenser, assuming the NTU of the unblocked condenser to be 0.5 and UA's known from parts (d) and (e).

(g) Finally, determine the reduction in the heat transfer rate of the condenser assuming that $C^* = 0$ and the same inlet temperature difference ΔT_{max}.

8 Heat Exchanger Surface Geometrical Characteristics

The objective of this chapter is to outline surface geometrical characteristics that are used in the determination of experimental j (or Nu) and f factors and in the design of various heat exchangers. If the surface geometries required are determined differently than those for the original correlations, the heat transfer and pressure drop computed can be significantly different from the real (or measured) values regardless of how highly accurate the original j and f data are. Important geometrical characteristics are: the heat transfer area (both primary and secondary, if any), minimum free-flow area, frontal area, hydraulic diameter, and flow length on each fluid side of the exchanger [the flow lengths could be different for heat transfer and pressure drop calculations e.g., see Eqs. (8.7) and (8.8)]. The ratio of free-flow area to frontal area is needed for the determination of entrance and exit pressure losses. Heat transfer surface area density is an important parameter used in heat exchanger calculations. For a finned surface, an appropriate length is needed for the fin efficiency determination. The foregoing geometrical characteristics are derived from basic geometric measurements of a heat exchanger and its surfaces. In this chapter, we take one set of basic dimensions known for each geometry and arrive at the geometrical characteristics for the following exchangers: tubular, tube-fin, plate-fin, and simple cylindrical passage regenerators; these are presented in Sections 8.1 through 8.4. All important geometrical characteristics associated with flow bypass and leakages for shell-and-tube heat exchangers are presented in Section 8.5 in terms of known geometrical parameters for segmental baffled exchangers.

8.1 TUBULAR HEAT EXCHANGERS

Geometrical characteristics are derived separately for the inline and staggered tube arrangements. Flow is idealized as being normal to the tube bank on the outside. Tubes are considered bare (without fins) in this section.

8.1.1 Inline Arrangement

The basic core geometry for an idealized single-pass crossflow tubular exchanger with an inline tube arrangement is shown in Fig. 8.1. The header (tubesheet) dimensions for this tube bank are considered as $L_2 \times L_3$ such that the overhangs of $X_\ell/2$ are idealized on each end in the L_2 dimension and of $X_t/2$ on each end in the L_3 dimension. Thus the core

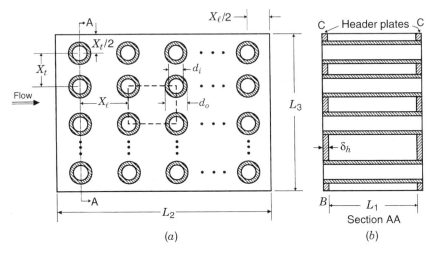

FIGURE 8.1 Tubular single-pass exchanger with an inline tube arrangement.

length for flow normal to the tube bank is L_2 and the noflow dimension is L_3. Thus the geometrical characteristics to be derived are for an infinite tube array.[†]

8.1.1.1 Tube Inside. Tubes have inside diameter d_i; length between headers L_1; total length, including header plates, $L_1 + 2\delta_h$; and a total number of tubes N_t, where

$$N_t = \frac{L_2 L_3}{X_t X_l} \tag{8.1}$$

The geometrical characteristics of interest for analyses for this geometry are straightforward:

$$\text{total heat transfer area } A = \pi d_i L_1 N_t \tag{8.2}$$

$$\text{total minimum free-flow area } A_o = \frac{\pi}{4} d_i^2 N_t \tag{8.3}$$

$$\text{core frontal area } A_{\text{fr}} = L_2 L_3 \tag{8.4}$$

$$\text{ratio of free flow to frontal area } \sigma = \frac{(\pi/4) d_i^2 N_t}{L_2 L_3} = \frac{(\pi/4) d_i^2}{X_t X_\ell} \tag{8.5}$$

$$\text{hydraulic diameter } D_h = d_i \tag{8.6}$$

$$\text{tube length for heat transfer} = L_1 \tag{8.7}$$

$$\text{tube length for pressure drop} = L_1 + 2\delta_h \tag{8.8}$$

$$\text{surface area density } \alpha_i = \frac{A}{V_{\text{total}}} = \frac{\pi d_i L_1 N_t}{L_1 L_2 L_3} \tag{8.9}$$

[†] An infinite tube array means an array having no end effects from the thermal hydraulic performance point of view.

The heat transfer area associated with the header plates on the tube side, represented by a plane through line BC (the leftmost plane) in Fig. 8.1b, is generally neglected. Note that different tube lengths are considered for heat transfer and pressure drop calculations. If there is a tube overhang beyond header plates or tubesheets, that length should be added for pressure drop calculations in Eq. (8.8), but that added length will not contribute to heat transfer from one fluid to the other fluid.

8.1.1.2 Tube Outside. The geometrical characteristics of the inline tube bank of Fig. 8.1 are summarized now. The total heat transfer area consists of the area associated with the tube outside surface and that with the two header plates:

$$A = \pi d_o L_1 N_t + 2 \left(L_2 L_3 - \frac{\pi d_o^2}{4} N_t \right) \tag{8.10}$$

The number of tubes N_t' in one row (in the X_t direction) is

$$N_t' = \frac{L_3}{X_t} \tag{8.11}$$

The minimum free-flow area A_o and frontal area A_{fr} are

$$A_o = (X_t - d_o) N_t' L_1 \tag{8.12}$$

$$A_{fr} = L_1 L_3 \tag{8.13}$$

Hence,

$$\sigma = \frac{A_o}{A_{fr}} = \frac{(X_t - d_o) N_t'}{L_3} = \frac{X_t - d_o}{X_t} \tag{8.14}$$

$$D_h = \frac{4 A_o L_2}{A} \tag{8.15}$$

where A_o and A are given by Eqs. (8.12) and (8.10), respectively.

$$\text{flow length for } \Delta p \text{ calculation} = L_2 \tag{8.16}$$

$$\text{heat exchanger total volume } V = L_1 L_2 L_3 \tag{8.17}$$

$$\text{surface area density } \alpha_o = \frac{A}{V} = \frac{\pi d_o}{p_t^2} \tag{8.18}$$

where A and V are given by Eqs. (8.10) and (8.17), respectively. The second equality in Eq. (8.18) is for a unit cell (shown by dashed lines in Fig. 8.1) for an inline arrangement.

It must be emphasized that the foregoing definition of the hydraulic diameter, Eq. (8.15), is used by Kays and London (1998) for tube banks. However, Žukauskas (1987) and other investigators use the tube outside diameter as the characteristic dimension in heat transfer and pressure drop correlations. In such correlations, the length L_2 may not be required and the tube bundle Δp is computed from Eq. (6.37) or (6.38) if the Euler number Eu or the Hagen number Hg is known, or other appropriate correlations. So it is essential to know that different researchers use different definitions of the characteristic length and other appropriate geometrical characteristics, and one must find out first the specific definition of the geometrical characteristics before using a particular correlation.

8.1.2 Staggered Arrangement

The geometrical characteristics are derived for the staggered arrangement of Fig. 8.2. In this case, any tube is halfway (in the X_t direction) between the two neighboring tubes in the next tube row, and the pattern of two transverse tube rows is repeated along the X_ℓ direction. However, if the transverse tube row pattern repeats after three tube rows along the X_ℓ direction, with the tubes in successive tube rows offset by $X_t/3$ distance (and thus tubes in every fourth tube rows arranged identical), it is referred to as a three-row echelon arrangement. An n-row echelon tube arrangement with $n \geq 3$ is possible as well. For geometry calculations for staggered tube arrangement, similar to the inline tube arrangement, we consider the overhangs $X_\ell/2$ on both sides of L_2, the overhangs $X_t/2$ on both sides of L_3, and half tubes at each end in the alternate tube row to simulate an infinite tube array. The total number of tubes in such a tube bundle is given by Eq. (8.1). If the half tubes are eliminated from the alternate intermediate tube rows, the number of tubes in the first row becomes L_3/X_t and in the second row, $L_3/X_t - 1$. The total number of tube rows then is L_2/X_ℓ, and the total number of tubes

$$N_t = \frac{L_3}{X_t} \frac{L_2/X_\ell + 1}{2} + \left(\frac{L_3}{X_t} - 1\right) \frac{L_2/X_\ell - 1}{2} \tag{8.19}$$

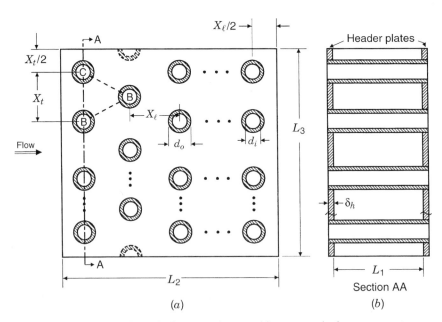

FIGURE 8.2 Tubular single-pass exchanger with a staggered tube arrangement.

8.1.2.1 Tube Inside. Tubes have inside diameter d_i, length between headers L_1, total length including header plates as $L_1 + 2\delta_h$ and the total number of tubes as N_t, as expressed by Eq. (8.1). The geometrical characteristics are identical to those for the inline arrangement given in Eqs. (8.2)–(8.9).

8.1.2.2 Tube Outside. The total heat transfer area consists of the area associated with the tube outside surface and that with the two header plates.

$$A = \pi d_o L_1 N_t + 2\left(L_2 L_3 - \frac{\pi d_o^2}{4} N_t\right) \tag{8.20}$$

The minimum free flow area occurs either at a plane through AA or at a plane through diagonals such as BB and BC of a unit cell of Fig. 8.2a. A unit cell for the analysis is shown in Fig. 8.3. From Fig. 8.3,

$$2a = X_t - d_o \tag{8.21}$$

$$b = \left[\left(\frac{X_t}{2}\right)^2 + X_\ell^2\right]^{1/2} - d_o = p_t - d_o \tag{8.22}$$

where the bracketed term is p_t for 30°, 60°, and 45° tube layouts of Table 8.1. Now define the minimum of $2a$ and $2b$ as c:

$$c = \begin{cases} 2a & \text{if } 2a < 2b \\ 2b & \text{if } 2b < 2a \end{cases} \tag{8.23}$$

The minimum free-flow area on the tube outside is then given by

$$A_o = \left[\left(\frac{L_3}{X_t} - 1\right)c + (X_t - d_o)\right]L_1 \tag{8.24}$$

Here the last term, $(X_t - d_o)L_1$, corresponds to the free-flow area between the last tube (at each end in the first row) and the exchanger wall.

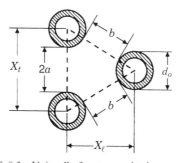

FIGURE 8.3   Unit cell of a staggered tube arrangement.

TABLE 8.1 Nomenclature and Geometrical Properties of Tube Banks Common in Shell-and-Tube Exchangers

	30° Triangular Staggered Array	60° Rotated Triangular Staggered Array	90° Square Inline Array	45° Rotated Square Staggered Array
Transverse tube pitch X_t	p_t	$\sqrt{3}\,p_t$	p_t	$\sqrt{2}\,p_t$
Longitudinal tube pitch X_ℓ	$\left(\dfrac{\sqrt{3}}{2}\right)p_t$	$\dfrac{p_t}{2}$	p_t	$\dfrac{p_t}{\sqrt{2}}$
Ratio of minimum free flow area to frontal area, $A_o/A_{\mathrm{fr}} = \sigma$	$\dfrac{p_t - d_o}{p_t}$	$\dfrac{\sqrt{3}\,p_t - d_o}{\sqrt{3}\,p_t}$ for $\dfrac{p_t}{d_o} \geq 3.732$ $\dfrac{2(p_t - d_o)}{\sqrt{3}\,p_t}$ for $\dfrac{p_t}{d_o} \leq 3.732$	$\dfrac{p_t - d_o}{p_t}$	$\dfrac{\sqrt{2}\,p_t - d_o}{\sqrt{2}\,p_t}$ for $\dfrac{p_t}{d_o} \geq 1.707$ $\dfrac{2(p_t - d_o)}{\sqrt{2}\,p_t}$ for $\dfrac{p_t}{d_o} \leq 1.707$

Other geometrical characteristics of interest are

$$A_{fr} = L_1 L_3 \tag{8.25}$$

$$\sigma = \frac{A_o}{A_{fr}} \tag{8.26}$$

$$D_h = \frac{4 A_o L_2}{A} \tag{8.27}$$

$$\text{flow length for } \Delta p \text{ calculation}^\dagger = L_2 \tag{8.28}$$

$$\text{heat exchanger volume } V = L_1 L_2 L_3 \tag{8.29}$$

$$\text{surface area density } \alpha_o = \frac{A}{V} \tag{8.30}$$

where A_o and A are given by Eqs. (8.24) and (8.20), respectively.

If we ignore all end effects and header plate surface area, α_o can be derived for a unit cell (shown by dashed lines) of Fig. 8.3 as follows:

$$A = \tfrac{1}{2} \pi d_o L_1 \qquad V = X_\ell X_t L_1 = (p_t \sin 60°) p_t L_1 = \frac{\sqrt{3}}{2} p_t^2 L_1 \tag{8.31}$$

and

$$\alpha_o = \frac{A}{V} = \frac{\pi d_o}{\sqrt{3} \, p_t^2} \tag{8.32}$$

For shell-and-tube exchangers, tube bundles with three specific staggered arrangements are commonly used, and are referred to as 30°, 45°, and 60° tube layouts, while the tube bundle with the inline arrangement is referred to as the 90° tube layout. The relationship between the tube pitch p_t and transverse and longitudinal pitches (X_t and X_ℓ) for these tube layouts is outlined in Table 8.1. Also summarized in this table is a ratio of the minimum free-flow area to the frontal area σ for these tube layouts. Note that Eq. (8.24) for A_o is valid for any staggered arrangement; its value in terms of σ in Table 8.1 is specific for those specific tube layouts.

8.2 TUBE-FIN HEAT EXCHANGERS

Geometrical characteristics are derived for two specific tube-fin exchangers: circular tubes having individual circular fins and circular tubes having flat plain fins.

8.2.1 Circular Fins on Circular Tubes

The basic core geometry of an idealized single-pass crossflow exchanger is shown in Fig. 8.4 for an inline tube arrangement. The finned tubes could also be in the staggered

---

† If the tube-side dimensionless pressure drop correlation is based on an average Euler number per tube row, there is no need to use or define L_2 for the Δp calculations.

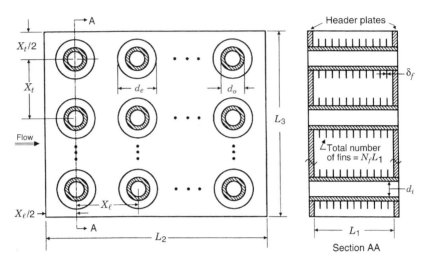

FIGURE 8.4   Circular-finned tubular exchanger. (From Shah, 1985.)

arrangement similar to those in Fig. 8.2. The total number of tubes in this exchanger is given by Eq. (8.1) for the inline or staggered arrangement.

8.2.1.1   Tube Inside. The core geometrical characteristics for the tube side applicable for both inline and staggered arrangements are identical to those of Eqs. (8.2)–(8.9).

8.2.1.2   Geometrical Characteristics for Tube Outside. The determination of geometrical characteristics for tube outside is somewhat complicated due to the presence of circular fins. It is idealized that the root of the circular fin has an effective diameter d_o and the fin tip has a diameter d_e. Depending on the manufacturing techniques, d_o may be the tube outside diameter or tube outside diameter plus the thickness of two collars made from the fin hole material for spacing fins evenly.

The total heat transfer area A consists of the area associated with the exposed tubes and header plates (primary surface area) A_p, and fins (secondary surface area) A_f. The primary surface area is the same as that given in Eq. (8.10) or (8.20) minus the area blocked by the fins:

$$A_p = \pi d_o \left(L_1 - \delta N_f L_1\right) N_t + 2\left(L_2 L_3 - \frac{\pi d_o^2}{4} N_t\right) \tag{8.33}$$

where δ is the fin thickness and N_f is the number of fins per unit length. The fin surface area is given by

$$A_f = \left[\frac{2\pi(d_e^2 - d_o^2)}{4} + \pi d_e \delta\right] N_f L_1 N_t \tag{8.34}$$

The factor 2 in the first term on the right-hand side is for two sides of a fin. The total heat transfer surface area is then

$$A = A_p + A_f \tag{8.35}$$

The minimum free-flow area for the inline arrangement is that area for a tube bank [Eq. (8.12)] minus the area blocked by the fins:

$$A_o = [(X_t - d_o)L_1 - (d_e - d_o)\delta N_f L_1] \frac{L_3}{X_t} \tag{8.36}$$

For the staggered tube arrangement, the minimum free-flow area could occur either through the front row or through diagonals similar to those of Fig. 8.3. A unit cell is shown in Fig. 8.5. The dimensions $2a$ and b as calculated by Eqs. (8.21) and (8.22) must be modified for the area blocked by the circular fins. We refer to these modified dimensions as $2a'$ and b'; note, they cannot be depicted in Fig. 8.5. They are given by

$$2a' = (X_t - d_o) - (d_e - d_o)\delta N_f \tag{8.37}$$

$$b' = \left[\left(\frac{X_t}{2}\right)^2 + X_\ell^2\right]^{1/2} d_o - (d_e - d_o)\delta N_f = (p_t - d_o) - (d_e - d_o)\delta N_f \tag{8.38}$$

where the bracketed term is p_t for 30°, 60° and 45° tube layouts of Table 8.1. Now define c' such that

$$c' = \begin{cases} 2a' & \text{if } 2a' < 2b' \\ 2b' & \text{if } 2b' < 2a' \end{cases} \tag{8.39}$$

The minimum free-flow area is then given by

$$A_o = \left[\left(\frac{L_3}{X_t} - 1\right)c' + (X_t - d_o) - (d_e - d_o)\delta N_f\right] L_1 \tag{8.40}$$

Other geometrical characteristics of interest are given by Eqs. (8.25)–(8.30) with appropriate values of A_o and A from the equations above.

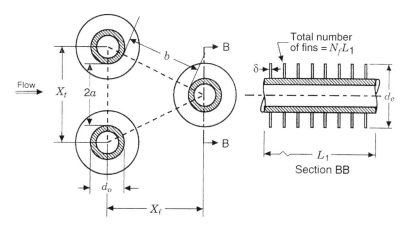

FIGURE 8.5   Unit cell of a staggered finned-tube arrangement. (From Shah, 1985.)

The fin efficiency of circular fins is calculated either from a formula in Table 4.5 or from Eq. (4.151) using $d_o/2$ and $d_e/2$ for r_o and r_e, respectively.

8.2.2   Plain Flat Fins on Circular Tubes

Conceptually, the fabrication of this exchanger is simple and amenable to mass production techniques. Proper holes are made into plain sheet metal (fins) of proper dimensions $L_2 \times L_3 \times \delta$. Tubes are then slipped into the properly stacked fins. Tubes are either expanded mechanically or are brazed. The basic core geometry of an idealized single-pass crossflow exchanger is shown in Fig. 8.6 for a staggered tube arrangement. The tubes could also be in an inline arrangement. The total number of tubes in this exchanger is given by Eq. (8.1) for the inline or staggered arrangement.

8.2.2.1   Tube Inside. The geometrical characteristics for the tube inside are the same as those for the preceding geometries, such as those given by Eqs. (8.2)–(8.9).

8.2.2.2   Tube Outside. The geometrical characteristics for the flow normal to the tubes and fins are similar to those for the circular fins on circular tubes except for some modifications due to the flat fin geometry. The total heat transfer area consists of the area associated with the exposed tubes and header plates (primary surface area) and the fins (secondary surface area). The primary surface area is the same as that given by Eq. (8.10) or (8.20) minus the area blocked by the fins:

$$A_p = \pi d_o \left(L_1 - \delta N_f L_1 \right) N_t + 2 \left(L_2 L_3 - \frac{\pi d_o^2}{4} N_t \right) \tag{8.41}$$

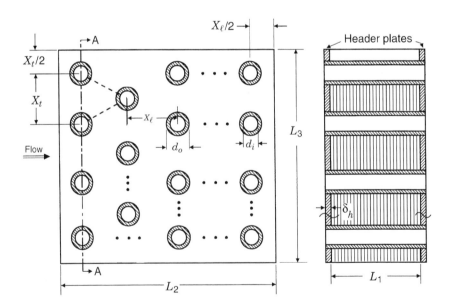

FIGURE 8.6   Flat fin and round tube exchanger.

The secondary area (fin surface) is

$$A_f = 2 \underbrace{\left[L_2 L_3 - \left(\frac{\pi d_o^2}{4} \right) N_t \right] N_f L_1}_{\text{fin surface area}} + \underbrace{2 L_3 \delta N_f L_1}_{\substack{\text{leading and trailing} \\ \text{edges area}}} \tag{8.42}$$

and the total heat transfer surface area

$$A = A_p + A_f \tag{8.43}$$

The minimum free-flow area for an inline arrangement is the area for a tube bank [Eq. (8.12)] minus the area blocked by the fins:

$$A_o = [(X_t - d_o)L_1 - (X_t - d_o)\, \delta N_f L_1] \frac{L_3}{X_t} \tag{8.44}$$

For the staggered tube arrangement, the minimum free-flow area could occur either through the front row or through the diagonals similar to those of Fig. 8.3 or 8.5. The dimensions $2a$ and b as calculated by Eqs. (8.21) and (8.22) must be modified for the area blocked by flat plain fins. These modified dimensions, referred to as $2a''$ and b'', are given by

$$2a'' = (X_t - d_o) - (X_t - d_o)\delta N_f \tag{8.45}$$

$$b'' = \left[\left(\frac{X_t}{2} \right)^2 + X_\ell^2 \right]^{1/2} - d_o - (X_t - d_o)\delta N_f = (p_t - d_o) - (X_t - d_o)\delta N_f \tag{8.46}$$

where the bracketed term is p_t for 30°, 60°, and 45° tube layouts of Table 8.1. Now define c'' such that

$$c'' = \begin{cases} 2a'' & \text{if } 2a'' < 2b'' \\ 2b'' & \text{if } 2b'' < 2a'' \end{cases} \tag{8.47} \\ \tag{8.48}$$

The minimum free flow area is then given by

$$A_o = \left[\left(\frac{L_3}{X_t} - 1 \right) c'' + (X_t - d_o) - (X_t - d_o)\delta N_f \right] L_1 \tag{8.49}$$

Other geometrical characteristics of interest are identical to those given by Eqs. (8.25)–(8.30) with appropriate values of A_o and A from the equations above.

For the determination of entrance (sudden contraction) and exit (sudden expansion) pressure losses, the area contraction and expansion ratio σ' is needed at the leading and trailing fin edges. It is given by

$$\sigma' = \frac{L_3 L_1 - L_3 \delta N_f L_1}{L_3 L_1} \tag{8.50}$$

The fin efficiency for flat fin geometry is determined by Eq. (4.155).

8.2.3   General Geometric Relationships for Tube-Fin Exchangers

In Sections 8.2.1 and 8.2.2 we showed how to evaluate A (A_p and A_f), A_o, A_{fr}, and D_h for the tube-fin surfaces on both fluid sides. Now we present how to evaluate α and σ, and the relationship between them, through definition of the hydraulic diameter for both individually finned tubes and flat fins on round or flat tubes. For these cases, the surface area density α is the ratio of the total transfer surface area A on one side of the exchanger to the total volume V of the exchanger as follows:

$$\alpha_1 = \frac{A_1}{V} \qquad \alpha_2 = \frac{A_2}{V} \qquad (8.51)$$

Similarly, the ratio of minimum free-flow area to frontal area σ for each fluid side is given by

$$\sigma_1 = \frac{A_{o,1}}{A_{fr,1}} \qquad \sigma_2 = \frac{A_{o,2}}{A_{fr,2}} \qquad (8.52)$$

The α's and σ's are related by the definition of the hydraulic diameter as

$$D_{h,1} = \frac{4\sigma_1}{\alpha_1} \qquad D_{h,2} = \frac{4\sigma_2}{\alpha_2} \qquad (8.53)$$

Note that the surface area density β has no meaning for tube-fin exchangers since the volumes occupied by each fluid side are not mutually independent. Hence, we refer to α's only for tube-fin exchangers.

8.3   PLATE-FIN HEAT EXCHANGERS

A large number of fin geometries are available for plate-fin heat exchangers. Some of the fin geometries are shown in Fig. 1.29. It is beyond the scope of this section to present derivations for geometrical characteristics for all these "corrugated" fins. As an illustration, geometrical characteristics are derived below for one side of a plate-fin heat exchanger having offset strip fins in Section 8.3.1 and corrugated louver fins in Section 8.3.2. In Section 8.3.3, a relationship is presented relating the α's and β's of the plate-fin surfaces.

8.3.1   Offset Strip Fin Exchanger

A schematic of a single-pass crossflow plate-fin exchanger core, employing offset strip fins on the fluid 1 side is shown in Fig. 8.7a. Offset strip fins are shown in Fig. 8.7b. The idealized fin geometry is shown in Fig. 8.7c.

The total heat transfer area consists of all surface area (primary and secondary) swept by fluid 1. The following four components are needed for calculating the primary surface area: (1) the plate area (Fig. 8.7a), (2) the fin base area that covers the plate (Fig. 8.7c), (3) the header bar area on the sides for fluid 1 near the ends of fins in the L_2 direction (Fig. 8.7a), and (4) header bars and plates exposed area of the blocked fluid 2 passages at fluid 1 core inlet and outlet faces (Fig. 8.7b). The secondary (fin) area consists of the fin height area (Fig. 8.7b), fin edge height area (Fig. 8.7c), and fin edge width area

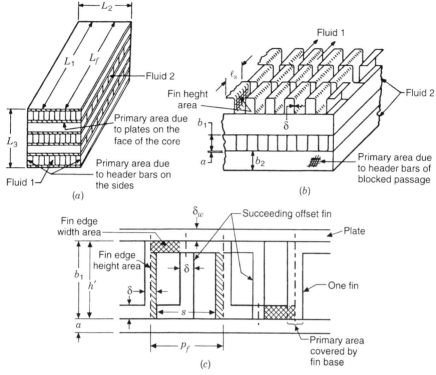

FIGURE 8.7   (a) Plate-fin exchanger; (b) offset strip fin geometry; (c) small section of an idealized offset strip fin geometry; the unit cell is bounded by b_1 and p_f. (From Shah, 1985.)

(Fig. 8.7c). The primary surface area is then the sum of components 1, 3, and 4 minus component 2.

These four components of the primary surface area are now derived:

$$\text{total plate area (Fig. 8.7a)} = 2L_1L_2N_p \tag{8.54}$$

where N_p is the total number of fluid 1 passages (in the L_3 direction).

$$\text{fin base area covering plates (Fig. 8.7c)} = 2\delta L_f n_f = 2\delta L_f N_f L_2 N_p \tag{8.55}$$

where L_f is the fin flow length, $n_f = N_f L_2 N_p$ is the total number of fins in the core if it is assumed that there are no fin offsets, and N_f is the number of fins per unit length (in the L_2 direction). Sometimes the fin flow length L_f is slightly shorter than the core flow length L_1 in an actual core, and hence this distinction has been made here:

$$\text{area of header bars on the side for fluid 1 (Fig. 8.7a, b)} = 2b_1 L_1 N_p \tag{8.56}$$

area of header bars and plates of fluid 2 at fluid 1 core inlet
and outlet faces (Fig. 8.7b)

$$= 2(b_2 + 2\delta_w)(N_p + 1)L_2 \tag{8.57}$$

Here, we have considered the total number of passages on the fluid 2 side as one more than that on the fluid 1 side. The total primary surface area on the fluid 1 side, from Eqs. (8.54)–(8.57), is then

$$A_{p,1} = 2L_1 L_2 N_p - 2\delta L_f n_f + 2b_1 L_1 N_p + 2(b_2 + 2\delta_w)(N_p + 1)L_2 \tag{8.58}$$

The three components of the secondary (fin) area are (refer to Fig. 8.7c)

$$\text{fin height area} = 2(b_1 - \delta)L_f n_f \tag{8.59}$$

The offset strip fins have leading and trailing edges at each strip, which contribute to the heat transfer area. The area associated with the edges is divided into two components: the area associated with the edge height and the area associated with the edge width. Because of the overlap of edge widths between two offset fins in the flow direction, only half of the edge width area is available at the front and half at the back edge of each fin, except for the edges at the front and back face of the core (refer to Fig. 8.7c):

$$\text{fin edge height area} = 2(b_1 - \delta)\delta n_{\text{off}} n_f \tag{8.60}$$

where n_{off} is the total number of offset strips in the flow (L_1) direction:

$$\text{fin edge width area} = (p_f - \delta)\delta(n_{\text{off}} - 1)n_f + 2p_f \delta n_f \tag{8.61}$$

where the last term of Eq. (8.61) represents the fin edge width area at the front and back face of the core. Thus, from Eqs. (8.59) through (8.61), the total secondary area on fluid 1 side is

$$A_{f,1} = 2(b_1 - \delta)L_f n_f + 2(b_1 - \delta)\delta n_{\text{off}} n_f + (p_f - \delta)\delta(n_{\text{off}} - 1)n_f + 2p_f \delta n_f \tag{8.62}$$

The total surface area on fluid 1 side is

$$A_1 = A_{p,1} + A_{f,1} \tag{8.63}$$

where $A_{p,1}$ and $A_{f,1}$ are given by Eqs. (8.58) and (8.62), respectively.

The free-flow area on fluid 1 side is given by the frontal area on fluid 1 side minus the area blocked by the fins at the entrance of the core on that side.

$$A_{o,1} = b_1 L_2 N_p - [(b_1 - \delta) + p_f]\delta n_f \tag{8.64}$$

Other geometrical characteristics of interest are

$$A_{\text{fr}} = L_2 L_3 \tag{8.65}$$

$$D_h = \frac{4A_o L}{A} \tag{8.66}$$

$$\sigma = \frac{A_o}{A_{\text{fr}}} \tag{8.67}$$

flow length for Δp calculation $= L_1$ \hfill (8.68)

heat exchanger volume between plates $V_p = (b_1 L_2 N_p) L_1$ \hfill (8.69)

$$\text{surface area density } \beta = \frac{A}{V_p} \tag{8.70}$$

It should be mentioned that if we would have ignored some secondary effects in geometry calculations as mentioned next, we could obtain the expressions for A, A_o, and D_h based on the unit cell of Fig. 8.7c. For the surface area based on the unit cell, the following effects in A_p and A_f calculated based on the full core need to be neglected: (1) the exposed surface area of the header bars on fluid 1 [the right-hand side of Eq. (8.56)], and (2) the exposed surface area of header bars and plates on fluid 2 [the right-hand side of Eq. (8.57)]. In addition, the exposed fin edge width area of the front and back (of the core) fin edges is considered the same as those areas for fins within the core [i.e., $2p_f \delta n_f$, the second term on the right-hand side of Eq. (8.61), is changed to $2(p_f - \delta)\delta n_f$]. In that case it can be shown that Eqs. (8.63) and (8.64) for the unit cell become

$$A_{\text{cell}} = (A_p + A_f)_{\text{cell}} = (2s\ell_s) + (2h'\ell_s + 2h'\delta + s\delta) = 2(s\ell_s + h'\ell_s + h'\delta) + s\delta \tag{8.71}$$

$$A_{o,\text{cell}} = sh' \tag{8.72}$$

where $s = p_f - \delta$ and $h' = b_1 - \delta$. The hydraulic diameter $D_h = 4A_o\ell_s/A$ for the unit cell is then given by

$$D_h = \frac{4A_{o,\text{cell}}\ell_s}{A_{\text{cell}}} = \frac{4sh'\ell_s}{2(s\ell + h'\ell_s + h'\delta) + s\delta} \tag{8.73}$$

This is the same expression as Eq. (7.126).

For the fin efficiency of the offset strip fin, it is assumed that the heat flow from both sides (plates) is uniform and the adiabatic plane occurs at the middle of the plate spacing. Hence,

$$\ell = \frac{b_1}{2} - \delta \tag{8.74}$$

The perimeter of the fin at a cross section is $(2\ell_s + 2\delta)$ and the cross section area is $\ell_s\delta$. Thus the value of $m\ell$ for the offset strip fin η_f of Eq. (4.146) is

$$m\ell = \left[\frac{2h}{k_f\delta}\left(1 + \frac{\delta}{\ell_s}\right)\right]^{1/2}\left(\frac{b_1}{2} - \delta\right) \tag{8.75}$$

Example 8.1   Determine the air-side core geometrical characteristics needed for data reduction of core 105 of Example 7.1. The following are the primary measurements, with most of them representing an arithmetic average of at least five individual measurements, for each item.

Airflow length $L_1 = 0.0532$ m   Total air-side passages $N_p = 46$
Core width $L_2 = 0.2148$ m   Number of fins per passage $n_f' = 204$

Core noflow height $L_3 = 0.2444$ m

Fin flow length $L_f = 0.0508$ m

Air-side plate spacing $b_1 = 1.91$ mm

Steam-side plate spacing $b_2 = 2.54$ mm

Fin offset length $\ell_s = 2.82$ mm

Fin thickness $\delta = 0.10$ mm

Plate thickness $\delta_w = 0.41$ mm

SOLUTION

Problem Data and Schematic: All pertinent geometrical data for the air side of a single-pass crossflow heat exchanger are provided above. See Fig. 8.7 as a schematic for this problem.

Determine: The air-side geometric characteristics, such as fin pitch p_f, total number of fins n_f, number of fin offsets n_{off}, total primary area A_p, total secondary (fin) area A_f, total heat transfer area A, surface area density β, minimum free-flow area A_o, frontal area A_{fr}, heat conduction area for wall thermal resistance A_w, and hydraulic diameter D_h.

Assumptions: The ideal fin geometry for the air side is shown in Fig. 8.7c. It is assumed that the total number of passages on the steam side is one more than that on the air side.

Analysis: Let us first calculate the fin pitch p_f, total number of fins in the core n_f, and number of fin offsets n_{off} before we use Eqs. (8.54)–(8.70) to determine the necessary geometrical characteristics.

$$\text{fin pitch } p_f = \frac{\text{core width } L_2}{\text{number of fins per passage } n_f'} = \frac{0.2148\,\text{m}}{204}$$

$$= 0.001053\,\text{m} = 1.05\,\text{mm}$$

$$\text{total number of fins } n_f = \left(\frac{\text{fins}}{\text{passage}}\right)(\text{number of passages}) = 204 \times 46 = 9384 \quad \textit{Ans.}$$

$$\text{number of fin offsets } n_{off} = \frac{\text{fin flow length } L_f}{\text{fin offset length } \ell_s} = \frac{0.0508\,\text{m}}{0.00282\,\text{m}} = 18 \qquad \textit{Ans.}$$

The total primary area A_p from Eq. (8.58) is given by

$$A_p = 2L_1 L_2 N_p - 2\delta L_f n_f + 2b_1 L_1 N_p + 2(b_2 + 2\delta_w)(N_p + 1)L_2$$

$$= 2 \times 0.0532\,\text{m} \times 0.2148\,\text{m} \times 46 - 2 \times (0.10 \times 10^{-3}\,\text{m}) \times 0.0508\,\text{m} \times 9384$$

$$+ 2 \times (1.91 \times 10^{-3}\,\text{m}) \times 0.0532\,\text{m} \times 46 + 2(2.54 + 2 \times 0.41) \times 10^{-3}\,\text{m}$$

$$\times (46 + 1) \times 0.2148\,\text{m}$$

$$= 1.0513\,\text{m}^2 - 0.0953\,\text{m}^2 + 0.0093\,\text{m}^2 + 0.0678\,\text{m}^2 = 1.0331\,\text{m}^2 \qquad \textit{Ans.}$$

The total secondary (fin) area A_f from Eq. (8.62) is

$$A_f = 2(b_1 - \delta)L_f n_f + 2(b_1 - \delta)\delta n_{\text{off}} n_f + (p_f - \delta)\delta(n_{\text{off}} - 1)n_f + 2p_f \delta n_f$$

$$= 2(1.91 - 0.10) \times 10^{-3}\,\text{m} \times 0.0508\,\text{m} \times 9384$$

$$+ 2(1.91 - 0.10) \times 10^{-3}\,\text{m} \times (0.10 \times 10^{-3}\,\text{m}) \times 18 \times 9384$$

$$+ (1.05 - 0.10)10^{-3}\,\text{m} \times (0.10 \times 10^{-3}\,\text{m}) \times (18 - 1) \times 9384$$

$$+ 2 \times 0.00105\,\text{m} \times (0.10 \times 10^{-3}\,\text{m}) \times 9384$$

$$= 1.7257\,\text{m}^2 + 0.0611\,\text{m}^2 + 0.0152\,\text{m}^2 + 0.0020\,\text{m}^2 = 1.8040\,\text{m}^2 \qquad Ans.$$

The total heat transfer area

$$A = A_p + A_f = 1.0331\,\text{m}^2 + 1.8040\,\text{m}^2 = 2.8371\,\text{m}^2 \qquad Ans.$$

$$\frac{A_f}{A} = \frac{1.8040\,\text{m}^2}{2.8371\,\text{m}^2} = 0.636$$

$$\beta = \frac{A}{L_1 L_2 (b_1 N_p)} = \frac{2.8371\,\text{m}^2}{0.0512\,\text{m} \times 0.2148\,\text{m} \times (1.91 \times 10^{-3}\,\text{m}) \times 46} = 2936.2\,\text{m}^2/\text{m}^3 \quad Ans.$$

The minimum free-flow area A_o from Eq. (8.64) is

$$A_o = b_1 L_2 N_p - [(b_1 - \delta) + p_f]\delta n_f$$

$$= (1.91 \times 10^{-3}\,\text{m}) \times 0.2148\,\text{m} \times 46 - [(1.91 - 0.10) + 1.05] \times 10^{-3}\,\text{m}$$

$$\times (0.10 \times 10^{-3}\,\text{m}) \times 9384$$

$$= 0.0162\,\text{m}^2$$

$$A_{\text{fr}} = L_2 L_3 = 0.2148\,\text{m} \times 0.2444\,\text{m} = 0.0525\,\text{m}^2 \qquad Ans.$$

$$\sigma = \frac{A_o}{A_{\text{fr}}} = \frac{0.0162\,\text{m}^2}{0.0525\,\text{m}^2} = 0.309$$

$$D_h = \frac{4A_{o,1}L_1}{A} = \frac{4 \times 0.0162\,\text{m}^2 \times 0.0532\,\text{m}}{2.8371\,\text{m}^2} = 0.00121\,\text{m} \qquad Ans.$$

The conduction area for wall thermal resistance

$$A_w = 2L_1 L_2 N_p = 2 \times 0.0532\,\text{m} \times 0.2148\,\text{m} \times 46 = 1.051\,\text{m}^2 \qquad Ans.$$

We compute the hydraulic diameter based on the unit cell approach [i.e., using Eqs. (8.71)–(8.73)]:

$$s = p_f - \delta = 1.05\,\text{mm} - 0.10\,\text{mm} = 0.95\,\text{mm}$$

$$h' = b_1 - \delta = 1.91\,\text{mm} - 0.10\,\text{mm} = 1.81\,\text{mm}$$

$$A_{\text{cell}} = 2(s\ell_s + h'\ell_s + h'\delta) + s\delta$$

$$= 2[(0.95\,\text{mm} \times 2.82\,\text{mm}) + (1.81\,\text{mm} \times 2.82\,\text{mm}) + (1.81\,\text{mm} \times 0.10\,\text{mm})]$$

$$+ (0.95\,\text{mm} \times 0.10\,\text{mm})$$

$$= 2(2.679 + 5.104 + 0.181)\,\text{mm}^2 + 0.0950\,\text{mm}^2 = 16.023\,\text{mm}^2$$

$$A_{o,\text{cell}} = sh' = 0.95\,\text{mm} \times 1.81\,\text{mm} = 1.7195\,\text{mm}^2$$

$$D_h = \left(\frac{4A_oL_1}{A}\right)_{\text{cell}} = \frac{4A_{o,\text{cell}}\ell_s}{A_{\text{cell}}} = \frac{4 \times 1.7195\,\text{mm}^2 \times 2.82\,\text{mm}}{16.023\,\text{mm}^2} = 1.21\,\text{mm} = 0.00121\,\text{m}$$

Discussion and Comments: Calculations for the geometrical characteristics are straightforward once the specific model for geometry is chosen. This example shows how the geometric characteristics are computed for surfaces reported by Kays and London (1998). Note that use of the cell approach for this test core provides the value of the hydraulic diameter, which is in excellent agreement with that which includes all secondary effects; however, this may not always be the case.

8.3.2   Corrugated Louver Fin Exchanger

In this section we present the geometrical characteristics of a louver fin (more precisely, a multilouver fin) of Fig. 1.29e or 7.29 on one fluid side of a plate-fin exchanger. This geometry is more complicated than the offset strip fin geometry, due to different shapes (including candy ribbon or omega shape), depending on the fin density and the applied pressure in brazing, and the size of the braze fillets (between the fin and primary surface), due to the brazing process (Sekulić et al., 2003). As a result, there are many variations in geometrical characteristics. It is beyond the scope of this book to go into details of the many possible geometric configurations. We use a simple idealized geometry of a unit cell without any braze fillets and ignore the surface area of edges of the cut louvers (i.e., as if there were no louvers or considering it as a plain corrugated fin). Ignoring the surface area of edges is a standard industrial practice; for high-fin-density thin fins, this area may be less than 2% of the uncut fin surface area. Following are the geometrical characteristics of a *unit cell* of Fig. 7.29.

primary (tube) surface area $A_{p,\text{cell}} = 2W_t(p_f - \delta) + 2p_f H_t$ (8.76)

fin surface area $A_{f,\text{cell}} = 2L_f[(b^2 + p_f^2)^{1/2} - \delta]$ (8.77)

total heat transfer surface area $A_{\text{cell}} = A_{p,\text{cell}} + A_{f,\text{cell}}$ (8.78)

free-flow area $A_{o,\text{cell}} = p_f b - \delta[(b^2 + p_f^2)^{1/2} - \delta]$ (8.79)

frontal area $A_{\text{fr,cell}} = p_f(b + H_t)$ (8.80)

ratio of free flow area to frontal area $\sigma = \dfrac{A_{o,\text{cell}}}{A_{\text{fr,cell}}}$ (8.81) hydraulic diameter $D_h = \dfrac{4A_{o,\text{cell}}L_f}{A_{\text{cell}}}$ (8.82)

cell volume $V_{\text{cell}} = p_f p_t L_f$ (8.83)

surface area density $\beta = \dfrac{A_{\text{cell}}}{V_{\text{cell}}}$ (8.84)

wall conduction area per unit cell $A_{w,\text{cell}} = W_t W$ (8.85)

For a full core, D_h and β are the same as above. To calculate the free-flow area A_o, the frontal area A_{fr}, and the primary and fin surface areas A_p and A_f for the full core, first compute the number of fins in the core as follows for known core width W (fin passage width, corresponds to L_2 in Fig. 8.7), the number of fin pitches or fins in the core width ($= W/p_f$), and the number of finned passages N_p. In this case, the total number of fins or unit cells in the core are given by

$$n_f = N_p \frac{W}{p_f} \qquad (8.86)$$

Then A_o, A_{fr}, A_p, A_f, A, and A_w for the full core are given by

$$
\begin{aligned}
A_o &= n_f A_{o,\text{cell}} & A_{\text{fr}} &= n_f A_{\text{fr,cell}} & A_p &= n_f A_{p,\text{cell}} \\
A_f &= n_f A_{f,\text{cell}} & A &= n_f A_{\text{cell}} & A_w &= (n_p - 1)A_{w,\text{cell}}
\end{aligned}
\qquad (8.87)
$$

Heat exchanger volume V_p between plates for this geometry can also be calculated using Eq. (8.69).

For the fin efficiency of the louver fin geometry, like the offset strip fin geometry, it is assumed that the heat flow from both sides (of plates or tubes representing the primary surface) is uniform and the adiabatic plane occurs at the middle point of the plate spacing. Hence,

$$\ell = \tfrac{1}{2}(b^2 + p_f^2)^{1/2} - \delta \qquad (8.88)$$

The perimeter of the fin at a cross section is $(2L_f + 2\delta)$, and the cross section for heat conduction through the fin is $L_f\delta$. Thus the value of $m\ell$ for the louver fin η_f of Eq. (4.146) is

$$m\ell = \left[\frac{2h}{k_f\delta}\left(1 + \frac{\delta}{L_f}\right)\right]^{1/2}\left[\frac{1}{2}(b^2 + p_f^2)^{1/2} - \delta\right] \qquad (8.89)$$

In practical applications it is assumed that the contact resistance between the fin and the primary surface is zero. In a real design this assumption is never fulfilled, but the influence of this effect is usually small and lumped into the experimental j factor of a test core (assuming that the same problem exists for the test core). Although the fin efficiency reduction due to the poor thermal contact may be significant in some local areas where the brazing is not good, the overall effect on the heat exchanger performance may be small (Zhao and Sakulić, 2001). This influential factor has to be considered independently for each particular manufacturing process.

The geometry used by Chang and Wang (1997) for the j and f correlations of Eqs. (7.127) and (7.129) neglected the inclination of the louver fins with respect to the primary surface. Hence, Eqs. (8.77) and (8.79) were modified for one term as follows:

$$(b^2 + p_f^2)^{1/2} - \delta \approx b \qquad (8.90)$$

This is an excellent approximation for many high-fin-density louver fin surfaces.

Example 8.2 Determine the air-side core geometrical characteristics of a corrugated multilouver fin and flat tube exchanger with the following measured geometrical parameters:

Fin flow length $L_f = 30.0$ mm	Core width $W = 225.0$ mm
Fin thickness $\delta = 0.10$ mm	Tube pitch $p_t = 10.0$ mm
Fin pitch $p_f = 1.00$ mm	Tube width $W_t = 28.0$ mm
Air-side plate spacing $b = 6.00$ mm	Tube height $H_t = 2.00$ mm
Number of fin passages $N_p = 20$	

SOLUTION

Problem Data and Schematic: All pertinent geometrical data for the air side of a flat tube and corrugated multilouver fin crossflow exchanger are provided above. The detailed geometry is shown in Fig. 7.29.

Determine: The air-side geometrical characteristics: the hydraulic diameter D_h, the ratio of free-flow area to frontal area σ, and the surface area density β; and for the full core, the minimum free-flow area, frontal area, primary surface area, fin surface area, total heat transfer surface area, and heat conduction area for wall thermal resistance.

Assumptions: The ideal fin geometry for the air side is shown in Fig. 7.29. The influence of braze fillets is neglected; note that the flat tube and corrugated multilouver fin crossflow exchangers are brazed.

Analysis: We first calculate D_h, σ, and β for the unit cell and then compute the core geometrical parameters using this information.

Primary (tube) surface area $A_{p,\,\mathrm{cell}} = 2W_t(p_f - \delta) + 2p_f H_t$

$$= 2 \times 28.0\,\mathrm{mm} \times (1.00 - 0.10)\,\mathrm{mm} + 2 \times 1.00\,\mathrm{mm}$$
$$\times 2.00\,\mathrm{mm} = 54.40\,\mathrm{mm}^2$$

Corrugated fin length $= [(b^2 + p_f^2)^{1/2} - \delta] = [(6.00^2 + 1.00^2)^{1/2}\,\mathrm{mm} - 0.10\,\mathrm{mm}]$

$$= 5.98\,\mathrm{mm}$$

Fin surface area $A_{f,\,\mathrm{cell}} = 2L_f[(b^2 + p_f^2)^{1/2} - \delta] = 2 \times 30.0\,\mathrm{mm} \times 5.98\,\mathrm{mm} = 358.80\,\mathrm{mm}^2$

Total heat transfer surface area $A_{\mathrm{cell}} = A_{p,\,\mathrm{cell}} + A_{f,\,\mathrm{cell}} = 54.40\,\mathrm{mm}^2 + 358.80\,\mathrm{mm}^2$

$$= 413.20\,\mathrm{mm}^2$$

Free-flow area $A_{o,\text{cell}} = p_f b - \delta[(b^2 + p_f^2)^{1/2} - \delta]$

$$= 1.00\,\text{mm} \times 6.00\,\text{mm} - 0.10\,\text{mm} \times 5.98\,\text{mm} = 5.402\,\text{mm}^2$$

Frontal area $A_{\text{fr,cell}} = p_f(b + H_t) = 1.00\,\text{mm}\,(6.00 + 2.00)\,\text{mm} = 8.00\,\text{mm}^2$

Ratio of free flow area to frontal area $\sigma = \dfrac{A_{o,\text{cell}}}{A_{\text{fr,cell}}} = \dfrac{5.402\,\text{mm}^2}{8.00\,\text{mm}^2} = 0.675$ *Ans.*

Hydraulic diameter $D_h = \dfrac{4A_{o,\text{cell}}L_f}{A_{\text{cell}}} = \dfrac{4 \times 5.402\,\text{mm}^2 \times 30.0\,\text{mm}}{413.20\,\text{mm}^2} = 1.569\,\text{mm}$ *Ans.*

Cell volume $V_{\text{cell}} = p_f p_t L_f = 1.00\,\text{mm} \times 10.0\,\text{mm} \times 30.0\,\text{mm} = 300.00\,\text{mm}^3$

Surface area density $\beta = \dfrac{A_{\text{cell}}}{V_{\text{cell}}} = \dfrac{413.20\,\text{mm}^2}{300.00\,\text{mm}^3} = 1.377\,\text{mm}^2/\text{mm}^3 = 1377\,\text{m}^2/\text{m}^3$ *Ans.*

Wall conduction area per unit cell $A_{w,\text{cell}} = W_t \times W = 28.0\,\text{mm} \times 225.0\,\text{mm}$

$$= 6300\,\text{mm}^2$$

The total number of fins in the exchanger using Eq. (8.86) is:

$$n_f = N_p \frac{W}{p_f} = 20\left(\frac{225.0\,\text{mm}}{1.00\,\text{mm}}\right) = 4500$$

Now all areas related to full core are computed from Eq. (8.87) as follows:

Core primary surface area $A_p = n_f A_{p,\text{cell}} = 4500 \times 54.40\,\text{mm}^2 = 244,800\,\text{mm}^2$

$$= 0.2448\,\text{m}^2$$ *Ans.*

Core fin surface area $A_f = n_f A_{f,\text{cell}} = 4500 \times 358.80\,\text{mm}^2 = 1,614,600\,\text{mm}^2$

$$= 1.6146\,\text{m}^2$$ *Ans.*

Core total surface area $A = n_f A_{\text{cell}} = 4500 \times 413.20\,\text{mm}^2 = 1,859,400\,\text{mm}^2 = 1.8594\,\text{m}^2$
    Ans.

Core free flow area $A_o = n_f A_{o,\text{cell}} = 4500 \times 5.402\,\text{mm}^2 = 24309\,\text{mm}^2 = 0.0243\,\text{m}^2$ *Ans.*

Core frontal area $A_{\text{fr}} = n_f A_{\text{fr,cell}} = 4500 \times 8.00\,\text{mm}^2 = 36000\,\text{mm}^2 = 0.0360\,\text{m}^2$ *Ans.*

Total wall conduction area $A_w = (N_p - 1)A_{w,\text{cell}} = 19 \times 6300\,\text{mm}^2 = 119,700\,\text{mm}^2$

$$= 0.1197\,\text{m}^2$$ *Ans.*

    Although industry practice is to ignore the edge area of the cut louvers, let us compute the approximate effect on the surface area. We will assume that the louver length is 80% of the louver corrugated fin length computed above as 5.98 mm, and the total number of louvers in the two banks for a fin is 12. In this case,

$$\text{louver edge area} = 0.8 \times 5.98\,\text{mm} \times 0.1\,\text{mm} \times 14 = 6.698\,\text{mm}^2$$

Since the total surface area for the cell is 413.20 mm^2, the edge area of 6.698 mm^2 is 1.6%. Hence our idealization of neglecting the edge area is good.

Discussion and Comments: The calculation procedure for computing geometrical characteristics of corrugated multilouver fin is straightforward. Here we adopted the cell approach as a starting point since it is accurate for the simplified geometry considered. It becomes clear why industry neglects the surface area of cut louver edges.

8.3.3   General Geometric Relationships for Plate-Fin Surfaces

In Sections 8.3.1 and 8.3.2 we showed how to evaluate A, A_o, A_{fr}, D_h, σ, V_p, β, and ℓ for the offset strip fins and louver fins on one side of a plate-fin exchanger. Now we show the general relationship between σ, β, and α for a plate-fin surface valid for *any* corrugated fin geometry.

If L_1 and L_2 are the flow lengths and N_p and $N_p + 1$ are the number of flow passages on the fluid 1 and 2 sides, respectively, the volume between plates on each side is

$$V_{p,1} = L_1 L_2 (b_1 N_p) \qquad V_{p,2} = L_1 L_2 b_2 (N_p + 1) \tag{8.91}$$

After including the volume occupied by the plates or parting sheets, the total volume of the exchanger is

$$V = [b_1 N_p + b_2 (N_p + 1) + 2\delta_w (N_p + 1)] L_1 L_2 \tag{8.92}$$

The heat transfer areas on each side are

$$A_1 = \beta_1 V_{p,1} \qquad A_2 = \beta_2 V_{p,2} \tag{8.93}$$

Here β_1 and β_2 are the surface area densities on each fluid side based on *unit volume between plates*. The ratio of minimum free-flow area to frontal area on the fluid 1 side, σ_1, is as follows after introducing the definitions of the hydraulic diameter and expressions for $V_{p,1}$ and V from Eqs. (8.91) and (8.92):

$$
\begin{aligned}
\sigma_1 &= \frac{A_{o,1}}{A_{\text{fr},1}} = \frac{A_{o,1} L_1}{A_{\text{fr},1} L_1} = \frac{A_1 D_{h,1}/4}{V} = \frac{V_{p,1} \beta_1 D_{h,1}/4}{V} \\
&= \frac{L_1 L_2 (b_1 N_p) \beta_1 D_{h,1}/4}{[b_1 N_p + b_2 (N_p + 1) + 2\delta_w (N_p + 1)] L_1 L_2} \\
&= \frac{b_1 N_p \beta_1 D_{h,1}/4}{b_1 N_p + b_2 (N_p + 1) + 2\delta_w (N_p + 1)} \approx \frac{b_1 \beta_1 D_{h,1}/4}{b_1 + b_2 + 2\delta_w} \qquad \text{for } N_p \gg 1 \tag{8.94}
\end{aligned}
$$

Here δ_w is the thickness of the parting sheets. The last approximate equality is for the case when $N_p \gg 1$ or the number of passages on each fluid side are the same. Similarly,

$$\sigma_2 = \frac{b_2 (N_p + 1) \beta_2 D_{h,2}/4}{b_1 N_p + b_2 (N_p + 1) + 2\delta_w (N_p + 1)} \approx \frac{b_2 \beta_2 D_{h,2}/4}{b_1 + b_2 + 2\delta_w} \tag{8.95}$$

The heat transfer surface area on one fluid side divided by the *total volume V of the exchanger*, designated as α_1, is obtained as follows by using the definitions of $D_{h,1}$

$(= 4A_{o,1}L_1/A_1)$ and σ_1 from Eq. (8.94):

$$\alpha_1 = \frac{A_1}{V} = \frac{A_1}{L_1 A_{\mathrm{fr},1}} = \frac{A_1/L_1}{A_{\mathrm{fr},1}} = \frac{4A_{o,1}/D_{h,1}}{A_{\mathrm{fr},1}} = \frac{4\sigma_1}{D_{h,1}} = \frac{b_1\beta_1}{b_1 + b_2 + 2\delta_w} \tag{8.96}$$

Similarly,

$$\alpha_2 = \frac{A_2}{V} = \frac{b_2\beta_2}{b_1 + b_2 + 2\delta_w} \tag{8.97}$$

These relationships between α and β will be useful in sizing a plate-fin heat exchanger, as will be shown in Sections 9.2.2.2 and 9.2.2.3.

8.4   REGENERATORS WITH CONTINUOUS CYLINDRICAL PASSAGES

Some of the continuous cylindrical passage geometries for the rotary regenerator are shown in Fig. 1.43. Geometrical characteristics of triangular passages (Fig. 1.43b) are derived below as an illustration.

8.4.1   Triangular Passage Regenerator

The ideal triangular passage model is shown in Fig. 8.8 along with the nomenclature to be used in the derivation. Note that most of the symbols for this subsection are local. In addition to the basic dimensions shown in Fig. 8.8, the total number of cells within the unit core face area is determined from the enlarged photographs of the core face and is designated as n_c with units of m^{-2} or ft^{-2}. Hence

$$n_c = \frac{\text{number of cells}}{\text{unit area}} \qquad \text{face area for one cell} = \frac{1}{n_c} = \frac{\text{face area}}{\text{unit cell}} \tag{8.98}$$

Note that the face or frontal area of one cell then is the reciprocal of n_c as designated in the second equation above. For the ideal model of Fig. 8.8a,

$$\text{face area for one cell} = \frac{\text{face area}}{\text{unit cell}} = \frac{1}{n_c} = dc \tag{8.99}$$

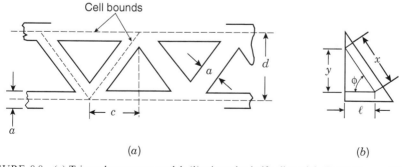

(a) $\qquad\qquad\qquad\qquad\qquad\qquad\qquad\qquad\qquad$ (b)

FIGURE 8.8   (a) Triangular passage model; (b) triangular half-cell model. (From Young, 1969.)

The heat transfer area associated with the unit cell is

$$A_{\text{cell}} = 2(\ell + x)L \tag{8.100}$$

where L is the regenerator flow length. From the geometry of Fig. 8.8b,

$$x = \frac{\ell}{\cos \phi}, \qquad \cos \phi = \frac{c}{(d^2 + c^2)^{1/2}} \tag{8.101}$$

so that

$$x = \ell \left[\left(\frac{d}{c} \right)^2 + 1 \right]^{1/2} \tag{8.102}$$

Substituting this into Eq. (8.100), we get

$$A_{\text{cell}} = 2L\ell \left\{ 1 + \left[\left(\frac{d}{c} \right)^2 + 1 \right]^{1/2} \right\} \tag{8.103}$$

The porosity σ is the ratio of free-flow to the frontal area for a cell:

$$\sigma = \frac{\ell y}{1/n_c} = \frac{\ell^2 \tan \phi}{1/n_c} = n_c \ell^2 \frac{d}{c} = \frac{\ell^2}{c^2} \tag{8.104}$$

where the last equality comes from Eq. (8.99). The heat transfer surface area density β is given by

$$\beta = \frac{A_{\text{cell}}}{V_{\text{cell}}} = \frac{2(\ell + x)L}{(1/n_c)L} = 2n_c(\ell + x) \tag{8.105}$$

Substituting the value of x from Eq. (8.102) and ℓ from Eq. (8.104) into Eq. (8.105) yields

$$\beta = 2n_c \left(\frac{\sigma/n_c}{d/c} \right)^{1/2} \left\{ 1 + \left[\left(\frac{d}{c} \right)^2 + 1 \right]^{1/2} \right\} \tag{8.106}$$

The hydraulic diameter and minimum free-flow area are

$$D_h = \frac{4\sigma}{\beta} \qquad A_o = \sigma A_{\text{fr}} \tag{8.107}$$

where A_{fr} is determined from the core dimensions.

Since the porosity σ is a critical parameter, it is generally determined from gravimetric measurements and then checked against the geometrically determined value. The gravimetric porosity is determined from the measurements of core mass M_w and core volume V of a small sample, and the known density ρ_w of the matrix material as follows:

$$\sigma = \frac{\text{void volume}}{\text{total volume}} = 1 - \frac{\text{solid volume}}{\text{total volume}} = 1 - \frac{M_w/V}{\rho_w} \tag{8.108}$$

TABLE 8.2 Surface Geometrical Properties for Some Idealized Flow Passages used in Compact Regenerators

Geometry	Cell Density N_c (cells/m²)	Porosity σ	Surface Area Density β (m²/m³)	Hydraulic Diameter D_h (m)
	—	0.37–0.39	$\dfrac{6(1-\sigma)}{b}$	$\dfrac{2b\sigma}{3(1-\sigma)}$
	$\dfrac{1}{(b+\delta)^2}$	$\dfrac{b^2}{(b+\delta)^2}$	$\dfrac{4b}{(b+\delta)^2}$	b
	$\dfrac{2}{\sqrt{3}(b+\delta)^2}$	$\dfrac{b^2}{(b+\delta)^2}$	$\dfrac{4b}{(b+\delta)^2}$	b
	$\dfrac{2}{\sqrt{3}(b+\delta)^2}$	$\dfrac{\pi b^2}{2\sqrt{3}(b+\delta)^2}$	$\dfrac{2\pi b}{\sqrt{3}(b+\delta)^2}$	b
	$\dfrac{1}{(b\alpha^*+\delta)(b+\delta)}$	$\dfrac{b^2\alpha^*}{(b\alpha^*+\delta)(b+\delta)}$	$\dfrac{2(1+\alpha^*)b}{(b\alpha^*+\delta)(b+\delta)}$	$\dfrac{2b\alpha^*}{1+\alpha^*}$
	$\dfrac{4\sqrt{3}}{(2b+2\delta)^2}$	$\dfrac{4b^2}{(2b+3\delta)^2}$	$\dfrac{24b}{(2b+3\delta)^2}$	$\dfrac{2b}{3}$

Source: Data modified from Mondt (1980).

The foregoing geometrical characteristics were used for ceramic regenerator cores of London et al. (1970). For some simple regenerator surfaces, the geometrical characteristics are summarized in Table 8.2 for completeness.

8.5 SHELL-AND-TUBE EXCHANGERS WITH SEGMENTAL BAFFLES

In this section, geometrical characteristics required for the rating and sizing of shell-and-tube exchangers with single segmental baffles are derived.

8.5.1 Tube Count

The total number of tubes in an exchanger is dependent on many geometrical variables: tube diameter, tube pitch and layout, the type of floating head, the number of tube passes, the thickness and position of pass dividers, the omission of tubes due to no-

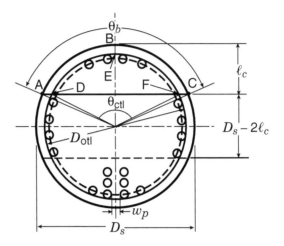

FIGURE 8.9   Nomenclature for basic baffle geometry relations for a single segmental exchanger (from Taborek, 1998).

tubes-in-window design or impingement plates, and the start of the drilling pattern relative to the shell inside diameter and pass dividers. For a fixed tubesheet design, the outermost tubes can be close to the shell inside diameter, or the diameter of the outer tube limit, D_{otl} (see Fig. 8.9), can be the largest followed by that for a split-ring floating head S (see Fig. 10.7), and D_{otl} being the smallest for a pull-through head T (see Fig. 10.5). For a U-tube bundle, some tubes are also lost near the centerline of the U-tube pattern because of the manufacturing limitations on the tube bend radius. Because of many variables involved, it is difficult to determine accurately the total number of tubes in an exchanger except for a direct count. As an alternative, the tube count may be determined approximately using published tabular values, such as those of Bell (1988), among others. For a specified diameter of the circle through the centers of the outermost tubes, D_{ctl}, the effect of the tube bundle type on the total number of tubes N_t is eliminated. Taborek (1998) provides an approximate expression for the tube count as follows in terms of D_{ctl}:

$$
N_t = \begin{cases} \dfrac{(\pi/4)D_{\text{ctl}}^2}{C_t p_t^2}(1 - \psi_c) & \text{single tube pass} \\[3mm] \dfrac{(\pi/4)D_{\text{ctl}}^2}{C_t p_t^2}(1 - \psi_n) & \text{multiple tube passes} \end{cases} \tag{8.109}
$$

where

$$
\psi_c = \begin{cases} 0 & \text{no impingement plate} \\[2mm] \dfrac{\theta_{\text{ctl}}}{2\pi} - \dfrac{\sin\theta_{\text{ctl}}}{2\pi} & \text{impingement plate on one side} \\[3mm] 2\left(\dfrac{\theta_{\text{ctl}}}{2\pi} - \dfrac{\sin\theta_{\text{ctl}}}{2\pi}\right) & \text{tube field removed on both sides} \end{cases} \tag{8.110}
$$

ψ_n is given in Fig. 8.10, $C_t = 0.866$ for 30° and 60° tube layouts and $C_t = 1.00$ for 45° and 90° tube layout. The angle θ_{ctl} in Eq. (8.110) is in radians and is given by Eq. (8.114).

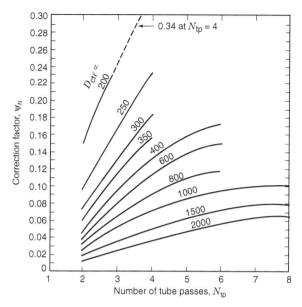

FIGURE 8.10   The correction factor ψ_n for estimation of number of tubes for tube bundles with the number of tube passses $n_p = 2 - 8$. (From Taborek, 1998.)

The accuracy in predicting tube count using Eq. (8.109) is 5% for single-tube-pass exchangers provided that large tubes are not used in relatively small shells. For multiple tube passes, the accuracy is approximately 10% for $D_s < 400\,$mm and 5% for larger shell diameters.

8.5.2   Window and Crossflow Section Geometry

The single segmental E shell exchanger is one of the most common exchangers used in the process, petroleum, and power industries. The geometrical information needed for rating such an exchanger by the Bell–Delaware method (discussed in Section 9.5.1) will be derived here. The original geometry of Bell (1988) is modified based on the suggestions made by Taborek (1998). As shown in Fig. 8.11a, b, and c, the shell side of an E shell exchanger can be divided into three sections: internal crossflow, window, and entrance and exit sections, respectively. We calculate the necessary geometrical characteristics for window and internal crossflow sections next. The effect of larger baffle spacings at the entrance and exit sections will be included by a correction factor when we discuss the thermal design of shell-and-tube heat exchangers in Section 9.5. In addition, in this section we compute various bypass and leakage flow areas needed for the thermal design of shell-and-tube heat exchangers.

8.5.2.1   Window Section. As shown in Fig. 8.9, the gross window area (i.e., without tubes in the window) or the area of a segment ABC corresponding to the window section is

$$A_{\text{fr},w} = \frac{\pi}{4} D_s^2 \left(\frac{\theta_b}{2\pi} - \frac{\sin \theta_b}{2\pi} \right) = \frac{D_s^2}{4} \left[\frac{\theta_b}{2} - \left(1 - \frac{2\ell_c}{D_s} \right) \sin \frac{\theta_b}{2} \right] \qquad (8.111)$$

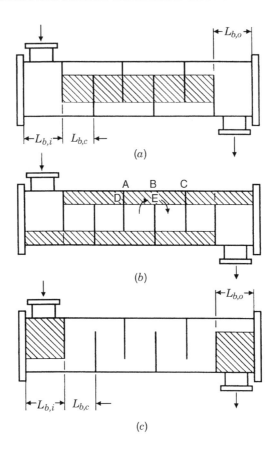

FIGURE 8.11   TEMA E shell exchanger: (*a*) internal cross flow sections, (*b*) window sections; (*c*) entrance and exit sections. (From Taborek, 1998.)

where θ_b is the angle in radians between two radii intersected at the inside shell wall with the baffle cut and is given by

$$\theta_b = 2\cos^{-1}\left(1 - \frac{2\ell_c}{D_s}\right)$$ (8.112)

To calculate the number of tubes in the window zone, we consider the tube field uniform within the shell cross section. This idealization is violated when there are tube pass lanes (in a multipass exchanger) or when tubes are removed due to impingement plates in the nozzle entry area. We will ignore this fact and assume that it causes negligible second-order effects. Then the fraction F_w of the number of tubes in one window section encircled by the centerline of the outer tube row (Fig. 8.9) is

$$F_w = \frac{\text{area of the segment } DEF}{\text{area of the circle with } D_{\text{ctl}}} = \frac{\theta_{\text{ctl}}}{2\pi} - \frac{\sin\theta_{\text{ctl}}}{2\pi}$$ (8.113)

where θ_{ctl} is the angle in radians between the baffle cut and two radii of a circle through the centers of the outermost tubes (see Fig. 8.9) as follows:

$$\theta_{ctl} = 2\cos^{-1}\left(\frac{D_s - 2\ell_c}{D_{ctl}}\right) \tag{8.114}$$

where $D_{ctl} = D_{otl} - d_o$. Consequently, the number of tubes in the window section is obtained from

$$N_{t,w} = F_w N_t \tag{8.115}$$

and the area occupied by tubes in the window section is

$$A_{fr,t} = \frac{\pi}{4} d_o^2 N_{t,w} = \frac{\pi}{4} d_o^2 F_w N_t \tag{8.116}$$

The net flow area in one window section is then

$$A_{o,w} = A_{fr,w} - A_{fr,t} \tag{8.117}$$

where the right-hand-side terms are evaluated from Eqs. (8.111) and (8.116).

By application of the conventional definition, the hydraulic diameter of the window section of a segmental baffle is

$$D_{h,w} = \frac{4A_{o,w}}{\mathbf{P}} = \frac{4A_{o,w}}{\pi d_o N_{t,w} + \pi D_s(\theta_b/2\pi)} \tag{8.118}$$

where θ_b is given by Eq. (8.112) and $\mathbf{P}$ is the wetted perimeter (of all tubes and the shell within the window region); the wetted perimeter of the baffle edge is usually neglected. This $D_{h,w}$ is used for shell-side pressure drop calculations in laminar flow ($\mathrm{Re} < 100$).

The final geometrical input required for the window section is the effective number of tube rows in crossflow needed for the heat transfer and pressure drop correlations. The fluid in the window section effectively makes a 180° turn while flowing from one internal crossflow section to another. In the window region, the fluid has both crossflow and longitudinal flow components of varying magnitudes as a function of the position. Based on the visual and experimental evidence (Bell, 1963), the effective distance of penetration for crossflow in the tube field in the baffle window is about $0.4\ell_{c,\mathrm{eff}}$ in the region AB in Fig. 8.11b while flowing away from the internal crossflow section, and about $0.4\ell_{c,\mathrm{eff}}$ in the region BC in Fig. 8.11b while flowing toward the internal crossflow section. Here $\ell_{c,\mathrm{eff}}$ is the distance between the baffle cut and D_{ctl} (Fig. 8.9). Hence, the number of effective tube rows in crossflow in the window section is

$$N_{r,cw} = \frac{0.8\ell_{c,\mathrm{eff}}}{X_\ell} = \frac{0.8}{X_\ell}[\ell_c - \tfrac{1}{2}(D_s - D_{ctl})] \tag{8.119}$$

8.5.2.2 Crossflow Section. The fraction F_c of the total number of tubes in the crossflow section is found from

$$F_c = 1 - 2F_w = 1 - \frac{\theta_{ctl}}{\pi} + \frac{\sin\theta_{ctl}}{\pi} \tag{8.120}$$

where the expression for F_w is obtained from Eq. (8.113). The number of tube rows $N_{r,cc}$ crossed during flow through one crossflow section between baffle tips may be obtained from a drawing or direct count or may be estimated from

$$N_{r,cc} = \frac{D_s - 2\ell_c}{X_\ell} \tag{8.121}$$

where X_ℓ is the longitudinal tube pitch summarized in Table 8.1 for various tube layouts.

The crossflow area at or near the shell centerline for one crossflow section may be estimated from

$$A_{o,\mathrm{cr}} = \left[D_s - D_{\mathrm{otl}} + \frac{D_{\mathrm{ctl}}}{X_t}(X_t - d_o) \right] L_{b,c} \qquad (8.122)$$

for 30° and 90° tube layout bundles. Here D_{ctl}/X_t denotes the number of $(X_t - d_o)L_{b,c}$ free-flow area in the given tube row. This equation is also valid for a 45° tube bundle having $p_t/d_o \geq 1.707$ and for a 60° tube bundle having $p_t/d_o \geq 3.732$. For 45° and 60° tube bundles having p_t/d_o lower than those indicated in the preceding line, the minimum free-flow area occurs in the diagonal spaces, and hence the term $(X_t - d_o)$ in Eq. (8.122) should be replaced by $2(p_t - d_o)$, or

$$A_{o,\mathrm{cr}} = \left[D_s - D_{\mathrm{otl}} + 2\frac{D_{\mathrm{ctl}}}{X_t}(p_t - d_o) \right] L_{b,c} \qquad (8.123)$$

for 45° and 60° tube bundles. If the tubes have circular fins, the area blocked by the fins should be taken into account as in Eq. (8.36). Hence, Eq. (8.122) modifies to

$$A_{o,\mathrm{cr}} = \left\{ D_s - D_{\mathrm{otl}} + \frac{D_{\mathrm{ctl}}}{X_t}[(X_t - d_o) - (d_e - d_o)\delta N_f] \right\} L_{b,c} \qquad (8.124)$$

which is valid for 30° and 90° tube bundles, 45° tube bundles having $p_t/d_o \geq 1.707$, and 60° tube bundles having $p_t/d_o \geq 3.732$. For circular finned tube bundles having 45° tube layout and $p_t/d_o \leq 1.707$ or 60° tube layout and $p_t/d_o \leq 3.732$, Eq. (8.124) modifies to

$$A_{o,\mathrm{cr}} = \left\{ D_s - D_{\mathrm{otl}} + 2\frac{D_{\mathrm{ctl}}}{X_t}[p_t - d_o) - (d_e - d_o)\delta N_f] \right\} L_{b,c} \qquad (8.125)$$

The number of baffles N_b is required to compute the total number of crossflow and window sections. It should be determined from the drawings or a direct count. Otherwise, compute it from the geometry of Fig. 8.11c as

$$N_b = \frac{L - L_{b,i} - L_{b,o}}{L_{b,c}} + 1 \qquad (8.126)$$

where $L_{b,c}$ is the central baffle spacing, and $L_{b,i}$ and $L_{b,o}$ are the baffle spacings in the inlet and outlet regions.

8.5.3 Bypass and Leakage Flow Areas

Flow area available for bypass streams C and F (see Fig. 4.19) associated with one crossflow section, normalized with respect to the crossflow open area at or near the shell centerline, is

$$F_{\mathrm{bp}} = \frac{A_{o,\mathrm{bp}}}{A_{o,\mathrm{cr}}} = \frac{(D_s - D_{\mathrm{otl}} + 0.5N_p w_p)L_{b,c}}{A_{o,\mathrm{cr}}} \qquad (8.127)$$

where N_p is the number of pass divider lanes through the tube field that are parallel to the crossflow stream B, w_p is the width of the pass divider lane (Fig. 8.9), and $A_{o,cr}$ is given by Eqs. (8.122)–(8.125). Since the tube field is on both sides of the pass divider bypass lane, the F stream is more effective in terms of heat transfer than is the C stream. Hence, the effective bypass lane width is considered as $0.5 w_p$, as indicated in Eq. (8.127).

Now let us determine the tube-to-baffle leakage area $A_{o,tb}$ for one baffle. The total number of tubes associated with one baffle is

$$N_{t,b} = N_t(1 - F_w) = N_t\left(\frac{1 + F_c}{2}\right) \tag{8.128}$$

where the value of F_w was substituted from the first equality of Eq. (8.120). If the diametral clearance (the difference between the baffle hole diameter d_1 and the tube outside diameter d_o) is δ_{tb} ($= d_1 - d_o$), the total tube-to-baffle leakage area for one baffle is

$$A_{o,tb} = \frac{\pi}{4}[(d_o + \delta_{tb})^2 - d_o^2]N_t(1 - F_w) \approx \frac{\pi d_o \delta_{tb} N_t(1 - F_w)}{2} \tag{8.129}$$

Finally, the shell-to-baffle leakage area for one baffle is associated with the gap between the shell inside diameter and the baffle. Note that this gap exists only within the sector ABC in Fig. 8.12. The shell-to-baffle leakage area

$$A_{o,sb} = \pi D_s \frac{\delta_{sb}}{2}\left(1 - \frac{\theta_b}{2\pi}\right) \tag{8.130}$$

where $\delta_{sb} = D_s - D_{baffle}$ and θ_b in radians is given by Eq. (8.112).

Example 8.3 Determine the shell-side geometrical characteristics (as outlined in Sections 8.5.2 and 8.5.3) of a 1–2 TEMA E shell-and-tube heat exchanger with a fixed tubesheet design and a 45° tube bundle with the following measured geometrical variables:

Shell-side inside diameter $D_s = 0.336\,\text{m}$ Number of sealing strip pairs $N_{ss} = 1$

Tube-side outside diameter $d_o = 19.0\,\text{mm}$ Total number of tubes $N_t = 102$

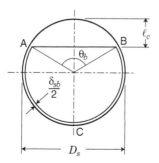

FIGURE 8.12   Single segmental baffle geometry showing shell-to-baffle diametral clearance δ_{sb}.

Tube-side inside diameter $d_i = 16.6\,\text{mm}$

Tube length $L = 4.3\,\text{m}$

Tube pitch $p_t = 25.0\,\text{mm}$

Tube bundle layout $= 45°$

Central baffle spacing $L_{b,c} = 0.279\,\text{m}$

Inlet baffle spacing $L_{b,i} = 0.318\,\text{m}$

Outlet baffle spacing $L_{b,o} = 0.318\,\text{m}$

Baffle cut $\ell_c = 86.7\,\text{mm}$

Transverse tube pitch $X_t = 35.4\,\text{mm}$

Longitudinal tube pitch $X_\ell = 17.7\,\text{mm}$

Width of bypass lane $w_p = 19.0\,\text{mm}$

Number of tube passes $n_p = 2$

Number of pass divider lanes $N_p = 2$

Diameter of the outer tube limit $D_{\text{otl}} = 0.321\,\text{m}$

Tube-to-baffle hole diametral clearance $\delta_{tb} = 0.794\,\text{mm}$

Shell-to-baffle diametral clearance $\delta_{sb} = 2.946$

SOLUTION

Problem Data and Schematic: All pertinent geometrical variables for the shellside are provided above. The detailed geometry is shown in Figs. 8.9, 8.11 and 8.12.

Determine: The shell-side geometrical characteristics: baffle cut angle, fraction of total number of tubes in the window section, the area for flow through the window section, number of effective crossflow rows in each window, fraction of total number of tubes in crossflow, the number of tube rows in one crossflow section, crossflow area at or near centerline, number of baffles, fraction of crossflow area available for flow bypass, tube-to-baffle leakage area for one baffle, and shell-to-baffle leakage area for one baffle.

Assumptions: The shell-and-tube heat exchanger is assumed to have the ideal geometrical characteristics summarized in Section 8.5.

Analysis: As outlined in the text, we compute geometrical characteristics for the window section, crossflow section, and bypass and leakage flow areas.

Window Section. Let us start the calculations with computing the angle θ_b from Eq. (8.112):

$$\theta_b = 2\cos^{-1}\left(1 - \frac{2\ell_c}{D_s}\right) = 2\cos^{-1}\left(1 - \frac{2 \times 86.7 \times 10^{-3}\,\text{m}}{0.336\,\text{m}}\right) = 2.131\,\text{rad} = 122°$$

Then the gross window area $A_{\text{fr},w}$ from Eq. (8.111) is

$$A_{\text{fr},w} = \frac{D_s^2}{4}\left[\frac{\theta_b}{2} - \left(1 - \frac{2\ell_c}{D_s}\right)\sin\frac{\theta_b}{2}\right]$$

$$= \frac{(0.336\,\text{m})^2}{4}\left[\frac{2.131}{2} - \left(1 - \frac{2 \times 0.0867\,\text{m}}{0.336\,\text{m}}\right)\sin\frac{122°}{2}\right] = 0.01813\,\text{m}^2$$

In order to calculate the fraction F_w of total tubes in the window section, first compute the baffle cut angle, using Eq. (8.114), as

$$\theta_{\rm ctl} = 2\cos^{-1}\left(\frac{D_s - 2\ell_c}{D_{\rm ctl}}\right) = 2\cos^{-1}\frac{0.336\,{\rm m} - 2\times 86.7\times 10^{-3}\,{\rm m}}{0.302\,{\rm m}} = 2.004\,{\rm rad} = 115^\circ$$

where $D_{\rm ctl} = D_{\rm otl} - d_o = 0.321\,{\rm m} - 19.0\times 10^{-3}\,{\rm m} = 0.302\,{\rm m}$. Now the fraction F_w of total tubes in the window section is given by Eq. (8.113) as

$$F_w = \frac{\theta_{\rm ctl}}{2\pi} - \frac{\sin\theta_{\rm ctl}}{2\pi} = \frac{2.004}{2\times\pi} - \frac{\sin(115^\circ)}{2\times\pi} = 0.1747$$

Consequently, the number of tubes in the window section, from Eq. (8.115), is

$$N_{t,w} = F_w N_t = 0.1747\times 102 = 17.8$$

The area occupied by tubes in the window section, Eq. (8.116), is

$$A_{fr,t} = \frac{\pi}{4}d_o^2 F_w N_t = \frac{\pi}{4}\times(0.0190\,{\rm m})^2\times 0.1747\times 102 = 0.00505\,{\rm m}^2$$

The net flow area in one window section is then, from Eq. (8.117),

$$A_{o,w} = A_{fr,w} - A_{fr,t} = 0.01813\,{\rm m}^2 - 0.00505\,{\rm m}^2 = 0.01308\,{\rm m}^2$$

The hydraulic diameter for the window section is given by Eq. (8.118) as

$$D_{h,w} = \frac{4A_{o,w}}{\pi d_o N_{t,w} + \pi D_s(\theta_b/2\pi)}$$

$$= \frac{4\times 0.01308\,{\rm m}^2}{\pi\times 0.0190\,{\rm m}\times 17.8 + \pi\times 0.336\,{\rm m}\times(2.131/2\pi)} = 0.03683\,{\rm m}$$

Finally, the number of effective tube rows in crossflow in each window is computed using Eq. (8.119) as

$$N_{r,cw} = \frac{0.8}{X_\ell}\left[\ell_c - \frac{1}{2}(D_s - D_{\rm ctl})\right]$$

$$= \frac{0.8}{17.7\times 10^{-3}\,{\rm m}}\left[86.7\times 10^{-3}\,{\rm m} - (1/2)(0.336\,{\rm m} - 0.302\,{\rm m})\right] = 3.15 \approx 3$$

Crossflow Section. The fraction F_c of the total number of tubes in the crossflow section is calculated from Eq. (8.120) as

$$F_c = 1 - 2F_w = 1 - 2\times 0.1747 = 0.6506$$

Next calculate the number of tube rows $N_{r,cc}$ crossed during flow through one crossflow section between the baffle tips [Eq. (8.121)] as

$$N_{r,cc} = \frac{D_s - 2\ell_c}{X_\ell} = \frac{0.336\,\text{m} - 2 \times 86.7 \times 10^{-3}\,\text{m}}{17.7 \times 10^{-3}\,\text{m}} = 9.19 \approx 9$$

The crossflow area for the 45° tube layout bundle with plain tubes at or near the shell centerline for one crossflow section can be calculated, using Eq. (8.123), as

$$A_{o,cr} = L_{b,c} \left[D_s - D_{\text{otl}} + 2\,\frac{D_{\text{ctl}}}{X_t} (p_t - d_o) \right]$$

$$= 0.279\,\text{m} \times \left[0.336\,\text{m} - 0.321\,\text{m} + 2 \times \frac{0.302\,\text{m}}{0.0354\,\text{m}} \times (0.0250\,\text{m} - 0.0190\,\text{m}) \right]$$

$$= 0.03275\,\text{m}^2$$

Now, compute the number of baffles from Eq. (8.126) as

$$N_b = \frac{L - L_{b,i} - L_{b,o}}{L_{b,c}} + 1 = \frac{4.3\,\text{m} - 0.318\,\text{m} - 0.318\,\text{m}}{0.279\,\text{m}} + 1 = 14.13 \approx 14$$

Bypass and Leakage Flow Areas. To calculate the fraction of crossflow area available for flow bypass, F_{bp} [Eq. (8.127)], we first have to calculate the magnitude of crossflow area for flow bypass:

$$A_{o,bp} = L_{b,c}(D_s - D_{\text{otl}} + 0.5 N_p w_p) = 0.279\,\text{m} \times [0.336\,\text{m} - 0.321\,\text{m} + 0.5 \times 2 \times 0.0190\,\text{m}]$$

$$= 0.00949\,\text{m}^2$$

Consequently,

$$F_{bp} = \frac{A_{o,bp}}{A_{o,cr}} = \frac{0.00949\,\text{m}^2}{0.03275\,\text{m}^2} = 0.2898$$

Tube-to-baffle leakage area is now given by Eq. (8.129) as follows:

$$A_{o,tb} = \frac{\pi d_o \delta_{tb} N_t (1 - F_w)}{2} = \frac{\pi \times 0.0190\,\text{m} \times 0.000794\,\text{m} \times 102 \times (1 - 0.1747)}{2}$$

$$= 0.001995\,\text{m}^2$$

Finally, the shell-to-baffle leakage area for one baffle [Eq. (8.130)] is

$$A_{o,sb} = \pi D_s \frac{\delta_{sb}}{2} \left(1 - \frac{\theta_b}{2\pi} \right) = \pi \times 0.336\,\text{m} \times \frac{0.002946\,\text{m}}{2} \left(1 - \frac{2.131}{2 \times \pi} \right) = 0.001027\,\text{m}^2$$

This concludes all geometrical characteristics needed for the thermal design/rating of a shell-and-tube heat exchanger using the Bell–Delaware method.

Discussion and Comments: The calculation procedure for computing geometrical characteristics of a shell-and-tube heat exchanger is tedious but straightforward. The same geometry has been utilized in the analysis of Example 9.4.

8.6 GASKETED PLATE HEAT EXCHANGERS

A large number of plate corrugation patterns have been developed worldwide. The chevron plate geometry (see Fig. 1.18b and c) is the most common in use today. We outline below the geometrical characteristics of the chevron plate gasketed PHE as an example by evaluating the actual heat transfer surface area due to corrugations. However, one of the common practices in industry is to ignore the effect of corrugations altogether and treat the chevron plates as if they were plain (uncorrugated) plates.

A plate with nomenclature used in the following geometrical derivations is shown in Fig. 7.28. The chevron corrugations increase surface area over the plain (uncorrugated) plate of the same outer (overall) dimensions. The ratio of the developed (actual) surface area of a chevron plate to the projected (for a plain or flat plate) is given by

$$\Phi = 1 + \frac{2}{\pi}\left(1 + \pi^2\alpha^{*2}\right)^{1/2}E(\alpha^*) \approx \frac{1}{6}\left(1 + \sqrt{1 + X^2} + 4\sqrt{1 + X^2/2}\right) \qquad (8.131)$$

where $\alpha^* = 2\mathbf{a}/\Lambda$, $E(\alpha^*)$ is the elliptical integral given in Table 7.4 with the formula for the elliptical duct, [i.e., $E(\mathrm{m})$], and $X = 2\pi\mathbf{a}/\Lambda$. While the expression with the first equality is exact, Martin (1996) used the last approximate formula in Eq. (8.131) using a three-point integration method. Thus the heat transfer surface area on one fluid side of a PHE is given by

$$A = 2\Phi(WL_h + 2\mathbf{a}L_h)N_p \approx 2\Phi WL_hN_p \qquad \text{since} \quad \mathbf{a} \ll W \qquad (8.132)$$

where W (width of the plate between gaskets) and L_h (length of the plate for heat transfer) are defined in Fig. 7.28 and N_p is the number of channels (passages) on the fluid side considered.

The free-flow area on one fluid side of a PHE is given by

$$A_o = 2\mathbf{a}WN_p \qquad (8.133)$$

Using the definition, the hydraulic diameter is given by

$$D_h = \frac{4A_oL_h}{A} = \frac{8\mathbf{a}WN_pL_h}{2\Phi WL_hN_p} = \frac{4\mathbf{a}}{\Phi} \qquad (8.134)$$

Another commonly used set of definitions for the heat transfer surface area, free-flow area, and characteristic dimension is based on the projected surface area (i.e., as if there were no corrugations). In that case,

$$A = 2WL_hN_p \qquad A_o = 2\mathbf{a}WN_p \qquad D_e = 4\mathbf{a} \qquad (8.135)$$

In this set of definitions, the characteristic dimension is referred to as an equivalent diameter, denoted by D_e. Thus,

$$D_e = \Phi D_h \qquad (8.136)$$

The typical range of Φ is from 1.15 to 1.25 with $\Phi \approx 1.22$ for $\Lambda/\mathbf{a} = 2$. This Φ can approach to 2 when $\Lambda/\mathbf{a}$ is as small as 2.46.

The total number of plates N_t in a PHE is related to the number of passes n_p and the number of channels per pass as $N_{c,p}$ as follows.

$$N_t = (n_p N_{c,p})_1 + (n_p N_{c,p})_2 + 1 \tag{8.137}$$

where the subscripts 1 and 2 refer to fluids 1 and 2. The number of thermal plates in this PHE is then $N_t - 2$. The overall height of the corrugation, $2\mathbf{a} + \delta_p$ (where δ_p is the plate thickness) in Fig. 7.28, represents the thickness of a fully compressed gasket since the plate corrugations are in metallic contact. It can be determined as the compressed plate pack length L_{pack} (see Section 1.5.2.1 for the definition) divided by the total number of plates N_t.

The plate length for heat transfer, L_h, and that for pressure drop, L_p, are related as follows where D_p is the port diameter.

$$L_p = L_h + D_p \tag{8.138}$$

SUMMARY

Heat transfer and pressure drop correlations are highly dependent upon the geometrical characteristics of a heat transfer surface. In this chapter, the geometrical characteristics are derived for determination of heat transfer and pressure drop correlations and exchanger performance (q and Δp) of the following exchangers: tubular, tube-fin, plate-fin, regenerative and plate heat exchangers. Also pertinent geometries are derived for computing the effects of leakage and bypass flows in segmental baffle shell-and-tube exchangers. It should be emphasized that if any heat transfer or friction factor correlation from the literature is used for design and analysis of any exchanger, the geometrical characteristics *must be* evaluated in the same manner as in the original source of the correlations.

REFERENCES

Bell, K. J., 1963, Final report of the cooperative research program on shell-and-tube heat exchangers, *Univ. Del. Eng. Exp. St. Bull.*, Vol. 5.

Bell, K. J., 1988, Delaware method for shell-side design, in *Heat Transfer Equipment Design*, R. K. Shah, E. C. Subbarao, and R. A. Mashelkar, eds., Hemisphere Publishing, Washington, DC, pp. 227–254.

Chang, Y. J., and C. C. Wang, 1997, A generalized heat transfer correlation for louver fin geometry, *Int. J. Heat Mass Transfer*, Vol. 40, pp. 533–544.

Kays, W. M., and A. L. London, 1998, *Compact Heat Exchangers*, reprint 3rd ed., Krieger Publishing, Malabar, FL.

London, A. L., M. B. O. Young, and J. H. Stang, 1970, Glass ceramic surfaces, straight triangular passages: Heat transfer and flow friction characteristics, *ASME J. Eng. Power*, Vol. 92, Ser. A, pp. 381–389.

Martin, H., 1996, A theoretical approach to predict the performance of chevron-type plate heat exchangers, *Chem. Eng. Process.*, Vol. 35, pp. 301–310.

Mondt, J. R., 1980, Regenerative heat exchangers: the elements of their design, selection and use, Rep. No. GMR-3396, Research Laboratories, GM Technical Center, Warren, MI,

Sekulić, D. P., A. J. Salazar, F. Gao, J. S. Rosen, and H. S. Hatchins, 2003, Local transient behavior of a compact heat exchanger core during brazing—equivalent zonal (EZ) approach, *Int. J. Heat Exchangers,* Vol. 4, No. 1, in print.

Shah, R. K., 1985, Compact heat exchangers, in *Handbook of Heat Transfer Applications,* 2nd ed. W. M. Rohsenow, J. P. Hartnett, and E. N. Ganić, eds., McGraw-Hill, New York., pp. 4-174 to 4-311.

Taborek, J., 1998, Shell-and-tube heat exchangers: single phase flow, in *Handbook of Heat Exchanger Design,* G. F. Hewitt, ed., Begell House, New York, pp. 3.3.3-1 to 3.3.11-5.

Young, M. B. O., 1969, Glass-ceramic, triangular and hexagonal pasage surfaces—heat transfer and flow friction characteristics, TR HE-2, Department of Mechanical Engineering, Stanford University, Stanford, California.

Zhao, H., and D. P. Sekulić, 2001, Brazed fin-tube joint thermal integrity vs. joint formation, *Proc. 2001 National Science Foundation Design, Service and Manufacturing Grantees and Research Conference,* Tampa, FL.; CD edition, University of Washington, Seattle, WA, DMII-998319.

Žukauskas, A. A., 1987, Convective heat transfer in cross flow, in *Handbook of Single-Phase Convective Heat Transfer,* S. Kakaç, R. K. Shah, and W. Aung, eds., Wiley, New York, Chap. 6.

REVIEW QUESTIONS

Where multiple choices are given, circle one or more correct answers. Explain your answers briefly.

8.1   In a given tube bank, the actual total length of the tube used is 3 m, the tubesheets are 6 mm thick, and the blanket insulation is 150 mm (Fig. RQ8.1). The effective length of a tube for heat transfer is:

(a) 3 m   (b) 2.988 m   (c) 2.838 m   (d) 2.688   (e) can't tell

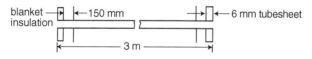

FIGURE RQ8.1

8.2   In Question 8.1, the effective tube length for pressure drop is:

(a) 3 m   (b) 2.988 m   (c) 2.838 m   (d) 2.688   (e) can't tell

8.3   In a tubular heat exchanger with an inline arrangement, the following geometric characteristics are of interest for in-tube heat exchanger calculations:

(a) heat transfer area related to the header plate on the tube side

(b) total heat transfer area

(c) core frontal area

(d) hydraulic diameter

8.4   The following surface area components are needed to evaluate primary surface area in a plate-fin heat exchanger:

(a) total plate area   (b) fin height area   (c) fin base area covering plate
(d) area of side header bars
(e) area of header bars and plates at the core inlet and outlet face

8.5 In an offset strip fin exchanger, the secondary area consists of the:

(a) fin height area   (b) fin edge height area
(c) plate area   (d) header surface area
(e) fin edge width area

8.6 For the same **b** and δ (see sketches in Table 8.2 for definitions), the cell density n_c of a regenerator of square passages compared to hexagonal passages is:

(a) higher   (b) the same
(c) lower   (d) can't tell

8.7 The total number of tubes in a shell-and-tube exchanger is dependent on the:

(a) tube diameter   (b) tube pitch
(c) number of tube passes   (d) tube length
(e) type of floating head   (f) tubesheet thickness

8.8 Which of the following arrangements have more tubes per pass for a specified shell inside diameter, diameter of the outer tube limit, tube outside diameter, and tube pitch?

(a) square   (b) rotated square
(c) triangular   (d) can't tell

8.9 The number of tube rows N_r in the flow direction in a crossflow zone with $D_s = 3$ m, $\ell_c = 0.675$ m, and $X_\ell = 50$ mm is approximately:

(a) 33   (b) 210   (c) 120   (d) 60   (e) none of these

8.10 In a no-tubes-in-window segmental baffle exchanger, the area of the window zone is:

(a) zero
(b) the area of the sector between the baffle tip and shell ID
(c) the area of the crossflow zone
(d) the bundle bypass area

9 Heat Exchanger Design Procedures

As mentioned in Chapter 2 and tabulated in Table 3.11, there are a large number of heat exchanger design problems, defined broadly as rating and sizing problems. In a rating problem, we determine the heat transfer rate and/or outlet temperatures and pressure drop performance of either an existing or an already designed heat exchanger. In a sizing problem, we design a heat exchanger; this involves the determination/selection of exchanger construction type, flow arrangement, tube/plate and fin material, and/or the physical size of an exchanger to meet the specified heat transfer and pressure drops within specified constraints. We discuss the selection of heat exchanger types, flow arrangements, and so on, in Chapter 10 and thermodynamic analysis and operating problems in Chapters 11 through 13. These aspects are equally or sometimes more important for the design of heat exchangers. We concentrate here, then, in a narrow sense on the determination of the physical size in a sizing problem for specific types of heat exchangers. Also, we determine heat transfer and pressure drop performance of a heat exchanger in a rating problem. Hence, in this chapter, stepwise solution procedures are presented separately for the rating (problem 12 in Table 3.11) and sizing (problem 2 or 4 in Table 3.11) of plate-fin, tube-fin, plate, and shell-and-tube heat exchangers. To present the solution procedures for rating and sizing, we use the theory, correlations, and geometrical properties summarized in Chapters 3, 6, 7, and 8.

Before presenting the procedures for rating and sizing problems, we start with how to evaluate the mean temperature of each fluid in a two-fluid heat exchanger. This mean temperature is needed to calculate fluid thermophysical properties for heat transfer and pressure drop calculations on each fluid side. Also note that an important assumption made in the heat exchanger design theory (see Section 3.2.1) is the uniformity of fluid/solid thermophysical properties. Fluid properties should be determined at fluid mean temperatures.

9.1 FLUID MEAN TEMPERATURES

The fluid properties are determined at the mean (flow length average) temperature on each fluid side in a heat exchanger. The single-phase fluid properties needed for heat transfer and pressure drop calculations are density, specific heat, viscosity, thermal conductivity, and Prandtl number. These properties are available for a large number of fluids in the literature; some of them are summarized in Appendix A. Now let us determine the appropriate fluid mean temperatures for various heat exchanger configurations.

For the ε-NTU or MTD method, we need to obtain a "single" temperature value to represent the temperature level of fluid through each fluid side of the exchanger. For counterflow and parallelflow exchangers, the fluid temperature varies in the flow direction as well as over the cross section of each flow passage (for example, as shown in Fig. 7.4). In a crossflow heat exchanger, it also varies in the flow direction of the other fluid. In more complex arrangements, the fluid temperature variation is generally both in the flow direction and across a given flow cross section. The temperature changes in the flow direction affect the fluid bulk properties. A scheme to correct the friction factor and convective heat transfer coefficient due to the temperature changes across the flow passage cross section was discussed in Section 7.6.1. If the fluid properties vary substantially within an exchanger, a flow length "average" temperature may not be adequate to determine the heat transfer rate and pressure drop accurately. In that case, refer to Section 4.2.3.2 for the stepwise calculations where arithmetic average temperatures for each fluid are used in each segment.

In the determination of the heat transfer rate through the exchanger, the true mean temperature difference is either used directly (as in the MTD method, $q = UA\,\Delta T_m = UAF\,\Delta T_{lm}$) or indirectly (as in the ε-NTU or P-NTU method). Note that $\Delta T_m = \Delta T_{lm}$ for counterflow and parallelflow exchangers, or $\Delta T_m = F\,\Delta T_{lm} \approx \Delta T_{lm}$ for a well-designed exchanger of any other flow arrangement. Here F is the log-mean temperature difference correction factor discussed in Section 3.7.2. Therefore, we use ΔT_{lm} to evaluate the log-mean average temperature of the C_{min} fluid as described below, when the temperature rise or drop of the C_{max} fluid is small. The average temperature of the C_{max} fluid in this case is the arithmetic average temperature. When the temperature drop of the hot fluid (ΔT_h) and the temperature rise of the cold fluid (ΔT_c) are of the same order of magnitude, the mean temperature on each fluid side is considered simply as the arithmetic average temperature. An approximate procedure is outlined next for evaluation of this temperature for two-fluid heat exchangers having some specific flow conditions/arrangements. The results are also summarized in Table 9.1.

TABLE 9.1   Approximate Mean Temperatures on the Hot- and Cold-Fluid Sides of a Two-Fluid Exchanger

C_{max} = Hot Fluid, C_{min} = Cold Fluid	C_{max} = Cold Fluid, C_{min} = Hot Fluid
$C^* < 0.5$ Case	
$T_{h,m} = \dfrac{T_{h,i} + T_{h,o}}{2}$	$T_{c,m} = \dfrac{T_{c,i} + T_{c,o}}{2}$
$T_{c,m} = T_{h,m} - \Delta T_{lm}$	$T_{h,m} = T_{c,m} + \Delta T_{lm}$
$\Delta T_{lm} = \dfrac{(T_{h,m} - T_{c,o}) - (T_{h,m} - T_{c,i})}{\ln[(T_{h,m} - T_{c,o})/(T_{h,m} - T_{c,i})]}$	$\Delta T_{lm} = \dfrac{(T_{h,i} - T_{c,m}) - (T_{h,o} - T_{c,m})}{\ln[(T_{h,i} - T_{c,m})/(T_{h,o} - T_{c,m})]}$
$C^* \geq 0.5$ Case	
$T_{h,m} = \dfrac{T_{h,i} + T_{h,o}}{2}$	$T_{c,m} = \dfrac{T_{c,i} + T_{c,o}}{2}$

9.1.1   Heat Exchangers with $C^* \approx 0$

Typical temperature distributions for both fluids are shown in Fig. 9.1. The flow length average temperatures for the case shown in Fig. 9.1a are

$$T_{c,m} = \frac{T_{c,i} + T_{c,o}}{2} \tag{9.1}$$

$$T_{h,m} = T_{c,m} + \Delta T_{\mathrm{lm}} \tag{9.2}$$

where

$$\Delta T_{\mathrm{lm}} = \frac{(T_{h,i} - T_{c,m}) - (T_{h,o} - T_{c,m})}{\ln\left[(T_{h,i} - T_{c,m})/(T_{h,o} - T_{c,m})\right]} \tag{9.3}$$

The flow length average temperatures for the case of Fig. 9.1b are

$$T_{h,m} = \frac{T_{h,i} + T_{h,o}}{2} \tag{9.4}$$

$$T_{c,m} = T_{h,m} - \Delta T_{\mathrm{lm}} \tag{9.5}$$

where

$$\Delta T_{\mathrm{lm}} = \frac{(T_{h,m} - T_{c,o}) - (T_{h,m} - T_{c,i})}{\ln[(T_{h,m} - T_{c,o})/(T_{h,m} - T_{c,i})]} \tag{9.6}$$

It can be shown theoretically that for $C^* = 0$, the foregoing method of computing the flow length average temperature is exact (Kays and London, 1998). In this case, the temperature of the $C_{\max}$ fluid remains truly constant along the flow length, in contrast to the small variations shown in Fig. 9.1.

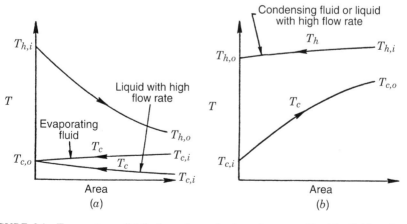

FIGURE 9.1   Temperature distributions in a heat exchanger with (a) $C_h/C_c \approx 0$ and (b) $C_c/C_h \approx 0$. (From Shah, 1981.)

9.1.2   Counterflow and Crossflow Heat Exchangers

Typical temperature distributions for these exchangers are shown in Figs. 1.50 (in the flow directions) and 1.54 (at inlet and outlet cross-sections). From Eqs. (3.192) and (3.199), the true mean temperature difference for these exchangers is

$$\Delta T_m = \begin{cases} \Delta T_{\mathrm{lm}} & \text{for counterflow} \\ F\,\Delta T_{\mathrm{lm}} & \text{for crossflow} \end{cases} \tag{9.7}$$

Since no simple relation is available for determining mean temperatures on both fluid sides for a counterflow exchanger for $C^* > 0$, we use the following approximation. When C^* is closer to zero (arbitrarily we limit $C^* < 0.5$), we recommend to compute the mean temperatures on each fluid side by Eqs. (9.1)–(9.6). For $C^* > 0.5$, again arbitrarily, we calculate the mean temperatures on both fluid sides as arithmetic mean values:

$$T_{h,m} = \tfrac{1}{2}(T_{h,i} + T_{h,o}) \qquad T_{c,m} = \tfrac{1}{2}(T_{c,i} + T_{c,o}) \tag{9.8}$$

For a "good" crossflow exchanger design, F generally varies between 0.80 and 0.99 and the resulting error in the average temperature, by considering $F = 1$, does not generally produce significant variations in the fluid properties, except possibly for oils. Hence, we evaluate the average temperature on each fluid side for a crossflow exchanger by considering it as a counterflow heat exchanger (i.e., using ΔT_{lm} where appropriate).

It should be emphasized that if we assume a value for exchanger effectiveness, the integral mean temperature could be calculated for those exchangers where a closed-form solution for the temperature distribution is available. In that case, the afore mentioned approximate procedure is not needed. However, in the lumped parameter approach (i.e., considering one value of the mean temperature along the complete flow length for a fluid as considered in the design theory of Chapter 3), the foregoing proposed approximate values of mean temperatures should be adequate. If the fluid properties vary significantly along the flow length, the ε-NTU or other methods outlined in Chapter 3 are not adequate. In that case, one should conduct numerical analysis as outlined in Section 4.2.3.2 by dividing the exchanger into sufficiently small elements so that arithmetic average temperatures can represent the mean temperatures reasonably accurately.

9.1.3   Multipass Heat Exchangers

Fluid temperatures are first determined after every pass so that the inlet and outlet fluid temperatures are known for every pass. Then *arithmetic averages* of the terminal temperatures are taken as the flow length average temperature for each fluid in each pass. This would constitute the first iteration. If needed, this procedure can be repeated with modified thermophysical/process properties/parameters.

The effectiveness ε_p or P_p of each pass in terms of overall effectiveness ε or P and C^* or R is given by Eq. (3.136) for an overall counterflow arrangement, and that for an overall parallelflow arrangement by Eq. (3.144). Mean outlet temperatures $T_{h,o}$ and $T_{c,o}$ for *each* pass are then determined from the definition of individual pass effectiveness ε_p; use Eq. (3.44) with ε replaced by ε_p, and the inlet and outlet temperatures as those of individual passes. Note that both inlet temperatures of each pass are known in a parallel-flow exchanger starting with the first pass. For an overall counterflow arrangement, use Eq. (3.44) modified as mentioned above to determine the mean outlet temperatures.

However, an iterative procedure will be required to determine the temperatures between passes. This is because the inlet temperatures of both fluids are not known for any pass in the beginning of calculation for a counterflow exchanger.

For heat exchangers with other flow arrangements that are not considered here, one should use one of the methods outlined above, such as the use of arithmetic or log-mean average temperature for a single-pass exchanger or for each pass of a multipass exchanger. Engineering judgment must be applied for the selection of a particular method for the determination of the flow length average temperature.

Now we outline the rating and sizing procedures for three major types of exchangers: extended surface (plate-fin and tube-fin), plate, and shell-and-tube exchangers. Because the regenerator rating and sizing are highly iterative and very complicated, it can be solved by a computer only. See Shah (1988b) for the detailed stepwise solution procedure for rating and sizing of rotary regenerators when the effects of leakage and bypass flows are neglected, and refer to Shah and Skiepko (1999) when these effects are included.

9.2 PLATE-FIN HEAT EXCHANGERS

In this section we consider rating and sizing of plate-fin exchangers (see Section 1.5.3.1 for a description) with "corrugated" fins on one or both sides. We use the ε-NTU method since it is used most commonly in industry for these types of exchangers.

9.2.1 Rating Problem

In this section we describe a step-by-step method for rating (problem 12 in Table 3.11) of *single-pass* counterflow and crossflow exchangers. A similar step-by-step method for a *two-pass* cross-counterflow exchanger is presented by Shah (1988a). The basic steps involved in the analysis of a rating problem are the determination of the following information: surface geometrical properties, fluid physical properties, Reynolds numbers, surface basic characteristics, corrections to the surface basic characteristics due to temperature-dependent properties, heat transfer coefficients and fin efficiencies, wall thermal resistance and overall thermal conductance, NTU, C^*, exchanger effectiveness, heat transfer rate, outlet temperatures, and pressure drops on each fluid side. These steps are outlined in detail now.

1. Determine the surface geometrical properties on each fluid side (see, e.g., Section 8.3). This includes the minimum free-flow area A_o, heat transfer surface area A (both primary and secondary), flow length L, hydraulic diameter D_h, heat transfer surface area density β, ratio of minimum free-flow area to frontal area σ, fin geometry (l, δ, etc.) for fin efficiency determination, and any specialized dimensions used for heat transfer and pressure drop correlations.

2. Compute the fluid mean temperature and fluid thermophysical properties on each fluid side. Since the outlet temperatures are not known for the rating problem, they are guessed initially. Unless it is known from past experience, assume exchanger effectiveness as 50 to 75% for most single-pass crossflow exchangers, or 75 to 85% for single-pass counterflow and two-pass cross-counterflow exchangers. For the assumed effectiveness, calculate the fluid outlet temperatures as follows:

$$T_{h,o} = T_{h,i} - \varepsilon \frac{C_{\min}}{C_h} (T_{h,i} - T_{c,i}) \tag{9.9}$$

$$T_{c,o} = T_{c,i} + \varepsilon \frac{C_{\min}}{C_c} (T_{h,i} - T_{c,i}) \tag{9.10}$$

Initially, assume that $C_c/C_h \approx \dot{m}_c/\dot{m}_h$ for a gas-to-gas exchanger, or $C_c/C_h \approx \dot{m}_c c_{p,c}/\dot{m}_h c_{p,h}$ for a gas-to liquid exchanger with approximate values of c_p's for the fluids in question. Determine fluid mean temperatures using the appropriate formulas from Table 9.1.

Once the mean temperatures are determined on each fluid side, obtain the fluid properties from thermophysical property books or handbooks (see Appendix A). The fluid properties needed for the rating problem are μ, c_p, k, Pr, and ρ. With this c_p, one more iteration may be carried out to determine $T_{h,o}$ and $T_{c,o}$ from Eq. (9.9) or (9.10) on the $C_{\max}$ side, and subsequently, T_m on the $C_{\max}$ side and refine fluid properties accordingly.

3. Calculate the Reynolds numbers (Re $= GD_h/\mu$) and/or any other pertinent dimensionless groups (using their definitions) needed to determine the nondimensional heat transfer and flow friction characteristics (e.g., j or Nu and f or Eu) of heat transfer surfaces on each fluid side of the exchanger. Subsequently, compute j or Nu and f factors. Correct Nu (or j) for variable fluid property effects in the second and subsequent iterations from the following equations:

$$\frac{\text{Nu}}{\text{Nu}_{\text{cp}}} = \left(\frac{T_w}{T_m}\right)^n \qquad \frac{f}{f_{\text{cp}}} = \left(\frac{T_w}{T_m}\right)^m \qquad \text{for gases} \tag{9.11}$$

$$\frac{\text{Nu}}{\text{Nu}_{\text{cp}}} = \left(\frac{\mu_w}{\mu_m}\right)^n \qquad \frac{f}{f_{\text{cp}}} = \left(\frac{\mu_w}{\mu_m}\right)^m \qquad \text{for liquids} \tag{9.12}$$

where the subscript cp denotes constant properties, and m and n are empirical constants presented in Tables 7.12 and 7.13. Note that T_w and T_m in Eq. (9.11) and Tables 7.12 and 7.13 are *absolute* temperatures and T_w is computed from Eq. (3.33).

4. From Nu or j, compute the heat transfer coefficients for both fluid streams from the following equations.

$$h = \text{Nu}\, \frac{k}{D_h} \quad \text{or} \quad h = jGc_p Pr^{-2/3} \tag{9.13}$$

Subsequently, determine the fin efficiency η_f and the extended surface efficiency η_o:

$$\eta_f = \frac{\tanh m\ell}{m\ell} \qquad m = \left(\frac{h\mathbf{P}}{k_f A_k}\right)^{1/2} \quad \text{or} \quad \left(\frac{2h}{k_f \delta}\right)^{1/2} \tag{9.14}$$

$$\eta_o = 1 - (1 - \eta_f)\frac{A_f}{A} \tag{9.15}$$

Here $\mathbf{P}$, A_k and δ are the wetted perimeter, fin cross-sectional area and fin thickness, respectively. Refer to Table 4.5 for the formulas for some additional fin geometries for a plate-fin exchanger. Also calculate the wall thermal resistance $\mathbf{R}_w = \delta_w / A_w k_w$. Finally, compute the overall thermal conductance UA from

$$\frac{1}{UA} = \mathbf{R}_h + \mathbf{R}_{h,f} + \mathbf{R}_w + \mathbf{R}_{c,f} + \mathbf{R}_c$$

$$= \frac{1}{(\eta_o hA)_h} + \frac{1}{(\eta_o h_f A)_h} + \mathbf{R}_w + \frac{1}{(\eta_o h_f A)_c} + \frac{1}{(\eta_o hA)_c} \qquad (9.16)$$

knowing the individual convective film resistances, wall thermal resistances, and fouling resistances, if any.

5. From the known heat capacity rates on each fluid side, compute $C^* = C_{min}/C_{max}$. From the known UA, determine $\mathrm{NTU} = UA/C_{min}$. Also calculate the longitudinal conduction parameter λ. With the known NTU, C^*, λ, and the flow arrangement, determine the exchanger effectiveness ε from either closed-form equations or tabular/graphical results (see Sections 3.4. and 4.1).

6. With this ε, finally compute the outlet temperatures from Eqs. (9.9) and (9.10). If these outlet temperatures are significantly different from those assumed in step 2, use these outlet temperatures in step 2 and continue iterating steps 2 through 6 until the assumed and computed outlet temperatures converge within the desired degree of accuracy. For a gas-to-gas exchanger, one iteration will probably be sufficient.

7. Finally, compute the heat duty from

$$q = \varepsilon C_{min}(T_{h,i} - T_{c,i}) \qquad (9.17)$$

and fluid outlet temperatures from Eqs. (9.9) and (9.10).

8. For the pressure drop calculations, first determine the fluid densities at the exchanger inlet and outlet (ρ_i and ρ_o) for each fluid. The mean specific volume on each fluid side is then computed from

$$\left(\frac{1}{\rho}\right)_m = \frac{1}{2}\left(\frac{1}{\rho_i} + \frac{1}{\rho_o}\right) \qquad (9.18)$$

Next, the entrance and exit loss coefficients, K_c and K_e, are obtained from Fig. 6.3 for known σ, Re, and the flow passage entrance geometry. The friction factor on each fluid side is corrected for variable fluid properties using Eq. (9.11) or (9.12). Here, the wall temperature T_w is computed from

$$T_{w,h} = T_{m,h} - (\mathbf{R}_h + \mathbf{R}_{h,f})q \qquad T_{w,c} = T_{m,c} + (\mathbf{R}_c + \mathbf{R}_{c,f})q \qquad (9.19)$$

where the various resistance terms are defined by Eq. (9.16). The core pressure drops on each fluid side are then calculated from

$$\Delta p = \frac{G^2}{2g_c\rho_i}\left[(1 - \sigma^2 + K_c) + 2\left(\frac{\rho_i}{\rho_o} - 1\right) + f\frac{L}{r_h}\rho_i\left(\frac{1}{\rho}\right)_m - (1 - \sigma^2 - K_e)\frac{\rho_i}{\rho_o}\right] \qquad (9.20)$$

We now demonstrate the aforementioned procedure by an example of a rating problem for a crossflow exchanger.

Example 9.1   A gas-to-air single-pass crossflow heat exchanger is contemplated for heat recovery from the exhaust gas to preheat incoming air in a solid oxide fuel cell cogeneration system. It has overall dimensions of 0.300 m × 0.300 m × 1.000 m as shown in Fig. E9.1A. Offset strip fins of the same geometry are employed on the gas and air sides; the geometrical properties and surface characteristics are provided in Fig. E9.1B. Both fins and plates (parting sheets) are made from Inconel 625 with $k = 18$ W/m · K. The plate thickness is 0.5 mm. The anode gas flows in the heat exchanger at 3.494 m^3/s and 900°C. The cathode air on the other fluid side flows at 1.358 m^3/s and 200°C. The inlet pressure of the gas is at 160 kPa absolute whereas that of air is at 200 kPa absolute. Determine the heat transfer rate, outlet fluid temperatures and pressure drops on each fluid side. Use the properties of air for the gas.

SOLUTION

Problem Data and Schematic: All necessary geometrical information for the core and the surfaces are provided in Figs. E9.1A and E9.1B, except that $\delta_w = 0.5$ mm.

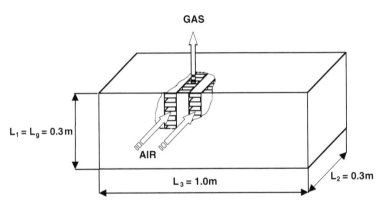

FIGURE E9.1A   Gas-to-air single-pass crossflow heat exchanger for the rating problem. (From Shah, 1981.)

Basic surface geometries on the gas and air sides along with their *j* and *f* data are given in Fig. E9.1B and are summarized below along with the operating conditions. The subscripts *g* and *a* are used for the gas and air sides, respectively.

$b_g = 2.49$ mm $D_{h,g} = 0.00154$ m $\beta_g = 2254$ m^2/m^3 $\delta_g = 0.102$ mm $\left(\dfrac{A_f}{A}\right)_g = 0.785$

$b_a = 2.49$ mm $D_{h,a} = 0.00154$ m $\beta_a = 2254$ m^2/m^3 $\delta_a = 0.102$ mm $\left(\dfrac{A_f}{A}\right)_a = 0.785$

$\dot{V}_g = 3.494$ m^3/s $T_{g,i} = 900$°C $p_{g,i} = 160$ kPa $k_f = k_w = 18$ W/m · K

$\dot{V}_a = 1.358$ m^3/s $T_{a,i} = 200$°C $p_{a,i} = 200$ kPa

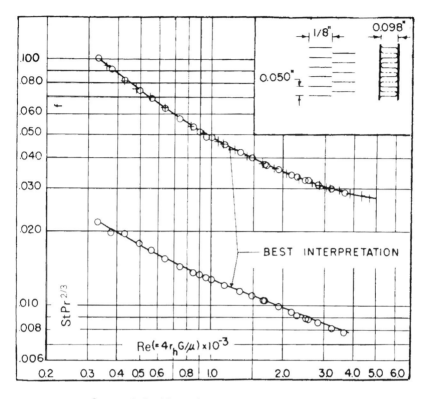

Gas and air-side surface
Fin density = 0.782 mm^{-1}
Plate spacing, b = 2.49 mm
Fin offset length ℓ_s = 3.18 mm
Hydraulic diameter, D_h = 0.00154 m
Fin metal thickness, δ = 0.102 mm
Fin area/total area, A_f/A = 0.785
Total heat transfer area/volume
between plates, β = 2254 m^2/m^3

FIGURE E9.1B   Surface basic characteristics for an offset strip fin surface $1/8$ -19.86. (From Kays and London, 1998.)

Determine: Heat duty and pressure drops (gas and air sides) for this exchanger.

Assumptions: The assumptions listed in Section 3.2.1 as applicable to a plate-fin exchanger are invoked.

Analysis: We follow the steps outlined in the preceding text, starting with the calculation of surface geometrical properties.

Surface Geometrical Properties. We assume that there are N_p passages for the gas and $(N_p + 1)$ passages for the air to minimize heat loss to the ambient.[†] The noflow height (stack height) is given by

$$L_3 = N_p b_g + (N_p + 1)b_a + (2N_p + 2)\delta_w$$

Therefore

$$N_p = \frac{L_3 - b_a + 2\delta_w}{b_g + b_a + 2\delta_w} = \frac{1000\,\text{mm} - 2.49\,\text{mm} - 2 \times 0.5\,\text{mm}}{2.49\,\text{mm} + 2.49\,\text{mm} + 2 \times 0.5\,\text{mm}} = 166.6 = 167$$

Here b is the fin height (plate spacing) and δ_w is the plate thickness. The frontal areas on the gas and air sides are

$$A_{\text{fr},g} = L_2 L_3 = 0.3\,\text{m} \times 1.0\,\text{m} = 0.3\,\text{m}^2$$
$$A_{\text{fr},a} = L_1 L_3 = 0.3\,\text{m} \times 1.0\,\text{m} = 0.3\,\text{m}^2$$

The heat exchanger volume between plates, on each fluid side, is

$$V_{p,g} = L_1 L_2 (b_g N_p) = 0.3\,\text{m} \times 0.3\,\text{m} \times (2.49 \times 10^{-3}\,\text{m}) \times 167 = 0.03742\,\text{m}^3$$
$$V_{p,a} = L_1 L_2 b_a (N_p + 1) = 0.3\,\text{m} \times 0.3\,\text{m} \times (2.49 \times 10^{-3}\,\text{m}) \times 168 = 0.03765\,\text{m}^3$$

The heat transfer areas A_g and A_a are

$$A_g = \beta_g V_{p,g} = 2254\,\text{m}^2/\text{m}^3 \times 0.03742\,\text{m}^3 = 84.345\,\text{m}^2$$

$$A_a = \beta_a V_{p,a} = 2254\,\text{m}^2/\text{m}^3 \times 0.03765\,\text{m}^3 = 84.863\,\text{m}^2$$

The minimum free-flow area is then calculated from the definition of the hydraulic diameter, $D_h = 4A_o L/A$:

$$A_{o,g} = \frac{(D_h A)_g}{4L_g} = \frac{0.00154\,\text{m} \times 84.345\,\text{m}^2}{4 \times 0.300\,\text{m}} = 0.1082\,\text{m}^2$$

$$A_{o,a} = \frac{(D_h A)_a}{4L_a} = \frac{0.00154\,\text{m} \times 84.863\,\text{m}^2}{4 \times 0.300\,\text{m}} = 0.1089\,\text{m}^2$$

Finally, the ratio of minimum free-flow area to frontal area is

$$\sigma_g = \frac{A_{o,g}}{A_{\text{fr},g}} = \frac{0.1082\,\text{m}^2}{0.3\,\text{m}^2} = 0.361 \qquad \sigma_a = \frac{A_{o,a}}{A_{\text{fr},a}} = \frac{0.1089\,\text{m}^2}{0.3\,\text{m}^2} = 0.363$$

Mean Temperatures and Fluid Properties. To determine the mean temperatures on each fluid side, we need to calculate C^*. Since the flow rates are specified as volumetric at inlet temperatures, let us first calculate the gas and air densities and then the mass flow rates.

[†] This is a common practice, whenever practical, to minimize the heat losses to ambient.

$$\rho_{g,i} = \frac{p_{g,i}}{\tilde{R}T_{g,i}} = \frac{160 \times 10^3 \, \text{Pa}}{287.04 \, \text{J/kg} \cdot \text{K} \times (273.15 + 900.0) \, \text{K}} = 0.4751 \, \text{kg/m}^3$$

$$\rho_{a,i} = \frac{p_{a,i}}{\tilde{R}T_{a,i}} = \frac{200 \times 10^3 \, \text{Pa}}{287.04 \, \text{J/kg} \cdot \text{K} \times (273.15 + 200.0) \, \text{K}} = 1.4726 \, \text{kg/m}^3$$

where $\tilde{R} = 287.04 \, \text{J/kg} \cdot \text{K}$ is the gas constant for air. Note that all temperatures are in kelvin. Hence the mass flow rates are

$$\dot{m}_g = \dot{V}_g \rho_g = 3.494 \, \text{m}^3/\text{s} \times 0.4751 \, \text{kg/m}^3 = 1.66 \, \text{kg/s}$$

$$\dot{m}_a = \dot{V}_a \rho_a = 1.358 \, \text{m}^3/\text{s} \times 1.4726 \, \text{kg/m}^3 = 2.00 \, \text{kg/s}$$

Hence, gas will be the $C_{\min}$ side since the change in the specific heat is not a strong function of temperature for air/gas. Now assume that $\varepsilon = 0.75$ for the crossflow exchanger. Then using the definition of the exchanger effectiveness [Eqs. (9.9) and (9.10)], we have

$$T_{g,o} = T_{g,i} - \varepsilon(T_{g,i} - T_{a,i}) = 900^{\circ}\text{C} - 0.75(900 - 200)^{\circ}\text{C} = 375.0^{\circ}\text{C}$$

$$T_{a,o} \approx T_{a,i} + \varepsilon \left(\frac{\dot{m}_g}{\dot{m}_a}\right)(T_{g,i} - T_{a,i}) = 200^{\circ}\text{C} + 0.75 \times \left(\frac{1.66 \, \text{kg/m}^3}{2.0 \, \text{kg/m}^3}\right)(900 - 200)^{\circ}\text{C}$$

$$= 635.8^{\circ}\text{C}$$

Note that we used $c_{p,a} \approx c_{p,g}$ as a first approximation for determining $T_{a,o}$. Since $C^* \approx \dot{m}_g/\dot{m}_a = 0.83$, we will use the arithmetic average temperature from [Eq. (9.8)] as the appropriate mean temperature on each fluid side.

$$T_{g,m} = \frac{(900.0 + 375.0)^{\circ}\text{C}}{2} = 637.5^{\circ}\text{C} = 910.65 \, \text{K}$$

$$T_{a,m} = \frac{(200.0 + 635.8)^{\circ}\text{C}}{2} = 417.9^{\circ}\text{C} = 691.05 \, \text{K}$$

In the absence of information on the composition of the gas, we treat both the gas and air as dry air. The properties of air are obtained from any source of thermophysical properties (see, e.g., Appendix A) as

	μ (Pa $\cdot$ s)	c_p (kJ/g $\cdot$ K)	Pr	Pr$^{2/3}$
Gas at 637.5°C	40.1×10^{-6}	1.122	0.731	0.811
Air at 417.9°C	33.6×10^{-6}	1.073	0.694	0.784

Mass Velocities, Reynolds Numbers, and j and f Factors

$$G_g = \frac{\dot{m}_g}{A_{o,g}} = \frac{1.66\,\text{kg/s}}{0.1082\,\text{m}^2} \qquad \text{Re}_g = \left(\frac{GD_h}{\mu}\right)_g = \frac{15.342\,\text{kg/m}^2 \cdot \text{s} \times 0.00154\,\text{m}}{0.0000401\,\text{Pa} \cdot \text{s}} = 589$$

$$= 15.342\,\text{kg/m}^2 \cdot \text{s}$$

$$G_a = \frac{\dot{m}_a}{A_{o,a}} = \frac{2.0\,\text{kg/s}}{0.1089\,\text{m}^2} \qquad \text{Re}_a = \left(\frac{GD_h}{\mu}\right)_a = \frac{18.365\,\text{kg/m}^2 \cdot \text{s} \times 0.00154\,\text{m}}{0.0000336\,\text{Pa} \cdot \text{s}} = 842$$

$$= 18.365\,\text{kg/m}^2 \cdot \text{s}$$

We get the j and f values from the curve fit of tabular values given in Kays and London (1998) as follows; the other sources are the graphical values from Fig. E9.1B or generalized correlations from the literature such as those given in Section 7.5.3.1.

	Re	j	f
Gas	589	0.0170	0.0669
Air	842	0.0134	0.0534

Since Reynolds numbers indicate the flow as laminar on both gas and air sides, the correction to the j factor is unity because $n = 0$ from Table 7.12. However, the correction to the f factor will not be unity since $m \neq 0$ from Table 7.12. We will determine this correction after calculating the wall temperature T_w.

Heat Transfer Coefficients and Fin Efficiency. We compute the heat transfer coefficient from the definition of the j factors as follows:

$$h_g = \left(\frac{jGc_p}{\text{Pr}^{2/3}}\right)_g = \frac{0.0170 \times 15.342\,\text{kg/m}^2 \cdot \text{s} \times (1.122 \times 10^3)\,\text{J/kg} \cdot \text{K}}{0.811} = 360.83\,\text{W/m}^2 \cdot \text{K}$$

$$h_a = \left(\frac{jGc_p}{\text{Pr}^{2/3}}\right)_a = \frac{0.0134 \times 18.365\,\text{kg/m}^2 \cdot \text{s} \times (1.073 \times 10^3)\,\text{J/kg} \cdot \text{K}}{0.784} = 336.81\,\text{W/m}^2 \cdot \text{K}$$

Now let us calculate the fin efficiency for air and gas sides. Since the offset strip fins are used on both gas and air sides, we will use Eq. (4.147) with L_f replaced by ℓ_s to take into account the strip edge exposed area.

$$m_g = \left[\frac{2h}{k_f\delta}\left(1 + \frac{\delta}{\ell_s}\right)\right]^{1/2} = \left[\frac{2 \times 360.83\,\text{W/m}^2 \cdot \text{K}}{18\,\text{W/m} \cdot \text{K} \times (0.102 \times 10^{-3})\,\text{m}}\left(1 + \frac{0.102\,\text{mm}}{3.175\,\text{mm}}\right)\right]^{1/2}$$

$$= 634.94\,\text{m}^{-1}$$

$$m_a = \left[\frac{2h}{k_f\delta}\left(1 + \frac{\delta}{\ell_s}\right)\right]^{1/2} = \left[\frac{2 \times 336.81\,\text{W/m}^2 \cdot \text{K}}{18\,\text{W/m} \cdot \text{K} \times (0.102 \times 10^{-3})\,\text{m}}\left(1 + \frac{0.102\,\text{mm}}{3.175\,\text{mm}}\right)\right]^{1/2}$$

$$= 615.37\,\text{m}^{-1}$$

$$\ell_a = \ell_g \approx b/2 - \delta = (2.49\,\text{mm}/2 - 0.102\,\text{mm}) = 1.143\,\text{mm} = 0.00114\,\text{m}$$

Thus,

$$\eta_{f,g} = \frac{\tanh(m\ell)_g}{(m\ell)_g} = \frac{\tanh(636.94\,\text{m}^{-1} \times 0.00114\,\text{m})}{636.94\,\text{m}^{-1} \times 0.00114\,\text{m}} = 0.8581$$

$$\eta_{f,a} = \frac{\tanh(m\ell)_a}{(m\ell)_a} = \frac{\tanh(615.37\,\text{m}^{-1} \times 0.00114\,\text{m})}{615.37\,\text{m}^{-1} \times 0.00114\,\text{m}} = 0.8657$$

The overall surface efficiencies are

$$\eta_{o,g} = \left[1 - (1 - \eta_f)\frac{A_f}{A}\right]_g = 1 - (1 - 0.8581) \times 0.785 = 0.8886$$

$$\eta_{o,a} = \left[1 - (1 - \eta_f)\frac{A_f}{A}\right]_a = 1 - (1 - 0.8657) \times 0.785 = 0.8946$$

It should be pointed out that the fin conduction length ℓ for the end passages on the airside will be b and not $(b/2 - \delta)$. This will result in lower fin efficiency for the end passages. However, its influence will be smaller on the weighted average fin efficiency considering all air passages. Hence, we have neglected it here. However, in a computer program, it can easily be incorporated.

Wall Resistance and Overall Conductance. For the $\mathbf{R}_w$ determination, the wall conduction area A_w is

$$A_w = L_1 L_2 (2N_p + 2) = 0.3\,\text{m} \times 0.3\,\text{m} \times (2 \times 167 + 2) = 30.24\,\text{m}^2$$

Therefore,

$$\mathbf{R}_w = \frac{\delta_w}{k_w A_w} = \frac{0.5 \times 10^{-3}\,\text{m}}{18\,\text{W/m} \cdot \text{K} \times 30.24\,\text{m}^2} = 9.186 \times 10^{-7}\,\text{K/W}$$

Since the influence of fouling is negligibly small for a gas-to-gas heat exchanger, we will neglect it. Then $1/UA$ from Eq. (9.16) is

$$\frac{1}{UA} = \frac{1}{(\eta_o hA)_h} + \mathbf{R}_w + \frac{1}{(\eta_o hA)_c}$$

$$= \frac{1}{0.8886 \times 360.83 \text{ W/m}^2 \cdot \text{K} \times 84.345 \text{ m}^2} + 9.186 \times 10^{-7} \text{ K/W}$$

$$+ \frac{1}{0.8946 \times 336.81 \text{ W/m}^2 \cdot \text{K} \times 84.863 \text{ m}^2}$$

$$= \frac{1}{27043.8 \text{ W/K}} + 9.186 \times 10^{-7} \text{ K/W} + \frac{1}{25,570.1 \text{ W/K}}$$

$$= 0.3698 \times 10^{-4} \text{ K/W} + 9.186 \times 10^{-7} \text{ K/W} + 0.3911 \times 10^{-4} \text{ K/W}$$

$$= 0.77009 \times 10^{-4} \text{ K/W}$$

$$UA = 12985 \text{ W/K}$$

NTU, Exchanger Effectiveness, and Outlet Temperatures. To determine NTU and ε, first calculate C_g and C_a:

$$C_g = (\dot{m}c_p)_g = 1.66 \text{ kg/s} \times (1.122 \times 10^3 \text{ J/kg} \cdot \text{K}) = 1863 \text{ W/K}$$

$$C_a = (\dot{m}c_p)_a = 2.00 \text{ kg/s} \times (1.073 \times 10^3 \text{ J/kg} \cdot \text{K}) = 2146 \text{ W/K}$$

$$C^* = \frac{C_{min}}{C_{max}} = \frac{C_g}{C_a} = \frac{1863 \text{ W/K}}{2146 \text{ W/K}} = 0.868$$

$$\text{NTU} = \frac{UA}{C_{min}} = \frac{12985 \text{ W/K}}{1863 \text{ W/K}} = 6.970$$

For NTU = 6.970 and $C^* = 0.868$, the effectiveness for the crossflow exchanger with both fluids unmixed from the expression of Table 3.3, is

$$\varepsilon = 0.8328$$

This effectiveness is higher than normally used for a crossflow exchanger. The reason for the selection of somewhat hypothetical example to yield high ε is to demonstrate how to take into account the effect of longitudinal conduction in the wall. We evaluate the decrease $\Delta\varepsilon$ in ε due to longitudinal heat conduction now. Let us first calculate the conduction cross-sectional area for longitudinal heat conduction in the wall.

$$A_{k,g} = 2N_p L_a \delta_w = 2 \times 167 \times 0.3 \text{ m} \times 0.5 \times 10^{-3} \text{ m} = 0.0501 \text{ m}^2$$

$$A_{k,a} = (2N_p + 2)L_g \delta_w = (2 \times 167 + 2) \times 0.3 \text{ m} \times 0.5 \times 10^{-3} \text{ m} = 0.0504 \text{ m}^2$$

Longitudinal conduction parameters on the gas and air sides are

$$\lambda_h = \lambda_g = \left(\frac{k_w A_k}{LC}\right)_g = \frac{18 \text{ W/m} \cdot \text{K} \times 0.0501 \text{ m}^2}{0.3 \text{ m} \times 1863 \text{ W/K}} = 0.0016$$

$$\lambda_c = \lambda_a = \left(\frac{k_w A_k}{LC}\right)_a = \frac{18 \text{ W/m} \cdot \text{K} \times 0.0504 \text{ m}^2}{0.3 \text{ m} \times 2146.4 \text{ W/K}} = 0.0014$$

Other parameters needed for determining $\Delta\varepsilon$ in addition to already determined $NTU = 6.970$ and $C_c/C_h = 0.868$ are

$$\frac{(\eta_o h A)_h}{(\eta_o h A)_c} = \frac{27,043.8 \text{ W/K}}{25,570.1 \text{ W/K}} = 1.06 \qquad \frac{\lambda_c}{\lambda_h} = \frac{0.0016}{0.0014} = 1.14$$

From the interpolation of tabular results of Chiou (1980) (see Table 4.1), it is found that $\Delta\varepsilon/\varepsilon \approx 0.002$. Thus $\Delta\varepsilon = 0.0017$, and actual exchanger effectiveness is

$$\varepsilon_{actual} = 0.8328 - 0.0017 = 0.8311$$

The heat transfer rate q is then

$$q = \varepsilon(T_{g,i} - T_{a,i})C_{min} = 0.8311 \times (900 - 200)°\text{C} \times 1863 \text{ W/K} = 1083.8 \times 10^3 \text{ W}$$

The outlet temperatures are then

$$T_{g,o} = T_{g,i} - \frac{q}{C_g} = 900°\text{C} - \frac{1083.8 \times 10^3 \text{ W}}{1863 \text{ W/K}} = 318.3°\text{C} = 591.5 \text{ K}$$

$$T_{a,o} = T_{a,i} + \frac{q}{C_a} = 200°\text{C} + \frac{1083.8 \times 10^3 \text{ W}}{2146 \text{ W/K}} = 705.0°\text{C} = 978.2 \text{ K}$$

Since these outlet temperatures are different from those assumed for the initial determination of the fluid properties, two more iterations were carried out with fluid properties evaluated at the new average temperatures. The values of C^*, NTU, ε, $T_{g,o}$ and $T_{a,o}$ were 0.857, 7.082, 0.8382, 314.4°C and 701.9°C, respectively, and after the third iteration were 0.857, 7.079, 0.8381, 314.5°C and 701.8°C respectively.

Pressure Drops. We use Eq. (9.20) to compute the pressure drop on each fluid side. The densities are evaluated using the perfect gas equation of state:

	T_i (K)	T_o (K)	ρ_i (kg/m³)	ρ_o (kg/m³)	ρ_m (kg/m³)
Gas	1173	591.5	0.4751	0.9424	0.6318
Air	473	978.2	1.4726	0.7123	0.9602

Note that we have also considered the outlet pressures as 160 kPa and 200 kPa for gas and air, respectively, since the pressure drop across the core is usually small and hence is neglected in the first iteration. The mean density in the last column is the harmonic mean value from Eq. (9.18).

Now let us determine K_c and K_e. Offset strip fins are used on gas and air sides. In such fin geometries, because of the frequent boundary layer interruptions, the flow is well mixed and is treated as having the Reynolds number very large ($\text{Re} \to \infty$). The aspect ratio of the rectangular passage, height/width $= 2.49/(1/0.615 - 0.15) = 1.15$. Since $\text{Re} = \infty$ curves for parallel plate and square passage geometries of Fig. 6.3 are identical, we could determine K_c and K_e from either geometry for $\sigma_a = 0.36$ as

$$K_c = 0.36 \qquad K_e = 0.42$$

Before we compute the pressure drop, we need to correct the values of the isothermal friction factors by the method of Section 7.6.1 to take into account the temperature-dependent properties. A review of Eq. (9.11) indicates that we need to calculate the fluid bulk mean temperatures and the wall temperature. The mean temperatures on the gas and air sides, based on the latest outlet temperatures, are

$$T_{g,m} = \frac{(900 + 318.3)^\circ C}{2} = 609.2^\circ C = 882.4 \, K$$

$$T_{a,m} = \frac{(200 + 705.0)^\circ C}{2} = 452.5^\circ C = 725.7 \, K$$

The thermal resistances on the gas and air sides are

$$\mathbf{R}_g = \frac{1}{(\eta_o hA)_h} = 0.3698 \times 10^{-4} \, K/W \qquad \mathbf{R}_a = \frac{1}{(\eta_o hA)_a} = 0.3911 \times 10^{-4} \, K/W$$

$$\frac{\mathbf{R}_g}{\mathbf{R}_a} = 0.946$$

If we neglect the wall resistance,

$$q = \frac{T_{g,m} - T_w}{\mathbf{R}_g} = \frac{T_w - T_{a,m}}{\mathbf{R}_a}$$

so that

$$T_w = \frac{T_{g,m} + (\mathbf{R}_g/\mathbf{R}_a)T_{a,m}}{1 + (\mathbf{R}_g/\mathbf{R}_a)} = \frac{609.2^\circ C + 0.946 \times 452.5^\circ C}{1 + 0.946} = 533.0^\circ C = 806.2 \, K$$

Since the gas is being cooled, using Eq. (9.11) and the exponent $m = 0.81$ from Table 7.12,

$$f = f_{cp} \left(\frac{T_w}{T_m}\right)^m = 0.0669 \left(\frac{806.2 \, K}{882.4 \, K}\right)^{0.81} = 0.0622$$

Since the air is being heated, using Eq. (9.11) and the exponent $m = 1.00$ from Table 7.12,

$$f = f_{cp} \left(\frac{T_w}{T_m}\right)^m = 0.0534 \left(\frac{806.2 \, K}{725.7 \, K}\right)^{1.00} = 0.0593$$

Now let us calculate the pressure drops using Eq. (9.20).

$$\Delta p_g = \frac{(15.342 \, kg/m^2 \cdot s)^2}{2 \times 1 \times 0.4751 \, kg/m^3} \left[(1 - 0.361^2 + 0.36) + 2 \left(\frac{0.4751 \, kg/m^3}{0.9424 \, kg/m^3} - 1 \right) \right.$$

$$\left. + \frac{0.0622 \times 0.3 \, m \times 0.4751 \, kg/m^3}{(0.00154/4) \, m \times 0.6318 \, kg/m^3} - (1 - 0.361^2 - 0.42) \left(\frac{0.4751 \, kg/m^3}{0.9424 \, kg/m^3} \right) \right]$$

$$\Delta p_a = \frac{(7.0484 \, kg/m^2 \cdot s)^2}{2 \times 1 \times 1.3827 \, kg/m^3} \left[(1 - 0.437^2 + 0.33) + 2 \left(\frac{1.3827 \, kg/m^3}{0.8194 \, kg/m^3} - 1 \right) \right.$$

$$\left. + \frac{0.0683 \times 0.6 \, m \times 1.3827 \, kg/m^3}{(0.002383/4) \, m \times 1.0290 \, kg/m^3} - (1 - 0.437^2 + 0.31) \frac{1.3827 \, kg/m^3}{0.9424 \, kg/m^3} \right]$$

$$= 247.71 \, \text{Pa} \times (1.2297 - 0.9917 + 36.4465 - 0.2267)$$

$$= 247.71 \, \text{Pa} \times 36.4578 = 9031 \, \text{Pa} = 9.031 \, \text{kPa}$$

$$\Delta p_a = \frac{18.365 \, \text{kg/m}^2 \cdot \text{s})^2}{2 \times 1 \times 1.4726 \, \text{kg/m}^3} \left[(1 - 0.363^2 + 0.36) + 2 \left(\frac{1.4726 \, \text{kg/m}^3}{0.7123 \, \text{kg/m}^3} - 1 \right) \right.$$

$$\left. + \frac{0.0593 \times 0.3 \, \text{m} \times 1.4726 \, \text{kg/m}^3}{(0.00154/4) \, \text{m} \times 0.9602 \, \text{kg/m}^3} - (1 - 0.363^2 - 0.42) \left(\frac{1.4726 \, \text{kg/m}^3}{0.7123 \, \text{kg/m}^3} \right) \right]$$

$$= 114.51 \, \text{Pa} \times (1.2282 + 2.1348 + 70.8660 - 0.9267)$$

$$= 114.51 \, \text{Pa} \times 73.3023 = 8394 \, \text{Pa} = 8.394 \, \text{kPa}$$

With these values of the pressure drops, the outlet pressures were recomputed, and with the corresponding values of the new outlet densities, the pressure drops were recalculated. The gas and airside pressure drops after the second iterations were 9.571 and 8.776 kPa, and after the third iterations 9.050 and 8.757 kPa, respectively.

Discussion and Comments: As shown above, the calculation procedure for rating of a crossflow exchanger is straightforward and demonstrates how all information presented in preceding chapters is integrated to obtain the performance of the exchanger. Note that this calculation assumes the validity of a number of assumptions. For example, the fin efficiency calculations were performed idealizing perfect brazing between the plates and fins. A manufactured heat exchanger would inevitably be characterized with a performance more or less different than predicted, depending on a designer's ability to incorporate relaxation of some of the assumptions into this procedure.

9.2.2 Sizing Problem

The sizing problem is more difficult. Many early decisions to choose the construction type and basic geometries on each fluid side are based on experience (including rules of thumb and engineering judgments), operating conditions, maintenance, manufacturing capability, and the expected life of the exchanger. Some of these issues were mentioned in Chapter 2. We discuss the selection of exchanger construction type, flow arrangement, surface geometries, and so on, in Chapter 10. With those inputs, the sizing problem then reduces to the determination of the core or exchanger dimensions for the specified heat transfer and pressure drop performance. One could, of course, reduce this problem to the rating problem by tentatively specifying the dimensions, then calculate the performance for comparison with the specified performance. This type of search for a solution is usually performed in the case of shell-and-tube exchangers and regenerators where one needs to take care of leakage and bypass flows in a very complex manner. However, leakage and bypass flows are not significant for plate-fin and tube-fin exchangers. The solution method can be made more straightforward, with fast convergence for these exchangers by reforming the surface characteristics input to include j/f vs. Re for surfaces on each fluid side, in addition to the separate j and f versus Re characteristics. This coupling of heat transfer and flow friction is now made in the derivation of the core mass velocity equation that has been proposed by Kays and London (1998). Once the core mass velocity is determined, the solution to the sizing problem is carried out itera-tively in a manner similar to the rating problem discussed in Section 9.2.1.

9.2.2.1 Core Mass Velocity Equation.

The coupling of heat duty and pressure drops is done by the core mass velocity equation as follows. From the required heat duty (and hence the exchanger effectiveness ε) and known heat capacity rates on each fluid side (known C^*), the overall NTU is determined for the selected exchanger flow arrangement. The overall conductance as a first approximation is given by

$$\frac{1}{UA} \approx \frac{1}{(\eta_o hA)_h} + \frac{1}{(\eta_o hA)_c} \tag{9.21}$$

Here we have neglected the wall and fouling thermal resistances. The overall NTU is related to individual side ntu_h and ntu_c as follows [see Eq. (3.67)]:

$$\frac{1}{NTU} = \frac{1}{ntu_h(C_h/C_{\min})} + \frac{1}{ntu_c(C_c/C_{\min})} = \frac{C^*}{ntu_g} + \frac{1}{ntu_a} \tag{9.22}$$

where the second equality is for $C_c = C_{\min}$.

From the known NTU, we need to determine ntu_h and ntu_c from this equation either from past experience or by guessing. If both fluids are gases or both fluids are liquid, one could consider that the design is "balanced" (i.e., the thermal resistances are distributed approximately equally on the hot and cold sides). In that case, $C_h \approx C_c$, and

$$ntu_h \approx ntu_c \approx 2NTU \tag{9.23}$$

Alternatively, if we have liquid on one side and gas on the other side, consider 10% thermal resistance on the liquid side:

$$0.10\left(\frac{1}{UA}\right) = \frac{1}{(\eta_o hA)_{\text{liq}}} \tag{9.24}$$

Substituting Eq. (9.24) into Eq. (9.21) with $C_c = C_{\text{gas}} = C_{\min}$ and $C_h = C_{\text{liq}}$, we can determine ntu_{gas} and with its subsequent substitution in Eq. (9.22), we get

$$ntu_{\text{gas}} = 1.11\ NTU \qquad ntu_{\text{liq}} = 10C^* \cdot NTU \tag{9.25}$$

The ntu on each fluid side is related to the Colburn factor j as follows by using Eq. (7.31) and (7.33):

$$ntu = \frac{\eta_o hA}{\dot{m}c_p} = \eta_o \frac{h}{Gc_p}\frac{A}{A_o} = \eta_o j \cdot Pr^{-2/3}\frac{A}{A_o} = \eta_o j \cdot Pr^{-2/3}\frac{L}{r_h} \tag{9.26}$$

The pressure drop on each fluid side is given by Eq. (9.20). Substituting L/r_h from Eq. (9.26) into Eq. (9.20) and simplifying, G reduces to

$$G = (2g_c\Delta p)^{1/2}\left[\frac{f}{j}\frac{\text{ntu}}{\eta_o}\text{Pr}^{2/3}\left(\frac{1}{\rho}\right)_m + 2\left(\frac{1}{\rho_o} - \frac{1}{\rho_i}\right) + (1 - \sigma^2 + K_c)\frac{1}{\rho_i}\right.$$

$$\left. -(1 - \sigma^2 - K_e)\frac{1}{\rho_o}\right]^{-1/2} \tag{9.27}$$

Equation (9.27) is a more generalized core mass velocity equation than that provided by Kays and London (1998), who considered only the first term in brackets in Eq. (9.27). Since the contribution of the last three terms in the brackets in Eq. (9.27) is generally very small, they can be neglected in light of other approximations already mentioned. In this case, Eq. (9.27) reduces to

$$G = \left[\frac{2g_c}{(1/\rho)_m \cdot \text{Pr}^{2/3}}\frac{\eta_o\,\Delta p}{\text{ntu}}\left(\frac{j}{f}\right)\right]^{1/2} \tag{9.28}$$

Equation (9.28) is referred to as the *core mass velocity equation*. The feature that makes this equation so useful is that the ratio j/f is a relatively weak function of the Reynolds number for most extended surfaces (see Fig. 2-41 of Kays and London, 1998, and Fig. E9.2). Thus, one can readily estimate a fairly accurate value of j/f in the operating range of Re. Also, for a "good design," the fin geometry is chosen such that η_o is in the range 70 to 90% and higher. Hence, $\eta_o \approx 80\%$ may be assumed for the first approximation in Eq (9.28), unless a better value is known from the past experience. All other information in Eq. (9.28) is known or evaluated from the problem specification. Thus, the first approximate value of G can be computed from Eq. (9.28). As a result, the iterative solution to the sizing problem converges relatively fast with this estimated value of G.

We will use Eq. (9.28) for the first iteration of a sizing problem, as described next for single-pass counterflow and crossflow exchangers.

9.2.2.2 Sizing of a Single-Pass Counterflow Exchanger. Now we outline a detailed procedure for arriving at core dimensions for a counterflow exchanger for specified heat transfer and pressure drop. In a single-pass counterflow heat exchanger of any construction, if the core dimensions on one side are fixed, the core dimensions for the other side (except for the passage height) are also fixed. Therefore, the design problem for this case is solved for the side that has more stringent pressure drop specification. This method is also applicable to the $C^* \approx 0$ exchanger, such as a gas-to-liquid or phase-changing fluid exchanger. In this case, the thermal resistance is primarily on the gas side and the pressure drop is also more critical on the gas side. As a result, the core dimensions obtained are based on the gas-side Δp and ntu$_{\text{gas}}$. The dimensions on the other side are then chosen such that the calculated pressure drop is within the specified Δp (i.e., $\Delta p_{\text{calculated}} \le \Delta p_{\text{specified}}$). Thus either for counterflow or for a $C^* = 0$ exchanger, the core dimensions are calculated for the side having the most stringent Δp. Following is a step-by-step procedure for the solution.

1. To compute the fluid mean temperature and the fluid thermophysical properties on each fluid side, determine the fluid outlet temperatures, for the specified heat duty, from the following equation considering fluid 1 as the hot fluid.

$$q = (\dot{m}c_p)_1(T_{1,i} - T_{1,o}) = (\dot{m}c_p)_2(T_{2,o} - T_{2,i}) \tag{9.29}$$

If the exchanger effectiveness is specified, use Eqs. (9.9) and (9.10) to compute outlet temperatures. For the first iteration, estimate the c_p values to determine the outlet temperatures from Eq. (9.29). Subsequently, determine the mean temperatures on both fluid sides using the procedure discussed in Section 9.1.2 or from Table 9.1. With these mean temperatures, determine the c_p's, and iterate one more time for the outlet temperatures if warranted. Subsequently, determine μ, c_p, k, Pr, and ρ on each fluid side.

2. Calculate C^* and ε (if q is given), and determine NTU from the ε-NTU expression, with tabular or graphical results for the flow arrangement selected [in this case, use Eq. (3.86) for counterflow]. The influence of longitudinal heat conduction, if any, is ignored in the first iteration since we don't yet know the exchanger size.

3. Determine ntu on each fluid side by the approximations discussed with Eqs. (9.23) and (9.25) unless it can be estimated differently (i.e., instead of 50:50% or 90:10% thermal resistance distribution) from past experience.

4. For the surfaces selected on each fluid side, plot j/f vs. Re curve from the given surface characteristics, and obtain an approximate mean value of j/f over the complete Reynolds number range; an accurate mean value of j/f is not necessary since we are making a number of approximations to get the first estimate of G. If fins are employed, assume that $\eta_o = 0.80$ unless a better value can be estimated based on the experience.

5. Evaluate G from Eq. (9.28) on each fluid side using the information from steps 1 through 4 and specified values of Δp.

6. Calculate Reynolds number Re, and determine j and f on each fluid side from the given design data for each surface. The design data may be in the form of graphs, curve fit to tabulated data or an empirical generalized equation.

7. Compute h, η_f, and η_o using Eqs. (9.13)–(9.15). For the first iteration, determine U_1 on the fluid 1 side from the following equation derived from Eq. (3.24):

$$\frac{1}{U_1} = \frac{1}{(\eta_o h)_1} + \frac{1}{(\eta_o h_f)_1} + \frac{\alpha_1/\alpha_2}{(\eta_o h_f)_2} + \frac{\alpha_1/\alpha_2}{(\eta_o h)_2} \tag{9.30}$$

where $\alpha_1/\alpha_2 = A_1/A_2$, $\alpha = A/V$, and V is the exchanger total volume, and subscripts 1 and 2 denote the fluid 1 and 2 sides. For a plate-fin exchanger, α's are related to β's by Eqs. (8.96) and (8.97) with β usually provided with the surface basic characteristics (see, e.g., Fig. E9.1B)

$$\alpha_1 = \frac{b_1 \beta_1}{b_1 + b_2 + 2\delta_w} \qquad \alpha_2 = \frac{b_2 \beta_2}{b_1 + b_2 + 2\delta_w} \tag{9.31}$$

Note that the wall thermal resistance in Eq. (9.30) is ignored in the first iteration since we do not yet know the size of the exchanger (i.e., A_w/A_1 is unknown). In second and subsequent iterations, compute U_1 from

$$\frac{1}{U_1} = \frac{1}{(\eta_o h)_1} + \frac{1}{(\eta_o h_f)_1} + \frac{\delta_w A_1}{k_w A_w} + \frac{A_1/A_2}{(\eta_o h_f)_2} + \frac{A_1/A_2}{(\eta_o h)_2} \tag{9.32}$$

where the necessary area ratios A_1/A_2 and A_1/A_w are determined from the geometry calculated in the preceding iteration.

8. Now calculate the core dimensions. In the first iteration, use NTU as computed in step 2. For subsequent iterations, calculate the longitudinal conduction parameter λ. With known ε, C^*, and λ, determine the correct value of NTU using Eq. (4.15) or (4.20) . Determine A_1 from NTU using U_1 from step 7 and known $C_{\min}$:

$$A_1 = \frac{\text{NTU} \cdot C_{\min}}{U_1} \tag{9.33}$$

and hence

$$A_2 = \frac{A_2}{A_1} A_1 = \frac{\alpha_2}{\alpha_1} A_1 \tag{9.34}$$

The free-flow area A_o from known $\dot{m}$ and G is given by

$$A_{o,1} = \left(\frac{\dot{m}}{G}\right)_1 \qquad A_{o,2} = \left(\frac{\dot{m}}{G}\right)_2 \tag{9.35}$$

so that

$$A_{fr,1} = \frac{A_{o,1}}{\sigma_1} \qquad A_{fr,2} = \frac{A_{o,2}}{\sigma_2} \tag{9.36}$$

where σ_1 and σ_2 are generally specified for the surface or can be computed from Eqs. (8.94) and (8.95) as follows using given geometrical properties:

$$\sigma_1 = \frac{b_1 \beta_1 D_{h,1}/4}{b_1 + b_2 + 2\delta_w} = \frac{\alpha_1 D_{h,1}}{4} \qquad \sigma_2 = \frac{b_2 \beta_2 D_{h,2}/4}{b_1 + b_2 + 2\delta_w} = \frac{\alpha_2 D_{h,2}}{4} \tag{9.37}$$

where the term after the second equality sign comes from the definition of α's from Eq. (9.31). In a single-pass counterflow exchanger, $A_{fr,1}$ and $A_{fr,2}$ must be identical, and those computed in Eq. (9.36) may not be identical. In this case, use the greater of $A_{fr,1}$ and $A_{fr,2}$. Finally, the core length L in the flow direction is determined from the definition of the hydraulic diameter of the surface employed on each fluid side.

$$L = \left(\frac{D_h A}{4A_o}\right)_1 = \left(\frac{D_h A}{4A_o}\right)_2 \tag{9.38}$$

The value of L calculated from either of the equalities will be the same, as can be shown using Eqs. (9.34), (9.36), and (9.37), and for $A_{fr,1} = A_{fr,2}$.

Once the frontal area is determined, any choice of exchanger width and height (product of which should be equal to the frontal area) will theoretically be a correct solution. If there are any constraints imposed on the exchanger dimension, select the frontal area dimensions accordingly. Also, from the header design viewpoint as well as from the flow distribution viewpoint, select the frontal area dimensions to make it the least prone to maldistribution; see Chapter 12 for a discussion of related issues.

9. Now compute the pressure drop on each fluid side, after correcting f factors for variable property effects, in a manner similar to step 8 of the rating problem (section 9.2.1).

10. If the values calculated for Δp are within the input specifications and close to them, the solution to the sizing problem is completed; finer refinements in the core dimensions, such as integer numbers of flow passages, may be carried out at this time. Otherwise, compute the new value of G on *both* fluid sides using Eq. (9.27), in which Δp is the input specified value, and f, K_c, K_e, and geometrical dimensions are from the previous iteration.

11. Iterate steps 6 through 10 until both heat transfer and pressure drops are met as specified. Probably, only one of the two pressure drops (whichever is the most critical) will be matched, the other will be lower than specified for a gas-to-gas exchanger. Only two or three iterations may be necessary to converge to the final size of the exchanger within 1% or the accuracy desired.

12. If the influence of longitudinal heat conduction is important, the longitudinal conduction parameter λ is computed from Eq. (4.13), and subsequently, NTU is computed iteratively from the ε formula/results of Section 4.1.2. This new value of NTU is then used in step 8 in the second and subsequent iterations.

9.2.2.3 Sizing of a Single-Pass Crossflow Exchanger. For a crossflow exchanger, determining the core dimensions on one fluid side (A_{fr} and L) does not fix the dimensions on the other fluid side. In such a case, the design problem is solved simultaneously on both fluid sides. The solution procedure follows closely that of Section 9.2.2.2 and is outlined next through detailed steps.

1. Determine G on each fluid side by following steps 1 through 5 of Section 9.2.2.2.

2. Follow steps 6 through 8 of Section 9.2.2.2 and compute A_1, A_2, $A_{o,1}$, $A_{o,2}$, $A_{fr,1}$, and $A_{fr,2}$.

3. Now compute the fluid flow lengths on each fluid side (see Fig. E9.1A) from the definition of the hydraulic diameter of the surface employed on each fluid side as follows:

$$L_1 = \left(\frac{D_h A}{4 A_o}\right)_1, \qquad L_2 = \left(\frac{D_h A}{4 A_o}\right)_2 \tag{9.39}$$

Since $A_{fr,1} = L_2 L_3$ and $A_{fr,2} = L_1 L_3$, we obtain

$$L_3 = \frac{A_{fr,1}}{L_2} \qquad L_3 = \frac{A_{fr,2}}{L_1} \tag{9.40}$$

Thus the noflow (or stack) height L_3 can be determined from the definition of either $A_{fr,1}$ or $A_{fr,2}$ and known L_2 or L_1 and should be identical. In reality, they may be slightly different because of the round-off error in calculations. In that case, consider an average value for L_3.

4. Now follow steps 9 and 10 of Section 9.2.2.2 to compute Δp on each fluid side. If Δp on one of two fluid sides does not match (i.e., too high compared to the specification), calculate new values of G on *both* fluid sides as mentioned in step 10 of Section 9.2.2.2.

5. Iterate steps 1 through 4 until both heat transfer and pressure drops are met as specified within the accuracy desired.

6. If the influence of longitudinal heat conduction is important, the longitudinal conduction parameter λ_h, λ_c, and other appropriate dimensionless groups are calculated based on the core geometry from the preceding iteration and input operating conditions with the procedure outlined in Section 4.1.4. Subsequently, NTU is computed iteratively from the ε results of Section 4.1.4. This new value of NTU is then used in step 8 of Section 9.2.2.2.

It should be emphasized that since we have not imposed any constraints on the exchanger dimensions, the procedure above will yield unique values of L_1, L_2, and L_3 for the surface selected such that theoretically the design will exactly meet the heat duty and pressure drops on both fluid sides.

Example 9.2 Consider the heat exchanger of the rating problem in Example 9.1. Design a gas-to-air single-pass crossflow heat exchanger operating at $\varepsilon = 0.8381$ having gas and air inlet temperatures as 900°C and 200°C respectively, and gas and air mass flow rates as 1.66 kg/s and 2.00 kg/s, respectively. The gas side and air side pressure drops are limited to 9.05 and 8.79 kPa, respectively. The gas and air inlet pressures are 160 kPa and 200 kPa absolute. The offset strip fin surface on the gas and air sides has the surface characteristics as shown in Fig. E9.1B. Both fins and plates (parting sheets) are made from Inconel 625 alloy (its thermal conductivity as 18 W/m · K). The plate thickness is 0.5 mm. Determine the core dimensions of this exchanger.

SOLUTION

Problem Data and Schematic: The following information is provided for the sizing of a crossflow exchanger.

$$\varepsilon = 0.8381 \qquad \Delta p_g = 9.05 \,\text{kPa} \qquad \Delta p_a = 8.79 \,\text{kPa}$$

Basic surface geometry parameters on the gas and air sides along with their j and f data are given in Fig. E9.1B. Geometry and operating parameters are:

$b_g = 2.49 \,\text{mm} \quad D_{h,g} = 0.00154 \,\text{m} \quad \beta_g = 2254 \,\text{m}^2/\text{m}^3 \quad \delta_g = 0.102 \,\text{mm} \quad \left(\dfrac{A_f}{A}\right)_g = 0.785$

$b_a = 2.49 \,\text{mm} \quad D_{h,a} = 0.00154 \,\text{m} \quad \beta_a = 2254 \,\text{m}^2/\text{m}^3 \quad \delta_a = 0.102 \,\text{mm} \quad \left(\dfrac{A_f}{A}\right)_a = 0.785$

$\dot{m}_g = 1.66 \,\text{kg/s} \quad T_{g,i} = 900°\text{C} \qquad p_{g,i} = 160 \,\text{kPa} \qquad k_f = k_w = 18 \,\text{W/m} \cdot \text{K}$

$\dot{m}_a = 2.0 \,\text{kg/s} \quad T_{a,i} = 100°\text{C} \qquad p_{a,i} = 200 \,\text{kPa}$

Determine: The length, width and height of this exchanger to meet specified exchanger effectiveness (heat duty) and pressure drops.

Assumptions: The assumptions listed in Section 3.2.1 applicable to a plate-fin exchanger are invoked. Neglect the effect of longitudinal heat conduction and treat the gas as air for fluid property evaluation.

Analysis: We will follow the steps outlined in Section 9.2.2.3 for the solution.

Outlet Temperatures. To determine outlet temperatures from the known ε, we first need to know which fluid side is the C_{min} side. Since $\dot{m}_a > \dot{m}_g$, the C_{min} side will be the gas side. Assuming that the specific heats of air and gas in the first iteration are the same $(c_{p,a} \approx c_{p,g})$, using Eqs. (9.9) and (9.10), we get

$$T_{g,o} = T_{g,i} - \varepsilon(T_{g,i} - T_{a,i}) = 900°\text{C} - 0.8381 \times (900 - 200)°\text{C} = 313.3°\text{C}$$

$$T_{a,o} \approx T_{a,i} + \varepsilon \frac{\dot{m}_g}{\dot{m}_a}(T_{g,i} - T_{a,i}) = 200°\text{C} + 0.8381 \times 0.83 \times (900 - 200)°\text{C} = 686.9°\text{C}$$

This value of $T_{a,o}$ will be refined after we determine the fluid properties.

Fluid Properties. Since $C^* \approx \dot{m}_g/\dot{m}_a = 0.83$, we will evaluate the fluid properties at the arithmetic mean temperatures.

$$T_{g,m} = \frac{900°\text{C} + 313.3°\text{C}}{2} = 606.7°\text{C} = 879.8\,\text{K}$$

$$T_{a,m} = \frac{200°\text{C} + 686.9°\text{C}}{2} = 443.5°\text{C} = 716.6\,\text{K}$$

The c_p values of gas and air at these temperatures are 1.117 and 1.079 J/kg · K, respectively (Raznjević, 1976). Hence, using Eq. (9.9), the correct $T_{a,o}$ will be

$$T_{a,o} = T_{a,i} + \varepsilon \left(\frac{\dot{m}_g c_{p,g}}{\dot{m}_a c_{p,a}} \right)(T_{g,i} - T_{a,i})$$

$$= 200°\text{C} + 0.8381 \left(\frac{1.66\,\text{kg/s} \times 1.117\,\text{kJ/kg} \cdot \text{K}}{2.00\,\text{kg/s} \times 1.079\,\text{kJ/kg} \cdot \text{K}} \right)(900°\text{C} - 200)°\text{C} = 704.1°\text{C}$$

Thus the refined value of $T_{a,m}$ is

$$T_{a,m} = \frac{704.1°\text{C} + 200°\text{C}}{2} = 452.0°\text{C} = 725.2\,\text{K}$$

The specific heat of air at 725.2 K is 1.081 kJ/kg · K, which has negligible difference from the previous value of 1.079 kJ/kg · K and hence there is no further need of iterations. The air properties at $T_{g,m} = 879.8$ K and $T_{a,m} = 725.2$ K from Appendix 1 are as follows.

	μ (Pa · s)	c_p (kJ/kg · K)	Pr	$\text{Pr}^{2/3}$
Gas at 879.8 K	39.3×10^{-6}	1.117	0.721	0.804
Air at 725.2 K	34.7×10^{-6}	1.081	0.692	0.782

The inlet and outlet gas densities are evaluated at 160 kPa and 150.95 (= 160 − 9.05) kPa, respectively. The inlet and outlet air densities are evaluated at 200 kPa and 191.21 (= 200 − 8.79) kPa, respectively. The mean densities are evaluated using Eq. (9.18):

	T_i (K)	T_o (K)	ρ_i (kg/m³)	ρ_o (kg/m³)	ρ_m (kg/m³)
Gas	1173.2	586.5	0.4751	0.8966	0.6212
Air	473.2	977.2	1.4726	0.6817	0.9319

C* and NTU. From the foregoing values of c_p and given flow rates, we evaluate

$$C_g = (\dot{m}c_p)_g = 1.66\,\text{kg/s} \times (1.117 \times 10^3)\,\text{J/kg} \cdot \text{K} = 1854\,\text{W/K}$$

$$C_a = (\dot{m}c_p)_a = 2.00\,\text{kg/s} \times (1.081 \times 10^3)\,\text{J/kg} \cdot \text{K} = 2162\,\text{W/K}$$

$$C^* = \frac{C_{\min}}{C_{\max}} = \frac{1854\,\text{W/K}}{2162\,\text{W/K}} = 0.858$$

Neglecting longitudinal heat conduction, NTU for a crossflow exchanger with both fluids unmixed for $\varepsilon = 0.8381$ and $C^* = 0.858$, from the expression of Table 3.3, is

$$\text{NTU} = 7.079$$

Now we need to estimate ntu_g and ntu_a from the overall NTU. The better the initial estimate, the closer will be the value of G as a first estimate. For a gas-to-gas heat exchanger, a good estimate would be equal resistances on each fluid side, considering a thermally balanced design. This would correspond to Eq. (9.23). Hence,

$$\text{ntu}_a = 2 \times \text{NTU} = 2 \times 7.079 = 14.16$$

Then neglecting the wall thermal resistance (since we do not know A_w yet), we get, from Eq. (9.22),

$$\text{ntu}_a = 2C^* \cdot \text{NTU} = 2 \times 0.858 \times 7.079 = 12.15$$

While $\text{ntu}_a = 12.15$ is a somewhat refined value, we could have taken $\text{ntu}_a = 14.16$ for the first iteration.

Core Mass Velocities. To determine G from Eq. (9.28), we need to estimate the values of j/f and η_o. Since j and f versus Re characteristics are specified for the surfaces on the gas and air sides, j/f versus Re curves are constructed as shown in Fig. E9.2. Since we do not know Re yet, an approximate average ("ballpark") value of j/f over the complete range of Re is taken for each surface from this figure as

$$(j/f)_g \approx 0.25 \qquad (j/f)_a \approx 0.25$$

Again, a more precise value of j/f is not essential since we are getting a first approximate value of G with a number of other approximations. In absence of any specific values of η_o, we will assume η_o on both the gas the air sides to be 0.80. Now substituting all values on the right-hand side of Eq. (9.28), we get

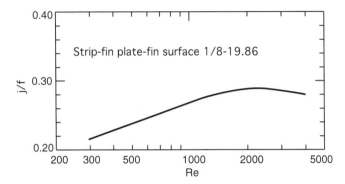

FIGURE E9.2 j/f vs. Re characteristics of surfaces of Fig. E9.1B. (From Shah, 1981).

$$G_g = \left[\frac{2g_c}{(1/\rho)_m \cdot \mathrm{Pr}^{2/3}} \frac{\eta_o\,\Delta p}{\mathrm{ntu}} \frac{j}{f}\right]^{1/2}_g = \left[\frac{2 \times 1 \times 0.8 \times (9.05 \times 10^3\,\mathrm{Pa}) \times 0.25}{(1/0.6212\,\mathrm{kg/m^3}) \times 0.8039 \times 14.16}\right]^{1/2}$$

$$= 14.06\,\mathrm{kg/m^2 \cdot s}$$

$$G_a = \left[\frac{2g_c}{(1/\rho)_m \cdot \mathrm{Pr}^{2/3}} \frac{\eta_o\,\Delta p}{\mathrm{ntu}} \frac{j}{f}\right]^{1/2}_a = \left[\frac{2 \times 1 \times 0.8 \times (8.79 \times 10^3\,\mathrm{Pa}) \times 0.25}{(1/0.9319\,\mathrm{kg/m^3}) \times 0.7823 \times 12.15}\right]^{1/2}$$

$$= 18.56\,\mathrm{kg/m^2 \cdot s}$$

Reynolds Numbers and j and f Factors. Compute the Reynolds number on each fluid side from its definition as

$$\mathrm{Re}_g = \left(\frac{GD_h}{\mu}\right)_g = \frac{14.06\,\mathrm{kg/m^2 \cdot s} \times 0.00154\,\mathrm{m}}{0.0000393\,\mathrm{Pa \cdot s}} = 551$$

$$\mathrm{Re}_a = \left(\frac{GD_h}{\mu}\right)_a = \frac{18.56\,\mathrm{kg/m^2 \cdot s} \times 0.00154\,\mathrm{m}}{0.0000347\,\mathrm{Pa \cdot s}} = 824$$

From Fig. E9.1B, (or the curve fit of j and f data), determine the j and f factors for these Reynolds numbers as follows:

	Re	j	f
Gas	551	0.0174	0.0695
Air	824	0.0135	0.0539

Since Reynolds numbers indicate the flow as laminar on both gas and air sides, the correction to the j factor for the temperature-dependent property effects is unity because $n = 0$ from Table 7.12.

Heat Transfer Coefficients, Fin Effectivenesses, and Overall Heat Transfer Coefficient. We compute the heat transfer coefficient from the definition of the j factor as follows:

$$h_g = \left(\frac{jGc_p}{\Pr^{2/3}}\right)_g = \frac{0.0174 \times 14.06\,\mathrm{kg/m^2 \cdot s} \times (1.117 \times 10^3)\,\mathrm{J/kg \cdot K}}{0.804} = 339.88\,\mathrm{W/m^2 \cdot K}$$

$$h_a = \left(\frac{jGc_p}{\Pr^{2/3}}\right)_a = \frac{0.0135 \times 18.57\,\mathrm{kg/m^2 \cdot s} \times (1.081 \times 10^3)\,\mathrm{J/kg \cdot K}}{0.782} = 346.36\,\mathrm{W/m^2 \cdot K}$$

Let us calculate m_g and m_a in order to calculate the fin efficiency on each fluid side. Since the offset fins are used on both gas and air sides, we use Eq. (4.147) with L_f replaced by ℓ_s to take into account strip edge exposed area.

$$m_g = \left[\frac{2h}{k_f \delta}\left(1 + \frac{\delta}{\ell_s}\right)\right]_g^{1/2} = \left[\frac{2 \times 339.88\,\mathrm{W/m^2 \cdot K}}{18\,\mathrm{W/m \cdot K} \times 0.102 \times 10^{-3}\,\mathrm{m}}\left(1 + \frac{0.102\,\mathrm{mm}}{3.175\,\mathrm{mm}}\right)\right]^{1/2}$$

$$= 618.17\,\mathrm{m^{-1}}$$

$$m_a = \left[\frac{2h}{k_f \delta}\left(1 + \frac{\delta}{\ell_s}\right)\right]_a^{1/2} = \left[\frac{2 \times 346.36\,\mathrm{W/m^2 \cdot K}}{18\,\mathrm{W/m \cdot K} \times 0.102 \times 10^{-3}\,\mathrm{m}}\left(1 + \frac{0.102\,\mathrm{mm}}{3.175\,\mathrm{mm}}\right)\right]^{1/2}$$

$$= 624.04\,\mathrm{m^{-1}}$$

$$\ell_a = \ell_g \approx \frac{b}{2} - \delta = (2.49\,\mathrm{mm}/2 - 0.102\,\mathrm{mm}) = 1.143\,\mathrm{mm} = 0.00114\,\mathrm{m}$$

Thus

$$\eta_{f,g} = \frac{\tanh(m\ell)_g}{(m\ell)_g} = \frac{\tanh(618.17\,\mathrm{m^{-1}} \times 0.00114\,\mathrm{m})}{618.17\,\mathrm{m^{-1}} \times 0.00114\,\mathrm{m}} = 0.8609$$

$$\eta_{f,a} = \frac{\tanh(m\ell)_a}{(m\ell)_a} = \frac{\tanh(624.04\,\mathrm{m^{-1}} \times 0.00114\,\mathrm{m})}{624.04\,\mathrm{m^{-1}} \times 0.00114\,\mathrm{m}} = 0.8592$$

The overall surface efficiencies with A_f/A values from Fig. E9.1B or input are

$$\eta_{o,g} = 1 - (1 - \eta_f)\frac{A_f}{A} = 1 - (1 - 0.8609) \times 0.785 = 0.8908$$

$$\eta_{o,a} = 1 - (1 - \eta_f)\frac{A_f}{A} = 1 - (1 - 0.8592) \times 0.785 = 0.8895$$

To calculate U_a from Eq. (9.30), we need to first calculate α_a and α_g using Eq. (9.31):

$$\alpha_a = \frac{(b\beta)_a}{b_a + b_g + 2\delta_w} = \frac{2.54\,\mathrm{mm} \times 2254\,\mathrm{m^2/m^3}}{2.54\,\mathrm{mm} + 2.54\,\mathrm{mm} + 2 \times 0.5\,\mathrm{mm}} = 941.6\,\mathrm{m^2/m^3}$$

$$\alpha_g = \frac{(b\beta)_g}{b_a + b_g + 2\delta_w} = \frac{2.54\,\mathrm{mm} \times 2254\,\mathrm{m^2/m^3}}{2.54\,\mathrm{mm} + 2.54\,\mathrm{mm} + 2 \times 0.5\,\mathrm{mm}} = 941.6\,\mathrm{m^2/m^3}$$

Hence,

$$\frac{A_a}{A_g} = \frac{\alpha_a}{\alpha_g} = \frac{941.6\,\text{m}^2/\text{m}^3}{941.6\,\text{m}^2/\text{m}^3} = 1.0$$

Thus U_g from Eq. (9.30), with no fouling, is

$$\frac{1}{U_g} = \frac{1}{(\eta_o h)_g} + \frac{\alpha_g/\alpha_a}{(\eta_o h)_a} = \frac{1}{0.8908 \times 339.88\,\text{W/m}^2 \cdot \text{K}} + \frac{1.0}{0.8895 \times 346.36\,\text{W/m}^2 \cdot \text{K}}$$

$$= 6.549 \times 10^{-3}\,\text{m}^2 \cdot \text{K/W}$$

$$U_g = 152.70\,\text{W/m}^2 \cdot \text{K}$$

Surface Area, Free Flow Area, and Core Dimensions. Since NTU $= 7.079$ and $C_{\min} = C_g = 1854\,\text{W/K}$,

$$A_g = \text{NTU}\,\frac{C_g}{U_g} = 7.079\left(\frac{1854\,\text{W/K}}{152.70\,\text{W/m}^2 \cdot \text{K}}\right) = 85.95\,\text{m}^2$$

From the specified $\dot{m}$ and computed G, the minimum free-flow area on the gas side is

$$A_{o,g} = \left(\frac{\dot{m}}{G}\right)_a = \frac{1.66\,\text{kg/s}}{14.06\,\text{kg/m}^2 \cdot \text{s}} = 0.1181\,\text{m}^2$$

The air flow length is then computed from the definition of the hydraulic diameter:

$$L_g = \left(\frac{D_h A}{4A_o}\right)_g = \frac{0.00154\,\text{m} \times 85.95\,\text{m}^2}{4 \times 0.1181\,\text{m}^2} = 0.280\,\text{m}$$

Since $A_g/A_a = 1$ and $A_g = 85.95\,\text{m}^2$, we get

$$A_a = A_g = 85.95\,\text{m}^2$$

Also,

$$A_{o,a} = \left(\frac{\dot{m}}{G}\right)_a = \frac{2.00\,\text{kg/s}}{18.56\,\text{kg/m}^2 \cdot \text{s}} = 0.1078\,\text{m}^2$$

and

$$L_a = \left(\frac{D_h A}{4A_o}\right)_a = \frac{0.00154\,\text{m} \times 85.95\,\text{m}^2}{4 \times 0.1078\,\text{m}^2} = 0.307\,\text{m}$$

To calculate the core frontal area on each fluid side, we first need to determine $\sigma = \alpha D_h/4$ as

$$\sigma_a = \frac{\alpha_a D_{h,a}}{4} = \frac{(941.6\,\text{m}^2/\text{m}^3) \times 0.00154\,\text{m}}{4} = 0.363$$

$$\sigma_g = \frac{\alpha_g D_{h,g}}{4} = \frac{(941.6\,\text{m}^2/\text{m}^3) \times 0.00154\,\text{m}}{4} = 0.363$$

Hence,

$$A_{fr,g} = \frac{A_{o,g}}{\sigma_g} = \frac{0.118\,\mathrm{m}^2}{0.363} = 0.3253\,\mathrm{m}^2 \qquad A_{fr,a} = \frac{A_{o,a}}{\sigma_a} = \frac{0.1078\,\mathrm{m}^2}{0.363} = 0.2970\,\mathrm{m}^2$$

Since $A_{fr,a} = L_g L_3$ or $A_{fr,g} = L_a L_3$, we get

$$L_3 = \frac{A_{fr,a}}{L_g} = \frac{0.2970\,\mathrm{m}^2}{0.280\,\mathrm{m}} = 1.061\,\mathrm{m} \quad \text{or} \quad L_3 = \frac{A_{fr,g}}{L_a} = \frac{0.3253\,\mathrm{m}^2}{0.307\,\mathrm{m}} = 1.060\,\mathrm{m}$$

The difference in two values of L_3 is due strictly to the round-off error.

Note that we obtained L_g, L_a, and L_3 as 0.280, 0.307, and 1.060. Thus, even with the very first iteration with very approximate value of G yielded the core dimensions that are within 6%.

Pressure Drops. We will now use Eq. (9.20) to determine the pressure drop on each fluid side. The entrance and exit loss coefficients will be the same as those determined during the rating problem.

Gas side: $K_c = 0.36$ $K_e = 0.42$

Air side: $K_c = 0.36$ $K_e = 0.42$

To correct f factors for the temperature-dependent property effects, let us first calculate T_w. The thermal resistances on the hot and cold fluid sides are

$$\mathbf{R}_g = \frac{1}{(\eta_o h A)_g} = \frac{1}{0.8908 \times 339.88\,\mathrm{W/m}^2 \cdot \mathrm{K} \times 85.95\,\mathrm{m}^2} = 3.843 \times 10^{-5}\,\mathrm{K/W}$$

$$\mathbf{R}_a = \frac{1}{(\eta_o h A)_a} = \frac{1}{0.8895 \times 346.36\,\mathrm{W/m}^2 \cdot \mathrm{K} \times 85.95\,\mathrm{m}^2} = 3.776 \times 10^{-5}\,\mathrm{K/W}$$

Therefore,

$$\frac{\mathbf{R}_g}{\mathbf{R}_a} = \frac{3.843 \times 10^{-5}\,\mathrm{K/W}}{3.776 \times 10^{-5}\,\mathrm{K/W}} = 1.018$$

Now

$$T_w = \frac{T_{g,m} + (\mathbf{R}_g/\mathbf{R}_a) T_{a,m}}{1 + (\mathbf{R}_g/\mathbf{R}_a)} = \frac{606.7°\mathrm{C} + 1.018 \times 452.0°\mathrm{C}}{1 + 1.018} = 528.7°\mathrm{C} = 801.8\,\mathrm{K}$$

Since the gas is being cooled, using Eq. (9.11) and the exponent $m = 0.81$ from Table 7.12, the corrected f factor is

$$f_g = \left[f_{cp} \left(\frac{T_w}{T_m} \right)^m \right]_g = 0.0695 \left(\frac{801.8\,\mathrm{K}}{879.8\,\mathrm{K}} \right)^{0.81} = 0.0645$$

Since the air is being heated, using Eq. (9.11) and the exponent $m = 1.00$ from Table 7.12, we have

$$f_a = \left[f_{cp} \left(\frac{T_w}{T_m} \right)^m \right]_a = 0.0539 \left(\frac{801.8 \text{ K}}{725.2 \text{ K}} \right)^{1.00} = 0.0596$$

The pressure drops, using Eq. (9.20), are

$$\Delta p_g = \frac{(14.06 \text{ kg/m}^2 \cdot \text{s})^2}{2 \times 1 \times 0.4751 \text{ kg/m}^3} \left[(1 - 0.363^2 + 0.36) + 2 \left(\frac{0.4751 \text{ kg/m}^3}{0.8966 \text{ kg/m}^3} - 1 \right) \right.$$

$$\left. + \frac{0.0645 \times 0.280 \text{ m} \times 0.4751 \text{ kg/m}^3}{(0.00154/4) \text{ m} \times 0.6212 \text{ kg/m}^3} - (1 - 0.363^2 - 0.42) \left(\frac{0.4751 \text{ kg/m}^3}{1.8966 \text{ kg/m}^3} \right) \right]$$

$$= 208.04 \text{ Pa} \times (1.2282 - 0.9402 + 35.8765 - 0.1123)$$

$$= 208.04 \text{ Pa} \times 36.0522 = 7500 \text{ Pa} = 7.50 \text{ kPa}$$

$$\Delta p_a = \frac{(18.56 \text{ kg/m}^2 \cdot \text{s})^2}{2 \times 1 \times 1.4726 \text{ kg/m}^3} \left[(1 - 0.363^2 + 0.36) + 2 \left(\frac{1.4726 \text{ kg/m}^3}{0.6817 \text{ kg/m}^3} - 1 \right) \right.$$

$$\left. + \frac{0.0596 \times 0.307 \text{ m} \times 1.4726 \text{ kg/m}^3}{(0.00154/4) \text{ m} \times 0.9319 \text{ kg/m}^3} - (1 - 0.363^2 + 0.42) \left(\frac{1.4726 \text{ kg/m}^3}{0.6817 \text{ kg/m}^3} \right) \right]$$

$$= 116.96 \text{ Pa} \times (1.2282 + 2.3204 + 75.0999 - 0.9682)$$

$$= 116.96 \text{ Pa} \times 77.6803 = 9085 \text{ Pa} = 9.09 \text{ kPa}$$

Since the air-side Δp is higher than specified, new values of G on *both* gas and air sides are determined again from Eq. (9.20), considering G as unknown.

$$9.05 \times 10^3 \text{ Pa} = \frac{G_g^2}{2 \times 1 \times 0.4751 \text{ kg/m}^3} \times [36.0522] \rightarrow G_g = 15.44 \text{ kg/m}^2 \cdot \text{s}$$

$$8.79 \times 10^3 \text{ Pa} = \frac{G_a^2}{2 \times 1 \times 1.4726 \text{ kg/m}^3} \times [77.6803] \rightarrow G_a = 18.26 \text{ kg/m}^2 \cdot \text{s}$$

Knowing the rating problem solution, we can see that this new value of G_g and G_a have almost converged to the true values. Repeating Steps 2 through 4 of section 9.2.2.3 with the new values of G yields the following results:

Iterations	NTU	L_g	L_a	L_3
Original	7.079	0.280	0.307	1.061
First	7.079	0.300	0.295	1.009
Second	7.079	0.300	0.297	1.003
Third	7.079	0.300	0.299	1.003

Thus with the iterations, the solution can be converged to the actual core dimensions within any desired accuracy.

Discussion and Comments: The foregoing method clearly indicates how fast a solution to the sizing problem will converge to the core dimensions that will meet the heat transfer and pressure drops on both sides for a crossflow exchanger when no constraints are imposed on the dimensions. However, with the imposed constraints on the core dimensions, the design will not meet the heat transfer and pressure drops specified on both fluid sides, and a constraint on the geometric parameters or operating condition variables must be relaxed. Also, in a sizing problem, one would like to find the optimum set of core/surface geometries, and/or operating conditions for the problem specification. In that case, it becomes an optimization problem. We discuss it in Section 9.6.

9.3 TUBE-FIN HEAT EXCHANGERS

Tube-fin exchangers are mostly used as single-pass crossflow or multipass cross-counter-flow exchangers. A heat pipe heat exchanger is effectively two tube-fin single-pass cross-flow exchangers placed side by side separated by a splitter plate and connected to each other by the same tubes, which are heat pipes. Fluids (usually air and exhaust gas) flow in counterflow (opposite) directions (crossflow to finned tubes). The design theory for heat pipe heat exchangers has been presented by Shah and Giovannelli (1988), and will not be discussed here.

The solution procedures for the rating and sizing problems for tube-fin exchangers with individually finned tubes or flat fins (see Fig. 1.31), either in single-pass crossflow or two-pass cross-counterflow, are identical to those for plate-fin exchangers described in detail in preceding sections. Hence, rather than repeating the same steps, only the differences are highlighted.

9.3.1 Surface Geometries

In this case, the surface area density α, a ratio of total transfer surface area A on one side of the exchanger to total volume V of the exchanger is used for heat transfer surfaces used in tube-fin exchangers. Hence, α, σ, and D_h are computed from Eqs. (8.51)–(8.53). Note that Eq. (9.31) and the first equality of Eq. (9.37) have no physical meaning for tube-fin exchangers. The heat transfer surface area density β does not have significance for the tube-fin exchangers.

9.3.2 Heat Transfer Calculations

All heat transfer equations, except for $\mathbf{R}_w$ and η_f, remain the same as those for plate-fin exchangers. The overall thermal resistance Eq. (9.16) should include a term for contact resistance if fins are wrapped tension wound or mechanically expanded onto the tubes. Also, the wall thermal resistance term should be for a tube. For a circular tube, it is as given in Eq. (3.26).

The fin efficiency for circular fins of Fig. 1.31a or flat fins of Fig. 1.31b is different from that for the straight fins [Eq. (9.14)]. For circular fins of Fig. 1.31a, the fin efficiency is given by Eq. (4.151), and an approximate formula, which does not involve Bessel functions, is given in Table 4.5. The fin efficiency of flat fins is obtained by an approximate method referred to as the *sector method*, discussed in Section 4.3.2.3. The details of how to evaluate the fin efficiency for this case are also presented there and hence are not repeated here.

9.3.3 Pressure Drop Calculations

The core pressure drop Eq. (9.20) for plate-fin exchangers needs to be modified for tube-fin exchangers for the tube outside, as discussed in Section 6.2.2.2. For individually finned tube exchangers, the entrance and exit pressure losses cannot readily be measured and evaluated separately. Hence, they are lumped into experimentally determined friction factors. In this case, the pressure drop is computed from Eq. (6.32) if f is the Fanning friction factor based on the hydraulic diameter. If, instead, the Euler number or Hagen number per tube row for the tube bank is used for the pressure drop evaluation, Eq. (6.33) should be used for the pressure drop calculation.

For continuous flat fins, the pressure drop components of Eq. (9.20) are all valid. However, while the entrance and exit pressure losses are evaluated based on the flow area at the leading and trailing edges of the fins, the core friction and momentum effect terms are based on G computed from the minimum free-flow area within the core. Thus, Eq. (6.36) should be used, instead of Eq. (9.20), for pressure drop evaluation of continuous flat fins on tubes.

9.3.4 Core Mass Velocity Equation

For a tube-fin exchanger with flat fins, if the flow friction and heat transfer correlations are based on the hydraulic diameter on the tube outside, the core mass velocity equation of Eq. (9.28) is also valid for the tube outside. For finned tubes, there are a number of different ways of correlating heat transfer and flow friction characteristics, such as that Nu is based on the tube outside diameter and the pressure drop is based on the Euler number or Hagen number. In that case, the core mass velocity equation should be derived for the specific cases depending on the nature of the correlations, such as using the Nu/Eu ratio instead of the j/f ratio.

9.4 PLATE HEAT EXCHANGERS

Plate heat exchangers can be designed with m passes on the fluid 1 side and n passes on the fluid 2 side, depending on flow rate imbalance, available pressure drop, and other design criteria. One of the most common flow arrangements is 1-pass 1-pass counterflow design, selected for reasonably balanced flow rates on hot and cold fluid sides. If the flow maldistribution within the PHE is ignored and if all plates have the same geometry, the rating of this exchanger is identical to that for the counterflow plate-fin exchanger described in Section 9.2.1 except that η_o and η_f are unity since there are no fins in a PHE. Unlike plate-fin exchangers, sometimes it is *not* possible to meet the pressure drop and heat transfer specified even on one side of a PHE. So let us first discuss this condition and the limiting cases for the design of a PHE; these limiting cases involve limitations imposed on the specified heat transfer and/or pressure drops. For a PHE with mixed plate design (i.e., having two different plate geometries), the channel-to-channel flow maldistribution must be taken into account for rating and sizing. Detailed analysis of flow maldistribution is given in Chapter 12. Here we present only a step-by-step method for rating a PHE with mixed plate design. We then very briefly discuss sizing a PHE.

9.4.1 Limiting Cases for the Design

Let us discuss how the design of a 1 pass–1 pass counterflow PHE differs from that of a pure counterflow plate-fin exchanger. For a plate-fin exchanger, the minimum free-flow area A_o and the surface area A on each fluid side are independent of each other. For example, for a specified (or selected) A_o, the surface area A could be varied by changing the fin density (this change has a minor effect on A_o). As a result, both heat transfer and pressure drop on one of the two fluid sides can be matched exactly. The design method for plate-fin exchangers then involves the coupling of specified NTU and Δp through a core mass velocity equation as presented by Eq. (9.28). This approach cannot be used for plate heat exchangers since A_o and A are not independent. Once A_o is fixed (i.e., the number of plates is selected), A is fixed automatically ($A = 4A_o L/D_e$) for a specified plate geometry[†] (the plate pattern, D_e, and L specified) because there is no secondary (or extended) surface. Hence, in most cases, it is not possible (unless mixed channels are used as discussed later) to match specified pressure drop and heat transfer identically even on one fluid side. The plate exchanger design in general is either pressure drop– or heat transfer–limited. In the pressure drop–limited design, the free-flow area is determined that satisfies the pressure drop limit; however, the corresponding surface area will be higher than that required to meet the heat duty. In the heat transfer–limited design, the surface area specified transfers the required heat duty; however, the corresponding free-flow area will be higher than that required to take advantage of the available pressure drop on either fluid side. Hence, the resulting pressure drops on both fluid sides will be lower than specified. This is explained further through Fig. 9.2 next.

A channel (a flow passage) in a PHE is made of two plates. Two given plates (with two different chevron angles) can be used to obtain three different channel types. For example, a plate type has two chevron angles β_{low} and β_{high}; three channel (plate) combinations are possible: β_{low} and β_{low}, β_{low} and β_{high}, and β_{high} and β_{high}. For discussion purposes, let us refer to them as channel types 1, 2, and 3, respectively. It is worth noting that there exists an apparent chevron angle β_{app} for a mixed-plate channel, which is approximately equal to $(\beta_{low} + \beta_{high})/2$. This approximate relationship can be unsatisfactory when a 90° plate is "mixed" with another plate having $(\beta < 90°)$ to form a channel. A more precise value of β_{app} can be obtained experimentally by testing the mixed-plate channels, determining their j (or Nu) and f vs. Re characteristics, comparing them with the data for one type of chevron plate of various β, and finding the β value of the closest match to the data. However, no such β_{app} values are available in the open literature.

For a given design, there exists an effective (or ideal) chevron angle β_{eff} that will meet the design criteria: the heat transfer specified, the pressure drop on one fluid side matched, and the pressure drop on the other fluid side lower than the value specified. In general, it is not possible to match the pressure drop on both sides because of the limited number of plate geometries available in a given size. In reality, the value of β_{eff} (consider it less than β_{app} for discussion purposes) indicates whether the heat exchanger should have channel types 1 and 2 or 2 and 3: In other words, if $\beta_{eff} < \beta_{app}$, channel types 1 and 2 should be used; and if $\beta_{eff} > \beta_{app}$, channel types 2 and 3 should be used in the

---

[†] There are a limited number of plate patterns available, due to the very high cost of tooling and of the press required to stamp the plates.

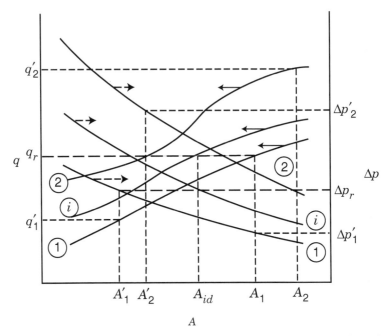

A

FIGURE 9.2   Heat-transfer-limited (channel type 1) and pressure-drop-limited (channel type 2) designs. Curves labeled 1 have $\beta_{\text{low}} - \beta_{\text{low}}$ plates; curves labeled 2 have $\beta_{\text{high}} - \beta_{\text{high}}$ plates. (From Shah and Focke, 1988.)

PHE. In both cases, a specific combination of one type of plates and mixed plates are chosen to meet the required q_r or Δp_r on one fluid side, depending on whether the design is heat transfer– or pressure drop–limited.

We now explain the concept of heat transfer– and pressure drop–limited designs and its relation to the concept of mixing the plates using Fig. 9.2. Consider two different channel types as possible candidates to meet the required heat duty q_r and pressure drops Δp_r for the two fluid streams. Generally, in a heat exchanger, one fluid stream has a more severe pressure drop restriction than the other stream. We consider only that fluid stream having the more severe pressure drop constraint,[†] designated as Δp_r (required or specified Δp). The heat transfer rate and pressure drop (on the more constrained side) as a function of the surface area (or the number of thermal plates) for these two channels are shown in Fig. 9.2 with solid line curves labeled as 1 and 2, with heat transfer rate and pressure drop scales on the left- and right-hand y axes, respectively. Also the specified q_r and Δp_r are shown in the same figure by horizontal long dashed lines. The following discussion assumes that β_{eff} lies between those given by channel types 1 and 2.

[†] Note that the fluid side having a severe pressure drop constraint does not necessarily have a lower pressure drop specified than that on the other fluid side; in fact, it could have higher pressure drop. All it means is that we have to ensure that the design pressure drop is equal or lower than the pressure drop specified.

Channel Type 1. As shown in Fig. 9.2, heat transfer of channel type 1 matches the required heat transfer q_r (see the intersection of dashed line q_r and solid line q for surface 1) with surface area A_1 and has pressure drop $\Delta p_1'$; thus, it does not utilize the available pressure drop since $\Delta p_1' < \Delta p_r$. If the entire specified pressure drop would have been utilized (consumed), the surface area required would be A_1', but the heat transfer rate of the exchanger would be only q_1', which is significantly lower than q_r. Hence, the exchanger design with channel type 1 is limited by the surface area A_1 to transfer the required q_r. Hence, it is designated as a *heat transfer–limited design.*

Channel Type 2. Pressure drop of channel type 2 matches the pressure drop requirement Δp_r with surface area A_2, but in doing so, it utilizes much more free-flow area (and hence surface area), which yields the heat transfer rate q_2'. This is higher than the value required. If the heat transfer would have been matched, the surface area required would have been A_2', but the resulting pressure drop would have been $\Delta p_2'$, which is significantly higher than Δp_r. Hence, the exchanger design with channel type 2 is limited by the surface area A_2 to meet the specified Δp_r constraint. Hence, it is designated as a *pressure drop–limited design.*

Mixed Channels. From Fig 9.2, it is clear that the ideal channel, designated as i, which meets the heat transfer and pressure drop specifications simultaneously, will require the ideal amount of the surface area A_{id}. Thus, a proper mixing (combination) of channel types 1 and 2 will yield the q and Δp curves as those indicated by i in Fig. 9.2. That design would enable the designer to satisfy the design heat duty and pressure drop on one of the two streams considered.

The use of two channel types in a given pass creates channel-to-channel flow maldistribution discussed in Section 9.4.2 and thereby a reduction in heat duty. Marriot (1977) reports that the effect of maldistribution of this type on q is typically less than 7%. In practical situations, a design based on mixed channels should be rated to quantify the effects of flow maldistribution (refer to Sections 12.1.2 and 12.1.3 for detailed discussion of flow maldistribution). If these effects are too severe, a pressure drop–limited design (using uniform channels giving a chevron angle higher than ideal) or a heat transfer–limited design (using uniform channels, giving a lower chevron angle) may be preferable to a mixed-channel design.

9.4.2   Uniqueness of a PHE for Rating and Sizing

Since 1 pass–1 pass counterflow PHE is the most common in application, its rating and sizing can be accomplished by using the methods described in Sections 9.2.1 and 9.2.2 for plate-fin exchangers, if the flow distribution is assumed uniform through all flow channels. However, due to the nature of exchanger construction, it leads to several flow maldistributions: within the channel, channel to channel, and manifold induced. These maldistributions are described further in some detail in Sections 12.1.2 and 12.1.3. To explain the rating procedure for a PHE with a mixed-plate design, we must consider, as a minimum, channel-to-channel flow maldistribution. This type of flow maldistribution occurs due to the presence of two different plate groups in a PHE. For example, consider two types of plates used in a PHE: part of an exchanger made up with all 30° chevron plates and the rest with alternating 30° and 60° chevron plates (i.e., having mixed-plate channels). In such a heat exchanger, in addition to having manifold-induced maldistribution in any plate group, the flow will be maldistributed among different plate groups, due to their different flow resistance (such as f vs. Re) characteristics. This can be quantified readily as shown below on a given fluid side if we imagine that the pressure drop across all channels (all plate groups) on a given fluid side is the same. Hence, we first summarize the theory as to how to determine different flow rates through two groups of plates. Subsequently, we show how to compute the heat transfer rate of this mixed-plate PHE.

Heat transfer rate (heat duty) for a PHE can be determined by idealizing the two different plate groups in a PHE exchanger as two exchangers in parallel coupling (Fig. 9.3). Hence, one needs to determine the individual mass flow rates through these two plate groups first. Consider the same core (frictional) pressure drop for each plate group, and neglect manifold and port pressure drops and momentum and elevation change effects; using the core frictional term only [the second term on the right-hand side of Eq. (6.44)], we get

$$\frac{f_I G_I^2}{D_{e,I}} = \frac{f_{II} G_{II}^2}{D_{e,II}} \tag{9.41}$$

where subscripts I and II denote plate groups I and II and D_e is the equivalent diameter [see Eq. (8.135) for the definition]. We consider that the friction factor can be represented as

$$f = a \cdot \mathrm{Re}^{-n} \tag{9.42}$$

Combining Eqs. (9.41) and (9.42), and noting that $\dot{m} = GA_o$, the ratio X of the mass flow rates through plate groups I and II can be presented as follows:

$$X = \frac{\dot{m}_I}{\dot{m}_{II}} = \left(\frac{a_{II}}{a_I}\right)^{1/(2-n)} \left(\frac{\mu_{II}}{\mu_I}\right)^{n/(2-n)} \left(\frac{D_{e,I}}{D_{e,II}}\right)^{(1+n)/(2-n)} \left(\frac{A_{o,I}}{A_{o,II}}\right) \tag{9.43}$$

where $A_{o,I}$ and $A_{o,II}$ are the total free-flow areas in plate groups I and II, respectively and can readily be calculated with a known number of plates or channels in each plate group. Then, from the mass balance, the total mass flow rate is

$$\dot{m} = \dot{m}_I + \dot{m}_{II} \tag{9.44}$$

Therefore, from Eqs. (9.43) and (9.44),

$$\dot{m}_I = \frac{X\dot{m}}{1+X} \qquad \dot{m}_{II} = \dot{m} - \dot{m}_I = \frac{\dot{m}}{1+X} \tag{9.45}$$

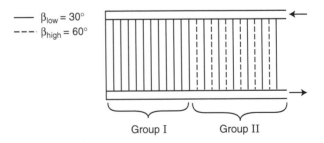

FIGURE 9.3   Idealized counterflow PHE with two plate groups in parallel.

Once the individual flow rates are determined, the pressure drop for each plate group can be determined from the last three terms on the right-hand side of Eq. (6.44). The manifold and port pressure drops then should be added to get the total pressure drop on each fluid side. If the f–Re correlation is about the same, ideally the total pressure drop on one fluid side of a PHE will be lower for the two different plate groups compared to that for only one plate group. This can readily be understood with an electric analogy that an electric circuit having two different resistances (such as 4 and 8 Ω) in parallel will have a lower electric potential than that of an electric circuit having two identical electrical resistances (6 and 6 Ω, a mean value for the two individual resistances of the first circuit) in parallel with the same total electric current, despite the fact that the sum of individual resistances is equal.

Once the total flow rates in each channel group are determined, the heat transfer analysis is straightforward by considering two exchangers in parallel, as shown in Fig. 9.3, corresponding to two plate groups. The temperature effectiveness of each plate group for a counterflow exchanger is given by

$$
P_{1,\mathrm{I}} = \begin{cases} \dfrac{1 - \exp[-\mathrm{NTU}_\mathrm{I}(1 - R_\mathrm{I})]}{1 - R_\mathrm{I}\exp[-\mathrm{NTU}_\mathrm{I}(1 - R_\mathrm{I})]} & \text{for } R_\mathrm{I} \neq 1 \\[4mm] \dfrac{\mathrm{NTU}_\mathrm{I}}{1 + \mathrm{NTU}_\mathrm{I}} & \text{for } R_\mathrm{I} = 1 \end{cases}
\tag{9.46}
$$

where

$$
\mathrm{NTU}_\mathrm{I} = \frac{(UA)_\mathrm{I}}{C_{1,\mathrm{I}}} = \frac{(UA)_\mathrm{I}}{(\dot{m}_1 c_{p,1})_1} \qquad R_\mathrm{I} = \frac{C_{1,\mathrm{I}}}{C_{2,\mathrm{I}}} = \frac{(\dot{m}_1 c_{p,1})_1}{(\dot{m}_1 c_{p,1})_2}
\tag{9.47}
$$

and

$$
\frac{1}{(UA)_\mathrm{I}} = \frac{1}{A_\mathrm{I}}\left(\frac{1}{h_1} + \hat{\mathbf{R}}_{1,f} + \frac{\delta_w}{k_w} + \hat{\mathbf{R}}_{2,f} + \frac{1}{h_2}\right)_\mathrm{I}
\tag{9.48}
$$

with $A_\mathrm{I} = A_{1,\mathrm{I}} = A_{2,\mathrm{I}} = A_{w,\mathrm{I}}$. Similarly, the temperature effectiveness $P_{1,\mathrm{II}}$ of the second plate group can be expressed in terms of NTU_II and $\mathbf{R}_\mathrm{II}$ defined in the same manner. The total exchanger heat duty is then given by

$$
q = q_\mathrm{I} + q_\mathrm{II} = \left(P_{1,\mathrm{I}} C_{1,\mathrm{I}} + P_{1,\mathrm{II}} C_{1,\mathrm{II}}\right)\left(T_{\mathrm{h},i} - T_{\mathrm{c},i}\right)
\tag{9.49}
$$

9.4.3 Rating a PHE

We now present a rating procedure or determination of heat transfer and pressure drop performance of a PHE that has two plate groups. In group I, all plates have the same chevron angle (such as β_{low}); in group II, two plate geometries (such as having β_{low} and β_{high}) are stacked alternately, thus having a mixed-plate pack (see Fig. 9.3). Since the performance of a given unit is to be determined, the following quantities are specified:

- Exchanger geometry (i.e., plate width and length, channel gap, number of plates, types of plates and how the mixing of plates is achieved in the given exchanger, etc.)

- Plate surface pattern with their heat transfer and pressure drop characteristics
- Flow arrangement of the two fluids (i.e., the number of passes on each fluid side and overall fluid flow direction)
- Mass flow rates, inlet temperatures, fluid physical properties, and fouling resistances for each fluid stream

With the foregoing known information, the following is a step-by-step rating procedure. This procedure is outlined for a PHE having two plate groups. If there is only one plate group, use the same procedure with all quantities for the plate group II ignored.

1. Calculate fluid properties (ρ, μ, k, and c_p) at the bulk mean temperature for each fluid side.
2. Compute $\dot{m}_\mathrm{I}$ and $\dot{m}_\mathrm{II}$ for both fluids from Eq. (9.45).
3. Determine Re for both fluids in each plate group.
4. Calculate h_h and h_c for both plate groups using the specified Nu or j vs. Re correlations.
5. Compute $(UA)_\mathrm{I}$ using Eq. (9.48). Similarly, compute $(UA)_\mathrm{II}$.
6. Calculate NTU_I and R_I using Eq. (9.47). Similarly, calculate NTU_II and R_II.
7. Determine $P_{1,\mathrm{I}}$ using Eq. (9.46). Similarly, determine $P_{1,\mathrm{II}}$.
8. Compute the heat duty q using Eq. (9.49).
9. Calculate f factors from Eq. (9.42).
10. Determine the combined channel pressure drops and other pressure drop components from Eq. (6.44) for both fluid sides.

Next we illustrate this procedure with one rating example.

Example 9.3 A 1 pass–1 pass counterflow water-to-water plate heat exchanger has 47 thermal plates or 48 fluid channels (24 channels for each fluid). On each fluid side, chevron plates of $\beta = 30°$ are used for 8 channels and 30° and 60° mixed chevron plates are used for 16 channels. Assume that $\beta_\mathrm{eff} = 39.8°$ and the following are empirical correlations for the Nusselt and Reynolds numbers based on D_e.

$$\mathrm{Nu} = 0.724\left(\frac{\beta}{30°}\right)^{0.646} \cdot \mathrm{Re}^{0.583} \cdot \mathrm{Pr}^{1/3}$$

$$f = \begin{cases} 0.80\mathrm{Re}^{-0.25} & \text{for } \beta = 30° \\ 3.44\mathrm{Re}^{-0.25} & \text{for } \beta = 30° \text{ and } 60° \text{ mixed plates} \end{cases}$$

The following process, geometry, and other information are provided.

Process Variables	Hot Fluid	Cold Fluid
Fluid type	Water	Water
Mass flow rate (kg/s)	18	10
Inlet temperature (°C)	40	20

Outlet temperature (°C)		30		38
Allowable pressure drop (kPa)		30		20

Plate geometry information:

Plate width W (m)	0.5	Plate length (height) L (m)	1.1
Port diameter D_p (m)	0.1	Channel spacing $2a$ mm	3.5
Equivalent diameter D_e (m)	7×10^{-3}	Projected area per plate A (m²)	0.55

Fluid properties [use the same constant properties (for simplicity) on both hot- and cold-fluid sides]:

Dynamic viscosity (Pa · s)	8.1×10^{-4}	Density (kg/m³)	995.4
Thermal conductivity (W/m · K)	0.619	Specific heat (J/kg · K) 4177 Pr = 5.47	

Additional information:

Total fouling resistance $= 4 \times 10^{-5}$ m² K · W   Plate wall thermal resistance
$$= 3 \times 10^{-6} \, \text{m}^2 \, \text{K} \cdot \text{W}$$

Determine heat transfer and pressure drop performance of this exchanger.

SOLUTION

Problem Data and Schematic: The detailed process, geometry, and other data are provided in the problem statement for the PHE.

Determine: The heat transfer and pressure drop on each fluid side for this exchanger.

Assumptions: The assumptions listed in Section 3.2.1 applicable to a PHE are invoked.

Analysis: We follow the steps outlined in Section 9.4.3 for the solution after we compute flow and heat transfer areas as follows:

$$A_{o,\text{I}} = A_{o,c,\text{I}} = A_{o,h,\text{I}} = W(2a)N_{c,\text{I}} = 0.5\,\text{m} \times (3.5 \times 10^{-3})\,\text{m} \times 8 = 0.014\,\text{m}^2$$

$$A_{o,\text{II}} = A_{o,c,\text{II}} = A_{o,h,\text{II}} = W(2a)N_{c,\text{II}} = 0.5\,\text{m} \times (3.5 \times 10^{-3})\,\text{m} \times 16 = 0.028\,\text{m}^2$$

$$A_{c,\text{I}} = A_{h,\text{I}} = 0.55\,\text{m}^2 \times 8 \times 2 = 8.8\,\text{m}^2$$

$$A_{c,\text{II}} = A_{h,\text{II}} = 0.55\,\text{m}^2 \times 16 \times 2 = 17.6\,\text{m}^2$$

Since we have one type of chevron plates ($\beta = 30°$) in group I of the PHE and mixed chevron plates ($\beta = 30°$ and $60°$) in group II of the PHE, let us first evaluate the flow distribution of each fluid in these sections using Eq. (9.43). For this equation, the ratio of dynamic viscosities and equivalent diameters will be unity from the problem statement. Hence, from Eq. (9.43), we get

$$X = \left(\frac{a_{\text{II}}}{a_{\text{I}}}\right)^{1/(2-n)} \left(\frac{\mu_{\text{II}}}{\mu_{\text{I}}}\right)^{n/(2-n)} \left(\frac{D_{e,\text{I}}}{D_{e,\text{II}}}\right)^{(1+n)/(2-n)} \left(\frac{A_{o,\text{I}}}{A_{o,\text{II}}}\right)$$

$$= \left(\frac{3.44}{0.80}\right)^{1/(2-0.25)} \times 1 \times 1 \times \frac{0.014\,\text{m}^2}{0.028\,\text{m}^2} = 1.151$$

where the values for a_j, $j = \text{I}$ or II, and n of Eq. (9.42) are given in the problem formulation. The mass flow rates from Eq. (9.45) are then

$$\dot{m}_{c,\mathrm{I}} = \frac{X\dot{m}_c}{1+X} = \frac{1.151 \times 10\,\mathrm{kg/s}}{1+1.151} \qquad \dot{m}_{c,\mathrm{II}} = \dot{m}_c - \dot{m}_{c,\mathrm{I}} = 10 - 5.351\,\mathrm{kg/s} = 4.649\,\mathrm{kg/s}$$
$$= 5.351\,\mathrm{kg/s}$$

$$\dot{m}_{h,\mathrm{I}} = \frac{X\dot{m}_h}{1+X} = \frac{1.151 \times 18\,\mathrm{kg/s}}{1+1.151} \qquad \dot{m}_{h,\mathrm{II}} = \dot{m}_h - \dot{m}_{h,\mathrm{I}} = 18 - 9.632\,\mathrm{kg/s} = 8.368\,\mathrm{kg/s}$$
$$= 9.632\,\mathrm{kg/s}$$

The mass velocities are then given by

$$G_{c,\mathrm{I}} = \frac{\dot{m}_{c,\mathrm{I}}}{A_{o,c,\mathrm{I}}} = \frac{5.351\,\mathrm{kg/s}}{0.014\,\mathrm{m}^2} \qquad G_{c,\mathrm{II}} = \frac{\dot{m}_{c,\mathrm{II}}}{A_{o,c,\mathrm{II}}} = \frac{4.649\,\mathrm{kg/s}}{0.028\,\mathrm{m}^2}$$
$$= 382.21\,\mathrm{kg/m}^2 \cdot \mathrm{s} \qquad\qquad = 166.04\,\mathrm{kg/m}^2 \cdot \mathrm{s}$$

$$G_{h,\mathrm{I}} = \frac{\dot{m}_{h,\mathrm{I}}}{A_{o,h,\mathrm{I}}} = \frac{9.632\,\mathrm{kg/s}}{0.014\,\mathrm{m}^2} \qquad G_{h,\mathrm{II}} = \frac{\dot{m}_{h,\mathrm{II}}}{A_{o,h,\mathrm{II}}} = \frac{8.368\,\mathrm{kg/s}}{0.028\,\mathrm{m}^2}$$
$$= 688.00\,\mathrm{kg/m}^2 \cdot \mathrm{s} \qquad\qquad = 298.86\,\mathrm{kg/m}^2 \cdot \mathrm{s}$$

The Reynolds numbers are determined from the definition as

$$\mathrm{Re}_{c,\mathrm{I}} = \frac{G_{c,\mathrm{I}}D_e}{\mu} = \frac{382.21\,\mathrm{kg/m}^2 \cdot \mathrm{s} \times 7 \times 10^{-3}\,\mathrm{m}}{8.1 \times 10^{-4}\,\mathrm{Pa} \cdot \mathrm{s}} = 3303$$

$$\mathrm{Re}_{c,\mathrm{II}} = \frac{G_{c,\mathrm{II}}D_e}{\mu} = \frac{166.04\,\mathrm{kg/m}^2 \cdot \mathrm{s} \times 7 \times 10^{-3}\,\mathrm{m}}{8.1 \times 10^{-4}\,\mathrm{Pa} \cdot \mathrm{s}} = 1435$$

$$\mathrm{Re}_{h,\mathrm{I}} = \frac{G_{h,\mathrm{I}}D_e}{\mu} = \frac{688.00\,\mathrm{kg/m}^2 \cdot \mathrm{s} \times 7 \times 10^{-3}\,\mathrm{m}}{8.1 \times 10^{-4}\,\mathrm{Pa} \cdot \mathrm{s}} = 5946$$

$$\mathrm{Re}_{h,\mathrm{II}} = \frac{G_{h,\mathrm{II}}D_e}{\mu} = \frac{298.86\,\mathrm{kg/m}^2 \cdot \mathrm{s} \times 7 \times 10^{-3}\,\mathrm{m}}{8.1 \times 10^{-4}\,\mathrm{Pa} \cdot \mathrm{s}} = 2583$$

Now calculate the heat transfer coefficients on the cold and hot sides for groups I and II using the given correlation for Nu written in terms of heat transfer coefficients.

$$h_{c,\mathrm{I}} = 0.724\left(\frac{k}{D_e}\right)\left(\frac{\beta_{\mathrm{I}}}{30^\circ}\right)^{0.646} \cdot \mathrm{Re}_{c,\mathrm{I}}^{0.583} \cdot \mathrm{Pr}^{1/3}$$

$$= 0.724\left(\frac{0.619\,\mathrm{W/m} \cdot \mathrm{K}}{7 \times 10^{-3}\,\mathrm{m}}\right)\left(\frac{30^\circ}{30^\circ}\right)^{0.646}(3303)^{0.583}(5.47)^{1/3}$$

$$= 12{,}701\,\mathrm{W/m}^2 \cdot \mathrm{K}$$

$$h_{c,\mathrm{II}} = 0.724 \left(\frac{0.619 \,\mathrm{W/m \cdot K}}{7 \times 10^{-3} \,\mathrm{m}} \right) \left(\frac{39.8^\circ}{30^\circ} \right)^{0.646} (1435)^{0.583} (5.47)^{1/3} = 9377 \,\mathrm{W/m^2 \cdot K}$$

$$h_{h,\mathrm{I}} = 0.724 \left(\frac{0.619 \,\mathrm{W/m \cdot K}}{7 \times 10^{-3} \,\mathrm{m}} \right) \left(\frac{30^\circ}{30^\circ} \right)^{0.646} (5946)^{0.583} (5.47)^{1/3} = 17{,}894 \,\mathrm{W/m^2 \cdot K}$$

$$h_{h,\mathrm{II}} = 0.724 \left(\frac{0.619 \,\mathrm{W/m \cdot K}}{7 \times 10^{-3} \,\mathrm{m}} \right) \left(\frac{39.8^\circ}{30^\circ} \right)^{0.646} (2583)^{0.583} (5.47)^{1/3} = 13{,}210 \,\mathrm{W/m^2 \cdot K}$$

The overall conductance (UA) for each group is computed from Eq. (9.48) as follows:

$$\frac{1}{(UA)_{\mathrm{I}}} = \frac{1}{A_{c,\mathrm{I}}} \left(\frac{1}{h_{c,\mathrm{I}}} + \hat{\mathbf{R}}_{c,f} + \frac{\delta_w}{k_w} + \hat{\mathbf{R}}_{h,f} + \frac{1}{h_{h,\mathrm{I}}} \right)$$

$$= \frac{1}{8.8 \,\mathrm{m^2}} \left(\frac{1}{12{,}701 \,\mathrm{W/m^2 \cdot K}} + 4 \times 10^{-5} \,\mathrm{m^2 \cdot K/W} \right.$$

$$\left. + \, 3 \times 10^{-6} \,\mathrm{m^2 \cdot K/W} + \frac{1}{17{,}894 \,\mathrm{W/m^2 \cdot K}} \right)$$

$$= 2.0184 \times 10^{-5} \,\mathrm{K/W}$$

or

$$(UA)_{\mathrm{I}} = 49{,}544 \,\mathrm{W/K}$$

Similarly,

$$\frac{1}{(UA)_{\mathrm{II}}} = \frac{1}{17.6 \,\mathrm{m^2}} \left(\frac{1}{9377 \,\mathrm{W/m^2 \cdot K}} + 4 \times 10^{-5} \,\mathrm{m^2 \cdot K/W} + 3 \times 10^{-6} \,\mathrm{m^2 \cdot K/W} \right.$$

$$\left. + \frac{1}{13{,}210 \,\mathrm{W/m^2 \cdot K}} \right)$$

$$= 1.2804 \times 10^{-5} \,\mathrm{K/W}$$

or

$$(UA)_{\mathrm{II}} = 78{,}103 \,\mathrm{W/K}$$

Next, determine NTU and R for groups I and II from their definitions.

$$\mathrm{NTU_I} = \frac{(UA)_{\mathrm{I}}}{(\dot{m}c_p)_{c,\mathrm{I}}} = \frac{49{,}544 \,\mathrm{W/K}}{5.351 \,\mathrm{kg/s} \times 4177 \,\mathrm{J/kg \cdot K}} = 2.217$$

$$\mathrm{NTU_{II}} = \frac{(UA)_{\mathrm{II}}}{(\dot{m}c_p)_{c,\mathrm{II}}} = \frac{78{,}103 \,\mathrm{W/K}}{4.649 \,\mathrm{kg/s} \times 4177 \,\mathrm{J/kg \cdot K}} = 4.022$$

$$R_{\mathrm{I}} = \frac{\dot{m}_{c,\mathrm{I}} c_p}{\dot{m}_{h,\mathrm{I}} c_p} = \frac{5.351 \,\mathrm{kg/s} \times 4177 \,\mathrm{J/kg \cdot K}}{9.632 \,\mathrm{kg/s} \times 4177 \,\mathrm{J/kg \cdot K}} = 0.556$$

$$R_{\mathrm{II}} = \frac{\dot{m}_{c,\mathrm{II}} c_p}{\dot{m}_{h,\mathrm{II}} c_p} = \frac{4.649 \,\mathrm{kg/s} \times 4177 \,\mathrm{J/kg \cdot K}}{8.368 \,\mathrm{kg/s} \times 4177 \,\mathrm{J/kg \cdot K}} = 0.556$$

The temperature effectiveness for the two groups of the counterflow exchanger, $P_{1,I}$ and $P_{1,II}$ are given by Eq. (I.1.1) of Table 3.6.

$$P_{1,I} = \frac{1 - \exp[-NTU_I(1 - R_I)]}{1 - R_I \exp[-NTU_I(1 - R_I)]} \frac{1 - \exp[-2.217 \times (1 - 0.556)]}{1 - 0.556 \times \exp[-2.217 \times (1 - 0.556)]} = 0.7906$$

$$P_{1,II} = \frac{1 - \exp[-NTU_{II}(1 - R_{II})]}{1 - R_{II} \exp[-NTU_{II}(1 - R_{II})]} \frac{1 - \exp[-4.022 \times (1 - 0.556)]}{1 - 0.556 \times \exp[-4.022 \times (1 - 0.556)]} = 0.9179$$

Finally, the heat transfer rate from the hot water to cold water in this exchanger is given by

$$
\begin{aligned}
q &= P_{1,I}\dot{m}_{c,I}c_p(T_{h,i} - T_{c,i}) + P_{1,II}\dot{m}_{c,II}c_p(T_{h,i} - T_{c,i}) \\
&= [0.7906 \times 5.351\,\text{kg/s} \times 4177\,\text{J/kg}\cdot\text{K} \times (40 - 20)\,\text{K}] \\
&\quad + [0.9179 \times 4.649\,\text{kg/s} \times 4177\,\text{J/kg}\cdot\text{K} \times (40 - 20)\,\text{K}] \\
&= 353.4 \times 10^3\,\text{W} + 356.5 \times 10^3\,\text{W} = 710\,\text{kW} \qquad\qquad Ans.
\end{aligned}
$$

To compute the pressure drop, the friction factors for the Reynolds numbers above can be computed from Eq. (9.42) as follows:

$$f_{c,I} = 0.80(3303)^{-0.25} = 0.1055 \qquad f_{c,II} = 3.44(1435)^{-0.25} = 0.5589$$

$$f_{h,I} = 0.80(5946)^{-0.25} = 0.0911 \qquad f_{h,II} = 3.44(2583)^{-0.25} = 0.4825$$

Now we compute the pressure drop associated within the plate pack on the cold and hot sides using Eq. (6.29) with given $\rho_m = \rho$:

$$\Delta p_{c,I} = \frac{4f_{c,I}LG_{c,I}^2}{2g_c\rho D_e} = \frac{4 \times 0.1055 \times 1.1\,\text{m} \times (382.21)^2}{2 \times 1 \times 995.4\,\text{kg/m}^3 \times 7 \times 10^{-3}\text{m}} = 4866\,\text{Pa}$$

$$\Delta p_{c,II} = \frac{4 \times 0.5589 \times 1.1\,\text{m} \times (166.04)^2}{2 \times 1 \times 995.4\,\text{kg/m}^3 \times 7 \times 10^{-3}\,\text{m}} = 4865\,\text{Pa}$$

$$\Delta p_{h,I} = \frac{4 \times 0.0911 \times 1.1\,\text{m} \times (688.00)^2}{2 \times 1 \times 995.4\,\text{kg/m}^3 \times 7 \times 10^{-3}\,\text{m}} = 13,615\,\text{Pa}$$

$$\Delta p_{h,II} = \frac{4 \times 0.4825 \times 1.1\,\text{m} \times (298.86)^2}{2 \times 1 \times 995.4\,\text{kg/m}^3 \times 7 \times 10^{-3}\,\text{m}} = 13,607\,\text{Pa}$$

Theoretically, $\Delta p_{c,I} = \Delta p_{c,II}$ and $\Delta p_{h,I} = \Delta p_{h,II}$. As found above, this is true within the round-off error margins. Thus we consider the following values for the pressure drop associated with this plate pack.

$$\Delta p_c = 4866\,\text{Pa} \qquad \Delta p_h = 13,615\,\text{Pa}$$

The other components of the pressure drop are the momentum effect, the elevation change effect, and the inlet and outlet manifolds and ports pressure drops. The first two effects are negligible for this case (no change in the density and negligible elevation change for 1.1-m-long plates) and are ignored. For the manifold and port pressure drop component, the corresponding mass velocities are

$$G_{c,p} = \frac{\dot{m}_c}{(\pi/4)D_p^2} = \frac{10\,\text{kg/s}}{(\pi/4)(0.1\,\text{m})^2} = 1273\,\text{kg/m}^2 \cdot \text{s}$$

$$G_{h,p} = \frac{\dot{m}_h}{(\pi/4)D_p^2} = \frac{18\,\text{kg/s}}{(\pi/4)(0.1\,\text{m})^2} = 2292\,\text{kg/m}^2 \cdot \text{s}$$

The manifold and port pressure drops for the cold and hot fluid sides are computed from the first term on the right-hand side of Eq. (6.44):

$$\Delta p_{c,p} = \frac{1.5G_{c,p}^2 n_p}{2g_c\rho_i} = \frac{1.5 \times (1273\,\text{kg/m}^2 \cdot \text{s})^2 \times 1}{2 \times 1 \times 995.4\,\text{kg/m}^3} = 1221\,\text{Pa}$$

$$\Delta p_{h,p} = \frac{1.5 \times (2292\,\text{kg/m}^2 \cdot \text{s})^2 \times 1}{2 \times 1 \times 995.4\,\text{kg/m}^3} = 3958\,\text{Pa}$$

where $n_p = 1$ represents the number of passes on the given fluid side.

Thus, the total pressure drops on the cold and hot fluid sides are

$$\Delta p_c = 4866\,\text{Pa} + 1221\,\text{Pa} = 6087\,\text{Pa} \qquad \Delta p_h = 13{,}615\,\text{Pa} + 3958\,\text{Pa} = 17{,}573\,\text{Pa} \quad \textit{Ans.}$$

Note that the pressure drops associated with the manifold and port on the cold and hot fluid sides are 20% $[= (1221/6087) \times 100]$ and 22.5% $[= (3958/17{,}573) \times 100]$, respectively, of the total pressure drop on individual fluid sides.

To compare the effect of mixed-plate performance to that for a single-plate geometry PHE, let us recalculate the performance of a similar PHE of the same number of plates (or channels), but all made from 30° chevron plates. The flow and surface areas on one side of that PHE are

$$A_{o,c} = A_{o,h} = W(2a)N_c = 0.5\,\text{m} \times 3.5 \times 10^{-3}\,\text{m} \times 24 = 0.042\,\text{m}^2$$

$$A_c = A_h = 0.55\,\text{m}^2 \times 24 \times 2 = 26.4\,\text{m}^2$$

The mass velocities, Reynolds numbers, and heat transfer coefficients on both fluid sides are as follows:

$$G_c = \frac{\dot{m}_c}{A_{o,c}} = \frac{10\,\text{kg/s}}{0.042\,\text{m}^2} = 238.10\,\text{kg/m}^2 \cdot \text{s} \qquad G_h = \frac{\dot{m}_h}{A_{o,h}} = \frac{18\,\text{kg/s}}{0.042\,\text{m}^2} = 428.57\,\text{kg/m}^2 \cdot \text{s}$$

$$\text{Re}_c = \frac{G_c D_e}{\mu} = \frac{238.10\,\text{kg/m}^2 \cdot \text{s} \times 7 \times 10^{-3}\,\text{m}}{8.1 \times 10^{-4}\,\text{Pa} \cdot \text{s}} = 2058$$

$$\text{Re}_h = \frac{G_h D_e}{\mu} = \frac{428.57\,\text{kg/m}^2 \cdot \text{s} \times 7 \times 10^{-3}\,\text{m}}{8.1 \times 10^{-4}\,\text{Pa} \cdot \text{s}} = 3704$$

$$h_c = 0.724\left(\frac{k}{D_e}\right)\mathrm{Re}_c^{0.583}\cdot\mathrm{Pr}^{1/3} = 0.724\left(\frac{0.619\,\mathrm{W/m}\cdot\mathrm{K}}{7\times10^{-3}\,\mathrm{m}}\right)(2058)^{0.583}(5.47)^{1/3}$$

$$= 9640\,\mathrm{W/m}^2\cdot\mathrm{K}$$

$$h_h = 0.724\left(\frac{k}{D_e}\right)\mathrm{Re}_h^{0.583}\cdot\mathrm{Pr}^{1/3} = 0.724\left(\frac{0.619\,\mathrm{W/m}\cdot\mathrm{K}}{7\times10^{-3}\,\mathrm{m}}\right)(3704)^{0.583}(5.47)^{1/3}$$

$$= 13{,}579\,\mathrm{W/m}^2\cdot\mathrm{K}$$

The overall thermal conductance UA is given by

$$\frac{1}{UA} = \frac{1}{A_c}\left(\frac{1}{h_c} + \hat{\mathbf{R}}_{c,f} + \frac{\delta_w}{k_w} + \hat{\mathbf{R}}_{h,f} + \frac{1}{h_h}\right)$$

$$= \frac{1}{26.4\,\mathrm{m}^2}\left(\frac{1}{9640\,\mathrm{W/m}^2\cdot\mathrm{K}} + 4\times10^{-5}\,\mathrm{m}^2\cdot\mathrm{K/W}\right.$$

$$\left. + 3\times10^{-6}\,\mathrm{m}^2\cdot\mathrm{K/W} + \frac{1}{13{,}579\,\mathrm{W/m}^2\cdot\mathrm{K}}\right) = 8.3476\times10^{-6}\,\mathrm{K/W}$$

or

$$UA = 119{,}790\,\mathrm{W/K}$$

Now we compute NTU, R, P, and q for this PHE.

$$\mathrm{NTU} = \frac{UA}{(\dot{m}c_p)_c} = \frac{119{,}790\,\mathrm{W/K}}{10\,\mathrm{kg/s}\times4177\,\mathrm{J/kg}\cdot\mathrm{K}} = 2.868$$

$$R = \frac{(\dot{m}c_p)_c}{(\dot{m}c_p)_h} = \frac{10\,\mathrm{kg/s}\times4177\,\mathrm{J/kg}\cdot\mathrm{K}}{18\,\mathrm{kg/s}\times4177\,\mathrm{J/kg}\cdot\mathrm{K}} = 0.556$$

$$P = \frac{1 - \exp[-\mathrm{NTU}(1-R)]}{1 - R\exp[-\mathrm{NTU}(1-R)]} = \frac{1 - \exp[-2.868\times(1-0.556)]}{1 - 0.556\times\exp[-2.868\times(1-0.556)]} = 0.8528$$

$$q = P(\dot{m}c_p)(T_{h,i} - T_{c,i}) = 0.8528\times10\,\mathrm{kg/s}\times4177\,\mathrm{J/kg}\cdot\mathrm{K}\times(40-20)\,\mathrm{K}$$

$$= 712\times10^3\,\mathrm{W} = 712\,\mathrm{kW}$$

Finally, the friction factors and the pressure drops are

$$f_c = a\,\mathrm{Re}_c^{-n} = 0.80(2058)^{-0.25} = 0.1188 \qquad f_h = 0.80(3704)^{-0.25} = 0.1025$$

$$\Delta p_{c,\mathrm{core}} = \frac{4f_c L G_c^2}{2g_c\rho D_e} = \frac{4\times0.1188\times1.1\,\mathrm{m}\times(238.10\,\mathrm{kg/m}^2\cdot\mathrm{s})^2}{2\times1\times995.4\,\mathrm{kg/m}^3\times7\times10^{-3}\,\mathrm{m}} = 2126\,\mathrm{Pa}$$

$$\Delta p_{h,\mathrm{core}} = \frac{4\times0.1025\times1.1\,\mathrm{m}\times(428.57\,\mathrm{kg/m}^2\cdot\mathrm{s})^2}{2\times1\times995.4\,\mathrm{kg/m}^3\times7\times10^{-3}\,\mathrm{m}} = 5944\,\mathrm{Pa}$$

Discussion and Comments: This example illustrates how to rate a PHE when it has a plate pack consisting of one type of plates in one group and mixed plates in another group, with the following results:

Group	$\dot{m}_c$ (kg/s)	$\dot{m}_h$ (kg/s)	q (kW)	$\Delta p_{c,\text{core}}$ (kPa)	$\Delta p_{h,\text{core}}$ (kPa)
I	5.351	9.632	353.4	4.866	13.615
II	4.649	8.368	356.5	4.865	13.607

We have also compared the performance of this exchanger with the one having all chevron plates of $\beta = 30°$. Following is the comparison of main results:

PHE	$\dot{m}_c$ (kg/s)	$\dot{m}_h$ (kg/s)	q (kW)	$\Delta p_{c,\text{core}}$ (kPa)	$\Delta p_{h,\text{core}}$ (kPa)
Mixed Plates	10	18	710	4.866	13.615
Single Plates	10	18	712	2.126	5.944

For the present problem, the friction factors for a given Re are over four times larger for the mixed-plate group II. This, in turn, reduces the flow to only 46.5% [$= 100 \times 8.368\,\text{kg/s}/18\,\text{kg/s}$] of the total flow in group II despite the flow area being double for group II than for group I. Hence, even though the heat transfer coefficient for the mixed-plate region is larger than that for the single-plate region at a given Re, the reduction in the flow reduces h, and as a result, the overall heat transfer is about the same (710 vs. 712 kW). Hence, it is important to keep in mind that the mixed-plate section should not have excessive friction factors which would otherwise defeat the advantage of having a mixed-plate section.

9.4.4 Sizing a PHE

When sizing a PHE, we have very little choice in the selection of plate dimensions, unlike plate-fin and tube-fin heat exchanger designs, because we cannot arbitrarily select a plate width W or plate length L. Instead, we should select from a relatively small pool of available plate sizes from any manufacturer. As the dies used for forming the plates are extremely expensive, each manufacturer offers only up to about 30 plate sizes. In selecting an appropriate plate size, we may compute the fluid velocity in the port and limit this value to a maximum of 6 m/s (20 ft/sec), as a rough rule of thumb. Further, the manifold and port pressure drops may not be allowed to exceed a certain percentage (typically 10% but up to 30% in rare cases) of the total pressure drop. Most plate sizes are generally available only in two chevron angles. However, by "mixing" plates of different chevron angles in various proportions, the designer is able to obtain considerable flexibility in β_{eff} for any PHE.

Two methods are published in the literature (Shah and Focke, 1988; Shah and Wanniarachchi, 1991) for sizing a PHE. However, we do not describe them here because (1) those procedures are quite involved, (2) engineers in the PHE industry use their own proprietary computer programs with their own data for j and f factors, and (3) it is easy to add or delete some plates if the designed PHE does not perform to the specifications.

Since sophisticated proprietary computer programs are available for rating a PHE which converge quickly, such programs have iterative rating schemes build into them to arrive at a size to meet the specified heat transfer and/or pressure drop; the size will depend on whether it is a heat transfer–limited design or a pressure drop–limited design, as discussed in Section 9.4.1.

9.5 SHELL-AND-TUBE HEAT EXCHANGERS

Accurate prediction of performance and design characteristics of conventional shell-and-tube heat exchangers is more difficult than that for plate-fin and tube-fin exchangers. This is due primarily to the complexity of shell-side flow conditions and the impact of that complexity on heat transfer performance. There are many variables associated with the geometry (i.e., baffles, tubes, front- and rear-end heads, etc.) in a shell-and-tube heat exchanger in addition to those for the operating conditions. So complete sizing (with a unique design) of a shell-and-tube exchanger is not possible as for a plate-fin exchanger, described earlier. As a result, the common practice is to presume the complete geometry of the exchanger and perform the rating of the exchanger to determine the tube (shell) length if the heat duty is given, or outlet temperatures if the length is given. In both cases, pressure drops are to be determined. Preliminary sizing (design) of a shell-and-tube heat exchanger is possible based on a number of approximations and the experience of past designs. Once the preliminary design is obtained, the design calculations are essentially a series of iterative *rating* calculations made on the preliminary design until a satisfactory design is achieved. In this section, we outline the basic steps of (1) a rating procedure with an example, and (2) a preliminary design and subsequent iteration technique for sizing of a shell-and-tube exchanger. It should be added that modern design practices are based almost exclusively on sophisticated commercial or proprietary computer software that takes into account many complex effects on the shell side that are beyond the simplified methods presented here.

In this section, we start with how to compute the shell-side heat transfer and pressure drop by taking into account various flow leakage, bypass, and other effects before providing the rating procedure. These effects are taken into consideration by a widely utilized method in the open literature referred to as the *Bell–Delaware method*. It was originally reported by Bell (1963) for rating of fully tubed segmentally baffled heat exchangers with plain tubes based on the experimental data obtained for an exchanger with geometrical parameters closely controlled. This method has been extended to rate low-finned-tube E shell, no-tubes-in-window E shell, and F shell heat exchangers (Bell, 1988b; Taborek, 1998).

9.5.1 Heat Transfer and Pressure Drop Calculations

Heat transfer and pressure drop calculations constitute the key part of the rating or design of an exchanger. Tube-side calculations are straightforward. The heat transfer coefficient is computed using available correlations for internal forced convection as presented in Section 7.4, and Eq. (9.20) is used for pressure drop calculations. The shell-side calculations, however, must take into consideration the effect of various leakage streams (A and E streams, Fig. 4.19) and bypass streams (C and F streams, Fig. 4.19) in addition to the main crossflow stream B through the tube bundle. Several methods have been in use over the years, but the most accurate method in the open literature is the

Bell–Delaware method. The set of correlations for calculating shell-side heat transfer coefficients and pressure drops discussed next constitutes the core of the Bell–Delaware method (Bell, 1988b).

9.5.1.1 Shell-Side Heat Transfer Coefficient.

In the Bell–Delaware method, the shell-side heat transfer coefficient h_s is determined using Eq. (4.169) by correcting the ideal heat transfer coefficient h_{id} for various leakage and bypass flow streams in a segmentally baffled shell-and-tube exchanger. The h_{id} is determined for pure crossflow in a rectangular tube bank assuming that the entire shell-side stream flow across the tube bank is at or near the centerline of the shell. It is computed from the Nusselt number correlations of Eq. (7.117) or other appropriate Nu or j vs. Re correlations, modified for property variation effects as outlined in Section 7.6.1. It is then corrected by five correction factors as follows:

$$h_s = h_{id} J_c J_\ell J_b J_s J_r \tag{9.50}$$

where

$J_c =$ correction factor for baffle configuration (baffle cut and spacing). It takes into account heat transfer in the window and leads to the average for the entire heat exchanger. It is dependent on the fraction of the total number of tubes in crossflow between baffle tips. Its value is 1.0 for an exchanger with no tubes in the windows and increases to 1.15 for small baffle cuts and decreases to 0.65 for large baffle cuts. For a typical well-designed heat exchanger, its value is near 1.0.

$J_\ell =$ correction factor for baffle leakage effects, including both tube-to-baffle and baffle-to-shell leakages (A and E streams) with heavy weight given to the latter and credit given to tighter constructions. It is a function of the ratio of the total leakage area per baffle to the crossflow area between adjacent baffles, and also of the ratio of the shell-to-baffle leakage area to tube-to-baffle leakage area. If the baffles are too close, J_ℓ will be lower, due to higher leakage streams. A typical value of J_ℓ is in the range 0.7 to 0.8.

$J_b =$ correction factor for bundle and pass partition bypass (C and F) streams. It varies from 0.9 for a relatively small clearance between the outermost tubes and the shell for fixed tubesheet construction to 0.7 for large clearances in pull-though floating head construction. It can be increased from about 0.7 to 0.9 by proper use of the sealing strips in a pull-through bundle.

$J_s =$ correction factor for larger baffle spacing at the inlet and outlet sections compared to the central baffle spacing. The nozzle locations result in larger end baffle spacing and lower velocities and thus lower heat transfer coefficients. J_s usually varies from 0.85 to 1.0.

$J_r =$ correction factor for any adverse temperature gradient buildup in laminar flows. This correction applies only for shell-side Reynolds numbers below 100 and fully effective for $Re_s < 20$; otherwise, it is equal to 1.

The combined correction factor, made up of five correction factors, in a well-designed shell-and-tube exchanger is about 0.6 (i.e., a reduction of 40% in the ideal heat transfer coefficient). The combined correction factor can be as low as 0.4. Comparison with a large amount of proprietary experimental data indicates that compared to measured values, the shell-side h_s predicted from Eq. (9.50) is from 50% too low to 200% too high, with a mean error of 15% low (conservative) at all Reynolds numbers.

TABLE 9.2   Correction Factors for the Heat Transfer Coefficient on the Shell Side by the Bell–Delaware Method

Correction Factors C's	Formulas for parameters for Correction Factors
$J_c = 0.55 + 0.72 F_c$	F_c given by Eq. (8.120)
$J_\ell = 0.44(1 - r_s) + [1 - 0.44(1 - r_s)]e^{-2.2 r_{lm}}$	$r_s = \dfrac{A_{o,sb}}{A_{o,sb} + A_{o,tb}}; \quad r_{lm} = \dfrac{A_{o,sb} + A_{o,tb}}{A_{o,cr}}$
	$A_{o,sb}$, $A_{o,tb}$, and $A_{o,cr}$ given by Eqs. (8.130), (8.129), and (8.125), respectively
$J_b = \begin{cases} 1 & \text{for } N_{ss}^+ \geq 1/2 \\ e^{-C r_b [1 - (2N_{ss}^+)^{1/3}]} & \text{for } N_{ss}^+ \leq 1/2 \end{cases}$	$r_b = \dfrac{A_{o,bp}}{A_{o,cr}}; \quad N_{ss}^+ = \dfrac{N_{ss}}{N_{r,cc}};$
	$C = \begin{cases} 1.35 & \text{for } \mathrm{Re}_s \leq 100 \\ 1.25 & \text{for } \mathrm{Re}_s > 100 \end{cases}$
$J_s = \dfrac{N_b - 1 + (L_i^+)^{(1-n)} + (L_o^+)^{(1-n)}}{N_b - 1 + L_i^+ + L_o^+}$	$L_i^+ = \dfrac{L_{b,i}}{L_{bc}}; \quad L_o^+ = \dfrac{L_{b,o}}{L_{bc}}$
	$n = \begin{cases} 0.6 & \text{for turbulent flow} \\ \frac{1}{3} & \text{for laminar flow} \end{cases}$
$J_r = \begin{cases} 1 & \text{for } \mathrm{Re}_s \geq 100 \\ (10/N_{r,c})^{0.18} & \text{for } \mathrm{Re}_s \leq 20 \end{cases}$	$N_{r,c} = N_{r,cc} + N_{r,cw}$; use Eqs. (8.121) and (8.119) for $N_{r,cc}$ and $N_{r,cw}$; For $20 < \mathrm{Re}_s < 100$, linearly interpolate J_r from two formulas

Source: Data from Taborek (1998).

These correction factors were determined from well-controlled experiments, and the results were presented graphically (Bell, 1963). Those correction factors have been curve fitted (Taborek, 1998) and are presented in Table 9.2.

The Bell–Delaware method can be used for a low-finned tube-bundle E shell, no-tubes-in-window E shell, and F shell exchangers. The modifications for these extensions are summarized briefly next.

- External low-finned tubes are used when the shell-side heat transfer coefficient is low, such as with viscous liquids. In this case, the ideal heat transfer coefficient for low-finned tubes is computed from the appropriate correlations, such as Eq. (7.135). Subsequently, the effective shell-side h_s is calculated from Eq. (9.50).

- The no-tubes-in-window design is used to minimize/eliminate the flow-induced tube vibration problem. In this case, the flow area $A_{o,w}$ through one window is given by $A_{fr,w}$ of Eq. (8.111) since $A_{fr,t} = 0$ in Eq. (8.117). The tube count for this exchanger is given by $N_t F_c$, where N_t is the number of tubes for a fully tubed exchanger. The fraction F_c of the total number of tubes in the crossflow section is given by Eq. (8.120). Also, J_c of Eq. (9.50) is unity and $N_{r,cw} = 0$ for the calculation of J_r from Table 9.2 for Eq. (9.50). The rest of the procedure remains the same.

- For the F shell exchanger, we have two-tube and two-shell passes by the use of a longitudinal baffle. If this baffle is not welded on both sides to the shell, there will be fluid leakage from the upstream to the downstream pass on the shell side due to the pressure difference. Also, there will be heat leakage across the baffle by heat conduction from the hotter to colder side of the shell-side pass. These effects may not be negligible in some cases. If we neglect these effects, the Bell–Delaware method remains identical except that all flow and surface areas need to be reduced by half compared to a single shell-side pass.

9.5.1.2 Shell-Side Pressure Drop. Similar to shell-side heat transfer, the shell-side pressure drop is also affected by various leakage and bypass streams in a segmentally baffled exchanger. The shell-side pressure drop has three components: (1) pressure drop in the central (crossflow) section, Δp_{cr}; (2) pressure drop in the window area, Δp_w; and (3) pressure drop in the shell-side inlet and outlet sections, Δp_{i-o} (see Section 6.4.2.2). It is assumed that each of the three components is based on the total flow rate, and that it can be calculated correcting the corresponding ideal pressure drops.

The ideal pressure drop in the central section, $\Delta p_{b,\mathrm{id}}$, assumes pure crossflow of the fluid across the ideal tube bundle. This pressure drop should be corrected for (1) the two leakage streams A and E in Fig. 4.19 using the correction factor ζ_ℓ, and (2) the bundle and pass partition bypass flow streams C and F in Fig. 4.19 using the correction factor ζ_b. The ideal window pressure drop, Δp_w, also has to be corrected for both baffle leakage streams. Finally, the ideal inlet and outlet section pressure drops, Δp_{i-o}, are based on an ideal crossflow pressure drop in the central section. These pressure drops should be corrected for bypass flow (correction factor ζ_b) and for uneven baffle spacing in inlet and outlet sections (correction factor ζ_s). Thus, the total shell-side pressure drop, from Eq. (6.43), is given as

$$\Delta p_s = \Delta p_{cr} + \Delta p_w + \Delta p_{i-o} = [(N_b - 1)\Delta p_{b,\mathrm{id}}\zeta_\ell + N_b \Delta p_{w,\mathrm{id}}]\zeta_\ell$$

$$+ 2\Delta p_{b,\mathrm{id}}\left(1 + \frac{N_{r,cw}}{N_{r,cc}}\right)\zeta_b\zeta_s \tag{9.51}$$

The formulas for $\Delta p_{b,\mathrm{id}}$ and $\Delta p_{w,\mathrm{id}}$ are given by Eqs. (6.37) and (6.39) respectively. The Hagen number for Eq. (6.37) is obtained from Eq. (7.109). Various correction factors of Eq. (9.51) are defined as follows:

ζ_ℓ = correction factor for tube-to-baffle and baffle-to-shell leakage (A and E) streams. This factor is related to the same effect as J_ℓ but is of different magnitude. Usually, $\zeta_\ell \approx 0.4$ to 0.5, although lower values are possible with small baffle spacing.

ζ_b = correction factor for bypass flow (C and F streams). It is different in magnitude from J_b and ranges from 0.5 to 0.8, depending on construction type and the number of sealing strips. The lower value will be typical of a pull-through floating head with one or two sealing strip pairs, and the higher value, if a fully tubed fixed-tubesheet design.

ζ_s = correction factor for inlet and outlet sections having different baffle spacing from that of the central section, in the range 0.5 to 2.

These correction factors, originally presented in graphical form (Bell, 1963, 1988b), are given in Table 9.3 in equation form by Taborek (1998).

The combined effect of pressure drop corrections reduces the ideal total shell-side pressure drop to 20 to 30% of the pressure drop that would be calculated for flow through the corresponding exchanger without baffle leakages and bundle bypass streams (i.e., $\Delta p_{s,\mathrm{actual}} \approx 0.2$ to $0.3\,\Delta p_{s,\mathrm{id}}$). Comparison with a large number of proprietary experimental data indicate that compared to measured values, the shell-side Δp_s

TABLE 9.3   Correction Factors for the Pressure Drop on the Shell Side by the Bell–Delaware Method

Correction Factors, ζ's	Formula for Parameters for Correction Factors
$\zeta_b = \begin{cases} \exp\{-Dr_b[1 - (2N_{ss}^+)^{1/3}]\} & \text{for } N_{ss}^+ < \frac{1}{2} \\ 1 & \text{for } N_{ss}^+ \geq \frac{1}{2} \end{cases}$	r_b and N_{ss}^+ defined in Table 9.2 $D = \begin{cases} 4.5 & \text{for } Re_s \leq 100 \\ 3.7 & \text{for } Re_s > 100 \end{cases}$
$\zeta_\ell = \exp[-1.33(1 + r_s)r_{1m}^p]$	r_s and r_{1m} defined in Table 9.2 $p = [-0.15(1 + r_s) + 0.8]$
$\zeta_s = \left(\dfrac{L_{b,c}}{L_{b,o}}\right)^{2-n'} + \left(\dfrac{L_{b,c}}{L_{b,i}}\right)^{2-n'}$	$n' = \begin{cases} 1.0 & \text{for laminar flow} \\ 0.2 & \text{for turbulent flow} \end{cases}$

Source: Data from Taborek (1998).

computed from Eq. (9.51) is from about 5% low (unsafe) at $Re_s > 1000$ to 100% high at $Re_s < 10$.

Despite the facts above, it should be emphasized that the window section contributes high pressure drop [compared to the other components of Eq. (9.51)] with insignificant contribution to heat transfer. This results in an overall lower heat transfer rate to pressure drop ratio for the segmental baffle exchanger than that for grid baffle and most newer shell-and-tube heat exchanger designs.

9.5.2   Rating Procedure

The following is a step-by-step rating procedure using the Bell–Delaware method (Bell, 1988b). For the rating problem, the detailed exchanger geometry is specified and we determine the heat duty, outlet temperatures as well as pressure drops on both fluid sides. We then describe the changes in the solution method if the exchanger length is to be determined.

1. Compute the surface geometrical characteristics on each fluid side. This includes shell-side flow areas in crossflow and window zones as well as all leakage flow areas and related information as detailed in Section 8.5. Also compute the tube-side flow area, surface area, ratio of free flow to frontal area, and other pertinent dimensions.

2. Calculate the fluid bulk temperature and fluid thermophysical properties on each fluid side. Since the outlet temperatures are not known for the rating problem, they are guessed initially. Unless it is known from past experience, assume the exchanger effectiveness as 50% for most single and multitube-pass shell-and-tube exchangers, or 60 to 75% for multishell-pass exchangers. For the assumed effectiveness, calculate the fluid outlet temperatures using Eqs. (9.9) and (9.10). Compute fluid mean temperatures on each fluid side, depending on the heat capacity ratio C^*, as outlined in Section 9.1. Subsequently, obtain the fluid properties (μ, c_p, k, Pr, and ρ) from thermophysical property books, handbooks, or Appendix A.

3. Calculate the Reynolds numbers ($Re = GD_h/\mu$) and/or any other pertinent dimensionless groups (from the basic definitions) needed to determine the nondimensional heat transfer and flow friction characteristics (e.g., j or Nu and f, Eu, or Hg)

of heat transfer surfaces on each fluid side of the exchanger. Subsequently, compute j or Nu and f, Eu or Hg factors. Correct Nu (or j) for variable fluid property effects in the second and subsequent iterations using Eqs. (9.11) and (9.12).

4. From Nu or j, compute the heat transfer coefficients for both fluid streams from the following equations:

$$h = \text{Nu}\frac{k}{D_h} \quad \text{or} \quad h = jGc_p \cdot \text{Pr}^{-2/3} \tag{9.52}$$

5. Compute various J correction factors for baffle configuration, flow leakage, flow bypass, unequal baffle spacing in the ends, and adverse temperature gradient on the shell side using Table 9.2. Determine the effective or actual shell-side heat transfer coefficient using Eq. (9.50).

6. Also calculate the tube-side heat transfer coefficient, wall thermal resistance, fouling resistances, and the overall heat transfer coefficient.

7. From the known heat capacity rates on each fluid side, compute $C^* = C_{min}/C_{max}$. From the known UA, determine NTU $= UA/C_{min}$. With the known NTU, C^*, and the flow arrangement, determine the exchanger effectiveness ε from either closed-form equations (see Table 3.6) or tabular/graphical results.

8. With this ε, finally compute the outlet temperatures from Eqs. (9.9) and (9.10). If these outlet temperatures are significantly different from those assumed in step 2, use these outlet temperatures in step 2 and continue iterating steps 2 through 8 until the assumed and computed outlet temperatures converge within the desired degree of accuracy.

9. For the pressure drop calculations, calculate the mean fluid densities on both fluid sides as follows: Use the arithmetic mean value for liquids and harmonic mean value for gases as given by Eq. (9.18). For the shell-side pressure drop, compute various correction factors using the formulas given in Table 9.3, the ideal crossflow and window zone pressure drops from Eqs. (6.37) and (6.39), and the shell-side total pressure drop from Eq. (6.43) or (9.51). For the tube-side pressure drop, determine the entrance and exit loss coefficients, K_c and K_e, from Fig. 6.3 for known σ, Re, and the flow passage entrance geometry. The friction factor on each fluid side is corrected for the variable fluid properties using Eq. (9.11) or (9.12). The core pressure drops on each fluid side are then calculated from Eq. (6.28) or (6.33).

If the heat duty and detailed exchanger geometry except for the exchanger (tube) length are given for the TEMA E exchanger, the tube length can be determined as follows by modifying the aforementioned detailed procedure. Follow step 1 except that the surface area is unknown. Since the heat duty is known, outlet temperatures are known, and as a result, the fluid properties mentioned in step 2 can be determined at mean temperatures in the exchanger. Follow steps 3 through 6 to compute the overall shell-side heat transfer coefficient U_s. Since all four temperatures are known, calculate the log-mean temperature difference ΔT_{lm} using the definition of Eqs. (3.172) and (3.173). Also compute the temperature effectiveness P_s and heat capacity rate ratio R_s from known four terminal temperatures using Eqs. (3.96) and (3.105) with the subscripts 1 and 2 replaced by s (shell side) and t (tube side). Next, determine the log-mean temperature difference correction factor F for known P_s, R_s and the exchanger type (flow arrangement). Finally compute the surface area on the shell side from the following equation.

$$A_s = \frac{q}{U_s F \Delta T_{lm}} \tag{9.53}$$

The required effective tube length of the exchanger is then calculated from $L = A_s / \pi d_o N_t$ and the number of baffles required by using Eq. (8.126).

We now illustrate the rating methodology with an example.

Example 9.4 Determine heat transfer rate, outlet fluid temperatures, and pressure drops on each fluid side for a TEMA E shell-and-tube heat exchanger with a fixed tubesheet and one shell and two tube passes. The tubes in the bundle are in 45° rotated square arrangement. The fluids are lubricating oil and seawater. Fouling factors for the oil and water sides are 1.76×10^{-4} and $8.81 \times 10^{-5} \, \text{m}^2 \cdot \text{K/W}$, respectively. The geometric dimensions and operating properties are provided as follows. Assume mean fluid temperatures to be 63°C and 35°C for oil and water, respectively.

Shell-side inside diameter $D_s = 0.336 \, \text{m}$
Tube-side outside diameter $d_o = 19.0 \, \text{mm}$
Tube-side inside diameter $d_i = 16.6 \, \text{mm}$
Tube pitch $p_t = 25.0 \, \text{mm}$
Tube bundle layout = 45°
Central baffle spacing $L_{b,c} = 0.279 \, \text{m}$
Inlet baffle spacing $L_{b,i} = 0.318 \, \text{m}$

Outlet baffle spacing $L_{b,o} = 0.318 \, \text{m}$

Baffle cut $\ell_c = 86.7 \, \text{mm}$ or 25.8%

Tube material
= admiralty (70% Cu, 30% Ni)

Number of sealing strip pairs $N_{ss} = 1$
Total number of tubes $N_t = 102$
Tube length $L = 4.3 \, \text{m}$
Width of bypass lane $w_p = 19.0 \, \text{mm}$
Number of tube passes $n_p = 2$
Number of pass partitions $N_p = 2$
Diameter of the outer tube limit
$\quad D_{otl} = 0.321 \, \text{m}$
Tube-to-baffle hole diametral clearance
$\quad \delta_{tb} = 0.794 \, \text{mm}$
Shell-to-baffle diametral clearance
$\quad \delta_{sb} = 2.946 \, \text{mm}$
Thermal conductivity of tube wall k_w
= 111 W/m · K

Operating conditions:

Oil flow rate $\dot{m}_{oil} = \dot{m}_s = 36.3 \, \text{kg/s}$
Oil inlet temperature $T_{s,i} = 65.6°\text{C}$
Oil side fouling factor
$\hat{R}_{o,f} = 0.000176 \, \text{m}^2 \cdot \text{W/K}$

Water flow rate $\dot{m}_{water} = \dot{m}_t = 18.1 \, \text{kg/s}$
Water inlet temperature $T_{t,i} = 32.2°\text{C}$
Water side fouling factor
$\hat{R}_{i,f} = 0.000088 \, \text{m}^2 \cdot \text{W/K}$

Fluid	Density ρ_s (kg/m³)	Specific heat c_p (J/kg·K)	Dynamic Viscosity μ (Pa·s)	Thermal Conductivity k (W/m²·K)	Prandtl Number Pr
Oil at 63°C	849	2094	64.6×10^{-3}	0.140	966
Seawater at 35°C	993	4187	0.723×10^{-3}	0.634	4.77

Use the Dittus–Boelter correlation [Eq. (7.80) in Table 7.6] for the tube-side heat transfer coefficient. Use the McAdams correlation [Eq. (7.72) in Table 7.6] for the tube-side friction factor. For the shell-side friction factor and Nusselt numbers, use the following correlations: $f_{id} = 3.5(1.33d_o/p_t)^b \cdot Re_s^{-0.476}$, where $b = 6.59/(1 + 0.14 Re_s^{0.52})$; $Nu_s = 1.04 Re_d^{0.4} Pr_s^{0.36}(Pr_s/Pr_w)^{-0.25}$.

SOLUTION

Problem Data and Schematic: The schematic of the 1–2 TEMA E shell-and-tube heat exchanger is given in Fig. 1.5b with characteristic heat exchanger zones and dimensions shown in Figs. 8.9 and 8.11. All major geometric dimensions, operating conditions, and thermophysical properties of the fluids are given in the problem statement.

Determine: This rating problem requires determination of the heat transfer rate, outlet temperatures, and pressure drops for each fluid.

Assumptions: The assumptions invoked in Section 3.2.1 applicable to a shell-and-tube exchanger are valid.

Analysis: We follow the solution procedure for this rating problem in several steps as outlined preceding this problem. First, all geometric characteristics of the shell side are determined as detailed in Example 8.3. Heat transfer coefficients on both the shell side (as outlined in Section 9.5.1) and the tube side are then calculated. Subsequently, the overall heat transfer coefficient and design parameters for the given operating point are computed. With all these data, the heat transfer rate and outlet temperatures are calculated in a straightforward manner. Finally, determination of pressure drops completes the procedure. Before we outline the details, some required geometrical characteristics are obtained from Example 8.3 as follows:

$A_{o,cr} = 0.03275\,\text{m}^2 \qquad F_c = 0.6506 \qquad A_{o,sb} = 0.001027\,\text{m}^2 \qquad A_{o,tb} = 0.001995\,\text{m}^2$

$A_{o,bp} = 0.00949\,\text{m}^2 \qquad N_{r,cc} = 9 \qquad N_b = 14 \qquad A_{o,w} = 0.01308\,\text{m}^2$

Shell-Side Heat Transfer Coefficient. We calculate the shell-side velocity, Reynolds number, ideal heat transfer coefficient, and then correct it for various leakage and bypass flow streams.

Shell-side mass velocity $G_s = \dfrac{\dot{m}_s}{A_{o,cr}} = \dfrac{36.3\,\text{kg/s}}{0.03275\,\text{m}^2} = 1108\,\text{kg/m}^2 \cdot \text{s}$

Shell-side Reynolds number $Re_s = \dfrac{G_s d_o}{\mu_s} = \dfrac{1108\,\text{kg/m}^2 \cdot \text{s} \times 0.0190\,\text{m}}{64.6 \times 10^{-3}} = 326$

Now we compute Nu_s from the given correlation with $Re_d = Re_s$. Note that we have not calculated T_w, so we cannot calculate Pr_w. So in this iteration, we consider $Pr_s = Pr_w$.

$$Nu_s = 1.04 Re_d^{0.4} \cdot Pr_s^{0.36} \left(\frac{Pr_s}{Pr_w}\right)^{0.25} = 1.04 \times (326)^{0.4} \times (966)^{0.36} = 125.0\,\text{W/m}^2 \cdot \text{K}$$

$$h_{id} = \frac{Nu_s k}{d_o}\left(\frac{\mu_w}{\mu_m}\right)^{-0.14} = \frac{125.0 \times 0.140\,\text{W/m}^2 \cdot \text{K}}{0.0190\,\text{m}}(1)^{-0.14} = 921.0\,\text{W/m}^2 \cdot \text{K}$$

Baffle cut and spacing effect correction factor

$$J_c = 0.55 + 0.72F_c = 0.55 + 0.72 \times 0.6506 = 1.018$$

To calculate the tube-to-baffle and baffle-to-shell leakage factor J_ℓ from Table 9.2, we need to calculate r_s and r_{lm} as follows:

$$r_s = \frac{A_{o,sb}}{A_{o,sb} + A_{o,tb}} = \frac{0.001027\,\text{m}^2}{0.001027\,\text{m}^2 + 0.001995\,\text{m}^2} = 0.3398$$

$$r_{lm} = \frac{A_{o,sb} + A_{o,tb}}{A_{o,cr}} = \frac{0.001027\,\text{m}^2 + 0.001995\,\text{m}^2}{0.03275\,\text{m}^2} = 0.0923$$

$$J_\ell = 0.44(1 - r_s) + [1 - 0.44(1 - r_s)]e^{-2.2r_{lm}}$$

$$= 0.44 \times (1 - 0.3398) + [1 - 0.44 \times (1 - 0.3398)]e^{-2.2 \times 0.0923} = 0.8696$$

Let us now calculate J_b using the formula from Table 9.2 after we determine $\mathbf{C}$ (for $Re_s = 326$), r_b, and N_{ss}^+ as follows:

$$\mathbf{C} = 1.25 \qquad r_b = \frac{A_{o,bp}}{A_{o,cr}} = \frac{0.00949\,\text{m}^2}{0.03275\,\text{m}^2} = 0.2898 \qquad N_{ss}^+ = \frac{N_{ss}}{N_{r,cc}} = \frac{1}{9} = 0.1111$$

$$J_b = \exp\{-\mathbf{C}r_b[1 - (2N_{ss}^+)^{1/3}]\} = \exp\{-1.25 \times 0.2898 \times [1 - (2 \times 0.1111)^{1/3}]\} = 0.8669$$

Now we compute L_i^+ and L_o^+ for determining unequal baffle spacing factor J_s from Table 9.2.

$$L_i^+ = \frac{L_{b,i}}{L_{b,c}} = L_o^+ = \frac{L_{b,o}}{L_{b,c}} = \frac{0.318\,\text{m}}{0.279\,\text{m}} = 1.14 \qquad n = 0.6 \text{ for turbulent flow } (Re_s > 100)$$

$$J_s = \frac{N_b - 1 + (L_i^+)^{(1-n)} + (L_o^+)^{(1-n)}}{N_b - 1 + L_i^+ + L_o^+} = \frac{14 - 1 + (1.14)^{0.4} + (1.14)^{0.4}}{14 - 1 + 1.14 + 1.14} = 0.9887$$

Finally, the adverse temperature gradient factor $J_r = 1$ for $Re_s = 326 > 100$.

Since all correction factors J are determined, the actual shell-side heat transfer coefficient is given by

$$h_s = h_o = h_{id}J_c J_\ell J_b J_s J_r = 921.0\,\text{W/m}^2 \cdot \text{K} \times 1.018 \times 0.8696 \times 0.8669 \times 0.9887 \times 1$$

$$= 698.8\,\text{W/m}^2 \cdot \text{K}$$

This heat transfer coefficient should be corrected for the fluid property variations as outlined in Section 7.6.1 once the wall temperature is calculated in the next iteration.

Tube-Side Heat Transfer Coefficient

Number of tubes per pass $N_{t,p} = \dfrac{N_t}{2} = \dfrac{102}{2} = 51$

Tube-side flow area per pass $A_{o,t} = \dfrac{\pi}{4} d_i^2 N_{t,p} = \dfrac{\pi}{4}(0.0166\,\text{m})^2 \times 51 = 0.01104\,\text{m}^2$

Tube-side Reynolds number $\text{Re}_t = \dfrac{\dot{m}_t d_i}{A_{o,t}\mu_t} = \dfrac{18.1\,\text{kg/s} \times 0.0166\,\text{m}}{0.01104\,\text{m}^2 \times (0.723 \times 10^{-3}\,\text{Pa}\cdot\text{s})} = 37{,}643$

Nusselt number $\text{Nu}_t = 0.024\text{Re}^{0.8} \cdot \text{Pr}^{0.4} = 0.024 \times (37{,}643)^{0.8}(4.77)^{0.4} = 205.2$

Heat transfer coefficient $h_t = h_i = \dfrac{(\text{Nu}k)_t}{d_i} = \dfrac{205.2 \times 0.634\,\text{W/m}\cdot\text{K}}{0.0166\,\text{m}} = 7837\,\text{W/m}^2\cdot\text{K}$

Overall Heat Transfer Coefficient. From Eq. (3.31a),

$$\frac{1}{U_o} = \frac{1}{h_o} + \hat{\mathbf{R}}_{o,f} + \frac{d_o \ln(d_o/d_i)}{2k_w} + \hat{\mathbf{R}}_{i,f}\frac{d_o}{d_i} + \frac{1}{h_i}\frac{d_o}{d_i}$$

$$= \frac{1}{698.8\,\text{W/m}^2\cdot\text{K}} + 0.000176\,\text{m}^2\cdot\text{K/W} + \frac{0.0190\,\text{m} \times \ln(0.0190\,\text{m}/0.0166\,\text{m})}{2 \times 111\,\text{W/m}\cdot\text{K}}$$

$$+ 0.000088\,\text{m}^2\cdot\text{K/W}\left(\frac{0.0190\,\text{m}}{0.0166\,\text{m}}\right) + \frac{1}{7837\,\text{W/m}^2\cdot\text{K}}\left(\frac{0.0190\,\text{m}}{0.0166\,\text{m}}\right)$$

$$= (0.001431 + 0.000176 + 0.0000116 + 0.0001007 + 0.000146)\,\text{m}^2\cdot\text{K/W}$$

$$= 0.0018653\,\text{m}^2\cdot\text{K/W}$$

or

$$U_o = 536.1\,\text{W/m}^2\cdot\text{K}$$

The unit thermal resistance $1/U_o$ in the calculation above indicates the individual unit thermal resistances as 76.7, 9.5, 0.6, 5.4, and 7.8%. Thus, the largest thermal resistance is on the shell side, and the fouling thermal resistances and wall thermal resistance are of the same order of magnitude as the tube-side thermal resistance.

Total tube outside heat transfer area

$$A_s = A_{t,o} = \pi L d_o N_t = \pi \times 4.3\,\text{m} \times 0.0190\,\text{m} \times 102 = 26.180\,\text{m}^2$$

$$C_{\min} = C_t = (\dot{m}c_p)_t = 18.1\,\text{kg/s} \times 4187\,\text{J/kg}\cdot\text{K} = 75{,}785\,\text{W/K}$$

$$C_{\max} = C_s = (\dot{m}c_p)_s = 36.3\,\text{kg/s} \times 2094\,\text{J/kg}\cdot\text{K} = 76{,}012\,\text{W/K}$$

$$C^* = \frac{C_{\min}}{C_{\max}} = \frac{75{,}785\,\text{W/K}}{76{,}012\,\text{W/K}} = 0.997 \approx 1$$

Number of heat transfer units

$$\text{NTU} = \frac{U_o A_{t,o}}{C_{\min}} = \frac{U_o A_{t,o}}{C_t} = \frac{536.1 \text{ W/m}^2 \cdot \text{K} \times 26.180 \text{ m}^2}{75,785 \text{ W/K}} = 0.1852$$

Heat exchanger effectiveness, using the formula from Table 3.3, is

$$\varepsilon = \frac{\sqrt{2}}{\sqrt{2} + \coth(\text{NTU}/\sqrt{2})} = \frac{\sqrt{2}}{\sqrt{2} + \coth(0.1852/\sqrt{2})} = 0.1555$$

Heat Transfer Rate and Exit Temperatures

Heat transfer rate

$$q = \varepsilon C_{\min}(T_{s,i} - T_{t,i}) = 0.1555 \times 75,785 \text{ W/K} \times (65.6 - 32.2)°\text{C}$$
$$= 393,600 \text{ W} = 393.6 \text{ kW}$$

Oil exit temperature

$$T_{s,o} = T_{s,i} - \varepsilon C^*(T_{s,i} - T_{t,i}) = 65.6°\text{C} - 0.1555 \times 0.997 \times (65.6 - 32.2)°\text{C}$$
$$= 60.4°\text{C}$$

Water exit temperature

$$T_{t,o} = T_{t,i} + \varepsilon(T_{s,i} - T_{t,i}) = 32.2°\text{C} + 0.1555 \times (65.6 - 32.2)°\text{C} = 37.4°\text{C}$$

Mean temperatures[†]:
$$T_{s,m} = \frac{T_{s,i} + T_{s,o}}{2} = \frac{(65.6 + 60.4)°\text{C}}{2} = 63.0°\text{C}$$

$$T_{t,m} = \frac{T_{t,i} + T_{t,o}}{2} = \frac{(32.2 + 37.4)°\text{C}}{2} = 34.8°\text{C}$$

Pressure Drop Calculations. To compute the idealized tube bundle pressure drop, we first calculate the ideal friction factor using the given formula:

$$f_{\text{id}} = 3.5 \left(1.33 \frac{d_o}{p_t}\right)^b \cdot \text{Re}_s^{-0.476} = 3.5 \left(\frac{1.33 \times 19.0 \text{ mm}}{25.0 \text{ mm}}\right)^{1.72} (326)^{-0.476} = 0.2269$$

where

$$b = \frac{6.59}{1 + 0.14\text{Re}_s^{0.52}} = \frac{6.59}{1 + 0.14 \times (326)^{0.52}} = 1.72$$

$$\Delta p_{b,\text{id}} = \frac{4 f_{\text{id}} G_s^2 N_{r,cc}}{2 g_c \rho_s} \left(\frac{\mu_w}{\mu_m}\right)^{0.25} = \frac{4 \times 0.2269 \times (1108 \text{ kg/m}^2 \cdot \text{s})^2 \times 9}{2 \times 1 \times 849 \text{ kg/m}^3} (1)^{0.25} = 5906 \text{ Pa}$$

To calculate the pressure drop in the crossflow section, we first compute the correction factors ζ_b and ζ_ℓ using the expressions from Table 9.3.

$$\zeta_b = \exp\{-Dr_b[1 - (2N_{ss}^+)^{1/3}]\} = \exp\{-3.7 \times 0.2898 \times [1 - (2 \times 0.1089)^{1/3}]\} = 0.6524$$

[†] The mean temperatures calculated (63°C and 35°C for oil and water, respectively) are the same as the values purposely assumed in the problem formulation for thermophysical properties. Consequently, no iterations are required for changes in thermophysical property values for this problem.

using $\mathbf{D} = 3.7$ for $\mathrm{Re}_s > 100$ and $r_b = 0.2898$ as calculated for J_b earlier.

$$\zeta_\ell = \exp[-1.33(1 + r_s)r_{\mathrm{lm}}^p] = \exp[-1.33 \times (1 + 0.3398)(0.0923)^{0.60}] = 0.6527$$

where p from the formula of Table 9.3 is given by

$$p = [-0.15(1 + r_s) + 0.8] = [-0.15 \times (1 + 0.3398) + 0.8] = 0.60$$

Hence, Δp_{cr} and Δp_w from Eq. (9.51) with $\Delta p_{w,\mathrm{id}}$ from Eq. (6.39a) are given by

$$\Delta p_{cr} = \Delta p_{b,\mathrm{id}}(N_b - 1)\zeta_b\zeta_\ell = 5906\,\mathrm{Pa} \times (14 - 1) \times 0.6524 \times 0.6527 = 32{,}694\,\mathrm{Pa}$$

$$\Delta p_w = N_b(2 + 0.6N_{r,cw})\frac{G_w^2}{2g_c\rho_s}\zeta_\ell = 14 \times (2 + 0.6 \times 3)\frac{(1754\,\mathrm{kg/m^2 \cdot s})^2}{2 \times 1 \times 849\,\mathrm{kg/m^3}} \times 0.6527$$

$$= 62{,}914\,\mathrm{Pa}$$

where

$$G_w = \frac{\dot{m}_s}{\left(A_{o,cr}A_{o,w}\right)^{1/2}} = \frac{36.3\,\mathrm{kg/s}}{(0.03275\,\mathrm{m^2} \times 0.01308\,\mathrm{m^2})^{1/2}} = 1754\,\mathrm{kg/m^2 \cdot s}$$

Next let us determine pressure drop in inlet–outlet sections using Eq. (9.51) after computing ζ_s from Table 9.3.

$$\zeta_s = \left(\frac{L_{b,c}}{L_{b,o}}\right)^{2-n'} + \left(\frac{L_{b,c}}{L_{b,i}}\right)^{2-n'} = \left(\frac{0.279\,\mathrm{m}}{0.318\,\mathrm{m}}\right)^{2-0.2} + \left(\frac{0.279\,\mathrm{m}}{0.318\,\mathrm{m}}\right)^{2-0.2} = 1.5803$$

where $n' = 0.2$

$$\Delta p_{i-o} = 2\Delta p_{b,\mathrm{id}}\left(1 + \frac{N_{r,cw}}{N_{r,cc}}\right)\zeta_b\zeta_s = 2 \times 5906\,\mathrm{Pa} \times \left(1 + \frac{3}{9}\right) \times 0.6524 \times 1.5803$$

$$= 16{,}237\,\mathrm{Pa}$$

Then

$$\Delta p_s = \Delta p_{cr} + \Delta p_w + \Delta p_{i-o} = (32{,}694 + 62{,}914 + 16{,}237)\,\mathrm{Pa} = 111{,}845\,\mathrm{Pa} = 112\,\mathrm{kPa}$$
$$(29.2\%) \quad (56.3\%) \quad (14.5\%)$$

Note that for this problem, the window section pressure drop is more than the crossflow section pressure drop, whereas the crossflow section provides most of heat transfer. Thus the window section results in excessive pressure drop with insignificant contribution to heat transfer.

Tube-Side Pressure Drop. From Eq. (7.72) in Table 7.6,

$$f = 0.046\mathrm{Re}_t^{-0.2} = 0.046 \times (37{,}643)^{-0.2} = 0.005593$$

From Fig. 6.3,

$$K_c = 0.3 \qquad K_e = 0.4 \qquad \text{for } \sigma = \frac{2(p_t - d_o)}{\sqrt{2}p_t} = \frac{2 \times (25.0 - 19.0)\,\text{mm}}{\sqrt{2} \times 25.0\,\text{mm}} = 0.34$$

$$
\begin{aligned}
\Delta p_t &= \frac{\dot{m}_t^2}{2 g_c \rho_t A_{o,t}^2} \left[\frac{4fL}{d_i} + (1 - \sigma^2 + K_c) - (1 - \sigma^2 - K_e) \right] n_p \\
&= \frac{(18.1\,\text{kg/s})^2}{2 \times 1 \times 993\,\text{kg/m}^3 \times (0.01104\,\text{m}^2)^2} \left(\frac{4 \times 0.005593 \times 4.3\,\text{m}}{0.0166\,\text{m}} + 0.3 + 0.4 \right) \times 2 \\
&= 17{,}582\,\text{Pa} = 17.58\,\text{kPa}
\end{aligned}
$$

Here since the inlet and outlet densities for water will not change appreciably and that information is not given, we have considered $\rho_i = \rho_o = \rho_m$.

Discussion and Comments: Despite the seemingly elaborate calculation procedure, this rating problem solution is straightforward. In principle, the calculations must be performed iteratively due to the unknown mean fluid temperatures. In this particular example, the mean temperatures were considered initially to be equal to the values calculated subsequently. To correct for the property variation, the shell-side Nu_s needs to be calculated again once the thermal resistances are known on both sides to compute T_w and hence Pr_w. The T_w can be computed using the same procedure as that outlined in Example 9.1.

9.5.3 Approximate Design Method

The objectives of the approximate design (sizing) method for a given service of a shell-and-tube heat exchanger are several-fold: (1) quick configuration and size estimation, (2) cost estimation, (3) plant layout, or (4) checking the results of a sophisticated computer program. The basis for this method is Eq. (3.184) rearranged as follows for the shell-side or tube outside surface area:

$$A_s = \frac{q}{U_o \, \Delta T_m} = \frac{q}{U_o F \, \Delta T_{\text{lm}}} \tag{9.54}$$

Here $U_o = U_s$ is the overall heat transfer coefficient based on tube outside or shell-side surface area. By approximately but rapidly estimating q, U_o, F, and ΔT_{lm}, one can arrive at the approximate surface area requirement and subsequently the size of the exchanger, as discussed next. Since this is a sizing procedure, either the heat duty and inlet temperatures are given or both inlet and outlet temperatures are specified. They are related by the energy balance of Eq. (3.5) as follows by considering the shell fluid as hot fluid:

$$q = \dot{m}_s c_{p,s} (T_{s,i} - T_{s,o}) = \dot{m} c_{p,t} (T_{t,o} - T_{t,i}) \tag{9.55}$$

The overall heat transfer coefficient U_o of Eq. (9.54) is calculated from Eq. (3.31a) as

$$\frac{1}{U_o} = \frac{1}{h_o} + \frac{1}{h_{o,f}} + \frac{d_o \ln(d_o/d_i)}{2k_w} + \frac{d_o}{h_{i,f} d_i} + \frac{d_o}{h_i d_i} \tag{9.56}$$

Here h_i, h_o, $h_{i,f}$, and $h_{o,f}$ (the subscripts i and o denote tube inside and tube outside or shell side) are selected from Table 9.4. It should be emphasized that the values given in this table are based on the usual velocities or nominally allowable pressure drops; allowance should be made on operating conditions that are quite unusual. Also, care should be exercised as noted in the appropriate footnotes of this table.

The log-mean temperature difference correction factor F should be estimated as follows. The correction factor $F = 1$ for a counterflow exchanger or if one stream changes its temperature only slightly in the exchanger. For a single TEMA E shell with an arbitrary even number of tube-side passes, the correction factor should be $F > 0.8$ if there is no temperature cross; a rough value would be $F = 0.9$ unless it can be determined from Fig. 3.13. Consider $F = 0.8$ when the outlet temperatures of the two streams are equal (thus avoiding the temperature cross). If $T_{s,o} < T_{t,o}$, there exists a temperature cross (we have assumed here that the shell fluid is hotter than the tube fluid) in the multipass exchanger; and in this case, multiple shells in series should be considered. They can be determined by the procedure outlined in Fig. 3.18.

For the known inlet temperatures and given or calculated outlet temperatures, compute the log-mean temperature difference ΔT_{lm} from its definition of Eqs. (3. 172) and (3. 173). Knowing all the parameters on the right-hand side of Eq. (9.54), the tube outside total surface area A_s (including the fin area, if any) can then be estimated from this equation.

9.5.3.1 *Exchanger Dimensions.* To relate the A_s above to the shell inside diameter and the effective tube length, we will use the information shown in Fig. 9.4 (Bell, 1998). It is generated for one of the commonly used fully tubed shell-and-tube heat exchangers that

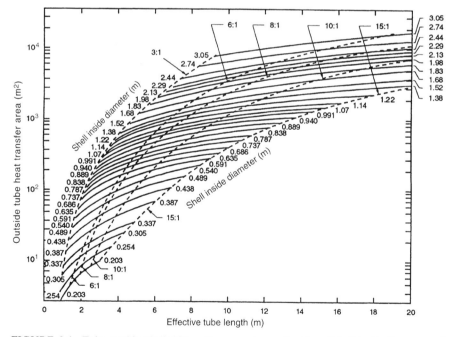

FIGURE 9.4   Tube outside (shell-side) surface area A_s as a function of shell inside diameter and effective tube length for a tube bundle having 19.05-mm ($\frac{3}{4}$-in.)-outside-diameter plain tubes, 23.8-mm ($\frac{15}{16}$-in.) equilateral triangular tube layout, single tube-side pass, and fully tubed exchanger with fixed tubesheets. (From Bell, 1998.)

TABLE 9.4 Typical Film Heat Transfer Coefficients and Fouling Factors for Shell-and-Tube Heat Exchangers

Fluid Conditions		h (W/m^2 · K)^{a,b}	Fouling Resistance (m^2 · K/W)^a
Sensible heat transfer			
Water^c	Liquid	5,000–7,500	1–2.5×10^{-4}
Ammonia	Liquid	6,000–8,000	0–1×10^{-4}
Light organics^d	Liquid	1,500–2,000	0–2×10^{-4}
Medium organics^e	Liquid	750–1,500	1–4×10^{-4}
Heavy organics^f	Liquid		
	Heating	250–750	2–10×10^{-4}
	Cooling	150–400	2–10×10^{-4}
Very heavy organics^g	Liquid		
	Heating	100–300	4–30×10^{-3}
	Cooling	60–150	4–30×10^{-3}
Gas^h	Pressure 100–200 kN/m^2 abs	80–125	0–1×10^{-4}
	Pressure 1 MN/m^2 abs	250–400	0–1×10^{-4}
	Pressure 10 MN/m^2 abs	500–800	0–1×10^{-4}
Condensing heat transfer			
Steam, ammonia	Pressure 10 kN/m^2 abs, no noncondensables^{i,j}	8,000–12,000	0–1×10^{-4}
	Pressure 10 kN/m^2 abs, 1% noncondensables^k	4,000–6,000	0–1×10^{-4}
	Pressure 10 kN/m^2 abs, 4% noncondensables^k	2,000–3,000	0–1×10^{-4}
	Pressure 100 kN/m^2 abs, no noncondensables^{i,j,k,l}	10,000–15,000	0–1×10^{-4}
	Pressure 1 MN/m^2 abs, no noncondensables^{i,j,k,l}	15,000–25,000	0–1×10^{-4}
Light organics^d	Pure component, pressure 10 kN/m^2 abs, no noncondensables^i	1,500–2,000	0–1×10^{-4}
	Pressure 10 kN/m^2 abs, 4% noncondensables^k	750–1,000	0–1×10^{-4}
	Pure component, pressure 100 kN/m^2 abs, no noncondensables	2,000–4,000	0–1×10^{-4}
	Pure component, pressure 1 MN/m^2 abs	3,000–4,000	0–1×10^{-4}
Medium organics^e	Pure component or narrow condensing range, pressure 100 kN/m^2 abs^{m,n}	1,500–4,000	1–3×10^{-4}
Heavy organics	Narrow condensing range, pressure 100 kN/m^2 abs^{m,n}	600–2,000	2–5×10^{-4}
Light multicomponent mixtures, all condensable^d	Medium condensing range, pressure 100 kN/m^2 abs^{k,m,o}	1,000–2,500	0–2×10^{-4}
Medium multicomponent mixtures, all condensable^e	Medium condensing range, pressure 100 kN/m^2 abs^{k,m,o}	600–1,500	1–4×10^{-4}
Heavy multicomponent mixtures, all condensable^f	Medium condensing range, pressure 100 kN/m^2 abs^{k,m,o}	300–600	2–8×10^{-4}
Vaporizing heat transfer^{p,q}			
Water^r	Pressure < 0.5 MN/m^2 abs, $\Delta T_{SH,max} = 25$ K	3,000–10,000	1–2×10^{-4}
	Pressure < 0.5 MN/m^2 abs, pressure < 10 MN/m^2 abs, $\Delta T_{SH,max} = 20$ K	4,000–15,000	1–2×10^{-4}
Ammonia	Pressure < 3MN/m^2 abs, $\Delta T_{SH,max} = 20$ K	3,000–5,000	0–2×10^{-4}
Light organics^d	Pure component, pressure < 2 MN/m^2 abs, $\Delta T_{SH,max} = 20$ K	1,000–4,000	1–2×10^{-4}
	Narrow boiling range,^{s} pressure < 2 MN/m^2 abs, $\Delta T_{SH,max} = 15$ K	750–3,000	0–2×10^{-4}
Medium organics^e	Pure component, pressure < 2 MN/m^2 abs, $\Delta T_{SH,max} = 20$ K	1,000–3,500	1–3×10^{-4}
	Narrow boiling range,^{s} pressure < 2 MN/m^2 abs, $\Delta T_{SH,max} = 15$ K	600–2,500	1–3×10^{-4}
Heavy organics^f	Pure component, pressure < 2 MN/m^2 abs, $\Delta T_{SH,max} = 20$ K	750–2,500	2–5×10^{-4}

| Heavy organics[g] | Narrow boiling range,[s] pressure < 2 MN/m^2 abs, $\Delta T_{SH,max} = 15\,K$ | 400–1,500 | 2–8×10^{-4} |
| Very heavy organics[h] | Narrow boiling range,[s] pressure < 2 MN/m^2 abs, $\Delta T_{SH,max} = 15\,K$ | 300–1,000 | 2–10×10^{-4} |

Source: Bell, K. J. (1998).

[a] Heat transfer coefficients and fouling resistances are based on area in contact with fluid. Ranges shown are typical, not all encompassing. Temperatures are assumed to be in normal processing range; allowances should be made for very high or low temperatures.

[b] Allowable pressure drops on each side are assumed to be about 50–100 kN/m^2 except for (1) low-pressure gas and two-phase flows, where the pressure drop is assumed to be about 5% of the absolute pressure; and (2) very viscous organics, where the allowable pressure drop is assumed to be about 150–250 kN/m^2.

[c] Aqueous solutions give approximately the same coefficients as water.

[d] Light organics include fluids with liquid viscosities less than about 0.5×10^{-3} N · s/m^2, such as hydrocarbons through C_8, gasoline, light alcohols and ketones, etc.

[e] Medium organics include fluids with liquid viscosities between about 0.5×10^{-3} and 2.5×10^{-3} N · s/m^2, such as kerosene, straw oil, hot gas oil, and light crudes.

[f] Heavy organics include fluids with liquid viscosities greater than 2.5×10^{-3} N · s/m^2, but not more than 50×10^{-3} N · s/m^2, such as cold gas oil, lube oils, fuel oils, and heavy and reduced crudes.

[g] Very heavy organics include tars, asphalts, polymer melts, greases, etc., having liquid viscosities greater than about 50×10^{-3} N · s/m^2. Estimation of coefficients for these materials is very uncertain and depends strongly on the temperature difference, because natural convection is often a significant contribution to heat transfer in heating, whereas congelation on the surface and particularly between fins can occur in cooling. Since many of these materials are thermally unstable, high surface temperatures can lead to extremely severe fouling.

[h] Values given for gases apply to such substances as air, nitrogen, carbon dioxide, light hydrocarbon mixtures (no condensation), etc. Because of the very high thermal conductivities and specific heats of hydrogen and helium, gas mixtures containing appreciable fractions of these components will generally have substantially higher heat transfer coefficients.

[i] Superheat of a pure vapor is removed at the same coefficient as for condensation of the saturated vapor if the exit coolant temperature is less than the saturation temperature (at the pressure existing in the vapor phase) and if the (constant) saturation temperature is used in calculating the MTD. But see note k for vapor mixtures with or without noncondensable gas.

[j] Steam is not usually condensed on conventional low-finned tubes; its high surface tension causes bridging and retention of the condensate and a severe reduction of the coefficient below that of the plain tube.

[k] The coefficients cited for condensation in the presence of noncondensable gases or for multicomponent mixtures are only for very rough estimation purposes because of the presence of mass transfer resistances in the vapor (and to some extent, in the liquid) phase. Also, for these cases, the vapor-phase temperature is not constant, and the coefficient given is to be used with the MTD estimated using vapor-phase inlet and exit temperatures, together with the coolant temperatures.

[l] As a rough approximation, the same relative reduction in low-pressure condensing coefficients due to noncondensable gases can also be applied to higher pressures.

[m] Absolute pressure and noncondensables have about the same effect on condensing coefficients for medium and heavy organics as for light organics. For large fractions of noncondensable gas, interpolate between pure component condensation and gas cooling coefficient.

[n] Narrow condensing range implies that the temperature difference between dew point and bubble point is less than the smallest temperature difference between vapor and coolant at any place in the condenser.

[o] Medium condensing range implies that the temperature difference between dew point and bubble point is greater than the smallest temperature difference between vapor and coolant, but less than the temperature difference between inlet vapor and outlet coolant.

[p] Boiling and vaporizing heat transfer coefficients depend very strongly on the nature of the surface and the structure of the two-phase flow past the surface in addition to all of the other variables that are significant for convective heat transfer in other modes. The flow velocity and structure are very much governed by the geometry of the equipment and its connecting piping. Also, there is a maximum heat flux from the surface that can be achieved with reasonable temperature differences between surface and saturation temperatures of the boiling fluid; any attempt to exceed this maximum heat flux by increasing the surface temperature leads to partial or total coverage of the surface by a film of vapor and a sharp decrease in the heat flux. Therefore, the vaporizing heat transfer coefficients given in this table are only for very rough estimating purposes and assume the sue of plain or low-finned tubes without special nucleation enhancement. $\Delta T_{SH,max}$ is the maximum allowable temperature difference between surface and saturation temperature of the boiling liquid. No attempt is made in this table to distinguish among the various types of vapor-generation equipment, since the major heat transfer distinction to be made is the propensity of the process stream to foul. Severely fouling streams will usually call for a vertical thermosiphon or a forced-convection (tube-side) reboiler for ease of cleaning.

[q] Subcooling heat load is transferred at the same coefficient as latent heat load in kettle reboilers, using the saturation temperature in the MTD. For horizontal and vertical thermosiphons and forced-circulation reboilers, a separate calculation is required for the sensible heat transfer area, using appropriate sensible heat transfer coefficients and the liquid temperature profile for the MTD.

[r] Aqueous solutions vaporize with nearly the same coefficient as pure water if attention is given to boiling-point elevation, if the solution does not become saturated, and if care is taken to avoid dry wall conditions.

[s] For boiling of mixtures, the saturation temperature (bubble point) of the final liquid phase (after the desired vaporization has taken place) is to be used to calculate the MTD. A narrow-boiling-range mixture is defined as one for which the difference between the bubble point of the incoming liquid and the bubble point of the exit liquid is less than the temperature difference between the exit hot stream and the bubble point of the exit boiling liquid. Wide-boiling-range mixtures require a case-by-case analysis and cannot be reliably estimated by these simple procedures.

has the following geometry: 19.05 mm ($\frac{3}{4}$ in.) outside diameter plain tubes, 23.8 mm ($\frac{15}{16}$ in.) equilateral triangular tube layout, single tube-side pass, and fixed tubesheets. In this figure, the effective tube length represents the actual tube length between tubesheets for the straight tube exchanger and the length between the tubesheet and the tangent line for the U-tube bundle. The solid black lines indicate the shell inside diameter. From the estimated value of A_s above, one can calculate a number of combinations of the effective tube length L_{eff} and the shell inside diameter D_s. The desired range of L_{eff}/D_s (shown by dashed lines in Fig. 9.4) is between 3 and 15, with a preferable range between 6 and 10. $L_{\text{eff}}/D_s \leq 3$ results in poor shell-side flow distribution and high Δp for the inlet and outlet nozzles. $L_{\text{eff}}/D_s \geq 15$ would be difficult to handle mechanically and would require a longer footprint for the tube bundle repair/removal.

We now explain how to use Fig. 9.4 for different tube diameters and layouts, tube-side multipass construction, and other tube bundle constructions. The effective tube-side surface area for geometry different from that for Fig. 9.4 will be designated as A'_s. The ordinate of Fig. 9.4 is then renamed A'_s. It is related to A_s calculated from Eq. (9.54) as

$$A'_s = A_s F_1 F_2 F_3 \tag{9.57}$$

Once we calculate the correction factors F_1, F_2, and F_3 as outlined next and A_s from Eq. (9.54), A'_s is computed from Eq. (9.57), and the combination of the effective tube length and the shell inside diameter is then determined from Fig. 9.4 as before. Let us describe how to calculate the correction factors.

F_1 = correction factor for the tube outside diameter and tube layout. $F_1 = 1$ for 19.05 mm tubes having a 23.8 mm 30° tube layout. For other d_o and p_t, obtain the value from Table 9.5.

F_2 = correction factor for the number of tube passes. $F_2 = 1$ for a one-tube-pass design. The value of F_2 for multiple tube passes (U-tube and floating head bundles) can be obtained from Table 9.6.

F_3 = correction factor for various rear-end head designs (shell construction) given in Table 9.7.

TABLE 9.5   Values of F_1 of Eq. (9.56) for Various Tube Diameters and Layouts

Tube Outside Diameter [in. (mm)]	Tube Pitch [in. (mm)]	Layout	F_1
$\frac{5}{8}$ (15.88)	$\frac{13}{16}$ (20.6)	→ ▽	0.90
$\frac{5}{8}$ (15.88)	$\frac{13}{16}$ (20.6)	→ ◇, □	1.04
$\frac{3}{4}$ (19.05)	$\frac{15}{16}$ (23.8)	→ ▽	1.00
$\frac{3}{4}$ (19.05)	$\frac{15}{16}$ (23.8)	→ ◇, □	1.16
$\frac{3}{4}$ (19.05)	1 (25.4)	→ ▽	1.14
$\frac{3}{4}$ (19.05)	1 (25.4)	→ ◇, □	1.31
1 (25.4)	$1\frac{1}{4}$ (31.8)	→ ▽	1.34
1 (25.4)	$1\frac{1}{4}$ (31.8)	→ ◇, □	1.54

Source: Data from Bell (1998).

TABLE 9.6   Values of F_2 for Various Numbers of Tube-Passes[a]

Inside Shell Diameter [in. (mm)]	F_2 for Number of Tube-Side passes			
	2	4	6	8
Up to 12 (305)	1.20	1.40	1.80	—
$13\frac{1}{4}$ to $17\frac{1}{4}$ (337 to 438)	1.06	1.18	1.25	1.50
$19\frac{1}{4}$ to $23\frac{1}{4}$ (489 to 591)	1.04	1.14	1.19	1.35
25 to 33 (635 to 838)	1.03	1.12	1.16	1.20
35 to 45 (889 to 1143)	1.02	1.08	1.12	1.16
48 to 60 (1219 to 1524)	1.02	1.05	1.08	1.12
Above 60 (above 1524)	1.01	1.03	1.04	1.06

Source: Data from Bell (1998).
[a] Since U-tube bundles must always have at least two passes, use of this table is essential for U-tube bundles estimation. Most floating-head buindles also require an even number of passes.

TABLE 9.7   Values of F_3 for Various Tube Bundle Constructions

Type of Tube bundle Construction	F_3 for Inside Shell Diameter [in. (mm)]				
	Up to 12 (305)	13–22 (330–559)	23–36 (584–914)	37–48 (940–1219)	Above 48 (above 1219)
Fixed tubesheet (TEMA L, M or N)	1.00	1.00	1.00	1.00	1.00
Split backing ring (TEMA S)	1.30	1.15	1.09	1.06	1.04
Outside packed floating head (TEMA P)	1.30	1.15	1.09	1.06	1.04
U-tube[a] (TEMA U)	1.12	1.08	1.03	1.01	1.01
Pull-through floating head (TEMA T)	—	1.40	1.25	1.18	1.15

Source: Data from Bell (1998).
[a] Since U-tube bundles must always have at least two passes, it is also essential to use Table 9.6 for this configuration.

9.5.4   More Rigorous Thermal Design Method

The more rigorous thermal design method includes all elements discussed in preceding sections. In a concise manner, the following is a step-by-step procedure for the design or sizing problem.

1. For given heat transfer duty and fluid streams inlet temperatures, compute the outlet temperatures using overall energy balances and the fluid mass flow rates specified or selected. If outlet temperatures are given, compute the heat duty requirement.

2. Select a preliminary flow arrangement (i.e., type of shell-and-tube heat exchanger) based on common industry practice (see Section 10.2.1 for selection criteria), mechanical integrity, and maintenance requirements.

3. Follow the approximate design method of Section 9.5.3 to arrive at a preliminary size for the exchanger. Select a shell inside diameter, tube diameter, length, pitch and layout, and baffle spacing. Calculate the number of tubes and number of passes.

4. Follow the rating procedure outlined in Section 9.5.2, which employs the Bell–Delaware method (see Section 9.5.1 for heat transfer coefficient calculations), or apply the stream analysis method (Taborek, 1998; see Section 4.4.1.4) or other available rating procedure.

5. Compare the heat transfer and pressure drop performance computed in step 4 with the values specified. If heat transfer is met and computed pressure drops are within specifications, the thermal design is finished. In that case, the mechanical design is pursued in parallel and series to thermal design to ensure structural integrity and compliance with applicable codes and standards. Also, a check for flow-induced vibration (and/or other operating problems) and a cost estimation are performed to finalize the design.

6. If heat transfer is not met or the computed pressure drop(s) are higher than the specifications, go to step 3 and select the appropriate shell-and-tube geometry, and iterate through step 5 until thermal, mechanical, and cost estimation criteria are met.

9.6 HEAT EXCHANGER OPTIMIZATION

In the preceding sections, rating problems for extended surface and shell-and-tube heat exchangers are presented, as is a sizing problem for an extended surface exchanger. For the sizing problem, no constraints were imposed on the design except for the pressure drops specified. The objective of that problem was to optimize the core dimensions to meet the heat transfer required for specified pressure drops.

Heat exchangers are designed for many different applications, and hence may involve many different optimization criteria. These criteria for heat exchanger design may be minimum initial cost, minimum initial and operating costs, minimum weight or material, minimum volume or heat transfer surface area, minimum frontal area, minimum labor (translated into a minimum number of parts), and so on. When a performance measure has been defined quantitatively and is to be minimized or maximized, it is called an *objective function* in a design optimization. A particular design may also be subjected to certain requirements, such as required heat transfer, allowable pressure drop, limitations on height, width and/or length of the exchanger, and so on. These requirements are called *constraints* in a design optimization. A number of different surfaces could be incorporated in a specific design problem, and there are many geometrical variables that could be varied for each surface geometry.[†] In addition, operating mass flow rates and temperatures could be changed. Thus, a large number of *design variables* are associated with a heat exchanger design. The question arises as to how one can effectively adjust these design variables within imposed constraints and come up with a design having an optimum objective function. This is what we mean by the optimum component

---

[†] For a shell-and-tube exchanger, the geometrical variables are those associated with the tube, baffles, shell, and front- and rear-end heads. For an extended surface exchanger, the geometrical variable associated with a fin are the fin pitch, fin height, fin thickness, type of fin, and other variables associates with each fin type.

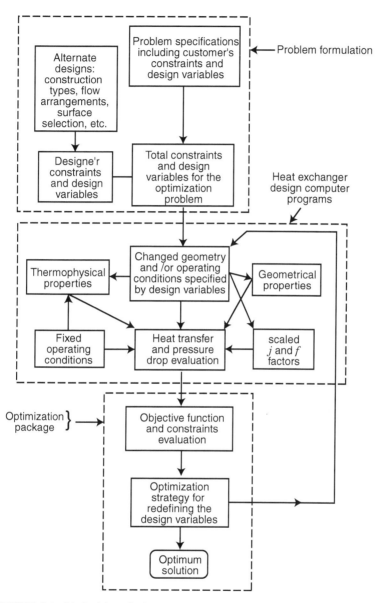

FIGURE 9.5   Methodology for heat exchanger optimization. (From Shah *et al.* 1978.)

design, sometimes also referred to as the most *efficient* design. If a heat exchanger is part of a system, it could also be optimized based on the system objective function by varying pertinent exchanger design variables as well as system variables in the optimization routine.

A complete mathematical component or system-based optimization of heat exchanger design is neither practical nor possible. Many engineering judgments based on experience are involved in various stages of the design. However, once the general configuration and surfaces are selected, an optimized heat exchanger design may be arrived at if the objective function and constraints can be expressed mathematically and if all the variables are changed automatically and systematically on some evaluation criteria basis.

A large number of optimization (search) techniques are available in the literature, and quite a lot of commercial optimization software is available. A typical design and optimization procedure for a heat exchanger is summarized here with the flowchart of Fig. 9.5 for completeness. The procedure is referred to as the *case study method*. In this method, each possible surface geometry and construction type is considered to be an alternative design, as indicated in Fig. 9.5. To make a legitimate comparison of these alternatives, each design must be optimized for the application specified. Thus there may be several independent optimized solutions satisfying the problem requirements. Engineering judgment, a comparison of objective function values, and other evaluation criteria are then applied to select a final optimum solution for implementation.

Assume a liquid-to-gas heat exchanger to be required for a specific application having minimum total cost. From the initial screening of surfaces (see Section 10.3), suppose that two plate-fin constructions (the louver-fin and strip-fin surfaces) and one flat-tube and wavy-fin construction appear to be promising for the gas side. Then, for this problem, there are three alternative designs that need to be optimized.

As shown in Fig. 9.5, first formulate the total number of constraints for the problem. This includes the customer's specified explicit constraints (such as fixed frontal area, the ranges of heat exchanger dimensions) and implicit constraints (such as required minimum heat transfer, allowable maximum pressure drop). Once the basic surface geometry for the design chosen is selected, the designer imposes some additional constraints, such as the minimum and maximum values for the fin height, fin thickness, fin pitch, fin thermal conductivity, flow length, number of finned passages, gas flow rate, and so on. The designer wants to vary all the design and operating variables within the ranges specified such that the exchanger will meet the required heat transfer, maximum pressure drop, and other constraints with minimum total cost.

To optimize the heat exchanger, the designer starts with one set of heat exchanger surface geometrical dimensions which may not even satisfy all or some of the constraints imposed. Subsequently, the various geometrical properties (such as heat transfer area, free-flow area, hydraulic diameter) and thermal properties are evaluated based on the input operating conditions. The heat transfer rate and pressure drop are then evaluated by the procedure outlined for the rating problem (see, e.g., Sections 9.2.1, 9.4.3, and 9.5.2). Next, the output from heat exchanger calculations is fed to the optimization computer program package, where the constraints and the objective function are evaluated. Subsequently, new values for the design variables are generated and heat exchanger calculations are repeated. The iterations are continued until the objective function is optimized (minimized or maximized as desired) within the accuracy specified and all constraints are satisfied. In some situations, it may not be possible to satisfy all constraints. Engineering judgment is then applied to determine whether or not the optimum design is satisfactory and which constraints to relax. One of the most important but least known inputs for the heat transfer and pressure drop evaluation is the magnitude for *scaled-up* or *scaled-down* (modified from the original) j and f factors. As soon as one of the surface geometrical dimensions is changed (such as the fin pitch, height, or thickness) but others may stay unchanged, the surface is no longer geometrically similar to the original surface for which experimental j and f data are available. In such cases, either theoretical or experimental correlations should be incorporated in the computer program to arrive at the scaled j and f factors for the new geometry. Some of these correlations are presented in Section 7.5. The designer must use his or her experience and judgment regarding the appropriate correlations to obtain the scaled j and f factors. In addition, care must be exercised to avoid excessive extrapolations.

A review of Fig. 9.5 indicates that heat exchanger design (rating) program and optimization software is needed for the optimization. In addition, a system simulation program is added if the heat exchanger optimization has to be done based on the system design approach.

Although the foregoing optimization procedure was outlined from a performance and cost point of view, the heat exchanger can also be optimized as a component or as part of a system based on a thermoeconomics point of view. This is discussed further in Section 11.6.

SUMMARY

The focus of this chapter is to provide step-by-step rating and sizing procedures for major types of heat exchangers. All the information and design theory outlined in previous chapters is applied and extended in formulating the design procedures discussed in this chapter. The detailed thermal and hydraulic design of heat exchangers outlined in this chapter is one of the major objectives of this book. After presenting how to determine the mean temperature on each fluid side in a heat exchanger, we have provided the rating and sizing of extended-surface (plate-fin and tube-fin), plate, and shell-and-tube heat exchangers in depth with an example. Subsequently, we have also provided a general approach to the optimization of a heat exchanger design.

REFERENCES

Bell, K. J., 1963, Final report of the cooperative research program on shell-and-tube heat exchangers, *Univ. Del. Eng. Stn. Bull.* No. 5.

Bell, K. J., 1988a, Overall design methodology for shell-and-tube exchangers, in *Heat Transfer Equipment Design*, R. K. Shah, E. C. Subbarao, and R. A. Mashelkar, eds., Hemisphere Publishing, Washington, DC, pp. 131–144.

Bell, K. J., 1988b, Delaware method for shell design, in *Heat Transfer Equipment Design*, R. K. Shah, E. C. Subbarao, and R. A. Mashelkar, eds., Hemisphere Publishing, Washington, DC, pp. 145–166.

Bell, K. J., 1998, Approximate sizing of shell-and-tube heat exchangers, in *Heat Exchanger Design Handbook*, G. F. Hewitt, exec. ed., Begell House, New York, Vol. 3, Sec. 3.1.4.

Chiou, J. P., 1980, The advancement of compact heat exchanger theory considering the effects of longitudinal heat conduction and flow nonuniformity, in *Compact Heat Exchangers: History, Technological Advancement and Mechanical Design Problems*, R. K. Shah, C. F. McDonald, and C. P. Howard, eds., Book G00183, American Society of Mechanical Engineers, New York.

Kays, W. M., A. L. London, 1998, *Compact Heat Exchangers*, reprint 3rd ed., Krieger Publishing, Malabar, FL.

Marriot, J., 1977, Performance of an Alfaflex plate heat exchanger, *Chem. Eng. Prog.*, Vol. 73, No. 2, pp. 73–78.

Raznjević, K., 1976, *Handbook of Thermodynamic Tables and Charts*, McGraw-Hill, New York.

Shah, R. K., 1981, Compact heat exchanger design procedures, in *Heat Exchangers: Thermal-Hydraulic Fundamentals and Design*, S. Kakaç, A. E. Bergles and F. Mayinger, eds., Hemisphere Publishing Corp., Washington, DC, pp. 495–536.

Shah, R. K., 1988a, Plate-fin and tube-fin heat exchanger design procedures, in *Heat Transfer Equipment Design*, R. K. Shah, E. C. Subbarao, and R. A. Mashelkar, eds., Hemisphere Publishing, Washington, DC, pp. 255–266.

Shah, R. K., 1988b, Counterflow rotary regenerator thermal design procedures, in *Heat Transfer Equipment Design*, R. K. Shah, E. C. Subbarao, and R. A. Mashelkar, eds., Hemisphere Publishing, Washington, DC, pp. 267–296.

Shah, R. K., and W. W. Focke, 1988, Plate heat exchangers and their design theory, in *Heat Transfer Equipment Design*, R. K. Shah, E. C. Subbarao, and R. A. Mashelkar, eds., Hemisphere Publishing, Washington, DC, pp. 227–254.

Shah, R. K., and A. D. Giovannelli, 1988, Heat pipe heat exchanger design theory, in *Heat Transfer Equipment Design*, R. K. Shah, E. C. Subbarao, and R. A. Mashelkar, eds., Hemisphere Publishing, Washington, DC, pp. 609–653.

Shah, R. K., and T. Skiepko, 1999, Influence of leakage distribution on rotary regenerator thermal performance, *Appl. Thermal Eng.*, Vol. 19, pp. 685–705.

Shah, R. K., and A. S. Wanniarachchi, 1991, Plate heat exchanger design theory, in *Industrial Heat Exchangers*, J.-M. Buchlin, ed., Lecture Series 1991-04, von Kármán Institute for Fluid Dynamics, Belgium.

Shah, R. K., K. A. Afimiwala, and R. W. Mayne, 1978, Heat exchanger optimization, *Heat Transfer 1978, Proc. 6th Int. Heat Transfer Conf.*, Vol. 4, pp. 185–191.

Taborek, J., 1998, Shell-and-tube heat exchangers: single phase flow, in *Handbook of Heat Exchanger Design*, G. F. Hewitt, ed., Begell House, New York, pp. 3.3.3-1 to 3.3.11-5.

REVIEW QUESTIONS

For each question, circle one or more correct answers. Explain your answers briefly.

9.1   The fluid mean temperature on each fluid side in a gas-to-gas multipass heat exchanger is generally computed as the:

(a) arithmetic mean temperature on one fluid side and log-mean average on the other fluid side in each pass

(b) arithmetic mean temperature on both fluid sides in each pass

(c) can't tell   (d) none of these

9.2   For sizing of a gas-to-gas plate-fin *counterflow* exchanger, we can determine the physical size of the exchanger (with no constraints imposed on the dimensions) such that:

(a) The pressure drops on both fluid sides will always be exactly matched.

(b) The critical pressure drop on one fluid side can be matched; the pressure drop on the other fluid side will be higher than the specified value.

(c) The critical pressure drop on one fluid side can be matched; the pressure drop on the other fluid side will be lower than the specified value.

(d) can't tell

9.3   In the following industrial heat exchangers, we can always meet both the heat transfer and pressure drop requirements on at least one fluid side during the design process:

(a) shell-and-tube exchanger   (b) gasketed plate exchanger

(c) plate-fin exchanger

9.4   The following streams do not contribute significantly to heat transfer on the shell side of a shell-and-tube heat exchanger:

(a) A   (b) B   (c) C

(d) E   (e) F

9.5   Various leakage and bypass streams on the shell side:

(a) increases heat transfer

(b) increases pressure drop

(c) decreases heat transfer and pressure drop

(d) none of these

9.6   A heat exchanger can be optimized *for a system* using:

(a) surfaces selected based on the screening methods

(b) performance evaluation criteria

(c) commercially available most sophisticated optimization software for heat exchanger optimization

(d) all of these   (e) none of these

PROBLEMS

9.1   A gas turbine–driven generator is to be installed in a power plant for peaking service. To obtain high system efficiency, the hot turbine exhaust gases (430°C and 102.7 kPa pressure) are used to preheat the combustion air, which leaves the compressor at 910 kPa and 175°C. The mass flow rate through the compressor and turbine are 24.3 and 24.7 kg/s, respectively.

We have selected an unmixed–unmixed crossflow heat exchanger having $\varepsilon = 0.75$. This exchanger has 2.0 m × 0.9 m frontal area on the air side and 2.0 m × 2.0 m frontal area on the gas side. The surface geometries and performance characteristics of the air- and gas-side surfaces are provided below.

Louvered Plate-Fin Surface 3/8-6.06 (Fig. 10-38, Kays and London, 1998)	Strip-Fin Plate-Fin Surface 1/2-11.94(D) (Fig. 10-64, Kays and London, 1998)
Fin density = 238.6 m^{-1}	Fin density = 470 m^{-1}
Plate spacing = 6.35 mm	Plate spacing = 6.02 mm
Louver spacing = 9.525 mm	Splitter symmetrically located
Louver gap = 1.4 mm	Strip length in flow direction = 12.7mm
Fin gap = 2.79 mm	Flow passage hydraulic
Flow passage hydraulic	diameter = 2.266 mm
diameter = 4.453 mm	Fin metal thickness = 0.152 mm
Fin metal thickness = 0.152 mm	Splitter metal thickness = 0.152 mm
Heat transfer surface area density	Heat transfer surface area density
$\beta = 840 \, m^2/m^3$	$\beta = 1521 \, m^2/m^3$
Fin area/total area = 0.640	Fin area (including splitter)/total area = 0.796

Re	j	f	Re	j	f
10,000	0.00551	0.0331	8,000	—	0.0123
8,000	0.00593	0.0340	7,000	0.00452	0.0126
6,000	0.00651	0.0354	6,000	0.00471	0.0131
5,000	0.00690	0.0363	5,000	0.00492	0.0137
4,000	0.00738	0.0375	4,000	0.00522	0.0146
3,000	0.00805	0.0394	3,000	0.00575	0.0162
2,500	0.00849	0.0406	2,000	0.00682	0.0198
2,000	0.00900	0.0426	1,500	0.00744	0.0231
1,500	0.00970	0.0461	1,200	0.00830	0.0265
1,200	0.0104	0.0496	1,000	0.00911	0.0306
1,000	0.0112	0.0532	800	0.01045	0.0347
800	0.0124	0.0587	600	0.01255	0.0429
600	0.0144	0.0682	500	0.01415	0.0493
500	0.0160	0.0755	400	0.0166	0.0592
			300	0.0205	0.0758

All materials are stainless steel with $k = 20.8\,\mathrm{W/m \cdot K}$. The following geometrical properties have been evaluated for the air and gas sides.

Geometrical Properties	Air Side	Gas Side
Minimum free-flow area A_o (m^2)	0.8247	1.5638
Heat transfer area A (m^2)	1469	2524
σ	0.455	0.398
L/r_h	1781	1614
A_w (m^2)		550.7

(a) At what temperatures will you evaluate fluid properties on each fluid side? Consider the following mean fluid properties: Air side: specific heat $1.04\,\mathrm{kJ/kg \cdot K}$, thermal conductivity $0.0431\,\mathrm{W/m \cdot K}$ and dynamic viscosity $0.283 \times 10^{-4}\,\mathrm{Pa \cdot s}$; these properties on gas side are: $1.06\,\mathrm{kJ/kg \cdot K}$, thermal conductivity $0.0473\,\mathrm{W/m \cdot K}$, and dynamic viscosity $0.305 \times 10^{-4}\,\mathrm{Pa \cdot s}$.

(b) Evaluate the quality of design by determining: (i) The percentage of the thermal resistance on the air and gas sides. Is this a thermally balanced design? Use η_f for air and gas sides as 0.674 and 0.773, respectively, and $\delta_w = 0.1\,\mathrm{mm}$. (ii) The relative pressure drops $\Delta p/p$ for each stream. Are the pressure drops balanced? What fraction of the total pressure drop is due to entrance/exit losses? Use the following densities: air-side inlet, outlet, and mean densities: 7.075, 4.962, and 5.833 kg/m^3; gas-side inlet, outlet, and mean densities: 0.5094, 0.6875, and 0.5852 kg/m^3.

9.2 A waste heat recovery heat exchanger is manufactured in small ($0.3\,\mathrm{m} \times 0.3\,\mathrm{m} \times 0.6\,\mathrm{m}$) modules. Each is a single-pass crossflow heat exchanger having fluids unmixed on both sides. The 0.6 m dimension corresponds to the noflow height. Assume that the surface on each fluid side is a plain triangular fin 11.1 of Kays and London (1998). The surface geometry and j and f data are given below.

Re	j	f	Re	j	f
10,000	0.00314	0.00878	2,000	0.00436	0.0129
8,000	0.00333	0.00923	1,500	0.00444	0.0149
6,000	0.00356	0.00971	1,200	0.00471	0.0169
5,000	0.00372	0.00991	1,000	0.00515	0.0190
4,000	0.00390	0.0103	800	0.00599	0.0228
3,000	0.00412	0.0112	600	0.00733	0.0294
2,500	0.00424	0.0119	500	0.00840	0.0350

Fin density $= 437\,\mathrm{m}^{-1}$, plate spacing $b = 6.35\,\mathrm{mm}$, flow passage hydraulic diameter $D_h = 3.081\,\mathrm{mm}$, fin metal thickness $= 0.15\,\mathrm{mm}$, material $=$ aluminum, heat transfer surface area density $\beta = 1204\,\mathrm{m}^2/\mathrm{m}^3$, and fin area/total area $= 0.756$. Assume the plate thickness to be $0.5\,\mathrm{mm}$. Try to accommodate an integer number of fins on each fluid side in an approximately $0.3\,\mathrm{m} \times 0.3\,\mathrm{m} \times 0.6\,\mathrm{m}$ envelope. On one fluid side of the exchanger, process air at $0.40\,\mathrm{m}^3/\mathrm{s}$ flows with the inlet temperature of $238°\mathrm{C}$. On the other fluid side, makeup air flows at a $0.26\,\mathrm{m}^3/\mathrm{s}$ flow rate at $40°\mathrm{C}$ inlet temperature. Both fluids are at atmospheric pressure at inlet. Determine the outlet temperatures on each fluid side as well as the pressure drop on each fluid side. Both the fin and plates are made of aluminum having a thermal conductivity of $190.4\ \mathrm{W/m \cdot K}$.

9.3 A finned-tube exchanger is designed to cool water from $52°\mathrm{C}$ to $38°\mathrm{C}$ with ambient air at $32°\mathrm{C}$ to be heated to $43°\mathrm{C}$. The water and air mass flow rates are 2.52 and 13.13 kg/s, respectively. The allowable air-side pressure drop is 149 Pa. Estimate the air-side frontal velocity to satisfy the desired performance. Consider $j/f = 0.30$, $\mathbf{R}_{\mathrm{air}}/\mathbf{R}_{\mathrm{tot}} = 0.7$, and counterflow performance. The finned tube exchanger has $A_o/A_{fr} = 0.56$ and $c_p = 1.005$ and $4.187\,\mathrm{kJ/kg \cdot K}$ for air and water, respectively. Use $\rho_m = 1.136\,\mathrm{kg/m^3}$, $\mathrm{Pr} = 0.705$, and $p_i = 101.4\,\mathrm{kPa}$ for air.

9.4 What will be the exit temperatures of the hot and cold streams and heat duty (heat transfer rate) of a plate heat exchanger if 2.52 kg/s of hot water enters at $104°\mathrm{C}$ and 4.04 kg/s of cold water enters at $16°\mathrm{C}$? These flow rates are expected at some part-load condition, and the plant engineer is interested in determining the temperatures to see how secondary cooling equipment downstream behaves. Use the following data for your solution: the plate exchanger is a 1 pass–1 pass counterflow exchanger (see Fig. 1.65a) with 14 flow channels for hot water and 14 flow channels for cold water (total 27 thermal plates). The effective width of plates is 0.457 m and the gasket thickness or gap between plates is 3.0 mm. Note that the hydraulic diameter will be equal to twice the gap dimension. The projected heat transfer area for each plate on one side is $0.28\ \mathrm{m}^2$. Use a fouling factor of $0.0002\,\mathrm{m}^2 \cdot \mathrm{K/W}$ for each fluid stream, a plate thickness of 0.9 mm, and wall thermal conductivity of $15.6\ \mathrm{W/m \cdot K}$. The j and f factors for the plate surface are given as $j = 0.2\mathrm{Re}^{-0.25}$ and $f = 0.6\mathrm{Re}^{-0.2}$. Also, the following properties are provided for water.

Property Hot	Hot Water at 88°C	Cold Water at 24°C
c_p (kJ/kg · K)	4.19	4.19
μ (Pa · s)	0.320×10^{-3}	0.922×10^{-3}
k (W/m · K)	0.68	0.61
ρ (kg/m³)	977	1001
Pr	1.97	6.33

9.5 Design a gas-to-gas two-fluid heat exchanger with both fluids assumed to be air. The inlet temperature, pressure, and mass flow rate of one airstream are 487 K, 490 kPa, and 21 kg/s, respectively, and it must leave the heat exchanger at a temperature of 619 K. The mass flow rate of the second airstream is the same as that of the first, and its inlet temperature and pressure are 690 K and 103 kPa, respectively. The pressure drops are limited to 4.900 kPa and 2.575 kPa for the cold and hot airstream. Additional data: Flow arrangement is a single-pass cross flow (unmixed–unmixed); heat transfer surface is a plain plate-fin surface with designation 19.86 (Kays and London, 1998). The heat transfer surface material is aluminum with a plate thickness of 1 mm. List all your assumptions and provide proper justification for each. If any additional assumption is needed, provide explicit reasons for it.

9.6 Analyze a three-pass two-fluid overall cross-counterflow heat exchanger of a gas-turbine plant with all three passes as equal unmixed–unmixed crossflow units (see the right-hand side of Fig. 1.58a as a schematic). The desired overall heat exchanger effectiveness must be 0.766 in order to reach the required plant effectiveness. The air mass flow rate is 21 kg/s and the inlet temperature 487 K. The mass flow rate of the gas is the same, but the inlet temperature is 690 K. Determine the outlet temperatures of both air and gas. Perform calculations involving determination of thermophysical properties of fluids at the true mean (integral) temperatures.

10 Selection of Heat Exchangers and Their Components

As described in Chapter 1, a variety of heat exchangers are available, and the question becomes which one to choose for a given application. In addition, for each type (core construction), either a large number of geometrical variables (such as those associated with each component of the shell-and-tube exchanger) or a large number of surface geometries (such as those for plate, extended surface, or regenerative exchangers) are available for selection. Again the question involves which set of geometries/surfaces will be most appropriate for a given application. There is no such thing as a particular heat exchanger or the selection of a particular heat transfer surface that is best (i.e., optimum) for a given application. Near-optimum heat exchanger designs involve many trade-offs, since many geometrical and operating variables are associated with heat exchangers during the selection process. For example, a cheaper exchanger may be obtained if one wants to give up some performance or durability. One can get higher performance if the exchanger is heavier or costs a little more. A heat exchanger can be made smaller if we accept a little lower performance or provide more pumping power for higher fluid flow rates. The heat exchanger design team must consider the trade-offs and arrive at the optimum exchanger for a given application to meet the design requirements and constraints.

In this chapter we discuss qualitative and quantitative criteria/methods used to select exchanger type and surface geometry for a given application for engineers not having prior design/operational experience. If one or more exchangers are already in service for similar applications, this prior experience is the best guide for the selection and design of a heat exchanger for a given application. We first describe important qualitative selection criteria for heat exchangers in two categories: (1) criteria based on important operating variables of exchangers in Section 10.1, and (2) general guidelines on major heat exchanger types in Section 10.2. Next, we describe some quantitative criteria for selection of extended heat exchanger surfaces (screening methods) in Section 10.3.1 and for selection of tubular exchangers (performance evaluation criteria) in Section 10.3.2. These quantitative criteria are energy-based (the first law of thermodynamics) for a heat exchanger as a component. Criteria based on the second law of thermodynamics can also be devised as indicated in Section 10.3.3, with the details presented in Section 11.7. Finally, selection based on cost criteria is presented briefly in Section 10.3.4 and discussed further in Section 11.6.6. Except for some qualitative discussion, it is not easy to present a method for system-based selection and optimization of a heat exchanger in a textbook since there are many systems in which heat exchangers are used, and each is different, depending on the process. Thus, the overall objective of this chapter is to provide a good understanding of heat exchanger selection in general as a component based on qualitative and quantitative criteria. However, system-based heat exchanger optimization is the current industrial practice.

10.1   SELECTION CRITERIA BASED ON OPERATING PARAMETERS

A large number of heat exchangers are described in Chapter 1, providing understanding of their functions, the range of operating parameters, reasons why they are used in certain applications, and so on. With such thorough understanding, one would have a good idea of which types of exchangers to use in given applications. Refer to Table 10.1 for operating conditions and principal features for many heat exchanger types. Here we highlight heat exchanger selection criteria based on major operating parameters.

10.1.1   Operating Pressures and Temperatures

The exchanger in operation must withstand the stresses produced by the operating pressure and the temperature differences between two fluids. These stresses depend on the inlet pressures and temperatures of the two fluids. The most versatile exchangers for a broad range of operating pressures and temperatures are shell-and-tube exchangers for medium- to high-heat duties and double-pipe exchangers for lower-heat duties. They can handle from high vacuum to ultrahigh fluid pressures [generally limited to 30 MPa (4350 psi) on the shell or annulus side and 140 MPa (20,000 psi) on the tube side]. Coupled with high pressures, shell-and-tube exchangers can withstand high temperatures, limited only by the materials used; however, the inlet temperature difference is limited to 50°C (120°F) from the thermal expansion point of view when the exchanger design allows only limited thermal expansion, such as in the E-shell design. These exchangers are used for gas, liquid, and phase-change applications.

For liquid–liquid or liquid–phase change applications, if the operating pressures and temperatures are modest (less than about 2.5 MPa and 200°C), gasketed or semiwelded plate exchangers should be considered. For somewhat higher pressures and temperatures, fully welded or brazed plate exchangers may be the choice, depending on other design criteria.

The plate-fin extended surface exchanger is designed for low-pressure applications, with the operating pressures on either side limited to about 1000 kPa (150 psig), except for cryogenics applications, where the operating pressure is about 9000 kPa gauge (1300 psig). The maximum operating temperature for plate-fin exchangers is below 650°C (1200°F) and usually below 150°C (300°F), to avoid the use of expensive materials. There is no limit on the minimum operating temperature; plate-fin exchangers are commonly used in cryogenic applications. Fins in a plate-fin exchanger act as a flow-mixing device for highly viscous liquids, and if properly designed, add surface area for heat transfer with a reasonably high fin efficiency. In a plate-fin exchanger, fins on the liquid side are used primarily for pressure containment and rigidity. Fins on the gas side are used for added surface area for heat transfer, with fin efficiencies usually greater than 80%.

The tube-fin exchanger is used to contain the high-pressure fluid on the tube side if only one fluid is at a high pressure. Fins on the liquid or phase-change side generally have "low" heights, to provide reasonably high fin efficiencies. Turbulators may be used within tubes for flow mixing. Tube-fin exchangers with or without shells are designed to cover the operating temperature range from low cryogenic temperatures to about 870°C (1600°F).

For ultrahigh temperature [870 to 2000°C (1600 to 3600°F)] and near-atmospheric pressures, as in high-temperature waste heat recovery, either rotary regenerators (870 to 1100°C) or fixed-matrix regenerators (up to 2000°C) are used.

10.1.2 Cost

Cost is a very important factor in the selection of the heat exchanger construction type. The cost per unit of heat transfer surface area is higher for a gasketed plate exchanger than for a shell-and-tube exchanger. However, from the total cost (capital, installation, operation, maintenance, etc.) point of view, PHEs are less expensive than shell-and-tube exchangers when stainless steel, titanium, and other higher quality alloys are used. Since tubes are more expensive than extended surfaces or a regenerator matrix, shell-and-tube (or broadly, tubular) exchangers are in general more expensive per unit of heat transfer surface area. In addition, the heat transfer surface area density of a tubular core is generally much lower than that of an extended surface or regenerative exchanger. Rotary regenerators made of paper or plastic are in general the least expensive per unit of heat transfer surface area.

10.1.3 Fouling and Cleanability

Fouling and cleanability are among the most important design considerations for liquid-to-liquid or phase-change exchangers and for some gas-to-fluid exchangers. Fouling should be evaluated for both design and off-design points. Periodic cleaning and/or replacement of some exchanger components depend on the fouling propensity of the fluids employed. In applications involving moderate to severe fouling, either a shell-and-tube or a gasketed plate heat exchanger is used, depending on the other operating parameters. In a shell-and-tube exchanger, the tube fluid is generally selected as the heavily fouling fluid since the tube side may be cleaned more easily. A plate heat exchanger is highly desirable in those relatively low temperature applications [< 300°C (575°F)] where severe fouling occurs on one or both sides, as plate disassembly, cleaning, and reassembly is a relatively easy task. For highly corrosive fluid heating or cooling applications, shell-and-tube exchangers are used exclusively, regardless of operating pressure and temperature conditions. Plate-fin exchangers usually have small hydraulic diameter passages and hence are more susceptible to fouling. They are also relatively difficult to clean and are not employed in even moderate fouling applications unless they can be cleaned chemically or thermally by baking (see Section 13.4).

The fouling and cleanability problem is not as severe for gas-to-gas exchangers as for liquid-to-liquid or phase-change exchangers, since in most applications gases are neither very dirty nor have the fouling propensity of water. Regenerators have self-cleaning characteristics because the hot and cold gases flow periodically in opposite directions through the same passage. Hence, they can tolerate moderate fouling. If the application has a potential for heavy fouling, a larger flow passage size is chosen, as in a fixed-matrix

TABLE 10.1 Principal Features of Several Types of Heat Exchangers

Type of Heat Exchanger	Compactness (m²/m³)	Stream Types[a]	Material[b]	Temperature Range (°C)	Maximum Pressure (bar)[c]	Cleaning Methods	Corrosion Resistance	Multistream Capability	Multipass Capability
					Feature				
Plate-and-frame (gaskets)	up to 200	Liquid–liquid, gas–liquid, two-phase	s/s, Ti, Incoloy, Hastelloy, graphite, polymer	−35 to +200	25	Mechanical[d]	Good[e]	Yes[f]	Yes
Partially welded plate	up to 200	Liquid–liquid, gas–liquid, two-phase	s/s, Ti, Incoloy, Hastelloy	−35 to +200	25	Mechanical,[d,g] chemical[h]	Good[e]	No	Yes
Fully welded plate (AlfaRex)	up to 200	Liquid–liquid, gas–liquid, two-phase	s/s, Ti, Ni alloys	−50 to +350	40	Chemical	Excellent	No	Yes
Brazed plate	up to 200	Liquid–liquid, two-phase	s/s	−195 to +220	30	Chemical[i]	Good[j]	No	No[k]
Bavex plate	200–300	Gases, liquids, two-phase	s/s, Ni, Cu, Ti, special steels	−200 to +900	60	Mechanical,[d,l] chemical	Good	In principle	Yes
Platular plate	200	Gases, liquids, two-phase	s/s, Hostelloy, Ni alloys	up to 700	40	Mechanical[d,m]	Good	Yes[n]	Yes
Compabloc plate	up to 300	Liquids	s/s, Ti Incoloy	up to 300	32	Mechanical[d]	Good	Not usually	Yes
Packinox plate	up to 300	Gases, liquids two-phase	s/s, Ti, Hastelloy, Inconel	−200 to +700	300	Mechanical[d,o]	Good	Yes[f]	Yes
Spiral	up to 200	Liquid–liquid, two-phase	c/s, s/s, Ti, Incoloy, Hastelloy	up to 400	25	Mechanical[d]	Good	No	No

676

Type	Area density or size	Duties[a]	Materials[b]	Temperature range (°C)	Max. pressure (bar)[c]	Cleaning method	Corrosion resistance			
Brazed plate-fin	800–1500	Gases, liquids, two-phase	Al, s/s, Ni alloy	Cryogenic to +650	90	Chemical	Good	Yes	Yes	Yes
Diffusion-bonded plate-fin	700–800	Gases, liquids, two-phase	Ti, s/s	up to 500	> 200	Chemical	Excellent	Yes	Yes	Yes
Printed-circuit	200–5000	Gases, liquids, two-phase	s/s, Ni, Ni alloys, Ti	−200 to +900	> 400	Chemical	Excellent	Yes	Yes	Yes
Polymer (e.g. channel plate)	450	Gas liquid[p]	PVDF[q] PP[r]	up to 150[s]	6	Water wash	Excellent	No	No	Not usually
Plate-and-shell	—	Liquids	s/s, Ti, (shell also in c/s)[t]	up to 350	70	Mechanical[d,o], Chemical[u]	Good	No	Yes	Yes
Marbond	up to 10,000	Gases, liquids, two-phase	S/s, Ni, Ni alloys, Ti	−200 to +900	> 400	Chemical	Excellent	Yes	Yes	Yes

Source: Data from Lancaster (1998).

[a] Two-phase includes boiling and condensing duties.

[b] s/s, stainless steel; Ti, titanium; Ni, nickel; Cu, copper. Alloys of these materials and other special alloys are frequently available.

[c] The maximum pressure capability is unlikely to occur at the higher operating temperatures, and assumes no pressure/stress-related corrosion.

[d] Can be dismantled.

[e] Function of gasket as well as plate material.

[f] Not common.

[g] On gasket side.

[h] On welded side.

[i] Ensure compatibility with copper braze.

[j] Function of braze as well as plate material.

[k] Not in a single unit.

[l] On tube side.

[m] Only when flanged access provided; otherwise, chemical cleaning.

[n] Five fluids maximum.

[o] On shell side.

[p] Condensing on gas side.

[q] Polyvinylidene difluoride.

[r] Polypropylene.

[s] PEEK (polyetheretherketone) can go to 250°C

[t] Shell may be composed of polymeric material.

[u] On plate side.

regenerator, so that the impact of fouling is reduced, or cleaning by one of the methods described in Section 13.4 may be employed.

10.1.4   Fluid Leakage and Contamination

Whereas in some applications, fluid leakage from one fluid side to the other fluid side is permissible within limits, in other applications fluid leakage is absolutely not allowed. Even in a good leak tight design, carryover and bypass leakages from the hot fluid to the cold fluid (or vice versa) occur in regenerators. Where these leakages and subsequent fluid contamination is not permissible, regenerators are not used. The choices left are either a tubular, extended surface, or some plate type heat exchangers. Gasketed plate exchangers have more probability of flow leakage than do shell-and-tube exchangers. Plate-fin and tube-fin exchangers have potential leakage problems at the joint between the corrugated fin passage and the header or at the tube-to-header joint. Where absolutely no fluid contamination is allowed (as in the processing of potable water), a double-wall tubular or shell-and-tube exchanger or a double-plate PHE is used.

10.1.5   Fluids and Material Compatibility

Materials selection and compatibility between construction materials and working fluids are important issues, in particular with regard to corrosion (see Section 13.5) and/or operation at elevated temperatures. While a shell-and-tube heat exchanger may be designed using a variety of materials, compact heat exchangers often require preferred metals or ceramics. For example, a requirement for low cost, light weight, high conductivity, and good joining characteristics for compact heat exchangers often leads to the selection of aluminum for the heat transfer surface. On the other side, plate exchangers require materials that are either used for food fluids or require corrosion resistance (e.g., stainless steel). In general, one of the selection criteria for exchanger material depends on the corrosiveness of the working fluid. In Table 10.2, a summary of some materials used for noncorrosive and/or corrosive services is presented. More details about the selection of materials are provided in specialized literature of TEMA (1999) and the ASME (1998) codes.

10.1.6   Fluid Type

A gas-to-gas heat exchanger requires a significantly greater amount of surface area than that for a liquid-to-liquid heat exchanger for a given heat transfer rate. This is because the heat transfer coefficient for the gas is $\frac{1}{10}$ to $\frac{1}{100}$ that of a liquid. The increase in surface area is achieved by employing surfaces that have a high heat transfer surface area density β. For example, fins are employed in an extended surface heat exchanger, or a small hydraulic diameter surface is employed in a regenerator, or small-diameter tubes are used in a tubular heat exchanger. Plate heat exchangers (of the type described in Section 1.5.2) are generally not used in a gas-to-gas exchanger application because they produce excessively high pressure drops. All prime surface heat exchangers with plain (uncorrugated) plates are used in some waste heat recovery applications. The fluid pumping power is generally significant and a controlling factor in designing gas-to-gas exchangers.

TABLE 10.2 Materials for Noncorrosive and Corrosive Service

Material	Heat Exchanger Type or Typical Service
Noncorrosive Service	
Aluminum and austenitic chromium–nickel steel	Any heat exchanger type, $T < -100°C$
$3\frac{1}{2}$ Ni steel	Any heat exchanger type, $-100 < T < -45°C$
Carbon steel (impact tested)	Any heat exchanger type, $-45 < T < 0°C$
Carbon steel	Any type of heat exchanger, $0 < T < 500°C$
Refractory-lined steel	Shell-and-tube, $T > 500°C$
Corrosive Service	
Carbon steel	Mildly corrosive fluids; tempered cooling water
Ferritic carbon–molybdenum and chromium–molybdenum alloys	Sulfur-bearing oils at elevated temperatures (above 300°C); hydrogen at elevated temperatures
Ferritic chromium steel	Tubes for moderately corrosive service; cladding for shells or channels in contact with corrosive sulfur bearing oil
Austenitic chromium–nickel steel	Corrosion-resistant duties
Aluminum	Mildly corrosive fluids
Copper alloys: admiralty, aluminum brass, cupronickel	Freshwater cooling in surface condensers; brackish and seawater cooling
High nickel–chromium–molybdenum alloys	Resistance to mineral acids and Cl-containing acids
Titanium	Seawater coolers and condensers, including PHEs
Glass	Air preheaters for large furnaces
Carbon	Severely corrosive service
Coatings: aluminum, epoxy resin	Exposure to sea and brackish water
Linings: lead and rubber	Channels for seawater coolers
Linings: austenitic chromium–nickel steel	General corrosion resistance

Source: Data from Lancaster (1998).

In liquid-to-liquid exchanger applications, regenerators are ruled out because of the associated fluid leakage and carryover (contamination). Fluid pumping power is, however, not as critical for a liquid-to-liquid heat exchanger as it is for a gas-to-gas heat exchanger.

In a liquid-to-gas heat exchanger, the heat transfer coefficient on the gas side is $\frac{1}{10}$ to $\frac{1}{100}$ of that on the liquid side. Therefore, for a "thermally balanced" design[†] (i.e., having hA of the same order of magnitude on each fluid side of the exchanger), fins are employed to increase the gas-side surface area. Thus, the common heat exchanger constructions used for a liquid-to-gas heat exchanger are the extended surface and tubular; plate-type and regenerative constructions are not used.

For phase-change exchangers, the condensing or evaporating fluid has a range of heat transfer coefficients that vary from low values approximating those for gas flows to high values approximating those for high liquid flows and higher. Therefore, the selection of

---

[†]A thermally balanced design usually results in an optimum design from the cost viewpoint since the cost of the extended surface per unit surface area is less than that of the prime surface, either tubes or plates.

exchanger type for phase-change exchangers parallels the guidelines provided for the gas or liquid side of the exchanger.

10.2 GENERAL SELECTION GUIDELINES FOR MAJOR EXCHANGER TYPES

A large number of heat exchangers are described in Chapter 1. That information, complemented by the material presented in this section, will provide a good understanding of the selection of heat exchanger types.

10.2.1 Shell-and-Tube Exchangers

More than 65% of the market share (in the late 1990s) in process and petrochemical industry heat exchangers is held by the shell-and-tube heat exchanger, for the following reasons: its versatility for handling a wide range of operating conditions with a variety of materials, design experience of about 100 years, proven design methods, and design practice with codes and standards. The selection of an appropriate shell-and-tube heat exchanger is achieved by a judicious choice of exchanger configuration, geometrical parameters, materials, and the "right" design. Next we summarize some guidelines on all these considerations qualitatively to provide the feel for the right design for a given application. The major components of a shell-and-tube exchanger are tubes, baffles, shell, front-end head, read-end head, and tubesheets. Depending on the applications, a specific combination of geometrical variables or types associated with each component is selected. Some guidelines are provided below. For further details on geometrical dimensions and additional guidelines, refer to TEMA (1999).

10.2.1.1 Tubes. Since the desired heat transfer in the exchanger takes place across the tube surface, the selection of tube geometrical variables is important from a performance point of view. In most applications, plain tubes are used. However, when additional surface area is required to compensate for low heat transfer coefficients on the shell side, low finned tubing with 250 to 1200 fins/m (6 to 30 fins/in.) and a fin height of up to 6.35 mm ($\frac{1}{4}$ in.) is used. While maintaining reasonably high fin efficiency, low-height fins increase surface area by two to three times over plain tubes and decrease fouling on the fin side based on the data reported.

The most common plain tube sizes have 15.88, 19.05, and 25.40 mm ($\frac{5}{8}$, $\frac{3}{4}$, and 1 in.) tube outside diameters. From the heat transfer viewpoint, smaller-diameter tubes yield higher heat transfer coefficients and result in a more compact exchanger. However, larger-diameter tubes are easier to clean and more rugged. The foregoing common sizes represent a compromise. For mechanical cleaning, the smallest practical size is 19.05 mm ($\frac{3}{4}$ in.). For chemical cleaning, smaller sizes can be used provided that the tubes never plug completely.

The number of tubes in an exchanger depends on the fluid flow rates and available pressure drop. The number of tubes is selected such that the tube-side velocity for water and similar liquids ranges from 0.9 to 2.4 m/s (3 to 8 ft/sec) and the shell-side velocity from 0.6 to 1.5 m/s (2 to 5 ft/sec). The lower velocity limit corresponds to limiting the fouling, and the upper velocity limit corresponds to limiting the rate of erosion. When sand and silt are present, the velocity is kept high enough to prevent settling.

The number of tube passes depends on the available pressure drop. Higher velocities in the tube result in higher heat transfer coefficients, at the expense of increased pressure drop. Therefore, if a higher pressure drop is acceptable, it is desirable to have fewer but longer tubes (reduced flow area and increased flow length). Long tubes are accommodated in a short shell exchanger by multiple tube passes. The number of tube passes in a shell generally range from 1 to 10 (see Fig. 1.61). The standard design has one, two, or four tube passes. An odd number of passes is uncommon and may result in mechanical and thermal problems in fabrication and operation.

10.2.1.2   Tube Pitch and Layout. The selection of tube pitch is a compromise between a close pitch (small values of p_t/d_o) for increased shell-side heat transfer and surface compactness, and an open pitch (large values of p_t/d_o) for decreased shell-side plugging and ease in shell-side cleaning. In most shell-and-tube exchangers, the ratio of the tube pitch to tube outside diameter varies from 1.25 to 2.00. The minimum value is restricted to 1.25 because the tubesheet ligament[†] may become too weak for proper rolling of the tubes and cause leaky joints. The recommended ligament width depends on the tube diameter and pitch; the values are provided by TEMA (1999).

Two standard types of tube layouts are the square and the equilateral triangle, shown in Fig. 10.1. The equilateral pitch can be oriented at 30° or 60° angle to the flow direction, and the square pitch at 45° and 90°.[‡] Note that the 30°, 45° and 60° arrangements are staggered, and 90° is inline. For the identical tube pitch and flow rates, the tube layouts in decreasing order of shell-side heat transfer coefficient and pressure drop are: 30°, 45°, 60°, and 90°. Thus the 90° layout will have the lowest heat transfer coefficient and the lowest pressure drop.

The square pitch (90° or 45°) is used when jet or mechanical cleaning is necessary on the shell side. In that case, a minimum cleaning lane of $\frac{1}{4}$ in. (6.35 mm) is provided. The square pitch is generally not used in the fixed tubesheet design because cleaning is not feasible. The triangular pitch provides a more compact arrangement, usually resulting in a smaller shell, and the strongest header sheet for a specified shell-side flow area. Hence, it is preferred when the operating pressure difference between the two fluids is large. If

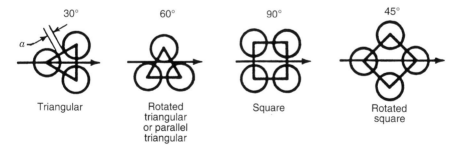

FIGURE 10.1   Tube layout arrangements.

[†] The ligament is a portion of material between two neighboring tube holes. The ligament width is defined as the tube pitch minus the tube hole diameter, such as the distance a shown in Fig. 10.1.

[‡] Note that the tube layout angle is defined in relation to the flow direction and is not related to the horizontal or vertical reference line. Refer to Table 8.1 for the definitions of tube layouts and associated geometrical variables.

designed properly, it can be cleaned from six directions instead of four as in the square pitch arrangement. When mechanical cleaning is required, the 45° layout is preferred for laminar or turbulent flow of a single-phase fluid and for condensing fluid on the shell side. If the pressure drop is limited on the shell side, the 90° layout is used for turbulent flow. For boiling applications, the 90° layout is preferred, because it provides vapor escape lanes. However, if mechanical cleaning is not required, the 30° layout is preferred for single-phase laminar or turbulent flow and condensing applications involving a high ΔT range[†] (a mixture of condensables). The 60° layout is preferred for condensing applications involving a low ΔT range (generally, pure vapor condensation) and for boiling applications. Horizontal tube bundles are used for shell-side condensation or vaporization.

10.2.1.3 Baffles. As presented in Section 1.5.1.1, baffles may be classified as either longitudinal or transverse type. Longitudinal baffles are used to control the overall flow direction of the shell fluid. Transverse baffles may be classified as plate baffles or grid baffles. Plate baffles are used to support the tubes, to direct the fluid in the tube bundle at approximately right angles to the tubes, and to increase the turbulence and hence the heat transfer coefficient of the shell fluid. However, the window section created by the plate baffles results in excessive pressure drop with insignificant contribution to heat transfer; flow normal to the tubes in crossflow section may create flow-induced vibration problems. The rod baffles, a most common type of grid baffles, shown in Fig. 1.11, are used to support the tubes and to increase the turbulence. Flow in a rod baffle heat exchanger is parallel to the tubes, and hence flow-induced vibration is virtually eliminated by the baffle support of the tubes. The choice of baffle type, spacing, and cut are determined largely by the flow rate, required heat transfer, allowable pressure drop, tube support, and flow-induced vibration. The specific arrangements of baffles in various TEMA shells are shown in Fig. 10.3.

Plate Baffles. Two types of plate baffles, shown in Fig. 1.10 are segmental, and disk and doughnut. Single and double segmental baffles are used most frequently. The single segmental baffle is generally referred to simply as a *segmental baffle*. The practical range of single segmental baffle spacing is $\frac{1}{5}$ to 1 shell diameter, although optimum could be $\frac{2}{5}$ to $\frac{1}{2}$. The minimum baffle spacing for cleaning the bundle is 50.8 mm (2 in.) or $\frac{1}{5}$ shell diameter, whichever is larger. Spacings closer than $\frac{1}{5}$ shell diameter provide added leakage[‡] that nullifies the heat transfer advantage of closer spacings. If the foregoing limits on the baffle spacing do not satisfy other design constraints, such as Δp_{max} or tube vibration, no-tubes-in-window or pure crossflow design should be tried.

The segmental baffle is a circular disk (with baffle holes) with one disk segment removed. The baffle cut varies from 20 to 49% (the height ℓ_c in Fig. 8.9 given as a percentage of the shell inside diameter), with the most common being 20 to 25%. At larger spacings, it is 45 to 50%, to avoid excessive pressure drop across the windows as compared to the bundle. Large or small spacings coupled with large baffle cuts are undesirable because of the increased potential of fouling associated with stagnant flow

---

[†] Here the ΔT range represents the difference in condensing temperature at the inlet minus condensing temperature at the outlet of an exchanger.
[‡] These are tube to baffle hole, baffle to shell, bundle to shell, and the tube pass partition leakages or bypasses described in Section 4.4.1.1.

areas. If fouling is a primary concern, the baffle cut should be kept below 25%. The baffle cut and spacing should be designed such that the flow velocity has approximately the same magnitude for the cross flow and window flow sections. Alternate segmental baffles are arranged 180° to each other, which cause shell-side flow to approach crossflow in the central bundle region[†] and axial flow in the window zone. All segmental baffles shown in Fig. 1.10 have horizontal baffle cuts. The direction of the baffle cut is selected as follows for shell-side fluids: Either horizontal or vertical for a single-phase fluid (liquid or gas), horizontal for better mixing for very viscous liquids, and vertical for the following shell-side applications: condensation (for better drainage), evaporation/boiling (for no stratification and for providing disengagement room), entrained particulates in liquid (to provide least interference for solids to fall out), and multishell pass exchanger, such as those in Fig. 1.62 and the F shell.

Since one of the principal functions of the plate baffle is to support the tubes, the terms *baffle* and *support plate* are sometimes used interchangeably. However, a support plate does not direct the fluid normal to the tube bank, it may be thicker than a baffle, it has less tube-to-baffle hole clearance, and it provides greater stiffness to the bundle. Support plates with single-segmental baffles are cut approximately at the centerline and spaced 0.76 m (30 in.) apart. This results in an unsupported tube span of 1.52 m (60 in.) because each plate supports half the number of tubes. The double-segmental baffle (Fig. 1.10), also referred to as a *strip baffle*, provides lower shell-side pressure drop (and allows larger fluid flows) than that for the single segmental baffle for the same unsupported tube span. The baffle spacing for this case should not be too small; otherwise, it results in a more parallel (longitudinal) flow (resulting in a lower heat transfer coefficient) with significant zones of flow stagnation. Triple-segmental baffles have flows with a strong parallel flow component, provide lower pressure drop, and permit closer tube support to prevent tube vibrations.

The lower allowable pressure drop results in a large baffle spacing. Since the tubes in the window zone are supported at a distance of two or more times the baffle spacing, they are most susceptible to vibration. To eliminate the possibility of tube vibrations and to reduce the shell-side pressure drop, the tubes in the window zone are removed and support plates are used to reduce the unsupported span of the remaining tubes. The resulting design is referred to as the *segmental baffle with no-tubes-in-window*, shown in Fig. 1.10. The support plates in this case are circular and support all the tubes. The baffle cut and number of tubes removed varies from 15 to 25%. Notice that low-velocity regions in the baffle corners do not exist, resulting in good flow characteristics and less fouling. Thus the loss of heat transfer surface in the window section is partially compensated for. However, the shell size *must* be increased to compensate for the loss in the surface area in the window zone, which in turn may increase the cost of the exchanger. If the shell-side operating pressure is high, this no-tubes-in-window design is very expensive compared to a similar exchanger having tubes in the window zone.

The disk-and-doughnut baffle is made up of alternate disks and doughnut-shaped baffles, as shown in Fig. 1.10. Generally, the disk diameter is somewhat greater than the half-shell diameter, and the diameter of the hole of the doughnut is somewhat smaller than the half-shell diameter. This baffle design provides a lower pressure drop compared to that in a single-segmental baffle for the same unsupported tube span and eliminates the tube bundle-to-shell bypass stream C. The disadvantages of this design are that (1) all the

---

[†] Various allowable clearances required for construction of a tube bundle with plate baffles are provided by TEMA (1999).

tie rods to hold baffles are within the tube bundle, and (2) the central tubes are supported by the disk baffles, which in turn are supported only by tubes in the overlap of the larger-diameter disk over the doughnut hole.

Rod Baffles. Rod baffles are used to eliminate flow-induced vibration problems. For certain shell-and-tube exchanger applications, it is desirable to eliminate the cross flow and have pure axial (longitudinal) flow on the shell side. For the case of high shell-side flow rates and low-viscosity fluids, the rod baffle exchanger has several advantages over the segmental baffle exchanger: (1) It eliminates flow-induced tube vibrations since the tubes are rigidly supported at four points successively; (2) the pressure drop on the shell side is about one-half that with a double segmental baffle at the same flow rate and heat transfer rate. The shell-side heat transfer coefficient is also considerably lower than that for the segmental baffle exchanger. In general, the rod baffle exchanger will result in a smaller-shell-diameter longer-tube unit having more surface area for the same heat transfer and shell-side pressure drop; (3) there are no stagnant flow areas with the rod baffles, resulting in reduced fouling and corrosion and improved heat transfer over that for a plate baffle exchanger; (4) since the exchanger with a rod (grid) baffle design has a counterflow arrangement of the two fluids, it can be designed for higher exchanger effectiveness and lower mean (or inlet) temperature differences than those of an exchanger with a segmental baffle design; and (5) a rod baffle exchanger will generally be a lower-cost unit and has a higher exchanger heat transfer rate to pressure drop ratio overall than that of a segmental baffle exchanger. If the tube-side fluid is controlling and has a pressure drop limitation, a rod baffle exchanger may not be applicable. Refer to Gentry (1990) for further details on this exchanger.

Impingement Baffles. Impingement baffles or plates are generally used in the shell side just below the inlet nozzle. Their purpose is to protect the tubes in the top row near the inlet nozzle from erosion, cavitation, and/or vibration due to the impact of the high-velocity fluid jet from the nozzle to the tubes. One of the most common forms of this baffle is a solid square plate located under the inlet nozzle just in front of the first tube row, as shown in Fig. 10.2. The location of this baffle is critical within the shell to minimize the associated pressure drop and high escape velocity of the shell fluid after the baffle. For this purpose, adequate areas should be provided both between the nozzle and plate and between the plate and tube bundle. This can be achieved either by omitting some tubes from the circular bundle as shown in Fig. 10.2 or by modifying the nozzle so that it has an expanded section (not shown in Fig. 10.2). Also, proper positioning of this plate in the first baffle space is important for efficient heat transfer.

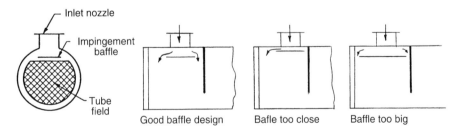

FIGURE 10.2   Impingement baffles at the shell-side inlet nozzle. (From Bell, 1998.)

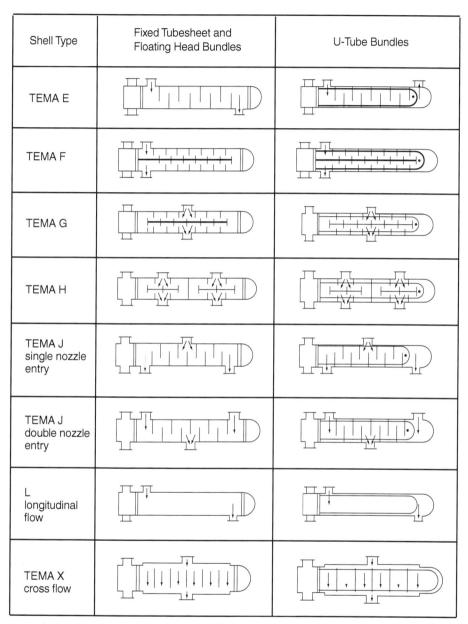

FIGURE 10.3   Shell-side flow arrangement for various shell types (Courtesy of Heat Transfer Research, Inc., College Station, Texas).

Enough space should be provided between the tip of the plate and the tubesheet and between the tip of the plate and the first segmental baffle. The most common cause of tube failure is improper location and size of the impingement plate.

10.2.1.4 Shells. Seven types of shells, as classified by TEMA (1999), are shown in Fig. 1.6; they are also shown in Fig. 10.3 with baffles. The E shell, the most common due to its low cost and relative simplicity, is used for single-phase shell fluid applications and for small condensers with low vapor volumes. Multiple passes on the tube side increase the heat transfer coefficient h (if corresponding more increased Δp is within allowed limits). However, a multipass tube arrangement can reduce the exchanger effectiveness or F factor compared to that for a single-pass arrangement (due to some tube passes being in parallelflow) if the increased h and NTU do not compensate for the parallel-flow effect. Two E shells in series (in overall counterflow configuration) may be used to increase the exchanger effectiveness ε.

As an alternative, a counterflow arrangement is desirable (i.e., high ε) for a two-tube-pass exchanger. This is achieved by the use of an F shell having a longitudinal baffle, resulting in two shell passes. However, a TEMA F shell is rarely used in practice because of heat leakage across the longitudinal baffle and potential flow leakage that can occur if the area between the longitudinal baffle and the shell is not sealed properly. Also, the F shell presents additional problems of fabrication and maintenance, and it is difficult to remove or replace the tube bundle. If one needs to increase the exchanger effectiveness, multiple shells in series are preferred over an F shell.

The TEMA G and H shells are related to the F shell but have different longitudinal baffles. Hence, when the shell-side Δp is a limiting factor, a G or H shell can be used; however, ε or F will be lower than that of a counterflow exchanger. The split-flow G shell has horizontal baffles with the ends removed; the shell nozzles are 180° apart at the midpoint of the tubes. The double-split-flow H shell is similar to the G shell, but with two inlet and two outlet nozzles and two longitudinal baffles. The G and H shells are seldom used for shell-side single-phase applications, since there is no advantage over E or X shells. They are used as horizontal thermosiphon reboilers, condensers, and other phase-change applications. The longitudinal baffle serves to prevent flashing of the lighter components of the shell fluid, helps flush out noncondensables, provides increased mixing, and helps distribute the flow. Generally, ΔT and Δp across longitudinal baffles are small in these applications, and heat transfer across the baffle and flow leakages at the sides have insignificant influence on the performance. The H shell approaches the cross-flow arrangement of the X shell, and it usually has low shell-side Δp compared to the E, F, and G shells. For high-inlet-velocity applications, two nozzles are required at the inlet, hence the H or J shell is used.

The divided-flow TEMA J shell has two inlets and one outlet or one inlet and two outlet nozzles (a single nozzle at the midpoint of the tubes and two nozzles near the tube ends). The J shell has approximately one-eighth the pressure drop of a comparable E shell and is therefore used for low-pressure-drop applications such as in a condenser in vacuum. For a condensing shell fluid, the J shell is used, with two inlets for the gas phase and one central outlet for the condensate and residue gases.

The TEMA K shell is used for partially vaporizing the shell fluid. It is used as a kettle reboiler in the process industry and as a flooded chiller (hot liquid in tubes) in the refrigeration industry. Usually, it consists of an overall circular-cross-section horizontal bundle of U tubes placed in an oversized shell with one or more vapor nozzles on the top side of the shell (see one vapor nozzle in Fig. 1.6) to reduce liquid entrainment. The tube

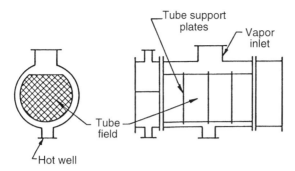

FIGURE 10.4   An X shell exchanger with omission of top tube rows for better flow distribution for a condensing application. (From Bell, 1998.)

bundle diameter ranges 50 to 70% of the shell diameter. The liquid (to be vaporized) enters from below near the tubesheet through the left-hand nozzle and covers the tube bundle. Pool and some convective boiling takes place on the shell side without forced flow of the vaporizing fluid outside the tubes on the shell side. The vapor occupies the upper space in the shell without the tubes. The large empty space in the shell acts as a vapor disengaging space; and if properly sized, almost dry vapor exits from the top nozzle, thus eliminating the need for an external vapor–liquid separator. Hence, it is commonly used, although it is more expensive to fabricate, particularly, for high-pressure applications. Generally, the kettle reboiler is considered as a pool boiling device; however, convective (flow) boiling prevails in the tube bundle.

For a given flow rate and surface area, the crossflow TEMA X shell has the lowest shell-side pressure drop of all (except for K) shell configurations. Hence, it is used for gas heating and cooling applications with or without finned tubes and for vacuum condensing applications. It is also used for applications having large shell flows. No transverse baffles are used in the X shell; however, support plates are used to suppress the flow-induced vibrations. Flow distributions on the shell side could be a serious problem unless proper provision has been made to feed the fluid uniformly at the inlet. This could be achieved by a bathtub nozzle, multiple nozzles, or by providing a clear lane along the length of shell near the nozzle inlet as shown in Fig. 10.4.

The type of shell described in Fig. 1.6 has either one or two shell passes in one shell. The cost of the shell is much more than the cost of tubes; hence, a designer tries to accommodate the required heat transfer surface in one shell. Three or four shell passes in a shell could be made by the use of longitudinal baffles.[†] Multipassing on the shell side with longitudinal baffles will reduce the flow area per pass compared to a single pass on the shell side in a single shell, resulting in a possibly higher shell-side pressure drop. Multiple shells in series are also used for a given application for the following reasons:

- They increase the exchanger effectiveness ε, or reduce the surface area for the same ε. For the latter case, a subsequent reduction in tubing cost may offset the cost of an additional shell and other components.

[†] Positive or tight sealing between the longitudinal baffles and the shell is essential to maintain the high exchanger effectivenesses predicted.

- For an exchanger requiring high effectiveness, multipassing is the only alternative.
- For part-load operation and where a spare bundle is essential, multiple shells (may be smaller in size) will result in an economical operation.
- Shipping and handling may dictate restrictions on the overall size or weight of the unit, resulting in multiple shells for an application.

In heat recovery trains and some other applications, up to six shells in series are commonly used. The limitation on the number of shells in such applications is the pressure drop limit on one of the fluid streams.

10.2.1.5   Front-End Heads. The front- and rear-end head types, as classified by TEMA (1999), are shown in Fig. 1.6. The front-end head is stationary, while the rear-end head can be either stationary or floating, depending on the allowed thermal stresses between the tubes and the shell. The major criteria for the selection of front- and rear-end heads are the thermal stresses, operating pressures, cleanability, hazards, and cost.

The front-end heads are primarily of two types, the channels and the bonnet. The bonnet head B is cast in one piece and has either a side- or an end-entering nozzle.[†] Although the bonnet head is less expensive, inspection and maintenance requires breaking the pipe joints and removing the bonnet. Hence, the bonnet head is generally used for clean tube-side fluids. The channel head can be removable, as in the TEMA A head, or can be integral with the tubesheet, as in TEMA C and N heads. There is a removable channel cover in these front-end heads for inspection and maintenance without disturbing the piping. The nozzles are side entering in these types. Notice that while the shell is welded onto the TEMA N head, it is flanged to the TEMA C head. In the TEMA N head, no mechanical joint exists (all welded joints) between the channel and tubesheet and between the tubesheet and the shell, thus eliminating leakage between the shell and the tubes. The TEMA D head has a special high-pressure closure and is used for applications involving 2100 kPa gauge (300 psig) for ammonia service and higher pressures for other applications.

10.2.1.6   Rear-End Heads. In a shell-and-tube exchanger, the shell is at a temperature different from that of the tubes because of heat transfer between the shell and tube fluids. This results in a differential thermal expansion and stresses among the shell, tubes, and the tubesheet. If proper provisions are not made, the shell or tubes can buckle, or tubes can be pulled apart or out of the tubesheet. Provision is made for differential thermal expansion in the rear-end heads. They may be categorized as fixed or floating rear-end heads, depending on whether there are no provisions or some provisions for differential thermal expansion. A more commonly used third design that allows tube expansion freely is the exchanger with U tubes having the front- and rear-end heads fixed; it is included in the floating rear-end-head category in the discussion to follow. The design features of shell-and-tube exchangers with various rear-end heads are summarized in Table 10.3.

A heat exchanger with a fixed rear-end head L, M, or N has a fixed tubesheet on that side. Hence, the overall design is rigid. The tube bundle-to-shell clearance is least among the designs, thus minimizing the bundle-to-shell bypass stream C. Any number of tube

---

[†] Notice that the nozzles on the front- and rear-end heads are for the tube fluid. The nozzles for the shell fluid are located on the shell itself.

TABLE 10.3   Design Features of Shell-and-Tube Heat Exchangers

Design Feature	Fixed Tubesheet	Return Bend (U-Tube)	Outside-packed Stuffing box	Outside-Packed Latern Ring	Pull-Through Bundle	Inside Split Backing Ring
TEMA Rear-Head Type:	L, M, N	U	P	W	T	S
Tube bundle removable	No	Yes	Yes	Yes	Yes	Yes
Spare bundles used	No	Yes	Yes	Yes	Yes	Yes
Provides for differential movement between shell and tubes	Yes, with bellows in shell	Tes	Yes	Yes	Yes	Yes
Individual tubes can be replaced	Yes	Yes[a]	Yes	Yes	Yes	Yes
Tubes can be chemically cleaned, both inside and outside	Yes	Yes	Yes	Yes	Yes	Yes
Tubes can be mechanicall cleaned on inside	Yes	With special tools	Yes	Yes	Yes	Yes
Tubes can be mechanically cleaned on outside	Yes	Yes[b]	Yes[b]	Yes[b]	Yes[b]	Yes[b]
Internal gaskets and bolting are required	No	No	No	No	Yes	Yes
Double tubesheets are practical	Yes	Yes	Yes	No	No	No
Number of tubesheet passes available	Any	Any even number	Any[c]	One or two[d]	Any[e]	Any[e]
Approximate diametral clearance (mm) (Shell ID, D_{otl})	11–18	11–18	25–50	15–35	95–160	35–50
Relative costs in ascending order, (least expensive = 1)	2	1	4	3	5	6

Source: Data from Shah (1995).
[a] Only those in outside rows can be replaced without special designs.
[b] Outside mechanical cleaning possible with square or rotated square pitch, or wide triangular pitch.
[c] Axial nozzle required at rear end for odd number of passes.
[d] Tube-side nozzles must be at stationary end for two passes.
[e] Odd number of passes requires packed or bellows at floating head.

passes can be employed. The TEMA L, M, and N rear-end heads are the counterparts of TEMA A, B, and N front-end heads. The major disadvantages of the fixed tubesheet exchanger are (1) no relief for thermal stresses between the tubes and the shell, (2) the impossibility of cleaning the shell side mechanically (only chemical cleaning is possible), and (3) the impracticality of replacing the tube bundle. Fixed tubesheet exchangers are thus used for applications involving relatively low temperatures [315°C (600°F) and lower] coupled with low pressures [2100 kPa gauge (300 psig) and lower]. As a rule of thumb, the fixed tubesheet design is used for an inlet temperature difference between the two fluids that is less than about 50 to 60°C (100°F). If an expansion bellows is used, this temperature difference can be increased to about 80 to 90°C (150°F). Expansion bellows are generally uneconomical for high pressures [> 4150 kPa gauge (600 psig)]. The fixed tubesheet exchanger is a low-cost unit ranked after the U-tube exchanger.

The differential thermal expansion can be accommodated by a floating rear-end head in which tubes expand freely within the shell, thus eliminating thermal stresses. Also, the tube bundle is removable for mechanical cleaning of the shell side. Basically, there are three types of floating rear-end heads: U-tube heads, internal floating heads (pull-through/split-ring heads), and outside packed floating heads.

In the U-tube bundle, the thermal stresses are significantly reduced, due to free expansion of the U-tubes, and the rear-end head has an integral cover which is the least expensive among rear-end heads. The exchanger construction is simple, having only one tubesheet and no expansion joints, and hence it is the lowest-cost design, particularly at high pressures. The tube bundle can be removed for shell-side cleaning; however, it is difficult to remove a U tube from the bundle except in the outer row, and it is also difficult to clean the tube-side bends mechanically. So a U-tube exchanger is used with clean fluids inside the tubes unless the tube side can be cleaned chemically. Flow-induced vibration can also be a problem for the tubes in the outermost row because of a long unsupported span, particularly in large-diameter bundles.

The next-simplest floating head is the pull-through head T shown in Fig. 10.5. On the floating-head side, the tubesheet is small, acts as a flange, and fits in the shell with its own bonnet head. The tube bundle can easily be removed from the shell by first removing the

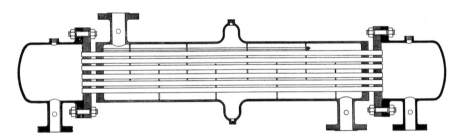

FIGURE 10.5   Two-pass exchanger (BET) with a pull-through (T) rear-end head. (Courtesy of Patternson-Kelley Co., Division of HARSCO Corporation, East Stroudsburg, Pennsylvania.)

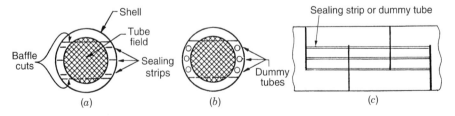

FIGURE 10.6   (a) Sealing strips; (b) dummy tubes or ties rods; (c) sealing strips, dummy tubes, or tie rods covering the entire length of the tube bundle.

front-end head. Individual tubes or the tube bundle can also be replaced if required. Due to the floating-head bonnet flange and bolt circle, many tubes are omitted from the tube bundle near the shell. This results in the largest bundle-to-shell circumferential clearance or a significant bundle-to-shell bypass stream C. So as not to reduce exchanger performance, sealing strips (or dummy tubes or tie rods with spacers) in the bypass area are essential, as shown in Fig. 10.6. They are placed in pairs every five to seven tube pitches between the baffle cuts. They force the fluid from the bypass stream back into the bundle. However, localized high velocities near the sealing strips could cause flow-induced tube vibration; hence, proper care must be exercised for the design. Since this design has the least number of tubes in a bundle for a given shell diameter compared to other floating-head designs, the shell diameter is somewhat larger, to accommodate a required amount of surface area. One of the ideal applications of the TEMA T head design is in the kettle reboiler, for which there is ample space on the shell side and the flow bypass stream C is of no concern.

The large bundle-to-shell clearance can be minimized by bolting the floating-head bonnet to a split backing ring (flange) as shown in Fig. 10.7. It is referred to as the TEMA S rear-end head. The shell cover over the tube floating head has a diameter larger than that of the shell. As a result, the bundle-to-shell clearances are reasonable and sealing strips are generally not required. However, both ends of the exchanger must be disassembled for cleaning and maintenance. In both TEMA S and T heads, the shell fluid is held tightly to prevent leakage to the outside. However, internal leakage is possible due to the failure of an internal hidden gasket and is not easily detectable. The TEMA T head has more positive gasketing between the two streams than does the S head. Both TEMA S and T head configurations are used for the tube-side multipass exchangers; the single-pass construction is not feasible if the advantages of the positive sealing of TEMA S and T heads are to be retained. The cost of TEMA S and T head designs is relatively high

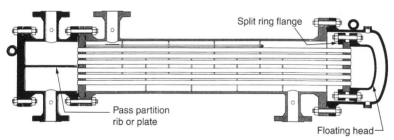

FIGURE 10.7 Two-pass exchanger (AES) with a split-ring (S) floating head. (Courtesy of Patternson-Kelley CO., Division of HARSCO Corporation, East Stroudsburg, Pennsylvania.)

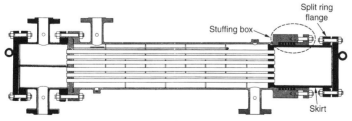

FIGURE 10.8 Two-pass exchanger (AEP) with an outside packed (P) floating head. (Courtesy of Patternson-Kelley Co., Division of HARSCO Corporation, East Stroudsburg, Pennsylvania.)

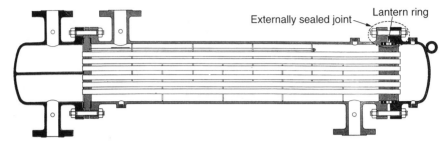

FIGURE 10.9   Two-pass exchanger (BEW) having a packed floating head (W) with lantern rings. (Courtesy of Patternson-Kelley Co., Division of HARSCO Corporation, East Stroudsburg, Pennsylvania.)

compared to U-tube or fixed-tubesheet units. The cost for the TEMA S head is higher than for the TEMA T head. The split backing ring floating head is used extensively in the petroleum industry for moderate operating pressures and temperatures. For very high operating pressures and temperatures, the TEMA S head design has a special test ring (TEMA, 1999).

In the outside-packed floating-head TEMA P design of Fig. 10.8, the stuffing box provides a seal against the skirt of the floating head and prevents shell-side fluid leakage to the outside. This skirt (and the tube bundle) is free to move axially against the seal to take thermal expansion into account. A split-ring flange near the end of the skirt seals the back end of the chamber. Because of the specific design of this floating head, any leak (from either the shell side or the tube side) at the gaskets is to the outside. Hence, the TEMA P head is generally not used with very toxic fluids. Also, the inlet and outlet nozzles must be located at the stationary end; hence, this design could have only an even number of tube passes. In this design, the bundle-to-shell clearance is large [about 38 mm (1.5 in.)]; as a result, sealing strips are required. The TEMA P head exchanger is more expensive than the TEMA W head exchanger.

The packed floating head with lantern ring or TEMA W head is shown in Fig. 10.9. Here a lantern ring rests on the machined surface of the tubesheet and provides an effective seal between the shell- and tube-side flanges. Vents are usually provided in the lantern ring to help locate any leaks in the seals before the shell-side and tube-side fluids mix. Although a single-pass design is possible on the tube side, generally an even number of tube passes is used. The TEMA W head exchanger is the lowest-cost design of all floating heads. Although its cost is higher than that of the U-tube bundle, this higher cost is offset by the accessibility to the tube ends (by opening both rear- and front-end heads) for cleaning and repair; consequently, this design is sometimes used in the petrochemical and process industries.

A large number of combinations of front- and rear-end heads with different shell types of Fig. 1.6 are possible, depending on the application and the manufacturer. Some common types of combinations result in the following shell-and-tube heat exchangers: AEL, AES, AEW, BEM, AEP, CFU, AKT, and AJW.

In light of the availability of different types of front- and rear-end heads, the tube bundle of a shell-and-tube exchanger may simply be classified as a straight-tube or U-tube bundle. Both have a fixed tubesheet at the front end. The U-tube bundle has a shell with a welded shell cover on the U-bend end. The straight tube bundle has either a fixed

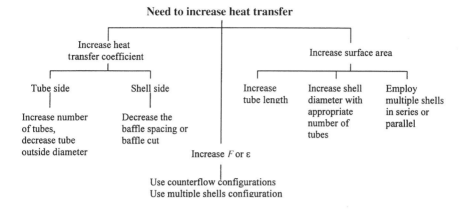

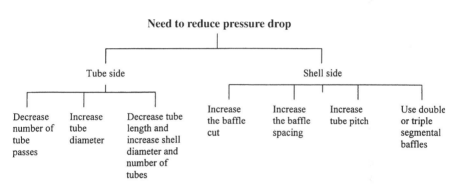

FIGURE 10.10   Influence of various geometrical parameters of a shell-and-tube exchanger on heat transfer and pressure drop.

tubesheet or a floating head at the rear end. The former is referred to as a fixed-tubesheet bundle; the latter is referred to as a floating-head bundle.

With this background, two flowcharts are presented in Fig. 10.10 to increase the overall heat transfer rate or decrease the pressure drop on either the tube or shell side during various stages of designing a shell-and-tube exchanger.

10.2.2   Plate Heat Exchangers

The chevron plate is the most common in PHEs. Hence, we will not discuss the reasoning behind why various other plate geometries have been used in PHEs. As described in Section 1.5.2.1, PHEs have a number of advantages over shell-and-tube heat exchangers, such as compactness, low total cost, less fouling, accessibility, flexibility in changing the number of plates in an exchanger, high q and ε, and low fluid residence time. Because of these advantages, they are in second place to shell-and-tube heat exchangers for market share in liquid-to-liquid and phase-change applications. The main reason for their limited versatility involves the pressure and temperature restrictions imposed by the gaskets. Replacing gaskets on one or both sides by laser welding of the plates (as in a welded PHE) increases both the operating pressure and temperature limits of the gasketed PHEs, and it allows the PHE to handle corrosive fluids compatible with the plate material. For low-heat duties (that translate into a total surface area up to 10 m^2), a more compact brazed PHE can replace the welded PHE, thus eliminating the frame, guide bars, bolts, and so on, of welded or gasketed PHEs. A variety of other PHEs have been developed for niche applications to cover specific operating conditions that cannot be handled by the above-described PHEs. Some of these PHEs are described briefly at the end of Section 1.5.2.2.

Although the shell-and-tube heat exchanger is versatile and can handle all kinds of operating conditions, it is not compact, not flexible in design, requires a large footprint, and is costly (total cost) compared to PHEs and other compact heat exchangers. Hence, PHEs and many other heat exchanger designs have been invented to replace shell-and-tube heat exchangers in individual narrow operating ranges. Refer to Sections 1.5.2 and 1.5.3 for details of these exchangers. Also, an excellent source of information on compact heat exchangers for liquid-to-liquid and phase-change applications is a recent monograph by Reay (1999).

10.2.3   Extended Surface Exchangers

10.2.3.1   Plate-Fin Exchanger Surfaces. The plate-fin construction is commonly used in gas–to–gas or gas–to–phase change exchanger applications where either the heat transfer coefficients are low or an ultrahigh exchanger effectiveness is desired. It offers high surface area densities [up to about 6000 m^2/m^3 (1800 ft^2/ft^3)] and a considerable amount of flexibility. The passage height on each fluid side could easily be varied and different fins can be used between plates for different applications. On each fluid side, the fin thickness and number of fins can be varied independently. If a corrugated fin (such as the plain triangular, louver, perforated, or wavy fin) is used, the fin can be squeezed or stretched to vary the fin pitch, thus providing added flexibility. The fins on each fluid side could easily be arranged such that the overall flow arrangement of the two fluids can result in crossflow, counterflow, or parallelflow. Even the construction of a multistream plate-fin exchanger is relatively straightforward with the proper design of inlet and outlet headers for each fluid (ALPEMA, 2000).

Plate-fin exchangers are generally designed for low-pressure applications, with operating pressures limited to about 1000 kPa gauge (150 psig). However, cryogenic plate-fin exchangers are designed for operating pressures of 8300 kPa (1200 psig). With modern manufacturing technology, they can be designed for very high pressures; for example, the gas cooler for an automotive air-conditioning system with CO_2 as the refrigerant has an operating pressure of 12.5 to 15.0 MPa (1800 to 2200 psig). The maximum operating temperatures are limited by the type of fin-to-plate bonding and the materials employed. Plate-fin exchangers have been designed from low cryogenic operating temperatures [−200°C (−400°F)] to about 800°C (1500°F). Fouling is generally not as severe a problem with gases as it is with liquids. A plate-fin exchanger is generally not designed for applications involving heavy fouling since there is no easy method of cleaning the exchanger unless chemical cleaning can be used. If an exchanger is made of small modules (stacked in the height, width, and length directions), and if it can be cleaned with a detergent, a high-pressure air jet, or by baking it in an oven (as in a paper industry exchanger), it could be designed for applications having considerable fouling. Fluid contamination (mixing) is generally not a problem in plate-fin exchangers since there is practically zero fluid leakage from one fluid side of the exchanger to the other.

Selection of a fin surface depends on the operating temperature, with references to bonding of the fins to plates or tubes and a choice of material. For low-temperature applications, a mechanical joint, or soldering or brazing may be adequate. Fins can be made from copper, brass, or aluminum, and thus maintain high fin efficiency. For high-temperature applications, only special brazing techniques and welding may be used; stainless steel and other expensive alloys may be used to make fins but with a possibly lower fin efficiency, due to their relatively lower thermal conductivities unless proper lower fin height is selected. Consequently, suitable high-performance surfaces may be selected to offset the potential reduction in fin efficiency unless the proper fin height is selected. Brazing would require added capital and maintenance cost of a brazing furnace, cost of brazing, and process expertise (Sekulić et al., 2003).

Cost is a very important factor in the selection of exchanger construction type and surface. The plate-fin surface in general is less expensive than a tube-fin surface per unit of heat transfer surface area. In most applications, one does not choose a high-performing surface, but rather, the least expensive surface, if it can meet the performance criteria within specified constraints. For example, if a plain fin surface can do the job for a specified application, the higher-performance louver or offset strip fin surface is not used because it is more expensive to manufacture.

We now discuss qualitatively the construction and performance behavior of plain, wavy, offset strip, louver, perforated, and pin fin plate-fin surfaces.

Plain Fin Surfaces. These surfaces are straight fins that are uninterrupted (uncut) in the fluid flow direction. Although triangular and rectangular passages are more common, any complex shape desired can be formed, depending on how the fin material is folded. Although the triangular (corrugated) fin (e.g., Fig. 1.29a, e and f) is less expensive, can be manufactured at a faster rate, and has the added flexibility of having an adjustable fin pitch, it is generally not structurally as strong as the rectangular fin (e.g., Fig. 1.29b and d) for the same passage size and fin thickness. The triangular fins can be made in very low to ultrahigh fin densities [40 to 2400 fins/m (1 to 60 fin/in.)].

Plain fins are used in applications where the allowed pressure drop is low and the augmented interrupted surfaces cannot meet the design requirement of allowed Δp for a desired fixed frontal area. Also, plain fins are preferred for very low Reynolds numbers applications. This is because with interrupted fins, when the flow approaches the fully developed state at such low Re, the advantage of the high h value of the interrupted fins is diminished while cost remains high, due to making interruptions. Plain fins are also preferred for high-Reynolds-number applications where the Δp for interrupted fins become excessively high.

Wavy Fin Surfaces. These surfaces also have uncut surfaces in the flow direction, and have cross-sectional shapes similar to those of plain surfaces (see Fig. 1.29c). However, they are wavy in the flow direction, whereas the plain fins are straight in the flow direction. The waveform in the flow direction provides effective interruptions to the flow and induces very complex flows. The augmentation is due to Görtler vortices, which form as the fluid passes over the concave wave surfaces. These are counterrotating vortices, which produce a corkscrewlike pattern. The heat transfer coefficient for a wavy fin is higher than that for an equivalent plain fin. However, the heat transfer coefficient for wavy fins is lower than that for interrupted fins such as offset or louver fins. Since there are no cuts in the surface, wavy fins are used in those applications where an interrupted fin might be subject to a potential fouling or clogging problem due to particulates, freezing moisture, bridging louvers due to condensate, and so on.

Offset Strip Fins. This is one of the most widely used enhanced fin geometries in plate-fin heat exchangers (see Fig. 1.29d). The fin has a rectangular cross section, and is cut into small strips of length ℓ_s. Every alternate strip is displaced (offset) by about 50% of the fin pitch in the transverse direction. In addition to the fin spacing and fin height, the major variables are the fin thickness and strip length in the flow direction. The heat transfer coefficients for the offset strip fins are 1.5 to 4 times higher than those of plain fin geometries. The corresponding friction factors are also high. The ratio of j/f for an offset strip fin to j/f for a plain fin is about 80%. Designed properly, the offset strip fin exchanger would require a substantially lower heat transfer surface area than that of plain fins at the same Δp. The heat transfer enhancement for an offset strip fin is caused mainly by the redeveloping laminar boundary layers for $Re \leq 10,000$. However, at higher Re, it acts as a rough surface (a constant value of f with decreasing j for increasing Re).

Offset strip fins are used in the approximate Re range 500 to 10,000, where enhancement over the plain fins is substantially higher. For specified heat transfer and pressure drop requirements, the offset strip fin requires a somewhat higher frontal area than a plain fin, but a shorter flow length and overall lower volume. Offset strip fins are used extensively by aerospace, cryogenic, and many other industries where higher heat transfer performance is required.

Louver Fins. Louvers are formed by cutting the metal and either turning, bending, or pushing out the cut elements from the plane of the base metal (see Fig. 1.29e). Louvers can be made in many different forms and shapes. The louver fin gauge is generally thinner than that of an offset strip fin. Louver pitch (also referred to as louver width) and louver angle (in addition, the fin spacing and fin height) are the most important geometrical parameters for the surface heat transfer and flow friction characteristics. On an absolute level, j factors are higher for louver fins than for the offset strip fin at the same Reynolds number, but the f factors are even higher than those for the offset strip fin geometry. Since the louver fin is triangular (or corrugated), it is generally not as strong as an offset strip fin; the latter fin has a relatively large flat area for brazing, thus providing strength. The louver fins may have a slightly higher potential for fouling than offset strip fins. Louver fins are amenable to high-speed mass production manufacturing technology, and as a result, are less expensive than offset strip fins and other interrupted fins when produced in very large quantities. The fin spacing desired can be achieved by squeezing or stretching the fin; hence it allows some flexibility in fin spacing without changes in tools and dies. This flexibility is not possible with the offset strip fin.

A wide range in performance can be achieved by varying the louver angle, width, and form. The operating Reynolds number range is 100 to 5000, depending on the type of louver geometry employed. Modern multilouver fins have higher heat transfer coefficients that those for offset strip fins, but with somewhat lower j/f ratios. However, the performance of a well-designed multilouver fin exchanger can approach that of an offset strip exchanger, possibly with increased surface compactness and reduced manufacturing cost. Multilouver fins (see Figs. 1.27, 1.28, and 1.29e) are used extensively by the automotive industry.

Perforated Fins. A perforated fin has either round or rectangular perforations with the size, shape, and longitudinal and transverse spacings as major perforation variables (see Fig. 1.29f). The perforated fin has either triangular or rectangular flow passages. When used as a plate-fin surface, it is generally brazed. The holes interrupt the flow and may increase h somewhat, but considerable surface area may also be lost, thus nullifying the advantage. Perforated fins are now used only in a limited number of applications. They are used as turbulators in oil coolers for mixing viscous oils, or as a high-Δp fin to improve flow distribution. Perforated fins were once used in vaporizing cryogenic fluids in air separation exchangers, but offset strip fins have now replaced them.

Pin Fins. These can be manufactured at very high speed continuously from a wire of proper diameter. After the wire is formed into rectangular passages (e.g., rectangular plain fins), the top and bottom horizontal wire portions are flattened for brazing or soldering with the plates. Pins can be circular or elliptical in shape. Pin fin exchanger performance is considerably lower, due to the parasitic losses associated with round pins in particular and to the inline arrangement of the pins (which results from the high-speed manufacturing techniques). The surface compactness achieved by pin fin geometry is much lower than that of offset strip or louver fin surfaces. Due to vortex shedding behind the round pins, noise- and flow-induced vibration may be a problem. Finally, the cost of a round wire is generally more than the cost of a flat sheet, so there may not be a material cost advantage. The potential application for pin fins is at very low Reynolds number (Re < 500), for which the pressure drop is of no major concern. Pin fins are used in electronic cooling devices with generally free convective flows over the pin fins.

10.2.3.2   Tube-Fin Surfaces.

When an extended surface is needed on only one fluid side (such as in a gas-to-liquid exchanger) or when the operating pressure needs to be contained on one fluid side, a tube-fin exchanger (see Section 8.2) may be selected, with the tubes being round, flat, or elliptical in shape. Also, when minimum cost is essential, a tube-fin exchanger is selected over a plate-fin exchanger since the fins are not brazed but are joined mechanically to the tubes by mechanical expansion. Flat or elliptical tubes, instead of round tubes, are used for increased heat transfer in the tube and reduced pressure drop outside the tubes; however, the operating pressure is limited compared to that for round tubes. Tube-fin exchangers usually have lower heat transfer surface compactness than a plate-fin unit, with a maximum heat transfer surface area density of about 3300 m^2/m^3 (1000 ft^2/ft^3).

A tube-fin exchanger may be designed for a wide range of tube fluid operating pressures [up to about 3000 kPa gauge (450 psig) or higher] with the other fluid being at low pressure [up to about 100 kPa (15 psig)]. The highest operating temperature is again limited by the type of bonding and the materials employed. Tube-fin exchangers are designed to cover the operating temperature range from low cryogenic temperatures to about 870°C (1600°F). Reasonable fouling can be tolerated on the tube side if the tubes can be cleaned. Fouling is generally not a problem on the gas side (fin side) in many applications; plain uninterrupted fins are used when "moderate" fouling is expected. Fluid contamination (mixing) of the two fluids is generally not a problem since there is essentially no fluid leakage between them. Since tubes are generally more expensive than extended surfaces, the tube-fin exchanger is in general more expensive. In addition, the heat transfer surface area density of a tube-fin core is generally lower than that of a plate-fin exchanger, as mentioned earlier.

The tube-fin construction is generally used in liquid-to-gas or phase-change fluid-to-gas heat exchanger applications with liquid, condensing fluid, or evaporating fluid on the tube side. Fins are generally used on the outside of tubes (on the gas side), although depending on the application, fins or turbulators may also be used inside the tubes. Round and flat tubes (rectangular tubes with rounded or sharp corners) are most common; however, elliptical tubes are also used. Round tubes are used for higher-pressure applications and also when considerable fouling is anticipated. Parasitic form drag is associated with flow normal to round tubes. In contrast, the flat tubes yield a lower pressure drop for flow normal to the tubes, due to lower form drag, and thus avoid the low-performance wake region behind the tubes. Also, the heat transfer coefficient is higher for flow inside flat tubes than for circular tubes, particularly at low Re. The use of flat tubes is limited to low-pressure applications, such as automotive radiators, unless the tubes are extruded with ribs inside (see the multiport tube in Fig. 1.27, also referred to as microchannels) or with integral fins outside.

Flat Fins on a Tube Array. This type of tube-fin geometry (shown in Fig. 1.31*b*) is most commonly used in air-conditioning and refrigeration exchangers in which high pressure needs to be contained on the refrigerant side. As mentioned earlier, this type of tube-fin geometry is not as compact (in terms of surface area density) as the plate-fin geometries, but its use is becoming widespread due to its lower cost. This is because the bond between the fin and tube is made by mechanically or hydraulically expanding the tube against the fin instead of soldering, brazing, or welding the fin to the tube. Because of the mechanical bond, the applications are restricted to those cases in which the differential thermal expansion between the tube and fin material is small, and preferably, the tube expansion is greater than the fin expansion. Otherwise, the loosened bond may have a significant thermal resistance.

Many different types of flat fins are available (see some examples in Fig. 1.33). The most common are the plain, wavy, and interrupted. The plain flat fins are used in those applications in which the pressure drop is critical (quite low), although a larger amount of surface area is required on the tube outside for the heat transfer specified than with wavy or interrupted fins. Plain flat fins have the lowest pressure drop than that of any other tube-fin surfaces at the same fin density. Wavy fins are superior in performance to plain fins and are more rugged. Wavy fins are used most commonly for air-conditioning condensers and other commercial heat exchangers. A variety of louver geometries are possible on interrupted flat fins. A well-designed interrupted fin would have even better performance than a wavy fin; however, it may be less rugged, more expensive to manufacture, and may have a propensity to clog.

Individually Finned Tubes. This tube-fin geometry (shown in Fig. 1.31*a*) is generally much more rugged than continuous fin geometry but has lower compactness (surface area density). Plain circular fins are the simplest and most common. They are manufactured by tension wrapping the fin material around a tube, forming a continuous helical fin or by mounting circular disks on the tube. To enhance the heat transfer coefficient on the fins, a variety of enhancement techniques have been used (see Fig. 1.32). Segmented or spine fins are the counterpart of the strip fins used in plate-fin exchangers. A segmented fin is generally rugged, has heavy-gauge metal, and is usually less compact than a spine fin. A studded fin is similar to a segmented fin, but individual studs are welded to the tubes. A slotted fin has slots in the radial direction; when radially slitted material is wound on a tube, the slits open, forming slots whose width increases in the radial direction. This fin geometry offers an enhancement over tension-wound plain fins; however, segmented or spine fins would yield a better performance. The wire loop fin is formed by spirally wrapping a flattened helix of wire around the tube. The wire loops are held to the tube by a tensioned wire within the helix or by soldering. The enhancement characteristic of small-diameter wires is important at low flows, where the enhancement of other interrupted fins diminishes.

10.2.4 Regenerator Surfaces

Regenerators, used exclusively in gas-to-gas heat exchanger applications, can have a higher surface area density (a more compact surface) than that of plate-fin or tube-fin surfaces. While rotary regenerators have been designed for a surface area density β of up to about 8800 m^2/m^3 (2700 ft^2/ft^3), the fixed-matrix regenerators have been designed for β of up to about 16,000 m^2/m^3 (5000 ft^2/ft^3). Regenerators are usually designed for low-pressure applications, with operating pressures limited to near-atmospheric pressures for rotary and fixed-matrix regenerators; an exception is the gas turbine rotary regenerator, having an inlet pressure of 615 kPa gauge or 90 psig on the air side. The regenerators are designed to cover an operating temperature range from low cryogenic to very high temperatures. Metal regenerators are used for operating temperatures up to about 870°C (1600°F); ceramic regenerators are used for higher temperatures, up to 2000°C (3600°F). The maximum inlet temperature for paper and plastic regenerators is 50°C (120°F).

Regenerators have self-cleaning characteristics because hot and cold gases flow in opposite directions periodically through the same passage. As a result, compact regenerators have minimal fouling problems and usually have very small hydraulic diameter passages. If severe fouling is anticipated, rotary regenerators are not used; fixed-matrix regenerators with large hydraulic diameter flow passages [50 mm (2 in.)] could be used for very corrosive/fouled gases at ultrahigh temperatures [925 to 1600°C (1900 to 2900°F)]. Carryover and bypass leakages from the hot fluid to the cold fluid (or vice versa) occur in the regenerator. Where this leakage and subsequent fluid contamination is not permissible, regenerators are not used. Hence, they are not used with liquids. The cost of the rotary regenerator surface per unit of heat transfer surface area is generally substantially lower than that of a plate-fin or tube-fin surface.

10.3 SOME QUANTITATIVE CONSIDERATIONS

As presented in Fig. 1.1, heat exchangers can be broadly classified according to construction as tubular, plate type, extended surface, and regenerative. A large variety of high-performance surfaces are used on the gas side of extended surface and regenerative exchangers. A large number of enhanced tube geometries are available for selection in tubular exchangers. For the general category of enhanced tubes, internally finned tubes, and surface roughness, Webb and Bergles (1983) have proposed a number of performance evaluation criteria (PEC) to assess the performance of enhanced surfaces compared to similar plain (smooth) surfaces. In plate-type exchangers (used primarily with liquids), although many different types of construction are available, the number of surface geometries used in modern exchangers is limited to high-performance chevron plate geometry, which is most commonly used in PHEs. As a result, in this chapter we focus on quantitative screening methods for

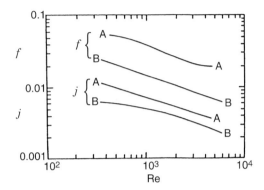

FIGURE 10.11 Comparison of surface basic characteristics of two heat exchanger surfaces. (From Shah, 1983.)

gas flows in compact heat exchangers and performance evaluation criteria (PECs) for tubular surfaces. For extended surfaces, particularly the plate-fin type, selection of the surfaces on both fluid sides are independent of each other, and generally, one fluid side is critical from the pressure drop requirement. Hence, we consider only one fluid side for the surface selection for plate-fin surfaces.

10.3.1 Screening Methods

Surface selection is made by comparing the performance of various heat exchanger surfaces and choosing the best under some specified criteria (objective function and constraints) for a given heat exchanger application. Consider the j and f characteristics of surfaces A and B in Fig. 10.11. Surface A has both j and f higher than those for surface B. Which is a better surface? This question is meaningless unless one specifies the criteria for surface comparison. If the pressure drop is of less concern, surface A will transfer more heat than surface B for the same heat transfer surface area for a given application. If the pressure drop is critical, one cannot in general say that surface A is better than surface B. One may need to determine, for example, the volume goodness factors for comparison (see Section 10.3.1.2); or one may even need to carry out a complete exchanger optimization after selecting the surface on the other fluid side of a two-fluid exchanger.

A variety of methods have been proposed in the literature for surface performance comparisons. These methods could be categorized as follows: (1) direct comparisons of j and f, (2) a comparison of heat transfer as a function of fluid pumping power, (3) miscellaneous direct comparison methods, and (4) performance comparisons with a reference surface. Over 30 such dimensional or nondimensional comparison methods have been reviewed critically by Shah (1978), and many more methods have been published since then.

It should be emphasized that most of these comparisons are for the surfaces only on one fluid side of a heat exchanger. When a complete exchanger design is considered that does not lend itself to having one fluid side as a strong side (i.e., having high $\eta_o hA$), the best surface selected by the foregoing methods may not be an optimum surface for a given application. This is because the selection of the surface for the other fluid side and its thermal resistance, flow arrangement, overall exchanger envelope, and other criteria (not necessarily related to the surface j and f vs. Re characteristics) influence the overall performance of a heat exchanger.

In addition, if the exchanger is considered as part of an (open or closed system), the exchanger surface (and/or other variables) may be selected based on the system as a whole rather than based on the optimum exchanger as a component. Current methods of surface selection for an optimum heat exchanger for a system include the use of sophisticated computer programs that take into account many possible effects. Such selection is not possible in simplified approaches presented in the open literature. We focus on considering simple but important quantitative screening methods for surface selection on the gas side of compact heat exchangers since these exchangers employ a large variety of high-performance surfaces.

The selection of a surface for a given application depends on exchanger design criteria. For a specified heat transfer rate and pressure drop on one fluid side, two important design criteria for compact exchangers (which may also be applicable to other exchangers) are the minimum heat transfer surface area requirement and the minimum frontal area requirement. Let us first discuss the significance of these criteria.

To understand the minimum frontal area requirement, let us first review how the fluid pressure drop and heat transfer are related to the flow area requirement, the exchanger flow length, and the fluid velocity. The fluid pressure drop on one fluid side of an exchanger, neglecting the entrance/exit and flow acceleration/deceleration losses, is given from Eq. (6.29) as

$$\Delta p = \frac{4fLG^2}{2g_c \rho D_h} \tag{10.1}$$

Since predominantly developed and/or developing laminar flows prevail in compact heat exchangers, the friction factor is related to the Reynolds number as follows (see Sections 7.4.1.1 and 7.4.2.1):

$$f = \begin{cases} C_1 \cdot \mathrm{Re}^{-1} & \text{for fully developed laminar flow} \\ C_2 \cdot \mathrm{Re}^{-0.5} & \text{for developing laminar flow} \end{cases} \tag{10.2}$$

where C_1 and C_2 are constants. Substituting Eq. (10.2) into Eq. (10.1) and noting that $\mathrm{Re} = GD_h/\mu$, we get

$$\Delta p \propto \begin{cases} LG & \text{for fully developed laminar flow} \\ LG^{1.5} & \text{for developing laminar flow} \end{cases} \tag{10.3}$$

Here $G = \dot{m}/A_o$. Therefore, for a specified constant flow rate $\dot{m}$, the pressure drop is proportional to the flow length L and inversely proportional to the flow area A_o or $A_o^{1.5}$.

The Nusselt number for laminar developed and developing temperature profiles and developed velocity profiles is given by (see Sections 7.4.1.1 and 7.4.3.1):

$$\mathrm{Nu} = \begin{cases} C_3 & \text{for thermally developed laminar flow} \\ C_4 (D_h \cdot \mathrm{Pr} \cdot \mathrm{Re}/L)^{1/3} & \text{for thermally developing laminar flow} \end{cases} \tag{10.4}$$

where C_3 and C_4 are constants. Thus, the heat transfer coefficient is independent of the mass flow rate $\dot{m}$ or mass velocity G for the thermally developed laminar flow and is proportional to $G^{1/3}$ for the thermally developing laminar flow. We have not considered simultaneously developing laminar flow (in which Nu $\propto G^{1/2}$) since thermally developing flow provides a conservative estimate of Nu.

Considering Δp and h simultaneously, a decrease in G will reduce Δp linearly without reducing h for fully developed laminar flow; for developing laminar flows, a reduction in G will reduce Δp as in Eq. (10.3), with a slight decrease in h as given by Eq. (10.4).

Now as discussed earlier, there are a variety of enhanced surfaces available for selection. An undesirable consequence of the heat transfer enhancement is an increase in the friction factor, which results in a higher pressure drop for an exchanger having a fixed frontal area and a constant flow rate. As noted in the preceding paragraph, reducing G can reduce the pressure drop in compact exchangers without significantly reducing the heat transfer coefficient h. For a specified constant flow rate, a reduction in G means an increase in the flow area A_o for constant $\dot{m}$ and approximately constant σ (the ratio of free-flow to frontal area). So as one employs the more enhanced surface, the required flow area (and hence frontal area) increases accordingly to meet the heat transfer and pressure drop requirements specified. Thus, one of the characteristics of highly compact surfaces is that the resulting shape of the exchanger becomes more like a pancake, having a large frontal area and a short flow length (e.g., think of the shape of an automotive radiator in contrast to a shell-and-tube exchanger; see also Example 10.3 to show the increase in free-flow area when employing a higher-performance surface). Hence, it is important to determine which of the compact surfaces will meet a minimum frontal area requirement.

The surface having the highest heat transfer coefficient at a specified flow rate will require the minimum heat transfer surface area. However, the allowed pressure drop is not unlimited. Therefore, one chooses the surface having the highest heat transfer coefficient for a specified fluid pumping power. The exchanger with the minimum surface area will have the minimum overall volume requirement.

From the foregoing discussion, two major selection criteria for compact surfaces with gas flows are (1) a minimum frontal area requirement and (2) a minimum volume requirement. For this purpose, the surfaces are evaluated based on the surface flow area and volume goodness factors. We discuss these comparison methods after the following example.

Example 10.1 Consider a gas turbine rotary regenerator (Fig. E10.1) having compact triangular flow passages operating at Re = 1000 and the pressure drop on the high-pressure air side as 10 kPa. Determine the change in heat transfer and pressure drop if this regenerator is operated at Re = 500. The following data are provided for the analysis:

$$j = \frac{3.0}{\text{Re}} \qquad f = \frac{14.0}{\text{Re}} \qquad \text{Pr} = 0.7 \qquad (hA)_h = (hA)_c \qquad \text{flow split} = 50{:}50$$

Ignore the effect of the flow area blockage by the hub and the radial seals to determine the required changes. Assume that the mass flow rate of air (and hence gas) does not change when Re is reduced. How would you achieve the reduction in Re when the L and D_h values of the regenerator surface are being kept constant?

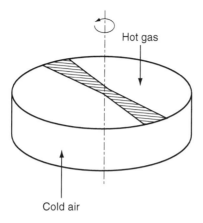

FIGURE E10.1

SOLUTION

Problem Data and Schematic: The data are given in the problem statement for j, f, Pr, and thermal conductances and flow split. Also, L and D_h are to be kept constant.

Determine: Determine the change in heat transfer and pressure drop for this regenerator when the air-side (and hence gas-side) Re is reduced from 1000 to 500. How is the reduction in Re achieved when the mass flow rate of the air is kept constant?

Assumptions: The flow is fully developed (thermally and hydrodynamically) laminar. The wall thermal resistance and fouling resistance are negligible. The fluid properties do not change with the change in Re; and with a flow split of 50:50, we expect the same effects on the air and gas sides with a change in Re.

Analysis: Since the flow is assumed to be fully developed laminar, Nu will be constant (see Section 7.4.1.1) and is given by its relationship to the j factor by Eq. (7.33) and input data $j \cdot \text{Re} = 3.0$ as

$$\text{Nu} = j \cdot \text{Re} \cdot \text{Pr}^{1/3} = 3.0(0.7)^{1/3} = 2.66$$

Using the definition of $\text{Nu} = h D_h / k$ and constant D_h for the change of Re from 1000 to 500, we get

$$\frac{\text{Nu}_2}{\text{Nu}_1} = \frac{h_1}{h_2} = 1$$

Here subscripts 1 and 2 denote the cases for $\text{Re} = 1000$ and 500, respectively. Thus, the heat transfer coefficient does not change with Re. This will also be the case for h on the gas side. Thus, UA and hence the heat transfer rate q will not change.

When Re $(= GD_h/\mu)$ is changed from 1000 to 500 without changing the mass flow rate and the regenerator geometry (L and D_h), we get

$$\frac{\text{Re}_2}{\text{Re}_1} = \frac{G_2}{G_1} = \frac{500}{1000} = \frac{1}{2}$$

Now let us evaluate the pressure drop. From Eq. (6.29),

$$\Delta p \propto \frac{L}{D_h^2} G(f \cdot \text{Re})$$

Since L, D_h, and $f \cdot \text{Re}$ are constant for this regenerator, we can evaluate Δp_2 with Re = 500 from

$$\frac{\Delta p_2}{\Delta p_1} = \frac{G_2}{G_1} = \frac{1}{2} \quad \text{or} \quad \Delta p_2 = 0.5 \times 10\,\text{kPa} = 5\,\text{kPa} \qquad \textit{Ans.}$$

Hence, the pressure drop will be reduced to 50% by decreasing Re by 50%. As we find above, the change in Re is accomplished by the corresponding change in G for constant $\dot{m} = GA_o$. This means the flow area A_o has to be doubled to reduce G by 50%, and hence the frontal area of the disk has to be doubled. By neglecting the effect of the area blockage by hub and radial seals, we obtain

$$\frac{\pi r_2^2}{\pi r_1^2} = \frac{A_{f,2}}{A_{f,1}} = \frac{A_{o,2}/\sigma}{A_{o,1}/\sigma} = 2$$

Therefore,

$$\frac{r_2}{r_1} = \sqrt{2} = 1.41$$

where σ is the ratio of free-flow area to frontal area, and r_2 and r_1 are disk radii for Re = 1000 and 500, respectively. Hence, the disk radius or diameter of this regenerator will need to be increased by 41%. $\qquad$ *Ans.*

Discussion and Comments: As discussed in Section 7.4.1.1, this problem clearly indicates that in fully developed laminar flow, the heat transfer coefficient is not reduced by reducing the fluid velocity ($u_m = G/\rho$), whereas the pressure drop is linearly reduced with the reduction in the flow velocity. Hence, a reduction in flow velocity without a reduction in the mass flow rate can be achieved by increasing the flow area and hence the frontal area of the exchanger.

10.3.1.1 Surface Flow Area Goodness Factor Comparison. London (1964) defined the ratio j/f as the surface flow area goodness factor. Using the definitions of j, Nu, f, and Re, we get the ratio j/f as

$$\frac{j}{f} = \frac{\text{Nu} \cdot \text{Pr}^{-1/3}}{f\,\text{Re}} = \frac{1}{A_o^2 \eta_o} \left(\frac{\text{Pr}^{2/3}}{2g_c \rho} \frac{\text{ntu} \cdot \dot{m}^2}{\Delta p} \right) \qquad (10.5)$$

The term in parentheses on right-hand side of Eq. (10.5) is dependent only on the operating parameters and is independent of the geometry and heat transfer surface involved. Equation (10.5) can be rearranged as

$$A_o^* = \frac{A_o}{[(\mathrm{Pr}^{2/3}/2g_c\rho)(\mathrm{ntu} \cdot \dot{m}^2/\Delta p)]^{1/2}} = \frac{1}{[\eta_o(j/f)]^{1/2}} \qquad (10.6a)$$

$$A_\mathrm{fr}^* = \frac{A_\mathrm{fr}}{[(\mathrm{Pr}^{2/3}/2g_c\rho)(\mathrm{ntu} \cdot \dot{m}^2/\Delta p)]^{1/2}} = \frac{1}{\sigma[\eta_o(j/f)]^{1/2}} \qquad (10.6b)$$

The left-hand sides of Eq. (10.6a) and (10.6b) are the dimensionless free-flow area A_o^* and frontal area A_fr^*, respectively. Equations (10.5) and (10.6) show the significance of j/f as being inversely proportional to A_o^2 (A_o is the surface minimum free-flow area) for specified operating conditions and η_o as constant. A surface having a higher j/f factor is good because it will require a lower free-flow area and hence a lower frontal area for the exchanger. The dimensionless j and f factors are independent of the length scale of the geometry (i.e., the hydraulic diameter).[†] Thus, the flow area A_o is independent of the hydraulic diameter, but dependent on the operating conditions ($\dot{m}$ and Δp), design condition (ntu), and fluid type (Pr). Note that for many compact surfaces, no significant variation is found in the j/f ratio over the reported test Reynolds number range, and hence A_o and A_fr are not a strong function of the surface type.

For fully developed laminar flow through simple geometries of Table 7.3, we find j_{H1}/f ranging from 0.263 for the equilateral triangular duct to 0.386 for the parallel-plate duct. Thus, the parallel-plate duct relative to the triangular duct has 47% (0.386/0.263 − 1) higher j/f. Then from Eq. (10.6), we get $A_{o,1}/A_{o,2} = (0.263/0.386)^{1/2} = 0.825$ (since $\eta_{o,1} = \eta_{o,2} = 1$), where the subscripts 1 and 2 are for the parallel-plate and triangular passages, respectively. Thus a parallel-plate exchanger would have a 17.5% smaller free-flow area requirement. In the free-flow area goodness factor comparison, no estimate of total heat transfer area or volume can be inferred. Such estimates may be derived from the core volume goodness factors described next.

10.3.1.2   Core Volume Goodness Factor Comparisons.

Two types of core volume goodness factor comparisons are suggested: h_std vs. E_std and $\eta_o h_\mathrm{std}\beta$ vs. $E_\mathrm{std}\beta$. In the first method, a comparison is made for surfaces having the same hydraulic diameter. In the second method, a comparison is made of the actual performance of surfaces having equal or different hydraulic diameters. The heat transfer rate per unit temperature difference and per unit surface area $[q/A(T_w - T_m)]^{[‡]}$, and the fluid pumping power due to friction per unit of surface area are expressed as

$$h = \frac{c_p\mu}{\mathrm{Pr}^{2/3}} \frac{1}{D_h} j \cdot \mathrm{Re} \qquad (10.7)$$

[†] As long as the surface geometrical similarity is maintained (i.e., the surface geometry is enlarged or reduced in size by changing the hydraulic diameter), and j and f vs. Re characteristics remain the same. Hence, the j/f ratio is independent of D_h at a given design Re.

[‡] The term heat transfer power is sometimes used to denote heat transfer rate q, watts.

$$E = \frac{\mathcal{P}}{A} = \frac{1}{2g_c} \frac{\mu^3}{\rho^2} \frac{1}{D_h^3} f \cdot \mathrm{Re}^3 \tag{10.8}$$

In either of the volume goodness factor comparisons, it is assumed that the surfaces under comparison will provide the same performance. It means that the following quantities are kept constant: (1) the same heat transfer rate, (2) the same pressure drop, (3) the same temperature difference between the wall and the fluid, and (4) the same fluid flow rate. Remember that we want to arrive at a "best" surface on one fluid side of the exchanger from the minimum-volume-requirement point of view. The heat transfer rate q and fluid pumping power $\mathcal{P}$ due to friction on one fluid side are

$$q = \eta_o h A (T_w - T_m) = \eta_o h \beta V (T_w - T_m) \tag{10.9}$$

$$\mathcal{P} = EA = E\beta V = \frac{\dot{m}\, \Delta p}{\rho} \tag{10.10}$$

where E is the fluid pumping power per unit surface area.

h_{std} vs. E_{std} *Comparisons.* For a rotary regenerator application in which all the surface is prime surface ($\eta_o = 1$), the comparison of performance is made at the same fluid pumping power due to friction per unit surface area (E) and the same compactness β or hydraulic diameter D_h ($= 4\sigma/\beta$) assuming that σ remains constant. The same D_h eliminates the scale or size variable of the passages. Then for the same q, $T_w - T_m$, and β, from Eq. (10.9),

$$h \propto \frac{1}{V} \tag{10.11}$$

Thus the higher $h^\dagger$ means a lower overall core volume requirement. Then the excellence of a particular surface geometry in terms of the core volume is characterized by a high position on a dimensional plot of h vs. E, as suggested by London and Ferguson (1949). Considering the gas turbine application, they evaluated all fluid properties for dry air at standard conditions ("std") of 260°C (500°F) and 1 atm pressure. However, for other applications, these standard conditions for fluid property evaluation could be changed to any conditions for the desired fluid for the h_{std} vs. E_{std} plot.

Since the regenerator is used for a gas-to-gas application, it is generally a thermally balanced heat exchanger. In this case, the thermal resistances on both sides are of the same order of magnitude and hence $UA \approx hA/2$. Thus the comparison of h_{std} with E_{std} is a realistic comparison for regenerators.

The h_{std} vs. E_{std} plot for fully developed flow with constant fluid properties through some constant cross-sectional ducts is presented in Fig. 10.12 for $D_h = 0.5\,\mathrm{mm}$ (0.0016 ft). From this figure, it is found that h_{std} varies from 256.4 to 700.6 $\mathrm{W/m^2 \cdot K}$, a factor of 2.7, with $h_{\mathrm{std}} = 264.7\,\mathrm{W/m^2 \cdot K}$ for the equilateral triangular duct.

When made for a fixed D_h, a plot of Fig. 10.12, clearly shows the influence of the passage shape. The parallel plate heat exchanger may prove impractical, but it is clear that there are several other configurations that possess significant advantages over the triangular and sine duct geometries. Based on this plot, the development and use of

---

$\dagger$ Note that the heat transfer coefficient h is the same as the heat transfer power per unit temperature difference and per unit surface area.

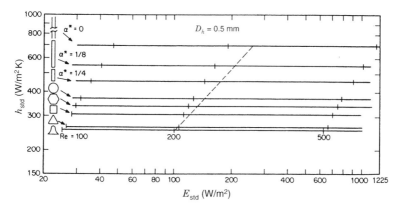

FIGURE 10.12   Theoretical laminar volume goodness factors for some simple duct geometries (From Shah, 1983.)

rectangular passage geometry is being continued for applications that involve fully developed laminar flows.

From Eqs. (10.7) and (10.8), it is evident that the dimensional h_{std} vs. E_{std} performance is strongly dependent on the length scale of the surface geometry (i.e., D_h). Thus, this comparison method reveals the benefit of increased performance (reduced surface area requirement) by going to a smaller D_h surface. This will also result in a much more compact surface.

For the foregoing reasons, the plot of h_{std} vs. E_{std} is recommended for selection of a heat exchanger surface for a new application for which there are no significant system or manufacturing constraints.

$\eta_o h_{std}\beta$ vs. $E_{std}\beta$ Comparisons.   The preceding method of comparison was for surfaces having the same D_h, σ, and η_o value (if there are any fins). When one wants to compare the performance of extended surfaces, for which j and f data are available, one may be interested in comparing the surfaces as they are. This is because we may not be able to manufacture a surface whose geometry is scaled up or scaled down. Such a comparison of actual surfaces could be made by a plot of $\eta_o h_{std}\beta$ vs. $E_{std}\beta$. Here β is the surface area density or compactness, $\eta_o h_{std}\beta$ represents the heat transfer power per unit temperature difference and unit core volume, and $E_{std}\beta$ represents the friction power expenditure per unit core volume. Note that this plot is modified from the $h_{std}\beta$ vs. $E_{std}\beta$ recommended by Kays and London (1998) by including the effect of overall fin efficiency η_o of the secondary surface. This effect is important for extended surface heat exchanger applications.

The foregoing variables, for a given set of surfaces, are evaluated from the following equations with fluid properties determined at some standard conditions:

$$\eta_o h_{std}\beta = \frac{c_p\mu}{Pr^{2/3}}\,\eta_o\,\frac{4\sigma}{D_h^2}\,j \cdot Re \qquad (10.12)$$

$$E_{std}\beta = \frac{\mu^3}{2g_c\rho^2}\,\frac{4\sigma}{D_h^4}\,f \cdot Re^3 \qquad (10.13)$$

where $\beta = 4\sigma/D_h$ and $\eta_o = 1 - (A_f/A)(1 - \eta_f)$. These equations have been derived from the definitions of j, f, and $\mathcal{P}$. From Eq. (10.9), $\eta_o h_{std}\beta \propto 1/V$ for a given q and $T_w - T_m$. Hence for constant $E_{std}\beta$, a surface having a high plot of $\eta_o h_{std}\beta$ vs. $E_{std}\beta$ is characterized as the best from the viewpoint of heat exchanger volume.

Example 10.2   Consider the rotary regenerator of Example 10.1 operating at Re $= 1000$ ($\Delta p = 10\,\text{kPa}$, $q =$ some given value). As noted in Table 7.3, the rectangular passage of aspect ratio $\frac{1}{8}$ has higher Nu and $f \cdot$ Re than those for the equilateral triangular flow passages. What would be the change in disk diameter, flow length, and volume of the regenerator if the flow passages are changed from triangular to rectangular with the *same* hydraulic diameter D_h, porosity σ, Δp, q, and air and gas mass flow rates. The following data are provided. Equilateral triangular: $j \cdot$ Re $= 3.0$ and $f \cdot$ Re $= 14.0$; rectangular: $j \cdot$ Re $= 5.2$ and $f \cdot$ Re $= 22.0$.

SOLUTION

Problem Data and Schematic: In addition to the data provided in Example 10.1, the following data are provided for the rectangular flow passages: $j \cdot$ Re $= 5.2$ and $f \cdot$ Re $= 22.0$. The hydraulic diameters of triangular and rectangular passages are identical. The regenerator sketch is shown in Fig. E10.1.

Determine: The change in the disk diameter, disk flow depth, regenerator core volume, and operating Reynolds number when changing the flow passages from triangular to rectangular.

Assumptions: The assumptions are the same as those in Example 10.1.

Analysis: Let us first evaluate the change in flow area due to the change in passage geometry. From Eq. (10.5),

$$\frac{A_{o,2}}{A_{o,1}} = \frac{(j/f)_1^{1/2}}{(j/f)_2^{1/2}} = \frac{(3.0/14.0)^{1/2}}{(5.2/22.0)^{1/2}} = 0.952$$

where the subscripts 1 and 2 are for triangular and rectangular flow passages. Thus, a rectangular passage exchanger will require 4.8% (0.048) smaller flow area and hence frontal area for the same porosity. This translates approximately into 2.4% ($1 - \sqrt{0.952}$) reduction in the disk diameter.

From Eqs. (10.11) and (10.12), the core volume ratio is

$$\frac{V_2}{V_1} = \frac{h_1}{h_2} = \frac{\text{Nu}_1}{\text{Nu}_2} = \frac{(j \cdot \text{Re} \cdot \text{Pr}^{1/3})_1}{(j \cdot \text{Re} \cdot \text{Pr}^{1/3})_2} = \frac{3.0}{5.2} = 0.577$$

Hence the reduction in the core volume is 42.3% ($1 - 0.577$).   *Ans.*

Since $V = LA_{fr} = LA_o/\sigma$, we get[†]

$$\frac{L_2}{L_1} = \frac{V_2\sigma}{A_{o,2}} \frac{A_{o,1}}{V_1\sigma} = \frac{V_2/V_1}{A_{o,2}/A_{o,1}} = \frac{0.577}{0.952} = 0.606$$

[†] As noted just before, Eq. (10.11) is valid for $\sigma_1 = \sigma_2 = \sigma$. Hence, we have considered σ here as constant. This means that the wall thickness for rectangular and triangular passages will be different. They can be calculated using the formulas of Table 8.2 by equating hydraulic diameters, and for the same σ computing **b** for rectangular passages for the known **b** for equilateral triangular passages.

Thus, the regenerator disk thickness or flow length will be reduced by 39.4% $(1 - 0.606)$.   *Ans.*

Finally, the operating Reynolds number will be changed as follows:

$$\frac{\text{Re}_2}{\text{Re}_1} = \frac{(\dot{m}D_h/A_o\mu)_2}{(\dot{m}D_h/A_o\mu)_1} = \frac{A_{o,1}}{A_{o,2}} = \frac{1}{0.952} = 1.05$$

Thus, the operating Reynolds number will increase by 5% or will be 1050 (1000×1.05).   *Ans.*

Discussion and Comments: As shown in this example, by going from triangular flow passages to rectangular flow passages of low aspect ratio, substantial savings in the regenerator volume and hence mass can be achieved along with lower packaging. However, it is a challenge to manufacture more difficult rectangular flow passages than triangular flow passages.

Example 10.3 Select an offset strip fin versus a plain fin on the air side of an exchanger for which the heat transfer rate and inlet temperatures are specified. The frontal area and surface area comparisons are to be done at the same airflow rates, hA, and fluid pumping powers. The j and f data for these fins at the same hydraulic diameter are provided in Fig. E10.3. The design Reynolds number for the plain fin is 3000. Determine the frontal area, surface area, flow length, and volume requirements for the offset strip fin compared to those for the plain fin.

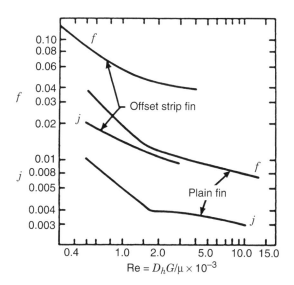

FIGURE E10.3   The j and f data comparison for an offset strip fin and plain fin at $D_h = 3.51$ mm. (From Webb, 1994.)

SOLUTION

Problem Data and Schematic: The j and f data are given in Fig. E10.3 for the offset strip fin and plain fin of Fig. 1.29d and b respectively. For the plain fin, Re $= 3000$.

Determine: For identical $\dot{m}$, hA, and $\mathcal{P}$, determine A_s/A_p and $A_{o,s}/A_{o,p}$ where the subscripts s and p denote the values for strip fin and plain fin surfaces, respectively.

Assumptions: Assume constant fluid properties and only one fluid side of the exchanger.

Analysis: Since the j and f factors are higher for the offset strip fin, the operating Reynolds number for the offset strip fin will be lower than that for the plain fin for the same $\dot{m}$, hA, $\mathcal{P}$, q, $T_{h,i}$, and $T_{c,i}$. This Reynolds number needs to be evaluated iteratively from the following relationship, which is derived from Eq. (10.5) with constant ntu, $\dot{m}$ and Δp, and, the definition of Re[$= (\dot{m}/A_o)D_h/\mu$ with $\dot{m}$ and D_h constant]:

$$\frac{\text{Re}_s}{\text{Re}_p} = \frac{A_{o,p}}{A_{o,s}} = \frac{(j/f)_s^{1/2}}{(j/f)_p^{1/2}} \tag{1}$$

From Fig. E10.3, the j and f factors at Re $= 3000$ for the plain fin are

$$j_p = 0.0038 \qquad f_p = 0.011$$

We need to assume a value of Re$_s$ such that the corresponding j_s and f_s satisfy Eq. (1) for the values of Re$_p$, j_p, and f_p above. Let us assume that Re$_p$/Re$_s$ = 1.20. Hence,

$$\text{Re}_s = \frac{3000}{1.20} = 2500$$

From Fig. E10.3, the j and f factors at Re $= 2500$ for the strip fin are

$$j_s = 0.010 \qquad f_s = 0.042$$

Substituting values of j's and f's in Eq. (1) yields

$$\frac{\text{Re}_s}{\text{Re}_p} = \frac{(0.010/0.042)^{1/2}}{(0.0038/0.011)^{1/2}} = 0.83$$

or Re$_p$/Re$_s$ = 1.20. Therefore, our guess of Re$_p$/Re$_s$ is correct (which was obtained based on iterations) and we don't need to iterate further. Otherwise, continue guessing the values of Re$_p$/Re$_s$ until Eq. (1) is satisfied.

From Eq. (10.8), the fluid pumping power $\mathcal{P}$ is given by

$$\mathcal{P} = \frac{1}{2g_c} \frac{\mu^3}{\rho^2} \frac{1}{D_h^3} f \cdot \text{Re}^3 \cdot A$$

Applying this equation, for equal pumping power ($\mathcal{P}_p = \mathcal{P}_s$) and $D_{h,p} = D_{h,s}$, we get

$$\frac{A_s}{A_p} = \frac{(f \cdot \text{Re}^3)_p}{(f \cdot \text{Re}^3)_s} = \frac{0.011}{0.042}(1.20)^3 = 0.453 \qquad \textit{Ans.}$$

where the numerical values computed earlier are substituted. The minimum free-flow area ratio, from Eq. (10.5), is given by

$$\frac{A_{o,s}}{A_{o,p}} = \frac{(j/f)_p^{1/2}}{(j/f)_s^{1/2}} = 1.20 \qquad\qquad Ans.$$

The offset strip fin will require 20% higher frontal area (for the same σ) and 54.7% $(1 - 0.453)$ less surface area than will the plain fin for this problem. Using the definition of the hydraulic diameter $D_h = 4A_oL/A$, we get

$$\frac{L_s}{L_p} = \frac{A_s}{A_p}\frac{A_{o,p}}{A_{o,s}}\frac{D_{h,s}}{D_{h,p}} = 0.453 \times \frac{1}{1.20} \times 1 = 0.378$$

since $D_{h,s} = D_{h,p}$ is given. Now if we assume the porosity σ is the same for the two surfaces, the volume ratio is given by

$$\frac{V_s}{V_p} = \frac{A_{o,s}}{A_{o,p}}\frac{L_s}{L_p} = 1.20 \times 0.378 = 0.454$$

Discussion and Comments: This example demonstrates that by employing higher-performing surface geometry, in general, one ends up with somewhat larger free-flow and frontal areas, but overall significantly lower surface area, smaller flow length, lower volume, and lower mass of the exchanger. Thus if packaging allows somewhat higher frontal area, the material cost of the exchanger surface will be lower with a higher-performing surface.

10.3.1.3 General Relationships for Compact Heat Exchanger Surfaces.

There is a continued technology drive toward higher compactness as practiced in the development of automotive, aerospace, and other compact heat exchangers. One of the common ways to describe the surface compactness is to characterize the surface with its hydraulic diameter as outlined in Section 1.4. In plate-fin exchangers, the heat transfer surface on each fluid side can often be selected independently of the shape and size of the resultant exchanger. There are other reasons (e.g., one fluid side is critical from a pressure drop requirement, a retrofit application, improved manufacturing process) why replacement of one heat transfer surface with the other may be an option. Hence, we provide general relationships that include geometry and surface performance characteristics (j, f, and related factors) when we change the hydraulic diameter of a compact surface. These relations will include the free-flow area, frontal area, volume, flow length surface area, and Reynolds number as functions of D_h and j, f, Re, η_o, β, and/or σ. Once the appropriate surface is selected, eventually we need to consider both fluid sides for the exchanger heat transfer performance and pressure drop calculations.

We now provide these relations based on the flow area and volume goodness factor comparisons provided in Sections 10.3.1.1 and 10.3.1.2, with all fluid properties as constant. These relationships will be useful for comparing two heat transfer surfaces on *one* fluid side of the exchanger (such as in plate-fin exchangers or rotary regenerators), if the surface on that fluid side can be selected or changed independently. They are not useful for the PHEs since the surfaces on both fluid sides cannot be selected independently (refer to Section 9.4.1). The following variables are kept constant: *fluid flow rate, heat transfer rate* (*heat duty*), and *pressure drop* (or *fluid pumping power*) for the derivation of these relationships.

The ratios of free-flow area and frontal areas on *one* fluid side for surface 1 to surface 2 are as follows using Eq. (10.6):

$$\frac{A_{o,2}}{A_{o,1}} = \left[\frac{\eta_{o,1}(j/f)_1}{\eta_{o,2}(j/f)_2}\right]^{1/2} \qquad \frac{A_{\mathrm{fr},2}}{A_{\mathrm{fr},1}} = \frac{\sigma_1}{\sigma_2}\frac{\eta_{o,1}}{\eta_{o,2}}\left[\frac{(j/f)_1}{(j/f)_2}\right]^{1/2} \tag{10.14}$$

Note that this equation is valid for *any* (the same or different) hydraulic diameter of the two surfaces and satisfies the heat transfer and pressure drop requirements specified for a given fluid flow rate (i.e., q, Δp, and $\dot{m}$ are constant for these relationships).

For an equal fluid pumping power requirement [i.e., $\mathcal{P} = $ constant for Eq. (10.10)] and equal heat duty q, from Eq. (10.9) we get the following equation for volume between plates on *one* fluid side with use of the definitions $\mathrm{Nu} = hD_h/k$, $D_h = 4\sigma/\beta$, and $\mathrm{Nu} = j \cdot \mathrm{Re} \cdot Pr^{1/3}$:

$$\frac{V_2}{V_1} = \frac{(\eta_o h\beta)_1}{(\eta_o h\beta)_2} = \frac{\eta_{o,1}}{\eta_{o,2}}\frac{\beta_1}{\beta_2}\frac{D_{h,2}}{D_{h,1}}\frac{\mathrm{Nu}_1}{\mathrm{Nu}_2} = \frac{\eta_{o,1}}{\eta_{o,2}}\frac{\sigma_1}{\sigma_2}\frac{(j \cdot \mathrm{Re})_1}{(j \cdot \mathrm{Re})_2}\left(\frac{D_{h,2}}{D_{h,1}}\right)^2 \tag{10.15}$$

Using D_h from $\mathrm{Re} = (\dot{m}/A_o)D_h/\mu$, substituting it in Eq. (10.15), and using Eq. (10.14), we get an alternative form of Eq. (10.15) as follows:

$$\frac{V_2}{V_1} = \frac{(f \cdot \mathrm{Re})_2}{(f \cdot \mathrm{Re})_1}\frac{\eta_{o,1}}{\eta_{o,2}}\frac{\sigma_1}{\sigma_2}\left(\frac{j_1}{j_2}\right)^2 \tag{10.16}$$

The flow length ratio on *one* fluid side can be computed from $L = V/A_{\mathrm{fr}}$ as follows using Eq. (10.15) and the definition of $\mathrm{Re} = (\dot{m}/\sigma A_{\mathrm{fr}})D_h/\mu$:

$$\frac{L_2}{L_1} = \frac{\eta_{o,1}}{\eta_{o,2}}\frac{j_1}{j_2}\frac{D_{h,2}}{D_{h,1}} \tag{10.17}$$

The heat transfer surface area ratio A_2/A_1 on *one* fluid side is then calculated from the definition $A = 4A_o L/D_h$ and $\mathrm{Re} \propto D_h/A_o$, Eqs. (10.14) and (10.17), as follows:

$$\frac{A_2}{A_1} = \frac{A_{o,2}}{A_{o,1}}\frac{L_2}{L_1}\frac{D_{h,1}}{D_{h,2}} = \frac{L_2}{L_1}\frac{\mathrm{Re}_1}{\mathrm{Re}_2} = \left[\frac{(j/f)_1}{(j/f)_2}\right]^{1/2}\left(\frac{\eta_{o,1}}{\eta_{o,2}}\right)^{3/2}\frac{j_1}{j_2} \tag{10.18}$$

Finally, the ratio of operating Reynolds numbers for the two surfaces can be determined from the definition $\mathrm{Re} = (\dot{m}/A_o)D_h/\mu$ and the use of Eq. (10.14) as follows:

$$\frac{\mathrm{Re}_2}{\mathrm{Re}_1} = \left[\frac{(j/f)_2}{(j/f)_1}\right]^{1/2}\frac{D_{h,2}}{D_{h,1}} \tag{10.19}$$

Since the relationships above are derived for constant q, Δp, $\dot{m}$, and $\mathcal{P}$, they are not included in the list of operational parameters. Based on Eqs. (10.14)–(10.18), we find that the ratios for A_o, A_{fr}, and A are independent of D_h, and the ratios for V and L are proportional to D_h for fully developed laminar flow. However, for other flows (such as turbulent, transition, and developing laminar flows), one more constraint (in addition to constant q, Δp, $\dot{m}$, $\mathcal{P}$) needs to be specified by keeping constant the left-hand side of one of the equations (10.14)–(10.19).

While the foregoing relationships are obtained for the case of keeping q, Δp and $\dot{m}$ constant, similar relationships can be obtained by keeping some variables constant and others varying from the following set: A_{fr}, V, L, A, D_h, Re, q, Δp, $\dot{m}$, $\mathcal{P}$, and so on. Cowell (1990) presents a number of such relationships.

Several important observations can be made from the foregoing relationships keeping q, Δp and $\dot{m}$ constant. These are useful when we have different surfaces for selection and we need to decide which to choose for a compact surface having *fully developed laminar flow*.

- Since there is no hydraulic diameter involved in the flow area ratio relationship of Eq. (10.14), the flow area on one fluid side is independent of D_h for fully developed laminar flow. Changing D_h without changing the wall thickness would result in a slight change in the porosity σ and hence in A_{fr}.
- From Eq. (10.15), $V \propto D_h/\sigma j$. Hence, the heat exchanger volume on one fluid side decreases with increasing porosity σ, increasing Colburn factor j and reducing hydraulic diameter D_h.
- Based on Eq. (10.17), the flow length on one fluid side decreases with increasing Colburn factor j and decreasing hydraulic diameter D_h.
- Based on Eq. (10.18), the heat transfer surface area on one fluid side decreases with increasing j, Re, or Nu $(= j \cdot \text{Re} \cdot \text{Pr}^{1/3})$, and decreasing D_h.

Note that j and f factors for any surface are independent of the hydraulic diameter as long as the surface is geometrically scaled up or down.

To emphasize the foregoing points, let us consider a case of a rotary regenerator having a 50:50% split for air and gas flows and having fully developed laminar flows, or one fluid side of a plate-fin exchanger with fully developed laminar flow. If the hydraulic diameter is reduced to one-half of the original value on one fluid side, one can show from Eqs. (10.14)–(10.19) that A_o and A remain constant, V reduces to $\frac{1}{2}$, L reduces to $\frac{1}{2}$, and Re reduces to $\frac{1}{2}$, all changes for constant q, Δp, and $\dot{m}$ on one fluid side.

10.3.2 Performance Evaluation Criteria

Webb (1994) has presented a number of performance evaluation criteria (PECs), shown in Table 10.4, to assess performance merits of enhanced heat transfer surfaces relative to plain surfaces in single-phase flow. When two fluid sides are not independent of each other, these PECs are applied in general; otherwise, use the methods of Section 10.3.1. These PECs have been developed for the following enhancement types: surface roughness, internally finned tubes, and enhanced tubes; and they can also be applied to plate-fin surfaces. These PECs are actually screening methods since they consider only the thermal and hydraulic performance and do not consider nonthermal performance considerations in the design and optimization of a heat exchanger. A PEC is established by selecting one of the operating variables for the performance objective subject to design constraints on the remaining variables. Operating variables considered are geometry (number of tubes n_t in a pass and tube

TABLE 10.4 Performance Evaluation Criteria for Enhanced Surfaces (with Constant d_i) for Single-Phase Heat Exchangers

PEC Case	Geometry	Fixed $\dot{m}$	Fixed $\mathcal{P}$	Fixed q	Fixed ΔT_{max}	Objective
FG-1a	n_t, L	×			×	Increase q
FG-1b	n_t, L	×		×		Decrease ΔT_{max}
FG-2a	n_t, L		×		×	Increase q
FG-2b	n_t, L		×	×		Decrease ΔT_{max}
FG-3	n_t, L			×	×	Decrease $\mathcal{P}$
FN-1	n_t		×	×	×	Decrease L
FN-2	n_t	×		×	×	Decrease L
FN-3	n_t	×		×	×	Decrease $\mathcal{P}$
VG-1		×	×	×	×	Decrease $n_t L$
VG-2a	$n_t L$	×	×		×	Increase q
VG-2b	$n_t L$	×	×	×		Decrease ΔT_{max}
VG-3	$n_t L$	×		×	×	Decrease $\mathcal{P}$

Source: Data from Webb (1994.)

length L), flow rate $\dot{m}$, fluid pumping power $\mathcal{P}$, heat transfer rate q, and fluid inlet temperature difference ΔT_{max}. For a given PEC, the ratio of the design objective for a surface of interest to that for a reference surface is then calculated as a function of a similar ratio of a design variable.

The heat transfer rate in an exchanger is given by

$$q = UA \, \Delta T_m = PR(T_{h,i} - T_{c,i}) = PR \, \Delta T_{max} \qquad (10.20)$$

Here P is the exchanger temperature effectiveness and R is the heat capacity rate ratio. Reviewing this equation, the design objectives for the use of enhanced surface may be as follows:

1. Increased UA for equal pumping power $\mathcal{P}$ and fixed geometry [frontal area (represented by n_t) and length L]. A higher UA means (a) higher q for a given ΔT_{max} or ΔT_m, and (b) lower ΔT_{max} or ΔT_m for a given q.

2. For a fixed flow area (i.e., fixed n_t in Table 10.4), (a) reduce the tube length L (and hence A) for equal q and $\dot{m}$ or $\mathcal{P}$, and (b) reduce $\mathcal{P}$ for equal q and $\dot{m}$. In all cases, ΔT_{max} or ΔT_m is fixed.

3. Reduce the surface area A and hence the volume and mass of the exchanger for fixed $\dot{m}$ and specified q (or ΔT_{max}) and $\mathcal{P}$.

Based on the foregoing objectives, three major categories of performance evaluation criteria are developed: Fixed geometry (FG), fixed flow area (FN), and variable geometry (VG) criteria. For fixed-geometry criteria, a plain surface is replaced by an enhanced surface of equal length, a retrofit application, resulting in higher q (or reduced ΔT_{max}) and higher $\mathcal{P}$. For fixed-flow-area criteria, either $\mathcal{P}$ (and L) is reduced at constant q and $\dot{m}$ or $\dot{m}$ (and L) is reduced at constant $\mathcal{P}$ and q, employing an enhanced surface. For variable-geometry criteria, the surface area A ($\propto n_t L$) is reduced and frontal area A_{fr} ($\propto n_t$) is increased for an enhanced surface for fixed q, $\dot{m}$, and $\mathcal{P}$. Several PECs are formulated based on these major criteria, as noted in various FG, FN, and VG cases in Table 10.4.

The advantages of PEC comparison methods are that (1) the designer can select his or her own criteria for comparison; (2) he or she can then compare the performance of a surface to that of a reference surface directly, and (3) he or she does not need to evaluate the fluid properties since they drop out in computing the performance ratios. The performance comparisons can include the effect of the thermal resistances of the wall and of fouling and convection on the other fluid side. The optimum surface selected by this method may not be optimum in a two-fluid heat exchanger when non-performance-related overall heat exchanger constraints are imposed. These aspects are considered in heat exchanger optimization, discussed in Section 9.6.

The algebraic relations for the PECs will now be summarized for comparing an enhanced tubular surface to the corresponding plain tube for a shell-and-tube heat exchanger having the tube length L per pass, tube diameters d_i and d_o, number of tubes n_t in each pass, and number of passes n_p. For this exchanger, the flow area and surface area on the tube side are given by

$$A_o = \frac{\pi}{4} d_i^2 n_t \qquad A = \pi d_i L n_t n_p \tag{10.21}$$

The frontal area A_{fr} on the tube side is then related to A_o for specified tube layout and pitches. Heat transfer and flow friction characteristics are needed for comparing the performance of an enhanced surface to the corresponding plain surface. We will use a subscript p to designate the quantities for the plain surface (except for c_p), and no subscript for the enhanced surface. From the definition of the j factor,

$$h = j G c_p \cdot \mathrm{Pr}^{-2/3} \tag{10.22}$$

To evaluate the heat duty q of the enhanced surface, we need to compare hA of the enhanced surface to $(hA)_p$ for the plain surface. From Eq. (10.22),

$$\frac{hA}{(hA)_p} = \frac{j}{j_p} \frac{G}{G_p} \frac{A}{A_p} \tag{10.23}$$

Note that here and in this section A_p represents the surface area of the *plain* surface. Using Eq. (6.30), the fluid pumping power ratio is given by

$$\frac{\mathcal{P}}{\mathcal{P}_p} = \frac{f}{f_p} \frac{A}{A_p} \left(\frac{G}{G_p}\right)^3 \tag{10.24}$$

A relationship between hA and $\mathcal{P}$ for enhanced and plain surfaces is obtained by eliminating the G/G_p term from Eqs. (10.23) and (10.24):

$$\frac{hA/(hA)_p}{(\mathcal{P}/\mathcal{P}_p)^{1/3}(A/A_p)^{2/3}} = \frac{j/j_p}{(f/f_p)^{1/3}} \tag{10.25}$$

This equation is used for a comparison of two heat transfer surfaces under various criteria of Table 10.4 when two of the three ratios on the left-hand side of Eq. (10.25) are given and the third is to be determined.

However, when comparing an enhanced surface with a plain surface in a heat exchanger, the ratio needed for heat transfer performance is $UA/(UA)_p$, which takes into account wall thermal resistance, fouling resistance, and convection resistance on the second fluid side. Using Eq. (3.20) or (3.24) for UA and $(UA)_p$, one can arrive at the following expression modified from Webb (1981):

$$\frac{UA}{(UA)_p} = \frac{1 + \mathbf{R}_p^*}{\dfrac{(hA)_{i,p}}{hA} + \mathbf{R}^*} = \frac{1 + \mathbf{R}_p^*}{\dfrac{j_p}{j}\dfrac{G_p}{G}\dfrac{A_p}{A} + \mathbf{R}^*} = \frac{1 + \mathbf{R}_p^*}{\dfrac{j_p}{j}\left[\dfrac{f}{f_p}\dfrac{\mathcal{P}_p}{\mathcal{P}}\left(\dfrac{A_p}{A}\right)^2\right]^{1/3} + \mathbf{R}^*} \tag{10.26}$$

where $\mathbf{R}^*$ and $\mathbf{R}_p^*$ are the total thermal resistances (excluding tube inside thermal resistance) for enhanced and plain surfaces, respectively, normalized with respect to the *plain* tube inside thermal resistance $[1/(hA)_{i,p}]$. They are given explicitly in Table 10.5. Note that $A_{i,p}$ is simply designated as A_p (plain tube inside surface area) in the last two equalities of Eq. (10.26) and throughout in Table 10.6.

In Eq. (10.26), $\mathbf{R}^*$ includes A_p/A. Hence, $\mathbf{R}^*$ can be presented as a function of A_p/A as follows (a formula alternative to that in Table 10.5) when A_p/A is not unity, but is a variable:

$$\mathbf{R}^* = \hat{\mathbf{R}}^* \frac{A_p}{A} = \left(\hat{\mathbf{R}}_o^* + \hat{\mathbf{R}}^* \frac{A_o}{A_w} + \hat{\mathbf{R}}_{o,f}^*\right)\frac{A_p}{A} = \left(\frac{h_{i,p}}{h_o} + \frac{h_{i,p}\delta_w}{k_w} + \frac{h_{i,p}}{h_{o,f}}\right)\frac{A_p}{A} \tag{10.27}$$

Using Eq. (3.94), the ratio of the heat transfer rate in an enhanced to the plain surface is given by

$$\frac{q}{q_p} = \frac{\dot{m}}{\dot{m}_p}\frac{P}{P_p}\frac{\Delta T_{\max}}{\Delta T_{\max,p}} \tag{10.28}$$

Here we have assumed the fluid c_p to be the same for enhanced and plain surfaces (and hence $C/C_p = \dot{m}/\dot{m}_p$), P is the temperature effectiveness, and $\Delta T_{\max} (= T_{h,i} - T_{c,i})$ is the inlet temperature difference. Since the temperature effectiveness P is dependent on NTU, the NTU of the enhanced surface is related to that for the plain surface by

$$\text{NTU} = \text{NTU}_p \frac{UA}{(UA)_p}\frac{\dot{m}_p}{\dot{m}} \tag{10.29}$$

TABLE 10.5 Dimensionless Thermal Resistances for Eq. (10.26)

Definition	Reference (Plain) Heat Exchanger	Enhanced Heat Exchanger
Outer surface thermal resistance	$\mathbf{R}_{o,p}^* = (hA)_{i,p}/(hA)_{o,p}$	$\mathbf{R}_o^* = (hA)_{i,p}/(hA)_o$
Wall thermal resistance	$\mathbf{R}_{w,p}^* = (hA)_{i,p}\delta_w/k_w A_w$	$\mathbf{R}_w^* = (hA)_{i,p}\delta_w/k_w A_w$
Fouling resistance	$\mathbf{R}_{f,p}^* = (hA)_{i,p}(\mathbf{R}_{i,f,p} + \mathbf{R}_{o,f,p})$	$\mathbf{R}_f^* = (hA)_{i,p}(\mathbf{R}_{i,f} + \mathbf{R}_{o,f})$
Combined resistance	$\mathbf{R}_p^* = \mathbf{R}_{o,p}^* + \mathbf{R}_{w,p}^* + \mathbf{R}_{f,p}^*$	$\mathbf{R}^* = \mathbf{R}_o^* + \mathbf{R}_w^* + \mathbf{R}_f^*$

TABLE 10.6 Algebraic Formulas to Evaluate PEC of Table 10.4[a]

Case	Determine:	Information or Constraints Given and Comments	Resultant Formulas to Compute the Objective of the PEC
FG-1a	$\dfrac{q}{q_p}$	$\dfrac{n_t}{n_{t,p}} = \dfrac{A_{\mathrm{fr}}}{A_{\mathrm{fr},p}} = 1 \qquad \dfrac{L}{L_p} = \dfrac{A}{A_p} = 1$ $\dfrac{\dot{m}}{\dot{m}_p} = \dfrac{G}{G_p} = 1 \qquad \dfrac{\Delta T_{\max}}{\Delta T_{\max,p}} = 1$ In this case, both q and $\mathcal{P}$ will go up for the enhanced surface	From Eq. (10.26), $$\dfrac{UA}{(UA)_p} = \dfrac{1+\mathbf{R}_p^*}{(j_p/j) + \mathbf{R}^*} \quad \text{since} \quad \dfrac{\mathcal{P}_p}{\mathcal{P}} = \dfrac{f_p}{f}$$ Compute NTU from Eq. (10.29) and subsequently $\mathcal{P}$ for the given flow arrangement. Then, from Eq. (10.28), $$\dfrac{q}{q_p} = \dfrac{\mathcal{P}}{\mathcal{P}_p}$$
FG-1b	$\dfrac{\Delta T_{\max}}{\Delta T_{\max,p}}$	$\dfrac{n_t}{n_{t,p}} = \dfrac{A_{\mathrm{fr}}}{A_{\mathrm{fr},p}} = 1 \qquad \dfrac{L}{L_p} = \dfrac{A}{A_p} = 1$ $\dfrac{\dot{m}}{\dot{m}_p} = \dfrac{G}{G_p} = 1 \qquad \dfrac{q}{q_p} = 1$ In this case, $\Delta T_{\max}$ will go down and $\mathcal{P}$ will go up for the enhanced surface	From Eq. (10.26), $$\dfrac{UA}{(UA)_p} = \dfrac{1+\mathbf{R}_p^*}{(j_p/j) + \mathbf{R}^*} \quad \text{since} \quad \dfrac{\mathcal{P}_p}{\mathcal{P}} = \dfrac{f_p}{f}$$ Compute NTU from Eq. (10.29) and subsequently, $\mathcal{P}$, for the given flow arrangement. Then, from Eq. (10.28), $$\dfrac{\Delta T_{\max}}{\Delta T_{\max,p}} = \dfrac{\mathcal{P}_p}{\mathcal{P}} \dfrac{\dot{m}_p}{\dot{m}}$$
FG-2a	$\dfrac{q}{q_p}$	$\dfrac{n_t}{n_{t,p}} = \dfrac{A_{\mathrm{fr}}}{A_{\mathrm{fr},p}} = 1 \qquad \dfrac{L}{L_p} = \dfrac{A}{A_p} = 1$ $\dfrac{\mathcal{P}}{\mathcal{P}_p} = 1, \qquad \dfrac{\Delta T_{\max}}{\Delta T_{\max,p}} = 1$	From Eq. (10.24) and (10.26), for this case $$\dfrac{G}{G_p} = \left(\dfrac{f_p}{f}\right)^{1/3} \quad \text{and} \quad \dfrac{UA}{(UA)_p} = \dfrac{1+\mathbf{R}_p^*}{(j_p/j)(f/f_p)^{1/3} + \mathbf{R}^*}$$ Solve G/G_p iteratively from the first equation knowing f_p vs. Re_p and f vs. Re characteristics. Compute j_p and j from the known j_p vs. Re_p and j vs. Re characteristics. Calculate $UA/(UA)_p$ from the foregoing equation. Then follow the procedure of FG-1a to get q/q_p from NTU.

continued over

TABLE 10.6 *Continued*

Case	Determine:	Information or Constraints Given and Comments	Resultant Formulas to Compute the Objective of the PEC
FG-2b	$\dfrac{\Delta T_{max}}{\Delta T_{max,p}}$	$\dfrac{n_t}{n_{t,p}} = \dfrac{A_{fr}}{A_{fr,p}} = 1 \quad \dfrac{L}{L_p} = \dfrac{A}{A_p} = 1$ $\dfrac{\mathcal{P}}{\mathcal{P}_p} = 1 \quad \dfrac{q}{q_p} = 1$	Compute $UA/(UA)_p$ as outlined for the FG-2a case. Compute NTU from Eq. (10.29) and subsequently, P, for the given flow arrangement. Then from Eq. (10.28), $$\frac{\Delta T_{max}}{\Delta T_{max,p}} = \frac{P_p}{P}\frac{\dot{m}_p}{\dot{m}}$$
FG-3	$\dfrac{\mathcal{P}}{\mathcal{P}_p}$	$\dfrac{n_t}{n_{t,p}} = \dfrac{A_{fr}}{A_{fr,p}} = 1 \quad \dfrac{L}{L_p} = \dfrac{A}{A_p} = 1$ $\dfrac{q}{q_p} = \dfrac{\Delta T_{max}}{\Delta T_{max,p}} = 1 \quad \dfrac{\dot{m}}{\dot{m}_p} = \dfrac{G}{G_p}$	From Eqs. (10.26), (10.24), and (10.28), $$\frac{UA}{(UA)_p} = \frac{1 + \mathbf{R}_p^*}{(j_p/j)[(f/f_p)(\mathcal{P}_p/\mathcal{P})]^{1/3} + \mathbf{R}^*} \quad (1)$$ $$\frac{\mathcal{P}}{\mathcal{P}_p} = \frac{f}{f_p}\left(\frac{G}{G_p}\right)^3 \quad (2)$$ $$\frac{q}{q_p} = 1 = \frac{G}{G_p}\frac{P}{P_p} \quad (3)$$ From these equations, compute G/G_p as follows: (1) Assume G/G_p. (2) Solve for $\mathcal{P}/\mathcal{P}_p$ from Eq. (2). (3) Compute $UA/(UA)_p$ from Eq. (1). (4) Calculate NTU from Eq. (10.28). (5) Determine P and P_p for the exchanger flow arrangement given. (6) compute q/q_p from Eq. (3) which would be different from unity until convergence. Iterate on G/G_p to obtain $q/q_p = 1$. Using converged value of G/G_p, calculate $\mathcal{P}/\mathcal{P}_p$ from Eq. (2).
FN-1	$\dfrac{L}{L_p}$	$\dfrac{n_t}{n_{t,p}} = \dfrac{A_{fr}}{A_{fr,p}} = 1 \quad \dfrac{A}{A_p} = \dfrac{L}{L_p}$ $\dfrac{\dot{m}}{\dot{m}_p} = \dfrac{G}{G_p} \quad \dfrac{\mathcal{P}}{\mathcal{P}_p} \quad \dfrac{q}{q_p} = \dfrac{\Delta T_{max}}{\Delta T_{max,p}} = 1$	From Eqs. (10.26), (10.24), and (10.28), $$\frac{UA}{(UA)_p} = \frac{1 + \mathbf{R}_p^*}{(j_p/j)(f/f_p)^{1/3}(L_p/L)^{2/3} + \mathbf{R}^*(L_p/L)}$$ $$\frac{\mathcal{P}}{\mathcal{P}_p} = 1 = \frac{f}{f_p}\frac{L}{L_p}\left(\frac{G}{G_p}\right)^3, \quad \frac{q}{q_p} = 1 = \frac{G}{G_p}\frac{P}{P_p}$$

continued over

718

Eliminating L_p/L from the foregoing first and second equations, we get

$$\frac{UA}{(UA)_p} = \frac{1 + \mathbf{R}_p^*}{(j_p/j)(f/f_p)(G/G_p)^2 + \mathbf{R}^*(f/f_p)(G/G_p)^3} \qquad (4)$$

From these equations, compute G/G_p as follows: (1) Assume G/G_p. (2) Solve for $UA/(UA)_p$ from Eq. (4). (3) Calculate NTU from Eq (10.29). (4) Determine P and P_p for the exchanger flow arrangement given. (5) Compute q/q_p from Eq. (3), which would be different from unity until convergence. Iterate on G/G_p to obtain $q/q_p = 1$. Using converged value of G/G_p, calculate L/L_p from the second $\mathcal{P}/\mathcal{P}_p$ equation above.

Since $G/G_p = 1$ the j and f factors are directly calculable. From Eq. (10.24), $\mathcal{P}/\mathcal{P}_p = (f/f_p)(L/L_p)$. From Eq. (10.28), $UA/(UA)_p = 1$ or $L/L_p = U_p/U$. Substitution of these equations and given information into Eq. (10.26) yields

$$\frac{U}{U_p} = \frac{1 + \mathbf{R}_p^*}{(j_p/j) + \mathbf{R}^*} \qquad \text{Then} \qquad \frac{L}{L_p} = \frac{U_p}{U}$$

All formulas and the procedure for the FN-2 case apply here. After computing the value of L/L_p, calculate $\mathcal{P}/\mathcal{P}_p$ from the equation

$$\frac{\mathcal{P}}{\mathcal{P}_p} = \frac{f}{f_p}\frac{L}{L_p}$$

FN-2 $\dfrac{L}{L_p}$

$$\frac{n_t}{n_{t,p}} = \frac{A_{\text{fr}}}{A_{\text{fr},p}} = 1 \qquad \frac{A}{A_p} = \frac{L}{L_p}$$

$$\frac{\dot{m}}{\dot{m}_p} = \frac{G}{G_p} = 1 \qquad \frac{q}{q_p} = \frac{\Delta T_{\max}}{\Delta T_{\max,p}} = 1$$

$$\text{Since} \quad \frac{q}{q_p} = \frac{P}{P_p} = 1 \Rightarrow \frac{\text{NTU}}{\text{NTU}_p} = 1$$

FN-3 $\dfrac{\mathcal{P}}{\mathcal{P}_p}$

Information here identical to that of the FN-2 case

continued over

TABLE 10.6 (Continued)

Case	Determine	Information and Constraints Given and Comments	Resultant Formulas to Compute the Objective of the PEC
VG-1	$\dfrac{n_t}{n_{t,p}}$	$\dfrac{\dot{m}}{\dot{m}_p}=1 \quad \dfrac{\mathcal{P}}{\mathcal{P}_p}=\dfrac{q}{q_p}=\dfrac{\Delta T_{\max}}{\Delta T_{\max,p}}=1$	Since, $P/P_p=1$, we get $\mathrm{NTU}/\mathrm{NTU}_p=1$ and hence $UA/(UA)_p=1$. Thus, Eqs. (10.26) and (10.24) reduce to
	$\dfrac{A_{\mathrm{fr}}}{A_{\mathrm{fr},p}}$	Hence, $P/P_p=1$	$\dfrac{UA}{(UA)_p}=1=\dfrac{1+\mathbf{R}_p^*}{(j_p/j)[(f/f_p)(A_p/A)^2]^{1/3}+\mathbf{R}^*(A_p/A)}$ (5)
	$\dfrac{L}{L_p}$		$\dfrac{\mathcal{P}}{\mathcal{P}_p}=1=\dfrac{f}{f_p}\dfrac{A}{A_p}\left(\dfrac{G}{G_p}\right)^3$ (6)
			Substitute A_p/A from Eq. (6) into Eq. (5). The resultant equation has only one unknown G/G_p since j/j_p and f/f_p are dependent on the value of G/G_p. Solve for G/G_p from this implicit equation. Then A/A_p is known from Eq. (6). Subsequently, compute $n_t/n_{t,p}$ (or $A_{\mathrm{fr}}/A_{\mathrm{fr},p}$) and L/L_p from Eq. (10.30).
VG-2a	$\dfrac{q}{q_p}$	$\dfrac{\dot{m}}{\dot{m}_p}=1 \quad \dfrac{L}{L_p}=\dfrac{A}{A_p}=1$ $\dfrac{\mathcal{P}}{\mathcal{P}_p}=1 \quad \dfrac{\Delta T_{\max}}{\Delta T_{\max,p}}=1$	The procedure outlined for the FG-2a case applies here. For this case, the mass flow rates are equal. This results in different n_t or frontal area because G's are different. In case FG-2a, n_t or frontal areas are the same and result in different mass flow rates. Otherwise, all formulas and the procedure of the FG-2a case are identical for this case.
VG-2b	$\dfrac{\Delta T_{\max}}{\Delta T_{\max,p}}$	$\dfrac{\dot{m}}{\dot{m}_p}=1 \quad \dfrac{L}{L_p}=\dfrac{A}{A_p}=1$ $\dfrac{\mathcal{P}}{\mathcal{P}_p}=1 \quad \dfrac{q}{q_p}=1$	All formulas and the procedure of the FG-2b case apply here. The comment of case VG-2a is also valid here.
VG-3	$\dfrac{\mathcal{P}}{\mathcal{P}_p}$	$\dfrac{A}{A_p}=1 \quad \dfrac{\dot{m}}{\dot{m}_p}=1 \quad \dfrac{q}{q_p}=\dfrac{\Delta T_{\max}}{\Delta T_{\max,p}}=1$ $n_t=\dfrac{A_{\mathrm{fr}}}{A_{\mathrm{fr},p}} \quad \dfrac{G}{G_p}\neq 1$	$U/A/(UA)_p=1$ since $q/q_p=P/P_p=1$. Then, we find that $G/G_p=(j/j_p)(1+\mathbf{R}_p^*)-\mathbf{R}^*$ from Eq. (10.26). Once G/G_p is calculated from the foregoing equation from the known j vs. Re correlations, f/f_p can be computed for known G/G_p. Subsequently, compute $\mathcal{P}/\mathcal{P}_p$ from Eq. (10.24) for $A/A_p=1$.

[a] When appropriate, it is assumed that the correlation for j vs. Re (and hence G) for the enhanced and plain surfaces have the same form [i.e., $j/j_p=(G/G_p)^n$ where n is an exponent]. Similarly, it is assumed that $f/f_p=(G/G_p)^m$, where m is some other exponent, and that $\mathbf{R}^*$ and $\mathbf{R}_p^*$ are known.

720

Thus, knowing the ratio $UA/(UA)_p$ from Eq. (10.26), the NTU of the enhanced surface can be calculated using Eq. (10.29) for a given flow rate, and subsequently, the temperature effectiveness for the given exchanger flow arrangement can be found from the formulas of Table 3.6. The heat transfer rate is then computed from Eq. (10.28).

Using Eq. (10.21) and $\dot{m} = A_o G$, the ratios of surface areas and flow rates for enhanced and plain tubular surface for the same d_i are given by

$$\frac{A}{A_p} = \frac{n_t}{n_{t,p}} \frac{L}{L_p} \qquad \frac{\dot{m}}{\dot{m}_p} = \frac{n_t}{n_{t,p}} \frac{G}{G_p} \qquad (10.30)$$

Using the foregoing equations, specific algebraic formulas for the PECs of Table 10.4 are summarized in Table 10.6. It should be emphasized that the formulas of Table 10.6 can also be used to compare two different surfaces 1 and 2 for the specific PEC by adding the subscript 1 to the enhanced surface values and replacing the subscript p by 2 as a base surface. Also, the values $\mathbf{R}^*$ and $\mathbf{R}_p^*$ in Eq. (10.26) are normalized with respect to the plain tube inside thermal resistance. They can be normalized consistently with respect to any fluid-side thermal resistance.

Example 10.4 Would a selection of the following heat transfer surface provide better performance than the existing design in which a given plain plate-fin surface (i.e., surface 11.1, Table 10-3, Fig. 10-26, Kays and London, 1998) is used? The comparison is based on the requirements of fixed (1) flow rate, (2) fluid pumping power, (3) heat transfer rate, and (4) inlet temperature difference between the hot- and cold-fluid streams. The argument is supposed to be valid for any Reynolds number. The new heat transfer surface has the following Colburn and Fanning friction factors:

$$j = \exp(a_0 + a_1 r + a_2 r^2 + a_3 r^3 + a_4 r^4 + a_5 r^5 + a_6 r^6)$$

and

$$f = \exp(b_0 + b_1 r + b_2 r^2 + b_3 r^3 + b_4 r^4 + b_5 r^5 + b_6 r^6)$$

where $r = \ln \mathrm{Re}$ and the numerical values of the coefficients are as follows:

Coefficient	Numerical Value	Coefficient	Numerical Value
a_0	0.1624564980E + 04	b_0	0.1242054696E + 04
a_1	−0.1404062382E + 04	b_1	−0.1077301312E + 04
a_2	0.4999486289E + 03	b_2	0.3855707180E + 03
a_3	−0.9400171748E + 02	b_3	−0.7291091700E + 02
a_4	0.9835078386E + 01	b_4	0.7674456065E + 01
a_5	−0.5428407378E + 00	b_5	−0.4263025064E + 00
a_6	0.1235104592E − 01	b_6	0.9766547339E − 02

SOLUTION

Problem Data and Schematic: Two surfaces should be compared under the fixed flow rate, pumping power, heat transfer rate, and inlet temperature difference between the hot- and cold-fluid streams. The entire range of applicable Re numbers should be considered. The correlations for Colburn and Fanning friction factors for a new heat transfer surface are provided in the problem formulation. The Colburn and Fanning factors for the plain plate-fin geometry 11.1 are given in Kays and London (1998) in tabular form. A curve fitting of these data (in the same form as the ones given in the problem formulation) leads to the following set of coefficients:

Coefficient	Numerical Value	Coefficient	Numerical Value
$a_{0,p}$	$-0.1305722226E + 05$	$b_{0,p}$	$-0.6188627536E + 04$
$a_{1,p}$	$0.1014346095E + 05$	$b_{1,p}$	$0.4958849985E + 04$
$a_{2,p}$	$-0.3268456896E + 04$	$b_{2,p}$	$-0.1647882636E + 04$
$a_{3,p}$	$0.5590860182E + 03$	$b_{3,p}$	$0.2906965749E + 03$
$a_{4,p}$	$-0.5355816661E + 02$	$b_{4,p}$	$-0.2872245480E + 02$
$a_{5,p}$	$0.2724931767E + 01$	$b_{5,p}$	$0.1507451536E + 01$
$a_{6,p}$	$-0.5753732235E - 01$	$b_{6,p}$	$-0.3283617030E - 01$

Determine: Determine which of the two heat transfer surfaces has better heat transfer/pressure drop performance.

Assumptions: The assumptions invoked in Section 3.2.1 are valid.

Analysis: We are given the information on only one fluid side of the exchanger in this problem. The problem formulation requires that the following relations be satisfied.

$$\frac{\dot{m}}{\dot{m}_p} = \frac{\mathcal{P}}{\mathcal{P}_p} = \frac{hA}{(hA)_p} = \frac{\Delta T_{max}}{\Delta T_{max,p}} = 1$$

where the symbols without a subscript denote the variables for the new surface and the symbols with the subscript p denote the plain plate-fin surface (surface 11.1 from Kays and London, 1998). This is the VG-1 PEC of Table 10.4 on one fluid side of the exchanger. According to Eq. (10.24) or Eq. (6) of Table 10.6, the pumping power ratio (which is equal to unity, as indicated above) must be equal to

$$\frac{\mathcal{P}}{\mathcal{P}_p} = 1 = \frac{f}{f_p} \frac{A}{A_p} \left(\frac{G}{G_p}\right)^3 \tag{1}$$

Similarly, from Eq. (10.25), we get

$$\frac{hA/(hA)_p}{(\mathcal{P}/\mathcal{P}_p)^{1/3}(A/A_p)^{2/3}} = \left(\frac{A_p}{A}\right)^{2/3} = \frac{j/j_p}{(f/f_p)^{1/3}} \tag{2}$$

From Eqs. (1) and (2), we can eliminate the surface area ratio, A/A_p. Hence, the ratio of mass velocities for the two surfaces must satisfy the relation

$$\frac{G}{G_p} = \left(\frac{f_p}{f} \frac{j}{j_p}\right)^{1/2}$$

Note that $G/G_p = \text{Re}/\text{Re}_p$. Therefore,

$$\frac{\text{Re}}{\text{Re}_p} = \left(\frac{f_p \cdot \text{Re}_p}{f \cdot \text{Re}} \frac{j \cdot \text{Re}}{j_p \cdot \text{Re}_p}\right)^{1/2} \tag{3}$$

This relation indicates that we can determine Re for which the new surface would satisfy the aforementioned constraints for any given Re_p for flow through the plain plate-fin surface.[†] The pair of Re numbers would then allow calculation of the corresponding Colburn and Fanning factors, and subsequently, the evaluation of heat transfer surface area reduction (or increase, if the new surface is not better than the old one) that would be accomplished with the proposed change of geometry. Therefore, we should determine these pairs of Re numbers, say for a low Re number range (say, around 500) and for a large Re number range (say, around 5000), and compare the heat transfer area changes. It should be noted that the determination of these Re numbers may be a tedious iterative job, in particular if one uses the graphical presentations of j and f factors as those given in Kays and London (1998). In our case, we perform the calculations numerically. This calculation leads to the following results. For $\text{Re}_p = 500$, a very little change in Re would occur (i.e., $\text{Re} = 505$), but $A/A_p = 0.93$. That means that the new surface would require only 7% less heat transfer area for the same performance as the old one. For $\text{Re}_p = 5000$, however, Re becomes 4870 and $A/A_p = 1.13$! The proposed new surface would require not less but 13% more heat transfer surface than the old one. So it can be concluded that the proposed solution does not bring a significant benefit for the low-Re range and may actually be worse for the high-Re range.

Discussion and Comments: The performance evaluation criteria may provide a useful tool for assessing the performance of a selected heat transfer surface as well as for the comparison of various solutions. In this case, the claim that a new design is better is not substantiated. Our conclusion is based on the use of a variable-geometry criterion (VG-1; see Table 10.6). It should be pointed out, though, that highly augmented heat transfer surfaces may reduce the required heat transfer area significantly (50% and more) compared with plain plate-fin heat transfer surfaces for the VG-1 criterion.

10.3.3 Evaluation Criteria Based on the Second Law of Thermodynamics

All performance evaluation criteria discussed so far are based exclusively on the first law of thermodynamics. These criteria were devised utilizing mass and energy balances without involving the thermodynamic quality of the energy flows. However, heat transfer and friction characteristics of heat transfer surfaces may easily be related to the quality level of energy flows defined by the second law of thermodynamics.[‡] That becomes very important in any system analysis, and feedback from

---

[†] For a given Re_p, j_p, and f_p are determined from the given surface data. Now one assumes a value of Re, determines corresponding j and f factors and computes Re from Eq. (3). If this Re does not match with the assumed Re, iterations are carried out with new values of Re, until Eq. (3) is satisfied, and corresponding j and f factors are computed to be used subsequently in Eq. (2).

[‡] The body of knowledge usually called the second law of thermodynamics analysis *always* involves both the first *and* second laws of thermodynamics. However, it is customary to name the product of such an analysis by indicating the second law of thermodynamics only (Bejan, 1988).

a system engineer may indicate a need for a change of design based on the second law of thermodynamics. This approach is considered in Section 11.7, with an additional performance evaluation criterion.

10.3.4   Selection Criterion Based on Cost Evaluation

The cost of an exchanger is usually an important selection criterion for a user. Let us assume that all other pertinent evaluation criteria, if not already incorporated in the cost evaluation/optimization routine, are satisfied. In that case, existing design options will be selected based on the most cost-effective design. So a methodology for cost estimation must be developed. Most heat exchanger manufacturers have their own proprietary methods for cost estimation. Some approaches to this problem have been reported in the literature in the past. We present here a simple procedure of the ESDU (1994) used in some industries.

Heat exchanger cost may be related either to the heat transfer surface area of an exchanger or to the heat exchanger duty required. In Fig. 10.13, this dilemma is presented schematically as two options that may be perceived from the perspective of a heat exchanger designer on one side and of a process engineer on the other. The simple logic implied by Fig. 10.13 has been used to define cost estimation evaluation.

The proposed methodology is based on empirical cost data compiled to evaluate all the various feasible heat exchanger types. The decision variable is the cost of a heat exchanger per unit of its thermal size, that is, per unit of the product UA ($= q/\Delta T_m$). In Fig. 10.13, this quantity is denoted as C_{UA}. An alternative solution would be the cost of an exchanger per unit heat transfer surface area, C_A of Fig. 10.13. The latter is less attractive because it does not explicitly take into account the heat transfer duty, or what is equivalent, the relevant thermal size of the exchanger, UA. It must be clear that the overall heat transfer coefficient, determined for a particular design, must influence the cost.

In Table 10.7, a *selection* of the cost data, represented by the values of C_{UA}, is compiled. This table is prepared for a particular exchanger purpose, namely, for an application in which the heat exchange is accomplished between gas as a hot fluid at

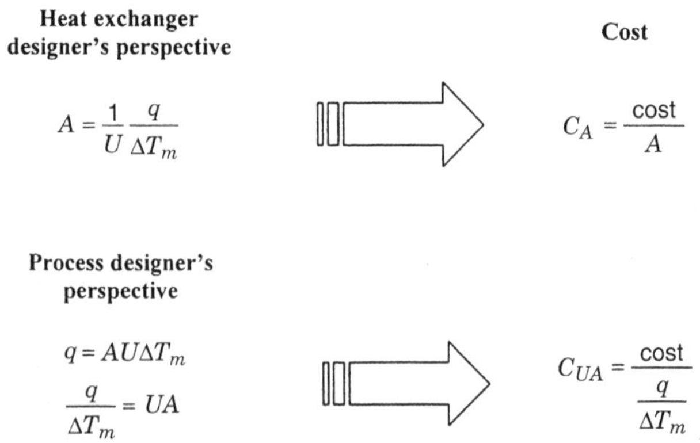

Heat exchanger designer's perspective

$$A = \frac{1}{U}\frac{q}{\Delta T_m}$$

$$C_A = \frac{\text{cost}}{A}$$

Cost

Process designer's perspective

$$q = AU\Delta T_m$$

$$\frac{q}{\Delta T_m} = UA$$

$$C_{UA} = \frac{\text{cost}}{\dfrac{q}{\Delta T_m}}$$

FIGURE 10.13   Cost of a heat exchanger vs. heat transfer area and/or heat duty.

TABLE 10.7   Cost Data C_{UA} vs UA for Various Heat Exchanger Types[a]

$q/\Delta T_m$ or UA (W/K)	C_{UA} [$/(W/K)]					
	Shell- and- Tube, $U = 484$ (W/m² · K)	Double Tube, $U = 484$ (W/m² · K)	Printed Circuit $U = 1621$ (W/m² · K)	Plate-Fin, $U = 491$ (W/m² · K)	Welded Plate	
					U (W/m² · K)	C_{UA} [$/(W · K)]
10^3	3.98	2.5	12	—	349	4.9
5×10^3	1.00	0.75	2.4	3.1	1187	1.22
3×10^4	0.29	0.31	0.6	0.513	1068	0.42
10^5	0.17	0.31	0.42	0.210	1112	0.28
10^6	0.106	0.31	0.28	0.115	1173	0.22

Source: Data from ESDU (1994).
[a] The hot fluid is medium-pressure gas and the cold fluid-treated water. The original ESDU cost data in the British pound are approximately in the US dollar value in 2000.

medium pressure (say, 20 bar) and cold fluid as treated water. Five different heat exchanger types may be used for this particular combination of working fluids (see Chapters 1 and 2 for a description of each type and the assessment of the feasibility of the possible selections): (1) shell-and-tube heat exchanger, (2) double-pipe heat exchanger, (3) printed-circuit heat exchanger, (4) plate-fin exchanger, and (5) welded plate exchanger. Depending on the magnitude of $q/\Delta T_m$, different cost C_{UA} values can be determined for each of the heat exchanger types. An extensive set of C_{UA} data for various heat exchangers is compiled by ESDU (1994), and partially summarized in Appendix D; they can be used for this purpose. From Table 10.7, it is clear that the cost of a heat exchanger per unit of its thermal size (i.e., per unit of UA), C_{UA}, decreases with an increase in the heat load ($q/\Delta T_m$) or heat exchanger size (UA).

The procedure for evaluation of a heat exchanger type based on the given cost criterion is as follows:

1. Estimate the heat duty q from a heat balance using Eq. (2.1).
2. Determine $q/\Delta T_m$ for the heat exchanger under consideration (a) by computing $\Delta T_m = F \Delta T_{lm}$ with ΔT_{lm} from Eq. (3.172) and the best estimate of F for (ESDU, 1994), or (b) from a known NTU and C_{min} using $q/\Delta T_m = C_{min} \cdot$ NTU [see Eqs. (3.12) and (3.59)].
3. Repeat step 2 for each heat exchanger type.
4. From empirical data (C_{UA} vs. $q/\Delta T_m$, see Appendix D), estimate the C_{UA} factor.
5. Calculate the cost of a particular heat exchanger type by multiplying C_{UA} and $q/\Delta T_m$.
6. Compare the costs for various heat exchanger types. If one of the types is much less expensive than the other (by a factor of 1.5 to 2.0 or more), that design should be selected. If the costs for all solutions are close to each other, a more detailed analysis of each individual cost must be performed.

The procedure outlined above is utilized in a simplified way in Example 2.4. A more elaborate analysis is the subject of Problem 10.8.

SUMMARY

Heat exchangers are designed for a variety of applications under varied operating conditions. As a result, an optimum heat exchanger will be different depending on the application. The most important selection criteria for a heat exchanger are summarized next.

- The heat exchanger must function as designed for performance, durability, and other criteria during its design life. As a result, the operating environment (i.e., pressure, temperature, fouling potential, fluid leakage and contamination, material compatibility, etc.), cost packaging, maintenance, and so on, are very important variables. Based on these operating and design conditions, an engineer can select an appropriate exchanger from Table 10.1 with additional considerations of the cost, manufacturability, and other requirements.
- A large number of geometric variables are associated with shell-and-tube exchangers. Considerable discussion is provided in Section 10.2.1 for the choice of specific geometrical variables. Similarly, for extended surface exchangers, a variety of fin geometries is available, and a qualitative discussion on the selection of particular geometries is presented in Section 10.2.3. Surface selection for plate heat exchangers and regenerators is discussed briefly in Sections 10.2.2 and 10.2.4.

In many applications, the heat exchanger operates in a system or a thermodynamic cycle. Therefore, quantitative criteria for component design and optimization have less meaning since the heat exchanger should be designed for optimum system performance. Hence, quantitative methods are presented in the text for screening various surfaces to select the most appropriate ones as components. In this regard, two categories of quantitative methods are summarized:

- The surface flow area and core volume goodness factor comparisons are presented to screen and arrive at higher-performing extended surfaces. Geometrical scaling laws are then summarized for a compact heat exchanger surface on one fluid side for changes in flow area, volume, surface area, and length of that surface for the case of constant q, Δp, and $\dot{m}$.
- Performance evaluation criteria are employed to compare the performance of an enhanced tubular to a plain tubular exchanger surface.

REFERENCES

ALPEMA, 2000, *The Standards of the Brazed Aluminum Plate-Fin heat Exchanger Manufacturer' Association (ALPEMA)*, 2nd Edition, AEA Technology plc, Didcot, Oxon, UK.

ASME, 1998, ASME Boiler and Pressure Vessel Code, 1998: Rules for Construction of Pressure Vessels, Sec. VIII, Div. 1, American Society of Mechanical Engineers, New York.

Bejan, A., 1988, *Thermodynamic Design in Advanced Engineering Thermodynamics*, Wiley, New York, pp. 594–669.

Bell, K. J., 1981, Construction features of shell-and-tube heat exchangers, in *Heat Exchangers: Thermal-Hydraulic Fundamentals and Design*, edited by S. Kakaç, A. E. Bergles and F. Mayinger, Hemisphere Publishing Corp., Washington, DC, pp. 721–763.

Bell, K. J., 1998, Approximate sizing of shell-and-tube heat exchangers, in *Heat Exchanger Design Handbook*, G. F. Hewitt, exec. ed., Begell House, New York, Vol. 3, Sec. 3.1.4.

Cowell, T. A., 1990, A general method for the comparison of compact heat transfer surfaces, *ASME J. Heat Transfer*, Vol. 112, pp. 288–294.

ESDU, 1994, Selection and costing of heat exchangers, *Engineering Science Data*, Item 92013, ESDU, Int., London, UK.

Gentry, C. G., 1990, RODbaffle heat exchanger technology, *Chem. Eng. Prog.* July, pp. 48–57.

Kays, W. M., and A. L. London, 1998, *Compact Heat Exchangers*, reprint 3rd ed., Krieger Publishing, Malabar, FL.

Lancaster, J. F., 1998, Materials of construction, in *Handbook of Heat Exchanger Design*, G. F. Hewitt, ed., Begell House, New York, Sec. 4.5.

London, A. L., 1964, Compact heat exchangers, Part 2, Surface geometry, *Mech. Eng.*, Vol. 86, June, pp. 31–34.

London, A. L., and C. K. Ferguson, 1949, Test results of high-performance heat exchanger surfaces used in aircraft intercoolers and their significance for gas-turbine regenerator design, *Trans. ASME*, Vol. 71, pp. 17–26.

Reay, D. A., 1999, *Learning from Experiences with Compact Heat Exchangers*, CADDET Analyses Series 25, Centre for the Analysis and Dissemination of Demonstrated Energy Technologies, Sittard, The Netherlands.

Sekulić, D. P., A. J. Salazar, F. Gao, J. S. Rosen, and H. S. Hutchins, 2003. Local transient behavior of compact heat exchanger core during brazing, *Int. J. Heat Exchangers*, Vol. 4, No. 1.

Shah, R. K., 1978, Compact heat exchanger surface selection methods, *Heat Transfer 1978, Proc. 6th Int. Heat Transfer Conf.*, Vol. 4, pp. 193–199.

Shah, R. K., 1983, Compact heat exchanger surface selection optimization and computer-aided thermal design, in *Low Reynolds Number Flow Heat Exchangers*, edited by S. Kakaç, R. K. Shah and A. E. Bergles, pp. 845–874, Hemisphere Publishing Corp., Washington, DC.

Shah, R. K., 1995, Heat exchangers, in *Encyclopedia of Energy Technology and the Environment*, A. Bisio and S. G. Boots, eds. Wiley, New York, Vol. 3, pp. 1651–1670.

TEMA, 1999, *Standard of the Tubular Exchanger Manufacturers Association*, 8th ed. Tubular Exchange Manufacturers Association, New York.

Webb, R. L., 1981, Performance evaluation criteria for use of enhanced heat transfer surfaces in heat exchanger design, *Int. J. Heat Mass Transfer*, Vol. 24, pp. 715–726.

Webb, R. L., 1994, *Principles of Enhanced Heat Transfer*, Wiley, New York.

Webb, R. L., and A. E. Bergles, 1983, Performance evaluation criteria for selection of heat transfer surface geometries used in low Reynolds number heat exchangers, in *Low Reynolds Number Flow Heat Exchangers*, S. Kakaç, R. K. Shah, and A. E. Bergles, eds., Hemisphere Publishing, Washington, DC, pp. 735–752.

REVIEW QUESTIONS

Where multiple choices are given, circle one or more correct answers. Explain your answers briefly.

10.1 For liquid-to-gas exchangers, the commonly used exchanger constructions are:

(a) shell-and-tube   (b) plate-type   (c) extended surface   (d) regenerators

10.2 For fouling fluids, the commonly used exchanger constructions are:

(a) shell-and-tube   (b) plate-type   (c) extended surface   (d) regenerators

10.3 Regenerators are exclusively used as:

(a) gas-to-liquid exchangers

(b) gas-to-gas exchangers

(c) condensing fluid-to-gas exchangers

(d) gas-to-evaporating fluid exchangers

10.4 Plate-fin exchangers are *commonly* used for the application having:

(a) 0 to 20 MPa operating pressures

(b) −200 to 540°C operating temperatures

(c) heavy fouling fluids   (d) highly corrosive fluids   (e) none of these

10.5   To cool concentrated hydrochloric acid with a heavy oil (low h), the following exchanger(s) would be feasible:

(a) plate exchanger with titanium plates

(b) plate-fin exchanger with fins on oil side

(c) TEMA AEW with oil in the shell

(d) TEMA CEN with oil in the shell.

10.6   In a shell-and-tube exchanger, low-finned tubes result in the following on the shell side:

(a) increase in the heat transfer coefficient

(b) increase in surface area

(c) increase in pressure containment and rigidity

(d) better flow mixing

(e) reduction in fouling

10.7   The tube-side turbulent flow heat transfer coefficient for a given flow rate is increased by:

(a) increasing the number of tubes with decreased tube length

(b) decreasing the number of tubes with increased tube length

(c) increasing the number of tube passes

(d) increasing the number of shell passes

(e) decreasing the tube gauge

10.8   The shell-side heat transfer coefficient is increased by:

(a) increasing the number of baffles   (b) decreasing the baffle cut

(c) increasing the tube pitch   (d) increasing number of tube passes

(e) increasing tube fluid velocity

10.9   In general, all shell-and-tube exchangers have transverse plate baffles except for the following TEMA shell:

(a) G   (b) J   (c) K   (d) X

10.10   The horizontal baffle cut is used for the following shell-side fluids:

(a) single-phase fluids   (b) condensing fluids

(c) evaporating fluids   (d) very viscous liquids   (e) slurries

10.11   Some consequences of using the transverse plate baffles are:

(a) reduce fouling on the shell side

(b) reduce pressure drop on the shell side

(c) minimize tube-to-tube temperature differentials

(d) eliminate flow-induced tube vibrations

(e) none of these

10.12   The use of rod baffles in shell-and-tube exchangers results in:

(a) increased shell-side heat transfer coefficient compared to a plate baffle exchanger at the same mean velocity

(b) reduced number of baffles in the exchanger compared to an equivalent segmental baffle exchanger

(c) increased rigidity of the tube bundle

(d) reduced shell-side fouling

(e) reduced shell-side pressure drop

10.13 Impingement baffles are used to:

(a) increase shell-side heat transfer coefficient

(b) support the tubes

(c) protect the tube damage in the inlet region

(d) provide nonuniform flow distribution at inlet on the shell side

10.14 The following shell types are commonly used for single-phase fluids on the shell side:

(a) E   **(b)** F   **(c)** G   **(d)** H   **(e)** J   **(f)** K   **(g)** X

10.15 The following shell types are commonly used for two-phase or multiphase fluids on the shell side:

(a) E   **(b)** F   **(c)** G   **(d)** H   **(e)** J   **(f)** K   **(g)** X

10.16 In a single tubesheet design, the *least* likelihood of leakage between the shell and tube fluids is with the following front-end heads:

(a) A   **(b)** B   **(c)** C   **(d)** N

10.17 The most important criteria for the selection of rear-end heads are:

(a) to control the fluid velocities

(b) operating pressures

(c) thermal stresses between tubes and shell

(d) shell- or tube-side cleaning requirement

(e) high shell-side heat transfer coefficient

(f) can't tell

10.18 The bundle-to-shell bypass stream C could be significant in the following rear-end head constructions, and as a result, sealing strips are usually required:

(a) L   **(b)** M   **(c)** N   **(d)** P   **(e)** S   **(f)** T   **(g)** U   **(h)** W

10.19 The flow-induced tube vibration could often be reduced in single-segmental exchangers by:

(a) adding sealing strips

(b) decreasing the number of tubes in the window area

(c) increasing the shell (and tube bundle) diameter

(d) decreasing the baffle cut

(e) none of these

10.20 Arrange the following shell types from high to low Δp on the shell side for a *liquid* at a specified flow rate (assume turbulent flow) and inlet temperature, for the same total surface area, and for single-segmental baffles/support plates at the same spacing:

(a) E   **(b)** F   **(c)** G   **(d)** H   **(e)** J   **(f)** X

10.21   Arrange the following rear-end heads in order of the most easy to the most difficult for cleaning and inspection of the shell side:

(a) M   (b) P   (c) S   (d) T   (e) U   (f) W

10.22   Arrange the following rear-end heads from the least-cost to highest-cost designs:

(a) P   (b) S   (c) T   (d) U   (e) W

10.23   Circle the following statements as true or false.

(a) T   F   Different constructions of shell types, front-end heads, and rear-end heads are available that cannot be identified clearly by the TEMA designation scheme.

(b) T   F   From the high- heat-transfer point of view, tubes with small diameters are preferred, but from the cleaning-requirement point of view, tubes of large diameters are preferred.

(c) T   F   Shell diameters of 4 m (160 in.) or tube lengths of 18.3 m (60 ft) in shell-and-tube exchangers are not possible because they do not represent values mentioned in TEMA standards.

(d) T   F   The commonly used ratio of the tube pitch to tube diameter in shell-and-tube exchangers is below 1.2.

(e) T   F   A square tube layout is generally used in a fixed tubesheet exchanger.

(f) T   F   A 60% baffle cut is used for single-segmental baffles in some applications.

(g) T   F   A bonnet head is generally used for ease in inspection and cleaning of the tubes.

(h) T   F   Improper location and size of the impingement baffle is a common cause of tube failure.

10.24   Circle the following statements as true or false for preferred tube layout when shell-side mechanical cleaning is required.

(a) T   F   A 60° tube layout is used for shell-side laminar flow.

(b) T   F   A 45° tube layout is used for shell-side turbulent flow.

(c) T   F   A 45° tube layout is used for condensing fluid on the shell side.

(d) T   F   A 60° tube layout is used for boiling fluid on the shell side.

10.25   Fill in the blanks:

(a) Flow-induced tube vibrations can be particularly serious in U-tube bundles due to _____ .

(b) The _____ design is used when thermal stresses in the tubes must be kept to a minimum.

(c) As an alternative to the floating heads or U-tubes, _____ can be used with E shell to allow increased inlet temperature difference between two fluids.

(d) Other than heat transfer, two of the main reasons for selecting tube orientation and pitch are _____ and _____ .

10.26   Double tubesheets are required in shell-and-tube exchangers to:

(a) relieve thermal stresses between the tubes and shell

(b) make overall design rigid

(c) prevent leakage from one fluid to the other fluid

10.27 For a specified heat transfer and pressure drop requirement, strip fins require the following compared to plain fins:

(a) larger frontal area   (b) larger frontal area and core volume

(c) shorter flow length but larger core volume

10.28 In a plain fin compact heat exchanger, the pressure drop on the side of concern is 500 Pa. If the flow is developing laminar, what is the approximate Δp if the flow length is doubled?

(a) 1000 Pa   (b) 2000 Pa   (c) 700 Pa   (d) can't tell

10.29 In fully developed laminar flow, increasing G will:

(a) increase Δp and h (b) decrease Δp and h

(c) increase Δp but h remains constant

10.30 Rotary regenerators:

(a) employ interrupted surfaces

(b) have surface area densities greater than 400 m^2/m^3

(c) are used for moderately fouling gases

(d) are more expensive per unit surface area compared to plate-fin and tube-fin surfaces

10.31 What type of fin is preferred in a plate-fin exchanger at high Re for low pressure drops?

(a) offset strip   (b) louver   (c) plain   (d) perforated

10.32 If the j and f characteristics for surface P are both 20% higher than those of surface Q, which surface would you select for your heat exchanger? Why?

(a) surface P   (b) surface Q   (c) can't tell

10.33 Usually, the maximum operating conditions for design of a metal plate-fin unit is 1000 kPa or 800°C due to:

(a) joining techniques between the fins and plates

(b) manufacturing technology limitations   (c) cost factors

10.34 Which compact heat exchanger has the highest heat transfer surface area density and the lowest cost per unit surface area?

(a) shell-and-tube   (b) plate   (c) plate-fin   (d) rotary regenerator

10.35 A long slender compact exchanger with triangular passages would have:

(a) a higher   (b) a lower   (c) the same

j/f factor than that of a similar unit with rectangular passages? Consider air flowing in both cases.

10.36 In a compact heat exchanger having fully developed laminar flow, the following relationships exist when we compare two surfaces. Consider q, $\dot{m}$, and Δp as given and constant for these comparisons in parts (a), (c), (d), and (f). Circle the following statements as true or false.

(a) T F   The flow length is inversely proportional to the hydraulic diameter and directly proportional to the j factor.

(b) T F   The flow areas for two surfaces are the same if the L/D_h ratio and f factors are the same for the given fluid, flow rate, and pressure drop.

(c) T   F   The flow area is practically independent of the surface hydraulic diameter.

(d) T   F   A compact heat exchanger having a 1-mm hydraulic diameter on one fluid side will have the identical performance (q, Δp, and $\dot{m}$) of a noncompact exchanger having a 20-mm hydraulic diameter on the same fluid side at the same Reynolds number.

(e) T   F   For a given compact surface, as one reduces the hydraulic diameter (truly scaled down geometry), the j factor also reduces in the same proportion at a specified Reynolds number.

(f) T   F   Reducing the hydraulic diameter on one fluid side, its volume and surface area reduce in the same proportion.

PROBLEMS

10.1   To select an appropriate shell type for a given application (having single-phase fluids on both sides), we want to evaluate all major shell types: E, F, G, H, J, and X. Because of the thermal stress considerations, we will use U tubes (two tube passes) in a single-shell exchanger with the design total $NTU_t = 1.4$ and $R_t = 0.8$. Consider the shell fluid mixed for all shell types except for the X shell, in which it is unmixed. For simple analysis purposes, consider the tube fluid as unmixed in each pass and mixed between passes for the X shell with an overall counterflow arrangement. Assume an overall counterflow arrangement for 1–2 TEMA G and H exchangers.

(a) Tabulate the exchanger effectiveness for the above shell types using the appropriate figures and equations from Chapter 3.

(b) For the identical tube fluid flow rate, will the pressure drop on the tube side be the same or different in the above types of shells? Why? Assume all fluid properties to be constant.

(c) Estimate the shell-side pressure drop for each shell type as a function of u_m and L, where u_m is the mean shell-side velocity in the E shell, and L is the shell length. Note that the shell-side flow rate is the same for all shell types and each has single-segmental baffles/support plates at the same spacing. The shell and tube diameters and number of tubes are the same for all shell types. The shell length is much greater than the shell diameter. Don't forget to add Δp's due to the fluid turning 180° on the shell side in some shell types.

(d) Select a shell type for **(i)** the highest heat transfer performance, and **(ii)** the lowest shell-side pressure drop.

10.2   You have designed two heat exchangers to do the same thermal "job" [i.e., they have the same q and total tube-side mass flow rate $\dot{m}_t$ (kg/s)]. A comparison of these heat exchangers shows the following:

Property	Smooth Tube	Rough Tube
Heat transfer area A (m^2)	13.62	9.80
Number of tubes N_t	100	120

Length of tubes L (m)	2.5	1.25
Tube inside diameter d_i (mm)	18	18
Tube-side friction factor f	0.008	0.016

Assume constant and identical fluid properties for both heat exchangers.

(a) Calculate $\mathcal{P}/\mathcal{P}_s$, where the subscript s is for the smooth tube.

(b) Calculate $\Delta p/\Delta p_s$.

(c) Suppose that $\mathcal{P}/\mathcal{P}_s$ is 0.8; how would you alter the rough tube exchanger geometry to make $\mathcal{P}/\mathcal{P}_s = 1$, keeping the same rough tube length? Give *quantitative* estimates considering no change in the friction factor.

(d) How would you expect UA to change for part (c)? Give explicit reasons whether it will increase, decrease, or will have no change.

10.3 An aircraft oil cooler has offset strip fins on the air side with 790 fins/m, a 0.15 mm fin thickness, a 0.25 mm plate thickness, and a 9.5 mm plate spacing. Air enters the heat exchanger with 5 m/s frontal velocity ($\nu = 1.579 \times 10^{-5}$ m^2/s, $\rho = 1.20$ kg/m^3). Assume that you have an option of using 3- and 12-mm strip lengths for the fin. Compare the performance of these two fin geometries.

(a) Calculate D_h, α^*, and σ. *Hint:* Draw a unit cell and use associated simple geometrical relationship.

(b) Compute j and f for each surface. *Hint:* Use Manglik and Bergles correlations, Eqs. (7.124) and (7.125).

(c) Determine h_2/h_1, E_2/E_1, and $(j/f)_2/(j/f)_1$. Here the subscript 1 refers to the 12 mm strip length.

(d) For both strip fins, compare the values of E and $\mathcal{P}$ for the same A_{fr} and the same hA. What would be the ratio of surface areas, A_2/A_1?

(e) What would be the approximate value of u_∞ for the 3 mm strip fin exchanger for the same hA and same $\mathcal{P}$? Assume that $f_{\text{old}} = f_{\text{new}}$ for the 3 mm strip fin. *Hint:* Use the functional relationship of $\mathcal{P}$ from Eq. (10.8). Knowing the new value of u_∞, compare it with the old value of u_∞ and estimate the new value of h and subsequently the new value of A for the same hA.

(f) Discuss the results of parts (d) and (e).

10.4 As a heat exchanger designer, you have designed a plain plate-fin heat exchanger to transfer the heat specified within the pressure drop allowed. The design point for the plain plate-fin corresponds to Re = 1000. You have an option to employ an offset strip fin of the same hydraulic diameter as an alternative. These surfaces are shown in Fig. 10.3.

(a) Determine the design Reynolds number for the offset strip fin for equal fluid pumping power per unit surface area (equal E).

(b) Determine h_2/h_1 for equal E, where subscripts 1 and 2 designate plain and offset strip fins, respectively. *Hint:* Use the definition of j and appropriate data.

(c) Based on the answer for part (b), how much reduction is achieved in the core volume on the one fluid side under consideration using the offset strip fin?

(d) Determine the area goodness factor ratio for these surfaces. Can you tell which surface will require a higher frontal area? Why?

10.5 Consider the triangular flow passage rotary regenerator of Example 10.1. Determine the change in the disk diameter, disk depth, and volume of the regenerator if the triangular flow passage hydraulic diameter is reduced by 50%.

10.6 A compact air-to-water heat exchanger is to be designed with an air-side mass flow rate of 0.83 kg/s. The required NTU for the exchanger is 1. We would like to design for an air-side Reynolds number (GD_h/μ) of 3000 for which $j = 0.018$. The following additional data are available: Airside: $D_h = 3.47$ mm, $\sigma = 0.48$, $\alpha = 558\,\text{m}^2/\text{m}^3$, $c_p = 1004.9\,\text{J/kg} \cdot \text{K}$, $\mu = 2.07 \times 10^{-5}\,\text{Pa} \cdot \text{s}$, $\text{Pr} = 0.7$, $\eta_o = 0.80$. Waterside: $h = 1.7\,\text{kW/m}^2 \cdot \text{K}$, $\alpha = 32.8\,\text{m}^2/\text{m}^3$, $\eta_o = 1$. Determine **(a)** the air-side frontal area, **(b)** the air-side heat transfer surface area, and **(c)** the air-side flow length. Neglect wall resistance and fouling on both sides. The air side is the minimum heat capacity rate side.

10.7 In automotive radiators, a corrugated multilouver fin (see Fig. 7.29) is used on the air side. Under consideration is replacing it with an offset strip fin geometry of Fig. 8.7. The following geometrical information is provided for the air-side fin geometry.

Fin Geometry	Multilouver Fin	Offset Strip Fin
Fin density (fins/m)	800	800
Louver pitch or offset strip length (mm)	1.25	1.25
Vertical fin height , (mm)	6	6
Louver cut length along the vertical height (mm)	4.8	6
Fin thickness (mm)	0.075	0.075
Louver angle (deg)	30	Not applicable
Tube width (mm)	24	24
Tube pitch (mm)	9.5	9.5
Tube height (mm)	3.5	3.5
Radiator core depth (mm)	24	24

Compute the change in the heat transfer surface and the air pumping power requirements by replacing the louver fin with the offset strip fin under the fixed flow area (FN-2) performance evaluation criterion. Assume the mean air velocity through both fin geometries to be 10 m/s. Use the appropriate correlations for the j and f factors from Chapter 7. Assume that the thermal resistances of the wall and the coolant are zero. Ignore fouling on both fluid sides. Use the following properties of air: $\rho = 1.058\,\text{kg/m}^3$, $c_p = 1.008\,\text{kJ/kg} \cdot \text{K}$, $k = 0.0288\,\text{W/m} \cdot \text{K}$, and $\mu = 20.4 \times 10^{-6}\,\text{Pa} \cdot \text{s}$.

10.8 A high viscosity liquid with $c_p = 1.9\,\text{kJ/kg} \cdot \text{K}$ and mass flow rate 0.6 kg/s enters a heat exchanger at 75°C having an inlet pressure of 3.1 MPa. This liquid is cooled to 35°C by water having an inlet temperature of 18°C and a mass flow rate of 2.1 kg/s. No significant fouling should be expected. Select the most feasible heat exchanger type using a cost estimate.

11 Thermodynamic Modeling and Analysis

The main objectives of this chapter are twofold: (1) to present and discuss important factors that affect heat exchanger performance, and (2) to introduce a basic analysis for the thermodynamic design and optimization of heat exchangers. A quest for answers regarding the first objective will help us to identify the important factors that affect heat exchanger effectiveness, to quantify the effects of these factors, and to provide guidelines for a qualitative assessment of the effectivenesses of the exchangers with different flow arrangements but with a given, identical design task. The second objective is to define a figure of merit for assessing the thermodynamic efficiency of a heat exchanger and to present an approach to thermoeconomic considerations.

In Section 11.1, the differences between a heat exchanger as a component and as part of a system are identified. In Section 11.2, a detailed modeling of a heat exchanger using energy balances only (i.e., the first law of thermodynamics) is provided for the determination of heat exchanger effectiveness and temperature distributions. In Section 11.3, a combined approach based on both the first and second laws of thermodynamics is introduced to quantify inherent irreversibilities in a heat exchanger. The most important source of irreversibility is heat transfer across the finite temperature differences, which is discussed first. Fluid mixing and fluid friction, as additional sources of irreversibility, are studied next. A temperature cross phenomenon is then discussed in detail in Section 11.4 by evaluating entropy generation in a 1–2 TEMA J shell-and-tube heat exchanger. Using all the analysis tools presented in the first four sections, a heuristic approach to an assessment of heat exchanger effectiveness is developed in Section 11.5. In Section 11.6, energy, exergy, and cost balances important for analysis and optimization of heat exchangers are presented. Finally, a thermodynamic criterion for evaluation/selection of heat transfer surfaces is summarized in Section 11.7.

11.1 INTRODUCTION

Traditionally, modeling of a heat exchanger is based on energy balances (i.e., on the consequences of both the first law of thermodynamics and the mass conservation principle), so only the concepts of heat transfer rate and enthalpy rate change would

735

suffice for such an analysis.[†] For an adiabatic heat exchanger (see the assumptions in Section 3.2.1), the enthalpy rate change of one fluid stream must be equal to the enthalpy rate change of the other, being at the same time equal to the exchanger heat transfer rate. This simple energy balance statement will be used in subsequent sections in the differential form to model spatial distributions of temperatures of both fluid streams. In Chapter 3, we also used the energy balances (both differential and overall), but only to determine the heat exchanger effectiveness without determining the temperature distributions.

Let us start with a review of the analysis of heat exchanger design methods discussed in Chapter 3 that relies on a relationship that can be presented in generalized form as follows:

$$
\begin{pmatrix} \text{heat transfer} \\ \text{rate } q \end{pmatrix} = \begin{pmatrix} \text{effectiveness/} \\ \text{correction} \\ \text{factor} \end{pmatrix} \times \begin{pmatrix} \text{heat capacity} \\ \text{rate or thermal} \\ \text{conductance} \end{pmatrix} \times \begin{pmatrix} \text{temperature} \\ \text{difference} \end{pmatrix}
$$

$$
= \begin{cases} \varepsilon C_{\min} \, \Delta T_{\max} & \text{in } \varepsilon\text{-NTU method} \\ P_1 C_1 \, \Delta T_{\max} & \text{in } P\text{-NTU method} \\ FUA \, \Delta T_{\mathrm{lm}} & \text{in LMTD method} \\ \psi UA \, \Delta T_{\max} & \text{in } \psi\text{-NTU method} \end{cases} \tag{11.1}
$$

Equation (11.1) is based on energy balances formulated as consequences of the first law of thermodynamics. It is important to note that each of the methods implied by Eq. (11.1) assumes the determination of either an effectiveness factor (heat exchanger effectiveness ε or temperature effectiveness P_1) or a correction factor (F or ψ) as a function of design parameters, e.g., NTU and C^*. We use the expression *factor* as a generic term to indicate a common first law of thermodynamics origin for both the effectiveness and correction factors. Of course, the correction factors do not have the same physical meaning as the effectiveness factors. Each relationship in Eq. (11.1) involves a temperature difference, either the maximum imposed temperature difference $\Delta T_{\max}$ or the logarithmic-mean temperature difference ΔT_{lm}. It has been demonstrated in Sections 3.3 and 3.5 that relationships between effectiveness factors and the pertinent design parameters can be devised from the heat transfer model of a heat exchanger. In some cases, these relationships can even be obtained without a detailed study of internal heat transfer interactions. Moreover, a designer who already has these relationships would be able to calculate the effectiveness for the given set of parameters and execute a design method procedure *without* a need to study temperature distributions of a selected flow arrangement as outlined in Sections 3.9 and 9.2 through 9.5. So one would treat the heat exchanger as a black box for determination of the overall heat transfer surface area or heat transfer performance of an exchanger as a component. In such a case, there is no need to know temperature distributions. For example, for any exchanger (for which the

---

[†] In our analysis, the enthalpy rate change is the rate change of enthalpy of a fluid stream caused by heat transfer interaction between the two fluid streams. From the thermodynamic point of view, note that the heat transfer rate is an energy interaction (not a change in the fluid property), while the enthalpy rate change represents a change of fluid property caused by existing heat interaction. This distinction is important for understanding of the interpretation of heat exchanger performance and will be emphasized in more detail later through the introduction of several advanced concepts of thermodynamics.

effectiveness relationships are already known or can be determined using the matrix formalism mentioned in Section 3.11.4), an engineer needs only a relationship between the effectiveness/correction factor and design parameters, without detailed insight into local temperature distributions. Thus, a misleading conclusion may be reached: that the only information a designer should possess concerning a flow arrangement is the relationship between an effectiveness factor and design parameters (e.g., P_1-NTU$_1$, ε-NTU, or F-P relationship).

Analysis presented so far does suffice for a design procedure for a heat exchanger with an already defined effectiveness relationship. However, a very important and still unanswered question should be addressed as well. Why does an effectiveness factor (say, heat exchanger effectiveness) have a high (or low) value for a given flow arrangement (especially for a complex one) compared to the corresponding value for another flow arrangement (for the same set of design parameters)? For example, we do know that a crossflow heat exchanger has less exchanger effectiveness than for a counterflow exchanger (for the same set of design parameters NTU and C^*). The only rational explanation that we can offer at this point (in addition to the intuitive ones) is that the effectiveness relationship for a crossflow exchanger simply provides a smaller numerical value for ε or P for the given heat capacity rate ratio and NTU than does ε or P for a counterflow exchanger. In addition, $\Delta\varepsilon/\Delta$NTU for a fixed heat capacity rate ratio C^* is different for counterflow than for crossflow, for NTU$_{min}$ < NTU < ∞. For NTU $\geq$ 4, this gradient is almost identical. For NTU$_{min}$ $\leq$ 0.4, all flow arrangements provide almost identical effectiveness values for a given set of design parameters [see Eq. (3.89)]. Why is that so? The reasoning will become clear when we present an astonishingly simple heuristic approach based on the second law analysis of exchanger flow arrangements. Also, we present a thermodynamic performance figure of merit, the efficiency of a heat exchanger from a system viewpoint. Consequently, these analysis tools will help us in assessing relative magnitudes of exchanger effectiveness for complex flow arrangements for the selection of an appropriate flow arrangement for a specified task. This understanding will also become valuable in finding an optimum heat exchanger design from a system viewpoint.

11.1.1 Heat Exchanger as Part of a System

Heat exchangers in numerous engineering applications are only one of many components of a system. Thus, the design of a heat exchanger is inevitably influenced by system requirements and should be based on system optimization rather than component optimization. An objective function for such system-based optimization is influenced by the main features of heat exchanger operation. For a given set of input data (e.g., flow rates and inlet temperatures), exchanger geometry, and other pertinent information, the output data (e.g., the outlet temperatures) will depend on heat transfer and fluid flow phenomena that take place within the boundaries of the heat exchanger. So even though one seeks a system optimum, in the process of determining that optimum, one must fully understand the features of the exchanger as a component.

Since heat exchangers are used in many systems, we do not attempt any specific system analysis or process integration. We discuss only the basic thermodynamic aspects. Despite exact mathematical/numerical results obtained through system-based optimization, the designer should know that the heat exchanger design (sizing) problem studied is a complex problem that *has no* single exact solution at all. In all but trivial cases, a designer must deal with uncertainty margins of the input data, in addition to numerous assumptions. Usually, a range of data (say, for a cost analysis in an optimization routine), and not a single set of parameters, must be considered. As shown in Fig. 2.1, for every case considered in the top left box of the problem specification, one arrives at an optimum solution at the end of the process of Fig. 2.1. Hence, there can be many (and not only one) optimum solution for a given exchanger sizing problem as provided by different heat exchanger manufacturers. With that in mind, the reader should understand the limitations of results obtained by modern computer software for design, optimization, and system integration.

11.1.2 Heat Exchanger as a Component

Before a system-based optimization can be carried out, a good understanding of the exchanger as a *component* must be gained. In addition to rating or sizing, this may include information about temperature distributions, local temperature differences, hot and cold spots, pressure drops, and sources of local irreversibilities—all as functions of possible changes of design and/or process variables and/or parameters.

The outlet state variables of the fluids depend on the efficiency of the heat transfer process influenced by fluid flow and heat transfer phenomena within the exchanger. A measure of this efficiency is not defined exclusively by heat exchanger or temperature effectivenesses because it gives relevant but limited information about heat exchanger performance since the influence of irreversibility, as discussed in Section 11.3 is not included. Thus, the key questions to be answered involve how to define exchanger efficiency, and how the heat transfer and fluid flow processes (manifested within the heat exchanger boundaries) affect the exchanger effectiveness and thermodynamic efficiency (see Section 11.6.5). To answer these questions, we first identify the important heat transfer/fluid flow phenomena in the operation of a heat exchanger having an arbitrary flow arrangement in Section 11.3.

Design of a heat exchanger as a component is to a large extent an engineering art. So, despite high sophistication in heat exchanger thermal modeling, some of the final decisions (in particular those related to optimization) are based on qualitative judgments due to nonquantifiable variables associated with exchanger manufacturing and other evaluation criteria. Still, analytical modeling—a very valuable tool—is crucial to *understanding* the relevant thermal–hydraulic phenomena and design options and various venues for design improvements. In structuring this chapter, special attention is devoted to a balanced use of both rigorous mathematical modeling (Sections 11.2 through 11.4) and qualitative analysis and heuristic judgments (Section 11.5). The results based on mathematical modeling, although elegant and transparent concerning the influences involved, always carry within them all consequences of numerous assumptions and often simplifications. So the primary purpose of our study in this chapter is to gain a good understanding of the factors that affect exchanger performance, not necessarily to provide new tools for design and system-based optimization of a heat exchanger.

11.2 MODELING A HEAT EXCHANGER BASED ON THE FIRST LAW OF THERMODYNAMICS

An important objective of the material presented in this section is to learn how to model a heat exchanger to determine temperature distributions. In Chapter 3, we focused on the determination of the exchanger efficiency factor ε, P, and F [see Eq. (11.1)], through various heat exchanger basic design methods; we did not pay any particular attention to fluid temperature distributions and their relationship to exchanger performance. Let us now consider the distribution of local temperatures and temperature differences in a heat exchanger having simple counterflow and parallelflow arrangements. Both these arrangements correspond to two limiting cases of the same *geometrical* situation: The two fluid streams are flowing in geometrically parallel orientation but in opposite or same directions to each other, thus providing the largest and smallest heat exchanger effectiveness values (see Figs. 3.7 and 3.8 and Table 3.3). It is assumed, by definition, that both fluids change their respective temperatures only in flow directions (i.e., the local temperature distribution is uniform for a fluid across a flow cross section). Subsequently, we consider a more complex situation with a cross flow of working fluids and the possibility of local mixing along the flow direction.

11.2.1 Temperature Distributions in Counterflow and Parallelflow Exchangers

In Fig. 11.1 a schematic of a counterflow heat exchanger is presented. The assumptions formulated in Section 3.2.1 are invoked here. In general, flow directions may be either in the positive (in Fig. 11.1, from left to right) or negative direction in relation to the axial coordinate (i.e., counterflow/parallelflow). The energy balances for the respective control volumes can be written as follows using the first law of thermodynamics and following rigorously the standard sign convention for heat/enthalpy rate flows across the control volume boundary (positive if entering into the system and negative if leaving the system).

For fluid 1 only (the elementary control volume in Fig. 11.1a):

$$\dot{i}_1(\dot{m}c_p)_1 T_1 - \dot{i}_1(\dot{m}c_p)_1 \left(T_1 + \frac{dT_1}{dx}\, dx \right) - U(T_1 - T_2)\, dA = 0 \qquad (11.2)$$

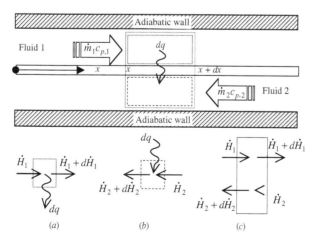

FIGURE 11.1   Energy balance control volumes for a counterflow arrangement. The control volumes are represented by rectangular areas. (*a*) Control volume for fluid 1 differential energy balance; (*b*) control volume for fluid 2 differential energy balance; (*c*) control volume for both fluids 1 and 2 differential energy balance. Enthalpy rates $\dot{H}_j = (\dot{m}c_p)_j T_j$, $j = 1$ or 2.

where $i_1 = +1$ or -1, for the same or opposite (positive or negative) direction of fluid 1 with respect to the positive direction of the x axis, respectively.

For fluid 2 only (the elementary control volume in Fig. 11.1b):

$$i_2(\dot{m}c_p)_2 T_2 - i_2(\dot{m}c_p)_2 \left(T_2 + \frac{dT_2}{dx} \, dx \right) + U(T_1 - T_2) \, dA = 0 \tag{11.3}$$

where $i_2 = +1$ or -1, for the same or opposite (positive or negative) direction of fluid 2 with respect to the positive direction of the x axis, respectively.

For both fluids 1 and 2 (the elementary control volume in Fig. 11.1c):

$$i_1 \left[(\dot{m}c_p)T - (\dot{m}c_p)\left(T + \frac{dT}{dx} \, dx \right) \right]_1 + i_2 \left[(\dot{m}c_p)T - (\dot{m}c_p)\left(T + \frac{dT}{dx} \, dx \right) \right]_2 = 0 \tag{11.4}$$

Note that an assumption of uniform distribution of the total heat transfer surface area A along the flow length L means that $dA/dx = A/L$; that is, $U \, dA = (UA/L) \, dx$. Let us, without restricting the model generality, fix the direction of fluid 1 to be in a positive axial direction ($i_1 = +1$) while $i_2 = \pm 1$ [i.e., $i_2 = -1$ for counterflow (see Fig. 11.1), or $i_2 = +1$ for parallelflow]). Rearranging Eqs. (11.2)–(11.4), we obtain

$$(\dot{m}c_p)_1 \frac{dT_1}{dx} = \frac{UA}{L} (T_2 - T_1) \tag{11.5}$$

$$i_2(\dot{m}c_p)_2 \frac{dT_2}{dx} = \frac{UA}{L} (T_1 - T_2) \tag{11.6}$$

$$(\dot{m}c_p)_1 \frac{dT_1}{dx} + i_2(\dot{m}c_p)_2 \frac{dT_2}{dx} = 0 \tag{11.7}$$

Note that only two of the three balance equations are sufficient to define the two temperature distributions. For example, either Eqs. (11.5) and (11.6) or Eqs. (11.5) and (11.7) can be utilized. Subsequently, distribution of the temperature difference along the heat exchanger can be determined.

To close the problem formulation, a set of boundary conditions is required at the heat exchanger terminal points. For a parallelflow exchanger, the inlet temperatures for both fluids are known at $x = 0$. For the counterflow exchanger, the known inlet temperatures are on the opposite sides of the exchanger, at $x = 0$ and $x = L$, respectively. Explicitly, these conditions are

$$T_1 = T_{1,i} \quad \text{at } x = 0 \qquad T_2 = T_{2,i} \text{ at } \begin{cases} x = 0 \text{ for parallelflow} \\ x = L \text{ for counterflow} \end{cases} \tag{11.8}$$

Equations (11.5)–(11.8) are made dimensionless with the following variables:

$$\Theta = \frac{T - T_{1,i}}{T_{2,i} - T_{1,i}} \qquad \xi = \frac{x}{L} \tag{11.9}$$

and design parameters NTU_1 and R_1, as defined by Eqs. (3.101) and (3.105). Hence,

$$\frac{d\Theta_1}{d\xi} + NTU_1(\Theta_1 - \Theta_2) = 0 \tag{11.10}$$

$$\frac{d\Theta_2}{d\xi} - i_2 NTU_1 R_1 (\Theta_1 - \Theta_2) = 0 \tag{11.11}$$

$$R_1 \frac{d\Theta_1}{d\xi} + i_2 \frac{d\Theta_2}{d\xi} = 0 \tag{11.12}$$

The boundary conditions are as follows:

$$\Theta_1 = 0 \quad \text{at } \xi = 0 \tag{11.13}$$

$$\Theta_2 = 1 \quad \text{at } \begin{cases} \xi = 0 & \text{for parallelflow} \\ \xi = 1 & \text{for counterflow} \end{cases} \tag{11.14}$$

The set of relationships given by Eqs. (11.5)–(11.8) or (11.10)–(11.14) define a mathematical model of the heat transfer process under consideration in terms of temperature distributions for both fluids. For example, one can solve Eqs. (11.5), (11.7), and (11.8) to obtain temperature distributions (as presented in Figs. 1.50 and 1.52). This can be done for virtually any combination of design parameters (NTU_1 and R_1) for both parallelflow and counterflow arrangements without a need for separate mathematical models.

Some mathematical aspects of the solution procedure and thermodynamic interpretation of the results will be addressed in the examples that follow. A rigorous and unified solution of the parallelflow heat exchanger problem defined above is provided in Example 11.1 (Sekulić, 2000). The relation between the heat exchanger and/or temperature effectiveness as a dimensionless outlet temperature of one of the fluids and a thermodynamic interpretation of these figures of merit is emphasized in Example 11.2. An approach to modeling more complex situations, such as a 1–2 TEMA J shell-and-tube heat exchanger or various crossflow arrangements, is left for an individual exercise (see Problems 11.1 through 11.10 at the end of the chapter).

Example 11.1 Determine temperature distributions of two parallel fluid streams in thermal contact. The fluid streams have constant mass flow rates and constant but different inlet temperatures. Show that a *unified* solution procedure can be formulated for both parallelflow and counterflow arrangements.

SOLUTION

Problem Data and Schematic: The two fluid streams flow in a parallel geometric orientation as presented in Fig. E11.1A. Both parallelflow and counterflow arrangements are considered (i.e., fluid 2 can be in either one of two opposite directions). Inlet temperatures, mass flow rates, and fluid properties are known. All the geometric characteristics of the flow passages are defined as well.

Determine: The local temperatures of both fluids as functions of the axial distance along the fluid flow direction.

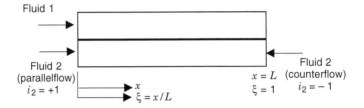

FIGURE E11.1A   Fluid flow orientations in counterflow and parallelflow heat exchangers.

Assumptions: It is assumed that thermal interaction between the fluids takes place under the assumptions described in Section 3.2.1

Analysis: Any two of three differential balances given by Eqs. (11.10)–(11.12) together with the boundary conditions given by Eqs. (11.13) and (11.14) describe the theoretical model for analysis. Let us define the model of this heat transfer process using Eqs. (11.10) and (11.12)–(11.14).

A general solution will be obtained utilizing the Laplace transforms method (Sekulić, 2000), although several other methods can be used as well (see Section 3.11). The rationale for using this particular method is that it can be applied efficiently to a number of more complex situations, such as for a crossflow arrangement with both fluids unmixed (see Problem 11.2).

Applying Laplace transforms to Eq. (11.10) yields

$$\mathscr{L}\left\{\frac{d\Theta_1}{d\xi} + \text{NTU}_1(\Theta_1 - \Theta_2)\right\}_{\xi \to s} = 0 \tag{1}$$

Using Laplace transforms rules, Eq. (1) is reduced to

$$s\bar{\Theta}_1(s) - \Theta_1(0) + \text{NTU}_1\left[\bar{\Theta}_1(s) - \bar{\Theta}_2(s)\right] = 0 \tag{2}$$

Variables $\bar{\Theta}_j(s)$, $j = 1$, 2, in Eq. (2) represent the Laplace transforms of the yet unknown temperature distributions $\Theta_j(\xi)$, $j = 1$, 2. A complex independent variable denoted as s replaces the original independent variable ξ. Knowing the inlet boundary condition at $\xi = 0$ [i.e., $\Theta_1(0) = 0$], Eq. (2) is solved with respect to the Laplace transform $\bar{\Theta}_1(s)$ as follows:

$$\bar{\Theta}_1(s) = \frac{\text{NTU}_1}{s + \text{NTU}_1} \bar{\Theta}_2(s) \tag{3}$$

In Eq. (3), an explicit form of $\bar{\Theta}_2(s)$ still has to be determined. This can be done involving the other differential equation of the mathematical model [i.e., Eq. (11.12)]. The same procedure as the one just outlined is applied. Hence,

$$\mathscr{L}\left\{R_1 \frac{d\Theta_1}{d\xi} + i_2 \frac{d\Theta_2}{d\xi}\right\} = 0 \tag{4}$$

and

$$R_1[s\bar{\Theta}_1(s) - \Theta_1(0)] + i_2[s\bar{\Theta}_2(s) - \Theta_2(0)] = 0 \tag{5}$$

Solving Eq. (5) for $\bar{\Theta}_2(s)$ and utilizing Eq. (3) for $\bar{\Theta}_1(s)$, we get

$$\bar{\Theta}_2(s) = \frac{s + \mathrm{NTU}_1}{s^2 + s\mathrm{NTU}_1(1 + i_2 R_1)} \Theta_2(0) \tag{6}$$

Substitute Eq. (6) into Eq. (3) to get $\bar{\Theta}_1(s)$ explicitly in terms of s.

Now, applying inverse Laplace transforms on Eqs. (3) and (6), we get (for $R_1 \neq 1$ if $i_2 = 1$)

$$\Theta_1(\xi) = \mathscr{L}^{-1}\{\bar{\Theta}_1(s)\} = \Theta_2(0) \frac{1 - e^{-\xi \mathrm{NTU}_1(1 + i_2 R_1)}}{1 + i_2 R_1} \tag{7}$$

and

$$\Theta_2(\xi) = \mathscr{L}^{-1}\{\Theta_2(s)\} = \Theta_2(0) \frac{1 + i_2 R_1 e^{-\xi \mathrm{NTU}_1(1 + i_2 R_1)}}{1 + i_2 R_1} \tag{8}$$

Parameter $\Theta_2(0)$ in Eqs. (7) and (8) depends on both design parameters (NTU_1, R_1) and the value of i_2. The value of $\Theta_2(0)$ can be determined for the parallelflow arrangement ($i_2 = +1$) directly from the boundary condition at the fluid 2 inlet [i.e., $\Theta_2(0) = \Theta_{2,i} = 1$]. For the counterflow arrangement (for $i_2 = -1$), the value of $\Theta_2(0)$ can be obtained by collocating Eq. (8) at the fluid 2 inlet (i.e., at $\xi = 1$), and solving for $\Theta_2(0)$. Consequently,

$$\Theta_2(0) = \frac{1 - R_1}{1 - R_1 e^{-\mathrm{NTU}_1(1 - R_1)}} \qquad \text{for } i_2 = -1 \tag{9}$$

By inspection of Eq. (9), a generalized algebraic expression for the parameter $\Theta_2(0)$ can be formulated to extend the validity of that equation to include parallelflow as follows:

$$\Theta_2(0) = \frac{1 - R_1}{1 - R_1 e^{-(1/2)(1 - i_2)\mathrm{NTU}_1(1 - R_1)}}$$

$$= \begin{cases} 1 & \text{for } i_2 = +1 \text{ (parallelflow)} \\ \dfrac{1 - R_1}{1 - R_1 e^{-(1 - R_1)\mathrm{NTU}_1}} & \text{for } i_2 = -1 \text{ (counterflow)} \end{cases} \tag{10}$$

In Eq. (10), one should first define the fluid stream direction parameter (i.e., $i_2 = \pm 1$), and then select the numerical values for design parameters.

Finally, combining Eqs. (7), (8), and (10), the general solution for temperature distributions for both parallelflow and counterflow exchangers can be written as follows:

$$\Theta_1(\xi) = \frac{1 - R_1}{1 + i_2 R_1} \frac{1 - e^{-\xi NTU_1(1+i_2 R_1)}}{1 - R_1 e^{-(1/2)(1-i_2)NTU_1(1-R_1)}}$$

$$\Theta_2(\xi) = \frac{1 - R_1}{1 + i_2 R_1} \frac{1 + i_2 R_1 e^{-\xi NTU_1(1+i_2 R_1)}}{1 - R_1 e^{-(1/2)(1-i_2)NTU_1(1-R_1)}} \tag{11}$$

Inserting $i_2 = +1$ for parallelflow and $i_2 = -1$ for counterflow into Eq. (11), we get the following temperature distributions:

Flow Arrangement	Flow Indicator i_2	$\Theta_1(\xi)$	$\Theta_2(\xi)$
Parallelflow	$+1$	$\dfrac{1 - e^{-\xi NTU_1(1+R_1)}}{1 + R_1}$	$\dfrac{1 + R_1 e^{-\xi NTU_1(1+R_1)}}{1 + R_1}$
Counterflow	-1	$\dfrac{1 - e^{-\xi NTU_1(1-R_1)}}{1 - R_1 e^{-NTU_1(1-R_1)}}$	$\dfrac{1 - R_1 e^{-\xi NTU_1(1-R_1)}}{1 - R_1 e^{-NTU_1(1-R_1)}}$

The temperature distributions for both parallelflow and counterflow arrangements and for several sets of parameters are presented in Fig. E11.1B.

All the results presented so far for $i_2 = -1$ require that $0 \le R_1 < 1$ (i.e., $R_1 \ne 1$). In the case of a balanced counterflow heat exchanger, $R_1 = 1$ and $i_2 = -1$, the original mathematical model given by the set of equations (11.10) and (11.12) transforms into

$$\frac{d\Theta_1}{d\xi} = \frac{d\Theta_2}{d\xi} = NTU_1(\Theta_2 - \Theta_1) \tag{12}$$

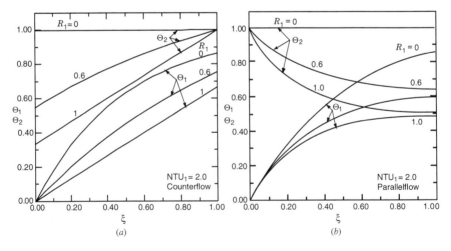

FIGURE E11.1B Temperature distributions in (a) counterflow and (b) parallelflow heat exchangers.

with the boundary conditions as given by Eqs. (11.13) and (11.14) for counterflow. The solution of this problem leads to linear dimensionless temperature distributions. The temperature distributions for fluids 1 and 2 are

$$\Theta_1(\xi) = \frac{NTU_1\xi}{1+NTU_1} \qquad \Theta_2(\xi) = \frac{NTU_1\xi + 1}{1+NTU_1} \tag{13}$$

Discussion and Comments: In this example, it has been shown how to find temperature distributions in a heat exchanger with parallel streams (in geometrical sense). This example demonstrates that both parallelflow and counterflow arrangements represent the two subproblems of a single heat transfer problem that differ only in the stream direction. Still, the character of temperature difference distributions in these two situations is radically different, as shown in Fig. 11.2. Note that this modeling is based on energy balances performed on two control volumes selected arbitrarily from a total of three balances (either for one or the other fluid, or for both fluids). All results of this modeling are the consequence of conservation principles.

11.2.2 True Meaning of the Heat Exchanger Effectiveness

In Section 3.3.1, heat exchanger effectiveness and the maximum possible heat transfer rate q_{max} are introduced by definition [see Eqs. (3.37) and (3.42)]. This has provided the basis for the formulation of heat exchanger effectiveness in terms of terminal temperatures of the fluids and their heat capacity rates as in Eq. (3.44). However, that approach requires a priori definition of a hypothetical infinite surface area of the heat exchanger. On the other hand, by knowing the temperature distributions of a given heat exchanger, we can devise the concept of heat exchanger effectiveness *without* invoking the concept of a hypothetical counterflow heat exchanger of infinite surface. We can show that the definition of heat exchanger effectiveness is obtained using the first law of thermodynamics only (Sekulić, 2000), without invoking explicitly the second law of thermodynamics. The true meaning of heat exchanger effectiveness as a dimensionless outlet temperature of the fluid stream having the smaller heat capacity rate is a direct consequence of this interpretation. Moreover, the maximum possible heat transfer rate

TABLE 11.1   Interpretations of the Meaning of Heat Exchanger Effectiveness

Traditional meaning	$\varepsilon = \dfrac{q}{q_{max}} = \dfrac{C_h(T_{h,i}-T_{h,o})}{C_{min}(T_{h,i}-T_{c,i})}$ $= \dfrac{C_c(T_{c,o}-T_{c,i})}{C_{min}(T_{h,i}-T_{c,i})}$	Based on a comparison of the actual heat transfer rate exchanged in the heat exchanger to that exchanged in an ideal, hypothetical heat exchanger having $UA \to \infty$	Defined utilizing the first law of thermodynamics explicitly, and the second law of thermodynamics implicitly
True meaning	$\varepsilon = \dfrac{T_{1,o}-T_{1,i}}{T_{2,i}-T_{1,i}}$ $C_1 = C_{min}$	Dimensionless outlet temperature of a fluid with smaller heat capacity rate, $C_1 < C_2$	The first law of thermodynamics is used explicitly

exchanged in a heat exchanger can subsequently be derived, not postulated. The two interpretations are summarized in Table 11.1.

In Example 11.2, these important thermodynamic consequences of the analysis of temperature distributions in a heat exchanger will be illustrated using both parallelflow and counterflow heat exchangers as an example. It should be reiterated that the interpretation given is universally valid, regardless of the complexity of the flow arrangement involved.

Example 11.2   Show that heat exchanger effectiveness and/or temperature effectiveness represent the nondimensional outlet temperature of one of the two fluid streams of a heat exchanger. Use the parallelflow and counterflow arrangements as examples. Demonstrate that heat exchanger effectiveness can be interpreted as a ratio of actual heat transfer rate to the heat transfer rate of a hypothetical exchanger with an infinitely large thermal size (NTU $\rightarrow \infty$), as emphasized in Chapter 3, but without invoking explicitly the second law of thermodynamics.

SOLUTION

Problem Data and Schematic: As presented in Example 11.1.

Determine: Demonstrate the equivalence between the heat exchanger effectiveness definition and the dimensionless outlet temperature of one of the fluids in a heat exchanger.

Assumptions: As invoked in Example 11.1.

Analysis: The outlet temperature of fluid 1 for parallelflow and counterflow can be obtained from Eq. (11) of Example 11.1 at $\xi = 1$ (the outlet of fluid 1).

$$\Theta_1(1) = \frac{1 - R_1}{1 + i_2 R_1} \frac{1 - e^{-\mathrm{NTU}_1(1 + i_2 R_1)}}{1 - R_1 e^{-(1/2)(1 - i_2)\mathrm{NTU}_1(1 - R_1)}} \tag{1}$$

Substituting $i_2 = +1$ for parallelflow, we obtain

$$\Theta_1(1) = P_1 = \frac{1 - e^{-\mathrm{NTU}_1(1 + R_1)}}{1 + R_1} \tag{2}$$

Substituting $i_2 = -1$ for counterflow, we obtain

$$\Theta_1(1) = P_1 = \frac{1 - e^{-\mathrm{NTU}_1(1 - R_1)}}{1 - R_1 e^{-\mathrm{NTU}_1(1 - R_1)}} \tag{3}$$

Equations (2) and (3) are identical to Eqs. (I.2.1) and (I.1.1) of Table 3.6. So the dimensionless outlet temperatures are equal to parallelflow (ε_{pf}) and counterflow (ε_{cf}) heat exchanger effectivenesses, respectively. We have obtained expressions for heat exchanger effectiveness without invoking the concept of an "ideal" heat exchanger. Hence, the true meaning of the effectiveness is simply the dimensionless outlet temperature of the fluid with the smaller heat capacity rate (note that $R_1 = C^*$ in this case). This conclusion is general and valid for any flow arrangement.

To devise a traditional definition of heat exchanger effectiveness, let us first determine the outlet temperature of a counterflow heat exchanger in the limit of an infinitely large heat exchanger with thermal size $NTU_1 \rightarrow \infty$:

$$\lim_{NTU_1 \to \infty} \Theta_1(1) = \lim_{NTU_1 \to \infty} \frac{1 - e^{-NTU_1(1-R_1)}}{1 - R_1 e^{-NTU_1(1-R_1)}} = \frac{1 - 0}{1 - 0} = 1 \tag{4}$$

Invoking the definition of the dimensionless temperature from Eq. (11.9), the following result can be obtained from Eq. (4):

$$\lim_{NTU_1 \to \infty} (T_{1,o}) = T_{2,i} \tag{5}$$

As indicated above, $R_1 = C^* \leq 1$, $C_1 \leq C_2$ by definition (i.e., fluid 1 has the smaller heat capacity rate, $C_1 = C_{min}$). In that case, from Eq. (3.103), $NTU_1 = NTU$.

Now the heat transfer rate in a countercurrent heat exchanger of $NTU_1 \rightarrow \infty$ is given as follows using Eq. (5):

$$\lim_{NTU \to \infty} q = \lim_{NTU \to \infty} [(\dot{m}c_p)_1(T_{1,o} - T_{1,i})] = (\dot{m}c_p)_1 |(T_{2,i} - T_{1,i})| = q_{max} \tag{6}$$

The actual heat transfer rate in a two-fluid single-phase heat exchanger of any flow arrangement is as follows:

$$q = (\dot{m}c_p)_1 |T_{1,o} - T_{1,i}| \tag{7}$$

Finally, dividing the right-hand side of Eq. (7) with q_{max} from Eq. (6), and comparing the result with the definition of the outlet temperature of fluid 1, one can obtain

$$\frac{q}{q_{max}} = \varepsilon = \frac{T_{1,o} - T_{1,i}}{T_{2,i} - T_{1,i}} = \Theta_1(1) = P_1 \tag{8}$$

This constitutes the proof required. Note that for a counterflow arrangement with $R_1 = C^* = 1$, the heat exchanger effectiveness becomes $\varepsilon = \Theta_1(1) = NTU/(1 + NTU)$; see Example 11.1 for the corresponding temperature distributions.

Discussion and Comments: The heat exchanger/temperature effectiveness has its true meaning as a dimensionless outlet temperature of the fluid with the smaller heat capacity rate (heat exchanger effectiveness, ε) or the dimensionless outlet temperature of a given fluid (temperature effectiveness, say P_1 for fluid 1). The numerical value of the heat exchanger effectiveness is between 0 and 1 and indicates how close the outlet temperature of one fluid can approach the inlet temperature of the other fluid. The traditional meaning of the heat exchanger effectiveness [although a thermodynamic interpretation based on Eq. (8) is perfectly valid and insightful] involves the concept of a hypothetical "infinitely large counterflow heat exchanger." Refer to Sekulić (2000) for a detailed discussion concerning an analysis of this approach; discussion of a concept of the thermodynamic efficiency for a heat exchanger is given in Section 11.6.5.

11.2.3 Temperature Difference Distributions for Parallelflow and Counterflow Exchangers

Let us now address the magnitude of the local temperature difference between fluids 1 and 2 ($\Delta T = |T_1 - T_2|$) in either a parallelflow or counterflow exchanger.[†] We need this information to better understand the influence of temperature distributions on ε, P, or F.

The distribution of local temperature differences for parallelflow and counterflow heat exchangers can be determined in a general form by utilizing the corresponding temperature distributions: for example, Eq. (11) of Example 11.1. The temperature difference distribution is as follows (see Problem 11.10 for details):

$$\Delta\Theta(\xi) = |\Theta_1(\xi) - \Theta_2(\xi)| = \frac{(1 - R_1)\exp[-\xi\mathrm{NTU}_1(1 + i_2 R_1)]}{1 - R_1 \exp[-(1/2)\mathrm{NTU}_1(1 - R_1)(1 - i_2)]} \qquad (11.15)$$

Note that Eq. (11.15) is valid for both parallelflow ($i_2 = +1$) and counterflow ($i_2 = -1$) arrangements for $0 \le R_1 < 1$. In Fig. 11.2, a graphical representation of Eq. (11.15) is given for both counterflow and parallelflow arrangements and for several

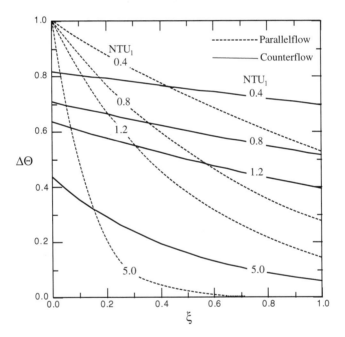

FIGURE 11.2 Temperature difference distributions for a counterflow exchanger and a parallelflow exchanger with $R_1 = 0.6$.

[†] The terms *parallelflow* and *counterflow* are used throughout the book following the practice by Kays and London (1998), which has been extensively used throughout the world for almost last five decades. As demonstrated in this section using a straightforward first law analysis, a more favorable terminology from a semantic point of view would be *unidirectional* and *bidirectional* flows (i.e., in both arrangements, fluid streams are flowing parallel in a geometric sense but oriented in the same or opposite directions).

values of NTU_1 and $R_1 = 0.6$. Refer to Sekulić (2000) for more detailed data. A study of these temperature difference distributions reveals the following conclusions.

- As we have learned in Chapter 3, counterflow is the best and parallelflow the worst flow arrangement from the effectiveness viewpoint for given NTU and C^* (or NTU_1 and R_1). For the same heat capacity rate ratio (say, $R_1 = 0.6$), and in particular at large NTU_1 values, the magnitude of temperature difference change along the flow direction (ξ) is substantially larger for parallelflow than it is for counterflow (compare corresponding curves in Fig. 11.2). The two temperature distributions are identical for $R_1 = 0$ (not shown in Fig. 11.2) and differ more at large heat capacity rate ratios.

- For a given NTU for counterflow, $\Delta\Theta$ features a more pronounced variation along ξ as R_1 (or C^*) decreases from 1 to 0. This means that there will be larger local temperature differences with decreasing R_1 (or C^*) if all other parameters remain the same. If the temperature difference distributions are close to each other or identical, they will have similar or identical effectiveness.

- If the temperature difference distributions differ substantially, the heat exchanger effectiveness will differ considerably as well.

In the limiting case of $R_1 = 1$, the temperature difference distribution for counterflow arrangement is uniform throughout the exchanger, $\Delta\Theta = \Theta_2 - \Theta_1 = 1/(1 + NTU_1)$, as derived in Example 11.1. For the same heat capacity rate ratio in the parallelflow arrangement, the *change* in local temperature difference is the largest possible. Comparing the corresponding heat exchanger effectiveness (see Figs. 3.7 and 3.8) for these conditions, one can easily conclude that the temperature effectiveness of these two flow arrangements differs the most.

The distribution of local temperature differences has a profound influence on exchanger effectiveness (either ε or P_1). A finite temperature difference between the two fluid streams is the driving potential for heat transfer, but large temperature differences lower the exchanger or temperature effectiveness, and ultimately may contribute to a lower system efficiency, as we demonstrate in subsequent sections.

11.2.4 Temperature Distributions in Crossflow Exchangers

A model of a crossflow arrangement with/without mixing provides a good example of how a simple geometric configuration of two fluid streams in thermal contact may lead to two-dimensional temperature fields with a very complex relationship between the design parameters. In addition, a study of temperature distributions within a crossflow heat exchanger illustrates how mixing may influence an outcome of the heat transfer process. This insight is very important for an assessment of the influence of mixing on a reduction in exchanger thermodynamic performance, as demonstrated in Section 11.5.

In Fig. 11.3, a schematic of the main features of the geometry and control volumes in a crossflow heat exchanger is presented. Fluids 1 and 2 flow perpendicular to each other over a heat transfer surface that separates them. This flow arrangement is discussed in Section 1.6.1.3 (see Fig. 1.53). All the assumptions of Section 3.2.1 are invoked here. Due to heat transfer across the heat transfer surface, both fluids will change their temperatures in either one or both flow directions, depending on the presence or absence of mixing, respectively.

Four distinct situations are possible with respect to mixing of the fluids, as emphasized in Section 1.6.1.3. These are also shown graphically in sketches for Eqs. (II.1)–(II.4) of Table 3.6. In Table 11.2, various temperature distributions are shown to be either one- or two-dimensional [$\theta_j = f(\chi \text{ or } \zeta)$ or $\theta_j = f(\chi, \zeta)$, $j = 1, 2$], depending on fluid mixing. Our goal now is to show how one can formulate the models and subsequently solve them to find these temperature fields and/or corresponding outlet temperatures. As a by-product of this analysis, one can easily devise temperature and/or heat exchanger effectiveness. In Tables 3.3 and 3.6, the formulas for the heat exchanger effectiveness are listed. Here we discuss the analytical models for determining both temperature fields and effectiveness. This will provide an insight into the influence of fluid mixing on heat exchanger effectiveness, discussed in Section 11.3.

Referring to Fig. 11.3, one can write energy balances for control volumes as follows:

$$\underbrace{d\dot{m}_1 c_{p,1} T_1}_{\substack{\text{fluid enthalpy rate} \\ \text{into the control volume}}} \quad \underbrace{- d\dot{m}_1 c_{p,1}\left(T_1 + \frac{\partial T_1}{\partial x}\, dx\right)}_{\substack{\text{fluid enthalpy rate out} \\ \text{of the control volume}}} - \underbrace{dq}_{\substack{\text{heat transfer rate} \\ \text{from fluid to wall}}} = 0 \qquad (11.16)$$

and

$$\underbrace{d\dot{m}_2 c_{p,2} T_2}_{\substack{\text{fluid enthalpy rate} \\ \text{into the control volume}}} + \underbrace{dq}_{\substack{\text{heat transfer rate in} \\ \text{from wall to fluid}}} \quad \underbrace{- d\dot{m}_2 c_{p,2}\left(T_2 + \frac{\partial T_2}{\partial y}\, dy\right)}_{\substack{\text{fluid enthalpy rate out} \\ \text{of the control volume}}} = 0 \qquad (11.17)$$

Note that $d\dot{m}_j c_{p,j} = dC_j$, $j = 1, 2$, in Eqs. (11.16) and (11.17) implies that constant thermophysical properties assumption is invoked. It is assumed that no mixing takes place on either side of the heat transfer surface. Consequently, both fluids will have two-dimensional temperature fields. In Eqs. (11.16) and (11.17), dq represents heat transfer by convection from the hot fluid to the wall; and in the steady-state formulation, that heat will be transferred by conduction through the wall and by convection to the cold fluid.

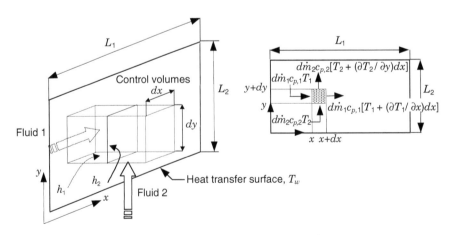

FIGURE 11.3   Energy balance control volumes for crossflow arrangements.

TABLE 11.2 Models of Crossflow Arrangements[a]

Information	Model			
	Unmixed–Unmixed	C_{max} Fluid Unmixed	C_{min} Fluid Unmixed	Mixed–Mixed
Temperature field	$\theta_1 = \theta_1(\chi, \zeta)$ $\theta_2 = \theta_2(\chi, \zeta)$	$\theta_1 = \theta_1(\chi)$ $\theta_2 = \theta_2(\chi, \zeta)$	$\theta_1 = \theta_1(\chi, \zeta)$ $\theta_2 = \theta_2(\zeta)$	$\theta_1 = \theta_1(\chi)$ $\theta_2 = \theta_2(\zeta)$
Differential equations	$\dfrac{\partial \theta_1}{\partial \chi} + \theta_1 = \theta_2$ $\dfrac{\partial \theta_2}{\partial \zeta} + \theta_2 = \theta_1$	$\dfrac{d\theta_1}{d\chi} + \theta_1 = \dfrac{1}{C^* \cdot \mathrm{NTU}} \displaystyle\int_0^{C^* \cdot \mathrm{NTU}} \theta_2 \, d\zeta$ $\dfrac{\partial \theta_2}{\partial \zeta} + \theta_2 = \theta_1$	$\dfrac{\partial \theta_1}{\partial \chi} + \theta_1 = \theta_2$ $\dfrac{d\theta_2}{d\zeta} + \theta_2 = \dfrac{1}{\mathrm{NTU}} \displaystyle\int_0^{\mathrm{NTU}} \theta_1 \, d\chi$	$\dfrac{d\theta_1}{d\chi} + \theta_1 = \dfrac{1}{C^* \cdot \mathrm{NTU}} \displaystyle\int_0^{C^* \cdot \mathrm{NTU}} \theta_2 \, d\zeta$ $\dfrac{d\theta_2}{d\zeta} + \theta_2 = \dfrac{1}{\mathrm{NTU}} \displaystyle\int_0^{\mathrm{NTU}} \theta_1 \, d\chi$
Independent variables	$0 \le \chi \le \mathrm{NTU}$ $0 \le \zeta \le C^* \cdot \mathrm{NTU}$	$0 \le \chi \le \mathrm{NTU}$ $0 \le \zeta \le C^* \cdot \mathrm{NTU}$	$0 \le \chi \le \mathrm{NTU}$ $0 \le \zeta \le C^* \cdot \mathrm{NTU}$	$0 \le \chi \le \mathrm{NTU}$ $0 \le \zeta \le C^* \cdot \mathrm{NTU}$
Boundary conditions	$\theta_1(0, \zeta) = 1$ $\theta_2(\chi, 0) = 0$	$\theta_1(0) = 1$ $\theta_2(\chi, 0) = 0$	$\theta_1(0, \zeta) = 1$ $\theta_2(0) = 0$	$\theta_1(0) = 1$ $\theta_2(0) = 0$
Schematic	 Fluid 1 → ↑ Fluid 2	 Fluid 1 → ↑ Fluid 2	 Fluid 1 → ↑ Fluid 2	 Fluid 1 → ↑ Fluid 2

[a] The dimensionless variables are defined as follows: $\theta_j = (T_j - T_{2,i})/(T_{1,i} - T_{2,i})$, $\chi = (x/L_1)\mathrm{NTU}$, and $\zeta = (y/L_2)C^* \cdot \mathrm{NTU}$. Horizontal and/or vertical lines in schematic figures symbolize flow of the respective fluid through the flow areas characterized with one-dimensionality (along the flow direction). The absence of lines in the direction of flow symbolizes fluid mixing in the direction transverse to the flow direction. Fluid 1 is assumed to have a smaller heat capacity rate C_{min}.

751

Using the rate equations for convection and conduction, dq from the hot fluid to the cold fluid can be expressed as follows:

$$dq = \underbrace{\eta_{o,1} h_1 (T_1 - T_{w,1}) \, dx \, dy}_{\text{convection from hot fluid to wall}} = \underbrace{k_w \left(\frac{T_{w,1} - T_{w,2}}{\delta_w} \right) dx \, dy}_{\text{conduction within the wall}} = \underbrace{\eta_{o,2} h_2 (T_{w,2} - T_2) \, dx \, dy}_{\text{convection from wall to cod fluid}}$$

$$(11.18)$$

Equation (11.18) indicates that there is neither energy generation nor axial conduction through the separating wall, as idealized in Section 3.2.1. The products $h_j \, \eta_{o,j} \, \Delta T_j$, $j = 1$, 2 of Eq. (11.18) represent the heat transfer rates exchanged per unit heat transfer area between fluids 1 or 2 and the wall separating the fluids. In the case where the thermal resistance of the separating wall is neglected, only convective terms exist in Eq. (11.18), with $T_{w,1} = T_{w,2} = T_w$ (i.e., the heat transfer surface has a uniform wall temperature orthogonal to the flow directions). Note that the balances presented by Eqs. (11.16)–(11.18) *do not* involve the overall heat transfer coefficient. They are formed by applying the thermodynamic convention for each of the thermal energy flow rates (positive if entering the system, otherwise negative). Note also that the assumptions of uniform distribution of the heat transfer area and uniform wall thermal resistance are invoked. Equation (11.18) is rewritten as follows:

$$T_1 - T_{w,1} = \frac{dq}{\eta_{o,1} h_1 \, dx \, dy} \qquad T_{w,1} - T_{w,2} = \frac{dq}{(k_w/\delta_w) \, dx \, dy} \qquad T_{w,2} - T_2 = \frac{dq}{\eta_{o,2} h_2 \, dx \, dy}$$

$$(11.19)$$

Adding up the temperature differences from three equations of Eq. (11.19), defining $dA = dx \, dy$, and using Eq. (3.18) for definition of the overall heat transfer coefficient U, but neglecting the fouling thermal resistances [since we did not include them in the formulation of Eq. (11.18), which we could have included readily if desired], we get

$$dq = U \, dA (T_1 - T_2) \qquad (11.20)$$

Substituting Eq. (11.20) into Eqs. (11.16) and (11.17) and simplifying, we can get the following partial differential equations:

$$\frac{\partial \theta_1}{\partial \chi} + \theta_1 = \theta_2 \qquad \frac{\partial \theta_2}{\partial \zeta} + \theta_2 = \theta_1 \qquad (11.21)$$

where $\theta_j = (T_j - T_{2,i})/(T_{1,i} - T_{2,i})$, with $j = 1$, 2, $\chi = (x/L_1)\text{NTU}$, and $\zeta = (y/L_2)C^* \cdot \text{NTU}$. The number of heat transfer units NTU is based on the heat transfer surface between the fluids defined (for the sake of clarity) as the product of L_1 and L_2 (see Fig. 11.3). Note that UA is distributed uniformly throughout the exchanger due to assumptions of uniformity of heat transfer surface and thermal resistances. The same holds for the heat capacity rates. Hence, NTU is based on any (hot or cold fluid side) heat transfer surface on which U is defined. The definition of dimensionless temperatures θ_j is complementary to the definition of the dimensionless temperature Θ of Eq. (11.9). This flexibility in defining dimensionless temperature allows an analyst to define the heat exchanger effectiveness as either a dimensionless outlet temperature of the fluid with smaller heat capacity rate or as its complementary value. Note also that both dimensionless temperatures θ_j are assumed to be locally dependent on both independent coordinates χ and ζ:

$$\theta_1 = \theta_1(\chi, \zeta) \qquad \theta_2 = \theta_2(\chi, \zeta) \tag{11.22}$$

where the ranges of independent variables take the values

$$0 \leq \chi \leq \text{NTU} \qquad 0 \leq \zeta \leq C^* \cdot \text{NTU} \tag{11.23}$$

Two boundary conditions (for uniform inlet temperatures) accompany the set of Eqs. (11.21):

$$\theta_1(0, \zeta) = 1 \qquad \theta_2(\chi, 0) = 0 \tag{11.24}$$

The set of equations Eqs. (11.21) and (11.24) represents the mathematical model of a crossflow heat exchanger. Four particular cases of crossflow (see Section 1.6.1.3, and Tables 3.6 and 11.2) differ from each other with respect to the presence or absence of fluid mixing on each fluid side within the heat exchanger core (see Problems 11.5 and 11.6). In Table 11.2, a summary containing all four models, deduced from Eqs. (11.21)–(11.24), is presented. Each of these models can be solved and closed-form analytical solutions can be obtained using various solution methods (see Section 3.11). The solution of the general unmixed–unmixed case is asked in Problem 11.2. A particular case of an unmixed–mixed crossflow arrangement is considered in detail in the following example. The mixed–mixed case is considered in Problem 11.7.

Example 11.3 Determine temperature difference fields in a heat exchanger with a mixed–unmixed crossflow arrangement. Assume that the fluid with the smaller heat capacity rate is mixed.

SOLUTION

Problem Data and Schematic: The flow arrangement under consideration corresponds to the model and schematic given in the third column of Table 11.2 ($C_{\max}$ fluid unmixed).

Determine: Temperature difference as a function of both axial and transverse coordinates (χ and ζ, as defined in Table 11.2).

Assumptions: The assumptions are as presented in Section 3.2.1.

Analysis: We first determine the temperature fields for both fluids followed by a temperature difference distribution relationship within the heat exchanger core. The analytical model consists of two equations, one partial and one ordinary differential equation, as presented in the third column of Table 11.2 (the details of the model development are the subject of Problem 11.6). Uniform temperatures are considered at inlets as corresponding boundary conditions. The solution to this analytical model will provide the desired temperature fields. Let us first solve the partial differential equation for fluid 2. Subsequently, using the temperature field solution for fluid 2 and replacing its explicit form in the ordinary differential equation for fluid 1, we will find the temperature distribution for fluid 1. Finally, the difference between the two fluid temperatures will provide the solution of the problem.

The solution of the partial differential equation for fluid 2 from Table 11.2 can be obtained as follows using the Laplace transforms technique:

$$\mathscr{L}\left\{\frac{\partial\theta_2}{\partial\zeta} + \theta_2\right\}_{\zeta\to s} = \mathscr{L}\{\theta_1\}_{\zeta\to s} \tag{1}$$

$$s\bar{\theta}_2(\chi,s) + \theta_2(\chi,0) + \bar{\theta}_2(\chi,s) = \frac{\theta_1(\chi)}{s} \tag{2}$$

where s represents a complex variable that replaces ζ. Rearranging Eq. (2) with $\theta_2(\chi,0) = 0$ from Table 2, we get

$$\bar{\theta}_2(\chi,s) = \frac{\theta_1(\chi)}{s(s+1)} \tag{3}$$

An inverse Laplace transform of Eq. (3) provides the temperature field for fluid 2:

$$\mathscr{L}^{-1}\{\theta_2(\chi,s)\}_{s\to\zeta} = \theta_2(\chi,\zeta) = \theta_1(\chi)(1 - e^{-\zeta}) \tag{4}$$

Note that the explicit form of $\theta_1(\chi)$ still has to be determined. The ordinary differential equation for fluid 1 of Table 11.2 can now be written as follows:

$$\frac{d\theta_1(\chi)}{d\chi} + \theta_1(\chi) = \frac{1}{C^* \cdot \mathrm{NTU}} \int_0^{C^* \cdot \mathrm{NTU}} \theta_1(\chi)(1 - e^{-\zeta})\, d\zeta \tag{5}$$

After determining the integral on the right-hand side of Eq. (5) and rearranging, the differential equation for the fluid 1 temperature distribution becomes

$$\frac{d\theta_1(\chi)}{d\chi} + \mathscr{K}\theta_1(\chi) = 0 \tag{6}$$

where $\mathscr{K} = [1 - \exp(-C^* \cdot \mathrm{NTU})]/(C^* \cdot \mathrm{NTU})$. The boundary condition for Eq. (6) is

$$\theta_1(0) = 1 \tag{7}$$

The solution of a simple problem defined by Eqs. (6) and (7) is

$$\theta_1(\chi) = e^{-\mathscr{K}\chi} \tag{8}$$

So the temperature field of fluid 2 can be obtained by introducing the temperature distribution of fluid 1 given by Eq. (8) into Eq. (4):

$$\theta_2(\chi,\zeta) = (1 - e^{-\zeta})e^{-\mathscr{K}\chi} \tag{9}$$

Finally, the relationship for the temperature difference distribution can easily be determined from Eqs. (8) and (9) as

$$\Delta\theta(\chi,\zeta) = \theta_1(\chi) - \theta_2(\chi,\zeta) = e^{-(\mathscr{K}\chi+\zeta)} \tag{10}$$

Discussion and Comments: As expected, the temperature distribution for unmixed fluid 2 is two-dimensional and that for mixed fluid 1 is one-dimensional, both dependent on $\mathscr{X}$, which in turn depends on NTU and C^*. Knowing the temperature distribution of the fluid with the smaller heat capacity rate (fluid 1 in this case), one can easily determine heat exchanger effectiveness (solve Problem 11.3 for understanding the details). Similarly, the analysis of a crossflow arrangement for the mixed fluid having a larger heat capacity rate can be performed, and heat exchanger effectiveness can subsequently be determined (see Problem 11.4). Even more complex situations with nonuniform inlet temperatures are the subject of Problems 11.8 and 11.9.

11.3   IRREVERSIBILITIES IN HEAT EXCHANGERS

Important phenomena that shape the heat transfer and flow characteristics within a heat exchanger are (1) heat transfer at finite temperature differences, (2) mixing and/or splitting of the fluid streams, and (3) fluid flow friction phenomena; additional phenomena when present are phase change, flow throttling, and so on. The first two phenomena influence temperature distributions, and the third the flow friction characteristics on each fluid side of a heat exchanger. Thermodynamics teaches us (Bejan, 1988) that these processes are accompanied by entropy generation, an indicator of undesirable thermodynamic irreversibilities that diminish the thermal performance. So thermodynamic irreversibility is an inevitable by-product of these processes and a principal cause of exchanger/system performance deterioration. Some of the irreversibilities associated with heat transfer and fluid flow are (Gregorig, 1965; Sontag and Van Wylen, 1982):

- Heat transfer across a finite temperature difference (including both heat transfer between the fluids and heat transfer across the heat exchanger boundary, i.e., heat leak and/or gain to/from surroundings)
- Mixing of "dissimilar" fluids (dissimilar with respect to p, T, and/or composition)
- Fluid friction and flow impact
- Phase change where initial conditions are not in equilibrium
- Flow throttling

In this section we focus on identification and quantitative evaluation of the three dominant irreversibilities in a heat exchanger: (1) irreversibility caused by a finite temperature difference, (2) irreversibility caused by fluid mixing, and (3) irreversibility caused by fluid friction. We evaluate these irreversibilities in terms of entropy generation. This analysis will assist us in assessing the quality of heat transfer and associated phenomena in heat exchangers that cannot be evaluated and explained by the analysis presented in Sections 3.3 through 3.8. This analysis requires simultaneous use of both the first and second laws of thermodynamics and introduction of the concept of exergy (see Section 11.6.4).

Let us first review very briefly the concepts of irreversibility, entropy, entropy generation, and exergy before we start the foundation of thermodynamic analysis of heat exchangers. For further details on these concepts and related thermodynamics background, refer to Bejan (1988) and Moran and Shapiro (1995).

Thermodynamic irreversibility (simply referred to as *irreversibility*) is a term used to describe a natural tendency of any real system not to be able to revisit the same sequence of states during a reverse change of state from the final to the initial state without additional energy interaction(s). An additional energy interaction is due to an absence of *reversibility* in a real world when thermal phenomena are involved. In practical terms, this means that the presence of irreversibilities is accompanied by thermodynamic losses, ultimately leading to poorer thermal performance than predicted by an idealized reversible process. In a narrower sense, the same term is used to describe the losses in energy terms $\dot{I}_{irr}$ caused by the presence of irreversibilities. The value of irreversibility cannot be negative and it is *not* a system property as discussed next.

Irreversibility can be expressed in energy terms as a product of *entropy generation* $\dot{S}_{irr}$ and a *temperature weighting factor* T_o (i.e., $\dot{I}_{irr} = T_o \dot{S}_{irr}$). It can be shown that in many engineering applications the weighting factor can be interpreted as the temperature of the surroundings, which is identified as a thermodynamic reference state for measuring the thermal energy potential of the system at hand. It should be noted that entropy generation is not a property of a system, while the *entropy* S is. Entropy is defined as a system property by a statement that its change in an ideal, reversible process must be equal to the transfer of an entity $\int dq/T$ that accompanies any heat transfer dq across the system boundary where the local temperature is T. Hence, this abstract system property indicates that heat transfer must be accompanied by an entropy change. As a consequence, a reversible adiabatic process can be identified by zero entropy change. If a process is not reversible (as with any heat transfer across a finite temperature difference), the situation is radically different. Entropy change ΔS is either equal (reversible process) or larger (irreversible process) than the entropy transfer ($\int dq/T$, nonproperty)[†] that accompanies heat transfer dq, the difference being attributed to entropy generation $\dot{S}_{irr}$ [see Eq. (11.36)]. The amount of entropy generation is the quantitative measure of the quality level of energy transfer. Entropy generation of zero corresponds to the highest quality of energy transfer and/or energy conversion (a reversible process), and entropy generation greater than zero represents poorer quality. All real processes are characterized by entropy generation greater than zero.

The concept of *exergy* or *available energy* $\mathscr{E}$ is introduced to describe the maximum available energy that can be obtained from a system in a given state. Each fluid stream that enters or leaves a heat exchanger carries exergy rate. Due to irreversible processes in a system (e.g., a heat exchanger), the available energy of a fluid decreases and the difference between the input and output exergy rates is equal to the lost exergy (lost available energy), which in turn is identical to the irreversibility in energy terms (a Guy–Stodola theorem, i.e., exergy destroyed = lost available energy = temperature weighting factor × entropy generation; that is, $\Delta\dot{\mathscr{E}} = \dot{W}_{lost} = T_o\dot{S}_{irr}$).

11.3.1   Entropy Generation Caused by Finite Temperature Differences

Temperature and temperature difference distributions within the heat exchanger influence thermodynamic irreversibility. Thermodynamics teaches us that a measure of the efficiency of any thermal process can be assessed by gaining an insight into the irreversibility level incurred in an associated heat transfer process (Bošnjaković, 1965). This irreversibility can be identified by determining the corresponding entropy generation. So there is a deeper rationale for turning our attention toward temperature differences within a heat exchanger to determine heat transfer performance behavior. The driving potential for heat transfer in a heat exchanger *is* the finite local temperature difference

---

[†] Note that the entropy is a system property and entropy transfer or entropy generation is *not* a system property.

between the fluids exchanging heat, and we should expect that it greatly influences the exchanger effectiveness as well. Consequently, it is plausible to conclude that these temperature differences are related to both the exchanger effectiveness and thermodynamic efficiency of an exchanger (one such thermodynamic figure of merit is defined in Section 11.6.5). How these temperature differences influence the irreversibility level is explained next in terms of entropy generation.

The thermodynamic irreversibility manifested within a heat exchanger as an adiabatic open system can be identified in terms of entropy generation by total entropy change (entropy measure of irreversibility, i.e., entropy generation rate $\dot{S}_{irr}$) of both fluid streams:

$$\dot{S}_{irr} = \Delta\dot{S} = \dot{m}_1\,\Delta s_1 + \dot{m}_2\,\Delta s_2 \tag{11.25}$$

Now we will evaluate $\dot{S}_{irr}$ only due to finite temperature differences, considering the fluids as pure simple single-phase compressible substances. Since $ds = d\mathbf{h}/T$ (where $\mathbf{h}$ is the specific enthalpy), the entropy rate change for fluid 1 in a heat exchanger operating under steady-state conditions for an ideal gas or an incompressible liquid is given by

$$\dot{m}_1\,\Delta s_1 = \int_i^o \left(\frac{\dot{m}\,d\mathbf{h}}{T}\right)_1 = \int_i^o \left(\frac{\dot{m}c_p\,dT}{T}\right)_1 = (\dot{m}c_p)_1 \ln\frac{T_{1,o}}{T_{1,i}} \tag{11.26}$$

Similarly, the entropy rate change for fluid 2 in the exchanger will be

$$\dot{m}_2\,\Delta s_2 = (\dot{m}c_p)_2 \ln\frac{T_{2,o}}{T_{2,i}} \tag{11.27}$$

Note that we *do not* need to distinguish at this point whether the fluid is hot or cold. Moreover, this distinction *is not* necessarily relevant for calculating the entropy measure of irreversibility. What matters, though, is that the two fluids have *different* temperatures. Hence, the concepts of hot and/or cold will not necessarily be used here. Consequently, Eq. (11.25) can be rewritten as follows:

$$\dot{S}_{irr} = \sum_{j=1}^{2} \dot{m}_j\,\Delta s_j = \dot{m}_1 c_{p,1} \ln\frac{T_{1,o}}{T_{1,i}} + \dot{m}_2 c_{p,2} \ln\frac{T_{2,o}}{T_{2,i}} \tag{11.28}$$

The two terms in Eq. (11.28) have the opposite signs since the fluids have different temperatures ($T_{1,i} \neq T_{2,i}$, i.e., either $T_{1,o} \leq T_{1,i}$ and $T_{2,o} \geq T_{2,i}$ or $T_{1,o} \geq T_{1,i}$ and $T_{2,o} \leq T_{2,i}$). Two important thermodynamic points have to be reiterated here. First, the fact that the two fluids have different temperatures is of far more importance than that one may conveniently be described as hot and the other as cold. This is because the hot/cold dichotomy is introduced *by convention*. In a heat exchanger, as will be demonstrated later (see Section 11.4.3), the same fluid may change the role of a hot/cold fluid side over some flow length! So, in this chapter, we will, as a rule, refer to a fluid as fluid 1 or fluid 2 whenever a general case has to be considered in which any of the two fluids may either be hot or cold. If a fluid is identified as having higher/lower temperature (such as in a particular given example), we denote as the hot fluid that has a temperature at its *inlet port* higher than the temperature of other fluid at its inlet port. Second, a more advanced thermodynamic analysis advocated in this chapter involves the concept of entropy; hence it involves not only temperature differences but also temperature ratios and the products of *absolute* temperature and entropy differences [see, e.g., Eq. (11.28) or (11.53)]. As a consequence, proper care must be taken regarding the use of *absolute temperatures* (K or °R) for all temperatures associated with entropy and also exergy later. To emphasize this fact, we use in this chapter, as a rule, temperatures on the absolute Kelvin (or Rankine) scale and *not* on the commonly used degree Celsius (or Fahrenheit) scale.

The heat transfer rate between the two fluid streams in thermal contact under adiabatic conditions is equal to the respective enthalpy rate changes (see Chapters 2 and 3):

$$q = \dot{m}_1 \, \Delta \mathbf{h}_1 = \dot{m}_2 \, \Delta \mathbf{h}_2 \qquad (11.29)$$

For better clarity, we consider fluid 1 as the hot fluid and fluid 2 as the cold fluid. Hence, the enthalpy rate changes for the hot and cold fluids are $\Delta \mathbf{h}_h = c_{p,h}(T_{h,i} - T_{h,o})$, $\Delta \mathbf{h}_h = c_{p,h}(T_{c,o} - T_{c,i})$ for Eq. (11.29). Changing the subscripts 1 and 2 of Eqs. (11.28) and (11.29) to h and c, combining them, and rearranging, we get,

$$\frac{\dot{S}_{\mathrm{irr}}}{q} = -\frac{1}{T_{h,\mathrm{lm}}} + \frac{1}{T_{c,\mathrm{lm}}} = \frac{T_{h,\mathrm{lm}} - T_{c,\mathrm{lm}}}{T_{h,\mathrm{lm}} T_{c,\mathrm{lm}}} \qquad (11.30)$$

where

$$T_{h,\mathrm{lm}} = \frac{T_{h,i} - T_{h,o}}{\ln(T_{h,i}/T_{h,o})} \qquad T_{c,\mathrm{lm}} = \frac{T_{c,o} - T_{c,i}}{\ln(T_{c,o}/T_{c,i})} \qquad (11.31)$$

Here $T_{h,\mathrm{lm}}$ represents the log-mean temperature of the hot fluid as defined using inlet and outlet temperatures $T_{h,i}$ and $T_{h,o}$. The $T_{c,\mathrm{lm}}$ is defined similarly. In contrast, the arithmetic mean temperatures of the hot and cold fluids are $T_{h,m} = (T_{h,i} + T_{h,o})/2$ and $T_{c,m} = (T_{c,i} + T_{c,o})/2$, and the log-mean temperature *difference* between hot and cold fluids in a heat exchanger is given by Eq. (3.172).

The entropy generation is related to the difference in the fluid temperatures. Equation (11.30) is written for the exchanger as a whole. On the local level, entropy generation is related to local temperature differences [such as Eq. (11.15)]. Hence, the *difference* between mean temperatures of two fluids [the numerator of Eq. (11.30)] directly influences the entropy measure of the irreversibility manifested within the heat exchanger. As a consequence, the heat exchanger irreversibility for a given heat transfer rate can be reduced by reducing temperature differences between the fluids, which in turn will increase the exchanger effectiveness ε.[†] A heat exchanger characterized by smaller temperature differences between the fluids generates a smaller irreversibility in a given system compared to a heat exchanger (for the same heat transfer rate) that has larger temperature differences between the fluids. Since the entropy measure of irreversibility is related directly to thermodynamic system efficiency (see Section 11.6.5), this statement leads to an anticipated conclusion about the possible detrimental influence of this source of irreversibility on the overall system efficiency.

Thermodynamic irreversibility represented by entropy generation as in Eq. (11.28) can be formulated in terms of heat exchanger thermal design parameters. Using the

---

[†] Note that the log-mean temperature difference ΔT_{lm} of Eq. (3.172) is proportional to $(T_{h,\mathrm{lm}} - T_{c,\mathrm{lm}})$. Hence, a smaller value of ΔT_{lm} means a larger value of ε for an exchanger.

definitions of heat exchanger effectiveness and heat capacity rate ratio Eqs. (3.44) and (3.56)] and considering $C_1 = C_{min}$, one can show that

$$\frac{T_{1,o}}{T_{1,i}} = 1 + \varepsilon(\vartheta^{-1} - 1) = 1 - P_1(\vartheta^{-1} - 1) \qquad \frac{T_{2,o}}{T_{2,i}} = 1 + C^*\varepsilon(\vartheta - 1) = 1 + R_1 P_1(\vartheta - 1)$$

$$(11.32)$$

where $\vartheta = T_{1,i}/T_{2,i}$ represents the inlet temperature ratio. Substituting these expressions in Eq. (11.28) results in

$$\frac{\dot{S}_{irr}}{C_{max}} = S^* = C^* \ln[1 + \varepsilon(\vartheta^{-1} - 1)] + \ln[1 + C^*\varepsilon(\vartheta - 1)] \qquad (11.33a)$$

$$\frac{\dot{S}_{irr}}{C_2} = S^* = R_1 \ln[1 + P_1(\vartheta^{-1} - 1)] + \ln[1 + R_1 P_1(\vartheta - 1)] \qquad (11.33b)$$

where $\dot{S}_{irr} \neq 0$ for $\vartheta \neq 1$ and $\dot{S}_{irr} = 0$ for $\vartheta = 1$. Note that normalizing $\dot{S}_{irr}$ by C_{max} or C_2, as indicated in the leftmost term of Eq. (11.33), is a matter of arbitrary choice. The entropy generation is equal to zero for the inlet temperature ratio equal to 1. Entropy generation [Eq. (11.33)] for different flow arrangements is different for the *same* inlet temperature ratio, heat capacity rate ratio C^* or R_1, and NTU (Sekulić, 1990b). This is because of different heat exchanger effectivenesses, $\varepsilon = \varepsilon$ (NTU, C^*), for different flow arrangements [and fixed values of NTU and C^* (or NTU$_1$ and R_1)]. It should be emphasized that the control volume for the entropy generation of Eq. (11.33) [see Eq. (11.25)] is drawn outside the exchanger core or matrix through inlet/outlet ports. Hence, the S^* expression of Eq. (11.33) is valid for an exchanger with *any flow arrangement* by employing its appropriate ε-NTU or P-NTU formula. Some additional features of Eq. (11.33) are discussed in Section 11.4.1.

11.3.2   Entropy Generation Associated with Fluid Mixing

Fluid mixing in a heat exchanger causes thermodynamic irreversibility and generates entropy, leading to a reduction in the thermodynamic efficiency of the heat transfer process, thus reducing the heat exchanger effectiveness. In general, the mixing of fluids that are dissimilar with respect to their composition and/or state variables is an irreversible process. These dissimilarities may be mechanical (pressure gradients), thermal (temperature gradients), and/or chemical (chemical potentials). Entropy generation associated with mixing depends on the degree of dissimilarity between mixing fluids.

The irreversibility associated with a mixing process is due to (1) a process of intermingling molecules of different substances, (2) energy interchange between the same or different substances or within the mixing substances, (3) heat transfer between surroundings and mixing substances, and (4) viscous dissipation effects.

For the analysis of many heat exchangers, thermal dissimilarities (due to temperature gradients on each fluid side in the transverse direction) of two or more fluid streams during mixing are of primary interest. For example, in a crossflow heat exchanger, heat transfer between two fluids causes the presence of local temperature nonuniformities in any given flow cross section. However, a fluid flowing through a nonpartitioned flow passage (i.e., a *mixed* fluid stream side) is characterized by an important feature. The unrestrained mixing attenuates temperature nonuniformities at a given cross section of

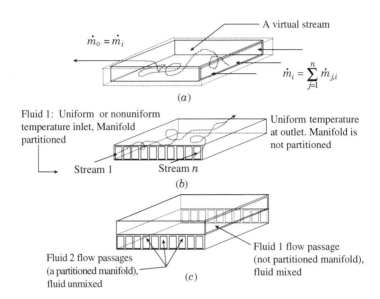

FIGURE 11.4   Flow passages with fluid mixing: (*a*) passage or duct with fluid mixing; (*b*) outlet manifold; (*c*) two adjacent flow passages with/without fluid mixing.

the mixed fluid and the available thermal potential is destroyed. This is certainly an irreversible phenomenon that leads to a corresponding entropy increase. A practical consequence of this thermodynamically detrimental process is the *equalization* of fluid local temperatures across the flow passage cross section, ultimately leading to a reduced heating/cooling manifested at the fluid exits. As a consequence, the respective temperature effectiveness (or the heat exchanger effectiveness) is also reduced.

Let us consider a fluid stream being mixed while flowing right to left through a duct, as shown in Fig. 11.4*a*. The objective of the analysis is to determine entropy generation associated with fluid mixing using a very simplified approach. This situation is often present in heat exchangers. For example, in the outlet header, mixing is accomplished between different streams of the same fluid ideally with no heat transfer with the surroundings as shown in Fig. 11.4*b*. In the heat exchanger core region, a mixed fluid exchanges heat with the other fluid, either mixed or unmixed. A significant simultaneous heat transfer and mixing (see Fig. 11.4*c* for the unmixed–mixed case) prevent us from determining the sole contribution of mixing to the total irreversibility for a control volume under consideration. That is the reason why we formulate a more general case but consider a mixing-only case as follows.

A mixed fluid (fluid 1 in Fig. 11.4*a*) flows through the passage while simultaneously being mixed in a direction transverse to the overall flow direction. For the sake of clarity (but with an inevitable loss of rigor), let us assume that resultant heat exchange between the fluid and the environment can be modeled as heat transfer between each of *n* virtual streams of the mixed fluid at the inlet that merge into one mixed stream at the outlet. Mass rate, energy/enthalpy rate, and entropy rate balances (continuity, energy, and entropy equations) for the control volume of the bulk flow of fluid 1 are as follows:

Continuity equation:
$$\dot{m}_o - \sum_{j=1}^{n} \dot{m}_{j,i} + \frac{dm_{cv}}{d\tau} = 0 \qquad (11.34)$$

Energy equation:
$$\dot{m}_o\mathbf{h}_o - \sum_{j=1}^{n} \dot{m}_{j,i}\mathbf{h}_{j,i} - \sum_{j=1}^{n} q_j + \frac{dE_{cv}}{d\tau} = 0 \qquad (11.35)$$

Entropy equation:
$$(\dot{m}s)_o - \sum_{j=1}^{n} (\dot{m}s)_{j,i} - \sum_{j=1}^{n} \left(\frac{q}{T}\right)_j + \frac{dS_{cv}}{d\tau} = \dot{S}_{irr} > 0 \qquad (11.36)$$

where q_j, $j = 1$, n, represent the equivalent heat transfer rates between virtual streams (having average individual bulk temperatures T_j along the respective flow paths) and the surroundings (the other fluid side). Note that $\dot{S}_{irr} > 0$ in Eq. (11.36) is a consequence of the second law of thermodynamics for a real system.

For a steady-state flow, Eqs. (11.34)–(11.36) reduce to[†]

$$\dot{m}_o = \sum_{j=1}^{n} \dot{m}_{j,i} \qquad (11.37)$$

$$\dot{m}_o\mathbf{h}_o = \sum_{j=1}^{n} \dot{m}_{j,i}\mathbf{h}_{j,i} + \sum_{j=1}^{n} q_j \qquad (11.38)$$

$$\dot{m}_o s_o = \sum_{j=1}^{n} (\dot{m}s)_{j,i} + \sum_{j=1}^{n} \left(\frac{q}{T}\right)_j + \dot{S}_{irr} \qquad (11.39)$$

Taking into account the change of local temperatures along the flow paths would, indeed, require writing the balances in a differential form and integrating them along the flow path. The form of these relationships would depend on actual heat transfer conditions. For isolating the mixing effect only, we consider an adiabatic mixing case.

The simplest physical situation of practical interest corresponds to the conditions encountered in the headers/manifolds or in the parts of a heat exchanger where the heat transfer q can be neglected (as is truly for example, in the exit zone of a TEMA J heat exchanger; Fig. 11.6). In such an adiabatic situation, Eqs. (11.38) and (11.39) can be simplified as follows:

$$\dot{m}_o\mathbf{h}_o = \sum_{j=1}^{n} \dot{m}_{j,i}\mathbf{h}_{j,i} \qquad (11.40)$$

$$\dot{S}_{irr} = \dot{m}_o s_o - \sum_{j=1}^{n} (\dot{m}s)_{j,i} = \sum_{j=1}^{n} \Delta(\dot{m}s)_j \qquad (11.41)$$

For a simple compressible substance, using the expression of Eq. (11.26) for $\dot{m}_j \, \Delta s_j$ for each stream, $\dot{S}_{irr}$ of Eq. (11.41) reduces to

$$\dot{S}_{irr} = \sum_{j=1}^{n} (\dot{m}c_p)_j \ln \frac{T_o}{T_{j,i}} \qquad (11.42)$$

[†] q_j is not shown in Fig. 11.4.

Hence, $\dot{S}_{irr} = 0$ for uniform inlet temperature ($T_{j,i} = T_i = T_o, j = 1, n$) and $\dot{S}_{irr} \neq 0$ when thermal dissimilarity is present (i.e., nonuniform temperature $T_{j,i} \neq T_o$). For example, mixing the two streams of fluid 1 at the exit of a 1–2 TEMA J shell-and-tube heat exchanger (see Fig. 11.6 where $T_o = T_{1,o}$, and $T_{j,i}$ are denoted as $T'_{1,o}$ and $T''_{1,o}$) is a source of irreversibility due to the fact that these outlet temperatures of the shell fluid (fluid 1) leaving the two zones A and B are not the same.

From Eq. (11.42), it is obvious that entropy generation caused by mixing would never be equal to zero if thermal dissimilarity is present while mixing the streams, even for a single fluid. The mixing process actually eliminates the presence of local temperature differences; hence it is an inherently irreversible process. This statement holds true regardless of the presence or absence of heat transfer to the environment (or other fluid) during mixing. Consequently, it is expected that a heat exchanger featuring mixing of either fluid within the heat exchanger core or within the header zones should have less effectiveness compared to an exchanger with the same design parameters but without mixing.

11.3.3   Entropy Generation Caused by Fluid Friction

The importance of fluid pressure drop and fluid pumping power $\mathcal{P}$ in the heat exchanger is discussed in Chapter 6. One of the important components of the fluid pressure drop in the heat exchanger is the fluid friction associated with flow over the heat transfer surface. We derive the irreversibility associated with this fluid friction in this section. Since the control volume is drawn at the inlet and outlet pipes/tanks, this analysis does take into account the contribution of both skin friction and form drag that is important in many exchangers.

To identify the irreversibility caused only by fluid friction, let us assume a fluid flowing through a flow passage of an arbitrary cross section. The flow is caused solely by the pressure difference between the two points along the fluid path. The entropy generated with such a flow is equal to the entropy change between the two points along the flow path, say between inlet and outlet. If the enthalpy change contribution to entropy change can be neglected (steady and adiabatic flow), the entropy change is as follows using the $T\,ds$ relationship: $d\mathbf{h} = 0 = T\,ds + v\,dp$ (where s and v are specific enthalpy and specific volume, respectively):

$$\int_i^o d\dot{S} = \int_i^o \dot{m}\,ds = -\int_i^o \dot{m}\,\frac{v}{T}\,dp \tag{11.43}$$

For an ideal gas flow, Eq. (11.43) reduces to

$$\dot{S}_{irr} = \Delta\dot{S} = -\dot{m}\tilde{R}\int_i^o \frac{dp}{p} = -\dot{m}\tilde{R}\ln\frac{p_o}{p_i} = -\dot{m}\tilde{R}\ln\left(1 - \frac{\Delta p}{p_i}\right) = \dot{m}\tilde{R}\ln\left(1 + \frac{\Delta p}{p_o}\right) \tag{11.44}$$

where pressure drop $\Delta p = p_i - p_o \geq 0$.

For an incompressible fluid (liquid) flow, under nonadiabatic conditions entropy generation caused by fluid friction is as follows as discussed by Roetzel in London and Shah (1983).

$$\dot{S}_{irr} = \frac{-\dot{m}\int_i^o v\,dp}{T_{lm}} = \frac{\Delta p}{\rho}\,\dot{m}\,\frac{\ln(T_o/T_i)}{T_o - T_i} \tag{11.45}$$

In both Eqs. (11.44) and (11.45), we have $\dot{S}_{irr} = 0$ for $\Delta p = 0$ and $\dot{S}_{irr} \neq 0$ for $\Delta p \neq 0$. Hence, the entropy generation caused by fluid friction is never equal to zero for $\Delta p > 0$, as in a heat exchanger. In a heat exchanger with two fluids, the irreversibility contribution of each of the two fluids has to be included [i.e., two terms of the form of Eq. (11.44) or (11.45) have to be calculated].

11.4 THERMODYNAMIC IRREVERSIBILITY AND TEMPERATURE CROSS PHENOMENA

In Section 11.3.1 we have demonstrated that heat transfer, accomplished at finite temperature differences in a heat exchanger, must be accompanied by entropy generation. This entropy generation is a function of heat exchanger design parameters [see Eq. (11.33) and Problem 11.11]. Let us now explore this relationship to relate thermodynamic performance of a heat exchanger to its heat transfer and design parameters.

11.4.1 Maximum Entropy Generation

Let us rewrite Eq. (11.33) in a symbolic form as a function of relevant design parameters as follows:

$$\frac{\dot{S}_{irr}}{C_{max}} = S^* = f(C^*, \varepsilon, \vartheta) = f(C^*, \text{NTU}, \vartheta, \text{flow arrangement}) \tag{11.46a}$$

$$\frac{\dot{S}_{irr}}{C_2} = S^* = f(R_1, P_1, \vartheta) = f(R_1, \text{NTU}_1, \vartheta, \text{flow arrangement}) \tag{11.46b}$$

The second equality in Eq. (11.46) is written by taking into account Eq. (3.50). So S^* is a function of the heat capacity rate ratio, NTU, inlet temperature ratio, and flow arrangement. In Fig. 11.5, this relationship is presented for counterflow and parallelflow arrangements for fluids having equal heat capacity rates ($C^* = 1$), and an inlet temperature ratio equal to 0.5 (i.e., $\vartheta = 0.5$). It can be shown that corresponding curves for numerous other flow arrangements will be located between the two limiting cases presented in Fig. 11.5 (Sekulić, 1990b). It should be emphasized that these curves (except for parallelflow) have at least one distinct maximum, as explained next. For an exchanger at small NTU values, when NTU $\to 0$, the magnitude of entropy generation will tend to zero (i.e., $S^* \to 0$). This is certainly an expected result because at NTU = 0 there is no heat transfer since $UA = 0$ despite the temperature potential for it (represented by the given inlet temperature difference). On the other side, if NTU $\to \infty$, the temperature differences along the heat exchanger tend to their minimum possible values (e.g., $\Delta T = 0$ for $C^* = 1$ for a counterflow exchanger). Consequently, S^* decreases to a limiting asymptotic value (equal to zero for $C^* = 1$ for a counterflow exchanger). Hence, a curve having minimum values at both ends will have at least one maximum value in between $0 < \text{NTU} < \infty$.

This analysis thus provides the following conclusions for many flow arrangements having only one maximum for S^* [including counterflow but excluding parallelflow (Sekulić, 1990a); see Section 11.4.3 for an exception; refer to Shah and Skiepko (2002) for other exceptions]:

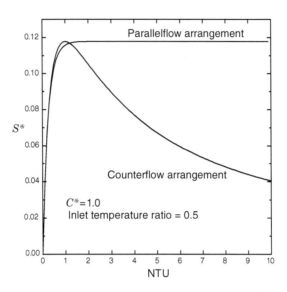

FIGURE 11.5   Entropy generation in parallelflow and counterflow exchangers with $C^* = 1$.

$$\frac{\partial S^*}{\partial \text{NTU}} = \begin{cases} > 0 & \text{at small NTU} & \lim_{\text{NTU} \to 0} S^* = 0 \\ = 0 & \text{at NTU} = \text{NTU}^* & S^* = S^*_{\text{max}} > 0 \text{ at NTU} = \text{NTU}^* \\ = 0 & \text{at large NTU} & \lim_{\text{NTU} \to \infty} S^* = S^*_{\text{min},\infty} \geq 0 \\ < 0 & \text{at NTU}^* < \text{NTU} < \infty \end{cases} \quad (11.47)$$

The most interesting feature of the heat transfer irreversibility behavior implied by Eq. (11.47) is an existence of at least one maximum of S^* for a finite-size heat exchanger at NTU* (or NTU$_1^*$).

Let us determine the value of NTU* and the corresponding effectiveness at that operating point (the same can be done for the number of transfer units defined as NTU$_1$ and the corresponding temperature effectiveness P_1). The entropy generation maximum is characterized by[†]

$$\frac{\partial S^*}{\partial \text{NTU}} = 0 \quad \text{at} \quad \text{NTU}\Big|_{S^*_{\text{max}}} = \text{NTU}^* \quad (11.48)$$

Performing the calculation as indicated in Eq. (11.48) on Eq. (11.33), one can show (see Problem 11.13) that the maximum of entropy generation corresponds to[‡]

[†] It can be shown that $\partial^2 S^*/\partial \text{NTU}^2 < 0$ at NTU = NTU* (Sekulić, 1990a).
[‡] Explicit expressions for NTU in terms of ε and C^* are available only for a limited number of flow arrangements as outlined in Table 3.4. For these arrangements, explicit formulas for NTU* (NTU at S^*_{max}) in terms of C^* can be obtained by substituting ε at S^*_{max} from Eq. (11.49). For the related formulas for countercurrent and crossflow exchangers, see Sekulić (1990b).

$$\varepsilon\bigg|_{S^*_{max}} = \frac{1}{1 + C^*} \quad \text{or} \quad P_1\bigg|_{S^*_{max}} = \frac{1}{1 + R_1} \qquad (11.49)$$

Note that Eq. (11.49) is identical to the corresponding relationships in Eq. (3.114). Thus, the number of heat transfer units at the maximum irreversibility in a heat exchanger is exactly the same as the limiting value of NTU ($= \text{NTU}^*$) at the onset of an external temperature cross. At that operating point, the outlet temperatures of both fluids are equal. Hence, beyond this NTU $> \text{NTU}^*$, there will be a temperature cross, and the hot-fluid outlet temperature will become lower than the cold-fluid outlet temperature. As defined in Sections 3.2.3 and 3.6.1.2, the *temperature cross* derives its name from fictitious or actual crossing of the temperature distributions of the hot and cold fluids in an exchanger. If there is no actual crossing of hot- and cold-fluid temperature distributions within an exchanger and $T_{c,o} > T_{h,o}$, we refer to it as an *external temperature cross* as found in the temperature distributions of a counterflow exchanger in Example 3.2 or the 1–2 TEMA E shell-and-tube heat exchanger of Fig. 3.17a for the high-NTU case (the solid lines). We will call it an *internal temperature cross* if there is an actual crossing of hot- and cold-fluid temperature distributions within an exchanger. There are two possibilities for the internal temperature cross: (1) $T_{c,o} > T_{h,o}$ (as in the 1–2 TEMA E exchanger of Fig. 3.17b at high NTU, or (2) $T_{c,o,\text{local}} > T_{h,o,\text{local}}$, where the subscript "local" means one of the multiple outlets on one or both fluid sides of an exchanger (see the temperature distributions in Fig. 11.7 for the 1–2 TEMA J shell-and-tube heat exchanger of Fig. 11.6). Note that there can be an external temperature cross without an internal temperature cross (such as in a counterflow exchanger, as shown in the temperature distribution of Example 3.2). Let us discuss the implications of the external and internal temperature crosses further in the following two subsections.

11.4.2 External Temperature Cross and Fluid Mixing Analogy

Results obtained in Section 11.4.1 [i.e., an interpretation of the physical meaning of Eq. (11.49), and the relation between the equalization of outlet temperatures and maximum entropy generation] lead to yet another interesting analogy. This analogy can be explained by studying Eq. (11.49) in a dimensional form.

First, let us reconfirm (see also Section 3.6.1.2) the statement of equality of the outlet temperatures at the operating point indicated by Eq. (11.49). By invoking definitions of exchanger effectiveness [Eq. (3.44)] and heat capacity rate ratio [Eq. (3.56)], Eq. (11.49) can be rewritten as follows:

$$\frac{T_{1,o} - T_{1,i}}{T_{2,i} - T_{1,i}} = \frac{1}{1 + (T_{2,i} - T_{2,o})/(T_{1,o} - T_{1,i})} \qquad (11.50a)$$

Simplifying this equation leads to

$$T_{1,o} = T_{2,o} = T_o \qquad (11.50b)$$

So Eq. (11.50) confirms that Eq. (11.49) corresponds to an equality of outlet temperatures of the two fluids of a heat exchanger. Now, using the conclusion reached by Eq. (11.50), let us rewrite Eq. (11.49) this time keeping the heat capacity rate ratio in its explicit form:

$$\frac{T_o - T_{1,i}}{T_{2,i} - T_{1,i}} = \frac{1}{1 + C_1/C_2} \tag{11.51a}$$

Therefore,

$$(C_1 + C_2)T_o = C_1 T_{1,i} + C_2 T_{2,i} \tag{11.51b}$$

Equation (11.51) clearly indicates an identical result as before, but this time it can be interpreted as having been obtained for a quite different physical situation of *adiabatic mixing* of the two fluid streams (with heat capacity rates C_1 and C_2 and inlet temperatures $T_{1,i}$ and $T_{2,i}$). Such imaginary mixing will lead to the outlet temperature T_o for the mixture of the two given streams ($C_1 + C_2$). From thermodynamics, we know that adiabatic mixing leads to total destruction of the available thermal energy potential (implied by the difference of the temperatures of the fluids) that exists at the onset of mixing. Thus, this process must be characterized by maximum entropy generation.

In conclusion, there is an analogy between a heat exchange process at the operating point that corresponds to the condition of an equalization of outlet temperatures and the adiabatic mixing process of the same two fluids. This analogy demonstrates that entropy generation in a heat exchanger at that operating point must be the maximum possible. This constitutes an additional physical explanation of the thermodynamic significance of an external temperature-cross operating point in a heat exchanger.

A clear thermodynamic meaning of the result given by Eq. (11.49) can easily be confirmed by comparing Eqs. (11.28) and (11.42), that is, comparing the entropy generations obtained for two completely different processes: (1) a heat transfer process in a heat exchanger characterized with equal outlet temperatures of the fluids involved [i.e., $T_{1,o} = T_{2,o} = T_o$; see Eq. (11.28)], and (2) a mixing process of the two fluids [$T_j = T_{j,i}, j = 1, 2$; see Eq. (11.42)]. The entropy generation rates for two physically quite different processes are found to be identical.

Analysis provided so far shows clearly how the facts related to the detrimental influence of fluid mixing on heat exchanger performance fit nicely into the consistent thermodynamic picture. Hence, the results for heat exchanger effectiveness presented in Chapter 3 (Tables 3.3 and 3.6) have deeper physical explanations. Let us now show how this kind of thermodynamic analysis can be used to understand the behavior of a relatively complex heat exchanger flow arrangement. Moreover, we demonstrate why such analysis has importance for practical design considerations.

11.4.3 Thermodynamic Analysis for 1–2 TEMA J Shell-and-Tube Heat Exchanger

It has been emphasized (see Section 3.6.1.2) that contrary to a general design requirement to transfer heat only from one fluid to the other, and not vice versa, in some heat exchangers, reverse heat transfer takes place. For example, consider the 1–2 TEMA J shell-and-tube heat exchanger of Fig. 11.6. The existence of a temperature cross leads to a situation in which an addition of more surface area in the second tube pass does not contribute a significant increase in heat transfer because of reverse heat transfer taking place in the second pass. Note that we reached this conclusion in Section 3.6.1.2 based on assumed (i.e., at that point not analytically derived) temperature distributions. In this

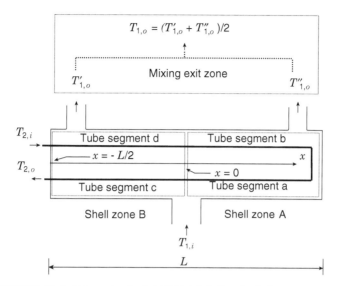

FIGURE 11.6   Schematic of a 1–2 TEMA J shell-and-tube heat exchanger.

chapter (see Section 11.4.2), we provided a thermodynamic interpretation of this un-desirable phenomenon through temperature distributions.

We elaborated in detail how one can determine temperature distributions and subse-quently evaluate the influence of local temperature differences and a mixing process on heat exchanger performance. Let us now show how the internal temperature cross can cause peculiar behavior in the P_1-NTU_1 or ε-NTU results. For a given heat capacity rate ratio, with increasing NTU_1, the temperature effectiveness P_1 reaches a maximum beyond which an increase in NTU_1 causes P_1 to decrease rather than increase mono-tonically as may be expected.[†] This behavior is illustrated in Fig. 3.16. Figures 11.7 and 11.8 summarize the results of a thermodynamic analysis of this exchanger. This analysis includes both the first *and* second laws of thermodynamics. The approach is straightfor-ward. It starts with a determination of temperature and heat transfer rate distributions obtained through an application of the first law of thermodynamics (Figs. 11.7 and 11.8*a*; model formulations in Problem 11.1). Subsequently, entropy generation is determined utilizing both the first and second laws of thermodynamics [see Section 11.3.1 and Eq. (11.33)].

Figure 11.7 shows three temperature distributions for $NTU_1 = 0.87$, 1.83, and 5.0 for $R_1 = 2$. Figure 11.8 provides data regarding the corresponding distribution of dimen-sionless heat transfer rates [the total, and those determined at different tube sections (a, b, c, and d) and shell zones (A and B) of the Fig. 11.6 exchanger] and entropy generations. To demonstrate this analysis, let us consider a numerical example from the results of Kmecko (1998).

Example 11.4   For a 1–2 TEMA J shell-and-tube heat exchanger, establish the number of temperature crosses and explain the meaning of the existence of a maximum effective-

---

[†]There are also different and unexpected P_1 vs. NTU_1 curve behaviors for n-pass n-pass ($n > 2$) plate heat exchangers and complex flow arrangements (Shah and Skiepko, 2002).

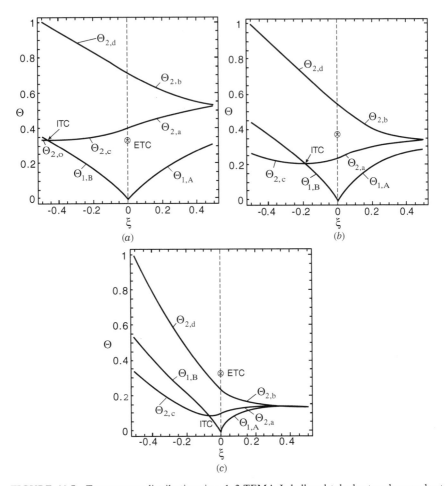

FIGURE 11.7 Temperature distributions in a 1–2 TEMA J shell-and-tube heat exchanger; heat capacity rate ratio $R_1 = 2$: (a) $NTU_1 = 0.87$; (b) $NTU_1 = 1.83$; (c) $NTU_1 = 5.00$. Note that the ξ scale at the abscissa is extended slightly beyond ± 0.5 (physical terminal ends) to show clearly the ends of the curves separate from the ordinates. (From Kmecko, 1998.)

ness at finite NTUs. The range of operating points under consideration is defined with $R_1 = 2$ and a range of NTU values between $NTU_1 = 0.87$ and 5.0. The inlet temperature ratio is equal to 2.

SOLUTION

Problem Data and Schematic: The schematic of a 1–2 TEMA J shell-and-tube heat exchanger is given in Fig. 11.6. The heat capacity rate ratio $R_1 = 2$. The range of NTU values is 0.87 to 5.0. The inlet temperature ratio $\vartheta = 2.0$.

Determine: The number of temperature crosses for this exchanger and explain the existence of maximum effectiveness at finite NTUs.

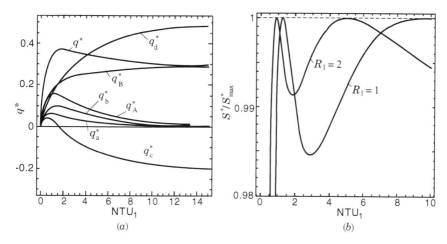

FIGURE 11.8 (*a*) Heat transfer rate ($R_1 = 2$), and (*b*) dimensionless entropy generation rate ($R_1 = 1$ and 2, $\vartheta = 2$) in a 1–2 TEMA J shell-and-tube heat exchanger. Note that the NTU_1 axis is extended slightly beyond 0 to show clearly the ends of the curves separate from the y axis. (From Kmecko, 1998.)

Assumptions: The assumptions are those listed in Section 3.2.1.

Analysis: The solution of the mathematical model of this heat exchanger (for the model formulation, study Problem 11.1) provides temperature distributions for both shell and tube fluids. The solution method may be the Laplace transforms technique, as discussed earlier in this chapter, or any other pertinent method as mentioned in Section 3.11. These distributions are given in graphical form in Fig. 11.7 for a fixed heat capacity rate ratio ($R_1 = 2$) and three values of NTU_1. The NTU_1 values correspond to the operating points for the following three specific cases: (1) the equalization of exit temperatures at small NTU (the first external temperature cross at $NTU_1 = 0.87$ at the exchanger outlets of two fluids ($T_{1,o} = T_{2,o}$), (2) the maximum temperature effectiveness ($P_{1.max}$ at $NTU_1 = 1.83$), and (3) the equalization of exit temperatures at a large NTU value (the second external temperature cross at $NTU_1 = 5.0$ at the exchanger outlet, $T_{1,o} = T_{2,o}$). In all three cases, we have internal temperature crosses.

The temperature distribution in Fig. 11.7*a* corresponds to an operating point $NTU_1 = 0.87$ and $R_1 = 2.0$. The fluid 2 (tube fluid) dimensionless temperature decreases along both tube passes (tube segments d–b–a–c in Fig. 11.6 and $\Theta_{2,d} \to \Theta_{2,b} \to \Theta_{2,a} \to \Theta_{2,c}$ in Fig. 11.7*a*) reaching the fluid exit with the dimensionless temperature $\Theta_{2,o}$ (its corresponding dimensional value is $T_{2,o}$). The outlet temperature of fluid 2 at that operating point is exactly equal to the mixed mean temperature of fluid 1 obtained after mixing both exit streams of fluid 1 [i.e., $T_{1,o} = (T'_{1,o} + T''_{1,o})/2$, or in dimensionless form $\Theta_{1,o} = (\Theta_{1,A,o} + \Theta_{1,B,o})/2$]. To emphasize, $\Theta_{2,o} = \Theta_{1,o}$ or $T_{2,o} = T_{1,o}$ for $NTU_1 = 0.87$ and $R_1 = 2$ for Fig. 11.7*a*. Note that fluid 1 (shell fluid) dimensionless temperature changes (increases) along both shell zones (A and B) following the distributions presented in Fig. 11.7*a* ($\Theta_{1,A}$ and $\Theta_{1,B}$). Therefore, this operating point corresponds to an equalization of exit temperatures, that is, an onset of an external temperature cross (ETC). In addition, an internal (or actual) temperature cross (ITC) is present in zone B at the ITC point $\Theta_{1,B} = \Theta_{2,c}$. Note that the temperature effectiveness of fluid 1 is $P_1 = \Theta_{1,o}$, and P_2 can then be calculated from $P_2 = P_1 R_1$.

If NTU_1 is increased from 0.87 to 1.83, a new operating point will be reached. Under that condition, temperature distributions will become as given in Fig. 11.7b. This figure reveals a peculiar increase of the fluid 2 dimensionless temperature ($\Theta_{2,c}$) in the second part of the tube segment c: instead of decreasing, the original hot tube fluid 2 dimensionless (or dimensional) temperature increases. This is due to reverse heat transfer taking place to the left of the ITC point in Fig. 11.7b. As expected, the resulting exit temperature of fluid 1 is higher than that for the operating point presented in Fig. 11.7a. This means that the temperature effectiveness of fluid 1 has increased from $P_{1,a} (= \Theta_{1,o,a})$ to $P_{1,b} (= \Theta_{1,o,b})$ (the subscripts a and b denote the cases from Fig. 11.7a and b, with $NTU = 0.87$ and 1.83, respectively). If we increase the number of transfer units even further, say from 1.83 to 5.0 (see Fig. 11.7c), the increase in the fluid 2 dimensionless temperature $\Theta_{2,c}$ in the tube segment c becomes very pronounced (see, e.g., $\Theta_{2,c}$ at $\xi = -0.5$ in Fig. 11.7c). The shell fluid temperature still continues to increase in both shell zones, but much more so in zone B, Θ_{1B} in Fig. 11.7c. However, $\Theta_{1,o} [= (\Theta_{1,A,o} + \Theta_{1,B,o})/2]$ of Fig. 11.7c is less than $\Theta_{1,o}$ of Fig. 11.7b. As a result, the temperature effectiveness of fluid 1 will become smaller than the corresponding value of Fig. 11.7b. This means that with an increase of NTU_1 starting from 1.83, the temperature effectiveness decreases from its maximum value at $NTU_1 = 1.83$ and $R_1 = 2.0$, as shown in Fig. 11.8a, in the $q^* (= P_1)$ vs. NTU_1 curve. To understand this peculiar behavior, let us now consider the distribution of heat transfer rates within the heat exchanger in various shell zones and tube segments, as presented in Fig. 11.8a.

It is interesting to notice from Fig. 11.8a that with an increase in NTU_1, an initial increase in total dimensionless heat transfer rate, $q^* = q/[(\dot{m}c_p)_1(T_{2,i} - T_{1,i})] = P_1$, is followed by a maximum at $NTU_1 = 1.83$ and decrease at higher NTU_1. Adding more heat transfer area beyond that for the maximum effectiveness decreases the heat transfer rate exchanged between the two fluids. How could that be possible? The components of this heat transfer rate are presented in the same diagram (Fig. 11.8a) for both the shell fluid (zones A and B) and the tube fluid (tube segments a to d) having appropriate subscripts with q^*. From this figure, it is clear that only the tube segment d (the inlet segment of the first tube pass) contributes to the heat exchanger performance by its positive dimensionless heat transfer rate slope in the direction of increasing NTU_1. The heat transfer contribution in tube segments a and b (and correspondingly, in the shell zone A) is diminishing rapidly, while the heat transfer in tube segment c becomes reversed and is increasing in the negative direction. Consequently, a large portion of the heat exchanger actually fails to contribute to the original design goal of increasing q with P_1 at large NTU_1 values, beyond $NTU_1 = 1.83$ for this case!

Discussion and Comments: The behavior of the analyzed heat exchanger can be interpreted in terms of entropy generation. For this exchanger, $\dot{S}_{irr}$ can be calculated using Eq. (11.33) for $\vartheta = T_{1,i}/T_{2,i} = 2.0$ and the P-NTU formula of Eq. (III.11) of Table 3.6 for given NTU_1 and $R_1 = 1$ and 2 as shown in Fig. 11.8b. For this exchanger, the entropy generation features two maximums and one local minimum. The *two* equalizations of outlet temperatures (ETCs, one at smaller and the other at larger NTU_1) correspond to two maximums in entropy generation for the case of $R_1 = 1$ and 2. In this situation, the first is at $NTU_1 = 0.87$ and the second at $NTU_1 = 5.0$, as shown in Fig. 11.8b. At the operating point characterized by the maximum temperature effectiveness, the heat exchanger operates at a local minimum entropy generation. Note that the existence of the second maximum at large NTU_1 is not of great practical importance per se (a 1–2 TEMA J shell-and-tube heat exchanger is never designed as a single-pass unit at such large NTU_1). Rather, it indicates that the entropy generation between $NTU_1 = 0.87$ and 5.0 may both decrease and increase. In other words, it makes no thermodynamic sense (or practical sense either) to design a 1–2 TEMA J heat exchanger for a large NTU (in the analyzed example, a large NTU means that $NTU > 1.83$).

The foregoing discussion demonstrates why study of a heat exchanger design should be accompanied by a study of internal heat transfer intricacies and a thermodynamic analysis. A very practical conclusion has been reached. The tube segments a and b are contributing virtually nothing to the heat exchange process at high NTUs, and the tube segment c has reverse heat transfer. This conclusion would never have been reached if study of this design were conducted looking at the heat exchanger as a black box. Of course, shell-and-tube heat exchangers are hardly ever designed for $NTU > 1.5$ with a single shell pass.

We find the following interesting observations from a detailed review of some ε-NTU curves of exchangers presented in Chapter 3 and many ε-NTU curves for other flow arrangements from Shah and Pignotti (1989).

- When there is an external temperature cross only (as in a counterflow exchanger) or only an external temperature cross can be found by modifying the original flow arrangement (such as modifying the flow arrangement of Fig. 3.17b to 3.17a), the exchanger effectiveness will continue to increase monotonically with increasing NTU.

- When an internal temperature cross cannot be eliminated [due to the exchanger geometry (such as the 1–2 TEMA J exchanger of Fig. 11.6) or even if by modifying the flow arrangement], the exchanger effectiveness will decrease with increasing NTU beyond the point of $S^*_{\min}$ at $NTU > 0$. For such exchangers, the $\dot{S}_{irr}$ versus NTU curve has at least two maximums, and the minimum value of S^* occurs between the two maximums (see, e.g., Fig. 11.8b).

  The following exchangers have an internal temperature cross *and* the exchanger effectiveness decreases with increasing NTU beyond the point of $S^*_{\min}$ at $NTU > 0$ based on the extensive P-NTU results of Shah and Pignotti (1989): a crossflow exchanger with both fluids mixed, 2–2, 2–3, and 2–4 overall parallelflow PHEs, and the following TEMA shell-and-tube heat exchangers: 1–3 E (two passes in parallelflow), 1–4 E, 1–2 G (overall parallelflow), 1–2 H (overall parallelflow), 1–2 J, and 1–4 J.

11.5 HEURISTIC APPROACH TO AN ASSESSMENT OF HEAT EXCHANGER EFFECTIVENESS

As demonstrated in previous sections, thermodynamics provides an insight into the relationship between the irreversibility level (entropy generation as an indicator of heat transfer quality performance) and the heat transfer and fluid flow features of a flow arrangement. For a detailed quantitative analysis, fluid temperature distributions within a heat exchanger need to be determined. However, one can use the conclusions reached through a qualitative study of temperature difference distributions, fluid mixing, and flow friction phenomena to assess the heat exchanger performance, even without a detailed quantitative analysis. We use this argument to show how a simple heuristic approach can be used to assess heat exchanger performance (Sekulić, 2003).

Our goal is to compare the pairs of flow arrangements having the same ϑ, NTU, and R. The objective is to predict through a heuristic analysis which of the two arrangements in a pair has better effectiveness without computing the effectiveness.

Let us compare two *single-pass* crossflow arrangements: (1) the arrangement with fluid 1 unmixed and connected in identical order between the rows A and B and fluid 2 split into two equal individually mixed streams as presented in Fig. 11.9a, and (2) the configuration with an inverted order coupling of fluid 1 shown in Fig. 11.9b. Thus the only difference between the two arrangements is the coupling of fluid 1 streams between the two rows of fluid 2. Let us also assume that the P_1-NTU$_1$ relationships for the two arrangements are not known but that the two exchangers have identical NTU$_1$ and R_1. The question is as follows: Which of the two arrangements has better performance?

To attempt to answer the question, let us recall that the presence of larger temperature differences and/or fluid mixing in an exchanger inevitably increases thermodynamic irreversibility (i.e., entropy generation), and correspondingly decreases the exchanger effectiveness, if all other conditions remain the same. As far as mixing in individual rows is concerned, the two arrangements are identical. However, the local temperature differences in the two flow arrangements are different due to overall flow configurations. Consequently, the two exchangers must have different effectivenesses.

Hence, let us compare qualitatively the magnitude of the local temperature differences in these two arrangements by concentrating on the two lateral streams S_1 and S_2 of fluid 1 in Fig. 11.9a and b. In the case of an *identical order coupling* (Fig. 11.9a), the stream S_1 leaves the first row A of the fluid 2 from a zone where fluid 2 enters that row, and subsequently enters the second row B of the same fluid again in the zone where fluid 2 enters that row (that is why we call this coupling *identical*). Stream S_2 leaves the first row A of fluid 2 with a temperature that has been changed less when compared to the change experienced by stream S_1 (due to the fact that fluid 2 has already experienced some heat transfer through row A before it meets stream S_2). Moreover, stream S_2 enters the second row B of fluid 2 in the zone where the exit of fluid 2 is located in that row. The heat transfer will be accomplished at established temperature differences, defined by the given operating and design conditions.

In the case of the *inverted order*, though (Fig. 11.9b), stream S_1 from the exit of the first row A of fluid 2, from a zone where fluid 2 enters the row, is led into the

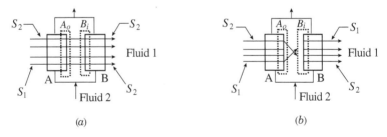

(a) (b)

FIGURE 11.9   Single-pass crossflow: fluid 2 split into two equal streams individually mixed and flowing through rows A and B: (a) fluid 1 unmixed and connected in identical order; (b) fluid 1 unmixed and connected in inverted order. The outlet and inlet zones of the two fluid 2 stream rows are denoted by A_o and B_i.

second row B of fluid 2 to the zone from which fluid 2 leaves that row. Similarly, stream S_2 is led from the first row A of fluid 2, from a zone where fluid 2 leaves the row, to the second row B of fluid 2, into the zone where fluid 2 enters the row. So stream S_1, the stream that has experienced a larger temperature change in the first row A of fluid 2 than stream S_2, will meet fluid 2 in two different zones of the second row B of fluid 2 for the two analyzed arrangements. As a consequence, in the case of the *identical order* coupling, the heat transfer between fluids 1 and 2 will be accomplished in the second row B of fluid 2 at smaller temperature differences at the entrance than that in the case of the *inverted order*. Based on this observation of the smaller temperature difference at the inlet of the second tube row or a pass, we could infer that it will have lower temperature difference irreversibility, which means higher exchanger effectiveness.

This conclusion has been reached without any knowledge about the corresponding effectiveness relationships; it is based on the qualitative analysis of the heat transfer process involved. Calculating the effectivenesses for the two arrangements at the same inlet operating conditions, using appropriate P-NTU relationships from Shah and Pignotti (1989), can easily provide a simple proof. For example, if $NTU_1 = 3$ and $R_1 = 0.8$, the temperature effectiveness of the identical order and inverted order arrangements are 0.7152 and 0.6668, respectively. The aforementioned conclusion has been verified for pairs of heat exchangers that differ only with respect to a particular source of irreversibility (a fluid coupling between the passes, mixing, finite temperature difference magnitude at a location where they may be at maximum).

A systematic analysis of the applicability of this heuristic approach can be conducted for other heat exchangers. For crossflow arrangements, it was shown that even in the cases for which the closed-form expressions of the effectiveness relationship does not exist, the simple heuristic approach usually works.

In Fig. 11.10, two two-pass arrangement pairs are compared.[†] For both the two-pass cross-parallelflow exchangers and the two-pass cross-counterflow arrangements, the identical order coupled passes (Fig. 11.10*a* and *c*) have larger effectiveness values compared to the corresponding inverted-order coupled passes (Fig. 11.10*b* and *d*).

Finally, let us consider an exchanger with the most complex two-pass cross-counterflow arrangement, with both fluids unmixed and connected either in identical (Fig. 11.11*a*) or inverted order (Fig. 11.11*b*). For these flow arrangements, there are no closed-form analytical solutions for effectiveness–NTU relationships. The solutions for $NTU_1 = 3$ and $R_1 = 0.8$ for the two cases are (1) both fluids coupled in identical order (Fig. 11.11*a*), $P_1 = 0.757$, and (2) both fluids coupled in inverted order (Fig. 11.11*b*), $P_1 = 0.736$. Consequently, the identical order arrangement has obviously higher effectiveness. We can reach the same conclusion without utilizing the complex semianalytical relationship if we simply use the heuristic approach described above. This is presented in Example 11.5.

Example 11.5 Provide a heuristic argument that a two-pass cross-counterflow arrangement with both fluids unmixed and connected in identical order has a larger fluid 1

[†] Note that the identical and inverted orders *between* rows/passes appear to be defined differently sometimes (see, e.g., Fig. 11.9*b* vs. Fig. 11.10*b*, both having inverted order). However, they are defined consistently as identical order if both fluid streams enter the second row/pass in the same order as they enter the first row/pass; if they enter the second row/pass in a different order (one of the streams enters from the other end of the row/pass), we refer to it as an inverted order. Review Sections 1.6.1.3 and 1.6.2.1 for further clarification.

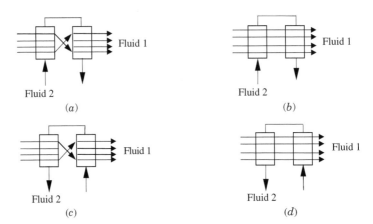

FIGURE 11.10   Two-pass cross-parallelflow and cross-counterflow arrangements: fluid 2 mixed: (*a*) two-pass cross-parallelflow arrangement with fluid 1 unmixed and coupled in identical order; (*b*) two-pass cross-parallelflow arrangement with fluid 1 unmixed and coupled in inverted order; (*c*) two-pass cross-counterflow exchanger with fluid 1 unmixed and coupled in identical order; (*d*) two-pass cross-counterflow with fluid 1 unmixed and coupled in inverted order.

temperature effectiveness than that of a two-pass cross-counterflow arrangement with both fluids unmixed and connected in inverted order.

SOLUTION

Problem Data and Schematic: Schematics of the two flow arrangements are given in Fig. 11.11.

Determine: Which of the two flow arrangements has the higher temperature effectiveness for the same operating point.

Assumptions: The assumptions are as discussed in Section 3.2.1.

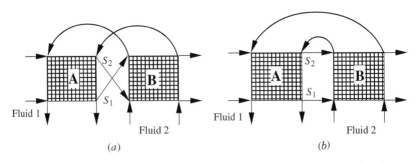

FIGURE 11.11   Two-pass cross-counterflow exchangers with both fluids unmixed and connected in (*a*) identical order and (*b*) inverted order.

Analysis: Assume that fluid 1 is the hot fluid. In that case, the fluid 1 stream S_1 leaves pass A with its temperature higher than the temperature of stream S_2. In the case of an identical order coupling (Fig. 11.11a), fluid 1 stream S_1 (the warmer of the two) is channeled to meet fluid 2 at the exit of pass B, at the point where the corresponding fluid 2 stream has already been heated by flowing through that pass. The fluid 1 stream S_2 (the colder of the two), however, is guided to meet fluid 2 at its inlet into pass B, where it has the lowest temperature (fluid 2 is assumed to be the cold fluid). So these temperature differences would be smaller for the identical order coupling (Fig. 11.11a) than for the case of an inverted order coupling (Fig. 11.11b). This is because in the inverted order coupling of Fig. 11.11b, streams S_1 and S_2 of fluid 1 are channeled to meet the corresponding streams of fluid 2 in pass B at locations characterized with opposite inlet/outlet sections when compared to pass A. The same reasoning can be conducted following the two bordering streams of fluid 2. So we should expect the identical order coupling to be more favorable than the inverted, the result already indicated above.

Discussion and Comments: A note regarding the presence of mixing is warranted at this point. It seems to be more obvious to assess two flow arrangements with respect to fluid mixing than with respect to local temperature differences. A good example is the sequence of decreasing effectiveness values for, say, single-pass crossflow arrangements given in Table 11.2, for the given NTU and R. For example, for the set of values $NTU_1 = 3$ and $R_1 = 0.8$, the temperature effectiveness P_1 takes the values 0.7355, 0.6791, 0.6655, and 0.6254, for unmixed–unmixed, unmixed (fluid 2)–mixed (fluid 1), unmixed (fluid 1)–mixed (fluid 2), and mixed–mixed flow arrangements, respectively. The increasing influence of mixing is obvious. Note that for unmixed–mixed and mixed–unmixed arrangements, fluid 1/fluid 2 is either mixed or unmixed, respectively, and because $R_1 = 0.8 < 1$, it is clear that mixing of the fluid with the larger heat capacity rate contributes more to performance deterioration, as expected. It should be mentioned that the heuristic approach might become more difficult to conduct in complex multipass heat exchangers. However, if performed correctly, it should provide at least guidance regarding the actual performance.

11.6 ENERGY, EXERGY, AND COST BALANCES IN THE ANALYSIS AND OPTIMIZATION OF HEAT EXCHANGERS

As emphasized in the introduction to this chapter, a heat exchanger is always part of a system. From the system point of view, heat exchanger design must be based on design specifications that are in full accord with an optimization objective defined for the system as a whole. The optimization objective may be formulated using energy rate and cost balances. When combined with the thermodynamic irreversibility analysis, the approach is called *thermoeconomics* (Bejan et al., 1995). Thus, a heat exchanger designer must be aware of (but not limited to) energy, cost, and exergy balances of the system. Therefore, this type of analysis requires not only energy balances based on the first law of thermodynamics, but also insights based on the second law of thermodynamics, as well as economic considerations. In this section, we address briefly a link between the thermal size of a heat exchanger (and accordingly, temperature distributions, including other related features of the heat exchanger performance) as well as energy, exergy, and cost balances. A more detailed analysis (and one that is less tightly related to the design theory of a heat exchanger) can be found in a number of books devoted to thermodynamic design and optimization of thermal systems, but such a discussion is beyond the scope of this book. Those interested in the subject should consult the literature provided by Linnhoff et al. (1982) and Bejan et al. (1995).

Heat exchanger optimization summarized in Section 9.6 was focused on optimization of a heat exchanger as a component. In this section, we discuss heat exchanger optimization when it is part of a system or when the system-imposed constraints during operation dictate the thermal design of a heat exchanger. However, it should be noted that we will not study the comprehensive system optimization, only some aspects of the analysis for system optimization with focus on the heat exchanger in the system. The system specified is first optimized based on a variety of criteria, including energy, entropy, economy criteria, and/or packaging. As a result, the performance and packaging requirements for individual heat exchangers in the system are deduced. Such optimization methods work and are usually straightforward when heat exchangers are part of a relatively simple system such as gas turbine or steam power plant, vapor compression air-conditioning, and so on, where we deal with a single or just a few working fluids to accomplish the system objectives. However, in the process industry, we deal with many process fluid streams to heat, cool, condense, vaporize, distill, concentrate, and so on, as mentioned in the beginning of Section 1.1. When a number of heat exchangers in a network are used to heat, cool, or change of phase of the process streams with available utilities, the analysis is often conducted based on the pinch analysis or *pinch technology* to ensure that all exchangers in a system meet the requirements of the process streams based on performance targets (Linnhoff et al., 1982). Practical thermodynamic performance targets must be defined before the actual design is carried out. These performance targets are either for energy performance of the system or for a realistic number of units to be used in a system. The actual heat exchanger performance and packaging requirements are deduced as a result of process integration. Subsequently, the methodology of Section 9.6 can be utilized for heat exchanger optimization as a component, which will also result in optimum individual heat exchangers in an optimum system. With the currently available very sophisticated commercial software for optimization, the methodologies of Section 9.6 and this section can be combined and the heat exchanger optimum dimensions and/or operating conditions can be obtained directly for an optimum system being characterized by the least cost, least energy consumption, or other set of criteria for optimization.

Now we concentrate on the analysis and optimization of a process industry application. First, we explain how the thermal size of a heat exchanger may become an object of optimization study based only on a first law of thermodynamics inquiry. In approaching that problem, we utilize energy rate balance considerations only. Subsequently, we extend our inquiry to include the second law of thermodynamics (exergy rate balance) and economic considerations (cost rate balance).

11.6.1   Temperature–Enthalpy Rate Change Diagram

Let us assume that the particular design requirements have been formulated for a process industry application. Since we are dealing here with a specific application, we identify the fluids as hot and cold instead of using more general terminology. The temperature of a hot-fluid stream in a process has to be reduced from $T_{h,i}$ to $T_{h,o}$ while simultaneously a cold-fluid stream temperature in the same process has to be increased from $T_{c,i}$ to $T_{c,o}$, as shown in Fig. 11.12.[†] We specify $T_{h,o} < T_{c,i}$ as shown in Fig. 11.12. Note that it is possible that in some applications $T_{c,o}$ may be desired to be even greater than $T_{h,i}$. For the sake of simplicity, both fluids are considered to be simple substances and have heat capacity rates C_h and C_c. In addition, cold and hot thermal sources (usually called *utilities*—the heat sources or sinks, or yet other process streams) are available if the required enthalpy changes cannot be accomplished utilizing the cold stream alone. Note that the general validity of this analysis is not restricted by adopting an assumption that the fluids are simple compressible substances within the given temperature ranges and away from the onset of phase-change phenomena.

Utilizing Eq. (2.1), and writing the outlet temperatures explicitly for the two streams leaving the process, one obtains

$$T_{j,o} = T_{j,i} \pm \frac{q}{C_j} = T_{j,i} \pm \frac{\Delta \dot{H}}{C_j} \qquad (11.52)$$

where $j = h$ or c. The enthalpy rate change, $\Delta \dot{H}$, equals the heat transfer rate q in the exchanger. Relationships defined by Eq. (11.52) can be interpreted as line segments in a $(T, \dot{H})$ graph, each with a slope equal to the reciprocal value of the corresponding heat capacity rate. These graphs are presented in Fig. 11.12 (for a selection of given inlet and outlet temperatures and heat capacity rates) for the three coupled exchangers shown underneath. They are cold utility, main, and hot utility exchangers.[‡] Note that equalization of heat capacity rates would lead to parallel temperature–enthalpy rate change lines. In Fig. 11.12, the temperature–enthalpy rate change lines are not parallel, therefore, the corresponding heat capacity rates are not equal. The greater the imbalance between the heat capacity rates, the more pronounced is the nonuniform temperature difference $(T_{h,\text{hex}} - T_{c,\text{hex}})$ distribution along the enthalpy rate change axis.

The situation presented in Fig. 11.12 can be interpreted as a general case situated between the two limiting designs—to accomplish the design goal (to reduce the hot-fluid enthalpy and to increase the cold-fluid enthalpy)—a designer may distribute the total heat load between (1) the main heat exchanger, (2) the cold utility heat exchanger, and (3) the hot utility heat exchanger, shown in Fig. 11.12. This may be done in such a way as to exploit the heat recovery in the overlapping region of the temperature–enthalpy rate change diagram (note: $T_{h,o} < T_{c,i}$ and $T_{c,o} < T_{h,i}$). An important question for the exchanger designer is the following: What would be the optimum thermal size, UA, of the main heat exchanger?

In a limiting case, the size of the main heat exchanger may be very large, say close to an infinitely large thermal size (i.e., NTU $\to \infty$). The outlet temperature of the hot fluid from that exchanger $(T_{h,o,\text{hex}})$ will become very close to the inlet temperature of the cold stream $(T_{c,i})$. That is, the minimum temperature difference between the fluids $\Delta T_{\min} = T_{h,o,\text{hex}} - T_{c,i,\text{hex}} = T_{h,o,\text{hex}} - T_{c,i}$ (the so-called "pinch") becomes very close to zero. Note that an increase in size of the main heat exchanger in the temperature–enthalpy diagram (Fig. 11.12) may easily be presented by shifting horizontally the

---

[†] It should be emphasized that the required heating and cooling of cold- and hot-fluid streams of Fig. 11.12 may need multiple (and not one) heat exchangers.

[‡] In Fig. 11.12, the main exchanger has the hot process stream on one fluid side and the cold process stream on the other fluid side, so that thermal energy is recovered from the hot process stream, heating the cold process stream (as desired) without any additional expense of utility streams. The cold utility exchanger cools the hot stream coming out of the main exchanger and the hot utility exchanger heats the cold stream coming out of the main exchanger. The utility streams are water, steam, or air in general.

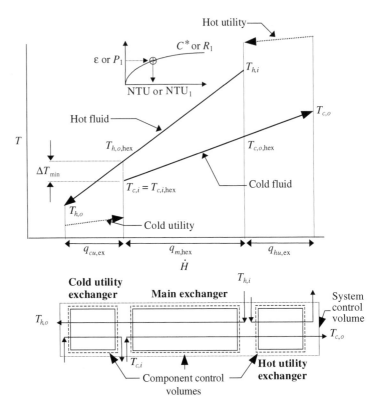

FIGURE 11.12 Heat exchanger temperature–enthalpy rate difference diagram. The energy balance control volumes (CVs) denote system component boundaries. The subscript "hex" denotes the main heat exchanger under consideration.

temperature curves toward each other until $\Delta T_{\min} = 0$ (note that $q_{m,\text{hex}}$ on the abscissa becomes the largest possible). This hypothetical design would reduce the requirement for utilities to a minimum ($q_{\text{hu,ex}}$ and $q_{\text{cu,ex}}$ become the smallest possible). Note that this idealized design cannot eliminate the need for either a hot or cold utility (the hot and cold fluid lines will not necessarily overlap entirely) for the given C_c and C_h, but it would reduce both utilities to the minimum. Still, this design is not a realistic one (i.e., it would not be a workable design) because it requires an infinitely large main heat exchanger.

Another limiting case would be to invest virtually nothing into the design of the main heat exchanger and to accomplish the required task by (1) cooling the hot fluid exclusively with a cold utility, and (2) heating the cold fluid with a hot utility. The temperature–enthalpy diagram in that case would feature the absence of the overlapping of the T–$\Delta \dot{H}$ lines, such as by shifting the T–$\dot{H}$ line for the cold fluid to the right until $T_{c,i}$ is vertically in the same location as $T_{h,i}$. This solution would require a maximum possible energy cost (no heat recovery at all) but at the same time would reduce the investment in the main heat exchanger to zero ($q_{m,\text{hex}} = 0$). For this example, it is assumed that hot and cold utilities would be available without an additional need for capital investment. This design is clearly workable (see Example 11.6 on the next page).

It is obvious that an optimum design must be somewhere between these two limiting cases. The objective functions may be, say, the total cost (involving the energy cost and number and size of the units required), the physical size of the exchanger, the exergy losses (see the following sections), and the like. Our intention in this section, however, is not to study thermoeconomic optimization and/or to extend the analysis considering a whole system to perform an integration of a heat exchanger network using either pinch analysis or exergy (or entropy generation) analysis. Our goal is less ambitious—to formulate the design optimization objective problem related to a single exchanger or a couple of them, and to show how an optimization problem arises.

In a more general case, most notably for the case of a heat exchanger network, the temperature–enthalpy rate change diagram becomes a valuable tool in constructing *composite curves*. Namely, instead of presenting temperature–enthalpy rate change lines for only two streams of a single heat exchanger, one composite T–$\dot{H}$ curve for all hot fluid streams and another such composite curve for all cold fluid streams within the corresponding temperature ranges can be depicted as well. Each such line has a slope related to the sum of all heat capacity rates of the involved streams within the given temperature range. This analysis, though, is beyond the scope of our interest. More details about the procedure of constructing the composite curves, very useful for the *pinch technology* method, can be found in the literature devoted to process synthesis and integration (e.g., Linnhoff et al., 1982; Gunderson and Naess, 1988; Sama, 1995a).

11.6.2   Analysis Based on an Energy Rate Balance

As emphasized in Chapter 9, an *optimum design*, or in terms of a practically achievable design solution, a *nearly optimum design*, is an engineering goal imposed by real-world requirements and constraints. For example, a heat exchanger positioned at a certain location in a large chemical engineering plant may fully satisfy the system purpose. Different integration and/or heat exchanger redesign may lead to a system that would be even more efficient or cost effective. However, the design requirements are, as a rule, conflicting propositions. For example, a smaller temperature difference in a heat exchanger leads to a larger system thermal efficiency but requires larger heat exchanger size and often more pumping power, which may, in turn, cause a significant system objective deterioration.

An optimization of a heat exchanger as a component integrated into a system may involve not only the traditional tools of thermal design (based almost exclusively on the first law of thermodynamics) but also an integrated approach known as *thermoeconomics*. This approach includes simultaneous analysis based on (1) energy rate balances, (2) exergy rate balances and/or entropy generation calculations, and (3) cost balances.

To illustrate a problem that involves determination of the physical size of an exchanger for which *only energy rate balances* are sufficient, we present Example 11.6. This example is based on a study of an optimum area allocation in a system of coupled heat exchangers as reported by Alt-Ali and Wilde (1980). This example is similar to the problem of Fig. 11.12, but without the main heat exchanger.

Example 11.6   A fluid stream with heat capacity rate of 52,740 W/K needs to be heated from 38°C (311 K) to 260°C (533 K). Two hot-fluid streams are available for accomplishing this task. These hot-fluid streams have the same heat capacity rates as that of the cold-fluid stream. The inlet temperatures of two hot fluids are 149°C (422 K) and 316°C

(589 K). The task can be accomplished either by using only one two-fluid heat exchanger (and utilizing only the stream with the higher temperature) or using two two-fluid heat exchangers connected in a two-stage array with the cold fluid flowing through both exchangers. The available two hot streams (the utilities) should be used in the two exchangers. Determine the optimal distribution of heat transfer surface areas of the two exchangers to get the minimum total heat transfer area sufficient to heat the cold stream to the desired temperature. Compare the two heat exchanger solution with the design having one heat exchanger. Assume constant and uniform overall heat transfer coefficients in both units to be equal to 454 $W/m^2 \cdot K$.

SOLUTION

Problem Data and Schematic: Two heat exchangers arranged in a two-stage array with only one cold stream flowing through both units is shown in Fig. E11.6. For a minimum surface area requirement, both exchangers are considered to be counterflow. The input data are given in Fig. E11.6.

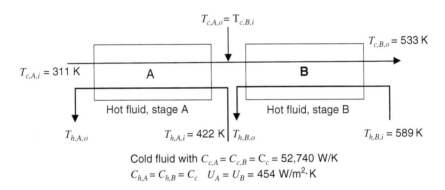

FIGURE E11.6   Two-stage heat exchanger system with operating conditions.

Determine: The heat transfer surface area of each of the two exchangers for a minimum total heat transfer surface area. Compare that solution to a single-exchanger solution.

Assumptions: All appropriate assumptions for modeling each of the exchangers of Section 3.2.1 are invoked. Additionally, it is assumed that both units have equal overall heat transfer coefficients.

Analysis: We need to formulate a functional relationship between total surface area A_{tot} and cold-fluid interstage temperature $T = T_{c,A,o} = T_{c,B,i}$ (shown in Fig. E11.6) for a minimum total surface area: Find

$$\min\{A_{tot} = f(T)\} \tag{1}$$

where

$$A_{tot} = A_A + A_B \quad \text{and} \quad T = T_{c,A,o} = T_{c,B,i} \tag{2}$$

Therefore, the problem is characterized with the following variables: (1) the interstage temperature T of the cold fluid is a decision variable, (2) the outlet temperatures ($T_{h,A,o}$ and $T_{h,B,o}$) of the two hot fluids are dependent variables, and (3) the independent variables (i.e., parameters of optimization) are: the inlet and outlet temperatures ($T_{c,A,i}$ and $T_{c,B,o}$) of the cold fluid, the inlet temperatures ($T_{h,A,i}$ and $T_{h,B,i}$) of the hot fluids, the heat capacity rates of both fluids, and the overall heat transfer coefficients of both units.

To define the objective function, $A_{\text{tot}} = f(T)$, first let us compute the surface areas A_A and A_B. To accomplish that task we have to utilize the first law of thermodynamics by formulating *energy rate balances* (i.e., equating the heat transfer rates and fluid enthalpy rates imposed by the first law of thermodynamics for each fluid side):

$$(UA)_j \, \Delta T_{\text{lm},j} = C_{c,j}(T - T_{c,j,i}) \tag{3}$$

where $j = A$ or B. Hence, the individual surface areas A_A and A_B from Eq. (3) are

$$A_A = \frac{C_{c,A}(T - T_{c,A,i})}{U_A \, \Delta T_{\text{lm},A}} \qquad A_B = \frac{C_{c,B}(T_{c,B,o} - T)}{U_B \, \Delta T_{\text{lm},B}} \tag{4}$$

As emphasized above, Eq. (3) is based on energy and mass conservation principles. Hence, all the subsequent results are the consequences of the first law of thermodynamics (energy equation) and the continuity equation only.

The total area is defined as follows:

$$A_{\text{tot}} = A_A + A_B = \frac{C_{c,A}(T - T_{c,A,i})}{U_A \, \Delta T_{\text{lm},A}} + \frac{C_{c,B}(T_{c,B,o} - T)}{U_B \, \Delta T_{\text{lm},B}} \tag{5}$$

where

$$\Delta T_{\text{lm},A} = \frac{(T_{h,A,i} - T) - (T_{h,A,o} - T_{c,A,i})}{\ln[(T_{h,A,i} - T)/(T_{h,A,o} - T_{c,A,i})]} \qquad \Delta T_{\text{lm},B} = \frac{(T_{h,B,i} - T_{c,B,o}) - (T_{h,B,o} - T)}{\ln[(T_{h,B,i} - T_{c,B,o})/(T_{h,B,o} - T)]} \tag{6}$$

In Eqs. (5) and (6), the outlet temperatures $T_{h,A,o}$ and $T_{h,B,o}$ are dependent on the interstage temperature T and can be found again from energy conservation principles. This time, however, we should formulate *enthalpy rate balances* for fluids in both units; that is, for units A and B,

$$C_{c,A}(T - T_{c,A,i}) = C_{h,A}(T_{h,A,i} - T_{h,A,o}) \qquad C_{c,B}(T_{c,B,o} - T) = C_{h,B}(T_{h,B,i} - T_{h,B,o}) \tag{7}$$

Therefore,

$$T_{h,A,o} = T_{h,A,i} - \frac{C_{c,A}}{C_{h,A}}(T - T_{c,A,i}) \qquad T_{h,B,o} = T_{h,B,i} - \frac{C_{c,B}}{C_{h,B}}(T_{c,B,o} - T) \tag{8}$$

Substitution of $T_{h,A,o}$ and $T_{h,B,o}$ of Eq. (8) in Eqs. (5) and (6) yields an objective function $A_{\text{tot}} = f(T)$, where T is a decision variable. Formally, to solve the problem, we should find an extremum of this function with respect to the decision variable (i.e., $\partial A_{\text{tot}}/\partial T = 0$). Performing this partial differentiation, we get (Alt-Ali and Wilde, 1980)

$$\frac{\partial A_{\text{tot}}}{\partial T} = \frac{C_{c,A}}{U_A} \frac{\Delta_A}{(\Delta_A - \delta_A)(\Delta_A - R_A \delta_A)} - \frac{C_{c,B}}{U_B} \frac{1}{\Delta_B - R_B \delta_B} = 0 \tag{9}$$

where

$$
\begin{array}{llll}
\delta_A = T - T_{c,A,i} & \delta_B = T_{c,B,o} - T & R_A = \dfrac{C_{c,A}}{C_{h,A}} & R_B = \dfrac{C_{c,B}}{C_{h,B}} \\
\Delta_A = T_{h,A,i} - T_{c,A,i} & \Delta_B = T_{h,B,i} - T & &
\end{array}
\tag{10}
$$

Temperature differences denoted as δ's represent the temperature differences between terminal temperatures of the cold fluid of concern for each of the two exchangers, while Δ's represent inlet temperature differences for the fluids in both heat exchangers, respectively. So the final results will depend not only on the individual temperature changes accomplished within the two exchangers, but also on the magnitudes of the inlet temperature differences. Rearranging Eq. (9) leads to

$$
\frac{C_{c,A}/U_A}{C_{c,B}/U_B} = \frac{\Delta_A - \delta_A}{\Delta_A}\frac{\Delta_A - R_A\delta_A}{\Delta_B - R_B\delta_B} = \frac{T_{h,A,i} - T_{\mathrm{opt}}}{T_{h,A,i} - T_{c,A,i}}\frac{T_{h,A,o} - T_{c,A,i}}{T_{h,B,o} - T_{\mathrm{opt}}}
\tag{11}
$$

where T_{opt} represents the value of the interstage temperature of the cold fluid for $\partial A_{\mathrm{tot}}/\partial T = 0$ (i.e., the optimum interstage temperature).

Finally, the optimum interstage temperature of the cold fluid T_{opt} can be obtained from Eq. (11) after introducing the definitions of Δ, δ, and R's as follows:

$$
T_{\mathrm{opt}} = T_{h,A,i} - \left[(T_{h,A,i} - T_{c,A,i})(T_{h,B,i} - T_{c,B,o})\frac{C_{c,A}/U_A}{C_{c,B}/U_B}\right]^{1/2}
\tag{12}
$$

For the given problem, inserting the numerical values of the variables in Eq. (12), we get

$$
T_{\mathrm{opt}} = 422\,\mathrm{K} - \left[(422 - 311)\,\mathrm{K}\,(589 - 533)\,\mathrm{K}\,\frac{(52{,}740\,\mathrm{W/K})/(454\,\mathrm{W/m^2 \cdot K})}{(52{,}740\,\mathrm{W/K})/(454\,\mathrm{W/m^2 \cdot K})}\right]^{1/2}
$$

$$
= 344\,\mathrm{K}
\tag{13}
$$

Now, replacing $T = T_{\mathrm{opt}} = 344\,\mathrm{K}$ into Eq. (4) and introducing other known variables, the heat transfer surface areas become $A_A = 48.1\,\mathrm{m}^2$ and $A_B = 395.5\,\mathrm{m}^2$. Total heat transfer area is $A_{\mathrm{tot}} = A_A + A_B = 444.6\,\mathrm{m}^2$. In a trivial case, if only one heat exchanger is used (with the hot fluid having an inlet temperature of 589 K), the heat transfer surface area can easily be calculated from Eq. (3) for $j = 1$. In such a case, this area would be $464.5\,\mathrm{m}^2$, which is 4.5% more than $A_{\mathrm{tot}} = 444.6\,\mathrm{m}^2$ for the optimum area for two exchangers.

Discussion and Comments: This example demonstrates how a proper distribution of heat transfer surface area can lead to an optimal design. This conclusion has been reached utilizing only energy and mass flow rate balances. The fact that a single heat exchanger would have a larger heat transfer surface area (compared to a two-heat-exchanger solution) does not mean that the solution with two exchangers would be the desired design solution. Cost (in addition to other considerations) may be a key decision factor for the best achievable design. Note that temperature differences between the fluids at the terminal points of both heat exchangers have an important role in the determination of the optimum solution. However, the temperature differences defined only on the basis of total area minimization (i.e., utilizing only the first law of thermodynamics and mass balances) do not necessarily minimize the entropy generation in the assembly or lead to the most economic solution. For this to be accomplished, a combined thermodynamic and economic analysis has to be performed as discussed in the following section.

11.6.3 Analysis Based on Energy/Enthalpy and Cost Rate Balancing

Let us now illustrate how cost rate balances may influence the design solution. Of course, the simultaneous use of energy balances is mandatory as well. This time, the objective function will be the total annual cost as a function of the cold-fluid exit temperature, keeping the heat duty fixed in a heat exchanger. Obviously, variation of the exit temperature will change exchanger ε or P_1 as well as cause a change in the coolant mass flow rate for fixed q. Thus, C^* or R_1 will change. This will yield a different NTU and hence A. Consequently, a search for the most favorable design based on an economic criterion would require resizing the exchanger.

Both economics and thermodynamics will influence the solution. It is important to notice that the corresponding thermodynamic part of the analysis would imply only the first law of thermodynamics. How that has to be accomplished can be illustrated best by using an example. The example represents a slightly modified problem presented by Peters and Timmerhaus (1980).

Example 11.7   Design a condenser in a distillation unit to operate at optimum total annual cost. The particular unit under consideration uses cooling water to condense vapor. It operates at a minimum total annual cost if water leaves the condenser at 52°C (325 K), and at the given inlet fluid conditions and with installed optimal heat transfer area. The outlet temperature of the cooling water, however, may increase to 57°C (330 K) due to changing environmental considerations with a corresponding decrease in the cooling-water mass flow rate and appropriate resizing of the exchanger. Assuming that all inlet variables (except for the mass flow rate of water) must stay unchanged, determine how large the change of total annual cost would be in excess of the established optimum value for the given design, if the mentioned changes take place. The additional information is as follows. The heat exchanger condenses $\dot{m} = 2000\,\text{kg/h}$ of vapor, which has an enthalpy of phase change of $4 \times 10^5\,\text{J/kg}$. Condensation occurs at 77°C (350 K). The inlet temperature of the cooling water is 17°C (290 K), and the specific heat of water at constant pressure is $4.2 \times 10^3\,\text{J/kg} \cdot \text{K}$. The overall heat transfer coefficient is $280\,\text{W/m}^2 \cdot \text{K}$. The distillation unit must operate for $\tau_o = 6500\,\text{hr/y}$. The unit cost of cooling water is $2 \times 10^{-5}\,\mathscr{C}/\text{kg}$, where $\mathscr{C}$ is the monetary unit. The unit cost for the heat exchanger per installed unit of heat transfer area is $300\,\mathscr{C}/\text{m}^2$. The annual cost of heat exchanger operation is 20% of the cost of installed heat exchanger area.

SOLUTION

Problem Data and Schematic: In the condenser, the cooling water increases its temperature along the flow path, and the condensing stream has a constant temperature, as shown in Fig. E11.7A. All pertinent data are provided in this figure.

Determine: The increase in total cost in excess of the optimum value when the exit temperature of the cooling water increases by 5 K.

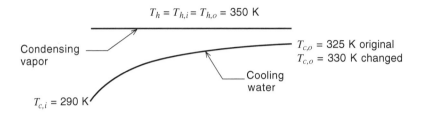

$$c_{p,w} = 4200 \text{ J/kg·K}, \quad \mathbf{h}_{\ell g} = 4 \times 10^5 \text{J/kg}, \quad U = 280 \text{ W/m}^2\text{·K} \quad \dot{m}_h = 2000 \text{ kg/h}$$
$$\tau_o = 6500 \text{ h/yr}, \quad C_w = 2 \times 10^{-5} \text{ } \mathscr{C}/\text{kg}, \quad C_A = 300 \text{ } \mathscr{C}/\text{m}^2, \quad c_A = 0.2/\text{yr}$$

FIGURE E11.7A Schematic of temperature distributions in a condenser.

Assumptions: It is assumed that Eq. (3.12) is valid. The mean temperature difference in that equation is assumed to be equal to the log-mean temperature difference determined using terminal temperature differences. Assume the total annual cost of heat exchanger operation to be $C_{\text{tot}} = C_w + C_A$, where C_w is the cost of the total amount of cooling water, and C_A is the fixed cost of the heat exchanger installed.

Analysis: Let us start with evaluating the heat transfer rate q:

$$q = \dot{m}_h \mathbf{h}_{\ell g} = (2000 \text{ kg/h})(4 \times 10^5 \text{ J/kg}) = 8 \times 10^8 \text{ J/h}$$

The total annual cost of the condenser is equal to

$$C_{\text{tot}} = C_w + C_A \tag{1}$$

Therefore,

$$C_{\text{tot}} = \tau_o C_w \dot{m}_2 + \mathbf{c}_A C_A A = \tau_o C_w \frac{q}{c_p(T_{c,o} - T_{c,i})} + \mathbf{c}_A C_A \frac{q}{U \, \Delta T_m}$$

$$= (6500 \text{ h/yr})(2 \times 10^{-5} \text{ } \mathscr{C}/\text{kg}) \left[\frac{8 \times 10^8 \text{ J/h}}{(4.2 \times 10^3 \text{ J/kg · K})(T_{c,o} - 290 \text{ K})} \right]$$

$$+ 0.2 \frac{1}{\text{yr}} \times 300 \, \mathscr{C}/\text{m}^2 \left[\frac{(8 \times 10^8 \text{ J/h})(1 \text{ h}/3600 \text{ s})}{(280 \text{ W/m}^2 \cdot \text{K}) \Delta T_m} \right]$$

$$= \frac{24{,}762}{(T_{c,o} - 290 \text{ K})} \frac{\mathscr{C}}{\text{yr}} + \frac{0.4762 \times 10^5}{\Delta T_m} \frac{\mathscr{C}}{\text{yr}} \tag{2}$$

where

$$\Delta T_m = \Delta T_{\text{lm}} = \frac{(T_h - T_{c,i}) - (T_h - T_{c,o})}{\ln \left(\dfrac{T_h - T_{c,i}}{T_h - T_{c,o}} \right)} = \frac{T_{c,o} - T_{c,i}}{\ln \left(\dfrac{T_h - T_{c,i}}{T_h - T_{c,o}} \right)} = \frac{T_{c,o} - 290 \text{ K}}{\ln \left[\dfrac{(350 - 290) \text{ K}}{350 \text{ K} - T_{c,o}} \right]} \tag{3}$$

The relationship of Eq. (2) is presented in Fig. E11.7B. The minimum value of total cost, $C_{tot,min} = 1898\ \mathscr{C}/yr$, is at $T_{c,o} = 325$ K, as suggested in the problem formulation. This can be confirmed easily by finding the first derivative of the objective function, Eq. (2), with respect to the coolant exit temperature, $T_{c,o}$ (i.e., $\partial C_{tot}/\partial T_{c,o} = 0$). This minimum point (see Fig. E11.7B), corresponds to the designed heat exchanger heat transfer area of $20\,\text{m}^2$ [calculated from the rate equation, Eq. (2.2)]. If the outlet temperature of the coolant increases to 330 K, the total cost will increase to $C_{tot} = 1927\ \mathscr{C}/yr$, a 1.5% higher value than the value at the optimal point. To keep the duty unchanged, the heat transfer area changes to approximately $22\,\text{m}^2$, and the coolant mass flow rate will change from 1.512 kg/s to 1.323 kg/s.

Discussion and Comments: In this example, the objective function is the total annual cost of heat exchanger operation. The optimum (minimum) total cost corresponds to the outlet coolant temperature equal to 325 K. Any change in this temperature incurred through a change of heat exchanger size (area) causes some increase in the total cost. At the same time, both the coolant mass flow rate and the exchanger heat transfer area must change to keep the heat load fixed. In this analysis, we utilized only the first law of thermodynamics to formulate energy balances (used implicitly for setting up the cost rate balance).

This example illustrates how an optimum design may lead to an economic penalty if a heat exchanger operates at off-design. In this case, only the exchanger surface area and the cooling water utility are considered for cost. The optimization problem required a minimum annual variable cost as a function of the exit temperature. Equations (2.1) and (2.2) defined the heat exchanger model. The temperature at optimum (i.e., at the minimum cost) is confirmed to be 325 K from Fig. E11.7B. The off-design condition requires a heat exchanger size change, as a result of a change in the exit temperature (for other variables constant except for the mass flow rate). Note that this economic optimum was not related in any way to a thermodynamic irreversibility minimization.

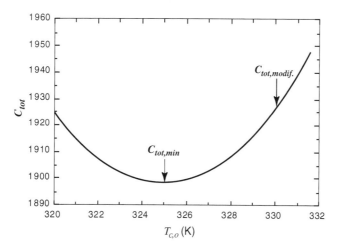

FIGURE E11.7B Total annual variable operating cost of a heat exchanger vs. coolant exit temperature.

11.6.4 Analysis Based on an Exergy Rate Balance

In previous sections, we demonstrated the utilization of both energy and cost rate balances. These analyses, though, have not considered the quality of energy rate flows. Under the term *energy rate flow*, we consider either the enthalpy rate or the heat transfer rate. Consequently, we should examine more closely the role of the irreversibility analysis for heat exchanger thermal design and optimization. In this section, we formulate the *exergy rate balance*. In the following two sections, we use this concept to define a thermodynamic figure of merit (thermodynamic efficiency of a heat exchanger) and to introduce the cost of irreversibility.

Let us assume that a heat exchanger represents a component of a process or power plant. The plant can be optimized for the total annual cost of operation. However, to find the cost of the related irreversibilities (see Section 11.6.6), we should define the irreversibility on an energy basis in a quantitative manner. This cost can be calculated as a sum of the *cost of compensation for the irreversibility* and the *capital investment*. Irreversibility may be measured by its energy measure—either the entropy generation multiplied by the appropriate weighting temperature factor (Ahrendts, 1980), or exergy (Kotas, 1995). So far in our analysis we measured irreversibilities in terms of entropy generation. As is known from thermodynamics (Moran and Shapiro, 1995), irreversibility can conveniently be defined in terms of energy rate units using the *exergy rate* as follows (excluding chemical exergy):

$$\dot{\mathscr{E}}_{j,k} = \dot{m}_j[(\mathbf{h}_{j,k} - \mathbf{h}_{\text{ref}}) - T_{\text{ref}}(s_{j,k} - s_{\text{ref}})] \tag{11.53}$$

The exergy rate represents the rate of the available energy of a given fluid stream with respect to the conveniently selected reference state. Each fluid stream (entering or leaving the heat exchanger; see Fig. 11.13) carries certain exergy, defined by Eq. (11.53). In this equation, the subscript j denotes either fluid 1 or fluid 2, or alternatively, the hot or cold fluid, k denotes inlet or outlet, and "ref" denotes a state of the respective fluid at the thermodynamic condition defined by a selected reference state (often, but not always, the state of the environment). Exergy does not obey a conservation principle because it includes not only the properties of a *thermodynamic system* at an exchanger terminal port, but also the reference thermodynamic state. In our case, the term *thermodynamic system* refers to whichever fluid stream exposed to heat transfer in a heat exchanger. In many applications, that state can simply be a state of the system in thermodynamic equilibrium with the environment. The usefulness of the concept of exergy is in providing a reference level regarding the maximum possible useful energy potential that would be available from a particular energy source with respect to the surroundings. The quality of energy flow rate can be interpreted by a simple example as follows. The quality of the same amount of available energy from a fluid at 500°C is higher than for the same fluid at 50°C if both have to be used (of course at different flow rates) to heat another fluid entering at 0°C in a heat exchanger. The first fluid stream would have a larger exergy rate and would therefore be able to transfer more heat over a wider temperature span. In other words, its potential to do work is high and its use at the low temperature level would be wasteful if there is a second stream at a lower temperature (< 500°C) that is available to do the same job.

Applied to a situation in a heat exchanger, the exergy balance can be closed only by introducing into the balance the *exergy destruction* $\mathscr{D}$. Consequently, the exergy rate balance for a heat exchanger (Fig. 11.13) may be written as follows:

$$\mathscr{D} = (\dot{\mathscr{E}}_{1,i} - \dot{\mathscr{E}}_{1,o}) + (\dot{\mathscr{E}}_{2,i} - \dot{\mathscr{E}}_{2,o}) \tag{11.54}$$

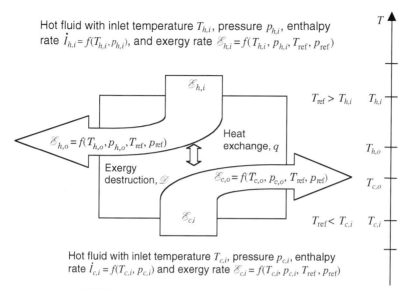

Hot fluid with inlet temperature $T_{h,i}$, pressure $p_{h,i}$, enthalpy rate $\dot{I}_{h,i} = f(T_{h,i}, p_{h,i})$, and exergy rate $\mathscr{E}_{h,i} = f(T_{h,i}, p_{h,i}, T_{ref}, p_{ref})$

$\mathscr{E}_{h,i}$

$\mathscr{E}_{h,o} = f(T_{h,o}, p_{h,o}, T_{ref}, p_{ref})$

Heat exchange, q

Exergy destruction, $\mathscr{D}$

$\mathscr{E}_{c,o} = f(T_{c,o}, p_{c,o}, T_{ref}, p_{ref})$

$\mathscr{E}_{c,i}$

$T_{ref} > T_{h,i} \qquad T_{h,i}$

$T_{h,o}$

$T_{c,o}$

$T_{ref} < T_{c,i} \qquad T_{c,i}$

Hot fluid with inlet temperature $T_{c,i}$, pressure $p_{c,i}$, enthalpy rate $\dot{I}_{c,i} = f(T_{c,i}, p_{c,i})$ and exergy rate $\mathscr{E}_{c,i} = f(T_{c,i}, p_{c,i}, T_{ref}, p_{ref})$

FIGURE 11.13   Exergy rate flows through a heat exchanger.

where the two expressions in parentheses on the right-hand side of the equality represent individual exergy destruction rates for each of the two fluids. The total exergy destruction rate given by Eq. (11.54), $\mathscr{D}$ in watts (Btu/hr), represents the quantitative energy measure of irreversibility. As demonstrated by Bosnjaković (1972), this irreversibility can also be calculated from entropy generation using the Gouy–Stodola theorem as follows:

$$\mathscr{D} = T_{ref}\dot{S}_{irr} \tag{11.55}$$

Equation (11.55) obviously implies a need for insight into the heat transfer thermodynamic intricacies (expressed in terms of entropy generation).

We started the analysis in this chapter with the argument that insight into temperature distributions must be gained and that the heat exchanger designer should understand how the exchanger performance depends on the most important sources of irreversibility, $\dot{S}_{irr}$. Now we can clearly see that this understanding may contribute to a reduction of exergy destruction in a system. A system engineer knows how to relate that destruction directly to the monetary value of the capital investment and operating costs and to optimize the system by providing appropriate changes in heat exchanger design. By writing exergy rate balances (or calculating entropy generation), energy rate measures of irreversibilities can be determined and subsequently their monetary value calculated (see Section 11.6.6).

11.6.5   Thermodynamic Figure of Merit for Assessing Heat Exchanger Performance

Let us now introduce a thermodynamic figure of merit[†] in the form of exergy efficiency for the performance of a heat exchanger in a system. The performance level may be

---

[†] The second law efficiency is defined in many different ways and they are in general referred to as the *thermodynamic figure of merit*.

defined using the concept of exergy. The exergy balance for a heat exchanger, based on Eq. (11.54), states simply that exergy entering a heat exchanger (carried in by both fluid streams) must be equal to the sum of exergy leaving the exchanger and the destruction of exergy caused by heat exchanger operation (as a consequence of irreversibilities), that is

$$\dot{\mathscr{E}}_i = \dot{\mathscr{E}}_o + \mathscr{D} \tag{11.56}$$

where

$$\dot{\mathscr{E}}_i = \dot{\mathscr{E}}_{1,i} + \dot{\mathscr{E}}_{2,i} \quad \text{and} \quad \dot{\mathscr{E}}_o = \dot{\mathscr{E}}_{1,o} + \dot{\mathscr{E}}_{2,o} \tag{11.57}$$

The exergy balance given by Eq. (11.56) can be rearranged as follows:

$$(\dot{\mathscr{E}}_{1,o} - \dot{\mathscr{E}}_{1,i}) + \mathscr{D} = \dot{\mathscr{E}}_{2,i} - \dot{\mathscr{E}}_{2,o} \tag{11.58}$$

In a particular case, when fluids 1 and 2 are identified as cold and hot fluids, respectively, Eq. (11.58) indicates that the exergy increase in the cold fluid plus the exergy destruction must be equal to the exergy decrease in the hot fluid utilized in this process. Equation (11.58) in such a case can be divided by the difference in the exergy rates on the right-hand side of the equality, to obtain

$$\frac{\dot{\mathscr{E}}_{c,o} - \dot{\mathscr{E}}_{c,i}}{\dot{\mathscr{E}}_{h,i} - \dot{\mathscr{E}}_{h,o}} + \frac{\mathscr{D}}{\dot{\mathscr{E}}_{h,i} - \dot{\mathscr{E}}_{h,o}} = 1 \tag{11.59}$$

Now define an *exergy efficiency* η of a heat exchanger as

$$\eta = \frac{\dot{\mathscr{E}}_{c,o} - \dot{\mathscr{E}}_{c,i}}{\dot{\mathscr{E}}_{h,i} - \dot{\mathscr{E}}_{h,o}} = 1 - \frac{\mathscr{D}}{\dot{\mathscr{E}}_{h,i} - \dot{\mathscr{E}}_{h,o}} \tag{11.60}$$

In general, the exergy efficiency η of a heat exchanger is defined as follows:

$$\eta = \begin{cases} \dfrac{\dot{\mathscr{E}}_{c,o} - \dot{\mathscr{E}}_{c,i}}{\dot{\mathscr{E}}_{h,i} - \dot{\mathscr{E}}_{h,o}} & \text{for } T_{c,i} \geq T_{\text{ref}} \\[4mm] \dfrac{\dot{\mathscr{E}}_{h,o} - \dot{\mathscr{E}}_{h,i}}{\dot{\mathscr{E}}_{c,i} - \dot{\mathscr{E}}_{c,o}} & \text{for } T_{c,i} \leq T_{\text{ref}} \end{cases} \tag{11.61}$$

The efficiency defined by Eq. (11.61), $\eta \leq 1$, takes care of different objectives of a heat exchanger used in a system. Namely, if the purpose of a heat exchanger is to increase the exergy rate (i.e., energy availability or energy quality) of a cold-fluid stream (at the expense of a decrease of exergy rate of the hot-fluid stream), η is calculated using the first expression of Eq. (11.61). An objective of heating a fluid by cooling another fluid is an example for the first expression of η of Eq. (11.61) (the heat source may be waste thermal energy, a heat exchanger in a heat pump, etc.). On the contrary, if the purpose of the exchanger is to cool the hot stream at or below the reference temperature (refrigeration for temperatures below the environment temperature), the second expression of Eq. (11.61) should be used.

The change of design parameters may lead to different behaviors of the heat exchanger effectiveness and thermodynamic efficiency for the same heat exchanger. Let us demonstrate this fact by an example.

Example 11.8   Assess the following two designs for a counterflow heat exchanger. The heat exchanger is to heat 1 kg/s of air from 227°C (500 K) and 3 bar to 307°C (580 K). The hot-fluid stream is also air at the inlet temperature of 327°C (600 K) and 2 bar. The two options available are as follows: The heat exchanger is designed for a hot-fluid mass flow rate of either 1.24 kg/s or 4.94 kg/s. Decide which design will yield better exergy efficiency in the system. The surroundings is at 300 K and 1 bar.

SOLUTION

Problem Data: Operating conditions: (a) cold fluid is air with $T_{c,i} = 500$ K, $T_{c,o} = 580$ K, $\dot{m}_c = 1$ kg/s, $p_{c,i} = 3$ bar; (b) hot fluid is air with $T_{h,i} = 600$ K, $p_{h,i} = 2$ bar, $\dot{m}_h = 1.24$ or 4.94 kg/s. The heat exchanger is a counterflow unit. The surroundings is at $T_{ref} = 300$ K, and $p_{ref} = 1$ bar.

Determine: Which of the two fluid flow rates will lead to a design with higher exergy efficiency?

Assumptions: The assumptions are as listed in Section 3.2.1. Air is assumed to behave as an ideal gas. Pressure drops are negligible, i.e., $p_{c,i} = p_{c,o}$ and $p_{h,i} = p_{h,o}$. Idealize the same overall heat transfer coefficient in both designs. Assume surroundings to be relevant for exergy definition (i.e., at reference conditions as follows: $T_{ref} = 300$ K and $p_{ref} = 1$ bar).

Analysis: A study of input data reveals the following. The heat capacity rates of the two fluids are not known, but we do know that they are not the same because the mass flow rates are different. The specific heats of both fluids are not known. Note that the specific heat of an ideal gas is a function of the temperature only. So, to determine the heat capacity ratio, we have to know the exit temperature of the hot fluid. However, the exit temperature of the hot fluid cannot be determined a priori because the specific heat is not known. However, both fluids are of the same type (i.e., air) and we can idealize that the specific heats at constant pressure of both fluids will not differ significantly from each other. The mass flow rate of the hot fluid is specified as larger than that for the cold fluid (i.e., $\dot{m}_h = 1.24$ kg/s or 4.94 kg/s $> \dot{m}_c = 1$ kg/s). Therefore, in both cases, the heat capacity rate of the hot fluid will be larger than the corresponding value for the cold fluid. Hence, cold air is the $C_{\min}$ fluid, and $\varepsilon = \varepsilon_c$. The heat exchanger effectiveness represents in this case a dimensionless outlet temperature of the cold fluid ($\varepsilon = \varepsilon_c$, i.e., the cold fluid has a smaller heat capacity outlet rate, $C_c = C_{\min}$; see Section 11.2.2). Since both terminal temperatures of the cold fluid are known and fixed, the heat exchanger effectiveness ε_c and heat transfer rate q for both designs are the same (a subject of Review Question 3.26). However, because the heat capacity rate ratios are not the same, the two designs must have different NTUs. Different NTUs for the same U and $C_{\min}$ must correspond to different heat transfer areas.

One can arrive at the same conclusion by noting that the heat transfer rate, the inlet conditions for both fluids, and the smaller heat capacity rate all are predetermined. From the definition of heat exchanger effectiveness based on its thermodynamic interpretation (see Example 11.2), it becomes clear that the heat exchanger effectiveness must be identical in both cases. Therefore, the change in NTU is due to different heat transfer surface areas.

Thus, the two designs considered would have the same ε and q but different A's. So the two exchangers would extract the same heat transfer rate from the hot fluid but at different temperature levels and with different local temperature distributions. Hence, the irreversibility level (responsible for thermodynamic performance) would not be the same. The question is: Which of the two designs (the one with the larger or the one with the smaller mass flow rate of the hot fluid) will provide higher exergy efficiency?

To be able to use one of the two definitions of the exergy efficiency of Eq. (11.61), one should recognize the purpose of the device. According to the problem formulation, the goal is to heat the cold fluid at the expense of the hot fluid. Therefore, the exergy efficiency is defined by the first of the two equations given by Eq. (11.61). In other words, the increase in the exergy rate of the cold fluid will be accomplished at the expense of a decrease in the exergy rate of the hot fluid:

$$\eta = \frac{\dot{\mathcal{E}}_{c,o} - \dot{\mathcal{E}}_{c,i}}{\dot{\mathcal{E}}_{h,i} - \dot{\mathcal{E}}_{h,o}} \tag{1}$$

where

$$\dot{\mathcal{E}}_{j,k} = \dot{m}_j [(\mathbf{h}_{j,k} - \mathbf{h}_{ref}) - T_{ref}(s_{j,k} - s_{ref})] \tag{2}$$

with

$$\mathbf{h}_{j,k} = \phi_{j,k}\,(air,\, T = T_{j,k}) \qquad s_{j,k} = \phi'_{j,k}\,(air,\, T = T_{j,k},\, p = p_{j,k}) \tag{3}$$

In Eqs. (2) and (3), $j = h$ or c and $k = i$ or o. The condition of the reference state (and values of stream properties at the condition of the surroundings) is denoted by "ref."

The design problem formulated in this example is a sizing problem in which both NTUs and one outlet temperature are unknown. The outlet temperature of the hot fluid in terms of $T_{c,i}$ can be determined from the definition of ε and C^* as follows:

$$T_{h,o} = T_{c,i} - (1 - C^*\varepsilon)(T_{c,i} - T_{h,i}) \tag{4}$$

The calculation of exergy efficiency must be performed numerically using Eqs. (1)–(4) because thermophysical properties depend on the temperature or on both temperature and pressure [see Eqs. (3) and (4)]. The results of the calculation of exergies and η are listed below. These results are obtained using EES software (2000). Air is treated as an ideal gas. C^* and ε are calculated from input data and subsequently, NTU is calculated for these values of ε and C^* for a counterflow exchanger.

$\dot{m}_h$ (kg/s)	$T_{h,o}$ (K)	C^*	ε	NTU	$\dot{\mathcal{E}}_{c,i}$ (kW)	$\dot{\mathcal{E}}_{c,o}$ (kW)	$\dot{\mathcal{E}}_{h,i}$ (kW)	$\dot{\mathcal{E}}_{h,o}$ (kW)	η
1.24	536	0.8	0.8	2.94	142.1	178.9	191.7	152.6	0.941
4.94	584	0.2	0.8	1.79	142.1	178.9	762.5	721.6	0.899

A review of the results above indicates that the exergy efficiency is larger for the design that requires larger NTU (smaller mass flow rate of the hot fluid). Both designs are characterized by the identical heat exchanger effectiveness.

Discussion and Comments: The results obtained confirm a statement made at the beginning of this chapter that heat exchanger (or temperature) effectiveness does not necessarily provide sufficient information about exchanger performance. The differences between the terminal temperatures on the hot fluid inlet side are the same for both exchangers but the temperature difference between the fluids at the cold fluid inlet side increases from 36 K (for a smaller mass flow rate or a larger NTU) to 84 K (for a larger mass flow rate or a smaller NTU). The results clearly demonstrate that smaller local temperature differences along the respective flow paths lead to larger exergy efficiency (as a consequence of the smaller entropy generation for the smaller $\dot{m}_h$ case). Moreover, in the case of the smaller exergy efficiency, the exit temperature of the hot fluid (584 K) is quite close to the cold-fluid outlet temperature (580 K). To understand the implication, review the S^* vs. NTU curve for $C^* = 1$ for a counterflow exchanger in Fig. 11.5. The trend for the S^* vs. NTU curve for other values of C^* is identical except for $C^* = 0$ and S^*_{max} is of the same order of magnitude for all C^* values (Sekulić, 1990) for a given ϑ. Hence, for this problem, one can see that the operating point approaches from the left to S^*_{max} at a high airflow rate, and the operating point for a low airflow rate will be considerably to the right of S^*_{max}. Thus, S^* and $\dot{S}_{irr}$ $(= S^* C_{max})$ will be higher for the high-airflow-rate case. So this high-airflow-rate case is an unfavorable design solution from the exergy point of view. An engineer's ultimate decision, though, will also depend on a number of additional considerations (pressure drop considerations, cost analysis, etc.). These considerations may include a trade-off between the physical size of the heat exchanger (cost of exchanger) and the mass flow rate of the hot fluid (operating cost), or alternately, the minimum cost of thermodynamic irreversibility, as discussed in the next subsection.

11.6.6   Accounting for the Costs of Exergy Losses in a Heat Exchanger

With the thermodynamic background developed so far, we can now point out how to relate an exergy flow rate to the cost rate. The heat exchanger transforms the total exergy rate input, carried in by both streams, into an exergy rate output that is always smaller than the one at the input due to the inevitable presence of irreversibilities (caused by heat transfer rate at finite temperature differences, fluid mixing, friction phenomena, etc.). Each exergy rate (j streams into and out of the heat exchanger) has an associated cost rate $\dot{C}_j$ as

$$\dot{C}_j = C_j \dot{\mathscr{e}}_j \qquad (11.62)$$

where C_j denotes the cost per unit of exergy ($\mathscr{C}/J$). The cost per unit of exergy of fluid flows that enter a heat exchanger has to be determined before an exergy costing can be performed. For example, if a particular fluid stream has to be utilized in a system, a designer must know the cost of the utilization of that stream. This cost has to be determined from an analysis located "upstream" from the location of the stream utilization (Bejan et al., 1995). Now we can write a cost rate balance for a heat exchanger:

$$\sum_{j=h,c} \dot{C}_{j,i} - \sum_{j=h,c} \dot{C}_{j,o} = \dot{Z}_{cap} + \dot{Z}_{op} \qquad (11.63)$$

In Eq. (11.63), Z_{cap} and Z_{op} denote the capital investment cost and operating expenses[†] of a heat exchanger that must be balanced by a difference between the sum of the cost rates of all inlet fluid flows and the cost rates of all outlet fluid flows. It should be noted that cost balancing as presented by Eq. (11.63) has to be performed for a heat exchanger as a component in a system. Note also that cost rates do not obey the conservation principle.

[†] For simplicity, we ignore here the installation cost, which can be comparable to capital and operating costs for some shell-and-tube heat exchangers and PHEs, and can be added as a third term on the right-hand side or can be included in the capital cost..

Before one attempts to perform an optimization, a careful utilization of a common-sense second law of thermodynamics approach to the design of the entire system has to be conducted (Sama et al., 1989; Sama, 1995b). This approach is based on a selection of thermodynamic rules such as the ones given in the summary of this chapter based on the results of Section 11.3 and this section for a heat exchanger design. If input cost rates of the form of Eq. (11.62) are known, and if both the capital investment and operating expenses are known, Eq. (11.63) can be used directly to determine the output cost rate of a stream. Once determined, a cost rate balance can be combined with the cost rate balances of other components and used to formulate an objective function (Bejan et al., 1995) for a system. Such an objective function combines thermodynamics and economics, exploiting the concepts of exergy and/or entropy generation for a system; that is the reason this branch of engineering is referred to as *thermoeconomics* (Bejan et al., 1995). Equations (11.62) and (11.63) clearly show how the cost rate balance can be related to the exergy rate balances (assigning a dollar sign to exergy flows) for a heat exchanger as a component of a system. Determination of cost per unit exergy flow is difficult and a single cost model could not be applied. This complex thermoeconomic topic is beyond the scope of our presentation.

A practical methodology to account for the cost of irreversibilities in a heat exchanger as a stand-alone unit has been developed by London (1982). This methodology has at its core an identification of the individual costs of irreversibilities incurred during heat exchanger operation, which is in fact related to both the capital (initial design) and oper-ating costs. This is because some small irreversibilities may be very expensive (such as airflow friction irreversibility for an automotive radiator), and some large irreversibilities may be less expensive (such as liquid-side friction irreversibility in a gas–liquid heat exchanger). London's approach does not use the exergy rates explicitly but does use the exergy destruction rate, referred to as the *energy measure of the irreversibilities*. Individual energy measures of irreversibility are determined by multiplying individual entropy generation rates with a temperature-weighting factor, T_{ref}. Total irreversibility destruction is related to total entropy generation through the relationship given by Eq. (11.55). Details of London's procedure are also presented by London and Shah (1983).

With the thermodynamic background developed so far, we can outline a procedure to evaluate in monetary terms the various irreversibilities in a heat exchanger as shown in Fig. 11.14. For example, five irreversibilities of significance in a heat exchanger may be the finite temperature difference between hot and cold fluids, fluid mixing at the exchanger ports (if applicable), pressure drops on the hot and cold sides, and irreversibility associated with heat leakage between the exchanger and the environment. The total energy measure of irreversibility normalized by the heat exchanger duty is then given by

$$\frac{\mathscr{D}}{q} = \frac{\dot{I}_{\text{irr}}}{q}\bigg|_{\Delta T} + \frac{\dot{I}_{\text{irr}}}{q}\bigg|_{\text{mixing}} + \frac{\dot{I}_{\text{irr}}}{q}\bigg|_{\Delta p,h} + \frac{\dot{I}_{\text{irr}}}{q}\bigg|_{\Delta p,c} + \frac{\dot{I}_{\text{irr}}}{q}\bigg|_{\text{leak}} \tag{11.64}$$

where $\dot{I}_{\text{irr},i} = T_{\text{ref}}\dot{S}_{\text{irr},i}$, with the subscript i representing individual irreversibilities of Eq. (11.64). These are summarized in Table 11.3 for an ideal gas or an incompressible liquid, all in measurable system operation quantities (temperatures, pressures, mass flow rates, etc.). These irreversibilities have an energy monetary value that is dependent on the system in which a particular heat exchanger is used. Assigning a monetary value to irreversibility is analogous to assigning a cost rate to exergy rates as in Eq. (11.62). Once the monetary values or costs of various irreversibilities are determined, the analyst

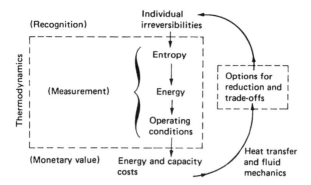

FIGURE 11.14   Thermodynamic optimization of a heat exchanger, including the development of trade-off factors. (From London, 1982.)

is in the position of deciding which particular irreversibilities are most costly and should first be reduced for a cost-effective heat exchanger. Hence, the industrial approach to the design of a heat exchanger is to reduce the most costly irreversibilities rather than reducing all irreversibilities in a heat exchanger [i.e., *not* to minimize $\mathscr{D}$ of Eq. (11.64), but to minimize only those irreversibilities on each fluid side that are the most expensive]. Note that reducing one irreversibility may increase or require the addition of another, or involve an increase in capital investment. These considerations lead to the development of trade-off factors, useful criteria for arriving at an optimum heat exchanger design as a component. Thus, coming back to Fig. 11.14, the design and optimization of a heat exchanger involves the process of evaluating various irreversibilities on an entropy basis and then converting them into an energy basis (either in the form of exergy flows or energy measures of irreversibility), assigning monetary values, minimizing the most costly irreversibilities and as a result developing trade-off factors, and continuing this process until the optimum heat exchanger is developed. To understand this process, London (1982) and London and Shah (1983) have provided a detailed example of a condenser in a thermal power plant with a clear demonstration of the accounting procedure above, including how to develop the trade-off factors. Problems 11.17 and 11.18 are based on this example. Even more intricate problems of second-law-based thermoeconomic optimization of exchangers (including both single- and two-phase heat exchangers) are proposed by Zubair et al. (1987), estimating the economic value of entropy generation in the heat exchanger caused by finite temperature differences and pressure drop. That method permits an engineer to trade the cost of entropy generation on each fluid side of the heat exchanger against its capital expenditure.

Now let us relate the cost rate analysis for a system [Eqs. (11.62) and (11.63)] with a similar analysis based on minimizing the most costly irreversibilities using the approach of Fig. 11.14 and the associated Eq. (11.64). The exergy cost analysis of a system, based on Eq. (11.63), includes *all* irreversibilities present individually within all components [say, a compressor, a condenser, an expansion device, an evaporator, and an accumulator/dehydrator in an automotive air-conditioning system, for which cost rates are calculated using Eq. (11.62)] if the fluid variables are determined at the boundaries of each component and each component is considered to be a black box. Hence, optimization of the system is performed by taking into account the irreversibility losses of *all* components without distinguishing various components of the total irreversibility in

TABLE 11.3 Normalized Energy Measure of Various Irreversibilities in a Heat Exchanger

Finite temperature difference

$$\frac{i_{irr}}{q}\bigg|_{\Delta T} = T_{ref}\left(\frac{1}{T_{c,lm}} - \frac{1}{T_{h,lm}}\right) \quad \text{where} \quad T_{j,lm} = \frac{T_{j,i} - T_{j,o}}{\ln(T_{j,i}/T_{j,o})} \quad j = c \text{ or } h$$

Fluid mixing (exchanger outlet into environment)

$$\frac{i_{irr}}{q}\bigg|_{mixing} = \frac{T_{ref}(\dot{m}c_p)}{q}\left[\left(\frac{T_{c,o}}{T_{ref}} - 1\right) - \ln\frac{T_{c,o}}{T_{ref}}\right]$$

Pressure drops on hot and cold fluid sides

$$\text{Hot fluid:} \quad \frac{i_{irr}}{q}\bigg|_{\Delta p,h} = \frac{T_{ref}}{T_{h,lm}}\frac{1}{q}\left(\frac{\dot{m}\Delta p}{\rho}\right)_h ; \quad \text{cold fluid:} \quad \frac{i_{irr}}{q}\bigg|_{\Delta p,c} = \frac{T_{ref}}{T_{c,lm}}\frac{1}{q}\left(\frac{\dot{m}\Delta p}{\rho}\right)_c$$

Heat leak

$$\text{To environment:} \quad \frac{i_{irr}}{q}\bigg|_{leak} = \frac{q_{leak}}{q}\left(1 - \frac{T_{ref}}{T_h}\right) ; \quad \text{from environment:} \quad \frac{i_{irr}}{q}\bigg|_{leak} = \frac{q_{leak}}{q}\left(\frac{T_{ref}}{T_c} - 1\right)$$

Source: Data from London and Shah (1983).

each component. Equation (11.64) is used to minimize the most costly irreversibilities to produce a cost-effective heat exchanger, and it does not necessarily lead to an optimum design of the system as a whole. So a system-based optimization would use an objective function based on Eq. (11.63), while a heat exchanger component optimization would be based on an analysis of individual irreversibilities. It should be noted that the cost rates in Eq. (11.63) can be calculated using Eq. (11.62) in such a way as to determine exergy rates [using Eq. (11.53)] by excluding from the entropy change all less costly irreversibility contributions. In such a way, the system-based optimization will include only the most costly irreversibilities in the same manner as would be done by using an individual irreversibility calculation for each component. However, the opposite is not possible; an optimization based on Eq. (11.64) only cannot provide an optimum system as a whole.

Example 11.9   Determine the annual cost of exergy destruction in a heat exchanger caused by finite temperature differences. The heat exchanger heats a cold fluid stream having a heat capacity rate of 5.8×10^6 W/K from 19°C (292 K) to 27°C (300 K). The hot fluid reduces its temperature from 30°C (303 K) to 25°C (198 K). The reference temperature (surroundings) is 17°C (290 K). The annual capital cost of the equipment involved is 10^3 $\mathscr{C}$/kW of energy at 10% annual interest rate. The average yearly energy cost is 150 $\mathscr{C}$/kW · yr.

SOLUTION

Problem Data and Schematic: The terminal temperatures of the fluids entering and leaving the heat exchanger and the cold-fluid heat capacity rate are as follows: $T_{c,i} = 292$ K, $T_{c,o} = 300$ K, $T_{h,i} = 303$ K, $T_{h,o} = 298$ K, and $C_c = 5.8 \times 10^6$ W/K. The reference surrounding temperature is $T_o = 290$ K (17°C). The annual capital cost of the equipment is $\mathbf{C}_{\mathrm{eqp}} = 1000$ $\mathscr{C}$/kW and the annual interest rate $\mathbf{r} = 0.1$/yr. The average annual cost of energy is $\mathbf{C}_q = 150$ $\mathscr{C}$/kW · yr. See Fig. 11.13 as a representative sketch.

Determine: The annual cost of exergy destruction in the exchanger caused by finite temperature differences.

Assumptions: All appropriate assumptions of Section 3.2.1 are invoked here.

Analysis: The exergy destruction is given by Eq. (11.55):

$$\mathscr{D} = T_{\mathrm{ref}} \dot{S}_{\mathrm{irr}} \tag{1}$$

Also, according to Eq. (11.64), taking into account *only* the finite temperature difference contribution

$$\frac{\mathscr{D}}{q} = \frac{\dot{I}_{\mathrm{irr}}}{q}\bigg|_{\Delta T} \tag{2}$$

In other words, the energy measure of irreversibility in the heat exchanger caused by finite temperature differences is equal to corresponding exergy destruction since no other irreversibility contributions are included. The entropy generation caused by finite temperature differences is given by Eq. (11.30):

$$\dot{S}_{\mathrm{irr}} = q\,\frac{T_{\mathrm{lm},h} - T_{\mathrm{lm},c}}{T_{\mathrm{lm},h}\,T_{\mathrm{lm},c}} \tag{3}$$

where

$$T_{\mathrm{lm},j} = \frac{T_{j,o} - T_{j,i}}{\ln(T_{j,o}/T_{j,i})} \tag{4}$$

Equation (4) provides the magnitudes of the log–mean temperatures of the two fluids (equal to the logarithmic mean between the inlet and outlet temperatures of each fluid) with $j = h$ and c.

Inserting problem data into Eq. (4), we get

$$T_{\mathrm{lm},h} = \frac{T_{h,o} - T_{h,i}}{\ln T_{h,o}/T_{h,i}} = \frac{(298 - 303)\,\mathrm{K}}{\ln(298/303)} = 300.5\,\mathrm{K}$$

$$T_{\mathrm{lm},c} = \frac{T_{c,o} - T_{c,i}}{\ln\left(T_{c,o}/T_{c,i}\right)} = \frac{(300 - 292)\,\mathrm{K}}{\ln(300/292)} = 296.0\,\mathrm{K}$$

Entropy generation is, from Eq. (3),

$$\dot{S}_{\mathrm{irr}} = C_c(T_{c,o} - T_{c,i})\left(\frac{1}{T_{\mathrm{lm},c}} - \frac{1}{T_{\mathrm{lm},h}}\right)$$

$$= (5.8 \times 10^6\,\mathrm{W/K}) \times (300 - 292)\,\mathrm{K}\left(\frac{1}{296.0\,\mathrm{K}} - \frac{1}{300.5\,\mathrm{K}}\right) = 2347\,\mathrm{W/K}$$

The exergy destruction caused by finite temperature differences, from Eq. (1), is

$$\mathscr{D} = T_{\mathrm{ref}}\dot{S}_{\mathrm{irr}} = 290\,\mathrm{K} \times 2347\,\mathrm{W/K} = 0.681 \times 10^6\,\mathrm{W}$$

Now, keeping Eq. (2) in mind and taking the cost data into account, we can calculate the cost of this exergy destruction:

$$C = \mathscr{D}(\mathbf{C}_{\mathrm{eqp}}\mathbf{r} + \mathbf{C}_q) = 0.681 \times 10^6\,\mathrm{W} \times (1 \times 0.1 + 0.15)(\mathscr{C}/\mathrm{W}\cdot\mathrm{yr}) = 0.17 \times 10^6\,\mathscr{C}/\mathrm{yr}$$

Discussion and Comments: This example illustrates how one can determine in a simplified way a cost value of the lost work (exergy destruction) caused by irreversibilities in a heat exchanger. In this example, only the finite temperature difference irreversibility contribution is considered. Problems 11.17 and 11.18 include all other relevant irreversibility contributions for a representative heat exchanger application.

11.7 PERFORMANCE EVALUATION CRITERIA BASED ON THE SECOND LAW OF THERMODYNAMICS[†]

All performance evaluation criteria for heat transfer surfaces presented in Section 10.3.2 are based on the first law of thermodynamics (i.e., energy and mass balances). In a different approach to a performance evaluation criterion, we introduce the *thermodynamic quality* of heat transfer and fluid flow processes to evaluate the heat transfer surface performance. Such an evaluation requires an assessment of the *irreversibility level* of heat transfer and fluid flow phenomena. Hence, a combined first *and* second law of

---

[†]The body of knowledge usually called the second law of thermodynamics analysis *always* involves both the first *and* second laws of thermodynamics. However, it is customary to name the product of such an analysis by indicating the second law of thermodynamics only.

thermodynamics analysis becomes necessary. Here, we emphasize only the possible formulation of a PEC that reflects these considerations. For details of this approach and a review of the literature, refer to Bejan (1988).

The *second law of thermodynamics performance evaluation criteria* are based on objective functions that include both heat transfer and pressure drop (fluid friction) irreversibilities and hence gauge the combined effect of these irreversibilities. Separation of the two irreversibilities can subsequently be performed if necessary. Note that heat transfer and pressure drop irreversibilities should be translated into costs separately because the unit costs of these irreversibilities are in general not equal.

The entropy rate balance for a duct/channel control volume (i.e., a flow passage of a heat exchanger) defines total entropy generation between the fluid inlet and outlet as follows:

$$\dot{S}_{irr} = \dot{S}_{irr,\Delta T} + \dot{S}_{irr,\Delta p} = \dot{S}_{irr,\Delta T}(1 + \Phi) \tag{11.65}$$

where the $\dot{S}_{irr,\Delta T}$ and $\dot{S}_{irr,\Delta p}$ terms denote the contributions to overall entropy generation incurred by either finite temperature difference between the fluid and the wall (ΔT) or pressure drop (Δp) (see Problem 11.19). The irreversibility distribution ratio, $\Phi(= \dot{S}_{irr,\Delta p}/\dot{S}_{irr,\Delta T})$ expresses, by definition, the trade-off between the two contributions. An explicit form of Eq. (11.65) depends on heat transfer and fluid flow conditions and idealizations (e.g., boundary conditions, free-flow area geometry, flow regime, selection of dimensionless parameters). One such expression in terms of dimensionless parameters for a constant-cross-section duct and constant-property fluid can be written as

$$S^* = \frac{N_q^2(D_h/L)}{4j \cdot Pr^{-2/3}} + \frac{2f \cdot Ec}{D_h/L} \tag{11.66}$$

where the dimensionless heat transfer rate N_q and Eckert number Ec are defined as $N_q = q/(\dot{m}c_p T_m)$ and $Ec = u^2/c_p T_m Jg_c$, respectively. Equation (11.66) represents a dimensionless form of the corresponding entropy generation expression derived by Bejan (1988). These dimensionless groups are conveniently defined using fluid bulk mean temperature T_m [physical modeling involving the derivation of Eq. (11.66) is the subject of Problem 11.20]. Note that the dimensionless group Ec is usually defined based on a temperature difference, not the bulk temperature. Such a definition can easily be introduced in Eq. (11.66), leading to the introduction of an additional temperature ratio parameter. For simplicity, the Ec number is defined as given above. Equation (11.66) assumes constant and known heat transfer and mass flow rates.

For a specified constant mass flow rate $\dot{m}$ and given duct length L, a change in the hydraulic diameter D_h causes a change in S^*. From the algebraic structure of Eq. (11.66) and Reynolds analogy (see Section 7.4.5), it becomes clear that the two terms on the right-hand side of Eq. (11.66) have opposite trends with respect to a change in Re (which affects j and f factors) or the hydraulic diameter. Thus, this objective function [i.e., Eq. (11.66)] may have an extremum (a minimum.) It has been recognized that what is good for the reduction of friction irreversibility (by decreasing surface area) is apparently bad for the reduction of finite temperature irreversibility (i.e., for an increase in exchanger effectiveness), and vice versa. An optimum trade-off between these two influences may exist. Consequently, such an objective function may be used as a basis for defining a new thermodynamic performance evaluation criterion. An optimum geometry (or flow regime) can ultimately be defined for a given selection of characteristic parameters. We can illustrate this approach by the following example.

Example 11.10   Thermoeconomic optimization of a large energy system requires mini-mization of irreversibility costs of a plant. Within the scope of that analysis, a heat exchanger designer must decide which geometry of a heat exchanger passage cross section would contribute the least to the overall irreversibility. The length and free-flow area of the duct representing the passage are known and fixed, as well as the fluid (air) inlet thermal state and mass flow rate. Temperature of the heating fluid is constant and equal to 100°C (373 K). The wall thermal resistance can be neglected. Duct geometry options include (1) square, (2) rectangular (aspect ratio $\alpha^* = \frac{1}{8}$), and (3) circular cross sections. Determine which cross section would be best from the point of view of irrever-sibility minimization. Compare the findings with an analysis of the magnitude of heat transfer and pressure drop for each of the duct shapes. The following data are available. Free-flow area of the cross section is 5×10^{-3} m², the duct length 5 m, the mass flow rate of air 5×10^{-2} kg/s, and the fluid inlet temperature 300 K. The thermophysical proper-ties of the air are as follows: density, 1.046 kg/m³; specific heat at constant pressure, 1.008 kJ/kg · K; dynamic viscosity, 2.025×10^{-5} Pa · s; Prandtl number, 0.702; and ther-mal conductivity, 2.91×10^{-2} W/m · K.

SOLUTION

Problem Data and Schematic: The duct geometries for this problem are shown below along with the input data for the problem.

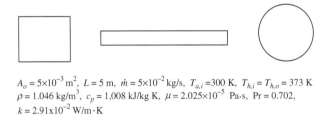

$A_o = 5 \times 10^{-3}$ m², $L = 5$ m, $\dot{m} = 5 \times 10^{-2}$ kg/s, $T_{a,i} = 300$ K, $T_{h,i} = T_{h,o} = 373$ K
$\rho = 1.046$ kg/m³, $c_p = 1,008$ kJ/kg K, $\mu = 2.025 \times 10^{-5}$ Pa·s, Pr $= 0.702$,
$k = 2.91 \times 10^{-2}$ W/m·K

FIGURE E11.10   Square, rectangular and circular duct geometries.

Determine: Which of the three cross-sectional shapes [square, rectangular ($\alpha^* = \frac{1}{8}$), or circular] will lead to minimum heat transfer and flow friction entropy generation?

Assumptions: The flow regime is assumed fully developed and the thermophysical proper-ties are constant. Heat transfer from the duct to the fluid is accomplished at a uniform rate across the assumed temperature difference between the wall and bulk mean tem-perature (determined as an arithmetic mean of the inlet and outlet temperatures). The temperature of the wall is uniform, constant, and equal to the temperature of the heating fluid (assume that the fluid is changing its phase at a constant temperature of 373 K). Consequently, the wall thermal resistance is neglected.

Analysis: We determine the dimensionless entropy generation associated with heat trans-fer and fluid friction, Eq. (11.66), for each duct and then compare the results to determine which geometry is most favorable. The calculation of heat transfer and pressure drop characteristics is summarized in Table E11.10. For details of some calculations for the square and rectangular ducts see Example 7.5. Entropy generation can subsequently be calculated in dimensionless form by Eq. (11.66), or in the corresponding dimensional

TABLE E11.10 Thermal and Hydraulic Characteristics and Entropy Generation of Various Ducts

Variable or Parameter	Equation	Square	Rectangular	Circular
D_h (m)	$4A_o/P$	7.071×10^{-2}	4.444×10^{-2}	7.979×10^{-2}
G (kg/m$^2 \cdot$ s)	$\dot{m}/A_o$	10	10	10
Re	GD_h/μ	34,919	21,946	39,402
Nu	$Nu_{sq} = Nu_{rect} = 0.024Re^{0.8} \cdot Pr^{0.4}$	89.78	61.92	94.77
	$Nu_{circ} = 0.023Re^{0.8} \cdot Pr^{0.4}$			
j	$j = St \cdot Pr^{2/3} = Nu \cdot Pr^{-1/3}/Re$	2.893×10^{-3}	3.174×10^{-3}	2.706×10^{-3}
h (W/m$^2 \cdot$ K)	kNu/D_h	36.94	40.55	34.56
A (m^2)	PL	1.4142	2.25	1.25
NTU	$hA/\dot{m}c_p$	1.038	1.83	0.87
ε	$1 - e^{-NTU}$	0.6458	0.8396	0.5810
q (W)	$q = \varepsilon \dot{m}c_p(T_w - T_{a,i})$	2,376	3,089	2,138
$T_{a,o}$(°C)	$T_{a,i} + q/\dot{m}c_p$	74.1	88.3	69.4
T_m(°C)	$(T_{a,i} + T_{a,o})/2$	50.6	57.7	48.2
f	$f_{rect} = 0.0791Re^{-0.25}(1.0875 - 0.1125\alpha^*)$	5.642×10^{-3}	6.976×10^{-3}	5.534×10^{-3}
	$f_{circ} = 0.00128 + 0.1143Re^{-1/3.2154}$			
Δp(Pa)	$\Delta p = 4fLG^2/2g_c\rho D_h$	76.3	150.0	66.3
N_q	$q/\dot{m}c_pT_m$	0.1456	0.1852	0.1320
Ec	$u^2/c_pT_mJg_c$	2.799×10^{-4}	2.738×10^{-4}	2.820×10^{-4}
$S_{\Delta T}^*$	$N_q^2(D_h/L)/4j \cdot Pr^{-2/3}$	2.043×10^{-2}	1.899×10^{-2}	2.021×10^{-2}
$S_{\Delta p}^*$	$2f \cdot Ec/(D_h/L)$	2.240×10^{-4}	4.292×10^{-4}	1.963×10^{-4}
S^*	$S_{\Delta T}^* + S_{\Delta p}^*$	2.065×10^{-2}	1.942×10^{-2}	2.041×10^{-2}

form (per unit of the duct length) as follows:

$$\frac{\dot{S}_{irr}}{L} = \frac{(q/L)^2 D_h}{4T_m^2 \dot{m}c_p \cdot St} + \frac{2\dot{m}^2 f}{\rho^2 A_o^2 c_p T_m Jg_c D_h}$$

From the last line of Table E11.10 it is obvious that the rectangular duct generates the minimum entropy. Therefore, the rectangular duct would be considered as most favorable from an entropy-generation point of view.

Discussion and Comments: To understand the influence of various sources of irreversibility on duct performance, the results of the analysis are summarized below. For the three duct geometries, the following quantities are compared: heat transfer rate, pressure drop, dimensionless entropy generation caused by temperature difference, dimensionless entropy generation caused by pressure drop, and total dimensionless entropy generation. The results are presented qualitatively in terms of the highest heat transfer rate, the lowest pressure drop, and the lowest entropy generations, all marked with the symbol ☺. The lowest heat transfer rate, the highest pressure drop, and the highest irreversibility are each marked with the symbol ☹. Finally, the medium values are marked ☺.

Variable or Parameter	Square	Rectangular	Circular
q	☺	☺	☹
Δp	☺	☹	☺
$S_{\Delta T}^*$	☹	☺	☺
$S_{\Delta p}^*$	☺	☹	☺
S^*	☹	☺	☺

It is obvious that the rectangular duct has the best performance in terms of both heat transfer rate and irreversibility level. The penalty for high heat transfer performance is paid by a very high pressure drop. It is interesting to note that the worst geometry from the heat-transfer-rate point of view is the circular duct, but from the entropy-generation point of view, it is a square duct. The reason for that is related to the different orders of magnitude of the entropy generation caused by temperature difference and pressure drop.

If the ultimate goal is to reach the minimum entropy generation for a given geometry, say a circular duct, an optimization based on the objective function given by Eq. (11.66) has to be performed. Solving $(\partial S^*)/\partial \text{Re}$ will lead to the determination of an optimum Re, or the optimum duct hydraulic diameter for a given mass flow rate. For the circular duct of this problem, this optimum diameter of the duct (the hydraulic diameter considered as the only degree of freedom, the other variables fixed) is found as very large: 135 mm.

The optimum trade-off between heat transfer and friction (pressure drop) irreversibilities may or may not exist. Also, the existence of a minimum of entropy generation is not always present for the range of parameters selected. It must also be emphasized that optimization based on a thermodynamic criterion cannot be a goal per se in a design effort related to an isolated heat exchanger. Usually, the ultimate goal is the minimization of cost within the framework of a system analysis.

SUMMARY

In this chapter, several interdisciplinary issues are discussed. These include (1) the use of energy and mass balances (first law of thermodynamics only) and mathematical modeling to obtain temperature distributions and temperature difference distributions in various flow arrangements, (2) the application of the first and second laws of thermodynamics (combined) to identify irreversibility sources that have a detrimental effect on heat exchanger performance, (3) a heuristic approach to the assessment of heat exchanger effectiveness, (4) thermodynamic analysis, including exergy and thermoeconomic accounting for heat exchanger optimization, and (5) performance evaluation criteria based on the minimization of entropy generation. Our goal has been to understand why certain heat exchanger designs would lead to a higher or lower effectiveness and/ or thermodynamic (exergy) efficiency than would a similar one, and to develop the skills needed for an approach to optimization of a heat exchanger as part of a system. Finite temperature differences, fluid mixing, and fluid friction are important irreversible phenomena associated with exchanger performance. Details on these irreversibilities are presented in the text.

The most important guidelines for the design of a heat exchanger as a component in a system are as follows. An approach to optimum design must be based on sound engineering judgment, along with utilization of a commercial or proprietary software (if any) with understanding. Such an approach should be performed utilizing not only the energy balances (implied by the first law of thermodynamics), but also entropy-generation balances (implied by the combined first and second laws of thermodynamics). An exergy balance or an entropy-generation calculation has to be accompanied by economic evaluations. The components of exergy balances or total entropy-generation rates should have assigned monetary values. Exergy cost balances may be used to define an objective function in a search for the optimum design of the system in which the analyzed heat exchanger is a component. The methodology outlined in Fig. 11.14 should be adopted for optimization of a heat exchanger in a system if the minimization of the most costly irreversibilities is considered. Trade-off factors are also developed in this methodology as part of the optimization procedure.

Based on the analysis presented throughout this chapter, a set of guidelines important for an assessment of a heat exchanger *as a component in a system* can be defined. These include the following:

- The thermodynamic driving potential (local temperature differences between the fluids) should be reduced as much as possible.
- Fluid mixing within the exchanger or at exchanger terminal ports (e.g., tanks, headers, etc.) has to be avoided whenever possible.
- Fluid streams in a heat exchanger network exchanging heat have to be matched beyond the temperature pinch (i.e., the point in the temperature enthalpy rate diagram where the composite curves are closest to each other). That means that one should not transfer heat across the pinch (cold and hot utilities should be used only below and above the pinch, respectively).
- Fluid streams have to be balanced as much as possible (i.e., C^* should have a value close to unity in an exchanger) for minimum irreversibilities.
- Fluid friction, throttling, and all the other inherently irreversible phenomena should be minimized.
- High temperatures (compared to the reference thermodynamic state) and large mass flow rates of the fluid streams should be avoided *if* a fluid stream having lower operational variables can be utilized (to minimize the exergy losses in the system). In other words, do not use the large thermal potentials if not needed. However, if heat recovery is desired from an available high-temperature stream, avoid dilution of that stream with a colder stream if possible. If the high-temperature stream can be used without adding cost for high-temperature materials, it is better to use the undiluted stream, since that will preserve more of the exergy in the high-exergy stream.

In conclusion, it should be noted that thermodynamic irreversibilities cause substantial deterioration of the performance level of a heat exchanger. They can never be eliminated, but should always be assessed, and if cost effective, should be minimized.

REFERENCES

Ahrendts, J., 1980, Reference states, *Energy*, Vol. 5, pp. 667–677.

Alt-Ali, M. A., and D. J. Wilde, 1980, Optimal area allocation in multistage heat exchanger systems, *ASME J. Heat Transfer*, Vol. 107, pp. 199–201.

Bejan, A., 1988, *Advanced Engineering Thermodynamics*, Wiley, New York.

Bejan, A., G. Tsatsaronis, and M. Moran, 1995, *Thermal Design and Optimization*, Wiley, New York.

Bosnjakovic, F., 1965, *Technical Thermodynamics*, Holt, Rinehart and Winston, New York (6th German ed., *Technische Thermodynamik*, Steinkopf, Dresden, 1972).

EES, 2000, *F_Chart Software*, Engineering Equation Solver, Middleton, WI.

Gregorig, R., 1965, Exergieverluste der Wärmeaustauscher (Exergy Losses in a Heat Exchanger), Teil 1, Reibung (Friction), *Chem. Ing. Techn.* Vol. 37, pp. 108–116; Teil 2, Endlichen Temperaturunterschiedes (Finite Temperature Differences), pp. 524–527.

Gunderson, T., and L. Naess, 1988, The synthesis of cost optimal heat exchanger networks: an industrial review of the state of the art, *Comput. Chem. Eng.*, Vol. 12, pp. 503–530.

Kays, W. M., and A. L. London, 1998, *Compact Heat Exchangers*, reprint 3rd ed., Krieger Publishing, Malabar, FL.

Kmecko, I., 1998, *Paradoxical Irreversibility of Enthalpy Exchange in Some Heat Exchangers*, M.S. thesis, University of Novi Sad, Novi Sad, Yugoslavia.

Kotas, T. J., 1995, *The Energy Method of Thermal Plant Analysis*, Krieger Publishing, Melbourne, FL.

Linnhoff, B., D. W. Townsend, D. Balard, G. F. Hewitt, B. E. A. Thomas, A. R. Guy, and R. H. Marshand, 1982, *User Guide on Process Integration for Efficient Use of Energy*. Institution of Chemical Engineers and Pergamon Press, Oxford.

London, A. L., 1982, Economics and the second law: an engineering view and methodology, *Int. J. Heat Mass Transfer*, Vol. 25, pp. 743–751.

London, A. L., and R. K. Shah, 1983, Costs of irreversibilities in heat exchanger design, *Heat Transfer Eng.*, Vol. 4, No. 2, pp. 59–73; discussion by W. Roetzel, in Vol. 5, No. 3–4, 1984, pp. 15, 17, and Vol. 6, No. 2, 1985, p. 73.

Moran, M. J., and H. N. Shapiro, 1995, *Fundamentals of Engineering Thermodynamics*, Wiley, New York.

Peters, M. S., and K. D. Timmerhaus, 1980, *Plant Design and Economics for Chemical Engineers*, McGraw-Hill, New York.

Sama, D. A., 1995a, Differences between second law analysis and pinch technology, *J. Energy Resour. Technol.*, Vol. 117, pp. 186–191.

Sama, D. A., 1995b, The use of the second law of thermodynamics in process design, *J. Energy Resour. Technol.*, Vol. 117, pp. 179–185.

Sama, D. A., S. Qian, and R. Gaggioli, 1989, A common-sense second law approach for improving process efficiencies, *Proc. Int. Symp. Thermodynamic Analysis and Improvement of Energy Systems*, Beijing, International Academic Publishing, Pergamon Press, New York, pp. 520–532.

Shah, R. K., and Skiepko, T., 2002, Entropy generation extrema and their relationship with heat exchanger effectiveness – number of transfer units behavior for complex flow arrangements, *Heat and Mass Transfer 2002, Proc. 5th ISHMT-ASME Heat Mass Transfer Conf.*, Tata McGraw-Hill., New Delhi, India, pp. 910–919.

Sekulić, D. P., 1990a, A reconsideration of the definition of a heat exchanger, *Int. J. Heat Mass Transfer*, Vol. 33, pp. 2748–2750.

Sekulić, D. P., 1990b, The second law quality of energy transformation in a heat exchanger, *ASME J. Heat Transfer*, Vol. 112, pp. 295–300.

Sekulić, D. P., 2000, A unified approach to the analysis of unidirectional and bi-directional parallel flow heat exchangers, *Int. J. Mech. Eng. Educ.*, Vol. 28, pp. 307–320.

Sekulić, D. P., 2003, A heuristic approach to an assessment of heat exchanger effectiveness, to be published in *Int. J. Mech. Eng. Education*, Vol. 31.

Shah, R. K., and A. Pignotti, 1989, *Basic Thermal Design of Heat Exchangers*, Report Int-8513531, National Science Foundation, Washington, DC.

Sontag, R. E., and G. J. Van Wylen, 1982, *Introduction to Thermodynamics*, Wiley, New York.

Zubair, S. M., P. V. Kadaba, and R. B. Evans, 1987, Second-law-based thermoeconomic optimization of two-phase heat exchangers, *ASME J. Heat Transfer*, Vol. 109, pp. 287–294.

REVIEW QUESTIONS

Where multiple choices are given, circle one or more correct answers. Explain your answers briefly.

11.1 Circle the following statements as true or false and provide detailed reasons.

  (a) T F   A workable solution of a heat exchanger design problem requires neither irreversibility analysis nor economic analysis.

  (b) T F   The true meaning of the concept of temperature effectiveness cannot be derived without invoking explicitly the second law of thermodynamics.

11.2 Optimization of a heat exchanger as an isolated component always makes sense.

  (a) true   **(b)** false   **(c)** The answer depends on the system application.

11.3 An ideal heat exchanger operates under the following conditions:

  (a) $\varepsilon = 1$ **(b)** $\dot{S}_{irr} = 0$ and $C^* = 1$

  (c) $\dot{S}_{irr} = 0$ and $C^* = 0$ **(d)** $\vartheta = 0$

11.4 Temperature distributions of the two fluids in a mixed–mixed crossflow arrangement are:

  (a) both two-dimensional   **(b)** both one-dimensional

  (c) one two-dimensional and the other one-dimensional

11.5 The existence of an internal temperature cross implies the existence of an external temperature cross.

  (a) always true   **(b)** always false

  (c) true only for some flow arrangements

11.6 A heat exchanger with equal exit temperatures of the two fluids is characterized with:

  (a) minimum entropy generation

  (b) maximum entropy generation

  (c) minimum heat exchanger effectiveness

  (d) maximum heat exchanger effectiveness

11.7 The concept of exergy is based on:

  (a) application of the first law of thermodynamics only

  (b) utilization of energy balances only

  (c) application of both the first and second laws of thermodynamics

  (d) application of the second law of thermodynamics only

11.8 Circle the following statements as true or false. Provide detailed reasons.

  (a) T F   Reduction of local temperature differences between the fluids in a heat exchanger always contributes positively to the decrease in the system total entropy generation.

  (b) T F   The presence of mixing in a heat exchanger causes deterioration of heat exchanger performance compared to that of an exchanger without mixing (all other design parameters remaining the same).

  (c) T F   An initially hot fluid in thermal contact with an initially cold fluid in a heat exchanger can become (locally, anywhere within the heat exchanger) colder than the other fluid.

(d) T F   For identical overall NTU and C^*, two identical exchangers in series coupling will yield an overall effectiveness higher than that for a similar pair of identical exchangers in parallel coupling.

(e) T F   In a parallel-coupled arrangement of two exchangers, the overall exchanger effectiveness will be higher if the C_{max} fluid is in the series (and C_{min} fluid in parallel) compared to the C_{min} fluid being in series.

(f) T F   An external temperature cross can occur for any flow arrangement only if there is an internal temperature cross.

(g) T F   The lower the mean temperature difference (MTD), the lower is the counterflow exchanger effectiveness.

(h) T F   Fluid mixing decreases irreversibility and results in a decrease in exchanger effectiveness.

(i) T F   As NTU increases, the irreversibility increases and hence the exchanger effectiveness increases in a counterflow exchanger.

(j) T F   For all exchanger flow arrangements and $C^* > 0$, as NTU increases, the exchanger effectiveness monotonically increases.

11.9 The total entropy on hot plus cold fluid sides in an industrial exchanger increases due to:

(a) heat transfer in the exchanger   **(b)** pressure drops in the exchanger

(c) leakage of hot fluid to cold fluid   **(d)** fouling

(e) all of these   **(f)** none of these

PROBLEMS

11.1 Consider a 1–2 TEMA J shell-and-tube heat exchanger. Define the analytical model (the set of differential equations and boundary conditions) that fully describes the temperature distributions for both tube and shell fluids. Assume that all fluid properties, process parameters, and heat exchanger dimensions are known.

11.2 The problem of predicting local temperatures and/or temperature differences along fluid stream paths in a plate-fin crossflow heat exchanger requires modeling of fluid temperature fields. Temperature fields for such a crossflow heat exchanger with both fluids unmixed throughout the exchanger core obey a model described with the set of equations as follows:

$$\frac{\partial \theta_1}{\partial \chi} + \theta_1 = \theta_2 \qquad \frac{\partial \theta_2}{\partial \zeta} + \theta_2 = \theta_1$$

with the boundary conditions

$$\theta_1(0, \zeta) = 1 \qquad \theta_2(\chi, 0) = 0$$

where dimensionless variables and the coordinates are as follows: $\theta_j = (T_j - T_{2,i})/(T_{1,i} - T_{2,i})$, $\chi = (x/L_1)\text{NTU}$, and $\zeta = (y/L_2)C^* \cdot \text{NTU}$ (see Section 11.2.4 for details). Determine temperature distributions θ_j as explicit functions of dimensionless coordinates χ and ζ.

11.3 Temperature distributions of both fluids in a crossflow heat exchanger with fluid 1 mixed throughout and fluid 2 unmixed are defined by the following relationships [see Example 11.3, Eqs. (8) and (9)]:

$$\theta_1(\chi) = e^{-\mathscr{K}\chi} \qquad \theta_2(\chi,\zeta) = (1 - e^{-\zeta})e^{-\mathscr{K}\chi}$$

where $\mathscr{K} = [1 - \exp(-C^* \cdot \mathrm{NTU})]/(C^* \cdot \mathrm{NTU})$. Show that the heat exchanger effectiveness of this heat exchanger is given by

$$\varepsilon = 1 - e^{-K/C^*} \qquad K = 1 - \exp(-C^* \cdot \mathrm{NTU})$$

11.4 Consider a crossflow heat exchanger with the smaller heat capacity rate fluid unmixed and the other fluid mixed throughout. Show that the heat exchanger effectiveness of this flow arrangement is given by $\varepsilon = [1 - \exp(-MC^*)]/C^*$, where $M = 1 - \exp(-\mathrm{NTU})$. Determine the numerical values for NTU that correspond to the maximum entropy generation of this exchanger for $C^* = 1$ and 0.1.

11.5 Using mass and energy balances, formulate a mathematical model of a general case of a crossflow heat exchanger. The inlet temperatures of either of the two fluids may be nonuniform at respective inlets. The formulation must be presented in dimensionless form. The model should consist of differential equations for determining temperature distributions and corresponding boundary conditions. Assume steady-state operation. Invoke traditional assumptions for heat exchanger analysis, except for nonuniformity of the inlet temperatures.

11.6 Formulate corresponding reduced mathematical models for determining temperature distributions of both fluids of a crossflow heat exchanger using the general model obtained in Problem 11.5. Consider the following particular cases:
(a) The fluid with the larger heat capacity rate is mixed; the other fluid is unmixed,
(b) The fluid with the smaller heat capacity rate is mixed; the other fluid is unmixed,
(c) Both fluids are mixed.

11.7 Determine the exact analytical solutions for temperature distributions of both fluid streams in a particular case of a mixed–mixed crossflow arrangement. The fluid inlet temperatures are uniform, and the heat exchanger operation is steady. All other traditional assumptions for heat exchanger theory are invoked.

11.8 Determine the exact analytical solutions for temperature distributions of both fluid streams in a particular case of an unmixed–mixed crossflow arrangement. The mixed fluid is the fluid with larger heat capacity rate. The unmixed fluid inlet temperature is nonuniform, and the heat exchanger operation is steady. All other traditional assumptions for heat exchanger theory are invoked. Reduce the general solution to a simplified one by assuming the uniformity of both inlet temperatures.

11.9 Determine the exact analytical solutions for temperature distributions of both fluid streams in a particular case of an unmixed–mixed crossflow arrangement. The mixed fluid is the fluid with smaller heat capacity rate. The unmixed fluid inlet temperature is nonuniform, and the heat exchanger operation is steady. All other traditional assumptions for heat exchanger theory are invoked. Reduce the general solution to a simplified one by assuming the uniformity of both inlet temperatures.

11.10 Derive the analytical expression for determining dimensionless temperature differences in a parallelflow/counterflow heat exchanger. Highlight important reasoning in the derivation with an explanation.

11.11 A two-fluid heat exchanger of an arbitrary flow arrangement has a specified number of transfer units NTU, and the heat capacity rate ratio C^*. The inlet temperatures of both fluids are known and their ratio is $\vartheta = T_{1,i}/T_{2,i}$. Derive the relationship between the entropy generated in this exchanger and its design parameters as given by Eq. (11.33), in which the pressure drops are neglected. Subsequently, calculate the entropy generation for a counterflow heat exchanger having $C^* = 1$ and an inlet temperature ratio of 0.5. Perform the calculations for NTU $= 1$, 5, and 10 and discuss the change in entropy generation with increased heat exchanger size. Finally, show how Eq. (11.28) would change if there is finite pressure drop.

11.12 Determine the entropy generation caused by fluid friction for fluid flow in a heat exchanger passage. Fluid is assumed to be an incompressible liquid described by mass density ρ and mass flow rate $\dot{m}$. The pressure drop along the flow length is $\Delta p = p_i - p_o$. The inlet and outlet temperatures are T_i and T_o, respectively.

11.13 The relationship between design parameters and entropy generated in a heat exchanger is defined by Eq. (11.33). This relationship is the subject of Problem 11.11. Show that the operating point corresponding to equal outlet temperatures must correspond to maximum entropy generation. Assume that the temperature effectiveness increases monotonically with NTU. Calculate the number of transfer units that correspond to the maximum entropy generation operating point for a counterflow exchanger with $C^* = 1$.

11.14 Compare the magnitude of the effectiveness of the two-pass cross-parallelflow and cross-counterflow arrangements presented in Fig. 11.10. Formulate a thermodynamic argument for comparison, and verify conclusions calculating the effectiveness for an arbitrarily selected set of design parameters, including NTU $= 1$ and $C^* = 1$.

11.15 Establish the rank of single-pass crossflow heat exchangers shown in Table P11.15. The rank is to be established by comparing the values of heat exchanger effectiveness using a heuristic approach in assessing irreversibility levels of the pairs of flow arrangements.

11.16 Establish the rank of two-pass crossflow heat exchangers shown in Table P11.16. The rank is to be established by comparing the values of heat exchanger effective-

TABLE P11.15 Single-Pass Crossflow Heat Exchanger

Schematic	Flow Arrangement
	Single-pass crossflow, fluid 1 unmixed and coupled in inverted order, fluid 2 split into two streams with equal mass flow rates, individually mixed.
	Single-pass crossflow, fluid 1 unmixed and connected in inverted order, fluid 2 split into three streams with equal mass flow rates, individually mixed
	Single-pass crossflow, fluid 1 unmixed and connected in inverted order, fluid 2 split into four streams with equal mass flow rates, individually mixed

ness using a heuristic approach in assessing irreversibility levels of the pairs of flow arrangements. Verify the conclusions by performing a numerical evaluation of heat exchanger effectiveness for each heat exchanger with $C^* = 0.8$ and NTU = 1, 2, and 3.

11.17 The condenser of a power plant operates under conditions as follows. The condensing water vapor enters the condenser at 30°C (303 K) with the quality (vapor fraction) $x < 1$. The pressure on the condensing side is 4 kPa. The pressure drop on the condensing side is 0.6 kPa. The cold-fluid (river water) inlet temperature is

TABLE P11.16 Two-Pass Crossflow Heat Exchanger

Schematic	Flow Arrangement
	Two-pass cross-counterflow, fluid 1 unmixed and coupled in inverted order fluid 2 mixed throughout
	Two-pass cross-counterflow, fluid 1 unmixed and coupled in identical order, fluid 2 mixed throughout
	Two-pass cross-parallelflow, fluid 1 unmixed and coupled in inverted order, fluid 2 mixed throughout
	Two-pass cross-parallelflow, fluid 1 unmixed and coupled in identical order, fluid 2 mixed throughout

17°C (290 K). The temperature of the water at the heat exchanger outlet is 27°C (300 K). The pressure drop on the coolant side of the heat exchanger (including manifolds and connecting pipes) is 50 kPa. It is assumed that the water pump operates with negligible losses. In addition, the exchanger (including the connecting piping) heat losses to the environment constitute approximately 2% of the condenser heat transfer rate. Determine the magnitude and relative importance of all irreversibilities associated with the operation of the heat exchanger.

11.18 Reconsider the analysis of Problem 11.17 (a steam electric power plant condenser), but now include the economic aspect of the entropy generation estimation. Assign a monetary value to the entropy-generation contributions in terms of busbar energy delivery costs. The relevant additional data are as follows:

Variable	Value
Rate of fuel consumption (MW)	1600
Boiler efficiency (%)	80
Net electric power delivered (MW)	700
Combined turbine/generator efficiency (%)	85
Combined motor/pump efficiency (%)	80
Capital cost of the equipment with 12% annual interest rate ($/kW)	700
Average energy cost ($/10^9 J)	3
Operation time (h/yr)	4000

11.19 Show that entropy generation caused by heat transfer and fluid friction associated with flow through a duct can be described by Eq. (11.65). Elaborate explicitly all the assumptions needed for derivation of this relationship.

11.20 A heat exchanger passage carries a constant property fluid. The cross-sectional area is A_o and the wetted wall perimeter is $\mathbf{P}$. The mass flow rate of the fluid is fixed. The heat transfer rate between the fluid and the wall is across a mean temperature difference ΔT, and it is considered to be constant along the flow direction of a short passage under consideration. Show that dimensionless entropy generation can be written in the form of Eq. (11.66) if the fluid represents a simple compressible substance with constant thermophysical properties. No phase change is present.

11.21 It can be shown that dimensionless entropy generation caused by heat transfer at finite temperature differences and fluid friction takes the following form for flow through an isothermal duct:

$$\frac{\dot{S}_{\text{irr}}}{\dot{m}c_p} = (\vartheta - 1)(1 - e^{-4j\text{Pr}^{-2/3}L/D_h}) + \ln \frac{(\vartheta - 1)e^{-4j\text{Pr}^{-2/3}L/D_h} + 1}{\vartheta}$$

$$+ \frac{1}{2}\frac{f}{j}\,\text{Pr}^{2/3} \cdot \text{Ec}\ln\frac{(\vartheta - 1)e^{-4j\text{Pr}^{-2/3}L/D_h} + 1}{\vartheta e^{-4j\text{Pr}^{-2/3}L/D_h}}$$

Using mass, energy, and entropy rate balances, show that this result is correct if the flow is assumed to be fully developed. Demonstrate that entropy generation may have a global minimum for a selected set of operating parameters. Perform the analysis for both air and water as working fluids.

12 Flow Maldistribution and Header Design

One of the common assumptions in basic heat exchanger design theory is that fluid be distributed *uniformly* at the inlet of the exchanger on each fluid side and throughout the core. However, in practice, *flow maldistribution*[†] is more common and can significantly reduce the desired heat exchanger performance. Still, as we discuss in this chapter, this influence may be negligible in many cases, and the goal of uniform flow through the exchanger is met reasonably well for performance analysis and design purposes.

Flow maldistribution can be induced by (1) heat exchanger geometry (mechanical design features such as the basic geometry, manufacturing imperfections, and tolerances), and (2) heat exchanger operating conditions (e.g., viscosity- or density-induced maldistribution, multiphase flow, and fouling phenomena). Geometry-induced flow maldistribution can be classified into (1) gross flow maldistribution, (2) passage-to-passage flow maldistribution, and (3) manifold-induced flow maldistribution. The most important flow maldistribution caused by operating conditions is viscosity-induced maldistribution and associated flow instability.

In this chapter, we consider geometry-induced flow maldistribution in Section 12.1 and operating condition–induced flow maldistribution in Section 12.2. Next, mitigation of flow maldistribution is discussed in Section 12.3. Finally, header design for compact heat exchangers is summarized in Section 12.4.

12.1 GEOMETRY-INDUCED FLOW MALDISTRIBUTION

One class of flow maldistribution, which is a result of geometrically nonideal fluid flow passages or nonideal exchanger inlet/outlet header/tank/manifold/nozzle design, is referred to as *geometry-induced flow maldistribution*. This type of maldistribution is closely related to heat exchanger construction and fabrication (e.g., header design, heat exchanger core fabrication including brazing in compact heat exchangers). This maldistribution is peculiar to a particular heat exchanger in question and cannot be influenced significantly by modifying operating conditions. Geometry-induced flow maldistribution is related to mechanical design-induced flow nonuniformities such as (1) entry conditions, (2) bypass and leakage streams, (3) fabrication tolerances,

---

[†] *Flow maldistribution* is defined as nonuniform distribution of the mass flow rate on one or both fluid sides in any of the heat exchanger ports and/or in the heat exchanger core. The term *ideal* fluid flow passage/header/heat exchanger would, as a rule, denote conditions of uniform mass flow distribution through an exchanger core.

(4) shallow bundle effects,[†] and (5) general equipment and exchanger system effects (Kitto and Robertson, 1989).

The most important causes of flow nonuniformities can be divided roughly into three main groups of maldistribution effects: (1) gross flow maldistribution (at the inlet face of the exchanger), (2) passage-to-passage flow maldistribution (nonuniform flow in neighboring flow passages), and (3) manifold-induced flow maldistribution (due to inlet/outlet manifold/header design). First, we discuss gross flow maldistribution. Subsequently, the passage-to-passage flow maldistribution is addressed, followed by a few comments related to manifold-induced flow maldistribution.

12.1.1 Gross Flow Maldistribution

The major feature of gross flow maldistribution is that nonuniform flow occurs at the macroscopic level (due to poor header design or blockage of some flow passages during manufacturing, including brazing or operation). The gross flow maldistribution does not depend on the local heat transfer surface geometry. This class of flow maldistribution may cause (1) a significant increase in the exchanger pressure drop, and (2) some reduction in heat transfer rate. To predict the magnitude of these effects for some simple exchanger flow arrangements, the nonuniformity will be modeled as one- or two-dimensional as follows, with some specific results.

Gross flow maldistribution can occur in one dimension across the free-flow area (perpendicular to the flow direction) as in single-pass counterflow and parallelflow exchangers, or it can occur in two or three dimensions as in single- and multipass crossflow and other exchangers.

Let us first model a one-dimensional gross flow maldistribution with an N-step inlet velocity distribution function. The heat exchanger will be represented by an array of N subunits, called *subexchangers*, having uniform flow throughout each unit but with different mass flow rates from unit to unit. The number of subexchangers is arbitrary, but it will be determined to be in agreement with the imposed flow maldistribution. The set of standard assumptions of Section 3.2.1 is applicable to each subexchanger. The following additional idealizations are introduced to quantify the influence of flow nonuniformity caused by gross flow maldistribution on each subexchanger and the exchanger as a whole.

1. Total heat transfer rate in a real heat exchanger is equal to the sum of the heat transfer rates that would be exchanged in N subexchangers connected in parallel for an idealized N-step inlet velocity distribution function.

2. The sum of the heat capacity rates of the respective fluid streams for all subexchangers is equal to the total heat capacity rates of the fluids for the actual maldistributed heat exchanger.

With these auxiliary assumptions, the temperature effectiveness of a counterflow/parallelflow heat exchanger can be calculated by modeling the heat exchanger as a parallel coupling of N subexchangers for a maldistributed fluid stream having N indivi-

[†] Use of an axial tube-side nozzle (if the depth of the head is insufficient to allow the jet from the nozzle to expand to the tubesheet diameter) may result in larger flow in central tubes and lead to flow maldistribution (Mueller, 1987).

dual uniform fluid streams (i.e., having an N-step function velocity distribution). The other fluid side is considered as having uniform flow distribution for such an analysis. If flow nonuniformity occurs on both fluid sides of a counterflow or parallelflow exchanger, the exchanger is divided into a sufficient number of subexchangers such that the flow distributions at the inlet on both fluid sides are uniform for each subexchanger. For all other exchangers, the solution can only be determined numerically, and the solutions of Sections 12.1.1.1 and 12.1.1.2 are not valid in that case.

12.1.1.1 Counterflow and Parallelflow Exchangers. In this section, we derive an expression for the exchanger effectiveness and hence heat transfer performance for counterflow and parallelflow exchangers having an N-step velocity distribution function on the fluid 1 side and perfectly uniform flow distribution on the fluid 2 side, as shown in Fig. 12.1a. Here fluid 1 can be either the hot or cold fluid, and in that case, fluid 2 will be the cold or hot fluid. Subsequently, we apply this analysis to a heat exchanger having a two-step velocity distribution function at the inlet.

Heat Transfer Analysis. Let us consider a counterflow exchanger with an N-step inlet distribution function of fluid 1, shown in Fig. 12.1a. The same analysis would be valid for a parallelflow heat exchanger. Fluid 2 is considered uniform. We may model this exchanger as an array of N subexchangers, each obeying the standard assumptions of Section 3.2.1. Hence,

$$q = \sum_{j=A}^{N} q_j \tag{12.1}$$

where q represents the total heat transfer rate, and $q_j, j = A, B, \ldots, N$, the fractions of heat transfer rate in N hypothetical subexchangers, each having uniform mass flow rates on both sides, as shown in Fig. 12.1b. The assumptions invoked, including the auxiliary ones introduced above, lead to the following results:

$$q = C_1\left|T_{1,i} - T_{1,o}\right| \quad \text{and} \quad q_j = C_{1,j}\left|(T_{1,i} - T_{1,o})_j\right| \qquad j = A, B, \ldots, N \tag{12.2}$$

$$P_1 = \frac{T_{1,i} - T_{1,o}}{T_{1,i} - T_{2,i}} \quad \text{and} \quad P_{1,j} = \frac{(T_{1,i} - T_{1,o})_j}{T_{1,i} - T_{2,i}} \qquad j = A, B, \ldots, N \tag{12.3}$$

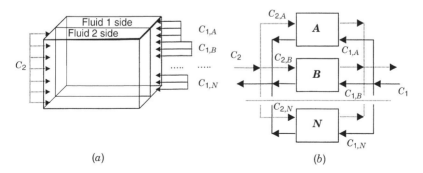

(a)   (b)

FIGURE 12.1   Idealized two-step function flow nonuniformity on fluid 1 side and uniform flow on fluid 2 side of a counterflow exchanger.

Substituting Eqs. (12.2) and (12.3) into Eq. (12.1) and rearranging, the expression for fluid 1 temperature effectiveness becomes

$$P_1 = \frac{1}{C_1} \sum_{j=A}^{N} C_{1,j} P_{1,j} \tag{12.4}$$

where

$$C_1 = \sum_{j=A}^{N} C_{1,j} \tag{12.5}$$

Note that Eqs. (12.4) and (12.5) are valid for a maldistributed fluid regardless of whether it is hot or cold, C_{min} or C_{max}, or denoted as 1 or 2. The subscript 1 in these equations may be replaced by a designator of the maldistributed stream, say as in P_{ms} and C_{ms}. The temperature effectiveness of fluid 1 for each of the subexchangers of Eq. (12.4) is computed knowing individual NTU and heat capacity rate ratio:

$$P_{1,j} = P_{1,j}\left(\mathrm{NTU}_{1,j}, \frac{C_{1,j}}{C_{2,j}}\right) \qquad j = A, B, \ldots, N \tag{12.6}$$

where $P_{1,j}(\cdot)$ on the right-hand side of Eq. (12.6) is computed for each exchanger using the expression provided in Table 3.6 as follows:

$$P_{1,j} = \begin{cases} \dfrac{1 - \exp[\mathrm{NTU}_{1,j}(1 - R_{1,j})]}{1 - R_{1,j}\exp[\mathrm{NTU}_{1,j}(1 - R_{1,j})]} & \text{counterflow} \\[4mm] \dfrac{1 - \exp[-\mathrm{NTU}_1(1 + R_{1,j})]}{1 + R_{1,j}} & \text{parallelflow} \end{cases} \tag{12.7}$$

Application of Eq. (12.6) requires the values of $\mathrm{NTU}_{1,j}$, $C_{1,j}$, and $C_{2,j}$ for each subexchanger. To determine these variables, we should invoke the standard assumptions of Section 3.2.1 to get the free-flow area and heat capacity rate ratio as

$$A_o = \sum_{j=A}^{N} A_{o,j} \tag{12.8}$$

$$\frac{C_{1,j}}{C_1} = \frac{\dot{m}_{1,j}}{\dot{m}_1} = \frac{u_{1,j}}{u_1}\left(\frac{A_{o,j}}{A_o}\right)_1, \qquad j = A, B, \ldots, N \tag{12.9}$$

Similarly, for fluid 2, we get

$$\frac{C_{2,j}}{C_2} = \frac{u_{2,j}}{u_1}\left(\frac{A_{o,j}}{A_o}\right)_2, \qquad j = A, B, \ldots, N \tag{12.10}$$

Note that the sets of relations given by Eq. (12.9) and (12.10) may be reduced by one equation each by utilizing Eq. (12.5) for fluid 1 (and similarly for fluid 2).

The number of heat transfer units and the capacity rate ratios for Eq. (12.7) can be determined from their definitions as follows:

$$\text{NTU}_{1,j} = \frac{UA_{1,j}}{C_{1,j}} = \frac{UA_1}{C_1} \frac{A_{1,j}}{A_1} \frac{C_1}{C_{1,j}} \tag{12.11}$$

$$R_{1,j} = \frac{C_{1,j}}{C_{2,j}} \qquad j = A, B, \ldots, N \tag{12.12}$$

Note also that $D_h = 4A_o L/A$ with D_h and L identical for all subexchangers $j = A, B, \ldots, N$ for the counterflow/parallelflow heat exchanger. $A_{1,j}/A_1$ of Eq. (12.11) is obtained then from the definitions of D_h as

$$\left(\frac{A_{o,j}}{A_o}\right)_1 = \left(\frac{A_j}{A}\right)_1 \qquad j = A, B, \ldots, N \tag{12.13}$$

In addition,

$$\left(\frac{A_{o,j}}{A_o}\right)_1 = \left(\frac{A_{o,j}}{A_o}\right)_2 \qquad j = A, B, \ldots, N \tag{12.14}$$

The reduction in the temperature effectiveness on the maldistributed side can be presented by the *performance (effectiveness) deterioration factor* as

$$\Delta P_1^* = \frac{P_{1,\text{ideal}} - P_1}{P_{1,\text{ideal}}} \quad \text{or} \quad \Delta\varepsilon^* = \frac{\varepsilon_{\text{ideal}} - \varepsilon}{\varepsilon_{\text{ideal}}} \tag{12.15}$$

where $P_{1,\text{ideal}}$ represents the temperature effectiveness for the case of having no flow maldistribution.

The influence of gross flow maldistribution is shown in Fig. 12.2 for a balanced ($C^* = 1$) counterflow heat exchanger in terms of $\Delta\varepsilon^*$ for a two-step inlet velocity distribution function (i.e., for two subexchangers). For a particular value of u_{max}/u_m

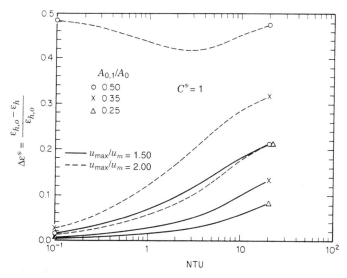

FIGURE 12.2 Performance deterioration factor $\Delta\varepsilon^*$ for a balanced heat exchanger, $C^* = 1$, $N = 2$. (From Shah, 1981.)

and given NTU, we find the greatest reduction in the heat exchanger effectiveness occurring when two-step function flow maldistribution occurs in equal flow areas (50 : 50%). The effect of flow maldistribution increases with NTU for a counterflow exchanger. Note that the reduction in the temperature effectiveness P_1 of Eq. (12.4), obtained using Eq. (12.15), is valid regardless of whether the maldistributed fluid is the hot, cold, C_{max}, or C_{min} fluid.

We can idealize an N-step function velocity distribution into an equivalent two-step function velocity distribution. Based on the analysis of passage-to-passage flow maldistribution presented in Section 12.1.2, it is conjectured that the deterioration in the exchanger effectiveness is worse for the two-step function velocity distribution. Hence, conservatively, any flow maldistribution can be reduced to a two-step function, and its effect can readily be evaluated on the exchanger effectiveness, which will represent the highest deterioration.

As indicated above, the discussion of the effect of the gross flow maldistribution in this section refers to a heat exchanger with counterflow arrangement and balanced flow ($C^* = 1$). Hence, the increased effect of flow maldistribution with increasing NTU is valid only for this special situation. If either $C^* \neq 1$ or if the flow arrangement is parallel-flow, the influence of flow maldistribution may decrease with increasing NTU. This can be determined from Eq. (12.15), assuming the validity of the appropriate effectiveness–NTU relationships for each subexchanger and for the heat exchanger as a whole. Also, the actual flow rate conditions in most practical cases would not correspond to a balanced heat exchanger case.

Pressure Drop Analysis. There is no rigorous theory available for predicting a change in the pressure drop due to flow maldistribution in the exchanger. This is because for nonuniform flow distribution, the static pressures at the core inlet and outlet faces will not be uniform, and hence, constant pressure drop across the core is not a valid assumption. The following is a suggested approximate procedure. This approach is not based on a rigorous modeling of the actual flow conditions and must be used very cautiously. Consider a two-step function velocity distribution at the core inlet on fluid 1 side as shown in Fig. 12.1a for $N = 2$. Subexchangers in Fig. 12.1b for $N = 2$ are in parallel. Using Eq. (6.28), evaluate the pressure drop Δp_j for a specific subexchanger which has the highest fluid velocity in the flow passages. Also compute $\Delta p_{uniform}$ for fluid 1 considering the flow as uniform at the core inlet in Fig. 12.1. Therefore, as a conservative approach, this largest Δp_j (i.e., Δp_{max}) will be the pressure drop on the fluid 1 side having imposed flow nonuniformity. The increase in pressure drop due to flow nonuniformity is then

$$(\Delta p)_{increase} = \Delta p_{max} - \Delta p_{uniform} \qquad (12.16)$$

It should be emphasized that the entrance and exit losses in addition to the core friction contribution will be higher (in the evaluation of Δp_{max}) than those for uniform flow.

If flow nonuniformity occurs on both sides of an exchanger, the procedure outlined above is applied to both sides, since the pressure drops on both sides of a two-fluid exchanger are relatively independent of each other, except for the changes in fluid density due to heat transfer in the core. Hence, the analysis above is applicable to any flow arrangement.

Example 12.1 A counterflow heat exchanger has a severe flow maldistribution due to poor header design. On the fluid 1 side, 25% of the total free-flow area has the flow velocity 50% larger than the mean flow velocity through the core as a whole. The number of heat transfer units of the heat exchanger is $NTU_1 = 3$. The total heat capacity rates through the exchanger are nearly the same (i.e., the heat exchanger is balanced). Determine the reduction in the temperature effectiveness of fluid 1 and an approximate increase in the pressure drop due to flow maldistribution. Assume fully developed laminar flow on both fluid sides (i.e., U remains constant).

SOLUTION

Problem Data and Schematic: A schematic of the heat exchanger under consideration is similar to that in Fig. 12.1 with only A and B subexchangers. The following data are known.

$$NTU_1 = 3 \qquad u_{1,A} = 1.5u_1 \qquad C_1 = C_2 \qquad A_{o,A} = 0.25A_{o,1}$$

Determine: The temperature effectiveness P_1 of the maldistributed fluid.

Assumptions: All the assumptions listed in Section 3.2.1 are valid with the exception of nonuniformity of the mass flow rate of fluid 1, which is assumed to have a two-step function velocity distribution. Even with this flow maldistribution, U based on A_1 or A_2 is assumed constant.

Analysis: The effectiveness of the maldistributed heat exchanger is given by Eq. (12.4) with $j = A$ and B as

$$P_1 = \frac{C_{1,A}}{C_1} P_{1,A} + \frac{C_{1,B}}{C_1} P_{1,B}$$

Let us determine all the parameters in this equation after determining the area ratios using Eqs. (12.13) and (12.14).

$$\left(\frac{A_{o,A}}{A_o}\right)_1 = \frac{A_{1,A}}{A_1} = 0.25 \qquad \left(\frac{A_{o,B}}{A_o}\right)_1 = \frac{A_{1,B}}{A_1} = 0.75$$

$$\left(\frac{A_{o,A}}{A_o}\right)_1 = \left(\frac{A_{o,A}}{A_o}\right)_2 = 0.25 \qquad \left(\frac{A_{o,B}}{A_o}\right)_1 = \left(\frac{A_{o,B}}{A_o}\right)_2 = 0.75$$

Also,

$$\frac{u_{1,A}}{u_1} = 1.5$$

where u_1 is the mean fluid velocity on the fluid 1 side.

The ratios of the heat capacity rates in the maldistributed subexchangers to the total capacity rate of fluid 1 are then given by Eq. (12.9) as

$$\frac{C_{1,A}}{C_1} = \frac{u_{1,A}}{u_1}\left(\frac{A_{o,A}}{A_o}\right)_1 = 1.5 \times 0.25 = 0.375 \qquad \frac{C_{1,B}}{C_1} = 1 - \frac{C_{1,A}}{C_1} = 1 - 0.375 = 0.625$$

Similarly, the heat capacity rate ratios of subexchangers to the total exchanger on fluid 2 side are given by Eq. (12.10) as

$$\frac{C_{2,A}}{C_2} = \left(\frac{A_{o,A}}{A_o}\right)_2 = \left(\frac{A_{o,A}}{A_o}\right)_1 = 0.25 \qquad \frac{C_{2,B}}{C_2} = 1 - \frac{C_{2,A}}{C_2} = 1 - 0.25 = 0.75$$

We will determine temperature effectivenesses, using Eq. (12.7), after calculating respective NTUs and R's. The NTUs from Eq. (12.11) are

$$\text{NTU}_{1,A} = \frac{UA_{1,A}}{C_{1,A}} = \frac{UA_1}{C_1} \frac{A_{1,A}}{A_1} \frac{C_1}{C_{1,A}} = 3 \times 0.25 \times \frac{1}{0.375} = 2.00$$

$$\text{NTU}_{1,B} = \frac{UA_{1,B}}{C_{1,B}} = \frac{UA_1}{C_1} \frac{A_{1,B}}{A_1} \frac{C_1}{C_{1,B}} = 3 \times 0.75 \times \frac{1}{0.625} = 3.60$$

The heat capacity rate ratios, required for effectiveness calculations, are computed using Eq. (12.12) as follows:

$$R_{1,A} = \frac{C_{1,A}}{C_{2,A}} = \frac{C_{1,A}}{C_1} \frac{C_2}{C_{2,A}} \frac{C_1}{C_2} = 0.375 \times \frac{1}{0.25} \times 1 = 1.50$$

$$R_{1,B} = \frac{C_{1,B}}{C_{2,B}} = \frac{C_{1,B}}{C_1} \frac{C_2}{C_{2,B}} \frac{C_1}{C_2} = 0.625 \times \frac{1}{0.75} \times 1 = 0.8333$$

Therefore, the temperature effectivenesses of subexchangers are given by Eq. (12.7) as

$$P_{1,A} = \frac{1 - e^{-\text{NTU}_{1,A}(1-R_{1,A})}}{1 - R_{1,A}e^{-\text{NTU}_{1,A}(1-R_{1,A})}} = \frac{1 - \exp[-2.00(1 - 1.50)]}{1 - 1.5\exp[-2.00(1 - 1.50)]} = 0.5584$$

$$P_{1,B} = \frac{1 - e^{-\text{NTU}_{1,B}(1-R_{1,B})}}{1 - R_{1,B}e^{-\text{NTU}_{1,B}(1-R_{1,B})}} = \frac{1 - \exp[-3.60(1 - 0.8333)]}{1 - 0.8333\exp[-3.60(1 - 0.8333)]} = 0.8315$$

The temperature effectiveness from Eq. (12.4) is

$$P_1 = \frac{C_{1,A}}{C_1} P_{1,A} + \frac{C_{1,B}}{C_1} P_{1,B} = 0.375 \times 0.5584 + 0.625 \times 0.8315 = 0.7291$$

The heat exchanger effectiveness of a balanced counterflow heat exchanger without any flow maldistribution on either fluid side would be

$$P_{1,\text{ideal}} = \frac{\text{NTU}_1}{1 + \text{NTU}_1} = \frac{3}{1 + 3} = 0.750$$

Finally, the quantitative measure of the reduction in the effectiveness due to maldistribution is [see Eq. (12.15)]

$$\Delta P_1^* = \frac{P_{1,\text{ideal}} - P_1}{P_{1,\text{ideal}}} = \frac{0.750 - 0.7291}{0.750} = 0.0279 \qquad \textit{Ans.}$$

Discussion and Comments: From the results, it becomes clear that a relatively large flow maldistribution on the fluid 1 side in this particular case causes a deterioration of the temperature effectiveness of approximately 2.8%. With all other parameters fixed, a heat exchanger with high NTU will suffer more pronounced effectiveness deterioration (see Fig. 12.2).

12.1.1.2   Crossflow Exchangers. A direct extension of the approach used for a counter-flow/parallelflow exchanger to that for a crossflow exchanger with different combinations of fluid mixing/unmixing on each fluid side is not necessarily straightforward. Only when flow nonuniformity is present on the unmixed fluid side with the other fluid side as mixed can a simple closed-form solution be obtained, as outlined next.

Mixed–Unmixed Crossflow Exchanger with Nonuniform Flow on the Unmixed Side. Let us consider a single-pass crossflow exchanger having the unmixed fluid (fluid 1) maldistributed. The inlet velocity distribution is represented with an *N*-step function (Fig. 12.3). Fluids 1 and 2 can be arbitrarily hot and cold, or vice versa.

The total heat transfer rate in the exchanger is given by

$$q = \sum_{j=A}^{N} q_j \tag{12.17}$$

where the q_j represent individual heat transfer rates/enthalpy rate changes as follows[†]:

$$q_A = P_{1,A} C_{1,A}(T_{1,i} - T_{2,i}) = C_2(T_{2M,A} - T_{2,i})$$
$$q_B = P_{1,B} C_{1,B}(T_{1,i} - T_{2M,A}) = C_2(T_{2M,B} - T_{2M,A})$$
$$\vdots \tag{12.18}$$
$$q_N = P_{1,N} C_{1,N}(T_{1,i} - T_{2M,N-1}) = C_2(T_{2M,N} - T_{2,o})$$

In Eq. (12.18), $T_{2M,j}$ ($j = A, B, \ldots, N$) represent the mixed mean temperatures of fluid 2 between the subexchangers. Note that the left-hand side of Eq. (12.17) can also be presented in the form

$$q = P_1 C_1(T_{1,i} - T_{2,i}) \tag{12.19}$$

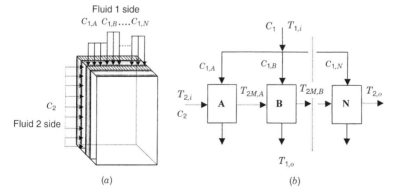

(a)   (b)

FIGURE 12.3   Idealized two-step function flow nonuniformity on the unmixed fluid 1 side and uniform flow on the mixed fluid 2 side of a crossflow exchanger.

[†] For the sake of simplified notation, the absolute value designators are omitted.

Our objective is to determine the relationship between the fluid 1 temperature effectiveness P_1 and the temperature effectiveness and heat capacity rates of the subexchangers of Fig. 12.3b. From Eq. (12.19), we get

$$P_1 = \frac{q}{C_1(T_{1,i} - T_{2,i})} \tag{12.20}$$

Replacing q in Eq. (12.20) with q from Eq. (12.17), and utilizing the relationships provided by Eq. (12.18), we get

$$P_1 = \frac{1}{C_1}\left(P_{1,A}C_{1,A} + P_{1,B}C_{1,B}\frac{T_{1,i} - T_{2M,A}}{T_{1,i} - T_{2,i}} + \cdots + P_{1,N}C_{1,N}\frac{T_{1,i} - T_{2M,N-1}}{T_{1,i} - T_{2,i}}\right) \tag{12.21}$$

Temperature difference ratios in Eq. (12.21) can be eliminated by manipulating relationships from Eq. (12.18) as follows:

$$\frac{T_{1,i} - T_{2M,A}}{T_{1,i} - T_{2,i}} = 1 - \frac{P_{1,A}C_{1,A}}{C_2}$$

$$\frac{T_{1,i} - T_{2M,B}}{T_{1,i} - T_{2,i}} = \left(1 - \frac{P_{1,A}C_{1,A}}{C_2}\right)\left(1 - \frac{P_{1,B}C_{1,B}}{C_2}\right)$$

$$\vdots$$

$$\frac{T_{1,i} - T_{2M,N-1}}{T_{1,i} - T_{2,i}} = \prod_{k=A}^{N-1}\left(1 - \frac{P_{1,k}C_{1,k}}{C_2}\right) \tag{12.22}$$

Combining Eqs. (12.22) and (12.21) and after rearrangement, we get

$$P_1 = \frac{1}{C_1}\left[P_{1,A}C_{1,A} + \sum_{j=B}^{N}P_{1,j}C_{1,j}\prod_{k=1}^{j-1}\left(1 - \frac{P_{1,k}C_{1,k}}{C_2}\right)\right] \tag{12.23}$$

To reemphasize, the fluid 1 side is unmixed and the fluid 2 side is mixed for the expression of a crossflow exchanger above. The temperature effectiveness of the ideal heat exchanger of Fig. 12.3 as a whole, and those of the subexchangers, can be expressed in terms of corresponding heat capacity rate ratios and numbers of transfer units as follows (see Table 3.6):

$$P_{1,\text{ideal}} = \frac{C_2}{C_1}\left\{1 - \exp\left[-\frac{C_1}{C_2}\left(1 - e^{-\text{NTU}_1}\right)\right]\right\} \tag{12.24}$$

$$P_{1,j} = \frac{(T_{1,i} - T_{1,o})_j}{T_{1,i} - T_{2,i}} = \frac{C_2}{C_{1,j}}\left\{1 - \exp\left[-\frac{C_{1,j}}{C_2}\left(1 - e^{-\text{NTU}_{1,j}}\right)\right]\right\} \qquad j = A, B, \ldots, N \tag{12.25}$$

where

$$\text{NTU}_1 = \frac{UA}{C_1} \qquad \text{NTU}_{1,j} = \frac{UA_j}{C_{1,j}} \qquad j = A, B, \ldots, N \tag{12.26}$$

When the flow maldistribution on the fluid 1 side is assumed to be a two-step velocity distribution function, Eq. (12.23) can be simplified to

$$P_1 = \frac{1}{C_1}\left[P_{1,A}C_{1,A} + P_{1,B}C_{1,B}\left(1 - \frac{P_{1,A}C_{1,A}}{C_2}\right)\right] \qquad (12.27)$$

Calculation of the temperature effectiveness of a maldistributed fluid stream using Eq. (12.23) or (12.27) is valid only if the maldistributed fluid side is the unmixed fluid side and the mixed side has uniform flow. However, if the maldistributed fluid is a mixed fluid and the unmixed fluid side has uniform flow, a closed-form solution cannot be obtained using this simplified approach. This is because T_{2M} temperatures for this case will not be uniform, and hence we cannot determine the effectiveness of this exchanger using the formula of Table 3.6 when the flow at the inlet of a subsequent subexchanger is not uniform. The problem becomes inevitably nonlinear, and no closed-form solution is available for this case; only a numerical solution is the option.

Note that Eqs. (12.6), (12.9), and (12.11) are also valid for this mixed–unmixed cross-flow exchanger, while Eqs. (12.10), (12.12), and (12.14) are not valid and we don't need them to determine the temperature effectiveness of this flow maldistributed case.

Unmixed–Unmixed Crossflow Exchangers. The two-dimensional flow maldistribution has been analyzed numerically only for an unmixed–unmixed crossflow exchanger. In a series of publications as summarized by Chiou (1980) and Mueller and Chiou (1988), Chiou has studied the effects of flow maldistribution on an unmixed–unmixed crossflow single-pass heat exchanger with flow maldistribution on one and both fluid sides. When flow maldistribution is present on only one fluid side, the following general conclusions have been obtained.

- For flow maldistribution on the $C_{\max}$ fluid side, the exchanger thermal performance deterioration factor $\Delta\varepsilon^*$ approaches a single value of 0.06 for all $C^* < 1$ when NTU approaches zero. The performance deterioration factor decreases as NTU increases. For a balanced heat exchanger ($C^* = 1$), the exchanger thermal performance deterioration factor increases continually with NTU.
- For flow maldistribution on the $C_{\min}$ fluid side, the thermal performance deterioration factor first increases and then decreases as NTU increases.
- If flow nonuniformities are present on both sides, the performance deterioration factor can be either larger or smaller than that for the case where flow nonuniformity is present on only one side, and there are no general guidelines about the expected trends.

A study of the influence of two-dimensional nonuniformities in inlet fluid temperatures (Chiou, 1982) indicates that there is a smaller reduction in exchanger effectiveness for the nonuniform inlet temperature than that for the nonuniform inlet mass flow rate. For various nonuniform flow models studied, the inlet nonuniform flow case showed a decrease in effectiveness of up to 20%; whereas for the nonuniform inlet temperature case, a decrease in effectiveness of up to 12% occurred, with even an increase in effectiveness for some cases of nonuniform inlet temperature. This occurs when the hotter portion of the inlet temperature is near the exit end of the cold fluid, whose inlet temperature is uniform. In a recent study, Ranganayakulu et al. (1996) obtained numerical solutions for the effects of two-dimensional flow nonuniformities on thermal performance and pressure drop in crossflow plate-fin compact heat exchangers.

Example 12.2 Analyze a crossflow heat exchanger with fluid 1 unmixed and fluid 2 mixed having pronounced maldistribution on fluid 1 side and $NTU_1 = 3$. The total heat capacity rates of the two fluids are nearly the same. Determine the temperature effectiveness of the maldistributed fluid 1 if 25% of the total free-flow area has the flow velocity 50% larger than the mean velocity through the core on the fluid 1 side corresponding to the uniform flow case.

SOLUTION

Problem Data and Schematic: A schematic of the heat exchanger under consideration is similar to that in Fig. 12.3 with only A and B subexchangers. The following data are known:

$$NTU_1 = 3 \qquad u_{1,A} = 1.5u_1 \qquad C_1 = C_2 \qquad A_{o,A} = 0.25A_{o,1}$$

Determine: The temperature effectiveness of fluid 1.

Assumptions: All the assumptions of Section 3.2.1 are valid here except for flow maldistribution on the fluid 1 side.

Analysis: Let us calculate the temperature effectiveness of fluid 1 under the idealized conditions of uniform mass flow rates on both fluid sides for a balanced unmixed–mixed crossflow heat exchanger using Eq. (12.24) as follows:

$$P_{1,\text{ideal}} = 1 - \exp[-(1 - e^{-NTU_1})] = 1 - \exp[-(1 - e^{-3})] = 0.6133$$

However, under given flow maldistribution conditions, this ideal effectiveness cannot be achieved. Thus, the temperature effectiveness should be calculated using Eq. (12.27). For this example, similar to Example 12.1, we can determine

$$\frac{C_{1,A}}{C_1} = 0.375 \qquad \frac{C_{1,B}}{C_1} = 0.625 \qquad NTU_{1,A} = 2.00 \qquad NTU_{1,B} = 3.60$$

Now $P_{1,A}$ and $P_{1,B}$ are computed using Eq. (12.25) as follows after incorporating $C_2 = C_1$:

$$P_{1,A} = \frac{C_1}{C_{1,A}} \left\{ 1 - \exp\left[-\frac{C_{1,A}}{C_1} \left(1 - e^{-NTU_{1,A}} \right) \right] \right\}$$

$$= \frac{1}{0.375} \left\{ 1 - \exp\left[-0.375 \left(1 - e^{-2.00} \right) \right] \right\} = 0.7385$$

$$P_{1,B} = \frac{C_1}{C_{1,B}} \left\{ 1 - \exp\left[-\frac{C_{1,B}}{C_1} \left(1 - e^{-NTU_{1,B}} \right) \right] \right\}$$

$$= \frac{1}{0.625} \left\{ 1 - \exp\left[-0.625 \left(1 - e^{-3.60} \right) \right] \right\} = 0.7288$$

Now, compute the temperature effectiveness of fluid 1 given by Eq. (12.27) as (with $C_1 = C_2$)

$$P_1 = \frac{1}{C_1}\left[P_{1,A}C_{1,A} + P_{1,B}C_{1,B}\left(1 - \frac{P_{1,A}C_{1,A}}{C_1}\right)\right] = P_{1,A}\frac{C_{1,A}}{C_1} + P_{1,B}\frac{C_{1,B}}{C_1}\left(1 - \frac{P_{1,A}C_{1,A}}{C_1}\right)$$

$$= 0.7385 \times 0.375 + 0.7288 \times 0.625 \times (1 - 0.7385 \times 0.375) = 0.6063 \qquad \textit{Ans.}$$

This actual effectiveness is, indeed, smaller than the one calculated for an idealized situation, 0.6063 vs. 0.6133. Finally, the fractional deterioration in the temperature effectiveness is given by Eq. (12.15) as

$$\Delta P^* = \frac{P_{1,\text{ideal}} - P_1}{P_{1,\text{ideal}}} = \frac{0.6133 - 0.6063}{0.6133} = 0.0114$$

Discussion and Comments: Deterioration in the temperature effectiveness caused by a relatively large flow maldistribution for this crossflow exchanger is 0.0114, much smaller than 0.0279 for the counterflow exchanger (see Example 12.1), for the same operating conditions. The results of Examples 12.1 and 12.2 emphasize the fact that flow maldistribution has the highest effect on a counterflow exchanger (since it has the highest ε for given NTU and C^*) compared to exchangers with other flow arrangements for similar operating parameters.

12.1.1.3 Tube-Side Maldistribution and Other Heat Exchanger Types.

Tube-side maldistribution in a 1–1 TEMA E shell-and-tube counterflow heat exchanger studied by Cichelli and Boucher (1956) led to the following major conclusions:

- For C_s/C_t small, say $C_s/C_t = 0.1$, the performance loss is negligible for large flow nonuniformities for $\text{NTU}_s < 2$.
- For C_s/C_t large, say $C_s/C_t > 1$, a loss can be noticed but diminishes for $\text{NTU}_s > 2$.
- $C_s/C_t = 1$ is the worst case at large NTU_s as can be found from Fig. 12.2.

Fleming (1966) and Chowdhury and Sarangi (1985) have studied various models of flow maldistribution on the tube side of a counterflow shell-and-tube heat exchanger. It is concluded that high-NTU heat exchangers are more susceptible to maldistribution effects. According to Mueller (1977), the well-baffled 1–1 counterflow shell-and-tube heat exchanger (tube side nonuniform, shell side mixed) is affected the least by flow maldistribution. Shell-and-tube heat exchangers, which do not have mixing of the uniform fluid [(1) tube side nonuniform, shell side unmixed; or (2) tube side uniform, shell side nonuniform in crossflow], are affected more by flow maldistribution.

According to Kutchey and Julien (1974), the radial flow variations of the mismatched air side and gas side reduce the regenerator effectiveness significantly.

12.1.2 Passage-to-Passage Flow Maldistribution

Compact heat exchangers with uninterrupted (continuous) flow passages, while designed for nonfouling applications, are highly susceptible to passage-to-passage flow maldistribution. That is because the neighboring passages are geometrically never identical, due to imperfect manufacturing processes. It is especially difficult to control the passage size precisely when small dimensions are involved [e.g., a rotary regenerator with $D_h = 0.5$ mm (0.020 in.)]. Since differently sized and shaped passages exhibit different flow resistances and the flow seeks a path of least resistance, a nonuniform flow through the matrix results. This phenomenon usually causes a slight reduction in pressure drop, while the reduction in heat transfer rate may be significant compared to that for nominal (average) size passages. The influence is of particular importance for continuous-flow passages at *low Re* (i.e., laminar flow) as found in compact rotary regenerators. For a theoretical analysis for passage-to-passage flow maldistribution, the actual nonuniform surface is idealized as containing large, small, and/or in-between size passages (in parallel) relative to the nominal passage dimensions. The models include (1) a two-passage model (London, 1970), (2) a three-passage model, and (3) an *N*-passage model (Shah and London, 1980). Although triangular and rectangular passage cross sections have been studied, similar analysis can be applied to any cross-sectional shapes of flow passages.

The analysis to follow can also be utilized for analyzing flow maldistribution in viscous oil cooler with constant-wall-temperature boundary conditions (i.e., condensation or vaporization taking place on the other fluid side). See Section 12.2.1 for further details. Let us first define the two-passage-model flow nonuniformity. From the methodological point of view, this approach is the most transparent and offers a clear idea of how the modeling of flow nonuniformity can be conducted. Also, the two-passage model predicts a more detrimental effect on heat transfer and pressure drop than that of an *N*-passage ($N > 2$) model.

12.1.2.1 Models of Flow Nonuniformity

Two-Passage Model.   Let us consider that a heat exchanger core characterizes flow non-uniformity due to two different flow cross sections differing in either (1) cross-section size of the same passage type, (2) different cross-sectional shapes of flow passages, or (3) a combination of both. The two most common types of idealized passage-to-passage nonuniformities are plate spacing and fin spacing, shown in Fig. 12.4a and b, respectively. For the analysis, the actual heat exchanger core will be assumed to be a collection of two (or more) distinct sets of uniform flow passages, passages 1 and passages 2

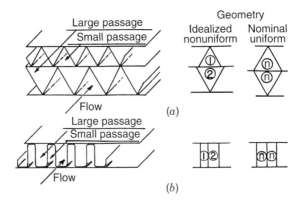

FIGURE 12.4 Two-passage nonuniformity model: (*a*) plate-spacing nonuniformity; (*b*) fin-spacing nonuniformity. Note that passages differ in size. The nominal size of the passage may be large, small, or in between, depending on how it is defined. (From London, 1968.)

(or N passages). Our objective here is to determine the reduction in heat transfer and pressure drop due to this passage-to-passage flow nonuniformity.

The following assumptions are invoked for setting up the model.

- Flow is hydrodynamically and thermally fully developed ($\mathrm{Nu} = \text{constant}$, $f \cdot \mathrm{Re} = \text{constant}$).
- Thermophysical properties of the fluids are constant and uniform.
- Entrance and exit pressure losses are negligible (the core friction component is dominant).
- Static pressures are constant and uniform across the cross section at the entrance and exit of this multipassage exchanger.
- The total flow rate through all nonuniform flow passages is identical to that going through all nominal flow passages.
- The lengths of all flow passages are the same.

The pressure drop for all flow passages (regardless of the size, shape, and distribution of flow passages) will be the same in the core based on the fourth assumption above:

$$(\Delta p)_j = (p_i - p_o)_j \tag{12.28}$$

where

$$\Delta p_j = f_j \left(\frac{4L}{D_h}\right)_j \left(\frac{\rho_m u_m^2}{2g_c}\right)_j \qquad j = 1, 2, n \tag{12.29}$$

where j denotes the flow passage type. Invoking the definitions of the Reynolds number and mass flow rate, Eq. (12.29) is regrouped as

$$\Delta p_j = \frac{2\mu L}{g_c \rho} \left[\frac{f \cdot \mathrm{Re}}{A_o D_h^2}\right]_j \dot{m}_j \tag{12.30}$$

For a two-passage model, applying Eq. (12.30) for $j = 1$ and 2 and taking the ratio and rearranging, we get

$$\frac{\dot{m}_1}{\dot{m}_2} = \frac{(f \cdot \mathrm{Re})_2}{(f \cdot \mathrm{Re})_1} \left(\frac{D_{h,1}}{D_{h,2}}\right)^2 \frac{A_{o,1}}{A_{o,2}} \tag{12.31}$$

since $\Delta p_1 = \Delta p_2$ from Eq. (12.28). Equation (12.31) provides the flow fraction distribution in the two types of flow passages. Normalizing flow rates with $\dot{m}_n$, hydraulic diameters with $D_{h,n}$, and free-flow areas with $A_{o,n}$, Eq. (12.31) becomes

$$\frac{\dot{m}_1/\dot{m}_n}{\dot{m}_2/\dot{m}_n} = \frac{(f \cdot \mathrm{Re})_2}{(f \cdot \mathrm{Re})_1} \left(\frac{D_{h,1}/D_{h,n}}{D_{h,2}/D_{h,n}}\right)^2 \frac{A_{o,1}/A_{o,n}}{A_{o,2}/A_{o,n}} \tag{12.32}$$

where $\dot{m}_n = \dot{m}_1 + \dot{m}_2$ and all variables with a subscript n denote nominal values (selected by the choice of an analyst), either the passage geometry 1, the passage geometry 2, or some nominal passage geometry in between (for normalization of D_h and A_o used in the equation) for a two-passage nonuniformity.

To compute the flow area ratios in Eq. (12.32), we maintain approximately the same frontal area of the heat exchanger core with actual and nominal flow passages. There are two choices for selecting the nominal passage geometry, and accordingly, the values of $A_{o,j}/A_{o,n}$ ($j = 1, 2$) will be different. They are as follows.

1. The number of flow passages for the nominal geometry is the sum of the number of flow passages for passage types 1 and 2 and the frontal area is the same. In this case,

$$\frac{A_{o,1}}{A_{o,n}} = \chi_1 \frac{\hat{A}_{o,1}}{\hat{A}_{o,n}} \qquad \frac{A_{o,2}}{A_{o,n}} = \chi_2 \frac{\hat{A}_{o,2}}{\hat{A}_{o,n}} \qquad (12.33)$$

where $\hat{A}_{o,1}$, $\hat{A}_{o,2}$, and $\hat{A}_{o,n}$ are the flow area for one passage of passage types 1, 2, and n, respectively, and χ_1 and χ_2 are the corresponding fractions of the number of passages of types 1 and 2.[†] This case applies when comparing sharp and rounded triangular (or any two similar passages), where the frontal area remains constant for the same total number of flow passages, regardless of which is the nominal flow passage. However, the free-flow area will be different for the nominal flow passages since the flow areas of sharp and rounded corner passages are different (see Example 12.3).

2. In the alternative case, the total number of flow passages for the nominal passages could be different from the actual number of flow passages for the same frontal area. This case applies when we compare the two-passage model (e.g., large and small rectangular or triangular passages with 50% : 50% or any other percent distribution) with the nominal passage geometry having approximately the same frontal area.[‡] In this case, the number of flow passages for the nominal passage geometry will be different from the sum of the number of flow passages of passage types 1 and 2. The flow area ratios $A_{o,1}/A_{o,n}$ and $A_{o,2}/A_{o,n}$ are given by

$$\frac{A_{o,1}}{A_{o,n}} = \frac{\chi_1 \hat{A}_{o,1}}{\chi_1 \hat{A}_{o,1} + \chi_2 \hat{A}_{o,2}} \qquad \frac{A_{o,2}}{A_{o,n}} = \frac{\chi_2 \hat{A}_{o,2}}{\chi_1 \hat{A}_{o,1} + \chi_2 \hat{A}_{o,2}} \qquad (12.34)$$

where the definitions of $\hat{A}_{o,1}$, $\hat{A}_{o,2}$, χ_1, and χ_2 are the same as defined above after Eq. (12.33). Note that we may use $\hat{A}_{o,j}/\hat{A}_{o,n}$ in Eq. (12.34) instead of $\hat{A}_{o,j}$, $j = 1$ or 2, since the fraction is primarily known.

The pressure drop ratio (the ratio of the pressure drop for either of the two passage types, either 1 or 2, to the nominal passage pressure drop) can be calculated, using Eq. (12.30), as

$$\frac{\Delta p_1}{\Delta p_n} = \frac{(f \cdot Re)_1}{(f \cdot Re)_n} \frac{\dot{m}_1}{\dot{m}_n} \left(\frac{D_{h,n}}{D_{h,1}}\right)^2 \frac{A_{o,n}}{A_{o,1}} \qquad (12.35)$$

[†] Note that $\chi_1 = 1 - \chi_2$. The parameter χ, by definition, represents a ratio of the number of the ith shaped passage to the total number of passages. If more than two passages are involved, the following relation holds: $\sum \chi_i = 1$. Also note that $(A_{o,1} + A_{o,2})/A_{o,n} \neq 1$ in general. This is because we have presumed the same frontal area, and as a result, the wall thickness is different for differently shaped passages (see Example 12.3).

[‡] The number of passages must be an integer, so the frontal area for a particular selection of passages may not necessarily be the same when compared to a two-passage model with a nominal passage model. However, in a compact heat exchanger with a very large number of flow passages, the difference will be negligible.

Note that the flow area of two nominal passages is the same as the two (large and small) passages of the nonuniform core (see Fig. 12.4 for two examples).

As we know, since the fluid seeks the path of least flow resistance, if we replace some nominal passages with different flow passages having larger and smaller flow areas, a larger fraction of the flow will go through the larger flow area passages. Then for a constant flow rate, the pressure drop (and hence heat transfer) will reduce for this exchanger with mixed passages.[†] This means that $\Delta p_1/\Delta p_n$ $(= \Delta p_2/\Delta p_n)$ will be less than unity. This gain (reduction) in the pressure drop due to passage-to-passage non-uniformity is

$$\Delta p_{\text{gain}} = 1 - \frac{\Delta p_1}{\Delta p_n} \tag{12.36}$$

Let us now determine a change in heat exchanger effectiveness due to passage-to-passage nonuniformity. Heat transfer through differently shaped passages would be different, which in turn would produce different temperature differences between fluids 1 and 2. Hence, one cannot consider different passages, let us say two passages A and B, in parallel to arrive at an effective h as the average of conductances h_A and h_B. To arrive properly at an effective value of h for a two-passage geometry heat exchanger, the passage geometrical properties, fluid physical properties, exchanger flow arrangement, and ε-NTU relationship must be considered. A procedure is outlined in the following subsections for the two most important cases of a two-passage geometry for a counter-flow exchanger with $C^* = 1$ (a rotary regenerator case) and an exchanger with $C^* = 0$ (an oil cooler case with constant wall temperature). Refer to Shah and London (1980) for an analysis of other flow arrangements.

For both these cases, the heat transfer results are presented in terms of the number of transfer units ntu_j for each type of passage on the maldistributed fluid side as follows:

$$\text{ntu}_j = \left(\frac{hA}{\dot{m}c_p}\right)_j = \left(\frac{\text{Nu}}{\text{RePr}}\frac{4L}{D_h}\right)_j = \left(\frac{4kL}{c_p}\frac{\text{Nu}A_o}{\dot{m}D_h^2}\right)_j \qquad j = 1, 2 \tag{12.37}$$

For the nominal passage, define ntu_n by Eq. (12.37) with $j = n$. The normalized $\text{ntu}_j/\text{ntu}_n$, based on Eq. (12.37), is

$$\frac{\text{ntu}_j}{\text{ntu}_n} = \frac{\text{Nu}_j}{\text{Nu}_n}\frac{\dot{m}_n}{\dot{m}_j}\left(\frac{D_{h,n}}{D_{h,j}}\right)^2\frac{A_{o,j}}{A_{o,n}} \tag{12.38}$$

Nu_j and Nu_n in this equation should be obtained from the results of Table 7.3 for the appropriate thermal boundary conditions for fully developed laminar flow. Note that for a counterflow exchanger with $C^* = 1$, the boundary conditions are Ⓗ1 or Ⓗ2, while the boundary condition is Ⓣ for the $C^* = 0$ case.

COUNTERFLOW HEAT EXCHANGER WITH $C^* = 1$. In this case, ε_j and ntu_j are related as follows using Eq. (3.85):

$$\varepsilon_j = \frac{\text{ntu}_j}{1 + \text{ntu}_j} \qquad j = 1, 2, n \tag{12.39}$$

[†] This is what we have shown through Eqs. (12.28)–(12.36) that the performance (q and Δp) of a continuous-flow-passage regenerator matrix will be *lower* when there are small and large passages in parallel compared to the performance of nominal (average)-size passages of the same shape.

where j depends on whether the heat exchanger unit considered has all uniform (nominal) passages, ε_n ($j = n$), or it refers to a maldistributed heat exchanger that consists of two subexchangers (with the effectivenesses ε_1 or ε_2 for the passage geometries $j = 1$ or $j = 2$, respectively). Note that since ntu_j is defined using a heat transfer coefficient (not the overall heat transfer coefficient U as in NTU), the heat exchanger effectiveness of Eq. (12.39) must be defined based on the passage wall temperature:

$$\varepsilon_j = \frac{T_{o,j} - T_i}{\bar{T}_w - T_i} = \frac{\mathrm{ntu}_j}{1 + \mathrm{ntu}_j} \qquad j = 1, 2, n \tag{12.40}$$

where $\bar{T}_w$ represents the mean wall temperature of the heat transfer surface, T_i is the inlet temperature of fluids in both subexchangers and nominal exchanger, and the distribution of $T_{w,j}$ vs. x are parallel to the distribution of T_j vs. x as shown in Fig. 1 of Shah and London (1980). The temperature $\bar{T}_w$ is assumed to be the same for both passage geometries at the inlet (thus leading to the same inlet temperature difference for both passage types). The ntu_j for Eq. (12.40) are computed from Eq. (12.38) for a specified value of ntu_n and known flow fraction distribution from Eq. (12.32).

The average effectiveness of the maldistributed heat exchanger can be calculated from the effectiveness of two subexchangers using a simple energy balance and assuming constant specific heat of the fluids as follows:

$$\dot{m}\varepsilon_{\mathrm{ave}} = \dot{m}_1\varepsilon_1 + \dot{m}_2\varepsilon_2 \tag{12.41}$$

It must be emphasized that the analysis presented here is for one fluid side of the exchanger (either the hot- or cold-fluid side of a rotary regenerator). To find the resultant effect on the exchanger performance, the effect of the other fluid side needs to be taken into account, as will be shown in Example 12.3. The effective ntu on one fluid side is then given by

$$\mathrm{ntu}_{\mathrm{eff}} = \frac{\varepsilon_{\mathrm{ave}}}{1 - \varepsilon_{\mathrm{ave}}} \tag{12.42}$$

The "cost" of the influence of passage-to-passage nonuniformity on ntu is defined as

$$\mathrm{ntu}_{\mathrm{cost}} = 1 - \frac{\mathrm{ntu}_{\mathrm{eff}}}{\mathrm{ntu}_n} \tag{12.43}$$

We need to compute $\mathrm{ntu}_{\mathrm{eff}}$ for both fluid sides and subsequently calculate $\mathrm{NTU}_{\mathrm{eff}}$ for the exchanger to determine a reduction in the exchanger effectiveness due to passage-to-passage nonuniformity, as shown for a specific exchanger in Example 12.3.

London (1970) determined $\mathrm{ntu}_{\mathrm{cost}}$ and Δp_{gain} for plate-spacing and fin-spacing type nonuniformities and concluded that the deviation in passage size causes a more severe reduction in the number of transfer units than does the pressure drop gain.

Specific results from the two-passage model for the passage-to-passage nonuniformity are presented in Fig. 12.5 for rectangular passages. This two-passage model consists of 50% of the flow passages large ($c_2 > c_n$) and 50% being small ($c_1 < c_n$) compared to the nominal passages, and the nominal aspect ratios $\alpha_n^* = 1, 0.5, 0.25,$ and 0.125. In Fig. 12.5a, a reduction in ntu is presented for the ⓗ and ⓣ boundary conditions and for a

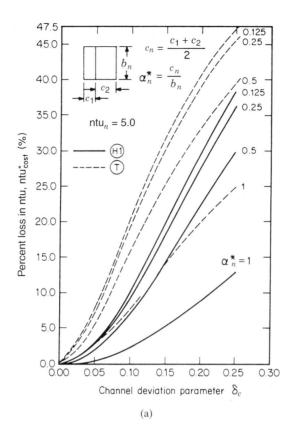

(a)

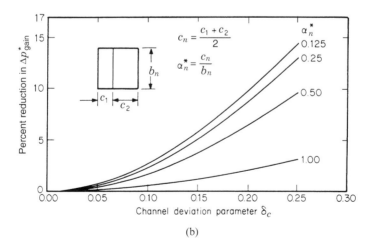

(b)

FIGURE 12.5 Deterioration factors for two-passage nonuniformities in rectangular passages: (a) percentage loss in ntu as a function of δ_c, α_n^*, and thermal boundary conditions; (b) percentage reduction in Δp as a function of δ_c and α_n^*. (From Shah, 1981.)

nominal (design) ntu_n of 5.0. Here, $\mathrm{ntu}_{\mathrm{cost}}$, a percentage loss in ntu, and the channel deviation parameter δ_c are defined as

$$\mathrm{ntu}_{\mathrm{cost}} = 1 - \frac{\mathrm{ntu}_{\mathrm{eff}}}{\mathrm{ntu}_n} \qquad \mathrm{ntu}^*_{\mathrm{cost}} = \left(1 - \frac{\mathrm{ntu}_{\mathrm{eff}}}{\mathrm{ntu}_n}\right) \times 100 \qquad \delta_c = 1 - \frac{c_1}{c_n} \qquad (12.44)$$

where $\mathrm{ntu}_{\mathrm{eff}}$ is the effective ntu for the two-passage model passage-to-passage non-uniformity, and ntu_n is the ntu for nominal (or reference) passages. It can be seen from Fig. 12.5a that a 10% channel deviation ($\delta_c = 0.10$, which is common for a highly compact surface) results in 10 and 21% reduction in $\mathrm{ntu}_{\mathrm{H1}}$ and $\mathrm{ntu}_{\mathrm{T}}$, respectively, for $\alpha_n^* = 0.125$ and $\mathrm{ntu}_n = 5.0$. In contrast, a gain in the pressure drop due to the passage-to-passage nonuniformity is only 2.5% for $\delta_c = 0.10$ and $\alpha_n^* = 0.125$, as found from Fig. 12.5b. Here $\Delta p^*_{\mathrm{gain}}$ is defined as

$$\Delta p^*_{\mathrm{gain}} = \left(1 - \frac{\Delta p_{\mathrm{actual}}}{\Delta p_{\mathrm{nominal}}}\right) \times 100 \qquad (12.45)$$

The following observation may be made from Fig. 12.5a and additional results presented by Shah and London (1980): (1) the loss in ntu is more significant for the ⊤ boundary condition than for the Ⓗ boundary condition; (2) the loss in ntu increases with higher values of nominal ntu; and (3) the loss in ntu is much more significant than the gain in Δp at a given δ_c.

N-Passage Model. The previous analysis was extended for an *N*-passage model by Shah and London (1980). In the *N*-passage model, there are *N* different-size passages of the same basic shape, either rectangular or triangular. The results of Fig. 12.5a and b for rectangular passages are also applicable to an *N*-passage model in which there are *N* different-size passages in a normal distribution about the nominal passage size with a proper definition of the channel deviation parameter δ_c as follows:

$$\delta_c = \left[\sum_{i=1}^{N} \chi_i \left(1 - \frac{c_i}{c_n}\right)^2\right]^{1/2} \qquad (12.46)$$

Here χ_i is the fractional distribution of the *i*th shaped passage. For $N = 2$ and $\chi_i = 0.5$, Eq. (12.46) reduces to Eq. (12.44) for δ_c.

Similar results are summarized in Fig. 12.6 for the *N*-passage nonuniformity model associated with equilateral triangular passages. In this case, the definition of the channel deviation parameter δ_c is modified to

$$\delta_c = \sum_{i=1}^{N} \left[\chi_i \left(1 - \frac{r_{h,i}}{r_{h,n}}\right)^2\right]^{1/2} \qquad (12.47)$$

where $r_{h,n}$ is the hydraulic radius of the nominal passages, $r_{h,i}$ is the hydraulic radius of the *i*th passage, and they are related for a two-passage model as follows: $2r_{h,n}^2 = r_{h,1}^2 + r_{h,2}^2$, but this particular case corresponds to an equilateral triangular passage. Qualitative trends of the results in Fig. 12.6 are similar to those in Fig. 12.5 for rectangular flow passages.

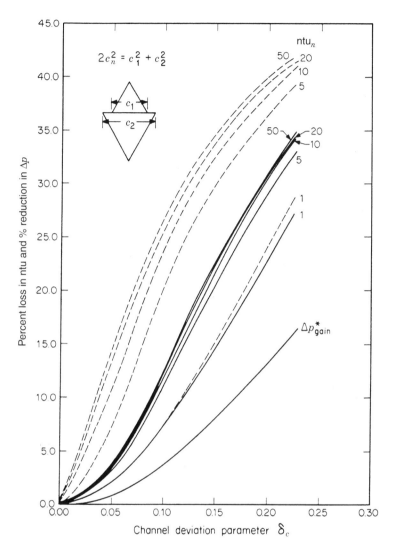

FIGURE 12.6 Percentage loss in ntu and percentage reduction in Δp as functions of δ_c for N-passage nonuniformities in equilateral triangular passages. (From Shah, 1985.)

Note that the percentage reduction in ntu and Δp vs. δ_c curves for $N = 2$ and $N > 2$ are *identical* (as shown in Figs. 12.5 and 12.6), except that the value of δ_c is higher for a two-passage model compared to the N-passage model for the same value of c_{max}/c_n. Hence, the two-passage model provides the highest deterioration in performance.

HEAT EXCHANGER WITH $C^* = 0$. In this case, ε_j and ntu_j are related as follows using Eq. (3.84):

$$\varepsilon_j = 1 - e^{-\mathrm{ntu}_j} \qquad j = 1, 2, n \tag{12.48}$$

where j depends on whether the heat exchanger unit considered has all nominal passages, ε_n ($j = n$), or it refers to a maldistributed heat exchanger that consists of two sub-exchangers (with the effectivenesses ε_1 or ε_2 for the passage geometries $j = 1$ or $j = 2$, respectively). The average effectiveness of the passage-to-passage maldistributed heat exchanger can be calculated using Eq. (12.41). Similar to the previous case, the cost of the influence of passage-to-passage nonuniformity on ntu is defined as follows:

$$\text{ntu}_{\text{cost}} = 1 - \frac{\text{ntu}_{\text{eff}}}{\text{ntu}_n} \tag{12.49}$$

where

$$\text{ntu}_{\text{eff}} = \ln \frac{1}{1 - \varepsilon_{\text{ave}}} \tag{12.50}$$

Refer to Shah and London (1980) for further details.

Example 12.3 A vehicular gas turbine counterflow rotary regenerator is made up of triangular flow passages. Due to brazing of the core, some of the flow passages became triangles with rounded corners. Hence, idealize that the matrix is made up of 50% passages having all three corners rounded and 50% passages having all three sharp corners. Determine:

(a) The flow fraction distribution in the two types of passages
(b) The change in pressure drop due to passage-to-passage nonuniformity. Does it represent a loss or a gain in comparison to all passages of ideal sharp corners?
(c) The change in the exchanger effectiveness due to the nonuniformity. Does it represent a loss or a gain?
(d) The subsequent change in ntu

The additional data are as follows:

Characteristic	Sharp Corner Passage	Rounded Corner Passage
$f \cdot \text{Re}$	13.333	15.993
Nu_{H1}	3.111	4.205
$\hat{A}_o / \hat{A}_{o,\Delta}$	1	0.868
$D_h / D_{h,\Delta}$	1	1.125

For this regenerator, the flow split gas : air = 50% : 50%. $\text{NTU}_n = 2.5$. Idealize negligible wall resistance, $C^* = 1$ and $C_r^* \to \infty$.

SOLUTION

Problem Data and Schematic: The passage-to-passage flow nonuniformity is caused by differing flow resistance of sharp and rounded corner flow passages shown in Fig. E12.3. All data for thermal and hydraulic characteristics of the flow passages are provided in the table above, and $\chi_1 = \chi_2 = 0.5$, $\text{NTU}_n = 2.5$, and $C^* = 1$.

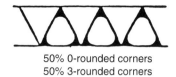

50% 0-rounded corners
50% 3-rounded corners

FIGURE E12.3

Determine: The influence of passage-to-passage flow nonuniformity on heat transfer and pressure drop.

Assumptions: The pressure drop is uniform across all heat exchanger flow passages. All the assumptions of Section 3.2.1 are valid except for the nonuniform flow through the regenerator due to differently shaped flow passages.

Analysis: The flow fraction distribution can be determined from the assumption of uniform pressure drop distribution across the heat exchanger core. We consider the ideal sharp corner flow passage as a nominal passage for this example. Therefore, using Eq. (12.33),

$$\frac{A_{o,1}}{A_{o,n}} = \chi_1 \frac{\hat{A}_{o,\mathrm{sharp}\,\triangle}}{\hat{A}_{o,\mathrm{sharp}\,\triangle}} = 0.5 \times 1 = 0.5 \qquad \frac{A_{o,2}}{A_{o,n}} = \chi_2 \frac{\hat{A}_{o,\mathrm{round}\,\triangle}}{\hat{A}_{o,\mathrm{sharp}\,\triangle}} = 0.5 \times 0.868 = 0.434$$

The ratio of mass flow rates through these passages, using Eq. (12.31), is given by

$$\frac{\dot{m}_1}{\dot{m}_2} = \frac{(f \cdot \mathrm{Re})_2}{(f \cdot \mathrm{Re})_1} \left(\frac{D_{h,1}}{D_{h,2}}\right)^2 \frac{A_{o,1}}{A_{o,2}} = \frac{15.993}{13.333}\left(\frac{1}{1.125}\right)^2 \frac{0.5}{0.434} = 1.092 \qquad (1)$$

From Eq. (1) we get

$$\frac{\dot{m}_1}{\dot{m}_n} = \frac{\dot{m}_1/\dot{m}_2}{1 + \dot{m}_1/\dot{m}_2} = \frac{1.092}{1 + 1.092} \;\Rightarrow\; \frac{\dot{m}_1}{\dot{m}_n} = 0.522 \quad \text{and} \quad \frac{\dot{m}_2}{\dot{m}_n} = 0.478 \qquad \textit{Ans.}$$

The ratio of the pressure drop for the sharp corner triangular passages to the nominal pressure drop can be determined as follows using Eq. (12.35):

$$\frac{\Delta p_{\mathrm{sharp}\,\triangle}}{\Delta p_n} = \frac{\Delta p_1}{\Delta p_n} = \frac{(f \cdot \mathrm{Re})_1}{(f \cdot \mathrm{Re})_n} \frac{\dot{m}_1}{\dot{m}_n}\left(\frac{D_{h,n}}{D_{h,1}}\right)^2 \frac{A_{o,n}}{A_{o,1}} = \frac{13.333}{13.333} \times 0.522 \times 1^2 \times \frac{1}{0.5} = 1.044$$

$$\textit{Ans.}$$

Similarly, ratio of the pressure drop within the rounded triangular passages to the pressure drop through the nominal passages should, ideally, be 1.044. We could have calculated this ratio using an analogous relationship to the one given by Eq. (12.35) by replacing the subscript 1 with 2.

$$\frac{\Delta p_{\text{round}\,\Delta}}{\Delta p_n} = \frac{\Delta p_2}{\Delta p_n} = \frac{(f \cdot \text{Re})_2}{(f \cdot \text{Re})_n} \frac{\dot{m}_2}{\dot{m}_n} \left(\frac{D_{h,n}}{D_{h,2}}\right)^2 \frac{A_{o,n}}{A_{o,2}}$$

$$= \frac{15.993}{13.333} \times 0.478 \times \left(\frac{1}{1.125}\right)^2 \times \frac{1}{0.434} = 1.044$$

The heat exchanger effectiveness due to the nonuniformity depends on the number of transfer units for the respective passages. The number of transfer units for sharp and rounded corner passages, normalized with respect to nominal passages, are determined using Eq. (12.38) as

$$\frac{\text{ntu}_j}{\text{ntu}_n} = \frac{\text{Nu}_j}{\text{Nu}_n} \frac{\dot{m}_n}{\dot{m}_j} \left(\frac{D_{h,n}}{D_{h,j}}\right)^2 \frac{A_{o,j}}{A_{o,n}}$$

where $j = 1$ or 2 (i.e., sharp-corner triangular passages or rounded-corner triangular passages, respectively). Utilizing the given and calculated data, we get

$$\frac{\text{ntu}_{\text{sharp}\,\Delta}}{\text{ntu}_n} = \frac{\text{ntu}_1}{\text{ntu}_n} = \frac{\text{Nu}_1}{\text{Nu}_n} \frac{\dot{m}_n}{\dot{m}_1} \left(\frac{D_{h,n}}{D_{h,1}}\right)^2 \frac{A_{o,1}}{A_{o,n}} = \frac{3.111}{3.111} \times \frac{1}{0.522} \times 1^2 \times 0.5 = 0.9579$$

$$\frac{\text{ntu}_{\text{round}\,\Delta}}{\text{ntu}_n} = \frac{\text{ntu}_2}{\text{ntu}_n} = \frac{\text{Nu}_2}{\text{Nu}_n} \frac{\dot{m}_n}{\dot{m}_2} \left(\frac{D_{h,n}}{D_{h,2}}\right)^2 \frac{A_{o,2}}{A_{o,n}} = \frac{4.205}{3.111} \times \frac{1}{0.478} \times \left(\frac{1}{1.125}\right)^2 \times 0.434 = 0.9697$$

For the given $\text{NTU}_n = 2.5$, we obtain $\text{ntu}_n = \text{ntu}_h = \text{ntu}_c = 5.0$ for $C^* = 1$ [see, for example, Eq. (9.23)]. With $\text{ntu}_n = 5$, we get

$$\text{ntu}_1 = 5 \times 0.9579 = 4.7893 \qquad \text{ntu}_2 = 5 \times 0.9697 = 4.8483$$

Consequently, the heat exchanger effectivenesses for the two types of passages for $C^* = 1$ would be

$$\varepsilon_j = \frac{\text{ntu}_j}{1 + \text{ntu}_j} \qquad j = 1, 2$$

Therefore, ε_1 for sharp-corner triangular passages and ε_2 for rounded-corner triangular passages are

$$\varepsilon_1 = \frac{4.7893}{1 + 4.7893} = 0.8277 \qquad \varepsilon_2 = \frac{4.8483}{1 + 4.8483} = 0.8290$$

The average heat exchanger effectiveness can be calculated from a simple energy balance, using Eq. (12.41), as

$$\varepsilon_{\text{ave}} = \frac{\dot{m}_1}{\dot{m}_n} \varepsilon_1 + \frac{\dot{m}_2}{\dot{m}_n} \varepsilon_2 = 0.522 \times 0.8277 + 0.478 \times 0.8290 = 0.8283$$

Therefore, ntu_{eff} from Eq. (12.43) is

$$\text{ntu}_{\text{eff}} = \frac{\varepsilon_{\text{ave}}}{1 - \varepsilon_{\text{ave}}} = \frac{0.8283}{1 - 0.8283} = 4.8241 = \text{ntu}_{\text{eff},h} = \text{ntu}_{\text{eff},c}$$

The effective NTU for this regenerator, using Eq. (5.54), is given by

$$NTU_{eff} = \frac{1}{1/ntu_{eff,h} + 1/ntu_{eff,c}} = \frac{1}{1/4.8241 + 1/4.8241} = 2.412$$

Subsequently, the regenerator effectiveness with the passage-to-passage flow non-uniformity is given by

$$\varepsilon_{eff} = \frac{NTU_{eff}}{1 + NTU_{eff}} = \frac{2.412}{1 + 2.412} = 0.7069$$

In contrast, the effectiveness of the heat exchanger with nominal uniform passages is

$$\varepsilon_n = \frac{NTU_n}{1 + NTU_n} = \frac{2.5}{1 + 2.5} = 0.7143$$

Thus, the loss in the regenerator effectiveness is

$$\Delta\varepsilon_{loss} = \frac{\varepsilon_n - \varepsilon_{ave}}{\varepsilon_n} \times 100 = \frac{0.7143 - 0.7069}{0.7069} \times 100 = 1.0\% \qquad Ans.$$

Discussion and Comments: It should be noted that in pressure drop analysis, sharp-corner passages are considered as nominal passages. Thus, the pressure drop of the matrix with nonuniform flow passages is 4.4% larger than that for the ideal matrix with uniform sharp corners. This is because the sharp-corner triangular passage (used for the comparison) has a lower $f \cdot Re$ than that for the rounded-corner triangular passage. In contrast, there is a reduction (1.0%) in regenerator effectiveness due to nonuniform flow because of the poor performance of the rounded-corner flow passages. The comparison was performed by comparing the performance of nonuniform passages with nominal sharp-corner passages which have a lower heat transfer coefficient and a lower friction factor. If we would have considered the rounded-corner triangular passages as nominal passages, there would have been a reduction in the pressure drop and a slight gain in heat exchanger effectiveness.

12.1.2.2 Passage-to-Passage Flow Nonuniformity Due to Other Effects.

Finally, passage-to-passage flow nonuniformity for very compact surfaces may be induced by brazing and/or fouling in addition to manufacturing imperfection. Both controlled atmosphere brazing and vacuum brazing have a negligible effect on j and f data if the plates/tubes/primary surface is clad and fins are unclad, and the ratio of the joint area to free-flow area is less than 10%. For ultracompact surfaces/flow passages, this ratio may not be small (i.e., flow area blockage and brazing-induced surface roughness may not be negligible, and accurate experimental j and f data are essential in this case). Gross blockage due to brazing may increase the pressure drop substantially. The influence of surface roughness induced by salt dip brazing (currently an outdated technology due to environmental concerns) is generally *nonnegligible* (i.e., can increase Δp considerably with only a slight increase in h or j) in highly compact surfaces (Shah and London, 1971). Controlled atmosphere brazing, a state-of-the-art manufacturing process for compact heat exchangers (Sekulić, 1999), provides a very uniform flow passage distribution due to uniform distribution of a re-solidified microlayer of cladding residue on a heat transfer surface and uniform fin area distribution (Sekulic et al., 2001).

12.1.3   Manifold-Induced Flow Maldistribution

Whereas manifolds are integral in plate heat exchangers due to construction features, manifolds are common and attached separately in many other applications. In the PHEs, the fluids enter and exit the manifolds laterally and flow within the core axially; here the axial direction is defined as the main direction of fluid flow within the PHE passages (see Fig. 12.7a and b). In other applications, the fluids enter and exit the core also axially, or a combination of axial and lateral entry and exit. In the PHEs, the manifolds are of two basic types: dividing flow and combining flow. In *dividing-flow manifolds*, fluid enters laterally and exits the manifold axially. The velocity *within* the manifold, parallel to the manifold axis, varies from the inlet velocity to zero value. Conversely, in *combining-flow*

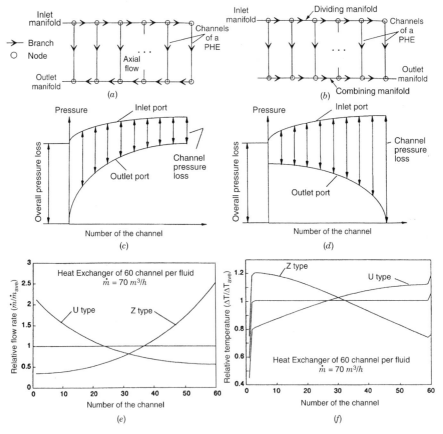

FIGURE 12.7   Manifold configurations: (*a*) U-flow or parallelflow configuration; (*b*) Z-flow, S-flow, or reverse-flow configuration. Pressure profile in (*c*) U-flow configuration, (*d*) Z-flow configuration. In these configurations, (*e*) typical flow distribution, and (*f*) typical temperature distribution. [Parts (e) and (f) from Thonon, 2002.]

manifolds, fluid enters axially from the PHE core and exits at the end of the manifold laterally, with the velocity *within* the exit manifold varying from zero to the outlet velocity. When interconnected by lateral branches, these manifolds result in parallel- and reverse-flow systems, or U- and Z-flow (or S-flow) arrangements, as shown in Fig. 12.7a and b. Because the inlet and outlet manifolds have the same effective-diameter pipes connected by the lateral branches, this construction has built-in inherent flow nonuniformity, as evidenced by the resultant typical pressure profiles as shown in Fig. 12.7c and d, and the mass flow rate distribution as shown in Fig. 12.7e.

Modeling of a manifold requires determination of both axial and lateral velocity and static pressure distributions. Available solutions of the manifold flow models may be either analytical (in simplified cases, Bajura and Jones, 1976; Edwards et al., 1984; Shen, 1992) or numerical (Majumdar, 1980; Thonon et al., 1991; Heggs and Scheidt, 1992) by considering the inlet and outlet manifolds connected by the flow channels in a PHE. The key problem in analytical modeling is the difficulty of identifying a relevant streamline on which to calculate energy and pressure losses and apply the Bernoulli equation. The state-of-the-art design procedures utilize commercial and/or proprietary CFD codes. The accurate modeling can be done only numerically and the reader is referred to the references noted in this section.

Such manifold-induced flow maldistribution has traditionally been analyzed by a sim- plified approach without explicitly including the flow resistance in the lateral branches. Bajura and Jones (1976) defined a set of generalized equations for manifold systems. The model consists of (1) pressure-flow equations, and (2) flow distribution equations, includ- ing the related boundary conditions. These equations are obtained by writing continuity and momentum equations for dividing and combining flow branch control volumes near branching points, and the discharge equation, which provides the relationship of the pressure differences between the manifold and the lateral branch. So this model is based on an application of first principles, using continuity and momentum equations for header flows and discharge equation for lateral flows. The experimental results for simple dividing and combining flows were in very good agreement with this theory.

Shen (1992) extended the work of Bajura and Jones (1976) numerically to include the effect of flow friction in the manifolds and the momentum losses associated with turning the flow in the lateral branches. Datta and Majumdar (1980) conducted a numerical analysis of both U- and Z-flow arrangements and found that the flow distribution within the heat exchanger core is dependent on the following three dimensionless groups: (1) the branch-to-manifold flow area ratio A_o^*, (2) the manifold friction parameter F, and (3) the lateral branch pressure loss coefficient K_b, all defined as follows:

$$A_o^* = \frac{N_c A_{o,b}^2}{A_{o,m}^2} \qquad F = \frac{\pi f_m L_m D_p}{N_c A_{o,b}^2} \qquad K_b = \frac{\Delta p}{\rho u_m^2 / 2 g_c} \tag{12.51}$$

Here N_c is the number of channels on one fluid side in a PHE, $A_{o,b}$ is the free-flow area for a branch or an exchanger and $A_{o,m}$ is the flow area of the manifold, D_p is the port or manifold diameter, and the subscript m is for manifold. Note that the pressure loss coefficient K_b of Eq. (12.51) is equal to the bracketed term of Eq. (6.28). The main conclusions of the influence of these parameters are as follows (Datta and Majumdar, 1980):

- The relative variation in the lateral flow distribution increases with increased A_o^*.
- A reverse-flow manifold provides relatively more uniform flow distribution than a parallel-flow manifold for otherwise identical conditions.

- In parallel- and reverse-flow manifolds, maximum flow occurs through the last port and first port, respectively.
- The effect of friction parameter is in general less significant than that of the area ratio A_o^*.

A few general conclusions from these studies for more uniform flow through manifold systems and some design guidelines for manifolds are as follows:

- Flow maldistribution is insignificant in PHEs with less than 20 flow channels on a given fluid pass.
- Flow maldistribution in the PHEs due to the manifold system (U- or Z-flow) increases with increasing flow rate, increasing the number of plates in a given pass and decreasing the liquid viscosity.
- In a U-flow manifold system, the maximum flow occurs through the first port, and in the Z-flow manifold system through the last port. Neither arrangement provides uniform flow through the PHE or lateral branches. However, flatter (relatively more uniform) flow distribution is obtained with the U-flow manifold system than with the Z-flow system (see Fig. 12.7e).
- To minimize flow maldistribution in a PHE, the flow area of the inlet manifold (area of the actual or simulated pipe before lateral branches) should be larger than the flow area of the lateral branches (heat exchanger core). The larger the port diameter, the more uniform flow through the heat exchanger core. Alternatively, flow maldistribution in a PHE plate pack (core) increases as the fraction of the total pressure drop in the manifold becomes significant.
- The flow area of a combining-flow manifold in Fig. 12.12b (the outlet manifold/ pipe in Fig. 12.7a and b) should be larger than that for the dividing-flow manifold in Fig. 12.12a (the inlet manifold/pipe in Fig. 12.7a and b) for a more uniform flow distribution through the core in the absence of heat transfer within the core. If there is heat transfer in lateral branches (core), the flow areas should be adjusted first for the density change and then the flow area of the combining manifold should be made larger than that calculated previously.
- Flow reversal is more likely to occur in a Z-flow system, which is subjected to poor flow distribution.
- Based on the limited tests, a 2-pass 2-pass Z-flow arrangement can be treated as if each pass were in a separate exchanger.

Thonon et al. (1991) and Heggs and Scheidt (1992), among others, have analyzed heat transfer in a PHE with U- and Z-flow arrangements having 60 channels (30 channels on each fluid side). They found that when both fluids enter the same end of the PHE with either a U- or a Z-flow arrangement, the reduction in the exchanger performance is small (ca. 2%) compared to the ideal uniform-flow case and may be neglected for practical purposes. Typical temperature distributions are shown in Fig. 12.7f. However, two fluids can enter at different ends in Z- or U-flow arrangements. Based on Fig. 12.7e, the Z-flow has a smaller flow rate/velocity at the entrance-end flow channels and a large flow rate/ velocity at the exit-end flow channels. When both fluids in the Z-flow arrangement enter from the different ends, there are two possibilities: (1) the flow rate in the end channels can be the largest, or (2) the flow rate in the end channels can be the smallest. A similar situation exists for the U-flow arrangement with fluids entering from the different ends. Heggs and Scheidt (1992) show that the effect of flow maldistribution on the PHE performance is severe (up to 15% for the case that they analyzed) when the two fluids enter the exchangers from different ends.

From the foregoing results, we find that if both fluids enter from the same end in a PHE, the manifold system has a significant negative impact on flow maldistribution and pressure drop, and a less degrading effect on overall heat transfer. However, if the fluids enter from different ends, both significant flow and temperature maldistribution can occur in a PHE. While in Section 1.5.2.1, we mentioned one of the advantages of PHEs as having the same residence time for all fluid particles on any fluid side, the foregoing results indicate that severe flow and temperature maldistributions can occur due to inherent construction features of a PHE, resulting in different residence times. This can have an impact on the use of a PHE in the chemical and food industries if the residence time or temperatures are to be controlled over the entire surface area.

12.2   OPERATING CONDITION–INDUCED FLOW MALDISTRIBUTION

Operating conditions (temperature level, temperature differences, multiphase flow conditions, etc.) inevitably influence thermophysical properties (viscosity, density, quality) and/or process characteristics (such as the onset of oscillations) of the exchanger fluids, which in turn may cause various flow maldistributions, both steady and transient in nature. We next summarize the influence of viscosity-induced flow maldistribution, common with oil flows, on exchanger performance. For flow maldistribution with phase change, refer to Hewitt et al. (1994) for details.

12.2.1   Viscosity-Induced Flow Maldistribution

Viscosity-induced flow instability and maldistribution are results of large changes in fluid viscosity within the exchanger as a result of different heat transfer rates in different tubes (flow passages). We discuss below two cases: (1) flow instability and associated flow maldistribution for liquid cooling when the wall temperature is kept constant (i.e., condensing or evaporating fluid on the other fluid side), and (2) a single-phase exchanger with viscous liquid on one fluid side and gas or liquid on the other fluid side in which an oil/viscous liquid is being heated or cooled.

12.2.1.1   Flow Instability with Liquid Coolers. A possibility for flow instability is present whenever one or more fluids are liquids in a heat exchanger and *if* the viscous liquid is being cooled. Flow maldistribution and flow instability are more likely in laminar flow ($\Delta p \propto \mu$) than in turbulent flow ($\Delta p \propto \mu^{0.2}$). Mueller (1974, 1987) has proposed a procedure for determining the pressure drop or mass flow rate (in a single-tube laminar flow cooler) above which the flow instability due to flow maldistribution is eliminated within a multitubular heat exchanger. Putnam and Rohsenow (1985) have investigated the flow instability phenomenon that occurs in noninterconnected parallel passages of laminar flow heat exchangers.

If a viscous liquid stream is cooled, depending on the liquid flow rate and the length of the tube, the liquid local bulk temperature T_m may or may not reach the wall temperature T_w along the flow length L (see Fig. 12.8a). If it reaches the constant wall temperature, the liquid temperature and hence its viscosity remains constant farther downstream,

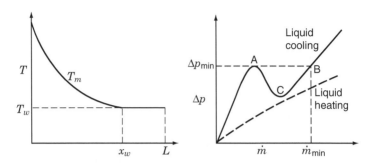

FIGURE 12.8 (*a*) Temperature distribution of a viscous liquid in the entrance region of a tube, and (*b*) pressure drop vs. mass flow rate for a single tube (flow passage) in laminar liquid flow cooling. (From Mueller, 1974.)

dependent on T_w [i.e., $\mu = \mu_w(T_w)$]. In the preceding region, the viscosity will be a function of the local bulk temperature, $\mu = \mu(T_m)$. The total pressure drop between the inlet and outlet of the flow passage could be approximated as a sum of the two terms that are based on the two viscosity regions: (1) a region between tube inlet and an axial location x_w below which μ is dependent on T_m, and (2) a region between the axial location x_w and the tube end in which μ is dependent only on the wall temperature T_w (and it is constant) as shown in Fig. 12.8*a*. Assume that one can define an average viscosity μ_{ave} for the region between the tube entrance and location x_w such that when used in the standard pressure drop equation it gives the true pressure drop for that region. From x_w to the tube exit ($x = L$), the pressure drop is calculated using the viscosity μ_w. The total pressure drop is the sum of the above-mentioned two pressure drops and behaves as shown by the solid line in Fig. 12.8*b* as a function of the flow rate (Mueller, 1974); it can be explained as follows. At any low flow rates $\dot{m}$, the pressure drop increases rapidly and almost linearly with $\dot{m}$ since the entire tube has the viscosity μ_w. As the flow rate increases, more of the tube has the fluid with viscosity μ_{ave} and less at μ_w. A maximum pressure drop for a tube is reached (at point A in Fig. 12.8*b*) when the point $x = x_w$ in Fig. 12.8*a* reaches $x = L$, and it depends on the ratio of the μ_{ave} to μ_w. The discharge temperature then starts to rise, the entire tube is then at a viscosity μ_{ave} which continuously decreases, and the pressure drop continuously decreases up to point C in Fig. 12.8*b*. With further increase in the flow rate $\dot{m}$, the pressure drop will again increase since the decrease due to the influence of μ_{ave} becomes smaller than the influence of the increase in $\dot{m}$. Note that if the exchanger contained more than one flow passage (i.e., for a multitubular exchanger), more than one flow rate would be possible for a given pressure drop in the operating range between points A and B in Fig. 12.8*b*. It is in this region that the flow instability is produced. There will be no flow maldistribution–induced instability in a multitubular cooler if the mass flow rate per tube $\dot{m}_m$ (assuming uniform flow distribution) is greater than $\dot{m}_{min}$ in Fig. 12.8*b*. The foregoing analysis is based on (1) fully developed laminar flow in the tube; (2) the viscosity is the only fluid property that can vary along the flow path; (3) the fact that only the frictional pressure drop contribution is significant; and (4) the wall temperature is considered constant and lower than the fluid inlet temperature.

Note that if a viscous liquid is being heated in a tube with constant wall temperature, the liquid viscosity will decrease along the tube length with increasing flow, however, at a rate lower than the increase in the flow rate. As a result, the pressure drop will increase monotonically with the liquid flow rate, as shown by a dashed-line curve for the liquid heating case in Fig. 12.8b. In this case, there will not be any flow instability as found for liquid cooling.

For gases, the viscosity increases with the temperature. Hence, flow maldistribution can occur with the constant-tube-wall-temperature case when heating the gas and *not* when cooling the gas, a phenomenon just opposite that for the liquids. With increasing temperature drop for liquids or the increasing temperature rise for gases, the flow maldistribution becomes more pronounced for the constant-wall-temperature boundary condition.

Example 12.4   A viscous liquid flows under steady, fully developed laminar flow conditions through a tubular heat exchanger having two tube rows connected to inlet and outlet pipes through lateral headers. The pressure drops within the headers are negligible. The dynamic viscosity of the liquid decreases exponentially with an increase in temperature while the other properties may be considered as being nearly constant. The temperature of the wall of the channels is either lower or higher than the temperature of liquid due to evaporating/condensing fluid stream on the tube outside. Thus, the wall temperature is uniform and constant along the tube length. The flow rate in each tube corresponds to the flow rate between that for points A and B in Fig. 12.8b. Determine which of the following conditions may exist in this heat exchanger: (1) different mass flow rates of the viscous liquid may be established in the two tube rows for the cases of both heating and cooling of the liquid, (2) a condition such as item (1) is possible only in the case of heating, or (3) such a condition is possible only in the case of cooling.

SOLUTION

Problem Data and Schematic: A liquid has the following property: $\mu = \mu(T)$; if $T_1 < T_2$, then $\mu_1 = \mu(T_1) > \mu_2 = \mu(T_2)$, and if $T_1 > T_2$, then $\mu_1 = \mu(T_1) < \mu_2 = \mu(T_2)$. Here subscripts 1 and 2 denote any two temperatures $T_1 \neq T_2$ of the liquid within the range of temperatures considered. The fluid flows through the two tube rows (A and B in Fig. E12.4A). The liquid temperature T_1 is either lower or higher than the uniform wall temperature T_w.

Determine: Whether or not two different mass flow rates may be established in the two tube rows connected to the same inlet and outlet headers under the conditions of either cooling and/or heating of the viscous liquid.

Assumptions: The flow is steady laminar. The pressure drops along the headers are negligible. Consequently, the pressure drops along the tube rows between the headers will be the same. The thermophysical properties of the liquid, except for the viscosity, are assumed to be constant. We assume that the liquid viscosity varies with the temperature as $\mu = C_1 \exp(C_2/T)$ where C_1 and C_2 are constants. Entrance and exit pressure drops are negligible. Thermal resistance on the tube side is controlling (i.e., the temperature of the wall is constant, that is, the heat transfer coefficient outside the channels is very large).

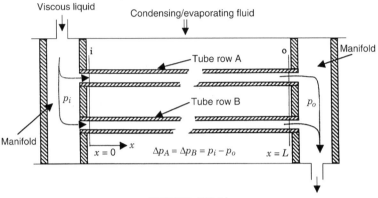

FIGURE E12.4A

Analysis: The pressure drop under fully developed laminar flow through tubes can be obtained from Eq. (6.67*a*) or directly from Eq. (13.2) by treating **P** independent of D_h as

$$\Delta p = \frac{1}{D_h^3} \left[\frac{1}{2g_c} \frac{\mu}{\rho} \frac{16L}{\mathbf{P}} \dot{m}(f \cdot \mathrm{Re}) \right] \tag{1}$$

Hence, we may conclude from Eq. (1) that the pressure drops depend on the products of viscosity and mass flow rate for each of the two tube rows:

$$\Delta p_j \propto (\bar{\mu}\dot{m})_j \tag{2}$$

where $\bar{\mu}_j$ ($j = $ A or B) represents an average viscosity between inlet and outlet of either tube row A or B.[†] For liquids, the dynamic viscosity may be approximated by an exponential decreasing function in terms of the local temperature (Mueller, 1987):

$$\mu_j = C_1 e^{C_2/T_j} \tag{3}$$

where C_1 and C_2 are constants, T is the local temperature, in K or °R and $j = $ A or B. Hence, to determine the flow distribution in two tube rows through Eq. (2), we need to know the corresponding temperature distributions for $\bar{\mu}_j$ evaluation.

With constant wall temperature, temperature distributions along the channels are given by [see Eq. (11) in Example 11.1]

$$\Theta_j = \frac{T_j - T_{j,i}}{T_w - T_{j,i}} = 1 - e^{-\xi \mathrm{NTU}_j} \propto 1 - e^{C_3(-x/\dot{m}_j)} \qquad j = \text{A or B} \tag{4}$$

where $\xi = x/L$, $\mathrm{NTU}_j = UA/(\dot{m}c_p)_j = hA/(\dot{m}c_p)_j = (3.66 \times k)A/(D_h \dot{m}c_p)_j$ and C_3 is a constant. In writing the last equality, assumptions that the fully developed laminar flow and controlling liquid side thermal resistance are utilized.

[†] Note that letters A and B in Figs. 12.8*b* and E12.4A are not related and denote different entities.

From Eq. (2) it is clear that pressure would differ along the flow direction at each location within tube rows A and B everywhere but at the terminal ports. Note that the pressure gradient along the flow $(\Delta p/L)$ is proportional to viscosity [see Eq. (1) above], which changes along the fluid flow direction as indicated by Eqs. (3) and (4). So even though the pressure drop between the inlet and outlet for each tube row is going to be the same, the local pressure distributions along the flow direction are not necessarily the same for the two tube rows [see Eq. (2)] in the flow instability region AB of Fig. 12.8*b* (specified input). From Eq. (2) it must be clear that the same pressure drop for both tube rows, $j = $ A and B, may be reached by having differing viscosities and differing mass flow rates in these two channels, at least in principle. We may say that if a larger mass flow rate and corresponding lower viscosity are established in one tube row, the same product of these two entities may be obtained for a lower mass flow rate and larger viscosity that may be established in the second tube row. The question is whether such conditions are possible for both heating and/or cooling. An answer would lead to a solution of the problem.

So, based on the relationships of Eqs. (2)–(4), the temperature and pressure distributions for a case of liquid cooling (i.e., $T_{A,i} = T_{B,i} > T_w$) are presented in Fig. E12.4B. From these plots we can conclude that if a viscous liquid is *cooled*, differing tube rows *may* have differing mass flow rates despite the same pressure drop in the flow instability region of Δp vs. $\dot{m}$ relationship. This is because the liquid *viscosity increases with a decrease in its temperature*. This means that flow misdistribution in different tube rows *will be present* in such a situation.

In the case of *heating* of the liquid, its *viscosity decreases with an increase in temperature*, and the situation is going to be different. Namely, the liquid with an assumed larger mass flow rate would have lower local temperatures than those of the liquid that would have the smaller mass flow rate based on Eq. (4). If that is the case, the viscosity along the flow direction for the fluid with a higher mass flow rate (lower temperatures) would have higher viscosities along the channel due to the relationship given by Eq. (3). The situation for the fluid with smaller mass flow rate (higher temperatures) would lead to the presence of lower viscosities along the flow direction. So, in both cases, both the mass flow rate and the viscosity are either increased or decreased for a considered tube row. However, to have an invariant pressure drop for both channels, the product of the mass flow rate and viscosity must stay invariant as shown by Eq. (2). That would be quite opposite from what would actually happen based on the analysis above. Therefore, such conditions could not be satisfied: It is not possible to have *heating* of a viscous liquid that would lead to maldistribution if the pressure drop must stay the same between the headers. That

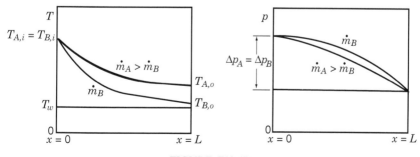

FIGURE E12.4B

means there is no instability region for a Δp vs. $\dot{m}$ relationship for viscous liquid heating (see the dashed curve in Fig. 12.8b).

Discussion and Comments: The analysis presented indicates that operating conditions may cause the flow maldistribution in a certain operating range of liquid flow rate. For example, a liquid with viscosity decreasing with an increase of temperature may feature differing mass flow rates in different tube rows within the heat exchanger, if the fluid is cooled and the pressure drops are kept the same for these channels at their respective terminal ports under certain operating conditions. If the fluid is heated, the mass flow rates in differing tubes must be approximately the same to keep the same pressure drops. It should be noted that this analysis was conducted for a viscous liquid that is characterized with a decreasing viscosity with increasing temperature. Gases may feature the opposite behavior (although the changes of the properties would be even smaller). Such a situation is considered in Problem 12.8.

Mueller (1974) proposed the following procedure to determine the minimum pressure drop, Δp_{min}, above which the flow maldistribution–induced instability would not be possible. The case considered is for a viscous liquid of a known inlet temperature being cooled as it flows through the length of a tube of known constant temperature T_w.

1. From viscosity data, determine the slope m of the curve $\ln \mu$ vs. $1/T$, where T is temperature on the absolute temperature scale.
2. With the known slope m and liquid viscosities at the inlet (μ_i) and wall (μ_w) temperatures, determine the average liquid viscosity (μ_{ave}) using Fig. 12.9. This figure is based on the assumption that the fluid temperature reaches the wall temperature within the tube. If the fluid exit temperature is still larger than the wall temperature, the average viscosity should be modified. The details are provided by Mueller (1974).
3. With these viscosities, determine x_w of Fig. 12.8 from

$$x_w = \frac{L}{2(1 - \mu_{ave}/\mu_w)} \tag{12.52}$$

4. Calculate the mass flow rate from

$$\text{Gz} = \frac{\dot{m}c_p}{kL} \approx 0.4 \tag{12.53}$$

The calculated mass flow rate corresponds to the *breakthrough mass flow rate* (Mueller, 1974). The maximum point A in Fig. 12.8b occurs when the actual mass flow rate through the tube approximately equals to the breakthrough value given by Eq. (12.53), and the ratio of viscosities must be less than 0.5.

5. Calculate the minimum pressure drop from

$$\Delta p_{min} = \frac{128}{\pi \rho g_c D_h^4} \dot{m}[x_w \mu_{ave} + (L - x_w)\mu_w] \tag{12.54}$$

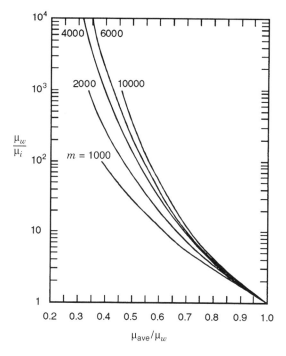

FIGURE 12.9   Viscosity ratio chart for various slopes of m of $\mu = C_1 e^{C_2/T}$. (From Mueller, 1974.)

If the pressure drop for the given design is found to be less than that calculated by Eq. (12.54), the fluid flow length should be increased (either increasing the duct length or considering a multipass design) to eliminate the flow instability. If the ratio of the average and wall temperature viscosities is larger than 0.5, the maximum pressure drop can occur at a flow rate larger than the breakthrough mass flow rate. If the exchanger pressure drop exceeds $\Delta p_{\min}$, flow instability will not be a problem. However, flow maldistribution is still possible, as considered in the next subsection.

When an exchanger has multiple (N) tubes, and the total flow rate on the tube side is less than $N\dot{m}_{\min}$ (where $\dot{m}_{\min}$ is defined in Fig. 12.8 per tube), different flow rates will establish in different tubes depending on their operating average temperature. Conceptually, the flow rate in individual tubes could be calculated based on the preceding method. Hence, one can visualize the flow maldistribution due to different fluid viscosity in different tubes. While Mueller (1974) provides a method for computing $\Delta p_{\min}$ or $\dot{m}_{\min}$ per tube and recommends operation of the multitubular exchanger at flow rates higher than $\dot{m}_{\min}$, Putnam and Rohsenhow (1985) provide a method for operation when $\dot{m} < \dot{m}_{\min}$ for a specific case.

12.2.1.2   Flow Maldistribution When No Flow Instability Present.   When the flow rate $\dot{m} > N\dot{m}_{\min}$ (in reality, each tube or flow passage should have the flow rate greater than $\dot{m}_{\min}$), there is no flow instability. However, there will be passage-to-passage flow

maldistribution, due to the viscosity change in different flow passages in parallel. In that case, the pressure drop in individual flow passages can be determined from Eq. (12.29). In this equation, f_j will be different in different flow passages, due to the viscosity change; this is because the change in the viscosity will result in $u_{m,j}$ and hence Re_j. For the analysis, consider a two-tube-row design with fully developed laminar flow in both tube rows. Now for the fully developed laminar flow case, we can use Eq. (12.30) for the two passages having different μ but the same $f \cdot \mathrm{Re}$ because the flow passage geometry of both tube rows is identical in the present case. From Eq. (12.30), we get the mass flow rate ratio for the two passages as

$$\frac{\dot{m}_1}{\dot{m}_2} = \frac{\mu_2}{\mu_1} \tag{12.55}$$

For turbulent flow, use $f \propto \mathrm{Re}^{-0.2}$ or a similar relationship in Eq. (12.29) to get the ratio $\dot{m}_1/\dot{m}_2$ in terms of the μ_2/μ_1 ratio (see, e.g., an equation for this ratio in Problem 12.6). In a similar manner, we can analyze developing laminar flow with the correct values for the f factors in Eq. (12.29).

Once the flow fraction distribution is found from Eq. (12.55), the pressure drop ratio for fully developed laminar flow case can be determined by Eq. (12.35), where the subscript n represents the case corresponding to the mean temperature for both tube rows (i.e., for the exchanger on that fluid side). For other flow types, calculate the pressure drop ratio from the following equation, derived from Eq. (12.29):

$$\frac{\Delta p_1}{\Delta p_n} = \frac{f_1}{f_n} \frac{\rho_{m,1}}{\rho_{m,n}} \left(\frac{u_{m,1}}{u_{m,n}} \right)^2 \tag{12.56}$$

The heat transfer case can be analyzed for the constant-wall-temperature boundary condition using Eq. (12.38) to determine ntu_j for the given ntu_n and the computed flow fraction distribution from Eq. (12.55). Subsequently, follow Eqs. (12.48)–(12.50) to determine the impact on the heat transfer performance due to flow maldistribution in a viscous cooler. Be sure to modify these equations as appropriate when the flow is turbulent or developing laminar flow on the flow maldistributed side.

12.3 MITIGATION OF FLOW MALDISTRIBUTION

Flow maldistribution in heat exchangers may have serious consequences on thermal and mechanical performance as elaborated in detail by Kitto and Robertson (1989). Flow maldistribution in a heat exchanger may be reduced through modifications in the existing design or taken into account by incorporating its effect in the design methodology. Most gross flow maldistributions may result in relatively minor heat transfer performance reduction, smaller than 5% for NTU < 4 for tube-side flow maldistribution in a shell-and-tube exchanger as reported by Mueller and Chiou (1988). At high NTUs (NTU > 10), the performance loss may be substantially larger. However, the increase in pressure drop is generally substantial with gross flow maldistribution. Where large temperature differences exist, the effect of gross flow maldistribution can result in excessive thermal stress in the heat transfer surface wall (Mueller, 1987). Passage-to-passage maldistribution may result in a significant reduction in heat transfer performance, particularly for laminar flow exchangers.

Any action to prevent flow maldistribution must be preceded by an identification of possible reasons that may cause the performance deterioration and/or may affect the mechanical characteristics of the heat exchanger. Possible consequences from the performance viewpoint are (1) deterioration in heat exchanger effectiveness and increase in pressure drop; (2) fluid "freezing," as in viscous flow coolers; (3) fluid deterioration; (4) enhanced fouling; and (5) mechanical and tube vibration problems due to flow instabilities, wear, fretting, erosion, and corrosion and mechanical failure. No generalized recommendations can be made for preventing the negative consequences of flow maldistribution. Most problems must be solved by intelligent designs and diagnosis on an individual basis. A few broad guidelines for shell-and-tube heat exchangers are:

- Gross flow maldistribution may be induced at inlet nozzles on the shell side. Placing an impingement perforated baffle about halfway to the tubesheet will break up the inlet jet stream (see Fig. 10.2).
- The shell inlet and exit baffle spaces are the regions prone to gross flow maldistribution. An appropriate design of the baffle geometry (e.g., the use of double segmental or disk-and-doughnut baffles) may reduce this maldistribution.
- Passage-to-passage flow maldistribution may be reduced by improved control of the manufacturing process (tolerances and gaps). For example, for brazed compact heat exchangers, use of the state-of-the-art controlled-atmosphere brazing, with improved temperature uniformity during brazing, good control of brazing parameters such as cladding ratio and flux amount, and proper fixturing, may significantly reduce the presence of passage-to-passage nonuniformities.
- Manifold-induced maldistribution may be controlled by careful control of the area ratio and lateral flow resistance. These parameters may be fixed in many systems by requirements other than these considerations. In such cases, the relative length of the manifold, the friction factors, and the orientation between the manifolds may be used as factors that may reduce flow maldistribution.
- Operating condition–induced flow maldistribution is difficult to control. For the laminar flow maldistribution, a design must be such as to allow sufficient pressure drop to prevent maldistribution or to resort to multipassing.

It should be noted that heat exchangers involving multiphase flows might be most prone to flow maldistribution. However, this type of heat exchanger is outside the scope of this book. For single-phase heat exchangers, good engineering judgment, involving the considerations discussed above, may greatly reduce any possible influence of flow maldistribution. In many cases, even though the overall influence of maldistribution on the heat exchanger performance is small, local phenomena caused by flow nonuniformity may be of great importance. These may cause increased corrosion, erosion, wear, fouling, and even material failure (Kitto and Robertson, 1989).

12.4 HEADER AND MANIFOLD DESIGN

Headers and manifolds are fluid distribution elements connecting the heat exchanger core and the inlet and outlet fluid flow lines. An inlet header is the transition duct joining the inlet face of the heat exchanger core or matrix to the inlet pipe for each fluid. Similarly, an outlet header joins the outlet face of the exchanger core to the outlet

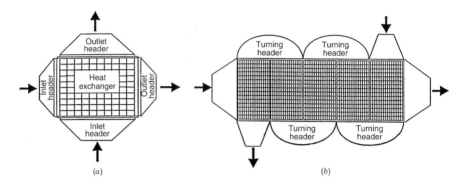

FIGURE 12.10 Typical compact heat exchangers with (a) normal headers and (b) turning headers.

(return) pipe. The header is variously referred to as a *tank*, *box*, or *distributor*. Manifolds have a bit more complex function, especially for compact heat exchangers. An incoming stream must be distributed uniformly into a heat exchanger core for the lowest core Δp and the highest achievable q. Basically, a manifold is a flow channel/duct with one (side or central) inlet and multiple sidewall outlets to the heat exchanger core, or vice versa. Compact heat exchangers having normal and turning headers are illustrated in Fig. 12.10 and oblique flow headers in Fig. 12.11. Examples of manifolds with flow distributions are shown in Fig. 12.12.

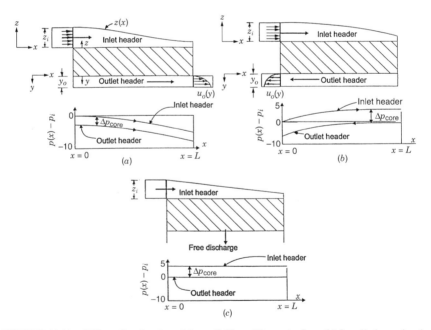

FIGURE 12.11 Oblique-flow headers: (a) parallelflow; (b) counterflow; (c) free discharge header. (From London et al., 1966.)

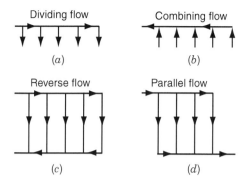

FIGURE 12.12   Major types of manifolds: (*a*) dividing-flow manifolds, (*b*) combining-flow manifolds. The corresponding inlet/outlet configurations (see also Fig. 12.7); (*c*) reverse-flow configuration, (*d*) parallelflow configuration.

Two important requirements may be identified for header and/or manifold design. They should be designed so that they result in (1) uniform distribution of the fluid stream within a heat exchanger core, and (2) minimal pressure drop within the header/manifold, since in general we do not get any heat transfer for that pressure drop expenditure.

The design of the inlet header is more critical. An area increase from the inlet pipe to the core face may be 5 to 50 times. It is impossible to maintain streamline flow in headers for such a large area enlargement. Hence, the flow is normally separated in an inlet header with either a completely detached or a singly attached jet on one wall. Flow separation results in increased pressure drop in the inlet header and nonuniform flow distribution at the core face. In addition to flow separation, the shape of inlet header could produce high-velocity regions, and this could lead to localized erosion at the core face (tube entrance from the tank), particularly for liquid flows.

The design of the outlet header should match that of the inlet header (or vice versa) so that the pressure drop across the core is uniform, resulting in the uniform flow distribution. To minimize pressure losses due to flow separation, area contraction in the outlet header should be smooth. Also, sharp turns in the outlet header should be avoided. Turns and bends create centrifugal forces, resulting in nonuniform pressure at the core face, which may lessen flow uniformity in the core.

From the foregoing viewpoints, the header design objective is to provide for acceptably uniform flow through the core with an acceptable header geometry and acceptably low pressure drops since this spent Δp in the headers is *not* associated with heat transfer between two fluids (the heat exchanger core is designed for this heat transfer).

In baffled shell-and-tube exchangers and in some multipass crossflow heat exchangers, good fluid mixing takes place on the shell side or within the core. Flow nonuniformity at the core inlet generally does not degrade the performance of the exchanger if there is good mixing within the core. The design of headers is important for those exchangers in which there is very little fluid mixing in the core, and for the gas side of a gas-to-fluid heat exchanger.

In the highly compact gas-to-fluid heat exchanger, the header design on the gas side is more difficult because the exchanger core shape is characterized by a large flow frontal area and a short flow length for the gas flow path. In such cases, either no header is used, as in an automobile radiator (on the air side), or an oblique flow header of the type mentioned in Section 12.4.1 (or a conical/tapered manifold of round cross section) is used. Normal-flow headers are also used. We now discuss these two types of headers (oblique flow and normal flow) for compact heat exchangers.

12.4.1 Oblique-Flow Headers

The design theory for oblique-flow headers for heat exchangers has been derived by London et al. (1968), based on the work of Perlmutter (1961). This theory is based on the study of flow conditions and corresponding header shape, assuming steady, constant-density, inviscid flow. Three single-pass header configurations are of particular interest, as shown in Fig. 12.11.

In an oblique-flow header, the fluid inlet flow direction with respect to the core face is at an angle different from $90°$ (i.e., normal flow, as in a normal-flow header). A special class of oblique-flow headers has an inlet flow direction parallel to the core face area. The main feature of this type of header is the minimization of header volume and flow separation. The three main types of oblique headers are:

1. *Parallelflow headers* (Fig. 12.11a). The fluid inlet and outlet are on opposite sides of the core and the fluid flows in the same directions through both headers.
2. *Counterflow headers* (Fig. 12.11b). The fluid inlet and outlet are on the same side of the core and the fluid flows in counterflow through the headers.
3. *Free discharge headers* (Fig. 12.11c). Only the inlet header exists and fluid discharges freely at the outlet without ducting.

In Table 12.1, the model predictions for pressure distributions, theoretical shapes, and pressure drops are compiled for the three types of oblique headers based on the results of London et al. (1968). The assumptions adopted are given in Example 12.5. A simple box configuration is considered for the outlet header (if the outlet header exists) from the construction and cost points of view; then inlet headers for parallelflow and counterflow oblique headers require a special shape to achieve the uniform flow distribution through the core; and the shapes are derived theoretically as summarized in Table 12.1. All geometrical characteristics and the notation are presented in Fig. 12.11.

If a designer has freedom to select the header type, a counterflow header would be the best option for the lowest Δp in headers, followed by the free discharge header and parallelflow header. Reviewing the results in Table 12.1, note that Δp for the inlet header is higher than that for the outlet header in all three cases, and Δp for the outlet headers is largely associated with the nonuniform velocity distribution at the exit shown in Fig. 12.11a and b.

In addition to the three configurations of oblique headers presented in Table 12.1 and Fig. 12.11, various other configurations are possible (such as headers at different incoming angles, or with turning vanes or guide vanes to turn the flow and minimize flow nonuniformity). No systematic studies are reported in the literature for such headers, due to many geometrical variations and lack of any available theory. One of the best alternatives is to conduct three-dimensional CFD analysis when accurate flow distribution and pressure drop predictions are required.

Example 12.5   Determine the geometry of the inlet header, z/y_o as a function of $X^* = x/L$, of the parallelflow oblique header of Fig. 12.11a (note that the outlet header

TABLE 12.1   Theoretical Shape, Pressure Profiles, and Pressure Drops of Oblique-Flow Headers[a]

Header Type		Theoretical Model[a]	Pressure Drop
Parallelflow header	Inlet header pressure	$\dfrac{p_i - p(X^*)}{H_i} = \dfrac{\pi^2}{4}(X^*)^2 \dfrac{\rho_i}{\rho_o}\left(\dfrac{z_i}{y_o}\right)^2$	Inlet: $\dfrac{\Delta p_i}{H_i} = 1 + \dfrac{\pi^2}{12}\dfrac{H_o}{H_i}$
	Inlet header profile[b]	$\dfrac{z}{y_o} = \dfrac{1 - X^*}{[(\pi^2/4)(\rho_i/\rho_o)(X^*)^2 + (y_o/z_i)^2]^{1/2}}$	Outlet: $\dfrac{\Delta p_o}{H_i} = 0.645\dfrac{H_o}{H_i}$
			Total: $\dfrac{\Delta p_t}{H_i} = 1 + 1.467\dfrac{H_o}{H_i}$
Counterflow header[c]	Inlet header pressure	$\dfrac{p(X^*) - p_i}{H_i} = \dfrac{\pi^2}{4}\dfrac{\rho_i}{\rho_o}\dfrac{z_i}{y_o}[1 - (1 - X^*)^2]$	Inlet: $\dfrac{\Delta p_t}{H_i} = \dfrac{1}{3}$
	Inlet header profile[b]	$\dfrac{z}{y_o} = \dfrac{1 - X^*}{\{(y_o/z_i)^2 - (\pi^2/4)(\rho_i/\rho_o)[1 - (1 - X^*)^2]\}^{1/2}}$	Outlet[d]: $\dfrac{\Delta p_o}{H_i} = 0.645\dfrac{H_o}{H_i}$
			Total: $\dfrac{\Delta p_t}{H_i} = 0.595$
Free discharge header	Inlet header profile	$\dfrac{z}{z_i} = 1 - X^*$	Inlet: $\dfrac{\Delta p_t}{H_i} = 1$
			Total: $\dfrac{\Delta p_t}{H_i} = 1$

$$H_i = \left(\dfrac{\rho u_m^2}{2g_c}\right)_i, \qquad H_o = \left(\dfrac{\rho u_m^2}{2g_c}\right)_o, \qquad X^* = \dfrac{x}{L}$$

Source: Data from London et al. (1968).

[a] The geometry of header designs is presented in Fig. 12.11. Subscripts i and o denote the entrance of the inlet header and the exit of the outlet header.

[b] The outlet header is a box header (required for a minimum inlet header loss for counterflow configuration and imposed for parallelflow configuration).

[c] For the parallelflow header, the inlet header dimension z_i can be either larger or smaller than the outlet dimension y_o; for counterflow, z_i must be $(2/\pi)y_o\sqrt{\rho_o/\rho_i}$ to assure the matching pressure profile needed for uniform flow distribution and for minimum header loss. The outlet header is also considered to be a box header.

[d] $H_o/H_i = 4/\pi^2$ for a counterflow header.

is a box type) for the case when inlet and outlet velocity heads are equal $(H_i = H_o)$ and $\rho_i/\rho_o = 1.4$. Use the inlet header profile given in Table 12.1 in which z/y_o is dependent on y_o/z_i, the ratio of outlet to inlet header height.

SOLUTION

Problem Data and Schematic: The inlet header is a parallelflow oblique header as shown in Fig. 12.11a with appropriate notations. The following data are given:

$$H_o = \left(\frac{\rho u_m^2}{2g_c}\right)_o = H_i = \left(\frac{\rho u_m^2}{2g_c}\right)_i \qquad \frac{\rho_i}{\rho_o} = 1.4$$

Determine: The inlet header geometry [i.e., $z/y_o = \phi(x/L)$], shown in Fig. 12.11a.

Assumptions: The following assumptions are made (London et al., 1968): (1) the inlet and outlet header fluid mass densities are individually constant, (2) the inlet header velocity distribution is uniform, (3) the flow velocity and pressure in the inlet header are functions of x only, (4) the inlet and outlet header fluid flows are inviscid, (5) the outlet header pressure is a function of x only, (6) the velocity is uniform through the core, and (7) the depth of the header (the third dimension) is unity.

Analysis: From Table 12.1, the geometry of an inlet parallelflow oblique header is given by

$$Z = \frac{z}{y_o} = \frac{1 - X^*}{[(\pi^2/4)(\rho_i/\rho_o)(X^*)^2 + (y_o/z_i)^2]^{1/2}} \tag{1}$$

Let us first demonstrate briefly how this equation may be derived (for details, see London et al., 1968). By applying the Bernoulli equation for a streamline through the inlet header, core, and outlet header (Fig. 12.11a), we can relate inlet pressure to a pressure at any x position as follows:

$$p_i - p(x) = \frac{\rho_i}{2g_c}(u^2 - u_i^2) \tag{2}$$

Invoking assumptions concerning constant fluid density, uniform inlet velocity, and uniform velocity v through the core (in the negative z direction in Fig. 12.11a), the continuity equation will have the following form[†]:

$$u_i z_i = uz(x) + vx = v(L - x) + vx = vL = u_o y_o \frac{\rho_o}{\rho_i} \tag{3}$$

[†] Note that x-direction mass flow rate through any cross section x plus the mass flow rate through the core in the y direction from $x = 0$ to x is equal to mass flow rate entering the inlet header at $x = 0$. Alternatively, x-direction mass flow rate through any cross section x is the same as the mass flow rate going through the core in the y direction from $x = x$ to L.

Note $v = v_m =$ constant; also $v_m L = u_i z_i$. Now it can be shown, based on the analysis of London et al. (1968), that the pressure drop described by Eq. (2) for $y =$ constant for a box type outlet header must be equal to $(\pi^2/4)H_o(x/L)^2$; that is,

$$p_i - p(x) = \frac{\rho_i}{2g_c}(u^2 - u_i^2) = \frac{\pi^2}{4}H_o(X^*)^2 \tag{4}$$

Invoking the definitions of H_o and H_i from Table 12.1, we get

$$\left(\frac{u_o}{u_i}\right)^2 = \frac{H_o}{H_i}\frac{\rho_i}{\rho_o} \tag{5}$$

Finally, combining Eqs. (3) and (4) to eliminate u from the second equality in Eq. (4) and keeping in mind the definition of velocity heads from Table 12.1 and as indicated above by Eq. (5), we obtain Eq. (1). A detailed derivation of Eq. (1) is the subject of Problem 12.9.

We will now derive the ratio y_o/z_i as a function of the velocity head ratio H_i/H_o from the continuity equation considering mass flow rates to be the same at the entrance of the inlet and outlet headers:

$$(\rho A_o u_m)_i = (\rho A_o u_m)_o$$

Therefore,

$$\frac{u_{m,i}}{u_{m,o}} = \frac{\rho_o}{\rho_i}\frac{A_{o,o}}{A_{o,i}} = \frac{\rho_o}{\rho_i}\frac{y_o}{z_i} \quad \text{or} \quad \left(\frac{u_{m,i}}{u_{m,o}}\right)^2 = \left(\frac{\rho_o}{\rho_i}\right)^2\left(\frac{y_o}{z_i}\right)^2 \tag{6}$$

From the definition of the velocity heads (see Table 12.1), we get

$$\frac{H_i}{H_o} = \frac{(\rho u_m^2/2g_c)_i}{(\rho u_m^2/2g_c)_o} \quad \text{or} \quad \frac{H_i}{H_o} = \frac{\rho_i}{\rho_o}\left(\frac{u_{m,i}}{u_{m,o}}\right)^2 \tag{7}$$

Now, combining Eqs. (6) and (7), we get

$$\frac{H_i}{H_o} = \frac{\rho_o}{\rho_i}\left(\frac{y_o}{z_i}\right)^2 \tag{8}$$

Substituting $(y_o/z_i)^2$ from Eq. (8) into Eq. (1), we get the geometry of the inlet header as

$$\frac{z}{y_o} = \frac{1 - X^*}{(\rho_i/\rho_o)[(\pi^2/4)(X^*)^2 + H_i/H_o]^{1/2}} \tag{9}$$

Since the inlet and outlet velocity heads must be equal (imposed by the problem formulation; i.e., $H_i = H_o$), Eq. (9) can be rewritten as follows using $\rho_i/\rho_o = 1.4$ as given

$$\frac{z}{y_o} = \frac{1 - X^*}{(\rho_i/\rho_o)[(\pi^2/4)(X^*)^2 + 1]^{1/2}} = \frac{1 - X^*}{1.4[(\pi^2/4)(X^*)^2 + 1]^{1/2}} \tag{10}$$

We can now calculate the coordinates of the inlet header shape z/y_o:

X^*	0	0.2	0.4	0.6	0.8	1.0
z/y_o	0.714	0.545	0.363	0.208	0.089	0

Discussion and Comments: The inlet header profile calculated demonstrates a need to design this header with a variable header cross section according to the functional relationship given by Eq. (10) for the case when the outlet header is box type and uniform flow is through the core. In this case, the inlet and outlet velocity heads would be equal, and the header loss would be, from Table 12.1, $\Delta p_t/H_i = 1 + 1.467(H_o/H_i) = 1 + 1.467 \times 1 = 2.47$ velocity heads.

12.4.2 Normal-Flow Headers

Normal-flow headers are characterized as having the flow direction perpendicular to the heat transfer core (see Fig. 12.10a). The design of a normal-flow header follows the design of a diffuser with a large increase in the free-flow area from the inlet pipe to the heat exchanger core face. This type of header design is qualitatively discussed by Wilson (1966). The pressure drop, flow separation, and recirculation (if any) depend on the diffuser geometry, which includes the type (two dimensional vs. three dimensional, rectangular vs. conical, etc.), included angle, aspect ratio (diffuser throat to length ratio), and flow type. For a heat exchanger, the diffuser (inlet normal header) is followed by the heat exchanger core having finite pressure drop. Hence, the design information for a diffuser having no downstream flow resistance will be conservative for a heat exchanger.

If the inlet header has a box configuration, a jet will be formed from the inlet pipe and will increase in diameter before impinging on the core face. So, to minimize the header volume and pressure losses, it is desirable to make the inlet header as a conical section (to match the spreading jet diameter reasonably) followed by a plenum chamber, instead of making it one large plenum chamber. For liquid flows, a plain or perforated baffle or a perforated plate is used in the inlet tank for distributing flow to all tubes in the *noflow* direction (see Section 1.6.1.3 for the definition of noflow direction). Such a design reduces header/tank volume and maintains an acceptably uniform flow distribution.

12.4.3 Manifolds

Two major types of manifolds, as noted in Section 12.1.3, are shown in Fig. 12.12a and b: dividing-flow manifolds and combining-flow manifolds. In dividing-flow manifolds, fluid enters either axially or laterally and exits the manifold laterally. The axial velocity within the manifold varies from the inlet velocity to the zero value. Conversely, in combining flow manifolds, fluid enters laterally and exits at the end of the manifold either axially or laterally.

Inlet and outlet manifolds in a heat exchanger can be arranged as (1) reverse-flow configuration and (2) parallelflow configuration, as shown in Fig. 12.12c and d, interconnected by lateral branches represented by flow passages in the heat exchanger core (see also Fig. 12.7).

Modeling a manifold requires determination of both axial and lateral velocity and static pressure distributions. Available solutions of the manifold flow models may be either analytical (in simplified cases, Bajura and Jones, 1976) or numerical (Majumdar,

1980). The key problem in analytical modeling is the difficulty in identifying a relevant streamline on which to calculate energy and pressure drop losses and apply the Bernoulli equation. The state-of-the-art design procedures utilize commercial and/or proprietary CFD codes. Still, simple analytical modeling has merit for assessment purposes (Shen, 1992).

SUMMARY

This chapter deals with several important flow maldistributions (geometry and operating conditions induced) in heat exchangers. Design theory is provided to determine quantitatively the influence of two major geometry-induced flow maldistributions: (1) gross flow maldistribution and (2) passage-to-passage flow maldistribution.

- Simple modeling for gross flow maldistribution is possible only for counterflow, parallelflow, and one crossflow unmixed–mixed arrangements. For other flow arrangements, the influence of flow maldistribution can be evaluated numerically. In many situations, gross flow maldistributions do not reduce heat transfer significantly but may cause a significant increase in pressure drop. Some specific gross flow maldistributions in an unmixed–unmixed crossflow exchanger *can* increase the exchanger effectiveness. This is mentioned in the paragraph just before Example 12.2.
- Compact heat exchangers with continuous-flow passages are highly susceptible to passage-to-passage flow maldistribution important in laminar flows. This maldistribution reduces the pressure drop slightly but may reduce heat transfer significantly. Usually, the pressure drop reduction is neglected in the design. The simplest model of passage-to-passage flow maldistribution is the two-passage model, where two different-size passages are in parallel. The two-passage model reduces exchanger effectiveness and hence heat transfer performance more than that for an N-passage ($N > 2$) model of passage-to-passage nonuniformity. Hence, for a conservative design, a two-passage model is most appropriate to determine the effect of passage-to-passage flow maldistribution.

Among operating condition–induced flow maldistributions, the most important is viscosity-induced flow maldistribution. It can induce flow instability in a multitube or multi-continuous passage exchanger if the flow rate is below some critical value ($\dot{m}_{min}$ in Fig. 12.8). If $\dot{m} > \dot{m}_{min}$, the viscosity-induced flow maldistribution problem reverts to the problem of passage-to-passage nonuniformity.

It should be mentioned that no generalized recommendations can be made for preventing negative influences of flow maldistribution. Each case should be considered separately.

Header and manifold design is very important in controlling the level of flow maldistribution within the core and reducing undesired pressure drop in headers and manifolds, particularly for compact heat exchangers. Specific information for designing the header cross-section profile is presented in Section 12.4 for oblique-flow headers.

REFERENCES

Bajura, R. A., and E. H. Jones, Jr., 1976, Flow distribution manifolds, *ASME J. Fluid Eng.*, Vol. 98, pp. 654–666.

Chiou, J. P., 1980, The advancement of compact heat exchanger theory considering the effects of longitudinal heat conduction and flow nonuniformity, in *Compact Heat Exchangers: History, Technological Advancement and Mechanical Design Problems*, R. K. Shah, C. F. McDonald, and C. P. Howard, eds., Book G00183, HTD-Vol. 10, American Society of Mechanical Engineers, New York, pp. 101–121.

Chiou, J. P., 1982, The effect of nonuniformities of inlet temperatures of both fluids on the thermal performance of crossflow heat exchanger, *Heat Transfer 1982, Proc. 7th Int. Heat Transfer Conf.*, Vol. 6, pp. 179–184.

Chowdhury, K., and S. Sarangi, 1985, The effect of flow maldistribution on multipassage heat exchanger performance, *Heat Transfer Eng.*, Vol. 6, No. 4, pp. 45–54.

Cichelli, M. T., and D. F. Boucher, 1956, Design of heat exchanger heads for low holdup, *AIChE Chem. Eng. Prog.*, Vol. 52, No. 5, pp. 213–218.

Datta, A. B., and A. K. Majumdar, 1980, Flow distribution in parallel and reverse flow manifolds, *Int. J. Heat Fluid Flow*, Vol. 2, pp. 253–262.

Fleming, R. B., 1966, The effect of flow distribution in parallel channels of counterflow heat exchangers, in *Advances in Cryogenic Engineering*, Vol. 12, K. D. Timmerhaus, ed., Plenum Press, New York, pp. 352–363.

Heggs, P. J., and H.-J. Scheidt, 1992, Thermal performance of plate heat exchangers with flow maldistribution, in *Compact Heat Exchangers for Power and Process Industries*, American Society of Mechanical Engineers, HTD-Vol. 201, pp. 87–93.

Hewitt, G. H., G. L. Shires, and T. R. Bott, 1994, *Process Heat Transfer*, CRC Press, Boca Raton, FL.

Kitto, J. B., and J. M. Robertson, 1989, Effects of maldistribution of flow on heat transfer equipment performance, *Heat Transfer Eng.*, Vol. 10, No. 1, pp. 18–25.

Kutchey, J. A., and H. L. Julien, 1974, The measured influence of flow distribution on regenerator performance, *SAE Trans.*, Vol. 83, SAE Paper 74013.

London, A. L., 1968, Laminar flow gas turbine regenerators – the influence of manufacturing tolerances, T.R. No. 69, Department of Mechanical Engineering, Stanford University, Stanford, California, 1968.

London, A. L., 1970, Laminar flow gas turbine regenerators: the influence of manufacturing tolerances, *ASME J. Eng. Power*, Vol. 92, Ser. A, pp. 45–56.

London, A. L., G. Klopfer, and S. Wolf, 1968, Oblique flow headers for heat exchangers – the ideal geometries and the evaluation of losses, T.R. No. 63, Department of Mechanical Engineering, Stanford University, Stanford, California.

London, A. L., G. Klopfer, and S. Wolf, 1968, Oblique flow headers for heat exchangers, *ASME J. Eng. Power*, Vol. 90, Ser. A, pp. 271–286.

Majumdar, A. K., 1980, Mathematical modeling of flows in dividing and combining manifolds, *Appl. Math. Model.*, Vol. 4, pp. 424–434.

Mueller, A. C., 1974, Criteria for maldistribution in viscous flow coolers, *Heat Transfer 1974, Proc. 5th Int. Heat Transfer Conf.*, Vol. 5, pp. 170–174.

Mueller, A. C., 1977, An inquiry of selected topics on heat exchanger design, *AIChE Symp. Ser. 164*, Vol. 73, pp. 273–287.

Mueller, A. C., 1987, Effects of some types of maldistribution on the performance of heat exchangers, *Heat Transfer Eng.*, Vol. 8, No. 2, pp. 75–86.

Mueller, A. C., and J. P. Chiou, 1988, Review of various types of flow maldistribution in heat exchangers, *Heat Transfer Eng.*, Vol. 9, No. 2, pp. 36–50.

Perlmutter, M., 1961, Inlet and exit header shapes for uniform flow through a resistance parallel to the main stream, *ASME J. Basic Eng.*, Vol. 83, pp. 361–370.

Putnam, G. R., and W. M. Rohsenow, 1985, Viscosity induced nonuniform flow in laminar flow heat exchangers, *Int. J. Heat Mass Transfer*, Vol. 28, pp. 1031–1038.

Ranganayakulu, C. H., K. N. Seetharamu, and K. V. Sreevatsan, 1996, The effects of inlet fluid flow nonuniformities on thermal performance and pressure drops in crossflow plate-fin compact heat exchanger, *Int. J. Heat Mass Transfer*, Vol. 40, pp. 27–38.

Sekulić, D. P., 1999, Behavior of aluminum alloy micro layer during brazing, in *Recent Res. Dev. Heat Mass Momentum Transfer*, Vol. 2, pp. 121–140.

Sekulić, D. P., C. Pan, F. Gao, and A. T. Male, 2001, Modeling of molten cladding flow and diffusion of Si across a clad-core interface of an aluminum brazing sheet, *DVS Berichte*, Vol. 212, pp. 204–219.

Shah, R. K., 1981, Compact heat exchangers, in *Heat Exchangers: Thermal-Hydraulic Fundamentals and Design*, S. Kakaç, A. E. Bergles, and F. Mayinger, eds., Hemisphere Publishing, Washington, DC, pp. 111–151.

Shah, R. K., 1985, Compact Heat Exchangers, in *Handbook of Heat Transfer Applications*, 2nd Ed., Eds. W. M. Rohsenow, J. P. Hartnett and E. N. Ganić, Chapter 4, Part III, pp. 4–174 to 4–311, McGraw-Hill, New York.

Shah, R. K., and A. L. London, 1971, *Influence of Brazing on Very Compact Heat Exchanger Surfaces*, Paper 71-HT-29, American Society of Mechanical Engineers, New York.

Shah, R. K., and A. L. London, 1980, Effects of nonuniform passages on compact heat exchanger performance, *ASME J. Eng. Power*, Vol. 102, Ser. A, pp. 653–659.

Shen, P. I., 1992, The effect of friction on flow distribution in dividing and combining flow manifolds, *ASME J. Fluids Eng.*, Vol. 114, pp. 121–123.

Thonon, B., 2002, Private communication, CAE-GRETh, Grenoble, France.

Thonon, B., P. Mercier, and F. Feidt, 1991, Flow distribution in plate heat exchanger and consequences on thermal and hydraulic performances, Proc. 18th Eurotherm Conference, Springer Verlag, Hamburg, Germany.

Wilson, D. G., 1966, *A Method of Design for Heat-Exchanger Inlet Headers*, Paper 66-WA/HT-41, Americal Society of Mechanical Engineers, New York.

REVIEW QUESTIONS

Where multiple choices are given, circle one or more correct answers. Explain your answers briefly.

12.1 The gross flow maldistribution is independent of:

(a) surface geometry

(b) passage-to-passage nonuniformity

(c) flow rate

(d) outlet header

(e) heat exchanger flow arrangement

12.2 The following are characteristic of gross flow maldistribution:

(a) increased total heat transfer

(b) increased core pressure drop on the maldistributed side

12.3 The following is a method of computing pressure drop on the side of gross flow maldistribution:

(a) a weighted average pressure drop based on apportioned flow rates $\dot{m}_i$

(b) the maximum pressure drop based on the largest $\dot{m}_i$ component

(c) an arithmetic average of Δp_i's   (d) a sum of Δp_i's

12.4 Flow nonuniformity in laminar flow surfaces can be caused by:

(a) large frontal areas and small core depths that keep flow velocity low but present fluid distribution problems

(b) manufacturing tolerances that are a significant fraction of the surface hydraulic diameter

(c) fouling (d) deposition of condensable substances

12.5 The temperature effectiveness of a maldistributed fluid stream of a counterflow heat exchanger can be determined using the following relationship:

(a) $P_i = \dfrac{1}{C_i} \sum_j C_{i,j} P_{i,j}$ (b) $P_i = \dfrac{1}{C_i} \prod_j C_{i,j} P_{i,j}$ (c) $P_i = \dfrac{1}{C_i} \sum_i C_{i,j} P_{i,j}$

where the subscript i denotes a fluid stream, and j denotes a subexchanger having the uniform fluid flow distribution.

12.6 Gross flow maldistribution increases irreversibilities caused by the following phenomena in a counterflow exchanger for heat transfer performance:

(a) temperature difference (b) fluid mixing (c) flow friction
(d) none of these (e) all of these

12.7 The following phenomena that cause irreversibilities in a counterflow exchanger are important for fluid pumping power requirement when gross flow maldistribution exists on the fluid side of interest:

(a) temperature difference (b) fluid mixing (c) flow friction
(d) none of these (e) all of these

Explain whether the irreversibility will increase or decrease.

12.8 Moderate gross flow maldistribution results in more total irreversibility in an exchanger than that for moderate temperature maldistribution.

(a) true (b) false
(c) It depends on the flow arrangement. (d) can't tell

Explain your reasoning from the irreversibility point of view.

12.9 Using the knowledge of irreversibilities gained in Chapter 11, explain why the temperature maldistribution increases exchanger effectiveness for an unmixed–unmixed crossflow exchanger when the hotter portion of the hot fluid maldistributed inlet temperature is near the exit end of the cold fluid, whose temperature is uniform.

12.10 The following are characteristics of passage-to-passage flow maldistribution on one fluid side of an exchanger:

(a) heat transfer and pressure drop generally unaffected

(b) a significant decrease in total heat transfer and also a slight decrease in pressure drop

(c) lower j and f factors for continuous-flow-passage surfaces

(d) lower j and f factors for interrupted surfaces

(e) poor header design

12.11 Passage-to-passage flow maldistribution:

(a) can be corrected by careful placement of turning vanes in the inlet and outlet headers

(b) results from core heat transfer surface nonuniformity for continuous-flow passages

    (c) is a critical design concern for all types of exchangers

    (d) can improve overall heat exchanger performance by promoting local turbulence

12.12 Brazing can affect exchanger performance:

    (a) very little at most because additional heat transfer surface created by the brazing roughness negates the effect of increased pressure drop due to the roughness

    (b) adversely when the flow passage geometry is affected appreciably by brazing roughness

    (c) positively because it significantly increases the j factor while slightly increasing the f factor

12.13 Gross blockage due to brazing may reduce:

    (a) the f factor greatly   (b) the f factor slightly

    (c) the j factor greatly   (d) the j factor slightly

12.14 Circle the following statements as true or false.

    (a) T F   For a given deviation in flow uniformity for a particular passage type of a compact heat exchanger, the gain in the pressure drop is greater than the reduction in NTU.

    (b) T F   Roughness introduced by brazing does not have any appreciable effect on flow maldistribution due to flow passage geometry if the passages are large.

    (c) T F   The problem of flow maldistribution due to fouling with the same fluid is more severe in heat exchangers of large-flow-passage geometry than those having small-flow-passage geometry.

12.15 Passage-to-passage flow maldistribution increases the following irreversibilities in a counterflow exchanger:

    (a) temperature difference   (b) fluid mixing   (c) flow friction

    (d) none of these   (e) all of these

12.16 The following irreversibilities in a counterflow exchanger are important for fluid pumping power requirements when passage-to-passage flow maldistribution exists on the fluid side of interest:

    (a) temperature difference   (b) fluid mixing   (c) flow friction

    (d) none of these   (e) all of these

    Explain whether the irreversibility will increase or decrease.

12.17 Manifold-induced flow maldistribution in a PHE increases with:

    (a) increasing relative pressure drop in the manifold compared to the exchanger core

    (b) increasing a large number of plates in a 1-pass 1-pass PHE

    (c) larger flow area of the dividing-flow manifold than for the combining-flow manifold

12.18 Viscosity-induced flow maldistribution (beyond the flow instability region) on one fluid side in an exchanger results in:

    (a) increase in heat transfer   (b) decrease in heat transfer

    (c) increase in Δp (d) decrease in Δp

12.19 The pressure drop is increased significantly compared to that for the uniform flow case if the following flow maldistributions exist on one fluid side of an exchanger:

(a) gross flow maldistribution   (b) viscosity-induced maldistribution

(c) manifold-induced maldistribution   (d) passage-to-passage maldistribution

12.20 The primary function of the headers/manifolds is:

(a) to avoid flow separation within the header/manifold

(b) to provide uniform flow distribution over the core face

(c) to result in the lowest possible pressure drop within the headers/manifolds

(d) to yield uniform temperature distribution at the core inlet

12.21 The design of the inlet header/manifold is more/less critical than the design of the outlet header/manifold:

(a) more   (b) less   (c) can't tell

12.22 Flow separation in the headers, caused by an area change in the free-flow area, leads to:

(a) increased pressure drop   (b) decreased pressure drop

(c) high-velocity regions   (d) localized erosion

12.23 The header/manifold design problem is more important in a:

(a) shell-and-tube heat exchanger   (b) compact heat exchanger

(c) spiral plate heat exchanger

12.24 Oblique-flow headers are characterized by:

(a) The inflow is orthogonal to the heat transfer core face.

(b) The inflow is parallel to the heat transfer core face.

(c) The inflow is at an angle different from $90°$ to the core face.

(d) The inlet and outlet headers are on the same side of the exchanger with side inlets.

(e) Only the inlet header is always present.

12.25 For specified heat transfer performance in an exchanger, important design considerations for normal and oblique flow headers are to:

(a) match inlet and outlet header designs

(b) minimize the core pressure drop

(c) allow nonuniform flow through the core to reduce the core pressure drop

(d) Any header design is acceptable as long as the heat exchanger core has perfectly manufactured flow passages.

12.26 Circle the following statements as true or false.

(a) T F   More uniform flow distribution through the core is achieved by a reverse-flow manifold system than by a parallelflow manifold system

(b) T F   More uniform flow distribution through the core is achieved if the flow area of the combining-flow header is smaller than that of the dividing-flow header.

(c) T F   Maintaining a larger pressure drop in an exchanger core than in the headers is important to provide uniform flow distribution through the core.

12.27 For a compact heat exchanger, header/manifold design is important because:

(a) turbulence must be minimized at the inlet

(b) of the awkward shape of the core, a relatively large frontal area, and a short flow length

(c) performance falls off sharply in a maldistributed situation

12.28 The header/tank on the coolant side in an automobile radiator is a:

(a) counterflow header   (b) oblique flow header

(c) normal flow header   (d) dividing/combining flow manifold

12.29 Rank the following fluids in order of least to most important for good header design:

(a) low-pressure gas   (b) water   (c) air   (d) oil

12.30 Which header configuration has the lowest pressure drop for the same inlet velocity and same inlet flow area?

(a) counterflow   (b) parallelflow   (c) free discharge   (d) none of these

12.31 For a free discharge header:

(a) The inlet header configuration is not important since the discharge can be considered as in an infinitely large reservoir.

(b) The outlet pressure varies linearly along the length of core discharge.

(c) The inlet header flow area decreases linearly along the length of the core inlet to ensure uniform flow distribution through the exchanger core.

(d) The total header pressure drop, exclusive of core pressure drop contribution, equals one velocity head.

(e) The pressure loss for the inlet header is higher than that for the outlet header.

12.32 Major functions of turning vanes in headers are to:

(a) promote turbulence and hence increase heat transfer

(b) protect header external walls from wear

(c) deflect any solid particles in the flow stream away from the core to prevent plugging

(d) improve flow distribution through the core

(e) stiffen the header duct to increase the natural frequency

PROBLEMS

12.1 On the fluid 1 side of a counterflow exchanger, 80% of the total free-flow area is fouled such that the velocity through that portion of the flow area constitutes only 60% of the mean velocity through the fluid 1 side core as a whole. The ratio of the heat capacity rates of fluids 1 and 2 for the heat exchanger as a whole is equal to 1. The number of transfer units of the heat exchanger is 3.5. Determine the temperature effectiveness of fluid 1 if flow on the fluid 2 side is uniform.

12.2 The crossflow unmixed (fluid 1) – mixed heat exchanger of Example 12.2 has to be analyzed for the influence of flow maldistribution in the entire range of possible nonuniformities of the mass flow rate on the fluid 1 side. Determine the change in the temperature effectiveness of the maldistributed fluid (fluid 1) if the X fraction of the total free-flow area on the fluid 1 side is characterized by a flow velocity larger by $Y\%$ than the mean velocity through the core. Consider the cases for which $X = \frac{1}{3}$, 0.5, $\frac{2}{3}$, and $Y = 25$, 50, and 75, respectively, for each of the X values.

12.3 Consider a gas turbine rotary regenerator made up of a deepfold surface (rectangular passages of $\alpha^* = 0.125$). Due to the manufacturing process, some of the passages are close to a trapezoidal ($\phi = 85°$) rather than a rectangular cross section. Assume that the matrix is made up of 50% rectangular and 50% trapezoidal passages, as shown in Fig. P12.3. Determine the mass flow fraction distribution and the reduction in heat exchanger effectiveness and Δp due to passage-

FIGURE P12.3

to-passage nonuniformity. Consider the design operating point as $ntu_n = 5$, $C_r^* > 5$, and $C^* = 1$. Following are $f \cdot Re$ and Nu_{H1} for fully developed flow and some pertinent geometry information:

Characteristic	Rectangular Passage	Trapezoidal Passage
Re	20.585	15.659
Nu_{H1}	6.490	3.256
$A_{o,\Box}/A_{o,\triangle}$	1	1
$D_{h,\Box}/D_{h,\triangle}$	1	0.9961

12.4 Consider a gas turbine rotary regenerator made up of rectangular passages of $\alpha^* = \frac{1}{6}$. Due to manufacturing imperfections, 70% of the passages are rectangular with $\alpha^* = \frac{1}{6}$, and 30% of the passages have $\alpha^* = \frac{1}{4}$. Determine:

(a) The free-flow area distribution $A_{o,j}/A_{o,n}$ with $j = 1, 2$.

(b) The flow fraction distribution in the two types of passages.

(c) The change in pressure drop due to passage-to-passage nonuniformity. Does it represent a loss or a gain in comparison to all passages of ideal $\alpha^* = \frac{1}{6}$ shape?

(d) The change in exchanger effectiveness due to the nonuniformity. Does it represent a loss or a gain?

(e) The change in ntu.

Use Fig. P12.4 and the following data for analysis: $ntu_n = 5$, $C_r^* > 5$, and $C^* = 1$.

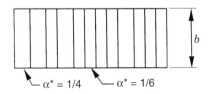

FIGURE P12.4

Characteristic	$\alpha^* = \frac{1}{4}$	$\alpha^* = \frac{1}{6}$
$f \cdot \mathrm{Re}$	18.233	19.702
$\mathrm{Nu_{H1}}$	5.331	6.049

Note:

$$D_h = \frac{2b\alpha^*}{1 + \alpha^*} \qquad \frac{A_{o,\alpha^*=1/4}}{A_{o,\alpha^*=1/6}} = 1.5 \text{ for one passage each only}$$

12.5 A crossflow regenerator for a gas turbine plant is characterized by the following data. The heat exchanger effectiveness is equal to 0.8, the air-side core relative pressure drop $\Delta p/p_i = 0.0042$ (for an inlet air pressure of 0.91 MPa), the inlet-to-outlet air density ratio is 1.5, and the mass flow rate of air is 25 kg/s. The frontal area of the air-side core is $2\,\mathrm{m^2}$. The inlet header air velocity u_i has to be 30 m/s and air density at the inlet $\rho_i = 7\,\mathrm{kg/m^3}$. Consider both parallelflow and counterflow header designs. Assume that the exit header has to be designed so as to have $H_o = H_i$ for a parallelflow header design. Determine the inlet header shape and header pressure losses.

12.6 Channel-to-channel flow maldistribution occurs in a plate heat exchanger (PHE) when two different plate groups are used. For example, consider a PHE with 47 thermal plates (24 channels for each fluid) having 8 channels with chevron plates of $\beta = 30°$ and 16 channels with chevron plates of $\beta = 39.8°$ for a $30°$–$60°$ mixed plate. Because of the turbulent flow in the channels, use a friction factor correlation of $f = a \cdot \mathrm{Re}^{-n}$, where a and n are constants. The channel-to-channel flow maldistribution for this exchanger can be derived as

$$\frac{\dot{m}_{\mathrm{I}}}{\dot{m}_{\mathrm{II}}} = \left(\frac{a_{\mathrm{II}}}{a_{\mathrm{I}}} \right)^{1/(2-n)} \left(\frac{\mu_{\mathrm{II}}}{\mu_{\mathrm{I}}} \right)^{n/(2-n)} \left(\frac{D_{e,\mathrm{I}}}{D_{e,\mathrm{II}}} \right)^{(1+n)/(2-n)} \frac{A_{o,\mathrm{I}}}{A_{o,\mathrm{II}}}$$

where $a_{\mathrm{I}} = 0.8$, $a_{\mathrm{II}} = 3.44$ and $n = 0.25$ and the subscripts I and II denote plate groups I and II.

Consider a water-to-water counterflow single-pass PHE with the aforementioned two groups of plates. Water flow rates on the hot and cold sides are 18 and 10 kg/s, respectively, and the inlet temperatures are 40 and 20°C, respectively. The total fouling resistance and wall resistance are given as 0.00004 and 0.000003 $\mathrm{m^2 \cdot K/W}$, respectively. The following is additional information.

Plate	Fluid Properties
Plate width $W = 0.05$ m	$\mu = 0.00081$ Pa $\cdot$ s
Plate height $L = 1.1$ m	$\rho = 995.4\,\mathrm{kg/m^3}$
Channel spacing $2\mathbf{a} = 0.0035$ m	$k = 0.619$ W/m $\cdot$ K
Equivalent diameter $D_e = 0.007$ m	$c_p = 4177$ J/kg $\cdot$ K
Projected area per plate $WL = 0.55\,\mathrm{m^2}$	Pr $= 5.47$

Consider the same fluid properties for water for each plate group and identical equivalent diameter for both plate groups.

(a) Determine the flow distribution of hot and cold fluids in plate groups I and II.

(b) Outline a step-by-step procedure to calculate the heat duty of this exchanger.

(c) Compute the heat duty of this PHE. Use the following equation to calculate the heat transfer coefficients:

$$h = 0.724 \frac{k}{D_h} \left(\frac{\beta}{30}\right)^{0.646} \cdot Re^{0.583} \cdot Pr^{1/3}$$

12.7 Consider an air-cooled tubular exchanger with engine oil flowing in the tubes. Because of the different amount of heat transfer taking place in the tubes, assume that oil flows at 300 K in 50% of the tubes and at 380 K in the remaining 50% of the tubes. The objective of this problem is to determine the influence of viscosity-induced flow maldistribution considering laminar flow in the tubes. Assume that $C^* = 1$ and $\text{ntu}_m = 1$. Since no results are presented in Table 3.6 or 3.3 on the ε-NTU relationship for a crossflow heat exchanger with a finite number of tubes and different NTUs associated with each tube, assume that the exchanger is counterflow.

(a) Determine the flow fraction distribution in the two types of passages.

(b) Calculate the change in Δp due to viscosity-induced flow maldistribution. Does it represent a loss or a gain in comparison to the "nominal" or base case with oil flow in both tubes at 340 K?

(c) Determine the change in exchanger effectiveness due to the flow nonuniformity. Does it represent a loss or a gain?

Use the following viscosity data for the oil and assume its density to be constant.

T (K)	300	340	380
μ (Pa · s)	0.486	0.053	0.014

12.8 A gas flows through the circular tubes (lateral branches) under laminar condition through the U-flow manifold configuration of Fig. 12.7. The two lateral headers are large enough to feature a negligible pressure drop along the fluid flow direction within each of the headers. For this gas, the dynamic viscosity increases linearly with temperature. The temperature of the tube wall is either lower or higher than the temperature of the gas flowing inside it, but it is uniform and constant along the flow length, due to high heat transfer on its outside surface. Determine which conditions will exist in this heat exchanger: (1) different mass flow rates of gas may be established in different tubes for both heating and cooling of the gas flowing through the tubes; (2) such a condition is possible only in the case of heating; or (3) such a condition is possible only in the case of cooling.

12.9 Derive an analytical expression that describes the geometry of the inlet parallel-flow oblique header given in Table 12.1. That expression should be written in the form of a relationship between the local dimensionless header wall coordinate $Z = z/y_o$ and the dimensionless coordinate in the flow direction $X^* = x/L$, as presented by Eq. (1) in Example 12.5. The main steps in the derivation are discussed in Example 12.5, but the complete derivation is to be performed for the solution of this problem.

13 Fouling and Corrosion

Fouling is an *accumulation* of undesirable material (deposits) on heat exchanger surfaces. Undesirable material may be crystals, sediments, polymers, coking products, inorganic salts, biological growth, corrosion products, and so on. This process influences heat transfer and flow conditions in a heat exchanger. Fouling is a synergistic consequence of transient mass, momentum and heat transfer phenomena involved with exchanger fluids and surfaces, and depends significantly on heat exchanger operation conditions. However, most manifestations of these various phenomena lead to similar consequences. In general, fouling results in a *reduction* in thermal performance, an *increase* in pressure drop, may *promote corrosion*, and may result in eventual failures of some heat exchangers.

Corrosion represents mechanical *deterioration* of construction materials of heat exchanger surfaces under the aggressive influence of flowing fluids and environment in contact. In addition to corrosion, other mechanically induced phenomena are important for heat exchanger design and operation, such as fretting (corrosion occurring at contact areas between metals under load subjected to vibration and slip).

Fouling and corrosion represent heat exchanger operation-induced effects and should be considered for both the design of a new heat exchanger and operation of an existing exchanger. In this chapter, we explain the impact of fouling and corrosion on heat transfer and pressure drop in Section 13.1. In Section 13.2, we present a detailed description of various fouling mechanisms and phenomenological considerations of fouling. The methodology to take into account the effect of fouling on exchanger performance and design is outlined in Section 13.3. Various techniques of prevention and mitigation of detrimental effects of fouling are summarized in Section 13.4. Finally, a brief account of the importance of corrosion, in particular its influence on heat exchanger operation and design practices, is provided in Section 13.5.

13.1 FOULING AND ITS EFFECT ON EXCHANGER HEAT TRANSFER AND PRESSURE DROP

Thermal fouling (in the presence of a temperature gradient) means accumulation of any undesirable deposition of a thermally insulating material (which provides added thermal resistance to heat flow) on a heat transfer surface occurring over a period of time.[†] This solid layer adds an additional thermal resistance to heat flow and also increases hydraulic resistance to fluid flow. Also, the thermal conductivity of fouling deposits is usually lower

---

[†] There are other types of fouling phenomena in nature (e.g., clogging of arteries) that are not of importance in heat exchanger design.

863

than that for the metals used for heat transfer surfaces. Fouling is an extremely complex phenomenon characterized by combined heat, mass, and momentum transfer under transient conditions. Liquid-side fouling occurs on the exchanger side where liquid is being heated, and gas-side fouling occurs on the gas cooling side; however, reverse examples can be found.

Fouling is very costly since it (1) increases capital costs due to the need to oversurface the heat exchanger and for cleaning; (2) increases maintenance costs resulting from cleaning, chemical additives, or troubleshooting; (3) results in loss of production due to shutdown or reduced capacity; and (4) increases energy losses due to reduced heat transfer, increased pressure drop, and dumping of dirty streams present. Gas-side fouling can also be a potential fire hazard in a fossil-fired exhaust environment, resulting in catastrophic lost production and repair costs. In some applications, increased pressure drop due to fouling may reduce gas flows affecting adversely heat transfer and increasing solvent concentration (such as during waste heat recovery from paint oven exhausts) which is not acceptable environmentally.

Fouling significantly reduces heat transfer with a relatively small increase in fluid pumping power in systems with liquid flows and high heat transfer coefficients. For systems having low heat transfer coefficients, such as with gases, fouling increases the fluid pumping power significantly with some reduction in heat transfer. Note that plugging will also increase pressure drop substantially but doesn't coat the surface and still may be considered as fouling in an application.

Let us first discuss only qualitatively the influence of a deposit on a heat transfer surface. We consider either fully developed laminar or turbulent flows. Using the results/correlations for laminar flow (Nu = constant; see Table 7.3) and turbulent flow [the Dittus–Boelter correlation, Eq. (7.79) in Table 7.6], we express the heat transfer coefficient as follows:

$$
h = \begin{cases}
\dfrac{\mathrm{Nu} \cdot k}{D_h} \quad \text{with } \mathrm{Nu} = \text{constant} & \text{for laminar flow} \\[4mm]
\dfrac{k}{D_h}\left[0.023\left(\dfrac{4\dot{m}}{\mathbf{P}\mu}\right)^{0.8} \cdot \mathrm{Pr}^{0.4}\right] & \text{for turbulent flow}
\end{cases}
\tag{13.1}
$$

Note that in the turbulent flow expression of Eq. (13.1), we substituted $\mathrm{Re} = \dot{m}D_h/A_o\mu = 4\dot{m}/\mathbf{P}\mu$ using the definition of D_h. Here $\mathbf{P}$ is the wetted perimeter of all flow passages in the exchanger. In general, we treat $\mathbf{P}$ as independent of D_h. For example, having a double number of 5 mm diameter tubes compared to a specified number of 10 mm diameter tubes in an exchanger will have the same total $\mathbf{P}$ but different D_h. A similar situation can exist for an extended surface. Using Eqs. (6.29) and (6.67b), we express the following pressure drop relationships:

$$
\Delta p = f\,\frac{4L}{D_h}\frac{G^2}{2g_c\rho} = \begin{cases}
\dfrac{1}{D_h^3}\left[\dfrac{1}{2g_c}\dfrac{\mu}{\rho}\dfrac{16L}{\mathbf{P}}\,\dot{m}(f\cdot\mathrm{Re})\right] & \text{for laminar flow} \\[4mm]
\dfrac{1}{D_h^3}\left[\dfrac{0.046\times 4L}{2g_c}\dfrac{\mu^{0.2}}{\rho}\left(\dfrac{4\dot{m}}{\mathbf{P}}\right)^{1.8}\right] & \text{for turbulent flow}
\end{cases}
\tag{13.2}
$$

In Eq. (13.2), $f\cdot\mathrm{Re}$ is approximately constant for fully developed laminar flow (a theoretical value for a circular tube is $f\cdot\mathrm{Re} = 16$), while the following relationship is applied for turbulent flow: $f = 0.046Re^{-0.2}$. The most important physical outcome of fouling is the flow cross section getting plugged and resulting in a reduced hydraulic diameter of flow passages. Therefore, for a given mass flow rate $\dot{m}$, fluid flow length L, heat transfer area $A (= \mathbf{P}L)$, and known fluid properties, one gets from Eqs. (13.1) and (13.2),

$$h \propto \frac{1}{D_h} \qquad \Delta p \propto \frac{1}{D_h^3} \tag{13.3}$$

The functional relationships given by Eq. (13.3) are obtained assuming total wetted perimeter as constant regardless of the change in the hydraulic diameter. In practice, when D_h for an exchanger changes, A may change as well as for a circular tube. In that case, since $\mathbf{P} = \pi d_i N_t = \pi D_h N_t$ for a tubular exchanger (N_t = total number of tubes), $\Delta p \propto 1/D_h^4$ from Eq. (13.2) for laminar flow and $\Delta p \propto 1/D_h^{4.8}$ for turbulent flow instead of $1/D_h^3$ of Eq. (13.3) for a tubular exchanger. Alternatively, the Δp expression for turbulent flow in a circular tube can be expressed as follows using Eq. (6.29) with the definitions $G = \dot{m}/A_o$, $A_o = (\pi/4)d_i^2$ and $D_h = d_i$:

$$\Delta p = f \frac{4L}{d_i} \frac{G^2}{2g_c \rho} = f \frac{4L}{d_i} \frac{\dot{m}^2}{(\pi/4)^2 d_i^4} \frac{1}{2g_c \rho} = \frac{32L\dot{m}^2}{\pi^2 g_c \rho} \frac{f}{d_i^5} \tag{13.4}$$

Substitution of $f = 0.046Re^{-0.2}$ will change the exponent of $\dot{m}$ in Eq. (13.4) to 1.8 and the exponent of d_i to 4.8. Also, the surface roughness change due to fouling on the f factors should be included as an additional effect (generally, we neglect the effect of surface roughness on the heat transfer coefficient for conservatism). Actual influences of fouling on the heat transfer coefficient and pressure drop are substantially more complex than those presented by Eqs. (13.3) and (13.4), due to inherently transient nature of fouling processes.

The pressure drop ratio $\Delta p_f/\Delta p_c$ of a fouled and a clean exchanger for a constant mass flow rate, from Eq. (13.4), is given by

$$\frac{\Delta p_f}{\Delta p_c} = \frac{f_f}{f_c} \left(\frac{D_{h,c}}{D_{h,f}}\right)^5 \tag{13.5}$$

If we consider that fouling does not affect the friction factor (i.e., $f_c \approx f_f$) and also consider that reduction in the tube inside diameter due to fouling is only 10 to 20%, the resulting pressure drop increase will be approximately 69% and 205%, respectively, according to Eq. (13.5), regardless of whether the fluid is liquid or gas [note that in contrast, $h \propto 1/D_h$, as shown in Eq. (13.1) or (13.3)]. This increased Δp can be translated into increased fluid pumping power using Eq. (6.1); and for liquids, the density being significantly higher than that for gases, a substantially higher Δp due to fouling can be allowed for liquids for a reasonable increase in liquid pumping power. Also, the equipment cost of fluid pumping power is lower for liquids than for gases for a given amount of pumping power.[†]

Now let us review the impact of fouling on exchanger heat transfer. As fouling will reduce the free-flow area and hence the passage D_h, it will increase the convection heat

---

[†] For example, for a midsized automobile, the cost of a 300-W fan for the radiator airflow was $35 to 40, compared to $20 to 25 for an equivalent power radiator coolant water pump in 2001.

transfer coefficient h [of Eq. (13.1)] between the fluid and heat transfer surface (which may be covered with a fouling layer), for two reasons: increased flow velocity with a reduction in the flow area, and increased surface roughness due to the fouling layer. Both these effects would increase the pressure drop substantially. Fouling layers (deposits) on one or both fluid sides increase thermal resistance to heat flow from the hot fluid to cold fluid by conduction through the fouling layers (see Fig. 3.4), which also have lower thermal conductivity. The added thermal resistances in general reduce the exchanger overall UA substantially compared to the increase in h due to fouling, as mentioned above.[†] Due to a large uncertainty, transient nature, variations in the fouling resistance ($\hat{\mathbf{R}}_f = 1/h_f$), and no accurate means of its measurement, the increase in h due to fouling is ignored or lumped into the reported values of fouling resistances. Hence, the heat transfer coefficients h_h and h_c for hot and cold fluids are determined for unfouled clean surfaces for the UA computation of fouled surfaces. From the overall thermal resistance Eq. (3.20) or (3.24), we find that fouling deposits will reduce UA and hence q more significantly in liquids than in gases. This is because liquids have h an order of magnitude higher than that for gases in general. To understand this, consider a process plant heat exchanger with clean $U = 1500 \, \text{W/m}^2 \cdot \text{K}$ or the overall unit thermal resistance $\hat{\mathbf{R}}_o = 6 \times 10^{-4} \, \text{m}^2 \cdot \text{K/W}$. If the fouling resistances $\hat{\mathbf{R}}_{f,h} + \hat{\mathbf{R}}_{f,c} = 3 \times 10^{-4}$ [a reasonable value from TEMA (1999)] are considered, 50% extra heat transfer area is chargeable to fouling since $\hat{\mathbf{R}}_{o,\text{new}} = (6 + 3) \times 10^{-4} \, \text{m}^2 \cdot \text{K/W}$ and $q = A \, \Delta T_m / \hat{\mathbf{R}}_{o,\text{new}}$. In contrast, for a gas-to-gas clean compact heat exchanger, consider $U_c = 300 \, \text{W/m}^2 \cdot \text{K}$ or $\hat{\mathbf{R}}_o = 3 \times 10^{-3} \, \text{m}^2 \cdot \text{K/W}$. For the same fouling resistances, $\hat{\mathbf{R}}_{f,h} + \hat{\mathbf{R}}_{f,c} = 3 \times 10^{-4} \, \text{m}^2 \cdot \text{K/W}$, the heat transfer surface area chargeable to fouling is only 10%.

Based on the foregoing discussion, fouling in liquids has a significant detrimental effect on heat transfer with some increase in fluid pumping power. In contrast, fouling in gases reduces heat transfer somewhat (5 to 10% in general) but increases pressure drop and fluid pumping power significantly (up to several hundred percent) from the cost point of view.

It should be emphasized that the same magnitude of a fouling factor (or fouling unit thermal resistance)[‡] can have a different impact on performance for the same or different applications. For example, the same fouling factor may represent heavy fouling in a clean service (such as a closed-loop refrigerant system) or low fouling in a dirty service (such as a refinery crude preheat train). As another example, the same fouling factor in two different plants may have radically different fouling rates because of different feedstocks, preprocessing, or equipment design.

13.2 PHENOMENOLOGICAL CONSIDERATIONS OF FOULING

As noted in Section 13.1, fouling is an extremely complex phenomenon characterized by a combined heat, mass, and momentum transfer under transient conditions. Fouling is affected by a large number of variables related to heat exchanger surface, operating conditions, and involved fluid streams. Despite the complexity of the fouling process, a general practice is to include the effect of fouling on the exchanger thermal performance

---

[†] For example, see the added thermal resistance terms $(1/\eta_o h_f A)$ for the hot and cold fluid sides in Eq. (3.20) or (3.24), which may reduce $1/UA$ more than the increase in h_h and/or h_c due to fouling, depending on their relative magnitudes in the equation.

[‡] The concept of fouling resistance introduced in Section 3.2.4 is explained further in Section 13.2.6.

by adding thermal resistances of fouling layers in the thermal circuit using empirical data, as explained through Fig. 3.4 and discussed further in Section 13.3. The problem, though, is that this simplified modeling approach does not (and cannot) reflect a real transient nature of the fouling process. The current practice is to use fouling factors or fouling unit thermal resistances from TEMA Standards (1999) (see Section 13.3 and Table 9.4 for tubular and shell-and-tube heat exchangers). However, probably a better approach would be to perform cost analysis for cleaning frequency by taking into account any initial overdesign (by including fouling resistances). This overdesign may provide added heat transfer performance initially due to larger surface area and flow area than required for a clean exchanger but will reduce the flow velocity and hence may accelerate initial fouling in some applications. Let us now consider in detail different types of fouling mechanisms, sequential events in fouling, and modeling of a fouling process as an example.

13.2.1 Fouling Mechanisms

There are six types of liquid-side fouling mechanisms: (1) *precipitation or crystallization fouling*, (2) *particulate fouling*, (3) *chemical reaction fouling*, (4) *corrosion fouling*, (5) *biological fouling*, and (6) *freezing (solidification) fouling*. Only biological fouling does not occur in gas-side fouling since there are in principle no nutrients in the gas flows. In reality, more than one fouling mechanisms is present in many applications and their synergistic effect makes the fouling even worse than predicted/expected with a single fouling mechanism present. Note that there are additional examples of fouling that may not fall in the foregoing categories, such as accumulation of noncondensables in a condenser. In addition, plugging will also increase pressure drop substantially, but doesn't coat the surface and still may be considered as fouling in applications. Refer to Melo et al. (1988) and Bott (1990) for a detailed study of fouling.

In *precipitation* or *crystallization fouling*, the dominant mechanism is the precipitation of dissolved salts in the fluid on the heat transfer surface when the surface concentration exceeds the solubility limit. Thus, a necessary prerequisite for an onset of precipitation is the presence of supersaturation. Precipitation of salts can occur within the process fluid, in the thermal boundary layer, or at the fluid–surface (fouling–film) interface. It generally occurs with aqueous solutions and other liquids of soluble salts which are either being heated or cooled. When the solution contains normal solubility salts (the salt solubility and concentration decrease with decreasing temperature such as wax deposits, gas hydrates and freezing of water/water vapor), the precipitation fouling occurs on the cold surface (i.e., by cooling the solution). For inverse solubility salts (such as calcium and magnesium salts), the precipitation of salt occurs with heating the solution. Precipitation/crystallization fouling is common when untreated water, seawater, geothermal water, brine, aqueous solutions of caustic soda, and other salts are used in heat exchangers. This fouling is characterized by deposition of divalent salts in cooling water systems. Crystallization fouling may occur with some gas flows that contain small quantities of organic compounds that would form crystals on the cold surface. If the deposited layer is hard and tenacious (as often found with inverse solubility salts such as cooling water containing hardness salts), it is often referred to as *scaling*. If it is porous and mushy, it is called *sludge*, *softscale*, or *powdery deposit*. The most important phenomena involved with precipitation or crystallization fouling include the following. Crystal growth during precipitation require formation of a primary nucleus. The mechanism controlling that process is nucleation, as a rule heterogeneous in the presence of impurities and on the heat transfer surface. Transfer of particulate solids to the fouled surface is accomplished by diffusion. Simultaneously with deposition, removal phenomena caused by shear stress are always present. Deposit mechanical integrity changes over time either by strengthening or by weakening it due to crystalization/recrystalization, temperature change, and so on. All these phenomena are controlled by numerous factors, the most dominant being local temperature and temperature gradient levels, composition of the fluid including concentration of soluble species.

Particulate fouling refers to the deposition of solids suspended in a fluid onto a heat transfer surface. If the settling occurs due to gravity, the resulting particulate fouling is called *sedimentation fouling*. Hence, particulate fouling may be defined as the accumulation of particles from heat exchanger working fluids (liquids and/or gaseous suspensions) on the heat transfer surface. Most often, this type of fouling involves deposition of corrosion products dispersed in fluids, clay and mineral particles in river water, suspended solids in cooling water, soot particles of incomplete combustion, magnetic particles in economizers, deposition of salts in desalination systems, deposition of dust particles in air coolers, particulates partially present in fire-side (gas-side) fouling of boilers, and so on. The particulate fouling caused by deposition of, for example, corrosion products is influenced by the following factors: metal corrosion process factors (at heat transfer surface), release and deposition of the corrosion products on the surface[†]; concentration of suspended particles, temperature conditions on the fouled surface (heated or nonheated), and heat flux at the heat transfer surface.

Chemical reaction fouling is referred to as the deposition of material (fouling precursors) produced by chemical reactions within the process fluid, in the thermal boundary layer, or at the fluid–surface (fouling–film) interface in which the heat transfer surface material *is not* a reactant or participant. However, the heat transfer surface may act as a catalyst as in cracking, coking, polymerization, and autoxidation. Thermal instabilities of chemical species, such as asphaltenes and proteins, can also induce fouling precursors. Usually, this fouling occurs at local hot spots in a heat exchanger, although the deposits are formed all over the heat transfer surface in crude oil units and dairy plants. It can occur over a wide temperature range from ambient to over $1000°C$ ($1800°F$) but is more pronounced at higher temperatures. Foulant deposits are usually organic compounds, but inorganic materials may be needed to promote the chemical reaction. This fouling mechanism is a consequence of an unwanted chemical reaction that takes place during the heat transfer process. Examples of chemical fouling include deposition of coke in petrochemical industries in cracking furnaces where thermal cracking of hydrocarbons is realized. This fouling mechanism is found in many applications of process industry, such as oil refining, vapor-phase pyrolysis, cooling of gas and oils, polymerization of process monomers, and so on. Furthermore, fouling of heat transfer surface by biological fluids may involve complex heterogeneous chemical reactions and physicochemical processes. The deposits from chemical reaction fouling may promote corrosion at the surface if the formation of the protective oxide layer is inhibited. All fouling deposits may promote corrosion.

In *corrosion fouling (in situ)*, the heat transfer surface itself reacts with the process fluid or chemicals present in the process fluid. Its constituents or trace materials are carried by the fluid in the exchanger, and it produces corrosion products that deposit on the surface. Hence, corrosion fouling could be considered as chemical reaction fouling

---

[†] It should be borne in mind that corrosion products may be soluble in a working fluid, and hence both precipitation and particulate fouling would usually occur concurrently.

in which heat transfer fouling affects the exchanger mechanical integrity, and the corrosion products add thermal resistance to heat flow from the hot fluid to the cold fluid. If corrosion products are formed upstream of the exchanger and then deposited on the heat transfer surface, the fouling mechanism refers to particulate or precipitation fouling, depending on whether the corrosion products are insoluble or soluble at the bulk fluid conditions. The interaction of corrosion and other types of fouling is the major concern for many industrial applications. Corrosion fouling is dependent on the selection of exchanger surface material and can be avoided with the right choice of materials (such as expensive alloys) if the high cost is warranted. Corrosion fouling is prevalent in many applications where chemical reaction fouling takes place and the protective oxide layer is not formed on the surface. Corrosion fouling is of significant importance in the design of the boiler and condenser of a fossil fuel–fired power plant. The important factors for corrosion fouling are the chemical properties of the fluids and heat transfer surface, oxidizing potential and alkalinity, local temperature and heat flux magnitude, and mass flow rate of the working fluid. It should be noted that although growth of corrosion influenced deposit has a detrimental effect on heat transfer, this influence is less important than fouling caused by particulate fouling of corrosion products formed elsewhere within the system. For example, fouling on the water side of boilers may be caused by corrosion products that originate in the condenser or feedtrain.

Biological fouling or *biofouling* results from the deposition, attachment, and growth of macro- or microorganisms to the heat transfer surface; it is generally a problem in water streams. In general, biological fouling can be divided into two main subtypes of fouling: microbial and macrobial. *Microbial fouling* is accumulation of microorganisms such as algae, fungi, yeasts, bacteria, and molds, and *macrobial fouling* represents accumulation of macroorganisms such as clams, barnacles, mussels, and vegetation as found in seawater or estuarine cooling water. Microbial fouling precedes macrobial deposition as a rule and may be considered of primary interest. Biological fouling is generally in the form of a biofilm or a slime layer on the surface that is uneven, filamentous, and deformable but difficult to remove. Although biological fouling could occur in suitable liquid streams, it is generally associated with open recirculation or once-through systems with cooling water. Since this fouling is associated with living organisms, they can exist primarily in the temperature range 0 to 90°C (32 to 194°F) and thrive in the temperature range 20 to 50°C (68 to 122°F). Biological fouling may promote corrosion fouling under the slime layer. Transport of microbial nutrients, inorganic salts, and viable microorganisms from the bulk fluid to the heat transfer surface is accomplished through molecular diffusion or turbulent eddy transport, including organic adsorption at the surface.

Freezing or *solidification fouling* is due to freezing of a liquid or some of its constituents, or deposition of solids on a subcooled heat transfer surface as a consequence of liquid–solid or vapor–solid phase change in a gas stream. Formation of ice on a heat transfer surface during chilled water production or cooling of moist air, deposits formed in phenol coolers, and deposits formed during cooling of mixtures of substances such as paraffin are some examples of solidification fouling (Bott, 1981). This fouling mechanism occurs at low temperatures, usually ambient and below depending on local pressure conditions. The main factors affecting solidification fouling are mass flow rate of the working fluid, temperature and crystallization conditions, surface conditions, and concentration of the solid precursor in the fluid.

Combined fouling occurs in many applications, where more than one fouling mechanism is present and the fouling problem becomes very complex with their synergistic effects. Some combined fouling mechanisms found in industrial applications are (Panchal, 1999):

- Particulate fouling combined with biofouling, crystallization, and chemical-reaction fouling
- Crystallization fouling combined with chemical-reaction fouling
- Condensation of organic/inorganic vapors combined with particulate fouling in gas streams
- Crystallization fouling of mixed salts
- Combined fouling by asphaltene precipitation, pyrolysis, polymerization, and/or inorganic deposition in crude oil
- Corrosion fouling combined with biofouling, crystallization, or chemical-reaction fouling

Some examples of the interactive effects of corrosion and fouling are as follows (Panchal, 1999):

- Microfouling-induced corrosion (MIC) (sustained-pitting corrosion)
- Under-deposit corrosion in petroleum and black liquor processing (concentration buildup of corrosion-causing elements)
- Simultaneous corrosion and biofouling in cooling water applications
- Fouling induced by corrosion products

It is obvious that one cannot talk about a single, unified theory to model the fouling process wherein not only the foregoing six types of fouling mechanisms are identified, but in many processes more than one fouling mechanism exists with synergistic effects. However, it is possible to extract a few variables that would most probably control any fouling process: (1) fluid velocity, (2) fluid and heat transfer surface temperatures and temperature differences, (3) physical and chemical properties of the fluid, (4) heat transfer surface properties, and (5) geometry of the fluid flow passage. The other important variables are concentration of foulant or precursor, impurities, heat transfer surface roughness, surface chemistry, fluid chemistry (pH level, oxygen concentration, etc.), pressure, and so on. For a given fluid–surface combination, the two most important design variables are the fluid velocity and heat transfer surface temperature. In general, higher flow velocities may cause less foulant deposition and/or more pronounced deposit erosion, but at the same time may accelerate corrosion of the surface by removing the heat transfer surface material. Higher surface temperatures promote chemical reaction, corrosion, crystal formation (with inverse solubility salts), and polymerization, but reduce biofouling and prevent freezing and precipitation of normal solubility salts. Consequently, it is frequently recommended that the surface temperature be maintained low.

13.2.2 Single-Phase Liquid-Side Fouling

Single-phase liquid-side fouling is most frequently caused by (1) precipitation of minerals from the flowing liquid, (2) deposition of various particles, (3) biological fouling, and (4) corrosion fouling. Other fouling mechanisms are also present. More important, though,

TABLE 13.1   Influence of Operating Variables on Liquid-Side Fouling[a]

Operating Variable	Precipitation	Freezing	Particulate	Chemical	Corrosion	Biological
Temperature	↑↓	↓	↑↓↔	↑↓	↑↓	↑↓↔
Velocity	↓↔	↑↓	↓	↓	↑↓↔	↑↓
Supersaturation	↑	↑	—	—	—	—
pH	↑	—	↑↓	—	↑↓	↑↓
Impurities	—	↓	—	—	—	—
Concentration	↑	↑	↑	—	—	—
Roughness	↑	↑	↑↔	—	↑↔	↑
Pressure	↔	↔	—	↑	↑	↑↓
Oxygen	↔	↔	—	↑	↑	↑↓

Source: Data from Cannas (1986).
[a] When the value of an operating variable is increased, it increases (↑), decreases (↓), or has no effect (↔) on the specific fouling mechanism listed. Dashes — indicate that no influence of these variables has been reported in the literature.

is the combined effect of more than one fouling mechanism present. The qualitative effects of some of the operating variables on these fouling mechanisms are shown in Table 13.1.

The quantitative effect of fouling on heat transfer can be estimated by utilizing the concept of fouling resistance and calculating the overall heat transfer coefficient under both fouling and clean conditions (see Section 13.3). An additional parameter for determining this influence, used frequently in practice, is the *cleanliness factor*. It is defined as a ratio of an overall heat transfer coefficient determined for fouling conditions to that determined for clean (fouling-free) operating conditions. The effect of fouling on the pressure drop can be determined by the reduced free-flow area due to fouling and the change in the friction factor, if any, due to fouling.

13.2.3   Single-Phase Gas-Side Fouling

Gas-side fouling may be caused by precipitation (scaling), particulate deposition, corrosion, chemical reaction, and freezing. Formation of hard scale from the gas flow occurs if a sufficiently low temperature of the heat transfer surface forces salt compounds to solidification. Acid vapors, high-temperature removal of an oxide layer by molten ash, or salty air at low temperatures may promote corrosion fouling. An example of particulate deposition is accumulation of plant residues. An excess of various chemical substances, such as sulfur, vanadium, and sodium, initiates various chemical reaction fouling problems. Formation of frost and various cryo-deposits are typical examples of freezing fouling on the gas side. An excellent overview of gas-side fouling of heat transfer surfaces is given by Marner (1990, 1996). Qualitative effects of some of the operating variables on gas-side fouling mechanisms are presented in Table 13.2.

13.2.4   Fouling in Compact Exchangers

Small channels associated with compact heat exchangers have very high shear rates, perhaps three to four times higher in a plate heat exchanger than in a shell-and-tube

TABLE 13.2   Influence of Operating Variables on Gas-Side Fouling[a]

Operating Variable	Particulate	Freezing	Chemical	Corrosion
Temperature	↑↓	↓	↑	↑↓↔
Velocity	↑↓↔	↓	↑↓↔	↑↔
Impurities	—	↓	—	—
Concentration	↑	↑	—	↑
Fuel-air ratio	↑	—	↑	—
Roughness	↑↔	—	—	↑↔
Oxygen	↔	↔	↑	—
Sulfur	—	—	↑	↑

Source: Data from Cannas (1986).
[a] When the value of an operating variable is increased, it increases (↑), decreases (↓), or has no effect (↔) on the specific fouling mechanism listed. Dashes — indicate that no influence of these variables has been reported in the literature.

exchanger. This reduces fouling significantly. However, small channel size creates a problem of plugging the passages. To avoid plugging, the particle size must be restricted by filtering or other means to less than one-third the smallest opening of heat exchanger passages. Even with this guideline, particulate fouling can occur and agglomerate, such as with waxy substances.

13.2.5   Sequential Events in Fouling

From the empirical evidence involving various fouling mechanisms discussed in Section 13.2.1, it is clear that virtually all these mechanisms are characterized by a similar sequence of events. The successive events occurring in most cases are the following: (1) initiation, (2) transport, (3) attachment, (4) removal, and (5) aging, as conceptualized by Epstein (1978). These events govern the overall fouling process and determine its ultimate impact on heat exchanger performance. In some cases, certain events dominate the fouling process, and they have a direct effect on the type of fouling to be sustained. Let us summarize these events briefly (Cannas, 1986).

Initiation of the fouling, the first event in the fouling process, is preceded by a *delay period* or *induction period* τ_d as shown in Fig. 13.1. The basic mechanism involved during this period is heterogeneous nucleation, and τ_d is shorter with a higher nucleation rate. The factors affecting τ_d are temperature, fluid velocity, composition of the fouling stream, and nature and condition of the heat exchanger surface. Low-energy surfaces (unwettable) exhibit longer induction periods than those of high-energy surfaces (wettable). In crystallization fouling, τ_d tends to decrease with increasing degree of supersaturation. In chemical reaction fouling, τ_d appears to decrease with increasing surface temperature. In all fouling mechanisms, τ_d decreases as the surface roughness increases due to available suitable sites for nucleation, adsorption, and adhesion.

Transport of species means transfer of a key component (such as oxygen), a crucial reactant, or the fouling species itself from the bulk of the fluid to the heat transfer surface. Transport of species is the best understood of all sequential events. Transport of species takes place through the action of one or more of the following mechanisms:

- *Diffusion:* involves mass transfer of the fouling constituents from the flowing fluid toward the heat transfer surface due to the concentration difference between the bulk of the fluid and the fluid adjacent to the surface.

- *Electrophoresis:* under the action of electric forces, fouling particles carrying an electric charge may move toward or away from a charged surface depending on the polarity of the surface and the particles. Deposition due to electrophoresis increases with decreasing electrical conductivity of the fluid, increasing fluid temperature, and increasing fluid velocity. It also depends on the pH of the solution. Surface forces such as London–van der Waals and electric double layer interaction forces are usually responsible for electrophoretic effects.

- *Thermophoresis:* a phenomenon whereby a "thermal force" moves fine particles in the direction of negative temperature gradient, from a hot zone to a cold zone. Thus, a high-temperature gradient near a hot wall will prevent particles from depositing, but the same absolute value of the gradient near a cold wall will promote particle deposition. The thermophoretic effect is larger for gases than for liquids.

- *Diffusiophoresis:* involves condensation of gaseous streams onto a surface.

- *Sedimentation:* involves the deposition of particulate matters such as rust particles, clay, and dust on the surface due to the action of gravity. For sedimentation to occur, the downward gravitational force must be greater than the upward drag force. Sedimentation is important for large particles and low fluid velocities. It is frequently observed in cooling tower waters and other industrial processes where rust and dust particles may act as catalysts and/or enter complex reactions.

- *Inertial impaction:* a phenomenon whereby "large" particles can have sufficient inertia that they are unable to follow fluid streamlines and as a result, deposit on the surface.

- *Turbulent downsweeps:* since the viscous sublayer in a turbulent boundary layer is not truly steady, the fluid is being transported toward the surface by turbulent downsweeps. These may be thought of as suction areas of measurable strength distributed randomly all over the surface.

Attachment of the fouling species to the surface involves both physical and chemical processes, and it is not well understood. Three interrelated factors play a crucial role in the attachment process: surface conditions, surface forces, and sticking probability. It is the combined and simultaneous action of these factors that largely accounts for the event of attachment.

- The properties of surface conditions important for attachment are the surface free energy, wettability (contact angle, spreadability), and heat of immersion. Wettability and heat of immersion increase as the difference between the surface free energy of the wall and the adjacent fluid layer increases. Unwettable or low-energy surfaces have longer induction periods than wettable or high-energy surfaces, and suffer less from deposition (such as polymer and ceramic coatings). Surface roughness increases the effective contact area of a surface and provides suitable sites for nucleation and promotes initiation of fouling. Hence, roughness increases the wettability of wettable surfaces and decreases the unwettability of the unwettable ones.

- There are several surface forces. The most important one is the London–van der Waals force, which describes the intermolecular attraction between nonpolar molecules and is always attractive. The electric double layer interaction force can be attractive or repulsive. Viscous hydrodynamic force influences the attachment of a particle moving to the wall, which increases as it moves normal to the plain surface.

- Sticking probability represents the fraction of particles that reach the wall and stay there before any reentrainment occurs. It is a useful statistical concept devised to analyze and explain the complicated event of attachment.

Removal of the fouling deposits from the surface may or may not occur simultaneously with deposition. Removal occurs due to the single or simultaneous action of the following mechanisms: shear forces, turbulent bursts, re-solution, and erosion.

- Shear forces result from the action of the shear stress exerted by the flowing fluid on the depositing layer. As the fouling deposit builds up, the cross-sectional area for flow decreases, thus causing an increase in the average velocity of the fluid for a constant mass flow rate and increasing the shear stress. Fresh deposits will form only if the deposit bond resistance is greater than the prevailing shear forces at the solid–fluid interface.

- Randomly distributed (about less than 0.5% at any instant of time) periodic turbulent bursts act as miniature tornadoes lifting deposited material from the surface. By continuity, these fluid bursts are compensated for by gentler fluid back sweeps, which promote deposition.

- The removal of the deposits by re-solution is related directly to the solubility of the material deposited. Since the fouling deposit is presumably insoluble at the time of its formation, dissolution will occur only if there is a change in the properties of the deposit, or in the flowing fluid, or in both, due to local changes in temperature, velocity, alkalinity, and other operational variables. For example, sufficiently high or low temperatures could kill a biological deposit, thus weakening its attachment to a surface and causing sloughing or re-solution. The removal of corrosion deposits in power-generating systems is done by re-solution at low alkalinity. Re-solution is associated with the removal of material in ionic or molecular form.

- Erosion is closely identified with the overall removal process. It is highly dependent on the shear strength of the foulant and on the steepness and length of the sloping heat exchanger surfaces, if any. Erosion is associated with the removal of material in particulate form. The removal mechanism becomes largely ineffective if the fouling layer is composed of well-crystallized pure material (strong formations); but it is very effective if it is composed of a large variety of salts each having different crystal properties.

Aging of deposits begins with attachment on the heat transfer surface, and refers to any changes the fouling material undergoes as time elapses. The aging process includes both physical and chemical transformations, such as further degradation to a more carbonaceous material in organic fouling, and dehydration and/or crystal phase transformations in inorganic fouling. A direct consequence of aging is change in the thermal conductivity of the deposits.[†] Aging may strengthen or weaken the fouling deposits.

[†] A common nonfouling example of aging is the transformation of fresh, soft, fluffy snow in an open field into hard, crystalline, yellowish ice after a week or so of exposure to the sun resulting differences in its material properties.

13.2.6   Modeling of a Fouling Process

Regardless of the type of fouling process, the principal characteristic feature of any type of fouling is that the net mass fouling rate (i.e., the change of the mass m of foulant deposited on the heat transfer surface for a given time, $dm/d\tau$, is a consequence of a net difference between the foulant deposit rate $\dot{m}_d$ and the foulant reentrainment rate $\dot{m}_r$:

$$\frac{\partial m(\mathbf{s}, \tau)}{\partial \tau} = \dot{m}_d(\mathbf{s}, \tau) - \dot{m}_r(\mathbf{s}, \tau) \tag{13.6}$$

In Eq. (13.6), $\mathbf{s}$ denotes symbolically the spatial dependence (say, x, y, and z) of the mass of foulant. Note that the mass m of the foulant deposited uniformly is given as a simple equation:

$$m = \rho_f A \delta_f \tag{13.7}$$

where ρ_f represents foulant mass density, A denotes heat transfer surface area covered with the foulant, and δ_f is the thickness of the foulant layer. In general, all three terms of Eq. (13.6) are spatially nonuniform and dependent on time. Equation (13.6) can conveniently be reformulated in terms of mass per unit heat transfer surface area, $M_A = m/A$, and for a uniform spatial distribution of deposit, it is

$$\frac{dM_A}{d\tau} = \dot{M}_{A,d} - \dot{M}_{A,r} \tag{13.8}$$

Equation (13.8) is a direct consequence of Eq. (13.6) after idealizing a uniform distribution of the fouling deposit over the surface A. Furthermore, mass per unit heat transfer surface (uniformly distributed along the heat transfer surface) can be written as

$$M_A = \rho_f \delta_f = \rho_f k_f \hat{\mathbf{R}}_f \tag{13.9}$$

where $\hat{\mathbf{R}}_f = \delta_f / k_f$, the fouling factor, represents fouling unit thermal resistance; it represents the thermal resistance of the layer of foulant deposited for a unit area of heat transfer surface. Concisely, we refer to this entity as *fouling resistance*. From the fouling factor definition, we obtain $\delta_f = k_f \hat{\mathbf{R}}_f$. Consequently,

$$\frac{dM_A}{d\tau} = \rho_f \frac{d\delta_f}{d\tau} = \rho_f k_f \frac{d\hat{\mathbf{R}}_f}{d\tau} \tag{13.10}$$

In Eq. (13.10), it is assumed that both mass density and thermal conductivity of the deposited layer are invariant with time. Combining Eqs. (13.8) and (13.10), we obtain

$$\frac{d\hat{\mathbf{R}}_f}{d\tau} = \hat{\dot{\mathbf{R}}}_d - \hat{\dot{\mathbf{R}}}_r \tag{13.11}$$

where $\hat{\dot{\mathbf{R}}}_j = \dot{M}_{A,j}/\rho_f k_f$ represents deposition ($j = d$) and removal ($j = r$) fouling resistance *rates*.

To solve either Eq. (13.8) or (13.11), one needs the explicit forms of either mass rates per unit heat transfer area (for both deposit or reentrainment process) or unit thermal resistances [the terms on the right-hand sides of Eqs. (13.8) and (13.11)]. A number of models for determining these variables have been developed; some of them are summarized in Table 13.3. Let us consider the model of Taborek et al. enlisted in that table as an illustration.

According to Taborek et al. [as reported by Epstein (1978)], the deposition and removal mass rates have the form

$$\dot{M}_{A,d} = c_1 \wp_1 \Omega^n \exp\left(-\frac{E}{\Re T_s}\right) \qquad \dot{M}_{A,r} = b_1 \tau_s \frac{m^i}{\psi} \tag{13.12}$$

where c_1 and b_1 are constants, $\wp_1$ is a deposition probability function related to the velocity and adhesion properties of the deposit, Ω^n is the water quality factor, E is the activation energy, $\Re$ is the universal gas constant, T_s is the absolute temperature of the deposit at the surface, τ_s denotes fluid shear stress at the deposit surface, ψ represents *removal resistance* of the deposit (scale strength factor), m is the foulant mass, and i is an exponent. Equation (13.12) can be represented in terms of deposition and removal thermal resistance rates of Eq. (13.11) in form as follows [as reported by Knudsen (1998)]:

$$\hat{\mathbf{R}}_d = c_2 \wp_1 \Omega^n \exp\left(-\frac{E}{\Re T_s}\right) \qquad \hat{\mathbf{R}}_r = b_2 \tau_s \frac{\delta_f}{\psi} = b_2 \tau_s \frac{k_f}{\psi} \hat{\mathbf{R}}_f \tag{13.13}$$

In Eq. (13.13), c_2 and b_2 are constants. It should be noted that both sets of equations [Eqs. (13.12) and (13.13)], are semiempirical, to include the variables that govern fouling. Introducing the expressions for mass per unit heat transfer surface area for both deposit and reentrainment processes (or their thermal resistance) into Eq. (13.8) or (13.11), we could integrate these governing equations and subsequently determine either deposited mass or their thermal resistance. These solutions have to fit empirical evidence that can be generalized as presented in Fig. 13.1.

In Fig. 13.1, four characteristic scenarios for the growth of the fouling resistance are presented (Knudsen, 1998). In this figure, τ_d is the delay period for the onset of fouling deposits for non-negligible $\hat{\mathbf{R}}_f$.

1. Linear characteristics (i.e., $\hat{\mathbf{R}}_f$ is linearly dependent on time) indicate that the deposition rate is constant and there is no reentrainment rate (or at least their difference is invariant in time). A linear fouling behavior is generally associated with the crystallization of a well-formed deposit consisting of a substantially pure salt that is largely uncontaminated by the presence of coprecipitated impurities. The strong bonds characterizing the structure of such deposits make removal mechanisms somewhat ineffective. If heat duty is kept constant linear fouling behavior is also often observed for reaction fouling.

2. Falling rate fouling normally occurs in situations where the deposition rate is decreasing but always greater than the removal rate. This type of fouling mechanism has been observed in crystallization fouling in a plate exchanger and also in particulate fouling.

3. The curve characterized by asymptotic behavior reflects the situation represented by the expression for $\hat{\mathbf{R}}_d$ of Eq. (13.13), which corresponds to fragile deposits exposed to shear stress of the flowing fluid. The asymptotic fouling growth model is often observed in cooling water heat exchangers. In these heat exchangers, the conditions leading to the formation of a scale layer of a weak, less coherent

TABLE 13.3 Deposition and Reentrainment Models[a]

Deposition Flux $\dot{M}_d$	Reentrainment Flux $\dot{M}_r$	Source[b]	Comments[c]
au_bC_b	$b_0\tau_s m$	Kern and Seaton (1959a,b)	a, deposition constant; u_b, fluid bulk velocity; C_b, bulk concentration; b_0, reentrainment constant; particulate and other fouling ($\dot{M}_d$); spalling (M_r); mass of foulant m given by Eq. (13.7)
$K_dC_b = k_mC_b$	$b_3\dot{M}_d$	Bartlett (BNWL)	K_d, deposition coefficient; k_m, mass transfer coefficient; b_3, reentrainment constant; convective mass transfer of depositable species ($\dot{M}_d$); erosion and bond fracture ($\dot{M}_r$)
K_dC_b	bm	Charlesworth (1970)	b, reentrainment coefficient; iron oxide deposition in flow boiling ($\dot{M}_d$); erosion, spalling (M_r)
$K_dC_b = \dfrac{C_b}{1/k_m + 1/Su_w}$	$b_4 m_{loose}$	Beal (1970, 1972, 1973)	S, sticking probability; u_w, particle velocity normal to wall; b_4, reentrainment coefficient; m_{loose}, loose deposit mass per unit heat transfer area; particle deposition by eddy and Brownian diffusion and inertial coasting ($\dot{M}_d$)/ erosion ($\dot{M}_r$)
	$b_5(C_s - C_b)m$	Burrill (1977)	b_5, reentrainment coefficient; C_s, surface concentration; dissolution ($\dot{M}_r$)
$c_1\wp_1\Omega^n e^{-(E/\Re T_s)}$	$b_1\tau_s \dfrac{m^i}{\psi}$	Taborek et al. (1972), Taborek and Ritter (1972)	See text for the description of symbols; cooling-water service ($\dot{M}_r$); spalling ($\dot{M}_r$);
$k_1 m$	$k_1 k_2 m^2$	Characklis (1990)	k_1 and k_2, constants; biofouling process
$K_m F C_m + km$	$b_6\tau_s m$	Panchal et al. (1997)	K_m, transport coefficient; F, fraction of microorganisms settled on the surface; C_m, concentration of microorganisms; k, rate constant; b_6, constant

[a] Specific symbols used in this table are all local and refer to the source cited for the units and detailed physical meaning. The list of models in this table illustrates the variety of models reported in the literature.

[b] D. Q. Kern and R. E. Seaton (1959a), *Brit. Chem. Eng.*, Vol. 4, p. 258; D. Q. Kern and R. E. Seaton (1959b), *Chem. Eng. Progr.*, Vol. 55, p. 71; J. W. Bartlett (1968), BNWL-676, UC-80, Reactor Technology, Battelle Northwest, Richland, Wash.; D. H. Charlesworth (1970), *Chem. Eng. Prog. Symp. Ser.*, Vol. 66, No. 104, p. 21; S. K. Beal (1970), *Nucl. Sci. Eng.*, Vol. 40, p. 1; S. K. Beal (1972), Paper 76-C, 65th Annual Meeting AIChE; S. K. Beal (1973), *Trans. Am. Nucl. Soc.*, Vol. 17, p. 163; K. A. Burrill (1977), *Can. J. Chem. Eng.*, Vol. 55, p. 54; J. Taborek et al. (1972), *Chem. Eng. Progr.*, Vol. 68, pp. 59, 69; J. Taborek and R. B. Ritter (1972), Paper 76-A, 65th Annual Meeting, AIChE, New York; W. G. Characklis (1990) Biofilm Process, in *Biofilms*, W. G. Characklis and K. C. Marshall, eds., Wiley, New York, pp. 195–231; C. B. Panchal et al. (1997), in *Fouling Mitigation of Industrial Heat-Exchange Equipment*, C. Panchal, ed., Begell House, New York, pp. 201–212.

[c] Specific fouling/deposition/removal types are mentioned for each model for $\dot{M}_d$ and $\dot{M}_r$.

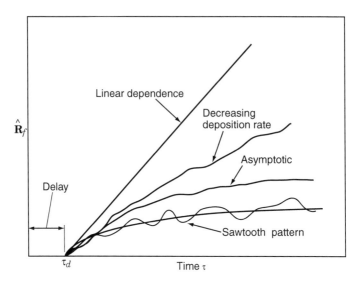

FIGURE 13.1 Time dependence of the fouling resistance.

structure are associated with simultaneous crystallization of salts of different crystal shapes or with the presence of suspended particles embedded in the crystalline structure. The growth of such deposits is expected to create internal stresses in the scale layer so that the removal processes become progressively more effective with deposit thickness. Such considerations lead to asymptotic scale thickness, at which the deposition is balanced by the scale removal mechanism.

4. Having a sawtooth pattern due to the aging process of the fouling deposits (decrease in strength and coherence) results in susceptibility to the removal process; this is found in corrosion fouling of copper tubes by seawater and in desalination evaporators.

Example 13.1 A fluid stream, rich in inert particles, flows through a tubular heat exchanger. The deposits form inside the tube surface due to particulate fouling. Assume that after a prolonged period of time, an asymptotic value of thermal resistance is reached at a level of $\hat{R}_{f,\tau\to\infty}$. Also consider that fouling resistance reaches 63% of its asymptotic value in 194 hours. Model this fouling process determining the relationship between fouling resistance and time. Assume the validity of the model given by Eqs. (13.11) and (13.13). How many hours of operation would be needed for the fouling resistance to reach within 90% of the asymptotic fouling resistance?

SOLUTION

Problem Data: Fouling takes place in a tubular heat exchanger. An asymptotic value of fouling resistance is $\hat{R}_{f,\tau\to\infty}$. It is known that 63% of this asymptotic value is reached in 194 hours.

Determine: The fouling process model determining the relationship between fouling resistance and time.

Assumptions: The assumptions invoked by the model of Taborek et al. (see Table 13.3), as presented in Section 13.2.6, are valid. That includes the fact that all parameters and variables for the problem are invariant in time.

Analysis: According to Eqs. (13.11) and (13.13), the model of fouling process is

$$\frac{d\hat{\mathbf{R}}_f}{d\tau} = c_2\wp_1\Omega^n \exp\left(-\frac{E}{\Re T_s}\right) - b_2\tau_s\frac{k_f}{\psi}\hat{\mathbf{R}}_f \tag{1}$$

with an initial condition

$$\hat{\mathbf{R}}_f = 0 \qquad \text{at} \quad \tau = 0 \tag{2}$$

The initial condition defined by Eq. (2) deserves an additional comment. In most fouling cases, fouling resistance is often noticed after a certain delay period (i.e., for $0 \leq \tau \leq \tau_d$, where τ_d represents the delay period for the onset of fouling deposits or a buildup of the fouling resistance; see Fig. 13.1). This is attributed to simultaneous influence of both initial nucleation of the deposited material on the heat transfer surface and its influence on heat transfer reduction due to lower thermal conductivity of the foulant material. Consequently, τ_d does not represent the delay of an actual fouling process, but it signifies a delay in reduction of the heat transfer rate due to fouling. In our analysis we treat $\tau_d = 0$.

A solution of the problem defined by Eqs. (1) and (2) can easily be found by using any of techniques for solving this linear, first-order ordinary differential equation. Let us introduce the set of new parameters defined as follows:

$$a = b_2\tau_s\frac{k_f}{\psi} \qquad b = c_2\wp_1\Omega^n e^{-E/\Re T_s} \tag{3}$$

Substituting a and b from Eq. (3) into Eq. (1), we get

$$\frac{d\hat{\mathbf{R}}_f}{d\tau} = b - a\hat{\mathbf{R}}_f \tag{4}$$

Integrating this linear first-order ordinary differential equation and simplifying, we get

$$\hat{\mathbf{R}}_f = \frac{b}{a} + Ce^{-a\tau} \tag{5}$$

The integration constant C in Eq. (5) can be determined by applying the initial condition of Eq. (2) to Eq. (5):

$$C = -\frac{c_2\wp_1\Omega^n e^{-E/\Re T_s}}{b_2\tau_s(k_f/\psi)} \tag{6}$$

Substituting the constants a, b, and C, Eq. (5) results in

$$\hat{\mathbf{R}}_f = \frac{c_2\wp_1\Omega^n e^{-E/\Re T_s}}{b_2\tau_s(k_f/\psi)}[1 - e^{-b_2\tau_s(k_f/\psi)\tau}] \tag{7}$$

Equation (7) represents the time history of the fouling thermal resistance. For large enough time ($\tau \to \infty$), an asymptotic value of thermal resistance, $\hat{\mathbf{R}}_{f,\tau \to \infty}$ is obtained:

$$\hat{\mathbf{R}}_{f,\tau \to \infty} = \lim_{\tau \to \infty} \hat{\mathbf{R}}_f = \frac{c_2 \wp_1 \Omega^n e^{-E/\Re T_s}}{b_2 \tau_s (k_f / \psi)} \tag{8}$$

Finally, Eq. (7) can be rearranged by using Eq. (8) in a more convenient form as follows:

$$\frac{\hat{\mathbf{R}}_f}{\hat{\mathbf{R}}_{f,\tau \to \infty}} = 1 - e^{-(\tau/\tau_c)} \tag{9}$$

where $\tau_c = \psi/b_2 \tau_s k_f$. Equation (9) is commonly referred as the *Kern–Seaton correlation*. Note that the right-hand side of Eq. (9) becomes 0.63 for $\tau = \tau_c$. This means that the fouling resistance reaches 63% of the asymptotic value for the time τ equal to the time constant τ_c, the value given in the problem formulation (i.e., $\tau_c = 194$ hours). Therefore, the number of hours of operation needed for the fouling thermal resistance to reach 90% of the asymptotic fouling resistance can be determined from Eq. (9) as follows:

$$0.9 = 1 - e^{-\tau/194} \qquad \text{therefore,} \quad \tau = 447 \, \text{h} \qquad \qquad Ans.$$

Discussion and Comments: In this example, we did not emphasize the influence of physical variables that are inherent in the original model [all the variables and constants introduced in Eq. (13.12)] since the values of these variables and constants are actually not known. Still, the model based on Eq. (9) describes quite well some fouling processes (e.g., particulate or crystallization fouling) having an asymptotic thermal resistance represented by a time constant. Note that when τ_c has a large value, Eq. (9) reduces to

$$\frac{\hat{\mathbf{R}}_f}{\hat{\mathbf{R}}_{f,\tau \to \infty}} = \frac{\tau}{\tau_c}$$

This equation represents the limiting value; here higher-order terms are neglected. Thus, in this case, the fouling resistance $\hat{\mathbf{R}}_f$ depends linearly on τ. For all cases for which the parameters of the differential equation given by Eq. (1) are constant (as implied by the set of assumptions in this example), the fouling process assumes the known deposit, defined fluid quality, and fixed-flow conditions. It should be added that many other models of fouling processes are available, as summarized by Epstein (1978). Most of them provide the expressions for $\hat{\mathbf{R}}_d$ and $\hat{\mathbf{R}}_r$ of Eq. (13.11). If these are known, the procedure for obtaining the time history of a fouling process will be similar to that demonstrated in this example after utilizing the expressions specified for deposition and removal fouling resistance rates and integrating the resulting differential equation. All these solutions, as a rule, can be treated at best as indicators of the fouling trend. However, the complexity of the process and involved nonlinearities (not included in the simple model discussed) prevent a reliable prediction.

13.3 FOULING RESISTANCE DESIGN APPROACH

Various practices relevant for heat exchanger design have been used to describe the influence of fouling on the thermal performance of a heat exchanger. The earliest one (around 1910) was the use of two combined coefficients, the *cleanliness coefficient C* and *material coefficient μ*, to correct the overall heat transfer coefficient U_c defined for a clean heat transfer surface. The resulting overall heat transfer coefficient for a fouled exchanger becomes $U_f = \mu C U_c$. The earliest values assigned to these coefficients, as reported by Somerscales (1990), were between 0.17 and 1.00 for the material coefficient and 0.5 and 1.0 for the cleanliness coefficient. Soon it became evident that introduction of a single coefficient, the cleanliness factor CF, would be more appropriate. Hence, an overall heat transfer coefficient under fouling conditions was predicted by the simple relation $U_f = \text{CF} \times U_c$, where CF < 1.00. CF values between 0.8 and 0.9 may be considered typical in the power industry.

Next we describe the modern practice of taking the influence of fouling into account.

13.3.1 Fouling Resistance and Overall Heat Transfer Coefficient Calculation

As we have introduced in Section 3.2.4, the overall thermal resistance for a heat exchanger involves a series of thermal resistances from the hot fluid to the cold fluid, including thermal resistances due to fouling on both fluid sides, as shown in Fig. 3.4. If the overall heat transfer coefficient is based on the fluid 1-side heat transfer surface area A_1, the following relation holds in the absence of fins on both fluid sides:

$$\frac{1}{U_1} = \left(\frac{1}{h_1} + \hat{\mathbf{R}}_{f,1}\right) + \frac{\delta_w}{k_w}\frac{A_1}{A_w} + \left(\frac{1}{h_2} + \hat{\mathbf{R}}_{f,2}\right)\frac{A_1}{A_2} \tag{13.14}$$

In Eq. (13.14), it is assumed that the wall thermal resistance is for a flat plate wall [see Eq. (3.31) for a tubular surface]. For a more general case, see Eq. (3.30), which includes the fin effects on both fluid sides. Equation (13.14) is further rearranged and simplified as

$$\frac{1}{U_1} = \frac{1}{h_1} + \hat{\mathbf{R}}_{f,1} + \hat{\mathbf{R}}_{f,2}\frac{A_1}{A_2} + \frac{\delta_w}{k_w}\frac{A_1}{A_w} + \frac{1}{h_2}\frac{A_1}{A_2} = \frac{1}{h_1} + \hat{\mathbf{R}}_f + \hat{\mathbf{R}}_w\frac{A_1}{A_w} + \frac{1}{h_2}\frac{A_1}{A_2} \tag{13.15}$$

Note that $\hat{\mathbf{R}}_f = \hat{\mathbf{R}}_{f,1} + \hat{\mathbf{R}}_{f,2}(A_1/A_2)$ represents the total fouling resistance, a sum of fouling resistances on both sides of the heat transfer surface, as shown. It should again be reiterated that the aforementioned reduction in the overall heat transfer coefficient due to fouling does not take into consideration the transient nature of the fouling process. According to Chenoweth (1990), use of the fouling resistance concept as represented by Eqs. (13.14) and (13.15) must be based on the following recommendations:

- Fouling resistances should reflect fouling alone and not uncertainties in the design of the heat exchanger.
- Appropriate values of fouling resistances should be based on operating experience and modified by economic considerations where possible.
- The buyer/user, not the manufacturer, should be responsible for selecting the fouling resistances because he or she may know his or her own application better.

- The effects of corrosion fouling and biofouling due to their complexity and a questionable predictability should always be controlled externally. That means that this control should be a system based on reducing or preventing fouling.

The current practice is to assume a value for the fouling resistance on one or both fluid sides as appropriate and to design a heat exchanger accordingly by providing extra surface area for fouling, together with a cleaning strategy (see Section 13.4.3).

The complexity in controlling a large number of internal and external factors of a given process makes it very difficult to predict the fouling growth as a function of time using deterministic (well-known kinetic) models. A more realistic fouling growth model can be devised by postulating fouling as a time-dependent random process[†] and analyzing using the probabilistic approaches (Zubair et al., 1997) in conjunction with the cleaning strategies as discussed in Section 13.4.3.

A note of caution is warranted at this point. There is an ongoing discussion among scholars and engineers from industry as to whether either fouling resistance or fouling rate concepts should be used as the most appropriate tool in resolving design problems incurred by fouling. One suggestion in resolving this dilemma would be that the design fouling-resistance values used for sizing heat exchangers be based on fouling-rate data and estimated cleaning-time intervals (Rabas and Panchal, 2000).

13.3.2   Impact of Fouling on Exchanger Heat Transfer Performance

In current practice, based on application and need, the influence of fouling on exchanger heat transfer performance can be evaluated in terms of either (1) required increased surface area for the same q and ΔT_m, (2) required increased mean temperature difference for the same q and A, or (3) reduced heat tranfer rate for the same A and ΔT_m.[‡] For these approaches, we now determine expressions for A_f/A_c, $\Delta T_{m,f}/\Delta T_{m,c}$ and q_f/q_c as follows.[§] In the first two cases, the heat transfer rate in a heat exchanger under clean and fouled conditions are the same. Hence,

$$q = U_c A_c \, \Delta T_m = U_f A_f \, \Delta T_m \qquad \text{for constant } \Delta T_m \tag{13.16}$$

Therefore,

$$\frac{A_f}{A_c} = \frac{U_c}{U_f} \tag{13.17}$$

According to Eq. (13.15), the relationships between overall heat transfer coefficients (based on tube outside surface area) and thermal resistances for clean and fouled conditions are defined as follows. For a clean heat transfer surface,

$$\frac{1}{U_c} = \frac{1}{h_{o,c}} + \hat{\mathbf{R}}_w \frac{A_o}{A_w} + \frac{1}{h_{i,c}} \frac{A_o}{A_i} \tag{13.18}$$

[†] The randomness in fouling process is due to time-dependent scatter in fouling resistance from a replicate to a replicate (repeat tests).

[‡] The first case is the design of an exchanger where an allowance for fouling can be made at the design stage by increasing surface area. The other two cases are for an already designed exchanger in operation, and the purpose is to determine the impact of fouling on exchanger performance.

[§] Throughout this chapter, the subscript c denotes a clean surface and f the fouled surface.

For a fouled heat transfer surface,

$$\frac{1}{U_f} = \frac{1}{h_{o,f}} + \hat{\mathbf{R}}_f + \hat{\mathbf{R}}_w \frac{A_o}{A_w} + \frac{1}{h_{i,f}} \frac{A_o}{A_i} = \frac{1}{h_{o,c}} + \hat{\mathbf{R}}_f + \hat{\mathbf{R}}_w \frac{A_o}{A_w} + \frac{1}{h_{i,c}} \frac{A_o}{A_i} \qquad (13.19)$$

Note that we have idealized that $h_{o,f} = h_{o,c}, h_{i,f} = h_{i,c}, A_{i,f} = A_{i,c} = A_i$, and $A_{o,f} = A_{o,c} = A_o$. Here A_o represents the tube outside surface area and *not* free-flow area in the exchanger. The difference between Eqs. (13.18) and (13.19) is

$$\hat{\mathbf{R}}_f = \frac{1}{U_f} - \frac{1}{U_c} \qquad (13.20)$$

It should be added that Eq. (13.20) is valid as long as clean overall heat transfer coefficients are constant. If this assumption is not satisfied, the right-hand side in Eq. (13.20) does not represent only the overall fouling resistance but a quantity that includes other influences on overall heat transfer coefficients in addition to fouling. In that case, the fouling assessment will be incorrect. Combining Eqs. (13.17) and (13.20), we get

$$\frac{A_f}{A_c} = U_c \hat{\mathbf{R}}_f + 1 \qquad (13.21)$$

Similarly, when q and A are the same and ΔT_m is different for clean and fouled exchangers, we have

$$q = U_c A_c \Delta T_{m,c} = U_f A_c \Delta T_{m,f} \qquad \text{for constant } A \qquad (13.22)$$

Hence,

$$\frac{\Delta T_{m,f}}{\Delta T_{m,c}} = \frac{U_c}{U_f} \qquad (13.23)$$

Combining Eqs. (13.23) and (13.20), we get

$$\frac{\Delta T_{m,f}}{\Delta T_{m,c}} = U_c \hat{\mathbf{R}}_f + 1 \qquad (13.24)$$

Finally, if one assumes that heat transfer area and mean temperature differences are fixed, heat transfer rates for the same heat exchanger under fouled and clean conditions are given by $q_f = U_f A \Delta T_m$ and $q_c = U_c A \Delta T_m$, respectively. Combining these two relationships with Eq. (13.20), we get

$$\frac{q_f}{q_c} = \frac{1}{U_c \hat{\mathbf{R}}_f + 1} \qquad (13.25)$$

Alternatively, Eq. (13.25) can be expressed as

$$\frac{q_c}{q_f} = U_c \hat{\mathbf{R}}_f + 1 \qquad (13.26)$$

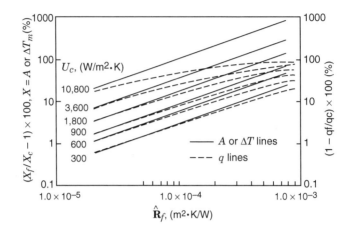

FIGURE 13.2   Percent change in heat transfer area, mean temperature difference, or heat duty vs. fouling unit thermal resistance for a fouled exchanger.

We find that the right-hand sides of Eqs. (13.21), (13.24), and (13.26) are the same. Equations (13.21), (13.24), and (13.25) are shown in Fig. 13.2 in terms of the percentage increase in A and ΔT_m and the percentage reduction in q for the fouled exchanger over that for the clean exchanger. From this figure, it is clear that fouling has a significant impact on the exchanger performance for high values of $\hat{\mathbf{R}}_f$ and/or U_c.

The cleanliness factor CF is related to the fouling resistance $\hat{\mathbf{R}}_f$ as

$$\mathrm{CF} = \frac{U_f}{U_c} = \frac{1}{1 + \hat{\mathbf{R}}_f U_c} \qquad (13.27)$$

Example 13.2   Overall heat transfer coefficient of a heat exchanger operating under clean conditions is calculated as $800\ \mathrm{W/m^2 \cdot K}$. Following industrial experience, the cleanliness factor for this exchanger is 0.7. Determine the magnitude of the corresponding fouling resistance.

SOLUTION

Problem Data: The following data are given: $U_c = 800\ \mathrm{W/m^2 \cdot K}$ and $\mathrm{CF} = 0.7$.

Determine: The fouling resistance $\hat{\mathbf{R}}_f$ of the deposit formed by this heat exchanger.

Assumptions: The convective heat transfer coefficients on the hot- and cold-fluid sides are the same for both fouled and clean heat transfer surfaces. Thermal resistance of the wall is unchanged under fouled conditions. The change in heat transfer surface areas due to fouling deposit formation is negligible. The fin efficiency is equal to unity. All idealizations adopted for heat exchanger design theory are valid (see Section 3.2.1).

Analysis: The relationship between fouling resistances and overall heat transfer coefficients for clean and fouled conditions is given by Eq. (13.20):

$$\hat{\mathbf{R}}_f = \frac{1}{U_f} - \frac{1}{U_c} \tag{1}$$

Using the definition of CF from $U_f = \mathrm{CF} \times U_c$, Eq. (1) reduces to

$$\hat{\mathbf{R}}_f = \frac{1}{\mathrm{CF} \times U_c} - \frac{1}{U_c} = \frac{1}{U_c}\frac{1 - \mathrm{CF}}{\mathrm{CF}} \tag{2}$$

Thus, substituting data given into Eq. (2), we get

$$\hat{\mathbf{R}}_f = \frac{1}{800}\left(\frac{1 - 0.7}{0.7}\right) = 5 \times 10^{-4}\,\mathrm{m}^2 \cdot \mathrm{K/W} \qquad\qquad Ans.$$

Discussion and Comments: In some industries (such as power industry), use of the cleanliness factor has been prevalent for assessing the influence of fouling. The reason for this is the practice of the industry and the difficulties associated with experimental determination of fouling thermal resistances (Somerscales, 1990). Equation (2) can be used to calculate fouling resistance or unit thermal resistance if the cleanliness factor is known (or vice versa) under the conditions governed by the above-mentioned assumptions.

Example 13.3   Determine how much will change the required heat transfer area of an exchanger under fouling conditions if the fouling resistance changes from $10^{-4}\,\mathrm{m}^2 \cdot \mathrm{K/W}$ to $10^{-3}\,\mathrm{m}^2 \cdot \mathrm{K/W}$. The heat transfer rate and mean temperature difference remain the same, and $U_c = 1000\,\mathrm{W/m}^2\,\mathrm{K}$. Consider no extended surface on either fluid side of the exchanger.

SOLUTION

Problem Data: The following data are given:

$$U_c = 1000\,\mathrm{W/m}^2 \cdot \mathrm{K} \qquad \hat{\mathbf{R}}_{f,1} = 10^{-4}\,\mathrm{m}^2 \cdot \mathrm{K/W} \qquad \hat{\mathbf{R}}_{f,2} = 10^{-3}\,\mathrm{m}^2 \cdot \mathrm{K/W}$$

$$q_c = q_f \qquad\qquad \Delta T_{m,c} = \Delta T_{m,f} \qquad\qquad \eta_{o,1} = \eta_{o,2} = 1$$

Determine: The change in heat transfer surface area required if the fouling resistance changes from $10^{-4}\,\mathrm{m}^2 \cdot \mathrm{K/W}$ to $10^{-3}\,\mathrm{m}^2 \cdot \mathrm{K/W}$.

Assumptions: The convection heat transfer coefficients are the same for fouled and clean heat transfer surfaces. The thermal resistance of the wall is unchanged under fouled conditions. Change in heat transfer surface areas due to deposit formation is negligible. All assumptions adopted for heat exchanger design theory are valid (see Section 3.2.1).

Analysis: The heat transfer rate and mean temperature differences in this exchanger under clean and fouled conditions are the same. Hence, from Eq. (13.21), we get

$$\frac{A_f}{A_c} = U_c\hat{\mathbf{R}}_f + 1 \tag{1}$$

From Eq. (1), it follows that the change in total fouling resistance from $\hat{\mathbf{R}}_{f,1}$ to $\hat{\mathbf{R}}_{f,2}$ causes a change in heat transfer area from $A_{f,1}$ to $A_{f,2}$ as follows:

$$\frac{A_{f,1}}{A_c} = U_c \hat{\mathbf{R}}_{f,1} + 1 \quad \text{and} \quad \frac{A_{f,2}}{A_c} = U_c \hat{\mathbf{R}}_{f,2} + 1 \tag{2}$$

From Eq. (2), inserting the data given, we get

$$\frac{A_{f,2}}{A_{f,1}} = \frac{U_c \hat{\mathbf{R}}_{f,2} + 1}{U_c \hat{\mathbf{R}}_{f,1} + 1} = \frac{10^3 \, \text{W/m}^2 \cdot \text{K} \times 10^{-3} \, \text{m}^2 \cdot \text{K/W} + 1}{10^3 \, \text{W/m}^2 \cdot \text{K} \times 10^{-4} \, \text{m}^2 \cdot \text{K/W} + 1} = 1.82 \qquad \textit{Ans.}$$

Thus an increase in the fouling resistance by a factor of 10 requires a surface area increase for this exchanger of 82%.

Discussion and Comments: This example clearly shows a significant increase in the surface area requirement for this exchanger when the total fouling resistance is increased by an order of magnitude. Inversely, a significant reduction in surface area can be achieved (by about one-half) if the total fouling resistance is reduced by an order of magnitude. Note that the result provides a direct information on how large percent change in heat transfer area would be compared to that for a clean heat exchanger for a given fouling resistance, as shown in Fig. 13.2.

13.3.3 Empirical Data for Fouling Resistances

Empirical data for fouling resistances have been obtained over many decades by industry since its first compilation by TEMA in 1941 for shell-and-tube heat exchangers. Selected data are summarized in Table 9.4 and hence are not repeated here. Many of the original values of TEMA fouling factors or fouling resistances established in 1941 for a typical exchanger service length of three months are still in use for a current typical service length of five years (Chenoweth, 1990)! TEMA fouling resistances are supposed to be representative values, asymptotic values, or those manifested just before cleaning to be performed. Chenoweth (1990) analyzed the current practice of customers' specifying fouling resistances on their specification sheets to manufacturers. He compiled the combined shell- and tube-side fouling resistances (by summing each side entry) of over 700 shell-and-tube heat exchangers and divided them into nine combinations of liquid, two-phase, and gas on each fluid side regardless of the applications. He then simply took the arithmetic average of total $\hat{\mathbf{R}}_f$ for each two-fluid combination value and plotted dimensional values in his Fig. I-3. His results are presented in Fig. 13.3 after normalizing with respect to the maximum combined shell- and tube-side $\hat{\mathbf{R}}_f$ of liquid–liquid applications so that the ordinate ranges between 0 and 1. For gases on both sides, the relative fouling resistance can be as high as 0.5 (the lowest value in Fig. 13.3) compared to liquids on both sides (relative total fouling resistance represented as 1.0 in Fig. 13.3). If liquid is on the shell side and gas on the tube side, the relative fouling resistance is 0.65. However, if liquid is on the tube side and gas on the shell side, it is 0.75. Since many process industry applications deal with liquids that are dirtier than gases, the general practice is to specify larger fouling resistances for liquids compared to those for the gases. Also, if fouling is anticipated on the liquid side of a liquid–gas exchanger, it is generally placed in the tubes for cleaning purposes and a larger fouling resistance is specified. These trends are clear from Fig. 13.3. It should again be emphasized that Fig. 13.3 indicates the current practice and has no scientific basis. Specification of larger fouling resistances for liquids (which

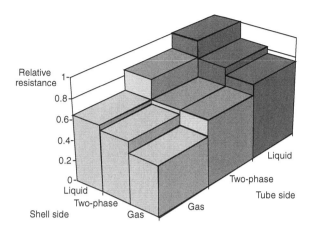

Relative resistance

FIGURE 13.3   Combined tube- and shell-side relative fouling resistances. (Based on data from Chenoweth, 1990.)

have higher heat transfer coefficients than those of gases) has even more impact on the surface area requirement for liquid–liquid exchangers than for gas–gas exchangers.

It should be reiterated that the recommended fouling resistances are believed to represent *typical* fouling resistances for design. Consequently, sound engineering judgment has to be made for each selection of fouling resistances, keeping in mind that actual values of fouling resistances in any application can be either higher or lower than the resistances calculated. Finally, it must be clear that fouling resistances, although recommended following the empirical data and a sound model, are still constant, independent of time, while fouling is a transient phenomenon. Hence, the value of $\hat{R}_f$ selected represents a correct value only at one specific time in the exchanger operation. As indicated by Chenoweth (1990): "... the new proposed (constant, independent of time) values reflect a careful review and the application of good engineering judgment by a group of knowledgeable engineers involved with the design and operation of shell-and-tube heat exchangers. ... It needs to be emphasized that the tables may not provide the applicable values for a particular design. They are only intended to provide guidance when values from direct experience are unavailable." With the use of finite fouling resistance, the overall U value is reduced, resulting in a larger surface area requirement, larger flow area, and reduced flow velocity which inevitably results in increased fouling. Thus, allowing more surface area for fouling in a clean exchanger may accelerate fouling initially.

Typical fouling resistances are roughly 10 times lower in plate heat exchangers than in shell-and-tube heat exchangers (Zubair and Shah, 2001). Some fouling resistances for PHEs are compared with those for shell-and-tube heat exchangers in Table 13.4.

TEMA (1999) presents fouling resistances for some gases used in process and petrochemical industries and Marner and Suitor (1987) summarize the literature data for gases used in many industries, as reported in Table 13.5.

Example 13.4   A heat exchanger with water-to-phase change fluid is designed keeping in mind that the fluid that changes its phase must be on the outside of the tube. The empirical data available reveal that an average heat transfer coefficient on the water side is 2715 W/m^2 K. On the tube outside, the heat transfer coefficient is 3200 W/m$^2 \cdot$ K.

TABLE 13.4   Liquid-Side Fouling Resistances for PHEs vs. TEMA Values

	$(m^2 \cdot K/kW)$	
Process Fluid	PHEs	$\hat{R}_f$-TEMA
Soft water	0.018	0.18–0.35
Cooling tower water	0.044	0.18–0.35
Seawater	0.026	0.18–0.35
River water	0.044	0.35–0.53
Lube oil	0.053	0.36
Organic solvents	0.018–0.053	0.36
Steam (oil bearing)	0.009	0.18

Source: Data from Panchal and Rabas (1999).

TABLE 13.5   Gas-Side Fouling Resistances $\hat{R}_f (m^2 \cdot K/kW)^a$

	Weierman (1982)	Zink (1981)	TEMA (1978)	Rogalski (1979)	Henslee and Bouge (1983)
Clean gas					
Natural gas	0.0881–0.528	0.176	—	—	—
Propane	0.176–0.528	—	—	—	—
Butane	0.176–0.528	—	—	—	—
Gas turbine	0.176	—	—	—	—
Average gas					
No. 2 oil	0.352–0.704	0.528	—	—	—
	0.264	—	—	0.528–6.69	—
	0.528	—	1.76	—	21.1–24.7
Dirty gas					
No. 6 oil	0.528–1.23	0.881	—	—	—
Crude oil	0.704–2.64	—	—	—	—
Residual oil	0.881–3.52	1.76	—	—	—
Coal	0.881–8.81	—	—	—	—
Miscellaneous					
Sodium-bearing waste	—	5.28	—	—	—
Metallic oxides	—	1.76	—	—	—
FCCU catalyst fines	—	1.41	—	—	—

Source: Data from Marner and Suitor (1987).
[a] R. C. Weierman (1982), JPL Publ. 82-67, Jet Propulsion Laboratory, California Institure of Technology, Pasadena, CA.; Jon Zink Co., Tulsa, OK (1981); R. D. Rogalski (1979), *SAE Trans.*, Vol. 88, pp. 2223–2239; S. P. Henslee and J. L. Bouge (1983), Report EGG-FM-6189, Idaho National Laboratory, Idaho Falls, ID.

Tubes are made of steel with thermal conductivity of $40 \, W/m \cdot K$. The tube outside diameter is 19 mm with 1.6 mm wall thickness. The asymptotic value of the fouling resistance on the water side is $4 \times 10^{-4} \, m^2 \cdot K/W$. There is no fouling on the tube outside. Based on past experience, the fouling phenomenon is of asymptotic nature for this exchanger, and the time constant for the fouling process is 280 hours for the Kern–Seaton model. Determine percentage unit thermal resistance distribution contributing to the overall unit thermal resistance for the following two cases: (a) after 280 hours of fouling initiation, and (b) for the asymptotic fouling condition.

SOLUTION

Problem Data: The following data are provided:

$h_i = 2715 \, \text{W/m}^2 \cdot \text{K}$ $h_o = 3200 \, \text{W/m}^2 \cdot \text{K}$ $k_w = 40 \, \text{W/m} \cdot \text{K}$ $d_o = 19 \, \text{mm}$

$\delta_w = 1.6 \, \text{mm}$ $\hat{\mathbf{R}}_{f,i} = 4 \times 10^{-4} \, \text{m}^2 \cdot \text{K/W}$ for $\tau \to \infty$ $\tau_c = 280 \, \text{h}$

Determine: The distribution of thermal resistances for two cases: (1) $\tau \to \infty$, and (2) $\tau = \tau_c$, where τ_c is the time constant for asymptotic fouling.

Assumptions: The set of assumptions introduced for heat exchanger analysis (see Section 3.1.2) and the Kern–Seaton model (see Section 13.2.5) is valid; the tube wall is thin.

Analysis: The overall unit thermal resistance for the exchanger, from Eq. (13.14), is

$$\frac{1}{U_o} = \left(\frac{1}{h_o} + \hat{\mathbf{R}}_{f,o}\right) + \frac{\delta_w}{k_w}\frac{A_o}{A_w} + \left(\frac{1}{h_i} + \hat{\mathbf{R}}_{f,i}\right)\frac{A_o}{A_i} = \frac{1}{h_o} + \frac{\delta_w}{k_w}\frac{A_o}{A_w} + \left(\frac{1}{h_i} + \hat{\mathbf{R}}_{f,i}\right)\frac{A_o}{A_i} \quad (1)$$

Let us first calculate the missing information (area ratios, $\hat{\mathbf{R}}_w$ and $\hat{\mathbf{R}}_{f,i}$) for this equation:

$$d_i = d_o - 2\delta_w = 19 \, \text{mm} - 2 \times 1.6 \, \text{mm} = 15.8 \, \text{mm}$$

$$\frac{A_o}{A_i} = \frac{\pi d_o L}{\pi d_i L} = \frac{19 \, \text{mm}}{15.8 \, \text{mm}} = 1.203 \qquad \frac{A_o}{A_w} \approx \frac{\pi d_o L}{\pi[(d_o + d_i)/2]L} = \frac{19 \, \text{mm}}{[(19 + 15.8)/2] \, \text{mm}} = 1.092$$

$$\hat{\mathbf{R}}_w = \frac{\delta_w}{k_w} = \frac{1.6 \times 10^{-3} \, \text{m}}{40 \, \text{W/mK}} = 4 \times 10^{-5} \, \text{m}^2 \, \text{K/W}$$

Since the asymptotic fouling resistance is given, the actual fouling resistance on the water side for $\tau = \tau_c = 280$ hours can be determined from Eq. (9) of Example 13.1 as follows:

$$\hat{\mathbf{R}}_{f,i} = \hat{\mathbf{R}}_{f,\tau \to \infty}(1 - e^{-\tau/\tau_c}) = 4 \times 10^{-4} \, \text{m}^2 \cdot \text{K/W} \times (1 - e^{-1}) = 2.53 \times 10^{-4} \, \text{m}^2 \cdot \text{K/W}$$

Now the individual unit thermal resistances of the last equality of Eq. (1), in absolute values and in percentages, are computed and summarized for this problem for $\tau = \infty$ and $\tau = \tau_c$:

Fouling Time	$\hat{\mathbf{R}}_{c,o} = \dfrac{1}{h_o}$ $(\text{m}^2 \cdot \text{K/W})$	$\hat{\mathbf{R}}_w \dfrac{A_o}{A_w}$ $(\text{m}^2 \cdot \text{K/W})$	$\hat{\mathbf{R}}_{f,i} \dfrac{A_o}{A_i}$ $(\text{m}^2 \cdot \text{K/W})$	$\hat{\mathbf{R}}_{c,i} = \dfrac{1}{h_i}\dfrac{A_o}{A_i}$ $(\text{m}^2 \cdot \text{K/W})$	Overall $1/U_o$ $(\text{m}^2 \cdot \text{K/W})$
$\tau \to \infty$	3.125×10^{-4}	4.368×10^{-5}	4.810×10^{-4}	4.429×10^{-4}	12.801×10^{-4}
$\tau = \tau_c$	3.125×10^{-4}	4.368×10^{-5}	3.042×10^{-4}	4.429×10^{-4}	11.033×10^{-4}
			Percentages		
$\tau \to \infty$	24	3	38	35	100
$\tau = \tau_c$	28	4	28	40	100

Discussion and Comments: From the results of this problem, it is obvious that fouling has quite a significant influence on the total unit thermal resistance $1/U_o$. If cleaning is not going to be performed, the fouling resistance will ultimately reach 38% of the total resistance, more than any other contribution. Note that the thermal resistance of the tube wall is an order of magnitude smaller, and hence our approximation of considering the thin wall of the tube for $\hat{R}_w$ determination is reasonable. Distribution of thermal resistances would be different if the foulant deposition is allowed to continue only up to the deposition time that equals the time constant. In that case, the fouling thermal resistance would be smaller compared to both the outside and inside tube convective resistances. Still, it would have the same order of magnitude.

13.4   PREVENTION AND MITIGATION OF FOULING

Ideally, a heat exchanger should be designed to minimize or eliminate fouling. For example, heavy-fouling liquids can be handled in a direct contact heat exchanger since heat and mass transfer takes place due to direct contact of the fluids over the "fill" or the surface in such an energy exchanger. The fill can get fouled without affecting energy transfer between the fluids in direct contact. In fluidized-bed heat exchangers, the bed motion scours away the fouling deposit. Gasketed plate-and-frame heat exchangers can easily be disassembled for cleaning. Compact heat exchangers are not suitable for fouling service unless chemical cleaning or thermal baking is possible. When designing a shell-and-tube heat exchanger, the following considerations are important in reducing or cleaning fouling. The heavy fouling fluid should be kept on the tube side for cleanability. Horizontal heat exchangers are easier to clean than vertical ones. Geometric features on the shell side should be such as to minimize or eliminate stagnant and low-velocity regions. It is easier to clean square or rotated square tube layouts mechanically on the shell side [with a minimum cleaning lane of $\frac{1}{4}$ in. (6.35 mm)] than to clean other types of tube layouts.

Some control methods are now summarized for specific types of fouling. Crystallization fouling can be controlled or prevented by preheating the stream so that crystallization does not occur. To control particulate fouling, use a filter or similar device to capture all particles greater than about 25% of the smallest gap size in the flow path. Eliminate any dead zones and low-velocity zones. Use back flushing, "puffing," or chemical cleaning, depending on the application. Chemical cleaning is probably the most effective cleaning method for chemical reaction fouling. For corrosion fouling, initial selection of corrosion-resistant material is the best remedy. For example, use proper aluminum alloy to prevent mercury corrosion in a plate-fin exchanger. Biofouling is usually easy to control with biocides, but must check compatibility with the exchanger construction materials. Chlorination aided by flow-induced removal of disintegrated biofilm is the most common mitigation technique.

General techniques to prevent or control fouling on the liquid or gas side are summarized briefly.

13.4.1   Prevention and Control of Liquid-Side Fouling

Among the most frequently used techniques for control of liquid-side fouling is the online utilization of chemical inhibitors/additives. The list of additives includes (1) dispersants to maintain particles in suspension; (2) various compounds to prevent polymerization and chemical reactions; (3) corrosion inhibitors or passivators to minimize corrosion; (4) chlorine and other biocide/germicides to prevent biofouling; and (5) softeners, poly-carboxylic acid and polyphosphates, to prevent crystal growth. Alkalis dissolve salts. Finally, filtration can be used as an efficient method of mechanical removal of particles. An extensive review of fouling control measures is provided by Knudsen (1998).

Mitigation of water fouling and the most recent review of the related issues are discussed extensively by Panchal and Knudsen (1998), where they suggest the following methods.

- *Chemical additives:* dispersants or coagulators for particulate fouling; dispersants, crystal modifiers, and chelating agents for crystallization fouling; inhibitors or surface filming for corrosion fouling; and biocides, biodispersants, and biostats for biofouling.

- *Process adjustments:* monitoring, modifications and replacements of devices, water flow reduction, and recirculation strategies.

- *Physical devices for cleaning:* sponge-ball cleaning and the use of reversing-flow shuttle brushes.

- *Utilization of enhanced heat transfer surfaces and devices:* It has sizable influence on fouling mitigation. The use of tube inserts (in particular, in refinery processes), such as wire mesh, oscillating wires, and rotating wires, is a standard method.

- *Various alternative devices and/or methods:* magnetic fields, radio-frequency, ultraviolet and acoustic radiation, and electric pulsation. Surface treatment and fluidized-bed designs are also used.

- The most frequently used technique for preventing water-side fouling is still the conventional water treatment. Strict guidelines have been developed for the quality of water for environmental concerns (Knudsen, 1998).

Heat transfer surface mitigation techniques can be applied either on- or offline. Online techniques (usually used for tube-side applications) include various mechanical techniques (flow-driven or power-driven rotating brushes, scrapers, drills, acoustic/mechanical vibration, air or steam lancing on the outside of tubes, chemical feeds, flow reversal, etc.). In some applications, flows are diverted in a bypass exchanger, and then the fouled exchanger is cleaned offline. Other offline techniques (without opening a heat exchanger) include chemical cleaning, mechanical cleaning by circulating particulate slurry, and thermal baking to melt frost/ice deposits. Offline cleaning with a heat exchanger opened or removed from the site include (1) high-pressure steam or water spray for a shell-and-tube heat exchanger, and (2) baking compact heat exchanger modules in an oven (to burn the deposits) and then rinsing. If fouling is severe, a combination of methods is required.

13.4.2 Prevention and Reduction of Gas-Side Fouling

The standard techniques for control and/or prevention of fouling on the gas side are (1) techniques for removal of potential residues from the gas, (2) additives for the gas-side fluid, (3) surface cleaning techniques, and (4) adjusting design up front to minimize fouling. Details regarding various techniques for gas-side fouling prevention, mitigation, and accommodation are given by Marner and Suitor (1987).

Control of gas (or liquid)-side fouling should be attempted before any cleaning method is tried. The fouling control procedure should be preceded by (1) verification of the existence of fouling, (2) identification of the feature that dominates the foulant accumulation, and (3) characterization of the deposit.

Some of the methods for mitigation of gas-side fouling are as follows:

- *Crystallization fouling* can be prevented if the surface temperature is kept above the freezing of vapors from the gaseous stream; the solidification can be minimized by keeping a "high" velocity of freezable species, having some impurities in the gas stream, and decreasing the foulant concentration, if possible.

- *Particulate fouling* can be minimized (1) by increasing the velocity of the gas stream if it flows parallel to the surface and decreasing the velocity if the gas flow impinges on the surface, (2) by increasing the outlet temperature of the exhaust gases from the exchanger above the melting point of the particulates, (3) by minimizing the lead content in gasoline or unburned hydrocarbons in diesel fuel, (4) by reducing the fuel–air ratio for a given combustion efficiency, and (5) by minimizing flow impact (e.g., flow over a staggered tube bank) or ensuring the narrowest dimension in the flow cross section, to three to four times the largest particle size anticipated.

- *Chemical reaction fouling* can be minimized (1) by maintaining the right temperature range in the exhaust gas within the exchanger, (2) by increasing or decreasing the velocity of the gaseous stream, depending on the application, (3) by reducing the oxygen concentration in the gaseous stream, (4) by replacing the coal with fuel oil and natural gas (in that order), and (5) by decreasing the fuel–air ratio.

- *Corrosion fouling* is strongly dependent on the temperature of the exhaust stream in the exchanger. The outlet temperature of the exhaust gas stream from the exchanger should be maintained in a very narrow range: above the acid dew point [above 150°C (300°F)] for sulfuric or hydrochloric acid condensation or below 200°C (400°F) for attack by sulfur, chlorine, and hydrogen in the exhaust gas stream. Since sulfur is present in all fossil fuels and some natural gas, the dew point of sulfur must be avoided in the exchanger, which is dependent on the sulfur content in the fuel (Shah, 1985). From the electrochemical condition of the metal surface, the corrosion rate increases with velocity up to a maximum value for an active surface and no sizable effect for a passive surface. The pH value has a considerable role in the corrosion fouling rate; the corrosion rate is minimum at a pH of 11 to 12 for steel surfaces. Low oxygen concentrations in the flue gases promote the fire-side corrosion of mild steel tubes in coal-fired boilers. Stainless steel, glass, plastic, and silicon are highly resistant to low-temperature corrosion [$T_{gas} < 260°C$ (500°F)], stainless steel and superalloys to medium-temperature corrosion [260°C (500°F) $< T_{gas} < 815°C$ (1500°F)], and superalloys and ceramic materials to high-temperature corrosion [$T_{gas} > 815°C$ (1500°F)]. Chrome alloys are suitable for high-temperature sulfur and chlorine corrosion, and molybdenum and chrome alloys protect against hydrogen corrosion.

13.4.3 Cleaning Strategies

An important element in mitigating a fouling problem is selection of a cleaning strategy (i.e., the cleaning-cycle period). The cleaning-cycle period is delineated by the operation of an exchanger until the performance reaches the minimum value acceptable. Subsequently, the exchanger must be cleaned by one of the methods summarized in Section 13.4.1 or 13.4.2. In the case of asymptotic fouling in a given application, generally no cleaning is necessary. If fouling rate data are available, ideally the exchanger can be optimized based on the life-cycle cost, and accordingly, the cleaning schedule can be established.

The cleaning-cycle period may also be determined based on a regular maintenance schedule during process shutdowns. In any case, the functional relationship between the operation time and fouling resistance (see Section 13.2.6) should be known at least partially. The importance of rational cleaning schedules based on such an understanding is critical when (Somerscales, 1990) (1) the allowable deviation from the process stream temperature is small compared to absolute values (steam power plant condensers), and (2) the cost of cleaning is a significant fraction of the operating cost.

Depending on the fouling process, the cleaning strategies for preventing maintenance are of two types: reliability-based and cost-based. There are three scenarios for reliability-based cleaning strategies: (1) maintenance restores exchanger performance (this is an idealized maintenance scheme, taking place at equal time intervals); (2) by decreasing the preventive maintenance time interval gradually (due to fixed degradation of performance after each maintenance interval), the exchanger performance is restored; (3) for the preceding case of a fixed degradation of performance after each preventive maintenance time interval, if the maintenance takes place at equal time intervals, it reduces the exchanger performance. The cost-based cleaning strategy includes the costs associated with online chemical cleaning, offline cleaning, additional fuel/power consumption due to fouling, and severity of the financial penalty associated with exchanger performance due to fouling.

It should be noted that operating a heat exchanger at the critical risk level of a system or component is important in some applications, such as in a heat exchanger network in a refinery. In this situation, an acceptable level of heat exchanger overall heat transfer coefficient will primarily govern the maintenance strategy. However, in some situations, heat exchangers are not in a network, or in a critical system; here, maintaining the exchanger at a higher reliability level r (or at a lower risk level p) implies more frequent maintenance intervals, which can often result in increasing operation and maintenance costs. It is thus important to note that in situations in which the cost of operation and maintenance is an important factor, along with exchanger reliability ($r = 1 - p$), maintenance decisions can be optimized by developing cost as a function of reliability (or risk level) and then searching for a minimum cost-based solution. This cost-optimized maintenance solution will also result in an optimal level of heat exchanger reliability (Zubair et al., 1997).

For further details on cleaning strategies, refer to Zubair et al. (1997) for shell-and-tube heat exchangers and Zubair and Shah (2001) for plate-and-frame heat exchangers.

13.5 CORROSION IN HEAT EXCHANGERS

Corrosion in an exchanger involves destruction of heat exchanger surfaces (construction materials, metals, and alloys), possibly caused by working fluids due to operating conditions (including stresses). Like fouling, corrosion is a complex transient phenomenon, affected by many variables, with a synergistic relationship as shown in Fig. 13.4. Corrosion can be classified according to such mechanisms as fretting corrosion, corrosion fatigue, and corrosion due to microorganisms. Alternatively, corrosion can be classified based on visual characteristics of the morphology of corrosion attack, as described in the following subsection. Corrosion can induce corrosion fouling as

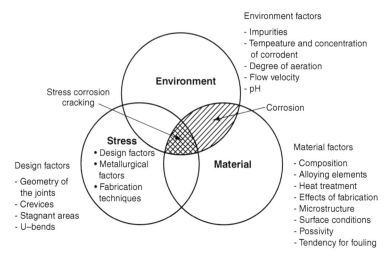

Environment factors

- Impurities
- Tempeature and concentration
  of corrodent
- Degree of aeration
- Flow velocity
- pH

Stress corrosion cracking

Corrosion

Design factors

- Geometry of
  the joints
- Crevices
- Stagnant areas
- U–bends

Stress
• Design factors
• Metallurgical
  factors
• Fabrication
  techniques

Environment

Material

Material factors

- Composition
- Alloying elements
- Heat treatment
- Effects of fabrication
- Microstructure
- Surface conditions
- Possivity
- Tendency for fouling

FIGURE 13.4   Factors influencing corrosion. (From Kuppan, 2000.)

described in Section 13.2.1, thus adding thermal resistance in the heat flow path and reducing heat transfer, increasing fluid pressure drop and pumping power, and increasing cost due to overdesign of the exchanger. The loss of material due to corrosion may result in crevices, holes, and/or partial removal of heat transfer surfaces, resulting in loss (leakage) of heat transfer fluids, some of which may be costly. If the fluid leaks outside, it may harm the environment if the fluid is corrosive or poisonous. If it leaks to the other fluid side, it may contaminate the other fluid and deteriorate its quality. Corrosion may add extra cost to the exchanger, due to the use of expensive material, maintenance, warranty, inventory of parts, and so on. Corrosion products carried downstream of the exchanger may corrode downstream components. Finally, corrosion may result in complete failure of an exchanger or partial failure in a tube-fin exchanger, due to corroding away fins, as in an automotive radiator.

There is no a single cause of corrosion and/or associated corrosion mechanisms. However, corrosion in general has clear electrochemical roots. Namely, different parts of a heat exchanger exposed to working fluids easily become polarized. The role of an electrolyte is usually taken by a working fluid (or sometimes by solid deposits or thick metal oxide scale) in the vicinity of or between parts made of different metals. If an external electrical circuit is established, metal surfaces involved take the role of either anode or cathode. Appearance of an electric current forces electrical particles (say, positively charged metal ions) to leave the metal on the anode end and enter the surrounding electrolyte. On the other end, a metal surface that plays the role of cathode serves as a site where electrical current escapes from the electrolyte. The presence of this mechanism opens the way for metal dissolution at the anode end of the established electrical circuit. This dissolution can be interpreted as a corrosion effect (if all other mechanisms are suppressed). This very simplified picture provides a background for many corrosion problems. Refer to Kuppan (2000) for further details.

Detailed study of corrosion phenomena is beyond the scope of this book. Due to the importance of corrosion in heat exchanger design and operation, we will address only the most important topics. A brief description of the main corrosion types is given first (keeping in mind their importance from a heat exchanger design point of view).

Subsequently, corrosion mechanisms are addressed, followed by a brief discussion of each mechanism. Possible locations of corrosion in a heat exchanger are emphasized as well. Finally, the most important guidelines for corrosion prevention are enlisted.

13.5.1   Corrosion Types

Corrosion types, important for heat exchanger design and operation, are as follows: (1) uniform attack corrosion, (2) galvanic corrosion, (3) pitting corrosion, (4) stress corrosion cracking, (5) erosion corrosion, (6) deposit corrosion, and (7) selective leaching, as categorized by Fontana and Greene (1978). Let us define each corrosion type briefly.

Uniform corrosion is a form of corrosion caused by a chemical or electrochemical reaction between the metal and the fluid in contact with it over the entire exposed metal surface. It occurs when the metal and fluid (e.g., water, acid, alkali) system and operating variables are reasonably homogeneous. It is usually easy to notice corroded areas attacked by uniform corrosion. All other forms of corrosion mechanisms discussed below cause localized corrosion.

Galvanic corrosion is caused by an electric potential difference between two electrically dissimilar metals in the system in the presence of an electrolyte (such as water in a heat exchanger). It occurs on the anode and does not affect the cathode (referred to as a *noble metal*).

Pitting corrosion is a form of localized autocatalytic corrosion due to pitting that results in holes in the metal. If anodes and cathodes rapidly interchange the sites randomly, uniform corrosion occurs, as in rusting of iron. If the anode becomes fixed on the surface, pitting corrosion takes place.

Stress corrosion is a form of corrosion that involves cracks on susceptible metals caused by the simultaneous presence of the tensile stress and a corrosive fluid medium.

Erosion corrosion is a form of surface corrosion due to erosion of the heat transfer surface due to a high-velocity fluid with or without particulates (e.g., fluid velocity greater than 2 m/s or 6 ft/sec for water flow over an aluminum surface) and subsequent corrosion of the exposed surface.

Crevice corrosion is a form of localized physical deterioration of a metal surface in crevices or under deposits in shielded areas (i.e., in stagnant fluid flow regions), often caused by deposits of dirt and corrosion products.

Selective leaching or *dealloying* is the selective removal of one metal constituent from an alloy by corrosion that leaves behind a weak structure.

13.5.2   Corrosion Locations in Heat Exchangers

Uniform (general) corrosion is not localized, and a surrounding corrosive medium affects the surface of the exposed metal prone to corrosion. Temperature, concentrations, oxidation, acidity, and so on, have a significant influence on the extent of this type of corrosion. Atmospheric corrosion and high-temperature gaseous corrosion are most probable in heat exchangers. This corrosion usually thins the heat transfer surface. Metals having 0.1 mm/yr surface thinning are considered excellent, those having 0.1 to 0.5 mm/yr satisfactory, and those having above about 1 mm/yr unsatisfactory for shell-and-tube heat exchangers (Kuppan, 2000).

As opposed to general corrosion, *galvanic corrosion* often attacks interfaces/contacts between tubes and baffles and/or tubesheets, contact between the baffle and shell, and joint areas (either welded, brazed, soldered, or mechanically joined). The likelihood of galvanic corrosion can easily be assessed knowing the position of the materials involved in the galvanic series summarized in Table 13.6. The metals next to each other in the galvanic series have little tendency to galvanic corrosion. In addition, metals closer to the anodic end of the galvanic series are more prone to corrosion, and the materials at the cathodic end are more stable. It can also occur in compact and other exchangers with water and other electrolytes in the circuit in which the exchanger is one of the components.

Pitting corrosion takes place when a protective surface film breaks down; these surface films are formed on the metal surface by reaction with an environment or during the surface treatment. The common metals exposed to this type of corrosion in descending order of nobility are aluminum, stainless steels, nickel, titanium, and their alloys. It is a very aggressive type of corrosion. Pitting corrosion is influenced by metallurgical and environmental factors, such as breakdown of protective coatings, inhomogeneities in the alloys, and inhomogeneities caused by joining processes. Consequently, the appearance of pitting corrosion is possible whenever such conditions are present. Pits caused by pitting corrosion are usually at places where the metal surface has surface deformities and scratches.

Stress corrosion cracking may be present at locations within the construction where the joint interaction of stress and a corrosive medium causes material deterioration. The presence of higher stress levels, increased temperature and concentration of a corrosive medium, and crack geometry may accelerate corrosion. For example, tube-to-tubesheet expanded joints may be prone to residual stresses, as well as thin-walled expansion joints and/or U-bends. Cold working parts and U-bends in shell-and-tube heat exchangers are locations where corrosion may take place in combination with existing stress.

TABLE 13.6 Galvani Series[a]

Mg (anodic; least noble)
Zn
Fe (galvanized)
Al 3004
Al 3003
Cast iron
SS 430 (active)
SS 304 (active)
Admiralty brass
Monel 400
SS 430 (passive)
SS 304 (passive)
Lead
Copper
Nickel
Inconel 825
Hastelloy C
Titanium
Graphite
Platinum (cathodic; most noble)

[a] In seawater at 25°C; materials listed in ascending order of nobility.

Erosion corrosion involves solid particle or liquid droplet impingement and cavitation. In shell-and-tube heat exchangers, impingement plates must be designed to prevent this type of erosion corrosion in tubes exposed to nozzle inlet flow. Erosion corrosion is more common at the inlet end of a heat exchanger flow passage or on the tube side.

Crevice corrosion is localized corrosion and may occur at metal-to-metal or metal-to-nonmetal joints (e.g., gasketed joints), or underneath biological growth or fouling deposits. In particular, areas prone to this type of corrosion are stagnant areas and complex geometric designs with sharp edges. This type of corrosion usually starts with an infiltration of a corrosive substance into a crack and/or small opening, such as clearances between rolled tubes and tubesheets, open welds, bolt holes, nut adjacent areas, gasket areas, or contacts between plates in a plate heat exchanger. Fouling and various deposits influence corrosion at shielded areas if the combination of fluid and heat exchanger surface material is inappropriate.

Selective leaching (parting) takes place depending on the combination of alloys selected and the presence of a corrosive substance in the surrounding medium. Some typical problems encountered in heat exchanger operation are related to (1) removal of Zn from brass in stagnant waters, (2) removal of Al from aluminum brass in acidic solutions, and (3) removal of Ni in Cu–Ni alloys under conditions of high heat flux. Such removal processes are referred to as *dezincification, dealuminumification, denickeli-fication,* and so on.

13.5.3 Corrosion Control

Corrosion control may be categorized as corrosion prevention and protection. In general, both prevention and protection should be planned.

Uniform corrosion can be suppressed by applying adequate inhibitors, coatings, or cathodic protection. *Galvanic corrosion* can be reduced by selecting dissimilar materials to be as close as possible to each other on the galvanic series list for the pairs of components in the system. In addition, insulation of dissimilar metals, application of coatings, addition of inhibitors, and installation of a third metal which is anodic to both metals in the galvanic contact may be used to minimize galvanic corrosion. *Pitting corrosion* is difficult to control. Materials that show pitting should be avoided in heat exchanger components. Adding inhibitors does not necessarily lead to efficient mitigation of the corrosion. The best prevention of *stress-corrosion* cracking is an appropriate selection of material, reduction of tensile stresses in the construction, elimination of critical environmental components (e.g., demineralization or degasification), cathodic protection, and addition of inhibitors. The selection of correct material less prone to erosion, making inlet flow more uniform (thus eliminating velocity spikes), and approximate maximum velocities for a working fluid may reduce *erosion* effects. For example, stainless steel 316 can sustain three times larger water velocity flowing inside tubes than can steel or cooper. Also, design modifications, coatings, and cathodic protection should be considered. The best prevention of *crevice corrosion* is a design in which the stagnation areas of the fluid flow and sharp corners are reduced to a minimum. Design should be adjusted for complete drainage, and if possible, welding should be used instead of rolling for tubes in tubesheets. Additives to an alloy, such as arsenic or tin, may reduce dezincification, thus solving the problem with *selective leaching* of brass.

Increased control of corrosion may be achieved through the following means.

- Use of corrosion-resistant and clad metal (bimetal) materials
- Use of fluids with corrosion inhibitors
- Good design, avoiding crevices, stagnant fluid zones, upgrading materials, having uniform and optimum fluid velocities (not too high or too low in the exchanger), using solid nonabsorbent gaskets (e.g., Teflon), minimizing tensile and residual stresses in exchanger surfaces, designing for desired startups and shutdowns, and so on
- Proper selection of construction metals from the point of proximity in the galvanic series
- Surface coatings, surface treatment, electrochemical protection, and so on
- Maintaining clean exchanger surfaces (no deposits) and fluids (use a filter in the flow circuit)
- Avoiding aluminum alloys if erosion corrosion cannot be prevented

SUMMARY

Fouling adds thermal resistance to heat transfer in a heat exchanger as well as increasing pressure drop. Fouling has a transient character, but for the purpose of thermal design considerations, it is often included into analysis through the concept of fouling unit thermal resistance, fouling resistance, or a fouling factor in steady-state thermal design analysis. The phenomena that control the fouling processes are complex in nature, and a comprehensive general theory cannot be defined. Most often, the influence of fouling is included through an overdesign (i.e., through an appropriate additional heat transfer surface allowance as compared to ideally clean conditions). In some applications, this overdesign accelerates fouling because of the lower-than-design value of the fluid velocity in the exchanger. Fouling has a significant detrimental effect on heat transfer in liquids and on pressure drop (fluid pumping power) in gases.

In this chapter, various fouling mechanisms are discussed that have been identified and investigated particularly with the emphasis on fouling in single-phase liquid and gas sides and in compact heat exchangers. After providing details on the sequential events in fouling, modeling of a fouling process is presented. From the heat exchanger design perspective, the effect of fouling on heat transfer performance is taken into account by the fouling resistance approach. The methodology of this approach is outlined followed by the impact of fouling on exchanger heat transfer performance and data on fouling resistances. Next we have covered the prevention, control and mitigation of liquid- and gas-side fouling and then cleaning strategies for liquid and gas-side fouling.

The influence of corrosion has to be taken into account not only during operation but also during design. Both prevention and protection must be included in design consideration. Basic information on corrosion types, locations in heat exchangers, and control ideas are presented in the text.

REFERENCES

Bott, T. R., 1981, Fouling due to liquid solidification, in *Fouling of Heat Transfer Equipment*, E. F. C. Somerscales and J. G. Knudsen, eds., Hemisphere Publishing, Washington, DC, pp. 201–226.

Bott, T. R., 1990, *Fouling Notebook*, Institution of Chemical Engineers, Rugby, UK.

Cannas, F. C., 1986, Fouling in heat exchanger design, M.S. Thesis, Mech. Eng. Dept., State Univ. New York, Buffalo.

Chenoweth, J. M., 1990, Final report of the HTRI/TEMA joint committee to review the fouling section of the TEMA standards, *Heat Transfer Eng.*, Vol. 11, No. 1, pp. 73–107.

Epstein, N., 1978, Fouling in heat exchangers, *Heat Transfer 1978, Proc. 6th Int. Heat Transfer Conf.*, Vol. 6, pp. 235–253; also in *Fouling of Heat Transfer Equipment*, E. F. C. Somerscales and J. G. Knudsen, eds., Hemisphere Publishing, Washington, DC, 1981, pp. 701–734.

Fontana, M. G., and N. D. Greene, 1978, *Corrosion Engineering*, McGraw-Hill, New York.

Knudsen, J. G., 1998, Fouling in heat exchangers, in *Heat Exchanger Design Handbook*, G. F. Hewitt, ed., Begell House, New York, Sec. 3.17.

Kuppan, T., 2000, *Heat Exchanger Design Handbook*, Marcel Dekker, New York.

Marner, W. J., 1990, Progress in gas-side fouling of heat transfer surfaces, *Appl. Mech. Rev.*, Vol. 43, No. 3, pp. 35–66.

Marner, W. J., 1996, Progress in gas-side fouling of heat transfer surfaces, *Appl. Mech. Rev.*, Vol. 49, No. 10, Pt. 2, pp. S161–S166.

Marner, W. J., and J. W. Suitor, 1987, Fouling with convective heat transfer, in *Handbook of Single-Phase Convective Heat Transfer*, S. Kakaç, R. K. Shah, and W. Aung, eds., Wiley, New York, Chap. 21.

Melo, L. F., T. R. Bott, and C. A. Bernardo, eds., 1988, *Advances in Fouling Science and Technology*, Kluwer Academic Publishers, Dordrecht, The Netherlands.

Panchal, C. B., 1999, Review of fouling mechanisms, mitigation of heat exchanger fouling and its economic and environmental impacts, *Proc. Engineering Foundation Conf.*, Banff, Alberta, Canada, July.

Panchal, C. B., and J. G. Knudsen, 1998, Mitigation of water fouling: technology status and challenges, in *Advances in Heat Transfer*, Vol. 31, J. P. Hartnett, T. F. Irvine, Jr., Y. I. Cho, and G. A. Greene, eds., Academic Press, San Diego, CA, pp. 437–474.

Rabas, T. J., and C. B. Panchal, 2000, Fouling rates, not fouling resistances, *Heat Transfer Eng.*, Vol. 21, No. 2, pp. 1–2.

Shah, R. K., 1985, Compact heat exchangers, in *Handbook of Heat Transfer Applications*, 2nd ed., W. M. Rohsenow, J. P. Hartnett, and E. N. Ganić, eds., McGraw-Hill, New York, p. 4-284.

Somerscales, E. F. C., 1990, Fouling of heat transfer surfaces: a historical review, *Heat Transfer Eng.*, Vol. 11, No. 1, pp. 19–36.

TEMA, 1999, *Standard of the Tubular Exchanger Manufacturers Association*, 8th ed., Tubular Exchanger Manufacturers Asscociation, New York.

Zubair, S. M., and R. K. Shah, 2001, Fouling in plate-and-frame heat exchangers and cleaning strategies, in *Compact Heat Exchangers and Enhancement Technology for the Process Industries 2001*, R. K. Shah, A. Deakin, H. Honda and T. M. Rudy, eds., Begell House, New York, pp. 553–565.

Zubair, S. M., A. K. Sheikh, M. A. Budair, and M. A. Badar, 1997, A maintenance strategy for heat-transfer equipment subject to fouling: a probabilistic approach, *ASME J. Heat Transfer*, Vol. 119, pp. 575–580.

REVIEW QUESTIONS

For each question circle one or more correct answers. Explain your answers briefly.

13.1   Fouling of a heat transfer surface in a heat exchanger may result in a(an):
   (a) decrease in heat transfer   **(b)** change in the local heat transfer coefficient
   (c) increase in pressure drop   **(d)** decrease in pressure drop
   (e) increase in pumping power

13.2   Fouling is costly for the following reasons:
   (a) an oversized exchanger   **(b)** periodic cleaning
   (c) reduced on-time for the system/process
   (d) reduced heat transfer   **(e)** more fluid pumping power

13.3   Fouling has generally significant influence on heat transfer in:
   (a) compact heat exchangers   **(b)** shell-and-tube heat exchangers
   (c) liquid side of a heat exchanger   **(d)** gas side of a heat exchanger

13.4   When thermal resistances are in series, the fouling has a significant influence on heat transfer in an exchanger having:
   (a) high U **(b)** low U **(c)** both of these   **(d)** can't tell

13.5   A change in the hydraulic diameter due to fouling influences the local heat transfer coefficient h in a heat exchanger with constant $\dot{m}$, L, A, and fluid properties as follows:
   (a) h increases linearly with an increase in D_h.
   (b) h is inversely proportional to D_h.   **(c)** h is inversely proportional to D_h^3.

13.6   A change in the hydraulic diameter due to fouling influences the pressure drop Δp in a heat exchanger with constant $\dot{m}$, L, A, and fluid properties as follows:
   (a) Δp is inversely proportional to D_h^5.
   (b) Δp is inversely proportional to D_h^3.
   (c) Δp is proportional to D_h^3.   **(d)** Δp does not depend on D_h.

13.7   The fouling resistance is:
   (a) an empirical factor equal to the ratio U_f/U_c
   (b) the unit thermal resistance of the fouling deposit
   (c) a ratio of fouled to clean pressure drops

13.8   Precipitation fouling involves:
   (a) formation of bioorganisms on a heat transfer surface
   (b) deposition of dissolved inorganic material from a fluid onto heat transfer surface
   (c) local deposition of corrosion products

13.9   Particulate fouling involves:
   (a) deposition of solids suspended in fluids on the heat transfer surface
   (b) local influence of corrosion   **(c)** phase-change phenomena

13.10 On the air side of a compact heat exchanger, 50% of the flow passages are large and 50% are small. Which passages would have more change in thermal performance when the flowing air is dirty and fully developed laminar flows are expected in both passages?

(a) passages with larger flow area   (b) passages with smaller flow area

(c) both of these

13.11 In general, deposit formed on the heat transfer surface due to fouling shows the time history character as follows:

(a) Deposit thermal resistance is hyperbolic with respect to time.

(b) Deposit thermal resistance changes linearly with time.

(c) Deposit thermal resistance changes exponentially with time.

13.12 The time constant in the Kern–Seaton correlation for time dependence of the fouling resistance has the following physical meaning:

(a) time required for the fouling resistance to reach 50% of its asymptotic value

(b) time required for the fouling resistance to reach its asymptotic value

(c) time required for the fouling resistance to reach 63% of the asymptotic value

(d) time required for the fouling resistance to reach 99% of the asymptotic value

13.13 Water-side fouling can be most efficiently mitigated by:

(a) utilizing chemical additives   (b) utilizing cleaning devices

(c) designing heat transfer surfaces with highly augmented heat transfer characteristics

13.14 Combined tube- and shell-side fouling resistance is the largest for the following combination of working fluids:

(a) vapor on both sides of a heat transfer surface

(b) vapor on one side and phase-change fluid on the other

(c) liquid on both sides

(d) phase-change fluids on both sides

13.15 Fouling process models are based on the concept of:

(a) exponential decrease of the deposition rate and linear increase of the removal rate

(b) deposition rate minus removal rate   (c) initiation plus aging minus removal

(d) initiation plus transport minus removal

13.16 Increasing the liquid-side velocity in a heat exchanger will definitely reduce the likelihood of fouling due to:

(a) precipitation   (b) freezing   (c) particulate fouling

(d) biological fouling   (e) chemical fouling

13.17 Increasing the liquid-side temperature will definitely reduce the likelihood of fouling due to:

(a) biological fouling   (b) freezing

(c) precipitation   (d) particulate fouling

13.18 In the case of gas-side chemical reaction fouling, the following influences are registered:

(a) Increase in the gas-side temperature decreases the likelihood of fouling.

(b) Increase in the oxygen level increases the likelihood of fouling

(c) Increase in the velocity may increase or decrease fouling.

13.19 Increased presence of sulfur on the gas side will definitely increase the likelihood of fouling due to:

(a) chemical reaction fouling   (b) corrosion fouling   (c) freezing fouling

(d) particulate fouling

13.20 If fouling takes place, h is going to be higher, and for the same heat transfer, the surface area should be:

(a) increased   (b) decreased   (c) unchanged

13.21 Corrosion fouling in a phosphoric acid condensation process may be prevented by the use of a:

(a) counterflow arrangement   (b) plate heat exchanger

(c) parallelflow arrangement

(d) Teflon coating over low-temperature exchanger surfaces

13.22 Corrosion fouling in a sulfuric acid condensation process may be prevented by use of a:

(a) counterflow arrangement   (b) plate heat exchanger

(c) parallelflow arrangement

(d) Teflon coating over low-temperature exchanger surfaces

13.23 Which materials are ideally prone to galvanic corrosion for contacting copper in a seawater solution?

(a) nickel   (b) lead   (c) zinc   (d) cast iron

13.24 Circle the following statements as true or false.

(a) T F Fouling resistance is a time-dependent quantity and has some finite value at time equal to zero.

(b) T F Attachment is associated with the delay period τ_d.

13.25 The delay period τ_d associated with initiation decreases with a(n):

(a) increase in the degree of supersatuartion in crystallization fouling

(b) decrease in surface roughness   (c) increase in the fluid viscosity

13.26 In a plate-fin heat exchanger with the fin spacing of 3 mm and gas velocity of 10 m/s, which of the following fuels would be likely to produce fouling exhaust gases?

(a) natural gas   (b) propane   (c) #2 oil   (d) coal

13.27 Chemical cleaning of foulant can be accomplished by:

(a) dilute acids   (b) steam   (c) chlorinated hydrocarbons

(d) hot air   (e) dispersants

13.28 Mechanical cleaning of foulant may be accomplished by:

    (a) sonic horns   **(b)** rotary soot blowers

    (c) an oven for thermal baking   **(d)** air cannons

PROBLEMS

13.1 A shell-and-tube heat exchanger has tubes with 19 mm outside diameter and 1.2 mm tube wall thickness. Water velocity in the tubes is 1.8 m/s. Determine an increase in the tube-side pressure drop per unit length due to fouling when the value of fouling resistance reaches $5.3 \times 10^{-4}\,\mathrm{m^2 \cdot K/W}$. Assume that the flow rate remains constant even when the tube is fouled. Use the following correlations for the Fanning friction factor for smooth (clean) and rough (fouled) tubes: (1) $f = 0.0014 + 0.125\mathrm{Re}^{-0.32}$ for a smooth tube, and (2) $f = 0.0035 + 0.264\mathrm{Re}^{-0.42}$ for a rough tube. The density and dynamic viscosity of water are $998\,\mathrm{kg/m^3}$ and $1.12 \times 10^{-3}\,\mathrm{Pa \cdot s}$, respectively. The thermal conductivity of the fouling deposit is $1.73\,\mathrm{W/m \cdot K}$.

13.2 Consider a simplified model of a heat exchanger with heat transfer areas equal on the hot and cold sides and having no fins. Assume that this heat exchanger has to be used alternately as (1) a gas-to-gas heat exchanger (fouling on the compressed air side), (2) steam-to-heavy fuel oil heat exchanger (fouling on the heavy fuel oil side), and (3) liquid-to-liquid heat exchanger (fouling on the ethylene glycol solution side). Determine what would be the required increase in heat transfer area for the various working fluids, comparing the fouled to the clean heat exchanger operation. Assume that only the thermal resistance of the foulant would change the overall heat transfer coefficient (the other conditions remaining unchanged). Assess the typical values of the overall heat transfer coefficients and fouling resistances for described physical situations. The heat transfer rate and mean temperature difference remain the same.

13.3 Using the same information as those given in Problem 13.2, determine what would be the change of heat transfer rate between clean and fouled operations. Assume the heat transfer area and the mean temperature difference of the fluid streams to be invariant.

13.4 A water–water gasketed plate heat exchanger has an overall heat transfer coefficient under fouled conditions of $4200\,\mathrm{W/m^2 \cdot K}$. Hot- and cold-fluid-side heat transfer coefficients are 15,000 and $14,000\,\mathrm{W/m^2 \cdot K}$, respectively. The plate thickness (stainless steel 316) is 0.6 mm, and the thermal conductivity is $17\,\mathrm{W/m \cdot K}$. Calculate the cleanliness factor CF and the total fouling resistance for this exchanger.

13.5 In the cement industry, large gas-to-air heat exchangers are used to cool hot exhaust gases leaving suspension preheaters. These gases are cooled prior to being vented through a baghouse to the environment. In such heat exchangers, the hot dirty gases are placed on the tube side. In one application, owing to static electricity, very small particles of cement dust are deposited on the inside surface of the tubes, thus reducing the effectiveness of the heat exchanger. In this particular case, the tubes are 6.45 m long, 76 mm in inside diameter, with a wall thickness of 3.2 mm. The tubes are made of carbon steel with a thermal conductivity of 43 W/m · K. The average outside tube wall temperature is 120°C and the average gas temperature is 280°C. For the clean exchanger, the gas-side velocity is 15.6 m/s, resulting in a gas-side convective heat transfer coefficient of 39.3 W/m² · K. If a 3.2 mm thick layer of cement dust ($k = 0.299$ W/m · K) is deposited on the inner surface of the tubes, determine **(a)** $\hat{R}_f$, the gas-side unit fouling resistance (m² · K/W), **(b)** the reduction in heat transfer due to the fouling layer, and **(c)** the increase in the pressure drop due to the fouling layer. Consider the mean density of the air and gas as 0.64 kg/m³ and the dynamic viscosity as 2.85 Pa·s. Assume the mass flow rate to be constant for clean and fouled exchangers. Use the following correlation for turbulent flow through the tube, and ignore entrance and exit losses as well as the momentum effect for pressure drop calculation.

$$f = 0.0014 + 0.125\mathrm{Re}^{-0.32}$$

Explicitly mention any additional assumptions that you need to make.

13.6   A double-pipe heat exchanger is used to condense steam at a rate of 113.7 kg/h at 50°C. Cooling water (brackish water) enters through the tubes at a rate of 0.9 kg/s at 10°C. The specific enthalpy of phase change $\mathbf{h}_{\ell g}$ of water is 2382.7 kJ/kg and the specific heat of water is 4.18 kJ/kg · K. The tube (25.4 mm OD and 22.1 mm ID) is made of mild steel ($k = 45$ W/m · K). The heat transfer coefficient on the steam side is 10,000 W/m² · K and that on the cooling water side 8000 W/m² · K. Inside and outside fouling resistances $\hat{R}_{f,i}$ and $\hat{R}_{f,o}$ are 0.176 and 0.088 m² · K/kW, respectively.

(a) Determine the outside surface area requirement for the plain tube for clean and fouled exchangers.

(b) Assume the plain tube to be replaced by a low-finned tube with fins inside tubes. In that case, the fin increases the surface area by a factor of 2.9. Assume 100% fin efficiency and identical heat transfer coefficients, wall resistances, and inside fouling resistances for both plain and finned tubes. What is the outside surface area requirement for clean and fouled heat exchanger for the same heat duty and ΔT_m?

(c) Analyze the case identical to part (b), but with the fins outside the tubes.

(d) Compare the results of parts (a), (b), and (c) and discuss them.

13.7   A shell-and-tube condenser is designed to condense a process stream at 120°C using water at 60 kg/s at the inlet temperature of 30°C. The water flows through the tubes. The copper tubes (thermal conductivity 401 W/m · K) have an OD and ID of 19 and 16 mm, respectively, and the tube length per pass is 2.5 m. There are a total of 800 tubes in the exchanger, with four tube passes on the tube side. Check through your calculations that the measured water outlet temperature under clean conditions is 89°C. Also determine the water velocity through the clean tubes. After six months of service, the exchanger was retested for fouling effects, and for the same water flow rate and inlet temperatures, the water outlet temperature was measured as 85°C. Compute the tube-side fouling resistance (factor) during the second test if the exchanger was clean originally, there was no fouling on the steam side, and the tube heat transfer surface area was the same under clean and fouled conditions. Explain what additional information you would need to compute the water velocity in the fouled tubes. Assume the condensing heat transfer coefficient to be 4000 W/m$^2 \cdot$ K. Assume the following properties for water: $\rho = 1000$ kg/m^3, $c_p = 4180$ J/kg $\cdot$ K, $k = 0.59$ W/m $\cdot$ K, and $\mu = 0.001$ Pa $\cdot$ s. *Hint:* Use the Dittus–Boelter correlation for the tube-side heat transfer coefficient.

13.8 A shell-side condenser with cooling tower water on the tube side is not performing satisfactorily, due to tube-side fouling. There is negligible shell-side fouling. Contemplated is replacement of the plain tubes with low-finned tubing having 1.18 fins/mm. This will increase the tube outside area by a factor of 2.9 over that of the plain tubes. The following are some design data provided.

Quantity	Plain Tube	Low-Finned Tube
Tube outside diameter d_o (mm)	25.4	25.4
Tube inside diameter d_i (mm)	22.9	22.9
Tube-side fouling resistance (m$^2 \cdot$ K/W)	0.00018	0.00018
Tube-side heat transfer coefficient (W/m$^2 \cdot$ K)	8517	8517
Shell-side heat transfer coefficient (W/m$^2 \cdot$ K)	8517	8517
Wall thermal conductivity k_w (W/m $\cdot$ K)	17.3	17.3
Area ratio:		
Outside/inside ($A_{p,o}/A_{p,i}$)	1.11	
Finned/bare tube outside [$(A_p + A_f)_o/A_{p,o,\text{bare}}$]		2.9
Finned/tube inside [$(A_p + A_f)_o/A_{p,i}$]		3.22

Assuming a fin efficiency of 100% in parts (a)–(c):

(a) Determine the overall heat transfer coefficient based on the shell-side surface area for plain and finned-tube exchangers.

(b) Compute the percentage increase in q by employing low-finned tubes. Assume that ΔT_{lm} remains constant.

(c) Discuss the results of part (b) in relation to the above-specified increase in surface area due to fins.

(d) Discuss qualitatively the results for part (b) if the fin efficiency would have been 90%.

APPENDIX A
Thermophysical Properties

TABLE A.1 Thermophysical Properties of Metals at 300 K

Metal	Density ρ (kg/m³)	Specific heat c (J/kg·K)	Thermal Conductivity k (W/m·K)	Metal	Density ρ (kg/m³)	Specific heat c (J/kg·K)	Thermal Conductivity k (W/m·K)
Aluminum				**Steel**			
Pure	2702	903	237	AISI 1010	7832	434	63.9
Duralumin[a]	2770	875	174	AISI 1042 (annealed)	7840	460	50
Coper				AISI 4130 (hardened)	7840	460	43
Pure	8933	385	401	AISI 302	8055	480	15.1
Bronze (90% Cu, 10% Al)	8800	420	52	AISI 304	7900	477	14.9
Brass (70% Cu, 30% Zn)	8530	380	111	AISI 316	8238	468	13.4
Iron				AISI 347	7978	480	14.2
Pure	7870	447	80.2	AISI 410	7770	460	25
4 C cast	7272	420	51	**Titanium**			
Inconel X-750	8510	439	11.7	Pure	4500	522	21.9
Nickel, pure	8900	444	91	Ti-6Al-4V	4420	610	5.8
Nichrome (80% Ni, 20% Cr)	8314	460	13	Ti-21Al-2Mn	4510	466	8.4
Hasteloy B	9240	381	12.2				

Sources: Data from A. F. Mills, *Heat and Mass Transfer*, Richard D. Irwin, Burr Ridge, IL, 1995; F. P. Incropera and D. P. DeWitt, *Introduction to Heat Transfer*, Wiley, New York, 2002; Y. S. Touloukian and C. Y. Ho. *Thermophysical Properties of Matter*, Vols. 1–9, Plenum Press, New York, 1972; American Society for Metals, *Metals Handbook*, Vol. 1, ASM, Metals Park, OH, 1961.

[a] 4.4% Cu, 1.0% Mg, 0.75% Mn, 0.4% Si, Al balance.

TABLE A.2 **Thermal Conductivity [k (W/m · K)] and Specific Heat [c (J/kg · K)] of Metals as a Function of Temperature**

Metal	Temperature (K)											
	200		300		400		500		600		800	
	c	k	c	k	c	k	c	k	c	k	c	k
Al pure	798	237	903	237	949	240	996	236	1033	231	1146	218
Cu pure	356	413	385	401	397	393	412	386	417	379	433	366
Bronze	785	42	420	52	460	52	500	55				
Brass	360	74	380	111	395	134	410	143	425	146		150
Iron Armco	384	81	447	73	490	66	530	59	574	53	680	42
Iron cast			420	51		44		39		36		27
AISI 1010			434	64	487	59	520	54	559	49	685	39
AISI 1042				52	500	50	530	48	570	45	700	37
AISI 4130				43	500	42	530	41	570	40	690	37
AISI 302			480	15	512	17	531	19	559	20	585	23
AISI 304	402	13	477	15	515	17	539	18	557	20	582	23
AISI 316			468	13	504	15	528	17	550	18	576	21
AISI 410		25	460	25		26		27		27		29
Inconel	372	10.3	439	11.7	473	13.5	490	15.1	510	17	546	20.5
Nichrome				13	480	14	500	16	525	17	545	21
Ti-pure	405	25	522	22	551	20	572	20	591	19	633	19

Sources: Data from A. F. Mills, *Heat and Mass Transfer,* Richard D. Irwin, Burr Ridge, IL, 1995; F. P. Incropera and D. P. DeWitt, *Introduction to Heat Transfer,* Wiley, New York, 2002; Y. S. Touloukian and C. Y. Ho, *Thermophysical Properties of Matter,* Vols. 1–9, Plenum Press, New York, 1972; American Society for Metals, *Metals Handbook,* Vol. 1, ASM, Metals Park, OH, 1961.

TABLE A.3 **Thermophysical Properties of Some Liquid Metals**

Liquid Metal	T (K)	ρ (kg/m^3)	k (W/m · K)	c_p (J/kg · K)	$\mu \times 10^4$ (Pa · s)
Potassium	400	814	45.5	800	4.9
	500	790	43.6	790	2.8
	600	765	41.6	780	2.1
	800	717	36.8	750	1.6
Sodium	500	900	79.2	1335	4.2
	600	868	74.7	1310	3.1
	800	813	65.7	1260	2.2
	1000	772	59.3	1255	1.8
Lithium	500	514	43.7	4340	5.31
	600	503	46.1	4230	4.26
	800	483	50.7	4170	3.10
	900	473	55.9	4160	2.47

Source: Data from A. F. Mills, *Heat and Mass Transfer,* Richard D. Irwin, Burr Ridge, IL, 1995.

TABLE A.4   Thermophysical Properties of Saturated Liquids

Liquid	T (K)	ρ (kg/m^3)	k (W/m · K)	c_p (J/kg · K)	$\mu \times 10^3$ (Pa · s)	$\beta \times 10^6$ (K^{-1})
Water	273.15	1000	0.569	4217	1.750	−68.05
	280	1000	0.582	4198	1.422	46.04
	290	999.0	0.598	4184	1.080	174.0
	300	997.0	0.613	4179	0.855	276.1
	310	993.0	0.628	4178	0.695	361.9
	320	989.1	0.640	4180	0.577	436.7
	330	984.3	0.650	4184	0.489	504.0
	340	979.4	0.660	4188	0.420	566.0
	350	973.7	0.668	4195	0.365	624.2
	400	937.2	0.688	4256	0.217	896
	500	831.3	0.642	4660	0.118	—
	600	648.9	0.497	7000	0.081	—
	647.3	315.5	0.238	∞	0.045	—
Engine oil	273	899.1	0.147	1796	3850	700
(unused)	280	895.3	0.144	1827	2170	700
	300	884.1	0.145	1909	486	700
	320	871.8	0.143	1993	141	700
	340	859.9	0.139	2076	53.1	700
	360	847.8	0.138	2161	25.2	700
	380	836.0	0.136	2250	14.1	700
	400	825.1	0.134	2337	8.74	700
	420	812.1	0.133	2427	5.64	700
Ethylene	273	1130.8	0.242	2294	65.1	650
glycol	290	1125.8	0.244	2323	42.0	650
	290	1118.8	0.248	2368	24.7	650
	300	1114.4	0.252	2415	15.7	650
	310	1103.7	0.255	2460	10.7	650
	320	1096.2	0.258	2505	7.57	650
	330	1089.5	0.260	2549	5.61	650
	340	1083.8	0.261	2592	4.31	650
	350	1079.0	0.261	2637	3.42	650
	360	1074.0	0.261	2682	2.78	650
	370	1066.7	0.262	2728	2.28	650
	373	1058.5	0.263	2742	2.15	650
Freon R-12	230	1528	0.068	881.6	0.457	1850
	240	1498	0.069	892.3	0.385	1900
	250	1470	0.070	903.7	0.354	2000
	260	1439	0.073	916.3	0.322	2100
	270	1407	0.073	930.1	0.304	2250
	280	1374	0.073	945.0	0.283	2350
	290	1341	0.073	960.9	0.265	2550
	300	1306	0.072	978.1	0.254	2750
	310	1269	0.069	996.3	0.244	3050
	320	1229	0.068	1015.5	0.233	3500

Source: Data from F. P. Incropera and D. P. DeWitt, *Introduction to Heat Transfer,* 5th ed., Wiley, New York, 2002.

TABLE A.5   Thermophysical Properties of Gases at Atmospheric Pressure

Gas	T (K)	ρ (kg/m^3)	k (W/m $\cdot$ K)	c_p (J/kg $\cdot$ K)	$\mu \times 10^3$ (Pa $\cdot$ s)
Air	100	3.5562	0.00934	1032	71.1
	150	2.3364	0.0138	1012	103.4
	200	1.7458	0.0181	1007	132.5
	250	1.3947	0.0223	1006	159.6
	300	1.1614	0.0263	1007	184.6
	350	0.9950	0.0300	1009	208.2
	400	0.8711	0.0338	1014	230.1
	450	0.7740	0.0373	1021	250.7
	500	0.6964	0.0407	1030	270.1
	600	0.5804	0.0469	1051	305.8
	700	0.4975	0.0524	1075	338.8
	800	0.4354	0.0573	1099	369.8
	900	0.3868	0.0620	1121	398.1
	1000	0.3482	0.0667	1141	424.2
Ammonia	300	0.6894	0.0247	2158	101.5
	320	0.6448	0.0272	2170	109
	340	0.6059	0.0293	2192	116.5
	360	0.5716	0.0316	2221	124
	380	0.5410	0.0340	2254	131
	400	0.5136	0.0370	2287	138
	420	0.4888	0.0404	2322	145
	440	0.4664	0.0435	2357	152.5
	460	0.4460	0.0463	2393	159
	480	0.4273	0.0492	2430	166.5
	500	0.4101	0.0525	2467	173
Steam	380	0.5863	0.0246	2060	127.1
	400	0.5542	0.0261	2014	134.4
	450	0.4902	0.0299	1980	152.5
	500	0.4405	0.0339	1985	170.4
	550	0.4005	0.0379	1997	188.4
	600	0.3652	0.0422	2026	206.7
	650	0.3380	0.0464	2056	224.7
	700	0.3140	0.0505	2085	242.6
	750	0.2931	0.0549	2119	260.4
	800	0.2739	0.0592	2152	278.6
	850	0.2579	0.0637	2186	296.9

Source: Data adapted and modified from F. P. Incropera, and D. P. DeWitt, *Introduction to Heat Transfer,* 5th ed., Wiley, New York, 2002.

TABLE A.6 Thermophysical Properties of R134a as Liquid and Vapour along the Saturation Line

T (K)	P (MPa)	Vapor				Liquid			
		ρ (kg/m³)	c_p (kJ/kg·K)	$\mu \times 10^3$ (Pa·s)	k (W/m·K)	ρ (kg/m³)	c_p (kJ/kg·K)	$\mu \times 10^3$ (Pa·s)	k (W/m·K)
273.2	0.293	14.43	0.90	10.7	0.0115	1295	1.34	271.1	0.0920
275	0.313	15.39	0.91	10.8	0.0117	1289	1.35	264.8	0.0912
280	0.373	18.23	0.93	11.0	0.0121	1272	1.36	248.5	0.0890
285	0.441	21.48	0.96	11.2	0.0126	1255	1.38	233.3	0.0868
290	0.518	25.19	0.98	11.4	0.0130	1237	1.39	219.2	0.0846
295	0.605	29.4	1.01	11.7	0.0135	1219	1.41	205.9	0.0825
300	0.703	34.19	1.04	11.9	0.0140	1200	1.43	193.3	0.0803
305	0.812	39.63	1.08	12.1	0.0145	1180	1.46	181.5	0.0782
310	0.933	45.79	1.12	12.4	0.0151	1160	1.48	170.2	0.0761
315	1.068	52.77	1.16	12.6	0.0157	1139	1.51	159.5	0.0739
320	1.217	60.71	1.21	12.9	0.0163	1117	1.54	149.3	0.0718
325	1.380	69.76	1.27	13.2	0.0170	1094	1.58	139.5	0.0696
330	1.560	80.09	1.34	13.6	0.0178	1069	1.63	130.0	0.0675
335	1.757	91.97	1.42	13.9	0.0187	1043	1.68	120.9	0.0653
340	1.972	105.7	1.52	14.4	0.0197	1015	1.75	111.9	0.0631
345	2.206	121.8	1.66	14.8	0.0209	984.7	1.84	103.2	0.0609
350	2.461	141	1.85	15.4	0.0225	951.3	1.96	94.5	0.0586
355	2.739	164.3	2.13	16.1	0.0246	913.8	2.14	85.8	0.0563
360	3.040	193.6	2.61	17.0	0.0274	870.1	2.44	76.8	0.0541
365	3.369	232.9	3.58	18.4	0.0318	816.3	3.04	67.2	0.0522
370	3.728	293.9	6.86	20.7	0.0407	740.3	5.11	55.8	0.0521
374	4.042	434.1	137.20	27.5	0.1018	587.9	101.70	38.9	0.0881

Source: Data from M. O. McLinden, S. A. Klein, E. W. Lemmon, and A. P. Peskin, *NIST Thermodyamic and Transport Properties of Refrigrants and Refrigerant Mixtures,* NIST Standard Reference Database 23, REFPROP Version 6.01, U.S. Department of Commerce, Technology Administration, National Institute of Standards and Technology, Physical and Chemical Properties Division, Boulder, CO, 1998.

APPENDIX B
ε-NTU Relationships for Liquid-Coupled Exchangers

The liquid-coupled indirect-transfer type exchanger system, also referred to as a *run-around coil system*, connects two direct-transfer type exchangers (recuperators) usually located apart by a circulating liquid as shown in Fig. B.1. Thus this system allows heat transfer between source and sink which are not closely located and/or must be separated. Such a system is commonly used in waste heat recovery applications such as HVAC and low-temperature process waste heat recovery, industrial dryers having inlet and exhaust ducts at the opposite end of the plant, drying of grains, and so on (Reay, 1979). It can also be used in waste heat recovery from a hot gas to cold air, where the inlet density difference could be very high (such as over fivefold). In such a case, complex gas ducting can be simplified through the use of a circulating liquid. This is also the case for a gas-to-gas exchanger, where the circulating liquid can simplify the gas ducting arrangement. Such a system would require two exchangers and theoretically would require a higher surface area for heat transfer from the hot to the cold fluid, about 10 to 20% (Kays and London, 1998). In addition, it will require additional components for the circulating liquid, adding cost and complexity.

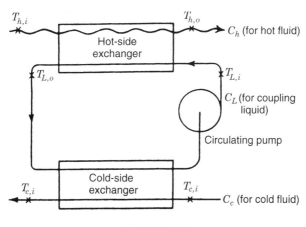

FIGURE B.1

TABLE B.1   Relationship between ε and ε_h and ε_c for Liquid-Coupled Exchangers

Heat Capacity Rate Criteria	ε Formula	Heat Capacity Rate Criteria	ε Formula
$C_L > C_c > C_h$	$\varepsilon = \dfrac{1}{\dfrac{1}{\varepsilon_h} + \dfrac{C_h/C_c}{\varepsilon_c} - \dfrac{C_h}{C_L}}$	$C_h > C_L > C_c$	$\varepsilon = \dfrac{1}{\dfrac{1}{\varepsilon_c} + \dfrac{C_c}{C_L}\left(\dfrac{1}{\varepsilon_h} - 1\right)}$
$C_L > C_h > C_c$	$\varepsilon = \dfrac{1}{\dfrac{1}{\varepsilon_c} + \dfrac{C_c/C_h}{\varepsilon_h} - \dfrac{C_c}{C_L}}$	$C_h = C_c = C > C_L$	$\varepsilon = \dfrac{C_L/C}{\dfrac{1}{\varepsilon_c} + \dfrac{1}{\varepsilon_h} - 1}$
$C_c > C_h > C_L$	$\varepsilon = \dfrac{1}{\dfrac{C_h}{C_L}\left(\dfrac{1}{\varepsilon_c} + \dfrac{1}{\varepsilon_h} - 1\right)}$	$C_h = C_c = C < C_L$	$\varepsilon = \dfrac{1}{\dfrac{1}{\varepsilon_c} + \dfrac{1}{\varepsilon_h} - \dfrac{C}{C_L}}$
$C_h > C_c > C_L$	$\varepsilon = \dfrac{1}{\dfrac{C_c}{C_L}\left(\dfrac{1}{\varepsilon_c} + \dfrac{1}{\varepsilon_h} - 1\right)}$	$C_c = C_h = C = C_L$	$\varepsilon = \dfrac{1}{\dfrac{1}{\varepsilon_c} + \dfrac{1}{\varepsilon_h} - 1}$
$C_c > C_L > C_h$	$\varepsilon = \dfrac{1}{\dfrac{1}{\varepsilon_h} + \dfrac{C_h}{C_L}\left(\dfrac{1}{\varepsilon_c} - 1\right)}$		

Source: Data from Kays and London (1998).

The analysis of individual exchangers in the liquid-coupled system is straightforward using the ε-NTU theory presented in Section 3.3. The individual effectiveness values for the overall system, the hot- and cold-side exchangers in Fig. B.1, are designated as ε, ε_h, and ε_c using the definition used in Section 3.3 (i.e., based on the $C_{\min}$ values for individual changers). For example, if $C_L > C_h > C_c$,

$$\varepsilon = \frac{C_c(T_{c,o} - T_{c,i})}{C_c(T_{h,i} - T_{c,i})} \qquad \varepsilon_h = \frac{C_h(T_{h,i} - T_{h,o})}{C_h(T_{h,i} - T_{L,i})} \qquad \varepsilon_c = \frac{C_c(T_{c,o} - T_{c,i})}{C_c(T_{L,o} - T_{c,i})} \qquad \text{(B.1)}$$

Note that ε_h and ε_c here are the exchanger effectiveness values of the hot and cold fluids, and they are *not* the temperature effectiveness values defined by Eqs. (3.51) and (3.52). The ε_h and ε_c are related to the overall effectiveness ε of the liquid-coupled exchangers as shown in Table B.1, where different formulas are presented depending on the relationships among the heat capacity rates of the hot fluid, cold fluid, and the circulating liquid. Note that in most applications, C_L is larger than C_h and C_c.

REFERENCES

Kays, W. M., and A. L. London, 1998, *Compact Heat Exchangers*, reprint 3rd ed., Krieger Publishing, Malabar, FL.

Reay, D. A., 1979, *Heat Recovery Systems*, E.&F.N. Spon, London.

APPENDIX C
Two-Phase Heat Transfer and Pressure Drop Correlations

Although the focus in this book is on single-phase flow heat exchanger design and analysis, there are situations when phase-change (condensation or vaporizing) fluid having negligible thermal resistance is on one fluid side of a two-fluid heat exchanger; the design and analysis for such an exchanger can be done using the slightly modified single-phase theory outlined in this book. However, we need to compute the heat transfer coefficient on the phase-change side even for this situation. Additionally, if one would like to estimate approximately the performance or size of the phase-change exchanger, it can be treated as a single-phase exchanger once the average heat transfer coefficient on the phase-change side is determined. Hence, in this appendix we provide some correlations for condensation and convective boiling. For the detailed information on the phase-change correlations and related phenomena, a comprehensive source is the handbook by Kandilkar et al. (1999). For completeness, we also provide a method to compute the pressure drop on the phase-change side and present it before the heat transfer correlations. Of course, many important topics of phase-change exchangers, such as the phase-change side not having the negligible thermal resistance, rating and sizing of the exchanger when phase change occurs on both fluid sides, flow maldistribution, and so on, are beyond the scope of this appendix and the book.

C.1 TWO-PHASE PRESSURE DROP CORRELATIONS

Due to the phase change during condensation or vaporization, the pressure gradient within the fluid changes along the flow path or axial length. The pressure drop in the phase-change fluid can then be computed by integrating the nonlinear pressure gradient along the flow path. In contrast, the pressure gradient is linear along the flow length (axial direction) in many single-phase flow applications, and hence we generally work directly with the pressure drop since there is no need to compute the pressure gradient in single-phase flow.

The total local pressure gradient in two-phase flow through a one-dimensional duct can be calculated as follows[†]:

$$\frac{dp}{dz} = \frac{dp_{fr}}{dz} + \frac{dp_{mo}}{dz} + \frac{dp_{gr}}{dz} \tag{C.1}$$

[†]Additional symbols used in this appendix are all defined here and are not included in the main nomenclature section.

where the three terms on the right-hand side correspond to the contributions by friction, momentum rate change, and gravity denoted by the subscripts fr, mo, and gr, respectively. The analysis that follows is based on a homogeneous model. The entrance and exit pressure loss terms of single-phase flow [see Eq. (6.28)] are lumped into the Δp_{fr} term since the information about these contributions is not available, due to the difficulty in measurements. The in-tube two-phase frictional pressure drop is computed from the corresponding pressure drop for single-phase flow as follows using the two-phase friction multiplier denoted as φ^2:

$$\left(\frac{dp}{dz}\right)_{fr} = f_{lo}\frac{4}{D_h}\frac{G^2}{2g_c\,\rho_l}\varphi_{lo}^2 \quad\text{where}\quad \varphi_{lo}^2 = \frac{(dp/dz)_{fr}}{(dp/dz)_{fr,lo}} \tag{C.2}$$

where f_{lo} is the single-phase Fanning friction factor (see Tables 7.3 through 7.8) based on the total mass flow rate as liquid and G is also based on the total mass flow rate as liquid; this means that the subscript "lo" indicates the two-phase flow considered as all liquid flow. The subscripts l and g in Eqs. (C.2) and (C.3) denote liquid and gas/vapor phases, respectively, and the subscript lo stands for entire two-phase flow as liquid flow.

Alternatively, $(dp/dz)_{fr}$ is determined using the liquid or vapor-phase pressure drop multiplier as follows.

$$\left(\frac{dp}{dz}\right)_{fr} = \left(\frac{dp}{dz}\right)_{fr,l}\varphi_l^2 = \left(\frac{dp}{dz}\right)_{fr,g}\varphi_g^2 \tag{C.3}$$

where

$$\varphi_l^2 = \frac{(dp/dz)_{fr}}{(dp/dz)_{fr,l}} \quad \varphi_g^2 = \frac{(dp/dz)_{fr}}{(dp/dz)_{fr,g}} \quad \left(\frac{dp}{dz}\right)_{fr,l} = \frac{4f_lG^2}{2g_c\rho_lD_h} \quad \left(\frac{dp}{dz}\right)_{fr,g} = \frac{4f_gG^2}{2g_c\rho_gD_h} \tag{C.4}$$

where the subscripts l and g denote liquid and gas/vapor phases. φ_{lo}^2 and φ_l^2 or φ_g^2 are functions of the parameter X (*Martinelli parameter*). φ_{go}^2 [defined similar to φ_{lo}^2 of Eq. (C.2), with the subscript lo replaced by go] is a function of Y (*Chisholm parameter*). The X and Y are defined as follows:

$$X^2 = \frac{(dp/dz)_{fr,l}}{(dp/dz)_{fr,g}} \quad Y^2 = \frac{(dp/dz)_{fr,go}}{(dp/dz)_{fr,lo}} \tag{C.5}$$

Here the subscript go means the total two-phase flow considered as all gas flow. The correlations to determine the two-phase frictional pressure gradient are presented in Table C.1 for various ranges of G and μ_l/μ_g (Kandlikar et al., 1999, p. 228).

The momentum pressure gradient can be calculated integrating the momentum balance equation (Collier and Thome, 1994), thus obtaining

$$\left(\frac{dp}{dz}\right)_{mo} = \frac{d}{dz}\left[\frac{G^2}{g_c}\left(\frac{x^2}{\alpha\rho_g} + \frac{(1-x)^2}{(1-\alpha)\rho_l}\right)\right] \tag{C.6}$$

where α represents the void fraction of the gas (vapor) phase (a ratio of volumetric flow rate of the gas/vapor phase divided by the total volumetric flow rate of the two-phase mixture), and x is the mass quality (a ratio of the mass flow rate of the vapor/gas phase

TABLE C.1   Frictional Multiplier Correlations Used for Determining the Two-Phase Frictional Pressure Gradient in Eq. (C.2)

Correlation	Parameters
Friedel correlation (1979) for $\mu_l/\mu_g > 1000$ and all values of G: $$\varphi_{lo}^2 = E + \frac{3.24FH}{Fr^{0.045} \cdot We^{0.035}}$$ Accuracy for annular flow: $\pm 21\%$ (Ould Dide et al., 2002	$E = (1-x)^2 + x^2 \dfrac{\rho_l \, f_{go}}{\rho_g \, f_{lo}}$ $F = x^{0.78}(1-x)^{0.24}$ $H = \left(\dfrac{\rho_l}{\rho_g}\right)^{0.91} \left(\dfrac{\mu_g}{\mu_l}\right)^{0.19} \left(1 - \dfrac{\mu_g}{\mu_l}\right)^{0.7}$ $Fr = \dfrac{G^2}{g d_i \rho_{hom}^2} \qquad We = \dfrac{G^2 d_i}{\rho_{hom}\sigma}$ $\dfrac{1}{\rho_{hom}} = \dfrac{x}{\rho_g} + \dfrac{1-x}{\rho_l} \qquad \sigma = \text{surface tension (N/m)}$
Chisholm correlation (1973) for $\mu_l/\mu_g > 1000$ and $G > 100\,kg/m^2 \cdot s$: $$\varphi_{lo}^2 = 1 + (Y^2 - 1)[Bx^{n^*}(1-x)^{n^*} + x^{1-n}]$$ $$n^* = \frac{2-n}{2}$$ Accuracy for annular flow: $\pm 38\%$ (Ould Didi et al., 2002)	Y defined in Eq. (C.4); $n = \frac{1}{4}$ (exponent in $f = C\,Re^n$) G = total mass velocity, $kg/m^2 s$ $B = \begin{cases} 4.8 & G < 500 \\ 2400/G & 500 \le G \le 1900 \\ 55/G^{1/2} & G \ge 1900 \end{cases} \Bigg\}$ for $0 < Y \le 9.5$ $B = \begin{cases} 520/(YG^{1/2}) & G \le 600 \\ 21/G & G > 600 \end{cases} \Bigg\}$ for $9.5 < Y \le 28$ $B = 15{,}000/(Y^2 G^{1/2})$ for $Y > 28$
Lockhart-Martinelli correlation (1949) for $\mu_l\mu_g > 1000$ and $G < 100\,kg/m^2 \cdot s$: $$\varphi_l^2 = \frac{d\rho/dz)_{fr}}{(dp/dz)_l} = 1 + \frac{c}{X} + \frac{1}{X^2}$$ $$\varphi_g^2 = \frac{(dp/dz)_{fr}}{(dp/dz)_g} = 1 + cX + X^2$$ Accuracy for annular flow: $\pm 29\%$ (Ould Didi et al., 2002)	Correlation constant by Chisholm (1967): $c = 20$ for liquid and vapor both turbulent $c = 10$ for liquid-turbulent, vapor-laminar $c = 12$ for liquid-laminar, vapor-turbulent $c = 5$ for liquid and vapor both laminar

divided by the total mass flow rate of the two-phase mixture). Equation (C.6) is valid for constant cross-sectional (flow) area along the flow length. For the homogeneous model, the two-phase flow behaves like a single phase and the vapor and liquid velocities are equal. A number of correlations for the void fraction α are given by Carey (1992) and Kandlikar et al. (1999). An empirical correlation for the void fraction whose general form is valid for several frequently used models is given by Butterworth (Carey, 1992) as

$$\alpha = \left[1 + A\left(\frac{1-x}{x}\right)^p \left(\frac{\rho_g}{\rho_l}\right)^q \left(\frac{\mu_l}{\mu_g}\right)^r\right]^{-1} \tag{C.7}$$

where the constants A, p, q, and r depend on the two-phase model and/or empirical data chosen. These constants for a nonhomogeneous model, based on steam–water data, are $A = 1$, $p = 1$, $q = 0.89$, and $r = 0.18$. For the homogeneous model, $A = p = q = 1$ and $r = 0$. For the Lockhart and Martinelli model, $A = 0.28$, $p = 0.64$, $q = 0.36$, and $r = 0.07$. For engineering design calculations, the homogeneous model yields the best results when the slip velocity between the gas and liquid phases is small (for bubbly or mist flows).

Finally, the pressure gradient due to the gravity (hydrostatic) effect is

$$\left(\frac{dp}{dz}\right)_{gr} = \pm\frac{g}{g_c}\sin\theta[\alpha\rho_g + (1-\alpha)\rho_l] \tag{C.8}$$

Note that the negative sign (i.e., the pressure recovery) stands for downward flow in inclined or vertical tubes/channels, and the positive sign (i.e., pressure drop) represents upward flow in inclined or vertical tubes/channels. And θ represents the angle of tube/channel inclination measured from the horizontal axis.

C.2 HEAT TRANSFER CORRELATIONS FOR CONDENSATION

Condensation represents a vapor–liquid phase-change phenomenon that usually takes place when vapor is cooled below its saturation temperature at a given pressure. The heat transfer rate per unit heat transfer surface area from the pure condensing fluid to the wall is given by

$$q'' = h_{con}(T_{sat} - T_w) \tag{C.9}$$

where h_{con} is the condensation heat transfer coefficient, T_{sat} is the saturation temperature of the condensing fluid at a given pressure, and T_w is the wall temperature. We summarize here the correlations for *filmwise* in-tube condensation, a common condensation mode in

TABLE C.2 Heat Transfer Correlations for Internal Condensation in Horizontal Tubes

Stratification Conditions	Correlation
Annular flow[a] (film condensation) (Shah, 1977), accuracy $\pm 14.4\%$ (Kandlikar et al., 1999)	$h_{loc} = 0.023\dfrac{k_l}{d_i}\cdot Re_l^{0.8}\cdot Pr_l^{0.4}\left[(1-x)^{0.8} + \dfrac{3.8x^{0.76}(1-x)^{0.04}}{(p_{sat}/p_{cr})^{0.38}}\right]$ $Re_l = \dfrac{Gd_i}{\mu_l}, \quad G = \text{total mass velocity (kg/m}^2\cdot\text{s)}$ $0.002 \leq p_{sat}/p_{cr} \leq 0.44 \quad\quad 11 \leq G \leq 1599\,\text{kg/m}^2\cdot\text{s}$ $21 \leq T_{sat} \leq 310°C, \quad 0 \leq x \leq 1, \quad\quad Pr_l > 0.5$ $3 \leq u_{vap} \leq 300\,\text{m/s, no limit on } q$ $7 \leq d_i \leq 40\,\text{mm} \quad\quad Re_l > 350 \text{ for circular tubes}$
Stratified flow (Carey, 1992), accuracy: $\pm 18\%$ (Ould Didi et al., 2002)	$h_m = 0.728\left[1 + \dfrac{1-x}{x}\left(\dfrac{\rho_g}{\rho_l}\right)^{2/3}\right]^{-3/4}\left[\dfrac{k_l^3\rho_l(\rho_l-\rho_g)g h_{lg}'}{\mu_l(T_{sat}-T_w)d_i}\right]^{1/4}$ where $\quad h_{lg}' = h_{lg} + 0.68c_{p,l}(T_{sat}-T_w)$

[a] Valid for horizontal, vertical, or inclined tubes.

most industrial applications. The two most common flow patterns for convective condensation are annular film flow in horizontal and vertical tubes and stratified flow in horizontal tubes. For annular film flow, the correlation for the local heat transfer coefficient h_{loc} [$h_{con} = h_{loc}$ in Eq. (C.9)] is given in Table C.2; and also for stratified flow, the correlation for mean condensation heat transfer coefficient $h_{con} = h_m$ is given in Table C.2. Shah et al. (1999) provide condensation correlations for a number of noncircular flow passage geometries.

C.3   HEAT TRANSFER CORRELATIONS FOR BOILING

Vaporization (boiling and evaporation) phenomena have been investigated and reported extensively in the literature. In this case, the heat transfer rate per unit heat transfer surface area from the wall to the pure vaporizing fluid is given by

$$q'' = h_{tp}(T_w - T_{sat}) \tag{C.10}$$

where h_{tp} is the two-phase heat transfer coefficient during the vaporization process. We present here a most general intube forced convective boiling correlation proposed by Kandlikar (1991). It is based on empirical data for water, refrigerants and cryogens. The correlation consists of two parts, the convective and nucleate boiling terms, and utilizes a fluid–surface parameter. The Kandlikar correlation for the two-phase heat transfer coefficient is as follows:

$$\frac{h_{tp}}{h_{lo}} = \text{larger of} \begin{cases} [0.6683\,\text{Co}^{-0.2} \cdot f_2(\text{Fr}_{lo}) + 1058\,\text{Bo}^{0.7} \cdot F_{fl}](1-x)^{0.8} \\ [1.136\,\text{Co}^{-0.9} \cdot f_2(\text{Fr}_{lo}) + 667.2\,\text{Bo}^{0.7} \cdot F_{fl}](1-x)^{0.8} \end{cases} \tag{C.11}$$

where

$$h_{lo} = \begin{cases} \dfrac{\text{Re}_{lo} \cdot \text{Pr}_l(f/2)(k_l/d_i)}{1.07 + 12.7(\text{Pr}^{2/3}-1)(f/2)^{0.5}} & 10^4 \le \text{Re}_{lo} \le 5 \times 10^6 \\[4mm] \dfrac{\text{Re}_{lo} \cdot \text{Pr}_l(f/2)(k_l/d_i)}{1.07 + 12.7(\text{Pr}^{2/3}-1)(f/2)^{0.5}} & 2300 \le \text{Re}_{lo} \le 10^4 \end{cases} \tag{C.12}$$

$$f_2(\text{Fr}_{lo}) = \begin{cases} (25\,\text{Fr}_{lo})^{0.3} & \text{for } \text{Fr}_{lo} < 0.04 \text{ in horizontal tubes} \\ 1 & \text{for vertical tubes and for } \text{Fr}_{lo} \ge 0.04 \text{ in horizontal tubes} \end{cases} \tag{C.13}$$

$$f = \frac{1}{[1.58\ln(\text{Re}_{lo}) - 3.28]^2} \tag{C.14}$$

Here h_{lo} is the single-phase heat transfer coefficient for the entire flow as liquid flow. Also, the convection number Co, the nucleate boiling number Bo, and the Froude number Fr for the entire flow as liquid are defined as follows:

$$\text{Co} = \left(\frac{\rho_g}{\rho_l}\right)^{0.5}\left(\frac{1-x}{x}\right)^{0.8} \qquad \text{Bo} = \frac{q''}{G h_{lg}} \qquad \text{Fr} = \frac{G^2}{\rho_l^2 g d_i} \tag{C.15}$$

TABLE C.3 F_{fl} Recommended by Kandlikar (1991)

Fluid	F_{fl}	Fluid	F_{fl}
Water	1.00	R-114	1.24
R-11	1.30	R-134a	1.63
R-12	1.50	R-152a	1.10
R-13B1	1.31	R-32/R-132 (60%–40% wt.)	3.30
R-22	2.20	Kerosene	0.488
R-113	1.30		

F_{fl} is a fluid–surface parameter and depends on the fluid and the heat transfer surface. F_{fl} values for several fluids in copper tubes are presented in Table C.3. F_{fl} should be taken as 1.0 for stainless tubes. This correlation is valid for either vertical (upward and downward) or horizontal intube flow. A mean deviation of slightly less than 16% with water and 19% with refrigerants has been reported by Kandlikar (1991).

Note that being fluid specific, F_{fl} cannot be used for other fluids (new refrigerants) and mixtures. It is also not accurate for stratified wavy flows and at high vapor qualities since it is not based on the onset of dryout. The Thome model (Kattan et al., 1998; Zrcher et al., 1999), based on a flow pattern map, is recommended for those cases.

REFERENCES

Carey, V. P., 1992, *Liquid-Vapor Phase Change Phenomena*, Taylor & Francis, Bristol, PA.

Chisholm, D., 1967, A theoretical basis for the Lockhart–Martinelli correlation for two-phase flow, *Int. J. Heat Mass Transfer*, Vol. 10, pp. 1767–1778.

Chisholm, D., 1973, Pressure gradients due to friction during the flow of evaporating two-phase mixtures in smooth tubes and channels, *Int. J. Heat Mass Transfer*, Vol. 16, pp. 347–358.

Collier, J. G., and J. R. Thome, 1994, *Convective Boiling and Condensation*, 3rd ed., McGraw-Hill, New York.

Friedel, L., 1979, Improved friction pressure drop correlations for horizontal and vertical two-phase pipe flow, European Two-Phase Flow Group Meeting, Ispra, Italy, Paper E2.

Hewitt, G. F., 1998, Gas–liquid flow, in *Handbook of Heat Exchanger Design*, G. F. Hewitt, ed., Begell House, New York, Sect. 2.3.2.

Kandlikar, S. G., 1991, Development of a flow boiling map for subcooled and saturated flow boiling of different fluids in circular tubes, *ASME J. Heat Transfer*, Vol. 113, pp. 190–200.

Kandlikar, S. G., M. Shoji, and V. K. Dhir, eds., 1999, *Handbook of Phase Change: Boiling and Condensation*, Taylor & Francis, New York.

Kattan, N., J. R. Thome, and D. Favrat, 1998, Flow boiling in horizontal tubes, Part 1; Development of a diabatic two-phase flow pattern map, *ASME J. Heat Transfer*, Vol. 120, pp. 140–147; Part 2; New heat transfer data for five refrigerants, *ASME J. Heat Transfer*, Vol. 120, pp. 148–155; Part 3; Development of a new heat transfer model based on flow patterns, *ASME J. Heat Transfer*, Vol. 120, pp. 156–165.

Ould Didi, M. B., N. Kattan, and J. R. Thome, 2002, Prediction of two-phase pressure gradients of refrigerants in horizontal tubes, *Int. J. Refrig.*, Vol. 25, pp. 935–947.

Shah, M. M., 1977, A general correlation for heat transfer during subcooled boiling in pipes and annuli, *ASHRAE Trans.*, Vol. 83, No. 1, pp. 205–215; also, M. M. Shah, 1982, Chart correlation for saturated boiling heat transfer: equations and further study, *ASHRAE Trans.*, Vol. 88, No. 1, pp. 185–196.

Shah, R. K., S. Q. Zhou, and K. Tagavi, 1999, The role of surface tension in film condensation in extended surface passages, *J. Enhanced Heat Transfer*, Vol. 6, pp. 179–216.

Zrcher, O., J. R. Thome, and D. Favrat, 1999, Evaporation of ammonia in a smooth horizontal tube: heat transfer measurements and predictions, *ASME J. Heat Transfer*, Vol. 121, pp. 89–101.

APPENDIX D
U and C_{UA} Values for Various Heat Exchangers

TABLE D.1 U and $C_{UA}(=C)$ Values for Shell-and-Tube Heat Exchangers[a]

$q/\Delta T$ (W/K)	Cold-Side Fluid	Parameter	Hot-Side Fluid								
			Low-Pressure Gas (< 1 bar)	Medium-Pressure Gas (20 bar)	High-Pressure Gas (150 bar)	Process Water	Low-Viscosity Organic Liquid	High-Viscosity Fluid	Condensing Steam	Condensing Hydrocarbon	Condensing Hydrocarbon with Inert Gas
1,000	Low-pressure gas (< 1 bar)	U [W/m²·K]	55	93	120	102	99	63	107	100	86
		C [£/(W/K)]	5.70	5.02	5.51	4.93	4.96	5.50	4.87	4.95	5.11
	Medium-pressure gas (20 bar)	U [W/m²·K]	93	300	350	429	375	120	530	388	240
		C [£/(W/K)]	5.02	4.18	4.81	4.03	4.09	4.76	3.95	4.07	4.28
	High-pressure gas (150 bar)	U [W/m²·K]	120	350	400	600	450	200	600	400	300
		C [£/(W/K)]	5.51	4.81	6.25	4.56	4.38	5.50	4.56	4.82	4.81
	Treated cooling water	U [W/m²·K]	105	484	600	938	714	142	1607	764	345
		C [£/(W/K)]	4.89	3.98	4.56	3.77	3.85	4.59	3.61	3.83	4.12
	Low-viscosity organic liquid	U [W/m²·K]	99	375	450	600	500	130	818	524	286
		C [£/(W/K)]	4.96	4.09	4.38	3.91	3.97	4.67	3.81	3.95	4.20
	High-viscosity liquid	U [W/m²·K]	68	138	200	161	153	82	173	155	214
		C [£/(W/K)]	5.39	4.61	5.50	4.46	4.51	5.16	4.42	4.50	4.33
	Boiling water	U [W/m²·K]	105	467	550	875	677	140	1432	722	336
		C [£/(W/K)]	4.89	3.99	4.91	3.79	3.87	4.60	3.64	3.85	4.13
	Boiling organic liquid	U [W/m²·K]	99	375	450	600	500	130	818	524	286
		C [£/(W/K)]	4.96	4.09	4.38	3.91	3.97	4.67	3.81	3.95	4.20
5,000	Low-pressure gas (< 1 bar)	U [W/m²·K]	55	93	120	102	99	63	107	100	86
		C [£/(W/K)]	2.11	1.63	2.26	1.58	1.59	1.95	1.55	1.59	1.68
	Medium-pressure gas (20 bar)	U [W/m²·K]	93	300	350	429	375	120	530	388	240
		C [£/(W/K)]	1.63	1.11	1.89	1.02	1.05	1.49	0.98	1.05	1.18
	High-pressure gas (150 bar)	U [W/m²·K]	120	350	400	600	450	200	600	400	300
		C [£/(W/K)]	2.26	1.89	2.25	1.10	1.46	1.93	1.10	1.45	1.45
	Treated cooling water	U [W/m²·K]	105	484	600	938	720	142	1607	764	345
		C [£/(W/K)]	1.56	1.00	1.10	0.88	0.91	1.41	0.83	0.90	1.07
	Low-viscosity organic liquid	U [W/m²·K]	99	375	450	600	500	130	818	524	286
		C [£/(W/K)]	1.59	1.05	1.46	0.95	0.99	1.46	0.89	0.98	1.13
	High-viscosity liquid	U [W/m²·K]	68	138	200	161	153	82	173	155	124
		C [£/(W/K)]	1.86	1.43	1.93	1.36	1.38	1.71	1.32	1.37	1.48

TABLE D.1 Continued

$q/\Delta T$ (W/K)	Cold-Side Fluid	Parameter	Low-Pressure Gas (< 1 bar)	Medium-Pressure Gas (20 bar)	High-Pressure Gas (150 bar)	Process Water	Low-Viscosity Organic Liquid	High-Viscosity Fluid	Condensing Steam	Condensing Hydrocarbon	Condensing Hydrocarbon with Inert Gas
30,000	Boiling water	U [W/m²·K]	105	467	550	875	677	140	1432	722	336
		C [£/(W/K)]	1.56	1.00	1.20	0.88	0.93	1.42	0.84	0.91	1.08
	Boiling organic liquid	U [W/m²·K]	99	375	450	600	500	130	818	524	286
		C [£/(W/K)]	1.59	1.05	1.46	0.95	0.99	1.46	0.89	0.98	1.13
	Low-pressure gas (< 1 bar)	U [W/m²·K]	55	93	120	102	99	63	107	100	86
		C [£/(W/K)]	1.11	0.76	1.06	0.73	0.74	0.99	0.71	0.73	0.80
	Medium-pressure gas (20 bar)	U [W/m²·K]	93	300	350	500	375	120	530	388	240
		C [£/(W/K)]	0.76	0.37	0.62	0.28	0.33	0.63	0.28	0.32	0.42
	High-pressure gas (150 bar)	U [W/m²·K]	120	350	400	600	450	200	600	400	300
		C [£/(W/K)]	1.06	0.62	0.94	0.40	0.53	0.73	0.40	0.62	0.62
	Treated cooling water	U [W/m²·K]	105	484	600	938	714	142	1607	764	345
		C [£/(W/K)]	0.71	0.29	0.40	0.23	0.25	0.56	0.19	0.24	0.34
	Low-viscosity organic liquid	U [W/m²·K]	99	375	450	600	500	130	818	524	286
		C [£/(W/K)]	0.74	0.33	0.53	0.27	0.38	0.59	0.24	0.28	0.38
	High-viscosity liquid	U [W/m²·K]	68	138	200	161	153	82	173	155	124
		C [£/(W/K)]	0.94	0.57	0.73	0.52	0.53	0.83	0.50	0.53	0.60
100,000	Boiling water	U [W/m²·K]	105	467	550	875	677	140	1432	722	336
		C [£/(W/K)]	0.71	0.29	0.49	0.23	0.25	0.56	0.20	0.25	0.35
	Boiling organic liquid	U [W/m²·K]	99	375	450	600	500	130	818	524	286
		C [£/(W/K)]	0.74	0.33	0.53	0.27	0.28	0.59	0.24	0.28	0.38
	Low-pressure gas (< 1 bar)	U [W/m²·K]	55	93	120	102	99	63	107	100	86
		C [£/(W/K)]	0.95	0.58	0.93	0.54	0.55	0.83	0.52	0.55	0.62
	Medium-pressure gas (20 bar)	U [W/m²·K]	93	300	350	429	375	120	530	388	240
		C [£/(W/K)]	0.58	0.23	0.35	0.18	0.20	0.47	0.16	0.19	0.27
	High-pressure gas (150 bar)	U [W/m²·K]	120	350	400	600	450	200	600	400	300
		C [£/(W/K)]	0.93	0.35	0.58	0.24	0.28	0.64	0.24	0.32	0.39
	Treated cooling water	U [W/m²·K]	105	484	600	938	714	142	1607	764	345
		C [£/(W/K)]	0.52	0.17	0.24	0.116	0.134	0.41	0.086	0.129	0.21

Hot-Side Fluid

Low-viscosity organic liquid	$U/(\text{W/m}^2 \cdot \text{K})$	99	375	450	609	500	130	818	524	286
	$C[\pounds/(\text{W/K})]$	0.55	0.20	0.28	0.145	0.162	0.44	0.125	0.158	0.24
High-viscosity liquid	$U/(\text{W/m}^2 \cdot \text{K})$	68	138	200	161	153	82	173	155	124
	$C[\pounds/(\text{W/K})]$	0.77	0.42	0.64	0.37	0.38	0.65	0.35	0.38	0.45
Boiling water	$U/(\text{W/m}^2 \cdot \text{K})$	105	467	550	875	677	140	1432	722	336
	$C[\pounds/(\text{W/K})]$	0.52	0.168	0.26	0.121	0.137	0.41	0.091	0.133	0.22
Boiling organic liquid	$U/(\text{W/m}^2 \cdot \text{K})$	99	375	450	600	500	130	818	524	286
	$C[\pounds/(\text{W/K})]$	0.55	0.20	0.28	0.146	0.162	0.44	0.125	0.158	0.24

Source: Selection and Costing of Heat Exchangers, ESDU Engineering Data 92013, ESDU International, London, 1994.

[a] A C_{UA} value for a $(q/\Delta T_m)$ between values of $(q/\Delta T_m)_j$, $j = 1, 2$, should be calculated by logarithmic interpolation:

$$C_{UA} = \exp\left\{\ln C_{UA,1} + \frac{\ln(C_{UA,1}/C_{UA,2})\ln[(q/\Delta T_m)/(q/\Delta T_m)_1]}{\ln[(q/\Delta T_m)_1/(q/\Delta T_m)_2]}\right\}$$

TABLE D.2   U and C_{UA} (=C) Values for Plate-Fin Heat Exchangers[a]

$q/\Delta T$ (W/K)	Cold-Side Fluid	Parameter	Low-Pressure Gas (< 1 bar)	Medium-Pressure Gas (20 bar)	High-Pressure Gas (150 bar)	Process Water	Low-Viscosity Hydrocarbon Liquid	High-Viscosity Hydrocarbon Liquid	Condensing Steam	Condensing Hydrocarbon	Condensing Hydrocarbon with Inert Gas
							Hot-Side Fluid				
5,000	Low-pressure gas (< 1 bar)	$U[\text{W}/\text{m}^2\cdot\text{K}]$	163	217	NA	NA	264	NUS	NUS	270	NUS
		$C[£/(\text{W}/\text{K})]$	3.10	3.10			3.10			3.10	
	Medium-pressure gas (20 bar)	$U[\text{W}/\text{m}^2\cdot\text{K}]$	217	325	NA	NA	377	NUS	NUS	402	NUS
		$C[£/(\text{W}/\text{K})]$	3.10	3.10			3.10			3.10	
	High-pressure gas (150 bar)	$U[\text{W}/\text{m}^2\cdot\text{K}]$	NA	NA	NA	NA	NA	NA	NA	NA	NA
		$C[£/(\text{W}/\text{K})]$									
	Treated cooling water	$U[\text{W}/\text{m}^2\cdot\text{K}]$	315	491	NA	NA	NUS	NUS	NUS	NUS	NUS
		$C[£/(\text{W}/\text{K})]$	3.10	3.10							
	Low-viscosity hydrocarbon	$U[\text{W}/\text{m}^2\cdot\text{K}]$	NUS	NUS	NA	NA	NUS	NUS	NUS	NUS	NUS
		$C[£/(\text{W}/\text{K})]$									
	High-viscosity hydrocarbon	$U[\text{W}/\text{m}^2\cdot\text{K}]$	NUS	NUS	NA	NUS	NUS	NUS	NUS	NUS	NUS
		$C[£/(\text{W}/\text{K})]$									
	Boiling water	$U[\text{W}/\text{m}^2\cdot\text{K}]$	NA	NA	NA	NA	NA	NA	NA	NA	NA
		$C[£/(\text{W}/\text{K})]$									
	Boiling hydrocarbon	$U[\text{W}/\text{m}^2\cdot\text{K}]$	270	402	NA	NA	453	NUS	NA	530	NUS
		$C[£/(\text{W}/\text{K})]$	3.10	3.10			3.10			3.10	
10,000	Low-pressure gas (< 1 bar)	$U[\text{W}/\text{m}^2\cdot\text{K}]$	163	217	NA	NA	264	NUS	NUS	270	NUS
		$C[£/(\text{W}/\text{K})]$	1.57	1.55			1.55			1.55	
	Medium-pressure gas (20 bar)	$U[\text{W}/\text{m}^2\cdot\text{K}]$	217	325	NA	NA	377	NUS	NUS	402	NUS
		$C[£/(\text{W}/\text{K})]$	1.55	1.55			1.55			1.55	
	High-pressure gas (150 bar)	$U[\text{W}/\text{m}^2\cdot\text{K}]$	NA	NA	NA	NA	NA	NA	NA	NA	NA
		$C[£/(\text{W}/\text{K})]$									
	Treated cooling water	$U[\text{W}/\text{m}^2\cdot\text{K}]$	315	491	NA	NA	NUS	NUS	NUS	NUS	NUS
		$C[£/(\text{W}/\text{K})]$	1.55	1.55							
	Low-viscosity hydrocarbon	$U[\text{W}/\text{m}^2\cdot\text{K}]$	NUS	NUS	NA	NA	NUS	NUS	NUS	NUS	NUS
		$C[£/(\text{W}/\text{K})]$									
	High-viscosity	$U[\text{W}/\text{m}^2\cdot\text{K}]$	NUS	NUS	NA	NUS	NUS	NUS	NUS	NUS	NUS

Wide data matrix (rotated table). Row labels carry units £/(W/m²·K) for U and £/(W/K) for C. Columns 1–8 correspond to fluid/service pairings; numeric entries give U and C, with NA (not applicable) and NUS (not usually selected).

Load	Service	Quantity	1	2	3	4	5	6	7	8
	hydrocarbon	$C[£/(W/K)]$	NA	NA	NA	NA	NA	NA	NA	NA
	Boiling water	$U[£/(W/m^2\cdot K)]$	NA	NA	NA	NA	NA	NA	NA	NA
		$C[£/(W/K)]$	NA	NA	NA	NA	NA	NA	NA	NA
30,000	Boiling hydrocarbon	$U[£/(W/m^2\cdot K)]$	270	402	NA	NA	453	NA	530	NA
		$C[£/(W/K)]$	1.55	1.55	NA	NA	1.55	NA	1.55	NA
	Low-pressure gas (< 1 bar)	$U[£/(W/m^2\cdot K)]$	163	217	NA	NUS	264	NUS	270	NUS
		$C[£/(W/K)]$	0.677	0.607	NA	NUS	0.574	NUS	0.579	NUS
	Medium-pressure gas (20 bar)	$U[£/(W/m^2\cdot K)]$	217	325	NA	NUS	377	NUS	402	NUS
		$C[£/(W/K)]$	0.607	0.551	NA	NUS	0.537	NUS	0.532	NUS
	High-pressure gas (150 bar)	$U[£/(W/m^2\cdot K)]$	NA	NA	NA	NA	NA	NA	NA	NA
		$C[£/(W/K)]$	NA	NA	NA	NA	NA	NA	NA	NA
	Treated cooling water	$U[£/(W/m^2\cdot K)]$	315	491	NA	NUS	NUS	NUS	NUS	NUS
		$C[£/(W/K)]$	0.560	0.513	NA	NUS	NUS	NUS	NUS	NUS
	Low-viscosity hydrocarbon	$U[£/(W/m^2\cdot K)]$	NUS	NUS	NA	NUS	NUS	NUS	NUS	NUS
		$C[£/(W/K)]$	NUS	NUS	NA	NUS	NUS	NUS	NUS	NUS
	High-viscosity hydrocarbon	$U[£/(W/m^2\cdot K)]$	NUS	NA	NUS	NUS	NA	NA	NA	NA
		$C[£/(W/K)]$	NA	NA	NA	NA	NA	NA	NA	NA
	Boiling water	$U[£/(W/m^2\cdot K)]$	NA	NA	NA	NA	NA	NA	NA	NA
		$C[£/(W/K)]$	NA	NA	NA	NA	NA	NA	NA	NA
100,000	Boiling hydrocarbon	$U[£/(W/m^2\cdot K)]$	270	402	NA	NA	453	NA	530	NA
		$C[£/(W/K)]$	0.579	0.532	NA	NA	0.527	NA	0.518	NA
	Low-pressure gas (< 1 bar)	$U[£/(W/m^2\cdot K)]$	163	217	NA	NUS	264	NUS	270	NUS
		$C[£/(W/K)]$	0.336	0.301	NA	NUS	0.280	NUS	0.273	NUS
	Medium-pressure gas (20 bar)	$U[£/(W/m^2\cdot K)]$	217	325	NA	NUS	377	NUS	402	NUS
		$C[£/(W/K)]$	0.301	0.245	NA	NUS	0.231	NUS	0.227	NUS
	High-pressure gas (150 bar)	$U[£/(W/m^2\cdot K)]$	NA	NA	NA	NA	NA	NA	NA	NA
		$C[£/(W/K)]$	NA	NA	NA	NA	NA	NA	NA	NA
	Treated cooling water	$U[£/(W/m^2\cdot K)]$	315	491	NA	NUS	NUS	NUS	NUS	NUS
		$C[£/(W/K)]$	0.250	0.210	NA	NUS	NUS	NUS	NUS	NUS
	Low-viscosity hydrocarbon	$U[£/(W/m^2\cdot K)]$	NUS	NUS	NA	NUS	NUS	NUS	NUS	NUS
		$C[£/(W/K)]$	NUS	NUS	NA	NUS	NUS	NUS	NUS	NUS
	High-viscosity hydrocarbon	$U[£/(W/m^2\cdot K)]$	NA	NA	NA	NA	NA	NA	NA	NA
		$C[£/(W/K)]$	NA	NA	NA	NA	NA	NA	NA	NA
	Boiling water	$U[£/(W/m^2\cdot K)]$	NA	NA	NA	NA	NA	NA	NA	NA
		$C[£/(W/K)]$	NA	NA	NA	NA	NA	NA	NA	NA
	Boiling hydrocarbon	$U[£/(W/m^2\cdot K)]$	270	402	NA	NA	453	NA	530	NA
		$C[£/(W/K)]$	0.273	0.227	NA	NA	0.216	NA	0.205	NA

Source: Selection and Costing of Heat Exchangers, ESDU Engineering Data 92013, ESDU International, London, 1994.

[a] A C_{UA} value for a $(q/\Delta T_m)$ between values of $(q/\Delta T_m)_j$, $j = 1, 2$ should be calculated by logarithmic interpolation:

$$C_{UA} = \exp\left\{\ln C_{UA,1} + \frac{\ln(C_{UA,1}/C_{UA,2})\ln[(q/\Delta T_m)/(q/\Delta T_m)_1]}{\ln[(q/\Delta T_m)_1/(q/\Delta T_m)_2]}\right\}$$

GENERAL REFERENCES ON OR RELATED TO HEAT EXCHANGERS

Ackerman, R. A., 1997, *Cryogenic Regenerative Heat Exchangers*, International Cryogenics Monograph Series, Plenum Publishing, New York.

Afgan, N., M. Carvalho, A. Bar-Cohen, D. Butterworth, and W. Roetzel, eds., 1994, *New Developments in Heat Exchangers*, Gordon & Breach, New York.

Afgan, N. H., and E. U. Schlünder, eds., 1974, *Heat Exchangers: Design and Theory Sourcebook*, McGraw-Hill, New York.

Andreone, C. F., and S. Yokell, 1997, *Tubular Heat Exchanger: Inspection, Maintenance and Repair*, McGraw-Hill, New York.

Apblett, W. R., Jr., ed., 1982, *Shell and Tube Heat Exchangers*, American Society for Metals, Metals Park, OH.

Au-Yang, M. K., 2001, *Flow-Induced Vibrations of Power Plants Components: A Practical Workbook*, ASME Press, New York.

Azbel, D., 1984, *Heat Transfer Applications in Process Engineering*, Noyes Publications, Park Ridge, NJ.

Bar-Cohen, A., M. Carvalho, and R. Berryman, eds., 1998, Heat exchangers for sustainable development, *Proc. Heat Exchangers for Sustainable Development*, Lisbon, Portugal.

Beck, D. S., and D. G. Wilson, 1996, *Gas Turbine Regenerators*, Chapman & Hall, New York.

Bhatia, M. V., and P. N. Cheremisinoff, 1980, *Heat Transfer Equipment*, Process Equipment Series, Vol. 2, Technomic Publishing, Westport, CT.

Bliem, C., et al., 1985, *Ceramic Heat Exchanger Concepts and Materials Technology*, Noyes Publications, Park Ridge, NJ.

Blevins, R. D., 1990, *Flow-Induced Vibration*, 2nd ed., Von Nostrand Reinhold, New York.

Bohnet, M., T. R. Bott, A. J. Karabelas, P. A. Pilavachi, R. Séméria, and R. Vidil, eds., 1992, *Fouling Mechanisms: Theoretical and Practical Aspects*, Eurotherm Seminar 23, Editions Européennes Thermique et Industrie, Paris.

Bott, T. R., 1990, *Fouling Notebook: A Practical Guide to Minimizing Fouling in Heat Exchangers*, Institution of Chemical Engineers, London.

Bott, T. R., 1995, *Fouling of Heat Exchangers*, Elsevier Science Publishers, Amsterdam, The Netherlands.

Bott, T. R., L. F. Melo, C. B. Panchal, and E. F. C. Somerscales, 1999, *Understanding Heat Exchanger Fouling and Its Mitigation*, Begell House, New York.

Bryers, R. W., ed., 1983, *Fouling of Heat Exchanger Surfaces*, Engineering Foundation, New York.

Bryers, R. W., ed., 1983, *Fouling and Slagging Resulting from Impurities in Combustion Gases*, Engineering Foundation, New York.

Buchlin, J. M., ed., 1991, *Industrial Heat Exchangers*, Lecture Series 1991–04, von Kármán Institute for Fluid Dynamics, Belgium.

Chen, S. S., 1987, *Flow-Induced Vibration of Circular Cylindrical Structures*, Hemisphere Publishing, Washington, DC.

Chisholm, D., ed., 1980, *Developments in Heat Exchanger Technology*, Vol. I, Applied Science Publishers, London.

Chisholm, D., ed., 1988, *Heat Exchanger Technology*, Elsevier Applied Science, New York.

Dragutinović, G. D., and B. S. Bačlić, 1998, *Operation of Counterflow Regenerators*, Vol. 4, Computational Mechanics Publications, WIT Press, Southampton, UK.

Dzyubenko, B. V., L.-V. Ashmantas, and M. D. Segal, 1999, *Modeling and Design of Twisted Tube Heat Exchangers*, Begell House, New York.

Foster, B. D., and J. B. Patton, eds., 1985, *Ceramic Heat Exchangers*, American Ceramic Society, Columbus, OH.

Foumeny, E. A., and P. J. Heggs, eds., 1991, *Heat Exchange Engineering*, Vol. 1; *Design of Heat Exchangers*, Ellis Horwood, London.

Foumeny, E. A., and P. J. Heggs, eds., 1991, *Heat Exchange Engineering*, Vol. 2; *Compact Heat Exchangers: Techniques for Size Reduction*, Ellis Horwood, London.

Fraas, A. P., and M. N. Ozisik, 1989, *Heat Exchanger Design*, 2nd ed., Wiley, New York.

Ganapathy, V., 1982, *Applied Heat Transfer*, PennWell Publishing, Tulsa, OK.

Ganapathy, V., 2002, *Industrial Boilers and Heat Recovery Steam Generators – Design, Applications, and Calculations*, Marcel Dekker, New York.

Garrett-Price, B. A., S. A. Smith, R. L. Watts, J. G. Knudsen, W. J. Marner, and J. W. Suitor, 1985, *Fouling of Heat Exchangers*, Noyes Publications, Park Ridge, NJ.

Gupta, J. P., 1986, *Fundamentals of Heat Exchanger and Pressure Vessel Technology*, Hemisphere Publishing, Washington, DC; also as *Working with Heat Exchangers*, in soft cover, Hemisphere Publishing, Washington, DC, 1990.

Hausen, H., 1983, *Heat Transfer in Counterflow, Parallel Flow and Cross Flow*, McGraw-Hill, New York.

Hayes, A. J., W. W. Liang, S. L. Richlen, and E. S. Tabb, eds., 1985, *Industrial Heat Exchangers*, American Society for Metals, Metals Park, OH.

Hesselgreaves, J. E., 2001, *Compact Heat Exchangers: Selection, Design, and Operation*, Elsevier Science, Oxford.

Hewitt, G. F., exec. ed., 1998, *Heat Exchanger Design Handbook*, three vols. (five parts), Begell House, New York; former publication: G. F. Hewitt, coord. ed., 1989, *Hemisphere Handbook of Heat Exchanger Design*, Hemisphere Publishing, New York.

Hewitt, G. F., G. L. Shires, and T. R. Bott, 1994, *Process Heat Transfer*, CRC Press and Begell House, Boca Raton, FL.

Hewitt, G. F., and P. B. Whalley, 1989, *Handbook of Heat Exchanger Calculations*, Hemisphere Publishing, Washington, DC.

Hryniszak, W., 1958, *Heat Exchangers: Applications to Gas Turbines*, Butterworth Scientific Publications, London.

Idelchik, I. E., 1994, *Handbook of Hydraulic Resistance*, 3rd ed., CRC Press, Boca Raton, FL.

Ievlev, V. M., ed., 1990, *Analysis and Design of Swirl-Augmented Heat Exchangers*, Hemisphere Publishing, Washington, DC.

Jakob, M., 1957, *Heat Transfer*, Vol. II, Wiley, New York.

Kakaç, S., ed., 1991, *Boilers, Evaporators, and Condensers*, Wiley, New York.

Kakaç, S., ed., 1999, *Heat Transfer Enhancement of Heat Exchangers*, Kluwer Academic Publishers, Dordrecht, The Netherlands.

Kakaç, S., A. E. Bergles, and E. O. Fernandes, eds., 1988, *Two-Phase Flow Heat Exchangers: Thermal Hydraulic Fundamentals and Design*, Kluwer Academic Publishers, Dordrecht, The Netherlands.

Kakaç, S., A. E. Bergles, and F. Mayinger, eds., 1981, *Heat Exchangers: Thermal-Hydraulic Fundamentals and Design*, Hemisphere Publishing, Washington, DC.

Kakaç, S., and H. Liu, 1998, *Heat Exchangers: Selection, Rating, and Thermal Design*, CRC Press, Boca Raton, FL.

Kakaç, S., R. K. Shah, and W. Aung, eds., 1987, *Handbook of Single-Phase Convective Heat Transfer*, Wiley, New York.

Kakaç, S., R. K. Shah, and A. E. Bergles, eds., 1983, *Low Reynolds Number Flow Heat Exchangers*, Hemisphere Publishing, Washington, DC.

Katinas, V., and A. Žukauskas, 1997, *Vibrations of Tubes in Heat Exchangers*, Begell House, New York.

Kays, W. M., and A. L. London, 1998, *Compact Heat Exchangers*, reprint 3rd edn., Krieger Publishing, Malabar, FL; first ed., National Press, Palo Alto, CA (1955); 2nd ed., (1964), 3rd ed., McGraw-Hill, New York (1984).

Kern, D. Q., 1950, *Process Heat Transfer*, McGraw-Hill, New York.

Kern, D. W., and A. D. Kraus, 1972, *Extended Surface Heat Transfer*, McGraw-Hill, Chaps. 9–12, pp. 439–641.

King, R., ed., 1987, *Flow Induced Vibrations*, BHRA Publication, London.

Kraus, A. D., 1982, *Analysis and Evaluation of Extended Surface Thermal Systems*, Hemisphere Publishing, Washington, DC.

Kraus, A. D., A. Aziz, and J. R. Welty, 2001, *Extended Surface Heat Transfer*, Wiley, New York.

Kröger, D. G., 1998, *Air-Cooled Heat Exchangers and Cooling Towers*, Tecpress, Uniedal, South Africa; also, Begell House, New York.

Kuppan, T., 2000, *Heat Exchanger Design Handbook*, Marcel Dekker, New York.

Lokshin, V. A., D. F. Peterson, and A. L. Schwarz, 1988, *Standard Handbook of Hydraulic Design for Power Boilers*, Hemisphere Publishing, Washington, DC.

Ludwig, E. E., 1965, *Applied Process Design for Chemical and Petrochemical Plants*, Vol. III, Gulf Publishing, Houston, TX, Chap. 10.

Manzoor, M., 1984, *Heat Flow through Extended Surface Heat Exchangers*, Springer-Verlag, Berlin.

Martin, M., 1992, *Heat Exchangers*, Hemisphere Publishing, Washington, DC.

Marto, P. J., and R. H. Nunn, eds., 1981, *Power Condenser Heat Transfer Technology*, Hemisphere Publishing, Washington, DC.

Marvillet, Ch., gen. ed., 1994, *Recent Developments in Finned Tube Heat Exchangers: Theoretical and Practical Aspects*, DTI Energy Technology, Danish Technological Institute, Taastrup, Denmark.

Marvillet, C., and R. Vidil, eds., 1993, *Heat Exchanger Technology: Recent Developments*, Eurotherm Seminar 33, Editions Européennes Thermique et Industrie, Paris.

McNaughton, K. J., ed., 1986, *The Chemical Engineering Guide to Heat Transfer*; Vol. 1; *Plant Principles*, Vol. 2; *Equipment*, Hemisphere Publishing, Washington, DC.

Melo, L. F., T. R. Bott, and C. A. Bernardo, eds., 1988, *Advances in Fouling Science and Technology*, Kluwer Academic Publishers, Dordrecht, The Netherlands.

Miller, D. S., 1990, *Internal Flow Systems*, 2nd ed., BHRA Fluid Engineering Series, Vol. 5, BHRA, Cranfield, UK.

Minton, P. E., 1986, *Handbook of Evaporator Technology*, Noyes Publications, Park Ridge, NJ.

Mori, Y., A. E. Sheindlin, and N. H. Afgan, eds., 1986, *High Temperature Heat Exchangers*, Hemisphere Publishing, Washington, DC.

Müller-Steinhagen, H., ed., 2000, *Heat Exchanger Fouling: Mitigation and Cleaning Technologies*, Publico Publications, Essen, Germany.

Palen, J. W., ed., 1987, *Heat Exchanger Sourcebook*, Hemisphere Publishing, Washington, DC.

Panchal, C. B., T. R. Bott, E. F. C. Somerscales, and S. Toyama, 1997, *Fouling Mitigation of Industrial Heat Exchange Equipment*, Begell House, New York.

Podhorsky, M., and H. Krips, 1998, *Heat Exchangers: A Practical Approach to Mechanical Construction, Design and Calculations*, Begell House, New York.

Putman, R. E., 2001, *Steam Surface Condensers: Basic Principles, Performance Monitoring and Maintenance*, ASME Press, New York.

Reay, D. A., 1979, *Heat Recovery Systems*, E&FN Spon, London.

Reay, D. A., 1999, *Learning from Experiences with Compact Heat Exchangers*, CADDET Analyses Series 25, Centre for the Analysis and Dissemination of Demonstrated Energy Technologies, Sittard, The Netherlands.

Rifert, V. G., 1998, *Condensation Heat Transfer Enhancement*, Computational Mechanics Publications, WIT Press, Southampton, UK.

Roetzel, W., P. J. Heggs, and D. Butterworth, eds., 1991, *Design and Operation of Heat Exchangers*, Springer-Verlag, Berlin.

Roetzel, W., and Y. Xuan, 1998, *Dynamic Behaviour of Heat Exchangers*, Vol. 3, Computational Mechanics Publications, WIT Press, Southampton, UK.

Saunders, E. A. D., 1989, *Heat Exchangers: Selection, Design and Construction*, Wiley, New York.

Schlünder, E. U., ed.-in-chief, 1982, *Heat Exchanger Design Handbook*, 5 vols., Hemisphere Publishing, Washington, DC.

Schmidt, F. W., and A. J. Willmott, 1981, *Thermal Energy Storage and Regeneration*, Hemisphere/McGraw-Hill, Washington, DC.

Shah, R. K., K. J. Bell, H. Honda, and B. Thonon, eds., 1999, *Compact Heat Exchangers and Enhancement Technology for the Process Industries*, Begell House, New York.

Shah, R. K., K. J. Bell, S. Mochizuki, and V. V. Wadekar, eds., 1997, *Compact Heat Exchangers for the Process Industries*, Begell House, New York.

Shah, R. K., A. W. Deakin, H. Honda, and T. M. Rudy, eds., 2001, *Compact Heat Exchangers and Enhancement Technology for the Process Industries 2001*, Begell House, New York.

Shah, R. K., and A. Hashemi, eds., 1993, *Aerospace Heat Exchanger Technology, 1993*, Elsevier Science, Amsterdam, The Netherlands.

Shah, R. K., A. D. Kraus, and D. Metzger, eds., 1990, *Compact Heat Exchangers: A Festschrift for A.L. London*, Hemisphere Publishing, Washington, DC.

Shah, R. K., and A. L. London, 1978, *Laminar Flow Forced Convection in Ducts*, Supplement 1 to *Advances in Heat Transfer Series*, Academic Press, New York.

Shah, R. K., and A. C. Mueller, 1985, Heat exchangers, in *Handbook of Heat Transfer Applications*, W. M. Rohsenow, J. P. Hartnett, and E. N. Ganić, eds., McGraw-Hill, New York, Chap. 4, pp. 1–312.

Shah, R. K., and A. C. Mueller, 1989, Heat exchange, in *Ullman's Encyclopedia of Industrial Chemistry*, Unit Operations II, Vol. B3, Chap. 2, VCH Publishers, Weinheim, Germany.

Shah, R. K., and D. P. Sekulić, 1998, Heat exchangers, in *Handbook of Heat Transfer*, W. M. Rohsenow, J. P. Hartnett, and Y. I. Cho, eds., McGraw-Hill, New York, Chap. 17.

Shah, R. K., E. C. Subbarao, and R. A. Mashelkar, eds., 1988, *Heat Transfer Equipment Design*, Hemisphere Publishing, Washington, DC.

Sheindlin, A. E., ed., 1986, *High Temperature Equipment*, Hemisphere Publishing, Washington, DC.

Singh, K. P., and A. I. Soler, 1984, *Mechanical Design of Heat Exchangers and Pressure Vessel Components*, Arcturus Publishers, Cherry Hill, NJ.

Smith, E. M., 1997, *Thermal Design of Heat Exchangers: A Numerical Approach: Direct Sizing and Stepwise Rating*, Wiley, New York.

Smith, R. A., 1987, *Vaporisers: Selection, Design, Operation (Designing for Heat Transfer)*, Wiley, New York.

Somerscales, E. F. C., and J. G. Knudsen, eds., 1981, *Fouling of Heat Transfer Equipment*, Hemisphere/McGraw-Hill, Washington, DC.

Soumerai, H., 1987, *Practical Thermodynamic Tools for Heat Exchanger Design Engineers*, Wiley, New York.

Stasiulevicius, J., and A. Skrinska, 1987, *Heat Transfer of Finned Tube Bundles in Crossflow*, Hemisphere Publishing, Washington, DC.

Sukhotin, A. M., and G. Tereshchenko, 1998, *Corrosion Resistance of Equipment for Chemical Industry Handbook*, Begell House, New York.

Sundén, B., and M. Faghri, eds., 1998, *Computer Simulation in Compact Heat Exchangers*, Computational Mechanics Publications, WIT Press, Southampton, UK.

Sundén, B., and P. J. Heggs, eds., 1998, *Recent Advances in Analysis of Heat Transfer for Fin Type Surfaces*, Computational Mechanics Publications, WIT Press, Southampton, UK.

Sundén, B., and R. M. Manglik, eds., 2001, *Plate and Frame Heat Exchangers*, Computational Mechanics Publication, WIT Press, Southampton, UK.

Taborek, J., G. F. Hewitt, and N. Afgan, eds., 1983, *Heat Exchangers: Theory and Practice*, Hemisphere/McGraw-Hill, Washington, DC.

Taylor, M. A., 1987, *Plate-Fin Heat Exchangers: Guide to Their Specification and Use*, HTFS, Harwell Laboratory, Oxon, UK.

Walker, G., 1990, *Industrial Heat Exchangers: A Basic Guide*, 2nd edn., Hemisphere Publishing, Washington, DC.

Webb, R. L., 1994, *Principles of Enhanced Heat Transfer*, Wiley, New York.

Willmott, A. J., 2001, *Dynamics of Regenerative Heat Transfer*, Taylor & Francis, New York.

Yokell, S., 1990, *A Working Guide to Shell-and-Tube Heat Exchangers*, McGraw-Hill, New York.

Žukauskas, A. A., 1989, *High Performance Single-Phase Heat Exchangers*, Hemisphere Publishing, Washington, DC. [This book has a misleading title. It should be *Forced Convection Heat Transfer*.]

Žukauskas, A., and R. Ulinskas, 1988, *Heat Transfer in Tube Banks in Crossflow*, Hemisphere Publishing, Washington, DC.

Žukauskas, A. A., R. Ulinskas, and V. Katinas, 1988, *Fluid Dynamics and Flow Induced Vibrations of Tube Banks*, Hemisphere Publishing, Washington, DC.

Reference Books by Title

Advances in Industrial Heat Exchangers, HEE96, Institution of Chemical Engineers, London, 1996.

ASME Boiler and Pressure Vessel Code, Sec. VIII, Div. 1, *Rules for Construction of Pressure Vessels*, American Society of Mechanical Engineers, New York, 1998.

Condensers: Theory and Practice, IChemE Symposium Series 75, Pergamon Press, Elmsford, NY, 1983.

Effectiveness N_{tu} Relationships for Design and Performance Evaluation of Multi-pass Crossflow Heat Exchangers, Engineering Sciences Data Unit Item 87020, ESDU International, McLean, VA, October 1987.

Effectiveness N_{tu} Relationships for Design and Performance Evaluation of Two-Stream Heat Exchangers, Engineering Sciences Data Unit Item 86018, ESDU International, McLean, VA, July 1986.

Shell-and-Tube Heat Exchangers for General Refinery Services, API Standard 660, 4th ed., American Petroleum Institute, Washington, DC, 1982.

Standard of the Tubular Exchanger Manufacturers Association, 8th edn., TEMA, New York, 1999.

Standard for Closed Feedwater Heaters, 4th ed., Heat Exchanger Institute, Cleveland, OH, 1984.

Standard for Power Plant Heat Exchangers, Heat Exchanger Institute, Cleveland, OH, 1980.

Standard for Steam Surface Condensers, 8th ed., Heat Exchanger Institute, Cleveland, OH, 1984.

Standards of the Brazed Aluminum Plate-Fin Heat Exchanger Manufacturers' Association (ALPEMA), 2nd ed., AEA Technology, Didcot, Oxon, UK, 2000.

Index

Printed in the USA/Agawam, MA
May 25, 2012

566166.039